second edition

CALCULUS

AND

ANALYTIC GEOMETRY

Sherman K. Stein

PROFESSOR OF MATHEMATICS

UNIVERSITY OF CALIFORNIA

DAVIS

McGraw-Hill Book Company

New York • St Louis • San Francisco

Auckland • Bogotá • Düsseldorf • Johannesburg

London • Madrid • Mexico • Montreal • New Delhi • Panama • Paris

São Paulo • Singapore • Sydney • Tokyo • Toronto

CALCULUS AND ANALYTIC GEOMETRY

4567890   DODO   7832109

This book was set in Times New Roman.   The editors were
A. Anthony Arthur, Alice Macnow, and Shelly Levine Langman;
the production supervisor was Thomas J. LoPinto.
New drawings were done by J & R Services, Inc.
R. R. Donnelley & Sons Company was printer and binder

Cover Illustration
The Pyramid at Sakkara illustrates the process of integration by step-like
intervals. Photo courtesy of the Bettmann Archive, Inc.

LIBRARY OF CONGRESS CATALOGING IN PUBLICATION DATA

Stein, Sherman K
    Calculus and analytic geometry.

    Includes index.
    1. Calculus   2.   Geometry, Analytic.   I.   Title.
QA303.S857   1977   515′.15   75-45189
ISBN 0-07-061008-8

To: Joshua, Rebecca, and Susanna

Arnold Toynbee, Experiences, Oxford University Press, pp. 12–13, 1969.

... at about the age of sixteen, I was offered a choice which, in retrospect, I can see that I was not mature enough, at the time, to make wisely. This choice was between starting on the calculus and, alternately, giving up mathematics altogether and spending the time saved from it on reading Latin and Greek literature more widely. I chose to give up mathematics, and I have lived to regret this keenly after it has become too late to repair my mistake. The calculus, even a taste of it, would have given me an important and illuminating additional outlook on the Universe, whereas, by the time at which the choice was presented to me, I had already got far enough in Latin and Greek to have been able to go farther with them unaided. So the choice that I made was the wrong one, yet it was natural that I should choose as I did. I was not good at mathematics; I did not like the stuff.... Looking back, I feel sure that I ought not to have been offered the choice; the rudiments, at least, of the calculus ought to have been compulsory for me. One ought, after all, to be initiated into the life of the world in which one is going to have to live. I was going to live in the Western World... and the calculus, like the fullrigged sailing ship, is one of the characteristic expressions of the modern Western genius.

CONTENTS

PREFACE

Our goal has been to provide the student and instructor with a readable, flexible text that covers the important topics of single and multivariable calculus as simply and as intuitively as possible.

The experiences of many users of the first edition have suggested most of the changes, large and small, that have been made in this, the second edition.

Organizational changes

Limits receive more attention than in the first edition, in the form of an extra section. The second derivative is introduced earlier, in time for graphing. L'Hôpital's rule also comes earlier, but only after the student has had adequate experience with limits. Partial derivatives, formerly developed in two chapters, are now covered in one. Tests for convergence of series, hyperbolic functions, integrals over surfaces, and determinants have been added, the last two strengthening the treatment of vector analysis.

To accommodate these changes, some topics had to go. The chapter on gravity and part of the chapter on traffic have been deleted. The latter chapter, which concerns primarily the Poisson distribution, now gives some attention to the idea of a mathematical model.

Pedagogical changes

I wrote this book for the student. After all, the student pays for the book and has to read it, while the instructor uses it primarily as a source of examples and exercises. I have strengthened the chapter summaries, which students have found helpful; they provide an emphasis and perspective that individual sections cannot. Also, I have added many asides in the margin to guide the student.

There are now more examples and exercises, both easy and difficult. Thus the instructor has the option of determining how demanding the course will be, and the student has ample exercises for review.

I believe that many students study calculus long before they are committed to a major or a career. Partly for this reason I have included many examples and exercises, by way of enrichment or perhaps for browsing, to offer the student a broader introduction to the way mathematics relates to the world. A few examples of these are the rotary engine, depletion of natural resources, present value of future income, and escape velocity.

I have added exercises for the calculator, some of which can be done even on a calculator that has only arithmetic operations. Some of these exercises reinforce a concept already presented, while others provide background for later developments, for instance, in limits and series.

Analytic geometry

The analytic geometry needed in calculus is amply presented in twelve sections that are inserted when needed (and are smoothly integrated into the flow of the course). Seven of these sections discuss such standard topics as the graph of a function, the slope of a line, polar coordinates, parametric equations, rectangular coordinates in space, and graphing surfaces. In addition, five of the sections focus on the geometry involved in setting up integrals: discussing such topics as how to find cross-sectional lengths and areas and how to describe plane and solid regions in coordinates. In particular, the geometric difficulties of setting up repeated integrals are separated from the computational aspects of the resulting integrals.

Differentiation

We reach the differentiation of the transcendental functions early for two reasons. First, they are far more important than polynomials in the applications that the student meets. Second, the arguments that obtain the derivative of the sine and logarithm functions reveal more clearly the idea of a limit than does the algebra that obtains the derivative of a polynomial (in which Δx can be set equal to 0 with impunity).

Applications

In addition to the standard geometric and physical applications, the text utilizes many illustrations from other fields. For instance, exercises and examples include the ideal lot problem, present value of future income, the multiplier effect, the homogeneous equation describing production, and compound interest. An entire chapter, which includes the Poisson distribution, is devoted to mathematical modeling.

Formal integration

While Chapter 9 covers all the standard techniques of formal integration, I recommend emphasizing Section 9.5, *Using a table of integrals*. If there is not enough time to cover the whole Chapter, perhaps Sections 9.6 to 9.9, which treat partial fractions and special techniques, could be played down. In any case, the integral of $1/(a^2 - x^2)$ should be mentioned. The instructor is urged to emphasize the many examples and exercises that concern nonintegrability in finite terms of such functions as

$$e^{x^2}, \quad (\sin x)/x, \quad \sqrt{\sin x}, \quad \sqrt{1 - x^3}, \quad \text{and} \quad \sqrt{x}\,\sqrt[3]{1 - x}.$$

I also suggest that formal integration be deemphasized in applications of the integral. For instance, as one exercise points out, the problem of finding the arc length along the curve $y = x^n$ for a positive integer n leads to an elementary integral only in case n is 1 or 2. One should not, like those who climb mountains "because they are there," evaluate integrals simply "because they are elementary." For similar reasons the computational tool of the repeated integral should not displace the important concept of a definite integral over a plane or solid region.

Multiple integrals

The evaluation of an integral over a plane or solid region by a repeated integral is presented intuitively, without any double or triple summations. The formulas for the use of rectangular or polar coordinates (in particular, the appearance of the "r" in the integrand) are made plausible. The evaluation of integrals over solids is separated by several chapters from the chapter devoted to the evaluation of two-dimensional integrals, thus giving the student a gradual introduction to repeated integration, in contrast to the customary abrupt single dose.

Exercises

The exercises are usually divided into three types set off by square boxes (■ or ■ ■). Those in the first group, which is the most numerous, are routine and help the student master the basic definitions and computations. Those in the second group may be longer or may require some departure from the text. Exercises in the last group either offer an alternative view, discuss theory, or are more difficult; they should be assigned sparingly, perhaps for review. There are far more exercises than can reasonably be worked during a typical course. The supply is generous to provide instructors with a convenient source of examples and test problems and students with ample review questions.

Summaries

Every chapter concludes with a substantial section called *Summary*, that provides the student with a survey of the chapter, including a list of terms and symbols, key facts, and guide quizzes. Taken together, they constitute a study guide; they are not intended to be covered in lectures. Frequently the exercises in these sections are not, strictly speaking, review. But, by developing new applications or alternative approaches, they give the student a chance to test his or her understanding of the material by using it.

Theory and proofs

I believe that the importance and excitement of introductory calculus lies in its applications, not in the formal foundations which are best supplied in advanced calculus and real analysis. Thus the idea of a limit is drawn from arithmetical computations; its formal definition in terms of ϵ and δ is relegated to an appendix. The logarithm is treated as the inverse of the exponential, thus giving the student a chance to review both functions; the more elegant definition of the logarithm as an integral, while pleasing to the instructor, strikes the typical freshman as unnatural. The chain rule for the derivative of $f \circ g$ is proved easily when $g'(x)$ is not 0, while an exercise disposes of the case when $g'(x)$ is 0. Moreover, when both a heuristic argument and a proof are presented, the instructor has the option of omitting the proof.

Duration

As a rule of thumb, one section corresponds to one class meeting, though some sections are longer and some shorter. Exclusive of summaries the text contains 137 sections. About 27 of these concern possibly expendable topics, namely natural growth (6.4), special integration techniques (9.6 to 9.9), Newton's method (11.4), angle between a line and a tangent (11.5), center of gravity (13.5), harmonic motion (15.3), error in estimating an integral (15.4), general binomial series (15.5), series for $f(x, y)$ (15.6), moment of a function (Chapter 16), models (Chapter 17), acceleration vector (20.3 to 20.5), Lagrange multipliers (20.8), and interchange of limits (Chapter 23). Thus about 110 sections cover the essentials of calculus through Stokes' theorem.

The first 18 chapters, which contain 108 nonsummary sections, cover calculus through series, partial derivatives, and multiple integrals. Of these, perhaps 90 are essential. The chapters and sections were written to give the instructor maximum flexibility. The map of the book on the next page shows how the chapters fit together.

Reading the book

I suggest that the student read a section carefully and completely before trying to do the exercises. In particular, it would be wise to try to work the examples in the text with the book closed, as a test of understanding.

Acknowledgments

I would like to thank those users of the first edition who participated in the McGraw-Hill survey on which I based many of the changes in this edition. In addition, Professors Konrad John Heuvers and Zane C. Motteler of Michigan Technological University, Charles V. Heuer of Concordia College, and Dale Varberg of Hamline University suggested various improvements.

Professors Frank Deane of Berkshire Community College, Paul C. Shields of the University of Toledo, and Henry Alder of University of California, Davis spent much time reviewing the manuscript.

My colleagues at Davis, whose suggestions were pretty close to those of the national survey, made countless recommendations. In particular, Professors Henry Alder, Curtis Fulton, and Donald Norton wrote me many a note to convey their thoughts based on their day-by-day teaching experience with the text.

Map of the book

The symbol $(a) \rightarrow (b)$ suggests a natural flow from Chapter (a) to Chapter (b), or that Chapter (a) is good background for Chapter (b). Whether (a) is a logical prerequisite for (b) will depend in some cases on the topics in (b) covered or omitted by the instructor.

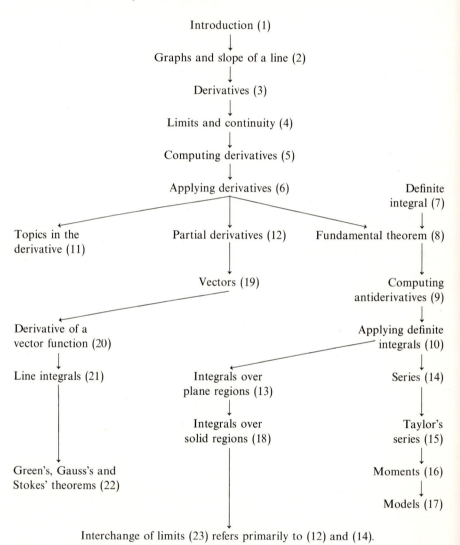

Interchange of limits (23) refers primarily to (12) and (14).

Sherman K. Stein

1

THE TWO MAIN CONCEPTS OF CALCULUS

If an object traveling at constant speed moves 6 feet in 2 seconds, its speed is easy to find:

$$\text{Speed} = \frac{\text{distance}}{\text{time}} = \frac{6}{2}$$
$$= 3 \text{ feet per second.}$$

However, suppose that the object travels at a varying speed. If we know how far it travels during any period of time, how can its speed at any given instant be found? For example, assume that a rock drops $16t^2$ feet in the first t seconds of its fall, as Galileo discovered. What is its speed t seconds after its release? This question, answered in Sec. 1.1, introduces the first of the two basic concepts of calculus, the derivative.

The question can be turned around: If an object travels at a varying speed and we know this speed at any instant, how can the distance it travels be found? For example, if the speed of a rocket is t^2 feet per second after t seconds, how far does it travel in the first 3 seconds? This question is explored in Sec. 1.2 but not completely answered until Chap. 7. It introduces the second basic concept of calculus, the definite integral.

1.1

How to find the varying speed from the distance

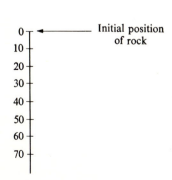

Initial position of rock

Suppose that a rock, starting from rest, falls $16t^2$ feet in t seconds. What is its speed after t seconds? For the sake of simplicity, consider a specific value of t, say $t = 2$. Let us find the speed after 2 seconds.

Start by introducing a vertical line to record the position of the rock, as in the accompanying figure.

The first half second, the rock falls $16(\frac{1}{2})^2 = 4$ feet. In the first second, the rock falls $16(1)^2 = 16$ feet. Thus, during the second half second the rock falls $16 - 4 = 12$ feet, which is three times as far as it falls during the first half second.

As the rock falls, its speed increases.

To find the rock's speed after it has been falling for 2 seconds, consider the distance it travels in a brief interval after the first 2 seconds of fall. If we observe only a single instant of time (precisely 2 seconds after the rock is dropped), we cannot hope to find the speed, which depends on distance traveled. In the same way, a photographer could not hope to compute the rock's speed from a single photograph. But from two photographs taken a very short time apart, he could at least estimate the speed of the rock.

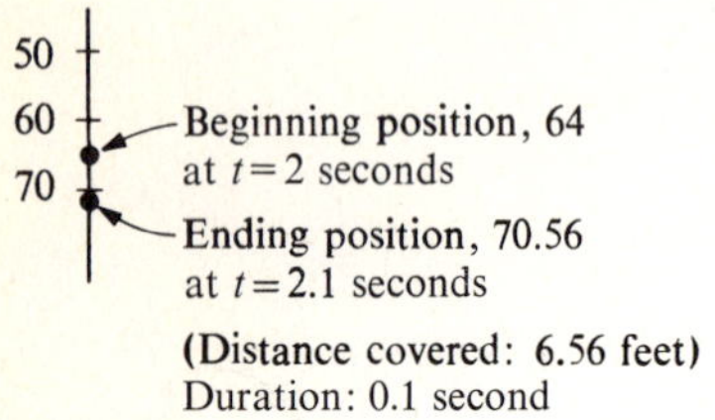

Take a specific short interval of time, say, from time $t = 2$ seconds to time 2.1 seconds, a duration of 0.1 second. To find the distance traveled, use the formula $16t^2$ and subtract:

$$16(2.1)^2 - 16(2)^2 = 16(4.41) - 16(4)$$
$$= 70.56 - 64$$
$$= 6.56 \text{ feet.}$$

These computations show that, during the interval of 0.1 second, the rock drops 6.56 feet.

A reasonable *estimate* of the speed of the rock at time $t = 2$ is therefore

$$\frac{\text{Distance}}{\text{Time}} = \frac{6.56}{0.1} = 65.6 \text{ feet per second.}$$

This may be thought of as the average speed during the brief interval from $t = 2$ to 2.1 seconds, a duration of 0.1 second.

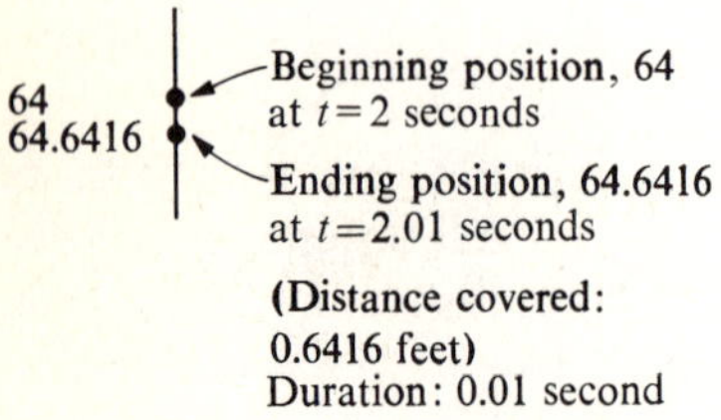

Use of a shorter time interval will presumably provide a more accurate estimate. Choose, say, the time interval from $t = 2$ to 2.01 seconds, a duration of only 0.01 second. The distance the rock falls during this shorter interval of time is

$$16(2.01)^2 - 16(2)^2 = 16(4.0401) - 16(4)$$
$$= 64.6416 - 64$$
$$= 0.6416 \text{ foot.}$$

The accompanying figure shows it pictorially.

Using this shorter time interval provides a more accurate estimate of the speed at time $t = 2$, namely

$$\frac{0.6416}{0.01} = 64.16 \text{ feet per second.}$$

The average speed during this shorter interval of time is 64.16 feet per second.

Rather than compute the average speed over shorter and shorter time intervals, let us use algebra to treat the *general* short interval of time, from time $t = 2$ to a time t_1 which is larger than 2. (Read t_1 as "t sub one.") This time interval has a duration of

$$t_1 - 2 \quad \text{seconds.}$$

During this time the rock falls

$$16t_1{}^2 - 16(2)^2 \quad \text{feet,}$$

and the quotient
$$\frac{16t_1{}^2 - 16(2)^2}{t_1 - 2}$$

50
60 — Beginning position, 64
70   at $t = 2$ seconds

Ending position, 70.56
at $t = 2.1$ seconds

(Distance covered: 6.56 feet)
Duration: 0.1 second

---

64   — Beginning position, 64
64.6416   at $t = 2$ seconds

Ending position, 64.6416
at $t = 2.01$ seconds

**(Distance covered:
0.6416 feet)**
Duration: 0.01 second

is an estimate of the speed after 2 seconds of fall. The closer t_1 is to 2, the more closely this estimate approximates the exact speed at time 2.

A little algebra changes the quotient

$$\frac{16t_1{}^2 - 16(2)^2}{t_1 - 2}$$

into an equivalent expression much easier to work with:

$$\frac{16t_1{}^2 - 16(2)^2}{t_1 - 2} = 16\frac{t_1{}^2 - (2)^2}{t_1 - 2}$$

$$= 16\frac{(t_1 + 2)(t_1 - 2)}{t_1 - 2} \quad \textit{factor}$$

$$= 16(t_1 + 2).$$

In short,

$$\frac{16t_1{}^2 - 16(2)^2}{t_1 - 2} = 16(t_1 + 2),$$

an algebraic identity that is valid whenever t_1 is different from 2.

Note for later reference: The step in which $t_1{}^2 - (2)^2$ is replaced by

$$(t_1 + 2)(t_1 - 2)$$

follows from the algebraic identity

$$c^2 - d^2 = (c + d)(c - d).$$

It is easy to see that, as t_1 gets closer and closer to 2, the expression

$$16(t_1 + 2)$$

approaches

$$16(2 + 2) = 64.$$

Therefore, it seems reasonable to claim that after 2 seconds of falling the rock has a speed of

$$64 \text{ feet per second.}$$

We have found the speed when $t = 2$ seconds. Before going on to find a formula for the speed at any time t, a warning should be posted.

Important warning: At no step in the reasoning was t_1 set equal to 2. When $t_1 = 2$, there is no physical sense in the notion "average speed from time 2 to time 2, a duration of 0 seconds." During such a time interval the rock moves a distance of 0 feet. The quotient

$$\frac{\text{Distance}}{\text{Time}}$$

in that case becomes the meaningless expression

$$\frac{0}{0}.$$

An argument similar to that used to find the speed at $t = 2$ holds for any

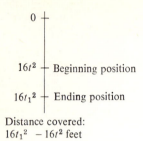

0 ⊢

$16t^2$ ⊢ Beginning position

$16t_1^2$ ⊢ Ending position

Distance covered:
$16t_1^2 - 16t^2$ feet

Duration: $t_1 - t$ seconds

time t during the descent of the rock. To find the speed at time t, consider a small interval from time t to time t_1. This interval has a duration of $t_1 - t$ seconds and is illustrated in the figure at the left. During this time the rock falls

$$16t_1^2 - 16t^2 \qquad \text{feet.}$$

Thus

$$\frac{16t_1^2 - 16t^2}{t_1 - t}$$

is a reasonable estimate of the speed at time t. This quotient equals

$$16\frac{t_1^2 - t^2}{t_1 - t} = 16\frac{(t_1 + t)(t_1 - t)}{t_1 - t}$$

$$= 16(t_1 + t).$$

When t_1 is near t, the quotient is near

$$16(t + t),$$

which is $\qquad\qquad\qquad\qquad 32t.$

Therefore the speed of the rock after t seconds is

$$32t \qquad \text{feet per second.}$$

When $t = 2$, the general formula for the speed of the rock, $32t$, gives the value $32(2) = 64$ feet per second, in accord with the earlier computation for that special case. The formula $32t$ shows that the speed is proportional to the time.

The preceding method, which finds the speed of the falling rock, may also be used to find the speed of an object whose motion is described by a formula other than $16t^2$. The basic approach is the same, though the specific algebraic details will be different. Exercises 8, 10, 12, and 13 illustrate this.

The procedure illustrated in this section is called *differentiation*. We say that differentiation of $16t^2$ yields $32t$, or that the *derivative* of $16t^2$ is $32t$.

Chapter 3 develops the ideas of this section in greater generality. It will be shown there that the speed of a moving object is just one of many applications of differentiation.

Exercises

1. The table below records how far a certain object travels up to the given time.

Time, seconds	Distance Traveled up to the Given Time, feet
1	4.2
1.01	4.3
1.1	5.7

(a) What is the average speed of the object from time $t = 1$ to 1.1?

(b) What is its average speed from time $t = 1$ to 1.01?

2. (a) How far does a rock that falls $16t^2$ feet in the first t seconds fall from time $t = 1$ second to $t = 1.01$ seconds?

(b) What is the average speed of the rock from time $t = 1$ second to time $t = 1.01$ seconds?

3. Find the speed of the rock of Exercise 2 at time $t = 1$ by considering the distance covered during the time interval from 1 second to t_1 seconds, where t_1 is larger than 1.

(a) How far does the rock fall during this time interval?

(b) How long is this time interval?

(c) What estimate of the speed after 1 second of fall is provided by this typical case?

(d) Letting t_1 approach 1, find the speed after 1 second of fall.

4. The formula developed in the text asserts that after t seconds, the speed of the rock is $32t$ feet per second. Using this formula, find its speed (in feet per second) when t is (a) 0, (b) 0.5, (c) 1, (d) 4.

5. A certain object moves t^2 feet in its first t seconds of motion. Find its speed at time t by considering short intervals of time from time t to t_1 and then letting t_1 approach t.

∎

6. To determine the speed of the rock at $t = 2$ seconds, the text considered its motion during a small interval of time that begins at $t = 2$. It could just as well have considered small intervals of time that end at $t = 2$, as this exercise shows.

(a) How far does the rock fall from time $t = 1.99$ seconds to time $t = 2$ seconds?

(b) How long is the time interval in (a)?

(c) Using (a) and (b), estimate the speed at time $t = 2$ seconds.

(d) Let t_1 be less than 2. Find the average speed of the rock from time t_1 to time 2 seconds.

(e) Letting t_1 approach 2, find the speed of the rock at time 2 seconds.

Exercise 8 concerns an object that travels t^3 feet in the first t seconds of its motion. It shows that the method of this section applies quite generally. Exercise 7 presents an algebraic identity that will be needed in Exercise 8.

7. By multiplying the two factors in parentheses, show that

$$c^3 - d^3 = (c^2 + cd + d^2)(c - d)$$

8. A certain object travels t^3 feet during its first t seconds of motion.

(a) How far does it travel up to time t_1?

(b) How far does it travel from time t to time t_1?

(c) How long is the time interval from time t to time t_1?

(d) Using (b), (c), and the identity from Exercise 7, show

that the average speed during the interval from time t to t_1 is

$$t_1^2 + t_1 t + t^2 \qquad \text{feet per second.}$$

(e) Use (d) to show that its speed at time t is $3t^2$ feet per second.

9. Use the formula in Exercise 8(e) to find the speed of the object in feet per second after (a) 0.5 second, (b) 1 second, (c) 2 seconds.

10. By multiplying the two factors in parentheses, show that

$$c^4 - d^4 = (c^3 + c^2 d + cd^2 + d^3)(c - d).$$

11. An object travels t^4 feet during its first t seconds of motion. Use the identity in Exercise 10 to show that its speed at time t is $4t^3$ feet per second.

12. An object travels t^5 feet during its first t seconds of motion. Find its speed at time t. (You will need to develop an algebraic identity for $c^5 - d^5$.)

∎ ∎

13. An object travels $\sqrt{t}$ feet in the first t seconds.

(a) What is its average speed from $t = 4$ seconds to $t = 4.1$ seconds? (Use the table of square roots in the appendix or a calculator.)

(b) Find its speed at time $t = 4$ seconds by the general method illustrated in the text. (The particular algebraic steps will, of course, be quite different.)

14. An exhausted snail, moving to the right along the x axis, at time t seconds, $t \geq 1$, reaches the point with coordinate $1 - 1/t$ feet.

(a) As t increases, what point is the snail approaching?

(b) Find its speed at time $t = 3$ seconds.

(c) Find its speed at any time $t \geq 1$.

15. Speed is the rate at which distance changes. The rate at which speed changes is called *acceleration*. For instance, an eight-cylinder car can accelerate from 50 to 80 miles per hour in 15 seconds. This is an *average acceleration* of 30 miles per hour in 15 seconds, or 2 miles per hour per second.

Consider an object that has a speed of t^2 feet per second at time t seconds.

(a) Find its average acceleration from time $t = 2$ to $t = 2.1$ seconds.

(b) Find its average acceleration from time $t = 2$ to t_1 seconds, where t_1 is larger than 2.

(c) Letting t_1 approach 2, find its acceleration at time $t = 2$.

1.2

How to find the distance from the varying speed

If an object travels at a constant speed of 1 foot per second, then in t seconds it travels t feet. It is simple to find the distance covered by an object moving at a constant speed: just use the formula

$$\text{Distance} = \text{speed} \cdot \text{time}.$$

But what if a rocket moves in such a way that after t seconds it is traveling at t^2 feet per second? How far does it travel in the first t seconds? This question is typical of the second main problem in calculus. Observe that this question is the opposite of that raised in Sec. 1.1. Here the varying speed of an object is given, and the distance it travels is sought. In Sec. 1.1 the distances were given, and the varying speeds were sought.

First, try to get a feel for the problem by estimating how far the rocket moves during its first 3 seconds of motion, that is, up to the time $t = 3$.

The rocket moves most slowly at the beginning of its flight, since its speed at time $t = 0$ is $0^2 = 0$ feet per second. It moves most quickly at the end, when $t = 3$, and its speed is $3^2 = 9$ feet per second. Since the entire time is 3 seconds, it follows that it moves at least

$$0 \cdot 3 = 0 \text{ feet}$$

and at most

$$9 \cdot 3 = 27 \text{ feet.}$$

This tells something, but not much, about the exact distance the rocket moves in the first 3 seconds.

Though the speed of the rocket continually increases, the speed changes very little during short intervals of time. So let us divide the time interval of 3 seconds into smaller intervals of time, say, into six intervals, each of which has a duration of $\frac{1}{2}$ second. Estimates of the distance covered during each of these short intervals, when added together, provide an estimate of the distance covered during the first 3 seconds.

This line segment represents the time during the first 3 seconds:

How far does the rocket move during the first half second, from time $t = 0$ to time $t = \frac{1}{2}$? Since its speed keeps changing, a precise answer cannot be given immediately. However, since we know that the speed is increasing, the greatest speed of the rocket during this time interval is $(\frac{1}{2})^2$ feet per second (since its speed is t^2 feet per second at any time t). The least speed is $0^2 = 0$ feet per second, at the beginning of this initial time interval. Therefore during the first half second it travels at least

$$0^2 \cdot \tfrac{1}{2} = 0 \text{ feet}$$

and at most
$$\left(\tfrac{1}{2}\right)^2 \cdot \tfrac{1}{2} = \tfrac{1}{8} \text{ foot.}$$

Similarly, during the second half second, from time $t = \tfrac{1}{2}$ to 1, the slowest it travels is $\left(\tfrac{1}{2}\right)^2$ feet per second, and the fastest is 1^2 feet per second. Therefore during the second half second it travels at least

$$\left(\tfrac{1}{2}\right)^2 \cdot \tfrac{1}{2} = \tfrac{1}{8} \text{ foot}$$

and at most
$$1^2 \cdot \tfrac{1}{2} = \left(\tfrac{2}{2}\right)^2 \cdot \tfrac{1}{2} = \tfrac{4}{8} \text{ foot.}$$

The remaining four half second intervals of time can be treated similarly. It follows that during the first 3 seconds the rocket travels at least

$$0 + \tfrac{1}{8} + \tfrac{4}{8} + \tfrac{9}{8} + \tfrac{16}{8} + \tfrac{25}{8} = \tfrac{55}{8} = 6.875 \text{ feet}$$

and at most

$$\tfrac{1}{8} + \tfrac{4}{8} + \tfrac{9}{8} + \tfrac{16}{8} + \tfrac{25}{8} + \tfrac{36}{8} = \tfrac{91}{8} = 11.375 \text{ feet.}$$

The table at left summarizes these computations. Thus, the distance covered during the first 3 seconds is between 6.875 and 11.375 feet. This certainly gives more information than did the first crude estimate, which was between 0 and 27 feet.

More accurate estimates can be obtained by dividing the 3-second time interval into much shorter ones: since the speed changes very little in very short time intervals, the estimates suggested by constant speed in each subinterval, namely,

$$\text{Distance} = \text{speed} \cdot \text{time}$$

becomes more accurate. For instance, when the 3-second interval is cut into 30 intervals, each of 0.1-second duration, a calculation similar to the preceding one shows that the object travels at least 8.565 feet and at most 9.455 feet.

The estimates discussed so far are recorded in this table:

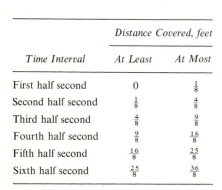

Time Interval	Distance Covered, feet	
	At Least	At Most
First half second	0	$\tfrac{1}{8}$
Second half second	$\tfrac{1}{8}$	$\tfrac{4}{8}$
Third half second	$\tfrac{4}{8}$	$\tfrac{9}{8}$
Fourth half second	$\tfrac{9}{8}$	$\tfrac{16}{8}$
Fifth half second	$\tfrac{16}{8}$	$\tfrac{25}{8}$
Sixth half second	$\tfrac{25}{8}$	$\tfrac{36}{8}$

Number of Time Intervals	Length of Each Time Interval, seconds	Lower Estimate, feet	Upper Estimate, feet
1	3	0.0	27.0
6	0.5	6.875	11.375
30	0.1	8.565	9.455

If the time interval of 3 seconds is divided into even smaller intervals, it seems reasonable to expect that the lower and upper estimates obtained will be even closer to the actual distance traveled.

It turns out that, as the time interval is divided into smaller and smaller intervals, both the lower and upper estimates get closer and closer to 9 feet. (Thus the exact distance traveled in 3 seconds is 9 feet.) However, the algebra that justifies this statement is a bit more complicated than that used in Sec. 1.1. For this reason, it is reserved until Chap. 7, which is concerned with the second main concept of calculus, the *definite integral.* In that chapter it will be shown that the definite integral of t^2 from 0 to 3 is 9.

Finding the total distance, knowing only the speed, is just one of many applications of the definite integral. (Exercises 9 and 10 present another illustration.)

Exercises

1. If during a 5-second time interval, a moving object never travels faster than 10 feet per second nor slower than 8 feet per second, what can be said about the total distance it travels during that time?

2. Find the lower and upper estimates for the distance that the rocket moves when you cut the 3-second time interval into nine equal intervals.

3. Find the upper and lower estimates of the distance the rocket in this section travels during the first 2 seconds,
   (a) cutting the time interval into two intervals of equal length;
   (b) cutting the time interval into four intervals of equal length;
   (c) cutting the time interval into six intervals of equal length. Incidentally, the exact distance covered is $\frac{8}{3}$ feet. This will be shown in Chap. 7.

The rocket discussed in this section had a speed of t^2 feet per second at time t. The next three exercises treat a simpler case, that of an object moving with a speed of t feet per second at time t.

4. A motorcycle travels t feet per second at time t.
   (a) Show that in the first half second, from $t = 0$ to $t = \frac{1}{2}$, it travels $\frac{1}{4}$ foot at most.
   (b) Show that during the second half second, from $t = \frac{1}{2}$ to $t = 1$, it travels at least $\frac{1}{4}$ but not more than $\frac{1}{2}$ foot.
   (c) From (a) and (b) show that during the first second the object travels at least $\frac{1}{4}$ foot but not more than $\frac{3}{4}$ foot.

5. This continues Exercise 4. Cut the one-second time interval into four equal intervals, and show that the motorcycle travels at least _____ foot and at most _____ foot. Show all work.

6. This continues Exercise 5. Cut the 1-second time interval into 10 equal intervals and show that the motorcycle travels at least _____ foot and at most _____ foot. Show all work. (In Chap. 7 it will be shown that the exact distance covered is $\frac{1}{2}$ foot.)

7. A bug travels at a speed of $\sqrt{t}$ feet per hour at time t hours, $t \geq 0$. Consider the distance it travels during the three hours from $t = 1$ to $t = 4$.
   (a) By dividing the time interval into three equal intervals, find a lower and an upper estimate of the distance the bug travels. (Use the square root table in the appendix or a calculator.)
   (b) By dividing the time interval into six equal intervals, find a lower and an upper estimate of the distance the bug travels.

8. A bug travels at a speed of $1/t$ feet per hour at time t hours, $t \geq 1$. Consider the distance it travels during the 2 hours from $t = 2$ to $t = 4$. Find a lower and an upper estimate of this distance by dividing the time interval into
   (a) two equal intervals;   (b) four equal intervals;
   (c) six equal intervals.
   (The table of reciprocals in the appendix or a calculator may be of use.)

9. The figure below, left shows the region below the curve $y = x^2$ from $x = 0$ to $x = 3$ and above the x axis. Use the five rectangles indicated in the figure to show that the area of the region is less than **11.88 square units**.

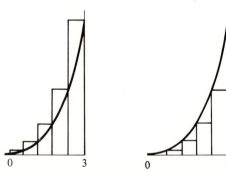

10. This exercise continues Exercise 9. Use the four rectangles shown in the figure above, right to show that the area of the region of Exercise 9 is more than **6.48 square units**.

11. (See Exercises 9 and 10.) Show that, if the rocket discussed in this section travels 9 feet, then the area of the region of Exercise 9 is 9 square units.

1.3

Summary

In Sec. 1.1 the speed of a falling rock at any instant is found. If the rock falls $16t^2$ feet in t seconds, its speed at time t is $32t$ feet per second. The method for finding this speed is somewhat indirect. The average speed over a short interval of time, from t to t_1, is calculated and then examined as t_1 approaches t. This method will be used in Chap. 3 to deal with concepts other than speed, such as slope of a curve, density of matter, and magnification. The method used to find the speed is an example of finding the derivative.

Section 1.2 examines a problem opposite that considered in Sec. 1.1. In this case a rocket moves with a speed of t^2 feet per second and the problem is to find how far it moves in a given interval of time, in particular, during the first 3 seconds. Though the precise answer must wait until Chap. 7, methods were presented for obtaining accurate estimates. The process of obtaining the exact answer is called integration.

The derivative, introduced in Sec. 1.1, is generally a measure of a rate of change; it will be defined and discussed in Chap. 3. The definite integral, on the other hand, expresses the total change if the varying rate of change is known. In Chap. 7 it will be used to compute areas and volumes of regions far more general than the triangles, rectangles, boxes, and spheres of high school geometry.

Guide quiz on Chap. 1

1. An object travels t^3 feet in t seconds.
   (a) How far does it travel during the time interval from 1 to 1.2 seconds?
   (b) What is its average speed during that time?
2. This table shows the speedometer readings of a car during a period when its speed was increasing.

Time, hours	0 (noon)	0.2	0.4	0.6	0.8	1 P.M.
Speed, miles per hour	30		38	44	49	52	54

Show that the car traveled (a) less than 47.4 miles, (b) more than 42.6 miles.
3. In the first t seconds an object travels $2t^3 + 5t$ feet. Find its speed at time $t = 2$ seconds. (Show all work.)
4. An object travels t^3 feet per second at time t. Find a lower and an upper estimate of the total distance it travels from time $t = 1$ to time $t = 2$, by breaking the time interval into
   (a) two equal intervals;
   (b) three equal intervals.
   (In Chap. 7 it will be shown that the exact distance traveled is $\frac{15}{4}$ feet.)

Algebra review quiz

This is a brief multiple-choice quiz on some essential algebra. Get help on any questions you miss.

1. The expression $2x(y + 3x)$ is equal to: (a) $2xy + 3x$, (b) $2xy + 6x$, (c) $2xy + 6x^2$, (d) $2xy + 9x^2$.
2. The expression $(a + b)^2$ is equal to: (a) $a^2 + b^2$, (b) $a^2 + b$, (c) $a^2 + ab + b^2$, (d) $a^2 + 2ab + b^2$.
3. The expression $(3 + x)(5 + y)$ is equal to: (a) $15 + xy$, (b) $15 + 3x + 5y + xy$, (c) $15 + 5x + 3y + xy$, (d) $5x + 3y$.
4. The solutions of the equation $x^3 - 3x = 0$ are: (a) 0, 3, -3; (b) 0, $\sqrt{3}$, $-\sqrt{3}$; (c) 1, 1, 27; (d) only 0.
5. The number $\sqrt{2} - 1$ is a solution of the equation: (a) $x^2 - 2 = 1$, (b) $x^2 - 2x - 1 = 0$, (c) $x^2 + 2x - 1 = 0$, (d) $x^2 + 2x + 1 = 0$.
6. If $1/x = 1/a + 1/b$, then x is equal to: (a) $a + b$, (b) $(a + b)/2$, (c) $ab/(a + b)$, (d) $(a + b)/(a^2 + b^2)$.
7. The simultaneous equations

$$y = 3x - 1$$
$$y = 5x - 2$$

have as solutions: (a) $x = \frac{1}{2}$, $y = \frac{3}{2}$; (b) $x = -\frac{3}{2}$, $y = -\frac{11}{2}$; (c) $x = \frac{1}{2}$, $y = \frac{1}{2}$; (d) $x = 5$, $y = 3$.
8. The number $-4[7 - 3(6 - 9)]$ equals: (a) 9, (b) -8, (c) 64, (d) -64.
9. The number $\sqrt{20} + \sqrt{45}$ simplifies to: (a) $\sqrt{65}$, (b) $13\sqrt{5}$, (c) $13 + \sqrt{5}$, (d) $5\sqrt{5}$.

10. If $(2x + 1)(x - 1) = 1$, then x is equal to: (a) $-\frac{1}{2}$ or 1, (b) $(1 \pm \sqrt{17})/4$, (c) $\frac{1}{2}$, (d) 0 and 2.

11. The expression $\sqrt{a + b}$ for positive a and b equals: (a) $\sqrt{a} + \sqrt{b}$, (b) $a^{1/2} + b^{1/2}$, (c) $\sqrt{a} + b$, (d) none of these.

12. The expression $\dfrac{x - 3}{x^2 - 9}$ simplifies to:

(a) $\dfrac{1}{x} + \dfrac{1}{3}$ (b) $\dfrac{1}{x + 3}$ (c) $\dfrac{1}{x} - \dfrac{1}{3}$ (d) none of these.

13. The expression $\dfrac{1}{a + b} + \dfrac{1}{a - b}$ equals:

(a) $\dfrac{2}{a^2 - b^2}$ (b) $\dfrac{2}{a^2 + b^2}$ (c) $\dfrac{2a}{a^2 - b^2}$ (d) $\dfrac{a}{a^2 - b^2}$.

14. The expression $\sqrt{x^2 + 9x^4}$ for positive x equals: (a) $x + 3x^2$, (b) $x\sqrt{1 + 9x^2}$ (c) $\sqrt{10x^3}$ (d) $x^2\sqrt{1 + 9x^2}$.

15. The value of $x^3 - 3x$ when $x = -1$ is: (a) 6, (b) 0, (c) 4, (d) 2.

2 FUNCTIONS, GRAPHS, AND THE SLOPE OF A LINE

In order to carry the ideas of Chap. 1 further, it is necessary to introduce certain concepts and terms. The most important of these are *function* and *graph*. The present chapter develops these two concepts, and in the third section explores a special case, the straight line. Then the slope of a line is defined. In Chap. 3, which defines the derivative, all these concepts will be needed.

2.1

Functions

The distance covered in the first t seconds by the falling rock of Sec. 1.1 is $16t^2$ feet. To each nonnegative number t, another number $16t^2$ is assigned; t is the time in seconds, and $16t^2$ is the distance in feet traveled during the first t seconds. The expression of distance in terms of time is an example of a *function*, which we now define.

DEFINITION *Function.* Let X and Y be sets. A *function* is a rule or method for assigning to each element x in X a unique element y in Y.

Recall that a set is any collection of objects or numbers, called *elements*. In calculus the elements are usually numbers, but, as in Examples 5 and 6 that follow, the elements of sets may be cities, persons, and so on.

The next three examples discuss some common and important functions.

EXAMPLE 1 *The doubling function.* Let X and Y both be the set of real numbers. Consider the function that assigns to each number x twice that number, $2x$. What does this function assign to 3? to -3? to 0?

SOLUTION Since the function always assigns to x the number $2x$, it assigns

$$2 \cdot 3 = 6 \text{ to } 3,$$
$$2 \cdot (-3) = -6 \text{ to } -3,$$

and

$$2 \cdot 0 = 0 \text{ to } 0. \quad \bullet$$

NOTATIONS   A function is usually denoted by the letter f. The element that f assigns to x is denoted

$$f(x),$$

which is read as "f of x."

For instance, if f is the doubling function, $f(x) = 2x$. Frequently the letters g and h are also used. Thus the doubling function may also be written as $h(x) = 2x$.

EXAMPLE 2   *The squaring function.* Let X and Y both be the set of real numbers. Let f be the function that assigns to each number x the square of x. Find $f(3), f(-3)$, and $f(0)$.

SOLUTION   The squaring function has the description

$$f(x) = x^2.$$

In particular, $$f(3) = 3^2 = 9,$$

$$f(-3) = (-3)^2 = 9,$$

and $$f(0) = 0^2 = 0. \bullet$$

DEFINITION $f(x)$ is called the *value* of f at x or the *image* of f at x or the *image* of x.

For instance, the value of the squaring function at 5 is 25.

EXAMPLE 3   *The cubing function.* Let f be the function whose value at x is x^3,

$$f(x) = x^3.$$

Find the value of f at $\frac{1}{2}$. What is the image of 2?

SOLUTION   The value of f at $\frac{1}{2}$ is $$f(\tfrac{1}{2}) = (\tfrac{1}{2})^3 = \tfrac{1}{8}.$$

The image of 2 is $$f(2) = 2^3 = 8. \bullet$$

In the case of the falling rock, the sets X and Y consist of the nonnegative numbers. The function that assigns to the time t the distance $16t^2$ that the rock falls in the first t seconds describes the motion of the rock.

EXAMPLE 4   If f is the cube root function, that is, $f(x) = \sqrt[3]{x}$, find $f(8), f(-1)$, and $f(0)$.

SOLUTION $$f(8) = \sqrt[3]{8} = 2,$$

$$f(-1) = \sqrt[3]{-1} = -1,$$

$$f(0) = \sqrt[3]{0} = 0. \bullet$$

In the following example, a function is given not by an algebraic formula but by a *table*.

EXAMPLE 5   A list that shows the population of each town in the United States is a function. In this case, X is the set of towns in the United States, and Y is the set of positive integers.   Such a list assigns to each town x its population $f(x)$.   For instance, the 1970 census shows

$$\begin{array}{ll} \text{New York} & 7{,}771{,}730 \\ \text{San Francisco} & 704{,}209. \end{array}$$

In the functional notation,

$$f(\text{New York}) = 7{,}771{,}730$$
$$f(\text{San Francisco}) = 704{,}209. \quad \bullet$$

A function may be described in words, as in the next example.

EXAMPLE 6   Assign to each time the population of the world at that time.   In the year 1977 the population was about 3.8 billion; thus $f(1977) = 3.8$ billion.   At the end of the last ice age the population was about 5 million; thus $f(-8000) = 5$ million.   The accompanying table lists several values of the function.

Year	-8000	0	1650	1810	1920	1966	1977
Population, billions	0.005	0.25	0.50	1	2	3	3.8

The accompanying chart shows gradual growth for thousands of years, the impact of medieval plagues, and then rapid growth in the past 2 centuries.

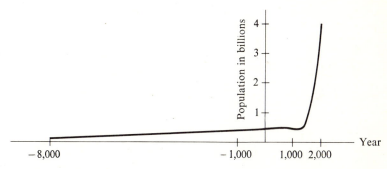

Though the values of the function are of course only integers, it is customary to treat the function as changing smoothly and the chart as a curve. $\bullet$

In Sec. 2.2, the charts (or graphs) of various functions given by formulas are discussed.

The next pair of examples will shed further light on the concept of a function.

EXAMPLE 7   *The absolute-value function.*   Define a function f as follows:

$$\text{Let } f(x) = \begin{cases} x & \text{if } x \text{ is positive} \\ -x & \text{if } x \text{ is negative} \\ 0 & \text{if } x \text{ is 0.} \end{cases}$$

This function is called the *absolute-value function.* Compute $f(3)$ and $f(-3)$.

SOLUTION
$$f(3) = 3 \quad \text{since 3 is positive,}$$
and
$$f(-3) = -(-3) = 3.$$

The absolute-value function assigns to each number its distance from 0. The absolute value of x is denoted $|x|$. Thus
$$|3| = 3 \qquad |-3| = 3. \ \bullet$$

EXAMPLE 8   Let $f(x) = \sqrt{x^2}$. Compute $f(3)$ and $f(-3)$.

SOLUTION
$$f(3) = \sqrt{3^2} = \sqrt{9} = 3$$
and
$$f(-3) = \sqrt{(-3)^2} = \sqrt{9} = 3. \ \bullet$$

Observe that the function in Example 8 is the same as the absolute-value function of Example 7, though the way in which it is given is different. Two functions that make the same assignments for every value of x will be considered to be the same function, even though the forms in which they are given may be different. As another example, the function that assigns to each number x the number $x^2 - 1$ is the same as the function that assigns to each number x the number $(x + 1)(x - 1)$.

The Domain of a Function   A function assigns to each element x in X an element $f(x)$ in Y. For our purposes, Y will usually be the set of real numbers and X some subset of the real numbers. The set X is called the *domain* of the function. Usually the domain will not be mentioned, especially when it is clear from the description of the function. For instance, if f is the function that assigns to a number its square root, the domain consists of all nonnegative numbers, since negative numbers do not have (real) square roots. If f is the "reciprocal" function, in which the value at x is given by the algebraic formula

$$f(x) = \frac{1}{x},$$

the domain consists of all numbers other than 0, since 0 does not have a reciprocal.

A function may be pictured as an "input-output" machine. When you insert an element x into the machine, an element which is called $f(x)$ falls out of the chute, as shown in the figure at the left.

When X and Y are sets of real numbers, we may also think of a function f as a complicated lens which projects x, on the slide, onto $f(x)$ on the screen.

When we insert x

The function f

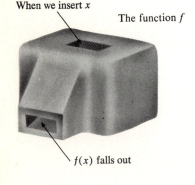

$f(x)$ falls out

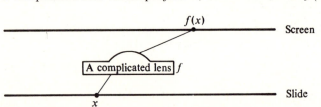

When dealing with any function f, it is important to keep in mind the difference between the function f and the number $f(x)$ that f assigns to an individual element x. However, in practice, we shall frequently describe a function f by the formula for $f(x)$. For instance, instead of speaking of "the squaring function," we may just say "the function x^2."

Exercises

1. Let f be the squaring function. Find $f(x)$ when x is (a) -4, (b) 4, (c) $\sqrt{2}$, (d) 0.
2. (a) Find $f(8)$ if f is the cube root function. (b) Find $g(8)$ if g is the absolute-value function. (c) Find $h(8)$ if h is the reciprocal function.
3. What is the domain of the function f if $f(x)$ is given by the formula (a) x^3? (b) $\sqrt[3]{x}$? (c) $\sqrt{x}$? (d) $\dfrac{3}{x}$? (e) $2x + 1$? (f) $\dfrac{1}{x(x+1)}$?

Exercises 4 to 8 present in the language of functions a quotient of the type that will be met in Sec. 3.1 and often in later chapters.

4. Let f be the squaring function $f(x) = x^2$. Compute decimally:
   (a) $f(3.01)$
   (b) $f(3)$
   (c) $f(3.01) - f(3)$
   (d) $\dfrac{f(3.01) - f(3)}{0.01}$
5. Let $f(t) = 5t^2$. Compute:
   (a) $f(1)$
   (b) $f(1.1)$
   (c) $f(1.1) - f(1)$
   (d) $\dfrac{f(1.1) - f(1)}{1.1 - 1}$.
6. Let $f(x) = 1/x$. Evaluate and express as a decimal:
   (a) $\dfrac{f(4) - f(3)}{4 - 3}$
   (b) $\dfrac{f(3.1) - f(3)}{3.1 - 3}$
   (c) $\dfrac{f(2.9) - f(3)}{2.9 - 3}$
7. Let $f(x) = \sqrt{x}$. Evaluate and express as a decimal:
   (a) $\dfrac{f(5) - f(4)}{5 - 4}$
   (b) $\dfrac{f(4.1) - f(4)}{4.1 - 4}$
   (c) $\dfrac{f(3.9) - f(4)}{3.9 - 4}$.
8. Evaluate $[f(x + h) - f(x)]/h$ and simplify using algebra if
   (a) f is the squaring function;
   (b) f is the cubing function;
   (c) f is the doubling function;
   (d) f is the reciprocal function.

9. Let f be the reciprocal function $1/x$. Compute as a decimal:
   (a) $f(5)$
   (b) $f(-2)$
   (c) $f(0.1)$
   (d) $f(2.5)$
   (e) $f(2)$
   (f) $\dfrac{f(2.5) - f(2)}{2.5 - 2}$.
10. Compute decimally $[f(2.5) - f(2)]/(2.5 - 2)$ if (a) f is the squaring function; (b) f is the cubing function; (c) f is the reciprocal function $1/x$.
11. Let f be the squaring function. Express as simply as possible:
    (a) $f(4) - f(3)$
    (b) $f(x) - f(3)$
    (c) $f(3.1) - f(3)$
    (d) $f(3 + 1) - f(2)$.
12. Let $f(x) = x^2$. Evaluate: (a) $f(2 + 0.5)$, (b) $f(2 - 0.1)$.
13. Let $f(x) = 1/x$. Evaluate: (a) $f(4)$, (b) $f(1)$, (c) $f(4 + 1)$, (d) $f(4 + 1) - f(4)$.
14. Let f be the reciprocal function; $f(x) = 1/x$. Express as simply as possible:
    (a) $f(3) - f(2)$
    (b) $f(2.25) - f(2)$
    (c) $f(t) - f(2)$
    (d) $f(x) - f(2)$
    (e) $f(2 + 1/2) - f(2)$
    (f) $f(2 - 2/5) - f(2)$
    (g) $f(2 + h) - f(2)$.
15. Let f be the square root function $f(x) = \sqrt{x}$ for $x \geq 0$. This function is tabulated in any handbook of mathematical tables and in the back of this text. Use the table to compute $\dfrac{f(x_1) - f(x)}{x_1 - x}$ when (a) $x_1 = 4.3$ and $x = 4$; (b) $x_1 = 4.2$ and $x = 4$; (c) $x_1 = 4.1$ and $x = 4$.

■

16. This table lists the number of aircraft owned by the scheduled air carriers in the United States on the last day of the indicated year.

Year	1950	1955	1960	1965	1968	1969	1970
Number of aircraft	1220	1409	1867	1806	2317	2363	2390

Make a chart of this function.

17. Let $f(t)$ be the number of new book titles published in the United States in the year t. This table lists a few values of the function

Year	1950	1955	1960	1965	1970	1972
New Titles, thousands	11	13	15	29	36	38

   (a) Plot the points for the 6 years cited.
   (b) Make a chart of the function.
18. What is the meaning of $[f(t_1) - f(t)]/(t_1 - t)$, where $t_1 > t$, and
   (a) $f(t)$ is the distance an object moves in the first t seconds?
   (b) $f(t)$ is the speed of an object at time t?
19. Let $f(x) = x^2$. Using algebra, simplify the quotient
$$\frac{f(t_1) - f(t)}{t_1 - t}.$$
20. Let $f(x) = 5x$. Show that
$$\frac{f(t_1) - f(t)}{t_1 - t} = 5.$$

   ■ ■

21. Let f be the squaring function.
   (a) Show that $f(1) = f(-1)$.
   (b) Compute $f(f(3))$. [First compute $f(3)$.]

(c) Show that $f(f(x)) = x^4$ for all x.
22. Let f be the function given by the formula $f(x) = x/(x-1)$.
   (a) What is the domain of f?
   (b) What is $f(2)$?
   (c) What is $f(-1)$?
   (d) What is $f(f(2))$? [First compute $f(2)$].
   (e) What is $f(f(3))$?
   (f) Show that $f(f(x)) = x$ if $x \neq 1$.
23. Let f be the function given by the formula $f(x) = 1/(1-x)$.
   (a) What is the domain of f?
   (b) What is $f(2)$? $f(f(2))$? $f(f(f(2)))$?
   (c) Show that $f(f(f(x))) = x$ (if $x \neq 1$ or $x \neq 0$).
24. Which of the following functions have the property that, for all choices of the numbers a and b in their domain, $f(a+b) = f(a) + f(b)$?
   (a) $f(x) = 3x$ (b) $f(x) = 3x + 2$
   (c) $f(x) = x^2$ (d) $f(x) = -x$
   (e) $f(x) = 1/x$ (f) $f(x) = |x|$
   Note: It is a common mistake to assume that *all* functions have the stated property.
25. Which of the following functions have the property that, for all choices of the numbers a and b in their domain, $f(ab) = f(a)f(b)$?
   (a) $f(x) = x^2$ (b) $f(x) = 1/x$
   (c) $f(x) = 3x$ (d) $f(x) = \sqrt{x}$
   (e) $f(x) = x + 4$ (f) $f(x) = |x|$

2.2

The table and the graph of a function

No matter how a function is described, a table can be formed for it, as in Examples 5 and 6 (the gazetteer and population illustrations) of the preceding section. In calculus, a function is usually described by a formula. Substitution of a few numbers in the formula provides a short table showing specific values of the function.

EXAMPLE 1 Let $f(x) = x^2$. Compute f at a few choices of x; make a short table for f.

SOLUTION

$$f(0) = 0$$
$$f(1) = 1 \qquad f(-1) = 1$$
$$f(2) = 4 \qquad f(-2) = 4$$
$$f(\tfrac{1}{2}) = \tfrac{1}{4} \qquad f(-\tfrac{1}{2}) = \tfrac{1}{4}$$

These seven cases are tabulated in this table:

x	0	$\frac{1}{2}$	1	$-\frac{1}{2}$	-1	-2	2
$f(x)$	0	$\frac{1}{4}$	1	$\frac{1}{4}$	1	4	4

Of course, the arithmetic is easier when the values of x are integers, but any number can be used for x. ●

In the table of Example 1 seven values are used. The more values of x you choose, the longer the table will be. In theory, the table has an entry for each value of x, but in practice only a few values of x are chosen.

The table in Example 1 can be used to draw a picture or chart of the squaring function. For instance, the information that

$$f(2) = 4$$

shows up at the end of the table above, in Example 1. This fact can be recorded by drawing the point whose x coordinate is 2 and whose y coordinate is 4, written briefly $(2, 4)$ as in the accompanying graph.

Of course, the squaring function is defined for all x, not just for the seven values that happened to be chosen. But the seven chosen suggest the general behavior of the function. A freehand sketch fills in the missing points and suggests the shape of the remainder of the curve when the absolute value of x is large. The curve is the familiar parabola, which is shaped like the profile of a headlight, as the figure at the left shows. The technical term for such a picture or chart is *graph*.

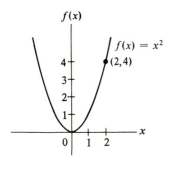

DEFINITION   *The graph of a function.*   Let f be a function such that the "inputs" x and the values or "outputs" $f(x)$ are numbers. The set of points of the form

$$(x, f(x))$$

is called *the graph of the function.*

When a function is graphed, it is customary to use the horizontal axis for the input x and the vertical axis for the output $f(x)$. To draw the graph of a function, begin by making a short table.

EXAMPLE 2   Sketch the graph of the absolute-value function,

$$f(x) = |x|.$$

SOLUTION   First make a short table.

x	0	1	-1	2	-2
$f(x) = |x|$	0	1	1	2	2

Then plot the corresponding points, one for each value, and sketch the graph, as in the accompanying figure. Note that the graph has a sharp corner at $(0, 0)$ and consists of two straight rays meeting at the point $(0, 0)$. ●

Sometimes the graph of a function may be given first, and then particular values read from it. The next example illustrates this.

EXAMPLE 3 An automatic recording device at a weather station produces a record of the temperature as a function of time during a 24-hour period.

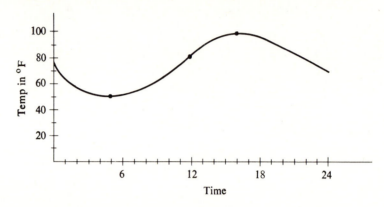

Describe the behavior of the temperature during the day.

SOLUTION Let $f(t) =$ temperature at time t. The lowest temperature occurs at 5 A.M., which corresponds to the lowest point on the graph. Reading off the graph (placing a ruler parallel to the temperature axis), we find that

$$f(5) = 50°.$$

At noon the temperature was 80°,

$$f(12) = 80°.$$

At 4 P.M. ($t = 16$), the temperature was a maximum,

$$f(16) = 100°.$$

The graph shows a cold morning followed by a hot afternoon, which starts cooling off at 4 P.M. ●

The next example shows how a graph can tell a good deal about a function at a glance.

EXAMPLE 4 Graph the reciprocal function given by the formula

$$f(x) = \frac{1}{x} \qquad (x \neq 0).$$

SOLUTION When $x = 0$, the function is not defined. For this reason, when making the table, it is wise to choose some values of x near 0. It is also wise to include some x's with large absolute value to get an idea of what happens to $f(x)$ when $|x|$ is large.

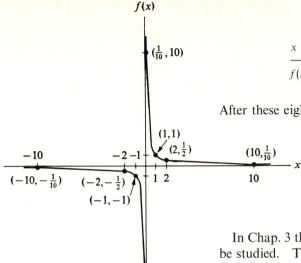

x	$\frac{1}{10}$	10	1	2	$-\frac{1}{10}$	-10	-1	-2
$f(x)$	10	$\frac{1}{10}$	1	$\frac{1}{2}$	-10	$-\frac{1}{10}$	-1	$-\frac{1}{2}$

After these eight points are plotted, it is easy to sketch the graph.

The graph comes in two identical pieces. Far to the right and to the left the graph gets closer and closer to the x axis without ever touching it. Also, when x approaches 0, the graph approaches the y axis. This graph is called a *hyperbola*. ●

In Chap. 3 the appearance of a graph in the vicinity of any of its points will be studied. The next example illustrates some of the computations that will be carried out then.

EXAMPLE 5   Sketch the graph of $f(x) = x^2$ in the vicinity of the point $(2, 4)$.

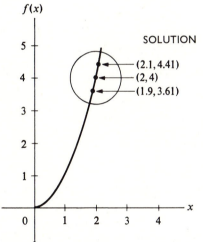

SOLUTION   Recall from Example 1 that $(2, 4)$ is in the graph since $f(2) = 4$. Select values of x near 2, and make a short table.

x	2	2.1	1.9
$f(x) = x^2$	4	$(2.1)^2 = 4.41$	$(1.9)^2 = 3.61$

Then plot the points $(2, 2^2) = (2, 4)$, $(2.1, (2.1)^2) = (2.1, 4.41)$, and $(1.9, (1.9)^2) = (1.9, 3.61)$. This section of the graph is almost a straight line, as can be seen in the accompanying figure. Of course, it is hard to be very accurate when the points plotted are so near each other. A larger scale which magnifies the graph would be more useful. ●

So far, all the curves considered have been the graphs of functions. However, not every curve is the graph of a function. For instance, the curve left, below is not the graph of a function. Whenever a function is graphed, a line parallel to the y axis meets the graph in at most one point. (This is not necessarily true of a line parallel to the x axis. Why not?) But in the diagram we are considering, a line through $(2, 0)$ and parallel to the y axis meets the curve in three points. [These points are $(2, 1)$, $(2, 3)$, and $(2, 5)$.] By the definition of a function, a function assigns to the number 2 only one number, not three numbers.

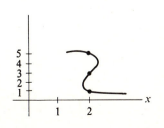

Exercises

1. Let $f(x) = 4x^2$.
(a) Fill in the table

x	3	2	1	0	-1	-2	-3
$f(x)$							

(b) Plot the seven points in (a) and graph f.

2. This exercise concerns the doubling function $f(x) = 2x$.
(a) Fill in this table.

x	2	1	0	-1	-2
$f(x)$					

(b) Plot the five points in (a) and graph f.

3. This exercise concerns the cubing function $f(x) = x^3$.
(a) Fill in this table.

x	0	$\frac{1}{2}$	1	2	$-\frac{1}{2}$	-1	-2
$f(x) = x^3$							

(b) Plot the seven points in (a) and graph f.

4. Let $f(x) = 2/x$.
(a) What is the domain of f?
(b) Make a table for f including values of x near 0 and large values of x.
(c) Graph f.

5. Graph the function f given by the formula

$$f(x) = -x.$$

6. A certain function f is described as follows: Start with x. Add 1. Then square. The result is $f(x)$.
(a) Compute $f(2)$.   (b) Fill in this table.

x	-2	-1	0	1	2
$f(x)$					

(c) Graph f.

7. A complicated lens projects from a linear slide to a linear screen as shown in the diagram, which indicates the paths of four of the light rays.

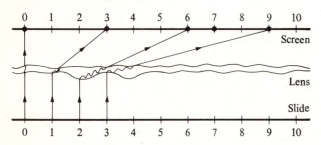

Let $f(x)$ be the image on the screen of x on the slide.
(a) What are $f(0), f(1), f(2), f(3)$?
(b) Fill in this table.

x	0	1	2	3
$f(x)$				

(c) Plot the four points in (b).

8. Graph $f(x) = x^2 + 1$.

9. Graph $f(x) = 4x + 1$.

10. Graph $f(x) = x^2 - x$.

11. Graph $f(x) = 1/(1 + x)$.

12. Graph $f(x) = (x - 1)^2$.

13. (a) For which x is $x^3 - 2x^2 = 0$?
(b) Graph $f(x) = x^3 - 2x^2$.

14. Graph $f(x) = x^4$, paying particular attention to values of x near 0.

15. Complete this table for the squaring function.

x	1	1.1	0.9
$f(x)$			

Then graph f in the vicinity of $(1, 1)$.

16. Graph $f(x) = x(x + 1)(x - 1)$.
(a) For which values of x is $f(x) = 0$?
(b) Where does the graph cross the x axis?
(c) Where does the graph cross the y axis?

17. Graph $y = (x - 3)^2 + 4$.

18. Let $f(x) = x + 1/x$.
(a) What is the domain of f?   (b) Graph f.
Note: It will be shown in Chap. 6 that for positive x the lowest point on the graph occurs when x is 1.

19. Let $f(x) = ax^2 + bx + c$, where a, b, and c are constants, a is positive, and $b^2 - 4ac$ is positive.
(a) Show that

$$ax^2 + bx + c = a\left(x + \frac{b}{2a}\right)^2 + \frac{4ac - b^2}{4a}.$$

(b) Show that the smallest value of $f(x)$ for all possible choices of x is $(4ac - b^2)/4a$.
(c) Show that the lowest point on the graph of f has an x coordinate that is the average of the x coordinates of the two points where the graph crosses the x axis.

20. Let $f(x) = ax^2 + bx + c$, where a, b, and c are constants and a is positive. How many points on the graph of f lie on the x axis if $b^2 - 4ac$ is (a) positive, (b) negative, (c) zero?

21. Graph $y = 2^x$, an important function in the study of population growth.

2.3

The slope of a line

This section is devoted to the graphs of functions given by a specially simple type of formula, namely

$$f(x) = ax + b,$$

where a and b are fixed numbers.

The first three examples illustrate the type of graph obtained from such a function, and also the influence of the numbers a and b on the graph.

We shall first look at the behavior of the graph as the value of a changes. In order to concentrate on a, first consider the case $b = 0$. The formula is then

$$f(x) = ax.$$

EXAMPLE 1   Graph the function given by the formula

$$f(x) = 2x.$$

(Here, $a = 2$ and $b = 0$.)

SOLUTION   To begin, make a table of values.

x	-1	0	1	2
$2x$	-2	0	2	4

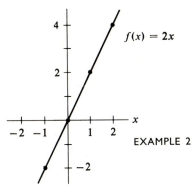

Then plot the four points in the table and sketch the graph, which is shown in the accompanying figure. The graph is a straight line, which passes through the origin $(0, 0)$ and is fairly steep. ●

The next example differs from Example 1 only in the replacement of a by 3 instead of by 2 in the formula for the function. Let us see what effect this change has on the graph.

EXAMPLE 2   Graph the function f given by the formula

$$f(x) = 3x.$$

SOLUTION   Begin again with a table of specific values.

x	-1	0	1	2
$3x$	-3	0	3	6

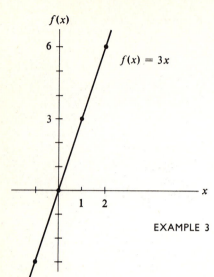

$f(x) = 3x$

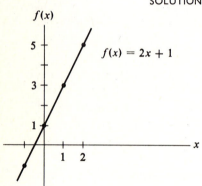

$f(x) = 2x + 1$

The graph then appears as in the accompanying figure. The graph is a straight line which passes through $(0, 0)$, but this line is steeper than the line in Example 1. ●

As Examples 1 and 2 suggest, for any fixed number a, the graph of

$$f(x) = ax$$

is a line which passes through $(0, 0)$. The next example shows what happens when the constant term b in the formula

$$f(x) = ax + b$$

is not 0.

EXAMPLE 3   Graph the function f given by the formula

$$f(x) = 2x + 1.$$

SOLUTION   As usual, a table of values is computed.

x	-1	0	1	2
$2x + 1$	-1	1	3	5

Then the graph is drawn as at the left. The straight line for this graph has exactly the same steepness as the line in Example 1 but, instead of passing through $(0, 0)$, this line meets the y axis one unit higher. The graph of

$$f(x) = 2x + 1$$

is simply the graph of $f(x) = 2x$

moved up by one unit. ●

As these three examples suggest, the graph of

$$f(x) = ax + b$$

is always a straight line. The number a determines how steep the line is. The number b determines where the line meets the y axis and is called the y *intercept*. Thus the graph of

$$f(x) = ax + b$$

is the graph of $f(x) = ax$

raised or lowered by the amount $|b|$; if b is negative, the graph is lowered and, if b is positive, the graph is raised. For instance, the graph of $f(x) = 2x - 5$ is the graph of $f(x) = 2x$ lowered by five units.

Let us examine these two ideas in detail. For convenience, denote $f(x)$ by y. The equation being considered is then

$$y = ax + b.$$

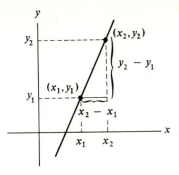

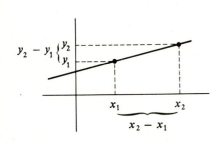

When $x = 0$,
$$y = a \cdot 0 + b$$
$$= 0 + b$$
$$= b.$$

As already mentioned, the number a determines how steep the graph of $y = ax + b$ is. Before this influence can be described in greater detail, it is necessary to introduce the notion of *slope*.

Consider two points (x_1, y_1) and (x_2, y_2) on a very steep line, shown in the accompanying figure. To the horizontal change from x_1 to x_2, that is, $x_2 - x_1$, corresponds a much greater vertical change $y_2 - y_1$. In short, the quotient

$$\frac{y_2 - y_1}{x_2 - x_1}$$

is large.

Next, consider two points (x_1, y_1) and (x_2, y_2) on a line that is not steep, a line that is almost horizontal, as in the drawing at the left. In this case, to the horizontal change $x_2 - x_1$ corresponds a much smaller vertical change $y_2 - y_1$. In short, the quotient

$$\frac{y_2 - y_1}{x_2 - x_1}$$

is small.

These two observations suggest that the quotient

$$\frac{y_2 - y_1}{x_2 - x_1}$$

is a reasonable measure of how steep a line is. But before it is safe to use it as a measure of the steepness of a line, it is important to know that the value of the quotient is the same for any two points (x_1, y_1) and (x_2, y_2) chosen on the line. The following theorem assures us that this is the case.

THEOREM   Consider the line
$$y = ax + b.$$

For any choice of distinct points (x_1, y_1) and (x_2, y_2) on the line, the quotient

$$\frac{y_2 - y_1}{x_2 - x_1}$$

equals a, the coefficient of x in the formula

$$y = ax + b.$$

PROOF   Choose x_1 and x_2. Then $y_1 = ax_1 + b$ and $y_2 = ax_2 + b$.

Thus
$$\frac{y_2 - y_1}{x_2 - x_1} = \frac{(ax_2 + b) - (ax_1 + b)}{x_2 - x_1}$$
$$= \frac{ax_2 - ax_1 + b - b}{x_2 - x_1}$$

$$= \frac{a(x_2 - x_1)}{x_2 - x_1}$$

$$= a.$$

This proves the theorem. ●

This theorem justifies the following definition.

DEFINITION   *Slope of a line.*   Consider a line $y = ax + b$ in the coordinate plane. Let (x_1, y_1) and (x_2, y_2) be two points on the line.   The quotient

$$\frac{y_2 - y_1}{x_2 - x_1}$$

is called the *slope of the line*.

As the proof of the theorem showed, the slope of the line $y = ax + b$ is simply a, the coefficient of x.   The number b has no influence on the slope, as Examples 1, 2, and 3 suggested.   For emphasis this observation is stated as a corollary.

COROLLARY   The slope of the line $y = ax + b$ is a. ●

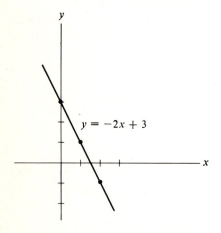

The slope of the line $y = -2x + 3$

is -2, the coefficient of x.   In this case, the slope is negative.   This means that, for any two points (x_1, y_1) and (x_2, y_2) on the line, the quotient

$$\frac{y_2 - y_1}{x_2 - x_1}$$

is negative.   In other words, if $x_2 - x_1$ is positive (x_2 is greater than x_1), then $y_2 - y_1$ is negative (y_2 is less than y_1).   This means that as you go *to the right* on this line, you move *downward*.   This fact will be evident when the line is graphed.   To graph a line, only two points are required.   For convenience choose $x = 0$ and $x = 1$.   But as a check, choose another value of x also, say $x = 2$.   The accompanying graph shows the result.

x	0	1	2
$-2x + 3$	3	1	-1

A line parallel to the y axis does not have a slope, since for any two points (x_1, y_1) and (x_2, y_2) on such a line $x_2 = x_1$ and the denominator of the fraction

$$\frac{y_2 - y_1}{x_2 - x_1}$$

is 0.   The fraction is meaningless.

A line parallel to the x axis, a horizontal line, does have a slope. For any two points (x_1, y_1) and (x_2, y_2) on such a line, $y_2 = y_1$; thus the fraction

$$\frac{y_2 - y_1}{x_2 - x_1}$$

is

$$\frac{0}{x_2 - x_1},$$

which is 0 (the fraction has meaning since the denominator is not 0). Thus a horizontal line has slope 0. This agrees with the corollary which asserts that the line $y = ax + b$ has slope a. A horizontal line with y intercept b has the equation

$$y = b,$$

or

$$y = 0x + b.$$

EXAMPLE 4 Let f be the function that assigns to every number x the number 3. Graph f.

SOLUTION A table of values for f looks like this.

x	-2	0	1	3
$f(x)$	3	3	3	3

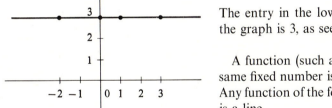

The entry in the lower row is always 3. The y coordinate of any point on the graph is 3, as seen in the drawing. The graph is a horizontal line. ●

A function (such as the one in Example 4) that assigns to all numbers the same fixed number is called a *constant function*. Its graph is a horizontal line. Any function of the form $f(x) = ax + b$ is called a *linear function*, since its graph is a line.

Exercises

1. Find the slope of the line through
   (a) $(2, 3)$ and $(4, 7)$ (b) $(1, 4)$ and $(2, 2)$
   (c) $(4, 6)$ and $(11, 6)$ (d) $(1, 5)$ and $(-3, 8)$.
2. Find the slope of the line through
   (a) $(2, 4)$ and $(3, 9)$
   (b) $(2, 4)$ and $(2.1, 2.1^2)$
   (c) $(2, 4)$ and $(1.9, 1.9^2)$
   (d) $(2, 4)$ and $(2 + h, (2 + h)^2)$
   (e) $(2, 4)$ and $(x_1, x_1{}^2)$.
3. Find the slope and y intercept of the lines given by the following equations:
   (a) $y = 3x + 2$ (b) $y = 3x - 2$
   (c) $y = 4x$ (d) $y = 5$
   (e) $y = -4x$ (f) $y = x + 2$.

4. Graph the lines given by the following equations:
   (a) $y = 3x$ (b) $y = 3x + 1$
   (c) $y = 3x - 1$.
5. Graph the lines given by the following equations:
   (a) $y = \dfrac{-x}{2}$ (b) $y = \dfrac{-x}{2} + 1$
   (c) $y = \dfrac{-x}{2} - 1$.
6. Describe the slope of each of the lines in the accompanying drawing as positive, negative, or 0.

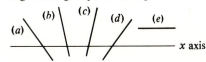

7. Write in the form $y = ax + b$ the line (*a*) whose slope is 5 and whose y intercept is -1; (*b*) whose slope is 1 and which passes through the point $(0, 2)$; (*c*) that is horizontal and passes through the point $(0, 4)$.

8. Write in the form $y = ax + b$ the equation of the line through $(1, 2)$ and $(0, 5)$.

9. Write in the form $y = ax + b$ the equation of the line through $(2, 4)$ and $(1, 5)$.

■

10. A line of slope 3 passes through the point $(1, 2)$. Write the equation of the line in the form $y = ax + b$.

11. Are the points $(1, -2)$, $(4, 5)$, and $(8, 13)$ on a line? Explain.

■ ■

12. Find the point where the lines $y = 2x + 1$ and $y = 5x - 3$ cross.

The next exercise provides a test for parallel lines.

13. Two distinct lines in the plane are said to be parallel if they do not intersect. (Any line is considered to be parallel to itself.)
    (*a*) Prove that, if a_1 is not equal to a_2, then the lines $y = a_1 x + b_1$ and $y = a_2 x + b_2$ are not parallel.

(*b*) Prove that, if a_1 equals a_2, then the lines $y = a_1 x + b_1$ and $y = a_2 x + b_2$ are parallel.

In short, this exercise shows that two lines are parallel if and only if they have the same slope.

14. The line $y = ax + b$ passes through the points $(1, 4)$ and $(2, -3)$. Find a and b.

15. Graph the lines $y = 3x$ and $y = -\frac{1}{3}x$. (Exercises 16 and 17 show that they make a right angle.)

16. In Appendix B it is shown that the distance from the point (x_1, y_1) to the point (x_2, y_2) is $\sqrt{(x_2 - x_1)^2 + (y_2 - y_1)^2}$. Recall from geometry that, in a triangle of sides a, b, and c, the sides of lengths a and b form a right angle if and only if $a^2 + b^2 = c^2$. Use the preceding information to show that the line through $(0, 0)$ and (x_1, y_1) is perpendicular to the line through $(0, 0)$ and (x_2, y_2) if and only if $x_1 x_2 + y_1 y_2 = 0$.

Exercise 17, together with Exercise 13, shows that two lines (neither of which is parallel to the axes) are perpendicular if and only if the product of their slopes is -1.

17. (See Exercise 16.) Let a_1 and a_2 be nonzero constants.
    (*a*) Show that the point $(1, a_1)$ lies on the line $y = a_1 x$ and that the point $(1, a_2)$ lies on the line $y = a_2 x$.
    (*b*) Show that the lines $y = a_1 x$ and $y = a_2 x$ are perpendicular if and only if the product of their slopes is -1, that is, $a_1 a_2 = -1$.

2.4

Summary

This chapter introduced first the idea of a function and then the table and graph of a function. It went on to treat in detail a special type of function described by the formula

$$f(x) = ax + b.$$

A function assigns to each element x in some set, called the *domain*, an element $f(x)$ in another set. (The two sets may be the same.) Through all the early chapters of this book both x and $f(x)$ will be numbers. The graph of such a function consists of all points $(x, f(x))$ in the coordinate plane. Thus $f(x)$ is the y coordinate, and x is the x coordinate of a point on the graph.

This is a list of some of the important functions discussed in the chapter.

Formula for f	Shape of Graph
$f(x) = x^2$	U-shape

Formula for f	Shape of Graph
$f(x) = |x|$	V-shape with vertex at $(0, 0)$
$f(x) = 1/x$	Hyperbola consisting of two parts, each approaching the axes
$f(x) = ax + b$	Line with slope a and y intercept b
$f(x) = b$	Horizontal line at height b

The slope of a line through the points (x_1, y_1) and (x_2, y_2) is defined as the quotient

$$\frac{y_2 - y_1}{x_2 - x_1}.$$

This quotient is independent of the points on the line. If it is positive, a pencil goes up when tracing the line to the right. If the quotient is negative, the pencil goes down when moving to the right.

KEY FACTS

The line through (x_1, y_1) and (x_2, y_2) has a slope given by the quotient

$$\frac{y_2 - y_1}{x_2 - x_1}.$$

A horizontal line has slope 0. The slope of a vertical line is not defined. The line $y = ax + b$ has slope a and y intercept b.

VOCABULARY AND SYMBOLS

function f	graph
absolute value $\lvert x \rvert$	domain
squaring function	table
square root function	slope
cubing function	y intercept
reciprocal function	constant function
value or image $f(x)$	linear function

Guide quiz on Chap. 2

1. Let $f(x) = \sqrt{x + 1}$.
   (a) What is the domain of f?
   (b) Make a table and graph of f.
2. Explain why the slope of the line $y = ax + b$ is a.
3. What are the slopes of the lines whose equations are the following?
   (a) $y = -\frac{1}{3}x + 2$;   (b) $y = -x$;
   (c) $y = -4$.
4. (a) Graph the function $f(x) = 3x^2 - 9x + 7$.
   (b) Does the graph meet the x axis?

The arithmetic in questions 5 and 6 will be important in the next chapter.

5. Let $f(x) = x^3 + 2x$. Compute:

   (a) $\dfrac{f(2) - f(1)}{2 - 1}$ (b) $\dfrac{f(1.1) - f(1)}{1.1 - 1}$.

6. Using algebra, express the quotient
   $$\frac{f(1 + h) - f(1)}{h}\qquad \text{where } h \text{ is not } 0$$
   as simply as possible if
   (a) $f(x) = x^2$ (b) $f(x) = 8x - 3$
   (c) $f(x) = 5/x$ (d) $f(x) = x^3 + 2x$.
7. Graph f if
   (a) $f(x) = \dfrac{x}{2} - \dfrac{1}{4}$ (b) $f(x) = x^4 - x^2$
   (c) $f(x) = \dfrac{1}{x^2 - 1}$.
8. If $(1, 4)$ is on the graph of a function f,
   (a) can $(5, 1)$ be on the graph?
   (b) can $(1, 5)$ be on the graph?
   (c) can $(5, 4)$ be on the graph?

Review exercises for Chap. 2

1. A function f is defined as follows: Start with x; add 2; square; take the reciprocal.
   (a) Write a formula for $f(x)$.
   (b) Graph f.
   (c) What is the domain of f?
2. Using algebra, express the quotient
   $$\frac{f(x_2) - f(x_1)}{x_2 - x_1}$$
   as simply as possible if (a) $f(x) = x^2 + 5$; (b) $f(x) = -1/x^2$; (c) $f(x) = 3x + 2$.
3. (a) Graph the functions $f(x) = 2x - 1/x$ and $f(x) = x$.
   (b) At what points do their graphs intersect?
4. Let $f(x) = \lvert x \rvert$. Compute:

   (a) $\dfrac{f(1) - f(0)}{1 - 0}$ (b) $\dfrac{f(0.1) - f(0)}{0.1 - 0}$

   (c) $\dfrac{f(-0.1) - f(0)}{-0.1 - 0}$

5. Let $f(x) = x^2$. Show that, for all numbers a and b,
   $$f(a + b) = f(a) + f(b) + 2ab.$$
6. Which of the following functions have the property that $f(-a) = f(a)$ for all numbers a in the domain?
   (a) $f(x) = 3x + 2$ (b) $f(x) = x^2$
   (c) $f(x) = x^3 + x$ (d) $f(x) = x^2 + x$
   (e) $f(x) = 1/x$ (f) $f(x) = x^4 - 5\lvert x \rvert$
7. Express in the form $y = ax + b$ the equation of the line

(*a*) that has slope $-\frac{1}{2}$ and passes through $(0, \frac{3}{2})$,

(*b*) that passes through the points $(0, -2)$ and $(2, 5)$.

8. Is the line through $(-2, 1)$ and $(3, 8)$ parallel to the line through $(3, -2)$ and $(10, 8)$?

9. A line has the equation $y = ax + b$. What can be said about a if the line

(*a*) is nearly vertical (almost parallel to the y axis)?

(*b*) is nearly horizontal (almost parallel to the x axis)?

(*c*) slopes downward as you move to the right?

(*d*) slopes upward as you move to the right?

10. Compute $f(2 + 0.5) - f(2)$ if (*a*) $f(x) = x^2 - 3x + 4$; (*b*) $f(x) = x/(x + 1)$.

11. Find $f(f(2))$ if (*a*) $f(x) = 1/x$; (*b*) $f(x) = -x$; (*c*) $f(x) = x^3$; (*d*) $f(x) = 3x - 2$.

12. A function is given by the graph in the right-hand column. Find (*a*) $f(0)$, (*b*) $f(1)$, (*c*) $f(2)$, (*d*) $f(3)$.

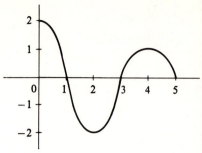

13. This is part of a table of a certain function f.

x	1	1.5	2	2.5
$f(x)$	2	3	4	6

Find (*a*) $f(1)$, (*b*) $f(f(1))$, (*c*) $f(1) + f(1.5)$.

3

THE DERIVATIVE

This chapter introduces one of the most important concepts of calculus, the derivative. The approach uses geometric and physical illustrations, but a few exercises show that the derivative has far more varied applications.

3.1

Four problems with one theme

This section discusses four problems which at first glance may seem unrelated. The first one, concerning speed, already appeared in Chap. 1. The other three involve slope, magnification, and density. A little arithmetic will quickly show that they are all just different versions of one mathematical idea, expressed in the problem of the falling rock of Sec. 1.1, which is repeated here as Problem 1.

PROBLEM I   A rock falls $16t^2$ feet in t seconds. What is its speed after 2 seconds?

The next problem involves the *tangent line* to a curve (see the figure in the margin). For the present, by the *tangent line* to a curve at a point P on the curve shall be meant the line through P that has the same direction as the curve at P. Later in the chapter this will be made more precise.

PROBLEM 2   What is the slope of the tangent line to the graph of $y = x^2$ at the point $P = (2, 4)$, as shown in the second diagram in the margin?

The following definition will be useful in stating the next problem, and more generally throughout the text.

DEFINITION   *Interval.* Let a and b be two numbers, with $a < b$. The interval $[a, b]$ consists of all real numbers x such that

$$a \leq x \leq b.$$

The interval $[a, b]$ extends from a to b on the number line.

The interval $[a, b]$

The third problem concerns magnification, a concept that occurs in everyday life. For instance, photographs can be blown up or reduced in size. The

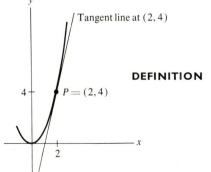

Tangent line at P

P

Tangent line at $(2, 4)$

$P = (2, 4)$

study of similar figures in geometry is related to the effects of such enlarging or reducing.

PROBLEM 3

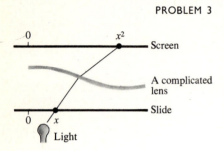

0 x^2 — Screen

A complicated lens

— Slide

0 x

Light

A light, two lines (a slide and a screen), and a complicated lens are placed as in the accompanying diagram. This arrangement projects the point on the bottom line whose coordinate is x to the point on the top line whose coordinate is x^2. For example, 2 is projected onto 4, and 3 onto 9. The projection of the interval $[2, 3]$ is $[4, 9]$, which is five times as long. The projection of the interval $[0, \frac{1}{3}]$ is $[0, \frac{1}{9}]$, which is only one-third as long. For large x the lens magnifies to a great extent; for x near 0 the lens markedly reduces. What is the magnification at $x = 2$?

The next problem is concerned with density, which is a measure of the heaviness of a material. Water has a density of 1 gram per cubic centimeter, while the density of lead is 11.3 grams per cubic centimeter. Air has a density of only 0.0013 gram per cubic centimeter. The density of an object may vary from point to point. For instance, the density of matter near the center of the earth is much greater than that near the surface. In fact, the *average* density of the earth is 5.5 grams per cubic centimeter, more than five times that of water.

The idea of density provides a concrete analog of several mathematical ideas and will be referred to frequently in later chapters. The next problem concerns a string of varying density. This density will be considered in terms of grams per linear centimeter, rather than grams per cubic centimeter. The matter is imagined as a continuous distribution, not composed of isolated molecules.

PROBLEM 4

The mass of the left-hand x centimeters of a nonhomogeneous string 10 centimeters long is x^2 grams. For instance, the left half has a mass of 25 grams, while the whole string has a mass of 100 grams. Clearly the right half is denser than the left half. What is the density, in grams per centimeter, of the material at $x = 2$?

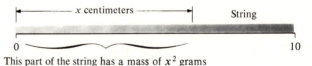

|← ------ x centimeters ------ →| String

0   10

This part of the string has a mass of x^2 grams

Now let us solve the four problems.

SOLUTION OF PROBLEM I (SPEED)

This problem was solved in Sec. 1.1. The speed is 64 feet per second. This answer was found by showing that the quotient

$$\frac{16t^2 - 16(2)^2}{t - 2}$$

approaches 64 when t approaches 2. ●

SOLUTION OF PROBLEM 2 (SLOPE)

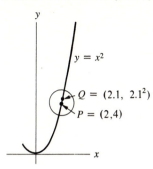

The line through P and Q

$(2.1, 2.1^2)$

$(2,4)$

Magnified view of
circled portion

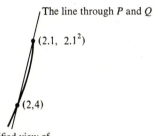

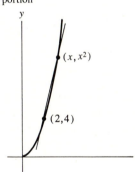

One approach consists of carefully drawing the curve $y = x^2$ and trying to draw the line tangent to the curve at $(2, 4)$. Though this is a reasonable method, its precision is limited (as some of the exercises will show). In this particular example, where the tangent line is so nearly vertical, a slight error in guessing the angle that the tangent line makes with the x axis can cause a large error in estimating the slope.

We shall choose a different approach, one which is perfectly accurate. Moreover this approach will reveal the similarity of this problem and Problem 1. As a start, compute the slope of a line that approximates the tangent line at $P = (2, 4)$. To do this, take a point Q near P on the curve $y = x^2$, say the point $Q = (2.1, 2.1^2)$, and compute the slope of the line passing through P and Q, as visualized in the accompanying figures. The slope of the line through P and Q is

$$\frac{2.1^2 - 2^2}{2.1 - 2},$$

which is

$$\frac{4.41 - 4}{0.1} = \frac{0.41}{0.1}$$

$$= 4.1.$$

Thus an estimate of the slope of the tangent line is 4.1. Note that in making this estimate there was no need to draw the curve.

To obtain a better estimate, we could repeat the process, using the line through $P = (2, 4)$ and $Q = (2.01, 2.01^2)$. Rather than do this, it is simpler to consider a *typical* point Q. That is, consider the line through $P = (2, 4)$ and $Q = (x, x^2)$ when x is near 2 (see the accompanying graph).

This line has slope

$$\frac{x^2 - 2^2}{x - 2},$$

which equals (by the algebra used in Sec. 1.1)

$$x + 2.$$

As x gets closer to 2, $x + 2$ approaches 4.

Thus the tangent line at $(2, 4)$ has slope 4.

Even though Problems 1 and 2 seem unrelated at first, their solutions turn out to be practically identical: The speed in Problem 1 is approximated by the quotient

$$\frac{16t^2 - 16(2)^2}{t - 2} = 16\frac{t^2 - 2^2}{t - 2},$$

and the slope in Problem 2 is approximated by the quotient

$$\frac{x^2 - 2^2}{x - 2}.$$

The only difference between the solutions is that the first quotient had a 16 and a t instead of an x. ●

SOLUTION OF PROBLEM 3
(MAGNIFICATION)

The lens projects the point having the coordinate 2 onto the point having the coordinate 2^2. More concisely, the image of 2 is $2^2 = 4$. The image of 3 is $3^2 = 9$; the image of 5 is $5^2 = 25$; and so on. Let us join some sample points to their images by straight lines:

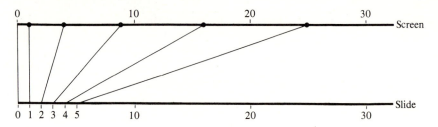

This diagram shows that the interval [2, 3] on the slide is magnified to become the interval [4, 9] on the screen, a fivefold magnification. Similarly, [3, 4] on the slide has as its image on the screen [9, 16], a sevenfold magnification. The magnifying power of the lens increases from left to right.

To estimate the magnification at 2 on the slide, examine the projection of a small interval in the vicinity of 2. Out of habit, let us see what the image of [2, 2.1] is on the screen. Since the image of 2 is 2^2 and the image of 2.1 is 2.1^2, the image of the interval [2, 2.1] of length 0.1 is the interval $[2^2, 2.1^2]$ of length

$$2.1^2 - 2^2 = 0.41.$$

The magnifying factor over the interval [2, 2.1] is

$$\frac{0.41}{0.1} = 4.1.$$

This number, 4.1, is an estimate of the magnification at $x = 2$.

You have probably guessed the next step. Rather than go on and consider the magnification of another specific interval, such as the interval [2, 2.01], we go directly to a typical interval with 2 as its left end.

The image of the interval [2, x], where x is greater than 2, is the interval $[2^2, x^2]$. Since [2, x] has length $x - 2$ and $[2^2, x^2]$ has length $x^2 - 2^2$, the magnification of the interval [2, x] is

$$\frac{x^2 - 2^2}{x - 2} = x + 2.$$

As already observed, when x approaches 2, this quotient approaches 4. Thus the magnification at 2 is 4. ●

SOLUTION OF PROBLEM 4 (DENSITY)

To estimate the density of the string 2 centimeters from its left end, examine the mass of the material in the interval [2, 2.1].

The material in the interval [2, 2.1] has a mass of $2.1^2 - 2^2$ gram, which equals

$$4.41 - 4 = 0.41 \text{ gram.}$$

Rather than make another estimate, consider the density in the typical small interval [2, x].   The mass in this interval is

$$x^2 - 2^2 \qquad \text{grams.}$$

The interval has length $x - 2$ centimeters.

Thus the density of matter in this interval is

$$\frac{x^2 - 2^2}{x - 2} = x + 2 \qquad \text{grams per centimeter.}$$

As x approaches 2, this quotient approaches 4, and we say that the density 2 centimeters from the left end of the string is 4 grams per centimeter. ●

From a mathematical point of view, the problems of finding the speed of the rock, the slope of the tangent line, the magnification of the lens, and the density along the string are the same.   Each leads to the same type of quotient as an estimate.   In each case the behavior of this quotient is studied as x (or t) approaches 2.

The underlying mathematical theme is explored in the next section, which introduces the derivative.

Exercises

1. This exercise is intended to emphasize the limitation of graphs in finding the slope of a tangent line.
(a) Draw the curve $y = x^2$ as carefully as you can.
(b) Draw as carefully as you can the tangent line at (4, 16).
(c) Using a ruler or the scale on your graph, estimate the slope of the line you drew in (b).
(d) Find the slope of the line through (4, 16) and the nearby point $(4.01, (4.01)^2)$.
(e) Find the slope of the line through (4, 16) and the nearby point $(3.99, (3.99)^2)$.
(f) How does your result in (c) compare with those in (d) and (e)?

2. What estimate do you obtain for
(a) the speed of the rock in Problem 1 when you use the time interval from 2 to 2.001 seconds?
(b) the slope of the tangent in Problem 2 when you use, as the second point on the curve, the nearby point $(2.001, (2.001)^2)$?
(c) the magnification in Problem 3 when you compute

the magnification of the interval [2, 2.001]?
(d) the density in Problem 4 when you compute the mass of material in the interval [2, 2.001] and divide by the length of that interval?

3. (a) Draw the curve $y = x^2$ as carefully as you can.
(b) Draw by eye the tangent line at (−1, 1).
(c) What is the slope of the line you drew in (b)?
(d) Examining the appropriate quotients, show that the tangent line at (−1, 1) has slope −2.

4. Find the density of the string in Problem 4 at a typical point x centimeters from the left end.   To do this, consider the mass in a short interval $[x, x_1]$ where $x_1 > x$.

5. Find the magnification of the lens in Problem 3 at the typical point x, by considering the magnification of the short interval $[x, x_1]$, where $x_1 > x$.

6. Find the slope of the tangent line to the curve $y = x^2$ of Problem 2 at the typical point $P = (x, x^2)$.   To do this, consider the slope of the line through P and the nearby point $Q = (x_1, (x_1)^2)$.

7. Estimate:

(a) the slope of the tangent line to $y = x^2$ at $P = (3, 9)$ using the line through P and the nearby point $Q = (3.1, (3.1)^2)$;

(b) the magnification of the lens in Problem 3 at 3 by computing the magnification of the interval $[3, 3.1]$;

(c) the density of the string in Problem 4, 3 centimeters from the left end, by considering the interval $[3, 3.01]$.

8. By what factor does the lens in Problem 3 magnify the interval (a) $[1, 1.1]$? (b) $[1, 1.01]$? (c) $[1, 1.001]$? (d) What is the magnification at 1?

9. An object travels $2t^2 + t$ feet in t seconds.

(a) Find its average speed during the interval of time $[1, t]$, where t is greater than 1.

(b) Letting t get closer and closer to 1, find the speed at time 1.

10. (a) Graph the curve $y = 2x^2 + x$.

(b) By eye, draw the tangent line to the curve at the point $(1, 3)$. Using a ruler, estimate the slope of this tangent line.

(c) Sketch the line that passes through the point $(1, 3)$ and the point $(x, 2x^2 + x)$.

(d) Find the slope of the line in (c).

(e) Letting x get closer and closer to 1, find the slope of the tangent line at $(1, 3)$. How close was your estimate in (b)?

11. (a) Using the method of the nearby point, find the slope of the curve $y = 3x^2 - x + 2$ at the point $P = (x, 3x^2 - x + 2)$.

(b) At which point on the curve is the tangent line horizontal?

12. (a) Sketch the curve $y = x + 1/x$.

(b) Estimate by eye the slope of the tangent line to the curve at the point $(1, 2)$.

(c) Using the method of the nearby point, find the slope of the tangent line at the point $(1, 2)$. How close was your estimate in (b)?

■

13. (a) The algebraic identity $c^3 - d^3 = (c^2 + cd + d^2)(c - d)$ will be needed in this exercise. Check it by multiplying out the right-hand side.

(b) Sketch the curve $y = x^3$ and estimate by eye the slope of the tangent line at the point $(0, 0)$.

(c) Using the method of the nearby point, find the slope of the tangent line to the curve at the point $(0, 0)$.

14. Using the method of the nearby point, find the slope of the tangent line to the curve $y = x^3$ at the point $(1, 1)$. The identity in Exercise 13(a) will be of aid.

15. A lens projects x on the slide onto x^3 on the screen. Find the magnification at $x = \frac{1}{2}$. The identity in Exercise 13(a) will be of aid.

The next two exercises show that the idea common to the four problems in this section also appears in biology and economics.

16. A certain bacterial culture has a mass of t^2 grams after t minutes of growth.

(a) How much does it grow during the time interval $[2, 2.01]$?

(b) What is its rate of growth during the time interval $[2, 2.01]$?

(c) What is its rate of growth when $t = 2$?

17. A thriving business has a profit of t^2 million dollars in its first t years. Thus, from time $t = 3$ to time $t = 3.5$ (the first half of its fourth year) it has a profit of $(3.5)^2 - 3^2$ million dollars. This gives an annual rate of

$$\frac{(3.5)^2 - 3^2}{0.5} = 6.5 \text{ million dollars per year.}$$

(a) What is its annual rate of profit during the time interval $[3, 3.1]$?

(b) What is its annual rate of profit during the time interval $[3, 3.01]$?

(c) What is its annual rate of profit after 3 years?

■ ■

18. Does the tangent line to the curve $y = x^2$ at the point $(1, 1)$ pass through the point $(6, 12)$?

19. An astronaut is traveling from left to right along the curve $y = x^2$. When she shuts off the motor, she will fly off along the line tangent to the curve at the point where she is at that moment. At what point should she shut off the motor in order to reach the point (a) $(4, 9)$? (b) $(4, -9)$?

20. (a) Sketch the curve $y = x^3 - x^2$.

(b) Using the method of the nearby point, find the slope of the tangent line to the curve at the typical point $(x, x^3 - x^2)$. The identity in Exercise 13(a) will be of use.

(c) Find all points on the curve where the tangent line is horizontal.

(d) Find all points where the tangent line has slope 1.

21. Answer the same questions as in Exercise 20 for the curve $y = x^3 - x$.

22. See Exercises 19 and 21. Where can the astronaut who is traveling from left to right along the curve $y = x^3 - x$ shut off the motor and pass through the point $(2, 2)$?

3.2

The derivative of a polynomial

The solution of the speed problem in Secs. 1.1 and 3.1 leads to the study of the quotient

$$\frac{16t^2 - 16(2)^2}{t - 2} = 16 \frac{t^2 - 2^2}{t - 2}$$

as t approaches 2. The slope, magnification, and density problems all lead to similar considerations of the quotient

$$\frac{x^2 - 2^2}{x - 2}.$$

These quotients arose from the particular formulas $16t^2$ and x^2 that had been given. But it is quite possible to study these four problems when those formulas are replaced with different ones. Example 1 illustrates this.

EXAMPLE I   Find the slope of the tangent line to the graph of $y = x^3$ at $P = (1, 1^3)$.

SOLUTION   The method that worked for the slope in Problem 2 of the preceding section can be used here. It is not necessary to draw the curve.

Choose a nearby point on the curve $Q = (x, x^3)$. The line through $P = (1, 1^3)$ and Q has the slope

$$\frac{x^3 - 1^3}{x - 1}.$$

To simplify the numerator $x^3 - 1^3$, use the algebraic identity

$$c^3 - d^3 = (c^2 + cd + d^2)(c - d).$$

When $\qquad\qquad c = x \qquad$ and $\qquad d = 1,$

this identity reduces to

$$x^3 - 1^3 = (x^2 + x \cdot 1 + 1^2)(x - 1).$$

Thus the slope of the line through P and Q is

$$\frac{(x^2 + x + 1)(x - 1)}{x - 1},$$

or $\qquad\qquad\qquad\qquad\qquad x^2 + x + 1.$

When x approaches 1, the slope of the line through P and Q therefore approaches

$$1 + 1 + 1 = 3.$$

The tangent line at $(1, 1^3)$ has slope 3.

This answers the question but, as a check, graph $y = x^3$ to see if the result is reasonable. First, make a short table for $y = x^3$:

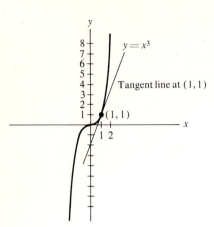

x	-2	-1	0	1	2
x^3	-8	-1	0	1	8

and graph the curve $y = x^3$ through the five points listed, as in the adjacent figure. The tangent line at $(1, 1)$ is steep and has a positive slope; hence a slope of 3 is reasonable. ●

The same algebra as that carried out in Example 1 is done when answering each of these questions:

An object moves t^3 feet in the first t seconds. What is its speed after 1 second?

A lens projects x on the slide onto x^3 on the screen. What is its magnification at $x = 1$?

The left-hand x centimeters of a string have a mass of x^3 grams. What is the density of the string at $x = 1$?

Identical algebraic steps obtain the answers to these three questions. The speed is 3 feet per second; the magnification is threefold; the density is 3 grams per centimeter.

Of course there is no need to restrict our attention to $x = 1$. In the next example, we calculate the slope of the tangent line at point $P = (x, x^3)$ on the curve $y = x^3$.

EXAMPLE 2   Find the slope of the tangent line to the curve $y = x^3$ at $P = (x, x^3)$.

SOLUTION   In this case denote the point near P by

$$Q = (x_1, x_1^3).$$

The slope of the line through P and Q is

$$\frac{x_1^3 - x^3}{x_1 - x}.$$

The identity $c^3 - d^3 = (c^2 + cd + d^2)(c - d)$

shows that $x_1^3 - x^3 = (x_1^2 + x_1 x + x^2)(x_1 - x).$

Hence the slope of the line through P and Q is

$$\frac{(x_1^2 + x_1 x + x^2)(x_1 - x)}{x_1 - x},$$

which equals $x_1^2 + x_1 x + x^2.$

As x_1 approaches x, this expression approaches

$$x^2 + xx + x^2 = 3x^2.$$

Hence the tangent line at (x, x^3) has slope $3x^2$. ●

Note that the formula $3x^2$ obtained in Example 2 agrees with the special case worked out in Example 1 $(x = 1)$. Observe also that $3x^2$ is 0 when $x = 0$.

This means that the tangent line at $(0, 0)$ to the curve is horizontal. It may seem strange that a tangent line can *cross the curve*, as this tangent line does. However, the basic property of a tangent line is that it indicates the direction of a curve at a point. In high school geometry, where only tangent lines to circles are considered, the tangent line never crosses the curve.

A Notation for "Approaches"   It is customary to abbreviate the phrase "x_1 approaches x" by the symbol

$$x_1 \to x.$$

Example 2 shows that, as $x_1 \to x$,

$$\frac{x_1^3 - x^3}{x_1 - x} \to 3x^2.$$

This is written more briefly as:

$$\lim_{x_1 \to x} \frac{x_1^3 - x^3}{x_1 - x} = 3x^2,$$

which is read as "the limit of

$$\frac{x_1^3 - x^3}{x_1 - x}$$

as x_1 approaches x is $3x^2$."

The notations

$$x_1 \to x \qquad \text{and} \qquad \lim_{x_1 \to x}$$

are used often in calculus. The next chapter explores the limit concept in greater detail.

The Derivative, Informally   Starting with the function x^3, we arrived at another function, $3x^2$. The function $3x^2$ is called the *derivative* of the function x^3. A more formal definition of the derivative is given in the next section. Computing the derivative requires finding out what happens to a certain quotient as $x_1 \to x$. When doing this algebra, we need not consider whether the derivative will be interpreted as speed, slope, magnification, or density.

With only a little more algebra, we shall now calculate the derivative of x^n for any positive integer n. The identity

This Identity Will Appear in Chaps. 4 and 7.

$$c^n - d^n = (c^{n-1} + c^{n-2}d + c^{n-3}d^2 + \cdots + d^{n-1})(c - d)$$

will be needed. To check it, multiply the two terms in parentheses on the right side, noticing that every product cancels except c^n and $-d^n$. The proof of the next theorem will use it in the form

$$x_1^n - x^n = (x_1^{n-1} + x_1^{n-2}x + x_1^{n-3}x^2 + \cdots + x^{n-1})(x_1 - x).$$

THEOREM   The derivative of x^n is nx^{n-1} for any positive integer n.

PROOF   Proceed as in Example 2, replacing 3 by n. It is necessary to find what happens to the quotient

$$\frac{x_1{}^n - x^n}{x_1 - x}$$

when $x_1 \to x$.   The identity

$$x_1{}^n - x^n = (x_1{}^{n-1} + x_1{}^{n-2}x + x_1{}^{n-3}x^2 + \cdots + x^{n-1})(x_1 - x)$$

shows that the quotient equals

$$\frac{(x_1{}^{n-1} + x_1{}^{n-2}x + x_1{}^{n-3}x^2 + \cdots + x^{n-1})(x_1 - x)}{x_1 - x},$$

which equals $x_1{}^{n-1} + x_1{}^{n-2}x + x_1{}^{n-3}x^2 + \cdots + x^{n-1}.$

As $x_1 \to x$, this approaches

$$x^{n-1} + x^{n-2}x + x^{n-3}x^2 + \cdots + x^{n-1}$$

$$= x^{n-1} + x^{n-1} + x^{n-1} + \cdots + x^{n-1}$$

$$= nx^{n-1} \qquad \text{since there are } n \text{ summands.}$$

This proves the theorem. ●

Direct application of this theorem yields, for instance:

The derivative of x^5 is $5x^{5-1} = 5x^4$.
The derivative of x^4 is $4x^{4-1} = 4x^3$.
The derivative of x^3 is $3x^{3-1} = 3x^2$.
The derivative of x^2 is $2x^{2-1} = 2x$.
The derivative of x^1 is $1x^0 = 1$ (in agreement with the fact that the line given by the formula $y = x$ has slope 1).

Here are some questions that can now be answered with the formula for the derivative of x^n.

Question: What is the slope of the tangent line to the curve $y = x^4$ at the point $(2, 2^4)$?   *Answer:* Since the derivative of x^4 is $4x^3$, the slope is $4(2)^3 = 32$.

Question: What is the speed at time $t = 3$ of an object that travels t^5 feet in the first t seconds?   *Answer:* Since the derivative of t^5 is $5t^4$, the speed when $t = 3$ is $5(3)^4 = 405$ feet per second.

Similar questions can now be answered for problems involving magnification and density.

In the next example the derivative of the function f given by the formula $f(x) = 5x^3$ is found.   It shows what effect the coefficient 5 has on the derivative.

EXAMPLE 3   Find the derivative of the function $5x^3$.

SOLUTION   In view of the definition of the derivative, it is necessary to determine

$$\lim_{x_1 \to x} \frac{5x_1{}^3 - 5x^3}{x_1 - x}.$$

As shown in Example 2,

$$x_1{}^3 - x^3 = (x_1{}^2 + x_1 x + x^2)(x_1 - x).$$

Hence
$$\frac{5x_1{}^3 - 5x^3}{x_1 - x} = \frac{5(x_1{}^3 - x^3)}{x_1 - x}$$

$$= \frac{5(x_1{}^2 + x_1 x + x^2)(x_1 - x)}{x_1 - x}$$

$$= 5(x_1{}^2 + x_1 x + x^2).$$

Thus
$$\lim_{x_1 \to x} \frac{5x_1{}^3 - 5x^3}{x_1 - x} = \lim_{x_1 \to x} 5(x_1{}^2 + x_1 x + x^2)$$

$$= 5(x^2 + xx + x^2)$$

$$= 5(3x^2)$$

$$= 15x^2.$$

The derivative of $5x^3$ is simply five times the derivative of x^3. ●

The argument used in Example 3 generalizes to show that the derivative of cx^n is cnx^{n-1} for any constant c and any positive integer n. For instance, the derivative of $-5x^2$ is $-5(2x) = -10x$.

The next example involves the derivative of a function given by a polynomial.

EXAMPLE 4   Find the derivative of the function f given by the formula

$$f(x) = x^3 - 5x^2 + 6x + 2.$$

SOLUTION   In this case, the quotient to be considered is

$$\frac{(x_1{}^3 - 5x_1{}^2 + 6x_1 + 2) - (x^3 - 5x^2 + 6x + 2)}{x_1 - x}$$

$$= \frac{x_1{}^3 - x^3 - 5(x_1{}^2 - x^2) + 6(x_1 - x) + 2 - 2}{x_1 - x}$$

$$= \frac{x_1{}^3 - x^3}{x_1 - x} - 5\frac{x_1{}^2 - x^2}{x_1 - x} + 6\frac{x_1 - x}{x_1 - x}$$

$$= \frac{x_1{}^3 - x^3}{x_1 - x} - 5\frac{x_1{}^2 - x^2}{x_1 - x} + 6.$$

As $x_1 \to x$, this last expression approaches (as a little algebra shows)

$$3x^2 - 5(2x) + 6 = 3x^2 - 10x + 6.$$

Thus the derivative of $x^3 - 5x^2 + 6x + 2$ is $3x^2 - 10x + 6$. ●

The Derivative of a Polynomial   Note that the constant term 2 of the polynomial in Example 4 contributes nothing to the derivative. The term $6x$ contributes the 6. The term x^3 contributes $3x^2$, which is the derivative of x^3. The term $-5x^2$ contributes its derivative $-10x$. This suggests that the derivative of any polynomial is simply the sum of the derivatives of its individual terms. This will be proved in general in Sec. 5.5. Thus, the derivative of

$$6x^4 + 2x^3 + 9x^2 - 7x + 4$$

is

$$24x^3 + 6x^2 + 18x - 7.$$

Exercises

1. What is the derivative of
   (a) x^2 when $x = 3$?   (b) x^4 when $x = -1$?
   (c) x^5 when $x = 2$?   (d) x^6 when $x = 0$?
   (e) $16x^3$ when $x = \frac{1}{2}$?   (f) $-\frac{1}{2}x^7$ when $x = 2$?
2. Using the definition of the derivative and carrying out all the necessary algebra, find the derivative of $6x^2 + 3x + 2$.

In Exercises 3 to 11 use the formula for the derivative obtained in this section.

3. Compute the derivative of
   (a) $5x^3 - x^2 + 2x + 5$ when $x = 1$;
   (b) $x^7 - x^6 + x + 3$ when $x = 0$;
   (c) $4x^{10} - 10x^4 + 8x$ when $x = -1$.
4. Find the derivative of $6x^5 - x^4 + 3x^2 + 2x + 5$ when $x = 1$.
5. Find the derivative of $2x^8 - 6x + 2$ when $x = -1$.
6. Find the derivative of $\frac{1}{2}x^7 - 4x$ when $x = 0$.
7. A particle travels $2t^4 + t^3$ feet in t seconds. Find its speed when $t = 2$.
8. Find the slope of the tangent line to the curve $y = -x^5 + 2x^3 + x$ at $(1, 2)$.
9. The left x centimeters of a string have a mass of $2x^5 + x$ grams. What is its density 3 centimeters from the left end?
10. A lens projects the point x on the slide to $x^6 + 3x^2 + x$ on the screen. Find its magnification at $x = 1$.
11. A certain object drops $16t^2 + 20t$ feet in t seconds.
    (a) Find its speed at time t.
    (b) What is its speed at the beginning of its motion, when $t = 0$?

■

12. (a) Show in detail that
    $$c^5 - d^5 = (c^4 + c^3d + c^2d^2 + cd^3 + d^4)(c - d).$$
    (b) Use (a) to show that the derivative of x^5 is $5x^4$.
13. (a) Graph the curve $y = x^5 - 5x$.
    (b) Where does the graph cross the x axis?
    (c) At what points is the tangent line horizontal?
14. (a) Sketch the graph of $y = 3x^2 + 5x + 6$.
    (b) By inspection of your graph, estimate the x coordinate of the point where the tangent line is horizontal.
    (c) Using the derivative, solve (b) precisely.
15. Without sketching the graph of $y = x^4$, draw the line that is tangent to it at the point $(\frac{1}{2}, \frac{1}{16})$.

■ ■

16. As $x \to 1$, what number does $(x^5 - 1)/(x^3 - 1)$ approach? Explain your answer.
17. Find the derivative of (a) $f(x) = (2x + 1)^2$; (b) $f(x) = (2x + 1)(3x - 1)$. *Warning:* Be careful.
18. (a) Find the derivative of $f(x) = x(x - 1)^2$.
    (b) At which points on the graph is the tangent line horizontal?
    (c) Graph $y = x(x - 1)^2$.
    (d) What do you think is the largest value of $x(x - 1)^2$ when x is in $[0, 1]$?
19. Let a, b, c, d, and e be constants, with a not 0.
    (a) At most how many points of the x axis can lie on the graph of $y = ax^4 + bx^3 + cx^2 + dx + e$?
    (b) At most how many horizontal tangent lines can there be on the graph?
20. Give two examples of a polynomial whose derivative is $x^3 - x^2$.

3.3

The derivative of a function

Section 3.2 obtained the derivative of functions given by the formula $f(x) = x^n$ and discussed the derivatives of polynomial functions. Derivatives can also be calculated for other functions, such as those given by the formulas

$$f(x) = \frac{1}{x}$$

and
$$f(x) = \sqrt{x}.$$

Let us define the derivative in general terms, and then find the derivatives of these two particular functions. The definition depends on the concept of a function and involves the type of computations carried out in the preceding section.

DEFINITION *The derivative of a function.* Let f be a function whose domain is part of or all the real numbers, and whose values are real numbers. Let x be some number in the domain of f. If

$$\lim_{x_1 \to x} \frac{f(x_1) - f(x)}{x_1 - x}$$

exists, this limit is called the derivative of f at x.

Let us see what the definition leads to for some specific functions other than those already studied.

EXAMPLE I Find the derivative of the reciprocal function f whose formula is

$$f(x) = \frac{1}{x}.$$

SOLUTION Since the domain of f excludes 0, consider only $x \neq 0$. The quotient

$$\frac{f(x_1) - f(x)}{x_1 - x}$$

becomes
$$\frac{1/x_1 - 1/x}{x_1 - x} \qquad x_1 \neq x$$

which equals
$$\frac{(x - x_1)/xx_1}{x_1 - x},$$

or
$$\frac{x - x_1}{x_1 - x} \frac{1}{xx_1}.$$

Since $x - x_1 = -(x_1 - x)$, the above product simplifies to

$$\frac{-1}{xx_1}.$$

What happens to this quotient as $x_1 \to x$? That is, what is

$$\lim_{x_1 \to x} \frac{-1}{xx_1}?$$

The limit is not hard to compute:

$$\lim_{x_1 \to x} \frac{-1}{x x_1} = -\frac{1}{xx} = -\frac{1}{x^2}.$$

This Formula for the Derivative of 1/x Is Worth Memorizing

Thus the derivative of the function $1/x$ is the function $-1/x^2$. ●

Is the result in Example 1 reasonable? Since x^2 is positive for all values of x, the slope of the tangent line, being $-1/x^2$, is negative. Moreover, when x is near 0, $-1/x^2$ is very large in size, meaning that the tangent lines are steep. When x is large, $-1/x^2$ is small and negative, indicating that the tangent lines are nearly horizontal but slanted downward.

Do these conclusions seem valid when checked against the graph of the function? The graph of $1/x$, copied from Sec. 2.2, is shown in the margin.

The conclusions are correct. The tangent lines all have negative slope; moreover, they are very steep when x is near 0 and become almost horizontal when $|x|$ is large.

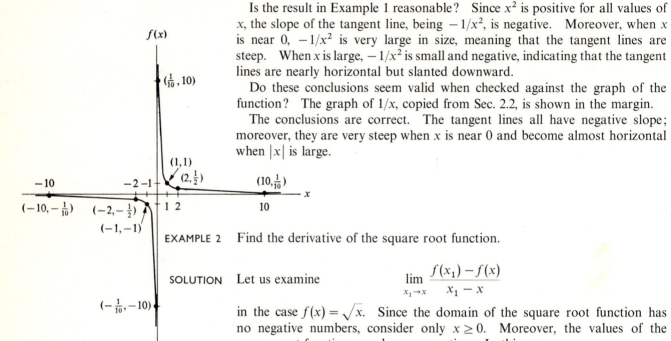

EXAMPLE 2   Find the derivative of the square root function.

SOLUTION   Let us examine

$$\lim_{x_1 \to x} \frac{f(x_1) - f(x)}{x_1 - x}$$

in the case $f(x) = \sqrt{x}$. Since the domain of the square root function has no negative numbers, consider only $x \geq 0$. Moreover, the values of the square root function are also nonnegative. In this case

$$f(x) = \sqrt{x} \qquad \text{and} \qquad f(x_1) = \sqrt{x_1}.$$

The quotient

$$\frac{f(x_1) - f(x)}{x_1 - x}$$

becomes

$$\frac{\sqrt{x_1} - \sqrt{x}}{x_1 - x}.$$

As $x_1 \to x$, what happens to this quotient? Both the numerator and the denominator approach 0. In order to avoid obtaining 0/0, which is meaningless, we must change the form of the quotient by the use of some algebra. The identity for $c^n - d^n$, which was of use in Sec. 3.2, is useless here. Instead, we multiply numerator and denominator by $\sqrt{x_1} + \sqrt{x}$, in order to obtain $x_1 - x$ in the numerator:

$$\frac{\sqrt{x_1} - \sqrt{x}}{x_1 - x} = \frac{(\sqrt{x_1} - \sqrt{x})(\sqrt{x_1} + \sqrt{x})}{x_1 - x} \frac{}{\sqrt{x_1} + \sqrt{x}}$$

$$= \frac{(\sqrt{x_1})^2 - (\sqrt{x})^2}{(x_1 - x)(\sqrt{x_1} + \sqrt{x})}$$

$$= \frac{x_1 - x}{(x_1 - x)(\sqrt{x_1} + \sqrt{x})}$$

$$= \frac{1}{\sqrt{x_1} + \sqrt{x}} \qquad x_1 \neq x.$$

The algebra is completed; now let us see what happens to the much simpler expression

$$\frac{1}{\sqrt{x_1} + \sqrt{x}}$$

as $x_1 \to x$. It will be convenient to consider two separate cases: $x > 0$ and $x = 0$. In the first case, $x > 0$: As $x_1 \to x$,

$$\frac{1}{\sqrt{x_1} + \sqrt{x}} \to \frac{1}{\sqrt{x} + \sqrt{x}} = \frac{1}{2\sqrt{x}}.$$

This Formula for the Derivative of $\sqrt{x}$ *Is Worth Memorizing*

Thus, when $x > 0$, the derivative of the square root function at the value x is

$$\frac{1}{2\sqrt{x}}.$$

In the second case, $x = 0$:

$$\frac{1}{\sqrt{x_1} + \sqrt{x}} = \frac{1}{\sqrt{x_1} + 0}$$

$$= \frac{1}{\sqrt{x_1}}.$$

As $x_1 \to 0$, the quotient $1/\sqrt{x_1}$ gets larger and larger; it does not have a limit. The derivative of the square root function is not defined at $x = 0$. ●

Is the result in Example 2 reasonable? It says that, when x is large, the slope of the tangent line at $(x, \sqrt{x})$ is near 0 [since $1/(2\sqrt{x})$ is near 0]. Let us draw the graph and see. First we make a brief table, as shown in the margin. With the aid of these six points, the graph is easy to sketch:

x	0	1	4	9	16	25
$\sqrt{x}$	0	1	2	3	4	5

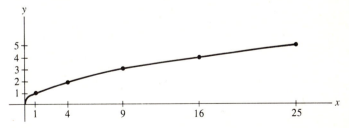

For points far to the right on the graph the tangent line is indeed almost horizontal, as the formula $1/(2\sqrt{x})$ suggests. When x is near 0, the derivative $1/(2\sqrt{x})$ is large. The graph gets steeper and steeper near $x = 0$.

You might have the impression from the functions examined so far in the

The graph of
the "absolute
value" function

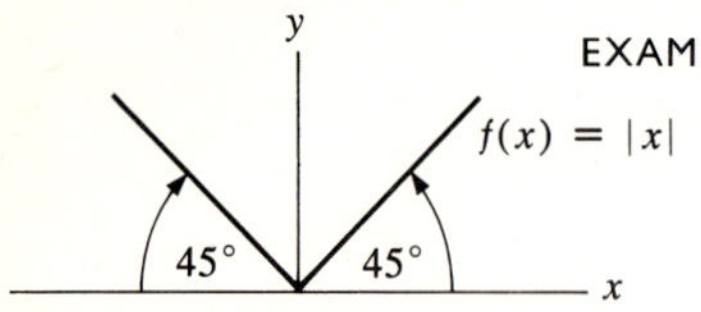

chapter that every function whose domain is the entire number line has a derivative for all values of x. This is not the case. The next example shows that the absolute-value function does not have a derivative at $x = 0$.

EXAMPLE 3   The graph of the function f whose formula is $f(x) = |x|$ was discussed in Sec. 2.2. Note the sharp bend at $(0, 0)$ in the accompanying illustration. This raises problems in speaking of the tangent line through $(0, 0)$. Show that this function does not have a derivative at $x = 0$.

SOLUTION   For positive values of x_1 near 0, $f(x_1) = x_1$, and the quotient

$$\frac{f(x_1) - f(x)}{x_1 - x}$$

is simply

$$\frac{x_1 - 0}{x_1 - 0},$$

or

$$1.$$

But x_1 can be negative, since we can approach 0 from either side. When x_1 is negative, $f(x_1) = -x_1$, and the quotient becomes

$$\frac{-x_1 - 0}{x_1 - 0},$$

or

$$-1.$$

Since the quotient approaches different values according to whether x_1 approaches 0 from the right or from the left, there is no one single number that the quotient

$$\frac{f(x_1) - f(0)}{x_1 - 0}$$

approaches as $x_1 \to 0$. Thus, the absolute-value function does not have a derivative at 0. In other words,

$$\lim_{x_1 \to 0} \frac{f(x_1) - f(0)}{x_1 - 0}$$

does not exist. ●

The function $|x|$ does have a derivative for all other values of x. This we indicate by saying that it is differentiable at any x except 0. More generally, we have the following definition.

DEFINITION   A function f is *differentiable at x* if

$$\lim_{x_1 \to x} \frac{f(x_1) - f(x)}{x_1 - x}$$

exists. If a function is differentiable at each x in its domain, we call the function *differentiable*.

The squaring function is differentiable. Indeed, most of the functions we need are differentiable. But neither the absolute-value function nor the square root function is differentiable at $x = 0$.

Usually, those functions with applications of importance are differentiable at all numbers in their domain, with at most isolated exceptions.

The graph of the typical function f is a curve. Let us draw $f(x)$ and $f(x_1)$ and then interpret the quotient

$$\frac{f(x_1) - f(x)}{x_1 - x}$$

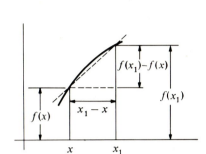

in terms of this curve, as shown in the figure in the margin. The numerator

$$f(x_1) - f(x)$$

corresponds to the vertical leg of a right triangle, and the denominator $x_1 - x$ corresponds to the horizontal leg. The quotient is the slope of the line through $(x, f(x))$ and $(x_1, f(x_1))$. This idea has already been used in the special case $f(x) = x^2$, when the slope of the tangent line to the graph of $y = x^2$ at $(2, 4)$ was found.

The x and x + h Notation

Sometimes it will be useful to label the same diagram as follows. Write the nearby number x_1 as

$$x + h,$$

where h is small and represents the difference $x_1 - x$. In this notation the preceding figure is labeled as in the accompanying figure. The quotient

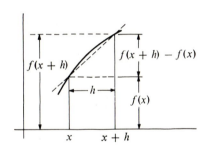

$$\frac{f(x_1) - f(x)}{x_1 - x}$$

is now written:

$$\frac{f(x + h) - f(x)}{h}.$$

The derivative at x is then

$$\lim_{h \to 0} \frac{f(x + h) - f(x)}{h}.$$

Though conceptually this is still the same definition of the derivative as that given earlier in this section, it leads to different algebraic steps. In Chap. 5, where the derivative of the trigonometric, exponential, and logarithmic functions are obtained, the notation using x and $x + h$ is more helpful than the notation using x and x_1.

To contrast the two notations, let us find the derivative of the squaring function, using x and $x + h$ instead of x and x_1. (We already know that the result should be $2x$.)

EXAMPLE 4   Compute $\displaystyle\lim_{h \to 0} \frac{f(x + h) - f(x)}{h},$

where f is the squaring function.

SOLUTION   Since the output of the squaring function is always the square of the input,
$$f(x + h) = (x + h)^2,$$
and
$$f(x) = x^2.$$

The quotient
$$\frac{f(x + h) - f(x)}{h} = \frac{(x + h)^2 - x^2}{h}$$

may be simplified as follows:
$$\frac{(x + h)^2 - x^2}{h} = \frac{x^2 + 2xh + h^2 - x^2}{h}$$
$$= \frac{2xh + h^2}{h}$$
$$= 2x + h.$$

The algebra in these steps is quite different from that of our previous work on the same problem. Instead of the identity $c^2 - d^2 = (c + d)(c - d)$, we used the identity $(c + d)^2 = c^2 + 2cd + d^2$ to expand $(x + h)^2$.

Finally, we must find what happens to $2x + h$ as $h \to 0$. As $h \to 0$, $2x + h \to 2x$. Consequently $\lim_{h \to 0} (2x + h) = 2x$. In short, the derivative is $2x$, as expected. ●

Now that we have the concept of the derivative, we are in a position to define *tangent line*, *speed*, *magnification*, and *density*, terms we have used only intuitively until now. These definitions are suggested by the similarity of the computations made in the four problems in Sec. 3.1.

DEFINITION   *Tangent line to a curve.* The tangent line to the graph of the function f at the point $P = (x, y)$ is the line through P that has a slope equal to the derivative of f at x.

The speed of a moving object will never be negative. For this reason we first define *velocity*, which may be positive or negative. Speed will be the absolute value of velocity.

DEFINITION   *Velocity and speed of a particle moving on a line.* The velocity at time t of an object, whose location on a line at time t is $f(t)$, is the derivative of f at time t. The speed of the particle is the absolute value of the velocity.

DEFINITION   *Magnification of a linear projector.* The magnification at x of a lens that projects the point x of one line onto the point $f(x)$ of another line is the derivative of f at x.

DEFINITION   *Density of material.* The density at x of material distributed along a line in such a way that the left-hand x centimeters have a mass of $f(x)$ grams is equal to the derivative of f at x.

The slope of a line was defined as the quotient $(y_2 - y_1)/(x_2 - x_1)$, where $P_1 = (x_1, y_1)$ and $P_2 = (x_2, y_2)$ are any distinct points on the line. Now it is possible to define the slope of a curve at a point on the curve.

DEFINITION   *Slope of a curve.* The slope of the graph of the function f at $(x, f(x))$ is the derivative of f at x.

There are many symbols for the derivative of the function f. The simplest is f' (read as "f prime"). Then the derivative of f at x is simply $f'(x)$. The long statement: "The derivative of the cubing function is three times the squaring function" is written simply

$$(x^3)' = 3x^2.$$

Similarly,

$$(x^2)' = 2x$$

$$\left(\frac{1}{x}\right)' = \frac{-1}{x^2}$$

and

$$(\sqrt{x})' = \frac{1}{2\sqrt{x}}.$$

The Derivative Records a Rate of Change.   The derivative of a function may always be thought of as a rate of change. If the function describes the position of a moving particle, then the derivative is the velocity of the particle; velocity describes the rate at which the position of the particle is changing. If the function describes the y value along a curve, then the derivative is the slope of the curve; the slope of a curve describes how rapidly y changes when x changes.

Exercises

In Exercises 1 to 6 find the derivative of the indicated function by examining

$$\lim_{x_1 \to x} \frac{f(x_1) - f(x)}{x_1 - x}.$$

1. $3\sqrt{x} + 4$
2. $1/\sqrt{x}$
3. $6x^2 + 3x + 2$
4. x^3
5. $1/x^2$
6. $2x + 3/x$
7. Let f be the function given by the formula $f(x) = x^2$. Compute, as a decimal,

$$\frac{f(x + h) - f(x)}{h}$$

when (*a*) $x = 3$ and $h = 0.1$; (*b*) $x = 3$ and $h = 0.01$; (*c*) $x = 3$ and $h = -0.1$.

In Exercises 8 and 9 find the derivative of the indicated function by examining

$$\lim_{h \to 0} \frac{f(x + h) - f(x)}{h}.$$

8. $3x^2 + 5/x$
9. $5 + 2x + 3\sqrt{x}$
10. A snail crawls $\sqrt{t}$ feet in t seconds. What is its speed when t is (*a*) $\frac{1}{9}$? (*b*) 1? (*c*) 4? (*d*) 9?
11. The left-hand x centimeters of a string have a mass of $\sqrt{x}$ grams. What is its density when x is (*a*) $\frac{1}{4}$? (*b*) 1? Is its density defined at $x = 0$?

■

12. In Example 2, which concerns the square root function, it was shown by examining the derivative that the graph of $y = \sqrt{x}$ is very steep near $(0, 0)$. Check this by filling

in the following table and using it to graph $y = \sqrt{x}$ for x in $[0, 1]$.

x	0	0.16	0.25	0.49	0.64	0.81	1.00
$\sqrt{x}$				0.7			

13. (a) Graph the function $1/x$.
   (b) Draw the line through $(1, 1)$ and $(1 + h, 1/(1 + h))$ for $h = 2$, $h = 1$, and $h = \frac{1}{4}$.
   (c) What is the slope of the tangent line to the graph at $(1, 1)$? (Example 1 has the general formula.)
   (d) Draw the tangent line in (c), making use of its slope. What angle does it make with the x axis?

14. Let f be the function whose value at x is $1/x^3$.
   (a) Express $[f(x_1) - f(x)]/(x_1 - x)$ as simply as possible.
   (b) Using (a), find the derivative of f.

15. Sketch a graph of $y = x^3$.
   (a) Why can the tangent line to this graph at $(0, 0)$ *not* be defined as "the line through $(0, 0)$ that meets the graph just once"?
   (b) Why can the tangent line to this graph at $(1, 1)$ *not* be defined as "the line through $(1, 1)$ that meets the graph just once"?
   (c) How is the tangent line at any point on the graph defined?

In the next exercise the magnification of a lens turns out to be negative. As will be noticed, this means that the lens switches right and left, that is, if x_1 is to the right of x the image of x_1 is to the left of the image of x.

16. A lens projects x of the slide onto $1/x$ on the screen.
   (a) How much does it magnify the interval $[0.5, 0.6]$?
   (b) What is its magnification at 0.5?
   (c) How much does it magnify the interval $[2, 2.1]$?
   (d) What is its magnification at 2?

■ ■

17. (a) Is $(x^3 + x^2)'$ equal to $(x^3)' + (x^2)'$?

(b) Is $(x^3 x^2)'$ equal to $(x^3)'(x^2)'$?
(c) Is $(x^3/x^2)'$ equal to $(x^3)'/(x^2)'$?

18. The height of a ball thrown straight up is $64t - 16t^2$ feet after t seconds.
   (a) Show that its velocity after t seconds is $64 - 32t$ feet per second.
   (b) What is its velocity when $t = 0$? $t = 1$? $t = 2$? $t = 3$?
   (c) What is its speed when $t = 0$? $t = 1$? $t = 2$? $t = 3$?
   (d) For what values of t is the ball rising? falling?

19. In the study of the seepage of irrigation water into soil, equations such as $y = \sqrt{t}$ are sometimes used. The equation says that the water penetrates $\sqrt{t}$ feet in t hours.
   (a) What is the physical significance of the derivative $1/(2\sqrt{t})$?
   (b) What does (a) say about the rate at which water penetrates the soil when t is large?

20. Let f be the exponential function given by the formula $f(x) = 10^x$.
   (a) Copy and complete this table.

x	-2	-1	0	1
10^x	0.01			10

   (b) Graph the function for x in $[-2, 1]$. (The same scale should be used for both axes.)
   (c) Using a ruler, draw what you think would be the tangent line at $(0, 1)$.
   (d) Using a ruler (preferably marked in centimeters), estimate the slope of the line you drew in (c).
   (e) Show that the derivative of 10^x at $x = 0$ is
   $$\lim_{h \to 0} \frac{10^h - 1}{h}.$$
   (f) The limit in (e) is far from obvious. What does (d) suggest as an estimate of this limit?
   (g) Call the limit in (e) simply c. Show that $(10^x)' = c 10^x$.

3.4

Summary

Four problems introduced the concept of the derivative: velocity of a moving particle, tangent line to a curve, magnification of a projector, and density of mass in a string.

The value of the derivative of the function f at x is defined as

$$\lim_{x_1 \to x} \frac{f(x_1) - f(x)}{x_1 - x}$$

or, what amounts to the same thing, as

$$\lim_{h \to 0} \frac{f(x + h) - f(x)}{h}.$$

For most functions of importance in calculus and its applications, this limit exists for all x in the domain of the function, with

perhaps a few exceptions.   For instance, the absolute-value function does not have a derivative at $x = 0$.

One notation for the derivative of the function f is f'.   The
derivative is a function, since it assigns to a number x the
derivative at x.   In particular it was shown that

$$(x^n)' = nx^{n-1} \qquad n = 1, 2, 3, 4, \ldots;$$

$$(\sqrt{x})' = \frac{1}{2\sqrt{x}};$$

$$\left(\frac{1}{x}\right)' = -\frac{1}{x^2}.$$

With the aid of the derivative concept, precise definitions have

been given for the tangent to a point on a curve, the slope of
a curve at a point, velocity (speed is the absolute value of
velocity), magnification, and density.

Just as velocity measures how swiftly distance changes, *the
derivative, in general, measures how quickly the value of a function
changes.*   If a slight change in the function's input causes a large
change in its output, the derivative will be large.   In general,
the derivative measures the rate at which a quantity is changing.

The table below is important and should be studied carefully.
The fifth and sixth rows are based on Exercises 16 and 17 of
Sec. 3.1.   The last row describes the general underlying concept.
The other rows are just individual interpretations of the
derivative.

If We Interpret x as	and f(x) as	then $\dfrac{f(x+h)-f(x)}{h}$ Is	and, as h Approaches 0, the Quotient Approaches
The abscissa of a point in the plane	The ordinate of that point	The slope of a certain line	The slope of a tangent line
Time	The location of a particle moving on a line	An average velocity over a time interval	The velocity at time x
A point on a linear slide	Its projection on a linear screen	An average magnification	The magnification at x
A location on a non-uniform string	The mass from 0 to x	An average density	The density at x
Time	Mass of a bacterial culture at time x	An average growth rate over a time interval	The growth rate at time x
Time	Total profit up to time x	An average rate of profit over a time interval	The rate of profit at time x
Just a number	A number depending on x	A quotient: the change in the output divided by the change in the input	The derivative evaluated at x (the rate of change of the function with respect to x)

VOCABULARY AND SYMBOLS

interval $[a, b]$	derivative: f'	velocity, speed
limit	differentiable at x	magnification
$x_1 \to x$, $\lim\limits_{x_1 \to x}$	differentiable	density
	slope of curve	

KEY FACTS

$(x^2)' = 2x$ $(x^3)' = 3x^2$

More generally $(x^n)' = nx^{n-1}$ $n = 1, 2, 3, 4, \ldots$

$$\left(\frac{1}{x}\right)' = -\frac{1}{x^2} \qquad x \neq 0$$

$$(\sqrt{x})' = \frac{1}{2\sqrt{x}} \qquad x \neq 0$$

$(cx^n)' = cnx^{n-1}$ $n = 1, 2, 3, 4, \ldots$

The derivative of a polynomial is the sum of the derivatives of its terms.

Guide quiz on Chap. 3

1. Define the derivative of a function.
2. Using the definition of the derivative, find the derivative of $x^2 + 3/x$ at $x = 2$.
3. Using formulas for derivatives developed in the text, compute the derivative of
   (a) $5x^3 - x^2 + 2$ at $x = 1$;
   (b) $1/x$ at $x = -1$;
   (c) $\sqrt{x}$ at $x = \frac{1}{4}$.
4. Let f be a differentiable function and let x_1 and x be numbers such that $x_1 > x$. Interpret $f(x_1) - f(x)$, $x_1 - x$, and their quotient $[f(x_1) - f(x)]/(x_1 - x)$ if
   (a) $f(x)$ is the height of a rocket x seconds after blast-off;

(b) $f(x)$ is the number of bacteria in a bacterial culture at time x;
(c) $f(x)$ is the mass of the left-hand x centimeters of a rod;
(d) $f(x)$ is the position of the image on the linear screen of the point x on the linear slide.

5. Find $(x^4)'$, using (a) the x and x_1 notation, (b) the x and $x + h$ notation.
6. Define: (a) tangent line to a curve, (b) velocity, (c) speed.
7. On a sketch of the graph of a typical function f,
   (a) show the line whose slope is $[f(x_1) - f(x)]/(x_1 - x)$;
   (b) show the tangent line at the point $(x, f(x))$.

Review exercises for Chap. 3

1. What is the derivative of
   (a) $x^5 - 3x^4 + 2x + 6$?   (b) $1/x$?
   (c) $\sqrt{x}$?   (d) $1/x^2$?
2. Graph $y = 2x^3 - 15x^2 + 36x$, showing in particular where the curve crosses the x axis and where the tangent line is horizontal.
3. Find the derivative of $(x^2 + 1)^2$.
4. Using the definition of the derivative, explain why $f(x) = |x - 1|$ is not differentiable at 1.
5. Let a_0, a_1, a_2, a_3 be fixed numbers. Using the x and x_1 notation, show that
   $$(a_0 x^3 + a_1 x^2 + a_2 x + a_3)' = 3a_0 x^2 + 2a_1 x + a_2.$$
6. Let f be the function given by the formula $f(x) = 3/(x + 1)$. Find f' at $x = 2$, using the x and x_1 notation.
7. Let f be the function whose value at x is $4x^2$.
   (a) Compute $[f(2.1) - f(2)]/0.1$.
   (b) What is the interpretation of the quotient in (a) if $f(x)$ denotes the total profit of a firm (in millions of dollars) in its first x years?

(c) What is the interpretation of the quotient in (a) if $f(x)$ denotes the height of the ordinate in the graph of $y = 4x^2$?
(d) What is the interpretation of the quotient in (a) if $f(x)$ is the distance a particle moves in the first x seconds?

8. This is a brief excerpt from the four-place table for the cube root function f, where $f(x) = \sqrt[3]{x}$.

x	62	63	64	65	66
$\sqrt[3]{x}$	3.9579	3.9791	4.0000	4.0207	4.0412

   (a) Use $x_1 = 65$ and $x = 64$ to estimate $(\sqrt[3]{x})'$ at 64.
   (b) Use $x_1 = 63$ and $x = 64$ to estimate $(\sqrt[3]{x})'$ at 64.
   (c) Use $x = 64$ and $h = -2$ to estimate $(\sqrt[3]{x})'$ at 64.
   (d) Use $x = 64$ and $h = 2$ to estimate $(\sqrt[3]{x})'$ at 64.

9. A lens projects the point x on the x axis onto the point x^3 on the linear screen.
   (a) How much does it magnify the interval $[2, 2.1]$?
   (b) How much does it magnify the interval $[1.9, 2]$?
   (c) What is its magnification at 2?

10. Express $[f(2 + h) - f(2)]/h$ as simply as possible if
    (a) $f(x) = 4x^2$;
    (b) $f(x) = 7/x$;
    (c) $f(x) = 3x + 5$;
    (d) $f(x) = 4$, for all x.
11. Express $[f(x_1) - f(1)]/(x_1 - 1)$ as simply as possible if
    (a) $f(x) = 5x^3$;
    (b) $f(x) = 3/x^2$;
    (c) $f(x) = 6x - 1$;
    (d) $f(x) = 4x^2 - x + 6$.
12. If the function f records the weight of a person (dependent on age), then we may think of the derivative f' as _____.
13. (a) If the function f records the trade-in value of a car (dependent on its age), then we may think of the derivative f' as _____.
    (b) When is the derivative in (a) negative? positive? Which is the more usual case?
14. (a) The left-hand x centimeters of a rod have a mass of $3x^4$ grams. What is its density at $x = 1$?
    (b) Devise a magnification problem which is mathematically equivalent to (a).
    (c) Devise a velocity problem which is mathematically equivalent to (a).
15. This is the chart of the temperature during a cold day.

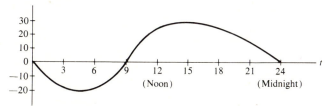

Temperature

(Noon)   (Midnight)

Let f be the function that assigns to time t the temperature at that time. Thus, reading the graph, we see that $f(12) = 25$.
    (a) For which t is $f(t) = 0$?
    (b) For which t is the derivative of f equal to 0?
    (c) For which t is $f(t) \geq 0$?
    (d) For which t is the derivative ≥ 0?
    (e) For which t is the derivative ≤ 0?
    (f) At what time was the temperature lowest? highest?
16. Express $[f(x + h) - f(x)]/h$ as simply as possible if
    (a) f is the squaring function and x is 5;
    (b) f is the reciprocal function and x is 2;
    (c) f is the cubing function and x is 4.
17. Let f be the function given by the formula $f(x) = 5x - 2$.
    (a) Express $[f(x + h) - f(x)]/h$ as simply as possible.
    (b) Use (a) to find $(5x - 2)'$.

18. (a) Consider the line of slope m that passes through the point (x_1, y_1). Show that, if (x, y) is on the line and distinct from (x_1, y_1), then

$$\frac{y - y_1}{x - x_1} = m.$$

    (b) Show that the graph of the equation $y = y_1 + m(x - x_1)$ is the line through (x_1, y_1) with slope m.
19. (See Exercise 18.) Find an equation of the tangent line to the curve $y = x^3 - 2x^2$ at $(1, -1)$.
20. (See Exercise 18.) Find an equation of the tangent line to the curve $y = 2x^4 - 6x^2 + 8$ at $(2, 16)$.

■ ■

21. Find $\lim\limits_{x \to 4} \dfrac{x^2 - 16}{\sqrt{x} - 2}$.
22. Find $\lim\limits_{h \to 0} \dfrac{f(x + 2h) - f(x)}{h}$ if (a) $f(x) = x^2$, (b) $f(x) = 1/x$.
23. (a) Using the formula for the derivative of a polynomial, find $(x^3 - 3x)'$.
    (b) For which x is the derivative in (a) positive? negative? zero?
    (c) Graph $y = x^3 - 3x$, plotting a few points and using the information in (b).
24. (a) Draw a freehand curve indicating a typical function f.
    (b) Label on it the points $P_0 = (x, f(x))$, $P_1 = (x + h, f(x + h))$, and $P_2 = (x - h, f(x - h))$.
    (c) Show that the slope of the line through P_1 and P_2 is

$$\frac{f(x + h) - f(x - h)}{2h}.$$

    (d) For a differentiable function, what do you think is the value of

$$\lim\limits_{h \to 0} \frac{f(x + h) - f(x - h)}{2h}?$$

    (e) Compute the limit in (d) if $f(x) = x^3$.

The remaining exercises suggest explorations with the aid of a hand-held calculator. They are included to offer an extra perspective on the limit concept, which will be the main topic of the next chapter.

25. What happens to the numbers $(0.99)^n$ as n assumes larger and larger positive integer values? Include $n = 5, 10, 20$ among your experiments.
26. Explore what happens to the quotient

$$\frac{2x^2 + x}{3x^2 + 5x}$$

    (a) when x is near 0 (for instance, $x = 0.1$), (b) when x is very large (for instance, $x = 20$).

27. If your calculator has a $\sqrt{x}$-key, examine what happens to $\sqrt{x^2 + x} - x$ when x is a very large positive number. Include $x = 10$ and $x = 100$.

28. (a) Find the values of $(1 + 1/n)^n$ for $n = 1, 2, 3, 4, 5$, and some larger positive integers. Also for $n = -2, -3$, and some larger negative integers.

    (b) Do the same for $(1 - 1/n)^n$.

(c) What do you think happens to the expressions in (a) and (b) as n gets larger?

29. If your calculator has a $\sqrt[n]{x}$-key, use it to estimate the derivative of $f(x) = \sqrt[3]{x}$ at $x = 8$.

30. If your calculator has a y^x-key, use it to help estimate the derivative of $f(x) = 2^x$ at $x = 3$.

4

LIMITS AND CONTINUOUS FUNCTIONS

The preceding chapter developed the idea of the derivative, whose definition depends on the concept of a limit. The present chapter explores this concept and the closely related notion of a continuous function. In addition it discusses two specific limits that will be needed in the next chapter, where the derivatives of all the functions needed in applications of calculus will be computed. One of these limits concerns exponentiation, the other, trigonometry. For this reason, a review of these topics is included.

4.1

Review of exponentiation

There are three basic operations that can be applied to a pair of numbers a and b:

$$\text{Addition, } a + b;$$

$$\text{Multiplication, } ab;$$

and $$\text{Exponentiation, } b^a.$$

This section reviews the third of these, exponentiation.

The numbers that concern us are called the *real numbers*. They may be thought of as marking points on the number line:

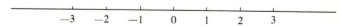

Every point corresponds to some real number. The points marked above correspond to the *integers*

$$\ldots, -3, -2, -1, 0, 1, 2, 3, \ldots.$$

Numbers to the right of 0 are called *positive*; those situated to the left of 0 are called *negative*.

A number that can be expressed as the quotient of two integers

$$\frac{m}{n} \quad \text{with} \quad n \neq 0$$

Definition of Rational Number   is called *rational*. We may always take n to be positive. Rational numbers with a given denominator are regularly spaced on the number line. This diagram shows a few rational numbers that have the denominator $n = 5$.

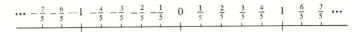

It is difficult to show on the number line the rationals that have denominator 1 million since they are so close to each other.

There are points on the number line that do not correspond to rational numbers. For instance, it is impossible to write $\sqrt{2}$ as the quotient of two integers (see Exercise 22). However, $\sqrt{2}$ is approximately 1.414, written

$$\sqrt{2} \doteq 1.414.$$

Thus $\sqrt{2}$ describes a point which lies between $\frac{7}{5} = 1.4$ and $\frac{8}{5} = 1.6$ in the accompanying diagram.

Definition of Irrational Number   A number that cannot be expressed as the quotient of two integers is called *irrational*. For instance, $\sqrt{2}$ is irrational.

An irrational number can be approximated by rational numbers as accurately as we please. For instance

$$1.414 < \sqrt{2} < 1.415,$$

since $(1.414)^2 = 1.999396$ and $(1.415)^2 = 2.002225.$

Definition of Base and Exponent   Now with this background, let us review the definition of the exponential b^x, where b is a fixed positive number, called the *base*. The number x is called the *exponent*. To keep things as simple as possible, let us review the definition of 2^x.

First of all, if x is a positive integer, 2^x is defined as the product of x of the 2s. For instance, $2^5 = 2 \cdot 2 \cdot 2 \cdot 2 \cdot 2 = 32;$

$$2^4 = 2 \cdot 2 \cdot 2 \cdot 2 = 16;$$

$$2^3 = 2 \cdot 2 \cdot 2 = 8;$$

$$2^2 = 2 \cdot 2 = 4;$$

and $2^1 = 2.$

Note that $\underbrace{2^2}_{2 \cdot 2} \cdot \underbrace{2^3}_{2 \cdot 2 \cdot 2} = \underbrace{2^{2+3}}_{2 \cdot 2 \cdot 2 \cdot 2 \cdot 2}.$

More generally, if x and y are positive integers,

$$2^{x+y} = 2^x \cdot 2^y.$$

Basic Law of Exponents

This *basic law of exponents* serves as a guide to defining 2^x when x is *not* a positive integer.

The Exponent 0

For instance, what should 2^0 be? If the basic law of exponents is to be true for all exponents, then in particular

$$2^{0+1} = 2^0 \cdot 2^1,$$

or

$$2^1 = 2^0 \cdot 2^1.$$

Since $2^1 = 2$, this equation says that

$$2 = 2^0 \cdot 2.$$

From this it follows that if the basic law of exponents is to hold and if 2^0 is to have meaning, then 2^0 must be 1; thus we define 2^0 to be 1,

$$2^0 = 1.$$

What should $2^{-1}, 2^{-2}, 2^{-3}, \dots$ be? To preserve the basic law of exponents, we must have, for instance,

$$2^3 \cdot 2^{-3} = 2^{3+(-3)}.$$

But $2^{3+(-3)} = 2^0 = 1$. Thus $2^3 \cdot 2^{-3} = 1$.

This shows that, if 2^{-3} is to have meaning, it must be the reciprocal of 2^3, that is,

$$2^{-3} = \frac{1}{2^3} = \frac{1}{8}.$$

Negative Integer Exponents

For this reason we define in general

$$2^{-n} \quad \text{to be} \quad \frac{1}{2^n}$$

for any positive integer n.

The exponential 2^x has now been defined for any *integer* x. Let us graph $y = 2^x$ for integer values of x, as in the accompanying figure.

It is tempting to draw a smooth curve through these points, but 2^x has not yet been defined if x is not an integer.

How should $2^{1/2}$ be defined? To preserve the basic law of exponents, it is necessary that

$$2^{1/2} \cdot 2^{1/2} = 2^{1/2+1/2};$$

thus

$$2^{1/2} \cdot 2^{1/2} = 2^1 = 2.$$

Hence $2^{1/2}$ is a solution of the equation

$$x^2 = 2.$$

Should it be the positive or the negative solution? The graph just sketched suggests that $2^{1/2}$ should be positive. Thus define $2^{1/2}$ to be $\sqrt{2}$, the square root of 2. Note that $2^{1/2} \doteq 1.4$, which fits nicely into the preceding graph.

Similarly, $2^{1/3}$ can be determined by the basic law of exponents:

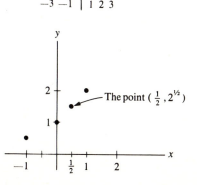

The point $(\tfrac{1}{2}, 2^{1/2})$

$$2^{1/3} \cdot 2^{1/3} \cdot 2^{1/3} = 2^{1/3 + 1/3 + 1/3}$$
$$= 2^1$$
$$= 2.$$

Thus $2^{1/3}$ must be a solution of the equation

$$x^3 = 2.$$

There is only one solution, and it is called the *cube root of* 2, denoted also $\sqrt[3]{2}$. Thus define

$$2^{1/3} = \sqrt[3]{2},$$

which is about 1.26.

Similarly, define $2^{1/n}$ for any positive integer n to be the positive solution of the equation $x^n = 2$, that is,

$$2^{1/n} = \sqrt[n]{2}.$$

Note that

$$(2^{1/n})^n = 2.$$

How should $2^{3/4}$ be defined? To preserve the basic law of exponents, we must have
$$2^{1/4} \cdot 2^{1/4} \cdot 2^{1/4} = 2^{3/4}.$$

In short,

$$2^{3/4} = (2^{1/4})^3.$$

Rational Exponents   In general, if m is an integer and n is a positive integer, define

$$2^{m/n} \text{ to be } (2^{1/n})^m.$$

So far, 2^x is defined for any rational number x.

How is 2^x defined when x is irrational? For instance, how can $2^{\sqrt{2}}$ be defined? One way is to consider the decimal expansion of $\sqrt{2}$, which begins

$$2 = 1.41421356 \cdots$$

The successive decimal approximations 1.4, 1.41, 1.414, ... are all rational numbers. For instance, $1.414 = 1{,}414/1{,}000$. Thus $2^{1.4}, 2^{1.41}, 2^{1.414}, \ldots$ have already been defined. This table lists some of their values:

x	1.4	1.41	1.414	1.4142	1.41421
2^x	$2.63901 \cdots$	$2.65737 \cdots$	$2.66474 \cdots$	$2.66511 \cdots$	$2.66513 \cdots$

Irrational Exponents   It can be shown that as the decimal approximation x gets closer and closer to $\sqrt{2}$, the number 2^x approaches a limit. This limit is called $2^{\sqrt{2}}$. Incidentally, $2^{\sqrt{2}} = 2.66514 \cdots$. The same approach provides a definition for b^x for any irrational exponent x. Thus we may now assume that b^x is defined for all real numbers x.

For any positive base b, b^x is defined similarly. For instance,

$$b^{-3} = \frac{1}{b^3}$$

$$b^{1/2} = \sqrt{b}$$

$$b^{1/3} = \sqrt[3]{b} \qquad \text{and so on.}$$

EXAMPLE 1   Evaluate $64^{1/3}$, $64^{2/3}$, $64^{1/2}$, and $64^{-2/3}$.

SOLUTION $64^{1/3}$ is the cube root of 64, $\sqrt[3]{64}$, which is 4.

$$64^{2/3} = (64^{1/3})^2 = 4^2 = 16;$$

$$64^{1/2} = \sqrt{64} = 8;$$

$$64^{-2/3} = (64^{1/3})^{-2} = 4^{-2} = \tfrac{1}{16} = 0.0625. \quad \bullet$$

EXAMPLE 2   Find a decimal approximation of $2^{1/3}$, the cube root of 2.

SOLUTION   The simplest way is to use the y^x key or $\sqrt[x]{y}$ key of a calculator. Another way is to look it up in a table of cube roots. Any handbook of mathematical tables contains such a table. A brief table is included in Appendix E, which shows the value 1.259921, accurate to six decimal places.

A slower but more instructive way is simply to experiment. For instance, since $1^3 = 1$ and since $1.5^3 = 3.375$, the cube root of 2 is located between 1 and 1.5:

$$1 < \sqrt[3]{2} < 1.5.$$

So try 1.2. Arithmetic shows that $1.2^3 = 1.728$, which is less than 2. Thus

$$1.2 < \sqrt[3]{2} < 1.5.$$

Let us see whether 1.3 is greater or smaller than $\sqrt[3]{2}$. Arithmetic shows that $1.3^3 = 2.197$, which is greater than 2. Therefore, $1.3 > \sqrt[3]{2}$ and it follows that

$$1.2 < \sqrt[3]{2} < 1.3.$$

From this we conclude that the decimal expansion of $\sqrt[3]{2}$ begins with 1.2;

$$2^{1/3} \doteq 1.2.$$

At least this gives some feel for the approximate size of $2^{1/3}$. $\bullet$

There are other useful laws of exponents in addition to the basic law of exponents,

$$b^{x+y} = b^x \cdot b^y.$$

They are (for positive a and b, and arbitrary x and y):

Other Laws of Exponents

$$(b^x)^y = b^{xy} \qquad \text{power of a power;}$$

$$(ab)^x = a^x b^x \qquad \text{power of a product;}$$

$$\left(\frac{a}{b}\right)^x = \frac{a^x}{b^x} \qquad \text{power of a quotient.}$$

Incidentally 0^x is defined only for positive x, $0^x = 0$.

Exercises

1. Evaluate: (a) $64^{1/2}$, (b) $64^{1/3}$, (c) $64^{-1/3}$, (d) $64^{2/3}$, (e) $64^{-5/6}$.

2. Evaluate: (a) $\sqrt{81}$, (b) $\sqrt{\sqrt{81}}$, (c) $81^{-1/4}$, (d) $81^{5/4}$.

3. Express in decimal notation: (a) 10^{-1},
   (b) $5 \cdot 10^{-2} + 3 \cdot 10^{-3}$, (c) $6 \cdot 10^2 + 6 \cdot 10^{-2}$.

4. (a) Copy and fill in this table.

x	2	1	0	-1	-2
3^x					

   (b) Use (a) to graph $y = 3^x$.

5. Evaluate: (a) 16^0, (b) $16^{1/4}$, (c) $16^{2/4}$, (d) $16^{3/4}$, (e) $16^{-1/4}$.

6. Use the data in Exercise 5 to graph $y = 16^x$.

7. Express in the form 2^x: (a) 4, (b) $\frac{1}{8}$, (c) $\sqrt{2}$, (d) 0.5, (e) 1, (f) $\sqrt[3]{2}$, (g) 0.25, (h) $\sqrt{8}$.

8. A table of square roots shows that $\sqrt{5} \doteq 2.236$. Which is larger, $\sqrt{5}$ or 2.236?

9. Express each of the following in the form b^x: (a) $(\sqrt[3]{b})^2$,
   (b) $1/b^2$, (c) $1/\sqrt{b}$, (d) $1/\sqrt[3]{b}$, (e) $(1/b)^5$, (f) $\sqrt{\sqrt[3]{b}}$.

10. Without using tables, show that $2.7 < 20^{1/3} < 2.8$.

11. Evaluate: (a) $(\sqrt{5})^4$, (b) $(2^{1/3})^6$, (c) $(1/3)^{-2}$, (d) $(\sqrt[3]{1/2})^{-6}$.

12. This exercise illustrates the law $(b^x)^y = b^{xy}$. Using the definition given in this section for b^x, compute both sides of each equation and check that $(b^x)^y = b^{xy}$ in each case.
    (a) $(b^3)^4 = b^{3 \cdot 4}$ (b) $(b^{1/2})^2 = b^{(1/2)(2)}$
    (c) $(b^{-1})^{-1} = b^{(-1)(-1)}$ (d) $(b^2)^{-1} = b^{2(-1)}$

13. This exercise illustrates the law $(ab)^x = a^x b^x$. Using the definition of exponentials given in this section, check that
    (a) $a^3 b^3 = (ab)^3$
    (b) $a^{-1} b^{-1} = (ab)^{-1}$
    (c) $a^{1/2} b^{1/2} = (ab)^{1/2}$.   *Hint for* (c): Square both sides.

14. Show that $2^{x+h} - 2^x = 2^x(2^h - 1)$.

■

15. Show that
$$\left(1 + \frac{h}{x}\right)^{1/h} = \left[\left(1 + \frac{h}{x}\right)^{x/h}\right]^{1/x}.$$

16. A common mistake is to assume that $(a + b)^x$ equals $a^x + b^x$. Find specific numbers a, b, and x for which the equality does *not* hold.

17. Show that $b/\sqrt[3]{b} = \sqrt[3]{b^2}$.

18. This exercise concerns the graph of $y = b^x$ when the base is between 0 and 1. For simplicity, take $b = \frac{1}{2}$.
    (a) Fill in the table below.

x	-2	-1	0	1	2
$(\frac{1}{2})^x$					

   (b) Use the table in (a) to plot the graph of $y = (\frac{1}{2})^x$.
   (c) Relative to the same axes as in (b), plot the graph of $y = 2^x$.
   (d) How is the graph in (b) related to the graph of $y = 2^x$?

19. We defined $b^{2/4}$ as $(b^{1/4})^2$ and $b^{1/2}$ as the square root of b. Prove that $b^{2/4} = b^{1/2}$.

■ ■

20. Let x be a positive number. For which x is (a) $x < x^2$?
    (b) $x^2 < x$?   (c) $x^2 < x^3$?   (d) $x^3 < x^2$?

21. For which positive x is (a) $\sqrt{x} < \sqrt[3]{x}$, (b) $\sqrt[3]{x} < \sqrt{x}$?

22. This exercise outlines a proof that $\sqrt{2}$ is not rational. It depends on the factorization of positive integers into primes.

    An integer larger than 1 is called a *prime* if it is not the product of smaller positive integers. The first 10 primes are 2, 3, 5, 7, 11, 13, 17, 19, 23, and 29. The following result from number theory will be needed: An integer larger than 1 either is prime or is uniquely expressible as the product of primes (the order in which the prime factors are written is disregarded). For instance,

$$2 \cdot 2 \cdot 3, \qquad 2 \cdot 3 \cdot 2, \qquad \text{and} \qquad 3 \cdot 2 \cdot 2$$

   are the only factorizations of 12 into primes; they are all considered to be the same factorization of 12.
   (a) Show that if the integer I is the product of d primes then I^2 is the product of $2d$ primes. (Thus I^2 is the product of an *even* number of primes.)
   (b) Show that for any positive integer n, $2n^2$ is the product of an *odd* number of primes.
   (c) If $\sqrt{2}$ can be written as the quotient of two integers, $\sqrt{2} = m/n$, then $2n^2 = m^2$. From (a) and (b) deduce a contradiction.

23. (See Exercise 22.) Prove that if a positive integer I has a rational square root, then I is the square of an integer.

24. (See Exercise 22.) (a) Prove that $\sqrt[3]{2}$ is not rational.
    (b) Determine which positive integers have rational cube roots.

The remaining exercises utilize a calculator.

25. Some 3,500 years ago the mathematicians of Babylon estimated $\sqrt{a}$ as follows. First they made a guess g_1. Then they computed

$$g_2 = \frac{g_1 + \dfrac{a}{g_1}}{2}.$$

   They then used g_2 as a better approximation of $\sqrt{a}$.

Repeating the process as often as desired, they produced a sequence g_2, g_3, ... of successive estimates of $\sqrt{a}$. Let $a = 5$ and $g_1 = 2$. Compute (a) g_2, (b) g_3.

26. Using a calculator and the method of Exercise 25, obtain a two-decimal estimate of (a) $\sqrt{7}$, (b) $\sqrt{19}$.

27. (See Exercise 25.) Prove that, if $g_1 > \sqrt{a}$, then (a) $g_2 > \sqrt{a}$ and (b) $g_2 < g_1$.

28. (See Exercise 25.) To estimate $\sqrt[3]{a}$ on a calculator that has only the four basic operations of arithmetic the following procedure can be used. Make a guess g_1 for $\sqrt[3]{a}$. Then choose

$$g_2 = \frac{2}{3} g_1 + \frac{1}{3} \frac{a}{g_1{}^2}.$$

(a) Show that, if g_1 happens to be $\sqrt[3]{a}$, then $g_2 = g_1$.

(b) Show that $g_2 = (2g_1{}^3 + a)/3g_1{}^2$ and therefore can be calculated without storage or pencil.

(c) Using $g_1 = 2$, estimate $\sqrt[3]{10}$. Calculate enough of g_2, g_3, ... so that the estimate stabilizes to two decimal places.

29. Recall that 0^0 was not defined. It might be hoped that if the positive number b and the number x are both close to 0, then b^x might be close to some fixed number. If that were so, it would suggest a definition of 0^0. The following computations using the y^x key show that, when b and x are both near 0, b^x is not near any specific number. Compute (a) $(0.001)^{0.001}$, (b) $(0.0000001)^{0.1}$. Consider also $(0.001)^0$ and $0^{0.001}$.

30. If a calculator has a square root key and a reciprocal key, how could you use them to calculate (a) $5^{1/4}$, (b) $5^{1/8}$, (c) $5^{3/8}$, (d) $5^{-1/4}$?

4.2

The number e

A question about interest on a bank account will introduce one of the most important numbers in calculus. Assume that the interest rate goes up and that banks pay 100 percent interest per year. Thus, 1 dollar left in a bank for 1 year earns 1 dollar interest and becomes

$$1 + 1 = 2 \text{ dollars}$$

at the end of the year.

If the bank compounds interest twice a year, then at the end of half a year there are

$$1 + \tfrac{1}{2} \quad \text{dollars}$$

in the account, and during the second half year this earns

$$(1 + \tfrac{1}{2})(\tfrac{1}{2}) \quad \text{dollars}$$

interest. Thus at the end of the year there are

$$(1 + \tfrac{1}{2}) + (1 + \tfrac{1}{2})(\tfrac{1}{2}) \quad \text{dollars}$$

or

$$(1 + \tfrac{1}{2})(1 + \tfrac{1}{2}) \quad \text{dollars}$$

or

$$(1 + \tfrac{1}{2})^2 \quad \text{dollars}$$

in the account. A little arithmetic shows that this is

$$2.25 \text{ dollars},$$

which is more than when the interest is computed just once a year.

A competing bank offers to compound interest three times a year. At the end of a year 1 dollar becomes

$$(1 + \tfrac{1}{3})^3 \qquad \text{dollars,}$$

which is about 2.37 dollars.

Another bank compounds monthly. At the end of a year 1 dollar becomes

$$(1 + \tfrac{1}{12})^{12} \qquad \text{dollars,}$$

which is about 2.61 dollars.

What happens to 1 dollar in 1 year if interest is compounded every week? Every day? Every hour? Every minute? Every second?

This financial question raises the purely mathematical question: What happens to

$$\left(1 + \frac{1}{n}\right)^n$$

when the integer n gets larger and larger?

When n is large, two influences affect the size of

$$\left(1 + \frac{1}{n}\right)^n.$$

Since $1 + (1/n)$ approaches 1 as n increases,

$$\left(1 + \frac{1}{n}\right)^n$$

is like a product of 1s, and so might stay near 1. On the other hand, $1 + (1/n)$ is multiplied many (n) times to obtain

$$\left(1 + \frac{1}{n}\right)^n.$$

For this reason

$$\left(1 + \frac{1}{n}\right)^n$$

might grow arbitrarily large.

n	1	2	3	12	20	100
$\left(1 + \dfrac{1}{n}\right)^n$	2	2.25	2.37	2.61	2.65	2.70

The computations already made, together with some tedious arithmetic, or use of a calculator, provide the data to two decimal accuracy shown in the margin.

It will be shown at the end of this section that as n increases so does $(1 + 1/n)^n$. (This agrees with the intuition of the competing banks.) It will also be shown that for all positive integers n the number $(1 + 1/n)^n$ is always less than 4. (This will assure the banks that they will not go bankrupt by compounding too frequently.) From these two assertions and the basic properties of real numbers, it follows that, as n increases, $(1 + 1/n)^n$ gets closer and closer to some fixed number which is not larger than 4. This specific number has a decimal expansion that begins $2.71828 \cdots$. It is called *e* in honor of the eighteenth-century Swiss mathematician Euler (pronounced "oiler").

Definition of e The above discussion is summarized in the brief equation

$$\lim_{n \to \infty} \left(1 + \frac{1}{n}\right)^n = e = 2.718 \cdots.$$

The symbol "$\lim_{n \to \infty}$" is read as "the limit as *n* goes to infinity of ...." The equation asserts that as *n* is chosen larger and larger, the expression $(1 + 1/n)^n$ approaches *e*.

When this mathematical discussion is translated into financial terms, it becomes apparent that no matter how often a bank that pays interest at the rate of 100 percent per year chooses to compound, 1 dollar left there for a year will never grow to more than 2.71828 dollars. Why the number *e* is important in calculus will become clear in the next chapter.

Note that, when *n* is large, $1/n$ is small. Also observe that *n* is the reciprocal of $1/n$. Thus when *x* is a small, positive number of the form

$$x = \frac{1}{n}$$

for some positive integer *n*, $(1 + x)^{1/x}$

is near *e*. This fact will be used in Sec. 4.4.

For use later in this chapter it will also be important to know what happens to

$$\left(1 + \frac{1}{n}\right)^n$$

when *n* is a *negative* integer of large absolute value. For instance, when $n = -4$,

$$\left(1 + \frac{1}{n}\right)^n = \left(1 + \frac{1}{-4}\right)^{-4}$$

$$= \left(1 - \frac{1}{4}\right)^{-4}$$

$$= \left(\frac{3}{4}\right)^{-4}$$

$$= \left(\frac{4}{3}\right)^{4}$$

$$= \frac{256}{81}$$

$$\doteq 3.16.$$

When $n = -10$, similar arithmetic shows that

$$\left(1 + \frac{1}{n}\right)^n \doteq 2.87.$$

It turns out that
$$\left(1 + \frac{1}{n}\right)^n \to e$$

as the negative integer n is chosen to be of larger and larger absolute value. (This is shown in Exercises 10 to 13.) The convenient notation for this fact is

$$\lim_{n \to -\infty} \left(1 + \frac{1}{n}\right)^n = e.$$

When n is negative and of large absolute value, $1/n$ is a small negative number. Thus, if x is a small negative number of the form $1/n$ for some (negative) integer n, it follows that $(1 + x)^{1/x}$ is near e. This fact will be needed in Sec. 4.4.

All that remains is to show that, as n increases, so does $(1 + 1/n)^n$ and also that $(1 + 1/n)^n$ remains always less than 4. The argument will rest on the first inequality in the following lemma. Both inequalities will be used, but for different purposes, in Chap. 7.

LEMMA   Let c and d be numbers such that $c > d \geq 0$. Let n be a positive integer. Then

$$c^{n+1} - d^{n+1} < (n + 1)c^n(c - d), \tag{1}$$

and

$$c^{n+1} - d^{n+1} > (n + 1)d^n(c - d). \tag{2}$$

PROOF   For simplicity take the case $n = 3$, hence $n + 1 = 4$. In this case

$$c^{n+1} - d^{n+1} = c^4 - d^4.$$

Now,

$$c^4 - d^4 = (c^3 + c^2 d + cd^2 + d^3)(c - d),$$

an identity that can be checked by multiplying out the right-hand side. Now, replacing d by c increases size. Thus the right-hand side is *less* than

$$(c^3 + c^2 \cdot c + c \cdot c^2 + c^3)(c - d),$$

which equals

$$4c^3(c - d).$$

Thus

$$c^4 - d^4 < 4c^3(c - d).$$

This establishes inequality (1) for $n = 3$. A similar argument proves it for any positive integer n.

To establish (2) for $n = 3$, start again with the same equality,

$$c^4 - d^4 = (c^3 + c^2 d + cd^2 + d^3)(c - d),$$

and note that the right-hand side is *greater* than

$$(d^3 + d^2 \cdot d + d \cdot d^2 + d^3)(c - d),$$

since replacing c by d decreases size. Thus

$$c^4 - d^4 > 4d^3(c - d).$$

This establishes inequality (2) for $n = 3$, and a similar argument proves it for any positive integer n. ●

The lemma provides quick proofs for the following two theorems.

THEOREM 1   For any positive integer n

$$\left(1 + \frac{1}{n}\right)^n < \left(1 + \frac{1}{n+1}\right)^{n+1}.$$

PROOF   Use inequality (1) with the particular choice,

$$c = 1 + \frac{1}{n} \quad \text{and} \quad d = 1 + \frac{1}{n+1}, \quad \text{for which } c > d.$$

Inequality (1) then becomes

$$\left(1 + \frac{1}{n}\right)^{n+1} - \left(1 + \frac{1}{n+1}\right)^{n+1} < (n+1)\left(1 + \frac{1}{n}\right)^n \left[\left(1 + \frac{1}{n}\right) - \left(1 + \frac{1}{n+1}\right)\right]$$

or $$\left(1 + \frac{1}{n}\right)\left(1 + \frac{1}{n}\right)^n - \left(1 + \frac{1}{n+1}\right)^{n+1} < (n+1)\left(1 + \frac{1}{n}\right)^n \left(\frac{1}{n(n+1)}\right).$$

Cancellation of the $(n+1)$'s on the right-hand side gives

$$\left(1 + \frac{1}{n}\right)\left(1 + \frac{1}{n}\right)^n - \left(1 + \frac{1}{n+1}\right)^{n+1} < \frac{1}{n}\left(1 + \frac{1}{n}\right)^n.$$

Collecting the terms and transposing yields the desired inequality

$$\left(1 + \frac{1}{n}\right)^n < \left(1 + \frac{1}{n+1}\right)^{n+1}.$$

This proves Theorem 1. ●

THEOREM 2   For every positive integer n,

$$\left(1 + \frac{1}{n}\right)^n < 4.$$

PROOF   Inequality (1) with

$$c = 1 + \frac{1}{2n} \quad \text{and} \quad d = 1, \quad \text{for which } c > d,$$

becomes

$$\left(1 + \frac{1}{2n}\right)^{n+1} - 1^{n+1} < (n+1)\left(1 + \frac{1}{2n}\right)^n \left[\left(1 + \frac{1}{2n}\right) - 1\right],$$

or $$\left(1 + \frac{1}{2n}\right)\left(1 + \frac{1}{2n}\right)^n - 1 < (n+1)\left(1 + \frac{1}{2n}\right)^n \left(\frac{1}{2n}\right),$$

or
$$\left(1 + \frac{1}{2n}\right)\left(1 + \frac{1}{2n}\right)^n - 1 < \left(\frac{1}{2} + \frac{1}{2n}\right)\left(1 + \frac{1}{2n}\right)^n.$$

Collecting like terms and transposing yields

$$\frac{1}{2}\left(1 + \frac{1}{2n}\right)^n < 1,$$

or
$$\left(1 + \frac{1}{2n}\right)^n < 2.$$

Squaring this last inequality shows that

$$\left(1 + \frac{1}{2n}\right)^{2n} < 4.$$

Now, $2n$ is an *even* integer. Thus Theorem 2 is established for any even integer. But, because the sequence $(1 + 1/n)^n$ increases as n increases, it follows that when n is odd Theorem 2 must also hold. This establishes Theorem 2. ●

While it is not important to memorize the proofs of Theorems 1 and 2, it is a good idea to remember the lemma on which they are based. It will be used later.

Exercises

1. Compute $(1 + 1/n)^n$ to two decimal places when (a) $n = 5$, (b) $n = -5$.
2. Compute $(1 + x)^{1/x}$ to two decimal places when x is equal to (a) $\frac{1}{2}$, (b) $\frac{1}{3}$, (c) $-\frac{1}{2}$, (d) $-\frac{1}{3}$. In Sec. 4.4 the behavior of $f(x) = (1 + x)^{1/x}$ for x near 0 will be studied.
3. Compute $(1 - 1/n)^n$ to two decimal places for n equal to (a) 2, (b) 3, (c) 4. [Exercise 14 shows that, when n is a large positive integer, $(1 - 1/n)^n$ is near $1/e \doteq 0.37$.]
4. (a) Define e. (b) Give its decimal representation to three places.
5. Find: (a) $\lim_{n \to \infty} (1 + 1/n)^{100}$, (b) $\lim_{n \to \infty} (1 + 1/2n)^n$. *Hint*: $(1 + 1/2n)^n = [(1 + 1/2n)^{2n}]^{1/2}$.
6. Verify inequality (1) with $c = 1.1$ and $d = 1$.
7. Prove the lemma for $n = 4$.

■

8. A certain bank pays interest at 5 percent per year compounded n times per year.
   (a) Show that at the end of a year an account of 1 dollar will grow to $(1 + 0.05/n)^n$ dollars.
   (b) Show that

$$\lim_{n \to \infty} \left(1 + \frac{0.05}{n}\right)^n = e^{0.05}.$$

9. Find

$$\lim_{n \to \infty} \left(\frac{n + 2}{n + 1}\right)^n.$$

The next four exercises show that if

$$\lim_{n \to \infty} \left(1 + \frac{1}{n}\right)^n = e$$

then, when *negative* integers n are chosen of larger and larger absolute value,

$$\left(1 + \frac{1}{n}\right)^n \to e.$$

10. Let n be a negative integer $n = -p$, where p is a positive integer. Show that

$$\left(1 + \frac{1}{n}\right)^n = \left(\frac{p}{p - 1}\right)^p.$$

11. Let $q = p - 1$, where p is defined in Exercise 10. Show that

$$\left(\frac{p}{p-1}\right)^p = \left(1 + \frac{1}{q}\right)^q \left(1 + \frac{1}{q}\right).$$

12. When q in Exercise 11 becomes larger, what happens to

(a) $\left(1 + \frac{1}{q}\right)^q$?

(b) $1 + \frac{1}{q}$?

(c) $\left(1 + \frac{1}{q}\right)^q \left(1 + \frac{1}{q}\right)$?

13. Combining Exercises 10 to 12, show that, when the *negative integer* n has a large absolute value, $(1 + 1/n)^n$ is near e.

■ ■

14. Prove that
$$\lim_{n \to \infty} \left(1 - \frac{1}{n}\right)^n = \frac{1}{e} \doteq 0.37.$$

15. Theorem 2 shows that for all positive integers n, $(1 + 1/n)^n < 4$. This exercise strengthens that result.

(a) Using inequality (1) with $c = 1 + 1/mn$ and $d = 1$, where m and n are positive integers and $m > 1$, show that

$$\left(1 + \frac{1}{mn}\right)^{mn} < \left(\frac{m}{m-1}\right)^m.$$

(b) Using $m = 6$ in (a), show that $e < 2.99$. This exercise requires the use of a calculator.

16. The number e is larger than $(1 + 1/n)^n$ for all positive integers n and, by Exercise 15(a), smaller than $[m/(m-1)]^m$ for all integers m greater than 1. Using values of n and m, such as 10, 20, and larger, obtain e with as much accuracy as time permits, perhaps two decimal places.

4.3

The limit of a real function

In Sec. 3.3, the derivative at x of the function f was defined as the limit of a certain quotient (if the limit exists). The concept of a limit, which occurs in the definition of a derivative, is of use throughout calculus. For this reason it is appropriate to pause and examine this concept in a variety of examples. After Example 4, the term *limit* will be defined.

EXAMPLE 1   Let $f(x) = 2x^2 + 1$. What happens to $f(x)$ as x is chosen closer and closer to 3?

SOLUTION   Let us make a table of the value of $f(x)$ for some choices of x near 3.

x	3.1	3.01	3.001	2.999	2.99	2.9
$f(x)$	20.22	19.1202	19.012002	18.988002	18.8802	17.82

When x is close to 3, $2x^2 + 1$ is close to $2(3)^2 + 1 = 19$. We say that "the limit of $2x^2 + 1$ as x approaches 3 is 19" and write

$$\lim_{x \to 3} (2x^2 + 1) = 19. \ \bullet$$

Example 1 presented no obstacle. The next example offers a slight challenge.

EXAMPLE 2   Let $f(x) = (x^3 - 1)/(x - 1)$. Note that this function is not defined when $x = 1$, for when x is 1, both numerator and denominator are 0. But we have every right to ask: How does $f(x)$ behave when x is *near* 1 but is *not* 1 itself?

SOLUTION   First make a brief table of values of $f(x)$ for x near 1, as shown in the margin. For instance,

x	1.1	1.01	0.9	0.99
$f(x)$	3.31	3.0301	2.71	2.9701

$$f(1.01) = \frac{(1.01)^3 - 1}{1.01 - 1} = \frac{1.030301 - 1}{0.01} = 3.0301.$$

There are two influences acting on the fraction $(x^3 - 1)/(x - 1)$ when x is near 1. *On the one hand, the numerator* $x^3 - 1$ *approaches* 0; *thus there is an influence pushing the fraction toward* 0. *On the other hand, the denominator* $x - 1$ *also approaches* 0; *division by a small number tends to make a fraction large.* How do these two opposing influences balance out?

The algebraic identity

$$x^3 - 1 = (x^2 + x + 1)(x - 1)$$

enables us to answer this question:

$$\lim_{x \to 1} \frac{x^3 - 1}{x - 1} = \lim_{x \to 1} \frac{(x^2 + x + 1)(x - 1)}{x - 1}$$

$$= \lim_{x \to 1} (x^2 + x + 1)$$

$$= 1^2 + 1 + 1$$

$$= 3.$$

The limit is 3, as the table may have suggested. ●

EXAMPLE 3   Consider the function f defined by

$$f(x) = \frac{x}{|x|}.$$

The domain of this function consists of every number except 0. For instance,

$$f(3) = \frac{3}{|3|} = \frac{3}{3} = 1,$$

and

$$f(-2) = \frac{-2}{|-2|} = \frac{-2}{2} = -1.$$

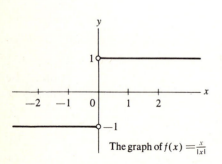

The graph of $f(x) = \frac{x}{|x|}$

When x is positive, $f(x) = 1$. When x is negative, $f(x) = -1$. This is visualized in the accompanying graph of f. The graph does not intersect the y axis, since f is not defined for $x = 0$. The hollow circles at $(0, 1)$ and $(0, -1)$ indicate that those points are not on the graph. What happens to $f(x)$ as $x \to 0$?

SOLUTION   As $x \to 0$ through positive numbers, $f(x) \to 1$, since $f(x) = 1$ for any positive number. As $x \to 0$ through negative numbers, $f(x) \to -1$, since $f(x) = -1$ for any negative number.

When x is near 0, it is *not* the case that $f(x)$ is near one specific number. Thus

$$\lim_{x \to 0} f(x)$$

does *not* exist, that is,

$$\lim_{x \to 0} \frac{x}{|x|}$$

does *not* exist. However, if $a \neq 0$,

$$\lim_{x \to a} f(x)$$

does exist, being 1 when a is positive and -1 when a is negative. Thus $\lim_{x \to a} f(x)$ exists for all a other than 0. ●

Whether a function f has a limit at a has nothing to do with $f(a)$ itself. In fact a might not even be in the domain of f. See, for instance, Examples 2 and 3. In Example 1, $a(=3)$ happened to be in the domain of f, but that fact did not influence the reasoning. It is only the behavior of $f(x)$ for x *near a* that concerns us.

Note: The derivative of a function f at x is defined as

$$\lim_{x_1 \to x} \frac{f(x_1) - f(x)}{x_1 - x}.$$

In this case x is fixed and plays the role of a in the preceding definition. The function whose behavior is being examined near x is the function that assigns to the number x_1 the value

$$\frac{f(x_1) - f(x)}{x_1 - x}.$$

If the function g is defined by the formula

$$g(x_1) = \frac{f(x_1) - f(x)}{x_1 - x},$$

then the derivative of f at x is

$$\lim_{x_1 \to x} g(x_1).$$

The next example illustrates this in a specific case.

EXAMPLE 4   Let $f(x) = x^3$ for each number x. Let x be fixed and define the function g by the formula

$$g(x_1) = \frac{f(x_1) - f(x)}{x_1 - x}$$

$$= \frac{x_1^3 - x^3}{x_1 - x}.$$

(This quotient is used in the definition of the derivative of f at x.) Find $\lim_{x_1 \to x} g(x_1)$.

SOLUTION
$$\lim_{x_1 \to x} g(x_1) = f'(x) = (x^3)' = 3x^2. \quad \bullet$$

These four examples provide a background for describing the limit concept which will be used throughout the text. The following definition will be of use in phrasing these descriptions and later theorems in the text.

DEFINITION   *Open interval.* Let a and b be two numbers, with $a < b$. The *open interval* (a, b) consists of all real numbers x such that

$$a < x < b.$$

The open interval (a, b) is obtained from the interval $[a, b]$ simply by the deletion of a and b. Sometimes, for clarity, the interval $[a, b]$ is called the *closed interval* $[a, b]$. The fact that the notation (a, b) is the same as that for the point (a, b) should cause no difficulty; the context will make it clear which meaning is intended.

The Solid-Dot and Hollow-Dot Notation

The closed interval $[a, b]$ is usually sketched as

and the open interval (a, b) as

In describing graphs or intervals a hollow dot indicates the absence of a point; a solid dot indicates the presence of a point.

A More Formal Definition of Limit Is to Be Found in Appendix C.

Consider now a function f and a number a which may or may not be in the domain of f. In order to discuss the behavior of $f(x)$ for x near a we must know that the domain of f contains numbers arbitrarily close to a. Note how this assumption is built into each of the following definitions.

DEFINITION   *Limit of $f(x)$ at a.* Let f be a function and a some fixed number. Assume that the domain of f contains open intervals (a, b) and (c, a) for some number $b > a$ and some number $c < a$. If, as x approaches a, both from the right and from the left, $f(x)$ approaches a specific number L, then L is called the limit of $f(x)$ as x approaches a. This is written

$$\lim_{x \to a} f(x) = L,$$

or   as $x \to a$, $f(x) \to L$.

In Example 1, we found $\lim_{x \to 3}(2x^2 + 1) = 19$, which illustrates this definition for $a = 3$ and $f(x) = 2x^2 + 1$. The fact that 3 happens to be in the domain of f is irrelevant. For Example 2, we found

$$\lim_{x \to 1} \frac{x^3 - 1}{x - 1} = 3,$$

which illustrates this definition for $a = 1$ and $f(x) = (x^3 - 1)/(x - 1)$. The fact that $f(x)$ is not defined for $x = 1$ did not affect the reasoning.

Example 3 concerns the behavior of $f(x) = x/|x|$ when x is near 0. As $x \to 0$, $f(x)$ does not approach a specific number. However, as x approaches 0 through positive numbers, $f(x) \to 1$. Also, as x approaches 0 through negative numbers, $f(x) \to -1$. This behavior illustrates the idea of a one-sided limit, which will now be defined.

DEFINITION   *Right-hand limit of $f(x)$ at a.*   Let f be a function and a some fixed number. Assume that the domain of f contains an open interval (a, b) for some number $b > a$. If, as x approaches a from the right, $f(x)$ approaches a specific number L, then L is called the right-hand limit of $f(x)$ as x approaches a. This is written

$$\lim_{x \to a^+} f(x) = L \qquad \text{or} \qquad \lim_{x \downarrow a} f(x) = L,$$

or   as $x \to a^+, f(x) \to L$ or   as $x \downarrow a, f(x) \to L.$

The left-hand limit is defined similarly. The only difference is the demand that the domain of f contain an open interval of the form (c, a), for some number $c < a$; $f(x)$ is examined as x approaches a from the left. The notations for the left-hand limit are

$$\lim_{x \to a^-} f(x) = L \qquad \text{or} \qquad \lim_{x \uparrow a} f(x) = L,$$

or   as $x \to a^-$, $f(x) \to L$ or   as $x \uparrow a$, $f(x) \to L.$

As Example 3 showed,

$$\lim_{x \to 0^+} \frac{x}{|x|} = 1 \qquad \text{and} \qquad \lim_{x \to 0^-} \frac{x}{|x|} = -1.$$

We could also write, for instance,

$$\text{as } x \to 0^+, \frac{x}{|x|} \to 1.$$

EXAMPLE 5   Let

$$f(x) = \frac{\sqrt{x^3 - 1}}{\sqrt{x^2 - 1}}.$$

This function is defined only for $x > 1$, since the square root of a negative number is not defined. What happens to $f(x)$ as x gets closer and closer to 1 from the right side of 1? Does $\lim_{x \to 1^+} f(x)$ exist?

SOLUTION   When x is near 1, both $x^3 - 1$ and $x^2 - 1$ are near 0. Hence $\sqrt{x^3 - 1}$ and $\sqrt{x^2 - 1}$ are near 0. Just as in Examples 2 and 4 there is an influence urging $f(x)$ to get small and an opposing influence urging $f(x)$ to get large. A little algebra will show how these two forces balance out:

$$f(x) = \frac{\sqrt{x^3 - 1}}{\sqrt{x^2 - 1}} \qquad x > 1$$

$$= \frac{\sqrt{(x^2 + x + 1)(x - 1)}}{\sqrt{(x + 1)(x - 1)}}$$

$$= \sqrt{\frac{x^2 + x + 1}{x + 1}}.$$

Now it is easy to see what happens to $f(x)$ as x approaches 1. It approaches

$$\sqrt{\frac{1^2 + 1 + 1}{1 + 1}} = \sqrt{\frac{3}{2}}.$$

In short, as x approaches 1, $f(x) \to \sqrt{\frac{3}{2}}$. ●

Note that, if both the right-hand and the left-hand limits of f exist at a and are equal, then $\lim_{x \to a} f(x)$ exists. But if the right-hand and left-hand limits are not equal, then $\lim_{x \to a} f(x)$ does not exist.

The next example reviews the three limit concepts.

EXAMPLE 6   Sketch the graph of a hypothetical function f that satisfies all these conditions:

1. The domain of f is the whole x axis.
2. $f(1) = 2$.
3. $\lim_{x \to 1} f(x) = 3$.
4. $\lim_{x \to 2^+} f(x) = 2$.

5. $\lim_{x \to 2^-} f(x)$ does not exist.
6. $f(2) = 1$.
7. $\lim_{x \to a} f(x)$ exists for all other values of a.

SOLUTION   Build up the graph freehand, first concentrating on $f(x)$ for x near 1 and then for x near 2. The following graph is just one of many possibilities.

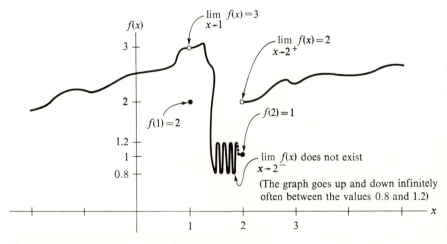

Note the use of solid dots and hollow dots. ●

Functions as wild as the one described in Example 6 will hardly be of major concern in calculus. However, they do serve to clarify the two definitions above, just as the notion of "sickness" illuminates the notion of "health."

One final point should be made. Consider

$$\lim_{x \to 3} \frac{1 + x^2}{1 + x^2}.$$

For all x, $(1 + x^2)/(1 + x^2) = 1$. Thus

$$\lim_{x \to 3} \frac{1 + x^2}{1 + x^2} = \lim_{x \to 3} 1$$

$$= 1.$$

More generally, if $f(x)$ is equal to a fixed number L for all x, then

$$\lim_{x \to a} f(x) = L.$$

In this case, it may seem a little strange to say that "the limit of L is L" or "L approaches L." In practice, though, this offers no difficulty.

Exercises

In Exercises 1 to 15 decide whether the limit exists and evaluate it if it does.

1. $\lim\limits_{x \to 3} \dfrac{x^2 - 3}{x + 2}$

2. $\lim\limits_{x \to 3} \dfrac{x^2 - 9}{x - 3}$

3. $\lim\limits_{x \to 0^+} \sqrt{x}$

4. $\lim\limits_{x \to 0^-} \sqrt{x}$

5. $\lim\limits_{x \to 3} \dfrac{(1/x) - \frac{1}{2}}{x - 2}$

6. $\lim\limits_{x \to 2} \dfrac{(1/x) - \frac{1}{2}}{x - 2}$

7. $\lim\limits_{h \to 1} \dfrac{(1 + h)^3 - 1}{h}$

8. $\lim\limits_{h \to 0} \dfrac{(1 + h)^3 - 1}{h}$

9. $\lim\limits_{x \to 2^-} \sqrt{2 - x}$

10. $\lim\limits_{x \downarrow 1} \dfrac{x - 1}{|x - 1|}$

11. $\lim\limits_{x \uparrow 1} \dfrac{x - 1}{|x - 1|}$

12. $\lim\limits_{x \to 1} \dfrac{x - 1}{|x - 1|}$

13. $\lim\limits_{x \to -2} \dfrac{x^3 + 8}{x + 2}$

14. $\lim\limits_{x \to 4^+} \sqrt{x - 4} + 2$

15. $\lim\limits_{x \to 0} \dfrac{\sqrt{x + 4} - 2}{x}$

In Exercises 16 to 21 the graphs of various functions are given. Describe those numbers a for which one or more of the limits $\lim_{x \to a^+} f(x)$, $\lim_{x \to a^-} f(x)$, and $\lim_{x \to a} f(x)$ exist; in each case indicate which of the limits exist.

16.

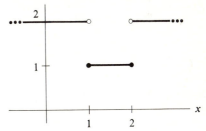

17.

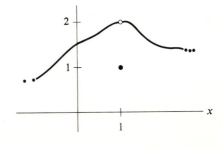

18.

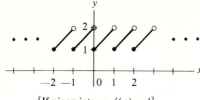

[If x is an integer, $f(x) = 1$]

19.

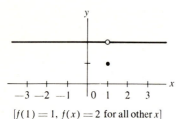

[$f(1) = 1$, $f(x) = 2$ for all other x]

20.

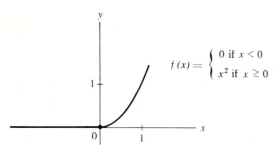

$$f(x) = \begin{cases} 0 & \text{if } x < 0 \\ x^2 & \text{if } x \ge 0 \end{cases}$$

21.

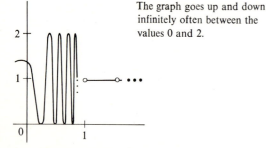

The graph goes up and down infinitely often between the values 0 and 2.

In Exercises 22 to 26 sketch the graph of a hypothetical function f that meets all the demands. Do not give a formula for the function.

22. The domain is $(0, 2)$; $\lim_{x \to a} f(x)$ exists at each $a \ne 1$; $\lim_{x \to 1^+} f(x) = 3$; $\lim_{x \to 1^-} f(x) = 4$; $\lim_{x \to 2^-} f(x)$ does not exist; $\lim_{x \to 0^+} f(x)$ exists.

23. The domain is $(0, 3)$; $\lim_{x \to a} f(x) = 1$ for each a; $f(1) = 3$.

24. The domain is the x axis; $f(1) = 3$; $\lim_{x \to 1^+} f(x) = 4$; $\lim_{x \to 1^-} f(x) = 3$; $\lim_{x \to a} f(x) = f(a)$ for all other a.

25. The domain is the x axis; as $x \to 1^+$, $f(x) \to 3$; as $x \to 1^-$, $f(x) \to 2$; $f(1) = 0$; $\lim_{x \to a} f(x)$ exists for all other a.

26. The domain is $[2, 3]$; $\lim_{x \to 2^+} f(x)$ does not exist; $\lim_{x \to 3^-} f(x) = 1$; $\lim_{x \to a} f(x)$ exists for all other a.

27. Define a certain function f as follows:

$$f(x) = \begin{cases} 1 & \text{if } x \text{ is an integer} \\ 0 & \text{if } x \text{ is not an integer} \end{cases}$$

(a) Graph f.
(b) Does $\lim_{x \to 3} f(x)$ exist?
(c) Does $\lim_{x \to 3.5} f(x)$ exist?
(d) For which numbers a does $\lim_{x \to a} f(x)$ exist?

28. (a) Graph the function f given by the formula

$$f(x) = x + |x|.$$

(b) For which numbers a does $\lim_{x \to a} f(x)$ exist?

29. Define f as follows:

$$f(x) = \begin{cases} x & \text{if } x \text{ is rational} \\ -x & \text{if } x \text{ is not rational} \end{cases}$$

(a) What does the graph of f look like?
(b) Does $\lim_{x \to 1} f(x)$ exist?
(c) Does $\lim_{x \to \sqrt{2}} f(x)$ exist?
(d) Does $\lim_{x \to 0} f(x)$ exist?
(e) For which numbers a does $\lim_{x \to a} f(x)$ exist?

■

30. The limit

$$\lim_{h \to 0} \frac{2^h - 1}{h}$$

will be determined in the next chapter; assume that this limit exists.

(a) Show that this limit is the derivative of the exponential function 2^x at $x = 0$.
(b) Sketch the graph of $y = 2^x$.
(c) Estimate the slope of the line through $(\frac{1}{2}, 2^{1/2})$ and $(0, 1)$, using the approximation $\sqrt{2} \doteq 1.4$.
(d) Estimate the slope of the line through $(-\frac{1}{2}, 2^{-1/2})$ and $(0, 1)$, using the approximation $\sqrt{2} \doteq 1.4$.
(e) What do (c) and (d) suggest about the numerical value of

$$\lim_{h \to 0} \frac{2^h - 1}{h}?$$

■ ■

This exercise requires the use of a calculator.

31. Explore the behavior of $(1/x)^x$ for x positive and near 0. What do you think $\lim_{x \downarrow 0} (1/x)^x$ is?

4.4

More on limits; e revisited

Certain properties of limits will be used often in subsequent chapters. In particular, let f and g be two functions, and assume that

$$\lim_{x \to a} f(x) = A \quad \text{and} \quad \lim_{x \to a} g(x) = B.$$

Then, as is proved in Appendix C,

$$\lim_{x \to a} [f(x) + g(x)] = A + B; \qquad \lim_{x \to a} [f(x) - g(x)] = A - B;$$

$$\lim_{x \to a} [f(x)g(x)] = AB; \qquad \lim_{x \to a} \frac{f(x)}{g(x)} = \frac{A}{B} \quad \text{if } B \neq 0.$$

These four basic theorems present no surprises; still, it is good to have them on record.

Sometimes it is useful to know how $f(x)$ behaves when x is a very large positive number. Example 1 serves as an illustration and introduces a variation on the theme of limits.

EXAMPLE 1 How does

$$f(x) = \frac{5x^2 + 2}{7x^3 + 3}$$

behave when x is a very large positive number?

SOLUTION As x gets large, the numerator $5x^2 + 2$ grows large, influencing the quotient to become large. On the other hand, the denominator is also getting large, influencing the quotient to become small. An algebraic device will reveal which influence dominates.

We have

$$f(x) = \frac{5x^2 + 2}{7x^3 + 3}$$

$$= \frac{x^2(5 + 2/x^2)}{x^3(7 + 3/x^3)}$$

$$= \frac{1}{x} \frac{(5 + 2/x^2)}{(7 + 3/x^3)}.$$

Now it is easier to see what happens to $f(x)$ for large x.

As x gets large, $1/x \to 0$, $2/x^2 \to 0$, and $3/x^3 \to 0$. Thus as x gets larger through positive values,

$$f(x) \to 0 \cdot \frac{5 + 0}{7 + 0},$$

or

$$f(x) \to 0. \quad \bullet$$

Rather than writing, "As x gets larger through positive values, $f(x)$ approaches the number L," it is customary to use the shorthand

$$\lim_{x \to \infty} f(x) = L.$$

How to Find the Limit of the
Quotient of Polynomials

As Example 1 suggests, to find the limit of the quotient of two polynomials as $x \to \infty$, first factor out of the numerator the highest power of x that appears there, and then factor out of the denominator the highest power of x that appears there.

The next example concerns a case in which $f(x)$ itself becomes arbitrarily large.

EXAMPLE 2   How does $f(x) = 1/x$ behave when x is near 0?

SOLUTION   The reciprocal of a small number x has a large absolute value.   For instance, when $x = 0.01$, $1/x = 100$; when $x = -0.01$, $1/x = -100$.   Thus, as x approaches 0 from the right, $1/x$, which is positive, becomes arbitrarily large. The notation for this is

$$\lim_{x \to 0^+} \frac{1}{x} = \infty.$$

As x approaches 0 from the left, $1/x$, which is negative, has arbitrarily large absolute values.   The notation for this is

$$\lim_{x \to 0^-} \frac{1}{x} = -\infty. \ \bullet$$

The many different types of limits all have the same flavor.   Rather than spell each out in detail, we list some of the typical cases that arise.   Formal definitions can be phrased in the style of Appendix C.   But the informal definitions are adequate for the problems that arise in introductory calculus.

Notation	In Words	Concept
$\lim_{x \to a} f(x) = L$	As x approaches a, $f(x)$ approaches L	Discussed in Sec. 4.2.
$\lim_{x \to \infty} f(x) = L$	As x approaches positive infinity, $f(x)$ approaches L	$f(x)$ is defined for all x beyond some number and, as x gets large through positive values, $f(x)$ approaches L.
$\lim_{x \to -\infty} f(x) = L$	As x approaches negative infinity, $f(x)$ approaches L	$f(x)$ is defined for all x to the left of some number and, as the negative number x takes on large absolute values, $f(x)$ approaches L.
$\lim_{x \to \infty} f(x) = \infty$	As x approaches infinity, $f(x)$ approaches positive infinity.	$f(x)$ is defined for all x beyond some number and, as x gets large through positive values, $f(x)$ becomes and remains arbitrarily large and positive.
$\lim_{x \to a^+} f(x) = \infty$	As x approaches a from the right, $f(x)$ approaches positive infinity.	$f(x)$ is defined in some open interval (a, b) and, as x approaches a from the right, $f(x)$ becomes and remains arbitrarily large and positive.
$\lim_{x \to a^+} f(x) = -\infty$	As x approaches a from the right, $f(x)$ approaches negative infinity.	$f(x)$ is defined in some open interval (a, b) and, as x approaches a from the right, $f(x)$ becomes negative and $|f(x)|$ becomes and remains arbitrarily large.

Other notations, such as

$$\lim_{x \to a} f(x) = -\infty, \qquad \lim_{x \to a^-} f(x) = \infty, \qquad \text{and} \qquad \lim_{x \to \infty} f(x) = -\infty$$

are defined similarly.

The next example illustrates how these ideas are of aid in graphing. At the same time it develops one of the most important limits in calculus,

$$\lim_{x \to 0} (1 + x)^{1/x}.$$

EXAMPLE 3   Let $f(x) = (1 + x)^{1/x}$.   Graph f.

SOLUTION   In order to keep the base $1 + x$ positive, consider only $x > -1$.   Furthermore, since the exponent in $(1 + x)^{1/x}$, $1/x$, is meaningless when $x = 0$, 0 is not in the domain of f.

Observe that for any integer n, whether positive or negative,

$$f\left(\frac{1}{n}\right) = \left(1 + \frac{1}{n}\right)^{1/(1/n)}$$

$$= \left(1 + \frac{1}{n}\right)^{n},$$

which is precisely the expression examined in Sec. 4.2.   Recall that

$$\lim_{n \to \infty} \left(1 + \frac{1}{n}\right)^{n} = e \doteq 2.718 \qquad \text{and that} \qquad \lim_{n \to -\infty} \left(1 + \frac{1}{n}\right)^{n} = e \doteq 2.718.$$

Thus, if x is a small number, positive or negative, of the form $1/n$, then

$$f(x) \text{ is near } e.$$

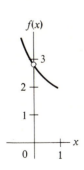

It can be shown that if x is near 0, even though x is not the reciprocal of an integer, still

$$f(x) \text{ is near } e.$$

(See Exercise 35.)   In symbols, therefore

$$\lim_{x \to 0} (1 + x)^{1/x} = e.$$

The graph of f for x near 0 appears in the margin.

All that remains is to see how $f(x)$ behaves when x is near -1 and also when x is positive and large.

Consider the first case, $x > -1$ but near -1.   Try $x = -0.99$.   We have

$$f(-0.99) = [1 + (-0.99)]^{1/-0.99}$$

$$= (0.01)^{-1/0.99}$$

$$= \frac{1}{(0.01)^{1/0.99}}.$$

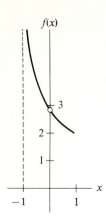

$f(x)$

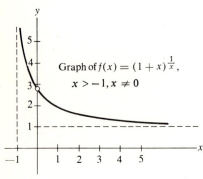

Graph of $f(x) = (1 + x)^{\frac{1}{x}}$,
$x > -1, x \neq 0$

Now, 0.01 is near 0, and the exponent, 1/0.99, is near 1. Thus

$$(0.01)^{1/0.99}$$

is near 0 and is positive. Thus $f(-0.99)$ is a large positive number. There are no opposing forces in this case; we may write

$$\lim_{x \to -1^+} (1 + x)^{1/x} = \infty.$$

The information gathered so far suffices to sketch the graph of f to the extent shown in the margin. It remains to see how $f(x)$ behaves for large positive values of x.

To get a feeling for the behavior of $(1 + x)^{1/x}$ when x is a large positive number, consider, say, $x = 10$. In this case

$$(1 + x)^{1/x} = (1 + 10)^{1/10}$$
$$= 11^{1/10}$$
$$= \sqrt[10]{11}.$$

The tenth root of 11 is slightly larger than 1. It is easy to show by the methods of Chap. 6 that

$$\lim_{x \to \infty} (1 + x)^{1/x} = 1.$$

The graph of f has the general character shown in the margin. ●

In Chap. 5 it will be useful to think of the assertion

$$\lim_{x \to 0} (1 + x)^{1/x} = e$$

as saying "$(1 + \text{small number } s)^{\text{reciprocal of } s}$ is close to e."

As an illustration,

$$\lim_{x \to 0} \left(1 + \frac{x}{2}\right)^{2/x} = e$$

since, when x is near 0, $2/x$ is the reciprocal of the small number $x/2$.

Exercises

In Exercises 1 to 26 find the indicated limits.

1. $\lim\limits_{x \to \infty} \dfrac{5x^2 + 2}{7x^2 + 3}$

2. $\lim\limits_{x \to \infty} \dfrac{6x^2 + 3}{4x + 1}$

3. $\lim\limits_{x \to -\infty} \dfrac{6x^3 + 2x^2 + 1}{5x^2 - 1}$

4. $\lim\limits_{x \to \infty} \dfrac{x^2 + (1/x)}{x^2 - (1/x)}$

5. $\lim\limits_{x \to 0} \dfrac{x^2 + (1/x)}{x^2 - (1/x)}$

6. $\lim\limits_{x \to 1^+} \dfrac{x}{x - 1}$

7. $\lim\limits_{x \to 1^-} \dfrac{x}{x - 1}$

8. $\lim\limits_{x \to \infty} \dfrac{x}{x + 1}$

9. $\lim\limits_{x \to -\infty} 2^x$

10. $\lim\limits_{x \to \infty} 2^x$

11. $\lim\limits_{x \to 0^+} (2 + x)^{1/x}$

12. $\lim\limits_{x \to 0^-} (2 + x)^{1/x}$

13. $\lim\limits_{x \to \infty} \left[\dfrac{1}{x - (x^2/2)} - \dfrac{1}{x}\right]$

14. $\lim\limits_{x \to 0} \left[\dfrac{1}{x - (x^2/2)} - \dfrac{1}{x}\right]$

15. $\lim\limits_{x \to \infty} \left[\dfrac{1}{x+1} - \dfrac{1}{x} \right]$

16. $\lim\limits_{x \to 0^+} \left[\dfrac{1}{x+1} - \dfrac{1}{x} \right]$

17. $\lim\limits_{x \to 0^-} \left[\dfrac{1}{x+1} - \dfrac{1}{x} \right]$

18. $\lim\limits_{x \to \infty} \dfrac{2^x}{3^x}$

19. $\lim\limits_{x \to 0^+} \dfrac{2^x + 3}{2^x - 1}$

20. $\lim\limits_{x \to 0^-} \dfrac{2^x + 3}{2^x - 1}$

21. $\lim\limits_{x \to \infty} \dfrac{2^x + 3}{2^x - 1}$

22. $\lim\limits_{x \to -\infty} \dfrac{2^x + 3}{2^x + 1}$

23. $\lim\limits_{x \to \infty} \left[\dfrac{1}{x-1} - \dfrac{1}{x^2 - 1} \right]$

24. $\lim\limits_{x \to 1^+} \left[\dfrac{1}{x-1} - \dfrac{1}{x^2 - 1} \right]$

25. $\lim\limits_{x \to 1^+} \dfrac{\sqrt{x^2 - 1}}{\sqrt{x^3 - 1}}$

26. $\lim\limits_{x \to \infty} \dfrac{x}{\sqrt{2x^2 + 1}}$

Exercises 27 to 30 concern $\lim_{x \to 0} (1 + x)^{1/x} = e$.

27. Evaluate $f(x) = (1 + x)^{1/x}$ to two decimal places for $x = \frac{1}{3}, \frac{1}{4}, \frac{1}{5}, -\frac{1}{3}, -\frac{1}{4}, -\frac{1}{5}$.

28. Find $\lim_{x \to 0} (1 + 3x)^{2/x}$.

29. Find $\lim_{x \to 0} (1 + x^2)^{1/x}$.

30. For a fixed nonzero x, find $\lim_{h \to 0} (1 + h/x)^{x/h}$.

■

In Exercises 31 to 33 sketch the graph of a hypothetical function that has the properties mentioned.

31. $f(1) = 2$, $\lim_{x \to 1^+} f(x) = \infty$, $\lim_{x \to 1^-} f(x) = \infty$.

32. $f(1) = 1$, $\lim_{x \to 1^+} f(x) = -\infty$, $\lim_{x \to 1^-} f(x) = \infty$.

33. $f(0) = 1$, $\lim_{x \to 0} f(x) = 2$, $\lim_{x \to \infty} f(x) = 1$.

34. Give an example of functions f and g such that we have $\lim_{x \to 0} f(x) = 0$ and $\lim_{x \to 0} g(x) = 0$ and (a) $\lim_{x \to 0} f(x)/g(x) = 3$; (b) $\lim_{x \to 0} f(x)/g(x) = 0$; (c) $\lim_{x \to 0} f(x)/g(x) = \infty$.

35. Let x be a positive number less than 1.
 (a) Show that there is a unique positive integer n such that

$$\frac{1}{n+1} < x \le \frac{1}{n}$$

or equivalently $\quad n + 1 > \dfrac{1}{x} \ge n$.

(b) Show that

$$(1 + x)^{1/x} \le \left(1 + \frac{1}{n} \right)^{1/x} < \left(1 + \frac{1}{n} \right)^{n+1}$$

$$= \left(1 + \frac{1}{n} \right)^n \left(1 + \frac{1}{n} \right).$$

(c) Similarly, show that

$$(1 + x)^{1/x} > \frac{[1 + 1/(n+1)]^{n+1}}{1 + 1/(n+1)}.$$

(d) Use (b) and (c) to show that, as $x \to 0$ through positive values,

$$(1 + x)^{1/x} \to e.$$

A similar argument can be applied to negative values of x.

■ ■

36. Sketch very carefully the graph of $f(x) = x^c$ for positive x near 0 if (a) $c = 1$, (b) $0 < c < 1$, (c) $c > 1$. For which c does the graph approach $(0, 0)$ horizontally? vertically?

The remaining exercises use the calculator to explore limits, some of which will be studied later.

37. Compute $(1 + x)^{1/x}$ for $x = 10, 20, 30$. As $x \to \infty$, does $(1 + x)^{1/x}$ seem to approach 1?

38. (a) Compute $x/2^x$ for $x = 3, 5, 10$, etc.
 (b) What do you think happens to $x/2^x$ as $x \to \infty$?

39. (a) Compute $x^5/2^x$ for $x = 3, 10, 20, 30$, etc.
 (b) What do you think happens to $x^5/2^x$ as $x \to \infty$?

40. How does $x^2/(1.01^x)$ behave for large x?

41. How does $[(2^x + 8^x)/2]^{1/x}$ behave as $x \to 0$?

42. (a) Compute $(1 + x)^{1/x}$ for $x = 0.012$ and $x = -0.012$.
 (b) What happens to $(1 + x)^{1/x}$ as $x \to 0$?

4.5

A review of trigonometry

In the next section an important limit involving a trigonometric function will be computed. The present section is included for the benefit of those whose trigonometry may be rusty.

The part of trigonometry most useful in calculus concerns three functions: sine, cosine, and tangent. The techniques for finding unknown sides and angles in triangles will seldom be needed. This review emphasizes the relation of angles to a circle in order to obtain the trigonometric functions quickly.

We can use a circle to measure the size of angles (rather than "degrees" which were introduced by Babylonian astronomers). To measure the size of an angle, such as the angle ABC shown in the accompanying figure, draw a circle with center at B.

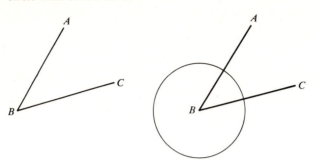

If the circle has radius r, and the angle intercepts an arc of length s, then the quotient

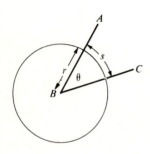

$$\frac{s}{r}$$

shall be the measure of the angle, and we say that the angle has a measure of s/r *radians*. It is frequently convenient to denote the measure of an angle by θ, and then write

$$\theta = \frac{s}{r}.$$

EXAMPLE 1 Find the radian measure of a right angle ABC.

SOLUTION Draw a circle of radius r and center at B and compute the quotient s/r. The circumference of a circle of radius r is $2\pi r$. The right angle intercepts a quarter of the circumference.

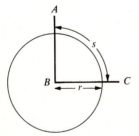

Thus

$$s = \frac{1}{4} 2\pi r = \frac{\pi r}{2},$$

and

$$\theta = \frac{s}{r} = \frac{\pi r/2}{r} = \frac{\pi}{2}.$$

So a right angle has the measure $\pi/2$ radians (about 1.57 radians). ●

The straight angle, which has 180°, consists of two right angles and therefore has a measure of π radians. This fact is helpful in translating from degrees to radians and from radians to degrees. One simply uses the proportion

$$\frac{\text{Degrees}}{180} = \frac{\text{radians}}{\pi}.$$

EXAMPLE 2 What is the measure in radians of the 30° angle?

SOLUTION   The proportion in this case becomes

$$\frac{30}{180} = \frac{\text{radians}}{\pi},$$

from which it follows that

$$\text{Radians} = \pi \frac{30}{180} = \frac{\pi}{6}. \quad \bullet$$

EXAMPLE 3   How many degrees are there in an angle of 1 radian?

SOLUTION   In this case the proportion is

$$\frac{\text{Degrees}}{180} = \frac{1}{\pi}$$

hence

$$\text{Degrees} = \frac{180}{\pi} \doteq \frac{180}{3.14} \doteq 57.3°.$$

So an angle of 1 radian is about 57.3°, a little less than 60°. ●

In the case of the *unit circle*, the circle whose radius is 1, the formula

$$\theta = \frac{s}{r}$$

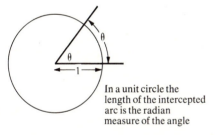

In a unit circle the length of the intercepted arc is the radian measure of the angle

becomes

$$\theta = \frac{s}{1} = s.$$

In that case the length of arc intercepted equals the measure of the angle in radians.

So far, an angle has been associated with each number θ in the interval $[0, 2\pi]$. Next, associate an angle with any *positive* number θ. For convenience, a unit circle will be used, and one arm of the angle will be placed along the positive x axis. To draw the second arm of the angle of θ radians, go around the unit circle in a counterclockwise direction a distance θ. The point P reached determines the second arm. For instance, if $\theta = 5\pi/2$ radians, it is necessary to travel clear around the circle once and then reach the point P above the center of the circle. In this case we obtain the right angle, also described by $\pi/2$ radians and shown in the accompanying drawing.

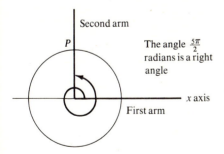

The angle $\frac{5\pi}{2}$ radians is a right angle

Every time we travel around the unit circle, we increase the measure of an angle by 2π radians. Thus the right angle of $\pi/2$ radians has an endless supply of descriptions:

$$\frac{\pi}{2}, \qquad \frac{\pi}{2} + 2\pi = \frac{5\pi}{2}, \qquad \frac{\pi}{2} + 4\pi = \frac{9\pi}{2}, \qquad \cdots.$$

To associate angles with the negative number θ, go *clockwise* around the unit circle through an angle $|\theta|$. For instance, to draw the angle $-\pi/2$, start at the point $(1, 0)$ and move along the unit circle clockwise through a right angle until

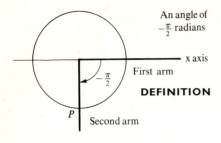

An angle of $-\frac{\pi}{2}$ radians

First arm

Second arm

The point P on the unit circle has coordinates $\cos \theta$ and $\sin \theta$

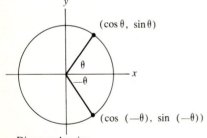

$P = (\cos \theta, \sin \theta)$

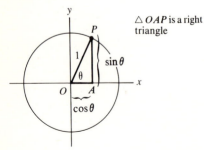

$\triangle OAP$ is a right triangle

Diagram showing that $\cos(-\theta) = \cos \theta$ and $\sin(-\theta) = -\sin \theta$

reaching the point P directly below the center of the circle. Note in the figure at the left that an angle of $-\pi/2$ radians coincides with an angle of $3\pi/2$ radians.

The two fundamental functions of trigonometry, sine and cosine, can now be defined.

DEFINITION *The sine and cosine functions.* For each number θ the sine of θ, denoted by $\sin \theta$, and the cosine of θ, denoted by $\cos \theta$, are defined as follows. Draw the angle of θ radians, whose first arm is the positive x axis and whose vertex is at $(0, 0)$. The second arm meets the unit circle whose center is at $(0, 0)$ in a point P. The x coordinate of P is called $\cos \theta$. The y coordinate of P is called $\sin \theta$.

EXAMPLE 4 Find $\cos \pi/2$ and $\sin \pi/2$.

SOLUTION If $\theta = \pi/2$, then the angle is a right angle, and $P = (0, 1)$. Hence

$$\cos \frac{\pi}{2} = 0 \qquad \text{and} \qquad \sin \frac{\pi}{2} = 1. \ \bullet$$

EXAMPLE 5 Find $\cos(-\pi)$ and $\sin(-\pi)$.

SOLUTION If $\theta = -\pi$, P is the point $(-1, 0)$.

Hence $\qquad \cos(-\pi) = -1 \qquad \text{and} \qquad \sin(-\pi) = 0. \ \bullet$

The trigonometric functions satisfy various identities. First of all, since a change of 2π in θ leads to the same point P in the circle,

$$\cos(\theta + 2\pi) = \cos \theta,$$

and $\qquad \sin(\theta + 2\pi) = \sin \theta.$

One says that the cosine and sine functions have period 2π. Second, the symmetry of the unit circle shows that

$$\cos(-\theta) = \cos \theta,$$

and $\qquad \sin(-\theta) = -\sin \theta.$

The numbers $\cos \theta$ and $\sin \theta$ are related by the equation

$$\cos^2 \theta + \sin^2 \theta = 1$$

[$\cos^2 \theta$ is short for $(\cos \theta)^2$]. To establish this, apply the pythagorean theorem to the right triangle $\triangle OAP$, shown in the accompanying figure.

With the aid of this relation between $\cos \theta$ and $\sin \theta$ the next two examples determine $\cos \pi/4$, $\sin \pi/4$, $\cos \pi/3$, and $\sin \pi/3$.

EXAMPLE 6 Find $\cos \pi/4$ and $\sin \pi/4$.

SOLUTION When the angle is $\pi/4$ ($45°$), a quick sketch shows that the cosine equals the sine:

$$\cos \frac{\pi}{4} = \sin \frac{\pi}{4}.$$

Thus
$$\cos^2 \frac{\pi}{4} + \cos^2 \frac{\pi}{4} = 1,$$

or
$$2 \cos^2 \frac{\pi}{4} = 1.$$

Consequently,
$$\cos^2 \frac{\pi}{4} = \frac{1}{2}.$$

Since $\cos \pi/4$ is positive,

$$\cos \frac{\pi}{4} = \sqrt{\frac{1}{2}}$$

$$= \frac{\sqrt{2}}{2}$$

$$(\doteq 0.707).$$

Thus
$$\cos \frac{\pi}{4} = \frac{\sqrt{2}}{2} = \sin \frac{\pi}{4}. \quad \bullet$$

EXAMPLE 7   Find $\cos \pi/3$ and $\sin \pi/3$.

SOLUTION   The angle $\pi/3$ ($60°$) is the angle in an equilateral triangle.   Place such a triangle in the unit circle as shown at the left.   Inspection of the figure shows that

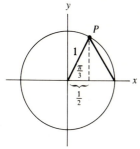

$$\cos \frac{\pi}{3} = \frac{1}{2}.$$

Then
$$\left(\frac{1}{2}\right)^2 + \sin^2 \frac{\pi}{3} = 1$$

$$\sin^2 \frac{\pi}{3} = \frac{3}{4}$$

$$\sin \frac{\pi}{3} = \frac{\sqrt{3}}{2}$$

$$(\doteq 0.866). \quad \bullet$$

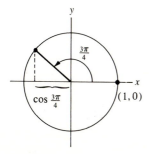

Once $\cos \pi/4$ is known, the cosine of multiples of $\pi/4$ can be found by sketching the unit circle.   For instance, to find $\cos 3\pi/4$ draw an angle of $3\pi/4$ radians, as in the accompanying figure.   It is clear that $\cos 3\pi/4$ is negative and that $|\cos 3\pi/4| = \sqrt{2}/2$.   Hence

$$\cos \frac{3\pi}{4} = -\frac{\sqrt{2}}{2}.$$

A similar method can be used to compute the cosine of multiples of $\pi/6$.   With the aid of such computations, this table for the cosine function can be obtained.

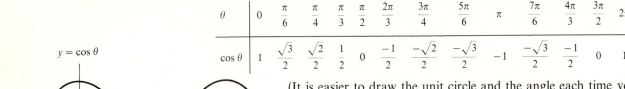

θ	0	$\dfrac{\pi}{6}$	$\dfrac{\pi}{4}$	$\dfrac{\pi}{3}$	$\dfrac{\pi}{2}$	$\dfrac{2\pi}{3}$	$\dfrac{3\pi}{4}$	$\dfrac{5\pi}{6}$	π	$\dfrac{7\pi}{6}$	$\dfrac{4\pi}{3}$	$\dfrac{3\pi}{2}$	2π
$\cos\theta$	1	$\dfrac{\sqrt{3}}{2}$	$\dfrac{\sqrt{2}}{2}$	$\dfrac{1}{2}$	0	$\dfrac{-1}{2}$	$\dfrac{-\sqrt{2}}{2}$	$\dfrac{-\sqrt{3}}{2}$	-1	$\dfrac{-\sqrt{3}}{2}$	$\dfrac{-1}{2}$	0	1

$y = \cos\theta$

(It is easier to draw the unit circle and the angle each time you need $\cos\theta$ than to memorize this table.) This table provides enough information to graph the cosine function. Since $\cos(\theta + 2\pi) = \cos\theta$, the graph consists of the portion from 0 to 2π endlessly repeated. It is sketched in the margin.

The graph of the sine function can be sketched in a similar manner.

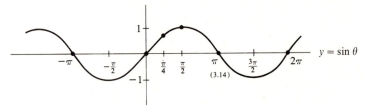

$y = \sin\theta$

Four important identities relate the cosine and sine of the sum and difference of two angles to the values of the cosine and sine of the angles:

*These Identities Are Obtained in
Exercises 20 to 24*

$$\cos(A + B) = \cos A \cos B - \sin A \sin B;$$
$$\sin(A + B) = \sin A \cos B + \cos A \sin B;$$
$$\cos(A - B) = \cos A \cos B + \sin A \sin B;$$
$$\sin(A - B) = \sin A \cos B - \cos A \sin B.$$

From these follow the "double angle" identities:

$$\cos 2\theta = \cos^2\theta - \sin^2\theta$$
$$\cos 2\theta = 2\cos^2\theta - 1$$
$$\cos 2\theta = 1 - 2\sin^2\theta$$
$$\sin 2\theta = 2\sin\theta\cos\theta$$
$$\sin^2\theta = \frac{1 - \cos 2\theta}{2}$$
$$\cos^2\theta = \frac{1 + \cos 2\theta}{2}.$$

The trigonometric function next in importance to the cosine and sine is the tangent function.

DEFINITION *The tangent function.* For each number θ that does not differ from $\pi/2$ or $-\pi/2$ by a multiple of 2π, the tangent of θ, denoted $\tan\theta$, is defined as follows:

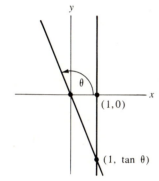

Draw the angle of θ radians whose first arm is the positive x axis and whose vertex is $(0,0)$. Let L be the line through $(1,0)$ parallel to the y axis. The line on the second arm of the angle meets the line L at a point Q. The y coordinate of Q is called $\tan \theta$.

Note from the diagram defining $\tan \theta$ that for θ near $\pi/2$ but less than $\pi/2$, $\tan \theta$ becomes very large. For θ slightly larger than $\pi/2$, $|\tan \theta|$ is large and $\tan \theta$ is negative. While $\cos \theta$ and $\sin \theta$ never exceed 1, $\tan \theta$ takes arbitrarily large values. Note that for

$$\frac{\pi}{2} < \theta < \pi$$

(a second-quadrant angle) $\tan \theta$ is negative, as the drawing in the margin shows. It follows from the definition that

$$\tan (\theta + \pi) = \tan \theta.$$

While cosine and sine repeat every 2π, the tangent function repeats more frequently, namely every π. The tangent has a period π. Tables for the functions $\cos \theta$, $\sin \theta$, and $\tan \theta$ are included in Appendix E; these functions are also available on many calculators.

The functions $\tan \theta$, $\sin \theta$, and $\cos \theta$ are related by the equation

$$\tan \theta = \frac{\sin \theta}{\cos \theta}.$$

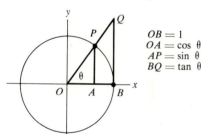

$OB = 1$
$OA = \cos \theta$
$AP = \sin \theta$
$BQ = \tan \theta$

This can easily be deduced from inspection of the accompanying diagram. By the similarity of triangles $\triangle OBQ$ and $\triangle OAP$,

$$\frac{\tan \theta}{1} = \frac{\sin \theta}{\cos \theta}$$

or

$$\tan \theta = \frac{\sin \theta}{\cos \theta}.$$

The following identities involving the tangent functions are of use:

$$\tan (A - B) = \frac{\tan A - \tan B}{1 + \tan A \tan B},$$

$$\tan (A + B) = \frac{\tan A + \tan B}{1 - \tan A \tan B},$$

$$\tan 2\theta = \frac{2 \tan \theta}{1 - \tan^2 \theta}.$$

It is easy to see that

$$\tan 0 = 0$$

and

$$\tan \frac{\pi}{4} = 1.$$

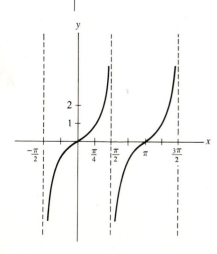

The accompanying diagram shows how the graph of the tangent function looks.

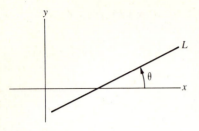

The tangent function provides another view of the slope of a line. Let a line L make an angle θ with the positive x axis. Then the slope of L is equal to $\tan \theta$. If the slope is known, then the angle can be estimated with the aid of a table of $\tan \theta$. For instance, when the slope is 2, the angle is about 1.1 radians or $63°$.

If the lengths of two sides of a triangle, a and b, and the angle between these sides θ are known, the length of the third side c is determined. The formula for finding c, the *law of cosines*, is

$$c^2 = a^2 + b^2 - 2ab \cos \theta.$$

This formula will be needed several times in the text. A proof is outlined in Exercise 28.

Exercises

1. What is the radian measure of the following angles? (*a*) $90°$, (*b*) $30°$, (*c*) $120°$, (*d*) $270°$, (*e*) $360°$.
2. How many degrees are there in an angle whose radian measure is (*a*) $3\pi/4$, (*b*) $\pi/3$, (*c*) $2\pi/3$, (*d*) 4π?
3. An angle intercepts an arc of 5 inches in a circle of radius 3 inches.
   (*a*) What is the measure of the angle in radians?
   (*b*) What is the measure of the angle in degrees?
4. How long an arc of a circle of radius 3 inches is intercepted by an angle of 0.5 radian?
5. (*a*) Express an angle of 3 radians in degrees.
   (*b*) Express an angle of 1 degree in radians.
6. How long an arc does an angle of 1.5 radians intercept in a circle of radius (*a*) 3 inches, (*b*) 4 inches, (*c*) 5 inches?
7. How would you draw an angle of 2 radians
   (*a*) With a protractor that shows angles in degrees?
   (*b*) With a string?
8. Find $\cos \pi/6$ and $\sin \pi/6$.
9. (*a*) Fill in this table by making a quick sketch on the unit circle:

θ	0	$\dfrac{\pi}{6}$	$\dfrac{\pi}{4}$	$\dfrac{\pi}{3}$	$\dfrac{\pi}{2}$	π	$\dfrac{3\pi}{2}$	2π
$\sin \theta$								

   (*b*) Graph the sine function.
10. Check the identity for $\cos (A + B)$ when $A = \pi/6$ and $B = \pi/3$.
11. Obtain the identity for $\cos 2\theta$ from the identity for $\cos (A + B)$.
12. Use a sketch of the angle in a unit circle to determine the sign (+ or −) of the function in each case: (*a*) $\sin \theta$,

$\pi < \theta < 2\pi$; (*b*) $\tan \theta$, $\pi < \theta < 3\pi/2$; (*c*) $\cos \theta$, $-\pi/2 < \theta < \pi/2$; (*d*) $\tan \theta$, $\pi/2 < \theta < \pi$.

13. What angle does a line make with the x axis if its slope is (*a*) 1, (*b*) $\frac{1}{2}$, (*c*) -1, (*d*) 2, (*e*) $\sqrt{3}$?
14. Consider an acute angle θ in any right triangle, as in the one at the left.

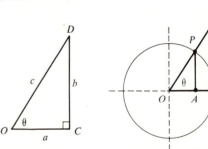

With the aid of the similar triangles in the second diagram, show that (*a*) $\cos \theta = a/c$, (*b*) $\sin \theta = b/c$, (*c*) $\tan \theta = b/a$. These formulas for cosine, sine, and tangent are sometimes taken as their definition for $0 < \theta < \pi/2$.

15. [See Exercise 14(*c*).] Use the triangle OCD below to

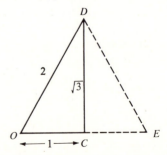

obtain $\cos \pi/3$, $\sin \pi/3$, and $\tan \pi/3$. ($\triangle ODE$ is equilateral.) A quick sketch of $\triangle OCD$ will remind you of the values of $\cos \pi/3$, $\sin \pi/3$, and $\tan \pi/3$ when you need them.

16. Use $\triangle OCD$ in Exercise 15 to compute $\cos \pi/6$, $\sin \pi/6$, and $\tan \pi/6$. *Hint:* See Exercise 14.

17. (See Exercise 14.) Solve for the length x.

 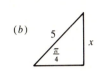

(a)   (b)   (c)

18. Use the pythagorean theorem to show that the distance d between the points (x_1, y_1) and (x_2, y_2) is given by the formula

$$d = \sqrt{(x_2 - x_1)^2 + (y_2 - y_1)^2}.$$

19. Use the formula in Exercise 18 to find the distance between (a) $(7, 3)$ and $(4, 7)$; (b) (x, y) and $(0, 0)$; (c) $(\cos \theta, \sin \theta)$ and $(1, 0)$.

20. This exercise outlines a proof that

$$\cos (A + B) = \cos A \cos B - \sin A \sin B.$$

It utilizes the distance formula of Exercise 18.
(a) In the two accompanying diagrams of the unit circle show that the line segments of $\overline{PQ}$ and $\overline{RS}$ have the same length.

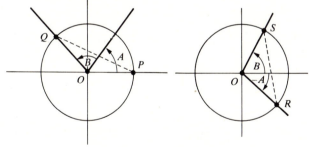

(b) Show that $Q = (\cos (A + B), \sin (A + B))$
           $S = (\cos B, \sin B)$
           $R = (\cos A, -\sin A)$.
(c) Using the coordinates in (b), show that

$$\overline{PQ}^2 = 2 - 2 \cos (A + B).$$

(d) Using the coordinates in (b), show that

$$\overline{RS}^2 = 2 - 2 \cos A \cos B + 2 \sin A \sin B.$$

(e) Deduce the identity for $\cos (A + B)$.

21. Replacing B by $-B$ in the identity for $\cos (A + B)$, obtain the identity

$$\cos (A - B) = \cos A \cos B + \sin A \sin B.$$

22. Using the identity for $\cos (A - B)$, show that

$$\cos \left(\frac{\pi}{2} - \theta \right) = \sin \theta.$$

23. With the aid of the identity in Exercise 22 and the identity for $\cos (A - B)$ obtain the identity

$$\sin (A - B) = \sin A \cos B - \cos A \sin B.$$

24. From the identity in Exercise 23 for $\sin (A - B)$ obtain the identity

$$\sin (A + B) = \sin A \cos B + \cos A \sin B.$$

25. From the identity for $\cos (A + B)$, show that
(a) $\cos 2\theta = \cos^2 \theta - \sin^2 \theta.$
(b) $\cos 2\theta = 2 \cos^2 \theta - 1 = 1 - 2 \sin^2 \theta.$
Then deduce that

(c) $\cos \theta = \pm \sqrt{\dfrac{1 + \cos 2\theta}{2}}.$

(d) $\sin \theta = \pm \sqrt{\dfrac{1 - \cos 2\theta}{2}}.$

26. Using the identities for $\cos (A - B)$ and $\sin (A - B)$, prove that

$$\tan (A - B) = \frac{\tan A - \tan B}{1 + \tan A \tan B}.$$

27. There are three more trigonometric functions:

$$\sec \theta = \frac{1}{\cos \theta}, \qquad \cot \theta = \frac{\cos \theta}{\sin \theta}, \qquad \csc \theta = \frac{1}{\sin \theta},$$

called *secant*, *cotangent*, and *cosecant*, respectively.
(a) Compute $\sec \theta$ for $\theta = \pi/3, \pi/4, \pi/6, 0$.
(b) Graph $y = \sec \theta$.

(c) Show that $\cot \theta = \dfrac{1}{\tan \theta}.$

(d) Show that $\sec^2 \theta = 1 + \tan^2 \theta$ and $\csc^2 \theta = 1 + \cot^2 \theta.$

28. This exercise outlines a proof of the law of cosines in the case $0 < \theta < \pi/2$. Consider the accompanying diagram.

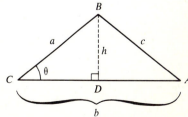

(a) Show that $\overline{CD} = a \cos \theta$ and $\overline{AD} = b - a \cos \theta$.
(b) Show that $a^2 - a^2 \cos^2 \theta = h^2 = c^2 - (b - a \cos \theta)^2$.
(c) From (b) deduce the law of cosines.
29. Show that, if a line makes an angle θ with the x axis, then its slope is $\tan \theta$. *Hint:* Consider the line through the origin and the point $(\cos \theta, \sin \theta)$.

The remaining exercises use a calculator to explore some limits that will be examined in the next section.
30. Use "sin x" to stand for "sine of an angle of x radians."

(a) Calculate $(\sin x)/x$ for $x = 1.5, 1, 0.1, 0.001$.
(b) What do you think happens to $(\sin x)/x$ as $x \to 0$?
31. Use "SIN x" to stand for "sine of an angle of x degrees."
(a) Calculate $(\text{SIN } x)/x$ for $x = 30, 20, 10, 5, 1$.
(b) What do you think happens to $(\text{SIN } x)/x$ as $x \to 0$?
32. Use "cos x" to stand for "cosine of an angle of x radians."
(a) Calculate $(1 - \cos x)/x^2$ for $x = 1, 0.5, 0.1$.
(b) What do you think happens to $(1 - \cos x)/x^2$ as $x \to 0$?

4.6

The limit of (sin θ)/θ as θ approaches 0

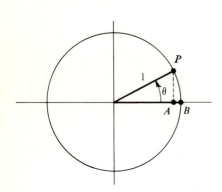

In the next chapter

$$\lim_{\theta \to 0} \frac{\sin \theta}{\theta}$$

will be needed to obtain the derivatives of the trigonometric functions. This limit will now be determined.

To begin, a sketch (shown in the margin) of a small angle in the unit circle suggests what the limit is. Since an angle is measured in radians, and the circle has radius 1, the length of the arc PB is θ. By the definition of $\sin \theta$, the length of $PA = \sin \theta$.

Thus

$$\frac{\sin \theta}{\theta} = \frac{\text{length of side } PA}{\text{length of arc } PB}.$$

When θ is small, so are the lengths of PA and PB. However, for small θ, PA looks so much like the arc PB that it seems likely that the quotient

$$\frac{\text{Length of side } PA}{\text{Length of arc } PB}$$

is near 1. This suggests that

$$\lim_{\theta \to 0} \frac{\sin \theta}{\theta} = 1.$$

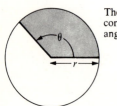

The shaded sector corresponds to the angle θ

We shall show that this guess is correct by a consideration of areas. First it will be necessary to develop a formula for the area of a sector of a circle subtended by an angle of θ radians, as in the accompanying figure. When the angle is 2π, the sector is the entire circle of radius r; hence it has an area of πr^2. Since the area of a sector is proportional to θ, it follows that

$$\frac{\text{Area of sector}}{\pi r^2} = \frac{\theta}{2\pi}.$$

From this equation it follows that

Formula for Area of a Sector

$$\text{Area of sector} = \frac{\theta}{2\pi} \pi r^2 = \frac{\theta r^2}{2}.$$

(This formula will be used in later chapters.   It is safer to memorize the proportion that led to it than the formula itself.   It is easy to forget the denominator 2 and also that the number π does *not* appear.)

The next theorem describes the behavior of $\sin\theta/\theta$ when θ is near 0.

THEOREM I   Let $\sin\theta$ denote the sine of an angle of θ radians.   Then

$$\lim_{\theta\to 0}\frac{\sin\theta}{\theta}=1.$$

PROOF   It will be enough to consider only $\theta > 0$, since

$$\frac{\sin(-\theta)}{-\theta}=\frac{-\sin\theta}{-\theta}=\frac{\sin\theta}{\theta}.$$

Moreover, it will be convenient to restrict θ to be less than $\pi/2$.

We shall compare the areas of three regions in the diagram at the left, $\triangle OAP$, sector OBP, and $\triangle OBC$.   Clearly

Area of $\triangle OAP <$ area of sector $OBP <$ area of $\triangle OBC$.   (1)

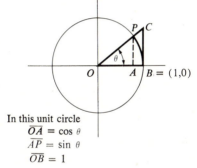

In this unit circle
$\overline{OA} = \cos\theta$
$\overline{AP} = \sin\theta$
$\overline{OB} = 1$

Now,   Area of $\triangle OAP = \frac{1}{2}$ base $\cdot$ altitude $=\frac{1}{2}\cos\theta\cdot\sin\theta$;

$$\text{Area of sector } OBP = \frac{\theta(1)^2}{2}=\frac{\theta}{2}.$$

We shall need to know the length $\overline{BC}$ in order to compute the area of $\triangle OBC$. Since $\triangle OAP$ is similar to $\triangle OBC$, their corresponding sides are proportional:

$$\frac{\overline{AP}}{\overline{OA}}=\frac{\overline{BC}}{\overline{OB}};$$

that is,

$$\frac{\sin\theta}{\cos\theta}=\frac{\overline{BC}}{1}.$$

Thus

$$\overline{BC}=\frac{\sin\theta}{\cos\theta}.$$

Hence   Area of $\triangle OBC = \frac{1}{2}\,\overline{OB}\cdot\overline{BC}=\frac{1}{2}\frac{\sin\theta}{\cos\theta}.$

Now the areas in inequalities (1) can all be expressed in terms of θ:

$$\frac{1}{2}\cos\theta\sin\theta<\frac{\theta}{2}<\frac{1}{2}\frac{\sin\theta}{\cos\theta}.$$

Hence, $\cos\theta\sin\theta<\theta<\dfrac{\sin\theta}{\cos\theta}.$

Dividing by the positive number $\sin\theta$ results in the inequalities

$$\cos \theta < \frac{\theta}{\sin \theta} < \frac{1}{\cos \theta}.$$

A glance at the unit circle shows that

$$\lim_{\theta \to 0} \cos \theta = 1$$

and hence

$$\lim_{\theta \to 0} \frac{1}{\cos \theta} = \frac{1}{1} = 1.$$

Thus, as $\theta \to 0$, both $\cos \theta$ and $1/\cos \theta$ approach 1.   Hence, $\theta/\sin \theta$, trapped between $\cos \theta$ and $1/\cos \theta$, must also approach 1.   Thus

$$\lim_{\theta \to 0} \frac{\theta}{\sin \theta} = 1.$$

If $\theta/\sin \theta \to 1$, so does its reciprocal $\sin \theta/\theta$.   Consequently,

$$\lim_{\theta \to 0} \frac{\sin \theta}{\theta} = 1,$$

as had been anticipated.   ●

We shall also need the limit

$$\lim_{\theta \to 0} \frac{1 - \cos \theta}{\theta}$$

in the next chapter.   It is not obvious what this limit is, if indeed it exists. As $\theta \to 0$, the numerator $1 - \cos \theta$ approaches 0; so does the denominator. The numerator influences the quotient to become small, while the denominator influences the quotient to become large.   The following theorem shows that the numerator is the stronger influence, causing the quotient to approach 0 as θ approaches 0.

THEOREM 2   Let $\cos \theta$ denote the cosine of an angle of θ radians.   Then

$$\lim_{\theta \to 0} \frac{1 - \cos \theta}{\theta} = 0.$$

PROOF

$$\frac{1 - \cos \theta}{\theta} = \frac{1 - \cos \theta}{\theta} \frac{1 + \cos \theta}{1 + \cos \theta}$$

$$= \frac{1 - \cos^2 \theta}{\theta(1 + \cos \theta)}$$

$$= \frac{\sin^2 \theta}{\theta(1 + \cos \theta)}$$

$$= \frac{\sin \theta}{\theta} \frac{\sin \theta}{1 + \cos \theta}.$$

Thus
$$\lim_{\theta \to 0} \frac{1 - \cos \theta}{\theta} = \lim_{\theta \to 0} \left(\frac{\sin \theta}{\theta} \frac{\sin \theta}{1 + \cos \theta} \right)$$

$$= 1 \frac{0}{1 + 1}$$

$$= 0.$$

Consequently,
$$\lim_{\theta \to 0} \frac{1 - \cos \theta}{\theta} = 0.$$

This implies that, when θ is small, $1 - \cos \theta$ is much smaller than θ. ●

Exercises

1. What is the area of a sector of a circle of (*a*) radius 3 and angle $\pi/2$? (*b*) radius 1 and angle θ? (*c*) radius 2 and angle θ?

2. What is the area of the sector of a circle of radius 6 inches subtended by an angle of (*a*) $\pi/4$ radians? (*b*) 3 radians? (*c*) 45°?

3. Use the values of $\sin \theta$, where angle is in radians, to fill in this table (to two decimals).

θ	$\dfrac{\pi}{2}$	$\dfrac{\pi}{4}$	$\dfrac{\pi}{6}$
$\dfrac{\sin \theta}{\theta}$			

In Exercises 4 to 12 find the indicated limit. Each can be solved with the aid of $\lim_{\theta \to 0} (\sin \theta)/\theta = 1$ and $\lim_{\theta \to 0} (1 - \cos \theta)/\theta = 0$.

4. $\lim\limits_{\theta \to 0} \dfrac{\sin 3\theta}{3\theta}$

5. $\lim\limits_{\theta \to 0} \dfrac{1 - \cos \theta}{\sin \theta}$

6. $\lim\limits_{x \to 0} \dfrac{\tan x}{x}$

7. $\lim\limits_{x \to 0} \dfrac{\sin 2x}{\sin 3x}$

8. $\lim\limits_{\theta \to 0} \dfrac{1 - \cos \theta}{\theta^2}$

9. $\lim\limits_{\theta \to 0^+} \dfrac{1 - \cos \theta}{\theta^3}$

10. $\lim\limits_{\theta \to \pi/2} \dfrac{\cos \theta}{(\pi/2) - \theta}$

11. $\lim\limits_{\theta \to 0} \dfrac{\sin 3\theta}{1 - \cos \theta}$

12. $\lim\limits_{x \to \infty} x \sin \dfrac{1}{x}$

13. Let $f(x) = \sin(1/x)$ for $x \neq 0$.
(*a*) Fill in this table.

x	$\dfrac{1}{3\pi}$	$\dfrac{1}{5\pi/2}$	$\dfrac{1}{2\pi}$	$\dfrac{1}{3\pi/2}$	$\dfrac{1}{\pi}$	100
$f(x)$						

(*b*) Graph f.
(*c*) Show that $\lim_{x \to 0} \sin(1/x)$ does not exist.
(*d*) Find $\lim_{x \to \infty} \sin(1/x)$.

14. Use the table of $\sin \theta$, θ in degrees, to fill in the table below.

θ (degrees)	90	45	30	10	5	1
$\dfrac{\sin \theta}{\theta}$						

As this exercise suggests, when angles are measured in degrees, the quotient $(\sin \theta)/\theta$, rather than approaching 1 as $\theta \to 0$, approaches a number whose decimal representation begins with 0.017. The next exercise shows that, when angles are measured in degrees,

$$\lim_{\theta \to 0} \frac{\sin \theta}{\theta} = \frac{\pi}{180}.$$

■

15. Let SIN θ denote the sine of an angle of θ degrees. Then
$$\text{SIN } \theta = \sin \frac{\pi\theta}{180},$$
where $\sin x$ denotes the sine of an angle of x radians.
(*a*) Show that
$$\frac{\text{SIN } \theta}{\theta} = \frac{\sin(\pi\theta/180)}{\pi\theta/180} \frac{\pi}{180}.$$

(*b*) Use the identity in (*a*) to show that
$$\lim_{\theta \to 0} \frac{\text{SIN } \theta}{\theta} = \frac{\pi}{180}.$$

16. (See Exercise 15.) When θ is near 0, $(\sin \theta)/\theta$ is near 1, if θ is the measure of an angle in radians. Thus $\sin \theta$ is approximately θ if the angle is measured in radians.
(*a*) Show that if θ measures an angle in degrees, then SIN θ is approximately $\theta\pi/180 \doteq 0.017\theta$.

(b) How accurate is this approximation when $\theta = 30°$? when $\theta = 10°$? $\theta = 5°$? $\theta = 1°$? (Use the table in Appendix E.)   The approximation SIN $\theta = 0.017\theta$ is frequently used in applications.

■ ■

17. Define the function f by the formula $f(x) = (\sin x)/x$.
   (a) What is the domain of f?
   (b) Fill in this table.

x	$\dfrac{\pi}{6}$	$\dfrac{\pi}{2}$	π	$\dfrac{3\pi}{2}$	2π
$f(x)$					

   (c) Graph f. Pay particular attention to x near 0 and also $|x|$ large.

18. (a) Find the derivative of $f(x) = \sin x$ at $x = 0$.
   (b) Sketch a graph of $y = \sin x$, using the same scales on both axes.
   (c) At what angle does the graph cross the x axis at $(0, 0)$?

19. (a) Define the derivative of a function, using the x and $x + h$ notation.
   (b) Using the definition in (a), show that $(\sin x)' = \cos x$. [The limits of $(\sin h)/h$ and $(1 - \cos h)/h$ as $h \to 0$ will come in handy.]

20. (See Exercise 19.)   Using the definition of the derivative, show that $(\cos x)' = -\sin x$.

The final two exercises use the calculator to explore some trigonometric limits.   The limits will be determined in Chap. 6.

21. Examine the behavior of $(\theta - \sin \theta)/\theta^3$ for θ near 0.

22. Examine the behavior of $(\cos \theta - 1 + \theta^2/2)/\theta^4$ for θ near 0.

4.7

Continuous functions

Graph of a differentiable function.

A small section of the graph around P resembles a short segment of the tangent line.

The differentiable functions have a derivative at any number x in their domain: At any point P on the graph of such a function there is a tangent line, or "the graph has a definite direction at every point P."   Very small sections of the graph around P closely resemble short segments of the tangent line to the curve at P, as is depicted in the accompanying figure.   For the most part, calculus is concerned with those functions whose graphs locally look like straight lines, for instance x^2, x^3, and $1/x$.   But there are important functions whose graphs are not that "nice," for instance, the absolute-value function $|x|$, since it is not differentiable at $x = 0$; its graph, sketched in Sec. 2.2, has a sharp corner at $(0, 0)$.

The next function is even less well behaved: Its graph has "jumps."

EXAMPLE I   Sketch the graph of the rounding-off function, defined as follows:

$$f(x) \begin{cases} = \text{ the integer nearest to } x \text{ if } x \\ \quad \text{is not midway between two} \\ \quad \text{consecutive integers} \\ = x + \tfrac{1}{2} \text{ if } x \text{ is midway between} \\ \quad \text{consecutive integers.} \end{cases}$$

SOLUTION   Begin by computing a few values of $f(x)$ where x is midway between two integers or nearly so:

$$f(2.4) = 2 \qquad f(2.5) = 3 \qquad f(2.6) = 3.$$

Note that $\lim_{x \to 2.5} f(x)$ does not exist, for when x is to the left of 2.5 and close to 2.5, $f(x) = 2$.   But when x is to the right of 2.5 and close to 2.5,

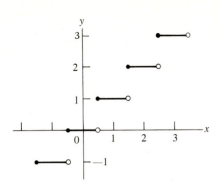

$f(x) = 3$. The graph "jumps" suddenly at 2.5. For $2.5 \leq x < 3.5$, $f(x) = 3$. This means that the graph of f has a horizontal line segment without its right-hand endpoint. In fact, the graph is composed of such segments, as shown in the diagram at the left.

The sudden change in the value of the function whenever x is midway between consecutive integers shows up as a break in the graph. ●

There is a wide class of functions that may not be as "smooth" as the differentiable functions, but at least they are not as "jumpy" as the function in Example 1. This class consists of the *continuous* functions. The simplest way to think of a continuous function—at least when its domain is an interval or the x axis—is one whose graph can be drawn without lifting the pencil from the paper. While this is certainly not a rigorous definition, it does convey the key property of a continuous function. In particular, it suggests, correctly, that the absolute-value function is continuous but that the rounding-off function of Example 1 is not continuous.

The precise definition of a continuous function f is easy to state if the domain of f is an open interval or the whole x axis. But functions with more peculiar domains present some technical problems. For instance, the domain of the function $f(x) = \sqrt{x}$ consists of all $x \geq 0$. Its graph has no "jumps," and the definition of continuous must be phrased in such a way that this function is continuous.

First let us define *continuity at a number a*. The definition is phrased in terms of limits, not "jumps" or pencils. In order to treat the various possible domains of interest, several definitions are required.

DEFINITION *A function being continuous at the number a if a is in an open interval in the domain.* Let a be a number in the domain of the function f, that is, $f(a)$ is defined. Assume that $f(x)$ is defined for all x in some open interval that includes a. Then the function f is *continuous at a* if

1. $\lim_{x \to a} f(x)$ exists, and
2. that limit is $f(a)$.

Note: The assertion that "$\lim_{x \to a} f(x)$ exists" implies that it is finite.

The function in Example 1 is continuous at any number a that is not midway between two integers. But at the other numbers, such as $a = \frac{1}{2}$, the function is not continuous, since

$$\lim_{x \to 1/2} f(x)$$

does not exist.

In the next example $\lim_{x \to a} f(x)$ exists, but f is not continuous at a.

EXAMPLE 2 Let

$$f(x) \begin{cases} = 2 & \text{if } x \text{ is an integer} \\ = 1 & \text{if } x \text{ is not an integer.} \end{cases}$$

Show that f is not continuous at 3.

SOLUTION   The graph of f is shown in the margin.

Is f continuous at $a = 3$?   First of all,

$$\lim_{x \to 3} f(x)$$

exists since, for x near 3, $f(x)$ is equal to 1; thus

$$\lim_{x \to 3} f(x) = 1.$$

But the value of the function at 3 is 2; hence

$$f(3) \neq \lim_{x \to 3} f(x)$$

so f is not continuous at 3.   However, at any number a that is not an integer, f is continuous.   ●

If a function whose domain is the x axis is continuous at every number a, it will be called a *continuous function*.   (The reader may stop and check that $f(x) = |x|$ is continuous.)   But this still leaves such functions as $\sqrt{x}$ or $\sqrt{1 - x^2}$ unsettled.   The domain of $\sqrt{1 - x^2}$ is the closed interval $[-1, 1]$. The graph of $\sqrt{1 - x^2}$ certainly has no "jumps": We would like it to be continuous.   The following definitions are phrased to cover such functions. Note that the two following definitions for "continuous at a" differ from the preceding one only in a condition about the domain of the function.   Frequently, they are described as defining *right continuity* and *left continuity*, but we will not use these special terms.

DEFINITION   *A function being continuous at the number a if a is a left endpoint.*   Let a be a number in the domain of the function f, that is, $f(a)$ is defined.   Assume that $f(x)$ is defined for all x in some closed interval $[a, b]$ but is *not* defined for any x in some open interval (c, a).   Then the function f is continuous at a if

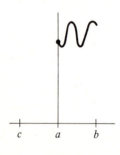

1. $\lim_{x \to a^+} f(x)$ exists, and
2. that limit is $f(a)$.

The diagram for this definition is in the margin.   According to this definition $\sqrt{x}$ is continuous at $a = 0$.   Also, $\sqrt{1 - x^2}$ is continuous at $a = -1$.   The next definition assures us that the function $\sqrt{1 - x^2}$ is also continuous at $x = 1$.

DEFINITION   *A function being continuous at the number a if a is a right endpoint.*   Let a be a number in the domain of the function f, that is $f(a)$ is defined.   Assume that $f(x)$ is defined for all x in some closed interval $[b, a]$ but is *not* defined for any x in some open interval (a, c).   Then the function f is continuous at a if

1. $\lim_{x \to a^-} f(x)$ exists, and
2. that limit is $f(a)$.

The diagram for this definition is in the margin.

At last we are in a position to define a continuous function precisely.

DEFINITION   *Continuous function.* A function that is continuous at every point in its domain is called continuous.

EXAMPLE 3   Show that $f(x) = \sqrt{1 - x^2}$ is a continuous function.

SOLUTION   The graph of $f(x) = \sqrt{1 - x^2}$ appears in the margin.   Is f continuous at every number in its domain $[-1, 1]$?

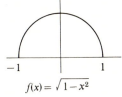

$f(x) = \sqrt{1-x^2}$

There are three cases to consider:

$$(1)\ -1 < a < 1, \qquad (2)\ a = -1, \qquad (3)\ a = 1.$$

In case (1), $\lim_{x \to a} f(x)$ clearly exists and equals $f(a)$. (A rigorous proof is given in advanced calculus courses.)   In case (2), $\lim_{x \to -1^+} f(x)$ exists; in fact, it is 0.   Note that $f(-1) = \sqrt{1 - (-1)^2} = 0$.   Thus f is continuous at $a = -1$. In a similar manner it is continuous at $a = 1$.   Thus f is continuous. ●

Though we may think of continuity in the broad terms of the general shape of a graph, the definition is really a "local" one, expressed in the behavior of the function in the vicinity of each of its points.

The next example reinforces the idea of continuity.

EXAMPLE 4   Is the *reciprocal* function f, given by $f(x) = 1/x$, continuous?

SOLUTION   At first glance, one might say, "No, there is a lot of trouble near $a = 0$."   It is true that $\lim_{x \to 0} f(x)$ does not exist, but that is irrelevant, for 0 is not in the domain of $1/x$.   The function $1/x$ is continuous at every number in its domain, and hence is said to be continuous. ●

Graph of reciprocal function, $f(x) = \frac{1}{x}$

Thus the technical term for a "nice, well-behaved function" is *continuous*. The graph of a continuous function may have sharp corners, but it does not have sudden "jumps."   If you know the values of a continuous function f near a number a in its domain, you can figure out its values at a:

$$f(a) = \lim_{x \to a} f(x).$$

Note that, if f is continuous at a, then

$$\lim_{x \to a} f(x) = f\left(\lim_{x \to a} x\right),$$

for

$$\lim_{x \to a} x = a.$$

Rigorous proofs that the functions commonly used in calculus, such as polynomials, exponentials, square roots, and trigonometric functions, are continuous depend on the precise definitions of limits to be found in Appendix C. An advanced calculus course devotes a good deal of time to them.   The interested reader will find in Appendix C a proof that any polynomial is continuous.

The equation

$$\lim_{x \to a} f(x) = f(a)$$

may also be written in the form

$$\lim_{x \to a} [f(x) - f(a)] = 0.$$

This observation is of use in proving the following theorem, which shows that a differentiable function is continuous, as might be expected.

THEOREM If f is differentiable at the number a, then it is continuous at a.

PROOF We wish to prove that $\lim_{x \to a} [f(x) - f(a)] = 0.$

Since f is differentiable at a, the derivative $f'(a)$, or

$$\lim_{x \to a} \frac{f(x) - f(a)}{x - a},$$

exists.   Then,

$$\lim_{x \to a} [f(x) - f(a)] = \lim_{x \to a} \frac{f(x) - f(a)}{x - a} (x - a)$$

$$= \lim_{x \to a} \frac{f(x) - f(a)}{x - a} \lim_{x \to a} (x - a)$$

$$= f'(a)0$$

$$= 0.$$

This concludes the proof. ●

Recall that the absolute-value function $|x|$ is continuous but not differentiable at 0.   Hence continuity does not imply differentiability.   Indeed, a function can be continuous throughout the entire x axis, yet not have a derivative anywhere!   Bolzano in 1834 concocted the first example of such a function.

Continuous functions (hence, by the theorem just proved, differentiable functions also) have an important property stated in the following theorem. The proof is omitted, since it would require too long a detour.

INTERMEDIATE-VALUE THEOREM Let f be continuous throughout $[a, b]$.   Let m be any number between $f(a)$ and $f(b)$.   [That is, $f(a) \le m \le f(b)$ if $f(a) \le f(b)$, or $f(a) \ge m \ge f(b)$ if $f(a) \ge f(b)$.] Then there is at least one number X in $[a, b]$ such that $f(X) = m$.

In ordinary English, the intermediate-value theorem reads: A continuous function defined on $[a, b]$ takes on all values between $f(a)$ and $f(b)$.   Pictorially, it asserts that a horizontal line of height m must meet the graph of f at least once if m is between $f(a)$ and $f(b)$, as seen in the accompanying figure.   In

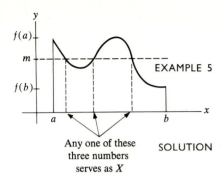

Any one of these
three numbers
serves as X

other words, when you move a pencil along the graph of a continuous function from one height to another, the pencil passes through all intermediate heights.

EXAMPLE 5 Use the intermediate-value theorem to show that the equation

$$x^4 + 4x - 6 = 0$$

has a solution in the interval $[1, 2]$.

SOLUTION The function $f(x) = x^4 + 4x - 6$, like any given by a polynomial, is continuous. Now

$$f(1) = 1^4 + 4 \cdot 1 - 6 = -1,$$

and

$$f(2) = 2^4 + 4(2) - 6 = 18.$$

The intermediate-value theorem, in the case $a = 1$ and $b = 2$, asserts that f takes on all values between -1 and 18 for x in $[1, 2]$. In particular, since we have $-1 < 0 < 18$, it follows that the equation

$$x^4 + 4x - 6 = 0$$

has a solution in the interval $[1, 2]$. ●

The intermediate-value theorem will be used on several occasions in later chapters in proofs of theorems.

Exercises

In each of Exercises 1 to 6 is presented the graph of a function whose domain is the interval $[0, 1]$. In each case indicate the values of a, if any, where the function is *not* continuous.

1.

2.

3.

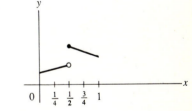

4.

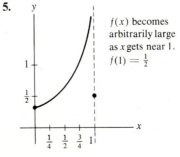

5.

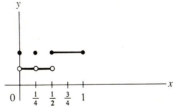

$f(x)$ becomes
arbitrarily large
as x gets near 1.
$f(1) = \frac{1}{2}$

6.

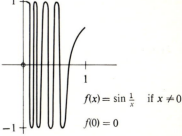

$f(x) = \sin \frac{1}{x}$ if $x \neq 0$

$f(0) = 0$

The graph oscillates infinitely often between the lines $y = 1$ and $y = -1$.

7. Let $f(x) = (\sin x)/x$ for $x \neq 0$. Is it possible to define $f(0)$ in such a way that f is continuous at 0?

8. Let $f(x) = 2x$ if $x > 1$, and let $f(x) = x^2$ if $x < 1$.
   (a) Sketch the graph of f.
   (b) Is it possible to define $f(1)$ in such a way that f is continuous?

9. Let $f(x) = (1 + x)^{1/x}$ for $x > -1$ but $x \neq 0$.
   (a) Is it possible to define $f(0)$ in such a way that f is continuous on the domain consisting of all $x > -1$?
   (b) Is it possible to define $f(-1)$ and $f(0)$ in such a way that f is continuous on the domain consisting of all $x \geq -1$?

10. Let $f(x) = (x^3 - 1)/(x - 1)$ if $x \neq 1$. Is it possible to define $f(1)$ in such a way that f is continuous?

Exercises 11 to 17 concern the intermediate-value theorem. Verify the intermediate-value theorem for the specified function f, the interval $[a, b]$, and the indicated value m. Find all X's in each case.

11. Function $3x + 5$; interval $[1, 2]$; $m = 10$.
12. Function $x^2 - 2x$; interval $[-1, 4]$; $m = 5$.
13. Function $\sin x$; interval $[\pi/2, 11\pi/2]$; $m = 0$.
14. Function $\cos x$; interval $[0, 5\pi]$; $m = 0$.
15. Function $\cos x$; interval $[0, 5\pi]$; $m = \frac{1}{2}$.
16. Function 2^x; interval $[0, 3]$; $m = 4$.
17. Function $x^3 - x$; interval $[-2, 2]$; $m = 0$.
18. Show that the equation $2^x - 3x = 0$ has a solution in the interval $[0, 1]$.
19. Show that the equation $x + \sin x = 1$ has a solution in the interval $[0, \pi/2]$.

■

20. Show that the equation $x^3 = 2^x$ has a solution in the interval $[1, 2]$. *Hint:* Consider the function $f(x) = x^3 - 2^x$.
21. Use the intermediate-value theorem to show that the equation $3x^3 + 11x^2 - 5x = 2$ has a solution.
22. (a) If a function whose domain is the x axis is continuous at a, is it necessarily differentiable at a?

(b) If a function is differentiable at a, is it necessarily continuous at a?

23. A continuous function f is defined for all x's on the x axis. If x is rational, $f(x) = x$. Must $f(x) = x$ when x is not rational?

24. Let $f(x) = \sin x$, $a = 0$, $b = 5\pi/2$.
   (a) Draw the graph of f from $x = 0$ to $x = 5\pi/2$.
   (b) For $m = 0$, find all the X's guaranteed by the intermediate-value theorem.
   (c) How many X's are there for various choices of m, $f(a) \leq m \leq f(b)$?

25. Let $f(x) = x^2$, $a = -1$, $b = 2$. Show that the intermediate-value theorem is valid in this case. Find, for each number m such that $f(a) \leq m \leq f(b)$, all corresponding X's.

26. (a) Draw the graph of a continuous function f, whose domain is the x axis, such that $f(0) = 3$ and $f(x)$ is an integer for each x.
   (b) How many such functions are there?

27. Let $f(x) = x$ if x is rational, and let $f(x) = -x$ if x is irrational. Where is f continuous?

28. Let $f(x) = 1/x$, $a = -1$, $b = 1$, $m = 0$. Note that $f(a) \leq m \leq f(b)$. Is there at least one X in $[a, b]$ such that $f(x) = m$? If so, find X; if not, does this imply that the intermediate-value theorem is sometimes false?

■ ■

29. Let f be a continuous function whose domain is the x axis and has the property that

$$f(x + y) = f(x) + f(y)$$

for all numbers x and y.
   (a) Show that there exists a constant c such that $f(x) = cx$ when x is an integer.
   (b) Show that $f(x) = cx$ when x is rational.
   (c) Show that $f(x) = cx$ for all x.

30. (a) Graph $f(x) = |\sin x|$.
   (b) Where is f continuous?
   (c) Where is f differentiable?

4.8

Summary

This chapter developed the idea of the limit, of which the derivative is a special case. After reviewing exponentiation, it dealt with

$$\lim_{n \to \infty} \left(1 + \frac{1}{n}\right)^n,$$

which is called e ($\doteq 2.718$). Then various types of limits of a function were introduced, such as

$$\lim_{x \to a} f(x) = L, \qquad \lim_{x \to a^+} f(x) = L, \qquad \lim_{x \to a^-} f(x) = L,$$

the latter two being the right-hand and left-hand limits.

In particular, $\lim\limits_{x \to 0} (1 + x)^{1/x} = e.$

The review of trigonometry emphasized the functions $\sin x$, $\cos x$, and $\tan x$, radian measure, and the area of a sector. It was shown that, if angles are measured in radians, then

$$\lim_{\theta \to 0} \frac{\sin \theta}{\theta} = 1 \quad \text{and} \quad \lim_{\theta \to 0} \frac{1 - \cos \theta}{\theta} = 0.$$

The chapter concluded with the notion of a continuous function. Though the definition is framed in terms of limits, the easiest way to think of a function continuous over an interval is that its graph can be sketched without lifting the pencil from the paper. It was proved that a differentiable function is necessarily continuous. However, there are continuous functions that are not differentiable [for instance, $f(x) = |x|$ is continuous, yet not differentiable at 0]. There are also everyday functions that are not continuous, for instance, the rounding-off function.

The intermediate-value theorem asserts that a continuous function defined on the closed interval $[a, b]$ takes on any value m between $f(a)$ and $f(b)$ at some X in $[a, b]$.

VOCABULARY AND SYMBOLS

b^n

b^{-n}

$b^{1/n}$

$b^{m/n}$

b^x

$\lim\limits_{n \to \infty} \left(1 + \dfrac{1}{n}\right)^n = e \doteq 2.718$

$\lim\limits_{x \to a} f(x)$

$\lim\limits_{x \to a^+} f(x), \ \lim\limits_{x \downarrow a} f(x)$

$\lim\limits_{x \to a^-} f(x), \ \lim\limits_{x \uparrow a} f(x)$

$\lim\limits_{x \to \infty} f(x)$

$\lim\limits_{x \to -\infty} f(x)$

radian

$\sin x, \cos x, \tan x$

continuous at a

continuous function

KEY FACTS

Concerning exponents:

$$b^{x+y} = b^x \cdot b^y$$
$$b^{-x} = \frac{1}{b^x}$$

$b^0 = 1 \qquad b \neq 0$

$b^{1/n} = \sqrt[n]{b}$

$b^{m/n} = (\sqrt[n]{b})^m$

$(b^x)^y = b^{xy}$ power of a power

$(ab)^x = a^x b^x$ power of a product

$\left(\dfrac{a}{b}\right)^x = \dfrac{a^x}{b^x}$ power of a quotient; $b \neq 0$

Concerning trigonometry:

$$\frac{\text{Degrees}}{180} = \frac{\text{radians}}{\pi}.$$

An angle of 1 radian is $\dfrac{180°}{\pi} \doteq 57.3°$.

An angle of $1°$ is $\dfrac{\pi}{180}$ radians $\doteq 0.017$ radian.

If θ is the measure of an angle in radians, and s is the arc intercepted on a circle of radius r,

$$\theta = \frac{s}{r} \quad \text{definition of } \theta,$$

$$s = r\theta,$$

and   Area of sector $= \dfrac{\theta r^2}{2}.$

$\left(\text{It is safer to remember the proportion: } \dfrac{\text{Area}}{\pi r^2} = \dfrac{\theta}{2\pi}.\right)$

$\cos^2 \theta + \sin^2 \theta = 1$

$\sin (A + B) = \sin A \cos B + \cos A \sin B$

$\cos (A + B) = \cos A \cos B - \sin A \sin B$

Law of cosines: $c^2 = a^2 + b^2 - 2ab \cos \theta$

$\tan (A - B) = \dfrac{\tan A - \tan B}{1 + \tan A \tan B}$

$\lim\limits_{\theta \to 0} \dfrac{\sin \theta}{\theta} = 1$ angle measured in radians

$\lim\limits_{\theta \to 0} \dfrac{1 - \cos \theta}{\theta} = 0$

Concerning continuous functions:

Every differentiable function is continuous. The intermediate-value theorem holds for continuous functions over a closed interval.

Guide quiz on exponents

1. Express or estimate as a decimal: (a) 8^{-1}, (b) $8^{1/2}$, (c) $8^{2/3}$, (d) 8^0, (e) $8^{-1/2}$, (f) $8^{-2/3}$.
2. Write in the form 3^x for suitable x: (a) $\sqrt{3}$, (b) $\frac{1}{3}$, (c) $\sqrt[3]{9}$, (d) $3^7/3^2$, (e) $3\sqrt{3}$, (f) $(3^5)^7$, (g) $1/\sqrt[2]{3}$, (h) $3^2 \cdot 9^4$.
3. By brute arithmetic find $7^{1/3}$ to one decimal place.

4. Write in the form b^x: (a) $(b^{1/2})^4$, (b) $(b\sqrt{b})$, (c) $\sqrt{b} \cdot \sqrt[3]{b}$, (d) $(1/\sqrt{b})^3$, (e) $(b\sqrt{b})/\sqrt[3]{b^2}$, (f) $(\sqrt{b^3})^5$.
5. (a) Compute $(\sqrt[3]{5})^{12}$ and $(\sqrt[4]{5})^{12}$.
   (b) Which is larger, $\sqrt[3]{5}$ or $\sqrt[4]{5}$?

Guide quiz on trigonometry

1. (a) How many degrees are there in an angle of 2 radians? Using a protractor, draw the angle.
   (b) With the aid of a unit circle of radius 10 centimeters, estimate cos 2 and sin 2 (angle in radians).
2. An angle of 1.3 radians is drawn with vertex at the center of a circle of radius 4 inches.
   (a) How long is the arc that the angle subtends?
   (b) What is the area of the sector determined by the angle?
3. (a) Graph $y = \cos \theta$.
   (b) How does the graph in (a) compare with the graph of $y = \sin \theta$?
4. Sketching a unit circle, fill in the opposite table with exact values.
5. (a) If $\cos \theta = 0.6$, what two possible values can $\sin \theta$ have?

(b) Draw the possible angles for which $\cos \theta = 0.6$.
6. Using the exact values for cos 45°, sin 45°, cos 30°, sin 30°: (a) and the identity for sin $(A + B)$, find sin 75°; (b) and the identity for sin $(A - B)$, find sin 15°; (c) and the identity relating $\cos \theta$ and $\cos 2\theta$, find $\cos 22\frac{1}{2}°$.
7. Graph $y = \tan x$.

θ (radians)	$\pi/2$		$3\pi/4$	$-\pi/4$	
θ (degrees)		120°		$-30°$	405°
$\cos \theta$					
$\sin \theta$					

Guide quiz on Chap. 4

1. (a) Define e.
   (b) Give the decimal expression for e to three places.
   (c) Evaluate $\lim_{h \to 0} \left(1 + \dfrac{h}{3}\right)^{3/h}$.

2. Prove that, if $c > d \geq 0$, then $c^5 - d^5 < 5c^4(c - d)$.
3. The graph of a function f appears below.

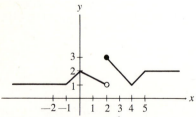

   (a) For which values of a does $\lim_{x \to a} f(x)$ not exist?
   (b) For which values of a does $\lim_{x \uparrow a} f(x)$ not exist?
   (c) For which values of a does $\lim_{x \downarrow a} f(x)$ not exist?
   (d) At which values of a is f not continuous?

(e) At which values of a is f not differentiable?
4. Evaluate the following limits, if they exist:
   (a) $\lim_{x \to \infty} \dfrac{5x^3 - 6x + 2}{10x^3 + 5}$
   (b) $\lim_{x \to 0} \dfrac{\tan 3x}{\tan 2x}$
   (c) $\lim_{x \to 1} \dfrac{3 \cos x - 3}{x}$
   (d) $\lim_{x \to 4} \dfrac{\sqrt{x} - 2}{x - 4}$
   (e) $\lim_{x \to 1^+} 2^{1/(x-1)}$
   (f) $\lim_{x \to -\infty} \left(1 + \dfrac{1}{x}\right)^{3x}$

5. Which limit in Exercise 4 can be interpreted as the derivative of some function at some number? Describe the function and the number.
6. As $\theta \to 0$, what happens to the quotients

$$\frac{\text{Length of } AP}{\text{Length of arc } BP}$$

and

$$\frac{\text{Length of } AB}{\text{Length of arc } BP},$$

where A, B, and P are shown in the accompanying diagram?

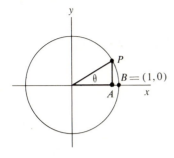

7. The velocity of a certain object at time t is $2^{-t} \sin^2 \pi t$ feet per second at time t.
   (a) Find its velocity at time $t = 0$ and at time $t = \frac{1}{2}$.
   (b) How do we know that at some time between $t = 0$ and $t = \frac{1}{2}$ the velocity was exactly $\frac{1}{2}$ foot per second?
8. (The three parts of this exercise are independent of each other.) Draw the graph of a function f whose domain is the x axis, is differentiable for all $x \neq 0$, and
   (a) is continuous at 0, but not differentiable at 0;
   (b) $\lim_{x \to 0} f(x)$ exists, but f is not continuous at 0;
   (c) $\lim_{x \to 0^+} f(x)$ and $\lim_{x \to 0^-} f(x)$ exist, but $\lim_{x \to 0} f(x)$ does not exist.

Review exercises for Chap. 4

1. The usual protractor shows angles in degrees. Make a radian protractor, showing in particular the angles whose radian measure is (a) π, (b) 1, (c) 2, (d) 3, (e) 1.5, (f) $\pi/2$, (g) $\pi/4$, (h) 0.5, (i) 0.1.
2. A ray forming an angle θ with the x axis meets the unit circle at the point $(\cos \theta, \sin \theta)$. Prove that it meets the circle of radius r and center $(0, 0)$ at the point $(r \cos \theta, r \sin \theta)$. *Hint:* Use similar triangles.
3. Sketch the graph of some hypothetical function f that has all these properties:
   (a) $\lim_{x \to a} f(x)$ exists at all a except $a = 1$;
   (b) f is not continuous at 1 and 2;
   (c) f is continuous everywhere else.
4. (a) What is the circumference of a circle of radius r?
   (b) What is the area of a circle of radius r?
   (c) What is the area of a sector of angle θ and radius r?
   (d) What is the length of arc of a circle of radius r if it subtends an angle θ?
5. Let f be a function whose domain is the x axis. True or false:
   (a) If f is continuous at 2, it is necessarily differentiable at 2.
   (b) If f is differentiable at 2, it is necessarily continuous at 2.
   (c) If f is continuous at 2, then $\lim_{x \to 2} [f(x) - f(2)] = 0$.
6. In the example in Sec. 4.2 the bank pays a fabulous 100 percent per year interest. More realistically, assume that it pays 6 percent per year interest.
   (a) Show that, if it compounds interest n times a year, 1 dollar grows to $(1 + 0.06/n)^n$ dollars at the end of 1 year.
   (b) Show that $\lim_{n \to \infty} (1 + 0.06/n)^n = e^{0.06}$.

7. In Sec. 4.6, it was shown that $\lim_{h \to 0} (1 - \cos h)/h = 0$.
   (a) Using the same technique, prove that
   $$\lim_{h \to 0} \frac{1 - \cos h}{h^2} = \frac{1}{2}.$$
   (b) From (a) deduce that a good estimate for $\cos h$, when h is small, is
   $$\cos h \doteq 1 - \frac{h^2}{2}.$$
   (c) From (b) deduce that for an angle measured in degrees
   $$\cos h \doteq 1 - 0.00015h^2.$$
   (d) How good is the estimate in (c) for $h = 10°$? $h = 5°$? $h = 30°$? Compare the results from the formula with those listed in the tables.
8. The law of cosines asserts that if θ is the angle opposite the side of length c in a triangle, then
   $$c^2 = a^2 + b^2 - 2ab \cos \theta.$$
   A triangle whose sides have length a, b, and c is placed in the xy plane as shown in the accompanying diagram.

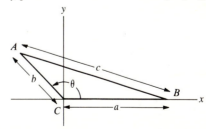

Use the diagram to prove the law of cosines, following the steps on the next page.

(a) Show that $B = (a, 0)$ and $A = (b \cos \theta, b \sin \theta)$.

(b) Using the distance formula, find c^2, the square of the distance between A and B.

■

9. Let $f(x) = x^2$ when x is rational, and $f(x) = x^4$ when x is not rational. Where is f continuous? Where is f differentiable?

10. Let $f(x) = x^2$ if x is rational and let $f(x) = x^3$ if x is not rational.

(a) Where is f continuous?

(b) Where is f differentiable?

11. Let a, b, and c be fixed numbers. Discuss the behavior of $x^3 + ax^2 + bx + c$

(a) when x is very large and positive.

(b) when $|x|$ is very large and x is negative.

(c) Prove that the equation $x^3 + ax^2 + bx + c = 0$ has at least one real root.

12. (a) Prove that a polynomial of odd degree, with real coefficients, has at least one real root.

(b) Give an example of a polynomial of degree 2 that has no real root.

13. Let $[x]$ denote the largest integer that is not greater than x. For instance, $[3.8] = 3$, $[3] = 3$, $[\sqrt{2}] = 1$, $[-\pi] = -4$. Let $f(x) = [x]$.

(a) Graph f.

(b) At which a (if any) is f *not* continuous?

14. Show that there is a positive number x such that $\tan x = 3 + \cos^2 x$.

■ ■

15. If $\lim_{n \to \infty} x^{2n}/(1 + x^{2n})$ exists, call it $f(x)$.

(a) Compute $f(\tfrac{1}{2})$, $f(2)$, $f(1)$.

(b) For which x is $f(x)$ defined? Graph $y = f(x)$.

(c) Where is f continuous?

16. Let f be a function defined on the x axis, with the property that

$$f(x + y) = f(x) + f(y)$$

for all x and y. Assume that $x > y$ implies that $f(x) \geq f(y)$. Prove that $f(x) = cx$ for some constant c.

17. (See Exercise 16.) Let f be a function defined on the x axis with the property that

$$f(x + y) = f(x) + f(y) \qquad \text{and} \qquad f(xy) = f(x)f(y)$$

for all x and y.

(a) Prove that $x > y$ implies that $f(x) \geq f(y)$.

(b) Prove that $f(x) = x$ for all x, or else that $f(x) = 0$ for all x.

18. (a) Let $f(x) = x \sin 1/x$ for $x \neq 0$. Define $f(0)$ in such a way that f is continuous on the x axis.

(b) Graph the resulting f. Pay special attention to x near 0 and x large.

19. Let f be a continuous function defined on the x axis. Assume that $f(f(f(x))) = x$ for all x. Prove that $f(x) = x$ for all x.

Review questions for Chaps. 1-4

1. (a) Define the derivative of a function f at a number x.

(b) List as many interpretations of the derivative as you can.

2. Using formulas developed in the book, compute the derivative of (a) $2x^4$ at $x = \tfrac{1}{2}$, (b) $x^3 + 5x$ at $x = -1$, (c) $\sqrt{x}$ at $x = 4$, (d) $1/x$ at $x = \sqrt{2}$.

3. Let $f(x) = x^3$. Evaluate:

(a) $f(x + h) - f(x)$ when $x = 2$ and $h = 0.5$;

(b) $f(t_1) - f(t)$ when $t_1 = 3.1$ and $t = 3$.

4. At time t seconds, a bobbing cork has a height of $3 \sin t$ feet relative to the water level. (Note that when $\sin t$ is negative the cork is submerged.) Find its speed at time $t = 0$.

5. What is the slope of the line (a) through the points $(-1, 4)$ and $(2, 6)$? (b) $y = 6x + 5$?

6. Graph $f(x) = (1 + x)^{1/x}$. Discuss the domain of f.

7. Graph $f(x) = (\sin x)/x$. Discuss the domain of f.

8. Graph $f(x) = (1 - \cos x)/x$. Discuss the domain of f.

9. (a) Differentiate $f(x) = 2x^3 + 3x^2 - 12x$.

(b) For which x is the derivative of f zero? positive? negative?

(c) With the aid of (b) graph f.

10. Using the definition of the derivative, find the derivative of

(a) $x^4 + 5x^3$ at $x = 1$;

(b) $1/x^2$ at $x = 2$.

11. A certain function f has a derivative equal to 3 at $x = 1$, that is, $f'(1) = 3$. What can be stated about the following:

(a) the slope of the graph of f?

(b) the speed of a particle that travels $f(t)$ feet in t seconds? About how far does the particle move during the time interval $[1, 1.01]$?

(c) the magnification of a lens that projects x onto $f(x)$? About how long is the image on the screen of the interval $[1, 1.01]$ on the slide?

(d) the density of a string whose left-hand x centimeters have a mass of $f(x)$ grams? About how much mass is there in the string within the interval $[1, 1.01]$?

12. (This concerns Sec. 1.2.) A particle has a speed of e^t feet per second at time t. Estimate the total distance it moves from time $t = 1$ to time $t = 3$, by dividing the time interval into 10 equal intervals and using the speed at
(a) the right end of each time interval;
(b) the left end of each time interval.
(Use the e^x table in Appendix E or a calculator.)

13. Using formulas obtained in the text, differentiate
(a) $6x^3 - 5x$, (b) $(1 + 3x^2)^2$, (c) $5/x$.

14. Using the definition of the derivative, find the derivative of $x^2/(1 + x)$.

15. (a) What is the domain of $f(x) = \sqrt{x^2 - 1}$?
(b) Graph f.
(c) Is f continuous?

16. (a) What is the domain of $f(x) = 1/\sqrt{x^2 - 1}$?
(b) Graph f.
(c) Is f continuous?

17. (a) What is the domain of $f(x) = \tan x$?
(b) Graph f.

18. Find: (a) $\lim_{x \to 0^+} 2^{x/|x|}$, (b) $\lim_{x \to 0^-} 2^{x/|x|}$, (c) $\lim_{x \to 1} 2^{x/|x|}$.

19. Find:

(a) $\lim\limits_{x \to \infty} \dfrac{\sqrt{x}}{\sqrt{4x + 2}}$
(b) $\lim\limits_{x \to \infty} \dfrac{8x^4 + 6x - 2}{2x^4 - 5x}$

(c) $\lim\limits_{x \to \infty} \left(\dfrac{5x^2 + 4x}{3x + 1} - \dfrac{5x + 8}{3} \right)$.

20. Let $y = f(x)$ be a differentiable function. What is the interpretation of "$f'(x)$ is positive at $x = a$" and "$f'(x)$ is negative at a"
(a) in terms of the graph of f?
(b) if $f(x)$ is the position at time x of a particle moving on the y axis?

■

21. Find $\lim\limits_{x \to 0} \dfrac{\tan x - \sin x}{x}$.

22. Find $\lim\limits_{x \to 0} \dfrac{\tan x - \sin x}{x^2}$.

23. Find (a) $\lim\limits_{x \to 0^+} \dfrac{2^{1/x} + 1}{2^{1/x} + 3}$ (b) $\lim\limits_{x \to 0^-} \dfrac{2^{1/x} + 1}{2^{1/x} + 3}$.

24. Let $f(x) = e^x$.
(a) Show that the derivative of f at $x = 0$ is

$$\lim_{h \to 0} \frac{e^h - 1}{h}.$$

(b) Estimate the limit of (a) by evaluating $(e^h - 1)/h$ for $h = 0.15, 0.10, 0.05, 0.01$ with the aid of the table of e^x in Appendix E or with a calculator.
(c) What do you think the limit in (a) is equal to? (It will be determined in the next chapter.)

25. Give an example of a function whose derivative is
(a) $3x^2$ (b) x^2 (c) $-1/x^2$ (d) $1/x^2$
(e) $9x^3 + 6x^2 - 6x + 2$ (f) $5/\sqrt{x}$.

■ ■

26. Find the derivative of $f(x) = \tan x$ at $x = \pi/6$. Use the definition of the derivative. *Hint:* The identity for $\tan(A + B)$ will come in handy.

27. Use the definition of the derivative to find $(\sqrt[3]{x})'$. *Hint:* The identity $c^3 - d^3 = (c^2 + cd + d^2)(c - d)$ will come in handy.

28. Let f be a continuous function such that $f(x)$ is in $[0, 1]$ when x is in $[0, 1]$. Prove that there is at least one number X in $[0, 1]$ such that $f(X) = X$.

29. Let f be a continuous function such that $f(f(x)) = x$ for all x. Prove that there is at least one number X such that $f(X) = X$.

30. If f is a function, then by a *chord of f* we shall mean a line segment whose ends are on the graph of f. Now let f be continuous throughout $[0, 1]$, and let $f(0) = f(1) = 0$.
(a) Explain why there is a horizontal chord of f of length $\frac{1}{2}$.
(b) Explain why there is a horizontal chord of f of length $1/n$, where $n = 1, 2, 3, 4, \ldots$.
(c) Must there exist a horizontal chord of f of length $\frac{2}{3}$?
(d) What is the answer to (c) if we also demand that $f(x) \geq 0$ for all x in $[0, 1]$?

31. A bank pays 5 percent interest per year.
(a) Show that, if it compounds interest n times per year, a deposit of 1 dollar grows to the value $(1 + 0.05/n)^{nt}$ dollars in t years,
(b) Show that $\lim_{n \to \infty} (1 + 0.05/n)^{nt} = e^{0.05t}$.
(c) Show that, if n in (a) is large, 1 dollar grows to 2 dollars in about 14 years. Use the e^x table in Appendix E. Notice that this is about 70 percent of the time it would take to double if interest were not compounded.

32. Find $\lim_{x \to \infty} (4^x + 3^x)^{1/x}$.

33. Find

$$\lim_{x \to 1} \frac{x^3 - 1}{x^2 - 1}.$$

34. A polynomial of degree $n \geq 0$ has the form $a_0 x^n + a_1 x^{n-1} + \cdots + a_n$, where the a_i's are constants and $a_0 \neq 0$. Let $P(x)$ be such a polynomial and let $Q(x) = b_0 x^m + b_1 x^{m-1} + \cdots + b_m$ be a polynomial of degree m. Find

$$\lim_{x \to \infty} \frac{P(x)}{Q(x)} \quad \text{if } (a) \ n = m, \ (b) \ n < m, \ (c) \ n > m.$$

35. The accompanying graph indicates a function which goes back and forth between the lines $y = x$ and $y = -x$ as $x \to \infty$.   Does $\lim_{x \to \infty} f(x)$ exist?

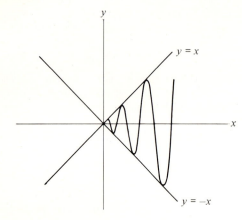

36. Find $\lim_{x \to \infty} (\sqrt{x^2 + 2x} - x)$.

37. Two citizens are arguing about

$$\lim_{x \to \infty} \left(\frac{3x^2 + 2x}{x + 5} - 3x \right).$$

The first claims, "For large x, $2x$ is small in comparison to $3x^2$, and 5 is small in comparison to x.   So $(3x^3 + 2x)/(x + 5)$ behaves like $3x^2/x = 3x$.. Hence the limit in question is 0." Her companion replies, "Nonsense.   After all,

$$\frac{3x^2 + 2x}{x + 5} = \frac{3x + 2}{1 + (5/x)},$$

which clearly behaves like $3x + 2$ for large x.   Thus the limit in question is 2, not 0."

Settle the argument.

38. Find

$$\lim_{x \to 0} \frac{\sqrt{3 + x} - \sqrt{3 - x}}{x}.$$

5

THE COMPUTATION OF DERIVATIVES

In Chap. 3 we defined the derivative of a function and computed the derivatives of various functions, showing that

$$(x^n)' = nx^{n-1} \qquad n = 1, 2, 3, \ldots,$$

$$(\sqrt{x})' = \frac{1}{2\sqrt{x}},$$

and

$$\left(\frac{1}{x}\right)' = -\frac{1}{x^2}.$$

In each case it was necessary to go back to the definition of the derivative and to examine

$$\lim_{x_1 \to x} \frac{f(x_1) - f(x)}{x_1 - x} \qquad \text{or} \qquad \lim_{h \to 0} \frac{f(x + h) - f(x)}{h},$$

whichever was algebraically more convenient. In view of the many applications of the derivative, this chapter develops efficient methods of computing the derivatives of the common functions, procedures which do not require going back to limits each time.

5.1

Some notations for derivatives

There are other common symbols for the derivative of the function f besides f', the one employed so far in this book. We shall illustrate them in the case that f is the squaring function,

$$y = f(x) = x^2.$$

The D Notation   First, there is the *D notation*, in which the derivative of f is denoted by $D(f)$. Thus

$$D(x^2) = 2x.$$

This is read, "The derivative of x^2 is $2x$," or, "The derivative of x^2 with respect to x is $2x$." The notation

$$D_x(x^2) = 2x$$

is also common.

The Differential Notation   Second, there is the *differential notation*, in which the derivative of f is denoted by

$$\frac{df}{dx}$$

(read as "dee f dee x" or "the derivative of f with respect to x") or

$$\frac{dy}{dx}$$

(read as "dee y dee x" or "the derivative of y with respect to x"). This notation was introduced by Leibnitz.

The notation dy/dx might tempt us to think of the derivative as a quotient of two numbers. Such a temptation should be resisted. The derivative is defined as a *limit* of a certain quotient, not as an actual quotient. (In the next chapter, the individual symbols dy and dx will be given meaning.) Generally it is best to think of the derivative as a measure of rate of change, of which velocity, slope, magnification, and density are specific examples.

In particular, when $y = x^2$,

$$\frac{dy}{dx} = 2x \qquad \text{or} \qquad \frac{d(x^2)}{dx} = 2x.$$

This notation will be used frequently in the text, and is standard in physics and chemistry.

If the squaring function is expressed with different letters, say, $x = t^2$, its derivative would be written

$$D(t^2) = 2t, \qquad D_t(t^2) = 2t, \qquad \frac{dx}{dt} = 2t, \qquad \text{or} \qquad \frac{d(t^2)}{dt} = 2t.$$

The Dot Notation   Third, there is the *dot notation*, in which the derivative of f is denoted

$$\dot{f}$$

(read as "f dot"). For instance, if $x = t^2$, then

$$\dot{x} = 2t.$$

This notation, introduced by Newton, is reserved for the study of motion, where $\dot{x}$ is then interpreted as velocity and t as time.

The D notation is of aid in stating some formulas; the $f'(x)$ notation is particularly appropriate if the argument is to be indicated.

These are further illustrations of the notations:

$$(x^5)' = 5x^4, \qquad D(x^5) = 5x^4, \qquad D_x(x^5) = 5x^4, \qquad D_t\left(\frac{1}{t}\right) = -\frac{1}{t^2},$$

$$\frac{d(x^5)}{dx} = 5x^4, \qquad \frac{d(t^5)}{dt} = 5t^4, \qquad \frac{d(\sqrt{z})}{dz} = \frac{1}{2\sqrt{z}}, \qquad \dot{t}^5 = 5t^4.$$

Exercises

In Exercises 1 to 13 use formulas already developed to find the indicated derivative.

1. $D(5x^3 + 6x - 2)$ at $x = -1$.
2. $d(t^3 - 5t)/dt$ at $t = 2$.
3. $\sqrt{t}$ at $t = 9$.
4. dy/dx if $y = \frac{1}{5}x^5$.
5. $D_u(6u^4)$.
6. $\dfrac{d(1/u)}{du}$.
7. du/dx if $u = 3x^5 - \frac{1}{2}x^2$.
8. dy/du if $y = \frac{1}{3}u^3$.
9. $D_u(y)$ if $y = 1/u$.
10. $D(s^3 - s^2)$.
11. $(x^5 + 4x^3)'$ at $x = \sqrt{2}$.
12. $f'(\sqrt{2})$ if $f(t) = t^5 + 4t^3$.
13. $(\frac{1}{4}t^4 + \frac{1}{3}t^3 - \frac{1}{2}t^2 + 5)'$.

■

14. Find $D(t^3)$ by calculating
$$\lim_{t_1 \to t} \frac{t_1^3 - t^3}{t_1 - t}.$$

15. Find $d(u^3)/du$ by calculating
$$\lim_{h \to 0} \frac{(u + h)^3 - u^3}{h}.$$

16. Let $y = f(v) = 2v^2 + 3v$. Find dy/dv by calculating

$$\lim_{v_1 \to v} \frac{f(v_1) - f(v)}{v_1 - v}.$$

17. Let $y = f(s) = (s^2 + s)^2$. Find dy/ds by calculating
$$\lim_{s_1 \to s} \frac{f(s_1) - f(s)}{s_1 - s}.$$

18. Let $x = f(t) = 5/t^3$. Find dx/dt by calculating
$$\lim_{t_1 \to t} \frac{f(t_1) - f(t)}{t_1 - t}.$$

■ ■

19. At time t seconds a certain object moving on the y axis has coordinate $y = -16t^2 + 80t$ meters.
    (a) Find dy/dt.
    (b) For which t is dy/dt positive? negative? zero?
    (c) What is the velocity when $t = 0$?

20. Let $y = f(x) = x + 4/x$. (a) Find dy/dx by calculating
$$\lim_{x_1 \to x} \frac{f(x_1) - f(x)}{x_1 - x}.$$
    (b) Graph f for $x > 0$. In particular, show where the slope is positive, negative, and zero.

21. Give an example of a function such that (a) $dy/du = u^3$; (b) $\dot{y} = 1/\sqrt{t}$; (c) $D_x(y) = 3x + 5$; (d) $D_t(y) = 5$. *Hint:* Use your experience with derivatives. All the answers have simple formulas.

5.2

The derivatives of a constant function, sine, and cosine

Later in this chapter it will be seen that the derivatives of all the functions most often used in applications of calculus can be computed if the derivatives of the following functions are known: the constant functions, the trigonometric functions sin x and cos x, and the logarithmic functions. In this section the derivatives of the constant functions and the two trigonometric functions sin x and cos x will be found. In each case it will, of course, be necessary to go back to the definition of the derivative of a function in order to obtain the formula for the derivative.

THEOREM I   The derivative of a constant function is 0, in symbols,

$$(c)' = 0 \quad \text{or} \quad \frac{dc}{dx} = 0.$$

PROOF Let c be a number and let f be the function that assigns to *any* input x the fixed output c. Thus, $f(x) = c$ for all numbers x. We have then

$$f(x + h) - f(x) = c - c = 0.$$

Thus
$$\frac{f(x + h) - f(x)}{h} = \frac{0}{h} \qquad h \neq 0$$

$$= 0.$$

Hence
$$\lim_{h \to 0} \frac{f(x + h) - f(x)}{h} = 0.$$

This shows that the derivative of any constant function is 0 for all x. ●

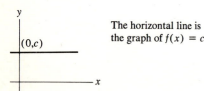

The horizontal line is the graph of $f(x) = c$

From two points of view, Theorem 1 is no surprise: Since the graph of $f(x) = c$ is a horizontal line, it coincides with each of its tangent lines, as can be seen in the accompanying figure. Also, if we think of x as time and $f(x)$ as the position of a particle, Theorem 1 implies that a stationary particle has a zero velocity.

In order to find $d(\sin x)/dx$ and $d(\cos x)/dx$, it will be necessary to make use of the limits

$$\lim_{h \to 0} \frac{\sin h}{h} \qquad \text{and} \qquad \lim_{h \to 0} \frac{1 - \cos h}{h},$$

where angles are measured in radians. In Sec. 4.6, (where the letter θ was used instead of h) it was found that

$$\lim_{h \to 0} \frac{\sin h}{h} = 1 \qquad \text{and} \qquad \lim_{h \to 0} \frac{1 - \cos h}{h} = 0.$$

THEOREM 2 The derivative of the sine function is the cosine function; symbolically,

$$(\sin x)' = \cos x \qquad \text{or} \qquad \frac{d(\sin x)}{dx} = \cos x.$$

PROOF The derivative at x of a function f is defined as

$$\lim_{h \to 0} \frac{f(x + h) - f(x)}{h}.$$

In this case f is the function "sine," and the limit under consideration is

$$\lim_{h \to 0} \frac{\sin (x + h) - \sin x}{h}.$$

Keep in mind that x is fixed while $h \to 0$. As $h \to 0$, the numerator approaches

$$\sin x - \sin x = 0,$$

while the denominator h also approaches 0. Since the expression 0/0 is meaningless, it is necessary to change the form of the quotient

$$\frac{\sin (x + h) - \sin x}{h}$$

before letting h approach 0.

Let us use the formula

$$\sin (A + B) = \sin A \cos B + \cos A \sin B$$

in the case $A = x$ and $B = h$, obtaining

$$\sin (x + h) = \sin x \cos h + \cos x \sin h.$$

Then the numerator, $\sin (x + h) - \sin x$, takes the form

$$\sin x \cos h + \cos x \sin h - \sin x = \sin x(\cos h - 1) + \cos x \sin h$$
$$= -\sin x(1 - \cos h) + \cos x \sin h.$$

Therefore

$$\lim_{h \to 0} \frac{\sin (x + h) - \sin x}{h} = \lim_{h \to 0} \frac{-\sin x(1 - \cos h) + \cos x \sin h}{h}$$

$$= \lim_{h \to 0} \left(-\sin x \frac{1 - \cos h}{h} + \cos x \frac{\sin h}{h} \right)$$

$$= (-\sin x)(0) + (\cos x)(1)$$

$$= \cos x.$$

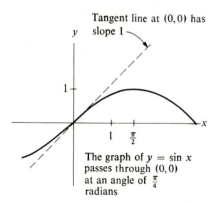

Tangent line at $(0,0)$ has slope 1

The graph of $y = \sin x$ passes through $(0,0)$ at an angle of $\frac{\pi}{4}$ radians

In short, the derivative of the sine function is the cosine function. This concludes the proof. ●

The formula obtained in Theorem 2 provides interesting information about the graph of $y = \sin x$. Since

$$\frac{d(\sin x)}{dx} = \cos x,$$

the derivative of the sine function when $x = 0$ is $\cos 0$, which is 1. This implies that the slope of the curve $y = \sin x$, when $x = 0$, is 1. Consequently, the graph of $y = \sin x$ passes through the origin at an angle of $\pi/4$ radians (45°). See the accompanying figure.

THEOREM 3   The derivative of the cosine function is the negative of the sine function; symbolically,

$$(\cos x)' = -\sin x. \ ●$$

We omit the proof, which is similar to that of Theorem 2. It makes use of the trigonometric identity

$$\cos (A + B) = \cos A \cos B - \sin A \sin B.$$

In Theorems 2 and 3 angles are measured in radians as is generally done in

calculus. What would the derivatives of sine and cosine be if angles were measured in degrees instead? The graph of $y = \sin x$, where angle is measured in degrees, has a slope near 0 for all x, as shown in the margin. Indeed the graph of $y = \sin x$ is now practically a horizontal line. We are tempted to change the vertical scale to stretch the graph vertically, but to do so would change the slopes, the very object of our inquiry.

If you look over the proof of Theorem 2, you will see that the key step is finding out that

$$\lim_{h \to 0} \frac{\sin h}{h} = 1.$$

If angle is measured in degrees, this limit is no longer 1. To see what the limit is, observe that an angle of h degrees is the same as an angle of $\pi h/180$ radians. Thus

$$\frac{\sin h^\circ}{h} = \frac{\sin (\pi h/180)}{h},$$

where, on the right side of the equation, the angle is measured in radians. Now

$$\frac{\sin (\pi h/180)}{h} = \frac{\sin (\pi h/180)}{\pi h/180} \frac{\pi}{180}.$$

Thus

$$\lim_{h \to 0} \frac{\sin h^\circ}{h} = \lim_{h \to 0} \frac{\sin (\pi h/180)}{\pi h/180} \frac{\pi}{180} = 1 \frac{\pi}{180} = \frac{\pi}{180}.$$

In short, when angles are measured in degrees,

$$\lim_{h \to 0} \frac{\sin h}{h} = \frac{\pi}{180} \doteq 0.017.$$

Even if angles are measured in degrees, it is still true that

$$\lim_{h \to 0} \frac{1 - \cos h}{h} = 0,$$

as the reader may check by examining the proof in Sec. 4.6 that this limit is 0 when angles are measured in radians.

Imitation of the steps in the proof of Theorem 2 shows that if the angle is measured in degrees, then

$$(\sin x)' = \frac{\pi}{180} \cos x \doteq 0.017 \cos x.$$

The great advantage of radian measure for angles is now evident: It makes the formula for the derivative of the sine function (indeed, of any trigonometric function) easier. There is no extra constant, such as $\pi/180$, to memorize. This goes back basically to the simplicity of

$$\lim_{h \to 0} \frac{\sin h}{h}$$

when angles are measured in radians; it is just 1.

Exercises

1. (a) Sketch a graph of the sine function for x in $[0, 2\pi]$. Use the same scale on both axes.
   (b) For which x in $[0, 2\pi]$ is the slope positive? negative? zero?
   (c) For which x in $[0, 2\pi]$ is the slope largest?
2. Evaluate the derivative of the sine function at (a) $\pi/4$, (b) $\pi/6$, (c) $\pi/2$, (d) π.
3. Compute the following:

   (a) $\dfrac{d(\sin x)}{dx}$ at $\dfrac{\pi}{3}$ (b) $\dfrac{d(\cos x)}{dx}$ at $\dfrac{\pi}{4}$

   (c) $\dfrac{d(x^3)}{dx}$ at -1 (d) $\dfrac{d(5)}{dx}$ at 4.7.

4. The left-hand x centimeters of a string $\pi/2$ centimeters long has a mass of $\sin x$ grams. What is its density at (a) its left end, (b) its center, and (c) its right end?
5. At time t, in seconds, a bobbing cork is $\sin t$ centimeters above the water. (When $\sin t$ is negative, the cork is below the water.)
   (a) Find its height above or below the water when $t = 0$, $\pi/2$, π, $3\pi/2$.
   (b) Find its velocity when $t = 0$, $\pi/2$, π, $3\pi/2$.
   (c) Find its speed at the times listed in (b).
6. (a) What is the slope of the tangent line to the curve $y = \sin x$ at $(\pi/3, \sqrt{3}/2)$?
   (b) Use (a) and a table of values of $\tan \theta$ to estimate the angle that the tangent line in (a) makes with the x axis.
7. Using the definition of derivative, prove that the derivative of the cosine function is minus the sine function.
8. Prove that when angles are measured in degrees

$$\lim_{h \to 0} \frac{1 - \cos h}{h} = 0.$$

   ■

9. Using the definition of the derivative, obtain the formula

$$(\cos x)' = -\frac{\pi}{180} \sin x$$

if angles are measured in degrees.

In Exercises 10 to 12 obtain the given formula by taking the limit of the appropriate quotient.

10. $D(\sin 3x) = 3 \cos 3x$.
11. $D(\cos 4x) = -4 \sin 4x$.
12. $\dfrac{d(3 \cos x + 2 \sin x)}{dx} = -3 \sin x + 2 \cos x$.
13. The height of the ocean surface above (or below) mean sea level is, say, $y = 2 \sin t$ feet at t hours.
    (a) Find the rate at which the tide is rising or falling at time t.
    (b) Is the surface rising most rapidly at low tide or when it is at mean sea level?

   ■ ■

14. Using the definition of the derivative as the limit of a certain quotient, show that

$$D(x \sin 2x) = 2x \cos 2x + \sin 2x.$$

15. Using the definition of the derivative as the limit of a certain quotient, show that

$$D(\sin x^2) = 2x \cos x^2.$$

16. (a) Using the definition of the derivative as the limit of a certain quotient, show that

$$D\left(\frac{1 - \cos 2x}{2}\right) = \sin 2x.$$

   (b) From (a) deduce that

$$D(\sin^2 x) = \sin 2x.$$

   (c) From (b) deduce that

$$D(\sin^2 x) = 2 \sin x \cos x.$$

   [Is this what you would expect $D(\sin^2 x)$ to be?]

5.3

A review of logarithms

In algebra and trigonometry, logarithms are used for computations. In calculus, the logarithm functions stand on their own as one of the most important classes of functions. This section reviews the logarithm functions and describes their basic properties; then Sec. 5.4 obtains their derivatives.

First recall the definition of a logarithm.

Consider the question $4^? = 16$;

which we read as "4 raised to what power equals 16?" The answer, whatever its numerical value might be, is called "the logarithm of 16 to the base 4." Since

$$4^2 = 16,$$

we say that "the logarithm of 16 to the base 4 is 2." The general definition of logarithm follows.

DEFINITION   *Logarithm.* If b and c are positive numbers and

$$b^x = c,$$

then the number x is the logarithm of c to the base b, and is written

$$\log_b c.$$

Any exponential equation $b^x = c$ may be translated into a logarithmic equation $x = \log_b c$, just as any English statement may be translated into French. This table illustrates some of these translations. Read it over several times, perhaps aloud, until you can, when covering a column, fill in the correct translation of the other column.

Exponential Language	*Logarithm Language*
$4^2 = 16$	$\log_4 16 = 2$
$7^0 = 1$	$\log_7 1 = 0$
$10^3 = 1{,}000$	$\log_{10} 1{,}000 = 3$
$10^{-2} = 0.01$	$\log_{10} 0.01 = -2$
$9^{1/2} = 3$	$\log_9 3 = \frac{1}{2}$
$8^{2/3} = 4$	$\log_8 4 = \frac{2}{3}$
$8^{-1} = \frac{1}{8}$	$\log_8 \frac{1}{8} = -1$
$5^1 = 5$	$\log_5 5 = 1$
$e^1 = e$	$\log_e e = 1$

Since $b^x = c$ is equivalent to $x = \log_b c$, it follows that

$$b^{\log_b c} = c.$$

EXAMPLE 1   Find $\log_5 125$.

SOLUTION   We are looking for an answer to the question

"5 to what power equals 125?"

or, equivalently, for a solution of the equation

$$5^x = 125.$$

Since $5^3 = 125$, the answer is 3.

$$\log_5 125 = 3. \ \bullet$$

EXAMPLE 2   Find $\log_{10} \sqrt{10}$.

SOLUTION   By the definition of $\log_{10} \sqrt{10}$,

$$10^{\log_{10} \sqrt{10}} = \sqrt{10}.$$

Now,

$$10^{1/2} = \sqrt{10}.$$

Thus

$$\log_{10} \sqrt{10} = \tfrac{1}{2}.$$

In words, we say, "The power to which we must raise 10 to get $\sqrt{10}$ is $\tfrac{1}{2}$." As the square root table shows, $\sqrt{10} \doteq 3.162$.

Thus

$$\log_{10} 3.162 \doteq \tfrac{1}{2} = 0.5. \; \bullet$$

In order to get an idea of the logarithm as a function, consider logarithms to the familiar base 10,

$$y = \log_{10} x.$$

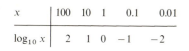

x	100	10	1	0.1	0.01
$\log_{10} x$	2	1	0	-1	-2

Begin with a table, as shown in the margin.   We must restrict ourselves to $x > 0$.   A negative number, such as -1, cannot have a logarithm, since there is no power of 10 that equals -1.   The domain of the function "log to the base 10" consists of the positive real numbers.

With the aid of the five points in the table, the graph is easy to sketch.   The graph lies to the right of the y axis.   Far to the right it rises slowly; not until x reaches 100 does the y coordinate reach 2.   When x is a small positive number, observe that $\log_{10} x$ is a negative number of large absolute value.   Observe also that as x increases so does $\log_{10} x$.   Though x is restricted to being positive, $\log_{10} x$ takes on all values, positive and negative.   Logarithms to the base 10 are called *common logarithms*.   A brief table of common logarithms is to be found in Appendix E and, in more detail, in any handbook of mathematical tables.   Many calculators have a $\log_{10}$-key, usually labeled "log."

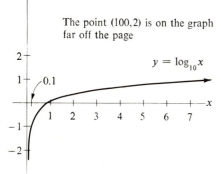

The point (100,2) is on the graph far off the page

$y = \log_{10} x$

Since each exponential equation $b^x = c$ translates into the corresponding logarithmic equation $x = \log_b c$, every property of exponentials must carry over to some property of logarithms.   For instance, let us find out what the basic law of exponents,

$$b^x \cdot b^y = b^{x+y},$$

implies about logarithms.

To do this, give b^x the name c, and b^y the name d; that is,

$$b^x = c \qquad \text{and} \qquad b^y = d.$$

The equation

$$b^x \cdot b^y = b^{x+y}$$

now reads

$$cd = b^{x+y},$$

and therefore

$$\log_b cd = x + y.$$

But since

$$x = \log_b c \qquad \text{and} \qquad y = \log_b d,$$

we have obtained

$$\log_b cd = \log_b c + \log_b d.$$

This equation is one of the fundamental properties of logarithms, namely that "the log of the product is the sum of the logs of the factors." The table below lists, without proof, the logarithmic equivalents of the properties of the exponential.

Exponents	*Logarithms*
$b^0 = 1$	$\log_b 1 = 0$
$b^{1/2} = \sqrt{b}$	$\log_b \sqrt{b} = \frac{1}{2}$
$b^1 = b$	$\log_b b = 1$
$b^{x+y} = b^x b^y$	$\log_b cd = \log_b c + \log_b d$
$b^{-x} = \dfrac{1}{b^x}$	$\log_b \left(\dfrac{1}{c}\right) = -\log_b c$
$b^{x-y} = \dfrac{b^x}{b^y}$	$\log_b \left(\dfrac{c}{d}\right) = \log_b c - \log_b d$
$(b^x)^y = b^{xy}$	$\log_b c^m = m \log_b c$

EXAMPLE 3   Use the last line in the preceding table to help simplify $\log_9 (3^7)$, $\log_5 \sqrt[3]{25^2}$, and $\log_{10} c^{x/h}$.

SOLUTION
$$\log_9 (3^7) = 7 \log_9 3 = 7(\tfrac{1}{2}) = \tfrac{7}{2}$$
$$\log_5 \sqrt[3]{25^2} = \log_5 (25)^{2/3} = \tfrac{2}{3} \log_5 25 = (\tfrac{2}{3})2 = \tfrac{4}{3}$$
$$\log_{10} c^{x/h} = \frac{x}{h} \log_{10} c. \;\bullet$$

The next example anticipates some computations in the next section where the derivative of the function $y = \log_b x$ is obtained.

EXAMPLE 4   Consider a positive number x and a number h such that $x + h$ is also positive. Show that
$$\frac{\log_b (x + h) - \log_b x}{h} = \log_b \left(1 + \frac{h}{x}\right)^{1/h}.$$

SOLUTION
$$\frac{\log_b (x + h) - \log_b x}{h} = \frac{\log_b [(x + h)/x]}{h} \qquad \log_b \frac{c}{d} = \log_b c - \log_b d$$

$$= \frac{1}{h} \log_b \left(1 + \frac{h}{x}\right) \qquad \text{algebra}$$

$$= \log_b \left(1 + \frac{h}{x}\right)^{1/h} \qquad m \log_b c = \log_b c^m. \;\bullet$$

The next example illustrates how the law of logarithms
$$\log_b c^m = m \log_b c,$$
combined with some arithmetic, can be used to estimate logarithms of specific numbers.

EXAMPLE 5   Estimate $\log_{10} e$.

SOLUTION   First of all, since $e < 3 < 10^{1/2}$, it follows that

$$\log_{10} e < \tfrac{1}{2}.$$

On the other hand,

$$2 < e;$$

hence

$$\log_{10} 2 < \log_{10} e.$$

To obtain information on $\log_{10} 2$, begin with the inequality

$$10^3 = 1{,}000 < 1{,}024 = 2^{10}.$$

Thus

$$\log_{10} 10^3 < \log_{10} 2^{10}$$

or

$$3 < 10 \log_{10} 2.$$

Hence

$$\tfrac{3}{10} < \log_{10} 2.$$

Combining this with the inequality $\log_{10} 2 < \log_{10} e$ shows that

$$\tfrac{3}{10} < \log_{10} e.$$

All told,

$$\tfrac{3}{10} < \log_{10} e < \tfrac{1}{2}.$$

More advanced methods show that

$$\log_{10} e \doteq 0.434. \quad \bullet$$

The final example shows how logarithms can be used to solve equations in which the unknown appears in an exponent.

EXAMPLE 6   Find x if

$$5 \cdot 3^x = 2.$$

SOLUTION   First take logarithms of both sides, using base 10:

$$\log_{10} (5 \cdot 3^x) = \log_{10} 2,$$

$$\log_{10} 5 + \log_{10} 3^x = \log_{10} 2, \qquad \text{log of a product}$$

$$\log_{10} 5 + x \log_{10} 3 = \log_{10} 2 \qquad \text{log of an exponential.}$$

Then solve this last equation for x:

$$x = \frac{\log_{10} 2 - \log_{10} 5}{\log_{10} 3}$$

$$= \frac{0.3010 - 0.6990}{0.4771} \qquad \text{from the table in Appendix E}$$

Thus

$$x = -0.8342. \quad \bullet$$

Exercises

1. Translate these equations into the language of logarithms.
   (a) $2^5 = 32$ (b) $3^4 = 81$
   (c) $10^{-3} = 0.001$ (d) $5^0 = 1$
   (e) $1,000^{1/3} = 10$ (f) $7 = \sqrt{49}$.

2. Let $f(x) = \log_2 x$.
   (a) Compute $f(x)$ for $x = \frac{1}{8}, \frac{1}{4}, \frac{1}{2}, 1, 2, 4$, and graph f.
   (b) What is $\lim_{x \to 0^+} f(x)$? $\lim_{x \to \infty} f(x)$?

3. Evaluate: (a) $\log_2 4$ and $\log_4 2$, (b) $\log_2 8$ and $\log_8 2$, (c) $\log_{10} 100$ and $\log_{100} 10$.

4. Assume that $\log_{10} 2 \doteq 0.30$ and $\log_{10} 3 \doteq 0.48$. From this information estimate
   (a) $\log_{10} 4$ (b) $\log_{10} 5$ Hint: $5 = \frac{10}{2}$
   (c) $\log_{10} 6$ (d) $\log_{10} 8$
   (e) $\log_{10} 9$ (f) $\log_{10} 1.5$ Hint: $1.5 = \frac{3}{2}$
   (g) $\log_{10} 1.2$ (h) $\log_{10} 1.33$.
   (i) $\log_{10} 20$ (j) $\log_{10} 200$
   (k) $\log_{10} 0.006$.

5. Translate these equations into the language of exponents.
   (a) $\log_2 7 = x$ (b) $\log_5 2 = s$
   (c) $\log_3 \frac{1}{3} = -1$ (d) $\log_7 49 = 2$.

6. Evaluate
   (a) $\log_4 2$ (b) $\log_5 25$
   (c) $\log_5 \frac{1}{25}$ (d) $\log_7 7$
   (e) $\log_2 4$.

7. Evaluate:
   (a) $\log_5 0.2$ (b) $\log_3 (3\sqrt{3})$
   (c) $\log_6 1$.

8. Evaluate:
   (a) $2^{\log_2 8}$ (b) $2^{\log_2 16}$
   (c) $10^{\log_{10} 100}$ (d) $3^{\log_3 7}$

9. Evaluate:
   (a) $\log_3 (\sqrt{3} \sqrt[3]{9})$ (b) $\log_3 (1/\sqrt{3})$
   (c) $\log_3 (\sqrt[3]{3}/\sqrt{3})$.

10. If $\log_4 A = 2.1$, find (a) $\log_4 A^2$, (b) $\log_4 64A$, (c) $\log_4 16/A$, (d) $\log_2 A$.

11. Find x if $2 \cdot 3^x = 7$.

12. Find x if $3 \cdot 5^x = 6^x$.

13. The common logarithm table can be used to find logarithms with other bases. For instance, $\log_3 2$ can be found as follows. Let $x = \log_3 2$.
    (a) Why is $3^x = 2$?
    (b) Show that $x \log_{10} 3 = \log_{10} 2$.
    (c) Using (b) and a common logarithm table, find $x = \log_3 2$.

■

14. (See Exercise 2.)

(a) Graph $g(x) = \log_4 x$.
(b) Show that, for a given x, $\log_4 x = (\frac{1}{2}) \log_2 x$.
(c) What does (b) imply about the graphs in Exercises 2(a) and 14(a)?

15. (a) Using the table of $\log_{10} x$ in Appendix E or a calculator, complete this table.

x	0.01	0.1	1	2	3	4	10	100
$\log_{10} x$								
$\dfrac{\log_{10} x}{x}$								

(b) Using (a), graph $y = (\log_{10} x)/x$.
(c) At what value of x do you think that $(\log_{10} x)/x$ is the largest?

16. Show that (a) $8^x = 2^{3x}$ and (b) $10^x = 2^{x \log_2 10}$. As (a) and (b) suggest, the exponential with a given base can be replaced by an exponential with any other base.

17. Prove that
$$\log_b x = \frac{\log_{10} x}{\log_{10} b}.$$

Hint: Take $\log_{10}$ of both sides of the equation $b^{\log_b x} = x$. This is the general formula for translating from $\log_{10}$ to $\log_b$.

18. (a) If $b^3 = c$, then we may write $3 = $ _____.
    (b) If $b^3 = c$, then we may write $b = $ _____.

19. Using the property $b^{-x} = 1/b^x$, deduce that $\log_b (1/c) = -\log_b c$.

20. Using the property $b^{x-y} = b^x/b^y$, deduce that $\log_b (c/d) = \log_b c - \log_b d$.

21. Prove that if $a, b > 0$, then $\log_b a \cdot \log_a b = 1$.

22. Using the property $(b^x)^y = b^{xy}$, deduce that $\log_b c^m = m \log_b c$.

23. Let a and b be fixed positive numbers.
    (a) Prove that
$$\frac{\log_b x}{\log_a x}$$
is independent of x; that is, the function f defined by
$$f(x) = \frac{\log_b x}{\log_a x}$$
is constant.
    (b) What does (a) imply about the relation between the graphs of $y = \log_b x$ and $y = \log_a x$?

24. Let A and k be constants. (a) Determine constants B and m such that, for all x, $A(2^{kx}) = B(3^{mx})$. (b) Express B and m in terms of A and k.

25. Let $y = A \cdot 10^{kx}$. (a) Show that the graph of $z = \log_{10} y$,

when z is considered a function of x, is a straight line. (b) What is the slope of the line in (a)?

26. The noise level of a sound with pressure p is measured in decibels by the formula

$$D = k \log \frac{p}{p_0},$$

where k and p_0 are constants and log denotes logarithm to some fixed base. The term is named in honor of Alexander Graham Bell. The noise in a subway car registers 95 decibels, conversational speech at 3 feet registers 60 decibels, and the noise in an average residence, 30 decibels.

(a) Show that if a certain barrier, such as a window or wall, cuts the pressure p of any sound in half, then it decreases the decibels of any sound by a fixed amount. (A hollow door decreases D by 15 decibels and a $\frac{1}{4}$-inch glass window by 30 decibels.)

(b) Solve for p in terms of D.

27. Prove that $\log_2 3$ is *not* a rational number.

The following two exercises require the use of a calculator.

28. What happens to $y = [(2^x + 5^x)/2]^{1/x}$ as $x \to 0$? Rather than explore this directly, examine instead the behavior of

$$\log_{10} y = \frac{1}{x} \log_{10}\left(\frac{2^x + 5^x}{2}\right).$$

Can you guess $\lim_{x \to 0} \log_{10} y$? $\lim_{x \to 0} y$?

29. (See Exercise 17.) Use the $\log_{10}$-key on a calculator to find $\log_3 5$.

5.4

The derivative of a logarithm function

In this section the derivative of the logarithm function, $\log_b x$, is obtained, and it is shown that in calculus the most convenient base for logarithms is e.

The graph of the function $y = \log_{10} x$ discussed in the preceding section suggests that, when x is large, $d(\log_{10} x)/dx$ should be near 0, since the graph is almost horizontal far to the right. When x is small, the graph is steep; the derivative should be very large. For any x in the domain of the function $\log_{10} x$, the derivative should be positive, since the tangent lines slope upward. After the formula for the derivative of the function $\log_{10} x$ is obtained in Theorem 1, these observations will be compared with the formula for the derivative.

In the proof of Theorem 1, two assumptions are made which are justified in advanced calculus, namely, that the function $\log_{10} x$ is continuous and that it is defined for all positive x. The number e, discussed in Secs. 4.2 and 4.4, will be needed; in particular, the fact that

$$(1 + \text{small number } s)^{1/s}$$

Review This Conclusion in Sec. 4.4 Before Going on. is approximately e is useful.

THEOREM 1 The derivative of the function $\log_{10} x$ is

$$\frac{\log_{10} e}{x}$$

for all positive numbers x. [The number e is defined as

$$\lim_{n \to \infty} \left(1 + \frac{1}{n}\right)^n.]$$

PROOF The function whose derivative is to be found in this case assigns to the number x the value

$$\log_{10} x.$$

It is therefore necessary to determine

$$\lim_{h \to 0} \frac{\log_{10} (x + h) - \log_{10} x}{h}.$$

Note that x is fixed, and that $h \to 0$.

As $h \to 0$, the numerator $\log_{10} (x + h) - \log_{10} x$ approaches 0, since the logarithm function is continuous. Hence both numerator and denominator approach 0 as $h \to 0$. Since the expression 0/0 is meaningless, it is not at all clear what happens to the quotient. Fortunately, the properties of logarithms permit us to rewrite the quotient in a form whose limit is easy to compute.

First, the identity

$$\log_b c - \log_b d = \log_b \frac{c}{d}$$

shows that the quotient equals

$$\frac{\log_{10} [(x + h)/x]}{h}.$$

Now,

$$\frac{x + h}{x} = 1 + \frac{h}{x}$$

so

$$\frac{\log_{10} [(x + h)/x]}{h} = \frac{\log_{10} [1 + (h/x)]}{h}.$$

What happens to

$$\frac{\log_{10} [1 + (h/x)]}{h}$$

as $h \to 0$? The numerator approaches $\log_{10} 1$, which is 0, and the denominator approaches 0. So the limit of the quotient is still far from obvious, since the expression 0/0 is meaningless.

However, the fact that h/x is small when h is small implies that

$$\left(1 + \frac{h}{x}\right)^{x/h} \text{ is near } e$$

when h is small. Thus

$$\lim_{h \to 0} \left(1 + \frac{h}{x}\right)^{x/h} = e.$$

This suggests the following steps:

$$\frac{\log_{10} (1 + h/x)}{h} = \frac{1}{h} \log_{10} \left(1 + \frac{h}{x}\right) \qquad \text{algebra}$$

$$= \log_{10} \left(1 + \frac{h}{x}\right)^{1/h} \qquad \log_b c^m = m \log_b c$$

$$= \log_{10}\left(1 + \frac{h}{x}\right)^{(x/h)(1/x)} \qquad \text{algebra}$$

$$= \log_{10}\left[\left(1 + \frac{h}{x}\right)^{x/h}\right]^{1/x} \qquad (b^x)^y = b^{xy}$$

$$= \frac{1}{x}\log_{10}\left(1 + \frac{h}{x}\right)^{x/h} \qquad \log_b c^m = m\log_b c.$$

After these changes, it is easy to find the limit. Since a logarithm function is continuous (and x is fixed),

$$\lim_{h\to 0}\frac{1}{x}\log_{10}\left(1 + \frac{h}{x}\right)^{x/h} = \frac{1}{x}\log_{10}\left[\lim_{h\to 0}\left(1 + \frac{h}{x}\right)^{x/h}\right]$$

$$= \frac{1}{x}\log_{10} e.$$

Thus $\log_{10} x$ has a derivative, and the derivative is

$$\frac{1}{x}\log_{10} e.$$

This concludes the proof. ●

Does this agree with the observations made in the opening paragraph about the graph of $y = \log_{10} x$? As mentioned in Example 5 of Sec. 5.3, $\log_{10} e \doteq 0.434$. Thus

$$\frac{d(\log_{10} x)}{dx} \doteq \frac{0.434}{x}.$$

Since x is positive, so is $0.434/x$, as expected. Also, when x is large, $0.434/x$ is small; when x is near 0, $0.434/x$ is large. Both these results agree with the predictions made about $d(\log_{10} x)/dx$.

The same argument that found that $(\log_{10} x)'$ equals $(\log_{10} e)/x$ applies to $\log_b x$ for any base b, and the conclusion is stated in Theorem 2.

THEOREM 2   The derivative of the function $\log_b x$ is

$$\frac{\log_b e}{x}.$$

Why e Is Used as a Base for Logarithms   Which is the best of all possible bases b to use? More precisely, for which base b does the formula

$$\frac{\log_b e}{x}$$

take the simplest form? Certainly not $b = 10$. It would be nice to choose the base b in such a way that

$$\log_b e = 1,$$

that is, b^1 must equal e. In this case b is e. The best of all bases to use for logarithms is e. The derivative of the $\log_e$ function is given by

$$\frac{d(\log_e x)}{dx} = \frac{\log_e e}{x} = \frac{1}{x}.$$

In this case there is no constant, such as 0.434, to memorize.

The Natural Logarithm   For this reason, the base e is preferred in calculus. We shall write $\log_e x$ as ln x, the *natural logarithm* of x. (Only for purposes of arithmetic, such as multiplying with the aid of logarithms, is base 10 preferable.) Most handbooks of mathematical tables include tables of $\log_{10} x$ (common logarithm) and ln x (natural logarithm). It would be helpful at this point to browse through both tables, shortened versions of which appear in Appendix E.

The Definition and Derivative of ln *x Is the Core of This Section.*   A simple equation summarizes much of this section:

$$\frac{d(\ln x)}{dx} = \frac{1}{x} \qquad x > 0.$$

It is well worth memorizing the following statement: The derivative of the natural logarithm function ln x is the reciprocal function $1/x$. The importance of this observation is emphasized by the following example.

EXAMPLE   Give an illustration of a function whose derivative is (*a*) $1/x^2$, (*b*) $1/x$.

SOLUTION   (*a*) The derivative of $1/x$ involves $1/x^2$, specifically,

$$D\left(\frac{1}{x}\right) = -\frac{1}{x^2}.$$

Thus

$$D\left(-\frac{1}{x}\right) = \frac{1}{x^2}.$$

(*b*) The derivative of ln x is $1/x$.

The contrast between (*a*) and (*b*) exhibits one of the uses of ln x in later chapters. ●

Exercises

1. What is the derivative of:
 (*a*) $\log_{10} x$?   (*b*) $\log_2 x$?
 (*c*) $\log_5 x$?   (*d*) $\log_e x$?
 (*e*) ln x?

2. Compute $(\log_{10} x)'$ at
 (*a*) $x = 2$ (*b*) $x = 3$
 (*c*) $x = \frac{1}{2}$.

3. Compute $(\ln x)'$ at
 (*a*) $x = 2$ (*b*) $x = 3$
 (*c*) $x = \frac{1}{2}$.

4. Using the estimates ln $2 \doteq 0.69$ and ln $3 \doteq 1.10$, estimate
 (*a*) ln $\frac{1}{2}$ (*b*) ln 6
 (*c*) ln 12   (*d*) ln $\frac{2}{3}$
 (*e*) ln 1.5   (*f*) ln 2^7
 (*g*) ln $\frac{1}{9}$.

5. (*a*) Graph $y = \ln x$. (Use the natural logarithm table.)
 (*b*) At what angle does the graph cross the x axis?

6. Give an example of a function whose derivative is
 (*a*) $4x^3$ (*b*) $4x^2$ (*c*) $4x$ (*d*) 4 (*e*) $\frac{4}{x}$ (*f*) $\frac{4}{x^2}$.

7. A lens projects $x > 0$ of a slide onto $\ln x$ on a screen.
  (*a*) What is its magnification at $x = 2$?
  (*b*) Approximately how long is the image of the interval $[2, 2.08]$?

8. (*a*) Using a ruler, estimate the slope of the graph of $y = \log_{10} x$ at $(1, 0)$. (This graph is shown in the preceding section.)
  (*b*) How does this compare with the value obtained from the formula $d(\log_{10} x)/dx = 0.434/x$?

9. This exercise shows how a table of common logarithms can be used to compute natural logarithms.
  (*a*) Taking natural logarithms of both sides of the equation
$$10^{\log_{10} x} = x,$$
  show that $\ln x = (\ln 10) \log_{10} x$.
  (*b*) Using the estimate $\ln 10 \doteq 2.3$ and the formula in (*a*), find $\ln 2$, $\ln 3$, and $\ln 7$.

10. In Exercise 21 of the preceding section it was shown that
$$\log_a b \cdot \log_b a = 1.$$
  Use this result and Theorem 2 to show that
$$\frac{d(\log_b x)}{dx} = \frac{1}{(\log_e b)x}.$$

■

11. At time $t > 0$ a particle moving on the x axis has the coordinate $x = \ln t$. (*a*) Find its speed when $t = 3$. (*b*) Using (*a*), estimate how far the particle moves during the time interval $[3, 3.06]$.

12. Show that there is a number x in $[1, 2]$ such that $\ln x = 1/x$.

13. Show that there is a number at which the functions $\ln x$ and $\sin x$ have the same derivatives.

14. The graph below shows the tangent line to the curve $y = \ln x$ at a point (x_0, y_0). Prove that AB has length 1, independent of the choice of (x_0, y_0).

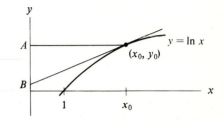

15. The derivative of $x \ln x$, by definition, equals
$$\lim_{x_1 \to x} \frac{x_1 \ln x_1 - x \ln x}{x_1 - x}.$$
  Rewrite the numerator as $x_1 \ln x_1 - x_1 \ln x + x_1 \ln x - x \ln x$, and show that
$$D(x \ln x) = 1 + \ln x.$$
  This exercise introduces ideas to be explored extensively in the next chapter.

16. See Exercise 15. (*a*) Show that
$$D\left(\frac{\ln x}{x}\right) = \frac{1 - \ln x}{x^2}.$$
  (*b*) Graph $y = (\ln x)/x$.
  (*c*) Where is the slope of the graph in (*b*) positive? negative? zero?
  (*d*) Which choice of x yields the largest value of $(\ln x)/x$?

5.5

The derivative of the sum, difference, and product of functions

Using algebraic, trigonometric, or logarithmic identities, we have found, for instance, that
$$\frac{d(x^2)}{dx} = 2x, \qquad \frac{d(x^3)}{dx} = 3x^2,$$
$$\frac{d(\sin x)}{dx} = \cos x, \qquad \text{and} \qquad \frac{d(\ln x)}{dx} = \frac{1}{x}.$$

If the derivative of $\qquad x^2 \cos x + x^3 \ln x$

is needed, must we go back to the definition of the derivative and study the limit of a quotient? The answer is "no," and the theorems of this section provide formulas for computing such derivatives.

So far, two notations for the definition of the derivative have been used:

$$\lim_{x_1 \to x} \frac{f(x_1) - f(x)}{x_1 - x} \quad \text{and} \quad \lim_{h \to 0} \frac{f(x + h) - f(x)}{h}.$$

We now introduce a third notation, which is very brief and quite useful when proving theorems about derivatives.

The Delta Notation The numerator

$$f(x + h) - f(x),$$

which is *the change in the output*, will be denoted by

$$\Delta f,$$

read as "delta f." Similarly, we denote the denominator h, which is *the change in the input*, by

$$\Delta x.$$

Both Δx and Δf can be positive or negative; Δf may be 0, but Δx, like h, is not permitted to be 0. Note that

$$\lim_{\Delta x \to 0} \frac{\Delta f}{\Delta x}$$

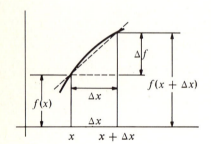

is the derivative of f at x. By the definition of Δf and Δx,

$$\Delta f = f(x + \Delta x) - f(x)$$

and

$$f(x + \Delta x) = f(x) + \Delta f.$$

On the graph of the function f, Δx and Δf denote the distances needed to form an estimate of the slope (see the accompanying figure). Note that if f is continuous at x, then $\lim_{\Delta x \to 0} \Delta f = 0$. In the proofs that follow it is easier to think and work in terms of Δf than the longer expression $f(x + \Delta x) - f(x)$.

We have one more preliminary matter to dispense with before turning to the theorems. If we have two functions f and g, we may form a new function $u = f + g$ as follows. For each number x that lies in domains of f and g, set $u(x) = f(x) + g(x)$. The graph of $f + g$ is obtained from the graphs of f and g by adding the corresponding y values for each value of x. Thus the function which assigns to x the value $x^2 + \cos x$ can be thought of as the sum of the squaring function and the cosine function. In a similar manner one may define $f - g$, fg (the product of two functions), and f/g (the quotient) where $g(x) \neq 0$.

The function whose value at x is $x^2 \cos x$ is the product of the squaring function and the cosine function. The function mentioned in the opening paragraph,

$$x^2 \cos x + x^3 \ln x,$$

is built up from the simpler functions x^2, x^3, $\cos x$, and $\ln x$ by forming sums and products of functions.

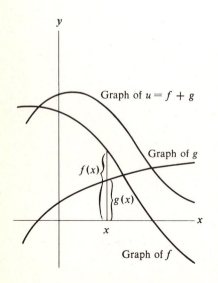

Graph of $u = f + g$

Graph of g

Graph of f

The three theorems of this section are used in the computation of almost any derivative.

THEOREM 1 If c is a constant and if f is differentiable, then cf is differentiable, and

$$\frac{d(cf)}{dx} = c\,\frac{df}{dx}.$$

PROOF Give the function cf the name u.

Then
$$u(x + \Delta x) = cf(x + \Delta x)$$

and
$$u(x) = cf(x).$$

Thus
$$\Delta u = u(x + \Delta x) - u(x) = cf(x + \Delta x) - cf(x)$$
$$= c[f(x + \Delta x) - f(x)]$$
$$= c \cdot \Delta f.$$

Then
$$\frac{du}{dx} = \lim_{\Delta x \to 0} \frac{\Delta u}{\Delta x}$$

$$= \lim_{\Delta x \to 0} \frac{c \cdot \Delta f}{\Delta x}$$

$$= \lim_{\Delta x \to 0} c\,\frac{\Delta f}{\Delta x}$$

$$= c \lim_{\Delta x \to 0} \frac{\Delta f}{\Delta x}.$$

Thus
$$\frac{d(cf)}{dx} = c\,\frac{df}{dx},$$

as was to be proved. ●

EXAMPLE 1 Use Theorem 1 to help compute $d(5x^3)/dx$, $D(6 \ln x)$, $(\sqrt{2} \sin x)'$, $d(\frac{3}{5} \cos x)/dx$.

SOLUTION
$$\frac{d(5x^3)}{dx} = 5\,\frac{d(x^3)}{dx} = 5 \cdot 3x^2 = 15x^2;$$

$$D(6 \ln x) = 6D(\ln x) = 6\,\frac{1}{x} = \frac{6}{x};$$

$$(\sqrt{2} \sin x)' = \sqrt{2}(\sin x)' = \sqrt{2} \cos x;$$

$$\frac{d(\frac{3}{5} \cos x)}{dx} = \frac{3}{5}\,\frac{d(\cos x)}{dx} = \frac{3}{5}(-\sin x) = \frac{-3}{5} \sin x. \quad ●$$

The next theorem asserts that if the functions f and g have derivatives at a certain value x, then so does their sum $f + g$, and

$$\frac{d(f+g)}{dx} = \frac{df}{dx} + \frac{dg}{dx}.$$

In other words, "The derivative of the sum is the sum of the derivatives." However, for the sake of clarity, it is stated much more briefly.

THEOREM 2   If f and g are differentiable functions, then so is $f + g$.   Its derivative is given by the formula

$$(f+g)' = f' + g'.$$

Similarly, $(f-g)' = f' - g'.$

PROOF   Give the function $f + g$ the name u.   That is,

$$u(x) = f(x) + g(x).$$

Then $u(x + \Delta x) = f(x + \Delta x) + g(x + \Delta x).$

So, $\Delta u = u(x + \Delta x) - u(x)$

$$= [f(x + \Delta x) + g(x + \Delta x)] - [f(x) + g(x)]$$
$$= [f(x + \Delta x) - f(x)] + [g(x + \Delta x) - g(x)]$$
$$= \Delta f + \Delta g.$$

Thus $u'(x) = \lim_{\Delta x \to 0} \dfrac{\Delta u}{\Delta x}$

$$= \lim_{\Delta x \to 0} \frac{\Delta f + \Delta g}{\Delta x}$$

$$= \lim_{\Delta x \to 0} \left(\frac{\Delta f}{\Delta x} + \frac{\Delta g}{\Delta x} \right)$$

$$= \lim_{\Delta x \to 0} \frac{\Delta f}{\Delta x} + \lim_{\Delta x \to 0} \frac{\Delta g}{\Delta x}$$

$$= f'(x) + g'(x).$$

Hence $f + g$ is differentiable, and

$$(f+g)' = f' + g'.$$

A similar argument applies to $f - g$. ●

EXAMPLE 2   Use Theorem 2 to help compute $d(x^2 + x^3)/dx$, $D(\ln x - x^5)$, and also $(\log_{10} x + \cos x)'$.

SOLUTION

$$\frac{d(x^2 + x^3)}{dx} = \frac{d(x^2)}{dx} + \frac{d(x^3)}{dx} = 2x + 3x^2;$$

$$D(\ln x - x^5) = D(\ln x) - D(x^5) = \frac{1}{x} - 5x^4;$$

$$(\log_{10} x + \cos x)' = (\log_{10} x)' + (\cos x)'$$

$$= \frac{\log_{10} e}{x} - \sin x. \quad \bullet$$

Theorems 1 and 2 justify the method described in Sec. 3.2 for differentiating any polynomial. The next example illustrates this.

EXAMPLE 3 Use Theorems 1 and 2 to differentiate $2x^4 - 6x^2 + x$.

SOLUTION

$$\frac{d(2x^4 - 6x^2 + x)}{dx} = \frac{d(2x^4 - 6x^2)}{dx} + \frac{d(x)}{dx}$$

$$= \left[\frac{d(2x^4)}{dx} - \frac{d(6x^2)}{dx} \right] + \frac{d(x)}{dx}$$

$$= (8x^3 - 12x) + 1$$

$$= 8x^3 - 12x + 1. \quad \bullet$$

The following theorem concerning the derivative of the product of two functions may be surprising, for it turns out that the derivative of the product is *not* the product of the derivatives. The formula is more complicated than that for the derivative of the sum, hence a little harder to apply. (It asserts that the derivative of the product is "the first function times the derivative of the second plus the second function times the derivative of the first.")

THEOREM 3 If f and g are differentiable functions, then so is fg. Its derivative is given by the formula

$$(fg)' = fg' + gf'.$$

Because the formula is more involved, it may be helpful to apply it in some examples before giving its proof.

EXAMPLE 4 Use Theorem 3 to differentiate $x^2 \sin x$.

SOLUTION

$$(x^2 \sin x)' = x^2(\sin x)' + \sin x(x^2)'$$

$$= x^2 \cos x + (\sin x)(2x),$$

usually rearranged as

$$x^2 \cos x + 2x \sin x. \quad \bullet$$

EXAMPLE 5 Use Theorem 3 to help differentiate $x^3 \cdot x^5$.

SOLUTION

$$\frac{d(x^3 \cdot x^5)}{dx} = x^3 \frac{d(x^5)}{dx} + x^5 \frac{d(x^3)}{dx}$$

$$\frac{d(x^3 \cdot x^5)}{dx} = x^3 \cdot 5x^4 + x^5 \cdot 3x^2$$

$$= 5x^7 + 3x^7$$

$$= 8x^7,$$

in agreement with the result in Chap. 3, $(x^8)' = 8x^7$. ●

PROOF OF THEOREM 3 Call the function fg simply u, that is,

$$u(x) = f(x)g(x).$$

Then
$$u(x + \Delta x) = f(x + \Delta x)g(x + \Delta x).$$

Rather than subtract directly, first write

$$f(x + \Delta x) = f(x) + \Delta f \quad \text{and} \quad g(x + \Delta x) = g(x) + \Delta g.$$

Then
$$u(x + \Delta x) = [f(x) + \Delta f][g(x) + \Delta g]$$
$$= f(x)g(x) + f(x)\,\Delta g + g(x)\,\Delta f + \Delta f\,\Delta g.$$

Hence
$$\Delta u = u(x + \Delta x) - u(x)$$
$$= f(x)g(x) + f(x)\,\Delta g + g(x)\,\Delta f + \Delta f\,\Delta g - f(x)g(x)$$
$$= f(x)\,\Delta g + g(x)\,\Delta f + \Delta f\,\Delta g,$$

and
$$\frac{\Delta u}{\Delta x} = f(x)\frac{\Delta g}{\Delta x} + g(x)\frac{\Delta f}{\Delta x} + \Delta f\frac{\Delta g}{\Delta x}.$$

As $\Delta x \to 0$, $\Delta g/\Delta x \to g'(x)$, $\Delta f/\Delta x \to f'(x)$, and, because f is differentiable (hence continuous), $\Delta f \to 0$. It follows that

$$\lim_{\Delta x \to 0} \frac{\Delta u}{\Delta x} = f(x)g'(x) + g(x)f'(x) + 0g'(x).$$

Therefore u is differentiable and

$$u' = fg' + gf'. \quad ●$$

To see how these theorems are used in practice, observe how, in the next example, the derivative is easily computed without any reference to limits.

EXAMPLE 6 Differentiate $x^2 \cos x + x^3 \ln x$.

SOLUTION
$$\frac{d(x^2 \cos x + x^3 \ln x)}{dx} = \frac{d(x^2 \cos x)}{dx} + \frac{d(x^3 \ln x)}{dx}$$

$$= \left[x^2\frac{d(\cos x)}{dx} + \cos x\frac{d(x^2)}{dx}\right] + \left[x^3\frac{d(\ln x)}{dx} + \ln x\frac{d(x^3)}{dx}\right]$$

$$= x^2(-\sin x) + (\cos x)(2x) + x^3\frac{1}{x} + (\ln x)(3x^2),$$

customarily rearranged to avoid the need for parentheses:

$$-x^2 \sin x + 2x \cos x + x^2 + 3x^2 \ln x. \ \bullet$$

Sometimes the formula for the derivative is simpler than the formula for the function, as the next example shows.

EXAMPLE 7   Differentiate $\cos x \sin x - x$.

SOLUTION

$$\frac{d(\cos x \sin x - x)}{dx} = \frac{d(\cos x \sin x)}{dx} - \frac{d(x)}{dx}$$

$$= \cos x \frac{d(\sin x)}{dx} + \sin x \frac{d(\cos x)}{dx} - 1$$

$$= \cos x \cos x + \sin x(-\sin x) - 1$$

$$= \cos^2 x - \sin^2 x - 1$$

$$= (1 - \sin^2 x) - \sin^2 x - 1$$

$$= -2 \sin^2 x. \ \bullet$$

The next example shows how to differentiate the product of three functions.

EXAMPLE 8   Differentiate $(x^2 + 1)(\sin x)(\ln x)$.

SOLUTION $D[(x^2 + 1)(\sin x \ln x)] = (x^2 + 1)D(\sin x \ln x) + (\sin x \ln x)D(x^2 + 1)$

$$= (x^2 + 1)[\sin x D(\ln x) + \ln x D(\sin x)]$$

$$+ \sin x \ln x(2x)$$

$$= (x^2 + 1)\left(\sin x \cdot \frac{1}{x} + \ln x \cos x \right) + 2x \sin x \ln x. \ \bullet$$

As shown in Exercise 29, for three functions f, g, and h.

$$(fgh)' = f'gh + fg'h + fgh'.$$

Exercises

In Exercises 1 to 27 differentiate the given function, using theorems developed in the text.

1. $\cos x + \sin x$
2. $2x^3 - 3 \cos x$
3. $8\sqrt{x}$
4. $x^3 + x$
5. $x^3 + 5 \sin x$
6. $5x^2 + \sqrt{2/x} - 15$
7. $\pi^2 + \sin x$
8. $6x - 7 \sin x$
9. $x + 1/x$
10. $5x^3 - 2x^2 + 6x + 4$
11. $x \sin x + \cos x$
12. $3 \cos x + 5 \sin x$
13. $4x^3 \sin x$
14. $8x^5 \ln x$
15. $3 \sin x \cos x$
16. $2x \cos x + (x^2 - 2) \sin x$
17. $5 \ln x - x \cos x$
18. $3x^2(1 + \sqrt{x})$
19. $(\ln x)^2$ (write as $\ln x \cdot \ln x$)
20. $\sin^3 x$ (write as $\sin x \cdot \sin x \cdot \sin x$)
21. $2 \ln x^5$ (first simplify)
22. $x \ln x - x$
23. $\dfrac{x^2 \ln x}{2} - \dfrac{x^2}{4}$
24. $\dfrac{x^3}{3} \ln x - \dfrac{x^3}{9}$

25. $x(\ln x)^2 - 2x \ln x + 2x$ **26.** $x^5\left(\dfrac{\ln x}{5} - \dfrac{1}{25}\right)$

27. $(\ln x)^3$

■

28. Prove that $(f - g)' = f' - g'$.
29. Use Theorem 3 to obtain the formula for the derivative of the product of three functions:

$$(fgh)' = f'gh + fg'h + fgh'.$$

30. Apply the formula in Exercise 29 to differentiate:
    (a) $x^2 \sin x \ln x$;   (b) $x^3 \cos x(1 + \sin x)$.
31. Give an example of a function whose derivative is

    (a) $x/5$ (b) $5/x$ (c) $x - \dfrac{3}{x}$

    (d) $\dfrac{x^2 + 2x - 1}{x}$ (e) $2 \cos x + 3 \sin x$

■ ■

32. (a) Graph $y = 2 \cos x + 3 \sin x$ for x in $[0, \pi/2]$.
    (b) By inspection of the graph, estimate the x coordinate of the point where the tangent line is horizontal.
    (c) Show that the x coordinate in (b) satisfies the equation $\tan x = \frac{3}{2}$.
    (d) Use the table of $\tan x$ (angle in radians) or a calculator to estimate x in (b).
33. (a) Assuming $(x^5)' = 5x^4$ and $(x)' = 1$, use Theorem 3 to obtain $(x^6)'$.
    (b) Show how one can use Theorem 3, together with $(x)' = 1$, to obtain, successively, $(x^2)' = 2x$, $(x^3)' = 3x^2$, and so on. This argument proves that $(x^n)' = nx^{n-1}$, when n is a positive integer.

5.6

The derivative of the quotient of two functions

The quotient f/g of two functions f and g is the function that assigns to the number x the number

$$\frac{f(x)}{g(x)}.$$

When $g(x) = 0$, the quotient is not defined. Thus the domain of the function f/g consists of those numbers x such that $f(x)$ and $g(x)$ are defined and $g(x)$ is not 0.

THEOREM 1   If f and g are differentiable functions, then so is

$$\frac{f}{g},$$

and $$\left(\frac{f}{g}\right)' = \frac{gf' - fg'}{g^2}.$$

PROOF   Denote the quotient function f/g by u, that is,

$$u(x) = \frac{f(x)}{g(x)}$$

and $$u(x + \Delta x) = \frac{f(x + \Delta x)}{g(x + \Delta x)}.$$

[Since we consider only values of x such that $g(x) \neq 0$ and g is continuous, for Δx sufficiently small, $g(x + \Delta x) \neq 0$.] Before computing Δu, write $f(x + \Delta x)$

as $f(x) + \Delta f$, and $g(x + \Delta x)$ as $g(x) + \Delta g$. Then

$$\Delta u = \frac{f(x) + \Delta f}{g(x) + \Delta g} - \frac{f(x)}{g(x)}.$$

Putting the right side over a common denominator, we obtain

$$\Delta u = \frac{g(x)[f(x) + \Delta f] - f(x)[g(x) + \Delta g]}{g(x)[g(x) + \Delta g]}$$

$$= \frac{g(x)f(x) + g(x)\,\Delta f - f(x)g(x) - f(x)\,\Delta g}{g(x)[g(x) + \Delta g]}.$$

Cancellation of $g(x)f(x)$ and $f(x)g(x)$ yields

$$\Delta u = \frac{g(x)\,\Delta f - f(x)\,\Delta g}{g(x)[g(x) + \Delta g]}.$$

The quotient, $\Delta u/\Delta x$ can be written as

$$\frac{\Delta u}{\Delta x} = \frac{g(x)\,\Delta f/\Delta x - f(x)\,\Delta g/\Delta x}{g(x)[g(x) + \Delta g]},$$

and we are now ready to examine $\lim_{\Delta x \to 0} \Delta u/\Delta x$. Since

$$\lim_{\Delta x \to 0} \frac{\Delta f}{\Delta x} = f'(x),$$

$$\lim_{\Delta x \to 0} \frac{\Delta g}{\Delta x} = g'(x),$$

and

$$\lim_{\Delta x \to 0} \Delta g = 0,$$

it follows that

$$\lim_{\Delta x \to 0} \frac{\Delta u}{\Delta x} = \frac{g(x)f'(x) - f(x)g'(x)}{g(x)[g(x) + 0]}.$$

Hence u is differentiable and

$$u' = \frac{gf' - fg'}{g^2}.$$

This proves the theorem. ●

EXAMPLE I   Differentiate

$$\frac{1 + x^2}{5 + x^3}.$$

SOLUTION

$$\left(\frac{1 + x^2}{5 + x^3}\right)' = \frac{(5 + x^3)(1 + x^2)' - (1 + x^2)(5 + x^3)'}{(5 + x^3)^2}.$$

Now,

$$(1 + x^2)' = (1)' + (x^2)' = 0 + 2x$$

and, similarly,

$$(5 + x^3)' = 3x^2.$$

Thus $$\left(\frac{1+x^2}{5+x^3}\right)' = \frac{(5+x^3)(2x) - (1+x^2)(3x^2)}{(5+x^3)^2},$$

which can be simplified. ●

A Suggestion for Using the Formula for the Derivative of a Quotient

Word of advice: When using the formula for $(f/g)'$, first write down the part

$$\frac{g}{g^2}.$$

In that way you will get the denominator correct and have a good start on the numerator. You may then go on to complete the numerator, *remembering that it has a minus sign.*

EXAMPLE 2   Compute $(x/\cos x)'$, showing each step in "slow motion."

SOLUTION   *Step 1* $\left(\dfrac{x}{\cos x}\right)' = \dfrac{\cos x \cdots\cdots}{\cos^2 x}$ first step in using the formula for $(f/g)'$.

Step 2 $\left(\dfrac{x}{\cos x}\right)' = \dfrac{\cos x(x)' - x(\cos x)'}{\cos^2 x}$ remember the minus sign.

Step 3 $\left(\dfrac{x}{\cos x}\right)' = \dfrac{(\cos x)(1) - x(-\sin x)}{\cos^2 x}.$

Step 4 $\left(\dfrac{x}{\cos x}\right)' = \dfrac{\cos x + x \sin x}{\cos^2 x}.$ ●

The formula for the derivative of the quotient has many consequences, as shown by the following corollaries. The symbol $(f/g)'$ beneath an equal sign is a shorthand justification for the equation. This is a handy device for explaining a computational step.

COROLLARY I   If n is a negative integer, $n = -1, -2, -3, \ldots$, then

$$(x^n)' = nx^{n-1}.$$

PROOF   Let $n = -m$, where m is a positive integer. Then

$$(x^n)' = (x^{-m})'$$

$$= \left(\frac{1}{x^m}\right)'$$

$$\underset{(f/g)'}{=} \frac{x^m(1)' - 1(x^m)'}{(x^m)^2}$$

$$= \frac{x^m \cdot 0 - 1 \cdot mx^{m-1}}{x^{2m}}$$

$$= \frac{-mx^{m-1}}{x^{2m}} = -mx^{m-1-2m}$$

$$= -mx^{-m-1} = nx^{n-1}.$$

This proves the corollary. ●

EXAMPLE 3   Use Corollary 1 to differentiate x^{-1}.

SOLUTION

$$\begin{aligned}(x^{-1})' &= -1x^{-1-1}\\ &= -1 \cdot x^{-2}\\ &= \frac{-1}{x^2}.\end{aligned}$$

Since $x^{-1} = 1/x$, this is simply the formula obtained in Chap. 3,

$$\left(\frac{1}{x}\right)' = \frac{-1}{x^2}. ●$$

EXAMPLE 4   Use the corollary to differentiate $1/x^3$.

SOLUTION

$$\begin{aligned}\left(\frac{1}{x^3}\right)' &= (x^{-3})'\\ &= -3x^{-4}\\ &= \frac{-3}{x^4}. ●\end{aligned}$$

The next corollary is quite useful. Though it is just a special case of the theorem, it is worth memorizing. It provides the formula for the derivative of the reciprocal of a function f, $1/f$, if f' is known.

COROLLARY 2

$$\left(\frac{1}{f}\right)' = \frac{-f'}{f^2}. ●$$

The proof is left to the reader.

EXAMPLE 5   Use Corollary 2 to differentiate $1/\cos x$.

SOLUTION

$$\begin{aligned}\left(\frac{1}{\cos x}\right)' &= \frac{-(\cos x)'}{(\cos x)^2}\\ &= \frac{-(-\sin x)}{\cos^2 x}\\ &= \frac{\sin x}{\cos^2 x}. ●\end{aligned}$$

The functions

$$\frac{\sin x}{\cos x}, \quad \frac{1}{\cos x}, \quad \frac{\cos x}{\sin x}, \quad \text{and} \quad \frac{1}{\sin x}$$

are of sufficient importance to have names.   Their names are:

$$\tan x = \frac{\sin x}{\cos x} \qquad \cot x = \frac{\cos x}{\sin x}$$

$$\sec x = \frac{1}{\cos x} \qquad \csc x = \frac{1}{\sin x}.$$

(The abbreviation "cot" is short for *cotangent*, "sec" is short for *secant*, "csc" is short for *cosecant*.)   With the aid of the theorem for $(f/g)'$ their derivatives are easy to compute.

The functions $\tan x$ and $\sec x$ are related by the equation

$$1 + \tan^2 x = \sec^2 x.$$

To show this, divide both sides of the equation

$$\cos^2 x + \sin^2 x = 1$$

by $\cos^2 x$, obtaining

$$\frac{\cos^2 x}{\cos^2 x} + \frac{\sin^2 x}{\cos^2 x} = \frac{1}{\cos^2 x},$$

or

$$1 + \tan^2 x = \sec^2 x.$$

A similar argument, using division by $\sin^2 x$ instead of $\cos^2 x$, shows that

$$1 + \cot^2 x = \csc^2 x.$$

COROLLARY 3 $(\tan x)' = \sec^2 x, \quad (\cot x)' = -\csc^2 x, \quad (\sec x)' = \sec x \tan x, \quad \text{and} \quad (\csc x)' = -\csc x \cot x.$

PROOF   We obtain only the formulas for $(\tan x)'$ and $(\sec x)'$.

$$(\tan x)' = \left(\frac{\sin x}{\cos x}\right)'$$

$$= \frac{\cos x(\sin x)' - \sin x(\cos x)'}{\cos^2 x}$$

$$= \frac{(\cos x)(\cos x) - \sin x(-\sin x)}{\cos^2 x}$$

$$= \frac{\cos^2 x + \sin^2 x}{\cos^2 x}$$

$$= \frac{1}{\cos^2 x}$$

$$= \sec^2 x.$$

To obtain the derivative of sec x begin by writing $(\sec x)' = (1/\cos x)'$, which, by Example 5, equals $\sin x/\cos^2 x$. This quotient can be written as

$$\frac{\sin x}{\cos x} \frac{1}{\cos x},$$

which is the same as $\tan x \sec x$. ●

Word of advice: Note that the "co" functions, cosine, cotangent, and cosecant, have the minus sign in the formula for their derivatives.

EXAMPLE 6   Differentiate $5x^{-2} + \tan x + \ln x$.

SOLUTION

$$\begin{aligned}
(5x^{-2} + \tan x + \ln x)' &= \underset{(f+g)'}{(5x^{-2} + \tan x)' + (\ln x)'} \\
&= \underset{(f+g)'}{(5x^{-2})' + (\tan x)' + (\ln x)'} \\
&= \underset{(cf)'}{5(x^{-2})' + (\tan x)' + (\ln x)'} \\
&= 5(-2x^{-3}) + \sec^2 x + \frac{1}{x} \\
&= \frac{-10}{x^3} + \sec^2 x + \frac{1}{x}.
\end{aligned}$$

This example also points out that the derivative of the sum of several functions is the sum of their derivatives. ●

Exercises

In Exercises 1 to 21 differentiate the given function using theorems developed in the text.

1. $\dfrac{2x + 1}{3x + 2}$

2. $\dfrac{(2x - 1)^2}{x^2 + 1}$

3. $\dfrac{\ln x}{x^2}$

4. $\dfrac{(\ln x)^2}{x}$

5. $\dfrac{4x^3 - 6x + 1}{x^2 + 1}(x^5 + 2x)$

6. $5 \csc x$

7. $\ln x \tan x$

8. $5 \sec x - \tan x$

9. $\dfrac{\sin x}{1 + \sec x}$

10. $(1 + \cos x)^2$

11. $\sin x \ln x \cos x$

12. $\dfrac{x}{\sin x}$

13. $\dfrac{\ln x}{x}$

14. $\dfrac{3 \cos x}{x^3}$

15. $\dfrac{1}{x^2 + 3x^5}$

16. $\dfrac{1}{x^7}$

17. $\dfrac{8}{x^8}$

18. $\frac{1}{2}x^{-4}$

19. $\dfrac{5 \tan x}{x^3}$

20. $x \sec x + \tan x$

21. $4 \csc x + 3 \cot x$

■

22. Differentiate (a) $x^4 - \tan x$, (b) $x^4 + \tan x$, (c) $x^4/\tan x$, (d) $x^4 \tan x$.

23. Prove Corollary 2, $\left(\dfrac{1}{f}\right)' = \dfrac{-f'}{f^2}$.

24. Prove that $(\cot x)' = -\csc^2 x$ and that $(\csc x)' = -\csc x \cot x$, as asserted in Corollary 3.

25. (*a*) Fill in this table.

x	0	$\dfrac{\pi}{6}$	$\dfrac{\pi}{4}$	$\dfrac{\pi}{3}$	π	$\dfrac{3\pi}{4}$	$\dfrac{7\pi}{4}$
$\sec x$							

    (*b*) What is the domain of $\sec x$?
    (*c*) Graph $y = \sec x$.

Exercise 26 illustrates the use of calculus in graphing. In Chap. 6 the ideas of this exercise will be extensively developed.

26. Let $f(x) = x + 3/x$. Consider only $x > 0$.
    (*a*) Compute $f(0.1)$, $f(1)$, and $f(10)$.
    (*b*) Find $\lim_{x \to \infty} f(x)$ and $\lim_{x \to 0^+} f(x)$.
    (*c*) Find $f'(x)$.
    (*d*) For which x is $f'(x)$ positive? negative? zero?

(*e*) Graph f. Show any points where the tangent line is horizontal.

■ ■

27. (*a*) Fill in this table.

x	$\dfrac{\pi}{6}$	$\dfrac{\pi}{3}$	$\dfrac{\pi}{4}$	$\dfrac{\pi}{2}$	$\dfrac{3\pi}{4}$
$\cot x$					

    (*b*) When x is near 0 and positive, what happens to $\cot x$?
    (*c*) When x is near 0 and negative, what happens to $\cot x$?
    (*d*) Graph $y = \cot x$.

28. (*a*) Using $\lim_{\theta \to 0} (\sin \theta)/\theta = 1$, prove that $\lim_{\theta \to 0} (\tan \theta)/\theta = 1$.
    (*b*) Using (*a*) and the identity $\tan (A + B) = (\tan A + \tan B)/(1 - \tan A \tan B)$, obtain a direct proof that $(\tan x)' = \sec^2 x$.

5.7

Composite functions

This chapter has already developed formulas for differentiating such functions as $1 + x^2$, $\sin x$, and $\ln x$. But how are the derivatives of $(1 + x^2)^{100}$, $\sin x^3$, and $\ln (1 + 2x)$ to be found? In the case of $(1 + x^2)^{100}$, we could multiply $1 + x^2$ times itself 100 times, obtaining a polynomial of degree 200 with 101 terms, and then compute the derivative of this polynomial by a procedure already available. Fortunately, in Sec. 5.8 a quicker way is developed. Moreover, by the procedure described there the derivatives of $\sin x^3$ and $\ln (1 + 2x)$ can be quickly obtained.

This section examines in detail the way such functions as $(1 + x^2)^{100}$, $\sin x^3$, and $\ln (1 + 2x)$ are built up from simpler functions and develops the concept of a *composite function* needed in the next section.

Note, for instance, that the function

$$y = (1 + x^2)^{100}$$

is built up by raising $1 + x^2$ to the one-hundredth power, that is,

$$y = u^{100}, \qquad \text{where} \quad u = 1 + x^2.$$

Similarly, the function $\qquad\qquad y = \sin x^3$

is built up by first cubing x and then taking the sine of the result,

$$y = \sin u, \qquad \text{where} \quad u = x^3.$$

Similarly, $\qquad\qquad\qquad\qquad y = \ln (1 + 2x)$

is built up as $\qquad y = \ln u,\qquad$ where $\quad u = 1 + 2x.$

The theme common to these three examples is spelled out in the following definition.

DEFINITION   *Composition of functions.* Let f and g be functions such that $g(x)$ is in the domain of f for each x in the domain of g.   The function that assigns to each x in the domain of g the value

$$f(g(x))$$

is called the *composition of f and g*, and is denoted h or $f \circ g$.   Thus, if

$$y = f(u) \qquad \text{and} \qquad u = g(x),$$

then $\qquad\qquad\qquad\qquad y = h(x),$

where $\qquad\qquad\qquad\qquad h(x) = f(g(x)),$

or $\qquad\qquad\qquad\qquad y = (f \circ g)(x).$

$(f \circ g$ is read as "f circle g.")

In ordinary English the definition says, "To compute $f \circ g$, first apply g, and then apply f to the result."

Thinking of functions as input-output machines, we may consider $f \circ g$ as the machine built by hooking the machine for f onto the machine for g, as shown in the margin.

The output of the g machine, $g(x)$, becomes the input for the f machine

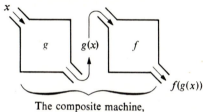

The composite machine, denoted h or $f\circ g$

EXAMPLE 1   If f is the function that squares and adds 1, that is, $f(x) = 1 + x^2$, and g is the cubing function, compute $f \circ g$ and $g \circ f$.

SOLUTION   By the definition of the composition of functions,

$$(f \circ g)(x) = f(g(x)) = f(x^3) = 1 + (x^3)^2,$$

while $\qquad (g \circ f)(x) = g(f(x)) = g(1 + x^2) = (1 + x^2)^3.$

Thus, if you hook up the f and g machines with f first, and then g, you obtain a machine with an effect different from that of the $f \circ g$ machine.   It is the "$g \circ f$ machine." ●

EXAMPLE 2   Let f be the function that raises each number to the one-hundredth power,

$$y = u^{100}.$$

Let g be the function that squares and adds 1,

$$u = 1 + x^2.$$

Compute the composite function $f \circ g$.

SOLUTION   In this case the composite function $h = f \circ g$ is given by the formula

$$y = u^{100}, \qquad \text{where} \quad u = 1 + x^2.$$

Hence
$$y = h(x) = (1 + x^2)^{100}.$$

This is one of the three functions cited in the opening paragraph of this section. In words, the function is given by these directions: Take a number, square it and add 1, then raise to the one-hundredth power. ●

EXAMPLE 3   Show that the other two functions mentioned in the opening paragraph, $\sin x^3$ and $\ln (1 + 2x)$, are also composite functions.

SOLUTION $y = \sin x^3$ is the composition of $y = \sin u$ with $u = x^3$; $y = \ln (1 + 2x)$ is the composition of $y = \ln u$ with $u = 1 + 2x$. ●

For our purposes, one of the most instructive ways of thinking of composite functions is in terms of projections from slides to screens. If g is interpreted as some complicated projection from a slide to a screen (see Sec. 3.1) and f as a projection from that screen to a second screen, then $h = f \circ g$ is the projection from the slide to the second screen.

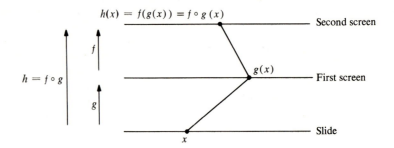

This point of view is helpful in the next section and in Sec. 22.4, where the slide and screen are no longer lines but, like ordinary slides and screens, regions in the plane.

Frequently a function is built up by the composition of more than two functions, as in the next example.

EXAMPLE 4   Let $y = (1 + \cos 2x)^3$. Show that this function can be obtained by the composition of three functions.

SOLUTION   First of all,
$$y = u^3, \qquad \text{where} \quad u = 1 + \cos 2x.$$

But $1 + \cos 2x$ is itself a composite function, built up as
$$u = 1 + \cos v, \qquad \text{where} \quad v = 2x.$$

Therefore, the function $(1 + \cos 2x)^3$ can be obtained in three steps:
$$y = u^3, \qquad \text{where} \quad u = 1 + \cos v, \qquad \text{and} \quad v = 2x. ●$$

Exercises

In Exercises 1 to 9 show that each of the given functions is composite by stating the formulas for the functions $y = f(u)$ and $u = g(x)$.

1. $\cos x^2$ **2.** $\sqrt{1 + x^2}$

3. $(\cos x)^3$ **4.** $\ln(3 + x^2)$

5. $3 + (\ln x)^2$ **6.** $(1 + \sin x)^{10}$

7. $\sec 5x$ **8.** $(1 + 5x^2)^{20}$

9. $\sin(1 + \ln x)$

In Exercises 10 to 12 express each of the given functions as the composition of three functions, as in Example 4.

10. $\sqrt{\tan 3x}$ **11.** $\sin^3 5x$

12. $\ln \cos 5x$.

13. Compute $(f \circ g)(x)$ and $(g \circ f)(x)$ if
  (a) f is the sine function and g is the absolute-value function;
  (b) f is the squaring function and g is the natural logarithmic function;
  (c) f is the squaring function and g is the cubing function.

14. Below are parts of the tables for two functions, f and g.

x	1	2	3	4
$f(x)$	3	4	1	2

x	1	2	3	4
$g(x)$	4	3	2	1

Compute the following:
  (a) $(f \circ g)(1)$ (b) $(g \circ f)(1)$
  (c) $(f \circ g)(2)$ (d) $(g \circ f)(2)$.

15. Below are the parts of the table for two differentiable functions f and g.

x	4	4.2
$f(x)$	7	7.6

x	3	3.1
$g(x)$	4	4.2

  (a) Use the data to estimate $g'(3)$, $f'(4)$, and $h'(3)$, where $h = f \circ g$.
  (b) How do you think $g'(3)$, $f'(4)$, and $h'(3)$ are related?

16. Write the formula for $y = h(x)$ if
  (a) $y = u^2$ and $u = 1 + 2x$;
  (b) $y = 1 + 2u$ and $u = x^2$;

  (c) $y = 3 \sin u$ and $u = 2 \sin x$;
  (d) $y = e^u$ and $u = \ln x$.

  ■

17. These are directions for computing two functions f and g:
  Directions for f: Square the number;
  Directions for g: Add 1 to the number.
  (a) What are the directions for $f \circ g$?
  (b) What are the directions for $g \circ f$?
  (c) What is the algebraic formula for $f \circ g$?
  (d) What is the algebraic formula for $g \circ f$?

18. Let h be the composition of $y = u^2$ and $u = 3x + 1$. Evaluate

$$\frac{h(x_1) - h(x)}{x_1 - x}$$

when (a) $x_1 = 3$ and $x = 2$; (b) $x_1 = 2.1$ and $x = 2$; (c) $x_1 = 2.01$ and $x = 2$.

19. A handbook of mathematical tables usually has a table called "Common Logarithms of Trigonometric Functions." [For instance, it may include $\log \sin(44°) = 9.84177 - 10$.] Then show that this table actually lists values of a composite function $f \circ g$, and that there are tables for f and g also.

20. Let $f(x) = -x$ for all x. Show that $(f \circ f)(x) = x$ for all x.

21. Determine $f \circ g$ and $g \circ f$ if (a) $f(x) = x^3$ and $g(x) = x^{1/3}$, (b) $f(x) = 10^x$ and $g(x) = \log_{10} x$.

  ■ ■

22. Let $f(x) = 1/(1 - x)$. What is the domain of f? $f \circ f$? $f \circ f \circ f$? Show that $(f \circ f \circ f)(x) = x$ for x in the domain of $f \circ f \circ f$.

23. Let $(f \circ g)(x) = x$ for all x. Assume that the domains of f and g consist of all real numbers. Also assume that every real number can be expressed in the form $g(x)$ for some number x. Show that $(g \circ f)(x) = x$.

24. Let $f(x) = x/(1 + x)$. Find $(f \circ f \circ f \circ f)(x)$.

5.8

The derivative of a composite function

This section develops a quick way for finding the derivative of composite functions such as $(1 + x^2)^{100}$, $\sin x^3$, and $\ln(1 + 2x)$.

We shall answer these two questions: If f and g are differentiable functions, is the composite function $h = f \circ g$ also differentiable? If so, what is its derivative?

Our experience with projectors suggests the answer. Let $h = f \circ g$. The effect of the first projector, which takes x on the slide to $g(x)$ on the first screen, is to magnify by a factor $g'(x)$. Then what is the effect of the second projector f, which takes $g(x)$ on the first screen to $f(g(x))$ on the second screen?

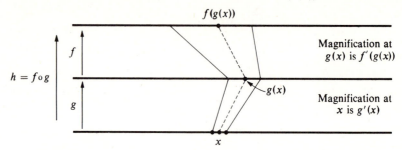

It magnifies by the factor f' evaluated at $g(x)$, that is, $f'(g(x))$. But when one magnification is followed by another, the total effect is the *product* of the two magnifications. For instance, a twofold magnification, followed by a threefold, is a sixfold magnification. We suspect, therefore, that $h'(x)$, the total magnification, is equal to $f'(g(x)) \cdot g'(x)$.

In the differential notation for derivatives, this formula becomes much shorter. Let

$$y = f(u), \qquad \text{where} \quad u = g(x),$$

and let

$$y = h(x)$$

be the composite function

$$h = f \circ g.$$

Then our suspicion is that the derivative of h is given by the formula

$$\frac{dy}{dx} = \frac{dy}{du}\frac{du}{dx}.$$

(This is a much more suggestive formula. If the symbol du had meaning by itself, it would look as if the du's cancel.)

In the D notation, this important formula, known as the *chain rule*, reads

$$D_x(y) = D_u(y) \cdot D_x(u).$$

Before proving that this conjecture is correct, let us test it in a specific case.

EXAMPLE 1   Let $$y = 5u \qquad \text{and} \qquad u = 3x.$$

Show that $$\frac{dy}{dx} = \frac{dy}{du}\frac{du}{dx}.$$

SOLUTION   The composite function h is given by the formula

$$y = 5u, \qquad \text{where} \quad u = 3x,$$

that is,

$$h(x) = 5(3x)$$
$$= 15x.$$

In this case, $\dfrac{dy}{dx} = 15,$ $\dfrac{dy}{du} = 5,$ and $\dfrac{du}{dx} = 3.$

It is true in this case that $\dfrac{dy}{dx} = \dfrac{dy}{du}\dfrac{du}{dx},$

since $15 = 5 \cdot 3.$ ●

With this background, we are ready to prove our conjecture.

THEOREM   *The chain rule.*   If f and g are differentiable functions, then so is $h = f \circ g$, and

$$h'(x) = f'(g(x)) \cdot g'(x).$$

More briefly, if $y = f(u)$ and $u = g(x)$ and, thus $y = h(x)$, then

$$\frac{dy}{dx} = \frac{dy}{du}\frac{du}{dx}.$$

(It is in the latter form that the chain rule is most easily memorized and applied.)

PROOF   To examine $h'(x)$, it is necessary to go back to the definition of the derivative,

$$h'(x) = \lim_{\Delta x \to 0} \frac{\Delta y}{\Delta x}.$$

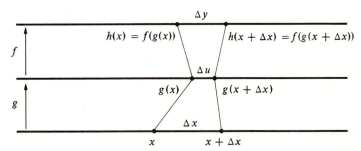

The computation will involve Δx, $\Delta u = \Delta g$, and Δy, shown in this diagram. That is, Δx, which is not 0, determines a number Δu, the change in u,

$$\Delta u = g(x + \Delta x) - g(x),$$

and a number Δy, the change in y,

$$\Delta y = h(x + \Delta x) - h(x).$$

It is important to note that, since g is differentiable, $\Delta u \to 0$ as $\Delta x \to 0$. However, it could happen that Δu is 0, even though Δx is not 0.   Since in a moment we shall wish to divide by Δu, we make an extra assumption, namely that

$$g'(x) \text{ is not } 0.$$

Since $g'(x) = \lim_{\Delta x \to 0} \dfrac{\Delta u}{\Delta x},$

this implies that, when Δx is sufficiently small, Δu is not 0.

With this assumption, which most functions satisfy at most values of x, the proof is short:

$$h'(x) = \lim_{\Delta x \to 0} \frac{\Delta y}{\Delta x}$$

$$= \lim_{\Delta x \to 0} \frac{\Delta y}{\Delta u} \frac{\Delta u}{\Delta x}$$

$$= \lim_{\Delta x \to 0} \frac{\Delta y}{\Delta u} \lim_{\Delta x \to 0} \frac{\Delta u}{\Delta x}$$

$$= \lim_{\Delta u \to 0} \frac{\Delta y}{\Delta u} \lim_{\Delta x \to 0} \frac{\Delta u}{\Delta x} \qquad \text{since } \Delta u \to 0 \text{ as } \Delta x \to 0$$

$$= \frac{dy}{du} \frac{du}{dx}.$$

[The special case, where $g'(x) = 0$, is discussed in Exercise 37.] This concludes the proof. ●

EXAMPLE 2   Use the chain rule to differentiate $(1 + x^2)^{100}$.

SOLUTION   Let $y = (1 + x^2)^{100}$, that is,

$$y = u^{100}, \qquad \text{where} \quad u = 1 + x^2.$$

The chain rule asserts that $\quad \dfrac{dy}{dx} = \dfrac{dy}{du} \dfrac{du}{dx}$

$$= \frac{d(u^{100})}{du} \frac{d(1 + x^2)}{dx}$$

$$= 100u^{99} \cdot 2x$$

$$= 100(1 + x^2)^{99} \cdot 2x$$

$$= 200x(1 + x^2)^{99}.$$

Thus $\qquad \dfrac{d(1 + x^2)^{100}}{dx} = 200x(1 + x^2)^{99}.$

The chain rule obtains a formula for the derivative of $(1 + x^2)^{100}$ without multiplying $1 + x^2$ 100 times. ●

EXAMPLE 3   Use the chain rule to differentiate $\sin x^3$.

SOLUTION   Let $y = \sin x^3$, that is, $\quad y = \sin u, \qquad \text{where} \quad u = x^3.$

The chain rule asserts that $\qquad \dfrac{dy}{dx} = \dfrac{dy}{du} \dfrac{du}{dx}$

$$= \frac{d(\sin u)}{du} \frac{d(x^3)}{dx}$$

$$= \cos u \cdot 3x^2$$

$$= \cos x^3 \cdot 3x^2$$

$$= 3x^2 \cos x^3,$$

that is,
$$\frac{d(\sin x^3)}{dx} = 3x^2 \cos x^3. \ \bullet$$

EXAMPLE 4   Use the chain rule to differentiate sin 2x.

SOLUTION   Let $y = \sin 2x$.   This is a composite function for

$$y = \sin u, \qquad \text{where} \quad u = 2x.$$

By the chain rule,
$$\frac{dy}{dx} = \frac{dy}{du} \frac{du}{dx}$$

$$= \frac{d(\sin u)}{du} \frac{d(2x)}{dx}$$

$$= (\cos u)(2)$$

$$= (\cos 2x)2$$

$$= 2 \cos 2x. \ \bullet$$

A *common* error in working with the chain rule, say, in Example 4, is to assert that $(\sin 2x)'$ is cos 2x, forgetting the extra 2.   As long as we are dealing with a composite function, we have the "product of two magnifications."   In this case, the projection from x to $2x$ already has a magnifying effect, namely 2.

EXAMPLE 5   Use the chain rule to differentiate $\ln (x^2 + 1)$.

SOLUTION   Let $y = \ln (x^2 + 1)$, hence

$$y = \ln u, \qquad \text{where} \quad u = x^2 + 1.$$

Thus
$$\frac{dy}{dx} = \frac{dy}{du} \frac{du}{dx}$$

$$= \frac{1}{u} 2x$$

$$= \frac{1}{x^2 + 1} 2x$$

$$= \frac{2x}{x^2 + 1}. \ \bullet$$

These examples show the "rhythm" of the chain rule in typical computations. The equation

$$\frac{dy}{dx} = \frac{dy}{du}\frac{du}{dx}$$

is read as "derivative of y with respect to x equals derivative of y with respect to u times derivative of u with respect to x." Optically, this says "total magnification of the two projections equals magnification of second projection times magnification of first projection."

The next example will be needed in Chap. 7.

EXAMPLE 6   Differentiate $\ln |x|$.

SOLUTION   Observe, first of all, that the domain of this function consists of all numbers other than 0. (The domain of $\ln x$, on the other hand, consists only of positive numbers.)

If x is positive, $|x| = x$, and $\ln |x| = \ln x$. Thus for positive x,

$$\frac{d(\ln |x|)}{dx} = \frac{d(\ln x)}{dx} = \frac{1}{x}.$$

This did not require use of the chain rule.

Next consider x negative. Then $|x| = -x$ and

$$y = \ln |x| = \ln (-x)$$

is composite,

$$y = \ln u, \qquad \text{where} \quad u = -x.$$

By the chain rule,

$$\frac{dy}{dx} = \frac{d(\ln u)}{du}\frac{d(-x)}{dx}$$

$$= \frac{1}{u}(-1)$$

$$= \frac{1}{-x}(-1)$$

$$= \frac{1}{x}.$$

Hence, for all x, whether positive or negative,

$$\frac{d(\ln |x|)}{dx} = \frac{1}{x}. \quad \bullet$$

The next example shows how to use the chain rule when a function is the composition of more than two functions.

EXAMPLE 7   Differentiate $\qquad\qquad\qquad y = \cos^2 3x.$

SOLUTION   Here $y = u^2$,   where $u = \cos 3x$.

But $\cos 3x$ is also a composite function,

$$u = \cos v, \qquad \text{where} \quad v = 3x.$$

All told, $y = \cos^2 3x$ is the composition of three functions,

$$y = u^2, \qquad \text{where} \quad u = \cos v, \qquad \text{and} \quad v = 3x.$$

In a sense, the magnifications of three projections must be computed and multiplied:

$$\frac{dy}{dx} = \frac{dy}{du}\frac{du}{dv}\frac{dv}{dx}$$

$$= 2u \cdot (-\sin v) \cdot 3$$

$$= 2 \cos v \cdot (-\sin 3x) \cdot 3$$

$$= 2 \cos 3x \cdot (-\sin 3x) \cdot 3$$

$$= -6 \cos 3x \sin 3x.$$

After some practice, such computations can be made without introducing any extra letters such as u and v. ●

Example 7 also indicates how the chain rule got its name: A function can be the composition of any number of functions, taken one after another like links in a chain.

The next example shows how the chain rule in combination with other formulas developed in this chapter can be used to find the derivatives of more complicated functions.

EXAMPLE 8   Compute $\dfrac{d(x^2 \sin^5 2x)}{dx}.$

SOLUTION   First of all, by the formula for the derivative of the product,

$$\frac{d(x^2 \sin^5 2x)}{dx} = x^2 \frac{d(\sin^5 2x)}{dx} + \sin^5 2x \frac{d(x^2)}{dx}.$$

The chain rule is needed for computing

$$\frac{d(\sin^5 2x)}{dx}.$$

Without all the details (that is, introduction of the letters u and v and exhibition of the various functions in detail), the computation looks like this:

$$\frac{d(\sin^5 2x)}{dx} = 5 \sin^4 2x \cdot \cos 2x \cdot 2$$

$$= 10 \sin^4 2x \cos 2x.$$

y	$\dfrac{dy}{dx}$
u^n	$nu^{n-1}\dfrac{du}{dx}$
$\sin u$	$\cos u\,\dfrac{du}{dx}$
$\cos u$	$-\sin u\,\dfrac{du}{dx}$
$\ln u$	$\dfrac{1}{u}\dfrac{du}{dx}$

Thus
$$\frac{d(x^2 \sin^5 2x)}{dx} = x^2(10\sin^4 2x \cos 2x) + \sin^5 2x \cdot (2x)$$
$$= 10x^2 \sin^4 2x \cos 2x + 2x \sin^5 2x. \;\bullet$$

As these examples suggest, the chain rule is the most important tool in the computation of derivatives.

The table on the left records a few special cases of the chain rule. They are used so often that they are worth memorizing. In each case u is a differentiable function of x.

Exercises

In Exercises 1 to 24 differentiate the given function.

1. $(2x^3 - 2x + 5)^4$
2. $(\sin x)^5$
3. $(\sin 3x)^5$
4. $\cos 2x$
5. $(1 + 2x)^5 \cos 3x$
6. $(t^2 + 1)^3(3t - 1)$
7. $(1 + \ln 2x)^3$
8. $5\cos^3 2x$

9. $\dfrac{\sin 3x}{(x^2 + 1)^5}$

10. $x\left(\dfrac{x^2}{1 + x}\right)^3$

11. $\cos 3x \sin 4x$

12. $\left(\dfrac{1 + 2x}{1 + 3x}\right)^4$

13. $\tan \sqrt{x}$

14. $x \sec 3x$

15. $\dfrac{\cot 5x}{1 + x^2}$

16. $[\ln (x^2 + 1)]^3$

17. $\log_{10}(\sin 3x)$

18. $\dfrac{(x^3 - 1)^3(x^{10} + 1)^4}{(2x + 1)^5}$

19. $\csc 3x^2$

20. $\dfrac{1}{1 + \sin^2 3x}$

21. $\sqrt{x^2 - 1}$

22. $\sqrt{5 - x^2}$

23. $\dfrac{1}{\sqrt{1 - x^2}}$

24. $\dfrac{x}{\sqrt{1 - x^2}}$

The functions in Exercises 25 to 32 have the property that their derivatives are much simpler than they themselves are. Check that this is so by differentiating them.

25. $\dfrac{x}{3} - \dfrac{5}{9}\ln(3x + 5)$

26. $\dfrac{2(9x - 2)}{135}\sqrt{(3x + 1)^3}$

27. $\dfrac{1}{3}\ln\left(\dfrac{\sqrt{2x + 3} - 3}{\sqrt{2x + 3} + 3}\right)$ (first simplify by laws of logarithms)

28. $\dfrac{x}{2}\sqrt{4x^2 + 3} + \dfrac{3}{4}\ln(2x + \sqrt{4x^2 + 3})$

29. $-\dfrac{1}{3}\cos 3x + \dfrac{1}{9}\cos^3 3x$

30. $\dfrac{3x}{8} - \dfrac{3\sin 10x}{80} - \dfrac{\sin^3 5x \cos 5x}{20}$

31. $\dfrac{1}{3}\ln(\tan 3x + \sec 3x)$

32. $\dfrac{1}{6}\tan^2 3x + \dfrac{1}{3}\ln \cos 3x.$

■

33. (a) Graph $y = \ln |x|$.
    (b) Check with a ruler that the slope for $x = -1$ is that given by the derivative $1/x$.

34. Let $y = (1 + x^2)^2$.
    (a) Find dy/dx by expanding $(1 + x^2)^2$.
    (b) Find dy/dx by the chain rule.

■ ■

35. We used the "magnification" interpretation of the derivative to guess the chain rule. The "slope" interpretation is not as illuminating. Let us see what the chain rule asserts about slopes.
    (a) Let f and g be differentiable functions, such that $g(1) = 2$ and $f(2) = 3$. Show that $(1, 2)$ is on the graph of g, $(2, 3)$ is on the graph of f, and $(1, 3)$ is on the graph of $f \circ g$.
    (b) If the slope of the graph of g at the point $(1, 2)$ is 5, and the slope of the graph of f at $(2, 3)$ is 7, what is the slope of the graph of $f \circ g$ at $(1, 3)$?

36. Compute $(\ln x^5)'$ (a) using the chain rule; (b) using the identity $\ln c^m = m \ln c$.

37. This exercise outlines the proof of the chain rule when $g'(x) = 0$.
    (a) Show that in this case it is sufficient to prove that
$$\lim_{\Delta x \to 0}\frac{\Delta y}{\Delta x} = 0.$$

(b) There are two types of Δx, those for which $\Delta u \neq 0$ and those for which $\Delta u = 0$. Show that, as $\Delta x \to 0$ through values of the first type, then $\Delta y/\Delta x \to 0$. *Hint:* Write $\Delta y/\Delta x = (\Delta y/\Delta u)(\Delta u/\Delta x)$.

(c) Show that, when Δx is of the second type, Δy is 0, hence

$$\frac{\Delta y}{\Delta x} = 0.$$

Thus, as $\Delta x \to 0$ through values of the second type, $\Delta y/\Delta x \to 0$.

(d) Combine (b) and (c) to show that, as $\Delta x \to 0$, $\Delta y/\Delta x \to 0$.

5.9

Inverse functions

In Sec. 5.10 we shall obtain formulas for the derivatives of such functions as 2^x, e^x, 10^x, $\sqrt[5]{x}$, and $x^{3/7}$. The concept of an *inverse function* will play an important part in the arguments there. In the present section this concept is introduced and illustrated.

Some functions have the property that different inputs always lead to different outputs. For instance, the doubling function has that property: If $2x_1 = 2x_2$, then x_1 and x_2 must be equal. However, the squaring function does *not* have that property. For instance,

$$(-3)^2 = 3^2,$$

yet -3 is different from 3. Functions with the property that different inputs lead to different outputs, such as the doubling function, are the ones of interest in this section. They are called *one-to-one*.

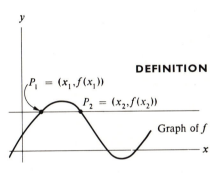

DEFINITION *One-to-one function.* A function f with domain X is called one-to-one if, whenever x_1 and x_2 are different elements in X, then the values $f(x_1)$ and $f(x_2)$ are different.

When f assigns numbers to numbers, which is the case of greatest interest in calculus, the property of being one-to-one can be described easily in terms of its graph. We describe this property first from a negative point of view: If some line parallel to the x axis meets the graph of f in *more* than one point, then f is *not* one-to-one. To see why, imagine that a horizontal line meets the graph of f in two different points, as in the accompanying figure. Since the line is parallel to the x axis, P_1 and P_2 have the same y coordinates, namely,

$$f(x_1) \quad \text{and} \quad f(x_2).$$

That is,

$$f(x_1) = f(x_2).$$

Thus f is not one-to-one.

Similar reasoning shows the following: If every line parallel to the x axis meets the graph of f in *at most* one point, then f is one-to-one.

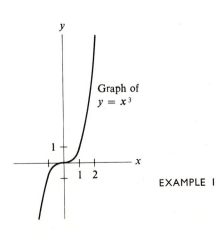

EXAMPLE 1 The cubing function $y = x^3$ is one-to-one. If $x_1{}^3 = x_0{}^3$, then $x_1 = x_0$. Observe that the graph of $y = x^3$ (see the figure in the margin) meets every horizontal line in exactly one point. ●

y

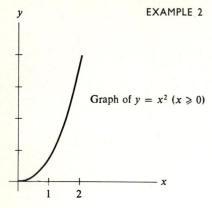

Graph of $y = x^2$ $(x \geqslant 0)$

1 2

x

EXAMPLE 2 The squaring function becomes one-to-one when its domain is restricted to nonnegative numbers. That is, if x_1 and x_2 are different *nonnegative* numbers, then

$$x_1^2 \neq x_2^2.$$

The graph of this function is shown in the accompanying figure. Note that any horizontal line below the *x* axis misses the graph, while all other horizontal lines meet it exactly once. ●

If a function is pictured as an input-output machine, then for a one-to-one function *f* it is possible to figure out the input, knowing the formula for *f* and the output. Another machine *g* can be designed which goes from the output of *f* back to the input.

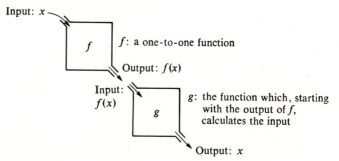

Input: *x*

f   *f*: a one-to-one function

Output: $f(x)$

Input: $f(x)$

g   *g*: the function which, starting with the output of *f*, calculates the input

Output: *x*

EXAMPLE 3 Let *f* be the cubing function and *g* be the cube root function. Thinking of these as input-output machines, hook them together and determine what happens when we start with the input 2 in the cubing machine *f*.

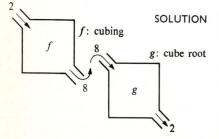

2

f   *f*: cubing

8   *g*: cube root

8   *g*

2

SOLUTION The output of the *f* machine is $2^3 = 8$. When 8 is fed into the cube root machine *g*, out comes $\sqrt[3]{8} = 2$, reversing the effect of *f* (as depicted in the accompanying figure). Therefore the cube root machine is just the opposite of the cubing machine. (All the wheels of the *g* machine run in the direction opposite those of the *f* machine.) ●

The relation between the cubing and cube root functions leads to the following general definition:

DEFINITION *Inverse function.* Let *f* be a one-to-one function. The function *g* that assigns to each value $f(x)$ the number *x* is called the *inverse* of *f*.

Let us look at some more examples of one-to-one functions and their inverses.

EXAMPLE 4 Determine the inverse of the "doubling" function *f* defined by $f(x) = 2x$.

SOLUTION If $y = 2x$, there is only one value of *x* for each value of *y*, and it is obtained by solving the equation $y = 2x$ for x, $x = \dfrac{y}{2}$.

Thus f is one-to-one and its inverse function g is the "halving" function: If y is the input in the function g, then the output is $y/2$.

For instance, $f(3) = 6$ and $g(6) = 3$. Thus $(3, 6)$ is on the graph of f, and $(6, 3)$ is on the graph of g. Since it is customary to reserve the x axis for inputs, we should write the formula for g, the "halving" function, as

$$g(x) = \frac{x}{2}.$$

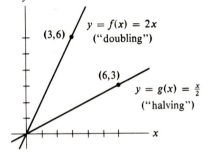

Thus f has the formula $\qquad y = 2x \qquad$ doubling,

and g has the formula $\qquad y = \dfrac{x}{2} \qquad$ halving.

The graphs of f and g are lines (see the accompanying figure); one has slope 2, and the other has slope $\frac{1}{2}$. ●

EXAMPLE 5 Find the inverse of the function $\log_{10} x$.

SOLUTION Let $y = \log_{10} x$. If we are told a value of $\log_{10} x$, we can figure out x. For instance, if $\log_{10} x = 2$, then $x = 10^2 = 100$. Thus the logarithmic function is one-to-one, and its inverse is the exponential function. In other words, if y is the logarithm of x to the base 10, then x is 10^y.

In order to graph the logarithm and exponential function, it is advisable to prepare brief tables first.

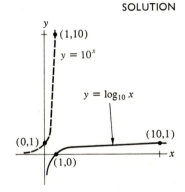

x	10	1	0.1
$\log_{10} x$	1	0	-1

x	1	0	-1
10^x	10	1	0.1

Graph the functions with the aid of the tables, as in the figure in the margin. ●

Note the relation between the two tables in Example 5. One is obtained from the other by switching inputs and outputs. This is the case for any one-to-one function and its inverse. If a and b are the numerical entries in a column for one function, then b and a are the entries in a column for the inverse function. Also, note the relation between the two graphs in Examples 4 and 5. One graph is obtained from the other by reflecting it around the line $y = x$. This can be done because, if (a, b) is on the graph of one function, then (b, a) is on the graph of the other. If you spin the paper around the line $y = x$, the point (b, a) is spun around to become the point (a, b), as you will note in the accompanying figure. This is the case for any one-to-one function and its inverse.

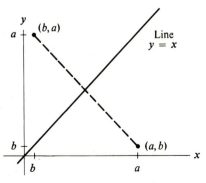

These examples are typical of the correspondence between a one-to-one function and its inverse. Perhaps the word *reverse* might be more descriptive than *inverse*. One final matter of notation: We have used the letter g to denote the inverse of f. It is common to use the symbol f^{-1} (read as "f

inverse") to denote the inverse function. We preferred to delay its use because its resemblance to the reciprocal notation might cause confusion. It should be clear by the examples that f^{-1} does *not* mean to divide 1 by f.

This table summarizes the work of the section.

Domain of f	Function f	Inverse Function f^{-1}	Domain of f^{-1}
x axis	cubing	cube root	x axis
$x \geq 0$	squaring	square root	$x \geq 0$
x axis	doubling	halving	x axis
$x > 0$	$\log_{10}$	exponential, with base 10	x axis

Exercises

The functions in Exercises 1 to 6 have inverse functions. In each case graph the given function and also the inverse function relative to the same axes.

1. $y = 3x$
2. $y = 3^x$
3. $y = \log_2 x$
4. $y = \sqrt[3]{x}$
5. $y = 2x + 1$
6. $y = x^5$
7. (a) Show that $y = x^4$ is not one-to-one.
   (b) When $y = x^4$ is considered on the domain $x \geq 0$, it is one-to-one. What is the formula for the inverse function?
8. On which of the following domains is the function $f(x) = \sin x$ one-to-one?
   (a) all real numbers
   (b) $[0, \pi]$
   (c) $[\pi/2, 3\pi/2]$
   (d) $[-\pi/2, \pi/2]$.
9. On which of the following domains is the function $f(x) = \cos x$ one-to-one?
   (a) all real numbers
   (b) $[0, \pi]$
   (c) $[\pi, 2\pi]$
   (d) $[-\pi/2, \pi/2]$.
10. Let $f(x) = \tan x$ for x in $(-\pi/2, \pi/2)$. Note that f is one-to-one. Evaluate:
    (a) $f^{-1}(1)$
    (b) $f^{-1}(\sqrt{3})$
    (c) $f^{-1}(0)$
    (d) $f^{-1}(-1)$
    (e) $f^{-1}(1/\sqrt{3})$
    (f) $f^{-1}(-\sqrt{3})$.
11. Let $f(x) = \sin x$ for x in $[-\pi/2, \pi/2]$. Evaluate:
    (a) $f^{-1}(1)$
    (b) $f^{-1}(\frac{1}{2})$
    (c) $f^{-1}(0)$
    (d) $f^{-1}(-\frac{1}{2})$
    (e) $f^{-1}(-\sqrt{3}/2)$
    (f) $f^{-1}(-1)$.
12. Let $f(x) = \cos x$ for x in $[0, \pi]$. Use a table of $\cos x$ (angle in radians) to estimate
    (a) $f^{-1}(0.9)$
    (b) $f^{-1}(0.8)$
    (c) $f^{-1}(0.7)$
    (d) $f^{-1}(0.1)$.
13. Let $y = (x + 1)/(x - 1)$. Show that this function is one-to-one and equals its inverse function.

In Exercises 14 to 19 differentiate the given function for practice.

14. $3 \cot 5x + 5 \csc 3x$
15. $\dfrac{\sqrt{4 - 9x^2}}{x}$
16. $\dfrac{1}{\sqrt{6 + 3x^2}}$
17. $(1 + x^2)^5 \sin 3x$
18. $\ln |\sin 2x|$
19. $\cos [\log_{10} (3x + 1)]$

In Exercises 20 to 22 first simplify the formula for the function by using laws of logarithms. Then differentiate the result.

20. $\ln \sqrt{\dfrac{1 + x^2}{1 + x^3}}$
21. $\ln \left[\dfrac{(5x + 1)^3(6x + 1)^2}{(2x + 1)^4} \right]$
22. $\ln \left(\dfrac{1}{6x^2 + 3x + 1} \right)$

■

23. Part of the table for a certain one-to-one function is shown below.

x	1	3	5
$f(x)$	-2	4	8

Relative to the same axes plot three points on the graph of f and three points on the graph of f^{-1}.

24. (a) A line of slope $m \neq 0$ is reflected around the line $y = x$. Find the slope of the resulting line.
    (b) A line of slope m is reflected around the x axis. Find the slope of the resulting line.
    (c) A line of slope m is reflected around the y axis. Find the slope of the resulting line.
25. At time t a moving particle has x coordinate $x = t^2 \sin 2\pi t$. Find its speed when $t = 0, \frac{1}{4}, \frac{1}{2}$, and 1.
26. (a) Show that the function $y = \sqrt[5]{1 + x^3}$ has an inverse

function by solving for x in terms of y. (b) What is the domain of the function $y = \sqrt[5]{1 + x^3}$? (c) What is the domain of the inverse function?

■ ■

27. Let $f(x) = -2x$ if $x \leq 0$ and let $f(x) = -x/2$ if $x > 0$. Show that $f = f^{-1}$.
28. Assume that $f(f(x)) = x$ for all x.
    (a) Show that f is one-to-one.
    (b) Show that $f^{-1} = f$.
29. Let X be the set of people and f the function that assigns to each person his or her father. Is f one-to-one?

30. Let X be the set of married men in a monogamous country. Let $f(x)$ be the wife of x.
    (a) Under what conditions is f one-to-one?
    (b) How would you describe f^{-1} if f is one-to-one?

Exercises 31 and 32 illustrate inverse functions on a calculator.
31. Read a positive number into the calculator. Then press the $\sqrt{x}$ key and x^2 key in that order. What number results? Try this for various initial inputs.
32. The function $y = \sin x$ is one-to-one for $0 \leq x \leq \pi/2$. How can you use your calculator to find x if $\sin x = 0.3$? (Some calculators have an "arc" key.) In Sec. 5.11 the inverses of the trigonometric functions are developed.

5.10

The derivative of b^x and x^a

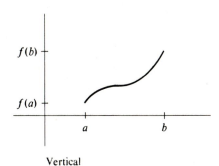

Vertical tangent line to graph of f^{-1}

Horizontal tangent line to graph of f

In this section we will use the idea of an inverse function to find the derivative of e^x and, from it, the derivative of b^x for any fixed positive base b and the derivative of x^a for any fixed exponent a.

Consider a one-to-one function f, differentiable at all x in the interval $[a, b]$. Since f is one-to-one, $f(a) \neq f(b)$ and therefore either $f(a)$ is less than $f(b)$ or $f(a)$ is greater than $f(b)$. In the following discussion of the graphs of f and f^{-1}, we take the case $f(a) < f(b)$. The other case, $f(a) > f(b)$, is similar. Since f is differentiable, it is continuous (this was proved in Sec. 4.7).

By the intermediate-value theorem of Sec. 4.7, for every number m between $f(a)$ and $f(b)$, the line

$$y = m,$$

which is parallel to the x axis, meets the graph of f. Moreover, it can be proved that the graph of f is a curve that rises as we trace it out from left to right, as in the accompanying figure. The domain of f^{-1} is $[f(a), f(b)]$, and f^{-1} assigns to each number in that interval a number in the interval $[a, b]$.

It is proved in advanced calculus that f^{-1} is continuous. Moreover f^{-1} is differentiable at any value $f(x)$ for which $f'(x)$ is not 0. (A horizontal tangent to the graph of f becomes, when the graph of f is reflected around the line $y = x$, a vertical tangent to the graph of f^{-1}.)

Our optical intuition will suggest the relation between the derivatives of f and f^{-1}. Think of f as the projection from a slide to the screen and f^{-1} a "reverse" projection that assigns to each point on the screen the point on the slide that projects onto it. Now compare the magnifications. If f magnifies by a certain ratio at x, say 3, then we expect f^{-1} to shrink by a factor 3 at $f(x)$, that is, "magnify" by a factor $\frac{1}{3}$. Intuitively, if f has a derivative at x, we expect f^{-1} to have a derivative at the number $f(x)$ and, in fact,

$$(f^{-1})'(f(x)) = \frac{1}{f'(x)}, \qquad \text{if} \quad f'(x) \neq 0.$$

It is easier to express this relation in the differential notation. Let $y = f(x)$ denote the given function and $x = f^{-1}(y)$ denote the inverse function. We shall assume that, if f is differentiable, then so is f^{-1}, if $f'(x) \neq 0$. With this assumption in mind, let us apply the chain rule to the special case

$$x = f^{-1}(y), \quad \text{where } y = f(x).$$

By the chain rule,
$$\frac{dx}{dx} = \frac{dx}{dy} \cdot \frac{dy}{dx},$$

or
$$1 = \frac{dx}{dy} \cdot \frac{dy}{dx}.$$

It is this equation that provides the basis of the proof of Theorem 1, which shows that the derivative of e^x is e^x. In the next section it is the key to obtaining the derivative of several other functions.

THEOREM I   *The derivative of e^x.* Let $y = e^x$ be the exponential function with base e. [Recall that e is defined as $\lim_{n \to \infty} (1 + 1/n)^n$.]

Then
$$\frac{d(e^x)}{dx} = e^x.$$

PROOF   Let $y = e^x$. This function is the inverse of the natural logarithmic function, that is,

$$y = e^x \quad \text{is equivalent to} \quad x = \ln y.$$

Since the derivative dx/dy exists, so does the derivative dy/dx, wherever dx/dy is not 0.

Since
$$x = \ln y,$$

we have
$$1 = \frac{dx}{dx} = \frac{dx}{dy} \cdot \frac{dy}{dx},$$

or
$$1 = \frac{d(\ln y)}{dy} \cdot \frac{dy}{dx}.$$

Hence
$$1 = \frac{1}{y} \cdot \frac{dy}{dx}.$$

Thus
$$\frac{dy}{dx} = y$$
$$= e^x.$$

In short,
$$\frac{dy}{dx} = e^x,$$

and the theorem is proved. ●

Theorem 1 shows that e^x equals its derivative. In the next chapter, it will be shown that the only functions that equal their own derivatives must be of the form Ae^x, where A is a constant.

EXAMPLE 1   Find the derivative of e^{3x}.

SOLUTION   Let $y = e^{3x}$.   Then $y = e^u$,   where $u = 3x$.

Thus
$$\frac{dy}{dx} = \frac{dy}{du}\frac{du}{dx} \qquad \text{chain rule}$$

$$= \frac{d(e^u)}{du}\frac{d(3x)}{dx}$$

$$= e^u \cdot 3 \qquad \text{Theorem 1}$$

$$= e^{3x} \cdot 3$$

$$= 3e^{3x}. \quad \bullet$$

The formula for the derivative of e^x is quite simple.   Let us next compute the derivative of 10^x.

EXAMPLE 2   Find the derivative of 10^x.

SOLUTION   Write 10 as a power of e: $10 = e^{\ln 10}$.

Then $10^x = (e^{\ln 10})^x$.

By the "power-of-a-power" rule,

$$(e^{\ln 10})^x = e^{(\ln 10)x}.$$

Thus $10^x = e^{(\ln 10)x}$.

Since $\ln 10$ is a constant, this problem is similar to Example 1.   Let $y = e^{(\ln 10)x}$.

This can be written as $y = e^u$, where $u = (\ln 10)x$.

Then
$$\frac{dy}{dx} = \frac{d(e^u)}{du}\frac{d[(\ln 10)x]}{dx} \qquad \text{chain rule}$$

$$= e^u \cdot \ln 10 \qquad \text{Theorem 1}$$

$$= e^{(\ln 10)x} \cdot \ln 10$$

$$= 10^x \cdot \ln 10$$

$$= (\ln 10) \cdot 10^x.$$

Thus
$$\frac{d(10^x)}{dx} = (\ln 10) \cdot 10^x.$$

Note that the derivative of 10^x is not equal to 10^x, but rather is proportional to 10^x, the constant of proportionality being $\ln 10 \doteq 2.3$. $\quad \bullet$

When Differentiating b^x, First Change the Base to e

From the point of view of calculus, 10 is as unfortunate a base for exponentials as it is for logarithms. It is advisable to remember the simple formula $(e^x)' = e^x$ and, if you need $(b^x)'$ for some other base b, write $b = e^{\ln b}$ and use the chain rule as in Example 2.

The importance of the exponential function e^x is further illustrated by the simplicity of the proof it provides for the following theorem, which generalizes the result obtained in Secs. 3.2 and 5.6 that $(x^n)' = nx^{n-1}$, when n is an integer, to the result $(x^a)' = ax^{a-1}$, when a is any real number, even irrational.

THEOREM 2   Let a be a fixed number.   Then

$$\frac{d(x^a)}{dx} = ax^{a-1}.$$

PROOF   Let $y = x^a$.   Since

$$x = e^{\ln x},$$

$$y = (e^{\ln x})^a.$$

Hence, by the power of a power rule,

$$y = e^{a \ln x}.$$

This can be written as $y = e^u$,   where $u = a \ln x$.

Hence

$$\frac{dy}{dx} = \frac{d(e^u)}{du}\frac{d(a \ln x)}{dx} \qquad \text{chain rule}$$

$$= e^u \frac{a}{x}$$

$$= e^{a \ln x} \frac{a}{x}$$

$$= x^a \frac{a}{x}$$

$$= ax^{a-1}.$$

This proves the theorem. ●

EXAMPLE 3   Compute the derivative of $x^{1/2}$.

SOLUTION

$$\frac{d(x^{1/2})}{dx} = \frac{1}{2}x^{1/2-1} \qquad \text{Theorem 2 with } a = \tfrac{1}{2}$$

$$= \frac{1}{2}x^{-1/2}$$

$$= \frac{1}{2x^{1/2}}.$$

This is in agreement with the formula obtained in Sec. 3.3,

$$\frac{d(\sqrt{x})}{dx} = \frac{1}{2\sqrt{x}}. \quad ●$$

EXAMPLE 4   Differentiate $\sqrt[3]{x^2}$.

SOLUTION

$$\frac{d(\sqrt[3]{x^2})}{dx} = \frac{d(x^{2/3})}{dx}$$

$$= \tfrac{2}{3}x^{2/3 - 1} \qquad \text{Theorem 2 with } a = \tfrac{2}{3}$$

$$= \tfrac{2}{3}x^{-1/3}$$

$$= \frac{2}{3x^{1/3}}$$

$$= \frac{2}{3\sqrt[3]{x}}. \quad \bullet$$

The next example concerns the derivative of x^x, an exponential in which neither the base nor the exponent is constant.

EXAMPLE 5   Compute $(x^x)'$.

SOLUTION   Let $y = x^x$.   To begin, introduce the base e by writing the base x as $e^{\ln x}$.

Then
$$x^x = (e^{\ln x})^x$$
$$= e^{x \ln x}$$

and, consequently,
$$y = e^{x \ln x}.$$

Then
$$y = e^u, \qquad \text{where} \quad u = x \ln x.$$

Hence
$$\frac{dy}{dx} = \frac{d(e^u)}{du}\frac{d(x \ln x)}{dx}$$

$$= e^u \left[x \frac{d(\ln x)}{dx} + \ln x \frac{dx}{dx} \right]$$

$$= e^{x \ln x} \cdot \left(x \frac{1}{x} + \ln x \right)$$

$$= x^x \cdot (1 + \ln x),$$

and
$$\frac{d(x^x)}{dx} = (1 + \ln x)x^x. \quad \bullet$$

The final two examples review several of the differentiation formulas and techniques discussed in the chapter.   The reader should *do* the differentiation and then compare the result with the steps shown.

EXAMPLE 6   Differentiate
$$f(x) = \frac{e^{ax}}{a^3}(a^2 x^2 - 2ax + 2),$$

where a is a nonzero constant.

SOLUTION First bring a^3, which is constant, to the front of the formula by writing the function as

$$f(x) = \frac{1}{a^3} e^{ax}(a^2x^2 - 2ax + 2).$$

Now differentiate:

$$f'(x) = \frac{1}{a^3} [e^{ax}(a^2x^2 - 2ax + 2)]' \quad \text{derivative of constant times a function}$$

$$= \frac{1}{a^3} [e^{ax}(a^2 \cdot 2x - 2a) + (a^2x^2 - 2ax + 2)ae^{ax}] \quad \begin{array}{l}\text{derivative of}\\ \text{product, chain rule,}\\ \text{etc.}\end{array}$$

$$= \frac{1}{a^3} [(2a^2x - 2a + a^3x^2 - 2a^2x + 2a)e^{ax}] \quad \text{collecting terms}$$

$$= x^2 e^{ax} \quad \text{after cancelling.} \quad \bullet$$

EXAMPLE 7   Differentiate $\qquad y = \dfrac{e^{ax}}{a^2 + b^2} (a \sin bx - b \cos bx),$

where a and b are nonzero constants.

SOLUTION $\qquad \dfrac{dy}{dx} = \dfrac{1}{a^2 + b^2} \dfrac{d[e^{ax}(a \sin bx - b \cos bx)]}{dx}$

$$= \frac{1}{a^2 + b^2} [e^{ax}(ab \cos bx + b^2 \sin bx) + (a \sin bx - b \cos bx)ae^{ax}]$$

$$= \frac{1}{a^2 + b^2} (e^{ax})(ab \cos bx + b^2 \sin bx + a^2 \sin bx - ab \cos bx)$$

$$= \frac{1}{a^2 + b^2} e^{ax}(a^2 + b^2) \sin bx$$

$$= e^{ax} \sin bx . \quad \bullet$$

Exercises

Differentiate the functions in Exercises 1 to 21.

1. $5e^{3x}$
2. 5^x
3. 2^{3x}
4. xe^x
5. e^{-x}
6. x^2e^{-x}
7. $e^x(x^2 - 2x + 2)$
8. $x - \ln(1 + e^x)$
9. $\dfrac{x^2}{1 + e^{3x}}$
10. $6x^{7/4}$
11. $x^{\sqrt{2}}$
12. $(\sqrt{2})^x$
13. $(\sqrt{2})^\pi$
14. $(1 + x)^x$
15. $\dfrac{e^{3x}(3x - 1)}{9}$
16. $\dfrac{10^x}{\ln 10}$

17. $x \sin 3x + \frac{1}{3} \cos 3x$
18. $\dfrac{e^{-x}}{5} (\sin 2x + 2 \cos 2x)$
19. $\frac{1}{10}[2x - \ln(5 + 3e^{2x})]$
20. $\ln[\ln(1 + x^2)]$
21. $x(\ln 5x)^2 - 2x \ln 5x + 2x$
22. Write each of the following functions in the form x^a and differentiate [parts (a) to (g)].

(a) $\sqrt{x^3}$ (b) $\sqrt[3]{x}$

(c) $\dfrac{1}{\sqrt{x}}$ (d) $x\sqrt[3]{x}$

(e) $\sqrt[3]{x^4}$

(f) $\dfrac{\sqrt[5]{x^6}}{x}$

(g) $xx^{\sqrt{2}}$

■

23. Let A and k be constants.   Show that the derivative of Ae^{kx} is proportional to Ae^{kx}.

24. (a) Graph $y = e^x$ and $y = -x$ on the same axes.
(b) Using the graphs in (a), graph $y = e^x - x$.
(c) Find all points on the graph in (b) where the tangent line is horizontal.

25. The formula $y = e^{-t}\sin t$ describes a decaying alternating current.   Consider t in $[0, 2\pi]$.
(a) Fill in this table and plot the resulting points.

t	0	$\dfrac{\pi}{2}$	π	$\dfrac{3\pi}{2}$	2π	$\dfrac{5\pi}{2}$	3π	$\dfrac{7\pi}{2}$	4π
$e^{-t}\sin t$									

(b) Graph $y = e^{-t}\sin t$ for t in $[0, 2\pi]$.
(c) Find all points on the graph in (b) where the tangent line is horizontal.

26. (a) Graph $y = x^{1/x}$ for $x > 0$.
(b) Find the points on the graph in (c) where the tangent line is horizontal.
(c) Where is dy/dx positive?   negative?
(d) What is the largest possible value of $x^{1/x}$?

27. Simplify the formulas for the following functions, using the   identities $\ln cd = \ln c + \ln d$, $\ln c^m = m\ln c$,   and $\ln c/d = \ln c - \ln d$.   Then differentiate.
(a) $\ln (x^2 + 3)^5$ (b) $\ln [(1 + e^{3x})(x^6)]$
(c) $\ln [\sqrt{1 + 2x}\,\sqrt[3]{1 + x^3}]$
(d) $\ln \dfrac{(x^3 - 2x)^5}{\sqrt{x^2 + 5}}$

■ ■

28. Using results of this section and the definition of the derivative, evaluate:
(a) $\lim\limits_{h\to 0} \dfrac{e^h - 1}{h}$ (b) $\lim\limits_{x\to 1} \dfrac{2^x - 2}{x - 1}$
(c) $\lim\limits_{h\to 0} \dfrac{10^h - 1}{h}$.

5.11

The derivative of the inverse trigonometric functions

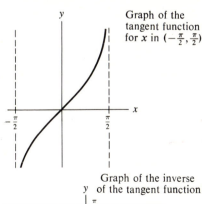

Graph of the tangent function for x in $(-\frac{\pi}{2}, \frac{\pi}{2})$

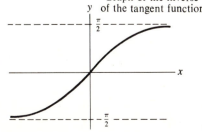

Graph of the inverse of the tangent function

The derivative of $\frac{1}{2}\ln (1 + x^2)$ is $x/(1 + x^2)$.   However, up to this point we have no function whose derivative is simply $1/(1 + x^2)$.   Nor do we have any function whose derivative is $1/\sqrt{1 - x^2}$ or $\sqrt{1 - x^2}$.   Such functions, which will be very useful in integral calculus, will be obtained in this section.   Surprisingly, they turn out to be inverse trigonometric functions.

First we will provide a function whose derivative is $1/(1 + x^2)$.

Consider the function $y = \tan x$ in the open interval $-\pi/2 < x < \pi/2$.   As x increases in the interval $-\pi/2 < x < \pi/2$, $\tan x$ increases.   Thus the function $\tan x$ is one-to-one if the domain is restricted to be between $-\pi/2$ and $\pi/2$.   The graph of $y = \tan x$ for $-\pi/2 < x < \pi/2$ appears in the margin.   Note that as $x \to \pi/2$ or $x \to -\pi/2$, $|\tan x|$ gets very large.

The graph of the inverse function g is obtained by spinning the graph of $y = \tan x$ around the line $y = x$.   As x gets large, note in the accompanying figure that $g(x) \to \pi/2$.

The inverse of the tangent function is called the *arctangent* function, and is written arctan x or $\tan^{-1} x$.   [This is not the reciprocal of $\tan x$, which is written $\cot x$, $1/\tan x$, or $(\tan x)^{-1}$ to avoid confusion.]   As an example, since $\tan \pi/4 = 1$,

$$\arctan 1 = \pi/4 \qquad \text{or} \qquad \tan^{-1} 1 = \pi/4.$$

Observe that the domain of the arctan function is the entire x axis, and that when $|x|$ is large arctan x is near $\pi/2$ or $-\pi/2$.

It is frequently useful to picture the tangent and arctangent functions in terms of the unit circle.

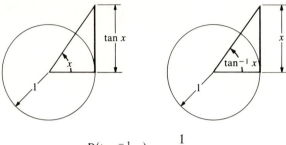

THEOREM 1

$$D(\tan^{-1} x) = \frac{1}{1 + x^2}.$$

PROOF   Let $y = \tan^{-1} x$.   The problem is to find dy/dx.   By the definition of the inverse tangent function, $x = \tan y$.   Note that $dx/dy = \sec^2 y$.

As in the preceding section,

$$\frac{dx}{dx} = \frac{dx}{dy} \cdot \frac{dy}{dx}$$

$$1 = \sec^2 y \frac{dy}{dx}.$$

Hence

$$\frac{dy}{dx} = \frac{1}{\sec^2 y}$$

$$= \frac{1}{1 + \tan^2 y}$$

$$= \frac{1}{1 + x^2}.$$

This completes the proof. ●

EXAMPLE 1   Find $D(\tan^{-1} \sqrt{x})$.

SOLUTION   Theorem 1 and the chain rule are needed.   Let

$$y = \tan^{-1} \sqrt{x};$$

then

$$y = \tan^{-1} u, \qquad \text{where } u = \sqrt{x}.$$

thus

$$\frac{dy}{dx} = \frac{dy}{du} \cdot \frac{du}{dx}$$

$$= \frac{d(\tan^{-1} u)}{du} \cdot \frac{d(\sqrt{x})}{dx}$$

$$= \frac{1}{1 + u^2} \cdot \frac{1}{2\sqrt{x}}$$

$$= \frac{1}{1 + x} \cdot \frac{1}{2\sqrt{x}}$$

$$= \frac{1}{2\sqrt{x}\,(1 + x)}. \; \bullet$$

We turn next to the inverse of the sine function.

The sine function is not one-to-one. For instance, $\sin(\pi/4) = \sqrt{2}/2 = \sin(3\pi/4)$. However, if the domain is restricted to $[-\pi/2, \pi/2]$ a one-to-one function results. The graph of $y = \sin x$ rises from $x = -\pi/2$ to $x = \pi/2$.

The inverse function is called the *arcsine* and is written $\arcsin x$ or $\sin^{-1} x$. Since $\sin \pi/2 = 1$, we have, for instance,

$$\arcsin 1 = \frac{\pi}{2} \quad \text{or} \quad \sin^{-1} 1 = \frac{\pi}{2}.$$

Both these latter equations say, "the angle whose sine is 1 is $\pi/2$."

We graph the sine and arcsine functions below.

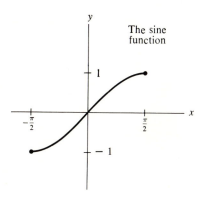

The sine function

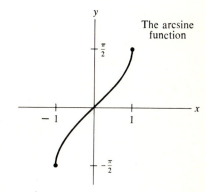

The arcsine function

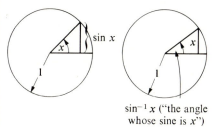

$\sin^{-1} x$ ("the angle whose sine is x")

It is also useful to visualize these two functions in terms of the unit circle; they are depicted in the accompanying figures. Note that

$$-\pi/2 \le \arcsin x \le \pi/2. \; \bullet$$

The proof of the next theorem is similar to that of Theorem 1.

THEOREM 2

$$D(\sin^{-1} x) = \frac{1}{\sqrt{1 - x^2}}.$$

PROOF Let $y = \sin^{-1} x$, hence $x = \sin y$. Thus

$$\frac{dx}{dx} = \frac{d(\sin y)}{dx}$$

$$\frac{dx}{dx} = \frac{d(\sin y)}{dy} \frac{dy}{dx}$$

or

$$1 = \cos y \frac{dy}{dx}.$$

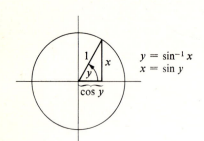

$y = \sin^{-1} x$
$x = \sin y$

$\overline{\cos y}$

Hence

$$\frac{dy}{dx} = \frac{1}{\cos y}.$$

To express $\cos y$ in terms of x, draw the unit circle and indicate on it that $x = \sin y$. Inspection of the accompanying figure shows that $\cos y$ is positive (since $-\pi/2 \le y \le \pi/2$) and

$$\cos^2 y + x^2 = 1.$$

Thus $\cos y = \sqrt{1 - x^2}$ the positive square root

and dy/dx takes the form

$$\frac{dy}{dx} = \frac{1}{\sqrt{1 - x^2}}$$

This proves that

$$\frac{d(\sin^{-1} x)}{dx} = \frac{1}{\sqrt{1 - x^2}}. \bullet$$

EXAMPLE 2 Find the derivative of $\sin^{-1} \dfrac{3x}{4}$.

SOLUTION Let $y = \sin^{-1} \dfrac{3x}{4}$. This is a composite function, with

$$y = \sin^{-1} u, \qquad \text{where} \quad u = \frac{3x}{4}.$$

Now

$$\frac{dy}{du} = \frac{1}{\sqrt{1 - u^2}} \qquad \text{and} \qquad \frac{du}{dx} = \frac{3}{4}.$$

Thus

$$\frac{dy}{dx} = \frac{dy}{du} \cdot \frac{du}{dx}$$

$$= \frac{1}{\sqrt{1 - u^2}} \cdot \frac{3}{4} = \frac{1}{\sqrt{1 - (3x/4)^2}} \cdot \frac{3}{4}$$

$$= \frac{1}{\sqrt{1 - 9x^2/16}} \cdot \frac{3}{4}$$

$$= \frac{\sqrt{16}}{\sqrt{16 - 9x^2}} \cdot \frac{3}{4}$$

$$= \frac{3}{\sqrt{16 - 9x^2}}. \bullet$$

EXAMPLE 3 Differentiate $x\sqrt{1 - x^2} + \sin^{-1} x$.

SOLUTION

$$\frac{d(x\sqrt{1-x^2}+\sin^{-1}x)}{dx}=\frac{d(x\sqrt{1-x^2})}{dx}+\frac{d(\sin^{-1}x)}{dx}$$

$$=\frac{xd(\sqrt{1-x^2})}{dx}+\sqrt{1-x^2}\frac{dx}{dx}+\frac{1}{\sqrt{1-x^2}}$$

$$=x\cdot\frac{1}{2}\cdot\frac{-2x}{\sqrt{1-x^2}}+\sqrt{1-x^2}+\frac{1}{\sqrt{1-x^2}}$$

$$=\frac{-x^2}{\sqrt{1-x^2}}+\sqrt{1-x^2}+\frac{1}{\sqrt{1-x^2}}$$

$$=\frac{1-x^2}{\sqrt{1-x^2}}+\sqrt{1-x^2}$$

$$=\sqrt{1-x^2}+\sqrt{1-x^2}$$

$$=2\sqrt{1-x^2}.$$

This is another illustration of a function whose derivative has a much simpler form than itself. ●

While $\tan^{-1}x$ and $\sin^{-1}x$ are the most frequently used inverse trigonometric functions, the inverse of the secant function is also important. Recall that in Sec. 5.6 the secant function was defined as

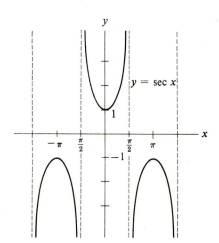

$y = \sec x$

$$\sec x=\frac{1}{\cos x}$$

Since $|\cos x|\le 1$, it follows that $|\sec x|\ge 1$. Also, when $\cos x$ is near 0, $\sec x$ takes on large values. The graph of $y=\sec x$ is shown in the margin. Since $\cos x$ has period 2π, so does $\sec x$. If x is restricted to being between 0 and π, $\sec x$ is one-to-one and therefore has an inverse function, denoted by $\sec^{-1}x$. Its graph appears in the margin. Observe that $\sec^{-1}x$ is defined only for $x\ge 1$ and $x\le -1$. For instance, to evaluate $y=\sec^{-1}2$, one would reason as follows:

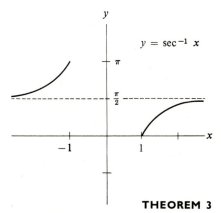

$y = \sec^{-1} x$

$$y=\text{angle whose secant is }2$$
$$=\text{angle whose cosine is }\tfrac{1}{2}$$
$$=\pi/3.$$

Thus

$$\sec^{-1}2=\frac{\pi}{3}.$$

THEOREM 3

$$D(\sec^{-1}x)=\frac{1}{|x|\sqrt{x^2-1}},\qquad\text{where}\quad|x|>1.$$

PROOF   Let $y=\sec^{-1}x$.

Then

$$x=\sec y,$$

and
$$\frac{dx}{dx} = \frac{dx}{dy} \cdot \frac{dy}{dx},$$

or
$$1 = \sec y \tan y \frac{dy}{dx}.$$

Hence
$$\frac{dy}{dx} = \frac{1}{\sec y \tan y}$$

$$= \frac{1}{x \tan y}.$$

All that remains is to express $\tan y$ in terms of x.

Since
$$\sec^2 y = 1 + \tan^2 y,$$

or
$$x^2 = 1 + \tan^2 y,$$

it follows that

$$\tan y = \pm\sqrt{x^2 - 1}, \qquad \text{where} \quad |x| > 1.$$

Which sign is to be chosen? If $x > 1$, $y = \sec^{-1} x$ is in the range $(0, \pi/2)$; thus $\tan y$ is positive. If $x < -1$, $y = \sec^{-1} x$ is in the range $(\pi/2, \pi)$; thus $\tan y$ is negative. But, in both cases $x \tan y$ is positive. Thus

$$\frac{dy}{dx} = \frac{1}{|x|\sqrt{x^2 - 1}}.$$

(Note that this derivative is positive, in agreement with the graph of $y = \sec^{-1} x$; its tangent lines slope upward). ●

EXAMPLE 4   Differentiate $y = \sec^{-1} 5x$.

SOLUTION   The chain rule is required.

Here
$$y = \sec^{-1} u, \qquad \text{where} \quad u = 5x.$$

Thus
$$\frac{dy}{dx} = \frac{1}{|u|\sqrt{u^2 - 1}} \cdot 5$$

$$= \frac{1}{|5x|\sqrt{25x^2 - 1}} \cdot 5$$

$$= \frac{1}{|5||x|\sqrt{25x^2 - 1}} \cdot 5$$

$$= \frac{1}{|x|\sqrt{25x^2 - 1}}. \; ●$$

The inverse of the remaining three trigonometric functions, cos x, cot x, and csc x, will not be needed. For the record, they may be defined as follows:

$$\cos^{-1} x = \frac{\pi}{2} - \sin^{-1} x,$$

$$\cot^{-1} x = \frac{\pi}{2} - \tan^{-1} x,$$

$$\csc^{-1} x = \sin^{-1} \frac{1}{x}.$$

The fact that

$$\frac{d(\cos^{-1} x)}{dx} = \frac{-1}{\sqrt{1 - x^2}}$$

is sometimes used.

Exercises

1. Draw a picture of a circle of radius 10 centimeters and use it, a centimeter ruler, and a protractor to estimate (a) $\tan^{-1} 1.5$, (b) $\tan^{-1} 0.7$, (c) $\tan^{-1} (-1.2)$, (d) $\sin^{-1} (0.4)$, (e) $\sin^{-1} (-0.5)$, (f) $\sin^{-1} (0.8)$. If your protractor reads degrees, turn your answer into radians by dividing by 57 (an approximation of $180/\pi$).

2. Evaluate (a) $\sin^{-1} 1$, (b) $\tan^{-1} 1$, (c) $\sin^{-1} (-\sqrt{3}/2)$, (d) $\tan^{-1} (-\sqrt{3})$, (e) $\sec^{-1} \sqrt{2}$.

In Exercises 3 to 8 evaluate the expression. A sketch of the unit circle or appropriate right triangle may help.

3. $\sin(\tan^{-1} 1)$
4. $\tan[\sin^{-1}(\sqrt{3}/2)]$
5. $\tan(\sec^{-1} 2)$
6. $\sin(\sin^{-1} 0.3)$
7. $\tan[\sin^{-1}(-\sqrt{2}/2)]$
8. $\sin(\tan^{-1} 0)$

In Exercises 9 to 20 differentiate the given functions.

9. $\sin^{-1} 5x$
10. $\sin^{-1} e^{-x}$
11. $\tan^{-1} 3x$
12. $\tan^{-1} \sqrt[3]{x}$

13. $\sec^{-1} 3x$
14. $-\frac{1}{3} \sin^{-1} \frac{3}{x}$

15. $\frac{x}{2} \sqrt{2 - x^2} + \sin^{-1}(x/\sqrt{2})$

16. $\sqrt{3x^2 - 1} - \tan^{-1} \sqrt{3x^2 - 1}$

17. $\frac{2}{5} \sec^{-1} \sqrt{3x^5}$

18. $\frac{1}{2} \left[(x - 3)\sqrt{6x - x^2} + 9 \sin^{-1} \frac{x - 3}{3} \right]$

19. $\sqrt{1 + x} \sqrt{2 - x} - 3 \sin^{-1} \sqrt{\frac{2 - x}{3}}$

20. $\sin^{-1} x - \sqrt{1 - x^2}$

■

In Exercises 21 to 23 differentiate the given functions. Note that quite different functions may have very similar derivatives.

21. (a) $\ln(x + \sqrt{x^2 - 9})$

(b) $\sin^{-1} \frac{x}{3}$

22. (a) $-\frac{1}{5} \ln \frac{5 + \sqrt{25 - x^2}}{x}$

(b) $-\frac{1}{5} \sin^{-1} \frac{5}{x}$

23. (a) $\ln \frac{\sqrt{2x^2 + 1} - 1}{x}$

(b) $\sec^{-1} x\sqrt{2}$

In Exercises 24 to 26 differentiate the given function. In each case the derivative has a short formula.

24. $x \sin^{-1} 3x + \frac{1}{3}\sqrt{1 - 9x^2}$.
25. $x(\sin^{-1} 2x)^2 - 2x + \sqrt{1 - 4x^2} \sin^{-1} 2x$.
26. $x \tan^{-1} 5x - \frac{1}{10} \ln(1 + 25x^2)$.
27. (a) Graph $y = \cos^{-1} x$.
    (b) Prove that $(\cos^{-1} x)' = -1/\sqrt{1 - x^2}$.

■ ■

28. Show that $\tan^{-1} \frac{1}{2} + \tan^{-1} \frac{1}{3} = \pi/4$.

5.12

Summary

In this chapter we obtained the formulas and procedures for computing the derivative of any function constructed from the functions x^a, the trigonometric functions and their inverses, exponentials and logarithms, by the operations of addition, subtraction, multiplication, division, and composition. These functions are called the *elementary functions*, and they are the functions of most interest in applications of calculus. As the various methods of this chapter show, *the derivative of an elementary function is again an elementary function.*

VOCABULARY AND SYMBOLS

$\dfrac{dy}{dx}, \dfrac{df}{dx}, D(f), y'$

composite function

$f \circ g$

chain rule

logarithm

one-to-one function

inverse of a one-to-one function

f^{-1}

$\sin^{-1} x$

$\tan^{-1} x$

$\sec^{-1} x$

KEY FACTS

The following table summarizes the chapter and calls attention to certain functions.

f	Derivative of f	Comment
Constant function (c)	0	
x^a	ax^{a-1}	
$\sqrt{x}$	$\dfrac{1}{2\sqrt{x}}$	
$\dfrac{1}{x}$	$\dfrac{-1}{x^2}$	
$\sqrt{1 + x^2}$	$\dfrac{x}{\sqrt{1 + x^2}}$	
$\sin x$	$\cos x$	
$\cos x$	$-\sin x$	Remember the minus sign
$\tan x$	$\sec^2 x$	
$\sec x$	$\sec x \tan x$	
$\csc x$	$-\csc x \cot x$	Not common
$\cot x$	$-\csc^2 x$	Not common
$\ln |x|$	$\dfrac{1}{x}$	
$\log_{10} |x|$	$\dfrac{\log_{10} e}{x}$	Not common
$\log_a |x|$	$\dfrac{\log_a e}{x}$	Not common
e^x	e^x	
e^{kx}	ke^{kx}	
e^{-x}	$-e^{-x}$	

f	Derivative of f	Comment
a^x	See comment	Do not memorize a formula for $d(a^x)/dx$. Write a^x as $e^{(\ln a)x}$ and differentiate
$\sin^{-1} x$	$\dfrac{1}{\sqrt{1-x^2}}$	
$\tan^{-1} x$	$\dfrac{1}{1+x^2}$	
$\sec^{-1} x$	$\dfrac{1}{|x|\sqrt{x^2-1}}$	

The chain rule, combined with some of the above formulas, yields the following rules where u is a differentiable function of x.

$$\frac{d(1/u)}{dx} = -\frac{1}{u^2}\frac{du}{dx}$$

$$\frac{d(\sin u)}{dx} = \cos u\,\frac{du}{dx}$$

$$\frac{d(\cos u)}{dx} = -\sin u\,\frac{du}{dx}$$

$$\frac{d(\tan u)}{dx} = \sec^2 u\,\frac{du}{dx}$$

$$\frac{d(\sec u)}{dx} = \sec u\,\tan u\,\frac{du}{dx}$$

$$\frac{d(\cot u)}{dx} = -\csc^2 u\,\frac{du}{dx}$$

$$\frac{d(\csc u)}{dx} = -\csc u\,\cot u\,\frac{du}{dx}$$

$$\frac{d(\ln |u|)}{dx} = \frac{1}{u}\frac{du}{dx}$$

$$\frac{d(e^u)}{dx} = e^u\,\frac{du}{dx}$$

$$\frac{d(\sin^{-1} u)}{dx} = \frac{1}{\sqrt{1-u^2}}\frac{du}{dx}$$

$$\frac{d(\tan^{-1} u)}{dx} = \frac{1}{1+u^2}\frac{du}{dx}$$

$$\frac{d(\sec^{-1} u)}{dx} = \frac{1}{|u|\sqrt{u^2-1}}\frac{du}{dx}$$

Techniques for finding derivatives:

$$(f+g)' = f' + g'$$

$$(cf)' = cf'$$

$$(fg)' = fg' + gf'$$

$$\left(\frac{1}{f}\right)' = \frac{-f'}{f^2}$$

$$\left(\frac{f}{g}\right)' = \frac{gf' - fg'}{g^2}.$$

When using the formula

$$\left(\frac{f}{g}\right)' = \frac{gf' - fg'}{g^2}$$

remember the minus sign in the numerator. It is advisable to write down first the g^2 in the denominator, and then immediately the g of gf' in the numerator. The rest of the numerator then fits into place.

And, most important of all the tools, the chain rule:

If $\qquad\qquad y = f(u)$

and $\qquad\qquad u = g(x),$

then $\qquad\qquad \dfrac{dy}{dx} = \dfrac{dy}{du}\dfrac{du}{dx}.$

Guide quiz on Chap. 5 (computations)

1. Differentiate $1/\sqrt{x}$ in the three following ways:
   (a) writing it in the form x^a and differentiating
   (b) using the formula $(1/f)' = -f'/f^2$
   (c) using the formula for $(f/g)'$.
2. Differentiate:
   (a) $(x^3 + x)^6$ (b) $\dfrac{(\sin 2x)^3}{x^4}$ (c) $x^3 \ln (x^2)$ (d) $\sec^5 3x$.
3. Differentiate:
   (a) $\sqrt{1 + x^3}$ (b) $\cos^2 3x$
   (c) e^{-3x} (d) $\sin^{-1} 3x$
   (e) 5^{x^2} (f) $\sqrt{x^3}$

4. Differentiate:
   (a) $\dfrac{e^x(\sin 2x - 2 \cos 2x)}{5}$
   (b) $x \sin^{-1} x + \sqrt{1 - x^2}$
   (c) $x \tan^{-1} x - \frac{1}{2} \ln (1 + x^2)$
   (d) $\sec^{-1} 3x \cdot \sec 3x$.
5. Evaluate:
   (a) $\tan^{-1}(-1)$,   (b) $\tan^{-1}\sqrt{3}$,   (c) $\sin^{-1}(\frac{1}{2})$,
   (d) $\sec^{-1}(-2)$,   (e) $\sec^{-1}(-1)$,   (f) $\sin^{-1}(-1)$.

Guide quiz on Chap. 5 (concepts)

1. Using the definition of the derivative and the values of the limits
$$\lim_{h \to 0} \frac{\sin h}{h} \quad \text{and} \quad \lim_{h \to 0} \frac{1 - \cos h}{h},$$
   show that $\dfrac{d(\cos x)}{dx} = -\sin x$.

2. (a) Prove that $\log_{10} x = (\log_{10} e) \ln x$, beginning with the equation
$$x = e^{\ln x}.$$
   (b) Use (a) and the formula for $(\ln x)'$ to find $(\log_{10} x)'$.
3. Evaluate
   (a) $\lim\limits_{x \to 2} \dfrac{e^x - e^2}{x - 2}$,   (b) $\lim\limits_{h \to 0} \dfrac{10^h - 1}{h}$.

4. Prove that (a) $(f + g)' = f' + g'$; (b) $(fg)' = fg' + gf'$.
5. (a) Show that, if $f(x) = x^3$ and $g(x) = x^2$, then $f \circ g = g \circ f$.
   (b) Show that, if $f(x) = x^3$ and $g(x) = 1 + x^2$, then $f \circ g \neq g \circ f$.
6. (a) Why does calculus use radian measure?
   (b) Why does calculus use the base e for logarithms?
   (c) Why does calculus use the base e for exponentials?
7. Let $f(x)$ be the mother of x and let $g(x)$ be the father of x.
   (a) What is $(f \circ f)(x)$?

(b) What is $(f \circ g)(x)$?
(c) What is $(g \circ f)(x)$?
(d) If $f(x_1) = f(x_2)$ and $g(x_1) = g(x_2)$, then what are x_1 and x_2 called?
(e) If $(f \circ f)(x_1) = (f \circ f)(x_2)$, but $f(x_1) \neq f(x_2)$, then what are x_1 and x_2 called? Assume $x_1 \neq x_2$.

8. What is the inverse of each of these functions?
   (a) $\ln x$ (b) e^x
   (c) x^3 (d) $3x$
   (e) $\sqrt[3]{x}$ (f) $\sin^{-1} x$
9. Let $y = x^{1/5}$. Using the fact that $x^{1/5}$ is the inverse of the function x^5, show that $dy/dx = \frac{1}{5}x^{-4/5}$.
10. This table records the values of f, g, f', and g' at $x = 1$ and $x = 2$.

x	$f(x)$	$g(x)$	$f'(x)$	$g'(x)$
1	2	2	4	5
2	5	7	9	3

Using the data in the table, determine, if possible
(a) $(f + g)'$ at $x = 2$; (b) $(f/g)'$ at $x = 1$; (c) $(f \circ g)'$ at $x = 1$; (d) $(g \circ f)'$ at $x = 1$; (e) $(f^2)'$ at $x = 2$.
11. Prove that (a) $D(\sin^{-1} x) = 1/\sqrt{1 - x^2}$; (b) $D(e^x) = e^x$;
(c) $D(\tan^{-1} x) = \dfrac{1}{1 + x^2}$.

Review exercises

Differentiate each of the functions in Exercises 1 to 23.

1. $6x^2 + 3x - 1$
2. $(x^2 + 1) \sin 2x$
3. $\dfrac{\sin^2 x}{\cos x}$
4. $\left(\dfrac{x^3 - 2x}{x^4}\right)^2$
5. $e^{-3x} \cos 2x$
6. $(2x + 1)^5$
7. (a) $\ln [x^3(1 - 2x)^5]$, (b) $\ln \sqrt[3]{\cos x}$. (First simplify by using the laws of logarithms.)

8. $\dfrac{1}{\sqrt{x}}$

9. $\dfrac{1}{3} \tan^{-1} \dfrac{x}{3}$

10. $\sqrt[3]{2x}$
11. $\sqrt{\sin x}$
12. $\cos \sqrt{x}$
13. $\ln (x + \sqrt{x^2 + 1})$

14. $x\sqrt{1 + 3x^2}$

15. $\dfrac{\sin^3 2x}{x^2 + x}$

16. $e^{-x} \tan x^2$
17. $e^{-x} \tan^{-1} x^2$
18. $\ln |\sec x + \tan x|$
19. $(e^{x^2})^3$
20. $5^{\cos x}$
21. $\sin^{-1} 2x$
22. $\csc 3x$
23. $\log_{10} (x^2 + 1)$

In Exercises 24 to 49, a, b, and c are constants. In each case verify that the derivative of the function in the left-hand column is the function in the right-hand column.

f	f'
24. $\frac{1}{2}[x\sqrt{x^2 + a^2} + a^2 \ln (x + \sqrt{x^2 + a^2})]$	$\sqrt{x^2 + a^2}$
25. $\dfrac{1}{2}\left(x\sqrt{a^2 - x^2} + a^2 \sin^{-1} \dfrac{x}{a}\right)$	$\sqrt{a^2 - x^2}$
26. $\ln (x + \sqrt{x^2 + a^2})$	$1/\sqrt{x^2 + a^2}$
27. $\frac{1}{2}[x\sqrt{x^2 - a^2} - a^2 \ln (x + \sqrt{x^2 - a^2})]$	$\sqrt{x^2 - a^2}$
28. $-\dfrac{1}{a} \sin^{-1} \dfrac{a}{x}$	$\dfrac{1}{x\sqrt{x^2 - a^2}}$
29. $\dfrac{1}{2a} \ln \dfrac{a + x}{a - x}$	$\dfrac{1}{a^2 - x^2}$
30. $\dfrac{1}{a} \tan^{-1} \dfrac{x}{a}$	$\dfrac{1}{a^2 + x^2}$
31. $\dfrac{x}{2}\sqrt{ax^2 + c} + \left(\dfrac{c}{2\sqrt{a}}\right) \ln (x\sqrt{a} + \sqrt{ax^2 + c})$	$\sqrt{ax^2 + c}$ $\quad a > 0$
32. $\dfrac{x}{2}\sqrt{ax^2 + c} + \dfrac{c}{2\sqrt{-a}} \sin^{-1} \left(x\sqrt{\dfrac{-a}{c}}\right)$	$\sqrt{ax^2 + c}$ $\quad a < 0$
33. $\sqrt{ax^2 + c} + \sqrt{c} \ln \dfrac{\sqrt{ax^2 + c} - \sqrt{c}}{x}$	$\dfrac{\sqrt{ax^2 + c}}{x}$ $\quad c > 0$
34. $\sqrt{ax^2 + c} - \sqrt{-c} \tan^{-1} \dfrac{\sqrt{ax^2 + c}}{\sqrt{-c}}$	$\dfrac{\sqrt{ax^2 + c}}{x}$ $\quad c < 0$

f	f'
35. $\dfrac{1}{\sqrt{b^2-4ac}}\ln\dfrac{2ax+b-\sqrt{b^2-4ac}}{2ax+b+\sqrt{b^2-4ac}}$	$\dfrac{1}{ax^2+bx+c}\qquad b^2>4ac$
36. $\dfrac{2}{\sqrt{4ac-b^2}}\tan^{-1}\dfrac{2ax+b}{\sqrt{4ac-b^2}}$	$\dfrac{1}{ax^2+bx+c}\qquad b^2<4ac$
37. $-\dfrac{2}{2ax+b}$	$\dfrac{1}{ax^2+bx+c}\qquad b^2=4ac$
38. $\dfrac{x}{2}-\dfrac{\sin 2ax}{4a}$	$\sin^2 ax$
39. $-\dfrac{1}{a}\cos ax+\dfrac{1}{3a}\cos^3 ax$	$\sin^3 ax$
40. $-\dfrac{1}{a}\ln\cos ax$	$\tan ax$
41. $\dfrac{x}{8}-\dfrac{\sin 4ax}{32a}$	$\sin^2 ax\cos^2 ax$
42. $\dfrac{1}{a}\tan ax-x$	$\tan^2 ax$
43. $\dfrac{1}{2a}\tan^2 ax+\dfrac{1}{a}\ln\cos ax$	$\tan^3 ax$
44. $-\dfrac{1}{2a}\cot^2 ax-\dfrac{1}{a}\ln\sin ax$	$\cot^3 ax$
45. $\dfrac{1}{a^2}\sin ax-\dfrac{1}{a}x\cos ax$	$x\sin ax$
46. $\dfrac{x^2}{4}-\dfrac{x\sin 2ax}{4a}-\dfrac{\cos 2ax}{8a^2}$	$x\sin^2 ax$
47. $\dfrac{1}{ab}[ax-\ln(b+ce^{ax})]$	$\dfrac{1}{b+ce^{ax}}$
48. $x(\ln ax)^2-2x\ln ax+2x$	$(\ln ax)^2$
49. $\dfrac{x^2}{2}\sin^{-1}ax-\dfrac{1}{4a^2}\sin^{-1}ax+\dfrac{x}{4a}\sqrt{1-a^2x^2}$	$x\sin^{-1}ax$

50. Show that $\sec^{-1}x=\cos^{-1}(1/x)$. (This is sometimes used as a definition for $\sec^{-1}x$.)

51. Let $f(x)=e^x+e^{-x}$. Find $(f^{-1})'$ at $e+e^{-1}$.

Express each limit in Exercises 52 to 55 as a derivative, and evaluate.

52. $\displaystyle\lim_{x_1\to 2}\dfrac{(1+x_1{}^2)^3-125}{x_1-2}$

53. $\displaystyle\lim_{x_1\to 3}\dfrac{\ln(1+2x_1)-\ln 7}{x_1-3}$

54. $\displaystyle\lim_{h\to 0}\dfrac{e^{(3+h)^2}-e^9}{h}$

55. $\displaystyle\lim_{\Delta x\to 0}\dfrac{\sin\sqrt{3+\Delta x}-\sin\sqrt{3}}{\Delta x}$

56. Show that the derivative of $2^{3x}5^{7x}$ is proportional to $2^{3x}5^{7x}$.

57. Give an example of a function whose derivative is (a) $1/x^3$, (b) $1/x^2$, (c) $1/x$.

58. Give an example of a function whose derivative is (a) e^x, (b) 10^x, (c) $1/(1 + x^2)$, (d) $x/(1 + x^2)$.

59. In each of these functions, y denotes a differentiable function of x. Express the derivative of each with respect to x in terms of y and dy/dx. (a) y^3, (b) $\cos y$, (c) e^y, (d) $1/y$.

60. (a) Compute the derivative of $y = \sqrt{9 + x^2}$ at $x = 4$.
(b) Find the slope of the tangent line to the curve $y = \sqrt{9 + x^2}$ at $(4, 5)$.
(c) Find the velocity of a particle at time $t = 4$ whose position on a line at time t is $y = \sqrt{9 + t^2}$.

61. Find $\lim\limits_{h \to 0} \dfrac{f(x + h) - f(x)}{h}$ at $x = 9$ if $f(x) = \tan^{-1}\sqrt{x}$.

62. Find $\lim\limits_{x_1 \to 2} \dfrac{f(x_1) - f(2)}{x_1 - 2}$ if $f(x)$ equals $\sqrt{1 + 4x}$.

63. Below is the graph of a function f. The function f has an inverse, which we shall call g. Estimate: (a) $f(1)$, (b) $f(3)$, (c) $g(1)$, (d) $g(2)$, (e) $g(3)$.

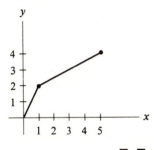

64. Let f be defined for all positive x. Assume that, for any positive x and y, $f(x/y) = f(x) - f(y)$. Also, assume that f has a derivative equal to 1 at $x = 1$. Prove that
(a) $f(1) = 0$;
(b) $f(1/x) = -f(x)$;
(c) $f(xy) = f(x) + f(y)$;
(d) f has a derivative for any positive x and it is equal to $1/x$.
(e) What familiar function f satisfies the equation $f(x/y) = f(x) - f(y)$?

65. Define $f(x)$ to be $x^2 \sin(1/x)$ if $x \neq 0$, and $f(0)$ to be 0.
(a) Show that f has a derivative at 0, namely 0.
$\left[\text{Investigate } \lim\limits_{h \to 0} \dfrac{f(h) - f(0)}{h}.\right]$
(b) Show that f has a derivative at $x \neq 0$.
(c) Show that the derivative of f is not continuous at $x = 0$.

66. Tell what is wrong with this alleged proof that $2 = 1$; Observe that $x^2 = x \cdot x = x + x + \cdots + x$ (x times). Differentiation with respect to x yields the equation $2x = 1 + 1 + \cdots + 1$ (x 1s). Thus $2x = x$. Setting $x = 1$ shows that $2 = 1$.

67. Show that the quotient $\log_2 3 / \log_2 7$ is irrational.

68. The graph below shows the tangent line to the curve $y = e^x$ at a point (x_0, y_0). Find the length of AB.

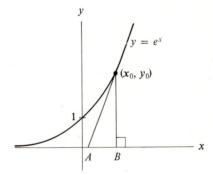

6

APPLICATIONS OF THE DERIVATIVE

In Sec. 1.1 the problem of finding the speed of a falling rock introduced the concept of the derivative informally. After the development of functions in Chap. 2, the derivative was formally defined in Chap. 3. Velocity, slope, magnification, and density are simply different interpretations of the derivative of a function, defined as the limit of a certain quotient. With the aid of certain limits developed in Chap. 4, methods were obtained in Chap. 5 for differentiating the "elementary" functions, built up by addition, subtraction, multiplication, division, and composition from the functions x^n, exponentials, logarithms, and the trigonometric functions. These tools will now be applied in a variety of ways, for instance, in graphing functions, in finding the maximum or minimum of a function, and in studying growth and decay in nature.

6.1

Rolle's theorem

The next two sections concern a fundamental property of the derivative, called the *law of the mean*, which will be the basis of much of this chapter. Setting the background for the law of the mean are several other theorems; some of these will also be needed later for other purposes.

The first theorem, whose proof is part of an advanced calculus course, says that a continuous function over an interval $[a, b]$ takes on a largest value and a smallest value.

MAXIMUM-VALUE THEOREM   Let f be continuous throughout $[a, b]$. Then there is at least one number X in $[a, b]$ for which f takes on a maximum value; that is, for some number X in $[a, b]$.

$$f(X) \geq f(x) \qquad \text{for all } x \text{ in } [a, b].$$

Similarly, f takes on a minimum value somewhere in the interval.

To persuade yourself that this theorem is plausible, imagine sketching the graph of a continuous function. As your pencil moves along the graph from some point on the graph to some other point on the graph, it passes through a highest point and also through a lowest point.

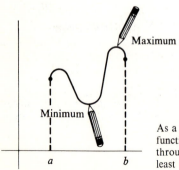

As a pencil runs along the graph of a continuous function from one point to another, it passes through at least one maximum point and at least one minimum point

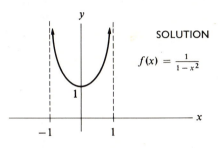

$f(x) = x^2$

EXAMPLE 1 Let $f(x) = x^2$ and $[a, b] = [-2, 2]$. Find all numbers X whose existence is guaranteed by the maximum-value theorem.

SOLUTION The maximum value of $f(x)$ for x in $[-2, 2]$ is 4. Either 2 or -2 serves as an X where the function takes on its maximum value, as the accompanying graph illustrates. ●

The next example shows the importance of the closed interval $[a, b]$ in the maximum-value theorem, namely, by an illustration of a function which is continuous throughout an open interval, yet has no maximum value in that open interval.

EXAMPLE 2 Let $f(x) = 1/(1 - x^2)$ and (a, b) be the open interval $(-1, 1)$. Show that f does not have a maximum value for x in (a, b).

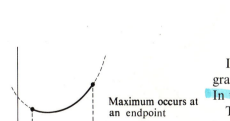

$f(x) = \dfrac{1}{1 - x^2}$

SOLUTION For x near 1, $f(x)$ gets arbitrarily large since the denominator $1 - x^2$ is close to 0. The graph of f, for x in $(-1, 1)$, looks like the drawing in the margin. This function is continuous throughout the open interval $(-1, 1)$, but there is no number X in $(-1, 1)$ at which f takes on a maximum value. ●

Now let us assume that the function f is differentiable. What can be said about $f'(X)$ if X is a number in $[a, b]$, where the function f takes its maximum value?

First, if X is neither a nor b, that is, X is in the open interval (a, b), the maximum would appear as in the figure below, left. It seems likely that a tangent to the graph at $(X, f(X))$ would be parallel to the x axis, in which case,

$$f'(X) = 0.$$

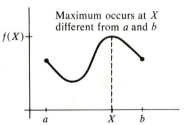

Maximum occurs at X
different from a and b

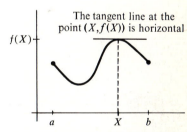

The tangent line at the point $(X, f(X))$ is horizontal

If the maximum occurs at an endpoint of the interval, a or b, as the graph in the margin shows, the derivative at such a point need not be 0. In this graph the maximum occurs at $X = b$, where $f'(X)$ is not 0.

The case in which the maximum occurs away from the ends of the interval is so important that we state it as a theorem and prove it without use of any pictures.

Maximum occurs at
an endpoint

THEOREM 1 Let f be a function defined on the closed interval $[a, b]$. If f takes on a maximum value at a number X in the open interval (a, b) and if $f'(X)$ exists, then
$$f'(X) = 0.$$

(The same holds for *minimum value*.)

PROOF We shall prove that $f'(X) \leq 0$ and that $f'(X) \geq 0$. From this it follows that $f'(X)$ must be 0.

Consider the quotient

$$\frac{f(X + h) - f(X)}{h}$$

used in defining $f'(X)$. Take h so small that $X + h$ is in the interval $[a, b]$. Since $f(X)$ is the maximum value of $f(x)$ for x in $[a, b]$,

$$f(X + h) \leq f(X).$$

Hence
$$f(X + h) - f(X) \leq 0.$$

Therefore, when h is positive,

$$\frac{f(X + h) - f(X)}{h} \quad \text{is negative or 0.}$$

Consequently, as $h \to 0$ through positive values,

$$\frac{f(X + h) - f(X)}{h},$$

being negative or 0, cannot approach a positive number. Thus

$$f'(X) = \lim_{h \to 0} \frac{f(X + h) - f(X)}{h} \leq 0.$$

If, on the other hand, h is negative, then the denominator of

$$\frac{f(X + h) - f(X)}{h}$$

is negative, and the numerator is still ≤ 0. Hence, for negative h,

$$\frac{f(X + h) - f(X)}{h} \geq 0,$$

(the quotient of two negative numbers being positive). Thus as $h \to 0$ through negative values, the quotient approaches a number ≥ 0. Hence $f'(X) \geq 0$.

Since $0 \leq f'(X) \leq 0$, $f'(X)$ must be 0, and the theorem is proved. ●

In Sec. 6.5 this theorem provides a method for finding the maximum of a function. It will be used in this section to establish another property of derivatives known as Rolle's theorem, named after a seventeenth-century mathematician.

DEFINITION *Chord of f.* If f is a function, then a line segment joining two points on the graph of f is called *a chord of f*.

Assume that a certain differentiable function f has a chord parallel to the x axis, as in the first diagram, next page. It seems reasonable that the graph will

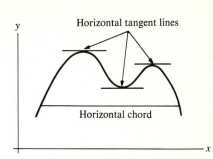

then have at least one horizontal tangent line. (In the case shown, there are three such lines tangent to the graph.) That this is no coincidence is the substance of the next theorem.

THEOREM 2   *Rolle's theorem.*   Let f be a continuous function on the closed interval $[a, b]$ and have a derivative at all x in the open interval (a, b). If $f(a) = f(b)$, then there is at least one number X in (a, b) such that $f'(X) = 0$.

PROOF   Since f is continuous, it has a maximum value M and a minimum value m for x in $[a, b]$. Certainly $m \leq M$.

   If $m = M$, f is constant, and $f'(x) = 0$ for all x in $[a, b]$. Then, any number x in (a, b) will serve as the desired X.

   If $m < M$, then the minimum and maximum cannot both occur at the ends of the interval a and b, since $f(a) = f(b)$. One of them, at least, occurs at a number X, $a < X < b$. And at that X, $f'(X)$ is 0. This proves Rolle's theorem. ●

EXAMPLE 3   Verify Rolle's theorem for the case $f(x) = x^2 - 2x + 5$ and $[a, b] = [0, 2]$.

SOLUTION   Note that $f(0) = 5 = f(2)$. Also, f is continuous and f' exists (even at a and b though this is not necessary to apply Rolle's theorem). According to Rolle's theorem, there is an X in $(0, 2)$ for which $f'(X) = 0$. It is easy to find such an X for this function, since $f'(x) = 2x - 2$. Setting $2x - 2 = 0$, we see that $X = 1$ (in this case X is unique). ●

EXAMPLE 4   Verify Rolle's theorem for the case $f(x) = \sqrt{1 - x^2}$ and $[a, b] = [-1, 1]$.

SOLUTION   Observe that $f(-1) = 0 = f(1)$, that f is continuous, and that $f'(x) = -x/\sqrt{1 - x^2}$, which is defined for all x in $(-1, 1)$. Rolle's theorem then guarantees that there is at least one number X in $(-1, 1)$ such that $f'(X) = 0$. We can find X by setting the derivative of f equal to 0:

$$\frac{-x}{\sqrt{1 - x^2}} = 0.$$

Thus $X = 0$ (and again happens to be unique). ●

The interpretation of the derivative as slope has led us to Rolle's theorem. The next example considers Rolle's theorem in terms of a particle moving on a line.

EXAMPLE 5 A particle moving on the y axis has the y coordinate $y = f(t)$ at time t, where f is a differentiable function. Initially, at time $t = a$, it has the coordinate $f(a)$. During a period of time it moves about on the line and finally, at time $t = b$, returns to its initial position, that is, $f(b) = f(a)$. What does Rolle's theorem assert for this case?

SOLUTION Rolle's theorem allows us to conclude that at some time t during the period of motion, the velocity $f'(t)$ had to be 0.

This is reasonable, for if the velocity were always positive, the particle would always be rising on the y axis and could not return to its initial position. Similarly, the velocity could not always be negative, or the particle would be falling all the time. When the particle shifts from a positive velocity to a negative velocity, say at the highest point in its journey, it momentarily has a velocity of 0. ●

The next example shows that it is necessary to assume in Rolle's theorem that f is differentiable throughout (a, b).

EXAMPLE 6 Can Rolle's theorem be applied to $f(x) = |x|$ in the interval $[a, b] = [-2, 2]$?

SOLUTION First, $f(-2) = 2 = f(2)$; second, f is continuous. The function, however, fails to be differentiable at 0. (See Example 3 in Sec. 3.3.) Not all of the hypotheses of Rolle's theorem hold; thus there is no need for the conclusion to hold. Indeed, there is no number X such that $f'(X) = 0$, as a glance at the graph of f in the margin shows. ●

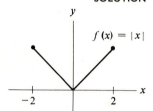

The final example illustrates Rolle's theorem with a function involving an exponential.

EXAMPLE 7 Verify Rolle's theorem for $f(x) = (x - x^2)e^{-x}$ in the interval $[0, 1]$.

SOLUTION Note that $f(0) = 0 = f(1)$ and that f is differentiable for all x, hence continuous. All the hypotheses of Rolle's theorem are satisfied. The theorem then asserts that there must be at least one number X in the open interval $(0, 1)$ such that $f'(X) = 0$.

To find X set $f'(x) = 0$:

$$(x - x^2)(-e^{-x}) + e^{-x}(1 - 2x) = 0$$

$$e^{-x}(-x + x^2 + 1 - 2x) = 0$$

$$x^2 - 3x + 1 = 0 \qquad \text{since } e^{-x} \text{ is never 0.}$$

$$x = \frac{3 \pm \sqrt{5}}{2} \qquad \text{by the quadratic formula.}$$

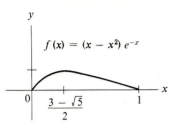

The solution with the plus sign lies outside the interval $(0, 1)$. The only number X in this case is

$$\frac{3 - \sqrt{5}}{2} \doteq \frac{3 - 2.24}{2}$$

$$= 0.38.$$

This result can be checked against the graph in the margin. ●

Exercises

1. Let $f(x) = \cos x$ for $0 \le x \le 2\pi$. Observe that $f(0) = f(2\pi)$.
(a) What does Rolle's theorem assert in this case?
(b) Find all values X that satisfy the conclusion of Rolle's theorem.

2. Let $f(x) = x^3 - x + 2$ for x in $[-1, 1]$. Observe that $f(-1) = f(1)$.
(a) What does Rolle's theorem assert in this case?
(b) Find all values X that satisfy the conclusion of Rolle's theorem.

3. Let $f(x) = x^4 - 2x^2 + 1$ for x in $[-2, 2]$. Observe that $f(-2) = f(2)$. Find all X guaranteed by Rolle's theorem.

4. (a) What is the maximum value of $\sin x$ for x in $[0, 4\pi]$?
(b) At which X in $[0, 4\pi]$ does the maximum occur?
(c) Verify Theorem 1 for the X's in (b).

5. (a) What is the maximum value of $\cos x$ for x in $[\pi/4, 7\pi/4]$?
(b) At which X in $[\pi/4, 7\pi/4]$ does the maximum occur?
(c) Does the derivative of the cosine function equal 0 at the X's in (b)?
(d) Compare (c) with Theorem 1. Is Theorem 1 false for the cosine function?

6. Let $f(x) = 5 + (x - x^2)e^x$.
(a) Show that $f(0) = f(1)$.
(b) Find all numbers X in $(0, 1)$ whose existence is guaranteed by Rolle's theorem.

7. At time t seconds a thrown ball has the height $f(t) = -16t^2 + 32t + 40$ feet.
(a) Show that after 2 seconds it returns to its initial height $f(0)$.
(b) What does Rolle's theorem imply about the velocity of the ball?
(c) Verify Rolle's theorem in this case by computing the number that it asserts exists.

8. Consider the function f given by the formula $f(x) = x^3 - 3x$.
(a) At which numbers x is $f'(x) = 0$?

(b) Use Theorem 1 to show that the maximum value of $x^3 - 3x$ for x in $[1, 5]$ occurs either at 1 or at 5.
(c) What is the maximum value of $x^3 - 3x$ for x in $[1, 5]$?

9. Let $f(x) = \sqrt[3]{x^2}$. Note that $f(1) = f(-1)$. Is there a number X in $(-1, 1)$ such that $f'(X) = 0$? If so, find it. If not, state which assumption in Rolle's theorem is not satisfied.

10. Differentiate (for practice):
(a) $x \tan^{-1} 5x - \frac{1}{10} \ln (1 + 25x^2)$
(b) xe^{-x^2}
(c) $\sqrt[3]{5 + 2x}$
(d) $x \sec^{-1} 2x - \frac{1}{2} \ln (2x + \sqrt{4x^2 - 1})$
(e) $\dfrac{\sin 3x}{\sqrt{1 + 4x}}$
(f) $\sqrt{1 + 3 \cos^2 5x}$

11. Let $f(x) = (x^2 - 4)/(x + \frac{1}{2})$. Observe that $f(2) = f(-2)$.
(a) What does Rolle's theorem say about f?
(b) For which values of x is $f'(x) = 0$?

12. Let $f(x) = e^{-x} \sin x$. Note that $f(0) = f(2\pi)$. Find all X in $(0, 2\pi)$ such that $f'(X) = 0$.

∎

13. Theorem 1 was proved for a maximum of f. Prove it for a minimum.

14. (a) Does the function x^2 have a maximum value when considered only on the open interval $(1, 2)$?
(b) Does the function x^2 have a minimum value when considered on the open interval $(-1, 2)$? On the open interval $(1, 2)$?

∎ ∎

15. Define a function f with domain the nonnegative numbers as follows. Let x be a nonnegative number. Express x as a nonnegative integer N plus a number $y, 0 \le y < 1$. Define $f(x)$ to be y. For instance, $f(3.5) = 0.5, f(4) = 0$, and $f(2.9) = 0.9$.

(a) Graph f.

(b) Does f have a maximum value for x in the closed interval [0, 1]?

(c) Does Theorem 1 apply to f?

16. Show that between any two roots of a polynomial there is at least one root of its derivative.

17. (a) Assume that every polynomial of degree 5 has at most five real roots. Use Rolle's theorem to prove that every polynomial of degree 6 has at most six real roots. *Hint:* Use Exercise 16.

(b) Use the technique of (a) to prove that a polynomial of degree n has at most n real roots.

18. Let $f(x) = x \sin x$.

(a) Show that $f(0) = f(\pi)$.

(b) Use Rolle's theorem to show that there is a number X in $(0, \pi)$ such that $\tan X = -X$.

(c) Graph the equations $y = -x$ and $y = \tan x$ to determine how many such numbers X there are. (Consider points where the two graphs intersect.)

19. Show that the equation $x^5 + 2x^3 - 2 = 0$ has exactly one solution in the interval [0, 1]. (Why does it have at least one? Use Rolle's theorem to show that there is at most one.)

20. Show that the equation $3 \tan x + x^3 = 2$ has exactly one solution in the interval $[0, \pi/4]$.

6.2

The law of the mean

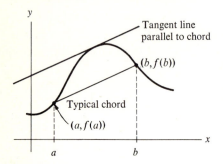

Tangent line parallel to chord

$(b, f(b))$

Typical chord

$(a, f(a))$

Rolle's theorem in Sec. 6.1 asserts that, if the graph of a function has a horizontal chord, then it has a tangent line parallel to that chord. The law of the mean is a generalization of Rolle's theorem, since it concerns any chord of f, not just horizontal chords.

In geometric terms the theorem asserts that, if you draw a chord for the graph of a well-behaved function (as in the accompanying figure), then somewhere above or below that chord the graph has at least one tangent line parallel to the chord.

Let us translate this geometric statement into the language of functions. Call the ends of the chord $(a, f(a))$ and $(b, f(b))$. The slope of the chord is then

$$\frac{f(b) - f(a)}{b - a},$$

while the slope of the tangent line at a typical point $(x, f(x))$ on the graph is

$$f'(x).$$

The law of the mean then asserts there is at least one number X in the open interval (a, b) such that

$$f'(X) = \frac{f(b) - f(a)}{b - a}.$$

THEOREM   *Law of the mean.* Let f be a continuous function on the closed interval $[a, b]$ and have a derivative at every x in the open interval (a, b). Then there is at least one number X in the open interval (a, b) such that

$$f'(X) = \frac{f(b) - f(a)}{b - a}.$$

We defer the proof of the law of the mean to the end of this section in order to emphasize first its several important consequences.

EXAMPLE 1 Verify the law of the mean for $f(x) = 2x^3 - 8x + 1$, $a = 1$, and $b = 3$.

SOLUTION
$$f(a) = f(1) = 2(1)^3 - 8(1) + 1 = -5$$
and
$$f(b) = f(3) = 2(3)^3 - 8(3) + 1 = 31.$$

According to the law of the mean there is at least one number X between $a = 1$ and $b = 3$ such that

$$f'(X) = \frac{31 - (-5)}{3 - 1} = \frac{36}{2} = 18.$$

Let us find X explicitly. Since $f'(x) = 6x^2 - 8$, we need to solve the equation

$$6x^2 - 8 = 18,$$

that is,
$$6x^2 = 26$$

$$x^2 = \frac{26}{6}.$$

The solutions are $\sqrt{\frac{13}{3}}$ and $-\sqrt{\frac{13}{3}}$. But only $\sqrt{\frac{13}{3}}$ is in $(1, 3)$. Hence there is only one number, namely $\sqrt{13/3}$, that serves as the X whose existence is guaranteed by the law of the mean. ●

The interpretation of the derivative as slope suggested the law of the mean. What does the law of the mean say when the derivative is interpreted, say, as velocity? This question is considered in Example 2.

EXAMPLE 2 A car moving on the x axis has the x coordinate $f(t)$ at time t. At time a its position is $f(a)$. At some later time b its position is $f(b)$. What does the law of the mean assert for this car?

SOLUTION
The quotient
$$\frac{f(b) - f(a)}{b - a}$$

equals
$$\frac{\text{Change in position}}{\text{Change in time}},$$

or "average velocity" for the interval of time $[a, b]$. The law of the mean asserts that at some time during this period the velocity of the car must equal its average velocity. To be specific, if a car travels 210 miles in 3 hours, then at some time its speedometer must read 70 miles per hour. ●

There are several ways of writing the law of the mean. For example, the equation

$$f'(X) = \frac{f(b) - f(a)}{b - a}$$

is equivalent to $\qquad f(b) - f(a) = (b - a)f'(X),$

hence to $\qquad f(b) = f(a) + (b - a)f'(X).$

In this form, the law of the mean asserts that $f(b)$ is equal to $f(a)$ plus a quantity that involves the derivative f'. The following important corollaries exploit this alternative view of the law of the mean.

COROLLARY I   If the derivative of a function is 0 throughout an interval, then the function is constant throughout that interval.

PROOF   Let a and b be any two numbers in the interval and let the function be denoted by f. To prove the corollary, it suffices to prove that

$$f(b) = f(a).$$

By the law of the mean, there is a number X between a and b such that

$$f(b) = f(a) + (b - a)f'(X).$$

But $f'(X) = 0$, since $f'(x)$ is 0 for all x in the given interval.   Hence

$$f(b) = f(a) + (b - a)(0),$$

which proves that $\qquad f(b) = f(a).$ ●

When Corollary 1 is interpreted in terms of motion, it is quite plausible.   It asserts that, if a particle has zero velocity for a period of time, then it does not move during that time.

EXAMPLE 3   Use Corollary 1 to show that $f(x) = \cos^2 3x + \sin^2 3x$ is a constant.   Find the constant.

SOLUTION $f'(x) = -6 \cos 3x \sin 3x + 6 \sin 3x \cos 3x = 0$.   Corollary 1 says that f is constant.   To find the constant, just evaluate f at some specific number, say at 0.   We have $f(0) = \cos^2 (3 \cdot 0) + \sin^2 (3 \cdot 0) = \cos^2 0 + \sin^2 0 = 1$.   Thus

$$\cos^2 3x + \sin^2 3x = 1$$

for all x.   This should be no surprise since, by the pythagorean theorem, $\cos^2 \theta + \sin^2 \theta = 1$. ●

COROLLARY 2   If two functions have the same derivatives throughout an interval, then they differ by a constant.   That is, if $f'(x) = g'(x)$ for all x in an interval, then there is a constant C such that $f(x) = g(x) + C$.

PROOF   Define a third function h by the equation

$$h(x) = f(x) - g(x).$$

Then $\qquad h'(x) = f'(x) - g'(x) = 0.$

Since the derivative of h is 0, Corollary 1 implies that h is constant, that is,

$$h(x) = C$$

for some fixed number C. Thus

$$f(x) - g(x) = C$$

or

$$f(x) = g(x) + C,$$

and the corollary is proved. ●

If two graphs have parallel tangent lines at all pairs of points with the same x coordinate, then one graph is obtainable from the other by raising or lowering it

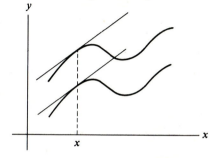

Is Corollary 2 plausible when the derivative is interpreted as slope? In this case, the corollary asserts that, if the graphs of two functions have the property that their tangent lines at points with the same x coordinate are parallel, then one graph can be obtained from the other by raising (or lowering) it by an amount C. If you sketch two such graphs (as in the accompanying diagram), you will see that the corollary is reasonable.

EXAMPLE 4   What functions have a derivative equal to $2x$ everywhere?

SOLUTION   One such function is x^2; another is $x^2 + 25$. For any constant C, $D(x^2 + C) = 2x$. Are there any other possibilities? Corollary 2 tells us there are not. For if f is a function such that $f'(x) = 2x$, then $f'(x) = (x^2)'$ for all x. Thus the functions f and x^2 differ by a constant, say C, that is,

$$f(x) = x^2 + C.$$

The only functions whose derivative is $2x$ are of the form $x^2 + C$. ●

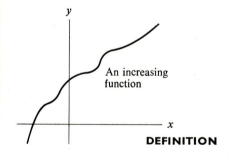

An increasing function

Corollary 1 asserts that if $f'(x) = 0$ for all x, then f is constant. What can be said about f if $f'(x)$ is *positive* for all x? In terms of the graph of f, this assumption implies that all the tangent lines slope upward. It is reasonable to expect that as we move from left to right on the graph, the y coordinate increases. In Corollary 3 this is proved.

DEFINITION   *Increasing (decreasing) function.* If $f(x_1) > f(x_2)$ whenever $x_1 > x_2$, for all such choices of x_1 and x_2 in an interval, then f is an *increasing* function in that interval. [If $f(x_1) < f(x_2)$ whenever $x_1 > x_2$, for all such choices of x_1 and x_2 in an interval, then f is a *decreasing* function in that interval.]

COROLLARY 3   If f is continuous on $[a, b]$ and has a positive derivative on the open interval (a, b), then f is increasing on the interval $[a, b]$.

PROOF   Take two numbers x_1 and x_2 such that

$$a \le x_2 < x_1 \le b.$$

By the law of the mean,

$$f(x_1) = f(x_2) + (x_1 - x_2)f'(X)$$

for some X between x_1 and x_2. Since $x_1 - x_2$ is positive, and since $f'(X)$ is assumed to be positive, it follows that

$$(x_1 - x_2)f'(X) > 0.$$

Thus $f(x_1) > f(x_2)$, and the corollary is proved. (The "decreasing" case is proved similarly.) ●

The next example applies Corollary 3.

EXAMPLE 5 Use Corollary 3 to show that $e^x > 1 + x$ if $x > 0$.

SOLUTION Consider the function f defined as follows:

$$f(x) = e^x - (1 + x)$$
$$= e^x - 1 - x.$$

The derivative of f is $\qquad f'(x) = e^x - 1.$

Since $e^x > 1$ when $x > 0$, f is an increasing function for $x \geq 0$ (that is, for any interval $[0, b]$). In particular, if $x > 0$, then $f(x) > f(0)$, that is,

$$e^x - (1 + x) > e^0 - (1 + 0) \qquad \text{if } x > 0.$$

Since $e^0 - (1 + 0) = 1 - 1 = 0$, this inequality asserts that

$$e^x - (1 + x) > 0$$

and therefore $\qquad\qquad e^x > 1 + x \qquad \text{if } x > 0$ ●

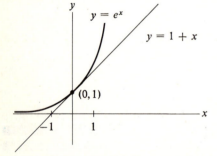

The inequality $e^x > 1 + x$ is easy to interpret in terms of the graphs of $y = e^x$ and $y = 1 + x$. The line $y = 1 + x$ is the tangent line to the curve $y = e^x$ at the point $(0, 1)$. [This is so because it passes through $(0, 1)$ and has a slope equal to 1, which is the derivative of e^x at $x = 0$.] As the graph in the margin suggests, e^x is larger than $1 + x$ for all $x \neq 0$. In the next section the inequality is needed only for the case $x > 0$.

Finally we present a proof of the law of the mean.

PROOF OF THE LAW OF THE MEAN

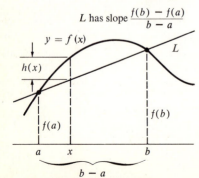

We shall prove the theorem by introducing a function to which Rolle's theorem can be applied. The chord through $(a, f(a))$ and $(b, f(b))$ is part of a line L, whose equation is, let us say, $y = g(x)$. Let $h(x) = f(x) - g(x)$, which represents the difference between the graph of f and the line L for a given x. It is clear from the picture that $h(a) = h(b)$, since both equal 0.

By Rolle's theorem, there is at least one number X in the open interval (a, b) such that

$$h'(X) = 0.$$

But $\qquad\qquad\qquad h'(X) = f'(X) - g'(X).$

Since $y = g(x)$ is the equation of the line through $(a, f(a))$ and $(b, f(b))$,

$$g'(x) = \frac{f(b) - f(a)}{b - a}$$

for all x.

Hence

$$0 = h'(X) = f'(X) - \frac{f(b) - f(a)}{b - a}.$$

In short

$$f'(X) = \frac{f(b) - f(a)}{b - a}$$

and the law of the mean is proved. ●

Exercises

In Exercises 1 to 7 find explicitly all values of X mentioned in the law of the mean for the given function and interval.

1. $f(x) = x^3 - x^2$, $a = -1$, $b = 2$.
2. $f(x) = x^3 - x$, $a = -1$, $b = 2$.
3. $f(x) = 5x$, $a = 1$, $b = 3$.
4. $f(x) = x^3$, $a = -2$, $b = 1$.
5. $f(x) = \cos x$, $a = 0$, $b = 4\pi$.
6. $f(x) = e^x$, $a = 2$, $b = 3$.
7. $f(x) = x \ln x$, $a = 1$, $b = e$.

The answers to Exercises 8 and 9 should be phrased in colloquial English. (Section 3.1 introduced the concepts of density and magnification.)

8. State the law of the mean in terms of density and mass. Let x be the distance from the left end of a string and $f(x)$ the mass of the string from 0 to x. When stated in these terms, does the law of the mean seem reasonable?

9. State the law of the mean in terms of a slide and a screen. Let x denote the position on the (linear) slide and $f(x)$ denote the position of the image on the screen. In optical terms, what does the law of the mean say?

10. Differentiate (for practice):

(a) $\sqrt{1 + x^5}$ (b) $\dfrac{\cos^2 3x}{\sin 2x}$

(c) $e^{-x} \sec 3x$ (d) $\sin^{-1} x^3$

(e) $\ln |\sec 2x|$.

11. Assume that f has a derivative for all x, that $f(3) = 7$, and that $f(8) = 17$. What can we conclude about f'?

12. Which of the corollaries to the law of the mean implies that (a) if two cars on a straight road have the same velocity at every instant, they remain a fixed distance apart? (b) if all the tangents to a curve are horizontal, the curve is a horizontal straight line? Explain in each case.

13. (a) Differentiate $\ln 3x$ and $\ln 2x$.
    (b) The derivatives in (a) are equal. According to Corollary 2, $\ln 3x$ differs from $\ln 2x$ by a constant. What is that constant?

14. (a) Differentiate $\cos^2 x$ and $-\sin^2 x$.
    (b) The derivatives in (a) are equal. According to Corollary 2, $\cos^2 x$ differs from $-\sin^2 x$ by a constant. What is that constant?

■

15. (a) Recall the definition of $g(x)$ in the proof of the law of the mean, and show that

$$g(x) = f(a) + \frac{x - a}{b - a}[f(b) - f(a)].$$

(b) Using (a), show that

$$g'(x) = \frac{f(b) - f(a)}{b - a}.$$

■ ■

16. Use the law of the mean to prove that $e^x > 1 + x$ for negative x.

Exercises 17 to 23 concern the notion of an increasing function.

17. Let f and g be two functions differentiable on (a, b) and continuous on $[a, b]$. Assume that $f(a) = g(a)$ and that $f'(x) < g'(x)$ for all x in (a, b). Prove that $f(b) < g(b)$.

18. (See Exercise 17.)
    (a) Show that $\tan x > x$ if $0 < x < \pi/2$.
    (b) What does (a) say about certain lengths related to the unit circle?

19. (See Exercise 17.) Show that, for $x > 0$,

$$\frac{x}{1 + x^2} < \tan^{-1} x < x.$$

20. (a) Use algebra to show that, for x in $(0, 1)$,

$$1 - x + x^2 - x^3 < \frac{1}{1+x} < 1 - x + x^2 - x^3 + x^4.$$

(b) (See Exercise 17.) Show that

$$x - \frac{x^2}{2} + \frac{x^3}{3} - \frac{x^4}{4}$$

$$< \ln(1+x) < x - \frac{x^2}{2} + \frac{x^3}{3} - \frac{x^4}{4} + \frac{x^5}{5}.$$

(c) Use (b) to estimate ln (1.2).

21. Generalize Exercise 20(b) to more summands.

22. (See Exercise 17.)

(a) Starting with the inequalities

$$0 < \cos x < 1,$$

which are valid for all x in $(0, \pi/2)$, obtain the inequalities

$$0 < \sin x < x$$

for the same interval

(b) Deduce from the inequalities obtained in (a) that

$$-1 < -\cos x < -1 + \frac{x^2}{2}$$

for x in $(0, \pi/2)$.

(c) Obtain, in order, the following inequalities for x in $(0, \pi/2)$:

$$-x < -\sin x < -x + \frac{x^3}{2 \cdot 3}$$

$$1 - \frac{x^2}{2} < \cos x < 1 - \frac{x^2}{2} + \frac{x^4}{4 \cdot 3 \cdot 2}$$

$$x - \frac{x^3}{3 \cdot 2} < \sin x < x - \frac{x^3}{3 \cdot 2} + \frac{x^5}{5 \cdot 4 \cdot 3 \cdot 2}.$$

23. Continuing Exercise 22, show that, for x in $(0, \pi/2)$,

(a) $1 - \dfrac{x^2}{2} + \dfrac{x^4}{4 \cdot 3 \cdot 2} - \dfrac{x^6}{6 \cdot 5 \cdot 4 \cdot 3 \cdot 2}$

$$< \cos x < 1 - \frac{x^2}{2} + \frac{x^4}{4 \cdot 3 \cdot 2}$$

(b) $x - \dfrac{x^3}{3 \cdot 2} + \dfrac{x^5}{5 \cdot 4 \cdot 3 \cdot 2} - \dfrac{x^7}{7 \cdot 6 \cdot 5 \cdot 4 \cdot 3 \cdot 2}$

$$< \sin x < x - \frac{x^3}{3 \cdot 2} + \frac{x^5}{5 \cdot 4 \cdot 3 \cdot 2}.$$

24. (a) Use the law of the mean to show that the function $f(x) = \sin x$ satisfies the inequality

$$|f(x_1) - f(x_2)| \le |x_1 - x_2|$$

for all numbers x_1 and x_2.

(b) Find all functions that satisfy the inequality

$$|f(x_1) - f(x_2)| \le |x_1 - x_2|^{1.01}$$

for all x_1 and x_2.

6.3

The relative sizes of e^x, x^n, and ln x

When x is large, so are 2^x and x^3, as the table indicates.

x	1	2	3	4	5	6	7	8	9	10	11	12
2^x	2	4	8	16	32	64	128	256	512	1,024	2,048	4,096
x^3	1	8	27	64	125	216	343	512	729	1,000	1,331	1,728

For $x = 2, 3, \ldots, 9$ the exponential 2^x is less than the power x^3. At $x = 10$, although they are almost equal, the exponential is now larger. By $x = 12$, the exponential is much larger than the power x^3. In view of the importance of the exponential functions b^x and the power functions x^n, the rate at which they increase in size for large x is worth investigating. Moreover, in Sec. 6.4 the growth of the exponential function will be of interest in the study of population size.

Recall from Example 5 in Sec. 6.2 that for $x > 0$

$$e^x > 1 + x.$$

The following theorem strengthens this result and is the basis of the present section.

THEOREM 1   If x is positive, then

$$e^x > 1 + x + \frac{x^2}{2}.$$

PROOF   Consider the function f defined by

$$f(x) = e^x - \left(1 + x + \frac{x^2}{2}\right).$$

Its derivative is $f'(x) = e^x - (1 + x).$

For $x > 0$ this derivative is positive, since $e^x > 1 + x$ for such x.   Hence f is an increasing function for $x \geq 0$ by Corollary 3 of Sec. 6.2.   In particular, if $x > 0$, it follows that

$$f(x) > f(0).$$

Hence $e^x - \left(1 + x + \dfrac{x^2}{2}\right) > e^0 - \left(1 + 0 + \dfrac{0^2}{2}\right) = 1 - 1 = 0$

and $e^x > 1 + x + \dfrac{x^2}{2}$

for $x > 0$.   This proves the theorem.   ●

All that will actually be needed from Theorem 1 in the proof of Theorem 2 is that e^x is larger than $x^2/2$ for positive x.

THEOREM 2 $$\lim_{x \to \infty} \frac{x}{e^x} = 0.$$

PROOF   Since e^x is larger than $x^2/2$, it follows that

$$\frac{x}{e^x} < \frac{x}{x^2/2} = \frac{2}{x}$$

for $x > 0$.   For large values of x, $2/x$ is a small positive number.   Since

$$0 < \frac{x}{e^x} < \frac{2}{x},$$

it follows that, as x gets large, x/e^x approaches 0, as the theorem claims.   ●

Note that though both the numerator and the denominator of x/e^x get large as $x \to \infty$, the denominator grows larger much faster, forcing the quotient toward 0.

Since e^x/x is the reciprocal of x/e^x, which approaches 0 as $x \to \infty$, it follows that

$$\lim_{x \to \infty} \frac{e^x}{x} = \infty.$$

The next theorem is simply a translation of Theorem 2 from exponentials to logarithms.

THEOREM 3
$$\lim_{x \to \infty} \frac{\ln x}{x} = 0.$$

PROOF   In the quotient
$$\frac{x}{e^x}$$

the numerator is the natural logarithm of the denominator. If e^x is denoted by t, the numerator is $\ln t$. Moreover, as $x \to \infty$, $t = \ln x \to \infty$. Thus

$$0 = \lim_{x \to \infty} \frac{x}{e^x} = \lim_{x \to \infty} \frac{\ln t}{t} = \lim_{t \to \infty} \frac{\ln t}{t}.$$

This proves the theorem. ●

Theorem 3 in turn is the basis of the next theorem, which asserts that any exponential b^x, $b > 1$, grows more rapidly than any fixed power of x. That is, b^x, for any fixed base greater than 1, grows more rapidly than x^a for any fixed exponent a.

THEOREM 4   Let b be greater than 1 and let a be a constant. Then

$$\lim_{x \to \infty} \frac{b^x}{x^a} = \infty.$$

PROOF   Let $y = b^x/x^a$. It will be easier to show that $\ln y$ becomes arbitrarily large as $x \to \infty$. Now,

$$\ln y = \ln \left(\frac{b^x}{x^a} \right)$$

$$= \ln b^x - \ln x^a$$

$$= x \ln b - a \ln x,$$

so
$$\ln y = x \left(\ln b - a \frac{\ln x}{x} \right).$$

Note that $\ln b$ is positive. Also, by Theorem 3, $\ln x/x \to 0$ as $x \to \infty$. Thus, when x is suitably large, $|(a \ln x)/x|$ is very close to 0 and will be less than, say, $\frac{1}{2} \ln b$. Thus, for suitably large x,

$$x \left(\ln b - a \frac{\ln x}{x} \right) > x(\ln b - \tfrac{1}{2} \ln b),$$

which is the same as

$$x \left(\ln b - a \frac{\ln x}{x} \right) > x \left(\frac{1}{2} \ln b \right),$$

or
$$\ln y > x(\tfrac{1}{2} \ln b).$$

Since $x(\frac{1}{2} \ln b) \to \infty$ as $x \to \infty$, it follows that $\ln y \to \infty$, hence $y \to \infty$. This proves the theorem. ●

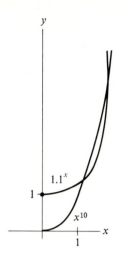

EXAMPLE   For large values of x, which is larger, $(1.1)^x$ or x^{10}?

SOLUTION   According to Theorem 4, with $b = 1.1$ and $a = 10$, $(1.1)^x$ becomes much larger than x^{10} for large x.   This may be surprising since, for small x, such as 2 or 3, x^{10} is much larger than $(1.1)^x$.   For instance,

$$2^{10} = 1{,}024 \qquad \text{while} \qquad (1.1)^2 = 1.21$$

and

$$3^{10} = 59{,}049 \qquad \text{while} \qquad (1.1)^3 = 1.331.$$

In spite of its slow start, $(1.1)^x$ overtakes x^{10} and eventually grows much more quickly.
The function $(1.1)^x$ catches up with x^{10} near $x = 685$.
The schematic graphs in the margin contrast the graphs of $y = (1.1)^x$ and $y = x^{10}$.   They cross twice for positive x, once just to the right of $x = 1$, and then again near 685. ●

Just as e^x grows faster than any fixed power of x, it turns out that $\ln x$ grows more slowly than any fixed power of x, that is,

$$\lim_{x \to \infty} \frac{\ln x}{x^a} = 0 \qquad a > 0.$$

See Exercise 21 for a proof.

Exercises

In Exercises 1 to 12 determine the given limits.

1. $\lim_{x \to \infty} x^3/e^x$

2. $\lim_{x \to \infty} x^5 e^{-x}$

3. $\lim_{x \to -\infty} x^5 e^{-x}$

4. $\lim_{x \to -\infty} x^4 e^{-x}$

5. $\lim_{x \to \infty} \dfrac{(1.01)^x}{x^{50}}$

6. $\lim_{x \to \infty} \dfrac{(\ln x)^2}{2^x}$

7. $\lim_{x \to \infty} \dfrac{\ln x}{x^3}$

8. $\lim_{x \to 0^+} \dfrac{\ln x}{x^3}$

9. $\lim_{x \to \infty} \dfrac{\log_{10} x}{x}$

10. $\lim_{x \to \infty} \dfrac{2^x}{e^x}$

11. $\lim_{x \to \infty} \dfrac{\log_2 x}{\log_3 x}$

12. $\lim_{x \to -\infty} 2^x e^{-x}$

13. Find $\lim_{x \to 0^+} x \ln x$.   *Hint:* Replace x by $1/t$.

14. Find $\lim_{x \to \infty} x^{1/x}$.   *Hint:* First find $\lim_{x \to \infty} \ln x^{1/x}$.

15. (a) Evaluate $(\log_{10} x)/x$ for $x = 10^2$, 10^3, 10^4.
(b) What is $\lim_{x \to \infty} (\log_{10} x)/x$?

16. Differentiate (for practice):

(a) $\dfrac{\sqrt{3x^2 - 1}}{e^{5x}}$

(b) $\sec \sqrt{x}$

(c) $(\cos 5x)^3$

(d) $\ln \sqrt[3]{(1 + \tan 2x)^2}$
(first simplify by laws of logarithms)

(e) $\sqrt{x} \tan^{-1} 2x$

(f) $x^3 \sin^{-1} e^{2x}$.

This exercise provides background experience for later work.

17. Give an example of a function whose derivative is:

(a) $\dfrac{1}{x^4}$

(b) $\sqrt[3]{x}$

(c) $\dfrac{3}{x}$

(d) $x^3 - 2x^2$

(e) $\dfrac{5}{1 + x^2}$

(f) $\dfrac{3x}{1 + x^2}$

(g) $\dfrac{1}{(1 + 3x)^2}$

(h) $\dfrac{1}{1 + 3x}$

(i) $\dfrac{1}{\sqrt{1 - 4x^2}}$.

■

18. This is a companion to Exercise 17 in Sec. 6.2.   Let f and g be two functions differentiable on (a, b) and continuous on $[a, b]$.   Assume that $f(b) = g(b)$ and that

$f'(x) < g'(x)$ for all x in (a, b). Prove that $f(a) > g(a)$.

19. See Exercise 18. Let $f(x) = (1 + x)^{1/x}$ for $x > -1$, $(x \neq 0)$, and $f(0) = e$. This function is graphed in Sec. 4.4. Show that $f(x)$ is a decreasing function. *Hint:* It may be simpler to show that $\ln f$ is a decreasing function. Be careful to consider the cases $x > 0$ and $-1 < x < 0$ separately.

20. Evaluate $\lim\limits_{x \to \infty} \dfrac{\ln x}{x^a}$ if $a < 0$.

21. Let a be a positive number. Introduce the variable $t = x^a$ to show that $\lim\limits_{x \to \infty} \dfrac{\ln x}{x^a} = 0$.

22. Prove that $\lim\limits_{x \to \infty} \dfrac{\log_b x}{x} = 0$ if $b > 0$.

23. Let $f(x) = 1/x$.
(a) Show that $f'(x)$ is negative for all x in the domain of f.
(b) If $x_1 > x_2$, is $f(x_1) > f(x_2)$?

Exercises 24 to 27 show how the law of the mean can be used to estimate the value of a function at some number if its value at a nearby number is known.

24. Show that, if $a \neq b$,

$$f(b) = f(a) + (b - a)f'(X)$$

for some number X between a and b.

Hint: Take the cases $b > a$ and $b < a$ separately. Both are an immediate consequence of the law of the mean.

25. Let $f(x) = \sqrt{x}$, $a = 100$ and $b = 105$ in Exercise 24.
(a) Deduce that

$$\sqrt{105} = 10 + \frac{5}{2\sqrt{X}} \quad \text{for some } X, \ 100 < X < 105.$$

(b) Since $100 < X < 121$, deduce that

$$10 + \frac{5}{2 \cdot 11} < \sqrt{105} < 10 + \frac{5}{2 \cdot 10}.$$

(c) From (b) deduce that

$$10.22 < \sqrt{105} < 10.25.$$

26. See Exercises 24 and 25. Use the method of Exercise 24 to estimate $\sqrt{67}$.

27. See Exercises 24 and 25. (a) Use the method of Exercise 25 to show that

$$\frac{\pi}{4} + \frac{0.1}{1 + (1.1)^2} < \tan^{-1} 1.1 < \frac{\pi}{4} + \frac{0.1}{1 + 1^2}.$$

(b) From (a) deduce that

$$0.835 < \tan^{-1} 1.1 < 0.836.$$

Exercises 28 to 33 develop a way of computing e^x for $0 \leq x \leq 1$. In Chap. 15 the result will be established for all x by a different method.

28. See the proof of Theorem 1. The notation $n!$ is short for the product $1 \cdot 2 \cdot 3 \cdots (n - 1)n$. For instance, $5! = 1 \cdot 2 \cdot 3 \cdot 4 \cdot 5$. Prove that, for $x > 0$,

(a) $e^x > 1 + x + \dfrac{x^2}{2!} + \dfrac{x^3}{3!}$;

(b) $e^x > 1 + x + \dfrac{x^2}{2!} + \dfrac{x^3}{3!} + \dfrac{x^4}{4!}$.

(c) Generalize (b).

29. See Exercise 28. Let $0 < x \leq 1$. Prove that
(a) $e^x < 1 + ex$;

(b) $e^x < 1 + x + \dfrac{ex^2}{2}$;

(c) $e^x < 1 + x + \dfrac{x^2}{2!} + \dfrac{ex^3}{3!}$;

(d) $e^x < 1 + x + \dfrac{x^2}{2!} + \dfrac{x^3}{3!} + e\dfrac{x^4}{4!}$.

(e) Generalize (d).

30. See Exercises 28 and 29. Recall that in Sec. 4.2 e was shown to be less than 4.
(a) Show that

$$1 + 1 + \tfrac{1}{2} + \tfrac{1}{6} + \tfrac{1}{24} < e < 1 + 1 + \tfrac{1}{2} + \tfrac{1}{6} + \tfrac{1}{24} + \tfrac{4}{120}.$$

(b) Use (a) to show that $2.708 < e < 2.742$.

A calculator is required in the remaining exercises.

31. (a) Show that the equation $b^x = x^a$ is equivalent to the equation

$$\frac{\ln x}{x} = \frac{\ln b}{a}.$$

(b) There are two solutions of the equation

$$(1.5)^x = x^5.$$

One is near 1. Use a calculator to estimate the other solution.

32. See Exercise 31. The equation $(1.2)^x = x^{20}$ has two positive solutions.
(a) Estimate the one near 1 to two decimal places.
(b) Estimate the other solution to the nearest integer.

33. Use the general method developed in Exercises 28 to 30 to estimate to three decimal places:
(a) $\sqrt{e} = e^{1/2}$.
(b) e.

6.4

Natural growth and decay

The change in the size of the world's population is determined by two basic variables: the birth rate and the death rate. If they are equal, the size of the population remains constant. But if one is larger than the other, the population either grows or shrinks.

The birth rate and death rate are determined by different variables, and therefore it would be sheer coincidence if they were equal. The birth rate is influenced, for instance, by age of marriage, infant mortality rate, age of weaning, attitudes toward birth control and abortion, woman's self-image, and the role of children in supporting aged parents. The death rate is influenced by such factors as public health, wars, nutrition, pollution, and stress. Since man has brought famines and epidemics under control, he has brought the death rate below the birth rate.

Consequently, for the past 2 centuries the world population has increased dramatically. See the graph in Example 6 of Sec. 2.1, which records the world population over a period of 10,000 years.

In the year 1977 the world population was 3.8 billion. It is forecast that, in the year 2000, if growth continues at the present rate, the population will be 6.0 billion.

How is this type of prediction made? As we shall see, the study of this type of population growth is based on the exponential function b^x, which was shown in Sec. 6.3 to grow faster than any power of x.

To treat the size of the population mathematically, let $f(t)$ denote the size of the population at time t. Actually, $f(t)$ is an integer, and the graph of f has "jumps" whenever someone is born or dies. However, assume that f is a "smooth" (differentiable) function that approximates the size of the population.

The derivative f' then records the rate of change of the population. If social, medical, and technological factors remain constant, then it is reasonable to expect the rate of growth $f'(t)$ to be proportional to the size of the population $f(t)$: A large population will produce more babies in a year than a small population. More precisely, there is a fixed number k, independent of time, such that

$$f'(t) = kf(t).$$

This equation involves the derivative of f and is therefore called a *differential equation*. This particular equation is the differential equation of *natural growth* (or *decay*, if k is negative). The constant k is called the *instantaneous growth rate*.

In order to predict the size of the population a number of years into the future, we must find an explicit formula for such a function f. One possible such function is

$$f(t) = Ae^{kt},$$

where A is a fixed number, since

$$f'(t) = (Ae^{kt})' = Ake^{kt} = kAe^{kt} = kf(t).$$

Thus, the function $f(t) = Ae^{kt}$ for any constant A satisfies the differential equation for natural growth.

But are there other functions f that satisfy the equation $f'(t) = kf(t)$? If there are, we must find them before we can make predictions. If there are no other such functions, then we must prove this beyond a shadow of a doubt. The next theorem shows that, under conditions of natural growth, the population must increase in accordance with the formula Ae^{kt}.

THEOREM Any function f that satisfies the differential equation

$$f'(t) = kf(t)$$

must be of the form $\qquad f(t) = Ae^{kt}$

for some constant A.

PROOF We prove first that, if $f(t)$ satisfies the above differential equation, the function

$$g(t) = \frac{f(t)}{e^{kt}}$$

is constant. To do this, we shall show that $g'(t) = 0$ for all t. Then Corollary 1 of Sec. 6.2 implies that g is constant.

The computation of g' is straightforward:

$$g'(t) = \frac{e^{kt}f'(t) - f(t)ke^{kt}}{(e^{kt})^2}.$$

But $f'(t) = kf(t)$. Hence

$$g'(t) = \frac{e^{kt}kf(t) - f(t)ke^{kt}}{(e^{kt})^2}.$$

Since the numerator of the quotient is 0,

$$g'(t) = 0$$

for all t. By Corollary 1 of Sec. 6.2, there is a constant which we call A, such that

$$g(t) = A$$

for all t. Thus $\qquad \dfrac{f(t)}{e^{kt}} = A$

for all t.

In short, $\qquad f(t) = Ae^{kt},$

and the theorem is proved. ●

COROLLARY   A function f that satisfies the differential equation

$$f'(t) = kf(t)$$

must be of the form

$$f(t) = Ab^t$$

for some constants A and b.

PROOF   By the preceding theorem, $f(t) = Ae^{kt}$ for some constant A.   Letting $b = e^k$, we have

$$f(t) = Ae^{kt}$$
$$= A(e^k)^t$$
$$= Ab^t.$$

This proves the corollary. ●

EXAMPLE 1   The size of the world population in 1977 was approximately 3.8 billion.   If it continues to grow at the rate of 2 percent a year, what will be the size of the population in the year 2000?   Find the instantaneous growth rate k and the constant b.

SOLUTION   Measure time t in years, with $t = 0$ corresponding to the year 1977.   Let $f(t)$ be the population at time t.   (Actually the population assumes only integer values.   The function f is a differentiable approximation to the size of the population.)

By the corollary, there are constants A and b such that

$$f(t) = Ab^t.$$

To find A, set $t = 0$:

$$f(0) = Ab^0 = A \cdot 1 = A.$$

Thus

$$A = 3.8 \text{ billion.}$$

How do we find b?   We know that in 1 year the population increases by 2 percent.   Consequently,

$$\frac{f(t+1)}{f(t)} = 1.02.$$

Thus

$$\frac{Ab^{t+1}}{Ab^t} = 1.02,$$

and cancellation yields

$$b = 1.02.$$

Thus b is determined.   Consequently,

$$f(t) = 3.8(1.02)^t \qquad \text{billion.}$$

Finding the size of the population in the year 2000, which is 23 years after the year 1977, requires evaluating $f(23)$.   Now,

$$f(23) = 3.8(1.02)^{23}$$
$$= (3.8)(1.577) \qquad \text{computation by calculator or logarithms}$$
$$= 6.0 \text{ billion.}$$

Finally, to determine the instantaneous growth rate k, recall that $b = e^k$. Thus

$$e^k = 1.02,$$

from which it follows that $k = \ln 1.02$

$$= 0.0198. \quad \bullet$$

As Example 1 shows, the constant b in the formula $f(t) = Ab^t$ is just the proportion

$$\frac{\text{Amount at end of a unit time interval}}{\text{Amount at beginning of a unit time interval}}.$$

Thus b is usually easy to determine experimentally: Simply find the relative increase in unit time and add 1. In Example 1 the relative increase in 1 year was 0.02, hence $b = 1.02$.

The instantaneous growth rate k is not directly determined by experiment. However, if b is known, k can be found by solving the equation

$$e^k = b,$$

that is, $k = \ln b$.

Note that it was necessary in Example 1 to compute $(1.02)^{23}$. This can be done easily with the aid of the y^x-key of a calculator. If a calculator is not handy, we could proceed as follows. First, use the natural logarithm table to find $k = 0.0198$. Thus $f(t) = 3.8e^{0.0198t}$ and, consequently,

$$f(23) \doteq 3.8e^{(0.0198)23}$$

$$\doteq 3.8e^{0.455}.$$

Then use an e^x table to find that

$$f(23) \doteq (3.8)(1.577)$$

$$\doteq 6.0 \text{ billion}.$$

Example 1 also illustrates another useful fact. The instantaneous growth rate $k = 0.0198$ is very close to the observed relative growth rate in unit time, which is 2 percent or 0.02. In the case of *small* growth rates the approximation

$$k \doteq \text{relative growth rate in unit time}$$

is very close. In other words, when k is small or, equivalently, b is near 1,

$$b \text{ is approximately } 1 + k.$$

(See either Exercise 18 or 19 for a justification of this assertion.) As a matter of fact, since population data are generally far from accurate, the use of $k = 0.02$ would be justified. Even the reported population of the United States, for instance, may be off by as much as 4 million; it is quite misleading to publish it to all nine digits.

The next example is an important illustration of natural growth in the field of biology.

EXAMPLE 2   At 1 P.M. a bacterial culture weighed 100 grams. At 4:30 P.M. it weighed 250 grams. Assuming that it grows at a rate proportional to the amount present, find at what time it will grow to 400 grams. What is the instantaneous growth rate?

SOLUTION   Measure time t with $t = 0$ corresponding to 1 P.M. Let $f(t)$ be the amount present after t hours. Thus there are constants A and k such that $f(t) = Ae^{kt}$.

The information given translates into the two equations

$$f(0) = 100$$

and

$$f(3.5) = 250.$$

Thus $A = 100$, and

$$Ae^{k(3.5)} = 250;$$

hence

$$100e^{k(3.5)} = 250,$$

or

$$e^{k(3.5)} = 2.5.$$

Thus

$$k(3.5) = \ln 2.5$$
$$\doteq 0.92,$$

and

$$k \doteq \frac{0.92}{3.5}$$
$$\doteq 0.26.$$

The instantaneous growth rate k is 0.26.

After how many hours will the bacteria weigh 400 grams? To find out, solve the equation

$$100e^{0.26t} = 400.$$

We have

$$e^{0.26t} = 4$$

or

$$0.26t = \ln 4$$
$$\doteq 1.39.$$

Thus

$$t \doteq \frac{1.39}{0.26}$$
$$\doteq 5.35 \text{ hours}$$
$$\doteq 5 \text{ hours and 21 minutes.}$$

Since $t = 0$ corresponds to 1 P.M., the bacterial culture will weigh 400 grams at about 6:21 P.M. ●

The first two examples concerned a quantity that *increases* at a rate proportional to the amount present. But a substance can *decrease* (decay) at a rate proportional to the amount present. In this case the same differential equation still holds,

$$f'(t) = kf(t),$$

but $f'(t)$ is negative, and therefore k is negative. The next example illustrates this type of decay.

EXAMPLE 3   Carbon 14 (chemical symbol, ^{14}C), one of the three isotopes of carbon, is radioactive and decays at a rate proportional to the amount present. In about 5,730 years only half of the material is left. (This is expressed by saying that its *half-life* is 5,730 years. The half-life is denoted by $t_{1/2}$.) Find the constant k in the formula Ae^{kt} in this case.

SOLUTION   Let $f(t) = Ae^{kt}$ be the amount present after t years. If A is the initial amount, then

$$f(5,730) = \frac{A}{2}.$$

Thus

$$Ae^{(k)(5,730)} = \frac{A}{2},$$

or

$$e^{5,730k} = \tfrac{1}{2}.$$

Thus

$$5,730k = \ln \tfrac{1}{2} = -\ln 2 \doteq -0.69.$$

Solving for k, we have

$$k \doteq \frac{-0.69}{5,730} \doteq -0.00012.$$

Thus

$$f(t) = Ae^{-0.00012t}.$$

Exercise 21 shows how this formula may be used to determine the age of fossils and other once-living things. ●

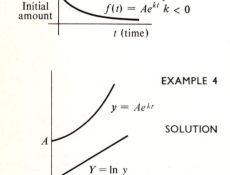

The graphs of natural growth and natural decay look quite different. The first gets arbitrarily steep as time goes on; the second approaches the t axis and becomes almost horizontal.

If a quantity is increasing subject to natural growth, its graph may quickly shoot off an ordinary size piece of paper. Witness, for instance, the population graph in Example 6 of Sec. 2.1. The next example describes a special way of graphing an equation of the form $y = Ae^{kt}$.

EXAMPLE 4   Let $y = Ae^{kt}$. Let $Y = \ln y$. Show that the graph of Y, considered as a function of t, is a straight line.

SOLUTION
$$\begin{aligned} Y &= \ln y \\ &= \ln Ae^{kt} \\ &= \ln A + \ln e^{kt} \\ &= \ln A + kt. \end{aligned}$$

The graph is thus a straight line with y intercept $\ln A$ and with slope k. The diagram in the margin contrasts the two graphs. ●

Special graph paper called *semilog*, based on the observation in Example 4, is available. It enables one to graph $Y = \ln y$ directly without having to compute logarithms. On the vertical axis the number y is placed a distance from the x axis proportional to $\ln y$, just as on a slide rule for multiplication.

Exercises

1. The weight of a certain bacterial culture after t hours is $10(3^t)$ grams.
   (a) What is the initial amount?
   (b) What is the instantaneous growth rate k?
   (c) What is the percent increase in any period of 1 hour?

2. Let $f(t) = 3(2^t)$.
   (a) Solve the equation $f(t) = 12$.
   (b) Solve the equation $f(t) = 5$.
   (c) Find k such that $f(t) = 3e^{kt}$.

3. Using a growth rate of 1 percent a year, determine how long it will take the population of the United States to double.

4. Using a growth rate of 1 percent a year, determine when the population of the United States will reach 1 billion. Assume that the population was 220 million in 1977.

5. Solve the equation $5^{0.3t} = 2$.

6. A bacterial culture grows from 100 to 400 grams in 10 hours.
   (a) How much was present after 3 hours?
   (b) How long will it take the mass to double? quadruple? triple?

7. See Example 1. When will the population of the world reach 20 billion if the present growth rate continues.

8. Assume that $f(t) = Ae^{kt}$, $f(0) = 4$, and $f(1) = 8$. Find
   (a) A (b) k (c) b.

9. See Example 3. How much carbon 14 remains after
   (a) 11,460 years? (b) 2,000 years?

10. Let $f(t) = Ae^{-ct}$, where c is positive. Note that this describes decay. Express the half-life $t_{1/2}$ in terms of c. That is, find t when $f(t) = A/2$.

11. Let $f(t) = 10^t$.
    (a) Show that $f'(t)$ is proportional to $f(t)$.
    (b) Does this violate the theorem that asserts that $f(t)$ must be of the form Ae^{kt} for suitable A and k?

12. The amount of a certain growing substance increases 10 percent every hour.
    (a) Find b. (b) Find k.

13. A disintegrating radioactive substance decreases from 12 grams to 11 grams in 1 day.
    (a) Find b. (b) Find k.

14. A radioactive substance disintegrates at the instantaneous rate of 0.05 gram per day when its mass is 10 grams.
    (a) How much of the substance will remain after t days if the initial amount is A?
    (b) What is its half-life?

15. Find all functions that equal their derivative.

16. It was once conjectured that the speed of a ball falling from rest is proportional to the distance s that it drops.
    (a) Show that if this conjecture were correct, s would grow exponentially as a function of time.
    (b) With the aid of (a) show that the speed would also grow exponentially.
    (c) Recalling that the initial speed is 0, show that (b) leads to an absurd conclusion.
    In fact, ds/dt is proportional to time t rather than to distance s.

17. This table lists the population of the United States during a 30-year period.

Year	1945	1950	1955	1960	1965	1970	1975
Population, millions	140	152	166	181	194	205	215

Does it seem to be of exponential character? Discuss.

18. Prove that, when k is small, $\ln(1 + k) \doteq k$. More precisely, prove that
$$\lim_{k \to 0} \frac{\ln(1+k)}{k} = 1.$$
Hint: Express the limit as the derivative of a certain function at a specific number.

19. Prove that, when k is small, $e^k \doteq 1 + k$. More precisely, prove that
$$\lim_{k \to 0} \frac{e^k - 1}{k} = 1.$$

20. A quantity is increasing according to the laws of natural growth. The amount present at time $t = 0$ is A. It will double when $t = 10$.
    (a) Express the amount in the form Ae^{kt} for suitable k.
    (b) Express the amount in the form Ab^t for suitable b.

■

21. If the carbon-14 concentration in the carbon from a plant or piece of wood of unknown age is half that of the carbon-14 concentration in a present-day live specimen, then it is about 5,730 years old. Show that, if A_c and A_u are the radioactivities of samples prepared from contemporary and from undated materials, respectively, then the age of the undated material is about $t = 8,300 \ln (A_c/A_u)$. (This method is dependable up to an age of about 70,000 years.) See Radiocarbon Dating in McGraw-Hill *Encyclopedia of Science and Technology*, McGraw-Hill, New York 1971.

Exercises 22 to 25 concern continuous compound interest.

22. Review Sec. 4.2 in which interest compounded n times per year is discussed.

    (a) A bank pays 5 percent interest per year with interest compounded n times per year. Show that in t years 1 dollar grows to

    $$\left(1 + \frac{0.05}{n}\right)^{nt} \qquad \text{dollars.}$$

    (b) Show that

    $$\lim_{n \to \infty} \left(1 + \frac{0.05}{n}\right)^{nt} = e^{0.05t}.$$

    If the bank pays continuous interest—compounding instantaneously—1 dollar grows to $e^{0.05t}$ dollars in t years.

    (c) Show that a deposit doubles in about 14 years. (At simple interest it would take $1/0.05 = 20$ years to double.)

23. See Exercise 22. Show that at continuous interest 1 dollar doubles in about 70 percent of the time it would at simple interest—for any fixed annual rate of interest.

24. See Exercises 22 and 23. Show that the doubling time for a population increasing at the rate of p percent a year is approximately $70/p$ years.

25. See Exercise 22. An interest rate of 8 percent per year, compounded continuously, equals what rate of simple interest (no compounding at all) over a duration of 1 year?

■ ■

26. A salesman, trying to persuade a tycoon to invest in Standard Coagulated Mutual Fund shows him the accompanying graph which records the value of a similar investment made in the fund in 1950. "Look! In the first 5 years the investment increased $1,000," the salesman observed, "but in the past 5 years it increased by $2,000. It's really improving. Look at the slope from 1970 to 1975."

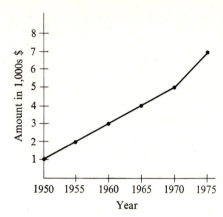

The tycoon replied, "Hogwash; in order to present an unbiased graph, you should use semilog paper. Though your graph is steeper from 1970 to 1975, in fact, the rate of return is less than from 1950 to 1955. Indeed, that was your best period."

    (a) If the percentage return on the accumulated investment remains the same over each 5-year period as the first 5-year period, sketch the graph.

    (b) Explain the tycoon's reasoning.

Exercises 27 to 30 present some simple differential equations.

27. Find all functions $y = f(t)$ that satisfy the differential equation

    $$\frac{dy}{dt} = kt \qquad k \text{ is constant.}$$

    The rate of change is proportional to the elapsed time.

28. Find all functions $y = f(t)$ such that

    $$\frac{dy}{dt} = k(y - A),$$

    where k and A are constants. *Hint:* first show that

    $$\frac{d(y - A)}{dt} = k(y - A).$$

    For negative k this is Newton's law of cooling; y is the temperature of some heated object at time t. The room temperature is A. The differential equation $dy/dt = k(y - A)$ says, "The object cools at a rate proportional to the difference between its temperature and the room temperature."

29. In many cases there is a clear-cut upper bound M on the extent to which a quantity may grow. It is reasonable in this case to assume, or take as a model, that

    $$\frac{dy}{dt} = ky(M - y).$$

This differential equation asserts, "The rate at which the quantity grows is proportional both to the amount present and the amount by which it can still increase." Let A be the amount when $t = 0$. Show that the function

$$y = f(t) = \frac{MA}{A + (M - A)e^{-Mkt}}$$

has the following properties:
(a) It satisfies the differential equation;
(b) $f(0) = A$;
(c) $\lim_{t \to \infty} f(t) = M$.
This type of growth is called *inhibited*, *logistic*, or *sigmoidal*.

30. Consider the differential equation for $y = f(x)$,

$$\frac{dy}{dx} = e^y.$$

(a) Show that y is an increasing function of x.
(b) The function f is increasing and differentiable. Thus x may be considered a function of y. Show that

$$\frac{dx}{dy} = e^{-y}.$$

(c) Find all functions that satisfy the differential equation.
(d) Is there any solution to the differential equation $dy/dx = e^y$ that has as its domain the entire x axis?

6.5

Using the derivative and limits when graphing a function

In this section it is shown how to use the derivative and limits to help sketch the graph of a function. Actually, most of the ideas have been illustrated in earlier sections. In addition, some important definitions will be introduced that will be of use not only in graphing but in many other applications as well. The opening example illustrates most of the ideas; the definitions will follow it.

EXAMPLE 1   Graph $y = f(x) = x^2 e^{-x}$.

SOLUTION   Observe that the domain of the function is the entire x axis.

Where does the graph meet the x axis? To find out, solve the equation $f(x) = 0$, that is,

$$x^2 e^{-x} = 0.$$

Since e^{-x} is never 0, the only solution is $x = 0$.

Next, where is the tangent line horizontal? To find out, solve the equation $f'(x) = 0$. Now,

$$f'(x) = \frac{d(x^2 e^{-x})}{dx}$$

$$= x^2 \frac{d(e^{-x})}{dx} + e^{-x} \frac{d(x^2)}{dx}$$

$$= -x^2 e^{-x} + 2x e^{-x},$$

so $$f'(x) = e^{-x}(2x - x^2).$$

Thus $f'(x) = 0$ only when $2x - x^2 = 0,$

That is, $x(2 - x) = 0.$

The solutions are $x = 0$ and $x = 2.$

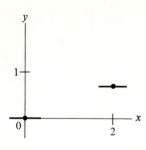

At these values of x the function has the values

$$f(0) = 0 \quad \text{and} \quad f(2) = 2^2 e^{-2} = 4e^{-2} \doteq 0.54.$$

The sketch in the margin incorporates the information gathered so far. A heavy, short, horizontal line records a point on the graph where the tangent line is horizontal.

Next, determine where f is increasing and where it is decreasing. By Sec. 6.2, this amounts to determining the sign of the derivative f'. As already shown,

$$f'(x) = e^{-x}(2x - x^2),$$

or

$$f'(x) = x(2 - x)e^{-x}.$$

Since e^{-x} is always positive, the sign of $f'(x)$ is determined by the sign of $x(2 - x)$. To determine the sign of $x(2 - x)$, make a little chart to keep track of the signs of the factors x and $2 - x$ for various ranges of x.

	$x < 0$	$0 < x < 2$	$x > 2$
Sign of x	$-$	$+$	$+$
Sign of $2 - x$	$+$	$+$	$-$
Sign of $x(2 - x)$	$-$	$+$	$-$

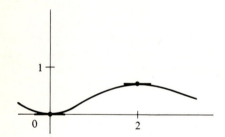

Thus $f'(x) = x(2 - x)e^{-x}$ is positive for x in $(0, 2)$ and negative for $x < 0$ or $x > 2$. Consequently, the function f is increasing for x in $(0, 2)$ and is decreasing elsewhere. This information is incorporated into the sketch in the margin.

To complete the graph, we must examine the behavior of $f(x)$ when $|x|$ is large. To do this, determine

$$\lim_{x \to \infty} x^2 e^{-x} \quad \text{and} \quad \lim_{x \to -\infty} x^2 e^{-x}.$$

By Theorem 4 of Sec. 6.3, $\quad \lim_{x \to \infty} x^2 e^{-x} = 0.$

Also we have $\quad \lim_{x \to -\infty} x^2 e^{-x} = \infty,$

since, when x is negative and $|x|$ is large, both x^2 and e^{-x} are large.

With this additional information, even though we have evaluated the function at only two places, 0 and 2, it is now possible to sketch the general behavior of the function:

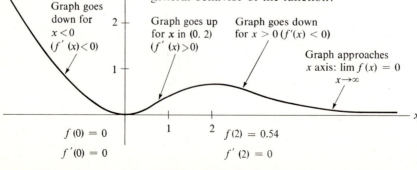

The graph to the left illustrates most of the concepts that will now be introduced. ●

DEFINITION   *Horizontal asymptote.*   If

$$\lim_{x \to \infty} f(x) = L \qquad \text{or if} \qquad \lim_{x \to -\infty} f(x) = L$$

the line $y = L$ is a *horizontal asymptote* of the graph of f.

In Example 1, because $\lim_{x \to \infty} x^2 e^{-x} = 0$, the line $y = 0$, that is, the x axis, is a horizontal asymptote.   A horizontal asymptote is simply a line parallel to the x axis which the graph of f approaches closer and closer, either far to the right or, perhaps, far to the left.   Actually, the definition permits the graph to coincide with the asymptote or to cross it infinitely often.   However, in most practical cases the graph sufficiently far from the origin does not meet the asymptote.

DEFINITION   *Vertical asymptote.*   If

$$\lim_{x \to a^+} f(x) = \infty \qquad (\text{or } -\infty)$$

or if

$$\lim_{x \to a^-} f(x) = \infty \qquad (\text{or } -\infty)$$

then the line $x = a$ is a *vertical asymptote* of the graph of f.

There is no vertical asymptote in Example 1; we will show that there is one in Example 2 at $a = 1$.   Asymptotes that are neither horizontal nor vertical are discussed in Exercises 82 to 86 of the summary Sec. 6.11.

DEFINITION   *Critical number* (or *critical point*).   A value of x at which $f'(x) = 0$ is called a critical number for the function f.   The corresponding point $(x, f(x))$ on the graph of f is a critical point on that graph.

At a critical point the tangent line is horizontal.   The curve in Example 1 has two critical points, $(0, 0)$ and $(2, 4e^{-2})$.

DEFINITION   *Relative maximum* (*local maximum*).   The function f has a relative maximum (or local maximum) at the number X if there is an open interval (a, b) around X such that $f(X) \geq f(x)$ for all x in (a, b) that lie in the domain of f.   A "local" or "relative minimum" is defined analogously.

DEFINITION   *Absolute maximum* (*global maximum*).   The function f has an absolute maximum (or global maximum) at the number X if $f(X) \geq f(x)$ for all x in the domain of f.   A global minimum is defined analogously.

The function in Example 1 has:

a relative (local) maximum at $x = 2$,
a relative (local) minimum at $x = 0$,
an absolute (global) minimum at $x = 0$,

but no absolute (global) maximum.   (However, if we restrict the domain of $f(x) = x^2 e^{-x}$ to $x \geq 0$, then the function will have a global maximum at $x = 2$.)

The following test for local maximum or local minimum is an immediate consequence of the fact that, when the derivative is positive, the function increases and, when it is negative, it decreases.

Derivative Test for Local Maximum (or Local Minimum) at $x = a$

Let f be a function and let a be a number. Assume that numbers b and c exist such that $b < a < c$ and

1. f is continuous on the open interval (b, c);
2. f is differentiable on the open interval (b, c), except possibly at a;
3. $f'(x)$ is positive for all $x < a$ in the interval and is negative for all $x > a$ in the interval.

Then f has a *local maximum* at a. A similar test, with "positive" and "negative" interchanged, holds for a *local minimum*.

Informally, the derivative test says, "If the derivative changes sign at a, then either you have a local minimum or a local maximum." To decide which it is, just make a crude sketch of the graph near $(a, f(a))$ to show on which side of a the function is increasing and on which side it is decreasing.

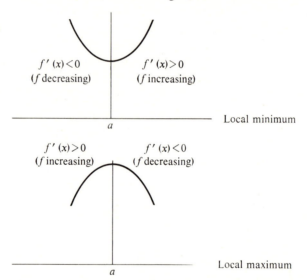

Note that in the derivative test there is no assumption that the derivative exists at $x = a$. Thus the test applies to $f(x) = |x|$, which has a local minimum at $x = 0$, where the derivative changes from -1 to 1. However, $f(x) = |x|$ is not differentiable at 0, as shown in Example 3 of Sec 3.3.

DEFINITION   *x intercept.* A value of x for which $f(x) = 0$ is called an *x intercept.*

The function in Example 1 has only one x intercept, namely $x = 0$. An x intercept indicates where the graph meets the x axis.

DEFINITION   *y intercept.* The value $f(0)$ is called the *y intercept.*

Most functions will not require all the preceding considerations.

General Procedure for Graphing a Function

	Calculations	Geometric Meaning
Domain	1. Find where $f(x)$ is defined.	Find horizontal extent of graph.
Intercepts	2. Find $f(0)$ and the values of x for which $f(x) = 0$.	Find where graph crosses the axes.
Critical Numbers	3. Find where $f'(x) = 0$.	Find where tangent line is horizontal.
	4. Compute $f(x)$ at all critical numbers.	Data needed for critical points.
Increasing, Decreasing	5. Find the values of x for which $f'(x)$ is positive and those for which $f'(x)$ is negative.	Find where graph goes up and where it goes down as pencil moves to the right.
Horizontal Asymptotes	6. Find $\lim_{x \to \infty} f(x)$ and $\lim_{x \to -\infty} f(x)$.	Find horizontal asymptotes or general behavior when $|x|$ is large.
Vertical Asymptotes	7. Find values of a where $\lim_{x \to a^+} f(x)$ or $\lim_{x \to a^-} f(x)$ is infinite.	Find vertical asymptotes (where the graph "blows up").
	8. Sketch the graph, showing intercepts, critical points, asymptotes, local and global maxima and minima.	

EXAMPLE 2   Graph
$$f(x) = \frac{2x - 5}{x - 1}.$$

SOLUTION   Note that at $x = 1$ the function is not defined.

Next observe that $f(0) = 5$ and that $f(x) = 0$ only when the numerator, $2x - 5$, is 0. Thus the y intercept is 5 and the only x intercept is $x = \frac{5}{2}$, the solution of the equation $2x - 5 = 0$. This information is recorded in the diagram in the margin.

Next, determine the critical numbers of f. To do this, compute $f'(x)$:

$$f'(x) = D\left(\frac{2x - 5}{x - 1}\right) = \frac{(x - 1) \cdot 2 - (2x - 5) \cdot 1}{(x - 1)^2}$$

$$= \frac{3}{(x - 1)^2}.$$

No Critical Numbers   Since the numerator is never 0, there are no critical numbers.

Where is the function increasing? Decreasing? Since $f'(x) = 3/(x - 1)^2$, the derivative is positive throughout the domain of the function. The func-
Always Increasing   tion is always increasing.

How does the function behave when $|x|$ is large? We have

$$\lim_{x \to \infty} \frac{2x - 5}{x - 1} = \lim_{x \to \infty} \frac{2 - 5/x}{1 - 1/x} = \frac{2}{1} = 2.$$

Similarly,
$$\lim_{x \to -\infty} \frac{2x - 5}{x - 1} = 2.$$

Horizontal Asymptote   Thus the line $y = 2$ is an asymptote of the graph both far to the right and far to the left. Since the function is *increasing*, the graph, for $|x|$ large, resembles the sketch in the margin.

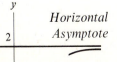

Are there any vertical asymptotes? In other words, are there any numbers *a* near which the function becomes arbitrarily large? Since

$$f(x) = \frac{2x - 5}{x - 1},$$

Vertical Asymptote   and $x - 1$ is small when x is near 1, near $a = 1$ the function "blows up." If x is near 1 and to the right of 1, the numerator is near $2 \cdot 1 - 5 = -3$, hence negative, and the denominator is a small positive number. Thus

$$\lim_{x \to 1^+} \frac{2x - 5}{x - 1} = -\infty.$$

Similarly,

$$\lim_{x \to 1^-} \frac{2x - 5}{x - 1} = \infty.$$

The line $x = 1$ is a vertical asymptote.

With this information, the graph can be completed.

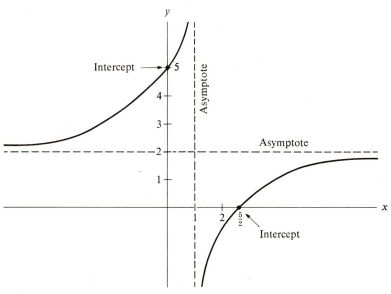

The next example illustrates the graphing of a polynomial.

EXAMPLE 3   Sketch the graph of $f(x) = 2x^3 - 3x^2 - 12x$.

SOLUTION   Note that $f(x)$, being a polynomial, is defined for all x.

Since $f(0) = 2 \cdot 0^3 - 3 \cdot 0^2 - 12 \cdot 0 = 0$, the y intercept is 0. To find the x intercepts it is necessary to solve the equation

$$f(x) = 0.$$

Intercepts   In the case of this function the equation can be solved easily:

$$2x^3 - 3x^2 - 12x = 0,$$

or $$x(2x^2 - 3x - 12) = 0.$$

Either $x = 0$ or $$2x^2 - 3x - 12 = 0.$$

The latter equation can be solved by the quadratic formula:

$$x = \frac{-(-3) \pm \sqrt{(-3)^2 - 4(2)(-12)}}{2(2)}$$

$$= \frac{3 \pm \sqrt{9 + 96}}{4}$$

$$= \frac{3 \pm \sqrt{105}}{4}.$$

These two solutions are approximately -1.8 and 3.3.

The intercepts are recorded in the graph in the margin.
When is $f'(x) = 0$?   We have

$$f'(x) = 6x^2 - 6x - 12$$
$$= 6(x^2 - x - 2)$$
$$= 6(x - 2)(x + 1).$$

Thus $f'(x) = 0$ when $$6(x - 2)(x + 1) = 0,$$

that is, when $$x = 2 \quad \text{or} \quad x = -1.$$

At these critical numbers the function has the values

$$f(2) = 2(2)^3 - 3(2)^2 - 12(2) = -20$$

and $$f(-1) = 2(-1)^3 - 3(-1)^2 - 12(-1) = 7.$$

The diagram in the margin records the data gathered so far.

Next, examine the sign of $f'(x)$ to determine where the function is increasing and where it is decreasing.   Recall that $f'(x) = 6(x - 2)(x + 1)$ and use this chart as an aid.

Sign of $x - 2$

Sign of $x + 1$

Sign of $6(x - 2)(x + 1)$

Thus the function is increasing for $x < -1$ and for $x > 2$; it is decreasing for $-1 < x < 2$.   The information gathered so far is recorded in the accompanying graph.

Finally, consider the behavior of $f(x) = 2x^3 - 3x^2 - 12x$ when $|x|$ is large. Since

$$\lim_{x \to \infty} f(x) = \lim_{x \to \infty} (2x^3 - 3x^2 - 12x)$$

Critical Numbers

(Vertical scale foreshortened)

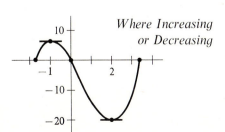

Where Increasing or Decreasing

$$\lim_{x \to \infty} f(x) = \lim_{x \to \infty} x^3 \left(2 - \frac{3}{x} - \frac{12}{x^2} \right)$$

$$= \infty,$$

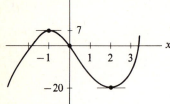

No Asymptotes

Graph of $f(x) = 2x^3 - 3x^2 - 12x$

The graph crosses the x axis at $x = 0$, $\frac{3 - \sqrt{105}}{4} \doteq -1.8$

and $\frac{3 + \sqrt{105}}{4} \doteq 3.3$

the graph does not have a horizontal asymptote as $x \to \infty$. Similar reasoning shows that it has no horizontal asymptote as $x \to -\infty$. In fact,

$$\lim_{x \to -\infty} f(x) = -\infty.$$

With this last information the curve can be sketched. The graph (with the y axis compressed) appears in the margin.

There is a local maximum at $x = -1$, a local minimum at $x = 2$, but no global maximum or minimum. ●

EXAMPLE 4 Graph $f(x) = 3x^4 - 4x^3$. Discuss relative maxima and minima.

SOLUTION To find the intercepts, note that $f(0) = 0$ and that $3x^4 - 4x^3 = 0$ when $x^3(3x - 4) = 0$, that is, when $x = 0$ and $x = \frac{4}{3}$. The derivative is

$$f'(x) = 12x^3 - 12x^2$$
$$= 12x^2(x - 1).$$

The critical numbers are the solutions of the equation

$$12x^2(x - 1) = 0,$$

namely 0 and 1.

How does the sign of $f'(x) = 12x^2(x - 1)$ behave when x is near 0? For $x < 0$, $12x^2$ is positive and $x - 1$ is negative; hence $12x^2(x - 1)$ is negative. For $0 < x < 1$, $12x^2$ is positive and $x - 1$ is still negative. Thus the sign of $f'(x)$ does *not* change as x passes through 0. In fact, since $f'(x)$ remains negative (except at 0), the function f is decreasing for $x \leq 1$. Thus there is no relative maximum or minimum at $x = 0$.

How does the sign of $f'(x) = 12x^2(x - 1)$ behave when x is near 1? The factor $12x^2$ remains positive, but $x - 1$ changes sign from negative to positive. Hence at $x = 1$ the function has a local minimum.

Writing $\qquad\qquad f(x) = 3x^4 - 4x^3$

$$= x^4 \left(3 - \frac{4}{x} \right).$$

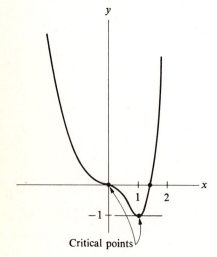

Critical points

shows that, when $|x|$ is large, $f(x)$ behaves like $3x^4$ (since $4/x$ is then near 0). Since $3x^4$ becomes arbitrarily large when x is large, the function has no global maximum. The accompanying sketch of the graph shows the x intercepts and the critical points. Note that at $x = 1$ a global minimum occurs. ●

In many applied problems we are interested in the behavior of a differentiable function just over some closed interval $[a, b]$. Such a function will have a global maximum by the maximum-value theorem of Sec. 6.1.

That maximum can occur either at an endpoint—a or b—or else at some number X in the open interval (a, b). In the latter case X must be a critical number, for $f'(X) = 0$ by Theorem 1 of Sec. 6.1.

The two accompanying graphs show some of the ways in which a relative or global maximum or minimum can occur for a function considered only on a closed interval $[a, b]$.

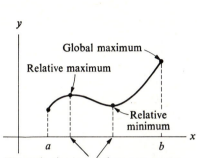

The derivative is 0 at these two numbers

The global maximum occurs at an end.
(The derivative need not be 0.)

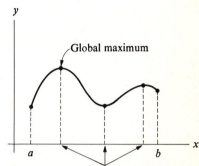

The derivative is 0 at these three numbers

The global maximum occurs at a number other than a or b. (The derivative is 0.)

The major point to keep in mind is that the maximum value of a differentiable function f on a closed interval occurs:

1. at an end of the interval, or
2. at a critical number [where $f'(x) = 0$].

EXAMPLE 5 Find the maximum value of $f(x) = x^3 - 3x^2 + 3x$ for x in $[0, 2]$.

SOLUTION First compute f at the ends of the interval, 0 and 2:

$$f(0) = 0 \quad \text{and} \quad f(2) = 2.$$

Next, compute $f'(x)$, which is $3x^2 - 6x + 3$. When is $f'(x) = 0$? When

$$3x^2 - 6x + 3 = 0,$$

or

$$3(x^2 - 2x + 1) = 0,$$

or

$$3(x - 1)^2 = 0.$$

Thus 1 is the only critical number, and it lies in the interval $[0, 2]$.

The maximum of f must therefore occur either at an endpoint of the interval (at 0 or 2) or at the only critical number, 1. It is necessary to calculate $f(1)$ to determine where the maximum occurs:

$$f(1) = (1)^3 - 3(1)^2 + 3(1)$$
$$= 1.$$

Since $f(0) = 0, f(2) = 2$, and $f(1) = 1$, the maximum value is 2, occurring at the endpoint 2.

The tangent line is horizontal
at the point (1, 1)

Now that the problem is solved, it may be instructive to sketch the graph of the function. Since

$$f'(x) = 3(x - 1)^2$$

is positive for all x other than 1, the function is increasing. The accompanying figure shows how the graph looks. Observe that the minimum occurs at 0. ●

This flowchart summarizes the method of finding the maximum of a differentiable function on a closed interval.

Flow Diagram of Procedure for Finding M, the Maximum of $f(x)$ for x in $[a, b]$

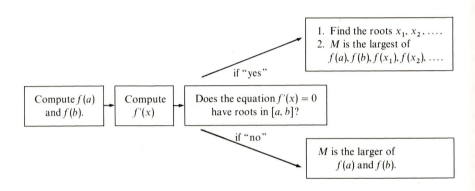

Exercises

In Exercises 1 to 24, graph the functions, showing x intercepts, y intercepts, critical points, local and global maxima and minima, and horizontal and vertical asymptotes.

1. xe^{-x}
2. $x^2 - 6x + 5$
3. x^3
4. $x/(x^2 + 1)$
5. $(3x^2 + 5)/(x^2 - 1)$
6. $e^{2x} + e^{-x}$
7. $x^3 + (1/x)$
8. $x + 2 \ln (1 + x^2)$
9. $(\ln x)/x^2$
10. $1/[(x - 1)^2(x - 2)]$
11. $e^{-x} \sin x$
12. $e^{-x} \sin \sqrt{3} x$
13. $\dfrac{\sin x}{1 + 2 \cos x}$
14. $2x^3 + 3x^2 - 6x$
15. $x^3 - 3x^2 + 3x$
16. $e^x - ex + 1$
17. $e^x + 2e^{-x}$
18. $x^4 - 4x$
19. $x^4 - 4x^3 + 4x^2$
20. $x^4 + 2x^3 - 3x^2$
21. $2x^{1/3} + x^{4/3}$
22. $(x^3 - 1)/(x^2 - 1)$
23. $\sqrt{x^2 - 1/x}$
24. $x^4 - 4x + 3$

Exercises 25 to 32 concern functions whose domain is restricted to a closed interval. In each find the maximum and the minimum value for the given function over the given interval.

25. $(\ln x)/x$; $[1, 10]$.
26. $4x - x^2$; $[0, 5]$.
27. $4x - x^2$; $[0, 1]$.
28. $2x^2 - 5x$; $[-1, 1]$.
29. $x^3 - 2x^2 + 5x$; $[-1, 3]$.
30. $x/(x^2 + 1)$; $[0, 3]$.
31. $e^{-x} \sin x$; $[0, 2\pi]$.
32. $(x + 1)/\sqrt{x^2 + 1}$; $[0, 3]$.

■

33. Differentiate (for practice):
(a) $\sqrt[5]{x^2}$
(b) $x^3 \sin^{-1} 3x$
(c) $e^{-2x} \cos 5x$
(d) $\sqrt{\dfrac{2x + 1}{3x + 2}}$
(e) $\csc^3 5x$
(f) $x^{\tan^{-1} 2x}$

6.6

The second derivative and motion; an application of higher derivatives

The Second Derivative

Velocity is the rate at which distance changes. The rate at which velocity changes is called *acceleration*. Thus if $y = f(t)$ denotes position at time t on a line, then the derivative $dy/dt = \dot{y}$ equals the velocity, and the derivative of the derivative, that is, $d(dy/dt)/dt$, equals the acceleration.

If $y = f(x)$, the *second derivative* $d(dy/dx)/dx$ is denoted by any of the following notations:

$$\frac{d^2y}{dx^2}, D^2y, y'', f'', D^2f, f^{(2)}, f^{(2)}_{(x)}, \frac{d^2f}{dx^2}, D_x^2y.$$

If $y = f(t)$, D^2y is also denoted $\ddot{y}$.

For instance, if $y = x^3$,

$$\frac{dy}{dx} = 3x^2$$

and

$$\frac{d^2y}{dx^2} = 6x.$$

Other ways of denoting the second derivative of this function are

$$D^2(x^3) = 6x,$$

$$\frac{d^2(x^3)}{dx^2} = 6x,$$

and

$$(x^3)'' = 6x.$$

This table lists dy/dx, the first derivative, and d^2y/dx^2, the second derivative, for a few common functions.

y	$\dfrac{dy}{dx}$	$\dfrac{d^2y}{dx^2}$
x^3	$3x^2$	$6x$
$\ln x$	$\dfrac{1}{x}$	$-\dfrac{1}{x^2}$
$\sin 3x$	$3\cos 3x$	$-9\sin 3x$
e^{2x}	$2e^{2x}$	$4e^{2x}$

The first derivative of f, $f'(x)$, is also denoted $f^{(1)}(x)$.

Most functions f in calculus can be differentiated repeatedly in the sense that Df exists, the derivative of Df, namely D^2f, exists, the derivative of D^2f exists, and so on. The derivative of the second derivative,

$$\frac{d\left(\dfrac{d^2y}{dx^2}\right)}{dx},$$

is called the *third derivative* and is denoted many ways, such as

$$\frac{d^3y}{dx^3},\ D^3y,\ y''',\ f''',\ f^{(3)},\ f^3(x),\ \frac{d^3f}{dx^3},\ D_x{}^3y.$$

The *fourth derivative* $f^{(4)}(x)$ is defined as the derivative of the third derivative and is represented by similar notations. Similarly, $f^{(n)}(x)$ is defined for $n = 5, 6, \ldots$. The derivatives $f^{(n)}(x)$ for $n \geq 2$ are called the *higher derivatives* of f.

Higher Derivatives

EXAMPLE 1   Compute $f^{(n)}(x)$ for $n \geq 1$ if $f(x) = e^{-3x}$.

SOLUTION

$$f^{(1)}(x) = \frac{df}{dx} = \frac{d(e^{-3x})}{dx} = -3e^{-3x};$$

$$f^{(2)}(x) = \frac{d(f^{(1)}(x))}{dx} = \frac{d(-3e^{-3x})}{dx} = 9e^{-3x};$$

$$f^{(3)}(x) = \frac{d(f^{(2)}(x))}{dx} = \frac{d(9e^{-3x})}{dx} = -27e^{-3x};$$

and so on. Each successive differentiation multiplies the coefficient of e^{-3x} by -3. Thus

$$f^{(n)}(x) = (-3)^n e^{-3x}. \ \bullet$$

The higher derivatives, $f^{(n)}(x)$, are of use in determining the rate at which the function $f(x)$ grows. This is discussed in Section 15.1. The second derivative has two main applications: first, to describe the rate of change of a rate of change, called acceleration; second, to help graph a function. This section is devoted to the first application, while Sec. 6.7 utilizes the second derivative to describe certain features of a graph.

Acceleration

The following example illustrates one way in which the second derivative may be used in the study of motion.

EXAMPLE 2   A falling rock drops $16t^2$ feet in the first t seconds. Find its velocity and acceleration.

SOLUTION   Place the y axis in the usual position, with 0 at the beginning of the fall and the part with positive values above 0. At time t the object has the y coordinate

$$y = -16t^2.$$

The velocity is $(-16t^2)' = -32t$ feet per second,

and the acceleration is

$$(-32t)' = -32 \text{ feet per second per second.}$$

The velocity changes at a constant rate. That is, the acceleration is constant. $\bullet$

Coordinate of rock at time t is $-16t^2$

The second derivative represents acceleration in other contexts, as the next example shows.

EXAMPLE 3   Translate into calculus this quotation from a news report: "The latest unemployment figures can be read as bearing out the forecast that the recession is nearing its peak. Though unemployment continues to increase, it is doing so at a slower rate than before."

SOLUTION   Let y be the number unemployed at time t. (We consider y to be a differentiable function of t that approximates the number unemployed.) As time changes, so does y;

$$y = f(t).$$

The rate of change in unemployment is the derivative

$$\frac{dy}{dt}.$$

The news that "unemployment continues to increase" is recorded by the inequality

$$\frac{dy}{dt} > 0.$$

There is optimism in the article. The rate of increase, dy/dt, is itself declining ("unemployment continues to increase ... at a *slower rate* than before"). The function dy/dt is decreasing. Thus its derivative

$$\frac{d\left(\dfrac{dy}{dt}\right)}{dt}$$

is negative:

$$\frac{d^2y}{dt^2} < 0.$$

In short, the bad news is that dy/dt is positive. But there is good news: d^2y/dt^2 is negative.

The promise that "the recession is nearing its peak" amounts to the prediction that soon dy/dt will be 0, and then switch sign to become negative. In short, a local maximum in the graph of $y = f(t)$ appears in the economists' crystal ball. ●

The remaining examples focus on the use of the second derivative in the study of motion.

EXAMPLE 4   In the simplest motion, no forces act on a moving particle. Assume that a particle is moving on the x axis and no forces act on it. Let its location at time t seconds be

$$x = f(t) \qquad \text{feet.}$$

If at time $t = 0$, $x = 3$ feet and the velocity is 5 feet per second, determine $f(t)$.

SOLUTION   The assumption that no force operates on the particle means that $d^2x/dt^2 = 0$. Call the velocity v. Then

$$\frac{dv}{dt} = \frac{d^2x}{dt^2} = 0.$$

Now, v is itself a function of time. Since its derivative is 0, v must be constant:

$$v(t) = C$$

for some constant C. Since $v(0) = 5$, the constant C must be 5.
    To find the position x as a function of time, note that

$$\frac{dx}{dt} = 5.$$

This equation implies that x must be of the form

$$x = 5t + K$$

for some constant K. Now, when $t = 0$, $x = 3$. Thus $K = 3$. In short, at any time t, the particle is at

$$x = 5t + 3. \ \bullet$$

The next example concerns the case in which the acceleration is constant, but not zero.

EXAMPLE 5   A ball is thrown straight up, with a speed of 64 feet per second, from a cliff 96 feet above the ground. Where is the ball t seconds later? When does it reach its maximum height? How high above the ground does the ball rise? When does the ball hit the ground? Assume that there is no air resistance and that the acceleration due to gravity is constant.

SOLUTION   Introduce a vertical coordinate axis to describe the position of the ball. It is more natural to call it the y axis, and so velocity is dy/dt, and acceleration is d^2y/dt^2. Place the origin at ground level and let the positive part of the y axis be above the ground.
    At time $t = 0$, the velocity $dy/dt = 64$, since the ball is thrown up at a speed of 64 feet per second. (If it had been thrown down, dy/dt would be -64.) As time increases, dy/dt decreases from 64 to 0 (when the ball reaches the top of its path and begins its descent) and continues to decrease through negative values as the ball falls down to the ground. Since v is decreasing, the acceleration dv/dt is negative. The (constant) value of dv/dt, obtained from experiments, is approximately -32 feet per second per second.
    From the equation

$$\frac{dv}{dt} = -32$$

Initial position of ball
Cliff
y
150
100
96' 50
0 ——— Ground level

it follows that
$$v = -32t + C,$$

where C is some constant. To find C, recall that $v = 64$ when $t = 0$. Thus

$$64 = -32(0) + C,$$

and $C = 64$. Hence $v = -32t + 64$ for any time t until the ball hits the ground. Now $v = dy/dt$, so

$$\frac{dy}{dt} = -32t + 64.$$

This equation implies that

$$y = -16t^2 + 64t + K,$$

where K is a constant. To find K, make use of the fact that $y = 96$ when $t = 0$.

Thus
$$96 = -16(0)^2 + 64(0) + K,$$

and $K = 96$.

We have obtained a complete description of the position of the ball at any time t while it is in the air:

$$y = -16t^2 + 64t + 96.$$

This, together with $v = -32t + 64$, provides answers to many questions about the ball's flight.

When does it reach its maximum height? When $v = 0$; that is, when $-32t + 64 = 0$, or when $t = 2$ seconds.

How high above the ground does the ball rise? Simply compute y when $t = 2$. This gives $-16(2)^2 + 64(2) + 96 = 160$ feet.

When does the ball hit the ground? When $y = 0$. Find y such that

$$y = -16t^2 + 64t + 96 = 0.$$

Division by -16 yields the simpler equation

$$t^2 - 4t - 6 = 0,$$

which has the solution

$$t = \frac{4 \pm \sqrt{16 + 24}}{2},$$

or simply
$$t = 2 \pm \sqrt{10}.$$

Since $2 - \sqrt{10}$ is negative, and the ball cannot hit the ground before it is thrown, the physically meaningful solution is $2 + \sqrt{10}$. The ball lands $2 + \sqrt{10}$ seconds after it is thrown; it is in the air for about 5.2 seconds.

The graphs of y, v, and speed, as functions of time, provide another perspective of the motion of the ball.

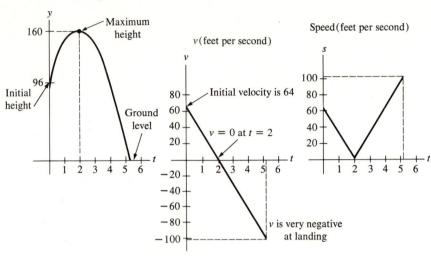

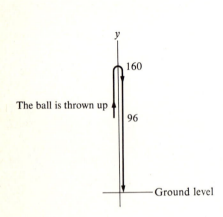

The ball is thrown up

Of course, the actual path of the ball is not restricted to a vertical line and looks somewhat as pictured in the margin. ●

In fact, the acceleration due to gravity is not constant. It varies inversely as the square of the distance from the center of the earth. However, it is almost constant if the particle moves only small distances—such as a few miles up; so, treating it as constant in practical engineering is justified. The next example concerns motion subject to an acceleration that is not constant. More complicated, it is included as a little detour for the interested reader, perhaps to be read later when reviewing. The mathematical basis for determining the escape velocity was developed in Newton's *Principia*, published in 1687. There he investigated not only the inverse square law but other laws of attraction, such as the *inverse cube*, which does not occur in nature.

EXAMPLE 6   Find the initial velocity that a payload must have in order that it will "coast to infinity" rather than fall back to earth. Assume that it is launched straight up.

SOLUTION   Begin by studying the motion of a projectile fired with an initial velocity of v_0 miles per second from the surface of the earth. For this purpose introduce a coordinate system whose origin is at the center of the earth. Let r denote the distance of the projectile from the center of the earth.

The velocity of the projectile is defined as

$$v = \frac{dr}{dt},$$

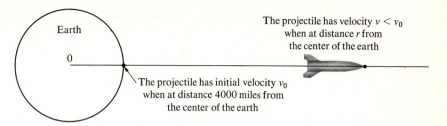

The projectile has velocity $v < v_0$ when at distance r from the center of the earth

Earth

0

The projectile has initial velocity v_0 when at distance 4000 miles from the center of the earth

and its acceleration as $d^2r/dt^2 = dv/dt$. Assume that the acceleration due to gravity is proportional to the force of gravity on the particle; that is,

$$\text{Acceleration} = \frac{dv}{dt} = \frac{-K}{r^2}, \qquad (1)$$

where K is some positive constant. [The negative sign in (1) reminds us that gravity slows the projectile down.]

Before analyzing (1) further, we determine K. At the surface of the earth, where $r = 4{,}000$ miles, the acceleration due to gravity is -32 feet per second per second, which is approximately -0.006 mile per second per second. Thus K satisfies the equation

$$-0.006 = \frac{-K}{4{,}000^2},$$

and

$$K = (4{,}000)^2(0.006). \qquad (2)$$

Now return to (1), which links velocity, time, and distance. It is possible to eliminate time from (1) by using the chain rule:

$$\frac{dv}{dt} = \frac{dv}{dr}\frac{dr}{dt} = \frac{dv}{dr}v.$$

Thus (1) now reads

$$v\frac{dv}{dr} = \frac{-K}{r^2}, \qquad (3)$$

an equation linking velocity and distance. Observe that (3) is equivalent to the equation

$$\frac{d(v^2/2)}{dr} = \frac{d(K/r)}{dr}.$$

Thus

$$\frac{v^2}{2} = \frac{K}{r} + C, \qquad (4)$$

where C is some constant. To determine C, again use information available at the surface of the earth, namely, $v = v_0$ when $r = 4{,}000$. From (4) it follows that

$$\frac{v_0{}^2}{2} = \frac{K}{4{,}000} + C,$$

and

$$C = \frac{v_0{}^2}{2} - \frac{K}{4{,}000}. \qquad (5)$$

Combining (2), (4), and (5) yields

$$\frac{v^2}{2} = \frac{K}{r} + \left(\frac{v_0{}^2}{2} - \frac{K}{4,000}\right)$$

$$= \frac{v_0{}^2}{2} + K\left(\frac{1}{r} - \frac{1}{4,000}\right)$$

$$= \frac{v_0{}^2}{2} + (4,000)^2(0.006)\left(\frac{1}{r} - \frac{1}{4,000}\right).$$

Hence

$$v^2 = v_0{}^2 + (4,000)^2(0.012)\left(\frac{1}{r} - \frac{1}{4,000}\right). \tag{6}$$

Equation (6) describes v as a function of r. If v in (6) is never 0, that is, if the payload never reaches a maximum distance from the earth, then the payload will not fall back to the earth. Thus, by (6), if v_0 is such that the equation

$$0 = v_0{}^2 + (4,000)^2(0.012)\left(\frac{1}{r} - \frac{1}{4,000}\right) \tag{7}$$

has no solution r, then v_0 is large enough to send the payload on an endless journey. To find such v_0, rewrite (7) as

$$(4,000)(0.012) - v_0{}^2 = (4,000)^2(0.012)\left(\frac{1}{r}\right). \tag{8}$$

If the left side of (8) is greater than 0, then there is a solution for r, and the payload will reach a maximum distance; if the left side of (8) is less than or equal to 0, then there is no solution for r. The smallest v_0 that satisfies the inequality $(4,000)(0.012) - v_0{}^2 \leq 0$ is

$$v_0 = \sqrt{(4,000)(0.012)} = \sqrt{48} \doteq 6.93 \text{ miles per second.}$$

The escape velocity is 6.93 miles per second, which is about 25,000 miles per hour. ●

In Sec. 20.3, it will be shown that the *orbit velocity* is equal to the escape velocity divided by $\sqrt{2}$. A satellite in orbit around the earth must have a speed of at least $6.93/\sqrt{2} \doteq 4.90$ miles per second. This is roughly 18,000 miles per hour.

Exercises

Exercises 1 to 9 concern the higher derivatives of a function.

1. Let $f(x) = x^3 - 2x^2 + 5x - 1$.

   (a) Compute $f^{(n)}(x)$ for $n = 1, 2, 3, 4$.

   (b) Find $f^{(n)}(x)$ for $n > 4$.

2. Let $y = \sin 3x$. Compute $\dfrac{dy}{dx}, \dfrac{d^2y}{dx^2}, \dfrac{d^3y}{dx^3}, \dfrac{d^4y}{dx^4},$ and $\dfrac{d^5y}{dx^5}$.

3. Find y', y'', and y''', if $y = \tan^{-1} 2x$.

4. Find $D(y)$, $D^2(y)$, $D^3(y)$, and $D^4(y)$, if $y = x \ln x - x$.

5. Find $f^{(n)}(x)$, for $n = 1, 2, 3$, if $f(x) = \sqrt{1 + 2x}$.

6. Find $D^n(y)$, for $n = 1, 2, 3$, if $y = e^{-x^2}$.

7. Find the first, second, and third derivatives of $\csc 2x$.

8. Find $\dfrac{d^3(\sin^{-1} 3x)}{dx^3}$

9. Find $D_t{}^3(e^{-t} \cos \pi t)$.

Exercises 10 to 13 concern the second derivative.

10. Translate into calculus the following news report about the leaning tower of Pisa. "The tower's angle from the vertical was increasing more rapidly." *Suggestion:* Let $\theta = f(t)$ be the angle of deviation from the vertical at time t.

    Incidentally, the tower, begun in 1174 and completed in 1350, is 179 feet tall and leans about 14 feet from the vertical. Each day it leans, on the average, another $\frac{1}{5,000}$ inch.

11. Show that, if $y = ae^x + be^{-x}$, where a and b are constants, then $d^2y/dx^2 = y$.

12. Show that, if $y = a \sin kt + b \cos kt$, where a and b are constants, then $D_t{}^2(y) = -k^2 y$. Incidentally, this describes the motion of a weight at the end of a spring, bobbing up and down, or any motion whose acceleration is proportional to the displacement from a certain point. Such motion is called *harmonic*.

13. Let $y = (t - 1)^{2/3}$. Show that y satisfies the differential equation

$$\frac{d^2y}{dt^2} = -\frac{2}{9}\frac{1}{y^2}.$$

This differential equation says that the acceleration of y is inversely proportional to the square of y. It describes the motion of an object falling straight toward the earth from space.

Exercises 14 to 22 concern motion in which the acceleration is constant.

14. Let $y = f(t)$ describe the motion on the y axis of an object whose acceleration has the constant value a. Show that

$$y = \frac{a}{2}t^2 + v_0 t + y_0,$$

where v_0 is the velocity when $t = 0$, and y_0 is the position when $t = 0$.

15. In Example 5 the origin of the y axis is at ground level. If the origin is located on top of the cliff, what would be the formulas for y and v as functions of t?

16. (a) How long after the ball in Example 5 is thrown does it pass by the top of the cliff?

    (b) What are its speed and velocity then?

17. If the ball in Example 5 had simply been dropped from the cliff, what would y be as a function of time? How long would the ball fall?

18. In view of the result of Exercise 17, interpret physically each of the three terms on the right side of the formula $y = -16t^2 + 64t + 96$.

19. What is a possible physical interpretation of the solution $2 - \sqrt{10}$ in Example 5?

20. A car accelerates with constant acceleration from 0 (rest) to 60 miles per hour in 15 seconds. How far does it travel in this period? Be sure to do your computations either all in seconds or all in hours; for instance, 60 miles per hour is 88 feet per second.

21. The reaction time of a driver is about 0.6 second. If a car can decelerate at 16 feet per second per second, find the total distance covered if the car is braked at (a) 60 miles per hour, (b) 30 miles per hour, (c) 20 miles per hour.

22. Show that a ball thrown straight up from the ground takes as long to rise as to fall back to its initial position. How does the velocity with which it strikes the ground compare with its initial velocity? Consider the same question for its speed.

■

Exercises 23 to 28 relate to Example 6, the launch of the payload.

23. If a payload is launched with the velocity of 7 miles per second, which is greater than its escape velocity, what happens to its velocity far out in its journey? In other words, determine $\lim_{r \to \infty} v$.

24. If we launch a payload with a speed of 6 miles per second, how far will it go from the center of the earth?

25. At what speed must we launch a payload if it is to reach the moon, 240,000 miles from the center of the earth? (Disregard the gravitational field of the moon.)

26. When a payload is launched with precisely the escape velocity, what happens to its velocity far out in its journey? In other words, determine $\lim_{r \to \infty} v$.

27. (Disregard air resistance.) In order to propel an object 100 miles straight up, what must the launching velocity be if it is assumed that (a) the force of gravity varies as in Example 6, (b) the force of gravity is constant as in Example 4?

28. Could it happen that a projectile shot straight out from the earth neither returns nor travels to "infinity" but approaches a certain finite limiting position?

29. A mass at the end of a spring oscillates. At time t seconds its position (relative to its position at rest) is $y = 6 \sin t$ inches.

(a) Graph y as a function of t.

(b) What is the maximum displacement of the mass?

(c) Show that its acceleration is proportional to its displacement y.

(d) Where is it when its speed is maximum?

(e) Where is it when the absolute value of its acceleration is maximum?

30. Show that if a particle at time t has the x coordinate $3e^{5t} + 6$, then its acceleration is proportional to its velocity.

31. (a) Let a and b be constants and $y = ax + b$. Show that $d^2y/dx^2 = 0$.

(b) Let f be a function defined for all x and having its second derivative equal to 0 for all x. Must $f(x)$ be of the form $ax + b$ for some constants a and b? Explain.

32. (a) Show that the second derivative of any polynomial of degree at most 2, $ax^2 + bx + c$, is constant.

(b) Let f be a function defined for all x and having a constant second derivative. Must $f(x)$ be of the form $ax^2 + bx + c$ for some constants a, `b, and c? Explain.

33. If as an object falls through the air, the air resistance is proportional to its velocity, then its height at time t is given by the formula

$$y = a(1 - e^{-kt}) - bt + c,$$

where a, b, c, and k, are positive constants.

(a) Find its velocity v at time t.

(b) As t gets large, what happens to v?

(c) Find its acceleration at time t.

(d) As t gets large, what happens to the acceleration?

■ ■

34. A particle moving through a liquid meets a "drag" proportional to its velocity; that is, its acceleration is proportional to its velocity.

(a) Show that there is a positive constant k such that

$$\frac{dv}{dt} = -kv.$$

(b) Show that there is a constant A such that

$$v = Ae^{-kt}.$$

(c) Show that there is a constant B such that

$$x = -\frac{1}{k} Ae^{-kt} + B.$$

(d) How far does the particle travel as t goes from 0 to ∞?

6.7

The geometric significance of the sign of the second derivative

Whether the first derivative is positive, negative, or zero tells a good deal about a function and its graph. This section will explore the geometric significance of the second derivative being positive, negative, or 0.

Assume that $f''(x)$ is positive for all x in the open interval (a, b). Since f'' is the derivative of f', it follows that f' is an increasing function throughout the interval (a, b). In other words, as x increases, the slope of the graph of $y = f(x)$ increases as we move from left to right on that part of the graph corresponding to the interval (a, b). The slope may increase from negative to positive values as in the first diagram. Or the slope may be positive

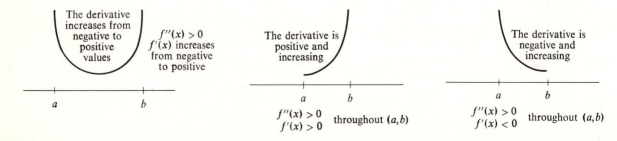

| The derivative increases from negative to positive values | $f''(x) > 0$ $f'(x)$ increases from negative to positive | The derivative is positive and increasing | The derivative is negative and increasing |

$a \qquad b$

$a \qquad b$
$f''(x) > 0$
$f'(x) > 0$ throughout (a,b)

$a \qquad b$
$f''(x) > 0$
$f'(x) < 0$ throughout (a,b)

throughout (a, b), and increasing, as in the second diagram. Or the slope may be negative throughout (a, b), and increasing. (See diagrams on page 210.)

As you drive along such a graph from left to right, your car keeps turning to the left.

Chord

Tangent line

A concave upward curve lies above its tangent lines and below its chords.

DEFINITION   *Concave upward.* A function f whose second derivative is positive throughout the open interval (a, b) is called *concave upward* in that interval.

Note that, when a function is concave upward, it is shaped like part of a cup (Concave *UP*ward).

It can be proved that where a curve is concave upward it lies *above its tangent lines* and *below its chords.* (See Exercises 34 to 36.)

If, on the other hand, $f''(x)$ is negative throughout (a, b), then f' is a decreasing function and the graph of f looks like part of the curve to the left.

$f''(x) < 0$ graph is concave downward

a ⊢——————⊣ b

DEFINITION   *Concave downward.* A function whose second derivative is negative throughout an open interval (a, b) is called *concave downward* in that interval.

Where a function is concave downward, it lies *below its tangent lines* and *above its chords.*

EXAMPLE 1   Where is the graph of $f(x) = x^3$ concave upward? Concave downward?

SOLUTION   We compute the second derivative. Since $D(x^3) = 3x^2$, $D^2(x^3) = 6x$.

Clearly $6x$ is positive for all positive x and negative for all negative x. The graph is concave upward if $x > 0$ and concave downward if $x < 0$. Note that the sense of concavity changes at $x = 0$. When you drive along this curve from left to right, your car turns to the right until you pass through $(0, 0)$. Then it starts turning to the left. ●

y

Concave upward

$y = x^3$

x

Concave downward

EXAMPLE 2   Examine the concavity of the curve $f(x) = xe^x$ and graph the curve.

SOLUTION   In this case,
$$f'(x) = (x + 1)e^x,$$
and
$$f''(x) = (x + 2)e^x.$$

For which x is $f''(x)$ positive? Negative?

Since e^x is positive for all x, the sign of $f''(x)$ is determined by the sign of $x + 2$. Clearly
$$x + 2 \text{ is positive when } x > -2$$
and
$$x + 2 \text{ is negative when } x < -2.$$

Thus $x = -2$ is an important number in determining the shape of the graph of $y = xe^x$.

Before graphing $f(x) = xe^x$, find those points where $f(x) = 0$ and where $f'(x) = 0$. Since $f(x) = xe^x$, $f(x) = 0$ only when $x = 0$. Since the derivative

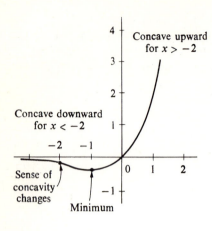

Concave upward
for $x > -2$

Concave downward
for $x < -2$

Sense of
concavity
changes

Minimum

$f'(x) = (x + 1)e^x$, it is 0 only when $x = -1$. Note that $f'(x)$ is positive if $x > -1$ and negative if $x < -1$. Hence at $x = -1$ the curve has a minimum. For convenience, tabulate $f(x)$ for $x = 0, -1, -2$, and a few other numbers.

x	xe^x
-2	$-2e^{-2} \doteq -0.3$
-1	$-1e^{-1} \doteq -0.4$
0	$0e^0 = 0$
1	$1e^1 \doteq 2.7$
2	$2e^2 \doteq 14.8$

Plotting the first four points and observing the sign of f' and of f'', we obtain the graph of $y = xe^x$ shown at the left. ●

The sense of concavity is a useful tool in sketching the graph of a function. Of special interest in both examples is the presence of a point on the graph where the sense of concavity changes. Such a point is called an *inflection point*.

DEFINITION *Inflection point.* Let f be a function and let a be a number. Assume that there are numbers b and c such that $b < a < c$ and

1. f is continuous on the open interval (b, c);
2. $f''(x)$ is positive for all $x < a$ in the interval and is negative for all $x > a$ in the interval, or vice versa.

Then the point $(a, f(a))$ is called an *inflection point* or *point of inflection*.

In short, at an inflection point the second derivative changes sign.
If the second derivative exists at an inflection point, it must be 0. But there can be an inflection point even if f'' is not defined there, as shown by the next example, which is closely related to Example 1. .

EXAMPLE 3 Examine the concavity of $y = x^{1/3}$.

SOLUTION Here

$$y' = \tfrac{1}{3}x^{-2/3},$$

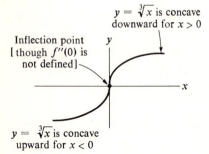

$y = \sqrt[3]{x}$ is concave
downward for $x > 0$

Inflection point
[though $f''(0)$ is
not defined]

$y = \sqrt[3]{x}$ is concave
upward for $x < 0$

and

$$y'' = \frac{1}{3}\frac{-2}{3}x^{-5/3}.$$

Neither y' nor y'' is defined at 0; however, the sign of y'' changes at 0. When x is negative, y'' is positive; when x is positive, y'' is negative. The concavity switches from upward to downward at $x = 0$.

Compare the graph at the left with the one in Example 1. Since the cubing function and the cube root function are inverses of each other, one graph is obtained from the other by reflection in the line $y = x$. ●

When graphing a function f, find where $f(x) = 0$, where $f'(x) = 0$, and where $f''(x) = 0$, if the solutions are easy.   Determine where $f'(x)$ is positive and where it is negative.   Determine also where $f''(x)$ is positive and where it is negative.   The following table contrasts the interpretations of the signs of f, f', and f''.   (It is assumed that f, $f^{(1)}$, and $f^{(2)}$ are continuous.)

	Is Positive	*Is Negative*	*Changes Sign*
Where the ordinate $f(x)$	The graph is above the x axis	The graph is below the x axis	The graph crosses the x axis
Where the slope $f'(x)$	The graph slopes upward	The graph slopes downward	The graph has a horizontal tangent and a relative maximum or minimum
Where $f''(x)$	The graph is concave upward (like a cup)	The graph is concave downward	The graph has an inflection point

Keep in mind that the graph can have an inflection point at x_0, even though the second derivative is not defined at x_0 (Example 3).   Similarly, a graph can have a maximum or minimum at x_0, even though the first derivative is not defined at x_0.   (Consider $f(x) = |x|$ at $x_0 = 0$.)

The second derivative is also useful in searching for relative maxima or minima.   For instance, let a be a critical number for the function f and assume that $f''(a)$ happens to be negative.   If f'' is continuous in some open interval that contains a, then $f''(a)$ remains negative for a suitably small open interval that contains a.   This means that the graph of f is concave downward near $(a, f(a))$, hence lies below its tangent lines.   In particular, it lies below the horizontal tangent line at the critical point $(a, f(a))$.   Thus the function has a *relative maximum* at the critical number a.   This observation suggests the following test for a relative maximum or minimum.

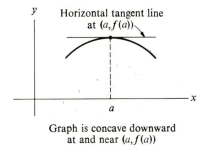

Horizontal tangent line at $(a, f(a))$

Graph is concave downward at and near $(a, f(a))$

THEOREM   *Second-derivative test for local maximum or minimum.*   Let f be a function with continuous derivative and second derivative.   Let a be a critical number for f, that is, $f'(a) = 0$.

If $f''(a) < 0$, f has a local maximum at a.
If $f''(a) > 0$, f has a local minimum at a.   ●

As Exercise 38 shows, the assumptions can be weakened.

EXAMPLE 4   Find all local maxima or minima of

$$f(x) = x^3 e^{-x}.$$

SOLUTION   Since f is differentiable, a local maximum or minimum can occur only at a critical number.

Now
$$f'(x) = (3x^2 - x^3)e^{-x}$$
$$= x^2(3 - x)e^{-x}.$$

The critical numbers are thus

$$x = 3 \quad \text{and} \quad x = 0.$$

Next, compute $f''(x)$:

$$f''(x) = (x^3 - 6x^2 + 6x)e^{-x}.$$

At the critical numbers the second derivative has the values

$$f''(3) = -9e^{-3}$$

and
$$f''(0) = 0.$$

Since $f''(3)$ is negative, there is a local maximum at 3. There is no need to determine whether $f'(x)$ changes sign at 3.

At the critical number 0, the second-derivative test gives no information, since it is neither positive nor negative. In this case we look more carefully at $f'(x)$ or $f(x)$ for x near 0.

When x is near 0 (but not 0), the derivative

$$f'(x) = x^2(3 - x)e^{-x}$$

is positive, since e^{-x} is positive, x^2 is positive, and $3 - x$ is positive. Thus f is an increasing function near 0. It has neither a local maximum nor minimum there.

Though it is not needed in the solution, a graph of the function helps clarify the behavior of the function at its two critical numbers.

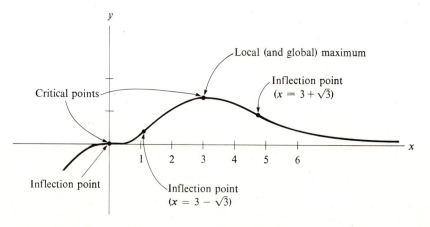

As Example 4 shows, if the second derivative is zero at a critical point, there may be neither a local maximum nor minimum there. In fact, just about anything could happen. To see this, consider the functions x^3, x^4, and $-x^4$ at $x = 0$. For each of them it is easy to check that 0 is a critical number and also that the second derivative is 0 there. However, the function x^3 has neither a local maximum nor local minimum there; x^4 has a local minimum; $-x^4$ has a local maximum. The reader may sketch their graphs in the vicinity of $(0, 0)$ to check these assertions.

Exercises

1. Let $f(x) = x^3 - 6x^2 - 15x$.
   (a) Where does f change sign?
   (b) Where does f' change sign?
   (c) Where does f'' change sign?
   (d) Compute $f(x)$ at these points and make a rough sketch of the graph of f.
2. Graph $y = x^3 - 3x^2$, showing where $f(x) = 0$, $f'(x) = 0$, $f''(x) = 0$, and where f' and f'' change sign.
3. Graph $y = x^2/2 + 1/x$, showing where y' and y'' change sign.
4. Let $f(x) = 2x^4 - 4x^3$.
   (a) Where does the graph slope upward?
   (b) Where is the graph concave upward?
   (c) Where does the ordinate change sign?
   (d) Where does the slope change sign?
   (e) Where are the inflection points, if any?
   (f) Sketch the curve with the aid of (a) through (e).
5. Let $y = 1/(1 + x^2)$. Where do y' and y'' change sign? Make a free hand graph showing these data.
6. Let $f(x) = \sin x$. For which values of x is $f(x) > 0$? $f'(x) > 0$? $f''(x) > 0$? Where do f, f', and f'' change sign? Show this information on a graph of $y = \sin x$.
7. Graph $y = e^{-x^2}$ and show its inflection points.
8. Let $y = x^4$.
   (a) Is y'' ever 0?
   (b) Does the graph have any inflection points? Explain.
   (c) Does the graph have any relative maxima or minima? Explain.
9. Graph $y = xe^{-x}$, showing where y' and y'' change sign.
10. Graph $y = x^4 + 2x^3$, showing the inflection points.
11. Graph $y = 3x^5 - 5x^4$, showing inflection points.
12. (a) Graph $y = xe^{x^2}$.
    (b) Does the graph in (a) have any inflection points?
13. The graph in the next column describes a function f.
    (a) Where does f change sign?
    (b) Where is $f(x) \geq 0$?
    (c) Where does f' change sign?
    (d) Where is $f'(x) \geq 0$?
    (e) Where does f'' change sign?
    (f) Where is $f''(x) \geq 0$?

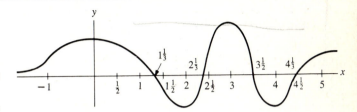

14. Let $f(x) = e^x \sin x$.
    (a) For which x is $f'(x) = 0$?
    (b) For which x is $f''(x) = 0$?
    (c) How many inflection points does the graph have?
15. Graph $y = \sin x + \cos x$, showing maxima, minima, and inflection points.
16. Graph $y = 1/(x^3 + 1)$ for $x > -1$, showing inflection points.
17. Graph $y = x^4 + \ln x$ for $x > 0$. Is y' ever 0? Is y'' ever 0? Show any inflection points.
18. Find the inflection points of $y = \tan^{-1} x$.

In Exercises 19 to 27 determine all relative maxima and minima of the given function. Use the second-derivative test, where it applies.

19. $y = x^2 e^{-x}$
20. $y = x^4 e^{-x}$
21. $y = x^5 e^{-x}$
22. $y = x^2 \ln x$
23. $y = x^4 + 2x^3$
24. $y = x^3 - 3x^2 + 3x$
25. $y = x + \dfrac{1}{x}$
26. $y = \dfrac{1}{x^2 + 2}$
27. $y = (x^2 - x)e^x$

■

28. Sketch the graph of a function f such that for all x:
    (a) $f(x) > 0, f'(x) > 0, f''(x) > 0$;
    (b) $f'(x) < 0, f''(x) < 0$.
    (c) Can there be a function such that $f(x) > 0, f'(x) < 0$, $f''(x) < 0$ for all x? Explain.
29. Let f be a function such that $f''(x) = (x - 1)(x - 2)$.
    (a) For which x is f concave upward?
    (b) For which x is f concave downward?

30. Find a specific function f whose second derivative is $(x - 1)(x - 2)$.

31. Sketch the graph of $y = f(x)$ near $(1, 1)$ if (a) $f(1) = 1$, $f'(1) = 0, f''(1) = -1$; (b) $f(1) = 1, f'(1) = -1, f''(1) = 1$. (Assume f'' is continuous.)

32. A certain function $y = f(x)$ has the property that

$$y' = e^y + 2y + x.$$

Show that at a critical number the function has a local minimum.

33. In natural growth, discussed in Sec. 6.4, the size of the population is unrestricted. In the theory of *inhibited growth* it is assumed that the growing quantity y approaches some limiting size M. Specifically one assumes that the rate of growth is proportional both to the amount present and to the amount left to grow:

$$\frac{dy}{dt} = ky(M - y).$$

Prove that the graph of y as a function of time has an inflection point when the amount y is exactly half the amount M.

■ ■

34. Let f be a function such that $f(0) = 0 = f(1)$ and $f''(x) \geq 0$ for all x in $[0, 1]$.
    (a) Using a sketch, explain why $f(x) \leq 0$ for all x in $[0, 1]$.
    (b) Without a sketch prove that $f(x) \leq 0$ for all x in $[0, 1]$.

35. (See Exercise 34.) Prove that if f is a function such that $f''(x) > 0$ for all x, then the graph of $y = f(x)$ lies below its chords; i.e., $f(ax_1 + (1 - a)x_2) < af(x_1) + (1 - a)f(x_2)$ for any a between 0 and 1, and for any x_1 and x_2.

36. Prove, without using a picture, that where the graph of f is concave upward it lies above its tangent

37. Prove (without referring to a picture) that if the graph of f lies above its tangent lines for all x in $[a, b]$, then $f''(x) \geq 0$ for all x in $[a, b]$. [The case $y = x^4$ shows that we should not try to prove that $f''(x) > 0$.]

38. The following theorem strengthens the second-derivative test: Let f be a function differentiable over an open interval that contains a. Assume that $f'(a) = 0$ and that $f''(a)$ exists and is negative. Prove that f has a local maximum at a. *Hint:* Show that the first derivative of f changes sign at a (from positive to negative as x increases).

6.8

Applied maximum and minimum problems

One of the most important applications of calculus is obtaining the most efficient design of a product. Frequently the problem of minimizing cost or maximizing the volume of a certain object reduces to maximizing or minimizing some function $f(x)$. In that case, the methods developed in Secs. 6.5 and 6.7 may be called on. They are the use of critical points, the first-derivative test, and the second-derivative test. Recall that, when maximizing or minimizing a function over a closed interval, it is essential to consider also the values of the function at the endpoints.

The five examples that follow are typical. The only novelty is the challenge of how to translate each problem into the terminology of functions.

A piece of cardboard

x x

x x

$12 - 2x$

$12 - 2x$

x x

x x

Tray

x

$12 - 2x$

$12 - 2x$

EXAMPLE I If we cut four congruent squares out of the corners of a square piece of cardboard 12 inches on each side, we can fold up the four remaining flaps to obtain a tray without a top. What size squares should be cut in order to maximize the volume of the tray?

SOLUTION Let us remove squares of side x, as shown in the two diagrams at the left. Folding on the dotted lines, we obtain a tray of volume

$$V(x) = (12 - 2x)^2(x) = 4x^3 - 48x^2 + 144x.$$

Since each side of the cardboard square has length 12 inches, the only values of x which make sense are those in $[0, 6]$. Thus, we wish to find the number x in the closed interval $[0, 6]$ that maximizes $V(x)$.

Notice that $V(x) = (12 - 2x)^2(x)$ is small when x is near 0 (that is, when we try to economize by making the height of the tray small) and small when x is near 6 (that is, when we try to economize by making the base small). We have a "two-influence" problem; to find the best balance between them, we use calculus.

The maximum value of $V(x)$ for x in $[0, 6]$ occurs either at 0, 6, or at a critical number [where $V'(x) = 0$]. Now, $V(0) = 0$ (the tray has height 0), and $V(6) = 0$ (the tray has a base of area 0). These are minimum values for the volume, certainly not the maximum volume, so the maximum must occur at some critical number in $(0, 6)$.

Next compute $V'(x)$:

$$
\begin{aligned}
V'(x) &= (4x^3 - 48x^2 + 144x)' \\
&= 12x^2 - 96x + 144 \\
&= 12(x^2 - 8x + 12) \\
&= 12(x - 6)(x - 2).
\end{aligned}
$$

The equation $\qquad 12(x - 6)(x - 2) = 0$

has two roots in $[0, 6]$, namely 2 and 6. The critical numbers are 2 and 6. As already remarked, the maximum does not occur at 0 or 6. Hence it occurs at $x = 2$. When $x = 2$, the volume is

$$
\begin{aligned}
V(2) &= [12 - 2(2)]^2(2) \\
&= 8^2 \cdot 2 \\
&= 128 \text{ cubic inches.}
\end{aligned}
$$

This is the largest possible volume and is obtained when the length of the cut is 2 inches.

As a matter of interest, let us graph the function V, showing its behavior for all x, not just for values of x significant to the problem. Note (in the accompanying figure) that at $x = 6$ the tangent is horizontal. ●

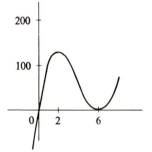

Only values of x in the portion above $[0, 6]$ correspond to physically realizable trays

EXAMPLE 2 A couple have enough wire to construct 100 feet of fence. They wish to use it to form three sides of a rectangular garden, one side of which is along a building. What shape garden should they choose in order to enclose the largest area?

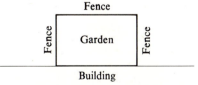

The three sides of the garden not along the building total 100 feet

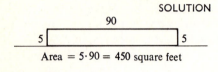

Area = 5·90 = 450 square feet

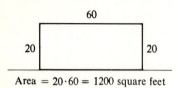

Area = 20·60 = 1200 square feet

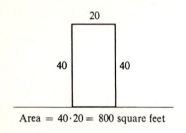

Area = 40·20 = 800 square feet

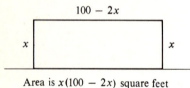

Area is x(100 − 2x) square feet

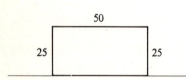

SOLUTION The three accompanying figures show some possible ways of laying out the 100 feet of fence. For convenience, let x denote the length of the side of the garden that is perpendicular to the building, and y the length of the side parallel to the building. Since 100 feet of fencing is available, $2x + y = 100$. When $x = 5$, the area is 450 square feet. When x has increased to 20, y has decreased to 60 feet, and the area is 1,200 square feet. When $x = 40$, $y = 20$, and the area is only 800 square feet. It is not immediately clear how to choose x to maximize the area of the garden. What makes the problem interesting is that, when you increase one dimension of the rectangle, the other automatically decreases. The area, which is the product of the two dimensions, is subject to two opposing forces, one causing it to increase, the other to decrease. This type of problem is easily solved with the aid of the derivative.

First of all, express the area A of the garden in terms of x and y:

$$A = xy.$$

Then use the equation $100 = 2x + y$ to express y in terms of x:

$$y = 100 - 2x.$$

Thus the area A is

$$A = x(100 - 2x) \qquad \text{square feet.}$$

Clearly $0 \le x \le 50$. Thus the problem now has become: Maximize $f(x) = x(100 - 2x)$ for x in $[0, 50]$.

In this case $f(x) = 100x - 2x^2$; hence

$$f'(x) = 100 - 4x.$$

Set the derivative equal to 0;

$$0 = 100 - 4x,$$

or

$$4x = 100;$$

hence

$$x = 25.$$

Thus 25 is the only critical number for the function. The maximum of f occurs either at 25 or at one of the ends of the interval, 0 or 50. Now,

$$f(0) = 0[100 - 2(0)] = 0,$$
$$f(50) = 50[100 - 2(50)] = 0,$$
$$f(25) = 25[100 - 2(25)] = 1,250.$$

Thus the maximum possible area is 1,250 square feet, and the fence should be laid out as shown in the figure in the margin. ●

Examples 1 and 2 illustrate the general procedure for solving applied maximum (or minimum) problems.

1. Name the various quantities in the problem by letters, such as x, y, A, V.
2. Express the quantity to be maximized in terms of one or more other letters.

3. By eliminating variables, express the quantity to be maximized as a function f of one variable.
4. Maximize the function f obtained in step 3.

Example 3 illustrates this procedure.

EXAMPLE 3   A package to be sent parcel post must have a combined girth (distance around) and length of at most 100 inches. What are the dimensions of the largest box that can be sent if its base is a square?

SOLUTION   First label the dimensions of a typical box with a square base. The side of the box is x; the height is y.

The quantity to be maximized is the volume; denote it by V. Observe that

$$V = \text{area of base} \cdot \text{height}$$
$$= x^2 y.$$

Since the problem is to find the largest permissible box, it may be assumed that

$$\text{Girth} + \text{length} = 100,$$

or $\qquad\qquad\qquad\qquad 4x + y = 100.$

This latter equation can be used to express y in terms of x or x in terms of y. Since it is easier to solve for y, rewrite the equation as

$$y = 100 - 4x.$$

Now the volume V can be expressed completely in terms of x:

$$V = x^2 y$$
$$= x^2(100 - 4x)$$
$$= 100x^2 - 4x^3.$$

The values of x that are of interest are in the interval $[0, 25]$.

Next compute the derivative of V, dV/dx, and find where it is 0:

$$\frac{dV}{dx} = \frac{d(100x^2 - 4x^3)}{dx}$$
$$= 200x - 12x^2.$$

Hence consider the equation

$$200x - 12x^2 = 0,$$

or $\qquad\qquad\qquad\qquad x(200 - 12x) = 0.$

The solutions of this equation are

$$x = 0 \qquad \text{and} \qquad x = \tfrac{200}{12} = 16\tfrac{2}{3}.$$

The maximum volume occurs either when $x = 16\frac{2}{3}$ or at an endpoint of the interval $[0, 25]$. But when $x = 0$ or $x = 25$, the box has a volume of zero (since either the base has zero area or the height is zero). Thus the maximum volume occurs when

$$x = 16\tfrac{2}{3} \text{ inches.}$$

The height of this largest box is

$$y = 100 - 4(16\tfrac{2}{3})$$
$$= 100 - 66\tfrac{2}{3}$$
$$= 33\tfrac{1}{3} \text{ inches.}$$

(The box is twice as high as it is wide.) ●

The next example minimizes a function defined for all positive numbers.

EXAMPLE 4   Of all the tin cans that enclose a volume of 100 cubic inches, which requires the least metal?

SOLUTION   Denote the radius of a can of volume 100 cubic inches by r, and its height by h. The can may be flat or tall. If the can is flat, the side uses little metal, but then the top and bottom bases are large. If the can is shaped like a mailing tube, then the two bases require little metal, but the curved side requires a great deal of metal. What is the ideal compromise between these two extremes?

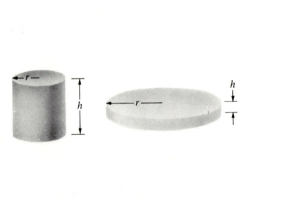

A tin can of volume 100 cubic inches; $\pi r^2 h = 100$. The can may be flat or long

The surface area S of the can is given by

$$S = 2\pi r^2 + 2\pi rh, \tag{1}$$

which accounts for the two circular bases and the side. Since the amount of metal in the can is proportional to S, it suffices to minimize S.

In the tin can under consideration the radius and height are related by the constraint

$$\pi r^2 h = 100. \tag{2}$$

In order to express S as a function of one variable, use Eq. (2) to eliminate either r or h. Choosing to eliminate h, we solve Eq. (2) for h:

$$h = \frac{100}{\pi r^2}.$$

Substitution into Eq. (1) yields

$$S = 2\pi r^2 + 2\pi r \frac{100}{\pi r^2},$$

or
$$S = 2\pi r^2 + \frac{200}{r}. \tag{3}$$

Equation (3) expresses S as a function of just one variable r. The domain of this function for our purposes is $r > 0$, since the tin can has a positive radius.

Compute dS/dr:

$$\frac{dS}{dr} = 4\pi r - \frac{200}{r^2}$$

$$= \frac{4\pi r^3 - 200}{r^2}. \tag{4}$$

This derivative is 0 only when

$$4\pi r^3 = 200, \tag{5}$$

that is, when
$$r = \sqrt[3]{\frac{50}{\pi}}.$$

Thus $r = \sqrt[3]{50/\pi}$ is the only critical number. Does it in fact provide a minimum?

First let us check by the second-derivative test. Differentiation of Eq. (4) yields

$$\frac{d^2S}{dr^2} = 4\pi + \frac{400}{r^3}.$$

When $r = \sqrt[3]{50/\pi}$, it follows that $r^3 = 50/\pi$. Thus, when $r = \sqrt[3]{50/\pi}$,

$$\frac{d^2S}{dr^2} = 4\pi + \frac{400\pi}{50}$$

$$= 12\pi.$$

Since d^2S/dr^2 is positive, and of course $dS/dr = 0$, we have a local minimum. But is it a global minimum?

The first derivative will enable us to answer this question. Recall that

$$\frac{dS}{dr} = \frac{4\pi r^3 - 200}{r^2}.$$

At the critical number the numerator is 0. If r is less than the critical number, the numerator, hence the quotient, is *negative*. If r is larger than the critical number, the quotient is positive. Thus the function decreases for $0 < r < \sqrt[3]{50/\pi}$ and increases for $r > \sqrt[3]{50/\pi}$. Thus the critical number indeed provides an absolute or global minimum.

The same conclusion could have been reached with the aid of the second derivative,

$$\frac{d^2S}{dr^2} = 4\pi + \frac{400}{r^3},$$

which is clearly positive for all positive r. A critical point on a curve that is concave upward everywhere is a global minimum for that curve. ●

The first four examples all had a geometric flavor. The final example describes an important and typical problem in the economics of warehousing. The interested reader may pursue this topic in C. R. Carr and C. W. Howe, *Quantitative Decision Procedures in Management and Economics*, McGraw-Hill, New York, 1964, in particular pp. 10–14.

EXAMPLE 5 *Ideal lot problem.* A firm sells A units of a certain item at a constant rate during the year. (We are assuming that there are no peak periods and no element of chance.) Goods purchased in a single order are delivered in one lot. If the firm orders all A units delivered at the beginning of the year, then it saves on *reorder* costs (such as secretarial work and delivery fees), but it incurs higher *carrying* costs (since the average inventory throughout the year is $A/2$). If it orders every day, it keeps the average inventory low, but then the reorder costs may become prohibitive. To arrive at the happy mean, the firm wishes to minimize cost as a function of the size of the order. How should the firm order?

SOLUTION Let $C(x)$ be the total annual carrying and reorder costs when the firm orders x units in each lot. It therefore places A/x orders per year. Let us assume that the cost of placing one order is made up of a fixed cost F (for instance, stationery) and a cost Px which is a linear function of the size of the order (such as packaging and shipping). Then the total reorder cost for the year is

$$(F + Px)\frac{A}{x}. \tag{6}$$

[Note from (6) that smaller lots increase reorder cost.] Assume that the annual carrying cost for one unit is I. When the lot size is x, the average inventory is $x/2$. Thus the carrying cost for 1 year is

$$I \cdot \frac{x}{2}. \tag{7}$$

[Note from (7) that smaller lots decrease annual carrying costs.] Combining (6) and (7) gives

$$C(x) = \frac{Ix}{2} + \frac{(F + Px)A}{x} = \frac{Ix}{2} + \frac{FA}{x} + PA. \tag{8}$$

To study the function C (for $x > 0$), examine dC/dx. We have

$$\frac{dC}{dx} = \frac{I}{2} - \frac{FA}{x^2} = \frac{Ix^2 - 2FA}{2x^2}. \qquad (9)$$

From (9) it follows that dC/dx is negative when x satisfies the inequality $Ix^2 < 2FA$, but positive when x satisfies $Ix^2 > 2FA$. When $Ix^2 = 2FA$, that is (since x is positive), when

$$x = \sqrt{\frac{2FA}{I}}, \qquad (10)$$

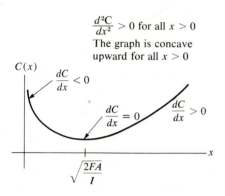

$\frac{d^2C}{dx^2} > 0$ for all $x > 0$

The graph is concave upward for all $x > 0$

$C(x)$

$\frac{dC}{dx} < 0$

$\frac{dC}{dx} = 0$ $\frac{dC}{dx} > 0$

$\sqrt{\frac{2FA}{I}}$

$dC/dx = 0$. Thus $C(x)$ decreases when x is less than $\sqrt{2FA/I}$ and increases when x is greater than $\sqrt{2FA/I}$. At $x = \sqrt{2FA/I}$, the function $C(x)$ reaches a minimum value. The graph looks like the one in the margin.

The function $C(x)$ has a global minimum, for $x > 0$, at $\sqrt{2FA/I}$. However, the firm may not, for practical reasons, be able to order an amount near this number. After all, it may wish to order at least once a year. Thus it would not order more than A units at a time; hence we may consider only $x \le A$. If A happens to be larger than $\sqrt{2FA/I}$, there is no restriction. But if A is less than $\sqrt{2FA/I}$, then the function $C(x)$ should be considered only on the domain $0 < x \le A$. A glance at the graph shows that in this case the minimum occurs at A: The firm should order once a year.

If A is larger than $\sqrt{2FA/I}$, the ideal lot size is $\sqrt{2FA/I}$, hence is proportional to $\sqrt{A}$. Then if A *quadruples*, the order size should *double*; hence the frequency should double. This is wiser than simply to quadruple the order size and keep the same frequency. ●

Exercises

1. (a) A tray is constructed, as in Example 1, by cutting four squares each with side x out of a rectangular piece of cardboard whose dimensions are 6 inches by 12 inches. Show that the volume of the tray is $V = f(x) = 4x^3 - 36x^2 + 72x$ cubic inches.
   (b) Graph V as a function of x for all x, not just for the x significant to the problem.
   (c) Why do we wish to examine $f(x)$ only for x in $[0, 3]$?
   (d) Using $f'(x)$, find the number in $[0, 3]$ that yields the maximum value of $f(x)$.

2. Find the maximum volume of a tray that can be made in the manner of Example 1 from a rectangular piece of paper 4 by 5 inches.

3. (a) Find the largest area of all rectangles whose perimeter is 4 inches.
   (b) Find the shortest perimeter of all rectangles whose area is 1 square inch.

4. (a) How should one choose two nonnegative numbers whose sum is 1 in order to maximize the sum of their squares?

   (b) To minimize the sum of their squares?

5. How should one choose two nonnegative numbers whose sum is 1 in order to maximize the product of the square of one of them and the cube of the other?

6. How should one choose two nonnegative numbers whose sum is 1 in order to maximize the sum of the square of one of them and the cube of the other?

7. The left-hand x centimeters of a string 12 centimeters long have a mass of $18x^2 - x^3$ grams.
   (a) What is its density x centimeters from the left-hand end?
   (b) Where is its density greatest?

8. An irrigation channel made of concrete is to have a cross section in the form of an isosceles trapezoid, three of

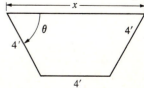

whose sides are 4 feet long. How should the trapezoid be shaped if it is to have the maximum possible area?
(a) Consider the area as a function of x and solve.
(b) Consider the area as a function of θ and solve.
(c) Do the two solutions agree? Explain.

9. The height of a ball t seconds after it is thrown straight up is $-16t^2 + 40t$ feet. (a) When does it reach its maximum height? (b) How high does it go?

10. If you have 100 feet of fence to enclose a rectangular garden, show that the fence should be laid out as a square to enclose the maximum area.

11. An optometrist knows that when the price of sunglasses is p, for $0 < p \leq 3$, then the number he can sell is $9 - p^2$ thousands.
(a) What is the maximum revenue if the optometrist has to set a single fixed price?
(b) How should the optometrist set the prices to achieve a maximum revenue if he can set two fixed prices, one of them 1 dollar higher than the other?

12. The cost of operating a certain truck (for gasoline, oil, and depreciation) is $(5 + s/10)$ cents per mile when it travels at a speed of s miles per hour. A truck driver earns $3.60 per hour. What is the most economical speed at which to operate the truck during a 600-mile trip?
(a) If you considered only the truck, would you want s to be small or large?
(b) If you considered only the expense of the driver's wages, would you want s to be small or large?
(c) Express cost as a function of s and solve.

13. The stiffness of a rectangular beam is proportional to the product of the width and the cube of the height of its cross section. What shape beam should be cut from a log in the form of a right circular cylinder of radius r in order to maximize its stiffness?

14. A printer is planning to produce 200,000 copies of an advertisement. It costs 2 dollars per hour to run his press, which can turn out 1,200 impressions per hour. A metal copy of set type costs 55 cents. How many copies of the type should the printer use on his press to minimize his cost? Note: The more copies of type used, the more advertisements he produces on each impression. However, if he makes too many copies of the type, the cost of the type will be large.

15. The *information content* or *entropy* of a binary source (such as a telegraph that transmits dots and dashes) whose two values occur with probabilities p and $1 - p$ is defined as $H(p) = -p \ln p - (1 - p) \ln (1 - p)$, where $0 < p < 1$. Show that H has a maximum at $p = \frac{1}{2}$. The practical significance of this result is that, for maximum flow of information per unit time, dots and dashes should, in the long run, appear in equal proportions.

16. (See Exercise 15.) Let p be fixed so that $0 < p < 1$. Define $M(q) = -p \ln q - (1 - p) \ln (1 - q)$. Show that $H(p) \leq M(q)$ for $0 < q < 1$ and that equality holds if and only if $p = q$.

17. When a tract of timber is to be logged, a main logging road is built from which small roads branch off as feeders. The question of how many feeders to build arises in practice. If too many are built, the cost of construction would be prohibitive. If too few are built, the time spent moving the logs to the roads would be prohibitive. The formula for total cost,

$$y = \frac{CS}{4} + \frac{R}{VS},$$

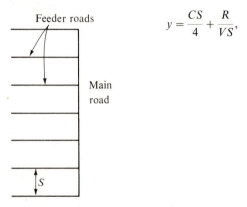

Feeder roads

Main road

is used in a logger's manual to find how many feeder roads are to be built. R, C, and V are known constants: R is the cost of road at "unit spacing"; C is the cost of moving a log a unit distance; V is the value of timber per acre. S denotes the distance between the regularly spaced feeder roads. Thus the cost y is a function of S, and the object is to find that value of S that minimizes y. The manual says, "To find the desired S set the two summands equal to each other and solve:

$$\frac{CS}{4} = \frac{R}{VS}."$$

Show that the method is valid.

18. A delivery service is deciding how many warehouses to set up in a large city. The warehouses will serve similarly shaped regions of equal area A and, let us assume, an equal number of people.
(a) Why would transportation costs per item presumably be proportional to $\sqrt{A}$?
(b) Assuming that the warehouse cost per item is inversely proportional to A, show that C, the cost of transportation and storage per item, is of the form $t\sqrt{A} + w/A$, where t and w are appropriate constants.
(c) Show that C is a minimum when $A = (2w/t)^{2/3}$.

19. On one side of a river 1 mile wide is an electric power

station; on the other side, s miles upstream, is a factory. It costs 3 dollars per foot to run cable over land, and 5 dollars per foot under water. What is the most economical way to run cable from the station to the factory?

(a) Using no calculus, what do you think would be (approximately) the best route if s were very small? if s were very large?

(b) Solve with the aid of calculus, and draw the routes for $s = \frac{1}{2}, \frac{3}{4}, 1$, and 2.

(c) Solve for arbitrary s.

20. (a) The crew from the power station of Exercise 19 has to inspect the connection at the factory. Their boat travels 9 miles per hour and their truck 15 miles per hour. What route should they take to reach the factory in the least time if they cross the river first?

(b) Compare the answer to (a) with the answer to Exercise 19.

21. Find the volume of the largest right circular cone that can be inscribed in a sphere of radius a.

22. What point on the line $y = 3x + 7$ is closest to the origin? (Instead of minimizing the distance, it is much more convenient to minimize the square of the distance. Doing so avoids square roots.)

23. Find the shape of the largest right circular cylinder that can be sent through the mails. The combined height and circumference can not be larger than 100 inches.

■

24. Of all right circular cones with fixed volume V, which shape has the least surface area, including the area of the base? (The area of the curved part of a cone of slant height l and radius r is $\pi r l$.)

25. Of all right circular cones with fixed surface area A (including the area of the base), which shape has the largest volume?

26. Of all cylindrical tin cans with circular bases made with a given amount of metal, what is the shape of the one that has the greatest volume? *Hint:* Call the given amount A, which is fixed.

27. A rectangular box-shaped house is to have a square floor. Three times as much heat per square foot enters through the roof as through the walls. What shape should the house be if it is to enclose a given volume and minimize heat entry? (Assume no heat enters through the floor.)

28. If the tin can in Example 4 is not to have a lid on top, what dimensions should it have to use the least metal?

29. If the top and bottom of the can in Example 4 are made of material twice as expensive per square inch as

the material used for the side, what are the most economical dimensions?

30. In Example 4 we found the radius for the most economical can.

(a) Knowing the radius, find the height.

(b) Show that the height of the most economical can equals its diameter.

31. Solve Example 4 by eliminating r rather than h.

32. A contractor who is removing earth from a large excavation can route trucks over either of two roads. There are 10,000 cubic yards of earth to move. Each truck holds 10 cubic yards. On one road the cost per truckload is $1 + 2x^2$ dollars, where x trucks use that road; the function records the cost of congestion. On the other road the cost is $2 + x^2$ dollars per truckload when x trucks use that road. How many trucks should be dispatched to each of the two roads?

33. The speed of traffic through the Lincoln Tunnel in New York City depends on the density of the traffic. Let S be the speed in miles per hour and let D be the density in vehicles per mile. The relation between S and D was seen to be approximated closely by the formula

$$S = 42 - \frac{D}{3},$$

for $D \le 100$.

(a) Express in terms of S and D the total number of vehicles that pass through the tunnel in an hour.

(b) What value of D will maximize the flow in (a)?

34. The base of a painting on a wall is a feet above the eye of an observer. The vertical side of the painting is b feet long. How far from the wall should the observer stand to maximize the angle that the painting subtends? *Hint:* It is more convenient to maximize $\tan \theta$ than θ itself.

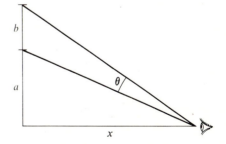

■ ■

35. The diagram shows two corridors meeting at a right angle. One has width 8; the other, width 27. Find the length of the longest pipe that can be carried horizontally from one hall, around the corner and into the other hall.

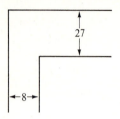

36. Two houses, A and B, are a distance p apart. They are distances q and r, respectively, from a straight road, and on the same side of the road. Find the length of the shortest path that goes from A to the road, and then on to the other house B.

(a) Use calculus.

(b) Use only elementary geometry. *Hint:* introduce an imaginary house C such that the midpoint of B and C is on the road and the segment BC is perpendicular to the road; that is, "reflect" B across the road to become C.

6.9

The differential

Let $f(x)$ denote the outdoor temperature measured in degrees at time x, measured in hours. At a certain time, say $x = 3$ P.M., the temperature is $70°$. What is a reasonable estimate of the temperature a quarter of an hour later? If nothing more is known, the safest estimate is $70°$. This estimate is essentially based on the assumption that the function is continuous: Its values do not fluctuate wildly. However, if in addition it is known that the temperature is increasing at 3 P.M. at the rate of $4°$ per hour, then a more accurate estimate can be made: A quarter of an hour later the temperature has probably increased about $1°$ and has become about $71°$. This more refined estimate is essentially based on the assumption that the temperature function is differentiable and that its derivative remains constant (or almost constant) during the 15-minute interval.

The present section discusses this use of the derivative to estimate the change in a function, and thereby the resulting value of a function. The reasoning will be geometric, utilizing the interpretation of the derivative as slope. The idea is that a very short piece of the graph around a point P, of a differentiable function, looks straight and closely resembles a short segment of the tangent line to the graph at P. This suggests that the tangent line can be used to estimate the change in the functional value caused by a small change in x.

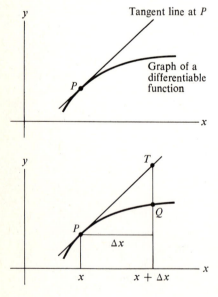

Let us look at the details. As the x coordinate changes from x to $x + \Delta x$, a point on the graph moves from P to Q, and a point on the tangent line to the graph at P from P to T. When Δx is small, the points T and Q are close together. Hence it seems reasonable that the y coordinate of T is a good estimate of $f(x + \Delta x)$, the y coordinate of Q, if Δx is small. Let us see if this is so in an example.

EXAMPLE 1   Let f be the square root function $f(x) = \sqrt{x}$. Let $x = 4$ and $\Delta x = 0.3$. In this case $P = (4, \sqrt{4}) = (4, 2)$. Compute the difference between the y coordinates of Q and T.

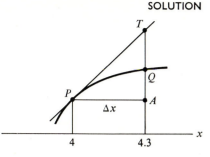

(Graph of $y = \sqrt{x}$ is distorted for clarity)

SOLUTION To aid the calculations, introduce the point A, at which the horizontal line through P meets the vertical line through T. Note in the accompanying figure that the y coordinate of Q is equal to $f(4) + \overline{AQ}$, where $\overline{AQ}$ is the change caused by moving from $x = 4$ to $x = 4.3$ on the graph; that is,

$$f(4.3) = f(4) + \overline{AQ} = f(4) + \Delta f.$$

(In this example $\overline{AQ}$ is positive; for other functions it may be negative.)
To compute $\overline{AT}$, observe that

$$\frac{\overline{AT}}{\overline{PA}} = \text{slope of tangent line at } P = f'(4).$$

Hence
$$\overline{AT} = f'(4) \cdot \overline{PA}$$
$$= f'(4) \cdot \Delta x$$
$$= f'(4) \cdot (0.3).$$

Since $f(x) = \sqrt{x}$, $$f'(x) = \frac{1}{2\sqrt{x}}.$$

Thus $$f'(4) = \frac{1}{2\sqrt{4}} = \frac{1}{4} = 0.25.$$

Consequently, $\overline{AT}$, the vertical change along the tangent line, is

$$(0.25)(0.3) = 0.075.$$

The y coordinate of T is therefore

$$2 + 0.075 = 2.075.$$

On the other hand, the y coordinate of Q is $\sqrt{4.3} \doteq 2.0736$. Thus, the y coordinates of Q and T differ by very little, approximately

$$2.0736 - 2.075 = -0.0014.$$

This example shows that the y coordinate of T is an excellent approximation to the y coordinate of Q. ●

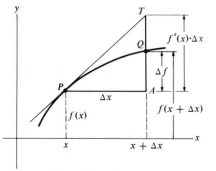

The method used in Example 1 applies to any differentiable function. The idea is to use vertical change along the tangent line to estimate vertical change along the graph.

The graph to the left shows the situation for any differentiable function: The change along the tangent line is $\overline{AT}$. Since

$$\frac{\overline{AT}}{\Delta x} = \text{slope of tangent line at } P$$

$$= f'(x),$$

it follows that $$\overline{AT} = f'(x) \cdot \Delta x.$$

Hence $f'(x) \cdot \Delta x$ is a good estimate of Δf, the change along the graph, when Δx is small.

The estimate $f'(x) \cdot \Delta x$ is of both practical and theoretical interest. For this reason it is given a name.

DEFINITION *Differential.* If f is a differentiable function, and x and Δx are numbers, the product $f'(x) \cdot \Delta x$ is called the differential of f at x. It is denoted by df. [If the notation $y = f(x)$ is used, the differential is also denoted dy.]

The differential df is a function of two variables x and Δx. For instance if $y = f(x) = \sqrt{x}$, then

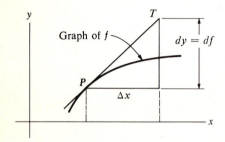

$$df = \frac{1}{2\sqrt{x}} \Delta x,$$

or, in the dy notation,

$$dy = \frac{1}{2\sqrt{x}} \Delta x.$$

The diagram in the margin shows Δx and $df = dy$. If the tangent lies below the graph, then the diagram appears as shown below (and df underestimates Δf).

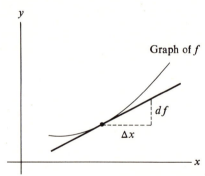

In Example 1, the value of the differential was computed when $x = 4$ and $\Delta x = 0.3$. Let us see how close df is to Δf in another example.

EXAMPLE 2 Let $y = f(x) = x^3$, $x = 5$, $\Delta x = 0.1$. Compute df and Δf and compare.

SOLUTION In this case $f'(x) = 3x^2$ and $df = 3x^2 \Delta x$. When $x = 5$ and $\Delta x = 0.1$,

$$df = 3(5)^2 \cdot (0.1) = 7.5.$$

Now, Δf is defined as $f(5 + \Delta x) - f(5)$, that is,

$$(5.1)^3 - 5^3 = 132.651 - 125 = 7.651.$$

Again, df, which is 7.5, is a good estimate of Δf, which is 7.651 ●

In what sense is the differential df a good estimate of the change Δf? After all, when x is small, both df and Δf are small. To claim that df differs very little from Δf when both are near 0, therefore, sheds no light. What is important is that their ratio is near 1 when Δx is small. This is proved in the following theorem.

THEOREM Let f be differentiable at a number x and assume that $f'(x) \neq 0$. Then

$$\lim_{\Delta x \to 0} \frac{\Delta f}{df} = 1.$$

PROOF Consider the quotient

$$\frac{\Delta f}{df} = \frac{\Delta f}{f'(x)\,\Delta x} = \frac{\Delta f}{\Delta x}\,\frac{1}{f'(x)}.$$

[Since $f'(x) \neq 0$, division is permissible.]   Thus

$$\lim_{\Delta x \to 0} \frac{\Delta f}{df} = \lim_{\Delta x \to 0} \frac{\Delta f}{\Delta x}\,\frac{1}{f'(x)} = f'(x)\,\frac{1}{f'(x)} = 1.$$

This concludes the proof. ●

EXAMPLE 3 The side of a cube is measured with an error of at most 1 percent. What percent error may this cause in calculating the volume of the cube?

SOLUTION Let x be the length of the cube and V its volume. Let dx denote the possible error in measuring x. The relative error

$$\frac{dx}{x}$$

is at most 0.01 in absolute value. That is, $|dx|/x \leq 0.01$.

The differential dV is an estimate of the actual error in calculating the volume. Thus

$$\frac{dV}{V}$$

is an estimate of the relative error in the volume.

Since

$$dV = d(x^3)$$
$$= 3x^2\,dx,$$

it follows that

$$\frac{dV}{V} = \frac{3x^2\,dx}{x^3}$$

$$= 3\,\frac{dx}{x}.$$

Therefore the relative error in the volume is about three times the relative error in measuring the side, hence at most 3 percent. ●

Function	Differential
$\sqrt{x}$	$\dfrac{\Delta x}{2\sqrt{x}}$
x^2	$2x\,\Delta x$
$\ln x$	$\dfrac{\Delta x}{x}$
e^x	$e^x\,\Delta x$
$\sin x$	$\cos x \cdot \Delta x$
x	Δx

The accompanying table exhibits the differentials of several important functions. In each case the differential is simply the derivative at x times Δx. Thus

$$d(\ln x) = \frac{1}{x}\,\Delta x,$$

$$d(e^x) = e^x\,\Delta x,$$

$$d(x) = 1\,\Delta x = \Delta x.$$

Notice that $d(x) = \Delta x$. For this reason it is customary to write Δx also as dx. The differential of f, then, is also written as

$$df = f'(x)\,dx,$$

or

$$dy = f'(x)\,dx.$$

The symbols dy and dx now have meaning individually. It is meaningful to divide both sides of the equation

$$dy = f'(x)\,dx$$

by dx, obtaining

$$dy \div dx = f'(x).$$

This is the origin of the symbol dy/dx for the derivative. It goes back to Leibniz at the end of the seventeenth century when dx denoted a number "vanishingly small," blasted by Bishop Berkeley in 1734 as "a ghost of a departed quantity."

It is easy to show that, if f and g are two differentiable functions, then

$$d(f + g) = df + dg.$$

To show this, use the definition of differentials:

$$\begin{aligned}
d(f + g) &= (f + g)'\,dx \\
&= (f' + g')\,dx \\
&= f'\,dx + g'\,dx \\
&= df + dg.
\end{aligned}$$

Similarly, $d(f - g) = df - dg$, $d(fg) = f\,dg + g\,df$, and

$$d\left(\frac{f}{g}\right) = \frac{g\,df - f\,dg}{g^2} \qquad g \neq 0.$$

It will be instructive to compare dy and $\Delta y\,(= \Delta f)$ for various values of $dx\,(= \Delta x)$ both large and small. Let us use the function x^3 and $x = 5$. Then $f'(x) = 3x^2 = 75$. Hence

$$dy = 75 \, dx,$$

while
$$\Delta y = (5 + dx)^3 - 5^3$$
$$= 125 + 75 \, dx + 15(dx)^2 + (dx)^3 - 125$$
$$= 75 \, dx + 15(dx)^2 + (dx)^3.$$

Note that for small dx, the terms $15(dx)^2$ and $(dx)^3$ become extremely small in comparison with $75 \, dx$.

This table contains information about the cubing function $y = x^3$ for various choices of dx, when $x = 5$.

dx	dy	Δy
3	225	387
2	150	218
1	75	91
0.1	7.5	7.651
0.01	0.75	0.751501
0	0	0
-1	-75	-61

$$dy = 75 \, dx;$$
$$\Delta y = (5 + dx)^3 - 5^3.$$

Note that, as the theorem asserts, the smaller dx is, the better dy approximates Δy. Keep in mind that it is usually easier to compute dy than Δy.

Exercises

Compute dy and Δy for the following functions and values of x and dx, and represent them on a graph of the function.

1. x^2 at $x = 1$ and $dx = 0.3$.

2. x^3 at $x = \frac{1}{2}$ and $dx = 0.1$.

3. x^3 at $x = 1$ and $dx = -0.1$.

4. $\sqrt{x}$ at $x = 9$ and $dx = 0.5$.

5. $\ln x$ at $x = 1$ and $dx = -0.2$.

6. $\sin x$ at $x = \pi/4$ and $dx = \pi/12$.

7. (a) Compute the differential of the function $1/x$ when $x = 1$ and $dx = 0.02$.

(b) Use (a) to show that $\frac{1}{1.02}$ is approximately 0.98.

8. (a) Compute the differential of the function $1/x$ when $x = 1$ and $dx = h$.

(b) Use (a) to show that when h is small $1 - h$ is a good estimate of $1/(1 + h)$.

9. (a) Compute the differential of the function $\sqrt{x}$ when $x = 1$ and $dx = h$.

(b) Use (a) to show that when h is small $1 + h/2$ is a good estimate of $\sqrt{1 + h}$.

(c) Compute $1 + h/2$ and $\sqrt{1 + h}$ when $h = 0.21$. By how much do they differ?

10. (a) Compute the differential of the function $\ln x$ at $x = 1$

and $dx = h$.

(b) Use (a) to show that $\ln (1 + h) \doteq h$ when h is small.

11. Fill in this table for the function $y = x^3$, and the indicated values of x and dx.

x	dx	dy	Δy	$\Delta y/dy$
3	1			
3	-0.5			
1	0.1			
2	-0.1			

In Exercises 12 to 20 express in terms of x and dx.

12. $d(e^{-x})$

13. $d[\ln (1 + x)]$

14. $d(\sqrt{1 + x^2})$

15. $d(\tan^{-1} x)$

16. $d(\sec x^2)$

17. $d(\sin^{-1} 3x)$

18. $d(\sec^{-1} 5x)$

19. $d(2^{x^2})$

20. $d(\cos 5x)$

21. The side of a square is measured with an error of at most 5 percent. Estimate the largest percent error this may induce in the measurement of the area.

∎

22. (a) Using differentials, show that $\log_{10}(1 + h) \doteq 0.434h$ when h is small.

(b) What is the percent error in using the estimate in (a) when computing $\log_{10}(1.2)$?

23. Let $f(x) = x^2$, the area of a square of side x.

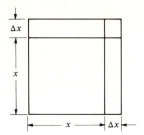

(a) Compute df and Δf in terms of x and Δx.

(b) In the square in the diagram, shade the part whose area is Δf.

(c) Shade the part of the square in (b) whose area is df.

24. Prove that, if f and g are two differentiable functions, then (a) $d(f - g) = df - dg$; (b) $d(fg) = f(dg) + g(df)$; (c) $d(f/g) = [(g(df) - f(dg))/g^2]$.

■ ■

25. Let y be a differentiable function of u, and u a differentiable function of x. Then $dy = D_u(y)\, du$ and $du = D_x(u)\, dx$. But y is a composite function of x, and one writes: $dy = D_x(y)\, dx$. Show that the two values of dy are equal.

6.10

L'Hôpital's rule

The problem of finding a limit has arisen in graphing a curve and will appear often in later chapters. For instance, in Chap. 10 it will be of importance in computing certain areas. Fortunately, there are some general techniques for computing a wide variety of limits. This section discusses one of the most important of these methods, l'Hôpital's rule, which concerns the limit of a quotient of two functions. If f and g are functions and a is a number such that

$$\lim_{x \to a} f(x) = 2 \quad \text{and} \quad \lim_{x \to a} g(x) = 3,$$

then

$$\lim_{x \to a} \frac{f(x)}{g(x)} = \frac{2}{3}.$$

This problem presents no difficulty; no more information is needed about the functions f and g. But if

$$\lim_{x \to a} f(x) = 0 \quad \text{and} \quad \lim_{x \to a} g(x) = 0,$$

then finding

$$\lim_{x \to a} \frac{f(x)}{g(x)}$$

may present a serious problem. For instance, Sec. 4.6 was dedicated to showing that

$$\lim_{\theta \to 0} \frac{\sin \theta}{\theta} = 1.$$

Here $f(\theta) = \sin \theta \to 0$ and $g(\theta) = \theta \to 0$ as $\theta \to 0$. The quotient $f(\theta)/g(\theta) \to 1$. In that same section it was proved that

$$\lim_{\theta \to 0} \frac{1 - \cos \theta}{\theta} = 0.$$

In this second limit, the numerator rushes toward 0 so much faster than the denominator that the quotient approaches 0. These two examples serve to point out that, if you know *only* that

$$\lim_{x \to a} f(x) = 0 \quad \text{and} \quad \lim_{x \to a} g(x) = 0,$$

you do not have enough information to determine

$$\lim_{x \to a} \frac{f(x)}{g(x)}.$$

Theorem 1 describes a general technique for dealing with the troublesome quotient

$$\frac{f(x)}{g(x)}$$

when

$$f(x) \to 0 \quad \text{and} \quad g(x) \to 0.$$

It is known as the *zero-over-zero case* of l'Hôpital's rule.

THEOREM I *L'Hôpital's rule.* Let *a* be a number and let *f* and *g* be differentiable over some open interval (a, b). Assume also that $g'(x)$ is not 0 for any *x* in that interval. If

$$\lim_{x \to a^+} f(x) = 0, \qquad \lim_{x \to a^+} g(x) = 0,$$

and if

$$\lim_{x \to a^+} \frac{f'(x)}{g'(x)} = L,$$

then

$$\lim_{x \to a^+} \frac{f(x)}{g(x)} = L.$$

Before worrying about *why* this theorem is true, we illustrate its use by an example.

EXAMPLE I Find

$$\lim_{x \to 1^+} \frac{x^5 - 1}{x^3 - 1}.$$

SOLUTION In this case

$$a = 1, \quad f(x) = x^5 - 1, \quad \text{and} \quad g(x) = x^3 - 1.$$

All the assumptions of l'Hôpital's rule are satisfied. In particular,

$$\lim_{x \to 1^+} (x^5 - 1) = 0 \quad \text{and} \quad \lim_{x \to 1^+} (x^3 - 1) = 0.$$

According to l'Hôpital's rule,

$$\lim_{x \to 1^+} \frac{x^5 - 1}{x^3 - 1} = \lim_{x \to 1^+} \frac{(x^5 - 1)'}{(x^3 - 1)'},$$

if the latter limit exists.

Now,

$$\lim_{x \to 1^+} \frac{(x^5 - 1)'}{(x^3 - 1)'} = \lim_{x \to 1^+} \frac{5x^4}{3x^2} \qquad \text{differentiation of numerator and of denominator}$$

$$= \lim_{x \to 1^+} \tfrac{5}{3}x^2 \qquad \text{algebra}$$

$$= \tfrac{5}{3}.$$

Thus

$$\lim_{x \to 1^+} \frac{x^5 - 1}{x^3 - 1} = \frac{5}{3}. \quad \bullet$$

A complete proof of Theorem 1 may be found in Exercises 92 and 93 of Sec. 6.11.   Let us pause long enough here to make the theorem plausible.

Argument for a Special Case of Theorem 1   To do so, consider the *special case* where f, f', g, and g' are all continuous throughout an open interval containing a.   Assume that $g'(x) \neq 0$ throughout the interval.   Since $\lim_{x \to a^+} f(x) = 0$ and $\lim_{x \to a^+} g(x) = 0$, it follows by continuity that $f(a) = 0$ and $g(a) = 0$.   Then

$$\lim_{x \to a^+} \frac{f(x)}{g(x)} = \lim_{x \to a^+} \frac{f(x) - f(a)}{g(x) - g(a)} \qquad \text{since } f(a) = 0 \text{ and } g(a) = 0$$

$$= \lim_{x \to a^+} \frac{\dfrac{f(x) - f(a)}{x - a}}{\dfrac{g(x) - g(a)}{x - a}} \qquad \text{algebra}$$

$$= \frac{f'(a)}{g'(a)} \qquad \text{by definition of } f'(a) \text{ and } g'(a)$$

$$= \frac{\displaystyle\lim_{x \to a^+} f'(x)}{\displaystyle\lim_{x \to a^+} g'(x)} \qquad f' \text{ and } g' \text{ are continuous}$$

$$= \lim_{x \to a^+} \frac{f'(x)}{g'(x)} \qquad \text{``limit of quotient'' property}$$

$$= L \qquad \text{by assumption.}$$

Consequently,

$$\lim_{x \to a^+} \frac{f(x)}{g(x)} = L.$$

The analogs of Theorem 1 hold if $x \to a^+$ is replaced by $x \to a^-$, or $x \to a$, or $x \to \infty$, or $x \to -\infty$.   Of course, corresponding changes are made in the hypotheses.

Sometimes it may be necessary to apply l'Hôpital's rule more than once, as in the next example.

EXAMPLE 2   Find

$$\lim_{x \to 0} \frac{\sin x - x}{x^3}.$$

SOLUTION   As $x \to 0$, both numerator and denominator approach 0.   By l'Hôpital's rule,

$$\lim_{x \to 0} \frac{\sin x - x}{x^3} = \lim_{x \to 0} \frac{\cos x - 1}{3x^2}.$$

But as $x \to 0$, both $\cos x - 1 \to 0$ and $3x^2 \to 0$.   So use l'Hôpital's rule again:

$$\lim_{x \to 0} \frac{\cos x - 1}{3x^2} = \lim_{x \to 0} \frac{-\sin x}{6x}.$$

Both $\sin x$ and $6x$ approach 0 as $x \to 0$.   Use l'Hôpital's rule yet another time:

$$\lim_{x \to 0} \frac{-\sin x}{6x} = \lim_{x \to 0} \frac{-\cos x}{6}$$

$$= -\tfrac{1}{6}.$$

So, after three applications of l'Hôpital's rule, we find that

$$\lim_{x \to 0} \frac{\sin x - x}{x^3} = -\frac{1}{6}. \ \bullet$$

Theorem 1 concerns the problem of finding the limit of $f(x)/g(x)$ when both $f(x)$ and $g(x)$ approach 0, the zero-over-zero case of l'Hôpital's rule.   But a similar problem arises when both $f(x)$ and $g(x)$ get arbitrarily large as $x \to a$ or as $x \to \infty$.   The behavior of the quotient $f(x)/g(x)$ will be influenced by how rapidly $f(x)$ and $g(x)$ become large.

Section 6.3 treated, for instance, the functions

$$f(x) = x \qquad \text{and} \qquad g(x) = e^x.$$

In that case

$$\lim_{x \to \infty} f(x) = \infty \qquad \text{and} \qquad \lim_{x \to \infty} g(x) = \infty,$$

while

$$\lim_{x \to \infty} \frac{f(x)}{g(x)} = \lim_{x \to \infty} \frac{x}{e^x} = 0.$$

Here the denominator gets large much faster than the numerator does.

On the other hand,

$$\lim_{x \to \infty} \frac{4x + 1}{2x} = \lim_{x \to \infty} \left(2 + \frac{1}{2x} \right)$$

$$= 2.$$

In this case, the numerator is increasing about twice as rapidly as the denominator.

The next theorem presents a form of l'Hôpital's rule that covers the case in which $f(x) \to \infty$ and $g(x) \to \infty$.   It is called the *infinity-over-infinity* case of l'Hôpital's rule.

THEOREM 2 *L'Hôpital's rule.* Let f and g be defined and differentiable for all x larger than some fixed number. Then, if

$$\lim_{x \to \infty} f(x) = \infty \qquad \text{and} \qquad \lim_{x \to \infty} g(x) = \infty,$$

and if

$$\lim_{x \to \infty} \frac{f'(x)}{g'(x)} = L,$$

it follows that

$$\lim_{x \to \infty} \frac{f(x)}{g(x)} = L.$$

First car:
position $f(t)$
velocity $f'(t)$

Second car:
position $g(t)$
velocity $g'(t)$

The proof of this is left to an advanced calculus course. However, it is easy to see why it is plausible. Imagine that $f(t)$ and $g(t)$ describe the location on the x axis of two cars at time t. Call the cars the f-car and the g-car. Their velocities are therefore $f'(t)$ and $g'(t)$. These two cars are on endless journeys. But let us assume that as time $t \to \infty$ the f-car tends to travel at a speed closer and closer to L times the speed of the g-car. That is, assume that

$$\lim_{t \to \infty} \frac{f'(t)}{g'(t)} = L.$$

No matter how the two cars move in the short run, it seems reasonable that in the *long run* the f-car will tend to travel about L times as far as the g-car; that is,

$$\lim_{t \to \infty} \frac{f(t)}{g(t)} = L.$$

Though not a proof, this argument does provide a perspective for a rigorous proof.

EXAMPLE 3 Use Theorem 2 to find

$$\lim_{x \to \infty} \frac{x}{e^x}.$$

SOLUTION By Theorem 2,

$$\lim_{x \to \infty} \frac{x}{e^x} = \lim_{x \to \infty} \frac{x'}{(e^x)'}$$

$$= \lim_{x \to \infty} \frac{1}{e^x} = 0.$$

This result was already obtained in Sec. 6.3. ●

There are many variations of l'Hôpital's rule beyond those already mentioned. For instance, in Theorem 2, ∞ could be replaced by $-\infty$ at either or both occurrences. Also, $x \to \infty$ could be replaced by $x \to a^+$ or $x \to a^-$, etc. Moreover, L may be replaced by ∞ or by $-\infty$. Of course, corresponding changes are called for in the hypotheses. We illustrate one case.

EXAMPLE 4 Find
$$\lim_{x \to 0^+} \frac{\ln x}{1/x}$$

SOLUTION In this case $f(x) = \ln x$ and $g(x) = 1/x$. Note that

$$\lim_{x \to 0^+} \ln x = -\infty \quad \text{and} \quad \lim_{x \to 0^+} \frac{1}{x} = \infty.$$

An analog of Theorem 2, with $x \to 0^+$, asserts that

$$\lim_{x \to 0^+} \frac{\ln x}{1/x} = \lim_{x \to 0^+} \frac{1/x}{-1/x^2}$$

$$= \lim_{x \to 0} (-x)$$

$$= 0.$$

Thus
$$\lim_{x \to 0^+} \frac{\ln x}{1/x} = 0.$$

In short, $\ln x$ gets large (in absolute value) much more slowly than does $1/x$ as $x \to 0^+$. ●

Many limit problems can be transformed to limits to which l'Hôpital's rule applies. For instance, the problem of finding

$$\lim_{x \to 0^+} x \ln x$$

does not seem to be related to l'Hôpital's rule, since it does not involve the quotient of two functions. But a little algebraic manipulation transforms it into a problem to which l'Hôpital's rule applies:

$$x \ln x = \frac{\ln x}{1/x}.$$

In the latter form the limit was determined in Example 4. Thus

$$\lim_{x \to 0^+} x \ln x = 0.$$

The remaining examples illustrate other limits that can be found by first relating them to limit problems to which l'Hôpital's rule applies.

EXAMPLE 5 Find
$$\lim_{x \to 0^+} x^x.$$

SOLUTION Since this limit involves an exponential, not a quotient, it does not fit directly into l'Hôpital's rule. But a little algebraic manipulation will change the problem to one covered by l'Hôpital's rule.

Let
$$y = x^x.$$

Then
$$\ln y = \ln x^x$$

$$= x \ln x.$$

By Example 4, solved by l'Hôpital's rule,

$$\lim_{x \to 0^+} \ln y = 0.$$

Thus

$$\lim_{x \to 0^+} y = 1.$$

In short,

$$\lim_{x \to 0^+} x^x = 1. \quad \bullet$$

The next example concerns the difference between two functions both of which become large.

EXAMPLE 6   Find

$$\lim_{x \to 0^+} \left[\frac{1}{x} - \frac{1}{\sin x} \right].$$

SOLUTION   As it stands, since it is not a quotient, the limit cannot be directly determined by l'Hôpital's rule.   But

$$\frac{1}{x} - \frac{1}{\sin x} = \frac{\sin x - x}{x \sin x}.$$

Now the problem is a natural one for l'Hôpital's rule.   Note that $\sin x - x \to 0$ and $x \sin x \to 0$ as $x \to 0^+$.

By l'Hôpital's rule,

$$\lim_{x \to 0^+} \frac{\sin x - x}{x \sin x} = \lim_{x \to 0^+} \frac{\cos x - 1}{x \cos x + \sin x}$$

Applying l'Hôpital's rule again,

$$= \lim_{x \to 0^+} \frac{-\sin x}{-x \sin x + 2 \cos x}$$

$$= \frac{-0}{0 + 2}$$

$$= 0.$$

Thus

$$\frac{1}{x} - \frac{1}{\sin x} \to 0$$

as $x \to 0^+$.   This means that the quantities $1/x$ and $1/\sin x$, though becoming large as $x \to 0^+$, get closer and closer to each other. $\bullet$

EXAMPLE 7   Find

$$\lim_{x \to -\infty} e^x x^3.$$

SOLUTION   As it stands, the problem does not involve a quotient.   Moreover, the answer is not immediate, since, as $x \to -\infty$, $e^x \to 0$ and $x^3 \to -\infty$.   The product of a small number and a large number can be anything.   The product, however, can be changed to a quotient:

$$\lim_{x \to -\infty} e^x x^3 = \lim_{x \to -\infty} \frac{x^3}{e^{-x}}.$$

As $x \to -\infty$, $x^3 \to -\infty$, and $e^{-x} \to \infty$; l'Hôpital's rule applies. Repeated application of l'Hôpital's rule yields the limit:

$$\lim_{x \to -\infty} \frac{x^3}{e^{-x}} = \lim_{x \to -\infty} \frac{3x^2}{-e^{-x}} \qquad \text{l'Hôpital's rule}$$

$$= \lim_{x \to -\infty} \frac{6x}{e^{-x}} \qquad \text{l'Hôpital's rule again}$$

$$= \lim_{x \to -\infty} \frac{6}{-e^{-x}} \qquad \text{l'Hôpital's rule yet again}$$

$$= 0,$$

since $e^{-x} \to \infty$ as $x \to -\infty$. ●

As a rule of thumb, l'Hôpital's rule applies when you need it, and not when you do not need it. It comes in handy generally if you run into limits that lead to expressions like

$$\frac{0}{0}, \frac{\infty}{\infty}, \frac{-\infty}{\infty}, \dots, \qquad \text{and} \qquad 0 \cdot \infty, 0°, \infty°, \infty - \infty,$$

which can often be reduced to the zero-over-zero case or the infinity-over-infinity case.

Exercises

Determine the limits (if they exist) in Exercises 1 to 36. In some exercises l'Hôpital's rule may not be of any aid, and using it carelessly may lead to a wrong answer.

1. $\lim_{x \to 0} \dfrac{xe^x}{e^x - 1}$

2. $\lim_{x \to 1} \dfrac{e^x + 1}{e^x - 1}$

3. $\lim_{x \to 2} \dfrac{x^3 - 8}{x^2 - 4}$

4. $\lim_{x \to 2} \dfrac{x^3 + 8}{x^2 + 2}$

5. $\lim_{x \to 0} \dfrac{\sin x^2}{x \sin x}$

6. $\lim_{x \to \pi/4} \dfrac{\cos x^2}{\cos x}$

7. $\lim_{x \to \pi/2} \dfrac{\sin x}{1 + \cos x}$

8. $\lim_{x \to 3^+} \dfrac{\ln (x - 3)}{x - 3}$

9. $\lim_{x \to 0} \dfrac{1 - \cos x}{x^2}$

10. $\lim_{x \to -\infty} \dfrac{5^x + 3^x}{4^x}$

11. $\lim_{x \to 0^+} \left(\dfrac{\sin x}{x}\right)^{1/x}$

12. $\lim_{x \to \infty} \dfrac{2^x}{3^x}$

13. $\lim_{x \to \infty} \dfrac{2x + 5 \cos x}{3x - 7 \sin x}$

14. $\lim_{x \to \infty} \dfrac{3x^2 + 6\sqrt{x}}{x^2 + 8\sqrt{x}}$

15. $\lim_{x \to \infty} \dfrac{x^2 + 5}{2x^2 + 6x}$

16. $\lim_{x \to 0^-} \dfrac{1 - \cos x}{x + \tan x}$

17. $\lim_{x \to 0^+} \dfrac{1 - \cos x}{x - \tan x}$

18. $\lim_{x \to \infty} \dfrac{e^{-x}}{x^2}$

19. $\lim_{x \to \infty} \dfrac{\cos x}{x}$

20. $\lim_{x \to \infty} \dfrac{\ln (x^2 + 1)}{\ln (x^2 + 8)}$

21. $\lim_{x \to 2} \dfrac{5^x + 3^x}{x}$

22. $\lim_{x \to \pi/2} \dfrac{\sin 2x}{\sin 3x}$

23. $\lim_{x \to 0} \dfrac{5^x - 3^x}{x}$

24. $\lim_{x \to \infty} e^{-x} \ln x$

25. $\lim_{x \to \infty} [(x^2 - 2x)^{1/2} - x]$

26. $\lim_{x \to 0} \left(\dfrac{1}{1 - \cos x} - \dfrac{2}{x^2}\right)$

27. $\lim_{x \to 0} (1 + 3x)^{2/x}$

28. $\lim_{x \to 0^+} x^{x^2}$

29. $\lim_{x \to 1^+} (x - 1) \ln (x - 1)$

30. $\lim_{x \to 0^+} \sin x \ln x$

31. $\lim_{x \to \infty} x \sin x$

32. $\lim_{x \to 0} (\sin x)^x$

33. $\lim_{x \to 0} (\cos x)^{1/x}$

34. $\lim_{x \to \pi/2^-} (1 - \sin x)^{\tan x}$

35. $\lim_{x \to \infty} x^{1/x}$

36. $\lim_{x \to \infty} (2^x - x^{10})^{1/x}$

■

37. Solve Example 1 without using l'Hôpital's rule.
38. Solve Example 4 with the aid of results in Sec. 6.3.

39. Find $\lim\limits_{x \to \infty} \dfrac{(x^3 - x^2 - 4x)^{1/3}}{x}$.

40. Find $\lim\limits_{x \to 0} \left(\dfrac{1 + 2^x}{2}\right)^{1/x}$.

■ ■

41. In the accompanying diagram, the unit circle is centered at the origin; BQ is a vertical tangent line; $\overline{BQ} = BP$. Prove that the x coordinate of R approaches -2 as $P \to B$.

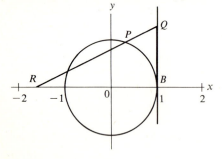

42. Find $\lim\limits_{x \to 0} \dfrac{\sin^2 2x - 4x^2}{(e^{x^2} - 1)^2}$.

This can be done by l'Hôpital's rule. An easier method is developed in Chap. 14.

43. In the accompanying diagram of a circle let $f(\theta) =$ area of triangle ABC and let $g(\theta) =$ area of the shaded region formed by deleting triangle OAC from the sector OBC. Clearly, $0 < f(\theta) < g(\theta)$.
    (a) What would you guess is the value of $\lim_{\theta \to 0} f(\theta)/g(\theta)$?
    (b) Find $\lim_{\theta \to 0} f(\theta)/g(\theta)$.

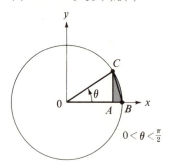

$0 < \theta < \frac{\pi}{2}$

6.11

Summary

This chapter presents some applications of the derivative. Sections 6.1 and 6.2 concern the law of the mean and some of its consequences, in particular the fundamental fact that, where f' is positive, f increases. Section 6.3 utilizes this fact to prove that e^x grows faster than $x^2/2$. From this some algebraic manipulations show that the exponential b^x, for fixed $b > 1$, grows faster when x is large than the power x^a for any fixed number a. For instance, $(1.1)^x$ is much larger than x^{10} when x is large.

Section 6.4 examines the differential equation of natural growth and decay,

$$f'(t) = kf(t),$$

showing that the only functions that satisfy it are of the form Ae^{kt} for suitable constants A and k.

The sign of the derivative $f'(x)$ was the focus of attention in Sec. 6.5, where it was used to help graph a function.

Higher derivatives were defined in Sec. 6.6, but only the second derivative was applied there, where it was used to analyze motion.

In Sec. 6.7 the second derivative was applied to graphs. In particular it tells where a graph is concave upward and where it is concave downward. It also provides a test for a relative maximum or minimum.

Section 6.8 applied the techniques of Secs. 6.5 and 6.7 to practical maximum and minimum problems.

The differential, introduced in Sec. 6.9, provides a quick way to approximate a small change in a function.

Finally, Sec. 6.10 discussed l'Hôpital's rule, which is an algorithm for finding limits of certain quotients and expressions that can be expressed in terms of a quotient.

KEY FACTS

A continuous function defined on a closed interval takes on a maximum (and a minimum) value.

If a differentiable function has a local maximum at some point in an open interval, its derivative must be zero at that point. (A similar statement holds for a minimum value.)

The law of the mean: If f is continuous on $[a, b]$ and differentiable on (a, b), then there is a number X in (a, b) such that

$$f'(X) = \frac{f(b) - f(a)}{b - a}.$$

In terms of graphs, this says that for any chord of f there exists at least one tangent line parallel to the chord. [This is a generalization of Rolle's theorem, which concerns only horizontal chords of f, that is, the case where $f(a) = f(b)$.]

When the law of the mean is rewritten in the form

$$f(b) = f(a) + f'(X)(b - a),$$

for some number X in (a, b), it can be used to obtain three useful corollaries:

A function whose derivative is 0 throughout an interval is constant.

Functions that have equal derivatives throughout an interval differ at most by a constant.

If a function has a positive derivative throughout an interval, the function is increasing. (If the derivative is negative throughout an interval, the function is decreasing.)

$\lim\limits_{x \to \infty} x^n/e^x = 0$, implying that e^x grows more rapidly than any fixed power of x.

$\lim\limits_{x \to \infty} x^a/b^x = 0$, if $b > 1$.

$\lim\limits_{x \to \infty} \ln x/x^a = 0$, implying that $\ln x$ grows more slowly than any fixed power of x.

The differential equation

$$f'(t) = kf(t)$$

describes natural growth for $k > 0$ and natural decay for $k < 0$. The function f must be of the form

$$f(t) = Ae^{kt}.$$

A is the amount present when $t = 0$. In short, the only solutions of the equation $dy/dt = ky$ are of the form $y = Ae^{kt}$ which can be written as $y = Ab^t$ by letting $b = e^k$.

When graphing a function, consider the sign of the derivative and of the second derivative, critical points, x intercepts, inflection points, asymptotes, and behavior of the function for $|x|$ large. Any relative or global maxima or minima should be pointed out. See also the table in Sec. 6.5, which outlines the procedure (except the use of the second derivative).

To find the maximum or minimum of a function f on a closed interval examine the value of the function at the endpoints of the interval and at critical numbers. See the flowchart description of the procedure in Sec. 6.5.

The second derivative of a function is the derivative of the first derivative. If the function describes the position of a particle moving on a line, then its second derivative is the acceleration, the rate at which velocity changes. The second derivative also determines the sense of concavity of a graph.

It is frequently possible to decide whether a critical point is a relative maximum or minimum by using the first or second derivative. If the first derivative changes sign or if the second derivative is not zero, the critical point is either a relative maximum or minimum.

The differential of a function f is defined as $f'(x) \Delta x$ and is denoted df. It is a function of x and Δx, useful in estimating Δf. Since $dx = \Delta x$, df is usually written $f'(x)\,dx$.

For determining the limit of a quotient $f(x)/g(x)$, l'Hôpital's rule, which comes in many forms, may be used. For instance, if $\lim_{x \to a^+} f(x) = \infty$ and $\lim_{x \to a^+} g(x) = \infty$,

and
$$\lim_{x \to a^+} \frac{f'(x)}{g'(x)} = L,$$

then
$$\lim_{x \to a^+} \frac{f(x)}{g(x)} = L.$$

Similar rules hold for $x \to \infty$, $x \to a^-$, or $f(x) \to 0$ and $g(x) \to 0$, etc.

VOCABULARY AND SYMBOLS

maximum-value theorem	asymptote
open interval (a, b)	x intercept
(closed) interval $[a, b]$	relative (local) maximum or
Rolle's theorem	minimum
chord of a function	(global) maximum or minimum
law of the mean	higher derivative, $D^n f$, $D^n y$,
increasing function	$d^n y/dx^n$, etc.
decreasing function	second derivative
differential equation	concave upward
natural growth and decay	concave downward
half life, $t_{1/2}$	inflection point
critical number	differential, df, dy, dx
critical point	l'Hôpital's rule

Guide quiz on Chap. 6

1. (a) State all the assumptions in Rolle's theorem.
   (b) State the conclusion of Rolle's theorem.
2. (a) State all the assumptions in the law of the mean.
   (b) State the conclusion of the law of the mean.

3. What does each of the following imply about the graph of a function?
   (a) As you move from left to right, $f(x)$ changes sign at a from positive to negative.

(b) As you move from left to right, $f'(x)$ changes sign at a from positive to negative.

(c) As you move from left to right, $f''(x)$ changes sign at a from positive to negative.

4. Use the law of the mean to show that a function whose derivative is zero throughout an interval is constant.

5. (a) Prove that $\tan x - x$ is an increasing function of x, when $0 \le x < \pi/2$.

(b) Deduce that $\tan x > x$ for x in $(0, \pi/2)$.

(c) From (b) obtain the inequality $x \cos x - \sin x < 0$, if x is in $(0, \pi/2)$.

(d) Prove that $\sin x/x$ is a decreasing function for x in $(0, \pi/2)$.

6. (a) Describe all functions whose derivative equals $\sin 3x$.

(b) How are you sure that you have found all possibilities in (a)?

7. List these functions in order of increasing size (when x is large):

$$(1.001)^x, \log_{10} x, x^{100}, 2^x.$$

8. (a) Let $f(x) = 5^{7x}6^{8x+3}$. Show that $f'(x)$ is proportional to $f(x)$.

(b) Does (a) contradict the theorem that asserts that the only functions whose derivative is proportional to the function are of the form Ae^{kt}?

9. (a) The area of a circle of radius r is πr^2. Use a differential to estimate the change in the area when the radius changes from r to $r + dr$.

(b) The circumference of a circle of radius r is $2\pi r$. Explain why $2\pi r$ appears in the answer to (a).

10. The graph of a certain function is shown below. List the x coordinates of (a) relative maxima, (b) relative minima, (c) critical points, (d) global maximum, (e) global minimum.

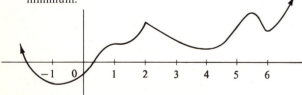

11. Show that, if $\dfrac{dy}{dx} = 3y^4$,

then $\dfrac{d^2y}{dx^2} = 36y^7$.

12. Let m be a number ≥ 1. Prove that $(1 + x)^m > 1 + mx$ for any $x > 0$. This is known as Bernoulli's inequality.

13. Graph $y = xe^{-x^2}$, showing critical points, asymptotes, and inflection points.

14. The half-life of radium is about 1,600 years (actually 1,590 years).

(a) From this, find k in the expression Ae^{kt}.

(b) How long does it take 75 percent of the radium to disintegrate?

(c) Solve (b) without using calculus.

(d) How long will it take for 90 percent of the radium to disintegrate?

(e) Without calculus, show that the answer to (d) is between 4,800 and 6,400 years.

15. A bacterial culture grows at a rate proportional to the amount present. From 9 to 11 A.M. it increases from 100 to 200 grams. What will be its weight at noon?

16. How should one choose two nonnegative numbers whose sum is 1 in order to minimize the sum of the square of one and the cube of the other?

17. A track of a certain length L is to be laid out in the shape of two semicircles at the ends of a rectangle. Find the relative proportion of the radius of the circle r and the length of the straight section x if the track is to enclose a maximum area. Discuss also the case of minimum area.

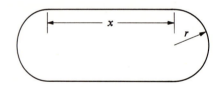

18. Using differentials, show that a good estimate (a) of $\sqrt{25 + dx}$ is $5 + dx/10$; (b) of $\sqrt{a^2 + dx}$ is $a + dx/2a$ (if a is positive). (c) Use the formula in (b) to estimate $\sqrt{65.6}$.

19. Find (a) $\lim\limits_{x \to 0} \dfrac{e^{x^2} - 1}{1 - \cos x}$

(b) $\lim\limits_{x \to 0} \dfrac{x - \tan^{-1} x}{x - \sin x}$

(c) $\lim\limits_{x \to 0} x^{1 - \cos x}$.

20. Find $D^3(f)$ if

(a) $f(x) = \csc 5x$ (b) $f(x) = \sin^{-1} x^2$

(c) $f(x) = \ln\sqrt{1 + x^2}$ (d) $f(x) = \sec 5x$

(e) $f(x) = e^{-\cos 2x}$ (f) $f(x) = 5x^3 + 2x + 1$.

Guide quiz on Chaps. 1 to 6 (computations)

1. Evaluate:

(a) $e^{\ln 2}$, $\ln e^8$, $\tan^{-1}\left(\tan\dfrac{\pi}{3}\right)$;

(b) $\sin\dfrac{3\pi}{2}$, $\tan\dfrac{\pi}{6}$, $\cos\dfrac{\pi}{3}$;

(c) $\sin^{-1}\frac{1}{2}$, $\tan^{-1}(-1)$, $\cos^{-1}\left(-\frac{1}{2}\right)$.

2. (a) About how many degrees are there in an angle of 1 radian?

(b) What is the area of a sector whose angle is θ radians?

(c) How is e defined?

(d) What is the decimal representation of e to three places?

3. Let y be a differentiable function of x, $y = f(x)$. If $f(3) = 5$ and $f'(3) = 4$, find $d(y^2)/dx$ when $x = 3$.

4. Graph $y = e^{-x^2}$.

5. Differentiate:

(a) e^{3x}

(b) 10^{-5x}

(c) $x^2\sin 3x$

(d) $\dfrac{\cos 5x}{x^2}$

(e) $x^{5/6}\sin^{-1}x$

(f) $\cot x^2$.

6. Differentiate:

(a) $x^5 - 2x + \ln(2x + 3)$

(b) $\sec^{-1}3x$

(c) $(x^7 + x^2)^{10}$

(d) $\dfrac{x^2}{1 + 3x}$.

7. Differentiate:

(a) $\ln\left|x + \sqrt{x^2 + 25}\right|$

(b) $(x - 3)\sqrt{6x - x^2} + 9\sin^{-1}\left(\dfrac{x - 3}{3}\right)$.

8. Give an example of a function whose derivative is equal to

(a) $\dfrac{3}{1 + x^2}$

(b) $\dfrac{5}{\sqrt{1 - x^2}}$

(c) $6e^{-x}$

(d) x^4

(e) $\dfrac{1}{\sqrt{1 - 4x^2}}$

(f) $\dfrac{1}{x^2}$

(g) $\dfrac{1}{x}$

(h) $\dfrac{x}{\sqrt{1 - x^2}}$.

9. Give an example of a function whose derivative is

(a) $3x^2 + 2x$

(b) $\cos x$

(c) $\cos 2x$

(d) $\dfrac{1}{\sqrt{x}}$

(e) $\dfrac{1}{2x + 1}$.

10. If one makes an error of about 2 percent in measuring the side of a cube, about what percent error will this cause in estimating its surface area? (Use differentials.)

11. In which cases below is it possible to determine $\lim_{x \to a} f(x)^{g(x)}$ without further information about the functions?

(a) $\lim_{x \to a} f(x) = 0$; $\lim_{x \to a} g(x) = 7$.

(b) $\lim_{x \to a} f(x) = 2$; $\lim_{x \to a} g(x) = 0$.

(c) $\lim_{x \to a} f(x) = 0$; $\lim_{x \to a} g(x) = 0$.

(d) $\lim_{x \to a} f(x) = 0$; $\lim_{x \to a} g(x) = \infty$.

(e) $\lim_{x \to a} f(x) = \infty$; $\lim_{x \to a} g(x) = 0$.

(f) $\lim_{x \to a} f(x) = \infty$; $\lim_{x \to a} g(x) = -\infty$.

12. In which cases below is it possible to determine $\lim_{x \to a} f(x)/g(x)$ without further information about the functions?

(a) $\lim_{x \to a} f(x) = 0$; $\lim_{x \to a} g(x) = \infty$.

(b) $\lim_{x \to a} f(x) = 0$; $\lim_{x \to a} g(x) = 1$.

(c) $\lim_{x \to a} f(x) = 0$; $\lim_{x \to a} g(x) = 0$.

(d) $\lim_{x \to a} f(x) = \infty$; $\lim_{x \to a} g(x) = -\infty$.

13. (a) State the assumptions in the zero-over-zero case of l'Hôpital's rule.

(b) State the conclusion.

14. In each case below decide if there is a function that meets all the conditions. If there is, sketch the graph of such a function. If there is none, indicate why not.

(a) $f'(x) > 0$ and $f''(x) < 0$ for all x.

(b) $f(x) > 0$ and $f''(x) < 0$ for all x.

(c) $f(x) > 0$ and $f'(x) > 0$ for all x.

15. (a) How do you go about finding the global maximum of a differentiable function on a closed interval?

(b) Describe two different tests for a relative (local) minimum at a critical point.

Guide quiz on Chaps. 1 to 6 (concepts)

1. (a) Using the definition of the derivative and the identity involving $c^4 - d^4$, show that $(x^4)' = 4x^3$.

(b) Using the fact that $(e^x)' = e^x$, and writing x^4 as a power of e, show that $(x^4)' = 4x^3$.

2. (a) Why are radians used in calculus to measure angles?
   (b) Why is e used in calculus as the base of logarithms?

3. Evaluate

   (a) $\lim\limits_{x \to \pi/4} \dfrac{\sin x}{x}$

   (b) $\lim\limits_{x \to \pi/4} \dfrac{\sin x - \sqrt{2/2}}{x - \pi/4}$

   (c) $\lim\limits_{h \to 0} \dfrac{e^{3+h} - e^3}{1 - h}$

   (d) $\lim\limits_{h \to 0} \dfrac{e^{3+h} - e^3}{h}$

   (e) $\lim\limits_{x \to 0} (1 + 3x)^{1/x}$

   (f) $\lim\limits_{x \to 1} \dfrac{\sin \pi x}{x - 1}$

   (g) $\lim\limits_{x \to \infty} x^{1/\log_2 x}$

   (h) $\lim\limits_{x \to 0} \dfrac{\cos \sqrt{x} - 1}{\tan x}$.

4. (a) Explain why each of these two proposed definitions of a tangent line is inadequate:
   (1) A line L is tangent to a curve at a point P if L meets the curve only at P.
   (2) A line L is tangent to a curve at a point P if L meets the curve at P and does not cross the curve at P.

(b) How is a tangent line defined?
(c) How is velocity defined?

5. If f and g are differentiable functions, with $f(0) = g(0)$ and $f'(x) \geq g'(x)$ for all x, is $f(x) \geq g(x)$ for all x?   for all positive x?   Explain.

6. If f is defined for all x, $f(0) = 0$, and $f'(x) \geq 1$ for all x, what is the most that can be said about $f(3)$?   Explain.

7. (a) Using differentials, show that $\sqrt[3]{8 + h} \doteq 2 + h/12$ when h is small.
   (b) What is the percent error when $h = 1$? $h = -1$?

8. For what value of the exponent a is the function x^a a solution to the differential equation

$$\frac{dy}{dx} = -y^2 ?$$

9. Find all functions f such that

$$\frac{d^2 f}{dx^2} = x.$$

Review exercises for Chaps. I to 6

In Exercises 1 to 18 find the derivatives of the functions.

1. (a) $\sqrt{x}$ (b) $\dfrac{1}{\sqrt{x}}$

2. (a) x^3 (b) $\dfrac{1}{x^3}$

3. $\dfrac{x}{3} - \dfrac{4}{9} \ln (3x + 4)$

4. (a) e^{x^2} (b) $\dfrac{e^{x^2}}{2x}$

5. (a) $e^{\sqrt{x}}$ (b) $2e^{\sqrt{x}}(\sqrt{x} - 1)$

6. (a) $x^2 e^{3x}$ (b) $x^2 e^{-3x}$

7. (a) $\sin^2 2x$ (b) $\sin^3 2x$

8. (a) $\ln (\sec 3x + \tan 3x)$ (b) $\ln (\cos 3x)$

9. (a) $\dfrac{1}{x^2 + 1}$ (b) $\dfrac{x}{x^2 + 1}$

10. (a) $\dfrac{1}{5} \tan^{-1} \dfrac{x}{5}$ (b) $\sin^{-1} \dfrac{x}{5}$

11. (a) 10^x (b) $\log_{10} x$

12. (a) $\dfrac{(3x + 1)^4}{12}$ (b) $\dfrac{12}{(3x + 1)^4}$

13. (a) $x^2 \ln x$ (b) $\dfrac{\ln x}{x^2}$

14. $\dfrac{x}{3} - \dfrac{4}{9} \ln |3x + 4|$

15. $x - 2 \ln (x + 1) - \dfrac{1}{x + 1}$

16. $\dfrac{1}{8} \left[\ln (2x + 1) + \dfrac{2}{2x + 1} - \dfrac{1}{2(2x + 1)} \right]$

17. $\sqrt{\dfrac{2}{15} (5x + 7)^3}$

18. (a) $x \cos 5x + \sin 5x$ (b) $\cos^2 5x$

In Exercises 19 to 22 first simplify by using laws of logarithms, and then differentiate.

19. $\ln \sqrt[3]{x}$

20. $\ln (\sqrt{4 + x} \sqrt[3]{x^2 + 1})$

21. $\ln \dfrac{10^x}{\sin x}$

22. $\ln (1 + \cos x)^2$

In Exercises 23 to 26 find the limit if it exists.

23. $\lim\limits_{x \to 0} \dfrac{\sin 2x}{e^{3x} - 1}$

24. $\lim\limits_{x \to \infty} \dfrac{(x^2 + 1)^5}{e^x}$

25. $\lim\limits_{x \to \infty} \dfrac{3x - \sin x}{x + \sqrt{x}}$

26. $\lim\limits_{x \to 0} \dfrac{1 - \cos^2 3x}{x^2 + x^3}$

In Exercises 27 to 29 compute du.

27. $u = x^5$

28. $u = \tan^{-1} 3x$

29. $u = -\cos 5x$

30. (a) What is the maximum value of the function $y = 3 \sin t + 4 \cos t$?

  (b) What is the maximum value of the function $y = A \sin kt + B \cos kt$, where A, B, and k are nonzero constants?

31. (a) Graph $y = \sqrt{x}$ for $0 \le x \le 5$.

  (b) Compute dy for $x = 4$ and $dx = 1$.

  (c) Compute Δy for $x = 4$ and $\Delta x = 1$.

  (d) Using the graph in (a), show dy and Δy.

32. Fill in this table.

Interpretation of $f(x)$	Interpretation of $f'(x)$
The y coordinate in a graph of $y = f(x)$	
Total distance traveled up to time x	
Projection by lens	
Size of population at time x	
Total mass of left x centimeters of string	
Velocity at time x	

33. Let F be a differentiable function and call its derivative f. Show that, if c and d are two numbers, then there is a number X between them such that

$$F(d) - F(c) = f(X)(d - c).$$

34. Below is the graph of a function f whose domain is the interval $[0, 4]$. [Dots indicate $f(1)$ and $f(2)$.]

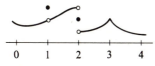

  (a) For which numbers a does $\lim_{x \to a} f(x)$ *not exist*?

  (b) At which numbers a is f not continuous?

  (c) At which numbers a is f not differentiable?

35. (a) Using differentials, show that

$$\tan^{-1}(1 + h) \doteq \frac{\pi}{4} + \frac{h}{2}$$

  when h is small.

  (b) Using $\pi \doteq 3.14$, estimate $\tan^{-1} 0.8$ with the aid of (a).

  (c) What is the percent error?

36. Using differentials, show that

$$\sin\left(\frac{\pi}{6} + h\right) \doteq \frac{1}{2} + \frac{\sqrt{3}}{2} h$$

for small h.

37. Using differentials, show that

$$\sin^{-1}\left(\frac{\sqrt{3}}{2} + h\right) \doteq \frac{\pi}{3} + 2h$$

when h is small.

38. Does every polynomial of even degree $n \ge 2$ have at least one critical point? at least one inflection point? a global maximum or minimum?

39. Does every polynomial of odd degree $n \ge 3$ have at least one critical point? at least one inflection point? a global maximum or minimum?

40. A window is made of a rectangle and equilateral triangle, as shown in the accompanying figure. What should the dimensions be to maximize the area of the window if its perimeter is prescribed?

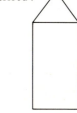

41. A wire of length l is to be cut into two pieces. One piece will be shaped into an equilateral triangle and the other into a square. How should the wire be cut in order to

  (a) minimize the sum of the areas of the triangle and square?

  (b) maximize the sum of the areas?

42. A square foot of glass is to be melted into two shapes. Some of it will be the thin surface of a cube. The rest will be the thin surface of a sphere. How much of the glass should be used for the cube and how much for the sphere if

  (a) their total volume is to be a minimum?

  (b) their total volume is to be a maximum?

43. Show that for $n \ge 3$ the curve $y = e^x$ meets the curve $y = x^n$ twice to the right of the y axis.

44. Consider the function $f(x) = x^3 + ax^2 + c$. Show that if $a < 0$ and $c > 0$, then f has exactly one negative root.

45. This is an excerpt from a news article published in 1975. There are 57,000 babies born daily in India, or 21 million a year. With eight million deaths, the annual population increase is 13 million. The population, nearly 570 million, is expected to reach a billion by the end of the century.

(a) Assuming natural growth, estimate the population of India in the year 2000.

(b) What is the instantaneous growth rate k?

(c) What is the percentage growth rate in 1 year?

46. (a) Copy and fill in this table:

x	y	$x + y$	$|x| + |y|$	$|x + y|$
2	3			
-2	3			
-2	-3			
2	-2			

(b) What relation is there between $|x + y|$ and $|x| + |y|$?

(c) For which pairs x and y does $|x + y| = |x| + |y|$?

47. Show that, if a quantity is growing or shrinking at a rate proportional to the amount present, then its changes in successive equal intervals of time form a geometric progression.

48. (a) Using differentials, show that $\log_{10}(1 + x)$ is approximately

$$x \log_{10} e \doteq 0.434x.$$

(b) Look up $\log_{10} 1.5$ and $\log_{10} 1.1$ in a table of common logarithms or use a calculator.

(c) What are the percent errors of the estimates in (a) for the two cases in (b)?

49. The acceleration of a particle at time t is e^{-t}. Find the most general form of its position function $x = f(t)$.

50. The derivative of a certain function f is 5 when x is 2.

(a) If $f(x)$ is the distance in feet that a rocket travels in x seconds, about how far does it travel from $x = 2$ to $x = 2.1$ seconds?

(b) If $f(x)$ is the projection of x on a slide, about how long is the projection of the interval $[2, 2.1]$?

(c) If $f(x)$ is the depth in feet that water penetrates the soil in the first x hours, how much does the water penetrate in the 6 minutes from 2 to 2.1 hours?

51. A certain function $y = f(x)$ has the property that

$$\frac{dy}{dx} = 3y^2.$$

Prove that $\dfrac{d^2y}{dx^2} = 18y^3$.

∎

52. Is this proposed proof of the law of the mean correct? *Proof:* Tilt the x and y axes until the x axis is parallel to the given chord. The chord is now "horizontal," and we may apply Rolle's theorem.

53. A ray of light travels from P to Q in such a way as to minimize the elapsed time.

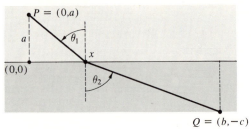

The speed of light is v_1 in the material above the x axis and v_2 in the material below it

Show that the path of light from P to Q is such that

$$\frac{\sin \theta_1}{\sin \theta_2} = \frac{v_1}{v_2}.$$

which is Snell's law of refraction.

54. Give an example of a function f such that $f(x)$ is defined for all x, f is differentiable and increasing, and yet the derivative of f is *not* positive for all x.

55. Graph $y = \dfrac{1}{x^2} + \dfrac{1}{x - 1}$.

56. A function f is *even* if $f(-x) = f(x)$ for all x in the domain of f. For instance, the cosine function is even, since $\cos(-x) = \cos x$. A function f is *odd* if $f(-x) = -f(x)$. For instance, the sine function is odd, since $\sin(-x) = -\sin x$. Which of the following functions are even? odd? neither?

(a) x^4, (b) x^3, (c) $x^3 + x^2$, (d) $|x|$, (e) $\tan x$, (f) $\sin 3x$, (g) e^x, (h) e^{x^2}.

57. (See Exercise 56.)

(a) If the point (x, y) lies on the graph of an even function, what other point also lies on the graph?

(b) If the point (x, y) lies on the graph of an odd function, what other point lies on the graph?

(c) How can you graph an even or odd function by doing only half the work?

58. Show that every function f is the sum of an even function and an odd function. (See Exercises 56.) *Hint:* First show that $g(x) = f(x) + f(-x)$ is even and $h(x) = f(x) - f(-x)$ is odd.

59. Approximately how large is $(1.03)^{100} - 1$?

(a) Estimate by using the differential of the function x^{100}.

(b) Estimate by relating the quantity $(1.03)^{100} - 1$ to an expression involving e.

60. If dy/dx is proportional to x^2, show that d^2y/dx^2 is proportional to x.

61. If dy/dx is proportional to y^2, show that d^2y/dx^2 is proportional to y^3.

62. Translate the following excerpt from a news article into the terminology of calculus.

> With all the downward pressure on the economy, the first signs of a slowing of inflation seem to be appearing. Some sensitive commodity price indexes are down; the over-all wholesale price index is rising at a slightly slower rate.

■ ■

63. Show that, if $f'(x) \neq 0$, then dy is a good approximation to Δy when dx is small, in the sense that

$$\lim_{dx \to 0} \frac{\Delta y - dy}{dx} = 0.$$

64. A swimmer stands at a point A on the bank of a circular pond of diameter 200 feet. He wishes to reach the diametrically opposite point B by swimming to some point P on the bank and walking the arc PB along the bank. If he swims 100 feet per minute and walks 200 feet per minute, to what point P should he swim in order to reach B in the shortest possible time?

65. Let f be differentiable everywhere. Assume that $f'(a) = 0$ and $f'(b) = 1$. Prove that there is a number X, with $a < X < b$, such that $f'(X) = \frac{1}{2}$. *Warning:* f' need not be continuous.

66. Let f have a derivative everywhere, such that $f' = f$ and $f(0) = 1$. In answering the following, do not make use of the explicit formula for f, $f(x) = e^x$, obtained in Sec. 6.4.

(a) Show that for any constant k we have

$$[f(x)f(k - x)]' = 0.$$

(b) In view of (a) what kind of function must the product $f(x) f(k - x)$ be?

(c) Prove that $f(x)f(k - x) = f(k)$ for all x.

(d) From (c), prove that $f(x + y) = f(x)f(y)$ for all x and y.

67. Let f have a derivative at all $x > 0$, with $f'(x) = 1/x$ and $f(1) = 0$.

(a) Prove that $f(xy) = f(x) + f(y)$ without referring to the function $\ln x$. *Hint:* Modify the technique used in Exercise 66.

(b) Without using (a), prove that $f(x) = \ln x$.

68. Let f have a derivative for all x.

(a) Is every chord of the graph of f parallel to some tangent to the graph of f?

(b) Is every tangent to the graph of f parallel to some chord of the graph of f?

69. An express subway train starts with a constant acceleration, then travels at its maximum speed, and finally slows down at a constant (negative) deceleration. It requires 120 seconds to go nonstop from 42d Street to 72d Street and 96 seconds to go from 72d Street to 96th Street.

(a) Sketch the graph of speed as a function of time in each case.

(b) What is the train's maximum speed? Assume that there are 20 blocks in a mile.

Exercises 70 to 73 concern various ways to "average" two numbers.

70. The most common way to average two positive numbers a and b is to find the so-called arithmetic mean, that is, form their sum and divide by 2, getting

$$\frac{a + b}{2}.$$

But there are other ways to average a and b. One such average is called the *logarithmic mean*, defined by

$$L(a, b) = \frac{a - b}{\ln a - \ln b} \qquad a, b > 0, a \neq b$$

Prove that $L(a, b)$ is between a and b. Incidentally, the logarithmic mean is used in the study of heat transfer and fluid mechanics.

71. The *power mean* is defined by the equation

$$M_p(a, b) = \left(\frac{a^p + b^p}{2}\right)^{1/p} \qquad a, b > 0, p \text{ fixed.}$$

It is used in the study of utility. The case $p = \frac{1}{3}$ is used in the study of gases.

(a) Show that, when $p = 1$, we obtain the arithmetic mean of a and b.

(b) Show that $M_2(a, b) \geq M_1(a, b)$.

(c) Show that $\lim_{p \to 0} M_p(a, b) = \sqrt{ab}$.

72. (See Exercise 71.) In the *American Mathematical Monthly*, vol. 81, October 1974, Harley Flanders presents the following proof of a theorem due to Tung-po Lin.

THEOREM

If $p = \frac{1}{3}$, then $L(x, y) < M_p(x, y)$ for any distinct $x, y > 0$.

PROOF

Consider $f(t) = \frac{3}{8} \ln t - \frac{t^3 - 1}{(t + 1)^3}$

for $t \geq 1$. A short calculation yields

$$f'(t) = \frac{3}{8} \frac{(t - 1)^4}{t(t + 1)^4}.$$

Hence $f'(t) > 0$ for $t > 1$. Since $f(1) = 0$, we conclude that $f(t) > 0$ for $t > 1$. Now let $0 < y < x$. Evaluate $f(t)$ at $t = x^{1/3}/y^{1/3}$. The result is

$$\left(\frac{\sqrt[3]{x} + \sqrt[3]{y}}{2}\right)^3 > \frac{x - y}{\ln x - \ln y}.$$

Fill in all the steps.

73. Let $0 < a < b$. Examine

$$\lim_{x \to \infty} \left(\frac{a^x + b^x}{2} \right)^{1/x}.$$

74. Find $\displaystyle\lim_{x \to 0} \frac{5^x - 3^x}{4^x - 2^x}$.

75. In R. P. Feynman, *Lectures on Physics*, Addison-Wesley, Reading, Mass., appears this remark. "Here is the quantitative answer of what is right instead of kT. This expression,

$$\frac{\hbar\omega}{e^{\hbar\omega/kT} - 1}$$

should of course approach kT as $\omega \to 0$ or as $T \to \infty$. See if you can prove it does—learn how to do the mathematics." Do the mathematics.

76. Assume that $\lim_{x \to \infty} f'(x) = 1$. Must the graph of $y = f(x)$ be asymptotic to a line of the form $y = x + k$ for some constant k? That is, must there exist a constant k such that $\lim_{x \to \infty} [f(x) - (x + k)] = 0$?

77. Let f and g be increasing functions such that

$$\lim_{x \to \infty} f(x) = \infty \qquad \text{and} \qquad \lim_{x \to \infty} g(x) = \infty.$$

Do any of these three assertions imply any of the others?

(a) $\displaystyle\lim_{x \to \infty} \frac{f(x)}{g(x)} = 1$

(b) $\displaystyle\lim_{x \to \infty} \frac{e^{f(x)}}{e^{g(x)}} = 1$

(c) $\displaystyle\lim_{x \to \infty} \frac{\ln f(x)}{\ln g(x)} = 1$

78. Let $f(x) = \left(\dfrac{e^x + 1}{2} \right)^{1/x}$, for $x > 0$.

(a) Find $\lim_{x \to 0^+} f(x)$.
(b) Find $\lim_{x \to \infty} f(x)$.
(c) Show that f is an increasing function for $x > 0$.
    Hint: Show that $\ln f$ increases.

79. Graph $f(x) = x + \sin 2x$.

80. Let $f(x) = (x - 1)^n(x - 2)$, where n is an integer, $n \geq 2$.
(a) Show that $x = 1$ is a critical number.
(b) For which values of n will $x = 1$ provide a relative maximum? a relative minimum? neither?

81. Let p and q be constants.
(a) Show that, if $p > 0$, the equation $x^3 + px + q = 0$ has exactly one real root.
(b) Show that, if $4p^3 + 27q^2 < 0$, the cubic equation $x^3 + px + q = 0$ has three distinct real roots.

Exercises 82 to 86 concern tilted asymptotes, asymptotes that are neither horizontal nor vertical.

82. The line $y = ax + b$ is called an asymptote to the graph of the function f if

$$\lim_{x \to \infty} [f(x) - (ax + b)] = 0,$$

or if

$$\lim_{x \to -\infty} [f(x) - (ax + b)] = 0.$$

(a) Show that

$$\lim_{x \to \infty} \left[\frac{2x^2 + 3x + 5}{x + 1} - (2x + 1) \right] = 0.$$

(b) Show that

$$\lim_{x \to -\infty} \left[\frac{2x^2 + 3x + 5}{x + 1} - (2x + 1) \right] = 0.$$

(c) Graph $f(x) = (2x^2 + 3x + 5)/(x + 1)$.

83. (a) Find a tilted asymptote to the graph of

$$f(x) = \frac{3x^2 + 2x + 3}{2x - 5}.$$

(b) Graph f.

84. (a) Find a tilted asymptote to the graph of

$$f(x) = \sqrt{x^2 + 2x}.$$

(b) Graph f.

85. Let a and b be positive constants. Let

$$y = f(x) = \frac{b}{a}\sqrt{x^2 - a^2}.$$

(a) Show that the graph of f has a tilted asymptote.
(b) Graph f.
This shows that the hyperbola $x^2/a^2 - y^2/b^2 = 1$ approaches the lines $y = \pm(b/a)x$ when $|x|$ is large.

86. Let $P(x)$ be a polynomial of degree $n \geq 1$ and let $Q(x)$ be a polynomial of degree $m \geq 1$. What relation must hold between m and n if the graph of $f(x) = P(x)/Q(x)$ has
(a) the x axis as asymptote?
(b) a horizontal asymptote other than the x axis?
(c) a tilted asymptote?

Exercises 87 to 90 present a result which may be quite surprising.

87. A projectile is fired directly up with an initial velocity v_0. Assume that the acceleration of gravity is constant and that the resistance of the air is proportional to the velocity of the projectile.
(a) Show that $dv/dt = -32 - kv$, where k is a positive constant.
(b) Show that the derivative of $v + 32/k$ is proportional to $v + 32/k$.

(c) Show that $v = (1/k)(ce^{-kt} - 32)$, where c is some constant.

(d) Show that $y = (c/k^2)(1 - e^{-kt}) - 32t/k$.

88. Show that the projectile in Exercise 87 reaches its maximum height $(c - 32)/k^2 - (32/k^2) \ln (c/32)$ at time $T = (1/k) \ln (c/32)$.

89. Prove that $(\ln x)^2 < 2(x - 1 - \ln x)$ for $x > 1$. *Hint:* Let $y = \ln x$, and write the inequality in terms of y.

90. (See Exercise 88.) In each of two experiments, a ball is thrown straight up. One experiment is carried out in a vacuum, the other in air. Both balls reach the same maximum height. Which takes longer going up?
(a) Guess the answer.
(b) Use calculus.

91. Linus proposes this proof for Theorem 1 in Sec. 6.10. Since

$$\lim_{x \to a^+} f(x) = 0 \qquad \text{and} \qquad \lim_{x \to a^+} g(x) = 0,$$

I will define $f(a) = 0$ and $g(a) = 0$. Next I consider $x > a$ but near a. I now have continuous functions f and g defined on the closed interval $[a, x]$ and differentiable on the open interval (a, x). So, using the law of the mean I conclude that there is a number X, $a < X < x$ such that

$$\frac{f(x) - f(a)}{x - a} = f'(X) \qquad \text{and} \qquad \frac{g(x) - g(a)}{x - a} = g'(X).$$

Since $f(a) = 0$ and $g(a) = 0$, these equations tell me that

$$f(x) = (x - a)f'(X) \qquad \text{and} \qquad g(x) = (x - a)g'(X).$$

Thus

$$\frac{f(x)}{g(x)} = \frac{f'(X)}{g'(X)}.$$

Hence

$$\lim_{x \to a^+} \frac{f(x)}{g(x)} = \lim_{x \to a^+} \frac{f'(X)}{g'(X)}$$
$$= L.$$

Alas, Linus made one error. What is it?

This exercise generalizes the law of the mean.

92. The proof of Theorem 1 in Sec. 6.10 to be outlined in Exercise 93 depends on the following *generalized law of the mean.*

Generalized law of the mean. Let f and g be two functions that are continuous on $[a, b]$ and differentiable on (a, b). Furthermore, assume

that $g'(x)$ is never 0 for x in (a, b). Then there is a number X in (a, b) such that

$$\frac{f(b) - f(a)}{g(b) - g(a)} = \frac{f'(X)}{g'(X)}.$$

(a) During a given time interval one car travels twice as far as another car. Use the generalized law of the mean to show that there is at least one instant when the first car is traveling exactly twice as fast as the second car.

(b) To prove the generalized law of the mean introduce a function h:

$$h(x) = f(x) - f(a) - \frac{f(b) - f(a)}{g(b) - g(a)} [g(x) - g(a)]$$

Show that $h(b) = 0$ and $h(a) = 0$. Then apply Rolle's theorem to h.

Remark: the function h is geometrically quite similar to the function h used in the proof of the law of the mean in Sec. 6.2. It is easy to check that $h(x)$ is the vertical distance between the point $(f(x), g(x))$ and the line through $(f(a), g(a))$ and $(f(b), g(b))$.

This exercise proves Theorem 1 of Sec. 6.10.

93. Assume the hypotheses of Theorem 1 of Sec. 6.10. Define $f(a) = 0$ and $g(a) = 0$.

Note that
$$\frac{f(x)}{g(x)} = \frac{f(x) - f(a)}{g(x) - g(a)}$$

and apply the generalized law of the mean from Exercise 92.

94. If $\lim_{t \to \infty} f(t) = \infty = \lim_{t \to \infty} g(t)$

and
$$\lim_{t \to \infty} \frac{\ln f(t)}{\ln g(t)} = 1,$$

must
$$\lim_{t \to \infty} \frac{f(t)}{g(t)} = 1?$$

95. If $\lim_{t \to \infty} f(t) = \infty = \lim_{t \to \infty} g(t)$

and
$$\lim_{t \to \infty} \frac{f(t)}{g(t)} = 3,$$

what can be said about

$$\lim_{t \to \infty} \frac{\ln f(t)}{\ln g(t)}?$$

(Do not assume that f and g are differentiable.)

7

THE DEFINITE INTEGRAL

Chapters 3 to 6 were devoted to the derivative, which was introduced in Sec. 1.1. Section 1.2 introduced the other main concept of the calculus, the definite integral. The next three chapters are devoted to this topic. Much of the work in Chaps. 2 to 6 will be of use in this development. It turns out that the derivative and the definite integral are so closely related that the computation of many definite integrals makes use of derivatives.

This chapter utilizes, for the sake of simplicity, geometric and physical illustrations, but a few exercises show that the definite integral has far more varied applications. As in Chap. 3, we begin with several problems that will turn out to be different forms of the same basic problem.

7.1

Estimates in four problems

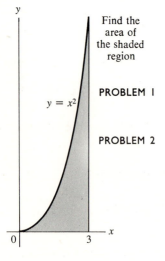

Find the area of the shaded region

$y = x^2$

Just as Chap. 3 introduced the derivative by four problems, this chapter introduces the definite integral by four problems. At first glance these problems may seem unrelated, but by the end of the section it will be clear that they represent one basic problem in various disguises.

PROBLEM 1 Find the area of the region bounded by the curve $y = x^2$, the x axis, and the vertical line $x = 3$.

PROBLEM 2 A thin nonuniform string 3 centimeters long is made of a material that is very light near one end and very heavy near the other end. In fact, at a distance of x centimeters from the left end it has a density of $5x^2$ grams per centimeter. Find the mass of the string.

Find the mass of the string

Nonuniform string

Light end   Heavy end

0   3

PROBLEM 3   An engineer drives a car whose clock and speedometer work, but whose odometer (mileage recorder) is broken. On a 3-hour trip out of a congested city into the countryside he begins at a snail's pace and, as the traffic thins, he gradually speeds up. Indeed, he notices that after traveling t hours his speed is $8t^2$ miles per hour. Thus after the first $\frac{1}{2}$ hour he is crawling along at 2 miles per hour, but after 3 hours he is traveling at 72 miles per hour. How far does the engineer travel in 3 hours?

PROBLEM 4   Find the volume of a tent with a square floor of side 3 feet, whose pole, 3 feet long, rises above a corner of the floor. The tent is shown in the left-hand figure below. It can also be thought of as the surface obtained when the

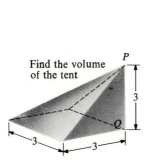

Find the volume of the tent

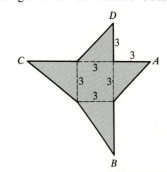

piece of paper illustrated is folded along the dotted lines and the free edges taped in such a way that A, B, C, and D come together (to become P).

We shall now estimate the answers to each of the four problems, showing that they all offer the same mathematical challenge.

The area of the six rectangles approximates the area under the curve

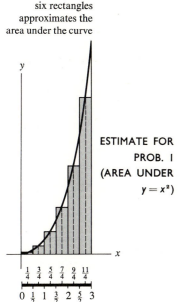

ESTIMATE FOR PROB. 1 (AREA UNDER $y = x^2$)   An estimate of the area can be made using a staircase of six rectangles, as in the figure to the left.

First break the interval from 0 to 3 into six smaller intervals, each of length $\frac{1}{2}$. Then above each small interval draw the rectangle whose height is that of the curve $y = x^2$ above the midpoint of that interval. The total area of the six rectangles is easily computed. This is equal to

$$\left(\tfrac{1}{4}\right)^2\left(\tfrac{1}{2}\right) + \left(\tfrac{3}{4}\right)^2\left(\tfrac{1}{2}\right) + \left(\tfrac{5}{4}\right)^2\left(\tfrac{1}{2}\right) + \left(\tfrac{7}{4}\right)^2\left(\tfrac{1}{2}\right) + \left(\tfrac{9}{4}\right)^2\left(\tfrac{1}{2}\right) + \left(\tfrac{11}{4}\right)^2\left(\tfrac{1}{2}\right),$$

which reduces to $\frac{286}{32}$, or decimally, to 8.9375.

We have *not* computed the area. The preceding computation provides an *estimate* of the area, 8.9375.

ESTIMATE FOR PROB. 2 (THE NONUNIFORM STRING)   Let us cut the string into six sections of equal length:

Nonuniform string

The density of the string in each of the six pieces varies less than it does over the whole length of the string. If the density were constant in a section, we would have, since

$$\text{Density} = \frac{\text{mass}}{\text{length}},$$

$$\text{Mass} = \text{density} \cdot \text{length}.$$

We shall treat each of the six sections as having throughout a constant density equal to that at its midpoint. To obtain an estimate of the mass of each of the six sections, let us multiply the density at the midpoint of the section by the length of the section.

The left section has a density of $5(\frac{1}{4})^2$ grams per centimeter at its midpoint, $\frac{1}{4}$, and thus has a mass of about $5(\frac{1}{4})^2(\frac{1}{2})$ gram. The next section, from $\frac{1}{2}$ to 1, has a density of $5(\frac{3}{4})^2$ at its midpoint, and thus has a mass of about $5(\frac{3}{4})^2(\frac{1}{2})$ grams. An estimate of the mass of each of the four other sections can be made similarly. An estimate of the total mass of the nonuniform string is then the sum

$$5(\tfrac{1}{4})^2(\tfrac{1}{2}) + 5(\tfrac{3}{4})^2(\tfrac{1}{2}) + 5(\tfrac{5}{4})^2(\tfrac{1}{2}) + 5(\tfrac{7}{4})^2(\tfrac{1}{2}) + 5(\tfrac{9}{4})^2(\tfrac{1}{2}) + 5(\tfrac{11}{4})^2(\tfrac{1}{2}).$$

This sum is five times the sum in Prob. 1, hence equals 5(8.9375), a little less than 45 grams. More important is the similarity in form between this sum and the sum used in the first problem.

ESTIMATE FOR PROB. 3
(THE BROKEN ODOMETER)

The speed during the 3-hour trip varies from 0 to 72 miles per hour. During shorter time intervals such a wide fluctuation will not occur.

As in the first two problems, cut the 3 hours of the trip into six equal intervals, each $\frac{1}{2}$ hour long, and use them to make an estimate of the total distance covered. Represent time by this line segment, cut into six parts of equal length.

To estimate the distance the engineer travels in the first $\frac{1}{2}$ hour, multiply his speed at $\frac{1}{4}$ hour by the duration of the first interval of time, $\frac{1}{2}$ hour. Since his speed at time t is $8t^2$, after $\frac{1}{4}$ hour his speed is $8(\frac{1}{4})^2$ miles per hour. Thus during the first $\frac{1}{2}$ hour the engineer travels about $8(\frac{1}{4})^2(\frac{1}{2})$ mile. During the second $\frac{1}{2}$ hour he travels about $8(\frac{3}{4})^2(\frac{1}{2})$ miles.

Making similar estimates for each of the other $\frac{1}{2}$-hour periods, we obtain this estimate for the length of the trip:

$$8(\tfrac{1}{4})^2(\tfrac{1}{2}) + 8(\tfrac{3}{4})^2(\tfrac{1}{2}) + 8(\tfrac{5}{4})^2(\tfrac{1}{2}) + 8(\tfrac{7}{4})^2(\tfrac{1}{2}) + 8(\tfrac{9}{4})^2(\tfrac{1}{2}) + 8(\tfrac{11}{4})^2(\tfrac{1}{2}).$$

This sum is eight times the sum in Prob. 1, hence equals

$$(8)(8.9375) = 71.5 \text{ miles.}$$

Keep in mind that this is only an estimate of the length of the trip.

ESTIMATE FOR PROB. 4
(THE TENT)

Observe that the cross section of the tent made by any plane parallel to the base is a square, as the marginal figure on the next page shows.

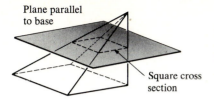

Plane parallel to base

Square cross section

This time we cut a vertical line, representing the pole, into six sections of equal length. Then we approximate the volume of each slab by a flat rectangular box, $\frac{1}{2}$ foot high. The cross section of the smallest box is obtained by passing a horizontal plane through the midpoint of the highest of the six sections. The remaining five boxes are determined in a similar manner, as shown here.

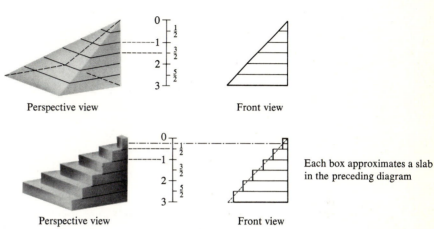

Perspective view   Front view

Perspective view   Front view

Each box approximates a slab in the preceding diagram

As the front view of the six boxes shows, the square cross section of the top box has a side equal to $\frac{1}{4}$ foot, the box below it has a side equal to $\frac{3}{4}$ foot, and so on until we reach the bottom box, whose side is $\frac{11}{4}$ feet.

Since the volume of a box is just the area of its base times its height, the total volume of the six boxes is

$$\left(\tfrac{1}{4}\right)^2\left(\tfrac{1}{2}\right) + \left(\tfrac{3}{4}\right)^2\left(\tfrac{1}{2}\right) + \left(\tfrac{5}{4}\right)^2\left(\tfrac{1}{2}\right) + \left(\tfrac{7}{4}\right)^2\left(\tfrac{1}{2}\right) + \left(\tfrac{9}{4}\right)^2\left(\tfrac{1}{2}\right) + \left(\tfrac{11}{4}\right)^2\left(\tfrac{1}{2}\right) \text{ cubic feet.}$$

This is the same sum that was met in estimating the area under the curve $y = x^2$. Thus the volume of the tent is estimated as 8.9375 cubic feet.

None of the four problems is yet solved; in each case all we have is an estimate. In the next section the precise answers will be found.

Exercises

1. To show the similarity of the four problems, the interval [0, 3] has been cut into six sections each time, and the midpoint of each section used to determine the cross section, density, speed, or area. Of course, we are free to cut the interval into more or fewer sections and to use a point other than the midpoint in each section. Furthermore, there is no need to restrict the sections to be of equal length.

(*a*) Estimate the area in Prob. 1 by using three sections, each of length 1, and the midpoint each time.

(*b*) Estimate the same area by using the same three sections, but now use the y coordinate of the point on $y = x^2$ above the right end of each section to determine the rectangle.

(*c*) Draw the three rectangles used in (*b*). Is their total area more or less than the area under the curve?

(*d*) Proceed as in (*b*) and (*c*), but use the point on $y = x^2$ above the left endpoint of each section.

2. Using your answers to Exercise 1, complete this sentence: The area in Prob. 1 is certainly less than _____ but larger than _____.

3. Cutting the interval from 0 to 3 into five sections of equal length, estimate the area in Prob. 1 by finding the sum of the areas of five rectangles whose heights are determined by (a) midpoints, (b) right endpoints, (c) left endpoints.

4. Cutting the interval of 3 hours into five periods of $\frac{3}{5}$ hour each, estimate the length of the engineer's trip in Prob. 3. For the approximate velocity in each period use the speedometer reading at (a) the middle of the period, (b) the end of the period, (c) the beginning of the period.

5. Estimate the mass of the string in Prob. 2 by cutting it into five sections of equal length. For an estimate of the mass of each of these sections use the density at (a) the midpoint of each section, (b) the right endpoint, (c) the left endpoint.

Exercises 6 and 7 are for review and preview.

6. Differentiate (for practice):

(a) $\sqrt{1 + \cot 3x}$

(b) $\dfrac{3x}{8} + \dfrac{3x \sin 4x}{32} + \dfrac{\cos^3 2x \sin 2x}{8}$

(c) $\dfrac{3}{8(2x + 3)^2} - \dfrac{1}{4(2x + 3)}$

(d) $\dfrac{1}{3(2x + 3)} - \dfrac{1}{9} \ln \left(\dfrac{2x + 3}{x} \right)$

(e) $\frac{1}{5} \ln (\csc 5x - \cot 5x)$

(f) $x \tan^{-1} 2x - \frac{1}{4} \ln (1 + 4x^2)$.

7. Give an example of a function whose derivative is the given function. This is background practice for Chaps. 8 and 9.

(a) $3x^2 + 2x - 1 + \dfrac{5}{x} + \dfrac{3}{x^2}$ (b) $\dfrac{x^2}{x^3 + 1}$

(c) $\dfrac{x^3 + 1}{x^2}$ (d) e^{3x}

(e) $\sin 2x$ (f) xe^x

■

8. A business which now shows no profit is to increase its profit flow gradually in the next 3 years until it reaches a rate of 9 million dollars per year. At the end of the first half year the rate is to be $\frac{1}{4}$ million dollars per year; at the end of 2 years, 4 million dollars per year. In general, at the end of t years, where t is any number between 0 and 3, the rate of profit is to be t^2 million dollars per year. Estimate the total profit during the next 3 years if the plan is successful. Use six intervals of equal length and midpoints as sampling points.

9. Make an estimate for each of the four problems, using in each case the accompanying partition into four sections. As the points where the cross section, density, or velocity is computed, use $\frac{1}{2}$, $\frac{3}{2}$, 2, and $\frac{14}{5}$ (one of these is in each of the four sections). See the figure in the next column.

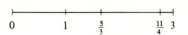

10. Estimate the area under $y = x^2$ and directly above the interval [1, 5] by the midpoint method with the aid of partition of [1, 5] into (a) four sections of equal length, (b) eight sections of equal length.

11. A right circular cone has a height of 3 feet and a radius of 3 feet, as shown in the figure below. Estimate its

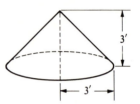

Right circular cone of height 3 feet and radius 3 feet

volume by the sum of the volumes of six cylindrical slabs, just as we estimated the volume of the tent with the aid of six rectangular slabs. In particular, (a) show with the aid of a diagram how the same partition and midpoints we used determine six cylinders, and (b) compute their total volume.

12. Carry out the analog of Exercise 3 for the cone of Exercise 11.

13. Estimate the area between the curve $y = x^3$, the x axis, and the vertical line $x = 6$ using a partition into (a) three sections of equal length and midpoints; (b) six sections of equal length and midpoints; (c) six sections of equal length and left endpoints; (d) same as (c) but right endpoints.

■ ■

14. Draw an accurate graph of $y = x^2$ and the six rectangles with heights equal to the ordinates of the midpoints that we used to estimate the area. Does each of these rectangles underestimate or overestimate the area under $y = x^2$ and above the base of the rectangle? (Form your opinion on the basis of your drawing.)

15. The kinetic energy of an object, for example, a bullet or car, of mass m and speed v is defined as $mv^2/2$ ergs. (Here mass is measured in grams and speed in centimeters per second.) Now, in a certain machine a uniform rod, 3 centimeters long and weighing 32 grams, rotates once per second around one of its ends. Estimate the kinetic energy of this rod by cutting it into six sections, each $\frac{1}{2}$ centimeter long, and taking as the "speed of a section" the speed of its midpoint.

The remaining two exercises introduce a theme which will be explored further in later sections. A calculator would come in handy. If a calculator is not available, use the table of reciprocals in Appendix E.

16. Let $A(t)$ denote the area of the region under the curve $y = 1/x$ and above the x axis from $x = 1$ to $x = t \geq 1$, as shown in the figure below.

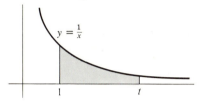

$y = \frac{1}{x}$

1   t

(a) Estimate $A(2)$ by cutting the interval $[1, 2]$ into 10 intervals of equal length and forming 10 rectangles,

each of width 0.1 and of a height equal to the value of $1/x$, where x is the left endpoint of the corresponding interval. Make a sketch to show that this estimate is larger than $A(2)$.

(b) Do the same as in (a), but this time use the right endpoints. Why is this estimate smaller than $A(2)$?

17. (a) Estimate the area under the curve $y = 1/x$ and above $[3, 6]$ by cutting the interval into 10 intervals of equal length and forming 10 rectangles, each of width 0.3 and of height equal to the value of $1/x$, where x is the left endpoint of the corresponding interval. Show that this estimate is too large.

(b) Do the same as in (a), but this time use right endpoints. Why is this estimate too small?

7.2

Precise answers to the four problems

The similarity of the sums formed in Sec. 7.1 to estimate the answers to the four problems suggests that the four problems are essentially just one problem. This is the case, and in this section the sums are studied closely to find out what the precise answers to the four problems are.

The estimates in the previous section were obtained by dividing the interval $[0, 3]$ into smaller sections. To minimize the arithmetic, partitions into only a few sections were considered, but partitions into a hundred, a thousand, or more pieces should provide better estimates. Let us describe how these estimates are made in general.

First the interval $[0, 3]$ is cut into many smaller intervals, which we shall call *sections*. These sections are formed by choosing endpoints x_0, x_1, x_2, ..., x_n, where $x_0 < x_1 < x_2 < \cdots < x_n$, with $x_0 = 0$ and $x_n = 3$. These $n + 1$ numbers determine n sections:

$$[x_0, x_1], [x_1, x_2], \ldots, [x_{n-1}, x_n].$$

The partition:

$x_0 = 0$ x_1 x_2 $\cdots$ x_{n-2} x_{n-1} $x_n = 3$

No matter what the choice of n may be, $x_0 = 0$ and $x_n = 3$. In Sec. 7.1 in all four estimates,

$$x_0 = 0, \qquad x_1 = \tfrac{1}{2}, \qquad x_2 = 1 \qquad x_3 = \tfrac{3}{2}, \qquad x_4 = 2, \qquad x_5 = \tfrac{5}{2}, \qquad x_6 = 3.$$

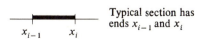

Typical section has ends x_{i-1} and x_i

x_{i-1} x_i

The first section has endpoints x_0 and x_1; the next has endpoints x_1 and x_2; the third has endpoints x_2 and x_3; and so on until we reach the nth section, which has endpoints x_{n-1} and x_n. A typical section, the ith section, has endpoints x_{i-1} and x_i, where i is 1, 2, 3, ..., or n. The n sections form what will be called a *partition* of the interval $[0,3]$.

Next pick a sampling point in each of the n sections—it may be the midpoint, an endpoint, or some other point. Call the point picked in the first section X_1; the point in the second section X_2, and so on. In general, the point chosen in the ith section $[x_{i-1}, x_i]$ is denoted by X_i.

$$X_1 \quad X_2 \quad \cdots \quad X_i \quad \cdots \quad X_{n-1} \quad X_n$$

$$x_0 = 0 \qquad x_1 \quad x_2 \qquad x_{i-1} \quad x_i \qquad x_{n-2} \quad x_{n-1} \quad x_n = 3$$

Typical section
in partition

Keep in mind that X_i can be any point in $[x_{i-1}, x_i]$, while x_{i-1} and x_i denote the two ends of the ith section.

Each choice of x's and X's provides an estimate for each of the four problems. For example, the typical estimating rectangle for the area in Prob. 1 is shown in the diagram at the left.

The sum of the areas of these typical rectangles,

$$X_1{}^2(x_1 - x_0) + X_2{}^2(x_2 - x_1) + \cdots + X_n{}^2(x_n - x_{n-1}),$$

is an estimate of the area in Prob. 1.

The same type of partition by the x's and choice of the X's provides a means of estimating the mass of the string.

$$X_1 \quad \cdots \quad X_i \quad \cdots \quad X_n$$

$$x_0 = 0 \qquad x_1 \qquad x_{i-1} \quad x_i \qquad x_{n-1} \quad x_n = 3$$

Typical section
in partition

Mass in typical section is roughly
$5X_i{}^2(x_i - x_{i-1})$ grams

Density Length of
at X_i section

Similarly, we can estimate the length of the trip in Prob. 3. Since time has been subdivided, let us use t's and T's instead of x's and X's.

$$T_1 \quad \cdots \quad T_i \quad \cdots \quad T_n$$

$$t_0 = 0 \quad t_1 \qquad t_{i-1} \quad t_i \qquad t_{n-1} \quad t_n = 3$$

Typical interval
of time

The distance the engineer travels during
a typical interval of time is roughly

$$8T_i{}^2(t_i - t_{i-1})$$

Speedometer reading Duration of ith
at time T_i time interval

Last, the same x's and X's (or t's and T's) provide an estimate of the volume in the tent.

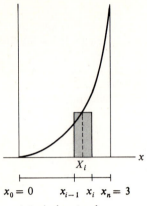

A typical rectangle has base $x_i - x_{i-1}$, height $X_i{}^2$, and area $X_i{}^2(x_i - x_{i-1})$

A typical box has thickness $x_i - x_{i-1}$, square cross section of side X_i and volume $X_i{}^2(x_i - x_{i-1})$

$0 = x_0$

$X_i \begin{cases} x_{i-1} \\ x_i \end{cases}$ We partition this vertical axis

$3 = x_n$

A brief list summarizes these results:

PROBLEM 1 A typical estimate of the area of the region under the curve $y = x^2$ and above the interval $[0, 3]$ is the sum

$$X_1^2(x_1 - x_0) + X_2^2(x_2 - x_1) + \cdots + X_n^2(x_n - x_{n-1}).$$

PROBLEM 2 A typical estimate of the mass of the string is the sum

$$5X_1^2(x_1 - x_0) + 5X_2^2(x_2 - x_1) + \cdots + 5X_n^2(x_n - x_{n-1}).$$

PROBLEM 3 A typical estimate of the distance the engineer travels is the sum

$$8T_1^2(t_1 - t_0) + 8T_2^2(t_2 - t_1) + \cdots + 8T_n^2(t_n - t_{n-1}).$$

PROBLEM 4 A typical estimate of the volume of the tent is the sum

$$X_1^2(x_1 - x_0) + X_2^2(x_2 - x_1) + \cdots + X_n^2(x_n - x_{n-1}).$$

When the typical sections $[x_{i-1}, x_i]$ are chosen shorter and shorter and n is chosen larger and larger, the sums in question should become more accurate estimates of the quantities sought, for when $[x_{i-1}, x_i]$ is short, the choice of X_i is quite restricted. Since the function x^2 is continuous, its values do not fluctuate much over short intervals.

As we consider finer and finer partitions of the interval $[0, 3]$ (and therefore larger and larger values of n), the sums in question should become more accurate estimates of the quantities we are seeking.

This is the mathematical challenge: Find what happens to sums of the form

$$X_1^2(x_1 - x_0) + X_2^2(x_2 - x_1) + \cdots + X_n^2(x_n - x_{n-1}) \tag{1}$$

as the little sections $[x_{i-1}, x_i]$ are chosen smaller and smaller. If we can find what number they approach, then all four problems will be solved in one stroke. For, if the sums approach the number c, say, then the area of the region under $y = x^2$ is c (square units), the string has a mass of $5c$ (grams), the engineer travels $8c$ (miles), and the tent has the volume c (cubic feet).

For this reason let us focus our attention on the sums

$$X_1^2(x_1 - x_0) + X_2^2(x_2 - x_1) + \cdots + X_n^2(x_n - x_{n-1}).$$

There is no need to compute any further specific estimates.

The typical summand, corresponding to the ith section, is

$$X_i^2(x_i - x_{i-1}).$$

Since

$$x_{i-1} \le X_i \le x_i,$$

the summand satisfies the inequalities

$$x_{i-1}^2(x_i - x_{i-1}) \le X_i^2(x_i - x_{i-1}) \le x_i^2(x_i - x_{i-1}).$$

(For the function x^2 is increasing for $x \ge 0$.)

These inequalities suggest that we should consider the smallest sums that can be formed using the given sections $[x_0, x_1], [x_1, x_2], \ldots, [x_{n-1}, x_n]$:

$$x_0^2(x_1 - x_0) + x_1^2(x_2 - x_1) + \cdots + x_{n-1}^2(x_n - x_{n-1}), \tag{2}$$

which *underestimate* the answer, and also the sums

$$x_1^2(x_1 - x_0) + x_2^2(x_2 - x_1) + \cdots + x_n^2(x_n - x_{n-1}), \tag{3}$$

which *overestimate* the answer. (Keep in mind that x^2 is an increasing function for $x \geq 0$.)

It is possible to think of these sums (2) and (3) geometrically, in terms of Prob. 1, the area of the region under $y = x^2$. Sums of type (2) correspond to rectangles whose height is the y coordinate at the left end of each section. The overestimating sum (3) corresponds to rectangles whose height is the ordinate at the right end of each section (see the two figures below).

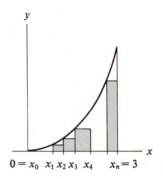

The area of the shaded rectangles equals the underestimating sum (2):

$$x_0^2(x_1 - x_0) + x_1^2(x_2 - x_1) + \ldots + \\ x_{n-1}^2(x_n - x_{n-1})$$

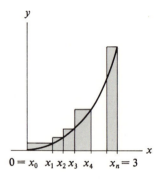

The area of the shaded rectangles equals the overestimating sum (3):

$$x_1^2(x_1 - x_0) + x_2^2(x_2 - x_1) + \ldots + \\ x_n^2(x_n - x_{n-1})$$

At this point the two inequalities that were developed in Sec. 4.2 will be needed. They are the following: If $c > d \geq 0$, and n is a positive integer, then

$$c^{n+1} - d^{n+1} < (n + 1)c^n(c - d)$$

and

$$c^{n+1} - d^{n+1} > (n + 1)d^n(c - d).$$

For convenience we write them on one line as

$$(n + 1)c^n(c - d) > c^{n+1} - d^{n+1} > (n + 1)d^n(c - d)$$

and then divide by $n + 1$, obtaining

$$c^n(c - d) > \frac{c^{n+1}}{n + 1} - \frac{d^{n+1}}{n + 1} > d^n(c - d).$$

The case we will need in a moment is $n = 2$, for which the inequalities become

$$c^2(c - d) > \frac{c^3}{3} - \frac{d^3}{3} > d^2(c - d). \tag{4}$$

Now we apply the inequalities (4) in each of the n sections, as follows.

When $c = x_1$ and $d = x_0$, inequalities (4) become

$$x_1{}^2(x_1 - x_0) > \frac{x_1{}^3}{3} - \frac{x_0{}^3}{3} > x_0{}^2(x_1 - x_0).$$

Similarly, when $c = x_2$ and $d = x_1$, inequalities (4) become

$$x_2{}^2(x_2 - x_1) > \frac{x_2{}^3}{3} - \frac{x_1{}^3}{3} > x_1{}^2(x_2 - x_1).$$

Continuing this process, we obtain n inequalities, of which the last one, obtained by letting $c = x_n$ and $d = x_{n-1}$, is

$$x_n{}^2(x_n - x_{n-1}) > \frac{x_n{}^3}{3} - \frac{x_{n-1}{}^3}{3} > x_{n-1}{}^2(x_n - x_{n-1}).$$

Adding these n inequalities shows that the overestimating sum,

$$x_1{}^2(x_1 - x_0) + x_2{}^2(x_2 - x_1) + \cdots + x_n{}^2(x_n - x_{n-1}),$$

is greater than

$$\frac{x_1{}^3}{3} - \frac{x_0{}^3}{3} + \frac{x_2{}^3}{3} - \frac{x_1{}^3}{3} + \cdots + \frac{x_n{}^3}{3} - \frac{x_{n-1}{}^3}{3}, \tag{5}$$

which, in turn, is greater than the underestimating sum

$$x_0{}^2(x_1 - x_0) + x_1{}^2(x_2 - x_1) + \cdots + x_{n-1}{}^2(x_n - x_{n-1}).$$

But look closely at the middle sum (5), in which many terms cancel:

$$\frac{x_1{}^3}{3} - \frac{x_0{}^3}{3} + \frac{x_2{}^3}{3} - \frac{x_1{}^3}{3} + \frac{x_3{}^3}{3} - \frac{x_2{}^3}{3} + \cdots + \frac{x_n{}^3}{3} - \frac{x_{n-1}{}^3}{3}.$$

Notice that $x_1{}^3/3$ cancels with $-x_1{}^3/3$, $x_2{}^3/3$ cancels with $-x_2{}^3/3$, and so on, as indicated by long parentheses. Because of cancellations, the sum (5) is simply

$$-\frac{x_0{}^3}{3} + \frac{x_n{}^3}{3},$$

since $-x_0{}^3/3$ and $x_n{}^3/3$ do not pair off with other terms. But $x_0 = 0$, and $x_n = 3$. So the middle sum is just

$$-\frac{0^3}{3} + \frac{3^3}{3} = 9.$$

This shows that the largest sum that can be formed with a given choice of

$$x_0, x_1, \ldots, x_n$$

is greater than 9, while the smallest sum that can be formed is less than 9.

This strongly suggests that the sums

$$X_1{}^2(x_1 - x_0) + X_2{}^2(x_2 - x_1) + \cdots + X_n{}^2(x_n - x_{n-1}), \tag{1}$$

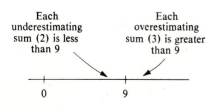

Each underestimating sum (2) is less than 9

Each overestimating sum (3) is greater than 9

0   9

trapped between sums (2) and (3), approach 9 as all sections $[x_{i-1}, x_i]$ are chosen smaller and smaller. To show this, it suffices to show that the difference between sums (3) and (2) approaches 0 when all the sections of a partition are short. This difference is

$$[x_1{}^2(x_1 - x_0) + x_2{}^2(x_2 - x_1) + \cdots + x_n{}^2(x_n - x_{n-1})]$$
$$- [x_0{}^2(x_1 - x_0) + x_1{}^2(x_2 - x_1) + \cdots + x_{n-1}{}^2(x_n - x_{n-1})]. \quad (6)$$

To show that (6) is small when all $x_i - x_{i-1}$ are small, collect similar terms in (6), obtaining

$$(x_1{}^2 - x_0{}^2)(x_1 - x_0) + (x_2{}^2 - x_1{}^2)(x_2 - x_1) + \cdots$$
$$+ (x_n{}^2 - x_{n-1}{}^2)(x_n - x_{n-1}). \quad (7)$$

We want to show that, when n is large and all the sections are short, that is, when $x_1 - x_0, x_2 - x_1, \ldots, x_n - x_{n-1}$ are all small, the difference (7), though it will have many summands, is near 0.

Call the length of the largest of all the n sections L. We want to show that, when L is small, so is the difference (6); equivalently, when L is small, so is the sum (7). In any case the sum (7) is less than or equal to

$$(x_1{}^2 - x_0{}^2)L + (x_2{}^2 - x_1{}^2)L + \cdots + (x_n{}^2 - x_{n-1}{}^2)L,$$

which equals

$$L[(x_1{}^2 - x_0{}^2) + (x_2{}^2 - x_1{}^2) + \cdots + (x_n{}^2 - x_{n-1}{}^2)]. \quad (8)$$

The product (8), due to many cancellations, is simply

$$L(-x_0{}^2 + x_n{}^2). \quad (9)$$

Since $x_n = 3$ and $x_0 = 0$, (9) is equal to

$$L(-0^2 + 3^2) = 9L.$$

When L is small, so is $9L$. Thus (9), and hence the difference (6), is small when all the sections $[x_{i-1}, x_i]$ are short.

Thus the underestimating sums (2) and the overestimating sums (3) are forced to be close to each other when all the sections $[x_{i-1}, x_i]$ are short. Since the number 9 is always between them, they both must approach 9 as the sections $[x_{i-1}, x_i]$ are chosen shorter and shorter.

This forces the more general sums (1) also to approach 9. With this mathematical result, all four problems are solved:

The area in Prob. 1 is 9 (square units).
The string in Prob. 2 has a mass of $5 \cdot 9 = 45$ (grams).
The engineer in Prob. 3 travels $8 \cdot 9 = 72$ (miles).
The tent in Prob. 4 has a volume 9 (cubic feet).

The method just used to find the area of the region under $y = x^2$ (and solve the other three problems) also works for the area of the region under $y = x^3$, $y = x^4$, ... It works for the functions x^n because of the existence of an algebraic identity involving $c^{n+1} - d^{n+1}$. Chapter 8 develops rapid general

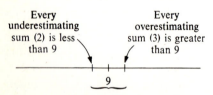

Every underestimating sum (2) is less than 9

Every overestimating sum (3) is greater than 9

But when all the sections $[x_{i-1}, x_i]$ are chosen shorter, and shorter, the difference between the overestimating and underestimating sums approaches 0

procedures for solving problems like the four discussed for functions other than x^n, such as e^x, $\sin x$, and more complicated functions.

Exercises

In Exercises 1 to 5 use the methods of this section to compute the indicated quantity.

1. The area of the region under $y = x^2$ that is above the interval $[0, 2]$.

2. The area of the region under $y = x^2$ that is above the interval $[2, 3]$.

3. The area of the region below the curve $y = x^2$ and above the interval (a) $[0, a]$, where a is some positive number; (b) $[0, b]$, where b is some positive number; (c) $[a, b]$, where $b > a \geq 0$.

4. The total mass of a string 2 centimeters long whose density x centimeters from the left end is x^2 grams per centimeter.

5. The mass in the right half of the string in Exercise 4.

6. If x_0, x_1, x_2, x_3, x_4, x_5 cut $[0, 3]$ into five sections of equal length, what is (a) x_0? (b) x_5? (c) $x_1 - x_0$? (d) $x_2 - x_1$?

7. An estimate of the area of the region under $y = x^2$ and above $[0, 3]$ is made using sections all of length less than 0.01. What can be said about the possible error in this estimate?

8. An estimate of the area of the region under $y = x^2$ and above $[2, 3]$ is made using sections all of length less than 0.01. What can be said about the possible error in this estimate?

Exercises 9 to 12 concern the curve $y = x^3$.

9. (a) Graph $y = x^3$ for x in $[0, 2]$.
   (b) Sketch a diagram showing typical $x_0 = 0$, x_1, ..., $x_n = 2$, sampling numbers X_i, and typical rectangles whose total area is an estimate of the area under $y = x^3$ and above $[0, 2]$.
   (c) What is the area of the rectangle corresponding to the ith section $[x_{i-1}, x_i]$?
   (d) What sum is an estimate of the area under the curve $y = x^3$ and above $[0, 2]$?

10. Show that any sum in Exercise 9(d) formed with $X_i = x_{i-1}$ is less than 4, while any such sum formed with $X_i = x_i$ is greater than 4. What is the area of the region in Exercise 9(d)?

■

11. Show that the area of the region under the curve $y = x^3$ and above $[a, b]$, where $0 \leq a < b$, is

$$\frac{b^4 - a^4}{4}.$$

12. Use the formula in Exercise 11 to find (a) the mass of a string 4 centimeters long whose density x centimeters from the left end is x^3 grams per centimeter; (b) the distance traveled in the third second of motion of an object that has a speed of t^3 feet per second t seconds after starting.

The next two exercises concern the curve $y = 1/x^2$.

13. (a) Show that, if $c > d > 0$, then

$$\frac{1}{c^2}(c - d) < \frac{1}{d} - \frac{1}{c} < \frac{1}{d^2}(c - d).$$

   (b) Use (a) and approximating sums to show that the area of the region under the curve $y = 1/x^2$ and above the interval $[a, b]$, where $b > a > 0$, is $1/a - 1/b$.

14. (See Exercise 13.) A tiring turtle travels at the velocity $1/t^2$ feet per second at time $t > 0$. How far does it travel (a) from time $t = 1$ to time $t = 2$ seconds? (b) from time $t = 2$ to time $t = 3$ seconds?

■ ■

15. Find the volume of a tent shaped like the one considered in the text but with a pole b feet long and square base of side b feet.

16. An approximating sum is formed and computed for the area in Problem 1 using a division of $[0, 3]$ into 100 sections of equal length. Show that no matter how the sampling point is chosen in each section the approximating sum differs from the area of the region under the curve by less than 0.27 square unit.

17. If you make three tents out of paper with the pattern given in Sec. 7.1, you will be able to fit them together easily to form a cube. If you have more geometric intuition than time, you might prefer to see that this is so by examining the diagram below. Use this information

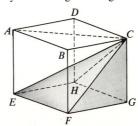

The three tents have
vertex C and base $GFEH$,
vertex C and base $ABFE$,
vertex C and base $DHEA$

to find the volume of the tent directly.   (This trick solves all four problems, since the answer to one determines the answers to the other three.)

The next two exercises develop the theme introduced in Exercises 16 and 17 of Sec. 7.1.

18. (a) Write out the typical approximating sum for the area under the curve $y = 1/x$ and above the interval $[1, 2]$, using left endpoints as the X_i and the typical partition $x_0 = 1, x_1, \ldots, x_n = 2$.

(b) Show that $3x_0, 3x_1, \ldots, 3x_n$ is a partition of the interval $[3, 6]$.   Show that if the left endpoints are used to form the approximating sum for the area

under $y = 1/x$ and above $[3, 6]$, then this sum has the same value as the one obtained in (a).

(c) Show that the area under the curve $y = 1/x$ and above $[1, 2]$ equals the area under the curve $y = 1/x$ and above $[3, 6]$.

19. (See Exercise 18.)   (a) Show that, if $A(t)$ is the area under the curve $y = 1/x$ and above $[1, t]$, then $A(2) = A(6) - A(3)$.   (b) Show that, if $x > 1$ and $y > 1$, then $A(x) = A(xy) - A(y)$.   (c) By (b), $A(xy) = A(x) + A(y)$ for x and y greater than 1.   What famous functions f have the property that $f(xy) = f(x) + f(y)$ for all positive x and y?

7.3

Summation notation

In Sec. 7.2 the sum

$$X_1{}^2(x_1 - x_0) + X_2{}^2(x_2 - x_1) + \cdots + X_n{}^2(x_n - x_{n-1})$$

was of great importance.   It takes a long time to write this sum out or to say it aloud.   Since such sums will be needed often, let us introduce a convenient notation for them, the so-called *sigma notation*, named after the Greek letter Σ, which corresponds to the S of Sum.

DEFINITION   *Summation notation.*   Let $a_1, a_2, \ldots, a_n$ be n numbers.   The sum $a_1 + a_2 + \cdots + a_n$ will be denoted by the symbol $\sum_{i=1}^{n} a_i$, which is read as "the sum of a sub i as i goes from 1 to n."

In the sigma notation the formula for the typical summand is given, as are directions where the summation starts and ends.

EXAMPLE 1   Write the sum $1^2 + 2^2 + 3^2 + 4^2$ in the sigma notation.

SOLUTION   Since the ith summand is the square of i and the summation extends from $i = 1$ to $i = 4$, we have

$$1^2 + 2^2 + 3^2 + 4^2 = \sum_{i=1}^{4} i^2.$$

Simple arithmetic shows that the sum is equal to 30.   ●

EXAMPLE 2   Compute $\sum_{i=1}^{3} 2^i$.

SOLUTION   This is short for the sum $2^1 + 2^2 + 2^3$, which is $2 + 4 + 8$, or 14.   ●

In the definition of the sigma notation the letter i (for "index") was used.

Any letter, such as j or k, would do just as well. Such an index is sometimes called a *summation index* or *dummy index*.

EXAMPLE 3 Compute $$\sum_{j=1}^{4} \frac{1}{j}.$$

SOLUTION This is short for $\frac{1}{1} + \frac{1}{2} + \frac{1}{3} + \frac{1}{4}$, which is approximately 2.083. ●

The summation notation has two properties which will be of use in coming chapters. First of all, if c is a fixed number, then

$$\sum_{i=1}^{n} ca_i = ca_1 + ca_2 + \cdots + ca_n = c(a_1 + a_2 + \cdots + a_n) = c\sum_{i=1}^{n} a_i.$$

Thus $$\sum_{i=1}^{n} ca_i = c\sum_{i=1}^{n} a_i.$$

This distributive rule is read as "a constant factor can be moved past Σ." Second,

$$\sum_{i=1}^{n} (a_i + b_i) = (a_1 + b_1) + (a_2 + b_2) + \cdots + (a_n + b_n)$$

$$= (a_1 + a_2 + \cdots + a_n) + (b_1 + b_2 + \cdots + b_n)$$

$$= \sum_{i=1}^{n} a_i + \sum_{i=1}^{n} b_i .$$

This is a direct consequence of the rules of algebra.

EXAMPLE 4 Compute $$\sum_{i=1}^{4} \left(i^2 + \frac{1}{i} \right).$$

SOLUTION This may be rewritten as $$\sum_{i=1}^{4} i^2 + \sum_{i=1}^{4} \frac{1}{i}.$$

By Examples 1 and 3, the sum, to three decimals, is $30 + 2.083 = 32.083$. ●

EXAMPLE 5 What is the value of $\sum_{i=1}^{5} 3$?

SOLUTION In this case $a_i = 3$ for each index i. Each summand has the value 3. Thus

$$\sum_{i=1}^{5} 3 = 3 + 3 + 3 + 3 + 3 = 15.$$

More generally, if c is a fixed number not depending on i, then $\sum_{i=1}^{n} c = cn$. ●

The next example shows how to interpret the sigma notation when the index does not start at 1.

EXAMPLE 6 Compute $\sum_{i=2}^{6} 5i$ (read as "the sum of $5i$ as i goes from 2 to 6").

SOLUTION This is short for $$5 \cdot 2 + 5 \cdot 3 + 5 \cdot 4 + 5 \cdot 5 + 5 \cdot 6,$$

which equals $5(2 + 3 + 4 + 5 + 6)$,

or 100. ●

A useful fact to note is that

$$\sum_{i=1}^{m} a_i + \sum_{i=m+1}^{n} a_i = \sum_{i=1}^{n} a_i.$$

For instance, $$\sum_{i=1}^{4} i^2 + \sum_{i=5}^{10} i^2 = \sum_{i=1}^{10} i^2.$$

The sigma notation provides brief expressions for the long sums needed in this chapter.   For instance, the sum

$$X_1^2(x_1 - x_0) + X_2^2(x_2 - x_1) + \cdots + X_n^2(x_n - x_{n-1})$$

may now be written as $$\sum_{i=1}^{n} X_i^2(x_i - x_{i-1}).$$

Note that this is the typical approximating sum for the area in Prob. 1 and the volume in Prob. 4.

How is the *underestimating* sum (2) in Sec. 7.2,

$$x_0^2(x_1 - x_0) + x_1^2(x_2 - x_1) + \cdots + x_{n-1}^2(x_n - x_{n-1}),$$

written in the sigma notation?

The *i*th summand is $x_{i-1}^2(x_i - x_{i-1})$.   The sum can therefore be written as

$$\sum_{i=1}^{n} x_{i-1}^2(x_i - x_{i-1}).$$

Clearly, care must be taken in denoting the subscripts, in this case, i and $i - 1$.

In Sec. 7.2 it was observed that the sum

$$\frac{x_1^3}{3} - \frac{x_0^3}{3} + \frac{x_2^3}{3} - \frac{x_1^3}{3} + \cdots + \frac{x_n^3}{3} - \frac{x_{n-1}^3}{3},$$

because of cancellations, equals $\dfrac{x_n^3}{3} - \dfrac{x_0^3}{3}$.

This is an instance of a "telescoping" sum, which Example 7 discusses.

EXAMPLE 7   Let b_0, b_1, b_2, b_3 be four numbers.   Form the three differences

$$a_1 = b_1 - b_0, \qquad a_2 = b_2 - b_1, \qquad a_3 = b_3 - b_2.$$

Show that $$\sum_{i=1}^{3} a_i = b_3 - b_0.$$

SOLUTION $$\sum_{i=1}^{3} a_i = a_1 + a_2 + a_3 = (b_1 - b_0) + (b_2 - b_1) + (b_3 - b_2).$$

Cancellations of b_1 and $-b_1$, of b_2 and $-b_2$ show that

$$a_1 + a_2 + a_3 = b_3 - b_0. \bullet$$

Example 7 easily can be generalized from four to any finite list of numbers. If $b_0, b_1, \ldots, b_n$ are $n + 1$ numbers and $a_i = b_i - b_{i-1}$, $i = 1, \ldots, n$, then

$$\sum_{i=1}^{n} a_i = \sum_{i=1}^{n} (b_i - b_{i-1}) = b_n - b_0.$$

Telescoping Sums   In short, the sum $\sum_{i=1}^{n} (b_i - b_{i-1})$ telescopes to $b_n - b_0$. In the special case $b_i = x_i^3/3$ this yields

$$\sum_{i=1}^{n} \left(\frac{x_i^3}{3} - \frac{x_{i-1}^3}{3} \right) = \frac{x_n^3}{3} - \frac{x_0^3}{3}.$$

In Sec. 7.1 an approximating sum involving six sections, each of length $\frac{1}{2}$, was used. The next example shows how this sum appears in the sigma notation.

EXAMPLE 8   Write the approximating sum

$$(\tfrac{1}{4})^2(\tfrac{1}{2}) + (\tfrac{3}{4})^2(\tfrac{1}{2}) + (\tfrac{5}{4})^2(\tfrac{1}{2}) + (\tfrac{7}{4})^2(\tfrac{1}{2}) + (\tfrac{9}{4})^2(\tfrac{1}{2}) + (\tfrac{11}{4})^2(\tfrac{1}{2})$$

in the sigma notation.

SOLUTION   The sum has six summands, all of the same general form. The 4 and 2 in the denominators are the same in each; the 1, 3, 5, 7, 9, 11 in the numerators are the changing part of the summands. The first summand has a 1, the second has a 3, the third has a 5, and so on. It is not hard to see that the ith summand has a $2i - 1$ in the numerator. Thus the ith summand is

$$\left(\frac{2i - 1}{4} \right)^2 \left(\frac{1}{2} \right),$$

and the sum is

$$\sum_{i=1}^{6} \left(\frac{2i - 1}{4} \right)^2 \left(\frac{1}{2} \right).$$

This may be simplified, as follows:

$$\sum_{i=1}^{6} \frac{(2i - 1)^2}{16} \frac{1}{2} = \frac{1}{32} \sum_{i=1}^{6} (2i - 1)^2. \bullet$$

Exercises

1. Evaluate: (a) $\displaystyle\sum_{i=1}^{3} i$ (b) $\displaystyle\sum_{i=1}^{4} 2i$ (c) $\displaystyle\sum_{d=1}^{3} d^2$.

2. Evaluate: (a) $\displaystyle\sum_{i=2}^{4} i^2$ (b) $\displaystyle\sum_{j=2}^{4} j^2$ (c) $\displaystyle\sum_{i=1}^{3} (i^2 + i)$.

3. Evaluate: (a) $\displaystyle\sum_{i=1}^{4} 1^i$ (b) $\displaystyle\sum_{k=2}^{6} (-1)^k$ (c) $\displaystyle\sum_{j=1}^{150} 3$.

4. Write in the sigma notation (do not evaluate):
   (a) $1 + 2 + 3 + 4 + \cdots + (n - 1) + n$;

(b) $1^2 + 2^2 + 3^2 + \cdots + 10^2$;

(c) $1 + 3 + 5 + 7 + 9 + 11 + 13$.

5. Write in the sigma notation (do not evaluate):

(a) $1 + 2 + 2^2 + 2^3 + \cdots + 2^{100}$;

(b) $x^3 + x^4 + x^5 + x^6 + x^7$;

(c) $\frac{1}{3} + \frac{1}{4} + \cdots + \frac{1}{102}$;

(d) $\frac{1}{2} + \frac{1}{3} + \cdots + \frac{1}{100}$;

(e) $\frac{1}{3} + \frac{1}{5} + \frac{1}{7} + \frac{1}{9} + \frac{1}{11}$;

(f) $\dfrac{1}{1^2} + \dfrac{1}{3^2} + \dfrac{1}{5^2} + \cdots + \dfrac{1}{101^2}$.

6. Write in the sigma notation:

(a) $x_1{}^2(x_1 - x_0) + x_2{}^2(x_2 - x_1) + \cdots + x_8{}^2(x_8 - x_7)$;

(b) $X_1{}^5(x_1 - x_0) + X_2{}^5(x_2 - x_1) + \cdots + X_n{}^5(x_n - x_{n-1})$.

7. Writing out each sum in longhand, show that

(a) $\displaystyle\sum_{i=1}^{3}(a_i - b_i) = \sum_{i=1}^{3}a_i - \sum_{i=1}^{3}b_i$;

(b) $\displaystyle\sum_{i=1}^{2}a_i b_i$ is *not* equal to $\displaystyle\sum_{i=1}^{2}a_i \cdot \sum_{i=1}^{2}b_i$;

(c) $\displaystyle\sum_{i=1}^{3}\left(\sum_{j=1}^{3}b_j\right)a_i = \sum_{j=1}^{3}\left(\sum_{i=1}^{3}a_i\right)b_j$.

8. Evaluate:

(a) $\displaystyle\sum_{i=1}^{100}(2^i - 2^{i-1})$.

(b) $\displaystyle\sum_{i=2}^{100}\left(\frac{1}{i} - \frac{1}{i-1}\right)$

(c) $\displaystyle\sum_{i=1}^{50}\left(\frac{1}{2i+1} - \frac{1}{2(i-1)+1}\right)$.

9. Let $x_0, x_1, \ldots, x_n$ determine a partition of $[0, 4]$. Evaluate:

(a) $\displaystyle\sum_{i=1}^{n}(x_i{}^3 - x_{i-1}{}^3)$

(b) $\displaystyle\sum_{i=1}^{n}(x_i{}^2 - x_{i-1}{}^2)$

(c) $\displaystyle\sum_{i=1}^{n}(5x_i - 5x_{i-1})$.

10. Writing out each sum in longhand, show that

(a) $\displaystyle\sum_{i=1}^{3}a_i = \sum_{j=1}^{3}a_j = \sum_{k=1}^{3}a_k$

(b) $\displaystyle\sum_{i=1}^{3}(a_i + 4) = 12 + \sum_{j=2}^{4}a_{j-1}$.

■

11. Let $S_n = 1 + 2 + 3 + \cdots + n = \sum_{i=1}^{n}i$. We outline a shortcut for computing this sum, illustrating the method in the case $n = 20$.

(a) $S_{20} = 1 + 2 + 3 + \cdots + 19 + 20$.
$S_{20} = 20 + 19 + 18 + \cdots + 2 + 1$.
Adding the two lines, show that
$2S_{20} = 21 + 21 + 21 + \cdots + 21 + 21$ (20 summands).

(b) From (a) deduce that $S_{20} = 21 \cdot 20/2 (= 210)$.

(c) Show that $S_n = [(n + 1)n]/2$.

(d) Use (c) to compute $\sum_{i=1}^{100}i$.

12. Write the expression

$$c^{n-1} + c^{n-2}d + c^{n-3}d^2 + \cdots + d^{n-1}$$

in the sigma notation.

13. (a) By inspection of the diagram below show, without using any arithmetic, that $1 + 3 + 5 + 7 = 4^2$.

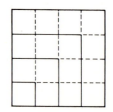

(b) Use the idea in (a) to find a short formula for the sum of the first n odd positive integers.

(c) Find a formula for the ith positive odd integer.

(d) Write the sum of the first n positive odd integers in sigma notation.

14. (a) Show that $2i + 1 = (i + 1)^2 - i^2$.

(b) Deduce from (a) a formula for the sum of the first n positive odd integers.

15. (a) Show that $i = [(i + 1)^2 - i^2 - 1]/2$.

(b) Use (a) to show that $\sum_{i=1}^{n}i = (n + 1)^2/2 - 1/2 - n/2$.

(c) From (b) obtain the formula in Exercise 11(c).

16. (a) Show that $(i + 1)^3 - i^3 = 3i^2 + 3i + 1$.

(b) Use (a) to show that

$$3\sum_{i=1}^{n}i^2 + 3\sum_{i=1}^{n}i + \sum_{i=1}^{n}1 = (n + 1)^3 - 1.$$

(c) Evaluate $\sum_{i=1}^{n}1$.

(d) Using Exercise 11 or 15, evaluate $\sum_{i=1}^{n}i$.

(e) Solving the equation in (b) for $\sum_{i=1}^{n}i^2$, obtain the formula

$$\sum_{i=1}^{n}i^2 = \frac{n(n+1)(2n+1)}{6}.$$

(f) Check the formula in (e) for the case $n = 5$.

■ ■

A calculator will come in handy in the remaining exercises.

17. Let $S_n = \sum_{i=1}^{n} i 2^{-i}$.

(a) Compute S_2, S_3, and S_4 to at least three decimal places.

(b) Compute S_{10} and S_n for some larger values of n.

(c) What do you think happens to S_n as $n \to \infty$? Does it become arbitrarily large or does it approach some number?

18. Let $S_n = \sum_{i=1}^{n} 1/[i(i+1)]$.

(a) Compute S_2, S_3, S_4, and S_5 to at least four decimal places.

(b) What do you think happens to S_n as $n \to \infty$?

19. Let $S_n = \sum_{i=n}^{2n} 1/i$. For example, $S_3 = \frac{1}{3} + \frac{1}{4} + \frac{1}{5} + \frac{1}{6}$.

(a) Compute S_1, S_2, S_3, and S_4 to at least three decimal places.

(b) What do you think happens to S_n as $n \to \infty$?

(c) Show that $S_n > \frac{1}{2}$ for all positive integers n.

20. Let $S_n = \sum_{i=1}^{n} 1/i^2$.

(a) Compute S_1, S_5, and S_{10}.

(b) What do you think happens to S_n as $n \to \infty$?

7.4

The definite integral over an interval

The four problems in Sec. 7.1 posed basically the same question: What happens to certain sums of products as partitions become very fine? In Sec. 7.2, with the aid of the two inequalities

$$c^2(c-d) > \frac{c^3}{3} - \frac{d^3}{3} > d^2(c-d),$$

it was possible to show that the sums in question,

$$\sum_{i=1}^{n} X_i^2 (x_i - x_{i-1}),$$

approach the number 9 when all the lengths $x_i - x_{i-1}$ are chosen smaller and smaller.

It is evident that the procedure of cutting up an interval into little sections, forming certain sums, and determining what happens to those sums when the sections are short is important. To emphasize this, we present some general situations in which this procedure is followed; each has already been illustrated by one of the four problems in Sec. 7.1. The presentation will be intuitive. A rigorous development of the area and volume cases below depends on theorems in advanced calculus.

SUMS AND THE AREA OF A PLANE REGION

Let S be some region in the plane whose area is to be found. Let L be a line in the plane which will be considered to be the x axis. Each line in the plane and perpendicular to L meets S in what shall be called a cross section. (If the line misses S, the cross section is empty.) Let the coordinate on L where the typical line meets L be x. The length of the typical cross section is denoted by $c(x)$. Assume that the lines that are perpendicular to L and that meet S intersect L in an interval whose ends are a and b. In Prob. 1 of Sec. 7.1, L is the x axis, $c(x) = x^2$, $a = 0$, and $b = 3$.

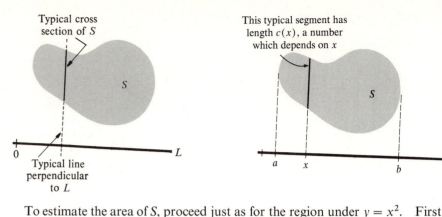

Typical cross section of S

Typical line perpendicular to L

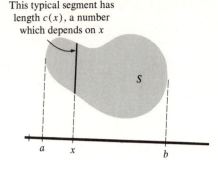

This typical segment has length $c(x)$, a number which depends on x

To estimate the area of S, proceed just as for the region under $y = x^2$. First cut $[a, b]$ into n sections by means of the numbers $x_0 = a, x_1, x_2, \ldots, x_n = b$.

$$x_0 = a < x_1 < x_2 < x_3 < \cdots < x_{n-1} < x_n = b$$

$x_0 = a \quad x_1 \quad x_2 \quad \cdots \quad x_{i-1} \quad x_i \quad \cdots \quad x_{n-1} \quad x_n = b$

Typical section $[x_{i-1}, x_i]$

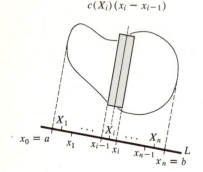

A typical rectangle has base $x_i - x_{i-1}$, height $c(X_i)$ and area $c(X_i)(x_i - x_{i-1})$

In each of these sections select a number at random. In the section $[x_0, x_1]$ select X_1, in $[x_1, x_2]$ select X_2, and so on. Simply stated, in the ith interval, $[x_{i-1}, x_i]$, select X_i.

With this choice of x's and X's, form a set of rectangles, whose typical member is shown in the diagram in the margin. Thus $\sum_{i=1}^{n} c(X_i)(x_i - x_{i-1})$ is an estimate of the area of S. As the lengths $x_i - x_{i-1}$ are chosen to be smaller and smaller, we would expect that for the sets S which one usually meets these sums tend toward the area of S.

In order to speak precisely about sums of this type, the following definitions are of use.

DEFINITION *Partition.* Let $[a, b]$ be an interval and let $x_0, x_1, \ldots, x_n$ be $n + 1$ numbers such that

$$a = x_0 < x_1 < x_2 < \cdots < x_n = b.$$

The n intervals

$$[x_{i-1}, x_i],$$

$i = 1, 2, \ldots, n$, are called a *partition of* $[a, b]$.

The intervals in a partition are called *sections*.

DEFINITION *Mesh.* The mesh of a partition is the length of the longest section (or sections) in the partition.

For instance, the partition used in Sec. 7.1 has mesh equal to $\frac{1}{2}$. The preceding discussion of the area of a plane region is summarized in this plausible assertion: As the mesh of the partitions approaches 0, the sum $\sum_{i=1}^{n} c(X_i)(x_i - x_{i-1})$ approaches the area of S.

SUMS AND THE MASS OF A STRING

A string is made of a material whose density may vary from point to point. (Such a string is called *nonuniform* or *nonhomogeneous*.) How would its total mass be computed if its density at each point is known?

Nonuniform string

First, place the string somewhere on the x axis and denote by $h(x)$ its density at x. The string occupies an interval $[a, b]$ on the axis. In a very small section the density is almost constant. So the mass of the ith section is approximately

$$h(X_i)(x_i - x_{i-1}),$$

where X_i is some point in $[x_{i-1}, x_i]$. Thus $\sum_{i=1}^n h(X_i)(x_i - x_{i-1})$ is an estimate of the mass of the string. And, what is more important, it seems plausible that, as the mesh of the partition approaches 0, the sum $\sum_{i=1}^n h(X_i)(x_i - x_{i-1})$ approaches the mass of the string. [The case when $a = 0, b = 3$, and $h(x) = 5x^2$ is Prob. 2 in Sec. 7.1.]

SUMS AND THE DISTANCE TRAVELED

An engineer takes a trip that begins at time a and ends at time b. Imagine that at any time t during the trip his speed is $v(t)$, depending on the time t. How far does he travel? [The case in which $a = 0, b = 3, v(t) = 8t^2$ is Prob. 3 in Sec. 7.1.]

First, cut the time interval $[a, b]$ into smaller intervals by a partition and estimate the trip's length by summing the estimates of the distance the engineer travels during each of the time intervals.

During a small interval of time, the velocity changes little. We thus expect to obtain a reasonable estimate of the distance covered during the ith time interval $[t_{i-1}, t_i]$ by observing the speedometer reading at some instant T_i in that interval, $v(T_i)$, and computing the product $v(T_i)(t_i - t_{i-1})$. Thus $\sum_{i=1}^n v(T_i)(t_i - t_{i-1})$ is an estimate of the length of the trip. Moreover, as the mesh of the partition approaches zero, the sum $\sum_{i=1}^n v(T_i)(t_i - t_{i-1})$ approaches the length of the trip.

SUMS AND THE VOLUME OF A SOLID REGION

Suppose that we wish to compute the volume of a solid S, and we happen to know the area $A(x)$ of each cross section made by planes in a fixed direction (see the figures on the next page). In the case of the tent in Prob. 4 of Sec. 7.1, $a = 0, b = 3$, and $A(x) = x^2$.

Every partition of $[a, b]$ and selection of X's provides an estimate of the volume of S, the sum of the volumes of slabs. A typical slab is shown in perspective and side views in the second and third figures on the next page.

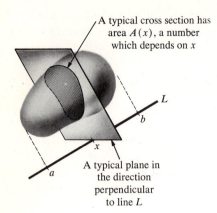

A typical cross section has area $A(x)$, a number which depends on x

A typical plane in the direction perpendicular to line L

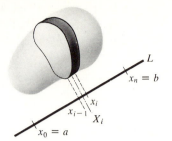

A typical slab is an irregular cylinder of cross-sectional area $A(X_i)$ and thickness $x_i - x_{i-1}$

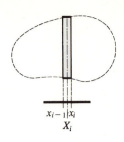

Thus $\sum_{i=1}^{n} A(X_i)(x_i - x_{i-1})$ is an estimate of the volume of the solid S. As the mesh of the partition shrinks, the slabs become thin, and the sum of their volumes becomes a more and more accurate estimate of the volume of S. As the mesh of the partition approaches 0, the sum $\sum_{i=1}^{n} A(X_i)(x_i - x_{i-1})$ approaches the volume of S.

The table below shows the similarities of these four general types of problems. To emphasize these similarities, denote the various function values $c(x)$, $h(x)$, $v(t)$, and $A(x)$ all by $f(x)$. Also, if the sums $\sum_{i=1}^{n} f(X_i)(x_i - x_{i-1})$ approach a number as the mesh approaches zero, denote that number by

$$\lim_{\text{mesh} \to 0} \sum_{i=1}^{n} f(X_i)(x_i - x_{i-1}),$$

read as "the limit of the sum $\sum_{i=1}^{n} f(X_i)(x_i - x_{i-1})$ as the mesh approaches 0."

Spend Some Time Examining This Table.   The Concepts It Describes Will Be Used Often.

$f(x)$	$\sum_{i=1}^{n} f(X_i)(x_i - x_{i-1})$	$\lim\limits_{\text{mesh} \to 0} \sum_{i=1}^{n} f(X_i)(x_i - x_{i-1})$
Variable length of cross section of set in plane	Approximation to area of the set in the plane	The area of the set in the plane
Variable density of string	Approximation to mass of the string	The mass of the string
Variable velocity	Approximation to the distance traveled	The distance traveled
Variable area of cross section of a solid	Approximation to the volume of the solid	The volume of the solid

The table suggests that one fundamental idea lies behind the solutions to these four "real-world" problems, so we should free ourselves from attachment to "area," "total mass," "distance traveled," and "volume." All that is needed to form the sums in question is an interval $[a, b]$ and some function f defined for every point in $[a, b]$.

We are now ready to define the definite integral of a function, the second fundamental concept of calculus. This definition is given informally and intuitively, using such expressions as "approach" and "shrinks toward 0." It is stated more precisely in Appendix C, but the present definition will be adequate for the theory and applications in this book.

DEFINITION   *The definite integral of a function f over an interval $[a, b]$.* If f is a function defined on $[a, b]$ and the sums $\sum_{i=1}^{n} f(X_i)(x_i - x_{i-1})$ approach a certain number as the mesh of partitions of $[a, b]$ shrinks toward 0 (no matter how the sampling number X_i is chosen in $[x_{i-1}, x_i]$), that certain number is called the *definite integral of f over* $[a, b]$.

The definite integral is also called the *definite integral of f from a to b* and the *integral of f from a to b*. The symbol for this number is $\int_a^b f(x)\, dx$. The symbol $\int$ comes from the letter S of Sum; the dx traditionally suggests a small section of the x axis and will be more meaningful and useful later. It is important to realize that area, mass, distance traveled, and volume are merely applications of the definite integral. (It is a mistake to link the definite integral too closely with one of its applications, just as it narrows our understanding of the number 2 to link it always with the idea of two fingers.) In advanced calculus it is proved that, if f is continuous, then

$$\lim_{\text{mesh} \to 0} \sum_{i=1}^{n} f(X_i)(x_i - x_{i-1})$$

exists; that is, a continuous function always has a definite integral. For emphasis we record this result, an important result in advanced calculus, as a theorem.

THEOREM *Existence of the definite integral.* Let f be a continuous function defined on $[a, b]$. Then the approximating sums

$$\sum_{i=1}^{n} f(X_i)(x_i - x_{i-1})$$

approach a single number as the mesh of the partition of $[a, b]$ approaches 0.

Though area is the most intuitive of the interpretations, physical scientists, if they want to think of the definite integral concretely, should think of it as giving total mass if we know the density everywhere. This interpretation carries through easily to higher dimensions; the area interpretation does not.

It is the concept of the definite integral that links the four problems of Sec. 7.1, which are summarized below.

PROBLEM 1 The area under the curve $y = x^2$ and above $[0, 3]$ equals the definite integral $\int_0^3 x^2\, dx$.

PROBLEM 2 The mass of the string equals the definite integral $\int_0^3 5x^2\, dx$.

PROBLEM 3 The distance that the engineer travels equals the definite integral $\int_0^3 8t^2\, dt$ (the t reminding us of time).

PROBLEM 4 The volume of the tent equals the definite integral $\int_0^3 x^2\, dx$.

Some examples will illustrate the use and theory of the definite integral.

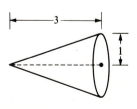

EXAMPLE 1 Find the volume of the right circular cone (in the accompanying drawing) of height 3 feet and radius 1 foot.

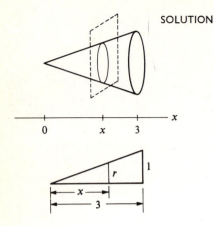

SOLUTION To begin, introduce an x axis parallel to the axis of the cone and compute the area $A(x)$ of the typical cross section of the cone made by a plane perpendicular to the x axis. The cross section is a circle whose radius depends on x. For instance, when $x = 0$, the radius is 0; when $x = 3$, the radius is 1. To find the radius r for any x, draw a side view of the cone. By similar triangles,

$$\frac{r}{1} = \frac{x}{3}.$$

Thus

$$r = \frac{x}{3}.$$

The area $A(x)$ of the typical cross section is

$$\pi r^2 = \pi\left(\frac{x}{3}\right)^2 = \frac{\pi}{9}x^2.$$

The volume of the cone is the definite integral of cross-sectional area:

$$\text{Volume of cone} = \int_0^3 \frac{\pi}{9}x^2\,dx.$$

The problem has now been translated into one concerning a definite integral. Once this has been done, we may forget the cone and try to evaluate the definite integral.

Fortunately, the labors of Sec. 7.2 do not have to be repeated here. As a beginning, the coefficient $\pi/9$ can precede the integral sign (as was noted in working Probs. 2 and 3 where the coefficients 5 and 8 appeared). Thus the volume of the cone is

$$\frac{\pi}{9}\int_0^3 x^2\,dx.$$

But we already know from Sec. 7.2 that

$$\int_0^3 x^2\,dx = 9.$$

The volume of the cone is therefore $(\pi/9)(9) = \pi$ cubic feet. ●

The next example treats a definite integral independently of any applications.

EXAMPLE 2 Let a and b be two numbers, $0 \leq a \leq b$. Compute the definite integral

$$\int_a^b x^2\,dx.$$

SOLUTION To do so, consider partitions of the interval $[a, b]$ instead of the interval $[0, 3]$ and choose $x_0, x_1, \ldots, x_n$ such that

$$a = x_0 < x_1 < \cdots < x_n = b.$$

Then examine the approximating sums

$$\sum_{i=1}^{n} X_i^2 (x_i - x_{i-1}),$$

when the mesh is small.

As in Sec. 7.2, observe that this sum is squeezed between the over-estimating sum $\sum_{i=1}^{n} x_i^2 (x_i - x_{i-1})$ and the corresponding underestimating sum $\sum_{i=1}^{n} x_{i-1}^2 (x_i - x_{i-1})$.

Next, as in Sec. 7.2, the inequalities

$$c^2(c - d) > \frac{c^3}{3} - \frac{d^3}{3} > d^2(c - d)$$

show that

$$\sum_{i=1}^{n} x_i^2 (x_i - x_{i-1}) > \sum_{i=1}^{n} \left(\frac{x_i^3}{3} - \frac{x_{i-1}^3}{3} \right) > \sum_{i=1}^{n} x_{i-1}^2 (x_i - x_{i-1}).$$

But the sum in the middle of the inequalities telescopes; its value is

$$\frac{x_n^3}{3} - \frac{x_0^3}{3}.$$

Since $x_n = b$ and $x_0 = a$, the middle sum is just

$$\frac{b^3}{3} - \frac{a^3}{3}.$$

Thus, for any partition of $[a, b]$,

$$\sum_{i=1}^{n} x_i^2 (x_i - x_{i-1}) > \frac{b^3}{3} - \frac{a^3}{3} > \sum_{i=1}^{n} x_{i-1}^2 (x_i - x_{i-1}).$$

All that remains, as in Sec. 7.2, is to show that the overestimating and underestimating sums get close to each other when the mesh of the partition is small.   The reasoning is unchanged, but it is presented again, to show what it looks like in the sigma notation.

The difference between the two sums is

$$\sum_{i=1}^{n} x_i^2 (x_i - x_{i-1}) - \sum_{i=1}^{n} x_{i-1}^2 (x_i - x_{i-1}) = \sum_{i=1}^{n} (x_i^2 - x_{i-1}^2)(x_i - x_{i-1}).$$

If the mesh is L, then

$$\sum_{i=1}^{n} (x_i^2 - x_{i-1}^2)(x_i - x_{i-1}) \le \sum_{i=1}^{n} (x_i^2 - x_{i-1}^2)L.$$

But

$$\sum_{i=1}^{n} (x_i^2 - x_{i-1}^2)L = L \sum_{i=1}^{n} (x_i^2 - x_{i-1}^2)$$

$$= L(x_n^2 - x_0^2)$$

$$= L(b^2 - a^2).$$

When L is small, the product $L(b^2 - a^2)$ is also small. The overestimating and underestimating sums must therefore approach the same number when the mesh $\to 0$, and the number is $b^3/3 - a^3/3$.

In short,

$$\int_a^b x^2\, dx = \frac{b^3}{3} - \frac{a^3}{3} \qquad 0 \le a < b.$$

Note that the computations were carried out without any thought of area, mass, distance, or volume. ●

EXAMPLE 3   Let f be the function whose value at any number x is 4; that is, f is the constant function given by the formula $f(x) = 4$. Use only the definition of the definite integral to compute

$$\int_1^3 f(x)\, dx.$$

SOLUTION   In this case a typical partition has $x_0 = 1$ and $x_n = 3$. The approximating sum

$$\sum_{i=1}^n f(X_i)(x_i - x_{i-1})$$

becomes

$$\sum_{i=1}^n 4(x_i - x_{i-1})$$

since, no matter how the sampling number X_i is chosen, $f(X_i) = 4$. Now

$$\sum_{i=1}^n 4(x_i - x_{i-1}) = 4 \sum_{i=1}^n (x_i - x_{i-1}) = 4(x_n - x_0).$$

Since $x_n = 3$ and $x_0 = 1$, it follows that all approximating sums have the same value, namely

$$4(3 - 1) = 8.$$

It does not matter whether the mesh is small or where the X_i are picked in each section. Thus, as the mesh $\to 0$,

$$\sum_{i=1}^n f(X_i)(x_i - x_{i-1})$$

approaches 8. Indeed, the sums are always 8. Thus

$$\int_1^3 4\, dx = 8. \quad ●$$

The definite integral $\int_1^3 4\, dx$ was found using only its definition. However, the result could have been anticipated by considering a string that occupies the interval $[1, 3]$ on the x axis and has the constant density 4 grams per centimeter throughout its length. Since the string is of uniform density, its mass is just

$$\text{Density} \cdot \text{length} = 4 \cdot 2 = 8 \text{ grams.}$$

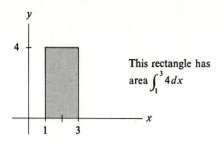

This rectangle has area $\int_1^3 4\,dx$

We could also have guessed the value of $\int_1^3 4\,dx$ by interpreting the definite integral as an area. To do so, draw a region in the plane whose cross section has length 4 for all lines perpendicular to the x axis. This region may be taken to be a rectangle of height 4 (and base coinciding with the interval $[1, 3]$), as in the figure at the left. Since the area of a rectangle is its base times its height, it follows again that $\int_1^3 4\,dx = 8$.

Similar reasoning shows that, for any constant function whose values are c,

$$\int_a^b c\,dx = c(b - a).$$

EXAMPLE 4   Compute $\int_a^b f(x)\,dx$, where $f(x) = x$.

SOLUTION   In this case $f(X_i) = X_i$, and one must estimate $\sum_{i=1}^{n} X_i(x_i - x_{i-1})$. As before, only the overestimating sums $\sum_{i=1}^{n} x_i(x_i - x_{i-1})$ and the underestimating sums $\sum_{i=1}^{n} x_{i-1}(x_i - x_{i-1})$ need to be considered. The inequalities

$$c(c - d) > \frac{c^2}{2} - \frac{d^2}{2} > d(c - d)$$

imply that

$$x_i(x_i - x_{i-1}) > \frac{x_i^{\,2}}{2} - \frac{x_{i-1}^{\,2}}{2} > x_{i-1}(x_i - x_{i-1});$$

hence

$$\sum_{i=1}^{n} x_i(x_i - x_{i-1}) > \sum_{i=1}^{n} \frac{x_i^{\,2}}{2} - \frac{x_{i-1}^{\,2}}{2} > \sum_{i=1}^{n} x_{i-1}(x_i - x_{i-1}).$$

Now

$$\sum_{i=1}^{n} \left(\frac{x_i^{\,2}}{2} - \frac{x_{i-1}^{\,2}}{2} \right)$$

is a telescoping sum whose value is

$$\frac{x_n^{\,2}}{2} - \frac{x_0^{\,2}}{2} = \frac{b^2}{2} - \frac{a^2}{2}.$$

Thus $\int_a^b x\,dx = b^2/2 - a^2/2$, since $b^2/2 - a^2/2$ is trapped between the upper and lower estimates. ●

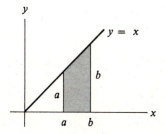

$y = x$

The result in Example 4 can be guessed by interpreting the definite integral as an area. The graph of $f(x) = x$ is a line; the area is that of the shaded trapezoid, seen in the accompanying figure. The trapezoid is the difference of a large triangle of base b and height b and a small triangle of base a and height a. Hence the trapezoid has area

$$\frac{1}{2}b^2 - \frac{1}{2}a^2 = \frac{b^2}{2} - \frac{a^2}{2}.$$

Formulas for   These formulas have been obtained in this section:
Some Integrals

$$\int_a^b 1\, dx = b - a;$$

$$\int_a^b x\, dx = \frac{b^2}{2} - \frac{a^2}{2};$$

$$\int_a^b x^2\, dx = \frac{b^3}{3} - \frac{a^3}{3}.$$

Also, as the reader may suspect,

$$\int_a^b x^3\, dx = \frac{b^4}{4} - \frac{a^4}{4}$$

and, in general, for any positive integer n,

$$\int_a^b x^n\, dx = \frac{b^{n+1}}{n+1} - \frac{a^{n+1}}{n+1}.$$

The reasoning in the last two cases is practically the same as for the cases $\int_a^b x\, dx$ and $\int_a^b x^2\, dx$. It is outlined in Exercises 15 and 16.

There are many other important definite integrals; for instance,

$$\int_a^b 2^x\, dx, \qquad \int_a^b \frac{1}{1 + x^2}\, dx, \qquad \text{and} \qquad \int_a^b \sqrt{1 - x^2}\, dx.$$

Chapters 8 and 9 develop the methods used for computing them.

Exercises

1. Find the mesh of each of these partitions of $[1, 6]$:
   (a) $x_0 = 1$, $x_1 = 2$, $x_2 = 3$, $x_3 = 4$, $x_4 = 5$, $x_5 = 6$;
   (b) $x_0 = 1$, $x_1 = 3$, $x_2 = 5$, $x_3 = 6$;
   (c) $x_0 = 1$, $x_1 = 4$, $x_2 = 4.5$, $x_3 = 5$, $x_4 = 5.5$, $x_5 = 6$.

2. A string occupying the interval $[a, b]$ has the density $f(x)$ grams per centimeter at x. Let $x_0, \ldots, x_n$ be a partition of $[a, b]$ and let $X_1, \ldots, X_n$ be sampling numbers. What is the physical interpretation of
   (a) $x_i - x_{i-1}$?
   (b) $f(X_i)$?
   (c) $f(X_i)(x_i - x_{i-1})$?
   (d) $\sum_{i=1}^{n} f(X_i)(x_i - x_{i-1})$?
   (e) $\int_a^b f(x)\, dx$?

3. A rocket moving with a varying speed travels at $f(t)$ miles per second at time t. Let $t_0, \ldots, t_n$ be a partition of $[a, b]$, and let $T_1, \ldots, T_n$ be sampling numbers. What is the physical interpretation of
   (a) $t_i - t_{i-1}$?
   (b) $f(T_i)$?
   (c) $f(T_i)(t_i - t_{i-1})$?
   (d) $\sum_{i=1}^{n} f(T_i)(t_i - t_{i-1})$?
   (e) $\int_a^b f(t)\, dt$?

4. Water is flowing into a lake at the rate of $f(t)$ gallons per second at time t. Answer the five questions in Exercise 3 for this interpretation of the function f.

5. Using the formulas developed in Examples 2 to 4, evaluate these definite integrals:

(a) $\int_5^7 3\,dx$ (b) $\int_2^8 x\,dx$ (c) $\int_4^6 x^2\,dx.$

6. Using formulas quoted after Example 4, evaluate:

(a) $\int_0^1 x^3\,dx$ (b) $\int_0^1 x^4\,dx$ (c) $\int_1^2 x^5\,dx.$

7. Estimate $\int_1^5 1/x\,dx$ using a partition of $[1, 5]$ into four sections of equal length and as the X_i (a) left endpoints, (b) right endpoints.

8. Using the formula for $\int_a^b x^2\,dx$, find:

(a) $\int_0^1 x^2\,dx$ (b) $\int_2^3 x^2\,dx$ (c) $\int_1^3 x^2\,dx.$

9. Using the method of Example 1, show that a right circular cone of height h and radius r has volume $\pi r^2 h/3.$

■

10. (a) Set up an appropriate definite integral $\int_a^b f(x)\,dx$ which equals the volume of the headlight in the figure below and whose cross section by a typical plane perpendicular to the x axis at x is a circle of radius $\sqrt{x/\pi}.$

(b) Evaluate the definite integral in (a) by a formula of this section.

11. A business firm has a rate of profit of x^3 million dollars per year after x years (x is any positive number, not necessarily an integer).
(a) What definite integral represents its profit during the first 2 years?
(b) Compute it with the aid of a formula from this section.
(c) What definite integral represents its profit during the third year?
(d) Compute it.

12. A driver makes a graph of his speed s as a function of time $s = f(t).$ Explain in detail why the area of the region below his graph and above $[a, b]$ is the distance he traveled from time a to time $b.$

13. Compute the area of the narrow region bounded by the line $y = x$, the curve $y = x^2 + x$, and the line $x = 6$ in the accompanying drawing.

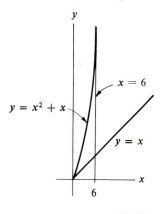

■ ■

14. Form an approximating sum for $\int_0^3 x\,dx$ as follows: Select a positive integer n and partition $[0, 3]$ into n sections, each of length $3/n.$ As sampling numbers choose X_i to be the right endpoint of the ith section.
(a) Draw the partition.
(b) What is the numerical value of x_0? x_1? x_2? x_i?
(c) Show that $A_n = \sum_{i=1}^n (3i/n)(3/n)$ is an estimate of $\int_0^3 x\,dx.$
(d) Exercise 11 of Sec. 7.3 contains the formula

$$\sum_{i=1}^n i = \frac{n(n+1)}{2}.$$

Use that formula to show that $A_n = \frac{9}{2}(1 + 1/n).$
(e) When n is large, what number does A_n approach?
(f) With the aid of (e), show that $\int_0^3 x\,dx = \frac{9}{2}.$

15. Modeling your argument after Example 2, show that

$$\int_a^b x^3\,dx = \frac{b^4}{4} - \frac{a^4}{4}.$$

You will need the inequalities

$$c^3(c - d) > \frac{c^4}{4} - \frac{d^4}{4} > d^3(c - d) \qquad \text{if } c > d \geq 0.$$

16. Carry out the analog of Exercise 15 for $\int_a^b x^m\,dx$, where m is any positive integer. (We call the exponent m, not n, since n will be needed to describe the number of sections in the partitions.)

A calculator or the tables in Appendix E will be useful in the remaining exercises.

17. (a) Sketch a graph of $y = \sin x$ for $0 \leq x \leq \pi/2.$

(b) Let A be the area of the region under the curve in (a) and above the interval $[0, \pi/2]$. Use elementary geometry to show that $\pi/4 < A < \pi/2$.

(c) Use a partition of $[0, \pi/2]$ into five sections of equal length and left endpoints to obtain an estimate of A that is too small.

(d) Use the same partition as in (c) and right endpoints to obtain an estimate of A that is too large.

In Chap. 8 it will be shown that $\int_0^{\pi/2} \sin x \, dx = 1$, hence that $A = 1$.

18. (a) Sketch a graph of $y = e^{-x}$ for $1 \le x \le 2$.

(b) Let A be the area under the graph in (a) and above

$[1, 2]$. Using a partition of $[1, 2]$ into 10 equal sections, obtain upper and lower estimates of A. By the methods of Chap. 8 it can be shown that $\int_1^2 e^{-x} \, dx = (e - 1)/e^2 \doteq 0.233$. Hence $A \doteq 0.233$.

19. (a) Sketch a graph of $y = 1/(1 + x^2)$ for $0 \le x \le 1$.

(b) Let A be the area under the graph in (a) and above $[0, 1]$.

(c) Use elementary geometry to show $\frac{3}{4} < A < 1$.

(d) Use a partition of $[0, 1]$ into five sections to obtain lower and upper estimates of A. In Chap. 8 it will be shown that $\int_0^1 1/(1 + x^2) \, dx = \pi/4$, hence that $A \doteq 0.7854$.

7.5

Summary

Four problems led to the concept of the definite integral: area of the region under a curve, total mass of a string of varying density, total distance traveled when the speed is varying, and volume of a certain tent.

All four problems required the same type of arithmetic: choosing partitions of some interval, sampling numbers, and then forming approximating sums,

$$\sum_{i=1}^{n} f(X_i)(x_i - x_{i-1}).$$

In Sec. 7.1 the interval was $[0, 3]$, and the function f was the squaring function.

The typical sum

$$\sum_{i=1}^{n} f(X_i)(x_i - x_{i-1})$$

is *not* the definite integral, any more than the quotient $[f(x + h) - f(x)]/h$ is the derivative. A definite integral $\int_a^b f(x) \, dx$ is the limit of the approximating sums as their mesh is chosen smaller and smaller:

$$\int_a^b f(x) \, dx = \lim_{\text{mesh} \to 0} \sum_{i=1}^{n} f(X_i)(x_i - x_{i-1}).$$

The table below shows at a glance why the definite integral is related to mass, area, volume, and distance. In each case consider only functions with nonnegative values. Study this table carefully. It records the core ideas of the chapter.

Function	Interpretation of Typical Summand	Approximating Sum	Definite Integral	Meaning of Definite Integral
Density	$f(X_i)(x_i - x_{i-1})$ is estimate of mass in $[x_i, x_{i-1}]$	$\sum_{i=1}^{n} f(X_i)(x_i - x_{i-1})$	$\int_a^b f(x) \, dx$	Mass
Length of cross section of a plane region by a line	$f(X_i)(x_i - x_{i-1})$ is area of an approximating rectangle	$\sum_{i=1}^{n} f(X_i)(x_i - x_{i-1})$	$\int_a^b f(x) \, dx$	Area
Area of cross section of a solid by a plane	$f(X_i)(x_i - x_{i-1})$ is volume of a thin approximating slab	$\sum_{i=1}^{n} f(X_i)(x_i - x_{i-1})$	$\int_a^b f(x) \, dx$	Volume
Speed	$f(T_i)(t_i - t_{i-1})$ is estimate of distance covered from time t_{i-1} to time t_i	$\sum_{i=1}^{n} f(T_i)(t_i - t_{i-1})$	$\int_a^b f(t) \, dt$	Distance
Just a function (no application in mind)	$f(X_i)(x_i - x_{i-1})$ is just a product of two numbers	$\sum_{i=1}^{n} f(X_i)(x_i - x_{i-1})$	$\int_a^b f(x) \, dx$	Just a number (no application in mind)

This table is summarized by saying, "The definite integral of density is mass, the definite integral of cross-sectional length is area, and so forth."

The inequalities

$$c^2(c-d) > \frac{c^3}{3} - \frac{d^3}{3} > d^2(c-d) \qquad c > d \geq 0$$

showed that

$$\int_0^3 x^2 \, dx = 9.$$

This mathematical fact, which is a statement about the behavior of approximating sums, simultaneously solved the four opening problems.

In a similar manner it was shown that

$$\int_a^b x^2 \, dx = \frac{b^3}{3} - \frac{a^3}{3}.$$

It was easy to show that, for any constant c, $\int_a^b c \, dx = c(b-a)$. The same type of reasoning that found the formula for $\int_a^b x^2 \, dx$ also would show that

$$\int_a^b x^n \, dx = \frac{b^{n+1}}{n+1} - \frac{a^{n+1}}{n+1}$$

for any positive integer n.

But just because there are formulas for computing some definite integrals, do not forget that a definite integral is a limit of sums. There are two reasons for keeping this fundamental concept clear:

1. In many applications in science the concept of the definite integral is more important than its use as a computational tool.
2. Many definite integrals cannot be evaluated by a formula. For instance, there is no formula for $\int_0^1 2^{x^2} \, dx$.

VOCABULARY AND SYMBOLS

sigma notation $\sum\limits_{i=1}^{n} a_i$

partition

section of a partition

sampling number X_i

mesh

definite integral $\int_a^b f(x) \, dx$

telescoping sum

$$\sum_{i=1}^{n} (b_i - b_{i-1}) = b_n - b_0$$

approximating sum

$$\sum_{i=1}^{n} f(X_i)(x_i - x_{i-1})$$

KEY FACTS

$$\int_a^b x^2 \, dx = \frac{b^3}{3} - \frac{a^3}{3}$$

$$\int_a^b c \, dx = c(b-a)$$

$$\int_a^b x^n \, dx = \frac{b^{n+1}}{n+1} - \frac{a^{n+1}}{n+1} \qquad n = 1, 2, 3, \ldots; b > a \geq 0.$$

Guide quiz on Chap. 7

1. Compute:

(a) $\sum\limits_{i=1}^{3} 2i^2$

(b) $\sum\limits_{i=1}^{50} [3i^4 - 3(i-1)^4]$

(c) $\sum\limits_{j=3}^{6} 5j$.

2. Using formulas of this chapter, evaluate:

(a) $\int_2^5 x^2 \, dx$

(b) $\int_0^2 5 \, dx$

(c) $\int_1^5 6x \, dx$

(d) $\int_1^2 x^4 \, dx$

3. (a) Write out in longhand the sum

$$\sum_{i=1}^{3} x_{i-1}^3 (x_i - x_{i-1}).$$

(b) What definite integral does the sum in (a) estimate if $x_0 = 2$, $x_1 = 3$, $x_2 = 3.5$, $x_3 = 4$?

(c) Is the sum in (a) smaller or larger than the definite integral in (b)?

4. Here is a brief table showing some values of the reciprocal function f, where $f(x) = 1/x$. (Entries are rounded off.)

x	$1/x$
1	1.000
1.2	0.833
1.4	0.714
1.6	0.625
1.8	0.556
2	0.500

(a) Using the table, graph $y = 1/x$ for x in $[1, 2]$.

(b) Consider the partition of $[1, 2]$ into five sections of equal length. Using the left end of each of the sections as sampling number X_i, evaluate the approximating sum

$$\sum_{i=1}^{5} f(X_i)(x_i - x_{i-1}).$$

(c) Does the sum in (b) over- or underestimate the definite integral $\int_1^2 1/x \, dx$?

(d) Like (b), except use the right end of each section.

(e) Does the sum in (d) over- or underestimate $\int_1^2 1/x \, dx$?

(f) Thinking of the definite integral in (e) as an area, show that $0.5 < \int_1^2 1/x \, dx < 0.75$.

5. A tiring turtle crawls slower and slower from 1 to 2 P.M. In fact, its speed at time t hours after noon is $1/t$ feet per hour.

(a) What definite integral equals the distance it travels from 1 to 2 P.M.?

(b) Use the results of Exercise 4 to estimate the definite integral in (a).

6. (a) Write in a paragraph or two (or describe to another student) what is meant by "the definite integral of ln x from 2 to 3." The description should not be phrased in terms of any application.

(b) Sketch a region in the plane whose area equals the definite integral mentioned in (a).

7. Evaluate $\sum_{i=1}^{5} f(X_i)(x_i - x_{i-1})$ if $f(x) = 1/x$, $x_0 = 1$, $x_1 = 1.2$, $x_2 = 1.4$, $x_3 = 1.6$, $x_4 = 1.8$, $x_5 = 2$, and (a) $X_i = x_{i-1}$; (b) $X_i = x_i$; (c) $X_i = (x_i + x_{i-1})/2$. (d) What definite integral is being estimated in parts (a), (b), and (c)?

8. Estimate $\int_1^2 e^{-x^2} \, dx$, using a partition into five sections of equal length and letting the sampling point X_i be the midpoint of the ith section. Incidentally, the methods developed in Chap. 8 for evaluating many definite integrals will *not* work for $\int_1^2 e^{-x^2} \, dx$.

Review exercises for Chap. 7

1. Evaluate:

(a) $\displaystyle \int_1^5 x^2 \, dx$

(b) $\displaystyle \int_2^6 3 \, dx$

(c) $\displaystyle \int_1^{10} x \, dx$

(d) $\displaystyle \int_0^2 x^3 \, dx.$

2. (a) Is the product $\int_0^1 x \, dx \int_0^1 x^2 \, dx$ equal to $\int_0^1 x^3 \, dx$?

(b) Is the sum $\int_0^1 x \, dx + \int_0^1 x^2 \, dx$ equal to $\int_0^1 (x + x^2) \, dx$?

(c) Is the sum $\int_0^1 x^2 \, dx + \int_1^2 x^2 \, dx$ equal to $\int_0^2 x^2 \, dx$?

3. A programmer of a computer is going to estimate $\int_0^1 \sqrt{1 + x^3} \, dx$ by using approximating sums. (There is no shortcut for evaluating it.) To do so, she divides the interval $[0, 1]$ into 50 sections of equal length. For the sampling point in each section she chooses the midpoint of that interval.

(a) What is the length of each section in the partition?

(b) What is x_0? x_1? x_i?

(c) What is X_1? X_2? X_i?

4. Write these sums without using the sigma notation:

(a) $\displaystyle \sum_{j=1}^{3} d^j$

(b) $\displaystyle \sum_{k=1}^{4} x^k$

(c) $\displaystyle \sum_{i=0}^{3} i2^{-i}$

(d) $\displaystyle \sum_{i=2}^{5} \frac{i+1}{i}$

(e) $\displaystyle \sum_{i=2}^{4} \left(\frac{1}{i} - \frac{1}{i+1} \right)$

(f) $\displaystyle \sum_{i=1}^{4} \sin \frac{\pi i}{4}.$

5. Write these sums without sigma notation and then evaluate:

(a) $\displaystyle \sum_{i=1}^{100} (2^i - 2^{i-1})$

(b) $\displaystyle \sum_{i=0}^{100} (2^{i+1} - 2^i)$

(c) $\displaystyle \sum_{i=1}^{100} \left(\frac{1}{i} - \frac{1}{i+1} \right).$

6. (a) Prove that $\sum_{i=1}^{n} ca_i = c \sum_{i=1}^{n} a_i$, where c is a fixed number.

(b) Write out in detail $\sum_{i=1}^{4} \left(\sum_{j=1}^{i} a_j \right).$

(c) Prove that $\sum_{i=1}^{m} a_i + \sum_{i=m+1}^{n} a_i = \sum_{i=1}^{n} a_i$.

7. The analog of Exercise 6(a) for definite integrals is $\int_a^b cf(x)\,dx = c\int_a^b f(x)\,dx$. Interpreting $f(x)$ as the speed of one moving object and $cf(x)$ as the speed of another always moving c times as fast as the first, interpret the equation $\int_a^b cf(x)\,dx = c\int_a^b f(x)\,dx$.

8. Interpret the equation

$$\int_a^b [f(x) + g(x)]\,dx = \int_a^b f(x)\,dx + \int_a^b g(x)\,dx$$

in terms of areas. Consider only functions f and g whose values are positive.

9. Write out in full:

(a) $\displaystyle\sum_{i=1}^{4} (-x)^i$

(b) $\displaystyle\sum_{k=1}^{4} \frac{x^k}{k}$

(c) $\displaystyle\sum_{k=1}^{4} \frac{(-x)^k}{k^2}$

10. Let n be a positive integer and f a function.
(a) Show that $\sum_{i=1}^{n} f(i/n)(1/n)$ is an approximating sum for the definite integral $\int_0^1 f(x)\,dx$.
(b) What is the length of the ith section of the partition in (a)?
(c) What is the mesh of the partition?
(d) Where does the sampling number X_i lie in the ith section?

11. Explain why $\frac{1}{100}\sum_{i=1}^{100} f(i/100)$ is an estimate of $\int_0^1 f(x)\,dx$. ∎

12. What definite integral is estimated by the following sums?

(a) $\displaystyle\sum_{i=1}^{200} \left(\frac{i}{100}\right)^3 \frac{1}{100}$

(b) $\displaystyle\sum_{i=1}^{100} \left(\frac{i-1}{100}\right)^4 \frac{1}{100}$

(c) $\displaystyle\sum_{i=101}^{300} \left(\frac{i}{100}\right)^5 \frac{1}{100}$.

13. (See Exercise 10.)
(a) Show that $1/n^3 \sum_{i=1}^{n} i^2$ is an approximation of $\int_0^1 x^2\,dx$.
(b) Compute the sum in (a) when $n = 4$.
(c) Compute $\int_0^1 x^2\,dx$.
(d) Find $\lim_{n\to\infty} 1/n^3 \sum_{i=1}^{n} i^2$. ∎ ∎

14. (a) Sketch the curve $y = \sqrt{x}$ for $0 \le x \le 1$.
(b) By considering areas and using the fact that $\int_0^1 x^2\,dx = \frac{1}{3}$, show that $\int_0^1 \sqrt{x}\,dx = \frac{2}{3}$.

15. Let f be a function of which we know only that $f(x)$ is between 3 and 5 for each x, that is, $3 \le f(x) \le 5$.
(a) How large can an approximating sum for $\int_0^4 f(x)\,dx$

be? How small?
(b) How large might $\int_0^4 f(x)\,dx$ be? How small?

16. (a) Show that, if $c > d > 0$, then

$$c^3 > cd\left(\frac{c+d}{2}\right) > d^3.$$

(b) Multiplying the inequalities in (a) by $(c - d)$ and dividing by $c^3 d^3$, show that

$$\frac{1}{d^3}(c-d) > \frac{1}{2d^2} - \frac{1}{2c^2} > \frac{1}{c^3}(c-d).$$

(c) With the aid of (b), show that, if $b > a > 0$, then

$$\int_a^b \frac{1}{x^3}\,dx = \frac{1}{2a^2} - \frac{1}{2b^2}.$$

(For convenience, assume that the definite integral exists.)

17. An unmanned satellite automatically reports its speed every minute. If a graph is drawn showing speed as a function of time during the flight, what is the physical interpretation of the area under the curve and above the time axis? Explain.

18. A number is *dyadic* if it can be expressed as the quotient of two integers m/n, where n is a power of 2. (These are the fractions into which an inch is usually divided.) Between any two numbers lies an infinite set of dyadic numbers and also an infinite set of numbers that are not dyadic. With this background we shall define a function f that does *not* have a definite integral over the interval $[0, 1]$, as follows.

$$\text{Let } f(x) = \begin{cases} 0 & \text{if } x \text{ is dyadic;} \\ 3 & \text{if } x \text{ is not dyadic.} \end{cases}$$

(a) Show that for any partition of $[0, 1]$ it is possible to choose sampling numbers X_i such that

$$\sum_{i=1}^{n} f(X_i)(x_i - x_{i-1}) = 3.$$

(b) Show that for any partition of $[0, 1]$ it is possible to choose sampling numbers X_i such that

$$\sum_{i=1}^{n} f(X_i)(x_i - x_{i-1}) = 0.$$

(c) Why does f not have a definite integral over the interval $[0, 1]$?

In this exercise $\int_0^1 e^x\,dx$ is evaluated.

19. Let a and $r \ne 1$ be real numbers and let m be a positive integer. In Sec. 14.2 it is proved that

$$a + ar + ar^2 + \cdots + ar^{m-1} = \frac{a(1 - r^m)}{1 - r}.$$

(The proof is short and can be read in a few minutes.) This formula will be needed in the present exercise.

(a) Graph $y = e^x$ for $0 \leq x \leq 1$.

(b) Let n be a positive integer. Sketch the partition $0 = x_0$, $x_1 = 1/n$, $x_2 = 2/n$, ..., $x_n = 1$, which yields n sections, each of length $1/n$. Let X_i be the right endpoint of the ith section. Show that the approximating sum for $\int_0^1 e^x \, dx$ formed with the indicated choice of the x_i and X_i has the value

$$\frac{1}{n} \frac{e^{1/n}(1 - e)}{1 - e^{1/n}}.$$

(c) Show that

$$\lim_{n \to \infty} \frac{1}{n} \frac{e^{1/n}(1 - e)}{1 - e^{1/n}} = e - 1.$$

Thus $\int_0^1 e^x \, dx = e - 1$.

20. Generalizing Exercise 19, show that $\int_a^b e^x \, dx = e^b - e^a$, where $b > a$.

This exercise concerns only material preceding Chap. 7.

21. Let f and g be functions defined on some open interval. Assume that they have only positive values and that they are differentiable and possess second derivatives. Let $F(x) = \ln f(x)$ and let $G(x) = \ln g(x)$.

(a) If F is concave upward, must f be concave upward?

(b) If f is concave upward, must F be concave upward?

(c) If f and g are concave upward, must $f + g$ be concave upward?

(d) If f and g are concave upward, must fg be concave upward?

(e) If F and G are concave upward, must $\ln fg$ be concave upward?

(f) If F and G are concave upward, must $\ln (f + g)$ be concave upward?

In each case explain your answer.

8

THE FUNDAMENTAL THEOREMS OF CALCULUS

The two basic concepts of calculus are the derivative and the definite integral. The derivative gives local information, such as slope and velocity, at a point or instant of time; the definite integral gives global information, such as total area or total distance. However, the two concepts are intimately related. This chapter will exploit their relation to develop a way of computing many definite integrals besides $\int_a^b 1\, dx$, $\int_a^b x\, dx$, and $\int_a^b x^2\, dx$, which were already computed in Chap. 7. It should be emphasized in advance that the method is limited; it does not provide a procedure for evaluating all definite integrals.

8.1

The first fundamental theorem of calculus

In Chap. 3 it was shown that *velocity is the derivative of the distance*. In Chap. 7 it was shown that *the definite integral of velocity is the change in distance*. These two facts suggest that there is a close relation between derivatives and definite integrals. In order to express this relation mathematically, let us introduce some mathematical symbols to describe these observations about velocity and change in distance.

Let x denote time, and let $F(x)$ denote the coordinate of a particle moving on a line. Then the velocity at time x is the derivative $F'(x)$. The change in distance of the moving particle from time a to time b is

Final coordinate − initial coordinate = $F(b) − F(a)$.

The assertion, "The definite integral of velocity is the change in distance," now reads mathematically.

$$\int_a^b F'(x)\, dx = F(b) − F(a).$$

This equation generalizes the statement, "Rate · time = distance," which is valid for a particle moving at a constant speed.

Recall also from Chap. 3 that *density is the derivative of mass*. In Chap. 7 it was shown that *mass is the definite integral of density*. Specifically, let x denote the distance of a point from the left end of a string whose density may not be constant.   Let $F(x)$ be the mass of the string situated to the left of x. Then $F'(x)$ is the density of the string at x.   The assertion, "The definite integral of density is mass," now reads

$$\int_a^b F'(x)\, dx = F(b) - F(a),$$

where a and b are the coordinates of the endpoints of a typical section of

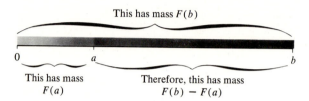

string. The equation relating F', $F(a)$, and $F(b)$ generalizes the simple statement, "Density · length = mass," which is valid for a homogeneous string.

These two physical arguments suggest that there is a single purely mathematical theorem in the background.   Moreover, computational evidence from Chap. 7 also points in the same direction for the particular function $F(x) = x^3/3$. It was proved there that

$$\int_a^b x^2\, dx = \frac{b^3}{3} - \frac{a^3}{3}.$$

Define the function F by the formula

$$F(x) = \frac{x^3}{3}.$$

Then

$$F'(x) = x^2,$$

and the equation

$$\int_a^b x^2\, dx = \frac{b^3}{3} - \frac{a^3}{3}$$

now reads

$$\int_a^b F'(x)\, dx = F(b) - F(a).$$

Thus three separate lines of reasoning all suggest the general and purely mathematical result:

$$\int_a^b F'(x)\, dx = F(b) - F(a),$$

that is, "The definite integral of the derivative of a function over an interval is simply the difference in the values of the function at the ends of the interval."

As it stands, the conjecture is not quite correct.   A function F may have such a wild derivative F' that $\int_a^b F'(x)\, dx$ does not exist; that is, the approximating sums do not approach a single number as the mesh of the partition approaches 0.   But, as mentioned in Sec. 7.4, it is proved in advanced calculus that if a function f is continuous on the interval $[a, b]$, then $\int_a^b f(x)\, dx$ *does* exist.   We shall accept this as an assumption, for it takes a long time to prove.

The observations about velocity, density, and $\int_a^b x^2\,dx$ all suggest the following theorem, whose proof will be presented in Sec. 8.3.

FIRST FUNDAMENTAL THEOREM OF CALCULUS

If f is continuous on $[a, b]$ and if f is the derivative of a function F, that is,

$$f = F',$$

then

$$\int_a^b f(x)\,dx = F(b) - F(a).$$

The symbol f is introduced for F' with a view toward applying this theorem to the computation of definite integrals. It says, "If you want to compute $\int_a^b f(x)\,dx$, search for a function F whose derivative is f, that is,

$$F' = f.$$

Then $\int_a^b f(x)\,dx$ is just $F(b) - F(a)$."

The following four examples will exhibit the power and the limitations of the fundamental theorem of calculus. (Generally the adjective "first" is omitted.)

EXAMPLE 1   Find the area of the region under the curve $y = \cos x$, above the x axis, and between $x = 0$ and $x = \pi/2$.

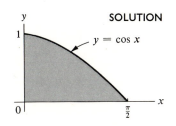

SOLUTION   As was shown in Sec. 7.4, area is the definite integral of the cross-sectional length $c(x)$. In this case $c(x) = \cos x$, and

$$\text{Area} = \int_0^{\pi/2} \cos x\,dx.$$

The fundamental theorem of calculus asserts that if we can find a function F such that

$$F'(x) = \cos x,$$

then the definite integral can be evaluated easily, as $F(\pi/2) - F(0)$. Now, in Chap. 5 it was shown that the derivative of the sine function is the cosine function. So let

$$F(x) = \sin x.$$

The fundamental theorem of calculus then says

$$\int_0^{\pi/2} \cos x\,dx = F\left(\frac{\pi}{2}\right) - F(0)$$

$$= \sin\frac{\pi}{2} - \sin 0$$

$$= 1 - 0$$

$$= 1.$$

The area in question is precisely 1 square unit. (This is reasonable, since the region is contained in a rectangle of height 1 and base $\pi/2$, and, in turn, contains a triangle of height 1 and base $\pi/2$. The rectangle has area $\pi/2 \doteq 1.57$ square units, and the triangle, $\pi/4 \doteq 0.79$ square unit.) ●

EXAMPLE 2   In Chap. 7 it was shown that

$$\int_a^b x^2 \, dx = \frac{b^3}{3} - \frac{a^3}{3},$$

when both a and b are nonnegative.   What does the fundamental theorem of calculus say about $\int_a^b x^2 \, dx$ for any a and b, $a < b$?

SOLUTION   The function F, defined by $F(x) = x^3/3$, has a derivative equal to x^2.   According to the fundamental theorem of calculus,

$$\int_a^b x^2 \, dx = F(b) - F(a) = \frac{b^3}{3} - \frac{a^3}{3}.$$

This holds even if a or b is negative.   ●

The next two examples form a fable whose moral should be remembered.

EXAMPLE 3   Compute $\int_0^1 e^{\sqrt{x}} \, dx$.

SOLUTION   To apply the fundamental theorem of calculus, it is necessary to find a function whose derivative is $e^{\sqrt{x}}$.   It happens that the derivative of $2e^{\sqrt{x}}(\sqrt{x} - 1)$ equals $e^{\sqrt{x}}$.   (Chapter 9 presents some methods for finding the formula for a function that has a prescribed derivative, when it is possible to do so.)   If

$$F(x) = 2e^{\sqrt{x}}(\sqrt{x} - 1),$$

the fundamental theorem of calculus asserts that

$$\int_0^1 e^{\sqrt{x}} \, dx \underset{\text{FTC}}{=} F(1) - F(0)$$
$$= [2e^{\sqrt{1}}(\sqrt{1} - 1)] - [2e^{\sqrt{0}}(\sqrt{0} - 1)]$$
$$= 2e^1(0) - 2e^0(-1)$$
$$= 0 + 2 = 2.$$

The letters FTC under the equality sign record the fact that this equation follows from the fundamental theorem of calculus.   ●

EXAMPLE 4   Encouraged by Example 3, try to compute $\int_0^1 e^{-x^2} \, dx$.

ATTEMPT AT SOLUTION   To apply the fundamental theorem of calculus, it is necessary to find a function F such that
$$F'(x) = e^{-x^2}.$$

As will be shown in Sec. 8.2, there is such a function F.   However, as Liouville proved in 1835, F is *not an elementary function*.   That is, F is not expressible in terms of polynomials, logarithms, exponentials, trigonometric functions, their inverses, or any composition of these functions.   We are therefore blocked, for the fundamental theorem of calculus is of use in computing $\int_a^b f(x)\,dx$ only if f is "nice" enough to be the derivative of an elementary function. ●

The moral of these last two examples is this: It is not easy to tell by glancing at f whether the desired F is elementary. After all, $e^{\sqrt{x}}$ looks more complicated than e^{-x^2}, yet it is the derivative of an elementary function, while e^{-x^2} is not.

If you differentiate an elementary function, you obtain an elementary function; this is a consequence of the work in Chap. 5 where it was shown, for instance, that $D(x^3) = 3x^2$, $D(\cos x) = -\sin x$, $D(\ln x) = 1/x$, $D(\tan^{-1} x) = 1/(1 + x^2)$, and so on. But if you start with an elementary function f and search for an elementary function F whose derivative is to be f, you may be frustrated—not because it may be hard to find F—but because it may be that no such F exists.

Incidentally $(1/\sqrt{2\pi}) \int_0^b e^{-x^2/2}\,dx$ is an important quantity in statistics. Most handbooks tabulate it (as a function of b) under the title "Area under the Normal Curve."

Exercises

In Exercise 1 differentiate the given functions. Keep a record of the results, for they will be useful in Exercises 2 to 13.

1. (a) e^{x^2} (b) $x^3/3$

(c) $\frac{1}{2}\sin^{-1} 2x$ (d) $-\dfrac{1}{2x^2}$

(e) $\dfrac{1}{2}\ln\dfrac{1-x}{1+x}$ (f) $\frac{1}{2}\tan^{-1} 2x$

(g) $x \sin x + \cos x$ (h) $-\cos x$
(i) $-e^{-x}$ (j) $\ln(\sec x + \tan x)$
(k) $\sqrt{1 + 2x}$ (l) $\ln x$.

2. Find the area of the region under the curve $y = e^{-x}$ and above $[0, 2]$.

3. Find the mass of a string 1 foot long if the density x feet from the left end is $1/(1 + x^2)$ grams per foot.

4. The velocity of a snail at time t hours, $t \geq 1$, is $1/t$ feet per hour. How far does the snail crawl during the 3 hours from time $t = 1$ to time $t = 4$?

5. Find the area of the shaded region in the accompanying diagram, if it is known that the length of the typical dotted cross section is $2xe^{x^2}$.

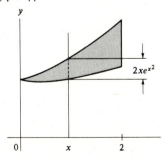

6. Find the area of the region under one arch of the sine curve $y = \sin x$ shown in the figure below.

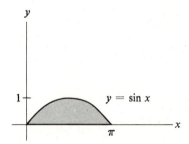

7. Find the area of the region under the curve $y = 1/x^3$ and above the interval $[2, 3]$.

8. Find the area of the region under the curve $y = 1/\cos x$ and above the interval $[0, \pi/3]$.

9. Find the area of the region under the curve $y = 1/(1 - x^2)$ and above the interval $[0, \frac{1}{2}]$.

10. The velocity of a particle at time t is $t \cos t$ feet per second. How far does it move from time $t = 0$ to time $t = \pi/2$ seconds?

11. Evaluate: $\displaystyle\int_0^4 \frac{1}{\sqrt{1 + 2x}}\,dx$.

12. Evaluate: $\displaystyle\int_0^{1/4} \frac{1}{\sqrt{1 - 4x^2}}\,dx$

13. In Example 1 $F(x) = \sin x$ was used to compute $\int_0^{\pi/2} \cos x \, dx$.
  (a) Show that the function $5 + \sin x$ also has a derivative equal to $\cos x$.
  (b) Use $F(x) = 5 + \sin x$ to evaluate $\int_0^{\pi/2} \cos x \, dx$.

14. As Example 4 indicated, the fundamental theorem of calculus is useless in evaluating $\int_0^1 e^{-x^2} \, dx$. Use the partition $x_0 = 0$, $x_1 = \frac{1}{3}$, $x_2 = \frac{2}{3}$, $x_3 = 1$ and the sampling numbers $X_1 = 0.3$, $X_2 = 0.4$, $X_3 = 0.8$ to estimate $\int_0^1 e^{-x^2} \, dx$.

■

In each of Exercises 15 to 17 there are three definite integrals. Two are easy to evaluate with the aid of the fundamental theorem of calculus, while the other, like $\int_0^1 e^{-x^2} \, dx$, cannot be evaluated by it. Decide which integrals can be evaluated by the fundamental theorem and evaluate them.

15. (a) $\displaystyle\int_0^{\pi/2} \sin 2x \, dx$ (b) $\displaystyle\int_0^{\pi/2} \sin x^2 \, dx$

  (c) $\displaystyle\int_0^1 (x^2/e^{x^3}) \, dx$.

16. (a) $\displaystyle\int_0^1 x^2 e^{x^3} \, dx$ (b) $\displaystyle\int_0^1 e^{x^3} \, dx$

  (c) $\displaystyle\int_0^{\pi/2} x \sin x^2 \, dx$.

17. (a) $\displaystyle\int_0^1 \sqrt[3]{x + x^2} \, dx$ (b) $\displaystyle\int_0^1 \sqrt[3]{1 + x} \, dx$

  (c) $\displaystyle\int_0^1 \frac{1}{\sqrt[3]{1 + x}} \, dx$.

■ ■

18. The region bordered by the curve $y = e^x$, the x axis, the y axis, and the line $x = 1$ is rotated around the x axis. Find the volume of the resulting solid.

19. A plane at a distance x from the center of a sphere of radius r, $0 \le x \le r$, meets the sphere in a circle.
  (a) Show that the radius of the circle is $\sqrt{r^2 - x^2}$.
  (b) Show that the area of the circle is $\pi r^2 - \pi x^2$.
  (c) Using the fundamental theorem, find the volume of the sphere.

20. There is no elementary function whose derivative is $\sqrt{\sin x}$. Consider the curve $y = \sqrt{\sin x}$ for x in $[0, \pi]$.
  (a) Set up a definite integral for the area of the region under the curve and above $[0, \pi]$.
  (b) Set up a definite integral for the volume of the solid obtained by revolving the region in (a) around the x axis.
  (c) Show that the fundamental theorem of calculus is useful in finding the volume in (b) but not in finding the area in (a).

8.2

The second fundamental theorem of calculus

In Sec. 8.1 interpretation of the derivative and definite integral in physical terms suggested the first fundamental theorem of calculus. A similar approach will now introduce the second fundamental theorem of calculus. The two theorems are closely related and will be proved in Sec. 8.3.

Let f denote the density, in grams per centimeter, of a string. That is, at a distance of t centimeters from the left end, the string has density $f(t)$ grams per centimeter. (The letter t rather than x is used since, in a moment, the letter x will be needed for another purpose.)

The total mass of the string from $t = a$ to $t = b$ is

$$\int_a^b f(t) \, dt \qquad \text{grams.}$$

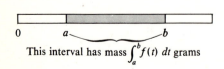

This interval has mass $\int_a^b f(t) \, dt$ grams

Now, keep a fixed and let b vary. To emphasize that b is not fixed, use the letter x instead of b to denote the right end of the interval $[a, b]$.

Now define a function G as follows.   For $x > a$, let

$$G(x) = \int_a^x f(t)\, dt.$$

Then $G(x)$ is the mass in the part of the string situated in the interval $[a, x]$. What is the derivative of G?   By the definition of the derivative,

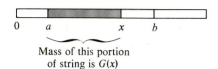

Mass of this portion
of string is $G(x)$

$$G'(x) = \lim_{\Delta x \to 0} \frac{G(x + \Delta x) - G(x)}{\Delta x}$$

$$= \lim_{\Delta x \to 0} \frac{\Delta G}{\Delta x}.$$

The numerator, $\qquad G(x + \Delta x) - G(x) = \Delta G,$

is the mass of the string in the interval $[x, x + \Delta x]$.

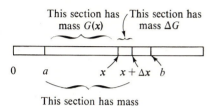

This section has mass $G(x)$

This section has mass ΔG

This section has mass $G(x + \Delta x)$

Thus the quotient $\qquad\qquad \dfrac{\Delta G}{\Delta x}$

is an estimate of the density at x.   Consequently,

$$\lim_{\Delta x \to 0} \frac{\Delta G}{\Delta x} = \text{density at } x.$$

But the density at x is given as $f(x)$.   Therefore we conclude that

$$G'(x) = f(x).$$

In short, $\qquad\qquad \dfrac{d(\int_a^x f(t)\, dt)}{dx} = f(x).$

Putting the above equation in words we would say, "The derivative of the definite integral of f from a to x is $f(x)$."

Let us check whether this is true in a particular case, where the formula for G is known, say when $f(t) = t^2$.

EXAMPLE I   Let $f(t) = t^2$ and let

$$G(x) = \int_a^x t^2\, dt.$$

Is $G'(x) = x^2$?

SOLUTION   In this case we may compute $G(x)$ explicitly:

$$G(x) = \int_a^x t^2\, dt \underset{\text{FTC}}{=} \frac{x^3}{3} - \frac{a^3}{3}.$$

Now $\qquad\qquad G'(x) = \dfrac{d(x^3/3 - a^3/3)}{dx} = \dfrac{3x^2}{3} - 0 = x^2.$

(Remember that a is constant.)   As expected, $G'(x)$ does equal x^2. ●

The assertion $$\frac{d(\int_a^x f(t)\,dt)}{dx} = f(x)$$

is the substance of the second fundamental theorem of calculus. It says, "The derivative of the definite integral of f with respect to the right end coordinate of the interval is simply f evaluated at that coordinate."

SECOND FUNDAMENTAL THEOREM OF CALCULUS

Let f be continuous on the interval $[a, b]$. Let

$$G(x) = \int_a^x f(t)\,dt$$

for $a < x \le b$. Then G is differentiable and its derivative is f; that is,

$$G'(x) = f(x).$$

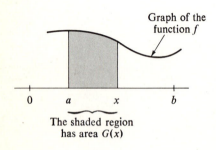

Graph of the function f

The shaded region has area $G(x)$

Considerations of density and mass suggested the second fundamental theorem. In view of its importance it is worthwhile to pause and see how considerations of area would also suggest the same conclusion.

Let f be a continuous function such that the $f(x)$ is positive for x in $[a, b]$. For x in $[a, b]$ let $G(x)$ be the area of the region under the graph of f and above the interval $[a, x]$, shown in the accompanying drawing. Let Δx be a small positive number. Then $G(x + \Delta x)$ is the area under the graph of f and above the interval $[a, x + \Delta x]$. Consequently,

$$\Delta G = G(x + \Delta x) - G(x)$$

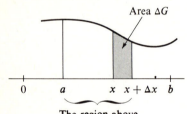

Area ΔG

The region above $[a, x + \Delta x]$ has area $G(x + \Delta x)$

is the area of the narrow strip shaded in the diagram at the left. When Δx is small, the narrow shaded strip above $[x, x + \Delta x]$ resembles a rectangle of base Δx and height $f(x)$, with area $f(x)\,\Delta x$. Therefore it seems reasonable that, when Δx is small,

$$\frac{\Delta G}{\Delta x} \text{ is approximately } f(x).$$

In short, it seems plausible that

$$\lim_{\Delta x \to 0} \frac{\Delta G}{\Delta x} = f(x).$$

Briefly, $$G'(x) = f(x).$$

Now $$G(x) = \int_a^x f(t)\,dt,$$

since area is the definite integral of the cross-sectional length. Thus considerations of area lead to the same formula as did considerations of mass, namely

$$\frac{d(\int_a^x f(t)\,dt)}{dx} = f(x).$$

This is again the second fundamental theorem.

As a consequence of this theorem, every continuous function is the derivative of some function. This is stated as a corollary for emphasis.

COROLLARY   Let f be continuous on an interval $[a, b]$.   Then f is the derivative of some function.

PROOF   Let $F(x) = \int_a^x f(t)\, dt$.   Then, by the second fundamental theorem,

$$F'(x) = f(x),$$

that is,

$$\frac{d(\int_a^x f(t)\, dt)}{dx} = f(x).$$

This proves the corollary. ●

EXAMPLE 2   Find a function whose derivative is e^{-x^2}.

SOLUTION   The function F defined by $F(x) = \int_0^x e^{-t^2}\, dt$

is such a function. ●

 The information in Example 2 is of no use in trying to apply the first funda-mental theorem to compute $\int_0^1 e^{-t^2}\, dt$.   While there are functions whose derivative is e^{-x^2}, there is no elementary function with this property.

EXAMPLE 3   Find $D_x(\int_0^{x^2} t^3\, dt)$.

SOLUTION   The upper limit of integration is x^2, not x.   The chain rule will be needed.   Let

$$y = \int_0^{x^2} t^3\, dt.$$

Then

$$y = \int_0^u t^3\, dt \qquad \text{where } u = x^2.$$

By the second fundamental theorem,

$$\frac{dy}{du} = u^3.$$

The chain rule then says that

$$\frac{dy}{dx} = \frac{dy}{du}\frac{du}{dx} = u^3 \cdot 2x$$

$$= (x^2)^3 2x = 2x^7. \; ●$$

Exercises

In Exercises 1 to 3 give an example of a function whose derivative is the given function.   Express your answer in terms of definite integrals.   Do not try to find a "formula" for the function.

1. $\sqrt{1 + x^3}$ **2.** $\sqrt[3]{1 + x^2}$ **3.** $\tan^{-1} x^2$

In Exercises 4 to 9 differentiate the given function.

4. $G(x) = \int_3^x t^5\, dt$

5. $G(x) = \int_4^x t^5\, dt$

6. $G(x) = \int_4^x \sec^3 t\, dt$

7. $G(x) = \int_1^x \ln t\, dt$

8. $G(x) = \int_0^x \sqrt{t}\, dt$ **9.** $G(x) = \int_3^x e^{t^3}\, dt$

In Exercises 10 to 15, find dG/dx. As in Example 3, the chain rule will be needed.

10. $G(x) = \int_1^{x^3} \frac{1}{t}\, dt$ **11.** $G(x) = \int_1^{\sqrt{x}} \sqrt{1 - t^2}\, dt$

12. $G(x) = \int_0^{1/x} e^{t^2}\, dt$

13. $G(x) = \int_0^{-x} (3t^2 + 4t)\, dt,\ x < 0$

14. $G(x) = \int_3^{\sin x} \sqrt{1 + t^2}\, dt$

15. $G(x) = \int_1^{x^2} f(t)\, dt$

■

16. Find $\displaystyle\lim_{\Delta x \to 0} \frac{\int_2^{3+\Delta x} x^3\, dx - \int_2^3 x^3\, dx}{\Delta x}$.

17. Find $\displaystyle\lim_{x_2 \to x_1} \frac{\int_0^{x_2} e^{t^2}\, dt - \int_0^{x_1} e^{t^2}\, dt}{x_2 - x_1}$.

18. Find $\displaystyle\lim_{h \to 0} \frac{\int_1^{x+h} \sqrt{\sin t}\, dt - \int_1^x \sqrt{\sin t}\, dt}{h}$.

■ ■

The remaining exercises continue Exercise 19 of Sec. 7.2.

19. (*a*) Let $A(x) = \int_1^x 1/t\, dt$ for $x > 1$. Find dA/dx.

(*b*) Show that $A(x) = \ln x$.

In some texts the logarithm of x to the base e is *defined* as $\int_1^x 1/t\, dt$. The exponential is then defined as the inverse of the logarithm function.

When doing Exercises 20 to 22, pretend that you never saw e before.

20. Let n be a positive integer.

(*a*) Show that

$$\int_1^{(1+1/n)^n} \frac{1}{x^{1-1/n}}\, dx = 1.$$

(*b*) Deduce that, as n increases, $(1 + 1/n)^n$ increases. *Hint:* Graph the curve $y = 1/x^{1-1/n}$.

21. Let m be a positive integer.

(*a*) Show that

$$\int_1^{(1-1/m)^{-m}} \frac{1}{x^{1+1/m}}\, dx = 1.$$

(*b*) Deduce that, as m increases, $(1 - 1/m)^{-m}$ decreases.

(*c*) Show that, for every pair of positive integers m and n,

$$\left(1 + \frac{1}{n}\right)^n < \left(1 - \frac{1}{m}\right)^{-m}.$$

Incidentally, Exercises 10 to 13 of Sec. 4.2 show that the two sides of this inequality have the same limit as $n \to \infty$ and $m \to \infty$.

22. Show that there is a number—call it "e"— such that

$$\int_1^e \frac{1}{t}\, dt = 1.$$

If logarithms are defined as integrals, this is how e is introduced into calculus.

8.3

Proofs of the two fundamental theorems of calculus

Geometric and physical intuition suggested the two fundamental theorems of calculus. It may be of value to give a mathematical proof for these two theorems, a proof which is independent of any particular interpretation of the definite integral. At least this will offer a chance to think of the definite integral as a purely mathematical concept, one which will be generalized in Chap. 13 from intervals to plane regions.

This section will first prove the second fundamental theorem, and then obtain the first fundamental theorem from it. A few definitions and preliminary lemmas will be needed.

DEFINITION   *The integral from a to b, where b is less than a.*   If b is less than a, then

$$\int_a^b f(x)\, dx = -\int_b^a f(x)\, dx.$$

EXAMPLE I   Compute $\int_3^0 x^2\,dx$, the integral from 3 to 0 of x^2.

SOLUTION   The symbol $\int_3^0 x^2\,dx$ is defined as

$$-\int_0^3 x^2\,dx.$$

As was shown in Chap. 7, $\int_0^3 x^2\,dx = 9$.   Thus

$$\int_3^0 x^2\,dx = -9. \quad \bullet$$

DEFINITION   *The integral from a to a.* $\int_a^a f(x)\,dx = 0$.

> *Remark:* The definite integral is defined with the aid of partitions.   Rather than permit partitions to have sections of length 0, it is simpler just to make the above definition.

The point of making these two definitions is that now the symbol $\int_a^b f(x)\,dx$ is defined for any numbers a and b and any continuous function f.   It is no longer necessary that a be less than b.

Three lemmas will be needed.

LEMMA I   If a, b, and c are numbers, and $\int_a^c f(x)\,dx$ and $\int_c^b f(x)\,dx$ exist, then $\int_a^b f(x)\,dx$ exists, and

$$\int_a^c f(x)\,dx + \int_c^b f(x)\,dx = \int_a^b f(x)\,dx.$$

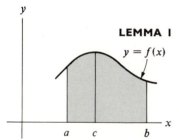

Lemma 1 asserts that the area from a to c plus the area from c to b is equal to the area from a to b

We omit the proof.   In the case that $a < c < b$ and $f(x)$ assumes only positive values, the equation asserts that the area of the region below the graph of f and above the interval $[a, b]$ is the sum of the areas of the regions below the graph and above the smaller intervals $[a, c]$ and $[c, b]$.   This is certainly plausible.

LEMMA 2   If $\int_a^b f(x)\,dx$ exists and if $m \le f(x) \le M$ for all x in $[a, b]$, then

$$m(b - a) \le \int_a^b f(x)\,dx \le M(b - a) \qquad \text{if} \quad a < b,$$

and

$$m(b - a) \ge \int_a^b f(x)\,dx \ge M(b - a) \qquad \text{if} \quad b < a.$$

PROOF   We prove this for the case $a < b$.   Since $m \le f(X_i) \le M$ if X_i is in $[a, b]$, then for any approximating sum based on a partition of $[a, b]$:

$$m \sum_{i=1}^n (x_i - x_{i-1}) \le \sum_{i=1}^n f(X_i)(x_i - x_{i-1}) \le M \sum_{i=1}^n (x_i - x_{i-1}).$$

But $\sum_{i=1}^n (x_i - x_{i-1}) = b - a$.   Therefore any approximating sum for $\int_a^b f(x)\,dx$ lies between $m(b - a)$ and $M(b - a)$.

Hence
$$\lim_{\substack{\text{mesh}\to 0}} \sum_{i=1}^{n} f(X_i)(x_i - x_{i-1})$$

is between $m(b - a)$ and $M(b - a)$. This concludes the proof. (The case $b < a$ can be deduced from the case $a < b$.) ●

LEMMA 3   Let a and b be numbers, and let f be a continuous function. Then there is a number X between a and b such that

$$\int_a^b f(x)\, dx = f(X)(b - a).$$

PROOF   Consider the case $a < b$. Let M be the maximum and m the minimum of $f(x)$ for x in $[a, b]$. [Recall the maximum- (and minimum-) value theorem of Sec. 6.1.] By Lemma 2,

$$m \le \frac{\int_a^b f(x)\, dx}{b - a} \le M.$$

By the intermediate-value theorem of Sec. 4.7 there is a number X in $[a, b]$ such that

$$f(X) = \frac{\int_a^b f(x)\, dx}{b - a},$$

and the lemma is proved. (The case $b < a$ can be obtained from the case $a < b$.) ●

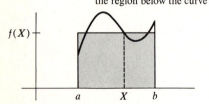

The area of the rectangle is the same as the area of the region below the curve

If $f(x)$ is positive, $\int_a^b f(x)\, dx$ can be interpreted as the area of the region below $y = f(x)$ and above $[a, b]$. Lemma 3 then asserts that there is a rectangle whose base is $[a, b]$ and whose height is $f(X)$ that has the same area as the region.

These three lemmas now provide a short proof of the second fundamental theorem, which asserts that the derivative of $G(x) = \int_a^x f(t)\, dt$ is $f(x)$. See Sec. 8.2 for its complete statement.

PROOF OF THE SECOND FUNDAMENTAL THEOREM

It is necessary to study the behavior of the quotient

$$\frac{\Delta G}{\Delta x} = \frac{G(x + \Delta x) - G(x)}{\Delta x}$$

as $\Delta x \to 0$ (x is fixed). The number Δx may be positive or negative.

Then
$$\Delta G = \int_a^{x + \Delta x} f(t)\, dt - \int_a^x f(t)\, dt.$$

By Lemma 1,
$$\int_a^{x + \Delta x} f(t)\, dt = \int_a^x f(t)\, dt + \int_x^{x + \Delta x} f(t)\, dt.$$

Hence
$$\Delta G = \int_a^x f(t)\, dt + \int_x^{x + \Delta x} f(t)\, dt - \int_a^x f(t)\, dt$$

$$= \int_x^{x+\Delta x} f(t)\, dt.$$

Now, by Lemma 3,

$$\int_x^{x+\Delta x} f(t)\, dt = f(X)((x + \Delta x) - x) \qquad \text{for some } X \text{ between } x \text{ and } x + \Delta x$$

$$= f(X)\, \Delta x.$$

Thus

$$\frac{\Delta G}{\Delta x} = f(X)$$

for some X between x and $x + \Delta x$. (Note that X depends on Δx and the fixed number x.)

Now it will be easy to compute

$$\lim_{\Delta x \to 0} \frac{\Delta G}{\Delta x}.$$

By the above reasoning, $\qquad \lim_{\Delta x \to 0} \dfrac{\Delta G}{\Delta x} = \lim_{\Delta x \to 0} f(X).$

Since f is continuous at x, and since X is between x and $x + \Delta x$, it follows that

$$\lim_{\Delta x \to 0} f(X) = f(x).$$

This proves that G is differentiable and that $G' = f$. ●

Remark: It may be illuminating to follow the various steps in this proof if we interpret "definite integral" as "area." If $f(x)$ is positive for all x in $[a, b]$, then $G(x)$ can be thought of as the area under the curve $y = f(t)$ from a to x. Then for $\Delta x > 0$, $G(x + \Delta x) - G(x) = \Delta G$ represents the area of a narrow strip above the interval $[x, x + \Delta x]$. Choose X in such a way that the rectangle with base Δx and height $f(X)$ has an area equal to the area of the shaded strip. Then $\Delta G/\Delta x = f(X)$, the height of the rectangle chosen as described. As Δx approaches 0, $f(X)$ approaches $f(x)$, since f is continuous.

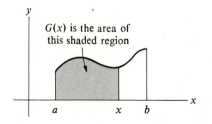

$G(x)$ is the area of this shaded region

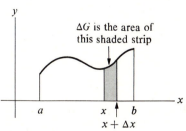

ΔG is the area of this shaded strip

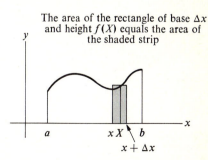

The area of the rectangle of base Δx and height $f(X)$ equals the area of the shaded strip

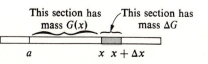

This section has mass $G(x)$. This section has mass ΔG

Remark: It may also be illuminating to interpret the proof of the second fundamental theorem in terms of density and mass. If we interpret f as density and $G(x) = \int_a^x f(t)\, dt$ as the mass in the interval $[a, x]$, then ΔG is the mass in the short interval $[x, x + \Delta x]$. Thus $\Delta G/\Delta x$ is the average density in the interval $[x, x + \Delta x]$ and equals the density somewhere in that interval,

$$\frac{\Delta G}{\Delta x} = f(X) \qquad \text{for some } X \text{ between } x \text{ and } x + \Delta x.$$

As $\Delta x \to 0$, $f(X) \to f(x)$.

Hence $\qquad\qquad\qquad\qquad\qquad\qquad G'(x) = f(x).$

EXAMPLE 2   Let f be continuous.   Find $\dfrac{d(\int_x^b f(t)\,dt)}{dx}$.

In other words, differentiate the definite integral with respect to the *lower* limit of integration.

SOLUTION   To do this, write $\qquad\qquad \displaystyle\int_x^b f(t)\,dt = -\int_b^x f(t)\,dt.$

Thus the derivative equals

$$\frac{d(-\int_b^x f(t)\,dt)}{dx} = \frac{-d(\int_b^x f(t)\,dt)}{dx} = -f(x).$$

Note the minus sign. ●

Remark: The minus sign in the answer in Example 2 makes sense if $f(x)$ is positive, $x < b$, and $\int_x^b f(t)\,dt$ is thought of as area (or mass).   As x increases, the area (or mass) *decreases*, and its derivative should be negative.

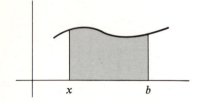

As x increases, the area of the shaded region decreases

The first fundamental theorem, which asserts that if $f = F'$, then $\int_a^b f(x) = F(b) - F(a)$, follows from the second as will now be shown.   (See Sec. 8.1 for its complete statement.)

PROOF OF THE FIRST FUNDAMENTAL THEOREM   Let $G(t) = \int_a^t f(x)\,dx$.   (The letter t is introduced to describe the interval of integration $[a, t]$, since x has already been used to describe the inputs of the function f.)   By the second fundamental theorem,

$$G'(x) = f(x).$$

Since it is assumed that

$$F'(x) = f(x),$$

it follows that the functions

$$G \text{ and } F$$

have the same derivative.   Thus they differ by a constant (see Corollary 2, Sec. 6.2),

$$G(t) = F(t) + C \qquad C \text{ constant.}$$

Then $\qquad\qquad G(b) - G(a) = [F(b) + C] - [F(a) + C]$

$$= F(b) - F(a).$$

But

$$G(a) = \int_a^a f(x)\, dx = 0,$$

and

$$G(b) = \int_a^b f(x)\, dx.$$

Hence

$$G(b) - G(a) = \int_a^b f(x)\, dx - 0$$

$$= \int_a^b f(x)\, dx.$$

Consequently,

$$\int_a^b f(x)\, dx = G(b) - G(a)$$

$$= F(b) - F(a),$$

and the theorem is proved. ●

Exercises

In Exercises 1 to 4 evaluate the expression.

1. $\displaystyle\int_3^3 x^3\, dx$

2. $\displaystyle\int_4^4 \sin x\, dx$

3. $\displaystyle\int_{\pi/2}^0 \sin x\, dx$

4. $\displaystyle\int_2^1 x^2\, dx$

5. (a) If f is continuous, and $3 \le f(x) \le 5$ for all x in $[2, 6]$, what can be said about $\int_2^6 f(x)\, dx$?

(b) Translate (a) into a statement about a string whose density is $f(x)$ grams per centimeter.

In Exercises 6 to 9 find at least one X whose existence is ensured by Lemma 3.

6. $\displaystyle\int_1^3 x\, dx = X(3 - 1)$

7. $\displaystyle\int_0^1 x^2\, dx = X^2(1 - 0)$

8. $\displaystyle\int_0^{2\pi} \cos x\, dx = \cos X(2\pi - 0)$

9. $\displaystyle\int_2^4 \frac{1}{x}\, dx = \frac{1}{X}(4 - 2)$

■

Exercises 10 to 12 provide background for the study of Chap. 9. In each case give a formula for the answer.

10. Find a function whose derivative is $e^x \sin 2x$.   *Hint:* Try $e^x(a \cos 2x + b \sin 2x)$ and find out what the constants a and b must be.

11. Find a function whose derivative is $\ln x$.   *Hint:* Try

$x \ln x$ as a first guess.   It will be close enough to suggest the answer.

12. Find a function whose derivative is $x \sin x^2$.

13. Letting $f(x)$ denote the velocity of an object at time x, interpret each step of the proof of the second fundamental theorem in terms of velocity and distance.

14. Verify that $\int_a^c f(x)\, dx + \int_c^b f(x)\, dx = \int_a^b f(x)\, dx$ when (a) $c = a$; (b) $c = b$; (c) $a = b$.

15. Find d^2y/dx^2 if $y = f(x) = \int_0^x (1 + t^2)\, dt$.

16. Find

$$\lim_{t \to 1} \frac{\int_1^t e^x\, dx}{\int_1^t e^{x^2}(1 + x)\, dx}.$$

17. Let $v(t)$ be the velocity of an object at time t moving on a straight path.   Its speed is $|v(t)|$.

(a) What is the physical interpretation of the area under the graph of $|v(t)|$ from t_1 to t_2?

(b) What is the physical significance of the slope of the graph of $v(t)$?

■ ■

18. The first fundamental theorem can be proved directly, without referring to the second fundamental theorem. Assume that f is continuous, $f = F'$, and $\int_a^b f(x)\, dx$ exists.   The steps are outlined as follows:

(a) Given x_{i-1} and x_i in $[a, b]$, show that there is a number X_i in $[x_{i-1}, x_i]$ such that

$$F(x_i) - F(x_{i-1}) = F'(X_i)(x_i - x_{i-1}).$$

(b) Given x_{i-1} and x_i in $[a, b]$, show that there is a number X_i in $[x_{i-1}, x_i]$ such that

$$f(X_i)(x_i - x_{i-1}) = F(x_i) - F(x_{i-1}).$$

(c) Let $x_0 = a, x_1, \ldots, x_n = b$ determine a partition of $[a, b]$ into n sections. Show that, if the sampling numbers X_i are chosen as in (b), then

$$\sum_{i=1}^{n} f(X_i)(x_i - x_{i-1}) = F(b) - F(a).$$

(d) Use (c) to show that

$$\lim_{\text{mesh} \to 0} \sum_{i=1}^{n} f(X_i)(x_i - x_{i-1})' = F(b) - F(a)$$

[even if the X_i are not chosen as in (b)]. This proves the first fundamental theorem directly.

8.4

Antiderivatives

The first fundamental theorem of calculus shows that it is of importance to consider functions whose derivatives are prescribed in advance. Since this concept is so useful, it will be given a name.

DEFINITION   *Antiderivative.* Any function F whose derivative is f is called an *antiderivative* of f. Sometimes F is called a *primitive* of f.

EXAMPLE I   Find two antiderivatives of $3x^2$.

SOLUTION   Since $(x^3)' = 3x^2$, x^3 is an antiderivative of $3x^2$. Also, $(x^3 + 21)' = 3x^2$. Hence $x^3 + 21$ is also an antiderivative of $3x^2$. ●

Does a continuous function f have at least one antiderivative? Yes, by the second fundamental theorem. If f is defined on $[a, b]$, then the function G described by

$$G(x) = \int_a^x f(t)\, dt$$

is an antiderivative of f. To express such a function G as an elementary function may be difficult or even impossible, even if the formula for f is quite simple.

As Example 1 suggests, if F is an antiderivative of f, then the function obtained by adding a constant to F is another antiderivative of f. That is, if $F' = f$, and C is a constant, then

$$(F + C)' = f.$$

The next theorem shows that any two antiderivatives of a function defined on an interval differ only by a constant.

THEOREM   If F and G are both antiderivatives of f on an interval $[a, b]$, then there is a constant C such that
$$F(x) = G(x) + C.$$

PROOF   The functions F and G have the same derivative f. By Corollary 2 of Sec. 6.2, they must differ by a constant. This proves the theorem. ●

Warning   Frequently an antiderivative is called an *integral* or *indefinite integral*. There is a danger that "integral" will become confused with "definite integral." It should be kept in mind that the definite integral $\int_a^b f(x)\, dx$ is defined as a number, a limit of certain sums, while an integral or antiderivative is a function. Sometimes, in applications, both the definite integral and the indefinite integral are called integrals; it takes a clear mind and mastery of the definitions to keep the ideas separate. The table of integrals in a mathematical handbook is primarily a table of antiderivatives (functions); it is usually followed by a short section that lists the values of a few common definite integrals (numbers).

NOTATION   An antiderivative of f is denoted $\int f(x)\, dx$.

Thus if $F' = f$, we write $F = \int f(x)\, dx$ and say "F is an antiderivative of f" or, for convenience, "$F(x)$ is an antiderivative of $f(x)$." For instance, we say "x^3 is an antiderivative of $3x^2$."

EXAMPLE 2   Evaluate $\int 3x^2\, dx.$

SOLUTION   Any function whose derivative is $3x^2$ serves as a value for $\int 3x^2\, dx$. Thus

$$x^3 + 17 = \int 3x^2\, dx$$

$$x^3 - 1 = \int 3x^2\, dx$$

$$x^3 = \int 3x^2\, dx.$$

Think of the equals sign in these equations as meaning "is a." ●

NOTATION $F(b) - F(a)$ is abbreviated to $F(x)\Big|_a^b$.

EXAMPLE 3   Evaluate $\displaystyle\int_1^2 \frac{1}{x^2}\, dx$ by the fundamental theorem of calculus.

SOLUTION
$$\int_1^2 \frac{1}{x^2}\, dx = \frac{-1}{x}\Big|_1^2 = \frac{-1}{2} - \frac{-1}{1} = \frac{-1}{2} + 1 = \frac{1}{2}. \;●$$

The fundamental theorem of calculus asserts that

$$\underbrace{\int_1^2 \frac{1}{x^2}\, dx}_{\text{The definite integral; a limit of sums.}} \qquad = \qquad \underbrace{\int \frac{1}{x^2}\, dx\; \Big|_1^2}_{\substack{\text{An antiderivative: a function; the value of this}\\ \text{function at 1 is then subtracted from its value}\\ \text{at 2.}}}$$

The symbols on the right and left of the equal sign are so similar that it is tempting to think that the equation is obvious or says nothing whatever. *Beware:*

That compact equation is in fact a terse statement of the (first) fundamental theorem of calculus.

DEFINITION *Integrand.* In the definite integral $\int_a^b f(x)\,dx$ and in the antiderivative $\int f(x)\,dx$, $f(x)$ is called the *integrand*.

The related processes of computing $\int_a^b f(x)\,dx$ and of finding an antiderivative $\int f(x)\,dx$ are both called *integrating* $f(x)$. Thus integration refers to two separate but related problems: computing a number $\int_a^b f(x)\,dx$ or finding a function $\int f(x)\,dx$. The fundamental theorem of calculus states that the second process may be of use in computing $\int_a^b f(x)\,dx$.

Exercises

Give two antiderivatives for each function in Exercises 1 to 12.

1. $\dfrac{1}{x}$

2. e^{4x}

3. $\dfrac{1}{\sqrt{1-x^2}}$

4. $3x^2 + 4x^3$

5. $2x^2$

6. $\dfrac{3}{1+x^2}$

7. $\dfrac{1}{\sqrt{x}}$

8. $\dfrac{1}{x^2}$

9. $\dfrac{5}{x\sqrt{x^2-1}}$

10. $\sec^2 3x$

11. $\tan 2x \sec 2x$

12. $\dfrac{x}{1+5x^2}$

Give at least one answer in Exercises 13 to 15.

13. $\int x^4\,dx = ?$

14. $\int \dfrac{1}{x^4}\,dx = ?$

15. $\int \sin 2x\,dx = ?$

16. Compute: (a) $(\int x^5\,dx)'$, (b) $(\int \sin x^5\,dx)'$.

17. What guarantees that

$$\int_0^{\pi/2} \sin x\,dx = \int \sin x\,dx \Big|_0^{\pi/2} \ ?$$

18. Compute: (a) $\int x^2\,dx \Big|_1^2$ (b) $\int \dfrac{1}{x}\,dx \Big|_1^2$

(c) $\int \dfrac{1}{1+x^2}\,dx \Big|_1^2$.

19. Compute: (a) $\int \cos x\,dx \Big|_0^{\pi/4}$ (b) $\int \dfrac{x}{1+x^2}\,dx \Big|_0^1$

(c) $\int \dfrac{1}{\sqrt[3]{x}}\,dx \Big|_1^2$.

20. Compute: (a) $x^2 \Big|_1^2$ (b) $x^2 \Big|_{-1}^1$ (c) $\sin x \Big|_0^{\pi}$

(d) $\cos x \Big|_0^{\pi}$.

21. Find $\int 1\,dx$. (This is usually written $\int dx$.)

22. Find: (a) $\int \dfrac{1+9x^2}{x}\,dx$ (b) $\int \dfrac{x}{1+9x^2}\,dx$.

23. (a) Is $\int x^2\,dx$ a function or a number?
(b) Is $\int x^2\,dx \big|_1^3$ a function or a number?
(c) Is $\int_1^3 x^2\,dx$ a function or a number?

24. (a) Which of these two numbers is defined as a limit of sums:

$$\int x^2\,dx \Big|_1^3 \quad \text{or} \quad \int_1^3 x^2\,dx ?$$

(b) Why are the two numbers in (a) equal?

25. Does the function e^{-x^2} have an antiderivative?

26. True or false:
(a) An elementary function has an elementary derivative.
(b) An elementary function has an elementary antiderivative.

■

In Exercises 27 and 28 verify the equation quoted from a table of antiderivatives (integrals). Just differentiate the alleged antiderivative and see whether you obtain the integrand. The number a is constant in each case.

27. $\int x^2 \sin ax\,dx = \dfrac{2x}{a^2}\sin ax + \dfrac{2}{a^3}\cos ax - \dfrac{x^2}{a}\cos ax$.

28. $\int x \sin^2 ax \, dx = \dfrac{x^2}{4} - \dfrac{x \sin 2ax}{4a} - \dfrac{\cos 2ax}{8a^2}.$

■ ■

29. For what values of θ does the function $\sin^2 \theta \sin 2\theta$ have a maximum?

30. Give an example of a function whose derivative is $\sin^2 \theta \sin 2\theta$.

31. Assume that the function f is defined for all x and has a continuous derivative. Assume that $f(0) = 0$ and that $0 < f'(x) \le 1$.
(a) Prove that

$$\left[\int_0^1 f(x) \, dx \right]^2 \ge \int_0^1 [f(x)]^3 \, dx.$$

Hint: Prove a more general result, namely that the inequality holds when the upper limit of integration, 1, is replaced by t, $0 \le t \le 1$.
(b) Give an example in which equality occurs in (a).

32. The accompanying graph shows a triangle ABC and a shaded region cut from the parabola by a horizontal line. Find the limit, as $x \to 0$, of the ratio between the area of the triangle and the area of the shaded region.

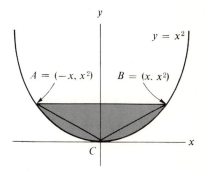

8.5

Summary

This chapter is the very core of calculus. It describes the close relation between the derivative, defined in Chap. 3, and the definite integral of Chap. 7. This relation is recorded in the two fundamental theorems of calculus.

First fundamental theorem: If f is continuous and is the derivative of the function F, then

$$\int_a^b f(x) \, dx = F(b) - F(a).$$

Second fundamental theorem: If f is continuous, then

$$\frac{d\left(\int_a^x f(t) \, dt \right)}{dx} = f(x).$$

The first fundamental theorem is also called the fundamental theorem and is abbreviated by the letters FTC. It provides a tool for computing many definite integrals. If f is the derivative of an elementary function F, then FTC is of use. But there are elementary functions, for instance, e^{-x^2}, which are not derivatives of elementary functions. In these cases it may be necessary to estimate the definite integral, say, by an approximating sum.

The FTC is *not* a theorem about area or mass. It is a theorem about the limit of the sums $\sum_{i=1}^n f(X_i)(x_i - x_{i-1})$. In many applications, it is first shown that a certain quantity (area, distance, volume, mass, etc.) is estimated by sums of that type, and then the FTC is called on. However, it may or may not be of use.

Any function whose derivative is the function f is called an antiderivative of f and is denoted $\int f(x) \, dx$. Any two antiderivatives of a function over an interval differ by a constant.

The second fundamental theorem implies that every continuous function is the derivative of some function. More specifically, it tells the rate of change of a definite integral as the interval over which the integral is computed is changed.

A glance back at Secs. 1.1 and 1.2 and the opening four problems in Secs. 3.1 and 7.1 and a review of the summaries of Chaps. 1 to 7 provide a perspective of the route followed. The related concepts of the derivative and definite integral provide the base for the topics covered throughout the rest of this book.

VOCABULARY AND SYMBOLS

First fundamental theorem of calculus = fundamental theorem of calculus = first fundamental theorem = fundamental theorem = FTC.

$$\int_a^b f(x) \, dx = -\int_b^a f(x) \, dx \qquad \text{if } b < a, \qquad \int_a^a f(x) \, dx = 0$$

second fundamental theorem of calculus = second fundamental theorem

VOCABULARY AND SYMBOLS (continued)

elementary function

integral

antiderivative $\int f(x)\, dx$

indefinite integral

integrand

$F(x) \Big|_a^b$

KEY FACTS

If f is continuous, $\int_a^b f(x)\, dx$ exists.   (This is assumed.)

If $f = F'$, then $\int_a^b f(x)\, dx = F(b) - F(a)$ (first fundamental theorem).

$\dfrac{d(\int_a^x f(t)\, dt)}{dx} = f(x)$ (second fundamental theorem).

$\displaystyle\int_a^b f(x)\, dx = \int_a^c f(x)\, dx + \int_c^b f(x)\, dx.$

Guide quiz on Chap. 8

1. True or false:
   (a) Every function that is continuous throughout $[a, b]$ has a definite integral over $[a, b]$.
   (b) Every function that is continuous for all x is the derivative of some function.
   (c) Every function that is differentiable for all x is the derivative of some function.

2. (a) Use the FTC to solve the opening problems in Chap. 7. That is, use it to compute $\int_0^3 x^2\, dx$, $\int_0^3 5x^2\, dx$, and $\int_0^3 8t^2\, dt$.
   (b) Give an example of a definite integral that cannot be evaluated by the FTC.

3. (a) State the two fundamental theorems of calculus.
   (b) Let $F(t)$ be the position on the x axis of a particle at time t. The particle is moving to the right with velocity $f(t)$ at time t. Interpret in terms of the moving particle the quantities

$$F(b) - F(a)$$

   and $$\int_a^b f(t)\, dt.$$

4. Two of these three antiderivatives are elementary. Evaluate those two, giving at least two answers for them.
   (a) $\displaystyle\int \left(x + \frac{3}{\sqrt{x}}\right) dx$ (b) $\displaystyle\int e^{x^3}\, dx$
   (c) $\displaystyle\int e^{-3x}\, dx.$

5. An integral table lists the formula

$$\int x^2 \cos ax\, dx = \frac{2x}{a^2}\cos ax - \frac{2}{a^3}\sin ax + \frac{x^2}{a}\sin ax$$

(a is a constant).   Verify it by differentiating the right side.

6. Find:
   (a) $\displaystyle\int \frac{1}{2x + 3}\, dx$ (b) $\displaystyle\int \frac{1}{(2x + 3)^2}\, dx$
   (c) $\displaystyle\int \frac{1}{\sqrt{2x + 3}}\, dx.$

7. Find the area of the region bordered by the curves $y = \sqrt{x}$, $y = 1/x$ and the line $x = 2$. (First sketch the region.)

8. (a) Compute $\dfrac{d(\int_1^x 3t^2 \sin 2t^3\, dt)}{dx}$.
   (b) Compute $\dfrac{d(\int_1^{x^3} \sin 2t\, dt)}{dx}$.
   (c) Using (a) and (b), show that

$$\int_1^x 3t^2 \sin 2t^3\, dt = \int_1^{x^3} \sin 2t\, dt.$$

9. Compute (a) $x^3 \Big|_{-1}^1$ (b) $x^4 \Big|_{-1}^1$
   (c) $\cos 2x \Big|_{-\pi/4}^{\pi/4}$ (d) $e^{-x} \Big|_0^1.$

Guide quiz on Chaps. 1 to 8

1. (a) Differentiate $1/x$.
   (b) Use (a) to find $\int 1/x^2\, dx$.
   (c) Find the area of the region under $y = 1/x^2$ and above $[1, 4]$.

(d) The curve $y = 1/x$ is rotated around the x axis. Find the volume of the solid enclosed by the resulting surface, between $x = 1$ and $x = 4$ (see the drawing below).

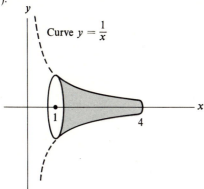

Curve $y = \frac{1}{x}$

2. (a) Differentiate $\sqrt{1 + x^2}$.
   (b) Use (a) to find an antiderivative of $x/\sqrt{1 + x^2}$.
   (c) Evaluate $\int_0^{\sqrt{3}} x/\sqrt{1 + x^2}\, dx$.
   (d) Find the area of the region under the curve

   $$y = \frac{x}{\sqrt{1 + x^2}}$$

   and above $[0, \sqrt{3}]$.
   (e) Find the distance an object travels from time $\sqrt{3}$ to time $\sqrt{8}$ if at time t its speed is

   $$\frac{t}{\sqrt{1 + t^2}} \quad \text{feet per second.}$$

3. Differentiate:
   (a) x^5
   (b) $\sqrt{x}$
   (c) $\ln\left(\dfrac{1 + x}{1 - x}\right)$
   (d) $\tan^{-1} x$.

4. Use the results of Exercise 3 (in some order) to find
   (a) $\displaystyle\int \frac{1}{1 - x^2}\, dx$
   (b) $\displaystyle\int \frac{1}{1 + x^2}\, dx$
   (c) $\displaystyle\int x^4\, dx$
   (d) $\displaystyle\int \frac{1}{\sqrt{x}}\, dx$.

5. Differentiate:
   (a) $\dfrac{1}{2}\left(x\sqrt{9 - x^2} + 9 \sin^{-1}\dfrac{x}{3}\right)$
   (b) $\dfrac{1}{8}\left[2x + 3 - 6 \ln (2x + 3) - \dfrac{9}{2x + 3}\right]$
   (c) $\tan 3x - 3x$
   (d) $\dfrac{1}{\sqrt{6}} \tan^{-1}\left(x\sqrt{\dfrac{2}{3}}\right)$ (continued)

(e) $-\frac{1}{3}\cos 3x + \frac{1}{9}\cos^3 3x$
(f) $\frac{1}{8}e^{-2x}(4x^2 + 4x + 2)$.

6. Use a differential to estimate $\sqrt[3]{1 + x}$ when x is small.

7. Find all functions $y = f(x)$ such that

   $$\frac{d^2y}{dx^2} = \sin 2x.$$

   How are you sure you have obtained all such functions?

8. Find:
   (a) $\displaystyle\lim_{x \to \pi/4} \frac{\sin 3x - \sin (3\pi/4)}{x - \pi/4}$;
   (b) $\lim_{\text{mesh} \to 0} \sum_{i=1}^n x_i^3(x_i - x_{i-1})$, where $x_0 = 2$, $x_1, \dots,$ $x_n = 3$ determines a partition of $[2, 3]$.

9. Define the differential and draw a picture showing df and Δf.

10. Graph $y = xe^{-x}/(x + 1)$, showing intercepts, critical points, relative maxima and minima, and asymptotes.

11. What is the fundamental theorem of calculus?

12. (a) Show that $y = Ab^x$, where A and b are constants, $b > 0$, satisfies the differential equation of natural growth.
    (b) Prove that any solution of the differential equation of natural growth must be of the form Ab^x for suitable constants A and b.

13. Find $\displaystyle\lim_{x \to \infty} \frac{5x^3 + 3/x}{10x^3 - x^2}$
    (a) with l'Hôpital's rule;
    (b) without it.

14. A window is to be made in the form of a rectangle topped by a semicircle, as shown below. The total perimeter is to be 20 feet. Find its maximum area.

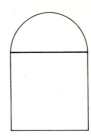

15. Consider the problem of finding the area under $y = e^{x^2}$ from $x = 0$ to $x = 1$.
    (a) Why is the FTC useless in determining this area?
    (b) Estimate $\int_0^1 e^{x^2}$ by utilizing a partition of $[0, 1]$ into the five sections $[0, 0.2], [0.2, 0.4], [0.4, 0.6], [0.6, 0.8], [0.8, 1]$, and as sampling numbers $X_1 = 0.1$, $X_2 = 0.3$, $X_3 = 0.5$, $X_4 = 0.7$, and $X_5 = 0.9$.

16. Fill in this table, indicating the interpretation of the definite integral.

f	$\int_a^b f(x)\,dx$
Velocity	
Density	
Cross-sectional length of region in plane	
Cross-sectional area of solid region	
(Varying) rate of flow of water through pipe in gallons per minute.	

17. Fill in this table, indicating the interpretation of the derivative.

$f(x)$	$f'(x)$
Position of moving particle at time x	
Mass in left x centimeters in string	
Area of region under some curve and above $[0, x]$	
Projection of x onto a linear screen	
y coordinate of graph of f	
Quantity of water in tank at time x	

18. One of the functions $\sqrt{x}\sqrt[3]{1-x}$ and $\sqrt{1-x}\sqrt[3]{1-x}$ has an elementary antiderivative F. Find F in the case for which an elementary antiderivative exists.

19. (a) Compute the derivative of $\ln (x^4 + 1)$ and $\tan^{-1} x^2$.
(b) Evaluate $\int_0^1 x^3/(x^4 + 1)\,dx$ and $\int_0^1 x/(x^4 + 1)\,dx$. Even though the integrands in (b) resemble each other, note the great difference in their antiderivatives.

Review exercises for Chaps. 1 to 8

1. (a) Differentiate $e^{-x}/(1 + 2x^2)$.
(b) What is the magnification at $x = 1$ of a projector that projects x onto $e^{-x}/(1 + 2x^2)$?
(c) What is the slope of the curve $y = e^{-x}/(1 + 2x^2)$ when $x = 1$?

2. Let $f(x) = x^2/(1 + x^4)$. Compute $\int_0^1 (df/dx)\,dx$.

3. (a) Differentiate $\ln (1 + x^2)$.
(b) Find the area of the region under the curve $y = x/(1 + x^2)$ and above the interval $[1, 2]$.

4. Prove that, if f and g are differentiable functions, then so is $f - g$, and $(f - g)' = f' - g'$.

5. Find an antiderivative and the derivative of (a) $1/\sqrt{x}$, (b) e^{-3x}, (c) $5/x$.

6. (a) Evaluate $(\int \tan x^3\,dx)'$.
(b) Evaluate $(\int_0^x \tan t^3\,dt)'$.
(c) Justify your answer in (a).
(d) Justify your answer in (b).

7. Differentiate: (a) $\ln (1 + e^{-3x})$, (b) $5 \sin x$, (c) $\sin 5x$.

8. In which parts of Exercise 7 did you need the chain rule?

9. What is the inverse of (a) the cubing function? (b) $\log_{10}$? (c) the square root function?

10. (a) Define $\sin x$ and $\cos x$ with the aid of the unit circle.
(b) Prove that $\sin^2 x + \cos^2 x = 1$.

11. (a) Define $\sec x$ and $\tan x$.
(b) Prove that $\sec^2 x = 1 + \tan^2 x$.

12. Compute:
(a) $\int \dfrac{3}{x^2}\,dx$ (b) $\int \dfrac{3}{1 + x^2}\,dx$ (c) $\int \dfrac{3x}{1 + x^2}\,dx$.

In Exercises 13 to 20 show by differentiating that

13. (a) $x \ln 3x - x = \int \ln 3x\,dx$;
(b) $\dfrac{x^2}{2} \ln 2x - \dfrac{x^2}{4} = \int x \ln 2x\,dx$.

14. (a) $\dfrac{\sin 3x}{3} = \int \cos 3x\,dx$;
(b) $\dfrac{\sin 3x}{9} - \dfrac{x \cos 3x}{3} = \int x \sin 3x\,dx$.

15. (a) $\dfrac{1}{5} \ln (\csc 5x - \cot 5x) = \int \dfrac{1}{\sin 5x}\,dx$;
(b) $\dfrac{-\cos 5x}{5} = \int \sin 5x\,dx$.

16. $\dfrac{6x - 4}{27} \sqrt{3x + 1} = \int \dfrac{x}{\sqrt{3x + 1}}\,dx$.

17. $\dfrac{5}{9(3x + 5)} + \dfrac{1}{9} \ln (3x + 5) = \int \dfrac{x}{(3x + 5)^2}\,dx$.

18. $\dfrac{1}{2}\left(x\sqrt{9 - x^2} + 9 \sin^{-1} \dfrac{x}{3}\right) = \int \sqrt{9 - x^2}\,dx$.

19. $\dfrac{1}{\sqrt{5}} \sec^{-1}\left(\sqrt{\dfrac{3}{5}}\,x\right) = \int \dfrac{1}{x\sqrt{3x^2 - 5}}\,dx$.

20. $\dfrac{x}{2\sqrt{3x^2 + 2}} = \int \dfrac{1}{(3x^2 + 2)^{3/2}}\,dx$.

21. Assuming that $\lim_{h\to 0} (\sin h)/h = 1$, prove, as a consequence, that $\lim_{h\to 0} (1 - \cos h)/h = 0$.

22. Find d^3y/dx^3 if $y = \int e^{-x^3}\, dx$.

23. (a) Assuming that $(\ln x)' = 1/x$, prove that $(e^x)' = e^x$.
    (b) Assuming that $(\sin x)' = \cos x$, prove that $(\sin^{-1} x)' = 1/\sqrt{1 - x^2}$.
    (c) Assuming that $(\tan x)' = \sec^2 x$, prove that $(\tan^{-1} x)' = 1/(1 + x^2)$.

24. A particle moving on the x axis has the x coordinate e^{-t} at time t.
    (a) What is its velocity at time $t = 2$?
    (b) What is its acceleration at time $t = 2$?

25. (a) Define e.
    (b) Give a decimal approximation of e.

26. Simplify by laws of logarithms, and then differentiate:
$$\ln \frac{\sqrt[3]{\sin 2x}}{\sqrt[4]{x}}.$$

27. Let f be a continuous function, such that $f(-x) = -f(x)$. [A function f such that $f(-x) = -f(x)$ is called *odd*.]
    (a) Show that $\int_{-1}^{1} f(x)\, dx = 0$.
    (b) Show that $\int_{-1}^{1} x^3 e^{x^2}\, dx = 0$.

28. Differentiate: (a) 5^{3x}, (b) $\log_{10} (1 + x^2)$.

29. Compute: (a) $d^2(x^5)/dx^2$, (b) $(\ln x^2)''$, (c) $D^3(x/(1 + x))$.

30. (a) Why is the differential df a function of two variables?
    (b) Show these variables on a graph of f.

31. Evaluate df when f is $5x + x^3$, $x = 2$, and $dx = 0.1$.

32. Find the minimum value of $x^2 - 6x + 11$ for x in $[0, 4]$.

33. Find the minimum value of $x^3 - 6x$ for x in $[0, 2]$.

34. Solve the equation $3 = 5e^{0.03t}$ for t.

35. What is the derivative of an antiderivative of (a) x^3, (b) $1/(1 + \sqrt{x})$?

36. (a) What is the mass of the left 3 centimeters of a string whose density x centimeters from the left is $4x^3$ grams per centimeter?
    (b) An object travels at the speed $4t^3$ feet per second after t seconds. How far does it move in the first 3 seconds?
    (c) The region bounded by the curve $y = x^{3/2}$, the x axis, and the line $x = 3$ is spun around the x axis. Find the volume of the resulting solid.

■

37. How often should a machine be overhauled? This depends on the rate $f(t)$ at which it depreciates and the cost A of overhaul. Denote the time interval between overhauls by T.
    (a) Explain why you would like to minimize $g(T) = [A + \int_0^T f(t)\, dt]/T$.
    (b) Find dg/dT.

(c) Show that when $dg/dT = 0$, $f(T) = g(T)$.
(d) Is this reasonable?

38. (a) Is there a relation between $\int f(x)\, dx$, $\int g(x)\, dx$, and $\int [f(x) - g(x)]\, dx$? Explain.
    (b) Is there a relation between $\int f(x)\, dx$, $\int g(x)\, dx$, and $\int [f(x)g(x)]\, dx$? Explain.

39. (a) Find the area of the region under the curve $y = e^{-x}$ from $x = 0$ to $x = b$.
    (b) As $b \to \infty$, what happens to the area in (a)?

40. One of the two definite integrals $\int_0^{\sqrt{\pi}} 2x \sin x^2\, dx$ and $\int_0^{\sqrt{\pi}} 2 \sin x^2\, dx$ is easy to evaluate; the other is not. Evaluate the easier one.

41. (a) Graph $y = e^{-x} \sin x$. (b) Graph $y = \dfrac{x + 1}{x^2 - 3x}$.

42. Evaluate:
    (a) $\displaystyle\lim_{h\to 0} \frac{(1 + h)^3 - (1 + h)^2}{2 + h}$
    (b) $\displaystyle\lim_{x\to\infty} \frac{2x^3 + e^{-x}}{x^3 - 1}$
    (c) $\displaystyle\lim_{x\to 0} \left(\frac{2^x + 8^x}{2}\right)^{1/x}$
    (d) $\displaystyle\lim_{t\to 0^+} \frac{\int_0^t \sqrt{\sin 3x}\, dx}{\int_0^t \sqrt{x}\, dx}$
    (e) $\displaystyle\lim_{t\to\infty} \frac{\int_1^t \sqrt[3]{x^6 + x + 1}\, dx}{\int_3^t \sqrt{x^4 + x^3 + 5}\, dx}$
    (f) $\displaystyle\lim_{t\to 2} \frac{\int_4^{t^2} e^{x^2}\, dx}{\int_8^{t^3} e^{x^3}\, dx}$.

In Exercises 43 to 47 differentiate the right side to verify these formulas from an integral table.

43. $\displaystyle\int x^n \tan^{-1} ax\, dx = \frac{x^{n+1}}{n + 1} \tan^{-1} ax - \frac{a}{n + 1}\int \frac{x^{n+1}\, dx}{1 + a^2x^2}$,
$$n \ne -1.$$

44. $\displaystyle\int \tan^{-1} ax\, dx = x \tan^{-1} ax - \frac{1}{2a} \ln (1 + a^2x^2)$.

45. $\displaystyle\int \sec^{-1} ax\, dx = x \sec^{-1} ax - \frac{1}{a} \ln (ax + \sqrt{a^2x^2 - 1})$.

46. $\displaystyle\int (\sin^{-1} ax)^2\, dx = x(\sin^{-1} ax)^2 - 2x$
$$+ \frac{2}{a}\sqrt{1 - a^2x^2} \sin^{-1} ax.$$

47. $\displaystyle\int x^n e^{ax}\, dx = \frac{1}{a} x^n e^{ax} - \frac{n}{a}\int x^{n-1} e^{ax}\, dx, n > 0$.

48. Let $A(t)$ be the area of the region under the curve $y = 1/(1 + x^2)$ and above the interval $[0, t]$. Find $\lim_{t\to\infty} A(t)$.

49. Let $a_n = n/(n + 1)^2 + n/(n + 2)^2 + \cdots + n/(n + n)^2$.
   (a) Compute a_n to three decimal places for $n = 1, 2$, and 3.
   (b) Rewrite a_n in such a way that it becomes an approximating sum to a certain definite integral.
   (c) With the aid of (b), find $\lim_{n \to \infty} a_n$.

50. As a stone is lowered into water, we record the volume of water it displaces. When x inches is submerged, the stone displaces $V(x)$ cubic inches of water. How can we find the area of the cross section of the stone made by the plane of the surface of the water when it is submerged to a depth of x inches? Assume we know $V(x)$ for all x.

51. A company is founded with a capital investment A. The plan is to have its rate of investment proportional to its total investment at any time. Let $f(t)$ denote the rate of investment at time t.
   (a) Show that there is a constant k such that $f(t) = k[A + \int_0^t f(x) \, dx]$ for any $t \geq 0$.
   (b) Differentiate the relation in (a), and with the aid of the equation you obtain, find the form of f.

52. What is the maximum slope of the curve $y = 1/(1 + x^2)$?

53. Let $A(t)$ be the area of the region under the curve $y = \sqrt{\ln x}$ from $x = 1$ to $x = t$. Find dA/dt.

54. Show that if $dy/dx = \sin y$, then $d^2y/dx^2 = \sin y \cdot \cos y$.

55. Find: (a) $D(\int_1^{x^3} e^{\sqrt{t}} \, dt)$, (b) $D(\int_{x^3}^2 \sin \sqrt{t} \, dt)$.

56. Find $D_x(\int_{x^2}^{x^3} \sqrt{1 + t^7} \, dt)$.

57. Give an example of a function that is not elementary but has an elementary derivative.

■ ■

58. Let f be a continuous function that has a derivative nowhere. (There are such functions.) Construct a function that has a derivative everywhere but a second derivative nowhere.

59. A man whose car has a vertical windshield drives a mile through a vertical rain consisting of drops that are uniformly distributed and falling at a constant rate. Should he go slow or fast in order to minimize the amount of rain that strikes the windshield?

60. Prove that if f is an increasing function on the interval $[a, b]$ and $f = F'$, then $\int_a^b f(x) \, dx$ exists and equals $F(b) - F(a)$. (Do not assume the fundamental theorem of calculus.)

61. Prove that if g is differentiable and $|g'(x)| < k$ for all x in $[a, b]$, then the function $f(x) = g(x) + kx$ is increasing on $[a, b]$.

62. (See Exercises 60 and 61.) Prove that if f is a differentiable function whose derivative f' is bounded over the interval $[a, b]$, and if $f = F'$, then $\int_a^b f(x) \, dx$ exists and equals $F(b) - F(a)$. (Do not assume that a continuous function has a definite integral.)

63. Prove that

$$2 \int_0^a f(u) \left[\int_0^u f(t) \, dt \right] du = \left[\int_0^a f(x) \, dx \right]^2.$$

Exercises 64 to 68 are based on, D. L. Meadows, J. Randers, W. W. Behrens, *The Limits to Growth*, Universe Books, New York, 1972.

64. Let $Y(t)$ be the amount of some natural resource that is consumed from time $t = 0$ to time $t > 0$. For our purposes $t = 0$ corresponds to the year 1970. Let $c(t)$ be the rate at which that resource is being consumed. Since population and industry are growing, $c(t)$ is an increasing function. Assume that $c(t)$ is increasing exponentially; that is, there are constants A and r such that $c(t) = Ae^{rt}$. Show that $Y(t) = A(e^{rt} - 1)/r$.

65. The amount of such natural resources as aluminum, natural gas, and petroleum is finite. The formula obtained in Exercise 64 enables one to estimate how long a given resource will last subject to various assumptions.
   Let R be the amount of a given resource remaining at time $t = 0$.
   (a) Show that if the rate of consumption remains constant, then the resource will last R/A years. This is called the *static index* and is denoted by s.
   (b) Show that, if the rate of consumption continues to grow exponentially, the resource will last
$$\frac{\ln (rs + 1)}{r} \quad \text{years.}$$
   This is called the *exponential index*. The letter s denotes the static index, defined in (a).

66. The static index of aluminum is 100 years (known global reserves are $1.17 \cdot 10^9$ tons). The projected growth rate for consumption is 6.4 percent a year. Thus $c(t) = A(1.064)^t$. However, $e^{0.064}$ is a very good approximation to 1.064 and is used in the cited reference to obtain the formula $c(t) = Ae^{0.064t}$.
   (a) Show that the exponential index is 31 years (approximately). Show that, if the global reserves are five times as large as quoted, the static index will be 500 years and the exponential index will be 55 years.
   (b) Show that, if $r = 0.032$ instead of 0.064, the static index doubles but the exponential index will be 45 years. (Assume known global reserves.)

67. The known global reserves of natural gas are $1.14 \cdot 10^{15}$ cubic feet. The static index is 38 years. If consumption continues to increase at 4.7 percent per year, show that the exponential index (a) is 22 years, (b) if global reserves are 5 times as large, is 49 years.

68. The known global reserves of petroleum are $455 \cdot 10^8$ barrels. The static index is 31 years. If the rate of consumption continues to increase at 3.9 percent per year, show that the exponential index is 20 years and, if reserves are five times as large, 50 years.

9

COMPUTING ANTIDERIVATIVES

In Chapter 5 it was shown how to compute the derivative of any elementary function. In Chap. 7 the computation of many definite integrals was reduced to the computation of antiderivatives. In the present chapter some techniques for finding antiderivatives will be developed.

The problem of computing antiderivatives differs from that for computing derivatives in two important aspects. First of all, some elementary functions, such as e^{x^2}, do not have an elementary antiderivative. Second, a slight change in the form of an integrand can cause a great change in the form of its antiderivative; for instance,

$$\int \frac{1}{x^2 + 1}\, dx = \tan^{-1} x \qquad \text{and} \qquad \int \frac{x}{x^2 + 1}\, dx = \frac{1}{2} \ln{(x^2 + 1)}.$$

A few moments of browsing through a table of antiderivatives (usually called a *table of integrals*) will yield many such examples. Such a table of integrals is included in Appendix E.

To be convenient, a table of antiderivatives should be short; it should not try to anticipate every antiderivative that may arise in practice. Sometimes it is necessary to transform a problem into one listed in the table, or else to solve it without the aid of the table. For this reason we present a few techniques for transforming problems and finding antiderivatives. It will sometimes be quicker to use these techniques than to thumb through the pages of the table, even though the table may list the answer.

For any constant C, $C + \tan^{-1} x$ is an antiderivative of $1/(x^2 + 1)$. Since we require only one antiderivative to apply the fundamental theorem of calculus, it suffices to take C to be 0. Hence C will usually be omitted.

In Chap. 5, in which we obtained formulas for differentiating all elementary functions, it was necessary to know what a derivative is—that it is defined as the limit of a certain quotient. But it is possible to become skillful in computing antiderivatives without knowing what a derivative is just by remembering the formulas for differentiation developed in Chap. 5.

Though the text contains a brief table of integrals, it is strongly recommended that the student purchase a standard handbook of mathematical tables, such as Burington's *Handbook of Mathematical Tables and Formulas*. Such a handbook also contains extensive lists of key facts from algebra, geometry, trigonometry, analytic geometry, statistics, etc. By the end of this chapter, the reader will be quite efficient in using such a handbook. The ubiquitous hand-held calculator may render its numerical tables obsolete, but it does not replace its many tables of mathematical facts and formulas.

9.1

Some basic facts

Every formula for a derivative provides a formula for an antiderivative. For instance, since

$$\frac{d(x^3)}{dx} = 3x^2,$$

it follows that

$$\int 3x^2 \, dx = x^3$$

or, more conveniently,

$$\int x^2 \, dx = \frac{x^3}{3}.$$

It is easy to show more generally that

$$\int x^a \, dx = \frac{x^{a+1}}{a+1} \qquad a \neq -1.$$

Just compute the derivative of $x^{a+1}/(a+1)$ and check that it is x^a.

The following miniature table of antiderivatives is included for convenience. Each formula may be checked by differentiation.

$$\int x^a \, dx = \frac{x^{a+1}}{a+1} \qquad \text{for} \quad a \neq -1 \qquad \int \frac{1}{x} \, dx = \ln |x|$$

$$\int e^x \, dx = e^x$$

$$\int \sin x \, dx = -\cos x \qquad \text{remember the minus sign}$$

$$\int \cos x \, dx = \sin x$$

$$\int \frac{1}{\sqrt{1-x^2}} \, dx = \sin^{-1} x$$

$$\int \frac{1}{1+x^2} \, dx = \tan^{-1} x$$

$$\int \frac{1}{|x|\sqrt{x^2-1}} \, dx = \sec^{-1} x$$

The following general formula is needed often enough to be included in this brief list:

$$\int \frac{f'}{f} \, dx = \ln |f|.$$

If the numerator is precisely the derivative of the denominator, then ln |denominator| is an antiderivative of the quotient.

The absolute value sign in $\ln |f|$ is a redundant nuisance if $f(x)$ is positive. In that case $\ln |f| = \ln f$. If $f(x)$ is negative, the absolute value sign is essential. After all, when $f(x)$ is negative, $\ln f(x)$ is meaningless, since the logarithm is defined only for positive numbers. To evaluate

$$\int_{-3}^{-2} \frac{1}{x} \, dx,$$

for instance, the absolute value is necessary. We have:

$$\int_{-3}^{-2} \frac{1}{x} \, dx \underset{\text{FTC}}{=} \ln |x| \Big|_{-3}^{-2}$$

$$= \ln |-2| - \ln |-3|$$

$$= \ln 2 - \ln 3$$

$$= \ln \tfrac{2}{3}$$

$$\doteq -0.41.$$

In addition, these two simple properties of antiderivatives help in many computations:

A Constant Can Be Moved Past the Integral Sign.

If c is constant, then

$$\int cf(x) \, dx = c \int f(x) \, dx.$$

To justify this assertion, it suffices to show that the derivative of $c \int f(x) \, dx$ is $cf(x)$. The details are simple:

$$\frac{d[c \int f(x) \, dx]}{dx} = \frac{cd[\int f(x) \, dx]}{dx}$$

$$= cf(x).$$

Thus $c \int f(x) \, dx$ is an antiderivative of $cf(x)$.

An Antiderivative of the Sum of Functions Can Be Found by Adding Antiderivatives of Each Function.

For two functions f and g

$$\int [f(x) + g(x)] \, dx = \int f(x) \, dx + \int g(x) \, dx.$$

Similarly,

$$\int [f(x) - g(x)] \, dx = \int f(x) \, dx - \int g(x) \, dx.$$

The proof of the first assertion depends on the fact that the derivative of a sum of functions is the sum of the derivatives. To show that

$$\int f(x) \, dx + \int g(x) \, dx$$

is an antiderivative of $f(x) + g(x)$, differentiate it:

$$\frac{d[\int f(x)\,dx + \int g(x)\,dx]}{dx} = \frac{d[\int f(x)\,dx]}{dx} + \frac{d[\int g(x)\,dx]}{dx}$$

$$= f(x) + g(x).$$

A few examples will show how these few formulas serve to integrate many functions.

EXAMPLE 1
Antiderivative of Any Polynomial.

Find

$$\int (2x^4 - 3x + 2)\,dx.$$

SOLUTION

$$\int (2x^4 - 3x + 2)\,dx = \int 2x^4\,dx - \int 3x\,dx + \int 2\,dx$$

$$= 2\int x^4\,dx - 3\int x\,dx + 2\int 1\,dx$$

$$= 2\frac{x^5}{5} - 3\frac{x^2}{2} + 2x. \bullet$$

EXAMPLE 2
Antiderivative of f'/f.

Find

$$\int \frac{4x^3}{x^4 + 1}\,dx.$$

SOLUTION   The numerator is precisely the derivative of the denominator.   Hence

$$\int \frac{4x^3}{x^4 + 1}\,dx = \ln |x^4 + 1|.$$

Since $x^4 + 1$ is always positive, the absolute-value sign is not needed, and

$$\int \frac{4x^3}{x^4 + 1}\,dx = \ln (x^4 + 1). \bullet$$

EXAMPLE 3
Antiderivative of x^a.

Find

$$\int \sqrt{x}\,dx.$$

SOLUTION

$$\int \sqrt{x}\,dx = \int x^{1/2}\,dx$$

$$= \frac{x^{1 + 1/2}}{1 + \frac{1}{2}}$$

$$= \tfrac{2}{3}x^{3/2}$$

$$= \tfrac{2}{3}(\sqrt{x})^3. \bullet$$

EXAMPLE 4   Find

$$\int \frac{1}{x^3}\,dx.$$

SOLUTION

$$\int \frac{1}{x^3}\,dx = \int x^{-3}\,dx$$

$$= \frac{x^{-3+1}}{-3+1}$$

$$= -\tfrac{1}{2}x^{-2}$$

$$= -\frac{1}{2x^2}. \ \bullet$$

EXAMPLE 5 Find

$$\int \left(3\cos x - 4\sin x + \frac{1}{x^2}\right) dx.$$

SOLUTION

$$\int \left(3\cos x - 4\sin x + \frac{1}{x^2}\right) dx = 3\int \cos x\,dx - 4\int \sin x\,dx + \int \frac{1}{x^2}\,dx$$

$$= 3\sin x + 4\cos x - \frac{1}{x}. \ \bullet$$

EXAMPLE 6 Find
Multiplying the Integrand by a Constant.

$$\int \frac{x}{1+x^2}\,dx.$$

SOLUTION If the numerator were $2x$, then the numerator would be the derivative of the denominator, and the antiderivative would be $\ln(1+x^2)$. But the numerator can be multiplied by 2, if we simultaneously divide by 2:

$$\int \frac{x}{1+x^2}\,dx = \frac{1}{2}\int \frac{2x}{1+x^2}\,dx.$$

This step depends on the fact that a constant can be moved past the integral sign:

$$\frac{1}{2}\int \frac{2x}{1+x^2}\,dx = \frac{1}{2}2\int \frac{x}{1+x^2}\,dx$$

$$= \int \frac{x}{1+x^2}\,dx.$$

Thus

$$\int \frac{x}{1+x^2}\,dx = \frac{1}{2}\int \frac{2x}{1+x^2}\,dx$$

$$= \tfrac{1}{2}\ln(1+x^2). \ \bullet$$

NOTATION Frequently, for the sake of brevity, the dx in $\int f(x)\,dx$ is combined with the integrand. For example,

$$\int \frac{x^2}{1+x^3}\,dx \quad \text{is written} \quad \int \frac{x^2\,dx}{1+x^3},$$

$$\int \frac{1}{1 + x} \, dx \quad \text{is written} \quad \int \frac{dx}{1 + x},$$

and $\quad \int 1 \, dx \quad$ is written $\quad \int dx.$

Exercises

In Exercises 1 to 27 compute the antiderivatives.

1. $\int x^3 \, dx$

2. $\int 5x^3 \, dx$

3. $\int x^{1/3} \, dx$

4. $\int \sqrt[3]{x^2} \, dx$

5. $\int \frac{3}{\sqrt{x}} \, dx$

6. $\int \frac{1}{\sqrt[4]{x}} \, dx$

7. $\int 5e^{-2x} \, dx$

8. $\int \frac{5}{1 + x^2} \, dx$

9. $\int \frac{6 \, dx}{x\sqrt{x^2 - 1}}$

10. $\int \sec^2 3x \, dx$

11. $\int \frac{x^3}{1 + x^4} \, dx$

12. $\int \frac{e^x}{1 + e^x} \, dx$

13. $\int \frac{\sin x}{1 + \cos x} \, dx$

14. $\int \cos 2x \, dx$

15. $\int \frac{1 + 2x}{x + x^2} \, dx$

16. $\int \frac{dx}{1 + 3x}$

17. $\int \left(3 \cos x - \frac{1}{\sqrt{1 - x^2}} \right) dx$

18. $\int \frac{dx}{x^5}$

19. $\int (1 + 3x)x^2 \, dx$

20. $\int \left(\sqrt{x} - \frac{2}{x} + \frac{x}{2} \right) dx$

21. $\int x^2 \sqrt{x} \, dx$

22. $\int \frac{1 + 2x}{1 + x^2} \, dx$

23. $\int \frac{1 + \sqrt{x}}{x} \, dx$

24. $\int \frac{\sin 2x}{\cos 2x} \, dx$

25. $\int (x^2 + 3)^2 \, dx$

26. $\int (1 + e^x)^2 \, dx$

27. $\int dx$

28. Differentiate (for practice)—all have short answers:

(a) $\dfrac{1}{a^3} \left[\dfrac{1}{2}(ax + b)^2 - 2b(ax + b) + b^2 \ln (ax + b) \right]$;

(b) $-\dfrac{x^2}{a\sqrt{ax^2 + c}} + \dfrac{2}{a^2} \sqrt{ax^2 + c}$;

(c) $\sqrt{p - x}\sqrt{q + x} + (p + q) \sin^{-1} \sqrt{\dfrac{x + q}{p + q}}$;

(d) $-\dfrac{1}{2a} \cot^2 ax - \dfrac{1}{a} \ln \sin ax$;

(e) $x(\ln ax)^2 - 2x \ln ax + 2x$.

■

29. Find all solutions of the differential equation

$$\frac{d^2 y}{dx^2} = (2x - 1)^5.$$

30. Find the minimum and maximum values, if any, of $f(x) = \ln x - 9x^3$ over (a) the interval $(0, 1)$, (b) the interval $[\frac{1}{2}, 1]$.

31. Find $\lim\limits_{x \to 0} \dfrac{e^{2x} - 1 - 2x}{1 - \cos 2x}$.

32. Find the area under the curve $y = \sin^2 x$ and above $[0, \pi]$. *Hint:* Write $\sin^2 x = (1 - \cos 2x)/2$.

33. Compute $\int x \, dx/(1 + x^4)$. *Hint:* The answer is an arctangent function.

■ ■

34. Jack claims that $\int 2 \cos \theta \sin \theta \, d\theta = -\cos^2 \theta$, while Jill claims that the answer is $\sin^2 \theta$. Who is right?

35. Jack's table of integrals shows that $\int \sec \theta \, d\theta = \ln (\sec \theta + \tan \theta)$, while Jill's table lists the integral as $\ln \tan (\theta/2 + \pi/4)$. Whose table is right?

This exercise concerns moving a nonconstant function past the integral sign.

36. (a) Give an example showing that $\int f(x) \, dx \int g(x) \, dx$ is not necessarily $\int f(x)g(x) \, dx$.

(b) If $\int f(x)g(x) \, dx = f(x) \int g(x) \, dx$, must f be a constant function? Assume that g is not the constant function 0.

9.2

The substitution technique

This method involves the introduction of a function that changes the form of the integrand, hopefully to a simpler integrand. Five examples will illustrate the mechanics of the method, known as *substitution*. The proof that it works will be given after these examples. Substitution is the most important tool in computing antiderivatives.

EXAMPLE 1 Find
$$\int \sin (x^2)\, 2x \; dx.$$

SOLUTION Note that $2x$ is the derivative of x^2. Introduce
$$u = x^2.$$
Then
$$du = 2x\, dx,$$
and
$$\int \sin (x^2) 2x \; dx = \int \sin u \; du.$$

Now it is easy to find $\int \sin u \; du$:
$$\int \sin u \; du = -\cos u.$$

Replacing u by x^2 in $-\cos u$ yields $-\cos (x^2)$. Thus
$$\int \sin (x^2) 2x \; dx = -\cos x^2.$$

This answer can be checked by differentiation (using the chain rule):
$$\frac{d(-\cos x^2)}{dx} = \sin x^2 \, \frac{d(x^2)}{dx}$$
$$= \sin (x^2) 2x. \; \bullet$$

EXAMPLE 2 Find
$$\int e^{x^5} 5x^4 \; dx.$$

SOLUTION Introduce $u = x^5$. Then
$$du = 5x^4\, dx,$$
and
$$\int e^{x^5} 5x^4 \; dx = \int e^u \; du$$
$$= e^u$$
$$= e^{x^5}. \; \bullet$$

EXAMPLE 3 Find
$$\int \sin^2 \theta \cos \theta \; d\theta.$$

SOLUTION   Note that $\cos \theta$ is the derivative of $\sin \theta$ and introduce

$$u = \sin \theta;$$

hence

$$du = \cos \theta \, d\theta.$$

Then

$$\int \underbrace{\sin^2 \theta}_{u^2} \underbrace{\cos \theta \, d\theta}_{du} = \int u^2 \, du$$

$$= \frac{u^3}{3}$$

$$= \frac{\sin^3 \theta}{3}. \quad \bullet$$

EXAMPLE 4   Find

$$\int (1 + x^3)^5 x^2 \, dx.$$

SOLUTION   The derivative of $1 + x^3$ is $3x^2$, which differs from x^2 in the integrand only by the constant factor 3.   So let

$$u = 1 + x^3;$$

hence

$$du = 3x^2 \, dx \qquad \text{and} \qquad \frac{du}{3} = x^2 \, dx.$$

Then

$$\int (1 + x^3)^5 x^2 \, dx = \int u^5 \frac{du}{3}$$

$$= \frac{1}{3} \int u^5 \, du$$

$$= \frac{1}{3} \frac{u^6}{6}$$

$$= \frac{(1 + x^3)^6}{18}.$$

It would be instructive to check this answer by differentiation. $\bullet$

In Example 4 note that, if the x^2 were not present in the integrand, the substitution method would not work.   To find $\int (1 + x^3)^5 \, dx$, it would be necessary to multiply out $(1 + x^3)^5$ first.

EXAMPLE 5   Compare the problems of finding these antiderivatives:

$$\int \frac{1}{\sqrt{1 + x^3}} \, dx \qquad \text{and} \qquad \int \frac{x^2}{\sqrt{1 + x^3}} \, dx.$$

SOLUTION   It turns out that the first antiderivative is *not* an elementary function, while the second is easy, because the x^2 is present.

Since x^2 differs from the derivative of $1 + x^3$ only by a constant factor 3, use the substitution $$u = 1 + x^3;$$

hence $$du = 3x^2 \, dx \quad \text{and} \quad \frac{du}{3} = x^2 \, dx.$$

Thus $$\int \frac{x^2 \, dx}{\sqrt{1 + x^3}} = \int \frac{1}{\sqrt{u}} \frac{du}{3} = \frac{1}{3} \int \frac{du}{\sqrt{u}} = \frac{1}{3} \int u^{-1/2} \, du$$

$$= \frac{1}{3} \frac{u^{1/2}}{\frac{1}{2}}$$

$$= \tfrac{2}{3} u^{1/2}$$

$$= \tfrac{2}{3}(1 + x^3)^{1/2}$$

$$= \tfrac{2}{3}\sqrt{1 + x^3}. \ \bullet$$

In the next example the choice of substitution is suggested not by something that can serve as du but rather by mathematical instinct.

EXAMPLE 6 Find $$\int \frac{x^2 + 1}{2x - 3} \, dx.$$

SOLUTION Since the denominator complicates the problem, try the substitution $u = 2x - 3$. We have

$$u = 2x - 3;$$

hence $$du = 2 \, dx, \quad dx = \frac{du}{2}, \quad \text{and} \quad x = \frac{u + 3}{2}.$$

Thus $$\int \frac{x^2 + 1}{2x - 3} \, dx = \int \frac{\left(\dfrac{u + 3}{2}\right)^2 + 1}{u} \frac{du}{2}$$

$$= \int \frac{u^2 + 6u + 13}{8u} \, du$$

$$= \int \left(\frac{u}{8} + \frac{3}{4} + \frac{13}{8u} \right) du$$

$$= \frac{u^2}{16} + \frac{3u}{4} + \frac{13}{8} \ln |u|$$

$$= \frac{(2x - 3)^2}{16} + \frac{3}{4}(2x - 3) + \frac{13}{8} \ln |2x - 3|. \ \bullet$$

The final example suggests that the substitution technique should always be tried as a last resort just to see what happens.

EXAMPLE 7   Try to find
$$\int \sqrt{1 - \tfrac{4}{5}\sin^2\theta}\, d\theta.$$

ATTEMPT AT SOLUTION   If the problem were that of computing
$$\int \sqrt{1 - \tfrac{4}{5}\sin^2\theta}\, \cos\theta\, d\theta,$$

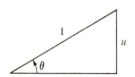

it would be easy, since $d(\sin\theta) = \cos\theta\, d\theta$, and the substitution $u = \sin\theta$ would work. Unfortunately, $\cos\theta$ is not part of the integrand.

Not seeing anything else to do, try the substitution $u = \sin\theta$ anyway. See the triangle in the margin. There is always a chance that it will transform the problem into one that is easier. We have

$$u = \sin\theta \qquad \text{and} \qquad du = \cos\theta\, d\theta,$$

thus
$$d\theta = \frac{du}{\cos\theta}.$$

$$= \frac{du}{\sqrt{1 - \sin^2\theta}} \qquad \begin{array}{l} \text{assume } -\pi/2 < \theta < \pi/2, \\ \text{so the positive square root} \\ \text{is appropriate.} \end{array}$$

$$= \frac{du}{\sqrt{1 - u^2}}.$$

Consequently,
$$\int \sqrt{1 - \tfrac{4}{5}\sin^2\theta}\, d\theta = \int \sqrt{1 - \tfrac{4}{5}u^2}\ \frac{du}{\sqrt{1 - u^2}}$$

$$= \int \sqrt{\frac{1 - \tfrac{4}{5}u^2}{1 - u^2}}\, du.$$

Though the integrand is now an algebraic instead of a trigonometric function, it still is tough. In fact, it is proved in advanced mathematics that the antiderivative we are seeking is not an elementary function. Thus, no substitution will solve the problem.

Incidentally, the definite integral
$$\int_0^{2\pi} \sqrt{1 - \tfrac{4}{5}\sin^2\theta}\, d\theta$$

appears in the calculation of the perimeter of the ellipse
$$\frac{x^2}{5} + \frac{y^2}{1} = 1,$$

as will be shown in the next chapter. An integral of the type
$$\int \sqrt{1 - k^2 \sin^2\theta}\, d\theta,$$

where $0 < k < 1$, is called an *elliptic integral*. ●

Example 7 concerns an elementary integrand with a nonelementary antiderivative. In the exercises in the text that involve the computation of an antiderivative, the integrands have the property that their antiderivatives are elementary. The few exercises that contain "impossible" antiderivatives are clearly so labelled. They serve as reminders of the limitations of the fundamental theorem of calculus.

General Description of the Substitution Technique   In each example the method is basically the same. In order to apply the substitution technique to find

$$\int f(x)\, dx,$$

look for a function $u = h(x)$ such that

$$f(x) = g(h(x))h'(x),$$

for some function g, or more simply

$$f(x)\, dx = g(u)\, du.$$

Then find the antiderivative of g,

$$\int g(u)\, du$$

and replace u by $h(x)$ in this antiderivative.

Of course, it is to be hoped that the problem of finding $\int g(u)\, du$ is easier than that of finding $\int f(x)\, dx$. If not, try another substitution or one of the methods presented in the rest of the chapter. It is important to keep in mind that there is no simple routine method for antidifferentiation of elementary functions. This is in contrast with the routine that exists for differentiation, which was presented in Chap. 5. Practice in integration pays off in the quick recognition of which technique is most promising.

The following argument justifies the method illustrated in the examples and shows that it depends on the chain rule.

SUBSTITUTION THEOREM   If $f(x) = g(h(x))h'(x)$

and

$$G(u) = \int g(u)\, du,$$

then

$$G(h(x)) = \int f(x)\, dx.$$

PROOF   It is necessary to show that

$$\frac{d[G(h(x))]}{dx} = f(x).$$

To do so, let $u = h(x)$ and use the chain rule:

$$\frac{d[G(h(x))]}{dx} = \frac{d[G(u)]}{du}\frac{du}{dx}$$

$$= g(u)h'(x)$$

$$= g(h(x))h'(x)$$

$$= f(x).$$

That is, $\qquad\qquad G(h(x)) = \int f(x)\, dx,$

which proves the theorem. ●

Exercises

In Exercises 1 to 15 *use the given substitution to find the anti-derivative.*

1. $\int (1 + 3x)^5 3\, dx; u = 1 + 3x.$

2. $\int \dfrac{x}{(1 + x^2)^3}\, dx; u = 1 + x^2.$

3. $\int e^{\sin\theta}\cos\theta\, d\theta; u = \sin\theta.$

4. $\int \dfrac{x}{\sqrt{1 + x^2}}\, dx; u = 1 + x^2.$

5. $\int \sqrt{1 + x^2}\, x\, dx; u = 1 + x^2.$

6. $\int \sin 2x\, dx; u = 2x.$

7. $\int \dfrac{e^{2x}}{(1 + e^{2x})^2}\, dx; u = 1 + e^{2x}.$

8. $\int e^{3x}\, dx; u = 3x.$

9. $\int \dfrac{e^{1/x}}{x^2}\, dx; u = 1/x.$

10. $\int \dfrac{1}{\sqrt{1 - 9x^2}}\, dx; u = 3x.$

11. $\int \dfrac{t\, dt}{\sqrt{2 - 5t^2}}; u = 2 - 5t^2.$

12. $\int \tan\theta \sec^2\theta\, d\theta; u = \tan\theta.$

13. $\int \dfrac{\sin\sqrt{x}}{\sqrt{x}}\, dx; u = \sqrt{x}.$

14. $\int \dfrac{(\ln x)^4}{x}\, dx; u = \ln x.$

15. $\int \dfrac{\sin(\ln x)}{x}\, dx; u = \ln x.$

In Exercises 16 to 30 *choose an appropriate substitution and find the antiderivative.*

16. $\int (1 - x^2)^5 x\, dx$

17. $\int \dfrac{x\, dx}{(x^2 + 1)^3}$

18. $\int \sqrt[3]{1 + x^2}\, x\, dx$

19. $\int \dfrac{\sin\theta}{\cos^2\theta}\, d\theta$

20. $\int \dfrac{e^{\sqrt{t}}}{\sqrt{t}}\, dt$

21. $\int \sin 3\theta\, d\theta$

22. $\int e^x \sin e^x\, dx$

23. $\int \dfrac{dx}{\sqrt{2x + 5}}$

24. $\int (x - 3)^{5/2}\, dx$

25. $\int \dfrac{dx}{(4x + 3)^3}$

26. $\int \dfrac{2x + 3}{x^2 + 3x + 2}\, dx$

27. $\int \dfrac{2x + 3}{(x^2 + 3x + 5)^4}\, dx$

28. $\int e^{2x}\, dx$

29. $\int \dfrac{dx}{\sqrt{x}(1 + \sqrt{x})^3}$

30. $\int x^4 \sin x^5\, dx$

■

In Exercises 31 to 45 find an antiderivative. Substitution may not be required, or may, in fact, not even work.

31. $\int \dfrac{dx}{e^x}$

32. $\int (3 + x^2)^2\, dx$

33. $\int (2e^x + 4e^{-x})^2 \, dx$

34. $\int \dfrac{x + 3x^2}{\sqrt{x}} \, dx$

45. $\int \dfrac{dx}{1 + 4x^2}$

■ ■

The antiderivatives in three of Exercises 46 to 51 are elementary. Compute those that are elementary.

35. $\int \dfrac{x}{1 + x^4} \, dx$

36. $\int \dfrac{x^3}{1 + x^4} \, dx$

37. $\int \dfrac{x}{\sqrt{1 - x^4}} \, dx$

38. $\int \dfrac{x \, dx}{1 + x}$

46. $\int x \cos x \, dx$

47. $\int \dfrac{\cos x}{x} \, dx$

39. $\int \sin^2 3x \, dx$

40. $\int \dfrac{e^x \, dx}{1 + e^{2x}}$

48. $\int \dfrac{x \, dx}{\ln x}$

49. $\int \dfrac{\ln x^2}{x} \, dx$

41. $\int \sqrt{3 + x^2} \, x^3 \, dx$

42. $\int \sqrt{1 + \cos 3\theta} \, d\theta$

50. $\int \sqrt{x - 1}\sqrt{x}\sqrt{x + 1} \, dx$

43. $\int (e^{2x} + 1)e^{-x} \, dx$

44. $\int \dfrac{dx}{\sqrt{1 - 5x^2}}$

51. $\int \sqrt{x - 1}\sqrt{x + 1} \, x \, dx$

9.3

Using a table of integrals

There are many methods of finding antiderivatives. The two most frequently used are substitution (which was described in Sec. 9.2) and searching through the 300 or 400 formulas of a table of integrals. Sections 9.1 to 9.4 dispose of practically all integration methods a student will need in this text or in subsequent applications of calculus. However, Secs. 9.5 to 9.8 will help the student feel more at home in a table of integrals. There is also a chance that he or she may someday have to call on one of the methods described there. Integration by partial fractions, which occupies Secs. 9.6 and 9.7, is of use in solving differential equations.

First of all, browse through your table of integrals. Notice how the formulas are grouped, For example the Burington tables are arranged as follows:

Group	*Formula Numbers*
1. Expressions containing $ax + b$	42–94
2. Expressions containing $ax^2 + c$, $ax^n + c$, $x^2 \pm p^2$, and $p^2 - x^2$	95–148
3. Expressions containing $ax^2 + bx + c$	149–174
4. Miscellaneous algebraic expressions	175–184
5. Expressions containing $\sin ax$	185–199
6. Expressions containing $\cos ax$	200–215
7. Expressions containing $\sin ax$ and $\cos ax$	216–246
8. Expressions containing $\tan ax$ or $\cot ax$	247–258
9. Expressions containing $\sec ax$ or $\csc ax$	259–268
10. Expressions containing $\tan ax$ and $\sec ax$ or $\cot ax$ and $\csc ax$	269–277
11. Expressions containing algebraic and trigonometric functions	278–297
12. Expressions containing exponential and logarithmic functions	298–341
13. Expressions containing inverse trigonometric functions.	342–363

Some tables, such as Burington's, use log instead of ln. It is understood that log is taken with respect to base e, *not* base 10.

EXAMPLE 1   Find $\int \sin^6 2x \, dx$ with the aid of the integral tables.

SOLUTION   The integrand involves sin $2x$, so turn to "Expressions containing sin ax." Here $a = 2$. There are formulas for $\int \sin ax \, dx$, $\int \sin^2 ax \, dx$, $\int \sin^3 ax \, dx$, $\int \sin^4 ax \, dx$, but for no higher power of sin ax. Instead, this formula is given:

$$\int \sin^n ax \, dx = \frac{-\sin^{n-1} ax \cos ax}{na} + \frac{n-1}{n} \int \sin^{n-2} ax \, dx.$$

This is a *recursion* formula. It reduces $\int \sin^n ax \, dx$ to $\int \sin^{n-2} ax \, dx$ recursively, knocking down the exponent by 2 at each stage. (Section 9.5 shows how this formula is found.)   We need the case $n = 6$, $a = 2$, which reads

$$\int \sin^6 2x \, dx = \frac{-\sin^5 2x \cos 2x}{6 \cdot 2} + \frac{5}{6} \int \sin^4 2x \, dx.$$

Now $\int \sin^4 ax \, dx$ is listed explicitly; there is no need to use the recursion for $n = 4$. The table includes a formula for $\int \sin^4 ax \, dx$, giving

$$\int \sin^4 2x \, dx = \frac{3x}{8} - \frac{3 \sin 2x}{32} - \frac{\sin^3 2x \cos 2x}{8}.$$

Combination of this with the equation relating $\int \sin^6 2x \, dx$ to $\int \sin^4 2x \, dx$ yields a formula for $\int \sin^6 2x \, dx$. The reader may complete the computation. ●

EXAMPLE 2   Find these three integrals with the aid of the integral tables:

$$\int \frac{dx}{x^2 - 5x + 3}, \tag{1}$$

$$\int \frac{dx}{x^2 - 2x + 3}, \tag{2}$$

$$\int \frac{dx}{4x^2 - 12x + 9}. \tag{3}$$

SOLUTION   Under "Expressions containing $ax^2 + bx + c$" are listed three different formulas for

$$\int \frac{dx}{ax^2 + bx + c}.$$

Many Tables Omit the Absolute Value Sign in ln $|u|$. *It Is Assumed that the User Will Supply It When Needed.*

The three separate cases are $b^2 > 4ac$, $b^2 < 4ac$, and $b^2 = 4ac$. The formulas listed are quite different:

$$\int \frac{dx}{ax^2 + bx + c} = \frac{1}{\sqrt{b^2 - 4ac}} \ln \frac{2ax + b - \sqrt{b^2 - 4ac}}{2ax + b + \sqrt{b^2 - 4ac}} \qquad b^2 > 4ac;$$

$$\int \frac{dx}{ax^2 + bx + c} = \frac{2}{\sqrt{4ac - b^2}} \tan^{-1} \frac{2ax + b}{\sqrt{4ac - b^2}} \qquad b^2 < 4ac;$$

$$\int \frac{dx}{ax^2 + bx + c} = -\frac{2}{2ax + b} \qquad b^2 = 4ac.$$

In (1), where the denominator is $x^2 - 5x + 3$, $a = 1$, $b = -5$, $c = 3$; hence $b^2 - 4ac = (-5)^2 - 4(1)(3) = 13 > 0$. Thus the first formula applies:

$$\int \frac{dx}{x^2 - 5x + 3} = \frac{1}{\sqrt{13}} \ln \frac{2x - 5 - \sqrt{13}}{2x - 5 + \sqrt{13}}.$$

In (2), where the denominator is $x^2 - 2x + 3$, $a = 1$, $b = -2$, $c = 3$; hence $b^2 - 4ac = (-2)^2 - 4(1)(3) = -8 < 0$. Thus the second formula applies:

$$\int \frac{dx}{x^2 - 2x + 3} = \frac{2}{\sqrt{8}} \tan^{-1} \frac{2x - 2}{\sqrt{8}}.$$

In (3), where the denominator is $4x^2 - 12x + 9$, $a = 4$, $b = -12$, $c = 9$; hence $b^2 - 4ac = (-12)^2 - 4(4)(9) = 144 - 144 = 0$. Thus the third formula applies:

$$\int \frac{dx}{4x^2 - 12x + 9} = \frac{-2}{8x - 12}.$$

The reader may check these three answers by simply differentiating them. ●

EXAMPLE 3 Find $\qquad\qquad \int \sqrt{\sin^2 \theta + 5} \cos \theta \, d\theta.$

SOLUTION A search through the expressions containing $\sin \theta$ and $\cos \theta$ will find no formula covering this case. However, note that $\cos \theta$ is the derivative of $\sin \theta$. This suggests the substitution

$$u = \sin \theta;$$

hence $\qquad\qquad\qquad du = \cos \theta \, d\theta.$

Then $\qquad\qquad \int \sqrt{\underbrace{\sin^2 \theta + 5}_{u^2 + 5}} \underbrace{\cos \theta \, d\theta}_{du} = \int \sqrt{u^2 + 5} \, du.$

Now look in the integral tables for $\int \sqrt{u^2 + 5} \, du$. If 5 is written as $(\sqrt{5})^2$, the integrand becomes "an expression in $u^2 + (\sqrt{5})^2$," so look among the "Expressions containing $x^2 \pm p^2$" and find (replacing x by u)

$$\int \sqrt{u^2 + p^2} \, du = \tfrac{1}{2}[u\sqrt{u^2 + p^2} + p^2 \ln (u + \sqrt{u^2 + p^2})].$$

In the problem considered, $p = \sqrt{5}$ and $p^2 = 5$. Thus

$$\int \sqrt{u^2 + 5} \, du = \int \sqrt{u^2 + (\sqrt{5})^2} \, du = \tfrac{1}{2}[u\sqrt{u^2 + 5} + 5 \ln (u + \sqrt{u^2 + 5})].$$

To find $\int \sqrt{\sin^2 \theta + 5} \cos \theta \, d\theta$, replace u by $\sin \theta$ in the formula just obtained. ●

EXAMPLE 4   Find

$$\int \frac{x^5 \, dx}{\sqrt{x^3 + 1}}.$$

SOLUTION   This integral is usually not listed in an integral table. Note that $x^5 \, dx = x^3 x^2 \, dx$ and that $x^2 \, dx$ is almost $d(x^3)$. So let $u = x^3$; hence

$$du = 3x^2 \, dx \qquad \text{and} \qquad x^2 \, dx = \frac{du}{3}.$$

Thus

$$\int \frac{x^5 \, dx}{\sqrt{x^3 + 1}} = \int \frac{x^3 x^2 \, dx}{\sqrt{x^3 + 1}}$$

$$= \int \frac{u}{\sqrt{u + 1}} \frac{du}{3}$$

$$= \frac{1}{3} \int \frac{u \, du}{\sqrt{u + 1}},$$

which is listed in the integral tables.

Alternatively, one could have substituted $u = x^3 + 1$. This leads to an easier integral, as the reader may check. Integral tables would not be needed. ●

There are many classes of functions not listed in the integral tables. For instance, there is no formula in the integral tables allowing us to find

$$\int \frac{5x^3 + x - 2}{x^3 - x^2 - x - 2} \, dx.$$

Sections 9.6 and 9.7 show how to integrate quotients of polynomials, the so-called *rational functions*. However, in practice, the integrands are much more likely to be polynomials and fairly simple expressions involving exponentials, logarithms, trigonometric functions, and square roots (especially $\sqrt{a^2 - x^2}$).

Exercises

Integrate with the aid of a table of integrals. For Exercises 1 to 45 the table in Appendix E suffices.

1. $\int (3x + 5)^5 \, dx$

2. $\int \frac{dx}{(3x + 7)^3}$

3. $\int \frac{x \, dx}{(2x - 5)^2}$

4. $\int \frac{dx}{x^2(3x - 2)}$

5. $\int \sqrt{5x - 7} \, dx$

6. $\int x\sqrt{2x + 3} \, dx$

7. $\int \dfrac{dx}{\sqrt{5x + 4}}$

8. $\int \dfrac{\sqrt{5x + 4}}{x}\, dx$

9. $\int \dfrac{dx}{x\sqrt{3x - 2}}$

10. $\int \dfrac{dx}{x\sqrt{3x + 2}}$

11. $\int \dfrac{dx}{9 + x^2}$

12. $\int \dfrac{dx}{x^2 + 5}$

13. $\int \dfrac{dx}{4 - x^2}$

14. $\int \dfrac{dx}{3 - x^2}$

15. $\int \dfrac{dx}{5x^2 + 3}$

16. $\int \dfrac{dx}{5x^2 - 3}$

17. $\int \dfrac{x\, dx}{3x^2 + 1}$

18. $\int \dfrac{dx}{x^2(3x + 1)}$

19. $\int \sqrt{x^2 - 1}\, dx$

20. $\int \sqrt{x^2 - 4}\, dx$

21. $\int \sqrt{x^2 - 3}\, dx$

22. $\int \dfrac{dx}{2x^2 + x + 3}$

23. $\int \dfrac{x\, dx}{2x^2 + x + 3}$

24. $\int \dfrac{dx}{\sqrt{x^2 + 3x + 5}}$

25. $\int \dfrac{dx}{\sqrt{-x^2 + 3x + 5}}$

26. $\int \sqrt{4x - x^2}\, dx$

27. $\int \sqrt{\dfrac{x + 3}{x - 2}}\, dx$

28. $\int \sin^2 3x\, dx$

29. $\int \sin^6 2x\, dx$

30. $\int \dfrac{dx}{1 - \sin 3x}$

31. $\int \cos^3 5x\, dx$

32. $\int \sin 3x \cos 2x\, dx$

33. $\int \sec 5x\, dx$

34. $\int x \sin 3x\, dx$

35. $\int x^2 \cos 3x\, dx$

36. $\int 2^x\, dx$

37. $\int x^2 e^{-x}\, dx$

38. $\int x^3 e^{2x}\, dx$

39. $\int e^{-3x} \sin 5x\, dx$

40. $\int x^2 \ln 3x\, dx$

41. $\int \sin^{-1} 3x\, dx$

42. $\int \tan^{-1} 5x\, dx$

43. $\int \sec^{-1} 4x\, dx$

44. $\int \sqrt{1 + \sin 3x}\, dx$

45. $\int \dfrac{dx}{1 + \sin 3x}$

■

After a substitution, evaluate with the aid of a table of integrals.

46. $\int \dfrac{x^5\, dx}{(5x^3 + 3)^2};$ let $u = x^3$

47. $\int x^2 \sqrt{5x^3 + 2}\, dx$

48. $\int \dfrac{x\, dx}{\sqrt{2x^2 + 3}}$

49. $\int \dfrac{\cos x\, dx}{4 + \sin^2 x}$

50. $\int \dfrac{\sin 2x\, dx}{9 - \cos^2 2x}$

51. $\int \sqrt{(1 + 3x)^2 - 5}\, dx$

52. $\int \dfrac{x^2}{\sqrt{1 - x^6}}\, dx$

53. $\int x\sqrt{x^4 + 9}\, dx$

54. $\int x\sqrt{9 - x^4}\, dx$

9.4

Substitution in the definite integral

Consider the steps involved when computing

$$\int_1^2 (x^3 + 1)^5 3x^2\, dx$$

by the fundamental theorem.
First, find an antiderivative,

$$\int (x^3 + 1)^5 3x^2\, dx.$$

To do so, introduce the substitution

$$u = x^3 + 1, \qquad du = 3x^2 \, dx.$$

Thus
$$\int (x^3 + 1)^5 3x^2 \, dx = \int u^5 \, du = \frac{u^6}{6} = \frac{(x^3 + 1)^6}{6}.$$

According to the fundamental theorem,

$$\int_1^2 (x^3 + 1)^5 3x^2 \, dx = \left. \frac{(x^3 + 1)^6}{6} \right|_1^2 = \frac{(2^3 + 1)^6}{6} - \frac{(1^3 + 1)^6}{6}$$

$$= \frac{9^6}{6} - \frac{2^6}{6}.$$

Note that the same result is obtained if we compute

$$\int_2^9 u^5 \, du,$$

where the ends of the interval of integration $[2, 9]$ are the values of u when x is an end of the interval of integration $[1, 2]$:

$$\int_2^9 u^5 \, du = \left. \frac{u^6}{6} \right|_2^9 = \frac{9^6}{6} - \frac{2^6}{6}.$$

A Shortcut for Evaluating Certain Definite Integrals. This suggests a shortcut for evaluating definite integrals when using a substitution: *When applying a substitution to the integrand, apply the same substitution to the limits of integration.* The proof that this works in general is given after Example 2.

EXAMPLE 1 Find
$$\int_0^{\pi/2} \sin^2 \theta \cos \theta \, d\theta.$$

SOLUTION Introduce the substitution

$$u = \sin \theta, \qquad du = \cos \theta \, d\theta.$$

When $\theta = 0$, u is 0; when θ is $\pi/2$, u is 1. Thus

$$\int_0^{\pi/2} \sin^2 \theta \cos \theta \, d\theta = \int_0^1 u^2 \, du = \left. \frac{u^3}{3} \right|_0^1$$

$$= \frac{1^3}{3} - \frac{0^3}{3}$$

$$= \tfrac{1}{3}.$$

Note that there is no need to express the antiderivative $u^3/3$ in terms of θ. ●

EXAMPLE 2 Find
$$\int_1^5 \sqrt{1 + 3x} \, dx.$$

SOLUTION   Let $u = 1 + 3x$, hence $du = 3\,dx$.   When $x = 1$, $u = 4$; when $x = 5$, $u = 16$.

Note the Change in Limits of   Thus $\displaystyle\int_1^5 \sqrt{1 + 3x}\,dx = \int_4^{16} \sqrt{u}\,\frac{du}{3} = \frac{1}{3}\int_4^{16}\sqrt{u}\,du = \frac{1}{3}\frac{u^{3/2}}{\frac{3}{2}}\Big|_4^{16}$
Integration

$$= \tfrac{2}{9}(16^{3/2} - 4^{3/2})$$

$$= \tfrac{2}{9}(64 - 8)$$

$$= \tfrac{112}{9}. \quad \bullet$$

The following theorem guarantees that the shortcut used in the examples is legitimate.

THEOREM   *Substitution in the definite integral.* Let f be a continuous function on the interval $[a, b]$, $u = h(x)$ be a differentiable function on the same interval, and g be a continuous function such that

$$f(x)\,dx = g(u)\,du;$$

that is, $$f(x) = g(h(x))h'(x).$$

Then $$\int_a^b f(x)\,dx = \int_A^B g(u)\,du,$$

where $A = h(a)$ and $B = h(b)$.

PROOF   Let $G(u)$ be an antiderivative of $g(u)$.   Then, by the substitution theorem in Sec. 9.2, $$G(h(x))$$

is an antiderivative of $f(x)$.   Then

$$\int_a^b f(x)\,dx \underset{\text{FTC}}{=} G(h(x))\Big|_{x=a}^{x=b}$$

$$= G(h(b)) - G(h(a))$$

$$= G(B) - G(A).$$

But $$\int_A^B g(u)\,du = G(B) - G(A)$$

by the fundamental theorem of calculus.   This proves that substitution in the definite integral is valid. $\bullet$

Exercises

Evaluate the definite integrals in Exercises 1 to 15 by the method of this section.

1. $\displaystyle\int_0^1 (e^x + 1)^3 e^x\,dx$

2. $\displaystyle\int_0^1 (x^4 + 1)^5 x^3\,dx$

3. $\displaystyle\int_0^{\pi/4} \sin 2\theta\,d\theta$

4. $\displaystyle\int_0^{\pi/2} \frac{\cos\theta\,d\theta}{\sqrt{1 + \sin\theta}}$

5. $\displaystyle\int_1^2 \frac{x^2\,dx}{2x^3 + 4}$

6. $\displaystyle\int_0^e \frac{\sqrt{\ln x}}{x}\,dx$

7. $\displaystyle\int_{2}^{3} x\sqrt{x^2 - 4}\, dx$

8. $\displaystyle\int_{\pi/6}^{\pi/4} \sin^3 2\theta \cos 2\theta\, d\theta$

9. $\displaystyle\int_{0}^{1} \frac{x^3\, dx}{(x^4 + 1)^2}$

10. $\displaystyle\int_{0}^{13} \frac{dx}{\sqrt[3]{1 + 2x}}$

11. $\displaystyle\int_{0}^{1} \frac{x^2\, dx}{1 + x^6}$

12. $\displaystyle\int_{1}^{2} \frac{e^{1/x}}{x^2}\, dx$

13. $\displaystyle\int_{1}^{e} \frac{\ln^3 x}{x}\, dx$

14. $\displaystyle\int_{0}^{1} \frac{dx}{(x + 1)^3}$

15. $\displaystyle\int_{0}^{2} \sqrt[3]{8 - x^2}\, x\, dx$

■

Exercises 16 to 22 offer a short review of a few topics in earlier chapters.

16. Find $\displaystyle\lim_{x\to 0} \left[\frac{1}{\ln(1 + x)} - \frac{1}{\sin x} \right]$.

17. Evaluate $\int_{-1}^{1} xe^{-x^4}\, dx$.
18. Prove that $D(f/g) = [gD(f) - fD(g)]/g^2$.
19. Show that $\lim_{a\to -1} \int_{1}^{t} x^a\, dx = \int_{1}^{t} (1/x)\, dx$ by evaluating both sides of the equation.
20. Show that the volume of a sphere of radius a is $4\pi a^3/3$.
21. Find the minimum value of the function $y = 3x^4 - 16x^3 + 30x^2 - 24x$.
22. Using the definition of the derivative, show that

$$D(\log_b x) = \frac{1}{x} \log_b e.$$

■ ■

23. Jill says, "$\int_0^\pi \cos^2 \theta\, d\theta$ is obviously positive." Jack claims, "No, it's zero. Just make the substitution $u = \sin \theta$, hence $du = \cos \theta\, d\theta$. Then I get

$$\int_{0}^{\pi} \cos^2 \theta\, d\theta = \int_{0}^{\pi} \cos \theta \cos \theta\, d\theta$$

$$= \int_{0}^{0} \sqrt{1 - u^2}\, du$$

$$= 0.$$

Simple." Who is right? The indentity $\cos^2 \theta = (1 + \cos 2\theta)/2$ permits the evaluation of the integral without substitution.

24. Jill asserts that $\int_{-2}^{1} 2x^2\, dx$ is obviously positive. "After all, the integrand is never negative and $-2 < 1$." "You're wrong again," Jack replies, "It's negative. Here are my computations. Let $u = x^2$; hence $du = 2x\, dx$. Then

$$\int_{-2}^{1} 2x^2\, dx = \int_{-2}^{1} x\, 2x\, dx$$

$$= \int_{4}^{1} \sqrt{u}\, du$$

$$= -\int_{1}^{4} \sqrt{u}\, du,$$

which is obviously negative." Who is right?

25. In the substitution theorems of Secs. 9.2 and 9.4 the functions f and g do not play interchangeable roles, even though in the equation

$$f(x)\, dx = g(u)\, du$$

the two sides look similar. In each of the two theorems, the assertion is, "If we can find $\int g(u)\, du$, then we can find $\int f(x)\, dx$." There is no mention of the reverse direction: "If we can integrate f, can we therefore integrate g?"

The answer is "yes" if the function $u = h(x)$, in addition to being differentiable, is one-to-one, and its derivative $h'(x)$ is not 0 in the interval considered. (Hence h is either an increasing or a decreasing function.)

To show this, let h^* denote the inverse of h. Then, starting from the equation

$$f(x) = g(h(x))h'(x),$$

prove that $\quad g(u) = f(h^*(u))(h^*)'(u).$

This is exactly the condition needed to show that from an antiderivative of f an antiderivative of g can be obtained.

9.5

Integration by parts

The chain rule is the basis for integration by substitution discussed in Sec. 9.2. The formula for the derivative of a product,

$$(uv)' = uv' + vu',$$

is the basis of a method of more restricted application known as *integration by parts*. If u and v are differentiable functions, we obtain by this method an antiderivative of uv' from an antiderivative of vu', and conversely.

THEOREM *Integration by parts.* If u and v are differentiable functions and $\int vu'\,dx$ is an antiderivative of vu', then

$$uv - \int vu'\,dx$$

is an antiderivative of uv'. In symbols,

$$\int uv'\,dx = uv - \int vu'\,dx$$

This Differential Form Is the Most Useful.

and, in the notation of differentials,

$$\int u\,dv = uv - \int v\,du.$$

PROOF Differentiate

$$uv - \int vu'\,dx$$

to see if the result is uv'. We have

$$\left(uv - \int vu'\,dx\right)' = (uv)' - \left(\int vu'\,dx\right)' \qquad \text{derivative of difference}$$

$$= (uv)' - vu' \qquad \text{definition of } \int vu'\,dx$$

$$= uv' + vu' - vu' \qquad \text{derivative of product.}$$

$$= uv'.$$

Thus $uv - \int vu'\,dx$ is an antiderivative of uv', as was to be shown. ●

EXAMPLE 1 Find

$$\int xe^{-x}\,dx.$$

SOLUTION To use the formula

$$\int u\,dv = uv - \int v\,du,$$

it is necessary to write

$$\int xe^{-x}\,dx$$

in the form

$$\int u\,dv.$$

Integration by Parts Frequently Involves Hope and Faith.

(The resulting $\int v\,du$, it is hoped, is easier to find than the original integral $\int u\,dv$.)

The integrand is so simple that there is not much choice. Try

$$u = x \qquad \text{and} \qquad dv = e^{-x}\,dx;$$

that is, break up the integrand this way:

$$\int \underset{u}{\underline{x}}\,\underset{dv}{\underline{e^{-x}\,dx}}.$$

Then find du and v.   Since $u = x$, it follows that $du = dx$.   Since $dv = e^{-x} dx$, we choose $v = -e^{-x}$.   (Of course v could be $-e^{-x} + C$ for any constant C, but choose the simplest v whose derivative is e^{-x}.)   Applying integration by parts yields

$$\int \underbrace{x}_{u} \underbrace{e^{-x} dx}_{dv} = \underbrace{x}_{u} \underbrace{(-e^{-x})}_{v} - \int \underbrace{(-e^{-x})}_{v} \underbrace{dx}_{du}.$$

Is $\int v \, du$ easier than the original integral, $\int u \, dv$?   Yes;

$$\int -e^{-x} dx = e^{-x}.$$

Hence

$$\int xe^{-x} dx = -xe^{-x} - e^{-x}$$

$$= e^{-x}(-x - 1).$$

The reader may check this by differentiation. ●

EXAMPLE 2   Find

$$\int x^2 e^{-x} dx.$$

SOLUTION   There are many ways to write the integral in the form $\int u \, dv$:

$$\int \underbrace{x^2}_{u} \underbrace{e^{-x} dx}_{dv}, \quad \int \underbrace{x}_{u} \underbrace{xe^{-x} dx}_{dv}, \quad \int \underbrace{xe^{-x}}_{u} \underbrace{x \, dx}_{dv}, \quad \int \underbrace{e^{-x}}_{u} \underbrace{x^2 dx}_{dv}.$$

The last one is to be avoided.   For in this case

$$v = \frac{x^3}{3} \quad \text{and} \quad du = -e^{-x} dx,$$

and the new integral $\displaystyle \int v \, du = \int \frac{x^3}{3}(-e^{-x}) \, dx$

is *harder* than the original; essentially, x^2 has been replaced by x^3.

The third way, $\int xe^{-x}x \, dx$, as the reader may check, runs into the same difficulty.   The first two choices actually *reduce*, rather than increase, the difficulty.   The first choice leads to the following arithmetic, since $du = 2x \, dx$ and $v = -e^{-x}$:

$$\int \underbrace{x^2}_{u} \underbrace{e^{-x} dx}_{dv} = \underbrace{x^2}_{u} \underbrace{(-e^{-x})}_{v} - \int \underbrace{(-e^{-x})}_{v} \underbrace{2x \, dx}_{du}$$

$$= -x^2 e^{-x} + 2 \int e^{-x} x \, dx.$$

The new integral is simpler than the first; x^2 has been replaced by x.   In fact, Example 1 already evaluated this integral,

$$\int xe^{-x} dx = e^{-x}(-x - 1).$$

Thus
$$\int x^2 e^{-x}\,dx = -x^2 e^{-x} + 2\int e^{-x}x\,dx$$

$$= -x^2 e^{-x} + 2[e^{-x}(-x-1)]$$
$$= -x^2 e^{-x} - 2xe^{-x} - 2e^{-x}$$
$$= -(x^2 + 2x + 2)e^{-x}. \quad \bullet$$

The method of Example 2 suggests how the formula in the integral tables for $\int x^n e^{ax}\,dx$ was found. It was obtained by an integration by parts:

$$\underbrace{\int \underbrace{x^n}_{u}\ \underbrace{e^{ax}\,dx}_{dv} = \underbrace{x^n}_{u}\ \underbrace{\frac{e^{ax}}{a}}_{v} - \int \underbrace{\frac{e^{ax}}{a}}_{v}\ \underbrace{nx^{n-1}\,dx}_{du};}$$

$$du = nx^{n-1}\,dx;$$

$$v = \frac{e^{ax}}{a}.$$

In this manner
$$\int x^n e^{ax}\,dx$$

is expressed in terms of
$$\int x^{n-1} e^{ax}\,dx.$$

The recursion lowers the exponent of x by 1 at each step. By a similar procedure, the recursion formula for

$$\int x^n \sin ax\,dx$$

can be obtained. (See Example 4 for another illustration.)

Moreover, Example 2 shows how integration by parts may be used to lower the degree of a polynomial that appears in an integrand.

The next example shows how to integrate any inverse trigonometric function. The resulting formula is in the integral tables.

EXAMPLE 3 Find $\int \tan^{-1} x\,dx$.

SOLUTION Recall that the derivative of $\tan^{-1} x$ is $1/(1+x^2)$, a much simpler function than $\tan^{-1} x$. This suggests the following approach:

$$\int \underbrace{\tan^{-1} x}_{u}\ \underbrace{dx}_{dv} = \underbrace{(\tan^{-1} x)}_{u}\underbrace{x}_{v} - \int \underbrace{x}_{v}\ \underbrace{\frac{1}{1+x^2}\,dx}_{du}$$

$$\left(du = \frac{dx}{1+x^2},\qquad v = x\right)$$

$$= x\tan^{-1} x - \int \frac{x}{1+x^2}\,dx.$$

It is easy to compute

$$\int \frac{x}{1+x^2}\, dx,$$

since the numerator is a constant times the derivative of the denominator:

$$\int \frac{x\, dx}{1+x^2} = \frac{1}{2}\int \frac{2x}{1+x^2}\, dx = \frac{1}{2}\ln(1+x^2).$$

Hence

$$\int \tan^{-1} x\, dx = x\tan^{-1} x - \tfrac{1}{2}\ln(1+x^2).$$

Observe that, in general,

$$\int f(x)\, dx = xf(x) - \int f'(x)x\, dx. \quad \bullet$$

EXAMPLE 4 Many formulas in a table of integrals express the integral of a function that involves the nth power of some expression in terms of the integral of a function that involves the $(n-1)$st or lower power of the same expression. These are reduction formulas. Usually they are obtained by an integration by parts. For instance, derive the formula

$$\int \sin^n x\, dx = -\frac{\sin^{n-1} x \cos x}{n} + \frac{n-1}{n}\int \sin^{n-2} x\, dx \qquad n \geq 2.$$

SOLUTION First write $\int \sin^n x\, dx$ as $\int \sin^{n-1} x \sin x\, dx$. Then let $u = \sin^{n-1} x$ and $dv = \sin x\, dx$. Thus

$$du = (n-1)\sin^{n-2} x \cos x\, dx \qquad \text{and} \qquad v = -\cos x.$$

Integration by parts yields

$$\int \underbrace{\sin^{n-1} x}_{u}\ \underbrace{\sin x\, dx}_{dv}$$

$$= \underbrace{(\sin^{n-1} x)}_{u}\underbrace{(-\cos x)}_{v} - \int \underbrace{(-\cos x)}_{v}\underbrace{(n-1)\sin^{n-2} x \cos x\, dx}_{du}.$$

But the integral on the right of the above equation is equal to

$$\int (1-n)\cos^2 x \sin^{n-2} x\, dx = (1-n)\int (1-\sin^2 x)\sin^{n-2} x\, dx$$

$$= (1-n)\int \sin^{n-2} x\, dx - (1-n)\int \sin^n x\, dx.$$

Thus

$$\int \sin^n x\, dx = -\sin^{n-1} x \cos x + (n-1)\int \sin^{n-2} x\, dx - (n-1)\int \sin^n x\, dx.$$

Rather than being dismayed by the reappearance of $\int \sin^n x\, dx$, collect like terms:

$$n\int \sin^n x\, dx = -\sin^{n-1} x \cos x + (n-1)\int \sin^{n-2} x\, dx,$$

from which the quoted formula follows.

Note that the recursion lowers the power of sin x by 2 at each step, $\sin^n x$ being replaced by $\sin^{n-2} x$. Contrast this with the recursion following Example 2, which replaces n by $n-1$. ●

EXAMPLE 5   Find $\int e^x \cos x \, dx$.

SOLUTION   Proceed as follows:

$$\int \underbrace{e^x}_{u} \underbrace{\cos x \, dx}_{dv} = \underbrace{e^x}_{u} \underbrace{\sin x}_{v} - \int \underbrace{\sin x}_{v} \underbrace{e^x \, dx}_{du};$$

$$du = e^x \, dx;$$

$$v = \sin x.$$

It may seem that nothing useful has been accomplished; cos x is replaced by sin x. But watch closely, as the new integral is treated to an integration by parts. Capital letters U and V, instead of u and v, are used to distinguish this computation from the preceding one.

$$\int \underbrace{e^x}_{U} \underbrace{\sin x \, dx}_{dV} = \underbrace{e^x}_{U} \underbrace{(-\cos x)}_{V} - \int \underbrace{(-\cos x)}_{V} \underbrace{e^x \, dx}_{dU} \quad (dU = e^x \, dx \quad V = -\cos x)$$

$$= -e^x \cos x + \int e^x \cos x \, dx.$$

Combining the two yields

$$\int e^x \cos x \, dx = e^x \sin x - \left(-e^x \cos x + \int e^x \cos x \, dx \right)$$

$$= e^x(\sin x + \cos x) - \int e^x \cos x \, dx.$$

This provides an equation for the unknown integral:

$$2 \int e^x \cos x \, dx = e^x(\sin x + \cos x).$$

Hence

$$\int e^x \cos x \, dx = \tfrac{1}{2}e^x(\sin x + \cos x). \quad ●$$

Since any table of integrals contains formulas covering these five examples, there is no need to memorize them.

Exercises

Using integration by parts, compute the integrals in Exercises 1 to 16.   Check by differentiation.

1. $\int x e^{2x} \, dx$

2. $\int x^2 e^{2x} \, dx$

3. $\int x^3 e^{2x} \, dx$

4. $\int \sin^{-1} x \, dx$

5. $\int x \sin x \, dx$

6. $\int x^2 \sin x \, dx$

7. $\int \ln (4 + x^2) \, dx$

8. $\int x \ln x \, dx$

9. $\int x(\ln x)^2 \, dx$

10. $\int x^5 e^{x^2} \, dx$

11. $\int e^x \sin x \, dx$ **12.** $\int e^{2x} \cos 3x \, dx$

13. $\int x \sec^2 3x \, dx$ **14.** $\int \ln(4+x) \, dx$

15. $\int e^{ax} \sin bx \, dx$,   where a and b are constants

16. $\int x^2 \cos x \, dx$

17. Use the recursion in Example 4 to find

(a) $\int \sin^2 x \, dx$ (b) $\int \sin^4 x \, dx$

(c) $\int \sin^6 x \, dx$.

18. Use the recursion in Example 4 to find

(a) $\int \sin^3 x \, dx$ (b) $\int \sin^5 x \, dx$.

■

19. Show that the third choice in Example 2 leads to a harder integral than the original one.

20. Find $\int x^3 \sin x \, dx$.   (See Exercise 6.)

21. By differentiating, check the answer in (a) Example 3, (b) Example 5.

22. Obtain this recursion formula, which is usually to be found in a table of integrals:

$$\int \sin^n ax \, dx =$$

$$-\frac{\sin^{n-1} ax \cos ax}{na} + \frac{n-1}{n} \int \sin^{n-2} ax \, dx.$$

23. Obtain this recursion formula ($m, n > 0$):

$$\int \sin^m x \cos^n x \, dx$$

$$= -\frac{\sin^{m-1} x \cos^{n+1} x}{m+n} + \frac{m-1}{m+n} \int \sin^{m-2} x \cos^n x \, dx.$$

24. Find $\int \ln(x+1) \, dx$ using

(a) $u = \ln(x+1)$, $dv = dx$, $v = x$;

(b) $u = \ln(x+1)$, $dv = dx$, $v = x+1$.

(c) Which is easier?

■ ■

Exercises 25 and 26 are related.

25. In a certain race, a car starts from rest and ends at rest, having traveled 1 mile in 1 minute. Let $v(t)$ be its velocity at time t, and $a(t)$ be its acceleration at time t. Show that

(a) $\int_0^1 v(t) \, dt = 1$;

(b) $\int_0^1 a(t) \, dt = 0$;

(c) $\int_0^1 ta(t) \, dt = -1$.

26. (Continuation of Exercise 25.)

(a) Show that at some time t we have $|a(t)| > 4$.

(b) Show graphically [drawing $v(t)$ as a function of time] that a race can be driven as in Exercise 25, but with $|a(t)| \leq 4.1$ for all t.

9.6

How to compute $\displaystyle\int \frac{dx}{(ax+b)^n}$, $\displaystyle\int \frac{dx}{(ax^2+bx+c)^n}$, **and** $\displaystyle\int \frac{x \, dx}{(ax^2+bx+c)^n}$

These particular integrals play an important role in Secs. 9.7 and 9.8. A table of integrals disposes of all of them. If instructor and student are so inclined, they may go directly to the exercises and solve them all with the aid of the integral table.

EXAMPLE 1   Find

$$\int \frac{1}{3x+4} \, dx.$$

SOLUTION   The numerator is a constant times the derivative of the denominator. This suggests the following steps:

$$\int \frac{1}{3x+4} \, dx = \frac{1}{3} \int \frac{3 \, dx}{3x+4} = \frac{1}{3} \ln|3x+4|. \; \bullet$$

EXAMPLE 2   Find

$$\int \frac{dx}{(3x+4)^7}.$$

SOLUTION   Since $3x+4$ has such a simple derivative, namely 3, make the substitution $u = 3x+4$; hence $du = 3\,dx$, and $du/3 = dx$.   Then

$$\int \frac{dx}{(3x+4)^7} = \int \frac{du/3}{u^7} = \frac{1}{3}\int u^{-7}\,du = \frac{1}{3}\frac{u^{-6}}{-6} = -\frac{u^{-6}}{18} = -\frac{1}{18(3x+4)^6}.\quad\bullet$$

Next consider integrals of the form $\int dx/(ax^2+bx+c)^n$, beginning with two examples where $b=0$ and $n=1$.

EXAMPLE 3   Find
$$\int \frac{dx}{4x^2+1}.$$

SOLUTION   This resembles
$$\int \frac{dx}{x^2+1} = \tan^{-1} x.$$

For this reason, make the substitution
$$u^2 = 4x^2,$$
that is,
$$u = 2x;$$
hence
$$du = 2\,dx \quad \text{and} \quad \frac{du}{2} = dx.$$

Then
$$\int \frac{dx}{4x^2+1} = \int \frac{1}{u^2+1}\frac{du}{2} = \frac{1}{2}\int \frac{du}{u^2+1} = \frac{1}{2}\tan^{-1} u = \frac{1}{2}\tan^{-1} 2x.\quad\bullet$$

EXAMPLE 4   Find
$$\int \frac{dx}{4x^2+9}.$$

SOLUTION   Again the motivation is provided by the fact that
$$\int \frac{dx}{x^2+1} = \tan^{-1} x.$$

This time choose u such that
$$9u^2 = 4x^2.$$

(This substitution is suggested by the equations
$$\frac{1}{4x^2+9} = \frac{1}{9u^2+9} = \frac{1}{9}\frac{1}{u^2+1}.)$$

In other words, choose u such that
$$3u = 2x;$$
hence
$$3\,du = 2\,dx \quad \text{and} \quad \tfrac{3}{2}\,du = dx.$$

Thus
$$\int \frac{dx}{4x^2+9} = \int \frac{1}{9u^2+9}\frac{3}{2}\,du = \frac{3}{18}\int \frac{du}{u^2+1} = \frac{1}{6}\tan^{-1} u = \frac{1}{6}\tan^{-1}\frac{2x}{3}.$$

(Note that only at the end is it necessary to solve for u; $u = 2x/3$.) $\bullet$

Completing the Square   Before presenting the next example we shall review a bit of algebra called *completing the square.*

The expression

$$x^2 + 6x$$

is the first two terms of the square of $x + 3$,

$$(x + 3)^2 = x^2 + 6x + 9.$$

Adding 9 to $x^2 + 6x$ completes the square of $x + 3$.

What must we add to $x^2 + 5x$ to complete the square? If $x^2 + 5x$ is the first two terms of

$$(x + k)^2 = x^2 + 2kx + k^2,$$

then $2k$ must be 5; that is, $k = \frac{5}{2}$,

and $k^2 = \left(\frac{5}{2}\right)^2.$

It is necessary to add $\left(\frac{5}{2}\right)^2$ to $x^2 + 5x$ to complete the square:

$$x^2 + 5x + \left(\frac{5}{2}\right)^2 = \left(x + \frac{5}{2}\right)^2.$$

The same reasoning establishes this rule:

To complete the square of $x^2 + bx$, add $(b/2)^2$, the square of half the coefficient of x.

EXAMPLE 5   Find

$$\int \frac{dx}{4x^2 + 8x + 13}$$

SOLUTION   First complete the square in the denominator $4x^2 + 8x + 13$, as follows:

$$4x^2 + 8x + 13 = 4(x^2 + 2x) + 13$$
$$= 4\left[x^2 + 2x + \left(\frac{2}{2}\right)^2\right] + 13 - 4\left(\frac{2}{2}\right)^2$$
$$= 4(x + 1)^2 + 9.$$

The integral now reads $\displaystyle\int \frac{dx}{4(x + 1)^2 + 9}$,

which resembles Example 4.

Make the substitution $u = x + 1$;

hence $du = dx.$

Then $$\int \frac{dx}{4(x + 1)^2 + 9} = \int \frac{du}{4u^2 + 9}.$$

By Example 4, $$\int \frac{du}{4u^2 + 9} = \frac{1}{6} \tan^{-1} \frac{2u}{3},$$

and therefore $$\int \frac{dx}{4(x + 1)^2 + 9} = \frac{1}{6} \tan^{-1} \frac{2(x + 1)}{3},$$

or $$\int \frac{dx}{4x^2 + 8x + 13} = \frac{1}{6} \tan^{-1} \frac{2(x + 1)}{3}. \quad \bullet$$

EXAMPLE 6   Find
$$\int \frac{x\,dx}{4x^2 + 8x + 13}.$$

SOLUTION   If the numerator were $8x + 8$, it would be the derivative of the denominator. The problem would then be covered by the formula

$$\int \frac{f'}{f}\,dx = \ln |f|.$$

This prompts the following maneuver:

$$\int \frac{x\,dx}{4x^2 + 8x + 13} = \frac{1}{8}\int \frac{8x\,dx}{4x^2 + 8x + 13}$$

$$= \frac{1}{8}\left(\int \frac{8x + 8}{4x^2 + 8x + 13}\,dx - \int \frac{8}{4x^2 + 8x + 13}\,dx \right)$$

($4x^2 + 8x + 13$ *Is Always Positive.*)
$$= \frac{1}{8}\left(\ln (4x^2 + 8x + 13) - \frac{8}{6}\tan^{-1}\frac{2(x + 1)}{3} \right).$$

Note that the result of Example 5 was used. ●

The next example should be contrasted with Example 4.

EXAMPLE 7   Find
$$\int \frac{dx}{4x^2 - 9}.$$

SOLUTION   The algebraic identity

$$\frac{1}{a^2 - b^2} = \frac{1}{2b}\left(\frac{1}{a - b} - \frac{1}{a + b} \right)$$

makes the problem easy to solve.   (Pause to verify that it is correct.)   In particular, when

$$a = 2x \quad \text{and} \quad b = 3,$$

it yields the identity

$$\frac{1}{(2x)^2 - 3^2} = \frac{1}{2 \cdot 3}\left(\frac{1}{2x - 3} - \frac{1}{2x + 3} \right).$$

Thus
$$\int \frac{dx}{4x^2 - 9} = \frac{1}{6}\int \frac{dx}{2x - 3} - \frac{1}{6}\int \frac{dx}{2x + 3}.$$

The original problem has been replaced by two integrals of the type discussed in Example 1.   Consequently,

$$\int \frac{dx}{4x^2 - 9} = \frac{1}{12}\ln |2x - 3| - \frac{1}{12}\ln |2x + 3|$$

$$= \frac{1}{12}\ln \left| \frac{2x - 3}{2x + 3} \right|. \ ●$$

Integrals of the form

$$\int \frac{dx}{(ax^2 + bx + c)^n} \quad \text{and} \quad \int \frac{x\,dx}{(ax^2 + bx + c)^n}$$

with $n > 1$ seldom occur.   The recursion formulas reducing n are developed in Exercises 30 and 31.

Happily, these formulas are in any standard table of integrals.   In fact all six examples are easily dealt with by use of a table of integrals, though it would probably be faster not to use a table in the first three or four examples.

Exercises

Compute the integrals in Exercises 1 to 15.

1. $\int \dfrac{dx}{3x - 4}$

2. $\int \dfrac{2\,dx}{3x + 6}$

3. $\int \dfrac{5\,dx}{(2x + 7)^2}$

4. $\int \dfrac{dx}{(4x + 1)^3}$

5. $\int \dfrac{dx}{x^2 + 9}$

6. $\int \dfrac{x\,dx}{x^2 + 9}$

7. $\int \dfrac{2x + 3}{x^2 + 9}\,dx$

8. $\int \dfrac{dx}{5x^2 + 1}$

9. $\int \dfrac{dx}{x^2 + 3}$

10. $\int \dfrac{dx}{2x^2 + 3}$

11. $\int \dfrac{x\,dx}{2x^2 + 3}$

12. $\int \dfrac{dx}{x^2 - 1}$

13. $\int \dfrac{dx}{x^2 - 4}$

14. $\int \dfrac{dx}{x^2 - 3}$

15. $\int \dfrac{dx}{2x^2 - 3}$

16. Complete the square in each case.
    (a) $x^2 + 6x$ (b) $x^2 + 7x$ (c) $x^2 - x$
17. Express in the form $k(x + c)^2 + m$ for suitable constants k, c, and m.
    (a) $2x^2 + 8x$ (b) $5x^2 + 10x$ (c) $3x^2 - 7x$

Compute the integrals in Exercises 18 to 26.

18. $\int \dfrac{dx}{9x^2 + 18x - 5}$

19. $\int \dfrac{dx}{4x^2 + 4x + 1}$

20. $\int \dfrac{x\,dx}{4x^2 + 4x - 1}$

21. $\int \dfrac{dx}{4x^2 + 4x + 3}$

22. $\int \dfrac{x\,dx}{4x^2 + 4x + 3}$

23. $\int \dfrac{x - 3}{2x^2 + 5}\,dx$

24. $\int \dfrac{dx}{x^2 + 4x + 7}$

25. $\int \dfrac{dx}{2x^2 + 4x + 7}$

26. $\int \dfrac{x\,dx}{2x^2 + 4x + 7}$

■

27. (a) Verify that

$$\int \frac{dx}{ax^2 + bx + c} = \frac{1}{\sqrt{b^2 - 4ac}} \ln \left| \frac{2ax + b - \sqrt{b^2 - 4ac}}{2ax + b + \sqrt{b^2 - 4ac}} \right|$$

$$b^2 > 4ac.$$

(b) Obtain the formula in (a) by the methods of this section.

28. (a) Verify that

$$\int \frac{dx}{ax^2 + bx + c} = \frac{2}{\sqrt{4ac - b^2}} \tan^{-1} \frac{2ax + b}{\sqrt{4ac - b^2}}$$

$$b^2 < 4ac.$$

(b) Obtain the formula in (a) by the methods of this section.

29. (a) Verify that

$$\int \frac{dx}{ax^2 + bx + c} = -\frac{2}{2ax + b} \qquad b^2 = 4ac.$$

(b) Obtain the formula in (a) by the methods of this section.

■ ■

30. Verify that the following recursion formula is correct by differentiating the right side of the equation.

$$\int \frac{dx}{(ax^2 + bx + c)^{n+1}} = \frac{2ax + b}{n(4ac - b^2)(ax^2 + bx + c)^n}$$

$$+ \frac{2(2n - 1)a}{n(4ac - b^2)} \int \frac{dx}{(ax^2 + bx + c)^n}.$$

31. Verify that the following recursion formula is correct by differentiating the right side of the equation.

$$\int \frac{x\, dx}{(ax^2 + bx + c)^{n+1}} = \frac{-(2c + bx)}{n(4ac - b^2)(ax^2 + bx + c)^n}$$

$$- \frac{b(2n - 1)}{n(4ac - b^2)} \int \frac{dx}{(ax^2 + bx + c)^n}.$$

32. Use the identity in Exercise 30 and the result of Example 3 to find

$$\int \frac{dx}{(4x^2 + 1)^2}.$$

33. Use the identity in Exercise 31 and the result of Example 5 to find

$$\int \frac{x\, dx}{(4x^2 + 8x + 13)^2}.$$

9.7

How to integrate rational functions: partial fractions

We shall not find

$$\int \frac{x^4 + x^3 - 3x + 5}{x^3 + 2x^2 + 2x + 1}\, dx$$

in any table of integrals. The integrand is a *rational function*—a quotient of two polynomials. This section, which is purely algebraic, shows how to integrate any rational function. The method depends on this result from advanced algebra: Every rational function can be expressed as a sum of a polynomial (which may be 0) and constant multiples of the three types of functions met in Sec. 9.6:

$$\frac{1}{(ax + b)^n}, \quad \frac{1}{(ax^2 + bx + c)^n}, \quad \text{and} \quad \frac{x}{(ax^2 + bx + c)^n}.$$

Since any polynomial and each of these three types of rational functions can be integrated, any rational function can be integrated. The only new question of interest is, "What is the method for expressing a rational function as a sum of these four types of simpler functions?" The resulting expression is called the *partial-fraction* representation of the rational function. The technique of partial fractions is outlined below. (Appendix D discusses it in more detail.)

To express A/B, where A and B are polynomials, as the sum of partial fractions, follow these steps:

1. *Make Degree of Numerator Less than Degree of Denominator.*

Step 1 If the degree of A is *not less* than the degree of B, divide B into A to obtain a quotient and a remainder: $A = QB + R$, where the degree of R is less than the degree of B or else $R = 0$. Then

$$\frac{A}{B} = Q + \frac{R}{B}.$$

Apply the remaining steps to R/B.

EXAMPLE I If

$$\frac{A}{B} = \frac{3x^3 + x}{x^2 + 3x + 5},$$

carry out step 1.

SOLUTION   Since the degree of the numerator is *not* less than the degree of the denominator, carry out a long division:

$$
\begin{array}{r}
3x - 9 \qquad \text{quotient} \\
x^2 + 3x + 5 \overline{\smash{\big)}\, 3x^3 + 0x^2 + x + 0} \\
\underline{3x^3 + 9x^2 + 15x} \\
-9x^2 - 14x + 0 \\
\underline{-9x^2 - 27x - 45} \\
13x + 45 \qquad \text{remainder.}
\end{array}
$$

Thus,
$$
\frac{3x^3 + x}{x^2 + 3x + 5} = 3x - 9 + \frac{13x + 45}{x^2 + 3x + 5}.
$$

(To check, just multiply both sides by $x^2 + 3x + 5$.)   ●

Similarly, in the case
$$
\frac{3x^2 + x}{x^2 + 3x + 5},
$$

a division would be carried out first.

Step 2   If the degree of A is *less* than the degree of B, then

2(a) *Factor Denominator.*

(a) Express B as the product of polynomials of degree 1 or 2, where the second-degree factors are *irreducible*. (Advanced mathematics guarantees that this is possible. A second-degree polynomial with real coefficients is *irreducible* if it is not the product of polynomials of degree 1 with real coefficients.) To find these factors of degrees 1 and 2, except in simple cases, may be quite difficult.

2(b) *List Summands of Form*

$$\frac{k_i}{(ax + b)^2}.$$

(b) If $ax + b$ appears exactly n times in the factorization of B, form the sum

$$
\frac{k_1}{ax + b} + \frac{k_2}{(ax + b)^2} + \cdots + \frac{k_n}{(ax + b)^n},
$$

where the constants $k_1, k_2, \ldots, k_n$ are to be determined later.

2(c) *List Summands of Form*

$$\frac{c_i x + d_i}{(ax^2 + bx + c)^i}$$

(c)   If $ax^2 + bx + c$ appears exactly m times in the factorization of B, then form the sum

$$
\frac{c_1 x + d_1}{ax^2 + bx + c} + \frac{c_2 x + d_2}{(ax^2 + bx + c)^2} + \cdots + \frac{c_m x + d_m}{(ax^2 + bx + c)^m},
$$

where the constants $c_1, c_2, \ldots, c_m$ and $d_1, d_2, \ldots; d_m$ are to be determined later.

2(d) *Find Constants*

$$c_i, d_i, k_i$$

(d) Determine the appropriate k's, c's, and d's defined in (b) and (c), such that A/B is equal to the sum of all the terms formed in (b) and (c) for all factors of B defined in (a).

Remark: The rational function

$$
\frac{c_i x + d_i}{(ax^2 + bx + c)^i}
$$

equals
$$\frac{c_i x}{(ax^2 + bx + c)^i} + \frac{d_i}{(ax^2 + bx + c)^i},$$

the sum of two functions that can be integrated by the methods of the preceding section. Combining the two latter quotients into one, as in (c), saves space.

Remark: If the polynomials $x + 2$ and $2x + 4 \, [= 2(x + 2)]$ both appear in the factorization, express one of them as a constant times the other. In step 2(a) no irreducible factor should simply be a constant times another irreducible factor.

EXAMPLE 2   Indicate the form of step 2 in the case in which
$$\frac{A}{B} = \frac{2x^2 + 3x + 3}{(x + 1)^3}.$$

SOLUTION   The denominator, if multiplied out, is a polynomial of degree larger than the degree of the numerator. In this case the denominator has only one first-degree factor, $x + 1$ (repeated three times), and no quadratic factors. Only step 2(b) applies, so that we write
$$\frac{2x^2 + 3x + 3}{(x + 1)^3} = \frac{k_1}{x + 1} + \frac{k_2}{(x + 1)^2} + \frac{k_3}{(x + 1)^3}.$$

The constants k_1, k_2, and k_3 can be found by the method illustrated in Examples 4 to 6, which follow. ●

EXAMPLE 3   What does the partial fraction decomposition of
$$\frac{A}{B} = \frac{2x^3 - 6x^2 + 2}{(x + 1)(x^2 + x + 1)^2}$$
look like?

SOLUTION   The denominator has degree 5 (if multiplied out), while the numerator has degree 3. Thus step 1 does not apply. The irreducible factors of B are $x + 1$, which appears only once, and $x^2 + x + 1$, which appears to the second power. Both parts (c) and (d) of step 2 apply, so that we write
$$\frac{2x^3 - 6x^2 + 2}{(x + 1)(x^2 + x + 1)^2} = \frac{k_1}{x + 1} + \frac{c_1 x + d_1}{x^2 + x + 1} + \frac{c_2 x + d_2}{(x^2 + x + 1)^2}. \quad ●$$

The next example shows how to find the promised constants in a simpler case, where there are fewer constants. Incidentally, these constants are always unique.

EXAMPLE 4   Express
$$\frac{4x - 7}{x^2 - 3x + 2}$$
as the sum of partial fractions.

SOLUTION   Beginning at step 2(a), factor $x^2 - 3x + 2$ as $(x - 1)(x - 2)$.   The denominator B in this case has only first-degree factors; hence step 2(c) will not apply. Since both $x - 1$ and $x - 2$ appear only once in the factorization, we have $n = 1$ in each case for step 2(b).

According to step 2(d) constants k and l exist such that

$$\frac{4x - 7}{x^2 - 3x + 2} = \frac{k}{x - 1} + \frac{l}{x - 2}. \tag{1}$$

To find k and l, multiply both sides of (1) by $x^2 - 3x + 2$, obtaining

$$4x - 7 = k(x - 2) + l(x - 1). \tag{2}$$

Equation (2) holds for all x, since it is an algebraic identity.   Thus it holds when x is replaced by any specific number.   To find two equations for the two unknowns k and l, we replace the x in (2) by two numbers.   Since $x - 1$ vanishes for x equal to 1, and since $x - 2$ vanishes for x equal to 2, for convenience replace x by 1 and by 2.   This gives

$$4(1) - 7 = k(1 - 2) + l(0) \qquad \text{setting } x = 1 \text{ in (2)};$$
$$4(2) - 7 = k(0) + l(2 - 1) \qquad \text{setting } x = 2 \text{ in (2)}.$$

These equations simplify to $-3 = -k$

and $1 = l.$

Hence $k = 3$ and $l = 1$, and

$$\frac{4x - 7}{x^2 - 3x + 2} = \frac{3}{x - 1} + \frac{1}{x - 2}. \ \bullet$$

EXAMPLE 5   Express

$$\frac{x^2 + 7x + 1}{(x + 2)^2(2x + 1)}$$

as the sum of partial fractions.

SOLUTION   The degree of the numerator is less than the degree of the denominator (which is 3).   Hence step 1 does not apply.   Since the denominator is already factored (a common occurrence in practice), step 2(a) is done.   We shall do parts (b) and (d) of step 2 simultaneously; step 2(c) does not apply, since the denominator B has no second-degree irreducible factors.

Since $x + 2$ appears twice in the factorization and $2x + 1$ once,

$$\frac{x^2 + 7x + 1}{(x + 2)^2(2x + 1)} = \frac{k_1}{x + 2} + \frac{k_2}{(x + 2)^2} + \frac{l}{2x + 1}. \tag{3}$$

To find the constants k_1, k_2, l, remove the denominators in (3) by multiplying both sides by $(x + 2)^2(2x + 1)$:

$$x^2 + 7x + 1 = k_1(x + 2)(2x + 1) + k_2(2x + 1) + l(x + 2)^2. \tag{4}$$

Three equations are needed to find the three unknowns k_1, k_2, and l. To obtain them, replace x in (4) by three different numbers in turn. Since $x + 2 = 0$ when $x = -2$, and $2x + 1 = 0$ when $x = -\frac{1}{2}$, replace x by -2 and then by $-\frac{1}{2}$. To obtain a third equation, use $x = 0$. Thus

$$-9 = -3k_2 \qquad \text{setting } x = -2 \text{ in (4)};$$

$$-\tfrac{9}{4} = \left(\tfrac{9}{4}\right)l \qquad \text{setting } x = -\tfrac{1}{2} \text{ in (4)};$$

$$1 = 2k_1 + k_2 + 4l \qquad \text{setting } x = 0 \text{ in (4)}.$$

Thus $k_2 = 3$, $l = -1$, and finally $k_1 = 1$. Replacing k_1, k_2, and l in (3) yields

$$\frac{x^2 + 7x + 1}{(x + 2)^2(2x + 1)} = \frac{1}{x + 2} + \frac{3}{(x + 2)^2} - \frac{1}{2x + 1} . \ \bullet$$

EXAMPLE 6 Express

$$\frac{x^4 + x^3 - 3x + 5}{(x + 1)(x^2 + x + 1)}$$

as the sum of partial fractions.

SOLUTION Since the degree of the numerator, 4, is at least as large as the degree of the denominator, 3, step 1 is applicable. Divide by the denominator, $(x + 1)(x^2 + x + 1) = x^3 + 2x^2 + 2x + 1$, as follows:

$$
\begin{array}{r}
x - 1 \qquad \text{quotient} \\
x^3 + 2x^2 + 2x + 1 \overline{\smash{)}\, x^4 + x^3 + 0x^2 - 3x + 5} \\
\underline{x^4 + 2x^3 + 2x^2 + x} \\
-x^3 - 2x^2 - 4x + 5 \\
\underline{-x^3 - 2x^2 - 2x - 1} \\
-2x + 6 \qquad \text{remainder.}
\end{array}
$$

Hence

$$\frac{x^4 + x^3 - 3x + 5}{(x + 1)(x^2 + x + 1)} = x - 1 + \frac{-2x + 6}{(x + 1)(x^2 + x + 1)}. \tag{5}$$

Next represent

$$\frac{-2x + 6}{(x + 1)(x^2 + x + 1)}$$

as a sum of partial quotients in accordance with step 2(b) and (c). Since $x + 1$ and $x^2 + x + 1$ are irreducible, there are constants k_1, c_1, and d_1 such that

$$\frac{-2x + 6}{(x + 1)(x^2 + x + 1)} = \frac{k_1}{x + 1} + \frac{c_1 x + d_1}{x^2 + x + 1}. \tag{6}$$

To find k_1, c_1, and d_1, multiply (6) by $(x + 1)(x^2 + x + 1)$, obtaining

$$-2x + 6 = k_1(x^2 + x + 1) + (c_1 x + d_1)(x + 1). \tag{7}$$

Let $x = -1$ (the root of $x + 1 = 0$); then let $x = 0$ and $x = 1$, which are easy numbers to work with, arriving at

$$8 = k_1 \qquad\qquad \text{setting } x = -1 \text{ in (7);}$$

$$6 = k_1 + d_1 \qquad \text{setting } x = 0 \text{ in (7);}$$

$$4 = 3k_1 + 2c_1 + 2d_1 \qquad \text{setting } x = 1 \text{ in (7).}$$

The first equation yields $k_1 = 8$, the second $d_1 = -2$, and the third $c_1 = -8$. Thus (6) takes the form

$$\frac{-2x + 6}{(x + 1)(x^2 + x + 1)} = \frac{8}{x + 1} - \frac{8x + 2}{x^2 + x + 1}. \tag{8}$$

Combining (5) and (8) shows that

$$\frac{x^4 + x^3 - 3x + 5}{(x + 1)(x^2 + x + 1)} = x - 1 + \frac{8}{x + 1} - \frac{8x + 2}{x^2 + x + 1}. \; \bullet \tag{9}$$

No integrations have been carried out in this section, which is purely algebraic. Sections 9.6 and 9.7 taken together, show how to integrate any rational function.

Exercises

In Exercises 1 to 6 indicate the form of the partial-fraction representation of the rational function listed, but do *not* find the constants k_i, c_i, and d_i.

1. $\dfrac{x^2 + 3x + 1}{(x + 1)^3}$

2. $\dfrac{5x + 6x^2}{(x + 1)^3}$

3. $\dfrac{x^3}{(x - 1)(x + 2)}$

4. $\dfrac{x + 3}{(x - 1)^2(x + 2)^2}$

5. $\dfrac{x^4 + 3x^2}{(x^2 + x + 1)^3}$

6. $\dfrac{x^7 - 1}{(x^2 + x + 1)^3(x + 1)^2}$

In Exercises 7 to 15, express the rational function in terms of partial fractions.

7. $\dfrac{x^2 + x + 3}{x + 1}$

8. $\dfrac{x}{x - 3}$

9. $\dfrac{2x^2 + 2x + 3}{2x + 1}$

10. $\dfrac{3x - 3}{x^2 - 9}$

11. $\dfrac{x + 4}{(x + 1)^2}$

12. $\dfrac{2x^2 + 9x + 12}{(x + 2)(x + 3)^2}$

13. $\dfrac{x^3}{x^2 + 3x + 5}$

14. $\dfrac{x^3}{x^2 - x - 6}$

15. $\dfrac{x^3}{x^2 - 4}$

In Exercises 16 to 24 compute the integral in the given exercise.

16. Exercise 7

17. Exercise 8

18. Exercise 9

19. Exercise 10

20. Exercise 11

21. Exercise 12

22. Exercise 13

23. Exercise 14

24. Exercise 15

■

25. Compute $\displaystyle\int \frac{dx}{(x - 1)x(x + 1)}$.

26. (*a*) If a is a constant other than 0, what is the partial-fraction decomposition of $1/(a^2 - x^2)$?
(*b*) Using (*a*), find $\int dx/(a^2 - x^2)$.

In Exercises 27 to 29 compute the integral.

27. $\displaystyle\int \frac{8 - 4x}{(x - 1)^2(x^2 + 1)} \, dx$

28. $\displaystyle\int \frac{x^4 \, dx}{(x^2 - 1)(x + 2)}$

29. $\displaystyle\int \frac{x^2 \, dx}{x^2 + 4x + 4}$

■ ■

30. Compute: (*a*) $\int (2x^3 + x)/(x^4 + x^2 + 1) \, dx$;
(*b*) $\int 1/(x^4 + x^2 + 1) \, dx$.
Hint: For (*b*) note that $x^4 + x^2 + 1 = (x^2 + x + 1) \times (x^2 - x + 1)$.

Exercises 31 to 33 provide a way of recognizing when a polynomial of degree 2 is irreducible.

31. Prove that if r is a root of the polynomial $P(x)$, that is, $P(r) = 0$, then $x - r$ divides P.

32. Write $2x^2 - 3x - 3$ as the product of two polynomials of degree 1.

33. Show that the polynomial $ax^2 + bx + c$ is irreducible if $b^2 - 4ac < 0$. Otherwise it can be factored.

34. This exercise describes a shortcut for finding the partial-fraction representation of

$$\frac{1}{(ax + b)(cx + d)} \quad (ax + b \text{ is not a constant times } cx + d.)$$

Start with the identity

$$a(cx + d) - c(ax + b) = ad - bc$$

and divide both sides by $(ax + b)(cx + d)$. For instance, to find the partial-fraction decomposition of

$$\frac{1}{(2x + 1)(3x + 5)},$$

start with $2(3x + 5) - 3(2x + 1) = 7.$

Division yields

$$\frac{2}{2x + 1} - \frac{3}{3x + 5} = \frac{7}{(2x + 1)(3x + 5)},$$

hence $\dfrac{1}{(2x + 1)(3x + 5)} = \dfrac{2}{7} \cdot \dfrac{1}{2x + 1} - \dfrac{3}{7} \cdot \dfrac{1}{3x + 5}.$

Obtain in this way the partial-fraction decomposition of

(a) $\dfrac{1}{(3x - 1)(4x + 5)}$ (b) $\dfrac{1}{(3x + 2)(3x + 5)}.$

9.8

How to integrate rational functions of sin θ and cos θ

A *polynomial* in $\sin \theta$ and $\cos \theta$ is simply a function built up from $\sin \theta$, $\cos \theta$, and real numbers by repeated additions, subtractions, and multiplications. For example,

$$5 \cos^3 \theta \sin \theta - \sqrt{2} \sin^2 \theta + \cos \theta + 11$$

is a polynomial in $\sin \theta$ and $\cos \theta$. The quotient of two such polynomials is called a *rational function* of $\sin \theta$ and $\cos \theta$. For example,

$$\frac{5 \sin^2 \theta \cos^3 \theta - \pi \cos \theta + 11}{2 \sin \theta + 3 \cos \theta}$$

is a rational function of $\sin \theta$ and $\cos \theta$. This section will provide a general technique for integrating any such function. We begin, however, with some rational functions of $\sin \theta$ and $\cos \theta$ that can be treated by special means. A few examples will illustrate these techniques.

How to Integrate $\sin^n \theta$ When n Is an Integer > 1.   Integration by parts in Sec. 9.5 provides a recursion for $\int \sin^n \theta \, d\theta$ if n is an integer > 1. Examples 1 and 2 provide some shortcuts for evaluating this type of integral.

EXAMPLE I   Find $\int \sin^2 \theta \, d\theta$ by using the identity

$$\sin^2 \theta = \frac{1 - \cos 2\theta}{2}.$$

SOLUTION
How to Find $\int \sin^2 \theta \, d\theta$.

$$\int \sin^2 \theta \, d\theta = \int \frac{1 - \cos 2\theta}{2} \, d\theta = \int \frac{1}{2} \, d\theta - \int \frac{\cos 2\theta}{2} \, d\theta = \frac{\theta}{2} - \frac{\sin 2\theta}{4}$$

$$= \frac{\theta}{2} - \frac{2 \sin \theta \cos \theta}{4} = \frac{\theta}{2} - \frac{\sin \theta \cos \theta}{2},$$

a result to be found in any table of integrals. ●

The particular definite integrals

$$\int_0^{\pi/2} \sin^2 \theta \, d\theta \quad \text{and} \quad \int_0^{\pi/2} \cos^2 \theta \, d\theta$$

occur so frequently that the following memory device is worth pointing out. A quick sketch of the graphs of $\sin^2 \theta$ and $\cos^2 \theta$ shows that the two definite integrals are equal.   Also, since

$$\sin^2 \theta + \cos^2 \theta = 1,$$

$$\int_0^{\pi/2} \sin^2 \theta \, d\theta + \int_0^{\pi/2} \cos^2 \theta \, d\theta = \int_0^{\pi/2} 1 \, d\theta = \frac{\pi}{2}.$$

How to Remember Hence

$$\int_0^{\pi/2} \sin^2 \theta \, d\theta = \frac{\pi}{4} = \int_0^{\pi/2} \cos^2 \theta \, d\theta.$$

$\int_0^{\pi/2} \sin^2 \theta \, d\theta$ and $\int_0^{\pi/2} \cos^2 \theta \, d\theta$.

In the next example a quick way is shown to find $\int \sin^n \theta \, d\theta$ if n is an *odd* positive integer.

EXAMPLE 2   Find

$$\int \sin^5 \theta \, d\theta.$$

SOLUTION   Recall that

$$d(\cos \theta) = -\sin \theta \, d\theta.$$

Thus

$$\int \sin^5 \theta \, d\theta = \int \sin^4 \theta \sin \theta \, d\theta$$

$$= -\int (1 - \cos^2 \theta)^2 \, d(\cos \theta).$$

Letting $u = \cos \theta$, we obtain then

$$\int \sin^5 \theta \, d\theta = -\int (1 - u^2)^2 \, du$$

$$= -\int (1 - 2u^2 + u^4) \, du$$

$$= -\left(u - \frac{2u^3}{3} + \frac{u^5}{5} \right)$$

$$= -\cos \theta + \frac{2 \cos^3 \theta}{3} - \frac{\cos^5 \theta}{5}. \quad \bullet$$

More generally, to find $\int \cos^m \theta \sin^n \theta \, d\theta$, where m and n are nonnegative integers and n is odd, pair one $\sin \theta$ with $d\theta$ to form $\sin \theta \, d\theta = -d(\cos \theta)$ and use the identity $\sin^2 \theta = 1 - \cos^2 \theta$ together with the substitution $u = \cos \theta$. The new integrand will be a polynomial in u.   A similar approach works on $\int \cos^m \theta \sin^n \theta \, d\theta$ if m is odd, as is illustrated by Example 3.

EXAMPLE 3   Find

$$\int \cos^3 \theta \sin^4 \theta \, d\theta.$$

SOLUTION   Pair one $\cos\theta$ with $d\theta$ to form

$$\cos\theta\, d\theta = d(\sin\theta),$$

which suggests the substitution $u = \sin\theta$:

$$\int \cos^3\theta \sin^4\theta\, d\theta = \int \cos^2\theta \sin^4\theta(\cos\theta\, d\theta)$$

$$= \int (1 - \sin^2\theta) \sin^4\theta\, d(\sin\theta)$$

$$= \int (1 - u^2)u^4\, du$$

$$= \int (u^4 - u^6)\, du$$

$$= \frac{u^5}{5} - \frac{u^7}{7}$$

$$= \frac{\sin^5\theta}{5} - \frac{\sin^7\theta}{7}. \quad \bullet$$

Similar pairing methods apply to the other trigonometric functions such as

$$\int \tan^6\theta \sec^4\theta\, d\theta.$$

[The integrand is actually a rational function of $\sin\theta$ and $\cos\theta$, namely

$$\left(\frac{\sin\theta}{\cos\theta}\right)^6 \left(\frac{1}{\cos\theta}\right)^4 = \frac{\sin^6\theta}{\cos^{10}\theta}.\bigg]$$

EXAMPLE 4   Find

$$\int \tan^6\theta \sec^4\theta\, d\theta.$$

SOLUTION   Recall that

$$d(\tan\theta) = \sec^2\theta\, d\theta.$$

So pair $\sec^2\theta$ with $d\theta$ to form $\sec^2\theta\, d\theta$. This suggests the substitution

$$u = \tan\theta \qquad du = \sec^2\theta\, d\theta.$$

Recall also that

$$\sec^2\theta = \tan^2\theta + 1.$$

Then

$$\int \tan^6\theta \sec^4\theta\, d\theta = \int \tan^6\theta \sec^2\theta \sec^2\theta\, d\theta$$

$$= \int u^6(u^2 + 1)\, du$$

$$= \int (u^8 + u^6)\, du$$

$$= \frac{u^9}{9} + \frac{u^7}{7}$$

$$= \frac{\tan^9 \theta}{9} + \frac{\tan^7 \theta}{7}. \quad \bullet$$

The substitution used in Example 4 can be applied to evaluate $\int \tan^m \theta \sec^n \theta \, d\theta$ for any number m and any positive *even* integer n. The reader should try $\int \tan^4 \theta \sec^3 \theta \, d\theta$ to see what complication arises when n is odd.

If m is a positive odd integer, $\int \tan^m \theta \sec^n \theta \, d\theta$ can be computed by pairing $\tan \theta \sec \theta$ with $d\theta$ to form $\tan \theta \sec \theta \, d\theta = d(\sec \theta)$. Example 5 illustrates this procedure.

EXAMPLE 5   Find $\int \tan^5 \theta \sec^3 \theta \, d\theta.$

SOLUTION   Recall that $\quad d(\sec \theta) = \tan \theta \sec \theta \, d\theta.$

and that $\quad \tan^2 \theta = \sec^2 \theta - 1.$

Let $\quad u = \sec \theta \quad$ and $\quad du = \sec \theta \tan \theta \, d\theta.$

Then $\quad \int \tan^5 \theta \sec^3 \theta \, d\theta = \int \tan^4 \theta \sec^2 \theta (\sec \theta \tan \theta \, d\theta)$

$$= \int (\sec^2 \theta - 1)^2 \sec^2 \theta (\sec \theta \tan \theta \, d\theta)$$

$$= \int (u^2 - 1)^2 u^2 \, du$$

$$= \int (u^6 - 2u^4 + u^2) \, du$$

$$= \frac{u^7}{7} - \frac{2u^5}{5} + \frac{u^3}{3}$$

$$= \frac{\sec^7 \theta}{7} - \frac{2 \sec^5 \theta}{5} + \frac{\sec^3 \theta}{3}. \quad \bullet$$

A slightly different trick disposes of $\int \tan^n \theta \, d\theta$, as the next example illustrates.

EXAMPLE 6   Obtain a recursion formula for $\int \tan^n \theta \, d\theta.$

SOLUTION   Keep in mind that $(\tan \theta)' = \sec^2 \theta$ and that $\tan^2 \theta = \sec^2 \theta - 1$. The steps are few:

Contrast This with the Recursion for $\int \sin^n \theta \, d\theta$ in Sec. 9.5, Which Is Based on Integration by Parts.

$$\int \tan^n \theta \, d\theta = \int \tan^{n-2} \theta \tan^2 \theta \, d\theta$$

$$= \int \tan^{n-2} \theta (\sec^2 \theta - 1) \, d\theta$$

$$= \int \tan^{n-2} \theta \sec^2 \theta \, d\theta - \int \tan^{n-2} \theta \, d\theta$$

$$= \frac{\tan^{n-1} \theta}{n-1} - \int \tan^{n-2} \theta \, d\theta.$$

Repeated application of this recursion eventually results in the case where $\tan \theta$ is raised to the zeroth power,

$$\int \tan^0 \theta \, d\theta = \int d\theta = \theta \qquad \text{if } n \text{ is even,}$$

or the case where $\tan \theta$ is raised to the first power,

$$\int \tan^1 \theta \, d\theta = \int \tan \theta \, d\theta = \int \frac{\sin \theta}{\cos \theta} \, d\theta = -\ln |\cos \theta| = \ln |\sec \theta|,$$

if n is odd. ●

The formulas

$$\int \sec \theta \, d\theta = \ln |\sec \theta + \tan \theta|$$

and

$$\int \sec \theta \, d\theta = \frac{1}{2} \ln \frac{1 + \sin \theta}{1 - \sin \theta}$$

are found in integral tables. Example 7 shows how to obtain them.

EXAMPLE 7   Find

$$\int \sec \theta \, d\theta.$$

SOLUTION   This is one way to obtain the formulas.

$$\int \sec \theta \, d\theta = \int \frac{1}{\cos \theta} \, d\theta$$

$$= \int \frac{\cos \theta}{\cos^2 \theta} \, d\theta$$

$$= \int \frac{\cos \theta}{1 - \sin^2 \theta} \, d\theta.$$

The substitution $u = \sin \theta$

and $du = \cos \theta \, d\theta$

transforms this last integral into the integral of a rational function:

$$\int \frac{du}{1 - u^2} = \frac{1}{2} \int \left(\frac{1}{1 + u} + \frac{1}{1 - u} \right) du$$

$$= \tfrac{1}{2}[\ln (1 + u) - \ln (1 - u)]$$

$$= \frac{1}{2} \ln \frac{1 + u}{1 - u}.$$

Since $u = \sin \theta$, $\qquad \frac{1}{2} \ln \frac{1 + u}{1 - u} = \frac{1}{2} \ln \frac{1 + \sin \theta}{1 - \sin \theta}.$

Note that $\sec \theta + \tan \theta$
Can Be Negative, But
$(1 + \sin \theta)/(1 - \sin \theta)$
Cannot Be.

The reader may check that this equals $\ln \lvert \sec \theta + \tan \theta \rvert$ by showing that

$$\frac{1 + \sin \theta}{1 - \sin \theta} = (\sec \theta + \tan \theta)^2. \ \bullet$$

The following table summarizes the techniques discussed and similar ones for other powers of trigonometric functions.

Integrand	Technique
$\sin^2 \theta$	Write $\sin^2 \theta$ as $\dfrac{1 - \cos 2\theta}{2}$.
$\cos^2 \theta$	Write $\cos^2 \theta$ as $\dfrac{1 + \cos 2\theta}{2}$.
$\sin^n \theta$ (n odd)	Write $\sin^n \theta \ d\theta = \sin^{n-1} \theta(\sin \theta \ d\theta)$ and use $u = \cos \theta$; hence $1 - u^2 = \sin^2 \theta$.
$\cos^m \theta \sin^n \theta$ (n odd)	Write $\cos^m \theta \sin^n \theta \ d\theta = \cos^m \theta \sin^{n-1} \theta(\sin \theta \ d\theta)$ and use $u = \cos \theta$; hence $1 - u^2 = \sin^2 \theta$.
$\cos^m \theta \sin^n \theta$ (m odd)	Write $\cos^m \theta \sin^n \theta \ d\theta = \cos^{m-1} \theta \sin^n \theta(\cos \theta \ d\theta)$ and use $u = \sin \theta$; hence $1 - u^2 = \cos^2 \theta$.
$\cos^m \theta \sin^n \theta$ (m and n positive even integers)	Replace $\cos^2 \theta$ by $\dfrac{1 + \cos 2\theta}{2}$ and $\sin^2 \theta$ by $\dfrac{1 - \cos 2\theta}{2}$. See Exercise 20.
$\tan^m \theta \sec^n \theta$ ($n \geq 2$ even)	Write $\tan^m \theta \sec^n \theta \ d\theta$ as $\tan^m \theta \sec^{n-2} \theta(\sec^2 \theta \ d\theta)$ and use $u = \tan \theta$; hence $1 + u^2 = \sec^2 \theta$.
$\tan^m \theta \sec^n \theta$ (m odd)	Write $\tan^m \theta \sec^n \theta \ d\theta$ as $\tan^{m-1} \theta \sec^{n-1} \theta(\tan \theta \sec \theta \ d\theta)$ and use $u = \sec \theta$; hence $u^2 - 1 = \tan^2 \theta$.
$\tan^n \theta$ (n an integer ≥ 2)	Write $\tan^n \theta = \tan^{n-2} \theta \tan^2 \theta =$ $\tan^{n-2} \theta \sec^2 \theta - \tan^{n-2} \theta$ and repeat.
$\tan \theta$	$\int \tan \theta \ d\theta = \ln \lvert \sec \theta \rvert.$
$\sec \theta$	$\int \sec \theta \ d\theta = \ln \lvert \sec \theta + \tan \theta \rvert.$
$\cot^m \theta \csc^n \theta$ ($n \geq 2$ even)	Write $\cot^m \theta \csc^n \theta$ as $\cot^m \theta \csc^{n-2} \theta(\csc^2 \theta \ d\theta)$ and use $u = \cot \theta$; hence $1 + u^2 = \csc^2 \theta$.
$\cot^m \theta \csc^n \theta$ (m odd)	Write $\cot^m \theta \csc^n \theta \ d\theta$ as $\cot^{m-1} \theta \csc^{n-1} \theta(\cot \theta \csc \theta \ d\theta)$ and use $u = \csc \theta$; hence $u^2 - 1 = \cot^2 \theta$.
$\tan^m \theta \sec^n \theta$ (m even, n a positive integer)	Replace $\tan^2 \theta$ by $\sec^2 \theta - 1$. See Exercise 24.

A General Method for Integrating Any Rational Function of cos θ *and* sin θ.

The next technique applies to *any rational function* of cos θ and sin θ. It depends on the fact that cos θ and sin θ can be expressed easily in terms of tan $(\theta/2)$.

Introduce the substitution $u = \tan(\theta/2)$, recorded in the right triangle at the left. Then, as the triangle shows,

$$\cos\frac{\theta}{2} = \frac{1}{\sqrt{1+u^2}}$$

and

$$\sin\frac{\theta}{2} = \frac{u}{\sqrt{1+u^2}}.$$

Thus

$$\cos\theta = \cos^2\frac{\theta}{2} - \sin^2\frac{\theta}{2} = \left(\frac{1}{\sqrt{1+u^2}}\right)^2 - \left(\frac{u}{\sqrt{1+u^2}}\right)^2 = \frac{1-u^2}{1+u^2}$$

and

$$\sin\theta = 2\sin\frac{\theta}{2}\cos\frac{\theta}{2} = 2\left(\frac{u}{\sqrt{1+u^2}}\right)\left(\frac{1}{\sqrt{1+u^2}}\right) = \frac{2u}{1+u^2}.$$

In addition, $d\theta$ can be expressed easily in terms of u and du, as will now be shown. Since $u = \tan(\theta/2)$,

$$du = \frac{1}{2}\sec^2\frac{\theta}{2}\,d\theta$$

$$= \frac{1}{2}\left(1 + \tan^2\frac{\theta}{2}\right)d\theta$$

$$= \tfrac{1}{2}(1 + u^2)\,d\theta.$$

Thus

$$d\theta = \frac{2\,du}{1+u^2}.$$

The substitution

$$u = \tan\frac{\theta}{2}$$

thus leads to

$$\cos\theta = \frac{1-u^2}{1+u^2}, \qquad \sin\theta = \frac{2u}{1+u^2}, \qquad \text{and} \qquad d\theta = \frac{2\,du}{1+u^2}.$$

Though Always Applicable, This Method Is Not Always the Most Convenient One to Use.

This substitution transforms any integral of a rational function of cos θ and sin θ into an integral of a rational function of u. The resulting rational function can then be integrated by the method of partial fractions.

EXAMPLE 8   Find

$$\int \frac{\sin^3\theta\cos\theta\,d\theta}{1 + 2\cos\theta}.$$

SOLUTION   Use the substitution $u = \tan(\theta/2)$. Then

$$\cos\theta = \frac{1-u^2}{1+u^2}, \qquad \sin\theta = \frac{2u}{1+u^2}, \qquad \text{and} \qquad d\theta = \frac{2\,du}{1+u^2}.$$

Hence
$$\int \frac{\sin^3\theta\cos\theta}{1+2\cos\theta}\,d\theta = \int \frac{\left(\dfrac{2u}{1+u^2}\right)^3 \dfrac{1-u^2}{1+u^2}}{1+2\dfrac{1-u^2}{1+u^2}}\,\frac{2\,du}{1+u^2},$$

which simplifies to
$$\int \frac{16u^3(1-u^2)}{(3-u^2)(1+u^2)^4}\,du.$$

This integral can be evaluated by the method of partial fractions, though it would be a tedious affair. ●

Exercises

In Exercises 1 to 18 compute the integrals.

1. $\int \sin^5\theta\,d\theta$

2. $\int \sin^2\theta\cos^3\theta\,d\theta$

3. $\int \tan^4\theta\sec^2\theta\,d\theta$

4. $\int \tan^5\theta\sec^3\theta\,d\theta$

5. $\int \sin^3\theta\cos^4\theta\,d\theta$

6. $\int \csc\theta\,d\theta$

7. $\int \cot\theta\,d\theta$

8. $\int \cos^2 3\theta\,d\theta$

9. $\int \tan^4\theta\,d\theta$

10. $\int \sec 3\theta\,d\theta$

11. $\int \cot^3\theta\csc^5\theta\,d\theta$

12. $\int \cot^4\theta\csc^4\theta\,d\theta$

13. $\int (\sin\theta+2\cos\theta)^2\,d\theta$

14. $\int (\tan\theta+2\cot\theta)^2\,d\theta$

15. $\int \cos^3\theta\,d\theta$

16. $\int_0^{\pi/2} \cos^2\theta\,d\theta$

17. $\int \cot^3\theta\csc^4\theta\,d\theta$

18. $\int \sin^3\theta\tan^2\theta\,d\theta$

19. Compute $\int \dfrac{d\theta}{1+\cos\theta}$ with the aid of the identity
$$\cos^2\frac{\theta}{2} = \frac{1+\cos\theta}{2}.$$

The method in Exercise 20 applies to $\int \sin^m\theta\cos^n\theta\,d\theta$ when m and n are positive even integers.

20. Compute $\int \sin^2\theta\cos^4\theta\,d\theta$ with the aid of the identities
$$\sin^2\theta = \frac{1-\cos 2\theta}{2}$$
and
$$\cos^2\theta = \frac{1+\cos 2\theta}{2}.$$

21. Use integration by parts to show that

$$\int \sin^n\theta\,d\theta = -\frac{\sin^{n-1}\theta\cos\theta}{n} + \frac{n-1}{n}\int \sin^{n-2}\theta\,d\theta.$$

Exercises 22 to 24 are related.

22. Use integration by parts to obtain the recursion
$$\int \sec^n a\theta\,d\theta = \frac{1}{a(n-1)}\frac{\sin a\theta}{\cos^{n-1}a\theta}$$
$$+ \frac{n-2}{n-1}\int \sec^{n-2}a\theta\,d\theta \qquad n>1.$$

23. Use the recursion in Exercise 22 to find
$$\int \sec^3\theta\,d\theta.$$

24. (See Exercise 22.)
(a) Show that $\int \tan^m\theta\sec^n\theta\,d\theta$, where m is a nonnegative even integer and n is a nonnegative integer, can be reduced to a sum of integrals of the type $\int \sec^n\theta\,d\theta$ by writing $\tan^2\theta = \sec^2\theta - 1$.
(b) Evaluate $\int \tan^2\theta\sec^3\theta\,d\theta$.

25. Show that
$$\frac{1}{2}\ln\frac{1+\sin\theta}{1-\sin\theta} = \ln|\sec\theta+\tan\theta|.$$

26. Find $\int \sin^2\theta\cos^2\theta\,d\theta$ by use of the identity $\sin 2\theta = 2\sin\theta\cos\theta$.

Exercises 27 to 31 concern the general method for integrating any rational function of $\sin\theta$ and $\cos\theta$.

27. Find $\int \sec\theta\,d\theta$ by the method that applies to any rational function of $\sin\theta$ and $\cos\theta$.

28. Transform $\int (1-3\cos\theta)/(1+4\sin\theta)\,d\theta$ to an integral of a rational function.

29. Transform $\int (\cos^3\theta-2\sin\theta)/(\sin\theta+\cos\theta)\,d\theta$ to an integral of a rational function.

30. Find $\int d\theta/(3+4\cos\theta)$.

31. Find $\int (3+\cos\theta)/(2-\cos\theta)\,d\theta$.

9.9

Trigonometric and algebraic substitutions

A rational function of x and $\sqrt{a^2 - x^2}$, or $\sqrt{a^2 + x^2}$, or $\sqrt{x^2 - a^2}$ can be integrated by using a trigonometric substitution. If the integrand is a rational function of x and

Case 1 $\quad \sqrt{a^2 - x^2}$; let $x = a \sin \theta$.

Case 2 $\quad \sqrt{a^2 + x^2}$; let $x = a \tan \theta$.

Case 3 $\quad \sqrt{x^2 - a^2}$; let $x = a \sec \theta$.

The motivation behind this general procedure is quite simple. Consider case 1, for instance. If you replace x in $\sqrt{a^2 - x^2}$ by $a \sin \theta$, you obtain

$$\sqrt{a^2 - x^2} = \sqrt{a^2 - (a \sin \theta)^2} = \sqrt{a^2(1 - \sin^2 \theta)}$$
$$= \sqrt{a^2 \cos^2 \theta}$$
$$= a \cos \theta,$$

if a and $\cos \theta$ are positive. The important point to observe is that *the square root sign disappears.* The general procedure is illustrated in Examples 1 to 3.

EXAMPLE 1   Compute

$$\int \frac{x^3 \, dx}{x + \sqrt{16 - x^2}}.$$

SOLUTION   Make the substitution $\qquad x = 4 \sin \theta$;

hence $\qquad dx = 4 \cos \theta \, d\theta$.

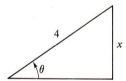

(See the triangle in the margin.)

Thus $\qquad \displaystyle\int \frac{x^3 \, dx}{x + \sqrt{16 - x^2}} = \int \frac{(4 \sin \theta)^3}{4 \sin \theta + \sqrt{16 - 16 \sin^2 \theta}} \, 4 \cos \theta \, d\theta$

$$= \int \frac{256 \sin^3 \theta \cos \theta}{4 \sin \theta + 4 \cos \theta} \, d\theta.$$

Since any rational function of $\sin \theta$ and $\cos \theta$ can be reduced to a rational function of $u = \tan (\theta/2)$ as shown in Sec. 9.8, the integral is solvable, though the entire computation would be long. ●

EXAMPLE 2   Compute $\int \sqrt{1 + x^2} \, dx$.

SOLUTION   The identity $\sec \theta = \sqrt{1 + \tan^2 \theta}$ suggests the substitution described in case 2:

$$x = \tan \theta;$$

hence $\qquad dx = \sec^2 \theta \, d\theta$.

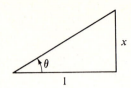

Thus
$$\int \sqrt{1 + x^2} \, dx = \int \sec \theta \sec^2 \theta \, d\theta = \int \sec^3 \theta \, d\theta,$$

which we rewrite as $\int \sec \theta \sec^2 \theta \, d\theta$ and subject to integration by parts with
$$u = \sec \theta, \qquad dv = \sec^2 \theta \, d\theta,$$
$$du = \sec \theta \tan \theta \, d\theta, \qquad v = \tan \theta.$$

This yields
$$\int \sec^3 \theta \, d\theta = \int \underbrace{\sec \theta}_{u} \underbrace{\sec^2 \theta \, d\theta}_{dv} = \underbrace{\sec \theta}_{u} \underbrace{\tan \theta}_{v} - \int \underbrace{\tan \theta}_{v} \underbrace{\sec \theta \tan \theta \, d\theta}_{du}$$

$$= \sec \theta \tan \theta - \int \tan^2 \theta \sec \theta \, d\theta$$

$$= \sec \theta \tan \theta - \int (\sec^2 \theta - 1) \sec \theta \, d\theta$$

$$= \sec \theta \tan \theta - \int \sec^3 \theta \, d\theta + \int \sec \theta \, d\theta.$$

Collecting $\int \sec^3 \theta \, d\theta$, which appears twice, yields
$$2 \int \sec^3 \theta \, d\theta = \sec \theta \tan \theta + \int \sec \theta \, d\theta.$$

By Example 7 in Sec. 9.8, $\int \sec \theta \, d\theta = \ln |\sec \theta + \tan \theta|$.

Thus
$$\int \sec^3 \theta \, d\theta = \frac{\sec \theta \tan \theta}{2} + \frac{1}{2} \ln |\sec \theta + \tan \theta|.$$

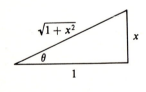

To express the antiderivative just obtained in terms of x rather than θ, it is necessary to express $\tan \theta$ and $\sec \theta$ in terms of x. Starting with the definition $x = \tan \theta$, find $\sec \theta$ by means of the relation $\sec \theta = \sqrt{1 + \tan^2 \theta} = \sqrt{1 + x^2}$. Thus,
$$\int \sqrt{1 + x^2} \, dx = \frac{x\sqrt{1 + x^2}}{2} + \frac{1}{2} \ln \left(\sqrt{1 + x^2} + x \right),$$

a formula listed in a table of integrals. ●

EXAMPLE 3   Compute
$$\int \frac{dx}{\sqrt{x^2 - 9}}.$$

SOLUTION   Let $x = 3 \sec \theta$, hence $dx = 3 \sec \theta \tan \theta \, d\theta$. (See the triangle in the margin.)

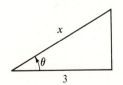

Thus
$$\int \frac{dx}{\sqrt{x^2 - 9}} = \int \frac{3 \sec \theta \tan \theta \, d\theta}{\sqrt{9 \sec^2 \theta - 9}}$$

$$= \int \frac{\sec \theta \tan \theta \, d\theta}{\tan \theta}$$

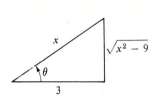

$$= \int \sec \theta \, d\theta$$

$$= \ln |\sec \theta + \tan \theta| \qquad \text{by Example 7 of Sec. 9.8.}$$

$$= \ln \left| \frac{x}{3} + \frac{\sqrt{x^2 - 9}}{3} \right| \qquad \text{by inspection of the triangle in the margin.}$$

Incidentally, $\ln |x + \sqrt{x^2 - 9}|$ is also an antiderivative of $1/\sqrt{x^2 - 9}$, since

$$\ln \left| \frac{x}{3} + \frac{\sqrt{x^2 - 9}}{3} \right| = \ln |x + \sqrt{x^2 - 9}| - \ln 3. \; \bullet$$

A rational function of x and $\sqrt{a^2 - b^2 x^2}$, or $\sqrt{a^2 x^2 - b^2}$, or $\sqrt{a^2 + b^2 x^2}$ can be treated by a similar procedure. For instance, to deal with a rational function of x and $\sqrt{a^2 - b^2 x^2}$ make the substitution suggested by

$$b^2 x^2 = a^2 \sin^2 \theta,$$

that is,

$$bx = a \sin \theta,$$

or

$$x = \frac{a}{b} \sin \theta \quad \text{and} \quad dx = \frac{a}{b} \cos \theta \, d\theta.$$

Then

$$\sqrt{a^2 - b^2 x^2} = \sqrt{a^2 - a^2 \sin^2 \theta}$$
$$= a \cos \theta.$$

A rational function of x and $a^2 - b^2 x^2$ is a rational function of x and therefore can be integrated by partial fractions. But in some special cases a trigonometric substitution may provide a quicker solution, as Example 4 illustrates.

EXAMPLE 4   Find
$$\int \frac{dx}{(4 - 9x^2)^2}.$$

SOLUTION   Assume that $9x^2 < 4$.

Let $9x^2 = 4 \sin^2 \theta$, choosing $3x = 2 \sin \theta$;

hence
$$dx = \tfrac{2}{3} \cos \theta \, d\theta.$$

Thus
$$\int \frac{dx}{(4 - 9x^2)^2} = \int \frac{\tfrac{2}{3} \cos \theta \, d\theta}{(4 - 4 \sin^2 \theta)^2}$$

$$= \frac{2}{3} \int \frac{\cos \theta \, d\theta}{16(1 - \sin^2 \theta)^2}$$

$$= \frac{1}{24} \int \frac{\cos \theta \, d\theta}{\cos^4 \theta}$$

$$= \frac{1}{24} \int \frac{1}{\cos^3 \theta} \, d\theta$$

$$\int \frac{dx}{(4 - 9x^2)^2} = \frac{1}{24} \int \sec^3 \theta \, d\theta.$$

Now, $\int \sec^3 \theta \, d\theta$ was found in Example 2.

If $9x^2 > 4$, then the substitution $9x^2 = 4 \sec^2 \theta$ is suggested, with the choice $3x = 2 \sec \theta$. ●

How to Integrate Rational   Integrals of rational functions of x and $\sqrt[n]{ax + b}$ can be transformed to
Functions of x and $\sqrt[n]{ax + b}$ integrals of rational functions by the algebraic substitution

$$u = \sqrt[n]{ax + b},$$

or equivalently

$$u^n = ax + b.$$

This substitution leads to

$$nu^{n-1} \, du = a \, dx$$

and

$$x = \frac{u^n - b}{a}.$$

Consequently, $\sqrt[n]{ax + b}$, x, and dx are easily expressed in terms of u and du. The radical sign is eliminated.   The following example illustrates the method.

EXAMPLE 5   Find

$$\int \frac{\sqrt[4]{1 + x}}{x^2} \, dx.$$

SOLUTION   Let

$$u = \sqrt[4]{1 + x}.$$

Then

$$u^4 = 1 + x$$

(hence $x = u^4 - 1$ and $4u^3 \, du = dx$).

The transformed integral is

$$\int \frac{u}{(u^4 - 1)^2} 4u^3 \, du,$$

which can be dealt with by partial fractions. ●

Exercises

In Exercises 1 to 9 make the appropriate substitution to remove the radical.   Do *not* find the resulting integral.

1. $\int \dfrac{\sqrt{4 - x^2}}{x^3} \, dx$

2. $\int \dfrac{\sqrt{x^2 + 2}}{4x + 1} \, dx$

3. $\int \dfrac{\sqrt{4x^2 + 2}}{4x + 5} \, dx$

4. $\int \dfrac{dx}{\sqrt[4]{x} + \sqrt{x}}$

5. $\int \dfrac{\sqrt{4x^2 - 1}}{x + \sqrt{4x^2 - 1}} \, dx$

6. $\int \dfrac{1 + \sqrt[3]{x}}{1 + \sqrt{x}} \, dx.$

   Hint: Let $u = \sqrt[6]{x}$.

7. $\int \dfrac{(\sqrt{1 - 9x^2})^3 + x}{\sqrt{1 - 9x^2}} \, dx$

8. $\int \dfrac{\sqrt[3]{1 - 5x} + 2}{\sqrt[3]{1 - 5x} + 3} \, dx$

9. $\int \dfrac{\sqrt{1 + 2x^2}}{1 + \sqrt{1 + 2x^2}} \, dx$

In Exercises 10 to 22 find the integral.

10. $\int \dfrac{\sqrt[3]{x + 8}}{x} \, dx$

11. $\int \dfrac{\sqrt{25 - 4x^2}}{x} \, dx$

12. $\int \dfrac{x^2 \, dx}{\sqrt{9 + x^2}}$

13. $\int \dfrac{x^3 \, dx}{\sqrt{4 + 9x^2}}$

14. $\displaystyle\int \frac{dx}{\sqrt[4]{x} + \sqrt{x}}$

15. $\displaystyle\int \frac{(9 - 16x^2)^{3/2}}{x^4}\,dx$

16. $\displaystyle\int x^3 \sqrt{9 - x^2}\,dx$

17. $\displaystyle\int \frac{dx}{\sqrt{16x^2 - 9}}$

18. $\displaystyle\int \frac{dx}{(4 + x^2)^{3/2}}$

19. $\displaystyle\int x^5(\sqrt{1 - 4x^2})^3\,dx$

20. $\displaystyle\int \frac{dx}{(x - 3)\sqrt{x + 3}}$

21. $\displaystyle\int \frac{(9x^2 - 4)^{3/2}}{x}\,dx$

22. $\displaystyle\int \frac{dx}{x^2 \sqrt{x^2 + 9}}$

■

In Exercises 23 to 28 transform the problem into one in which the integrand is a rational function of convenient trigonometric functions. Do not evaluate the resulting integral.

23. $\displaystyle\int \frac{x^3\,dx}{(1 - 4x^2)^3},\ x^2 < \tfrac{1}{4}$

24. $\displaystyle\int \frac{x^3\,dx}{(1 - 4x^2)^3},\ x^2 > \tfrac{1}{4}$

25. $\displaystyle\int (4 + 9x^2)^5\,dx$

26. $\displaystyle\int \frac{x^6\,dx}{(9x^2 - 4)^5},\ 9x^2 > 4$

27. $\displaystyle\int \frac{x + x^3}{(1 - 2x^2)^5}\,dx,\ x^2 < \tfrac{1}{2}$

28. $\displaystyle\int \frac{x^3\,dx}{\sqrt{25 - 4x^2}}$

■ ■

29. Show that any rational function of $\sqrt{x}$ and $\sqrt{x + 1}$ can be integrated by using the substitution $x = \tan^2 \theta$.

30. Show that any rational function of x, $\sqrt{x + a}$, and $\sqrt{x + b}$ can be integrated by introducing the substitution defined by

$$x + a = \frac{1}{4}(a - b)\left(t + \frac{1}{t}\right)^2,$$

hence $\qquad x + b = \frac{1}{4}(a - b)\left(t - \frac{1}{t}\right)^2.$

However, it is not the case that every rational function of $\sqrt{x + a}$, $\sqrt{x + b}$, and $\sqrt{x + c}$ has an elementary integral. For instance,

$$\int \frac{1}{\sqrt{x}\sqrt{x - 1}\sqrt{x + 1}}\,dx = \int \frac{1}{\sqrt{x^3 - x}}\,dx$$

is not an elementary function.

31. Show that every rational function of x and

$$\sqrt[n]{\frac{ax + b}{cx + d}}$$ has an elementary antiderivative.

9.10

Summary

Method	Description
Substitution	Introduce $u = h(x)$. If $f(x)\,dx = g(u)\,du$, then $\int f(x)\,dx = \int g(u)\,du$.
Substitution in the definite integral	If, in the above substitution, $u = A$ when $x = a$, and $u = B$ when $x = b$, then $\displaystyle\int_a^b f(x)\,dx = \int_A^B g(u)\,du$.
Table of integrals	Obtain and become familiar with a table of integrals. Substitution, together with integral tables, will usually be adequate.
Integration by parts	$\int u\,dv = uv - \int v\,du$. Choose u and v so $u\,dv = f(x)\,dx$, with the hope that $\int v\,du$ is easier than $\int u\,dv$.
Partial fractions (applies to any rational function of x)	This is an algebraic method. Write the integrand as a sum of a polynomial (if the degree of the numerator is greater than or equal to the degree of the denominator) plus terms of the type $\dfrac{k_i}{(ax + b)^n}$, $\dfrac{c_i x + d_i}{(ax^2 + bx + c)^n}$. A table of integrals treats the integrals of these two types. (For the first type, use the substitution $u = ax + b$.)

Method	Description
To integrate certain powers of trigonometric functions	There are many special cases. For instance $\displaystyle\int \cos^k \theta \sin^{2n+1} \theta \, d\theta$; rewrite as $\displaystyle\int \cos^k \theta (1 - \cos^2 \theta)^n \sin \theta \, d\theta$ and make the substitution $u = \cos\theta$. Keep in mind that $d(\sin\theta) = \cos\theta \, d\theta$; $d(\cos\theta) = -\sin\theta \, d\theta$; $d(\sec\theta) = \sec\theta\tan\theta \, d\theta$; $d(\tan\theta) = \sec^2\theta \, d\theta$. Also, trigonometric identities, such as $\sin^2\theta = (1 - \cos 2\theta)/2$ and $\cos^2\theta = (1 + \cos 2\theta)/2$.
To integrate any rational function of $\cos\theta$ and $\sin\theta$	Let $u = \tan(\theta/2)$. Then $$\cos\theta = \frac{1 - u^2}{1 + u^2},$$ $$\sin\theta = \frac{2u}{1 + u^2}, \quad d\theta = \frac{2\,du}{1 + u^2},$$ and the new integrand is a rational function of u.
To integrate rational functions of x and one of $$\sqrt{a^2 - x^2}, \sqrt{a^2 + x^2}, \sqrt{x^2 - a^2},$$ $$a^2 - x^2, a^2 + x^2$$	For $\sqrt{a^2 - x^2}$ or $a^2 - x^2$, let $x = a\sin\theta$. For $\sqrt{a^2 + x^2}$ or $a^2 + x^2$, let $x = a\tan\theta$. For $\sqrt{x^2 - a^2}$, let $x = a\sec\theta$. (Recall the two right triangles shown to the right.)
To integrate rational functions of x and $\sqrt[n]{ax + b}$	Let $u = \sqrt[n]{ax + b}$, hence $u^n = ax + b$, $nu^{n-1}\,du = a\,dx$, and $x = (u^n - b)/a$. The new integrand is a rational function of u.

The fundamental theorem of calculus, proved in Chap. 7, raised the problem of finding antiderivatives. Now, some very simple and important functions do not have elementary antiderivatives; for instance,

$$\int \frac{\sin x \, dx}{x}, \quad \int e^{x^2} \, dx, \quad \int \frac{1}{\ln x} \, dx, \quad \int x \tan x \, dx,$$

$$\int \frac{\ln x}{x + 1} \, dx, \quad \int \sqrt{1 - \frac{\sin^2 x}{4}} \, dx, \quad \text{and} \quad \int \sqrt[3]{x - x^2} \, dx$$

are not elementary. If the definite integral

$$\int_0^1 e^{x^2} \, dx$$

is needed, an estimate must be made, for example, by using an approximating sum. (After all, $\int_0^1 e^{x^2} \, dx$ is defined as a limit of approximating sums; just choose a sum with small mesh.) The "elliptic integral"

$$\int_0^{\pi/2} \sqrt{1 - k^2 \sin^2 x} \, dx,$$

frequently used in engineering, is tabulated for various values of k to four decimal places in most handbooks.

Fortunately many commonly used functions do have elementary antiderivatives, and this chapter has presented a few methods for finding antiderivatives in these cases. The two most important techniques are substitution and use of integral tables. The preceding table summarizes the various methods discussed.

Guide quiz on Chap. 9

In Problems 1 to 18 evaluate the integral.

1. $\displaystyle\int \frac{x^3 \, dx}{1 + x^4}$

2. $\displaystyle\int \sqrt{4 - 9x^2} \, dx$

3. $\displaystyle\int \frac{dx}{x^4 - 1}$

4. $\displaystyle\int \tan^5 2x \sec^2 2x \, dx$

5. $\displaystyle\int \frac{x^4 \, dx}{x^4 - 1}$

6. $\displaystyle\int \frac{dx}{\sqrt{9x^2 + 16}}$

7. $\int \dfrac{dx}{2\sqrt{x} - \sqrt[4]{x}}$

8. $\int \dfrac{dx}{(x-3)\sqrt{x+3}}$

13. $\int \dfrac{x\,dx}{1+x^4}$

14. $\int \dfrac{dx}{x\sqrt{4+x^2}}$

9. $\int \dfrac{dx}{3-x^2}$

10. $\int \sin^5 2x\,dx$

15. $\int \dfrac{x^2\,dx}{\sqrt{x^2-9}}$

16. $\int (9-x^2)^{3/2}\,dx$

11. $\int e^x \cos 2x\,dx$

12. $\int \dfrac{dx}{\sin^5 3x}$

17. $\int \dfrac{dx}{3+\cos x}$

18. $\int \dfrac{dx}{x^2\sqrt{x^2+25}}$

Review exercises for Chap. 9

1. (a) By an appropriate substitution, transform this definite integral into a simpler definite integral.

$$\int_0^{\pi/2} \sqrt{(1+\cos\theta)^3}\,\sin\theta\,d\theta.$$

(b) Evaluate the new definite integral in (a).

2. Two of these antiderivatives are elementary functions; evaluate them.

(a) $\int \ln x\,dx$

(b) $\int \dfrac{\ln x\,dx}{x}$

(c) $\int \dfrac{dx}{\ln x}$.

3. Evaluate:

(a) $\int_1^2 (1+x^3)^2\,dx$

(b) $\int_1^2 (1+x^3)^2 x^2\,dx$.

4. Compute with the aid of a table of integrals:

(a) $\int \dfrac{e^x\,dx}{5e^{2x}-3}$

(b) $\int \dfrac{dx}{\sqrt{x^2-3}}$.

5. Compute:

(a) $\int \dfrac{dx}{x^3}$

(b) $\int \dfrac{dx}{\sqrt{x+1}}$

(c) $\int \dfrac{e^x\,dx}{1+5e^x}$.

6. Compute:

$$\int \dfrac{5x^4 - 5x^3 + 10x^2 - 8x + 4}{(x^2+1)(x-1)}\,dx.$$

7. Compute:

$$\int \dfrac{x^3\,dx}{(1+x^2)^4}$$

in two different ways:
(a) by the substitution $u = 1 + x^2$;
(b) by the substitution $x = \tan\theta$.

8. Transform the definite integral

$$\int_0^3 \dfrac{x^3}{\sqrt{x+1}}\,dx$$

to another definite integral in two different ways (and evaluate):
(a) by the substitution $u = x + 1$;
(b) by the substitution $u = \sqrt{x+1}$.

9. Compute $\int x^2 \ln(1+x)\,dx$

(a) without an integral table;
(b) with an integral table.

10. Find $\int \dfrac{x\,dx}{\sqrt{9x^4+16}}$

(a) without an integral table;
(b) with an integral table.

11. Compute $\int \dfrac{\sin\theta\,d\theta}{1+\sin^2\theta}$

(a) by using the substitution that applies to any rational function of $\cos\theta$ and $\sin\theta$;
(b) by writing $\sin^2\theta$ as $1 - \cos^2\theta$ and using the substitution $u = \cos\theta$.

12. (a) Without an integral table, evaluate

$$\int \sin^5\theta\,d\theta \qquad \text{and} \qquad \int \tan^6\theta\,d\theta.$$

(b) Evaluate them with an integral table.

13. Two of these three antiderivatives are elementary. Find them.

(a) $\int \sqrt{1 - 4\sin^2\theta}\,d\theta$

(b) $\int \sqrt{4 - 4\sin^2\theta}\,d\theta$

(c) $\int \sqrt{1 + \cos\theta}\,d\theta$.

14. (a) Transform the definite integral $\int_{-1}^{4} (x+2)/\sqrt{x+3}\,dx$ into another definite integral, using the substitution $u = x + 3$.

(b) Evaluate the integral obtained in (a).

15. The fundamental theorem can be used to evaluate one of these definite integrals, but not the other. Evaluate one and estimate the other.

(a) $\int_0^1 \sqrt[3]{x} \sqrt{x} \, dx$ (b) $\int_0^1 \sqrt[3]{1-x} \sqrt{x} \, dx$.

16. Verify that the following factorizations into irreducible polynomials are correct.
(a) $x^3 - 1 = (x - 1)(x^2 + x + 1)$;
(b) $x^4 - 1 = (x - 1)(x + 1)(x^2 + 1)$;
(c) $x^3 + 1 = (x + 1)(x^2 - x + 1)$;
(d) $x^4 + 1 = (x^2 + \sqrt{2} x + 1)(x^2 - \sqrt{2} x + 1)$.

In Exercises 17 to 24 express as a sum of partial fractions. (Do not integrate.)

17. $\dfrac{2x^2 + 3x + 1}{x^3 - 1}$

18. $\dfrac{x^4 + 2x^2 - 2x + 2}{x^3 - 1}$

19. $\dfrac{2x - 1}{x^3 + 1}$

20. $\dfrac{x^4 + 3x^3 - 2x^2 + 3x - 1}{x^4 - 1}$

21. $\dfrac{x^3 - (1 + \sqrt{2})x^2 + (1 - \sqrt{2})x - 1}{x^4 + 1}$

22. $\dfrac{2x + 5}{x^2 + 3x + 2}$

23. $\dfrac{5x^3 + 11x^2 + 6x + 1}{x^2 + x}$

24. $\dfrac{5x^3 + 6x^2 + 8x + 5}{(x^2 + 1)(x + 1)}$

25. For which values of the nonnegative integers m and n are the following integrations especially simple?

(a) $\int \sin^m x \, dx$ (b) $\int \sec^n x \, dx$

(c) $\int \sin^m x \cos^n x \, dx$ (d) $\int \sec^m x \tan^n x \, dx$

(e) $\int \cot^m x \csc^n x \, dx$.

26. (a) Develop the reduction formula relating

$$\int \sin^n x \, dx \quad \text{to} \quad \int \sin^{n-2} x \, dx.$$

(b) If n is odd, what technique may be used for finding $\int \sin^n x \, dx$?

In Exercises 27 to 32 use an appropriate substitution to obtain an integrand that is a rational function. Do not evaluate.

27. $\int \dfrac{(\sqrt{4 - x^2})^3 + 1}{[(4 - x^2)^3 + 5]\sqrt{4 - x^2}} \, dx$

28. $\int \dfrac{x + \sqrt[3]{x} - 2}{x^2 - \sqrt[3]{x} - 2} \, dx$

29. $\int \dfrac{(x^2 - 5)^7}{x^2 + 3 + \sqrt{x^2 - 5}} \, dx$

30. $\int \dfrac{\cos^2 \theta + \sin \theta}{1 - \sin \theta \cos \theta} \, d\theta$

31. $\int \dfrac{3 \tan^2 \theta + \sec \theta + 1}{2 + \tan \theta + \cos \theta} \, d\theta$

32. $\int \dfrac{(4 + x^2)^{1/2}}{5 + (4 + x^2)^{3/2}} \, dx$

In Exercises 33 to 145 find the integral. Consult with your instructor whether to use integral tables.

33. $\int \dfrac{\cos x \, dx}{\sin^3 x - 8}$ 34. $\int \dfrac{dx}{\sqrt{2 + \sqrt{x}}}$

35. $\int \dfrac{\sqrt{x^2 + 1}}{x^4} \, dx$ 36. $\int \dfrac{\sin x \, dx}{1 + 3 \cos^2 x}$

37. $\int \dfrac{\sin x \, dx}{3 + \cos x}$ 38. $\int x\sqrt{x^4 - 1} \, dx$

39. $\int x^2\sqrt{x^3 - 1} \, dx$ 40. $\int \sin \sqrt{x} \, dx$

41. $\int \dfrac{dx}{(4 + x^2)^2}$ 42. $\int (\sqrt[3]{x} + \sqrt[3]{x + 1}) \, dx$

43. $\int \sin^2 3x \cos^2 3x \, dx$ 44. $\int \sin^3 3x \cos^2 3x \, dx$

45. $\int \tan^4 3\theta \, d\theta$ 46. $\int \dfrac{x^2 \, dx}{x^4 - 1}$

47. $\int \dfrac{x^4 + x^2 + 1}{x^3} \, dx$ 48. $\int \dfrac{3 \, dx}{\sqrt{1 - 5x^2}}$

49. $\int 10^x \, dx$ 50. $\int \dfrac{x^3}{(x^4 + 1)^3} \, dx$

51. $\int \dfrac{x \, dx}{(x^4 + 1)^2}$ 52. $\int \cos^3 x \sin^2 x \, dx$

53. $\int \cos^2 x \, dx$ 54. $\int x\sqrt{x + 4} \, dx$

55. $\int x\sqrt{x^2 + 4}\, dx$

56. $\int \dfrac{x + 2}{x^2 + 1}\, dx$

57. $\int \dfrac{x^2\, dx}{1 + x^6}$

58. $\int \sqrt[3]{4x + 7}\, dx$

59. $\int x^2 \sin x^3\, dx$

60. $\int \dfrac{\ln x^4}{x}\, dx$

61. $\int x^4 \ln x\, dx$

62. $\int \dfrac{\tan^{-1} 3x}{1 + 9x^2}\, dx$

63. $\int \dfrac{e^{\sqrt{x}}}{\sqrt{x}}\, dx$

64. $\int \sin(\ln x)\, dx$

65. $\int \ln(x^3 - 1)\, dx$

66. $\int \tan x\, dx$

67. $\int \dfrac{x\, dx}{\sqrt{(x^2 + 1)^3}}$

68. $\int \dfrac{2 + \sqrt[3]{x}}{x}\, dx$

69. $\int \dfrac{dx}{\sqrt{(x + 1)^3}}$

70. $\int \dfrac{2x + 3}{x^2 + 3x + 5}\, dx$

71. $\int \dfrac{3\, dx}{x^2 + 4x + 5}$

72. $\int \dfrac{3\, dx}{x^2 + 4x - 5}$

73. $\int \dfrac{x\, dx}{1 + \sqrt[3]{x}}$

74. $\int \ln\sqrt{2x - 1}\, dx$

75. $\int \dfrac{x^7\, dx}{\sqrt{x^2 + 1}}$

76. $\int x^3 \tan^{-1} x\, dx$

77. $\int \dfrac{\tan^{-1} x}{x^2}\, dx$

78. $\int \dfrac{dx}{x^3 + 4x}$

79. $\int e^x \sin 3x\, dx$

80. $\int \sqrt{1 - \cos x}\, dx$

81. $\int \dfrac{dx}{(4 - x^2)^{3/2}}$

82. $\int x^{1/4}(1 + x^{1/5})\, dx$

83. $\int \dfrac{x\, dx}{x^4 - 2x^2 - 3}$

84. $\int \sin^{-1}\sqrt{x}\, dx$

85. $\int \dfrac{x^2\, dx}{\sqrt[3]{x - 1}}$

86. $\int \ln(4 + x^2)\, dx$

87. $\int \dfrac{\sqrt{x^2 + 4}}{x}\, dx$

88. $\int \sqrt{\tan\theta}\, \sec^2\theta\, d\theta$

89. $\int \sec^5\theta \tan\theta\, d\theta$

90. $\int \tan^6\theta\, d\theta$

91. $\int \dfrac{dx}{x\sqrt{x^2 + 9}}$

92. $\int (e^x + 1)^2\, dx$

93. $\int \dfrac{(1 - x)^2}{\sqrt[3]{x}}\, dx$

94. $\int (1 + \sqrt{x})x\, dx$

95. $\int \sin^2 2x \cos x\, dx$

96. $\int (e^{2x})^3 e^x\, dx$

97. $\int \left(e^x - \dfrac{1}{e^x}\right)^2 dx$

98. $\int \dfrac{dx}{(\sqrt{x} + 1)(\sqrt{x})}$

99. $\int x \sin^{-1} x^2\, dx$

100. $\int x \sin^{-1} x\, dx$

101. $\int \dfrac{dx}{e^{2x} + 5e^x}$

102. $\int \dfrac{e^x\, dx}{1 - 6e^x + 9e^{2x}}$

103. $\int (2x + 1)\sqrt{3x + 2}\, dx$

104. $\int \dfrac{2x^3 + 1}{x^3 - 4x^2}\, dx$

105. $\int \dfrac{x^2\, dx}{(x - 1)^3}$

106. $\int \dfrac{dx}{\sqrt{9 + x^2}}$

107. $\int \dfrac{e^x + 1}{e^x - 1}\, dx$

108. $\int \dfrac{dx}{4x^2 + 1}$

109. $\int (1 + 3x^2)^2\, dx$

110. $\int \dfrac{x\, dx}{x^3 + 1}$

111. $\int \dfrac{x^3\, dx}{x^3 + 1}$

112. $\int \dfrac{x^2\, dx}{\sqrt{2x + 1}}$

113. $\int \dfrac{dx}{\sqrt{2x + 1}}$

114. $\int (x + \sin x)^2\, dx$

115. $\int \dfrac{x\, dx}{x^4 - 3x^2 - 2}$

116. $\int \dfrac{x^3\, dx}{x^4 - 1}$

117. $\int \dfrac{e^x\, dx}{1 + e^{2x}}$

118. $\int \dfrac{dx}{x^2 + 5x - 6}$

119. $\int \dfrac{dx}{x^2 + 5x + 6}$

120. $\int \dfrac{x\, dx}{2x^2 + 5x + 6}$

121. $\int \dfrac{4x + 10}{x^2 + 5x + 6}\, dx$

122. $\int \sqrt{4x^2 + 1}\, dx$

123. $\int \dfrac{dx}{2x^2 + 5x + 6}$

124. $\int \sqrt{-4x^2 + 1}\, dx$

125. $\int \dfrac{dx}{2x^2 + 5x - 6}$

126. $\int \dfrac{dx}{2 + 3\sin x}$

127. $\int \dfrac{dx}{\sin^2 x}$

128. $\int \dfrac{dx}{3 + 2 \sin x}$

129. $\int \dfrac{dx}{\sin^4 x}$

130. $\int \ln (x^2 + 5)\, dx$

131. $\int \dfrac{dx}{\sqrt{x - 2}\sqrt{3 - x}}$

132. $\int x^3 e^{-5x}\, dx$

133. $\int \sqrt{(1 + 2x)(1 - 2x)}\, dx$

134. $\int x \sin 3x\, dx$

135. $\int \dfrac{2x\, dx}{\sqrt{x^2 + 1}}$

136. $\int \dfrac{2\, dx}{\sqrt{x^2 + 1}}$

137. $\int \dfrac{x^4 + 4x^3 + 6x^2 + 4x - 3}{x^4 - 1}\, dx$

138. $\int \dfrac{x^3 + 6x^2 + 11x + 5}{(x + 2)^2(x + 1)}\, dx$

139. $\int \dfrac{-3x^2 - 11x - 11}{(x + 2)^2(x + 1)}\, dx$

140. $\int \dfrac{x^2 - 3x}{(x + 1)(x - 1)^2}\, dx$

141. $\int \dfrac{12x^2 + 2x + 3}{4x^3 + x}\, dx$

142. $\int \dfrac{6x^3 + 2x + \sqrt{3}}{1 + 3x^2}\, dx$

143. $\int \dfrac{-6x^3 - 13x - 3\sqrt{3}}{1 - 3x^2}\, dx$

144. $\int \dfrac{x\, dx}{\sqrt{1 - 9x^2}}$

145. $\int \dfrac{dx}{\sqrt{1 - 9x^2}}$

146. Complete the square in
   (a) $x^2 + 2x + 1$,
   (b) $x^2 + 2x + 2$,
   (c) $x^2 - 2x + 2$.

147. Compute:
   (a) $\int \dfrac{dx}{x^2 + 4x + 3}$
   (b) $\int \dfrac{dx}{x^2 + 4x + 4}$
   (c) $\int \dfrac{dx}{x^2 + 4x + 5}$
   (d) $\int \dfrac{dx}{x^2 + 4x - 2}$.

148. Compute $\int \dfrac{x^3\, dx}{(x - 1)^2}$

   (a) using partial fractions;
   (b) using the substitution $u = x - 1$.
   (c) Which method is easier?

149. Compute:

   (a) $\int \sec^5 x\, dx$

   (b) $\int \sec^5 x \tan x\, dx$

   (c) $\int \dfrac{\sin x}{(\cos x)^3}\, dx$.

150. (a) Compute $\int \dfrac{x^{2/3}\, dx}{x + 1}$.

   (b) What does a table of integrals say about $\int \dfrac{x^{2/3}\, dx}{x + 1}$?

151. Find $\int x^3/\sqrt{1 + x^2}\, dx$ with the aid of a table of integrals.

152. Compute $\int x\sqrt[3]{x + 1}\, dx$ using
   (a) the substitution $u = \sqrt[3]{x + 1}$;
   (b) the substitution $u = x + 1$.

153. Transform $\int x^2/\sqrt{1 + x}\, dx$ by each of the substitutions
   (a) $u = \sqrt{1 + x}$;
   (b) $u = 1 + x$;
   (c) $x = \tan^2 \theta$.
   (d) Solve the easiest of the resulting problems.

■

154. The following is a quote from an article on the management of energy resources:

   Let $u(t)$ be the rate at which water flows through the turbines at time t. Then $U(t) = \int_0^t u(s)\, ds$ is the total flow during the time $[0, t]$. Let $z(t)$ represent the rate of demand for electric energy at time t, as measured in equivalent water flow. Then the cost incurred during this period of time is $\int_0^t c(z(s) - u(s))\, ds$, where c is a cost function for imported energy.

   Explain why each integral above makes sense.

155. Transform the problem of finding $\int x^3/\sqrt{1 + x^2}\, dx$ to a different problem, using
   (a) integration by parts with $dv = (x\, dx)/\sqrt{1 + x^2}$;
   (b) the substitution $x = \tan \theta$;
   (c) the substitution $u = \sqrt{1 + x^2}$.

156. Two of these three integrals are elementary. Evaluate them.

   (a) $\int \sin^2 x\, dx$

   (b) $\int \sin x^2\, dx$

   (c) $\int \sin \sqrt{x}\, dx$.

157. Compute $\int x\sqrt{1 + x}\, dx$ in three ways:
   (a) let $u = \sqrt{1 + x}$;
   (b) let $x = \tan^2 \theta$;
   (c) by parts, with $u = x$, $dv = \sqrt{1 + x}\, dx$.

158. (a) Which is a more convenient method for finding $\int x^2/(x - 3)^{10}\, dx$, partial fractions or the substitution $u = x - 3$?
   (b) Find $\int x^2/(x - 3)^{10}\, dx$.

159. Assuming that $\int e^x/x\, dx$ is not elementary (a theorem of Liouville), prove that $\int 1/\ln x\, dx$ is not elementary.

160. (a) Explain why $\int x^m e^x\, dx$ is an elementary function for any positive integer m.

(b) Explain why $\int x^m (\ln x)^n\, dx$ is an elementary function for any positive integers m and n. *Hint:* Make the substitution $u = \ln x$.

161. (a) Prove the trigonometric identity

$$\sin mx \sin nx = \tfrac{1}{2}\{\cos [(m-n)x] - \cos [(m+n)x]\}$$

by expanding the right side.

(b) Use it to compute $\int \sin 2x \sin 3x\, dx$.

■ ■

162. Liouville proved that, if f and g are rational functions and if $\int e^{f(x)} g(x)\, dx$ is an elementary function, then $\int e^{f(x)} g(x)\, dx$ can be expressed in the form $e^{f(x)} w(x)$, where $w(x)$ is a rational function. With the aid of this result, prove that $\int e^x/x\, dx$ is not an elementary function. *Hint:* Assume $[e^x(p/q)]' = e^x/x$, where p and q are relatively prime polynomials. Write $q = x^i r$, where $i \geq 0$ and x does not divide the polynomial r.

Exercises 163 to 166 concern Wallis' formula for π.

163. (a) Using the reduction formula of Example 4 of Sec. 9.5, prove that

$$\int_0^{\pi/2} \sin^n x\, dx = \frac{n-1}{n} \int_0^{\pi/2} \sin^{n-2} x\, dx.$$

(b) Let $I_n = \int_0^{\pi/2} \sin^n x\, dx$. Show that $I_0 = \pi/2$ and $I_1 = 1$.

(c) With the aid of (a) and (b), show that

$$I_7 = \frac{6}{7} \frac{4}{5} \frac{2}{3} \quad \text{and} \quad I_6 = \frac{5}{6} \frac{3}{4} \frac{1}{2} \frac{\pi}{2}.$$

(d) With the aid of (a) and (b), show that

$$I_{2n} = \frac{2n-1}{2n} \frac{2n-3}{2n-2} \cdots \frac{3}{4} \frac{1}{2} \frac{\pi}{2};$$

and

$$I_{2n+1} = \frac{2n}{2n+1} \frac{2n-2}{2n-1} \cdots \frac{4}{5} \frac{2}{3}.$$

164. (See Exercise 163.)

(a) Show that

$$\frac{I_7}{I_6} = \frac{6}{7} \frac{6}{5} \frac{4}{5} \frac{4}{3} \frac{2}{3} \frac{2}{1} \frac{2}{\pi}.$$

(b) Show that

$$\frac{I_{2n+1}}{I_{2n}} = \frac{2n}{2n+1} \frac{2n}{2n-1} \frac{2n-2}{2n-1} \cdots \frac{2}{3} \frac{2}{1} \frac{2}{\pi}.$$

(c) Show that

$$\frac{2n}{2n+1} I_{2n} < \frac{2n}{2n+1} I_{2n-1} = I_{2n+1} < I_{2n},$$

and thus

$$\lim_{n \to \infty} \frac{I_{2n+1}}{I_{2n}} = 1.$$

(d) From (b) and (c) deduce that

$$\lim_{n \to \infty} \frac{2 \cdot 2\; 4 \cdot 4\; 6 \cdot 6}{1 \cdot 3\; 3 \cdot 5\; 5 \cdot 7} \cdots \frac{(2n)(2n)}{(2n-1)(2n+1)} = \frac{\pi}{2}.$$

This is Wallis' formula, usually written in shorthand

as

$$\frac{2 \cdot 2\; 4 \cdot 4\; 6 \cdot 6}{1 \cdot 3\; 3 \cdot 5\; 5 \cdot 7} \cdots = \frac{\pi}{2}.$$

165. (See Exercise 164.)

(a) Show that Wallis' formula is equivalent to

$$\left(1 - \frac{1}{2^2}\right)\left(1 - \frac{1}{4^2}\right)\left(1 - \frac{1}{6^2}\right) \cdots = \frac{2}{\pi}.$$

(b) Show that

$$\left(1 - \frac{1}{2^2}\right)\left(1 - \frac{1}{3^2}\right)\left(1 - \frac{1}{4^2}\right) \cdots = \frac{1}{2}.$$

(c) From (a) and (b) deduce that

$$\left(1 - \frac{1}{3^2}\right)\left(1 - \frac{1}{5^2}\right)\left(1 - \frac{1}{7^2}\right) \cdots = \frac{\pi}{4}.$$

166. Let $n!$ denote the product of the integers from 1 to the positive integer n. For instance, 5! equals the product $1 \cdot 2 \cdot 3 \cdot 4 \cdot 5 = 120$.

(a) Show that $2 \cdot 4 \cdot 6 \cdot 8 \cdots 2n = 2^n n!$.

(b) Show that $1 \cdot 3 \cdot 5 \cdots (2n-1) = (2n)!/(2^n n!)$.

(c) From Exercise 164(d) deduce that

$$\lim_{n \to \infty} \frac{(n!)^2 4^n}{(2n)! \sqrt{2n+1}} = \sqrt{\frac{\pi}{2}}.$$

Exercise 167 is the basis of the remaining exercises.

167. Let p and q be rational numbers. Prove that $\int x^p (1-x)^q\, dx$ is an elementary function

(a) if p is an integer (*Hint:* if $q = s/t$, let $1 - x = v^t$);

(b) if q is an integer;

(c) if $p + q$ is an integer. Chebyshev proved that these are the only cases for which the antiderivative in question is elementary. In particular,

$$\int \sqrt{x} \sqrt[3]{1-x}\, dx \quad \text{and} \quad \int \sqrt[3]{x - x^2}\, dx$$

are not elementary.

168. Deduce from Exercise 167 that $\int \sqrt{1-x^3}\, dx$ is not elementary.

169. Deduce from Exercise 167 that $\int (1-x^n)^{1/m}\, dx$, where m and n are positive integers, is elementary if and only if $m = 1$ and n is arbitrary, $n = 1$ and m is arbitrary, or $m = 2 = n$.

170. Deduce from Exercise 167 that $\int \sqrt{\sin x}\, dx$ is not elementary. *Hint:* Let $u = \sin^2 x$.

171. Deduce from Exercise 167 that $\int \sin^a x\, dx$, where a is rational, is elementary if and only if a is an integer.

172. Deduce from Exercise 167 that $\int \sin^p x \cos^q x\, dx$, where p and q are rational, is elementary if and only if p or q is an odd integer or $p + q$ is an even integer.

10

COMPUTING AND APPLYING DEFINITE INTEGRALS OVER INTERVALS

This chapter applies the ideas and techniques of Chaps. 7 to 9 to problems mainly about area, volume, arc length, and surface area. The purpose of the chapter is to develop skill in (1) recognizing when a quantity can be represented as a definite integral, and (2) evaluating definite integrals, by means of the fundamental theorem of calculus. It is pointless to memorize the various formulas that will be obtained. Understanding of the basic ideas should make it possible to reconstruct any formula when needed. After all, the areas, arc lengths, and volumes that are of interest have already been computed and have long been listed in mathematical handbooks. It is important, though, to understand why such quantities can be represented as definite integrals.

10.1

How to compute the cross-sectional length $c(x)$

Section 7.4 showed that the area of a region in the plane is

$$\int_a^b c(x)\, dx,$$

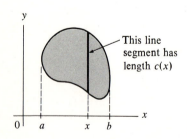

This line segment has length $c(x)$

where $c(x)$ is the length of the typical intersection of a line perpendicular to the x axis with the given region. See the accompanying figure and note that x need not refer to the x axis of the xy plane; it may refer to any conveniently chosen line in the plane. It may even refer to the y axis; in this case the cross-sectional length would be noted $c(y)$. If these facts seem vague, review the part of Sec. 7.4 labeled "Sums and the Area of a Plane Region."

To compute an area:

1. Find a, b, and the cross-sectional length $c(x)$.
2. Evaluate $\int_a^b c(x)\, dx$ by the fundamental theorem of calculus if $\int c(x)\, dx$ is elementary and simple enough to find.

Chapter 9 showed how to accomplish step 2. A few examples will show how to find a, b, and $c(x)$, a procedure known as *setting up a definite integral*.

Two basic geometric facts are of use in finding $c(x)$ in many problems:

THE PYTHAGOREAN THEOREM   In a right triangle whose legs have lengths a and b, and whose hypotenuse has length c, $c^2 = a^2 + b^2$.

CORRESPONDING PARTS OF SIMILAR TRIANGLES ARE PROPORTIONAL   If a, b, c are the lengths of the sides of one triangle and a', b', c' are the lengths of the corresponding sides of a similar triangle, then $a'/a = b'/b = c'/c$.

The following examples illustrate the use of these two geometric facts. In order to concentrate on the geometric aspects of setting up the appropriate definite integrals, the evaluations of the definite integrals for area are delayed until Sec. 10.3.

EXAMPLE 1   Find the cross-sectional length $c(x)$ if R is a circle of radius 5. Set up a definite integral for the area of R.

SOLUTION   We are free to put the axis anywhere in the plane. Place it in such a way that its origin is below the center of the circle. This provides the simplest formula for $c(x)$, the length of typical chord AB perpendicular to the x axis. This is illustrated at the left.

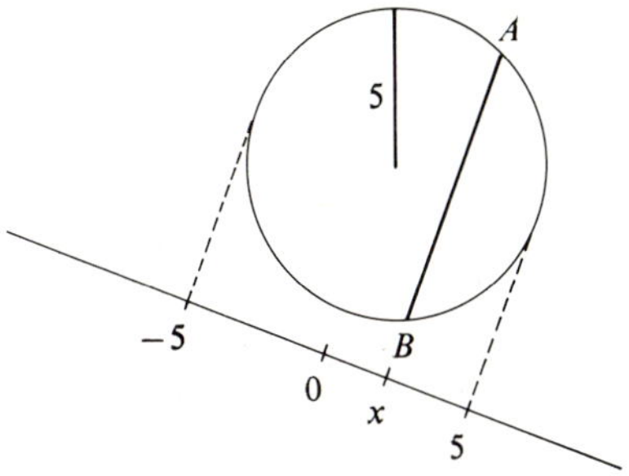

Note that $a = -5$ and $b = 5$. To get a feel for $c(x)$ in this case, note by glancing at the diagram that

$$c(-5) = 0, \qquad c(0) = 10, \qquad \text{and} \qquad c(5) = 0.$$

$c(x) = $ length of chord AB

To find $c(x)$ for any x in the interval $[-5, 5]$, draw the line through the center of the circle and parallel to the x axis. It meets the segment AB at a point M. Call the center of the circle C. Also draw the segment AC, a radius of the circle. Then

$$c(x) = \overline{AB} = 2\overline{AM}.$$

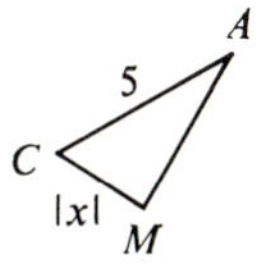

To find $\overline{AM}$, use the right triangle $\triangle ACM$. One side CM has length $|x|$, while the hypotenuse has length 5. Hence $5^2 = |x|^2 + \overline{AM}^2$. Since $|x|^2 = x^2$, this equation gives a simple formula for $\overline{AM}^2$,

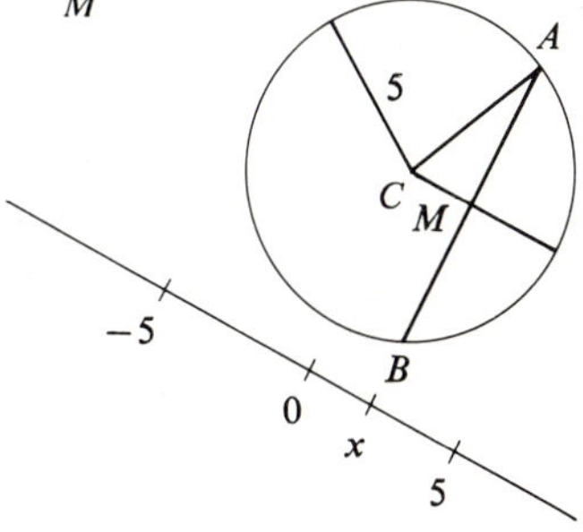

$$\overline{AM}^2 = 5^2 - x^2 = 25 - x^2.$$

Hence

$$\overline{AM} = \sqrt{25 - x^2}$$

and, since $c(x) = 2\overline{AM}$,

$$c(x) = 2\sqrt{25 - x^2}.$$

Thus the area of the circle is

$$\int_{-5}^{5} 2\sqrt{25 - x^2} \, dx. \ \bullet$$

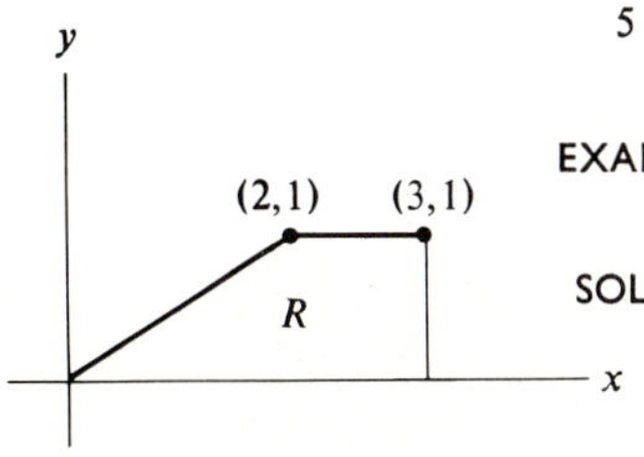

EXAMPLE 2   Set up a definite integral for the area of the trapezoid R shown in the diagram.

SOLUTION   Consider cross sections of R by lines parallel to the y axis. Let $c(x)$ be the length of a typical cross section. Observe that x is in the interval $[0, 3]$. The cross section changes character at $x = 2$; if $x \geq 2$, the cross-section function is constant, and $c(x) = 1$; if $x \leq 2$, a little work is required to find $c(x)$. The diagram in the margin shows the typical cross section $B'C'$ for x in $[0, 2]$. Note that

$$c(x) = \overline{B'C'}.$$

Since $\triangle AB'C'$ is similar to $\triangle ABC$,

$$\frac{x}{2} = \frac{\overline{B'C'}}{1}.$$

Thus

$$\overline{B'C'} = \frac{x}{2}.$$

Consequently, for x in the interval $[0, 2]$,

$$c(x) = \frac{x}{2}.$$

Both cases together provide a formula for $c(x)$:

$$c(x) = \begin{cases} \dfrac{x}{2} & \text{if } 0 \leq x \leq 2 \\ 1 & \text{if } 2 \leq x \leq 3. \end{cases}$$

The area of the trapezoid is therefore

$$\int_0^3 c(x)\,dx = \int_0^2 \frac{x}{2}\,dx + \int_2^3 1\,dx. \ \bullet$$

The next example shows how to find $c(x)$ if the border of R is described by formulas.

EXAMPLE 3   Set up a definite integral for the area of the region R shown in the accompanying diagram.

SOLUTION   Consider cross sections parallel to the y axis. Observe that x runs from 0 to 2. What is $c(x)$ for each x? For instance,

$$c(0) = 0 \qquad \text{and} \qquad c(2) = 7.$$

The typical cross-sectional segment perpendicular to the x axis has one end A on the line through $(0, 0)$ and $(2, -3)$, the other end B on the parabola $y = x^2$, and meets the x axis at point C.

Then

$$c(x) = \overline{AB} = \overline{AC} + \overline{CB}.$$

The length $\overline{CB}$ is given by

$$\overline{CB} = x^2.$$

Though $\overline{AC}$ could be found by the method of similar triangles, let us use analytic geometry. The point $A = (x, y)$ is on the line through the points

$$(0, 0) \qquad \text{and} \qquad (2, -3).$$

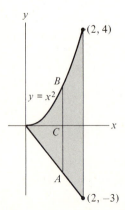

Hence
$$\frac{y - 0}{x - 0} = \frac{-3 - 0}{2 - 0},$$

or
$$\frac{y}{x} = -\frac{3}{2}.$$

Thus the y coordinate of A is $-\frac{3}{2}x$.

The length $\overline{AC}$, being positive, is $\frac{3}{2}x$. Consequently,

$$c(x) = \overline{AC} + \overline{CB} = x^2 + \tfrac{3}{2}x.$$

The area of R is therefore represented by the definite integral

$$\int_0^2 (x^2 + \tfrac{3}{2}x)\, dx,$$

easily evaluated by the fundamental theorem. (As the reader may check, the result is $\frac{17}{3}$.) ●

Remark: If the region R is bordered by the curves $y = f_1(x)$ and $y = f_2(x)$, $f_1(x) \geq f_2(x)$, for all x in the interval $[a, b]$, then

$$c(x) = f_1(x) - f_2(x).$$

So the region between the two curves for x in $[a, b]$ has area

$$\int_a^b [f_1(x) - f_2(x)]\, dx.$$

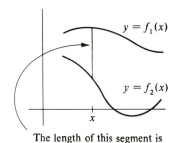

$y = f_1(x)$

$y = f_2(x)$

x

The length of this segment is
$c(x) = f_1(x) - f_2(x)$

EXAMPLE 4   Set up a definite integral for the area of the region R bounded by $y = \sqrt{x}$, the x axis, and the line $x = 4$. Use cross sections parallel to the x axis (see the first figure below in the margin).

SOLUTION   First find $c(y)$, the length of a typical cross section of R by a line perpendicular to the y axis. Though the letter has switched from x to y, the idea is the same. First determine the interval of y values, and then find $c(y)$. To begin, label several points in the diagram, shown below. The largest value of y in this region is the y coordinate of point C. Since C lies on the curve $y = \sqrt{x}$, $C = (4, \sqrt{4}) = (4, 2)$. Hence y goes from 0 to 2.

Now find $c(y)$. Inspection of the diagram shows that $c(y) = \overline{AB} = (x$ coordinate of $B) - (x$ coordinate of $A) = 4 - (x$ coordinate of $A)$. How do we obtain the x coordinate of A? Since A lies on the curve $y = \sqrt{x}$, the coordinates of A satisfy the equation

$$y = \sqrt{x}.$$

To express x in terms of y, solve for x:

$$y^2 = x,$$

hence
$$x = y^2.$$

Thus
$$c(y) = 4 - (x \text{ coordinate of } A) = 4 - y^2.$$

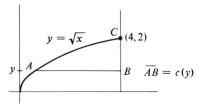

$y = \sqrt{x}$ C (4, 2)

y A B $\overline{AB} = c(y)$

The definite integral for the area of the region is therefore

$$\int_0^2 (4 - y^2)\, dy.$$

(The reader may easily compute it by the fundamental theorem. The result is $\frac{16}{3}$.) ●

Exercises

1. Let R be the region bordered by $y = x^2$ and $y = x^3$.
   (a) Draw R.
   (b) Find the length of the vertical cross section $c(x)$.
   (c) Find the length of the horizontal cross section $c(y)$.
2. In Example 1 the origin of the x axis was chosen below the center of the circle. If, instead, it is chosen as shown in the accompanying diagram, what would a, b, and the vertical cross section $c(x)$ be?

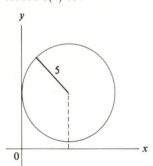

3. Find the vertical cross-sectional length $c(x)$ if x denotes the x of the x axis and R is the shaded region shown in the figure below.

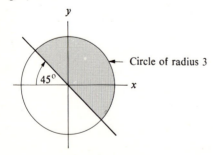

4. Find the horizontal cross-sectional length $c(y)$ for the region in Exercise 3.
5. Find the horizontal cross-sectional length $c(y)$ for the region in Example 2.
6. Find the horizontal cross-sectional length $c(y)$ for the region in Example 3.

7. Find the vertical cross-sectional length $c(x)$ for the region between $y = e^x$, $y = 1 + x$, and $x = 2$.
8. Find the horizontal cross-sectional length $c(y)$ for the region in Exercise 7.
9. Let R be the triangle whose vertices are $(1, 0)$, $(3, 2)$, and $(4, 1)$. Find $\overset{\circ}{c}(x)$, the vertical cross-sectional length.
10. (a) Find the horizontal cross-sectional length $c(y)$ for the region in Exercise 9, using similar triangles.
    (b) Which has the easier description, $c(y)$ in (a) or $c(x)$ in Exercise 9?
11. Let R be the region between $y = x^2$ and $y = x + 1$.
    (a) Draw R.
    (b) Find $c(x)$, the vertical cross-sectional length.
    (c) Find $c(y)$, the horizontal cross-sectional length.

■

12. (a) Draw the region bordered by $y = \sin x$ and the x axis for x in $[0, \pi]$.
    (b) Find $c(x)$, the vertical cross-sectional length.
    (c) Find $c(y)$, the horizontal cross-sectional length.
13. (a) Draw the region bordered by $y = x/2$, $y = x - 1$, and the x axis.
    (b) Find $c(x)$, the vertical cross-sectional length.
    (c) Find $c(y)$, the horizontal cross-sectional length.
14. (a) Draw the region bordered by $y = \ln x$, $y = x/e$, and the x axis.
    (b) Find $c(x)$, the vertical cross-sectional length.
    (c) Find $c(y)$, the horizontal cross-sectional length.
15. Compute the horizontal cross-sectional length $c(y)$ for the triangle with each of these three choices of the origin and direction of the y axis.

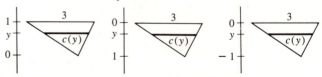

Hint: Use similar triangles in each case. Note that a wise choice of coordinate system can simplify the cross-section function.

16. (a) Guess what the horizontal cross-sectional length $c(y)$ is in the isosceles trapezoid depicted below. *Hint:* What is the simplest formula that gives $c(0) = 8$ and $c(2) = 14$?

(b) Using similar triangles, find $c(y)$.

17. (a) Draw the region R inside the ellipse

$$\frac{x^2}{4} + \frac{y^2}{9} = 1.$$

(b) Find $c(x)$, the vertical cross-sectional length.
(c) Find $c(y)$, the horizontal cross-sectional length.

18. Let R be the region cut from a circle of radius 5 by two parallel lines, each a distance 3 from the center.

(a) Choosing an x axis parallel to the lines, and origin beneath the center, find $c(x)$, the length of cross sections perpendicular to the x axis.
(b) Choosing an x axis perpendicular to the lines, and the origin beneath the center, find $c(x)$, the length of cross sections perpendicular to the x axis.

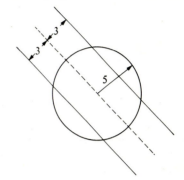

10.2

How to compute the cross-sectional area $A(x)$

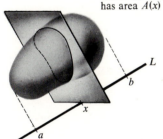

This plane region inside the solid has area $A(x)$

Section 7.4 showed that the volume of a solid is $\displaystyle\int_a^b A(x)\,dx$,

where $A(x)$ is the area of the intersection with the solid of a typical plane perpendicular to the x axis. If there is any doubt about this, review the part of Sec. 7.4 labeled "Sums and the Volume of a Solid Region."

This section shows how to find $A(x)$ in specific cases, and how to set up definite integrals for volumes. In Sec. 10.3 definite integrals for volumes will be evaluated by the methods of Chap. 9.

In addition to the pythagorean theorem and the properties of similar triangles, formulas for the areas of familiar plane figures will be needed:

Figure	Area
Circle of radius r	πr^2
Square of side s	s^2
Equilateral triangle of side s	$\dfrac{\sqrt{3s^2}}{4}$
Triangle of height h and base b	$\dfrac{bh}{2}$
Trapezoid of height h and parallel bases b and B	$h\left(\dfrac{b+B}{2}\right)$
Sector of circle of radius r and angle θ (in radians)	$\dfrac{r^2\theta}{2}$

Also keep in mind that, if corresponding sides of similar figures have the ratio k, then their areas have the ratio k^2, that is, the area is proportional to the square of the lengths of corresponding sides.

A few examples will show what may be involved in finding the cross-sectional area $A(x)$.

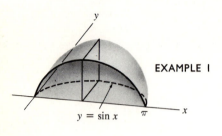

y = sin x

EXAMPLE I Set up a definite integral for the volume of the solid whose base is the region bounded by the x axis and the arch of the curve $y = \sin x$ from $x = 0$ to $x = \pi$ and for which each plane section perpendicular to the x axis is a square whose base lies in the region.

SOLUTION Since the cross section is a square whose side is $\sin x$, its area is $\sin^2 x$. Thus

$$A(x) = \sin^2 x.$$

The definite integral for the volume is therefore

$$\int_0^\pi \sin^2 x \, dx.$$

(By the shortcut after Example 1 of Sec. 9.8, this definite integral has the value $\pi/2$.) ●

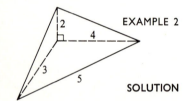

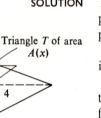

Triangle T of area $A(x)$

EXAMPLE 2 Set up a definite integral for the volume of a solid triangular pyramid whose base is a right triangle of sides 3, 4, and 5. The altitude of the pyramid is above the vertex of the right angle and has length 2.

SOLUTION There are three convenient directions in which to define cross sections by planes, namely parallel to each of the three right-triangular faces. Choose planes perpendicular to the altitude, that is, parallel to the base.

Introduce an x axis perpendicular to the base triangle and with origin in the plane of that triangle.

The typical cross section is a triangle T. Moreover, this triangle is similar to the base triangle. Since the dimensions of T are proportional to the distance from T to the top vertex, $A(x)$ is proportional to the square of the distance from T to the top vertex. This distance is $2 - x$. Therefore there is a constant k such that

$$A(x) = k(2 - x)^2.$$

To find k, notice that, when $x = 0$, T is the base triangle, of area $\frac{1}{2}(3)(4) = 6$. Hence

$$A(0) = k(2 - 0)^2 = 6.$$

Thus

$$4k = 6$$

and

$$k = \tfrac{6}{4} = \tfrac{3}{2}.$$

Consequently

$$A(x) = \tfrac{3}{2}(2 - x)^2.$$

Thus the volume of the pyramid is

$$\int_0^2 \tfrac{3}{2}(2 - x)^2 \, dx. \; \bullet$$

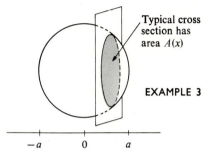

Typical cross section has area $A(x)$

Incidentally, in Exercise 3 the coordinate system on the x axis is chosen in such a way that 0 corresponds to the plane through the top vertex and 2 to the plane of the base triangle; $A(x)$ then has a formula simpler than that just obtained in this example.

EXAMPLE 3   Consider a solid sphere of radius a.   Place an x axis in such a way that its origin is beneath the center of the sphere.   Find a formula for $A(x)$, the area of the cross section of the sphere by a plane perpendicular to the x axis at the point whose coordinate is x.   Set up a definite integral for the volume of the sphere.

SOLUTION   The typical cross section is a circle.   Since the area of a circle of radius r is πr^2, all that remains is to find r^2 in terms of x.

To accomplish this, draw a side view of the sphere and the typical cross section, showing r and x clearly.

By the pythagorean theorem,

$$a^2 = |x|^2 + r^2;$$

hence

$$a^2 = x^2 + r^2,$$

and

$$r^2 = a^2 - x^2.$$

Consequently, $$A(x) = \pi r^2 = \pi(a^2 - x^2).$$

The volume of the sphere of radius a is therefore

$$\int_{-a}^{a} \pi(a^2 - x^2) \, dx,$$

an integral easy to evaluate by the fundamental theorem. $\bullet$

The next example is of special interest, because there are several directions in which to choose cross sections, some convenient and some inconvenient.

EXAMPLE 4   A drinking glass is a right circular cylinder of radius a and height h.   It is tilted until the water level bisects the base and touches the rim, as shown in the accompanying figure.   The water occupies a region R in space.   Find the area of plane cross sections for this region.

Radius of base is a; AB is a diameter

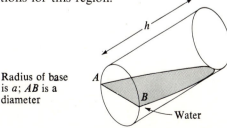

SOLUTION   Here are some of the possibilities for the direction to use for plane cross sections:

1. Planes parallel to the surface of the water.
2. Planes parallel to the base of the glass.
3. Planes perpendicular to the diameter AB.
4. Planes parallel to the plane through AB and the axis of the cylinder.

As each is considered, the reader should make his own sketch in order to develop his geometric intuition. (Little that is done in high school or in analytic geometry develops a feeling for space; topics there are usually at most two-dimensional.) Even better, saw a piece out of a cylinder of wood to represent the water.

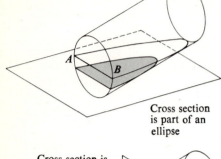

Cross section is part of a disk

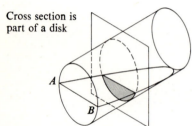

Cross section is part of an ellipse

1. *Planes parallel to the surface of the water.* In this case, each cross section is part of an ellipse. (See Appendix B for a discussion of the ellipse, the oval curve obtained when sawing a right circular cylinder, for instance, a broomstick, at an angle.) The highest cross section, the one through AB, is exactly half of an ellipse, but the lower ones are smaller portions of ellipses. (See the first figure in the margin.) This is not a good choice, since the area of a portion of an ellipse is hard to compute.

2. *Planes parallel to the base of the glass.* In this case each cross section is part of a disk. This is a bit easier than case 1, but still not the most convenient choice.

3. *Planes perpendicular to the diameter AB.* In this case the cross sections are right triangles. (In spite of all the curves in the figure, the cross sections have a border made up only of line segments.) Since the area of a right triangle is easy to compute, let us find $A(x)$ in this case.

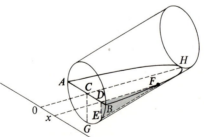

Cross section is a right triangle *DEF*

Introduce an x axis parallel to the diameter AB, and locate the origin 0 to correspond with the midpoint of the diameter. Next, label key points, as shown in the figures above and to the left.

The area of the triangle is

$$A(x) = \tfrac{1}{2} \text{ base} \cdot \text{altitude} = \tfrac{1}{2}\overline{DE} \cdot \overline{EF}.$$

First find $\overline{DE}$ in terms of x, using the pythagorean theorem:

$$a^2 = \overline{CD}^2 + \overline{DE}^2 = x^2 + \overline{DE}^2.$$

(Use right triangle $\triangle\, CDE$, whose hypotenuse CE is dotted in lightly.)

Thus $\overline{DE} = \sqrt{a^2 - x^2}.$

View of base

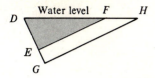

Side view showing cross section
of area $A(x)$ (C and D overlap)

This expresses $\overline{DE}$ in terms of x.   Next consider $\overline{EF}$.
The side view shows two similar triangles,

$$\triangle DEF \qquad \text{and} \qquad \triangle DGH.$$

Thus there is this equality relating corresponding parts:

$$\frac{\overline{DE}}{\overline{DG}} = \frac{\overline{EF}}{\overline{GH}}.$$

Since $\qquad \overline{DE} = \sqrt{a^2 - x^2}, \qquad \overline{DG} = a, \qquad$ and $\qquad \overline{GH} = h,$

the above equality relating corresponding parts becomes

$$\frac{\sqrt{a^2 - x^2}}{a} = \frac{\overline{EF}}{h}.$$

Thus $\qquad \overline{EF} = \frac{h}{a}\sqrt{a^2 - x^2}.$

The area of the typical triangle then is

$$A(x) = \tfrac{1}{2}\,\overline{DE} \cdot \overline{EF}$$

$$= \frac{1}{2}\sqrt{a^2 - x^2} \cdot \frac{h}{a}\sqrt{a^2 - x^2}$$

$$= \frac{h(a^2 - x^2)}{2a}.$$

(The volume of the solid is therefore

$$\int_{-a}^{a} \frac{h(a^2 - x^2)}{2a}\,dx,$$

an integral easily evaluated by the fundamental theorem.)

4. *Planes parallel to the plane through AB and the axis of the cylinder.* In this case the cross sections are rectangles.   All four sides are straight. (Even the side at the water surface is straight, for it is the intersection of two planes, the plane of the water surface and the plane that defines the cross section.)   This case, like case 3, is not hard, for the area of a rectangle is easy to compute. ●

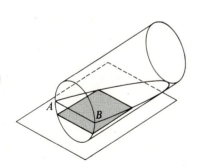

Exercises

1. (a) Compute the areas of a square of side s and a square of side ks.
   (b) Find the quotient of the second area divided by the first area.
   (c) What principle does (b) illustrate?

2. The base of a certain solid is the region in the xy plane bounded by the curve $y = x^2$, the x axis, and the line $x = 2$. Each section cut by a plane perpendicular to the x axis is an equilateral triangle.
   (a) Draw the solid.

(b) Find $A(x)$, the area of a cross section of the solid by a typical plane perpendicular to the x axis.

(c) Set up the definite integral for the volume of the solid.

3. Find $A(x)$ and set up the definite integral for Example 2 if the origin of the x axis, instead of being in the plane of the base triangle, is level with the top vertex, and the positive part of the x axis is downward.

4. Consider the pyramid in Example 2. Take cross sections parallel to the right triangular face whose legs have length 2 and 3. Compute the typical cross-sectional area $A(x)$ if the x axis, perpendicular to the above face, (a) has its origin in the plane of the face mentioned; (b) has its origin at the vertex that is not in the face mentioned. (c) In each case above set up the definite integral for the volume.

5. A right circular cone has height h and radius a. Consider cross sections parallel to the base. Introduce an x axis such that $x = 0$ corresponds to the vertex of the cone and $x = h$ corresponds to the base.

(a) Draw the cone, the x axis, and the typical cross section.

(b) Compute $A(x)$, the typical cross-sectional area for planes perpendicular to the x axis.

(c) Set up the definite integral for the volume.

6. Consider the cone in Exercise 5. (a) Draw a typical cross section of it made by a plane parallel to the axis of the cone. (b) Is this cross section a triangle, part of a parabola, or part of a hyperbola?

7. Using a sharp pencil and a ruler, draw clearly the four types of cross sections discussed for the water in the drinking glass of Example 4.

8. (a) Make a physical model of the solid in Example 4, either by carving soap, modeling clay, or sawing a dowel.

(b) Mark in the model the outlines of the four types of cross sections discussed in Example 4.

9. Compute $A(x)$ for case 4, the rectangular cross sections of Example 4. Place the x axis perpendicular to the diameter AB and parallel to the base of the glass. Let $x = 0$ correspond to a plane through AB.

(a) Find the width of the typical rectangle, that is, the length of the side parallel to AB.

(b) Find the length of the typical rectangle.

(c) Compute $A(x)$.

(d) Use (c) to set up a definite integral for the volume of the solid.

(e) Evaluate the integral in (d).

10. A drinking glass is a right circular cylinder of radius a and height h. It is tilted until the water level just covers the base and touches the rim.

(a) Draw the water.

(b) Draw the typical cross sections corresponding to the four cases in Example 4.

11. (See Exercise 10.) Compute $A(x)$, the cross-sectional area, using trapezoidal cross sections. (Place the origin of the x axis on the axis of the cylinder.)

12. (See Exercise 10.) Compute the typical cross-sectional area $A(x)$, using rectangular cross sections. (Place the origin of the x axis on the axis of the cylinder.)

13. Without using any calculus, find the volume of the solid in Exercise 10.

■

14. A lumberjack saws a wedge out of a cylindrical tree of radius a. His first cut is parallel to the ground and stops at the axis of the tree. His second cut makes an angle θ with the first cut and meets it along a diameter. Draw the solid.

15. (See Exercise 14.) Place the x axis in such a way that the cross sections of the wedge in Exercise 14 are triangles.

(a) Draw the typical cross section.

(b) Find the area of a typical cross section made by a plane at a distance x from the axis of the tree.

16. (See Exercise 14.) Place the x axis in such a way that the sections of the wedge in Exercise 14 are rectangles.

(a) Draw a typical cross section.

(b) Find its area if it is made by a plane at a distance x from the axis of the cylinder.

17. The circle in the xy plane whose center is $(4, 0)$ and whose radius is 3 is rotated around the y axis, producing a doughnut-shaped solid. Find the area of a cross section of this solid made by a plane perpendicular to the y axis and passing through the point $(0, y)$.

18. A drill of radius 3 inches bores a hole through a sphere of radius 5 inches, passing symmetrically through the center of the sphere.

(a) Draw the part of the sphere removed by the drill.

(b) Find $A(x)$, the area of a cross section of the region in (a) made by a plane perpendicular to the axis of the drill and at a distance x from the center of the sphere.

10.3

Computing area and volume by cross sections

This section, which contains no new ideas, simply combines techniques already developed in order to compute areas and volumes.

To compute the area of a region in the plane, follow these steps:

1. Choose a direction for cross sections, compute the cross-sectional length $c(x)$, and find the interval $[a, b]$ through which x ranges. (This is the substance of Sec. 10.1.)

2. Express the area of the region as the definite integral

$$\int_a^b c(x)\, dx.$$

(This was discussed in Sec. 7.4.)

3. Evaluate $\int_a^b c(x)\, dx$ by the fundamental theorem of calculus. (Techniques for doing this were discussed in Chap. 9.) If the antiderivative of the function $c(x)$ is not elementary, then it is necessary to estimate the definite integral, and hence the area.

A similar approach works for the computation of volumes, the main difference being that cross sections are made by planes rather than by lines. The resulting integral is

$$V = \int_a^b A(x)\, dx.$$

The exercises offer ample opportunity to apply these techniques.

Exercises

Exercises 1 to 24 concern area.

1. Find the area of the circle of radius 5 discussed in Example 1 of Sec. 10.1. (This is circular reasoning, since the formula for the area of a circle was used to find the derivative of $\sin x$, hence of $\sin^{-1} x$, which is needed in order to apply the fundamental theorem of calculus.)

2. Find the area of the region below $y = x^2$ and above, $y = -\frac{3}{2}x$, between $x = 0$ and $x = 2$. (See Example 3 of Sec. 10.1.)

3. Find the area of the region in Example 3 of Sec. 10.1, using horizontal cross sections.

4. Find the area of the region below $y = \sqrt{x}$, above the x axis, and to the left of $x = 4$. (See Example 4 of Sec. 10.1.)
   (a) Use vertical cross sections.
   (b) Use horizontal cross sections.

5. Find the area of the region described in Exercise 7 of Sec. 10.1.

6. Find the area of the region described in Exercise 7 of Sec. 10.1, using horizontal cross sections.

7. Find the area of the region described in Exercise 11 of Sec. 10.1, using (a) vertical cross sections, (b) horizontal cross sections.

8. Find the area of the region described in Exercise 13 of Sec. 10.1, using (a) vertical cross sections, (b) horizontal cross sections.

9. Find the area of the region inside the ellipse in Exercise 17 of Sec. 10.1.

10. Find the area of the region in Exercise 18 of Sec. 10.1.

11. Find the area of the region bordered by $y = 1/x^2$, the x axis, and the lines $x = 1$ and $x = 3$, using (a) vertical cross sections, (b) horizontal cross sections.

12. Find the area of the region between $y = 4x$ and $y = 2x^2$, using (a) vertical cross sections, (b) horizontal cross sections.

13. Find the area of the region bounded by the curves $y = 2^x$ and $y = 1 + x$.

14. Find the area of the region bounded by the curves $y = \ln x$, $y = x/e$, and $y = 0$.

15. Find the area of the region bounded by $y = (\pi/2) \sin x$ and $y = x$, and to the right of the y axis.

16. Find the area of the region bounded by $y = x^2$ and $y = 8 - x^2$.

17. (a) Sketch the region in the plane bounded by the curve $y = x^2$ and the line $y = 4$.
    (b) Find its area, using vertical cross sections.
    (c) Find its area, using horizontal cross sections.

18. (a) Sketch the region in the plane bounded by the curve $x = y^2$, the y axis, and the lines $y = 1$ and $y = 2$.
    (b) Find its area, using horizontal cross sections.
    (c) Find its area, using vertical cross sections.

19. (a) Sketch the region bounded by the curve $y = 1/(1 + x^2)$, the x axis, and the lines $x = 1$ and $x = -1$.
    (b) Find the area of this region, using vertical cross sections.
    (c) Find its area, using horizontal cross sections.

20. Find the area of the shaded region in the figure shown below.

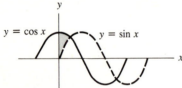

21. Find the area of the shaded region in the figure shown below (a) without calculus, (b) using vertical cross sections, (c) using horizontal cross sections.

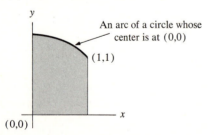

22. Find the area between the curves $y^2 = x$ and $y = x - 2$, using (a) horizontal cross sections, (b) vertical cross sections.

23. Sketch the region common to two circles of radius 1 whose centers are a distance 1 apart. Find the area of this region, using (a) vertical cross sections, (b) horizontal cross sections, (c) only elementary geometry, but no calculus.

24. (a) What fraction of the rectangle whose vertices are $(0, 0)$, $(a, 0)$, (a, a^4), $(0, a^4)$, with a positive, is occupied by the region under the curve $y = x^4$ and above $[0, a]$?
    (b) Repeat part (a), but with 4 replaced throughout by $\frac{1}{4}$.

Exercises 25 to 43 concern volume.

25. Find the volume of the solid described in Example 1 of Sec. 10.2.

26. Find the volume of a sphere of radius a. (See Example 3 of Sec. 10.2.)

27. Find the volume of a right circular cone of height h and radius a.

28. Find the volume of water discussed in Example 4 of Sec. 10.2, using the triangular cross sections of Case 3.

29. Find the volume of water discussed in Example 4 of Sec. 10.2, using the rectangular cross sections of Case 4.

Exercise 30 provides an easy way to remember the volume of a cone or pyramid.

30. A solid is formed in the following manner. A plane region R is given, and a point P not in that plane. The solid consists of all line segments, one of whose ends is P and whose other end is in R. If R has area A, and P is a distance h from the plane of R, show that the volume of the solid is $Ah/3$.

31. The base of a solid is a circle of radius 3. Each plane perpendicular to a given diameter meets the solid in a square, one side of which is in the base of the solid. (See the accompanying figure.) Find its volume.

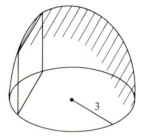

32. The region bordered by $y = e^x$, $y = 1$, and $x = 1$ is rotated around the x axis. Find the volume of the resulting solid.

33. The region in the preceding exercise is revolved around the y axis. Find the volume of the resulting solid.

34. The base of a solid S is the region bordered by $y = 1 - x^2$ and $y = 1 - x^4$. Cross sections of S by planes that are perpendicular to the x axis are squares. Find the volume of S.

35. The region between the curve $y = x^3$, the x axis, $x = 2$, and $x = 3$ is rotated around the y axis. Find the volume of the resulting solid.

36. The region between the curves $y = 2^x$ and $y = x + 1$ is rotated around the x axis. Find the volume of the resulting solid.

37. Find the volume of the solid whose cross sections

perpendicular to the x axis consist of circles whose centers are in the xy plane and whose intersection with the xy plane is the line segment from the curve $y = \sqrt{x}$ to the line $y = x$.

38. Find the volume of the solid whose base is the disk of radius 5 and whose cross sections perpendicular to the x axis are equilateral triangles. See the figure below.

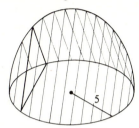

39. The region bounded by $y = \ln x$, $y = 0$, and $x = e$ is rotated around the x axis. Find the volume of the resulting solid.

40. The region bounded by $y = x^3$ and $y = x^2$ is rotated around the x axis. Find the volume of the resulting solid.

41. The same as Exercise 40, except that the region is rotated around the y axis.

42. What fraction of the volume of a sphere is contained between parallel planes that trisect the diameter to which they are perpendicular?

■

43. Find the volume of one octant of the region common to two right circular cylinders of radius 1, whose axes intersect at right angles, as shown in the accompanying figure.

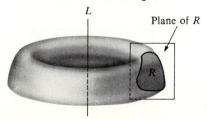

Quadrant of a circle
A
Quadrant of a circle
B
1
1

Hint: Find the cross sections by planes perpendicular to the line AB.

■ ■

44. Let f be an increasing function with $f(0) = 0$, and assume that it has an elementary antiderivative. Then f^{-1} is an increasing function, and $f^{-1}(0) = 0$. Prove that if f^{-1} is elementary, then it also has an elementary antiderivative. *Hint:* Observe that the sum of the areas of I and II in the figure below is $tf(t)$.

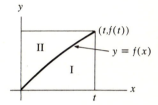

45. Show that the shaded area in the accompanying drawing is $\frac{2}{3}$ the area of the parallelogram $ABCD$. This is an illustration of a theorem of Archimedes concerning sectors of parabolas.

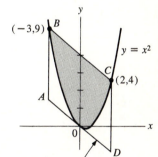

$(-3,9)$ B
$y = x^2$
C
$(2,4)$
A
0
D

This line is parallel to the line through $(2,4)$ and $(-3,9)$ and is tangent to the curve $y = x^2$

10.4

Computing the volume of a solid of revolution by the shell technique

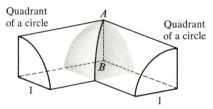

L
Plane of R
R

Let R be a region in the plane and L a line in the plane (see the figure to the lĕft). Assume that L does not meet R at all, or that L meets R only at points of the boundary. Consider the solid formed by revolving R about L. If L does not meet R, the solid is shaped like a doughnut. What definite integral equals the volume of this solid?

If each line in the plane of R and perpendicular to L meets R in a line segment, then it is not hard to find the volume of the solid by taking cross sections by planes perpendicular to L. Each such cross section is a ring, the region between two concentric circles. This area is just the difference between the areas of the inner and outer circles. The volume can be computed by the integration of cross-sectional area. (In fact there are some exercises doing this in Sec. 10.3.)

In this section a second method is presented for computing the volume of a solid of revolution. This approach will play an important theoretical role in Chap. 16, which is the main justification for including this alternative method, known as the *cylindrical-shell technique*.

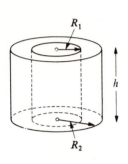

The key tool is the formula for the volume of a cylindrical shell. Assume that its inner radius is R_1, its outer radius R_2, and its height h. (See the figure in the margin.) The volume of this cylindrical shell equals

$$\text{Area of base} \cdot \text{height} = (\pi R_2{}^2 - \pi R_1{}^2)h$$
$$= \pi(R_2 + R_1)(R_2 - R_1)h$$
$$= 2\pi \frac{R_2 + R_1}{2} h(R_2 - R_1).$$

Thus the volume of the shell is

$$2\pi \cdot \text{average of the inner and outer radii} \cdot \text{height} \cdot \text{thickness}.$$

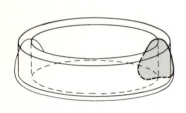

To develop this alternative method, look at a solid of revolution in a new way. Instead of approximating its volume by parallel horizontal slabs, approximate it by concentric cylindrical shells, as suggested in the diagram at the left, which shows one of the shells.

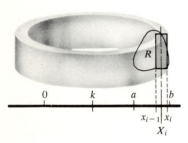

Introduce an axis perpendicular to L on which to record the shells. The position of the origin 0 is optional. Assume that L lies to the left of R and cuts the x axis at k, and that R lies above the interval $[a, b]$. With the partition

$$a = x_0, x_1, \ldots, x_n = b$$

are associated n concentric shells: the ith shell is formed by selecting X_i to be the midpoint, $(x_{i-1} + x_i)/2$, of the ith interval; form the rectangle above the interval $[x_{i-1}, x_i]$ whose height is $c(X_i)$, the cross section of R by the line parallel to L and meeting the x axis at X_i. The ith shell is obtained by revolving this rectangle around the line L.

The shell has thickness $x_i - x_{i-1}$

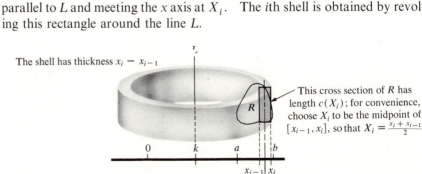

This cross section of R has length $c(X_i)$; for convenience, choose X_i to be the midpoint of $[x_{i-1}, x_i]$, so that $X_i = \frac{x_i + x_{i-1}}{2}$

The volume of the typical shell is

$2\pi \cdot$ average of inner and outer radii $\cdot$ height $\cdot$ thickness

$$= 2\pi \left[\frac{(x_{i-1} - k) + (x_i - k)}{2} \right] c(X_i)(x_i - x_{i-1})$$

$$= 2\pi \left(\frac{x_{i-1} + x_i}{2} - k \right) c(X_i)(x_i - x_{i-1})$$

$$= 2\pi(X_i - k)c(X_i)(x_i - x_{i-1}).$$

Now make the assumption that, when the shells are thin, the total volume of these shells is a good estimate of the volume of the solid of revolution. That is, assume that

$$\lim_{\text{mesh} \to 0} \sum_{i=1}^{n} 2\pi(X_i - k)c(X_i)(x_i - x_{i-1}) = \text{volume of solid}.$$

But this limit is the definite integral

$$\int_a^b 2\pi(x - k)c(x)\, dx.$$

FORMULA FOR VOLUME BY SHELL TECHNIQUE   This is the formula for computing volumes by the shell technique: If $x - k$ is denoted $R(x)$, the radius of the shell, then

$$\text{Volume} = \int_a^b 2\pi R(x)c(x)\, dx.$$

Memory aid: The expression $2\pi R(x)c(x)\, dx$ is the volume of a flat box of height dx, width $c(x)$, and length $2\pi R(x)$, the circumference of a circle of radius $R(x)$. The flat box shown below is approximately that solid formed when the typical shell is cut along a line parallel to L and laid flat.

These Two Figures Show the Essence of the Shell Technique at a Glance.   Study Them Carefully.

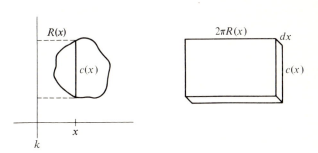

EXAMPLE 1   The region between the curves $y = x^2$ and $y = x^3$ is revolved around the y axis. Find the volume of the solid of revolution produced.   (It is shaped like a bowl.)

SOLUTION   Use the given x axis to set up the integral $\int_a^b 2\pi R(x)c(x)\,dx$.   Since the y axis is the line L about which the region is revolved, $R(x) = x$.   Also, $c(x) = x^2 - x^3$.

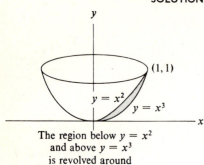

The region below $y = x^2$
and above $y = x^3$
is revolved around
the y axis

Thus

$$\text{Volume} = \int_0^1 2\pi x(x^2 - x^3)\,dx$$

$$= \int_0^1 2\pi(x^3 - x^4)\,dx$$

$$\underset{\text{FTC}}{=} 2\pi\left[\frac{x^4}{4} - \frac{x^5}{5}\right]_0^1$$

$$= 2\pi\left[\left(\frac{1^4}{4} - \frac{1^5}{5}\right) - \left(\frac{0^4}{4} - \frac{0^5}{5}\right)\right]$$

$$= 2\pi\left(\tfrac{1}{4} - \tfrac{1}{5}\right)$$

$$= \frac{2\pi}{20} = \frac{\pi}{10}.\ \bullet$$

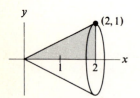

EXAMPLE 2   The triangle whose vertices are $(0, 0)$, $(2, 0)$, and $(2, 1)$ is revolved around the x axis.   Find the volume of the resulting cone.

SOLUTION   In this case the shells are formed by partitions of the y axis, not the x axis.   The formula for the volume, written with y replacing x, is

$$\int_a^b 2\pi R(y)c(y)\,dy.$$

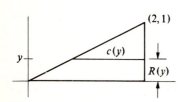

To determine $R(y)$ and $c(y)$, look closely at the given triangle, not at the solid of revolution.   Inspection of the accompanying diagram shows that $R(y) = y$.   To find $c(y)$, make use of the equation of the line through $(0, 0)$ and $(2, 1)$, namely $y = x/2$, or $x = 2y$.   Inspection of the diagram then shows that

$$c(y) = 2 - 2y.$$

The interval of integration is $[0, 1]$.   Thus

$$\text{Volume} = \int_0^1 2\pi y(2 - 2y)\,dy$$

$$= \int_0^1 (4\pi y - 4\pi y^2)\,dy$$

$$\underset{\text{FTC}}{=} \left[2\pi y^2 - \frac{4\pi y^3}{3}\right]_0^1$$

$$= \left[2\pi(1)^2 - \frac{4\pi(1)^3}{3}\right] - \left[2\pi(0)^2 - \frac{4\pi(0)^3}{3}\right]$$

$$= 2\pi - \frac{4\pi}{3}$$

$$= \frac{2\pi}{3}. \; \bullet$$

EXAMPLE 3 The region below $y = 1 + \sin x$, above the x axis, and situated between $x = 0$ and $x = 2\pi$ is revolved around the y axis.   Find the volume of the resulting solid of revolution.

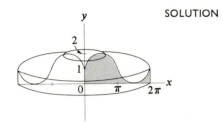

SOLUTION (Note that cross sections by planes perpendicular to the y axis would be quite messy.   For y between 1 and 2, the cross section is one ring, whose radii would be expressed in terms of the function $\sin^{-1}$.   For y between 0 and 1, the cross section consists of two pieces.)

The method of concentric shells is easy.

$$\text{Volume} = \int_a^b 2\pi R(x) c(x) \, dx = \int_0^{2\pi} 2\pi x (1 + \sin x) \, dx$$

$$= \int_0^{2\pi} 2\pi x \, dx + \int_0^{2\pi} 2\pi x \sin x \, dx.$$

Both integrals can be evaluated with the aid of the fundamental theorem of calculus.   (The second requires an integration by parts or an integral table.)

$$\int_0^{2\pi} 2\pi x \, dx = \pi x^2 \Big|_0^{2\pi} = \pi (2\pi)^2 - \pi (0)^2 = 4\pi^3,$$

and $$\int_0^{2\pi} 2\pi x \sin x \, dx = 2\pi (\sin x - x \cos x) \Big|_0^{2\pi}$$

$$= 2\pi [(\sin 2\pi - 2\pi \cos 2\pi) - (\sin 0 - 0 \cos 0)]$$

$$= 2\pi \{[0 - 2\pi(1)] - (0 - 0)\}$$

$$= -4\pi^2.$$

Thus the volume is $4\pi^3 - 4\pi^2$. $\bullet$

Exercises

In Exercises 1 to 6 draw a neat, clear picture of the typical shell, and set up by the shell method an integral for the volume of the indicated solid of revolution.   *Do not evaluate the integral.*

1. The region below $y = x^2$, above the x axis, between $x = 0$ and $x = 1$ is revolved around the y axis.
2. The region below $y = e^x$, above $y = 1$, between $x = 0$ and $x = 2$ is revolved around the y axis.
3. The region below $y = 1$, above $y = \sqrt{x}$, and to the right of $x = 0$ is revolved about the x axis.
4. The region in Exercise 1 is revolved around the line $x = -1$.

5. The region in Exercise 1 is revolved around the line $x = 2$.
6. The region in Exercise 3 is revolved around the line $y = 1$.

In Exercises 7 to 13 find the volume of the solid of revolution described.   (Use the concentric-shell method.)

7. The region between $y = x^2$, the x axis, and the line $x = 1$ is revolved around (a) the y axis, (b) the x axis.
8. The region between $y = x^3$, the x axis, and the line $x = 1$ is revolved around (a) the x axis, (b) the line $y = -2$.
9. The region between $y = \sin x$, the x axis, the y axis, and

the line $x = \pi/2$ is revolved around (a) the y axis, (b) the x axis.

10. The region between $y = e^x$, $y = 1$, and $x = 1$ is revolved around (a) the x axis, (b) the line $y = 1$, (c) the y axis, (d) the line $x = 1$.

11. The region between $y = e^{x^2}$, the x axis, $x = 0$, and $x = 1$ is revolved around the y axis. (It is interesting to note that the fundamental theorem of calculus is of no use in evaluating the area of this region.)

12. The region between $y = \sqrt{1 + x^2}$, $y = 1$, and $x = 1$ is revolved around the line $y = 1$.

13. The region between $y = \ln x$, the x axis, and $x = e$ is revolved around (a) the y axis, (b) the line $y = 1$, (c) the line $y = -1$.

14. A circle of radius 3 is revolved about a line in the plane of the circle and a distance 5 units away from the center of the circle. Find the volume of the doughnut (torus) produced.

15. Find the volume of a sphere of radius a by the shell method.

16. Find the volume of a right circular cone of radius a and height h by the shell method.

■

17. Solve Example 1 by the method of parallel cross sections.
18. Solve Example 2 by the method of parallel cross sections.
19. Solve Exercise 9 by the method of parallel cross sections.
20. A circle of radius a is revolved around a line a distance b from its center, $b > a$. Show that the volume of the solid produced is equal to the product of the area of the circle times the distance its center moves in the revolution.

■ ■

21. Let a and b be positive numbers and $y = f(x)$ be a decreasing differentiable function of x, such that $f(0) = b$ and $f(a) = 0$. Prove that $\int_0^a 2xy \, dx = \int_0^b x^2 \, dy$, (a) by considering the volume of a certain solid, (b) by integration by parts.

22. Let f be a decreasing differentiable function, positive for x in the interval $[0, 1]$. Let R be the region below $y = f(x)$ and above $[0, 1]$. Let V_1 be the volume of the solid obtained by revolving R about the y axis, and V_2 be the volume of the solid obtained by revolving R about the x axis. (a) Prove that

$$V_1 \le 2\sqrt{\frac{\pi}{3}} \sqrt{V_2}.$$

(b) When does equality hold in (a)? *Hint:* Look up Schwarz' inequality in Exercise 12 of the next section.

Exercise 23 shows that from a computational viewpoint the shell technique is not more effective than the cross-section technique.

23. Let f in Exercise 21 be elementary. (a) Show that, if $x^2 f'$ has an elementary integral, so does xf, and conversely. (b) Consider the solid obtained by rotating the region bounded by the curve $y = f(x)$, the x axis, and the y axis around the y axis. Show that its volume expressed by the shell technique involves an elementary integral only when its volume by the cross-section technique involves an elementary integral.

10.5

The average of a function over an interval

The average or, more precisely, the arithmetic mean of n numbers is simply their sum divided by n. Thus the average of 2, 3, and 7 is $(2 + 3 + 7)/3$, or 4. How can we define the average of the function x^2, for x in the interval $[0, 3]$? It makes no sense to find the sum of the squares of all the numbers from 0 to 3 and then divide by the number of those quantities, since the interval $[0, 3]$ contains an infinite set of numbers.

To define the average of a function f over an interval $[a, b]$, proceed as follows. Pick an integer n and then n equally spaced points in the interval $[a, b]$. Call these points $x_1, x_2, \ldots, x_n$, with $x_1 = a + (b - a)/n$ and $x_n = b$.

Then

$$\frac{f(x_1) + f(x_2) + \cdots + f(x_n)}{n} \tag{1}$$

would be a reasonable estimate of the average value of f over $[a, b]$. And, as n increases, this quotient becomes a better estimate of the average.

The sum (1) resembles the approximating sum for a definite integral. A little algebra relates it to such a sum, as follows: first of all,

$$\frac{f(x_1) + f(x_2) + \cdots + f(x_n)}{n}$$

$$= \frac{1}{b-a}\left[f(x_1)\frac{b-a}{n} + f(x_2)\frac{b-a}{n} + \cdots + f(x_n)\frac{b-a}{n} \right]. \tag{2}$$

To make the relation between (1) or (2) and $\int_a^b f(x)\, dx$ more evident, note that, for each $i = 1, 2, \ldots, n$ (if x_0 is taken to be a),

$$x_i - x_{i-1} = \frac{b-a}{n}.$$

Hence the right side of (2) is equal to

$$\frac{1}{b-a}\left[f(x_1)(x_1 - x_0) + f(x_2)(x_2 - x_1) + \cdots + f(x_n)(x_n - x_{n-1}) \right]. \tag{3}$$

The bracketed expression in (3) is an approximation of $\int_a^b f(x)\, dx$. This observation suggests the following definition of the average of f over $[a, b]$ in terms of the definite integral.

DEFINITION   *Average value of a function over an interval.* The average value of f over $[a, b]$ is the quotient

$$\frac{\int_a^b f(x)\, dx}{b-a}.$$

If $f(x)$ is positive for x in $[a, b]$, there is a simple geometric interpretation of the average of the function over the interval. Call the average A; then

$$A = \frac{\int_a^b f(x)\, dx}{b-a} \qquad \text{or} \qquad A(b-a) = \int_a^b f(x)\, dx.$$

Now,

$$\int_a^b f(x)\, dx$$

is the area of the region below the graph of f and above the interval $[a, b]$. The equation

$$A(b-a) = \int_a^b f(x)\, dx$$

asserts that A, the average value of the function, is the height of a rectangle whose base is $(b-a)$ and whose area is equal to the area of the region under the graph of f.

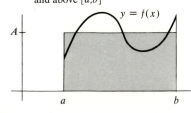

The height of this rectangle is the average value A of f over $[a,b]$, and the area is the same as the area of the region below the curve $y = f(x)$ and above $[a,b]$

EXAMPLE 1   Compute the average value of sin x for x in the interval $[0, \pi]$.

SOLUTION   According to the definition of the average value of a function, this average is

$$\frac{\int_0^\pi \sin x \, dx}{\pi - 0}.$$

Now,
$$\int_0^\pi \sin x \, dx \underset{\text{FTC}}{=} -\cos x \Big|_0^\pi$$
$$= (-\cos \pi) - (-\cos 0)$$
$$= -(-1) - (-1)$$
$$= 1 + 1 = 2.$$

Hence the average value of sin x for x in $[0, \pi]$ is

$$\frac{2}{\pi},$$

which is about 0.64. ●

The next two examples show that "average" means "average of a function." It is not enough to ask, "What is the average velocity?" We must ask, "What is the average of velocity with respect to time?" or "What is the average of velocity with respect to distance?" These averages are sometimes called the *time average* and the *distance average*, respectively. The answers will usually be different.

EXAMPLE 2   If we travel 30 miles per hour for 1 hour and then 50 miles per hour for another hour, what is the average of our velocity with respect to time?

SOLUTION   Denote the velocity at time t by $f(t)$. The average velocity with respect to time is defined as

$$\frac{\int_0^2 f(t) \, dt}{2 - 0}.$$

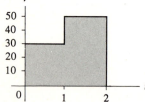

The simplest way to compute $\int_0^2 f(t) \, dt$ is to interpret it as the area of the shaded region shown in the diagram at the left. Since the shaded region has area 80, $\int_0^2 f(t) \, dt = 80$. Thus

$$\text{Average velocity with respect to time} = \frac{\int_0^2 f(t) \, dt}{2 - 0} = \frac{80}{2} = 40 \text{ miles per hour.} \ ●$$

The notion of average velocity with respect to time coincides with the ordinary notion of average velocity, which is defined as total change in position divided by total time. To see this, consider for simplicity the case in which the velocity $v(t)$ is positive. Then

$$\text{Average velocity with respect to time} = \frac{\int_a^b v(t) \, dt}{b - a} = \frac{\text{total distance}}{\text{total time}}.$$

EXAMPLE 3 If we travel 30 miles per hour for 30 miles and then 50 miles per hour for another 50 miles, what is the average of our velocity with respect to distance? (Note that this is the same trip as in Example 2.)

SOLUTION Now the velocity is considered as a function of distance. Denote the velocity after x miles of travel by $g(x)$. The average with respect to distance is therefore .

$$\frac{\int_0^{80} g(x)\, dx}{80 - 0}.$$

To compute the definite integral, interpret it as the area of the shaded region in the accompanying diagram. The area is $30 \cdot 30 + 50 \cdot 50 = 3{,}400$. Therefore,

$$\text{Average velocity with respect to distance} = \frac{\int_0^{80} g(x)\, dx}{80 - 0}$$

$$= \frac{3{,}400}{80} = 42.5 \text{ miles per hour.} \; \bullet$$

The reader should pause and explain to his own satisfaction why the average with respect to distance "ought to be" larger than the average with respect to time; Exercises 12 and 13 will answer this question formally, but a practical qualitative explanation is more valuable.

The moral of these two examples is this: When speaking of the average value of a quantity, it is necessary to indicate the variable with respect to which the average is to be computed.

Exercises

In Exercises 1 to 6 find the average of the given function over the given interval.

1. x^2, $[-1, 3]$

2. $\sin x$, $[0, 2\pi]$

3. $\ln x$, $[1, e]$

4. $\dfrac{1}{1 + x^2}$, $[0, 1]$

5. $\dfrac{1}{1 + x}$, $[1, 2]$

6. $\sin^2 x$, $\left[0, \dfrac{\pi}{2}\right]$

7. A person travels 20 miles per hour for $\frac{1}{2}$ hour, 30 miles per hour for 2 hours, and 40 miles per hour for $\frac{1}{2}$ hour.
(a) Compute the average of velocity with respect to time.
(b) Compute the average of velocity with respect to distance.

8. In the first t seconds a falling body drops $16t^2$ feet. Let its position relative to the y axis at time t be $16t^2$. (We aim the positive part of the axis downward.)
(a) Compute velocity as a function of time. (Note that it is positive.)

(b) Compute velocity as a function of y.
(c) Find the average of velocity with respect to time during the first t seconds.
(d) Find the average of velocity with respect to distance during the first t seconds.

■

9. Find the average length of the vertical cross section of the circle of radius r pictured below
(a) as a function of x,
(b) as a function of θ,
(c) as a function of y.
(d) Why would you expect the answer to (b) to be less than the answer to (a)?

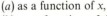
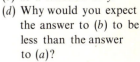

10. Prove that the average velocity with respect to time is equal to the change in position divided by the elapsed time.

11. A certain function f is defined throughout $[0, 2]$, but only limited data concerning it are available from experiments. It is known that $f(0) = 3, f(\frac{1}{4}) = \frac{7}{2}, f(\frac{1}{2}) = 4$, $f(1) = 6$, $f(2) = 8$. What is a sensible estimate of the average value of f over $[0, 2]$?

■ ■

12. This exercise obtains a famous inequality, known as the Schwarz inequality:

$$\int_A^B f(x)g(x)\,dx \le \left(\int_A^B [f(x)]^2\,dx\right)^{1/2} \left(\int_A^B [g(x)]^2\,dx\right)^{1/2}.$$

(a) Prove that if the quadratic polynomial $at^2 + bt + c = 0$ has at most one (real) root, then $b^2 - 4ac \le 0$. (When does it have exactly one root?)

(b) Let f and g be continuous functions. Define a third function h as follows:

$$h(t) = \int_A^B [tf(x) - g(x)]^2\,dx.$$

Show that h is a quadratic polynomial in t, $at^2 + bt + c$.

(c) Express the coefficients a, b, and c of h in terms of f and g.

(d) Combining (a), (b), and (c), derive the Schwarz inequality.

(e) When does equality occur in the Schwarz inequality?

In Exercise 13 it is shown that the average of velocity with respect to time is not larger than the average of velocity with respect to distance.

13. Assume for convenience that the velocity $v = dx/dt$ is positive, hence that speed equals velocity. Assume that from time a to time b the object moves from x_1 to x_2 on the x axis.

(a) Show that the average of velocity with respect to distance equals $\int_{x_1}^{x_2} v(x)\,dx / \int_a^b v(t)\,dt$.

(b) Show that $\int_{x_1}^{x_2} v(x)\,dx = \int_a^b [v(t)]^2\,dt$.

(c) Using the Schwarz inequality from Exercise 12, show that the average of velocity with respect to time is not larger than the average of velocity with respect to distance.

14. Let f be a continuous function such that $f(x)$ is always positive. Prove that

$$\int_a^b f(x)\,dx \int_a^b \frac{1}{f(x)}\,dx \ge (b - a)^2.$$

Hint: See Exercise 12.

15. Compute

$$\lim_{b \to a} \frac{\int_a^b f(x)\,dx}{b - a},$$

where f is a continuous function.

16. Differentiate (for practice):

(a) $\ln \sec 3x$

(b) $\dfrac{e^{x^2}}{\cos 2x}$

(c) $\dfrac{\sin^{-1} 3x}{\tan 2x}$

(d) $3 \cot 2x \ln \sqrt[3]{1 + 2x}$

(e) $\left(\dfrac{1 + x^2}{1 - x^2}\right)^3$

(f) $\sec^{-1} 3x \sec 3x$.

10.6

Improper integrals

Consider the volume of the solid obtained by revolving about the x axis the region bordered by $y = 1/x$ and the x axis, to the right of $x = 1$. The typical

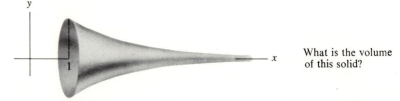

What is the volume of this solid?

cross section made by a plane perpendicular to the x axis is a circle of radius $1/x$. We might therefore be tempted to say that the volume is $\int_1^\infty \pi(1/x)^2\,dx$. Unfortunately, the symbol $\int_a^\infty f(x)\,dx$ has not been given any meaning so far in this book. The definition of the definite integral involves sums of the form

$\sum_{i=1}^{n} f(X_i)(x_i - x_{i-1})$. If a section in the partition has infinite length, such a sum is meaningless.

It does make sense, however, to examine the volume of that part of the solid from $x = 1$ to $x = b$, where b is some number greater than 1, and then to determine what happens to this volume as $b \to \infty$. In other words, consider $\lim_{b \to \infty} \int_1^b \pi(1/x)^2 \, dx$. Now,

$$\int_1^b \pi\left(\frac{1}{x}\right)^2 dx \underset{\text{FTC}}{=} -\frac{\pi}{x}\Big|_1^b$$

$$= -\frac{\pi}{b} - \left(-\frac{\pi}{1}\right)$$

$$= \pi - \frac{\pi}{b}.$$

Thus $\lim_{b \to \infty} \int_1^b \pi/x^2 \, dx = \pi - 0 = \pi$. The volume of the endless solid is finite. This approach suggests a way to give meaning to the symbol $\int_a^\infty f(x) \, dx$.

DEFINITION *Convergent improper integral $\int_a^\infty f(x) \, dx$.* Let f be continuous for $x \geq a$. If $\lim_{b \to \infty} \int_a^b f(x) \, dx$ exists, the function f is said to have a convergent improper integral from a to ∞. The value of the limit is denoted by $\int_a^\infty f(x) \, dx$.

It was shown above that $\int_1^\infty \pi(1/x)^2 \, dx$ is a convergent improper integral, with value π.

DEFINITION *Divergent improper integral $\int_a^\infty f(x) \, dx$.* Let f be a continuous function. If $\lim_{b \to \infty} \int_a^b f(x) \, dx$ does not exist, the function f is said to have a divergent improper integral.

EXAMPLE I Determine the area of the region below $y = 1/x$, above the x axis, and to the right of $x = 1$.

SOLUTION The area in question is given by

$$\int_1^\infty \frac{1}{x} \, dx = \lim_{b \to \infty} \int_1^b \frac{1}{x} \, dx \underset{\text{FTC}}{=} \lim_{b \to \infty} (\ln b - \ln 1) = \lim_{b \to \infty} \ln b.$$

How does $\ln b$ behave as b increases without bound? First of all, since $D(\ln t) = 1/t$, the function $\ln t$ is increasing. Moreover, since $\ln e = 1$, $\ln e^2 = 2, \ldots, \ln e^n = n$, it follows that $\ln b$ becomes arbitrarily large. Thus $\lim_{b \to \infty} \ln b = \infty$. These observations are summarized in the following statements: The area is infinite, $\int_1^\infty 1/x \, dx = \infty$ or, simply, $\int_1^\infty 1/x \, dx$ is a divergent improper integral. ●

The improper integral $\int_{-\infty}^b f(x) \, dx$ is defined similarly, by considering

$$\int_t^b f(x) \, dx$$

for negative values of t of large absolute value.

The Improper Integral   If

$$\lim_{t \to -\infty} \int_t^b f(x)\, dx$$

$$\int_{-\infty}^b f(x)\, dx \quad \text{exists, it is denoted} \qquad \int_{-\infty}^b f(x)\, dx.$$

In such a case, the improper integral $\int_{-\infty}^b f(x)\, dx$ is said to be convergent.   If

$$\lim_{t \to -\infty} \int_t^b f(x)\, dx$$

does not exist, then the improper integral

$$\int_{-\infty}^b f(x)\, dx$$

is said to be divergent.

The Improper Integral   To deal with improper integrals over the entire x axis, define

$$\int_{-\infty}^{\infty} f(x)\, dx$$

$$\int_{-\infty}^{\infty} f(x)\, dx$$

to be the sum $\displaystyle \int_{-\infty}^0 f(x)\, dx + \int_0^{\infty} f(x)\, dx,$

which will be called convergent if both

$$\int_{-\infty}^0 f(x)\, dx \qquad \text{and} \qquad \int_0^{\infty} f(x)\, dx$$

are convergent.   [If at least one of the two is divergent, $\int_{-\infty}^{\infty} f(x)\, dx$ will be called divergent.]

EXAMPLE 2   Determine the area of the region bounded by the curve $y = 1/(1 + x^2)$ and the x axis

SOLUTION   The area in question equals

$$\int_{-\infty}^{\infty} \frac{dx}{1 + x^2}.$$

Now, $\displaystyle \int_0^{\infty} \frac{dx}{1 + x^2} = \lim_{b \to \infty} \int_0^b \frac{dx}{1 + x^2}$

$$= \lim_{b \to \infty} (\tan^{-1} b - \tan^{-1} 0)$$

$$= \frac{\pi}{2}.$$

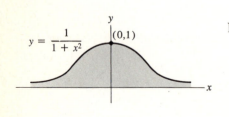

$y = \dfrac{1}{1 + x^2}$ $(0,1)$

By symmetry, $\displaystyle \int_{-\infty}^0 \frac{dx}{1 + x^2} = \frac{\pi}{2}.$

Hence
$$\int_{-\infty}^{\infty} \frac{dx}{1+x^2} = \frac{\pi}{2} + \frac{\pi}{2},$$

and the area in question is π. ●

EXAMPLE 3   Is $\int_0^\infty \cos x \, dx$ convergent or divergent?

SOLUTION   It is necessary to consider

$$\lim_{b\to\infty} \int_0^b \cos x \, dx = \lim_{b\to\infty} \sin x \Big|_0^b$$

$$= \lim_{b\to\infty} \sin b.$$

But $\lim_{b\to\infty} \sin b$ does not exist.   Thus $\int_0^\infty \cos x \, dx$ is divergent. ●

There is a second type of improper integral, in which the function is un-bounded in the interval $[a, b]$.   If $f(x)$ becomes arbitrarily large in the interval $[a, b]$, then arbitrarily large approximating sums $\sum_{i=1}^n f(X_i)(x_i - x_{i-1})$ can be obtained no matter how fine the partition may be by choosing an X_i that makes $f(X_i)$ large.   The next example shows how to get around this difficulty.

EXAMPLE 4   Determine the area of the region bounded by $y = 1/\sqrt{x}$, $x = 1$, and the coordin-ate axes shown in the figure in the margin.

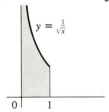

$y = \frac{1}{\sqrt{x}}$

SOLUTION   First of all, resist the temptation to write that area $= \int_0^1 1/\sqrt{x} \, dx$, for $\int_0^1 1/\sqrt{x} \, dx$ does not exist according to the definition of the definite integral given in Chap. 7, since its integrand is unbounded in $[0, 1]$.   (Note also that the integrand is not defined at 0.)   Instead consider the behavior of $\int_t^1 1/\sqrt{x} \, dx$ as t approaches 0 from the right.   Since

$$\int_t^1 \frac{1}{\sqrt{x}} \, dx = 2\sqrt{x} \Big|_t^1$$

$$= 2\sqrt{1} - 2\sqrt{t}$$

$$= 2(1 - \sqrt{t}),$$

it follows that, as t shrinks toward 0 from the right, $\int_t^1 1/\sqrt{x} \, dx$ approaches 2. The area in question is said to be 2.

The reader should check and see that this is the same value for the area that can be obtained by taking horizontal cross sections and evaluating an improper integral from 0 to ∞. ●

The reasoning in Example 4 motivates the definition of the second type of improper integral, in which the function rather than the interval is unbounded.

DEFINITION   *Convergent and divergent improper integrals $\int_a^b f(x)\,dx$.*   Let f be continuous at every number in $[a, b]$ except a, and become arbitrarily large for values in (a, b).   If $\lim_{t \to a^+} \int_t^b f(x)\,dx$ exists, the function f is said to have a convergent improper integral from a to b.   The value of the limit is denoted $\int_a^b f(x)\,dx$.   If $\lim_{t \to a^+} \int_t^b f(x)\,dx$ does not exist, the function f is said to have a divergent improper integral from a to b; in brief, $\int_a^b f(x)\,dx$ is not defined.

In a similar manner, if f is unbounded only near b and to the left of b, define $\int_a^b f(x)\,dx$ as $\lim_{t \to b^-} \int_a^t f(x)\,dx$.

Example 4 is summarized in the statement, "The improper integral $\int_0^1 1/\sqrt{x}\,dx$ is convergent and has the value 2."

It may happen that a function behaves well everywhere in the interval $[a, b]$ except at the number c, distinct from a and b, where it may be infinite.   In that case $\int_a^b f(x)\,dx$ makes no sense.   In such a case consider $\int_a^c f(x)\,dx$ and $\int_c^b f(x)\,dx$.   If both exist, then the integral $\int_a^b f(x)\,dx$ is said to be convergent and have the value $\int_a^c f(x)\,dx + \int_c^b f(x)\,dx$.   More generally, if a function f has both an infinite range of values and points where it becomes infinite, break the entire integral into the sum of integrals each of which has *only one* of the two basic "troubles," either an infinite range or an endpoint where the function is infinite.   For instance, the improper integral $\int_{-\infty}^\infty 1/x^2\,dx$ is troublesome for four reasons: $\lim_{x \to 0^-} 1/x^2 = \infty$, $\lim_{x \to 0^+} 1/x^2 = \infty$, and the range extends infinitely to the left and also to the right.   To treat the integral, write it as the sum of four improper integrals of the two basic types:

$$\int_{-\infty}^\infty \frac{1}{x^2}\,dx = \int_{-\infty}^{-1} \frac{1}{x^2}\,dx + \int_{-1}^0 \frac{1}{x^2}\,dx + \int_0^1 \frac{1}{x^2}\,dx + \int_1^\infty \frac{1}{x^2}\,dx.$$

All four of the integrals on the right have to be convergent for $\int_{-\infty}^\infty 1/x^2\,dx$ to be convergent.   As a matter of fact, only the first and last are.   So $\int_{-\infty}^\infty 1/x^2\,dx$ is divergent.

Just as substitution in a definite integral is valid as long as the same substitution is applied to the limits of integration, substitution in improper integrals is also permissible.

The final two examples illustrate the use of improper integrals in physics and economics.   Intended for enrichment and perspective, they are not essential for understanding the concept of improper integrals.   The reader is invited to read them, if not the first time through the section, then perhaps when reviewing it.

EXAMPLE 5   How much work is required to lift a 1-pound payload from the surface of the earth to "infinity"?   If this work should turn out to be infinite, then it would be impossible to send rockets off on unlimited orbits.   Fortunately, it is finite, as will now be shown.

SOLUTION   The work W necessary to lift an object a distance x against a constant force F is the product of force times distance,

$$W = F \cdot x.$$

Since the gravitational pull of the earth on the payload *changes* with distance from the earth, an (improper) integral will be needed to express the total work required to lift the load to "infinity."

The mass of the payload weighs 1 pound at the surface of the earth. The farther it is from the center of the earth, the less it weighs, for the force of the earth on the mass is inversely proportional to the square of the distance of the mass from the center of the earth. Thus the force on the payload is given by k/r^2, where k is a constant which we shall determine in a moment, and r is the distance from the payload to the center of the earth. When $r = 4{,}000$ (miles), the force is 1 pound; thus

$$1 = \frac{k}{4{,}000^2}.$$

From this it follows that $k = 4{,}000^2$, and therefore the gravitational force on a 1-pound mass is, in general, $(4{,}000/r)^2$ pounds. As the payload recedes from the earth, it loses weight (but not mass), as recorded on the diagram shown.

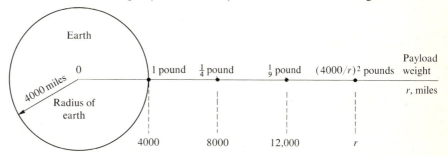

The work done in lifting the payload from point r to point $r + \Delta r$ is approximately

$$\underbrace{\left(\frac{4{,}000}{r}\right)^2}_{\text{Force}} \underbrace{(\Delta r)}_{\text{Distance}}.$$

Hence the total work required to move the 1-pound mass from the surface of the earth to infinity is given by the improper integral $\int_{4{,}000}^{\infty} (4{,}000/r)^2 \, dr$. Now

$$\int_{4{,}000}^{\infty} \left(\frac{4{,}000}{r}\right)^2 dr = \lim_{b \to \infty} \int_{4{,}000}^{b} \left(\frac{4{,}000}{r}\right)^2 dr = \lim_{b \to \infty} \frac{-4{,}000^2}{r} \Big|_{4{,}000}^{b}$$

$$= \lim_{b \to \infty} \left[\frac{-4{,}000^2}{b} + \frac{4{,}000^2}{4{,}000} \right] = 4{,}000 \text{ mile-pounds.}$$

The total work is finite because the improper integral is convergent. It is just as if the payload were lifted 4,000 miles against a constant gravitational force equal to that at the surface of the earth. ●

EXAMPLE 6   *Present value of future income.* Both business and government frequently face the question: "What is 1 dollar t years in the future worth today?" Implicit in this question are such considerations as the present value of a business being dependent on its future profit and the cost of a dam being weighed against its future revenue." Determine the present value of a business whose rate of profit t years in the future is $f(t)$ dollars per year.

SOLUTION To begin the analysis, assume that the annual interest rate r remains constant and that 1 dollar deposited today is worth e^{rt} dollars t years from now. This assumption corresponds to continuously compounded interest or to natural growth. Thus A dollars today will be worth Ae^{rt} dollars t years from now. What is the present value of the promise of 1 dollar t years from now? In other words, what amount A invested today will be worth 1 dollar t years from now? To find out, solve the equation

$$Ae^{rt} = 1$$

for A. The solution is $\qquad A = e^{-rt}.$ $\qquad$ (1)

Now consider the present value of the future profit of a business (or future revenue of a dam). Assume that the profit flow t years from now is $f(t)$ dollars per year. This rate may vary within the year; consider f to be a continuous function of time. The profit in the small interval of time Δt, from time t to time $t + \Delta t$, would be approximately $f(t) \Delta t$. The total future profit $F(T)$ from now, when $t = 0$, to some time T in the future is therefore

$$F(T) = \int_0^T f(t) \, dt. \qquad (2)$$

But the *present value* of the future profit is *not* given by (2). It is necessary to consider the present value of the profit earned in a typical short interval of time from t to $t + \Delta t$. According to (1) its present value is approximately

$$e^{-rt}f(t) \, \Delta t.$$

Hence the present value of future profit from $t = 0$ to $t = T$ is given by a definite integral:

$$\int_0^T e^{-rt}f(t) \, dt. \qquad (3)$$

The present value of all future profit is therefore the improper integral $\int_0^\infty e^{-rt}f(t) \, dt$. (This is a convenient idealization; in practice, an economist usually restricts the future to a finite period.)

To see what influence the interest rate r has, denote by $P(r)$ the present value of all future revenue when the interest rate is r; that is,

$$P(r) = \int_0^\infty e^{-rt}f(t) \, dt. \qquad (4)$$

If the interest rate r is raised, then according to (4) the present value of a business declines. An investor choosing between investing in a business or placing his money in a bank account finds the bank account more attractive when r is raised.

Laplace transform $\qquad$ Equation (4) assigns to a profit function f (which is a function of t) a present-value function P, which is a function of r, the interest rate. In the theory of differential equations P is called the *Laplace transform of f*. ●

Exercises

In Exercises 1 to 21 determine whether the given improper integral is convergent or divergent. Evaluate the convergent ones.

1. $\int_1^\infty \dfrac{dx}{x^3}$

2. $\int_0^\infty \dfrac{dx}{x^3}$

3. $\int_0^\infty e^{-x}\, dx$

4. $\int_0^\infty e^{-x} \sin 3x \, dx$

5. $\int_{-\infty}^\infty \dfrac{x\, dx}{1 + x^4}$

6. $\int_1^\infty x^{-1.01}\, dx$

7. $\int_0^1 x^{-1.01}\, dx$

8. $\int_0^1 x^{-0.99}\, dx$

9. $\int_1^\infty x^{-0.99}\, dx$

10. $\int_0^e x \ln x \, dx$

11. $\int_{-1}^1 \dfrac{dx}{\sqrt{1 - x^2}}$

12. $\int_0^1 \dfrac{dx}{\sqrt{1 - x}}$

13. $\int_0^4 \dfrac{dx}{\sqrt[3]{x - 2}}$

14. $\int_0^\infty \dfrac{dx}{\sqrt{x + 1}}$

15. $\int_0^\infty (1 - x)e^{-x}\, dx$

16. $\int_0^\infty x^{-1/3}\, dx$

17. $\int_0^1 \ln x \, dx$

18. $\int_1^\infty e^{-\sqrt[3]{x}} x^{-2/3}\, dx$

19. $\int_0^\infty \dfrac{dx}{(x - 1)^2}$

20. $\int_0^\infty x \ln x \, dx$

21. $\int_0^\infty e^{-x} \sin (2x + 1) \, dx$

22. Let R be the region to the right of the y axis, below $y = e^{-x}$ and above the x axis.
(*a*) Find the area of R.
(*b*) Find the volume of the region obtained by revolving R about the x axis.
(*c*) Find the volume of the region obtained by revolving R about the y axis.

■

Frequently one can determine whether an improper integral $\int_a^b f(x)\, dx$ is convergent even though f does not have an

elementary antiderivative. For inst. nce, if $0 \le f(x) \le g(x)$ and $\int_a^b g(x)\, dx$ is convergen., then it can be proved that $\int_a^b f(x)\, dx$ is convergent. Use this principle in Exercises 23 to 28.

23. Plankton are small football-shaped organisms. The resistance they meet when falling through water is proportional to the integral

$$\int_0^\infty \frac{dx}{\sqrt{(a^2 + x)(b^2 + x)(c^2 + x)}},$$

where a, b, and c describe the dimensions of the plankton. Is this improper integral convergent or divergent?

24. The function $f(x) = (\sin x)/x$ for $x \ne 0$ and $f(0) = 1$ occurs often in communication theory. Show that the energy E of the signal represented by f is finite, where

$$E = \int_{-\infty}^\infty [f(x)]^2\, dx.$$

25. Show that $\int_1^\infty e^{-x^2}\, dx$ is convergent by showing that it is smaller than $\int_1^\infty e^{-x}\, dx$.

26. Show that $\int_1^\infty 1/\sqrt{1 + x^3}\, dx$ is convergent, but do not try to evaluate it.

27. Show that $\int_0^\infty e^{-t} t^{-1/2}\, dt$ is convergent.

■ ■

Exercise 28 is used in Exercise 29.

28. From the fact that $\int_0^{\pi/2} \ln x \, dx$ is convergent deduce that $\int_0^{\pi/2} \ln \sin x \, dx$ is convergent. *Hint:* Show that, for $0 < x \le \pi/2$, $\sin x \ge 2x/\pi$.

29. (See Exercise 28.) Let $A = \int_0^{\pi/2} \ln \sin \theta \, d\theta$.
(*a*) Show that $A = \int_0^{\pi/2} \ln \cos \theta \, d\theta$.
(*b*) Show that $\int_0^\pi \ln \sin \theta \, d\theta = 2A$.
(*c*) Show that

$$\int_0^\pi \ln \sin \theta \, d\theta = \int_0^\pi \ln 2 \, d\theta + \int_0^\pi \ln \sin \frac{\theta}{2} \, d\theta$$

$$+ \int_0^\pi \ln \cos \frac{\theta}{2} \, d\theta.$$

(*d*) From (*c*) deduce that $2A = \pi \ln 2 + 4A$.
(*e*) Show that $A = -(\pi/2) \ln 2$.

30. Find the error in the following computations. The substitution $x = y^2$, $dx = 2y \, dy$, yields

$$\int_0^1 \frac{1}{x}\, dx = \int_0^1 \frac{2y}{y^2}\, dy = \int_0^1 \frac{2}{y}\, dy = 2\int_0^1 \frac{1}{y}\, dy = 2\int_0^1 \frac{1}{x}\, dx.$$

Hence

$$\int_0^1 \frac{1}{x}\, dx = 2 \int_0^1 \frac{1}{x}\, dx,$$

from which it follows that $\int_0^1 1/x\, dx = 0$.

31. (a) Show that, if $\int_{-\infty}^{\infty} f(x)\, dx$ is convergent, it equals $\lim_{L\to\infty} \int_{-L}^{L} f(x)\, dx$, but that $\lim_{L\to\infty} \int_{-L}^{L} f(x)\, dx$ may be finite while $\int_{-\infty}^{\infty} f(x)\, dx$ is a divergent improper integral.
(b) From (a), explain why $\lim_{L\to\infty} \int_{-L}^{L} f(x)\, dx$ would not be a good definition for $\int_{-\infty}^{\infty} f(x)\, dx$.

32. Find the error in the following computations. Using the substitution $u = 1/x$, $du = -1/x^2\, dx$, we have

$$\int_{-1}^{1} \frac{1}{1+x^2}\, dx = \int_{-1}^{1} \frac{1}{1+1/u^2}\left(-\frac{1}{u^2}\, du\right)$$

$$= -\int_{-1}^{1} \frac{1}{1+u^2}\, du.$$

Thus $\int_{-1}^{1} 1/(1+x^2)\, dx$, being equal to its negative, is 0.

Exercises 33 and 34 refer to Example 5.

33. How much work is done in lifting the 1-pound payload the first 4,000 miles of its journey to infinity?

34. (a) If the force of gravity were of the form $k/r^{1.01}$, could a payload be sent to infinity? (b) If the force of gravity were of the form k/r, could a payload be sent to infinity?

Exercises 35 and 36 refer to Example 6.

35. If the profit flow remains constant, say $f(t) = k > 0$, the total future profit is obviously infinite. Show that the present value is k/r, which is finite.

36. Find the Laplace transform of (a) $f(t) = k$, (b) $f(t) = t$, (c) $f(t) = \sin t$.

10.7

Polar coordinates

Rectangular coordinates are only one of the ways to locate points in the plane by pairs of numbers. In this section another system is described, called *polar coordinates*, which will be used later in this chapter and in subsequent chapters.

The rectangular coordinates x and y locate a point P in the plane as the intersection of a vertical line and a horizontal line. Polar coordinates locate a point P as the intersection of a circle and a ray from the center of that circle. They are defined as follows.

Select a point in the plane and a ray emanating from this point. The point is called the *pole*, and the ray the *polar axis*. Measure positive angles θ counterclockwise from the polar axis and negative angles clockwise. Now let r be a number. To plot the point P that corresponds to the pair of numbers r and θ proceed as follows:

If r is positive, P is the intersection of the circle of radius r whose center is at the pole and the ray of angle θ emanating from the pole.
If r is 0, P is the pole, no matter what θ is.
If r is negative, P is at a distance $|r|$ from the pole on the ray directly opposite the ray of angle θ.

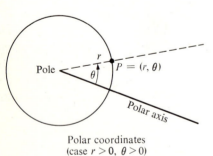

Polar coordinates
(case $r > 0$, $\theta > 0$)

In each case P is denoted (r, θ), and the pair r and θ are called polar coordinates of P. Note that the point (r, θ) is on the circle of radius $|r|$ whose center is the pole. Observe that the pole is the midpoint of the points (r, θ) and $(-r, \theta)$.

EXAMPLE 1   Plot the points $(3, \pi/4)$, $(2, -\pi/6)$, $(-3, \pi/3)$ in polar coordinates.

SOLUTION   To plot $(3, \pi/4)$, go out a distance 3 on the ray of angle $\pi/4$ (shown below).

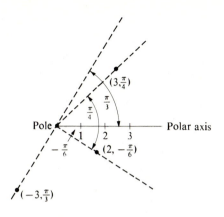

To plot $(2, -\pi/6)$, go out a distance 2 on the ray of angle $-\pi/6$. To plot $(-3, \pi/3)$, draw the ray of angle $\pi/3$, and then go a distance 3 in the *opposite* direction from the pole. ●

The Relation between Polar and Rectangular Coordinates

It is customary to have the polar axis coincide with the positive x axis as in the diagram in the margin. In that case, inspection of the diagram shows the following relation between the rectangular and polar coordinates of the point $P = (x, y) = (r, \theta)$:

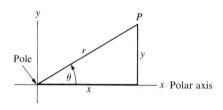

$$x = r \cos \theta, \qquad y = r \sin \theta,$$

and

$$r^2 = x^2 + y^2, \qquad \tan \theta = \frac{y}{x}.$$

For instance, the point whose rectangular coordinates are $(1, 1)$ has polar coordinates $(\sqrt{2}, \pi/4)$, or $(\sqrt{2}, \pi/4 + 2\pi)$, or $(\sqrt{2}, \pi/4 + 4\pi)$, or $(-\sqrt{2}, \pi/4 + \pi)$, and so on.

Just as we may graph the set of points (x, y), where x and y satisfy a certain equation, so may we graph the set of points (r, θ), where r and θ satisfy a certain equation. It is important, however, to keep in mind that, although each point in the plane is specified by a unique ordered pair (x, y) in rectangular coordinates, there are many ordered pairs (r, θ) in polar coordinates which specify each point.

EXAMPLE 2   Graph the equation $r = 2 \cos \theta$.

SOLUTION   First make a table, choosing convenient values of θ. Then sketch the points listed in the table.

θ	0	$\dfrac{\pi}{4}$	$\dfrac{\pi}{3}$	$\dfrac{\pi}{2}$	$\dfrac{3\pi}{4}$	π	$\dfrac{3\pi}{2}$	2π
$r = 2 \cos \theta$	2	$\sqrt{2} \doteq 1.4$	1	0	$-\sqrt{2} \doteq -1.4$	-2	0	2

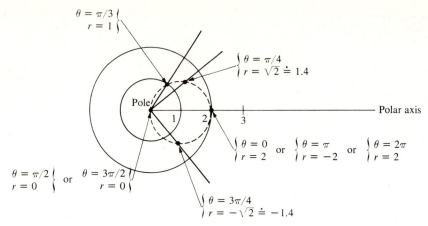

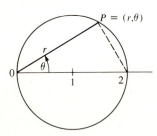

The dotted curve is the graph of $r = 2 \cos \theta$.

A little geometry and trigonometry show that the graph of $r = 2 \cos \theta$ is a circle. To see this, recall that an angle inscribed in a semicircle is a right angle, and consider a typical point $P = (r, \theta)$ for $r > 0$ on the circle shown in the diagram at the left. Since $\cos \theta =$ adjacent side/hypotenuse in the triangle, $\cos \theta = r/2$ or $r = 2 \cos \theta$. ●

EXAMPLE 3   Transform the equation $r = 2 \cos \theta$ into rectangular coordinates.

SOLUTION   Since $r^2 = x^2 + y^2$ and $r \cos \theta = x$, first multiply the equation $r = 2 \cos \theta$ by r, obtaining

$$r^2 = 2r \cos \theta.$$

Hence $$x^2 + y^2 = 2x.$$

Example 2 showed that this describes a circle of radius 1 and center $(1, 0)$. This could also be shown by rewriting the equation $x^2 + y^2 = 2x$ in the form

$$(x - 1)^2 + (y - 0)^2 = 1^2.$$

This equation says that the distance from (x, y) to $(1, 0)$ is 1. ●

EXAMPLE 4   Transform the equation $y = 2$, which describes a horizontal straight line, into polar coordinates.

SOLUTION   Since $y = r \sin \theta$, $$r \sin \theta = 2,$$

or $$r = \frac{2}{\sin \theta}.$$

This is far more complicated than the original equation but is still sometimes useful. ●

Generally, rectangular coordinates are best for describing straight lines, while polar coordinates are best for describing circles.

EXAMPLE 5   Graph $r = 1 + \cos \theta$.

SOLUTION   Begin by making a table.

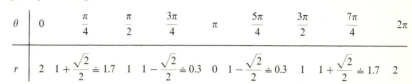

θ	0	$\dfrac{\pi}{4}$	$\dfrac{\pi}{2}$	$\dfrac{3\pi}{4}$	π	$\dfrac{5\pi}{4}$	$\dfrac{3\pi}{2}$	$\dfrac{7\pi}{4}$	2π
r	2	$1+\dfrac{\sqrt{2}}{2} \doteq 1.7$	1	$1-\dfrac{\sqrt{2}}{2} \doteq 0.3$	0	$1-\dfrac{\sqrt{2}}{2} \doteq 0.3$	1	$1+\dfrac{\sqrt{2}}{2} \doteq 1.7$	2

(2, 0)

The last point is the same as the first.   The graph begins to repeat itself.   This heart-shaped curve, shown at the left, is called a *cardioid*.   ●

Spirals turn out to be quite easy to describe in polar coordinates.   This will be illustrated by the graph of $r = 2\theta$ in the next example.

EXAMPLE 6   Graph $r = 2\theta$.

SOLUTION   First make a table.

θ	0	$\dfrac{\pi}{2}$	π	$\dfrac{3\pi}{2}$	2π	$\dfrac{5\pi}{2}$	$\cdots$
r	0	π	2π	3π	4π	5π	$\cdots$

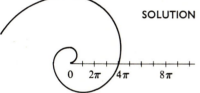

Increasing θ by 2π does *not* produce the same value of r.   As θ increases, r increases.   The graph for $\theta \geq 0$ is an endless spiral, going infinitely often around the pole.   ●

Polar coordinates are also convenient for describing loops arranged like the petals of a flower, as Example 7 shows.

EXAMPLE 7   Graph $r = \sin 3\theta$.

SOLUTION   As θ increases from 0 through $\pi/3$, 3θ increases from 0 through π.   Thus, r, which is $\sin 3\theta$, goes from 0 up to 1, then back to 0, for θ in $[0, \pi/3]$.   This gives one loop of the three loops making up the graph of $r = \sin 3\theta$.   For θ in $[\pi/3, 2\pi/3]$, $r = \sin 3\theta$ is negative (or 0).   This yields the lower loop.   For θ in $[2\pi/3, \pi]$, r is again positive, and we obtain the upper left loop.   Further choices of θ lead only to repetition of the loops already shown.   ●

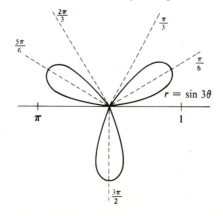

Exercises

1. Plot the points whose polar coordinates are:
   (a) $(1, \pi/6)$
   (b) $(2, \pi/3)$
   (c) $(2, -\pi/3)$
   (d) $(-2, \pi/3)$
   (e) $(2, 7\pi/3)$
   (f) $(0, \pi/4)$.
2. Find the rectangular coordinates of the points in Exercise 1.
3. Give at least three pairs of polar coordinates (r, θ) for the point $(3, \pi/4)$,
   (a) with $r > 0$,
   (b) with $r < 0$.
4. Find polar coordinates (r, θ), with $0 \le \theta < 2\pi$ and r positive, for the point whose rectangular coordinates are
   (a) $(\sqrt{2}, \sqrt{2})$
   (b) $(-1, \sqrt{3})$
   (c) $(-5, 0)$
   (d) $(-\sqrt{2}, -\sqrt{2})$
   (e) $(0, -3)$
   (f) $(1, 1)$.

In Exercises 5 to 9 transform the equation into one in rectangular coordinates.

5. $r = \sin \theta$
6. $r = \csc \theta$
7. $r = 3/(4 \cos \theta + 5 \sin \theta)$
8. $r = 4 \cos \theta + 5 \sin \theta$
9. $r = \sin 2\theta$.
   Hint: use the identity for $\sin 2\theta$.

In Exercises 10 to 15 transform the equation into one in polar coordinates.

10. $y = x^2$
11. $x + 2y = 3$
12. $x^2 + y^2 = 5$
13. $xy = 1$
14. $x^2 + y^2 = 4x$
15. $x = -2$

In Exercises 16 to 21 graph the given equations.

16. $r = 1 + \sin \theta$
17. $r = 3 + 2 \cos \theta$
18. $r = 1/\theta, \theta > 0$
19. $r = e^\theta, \theta \ge 0$
20. $r = \cos 3\theta$
21. $r = \cos 2\theta$

22. (a) Sketch the curves $r = 1 + \cos \theta$ and $r = 2 \cos \theta$ relative to the same polar axis.
    (b) Where do the curves in (a) intersect? (There are two intersections. Note that they are not both obtained by setting $1 + \cos \theta$ equal to $2 \cos \theta$.)
23. (a) Sketch the curves $r = \sin \theta$ and $r = \cos 2\theta$ relative to the same polar axis.
    (b) Where do the curves in (a) intersect? (There are four intersections.)

The curve $r = 1 + a \cos \theta$ (or $r = 1 + a \sin \theta$) is called a limaçon. Its shape depends on the choice of the constant a. For $a = 1$ we have the cardioid of Example 5. Exercises 24 and 25 concern other choices of a.

24. Graph $r = 1 + 2 \cos \theta$. (If $|a| > 1$, the graph of $r = 1 + a \cos \theta$ crosses itself and forms a loop.)
25. Graph
$$r = 1 + \tfrac{1}{2} \cos \theta.$$

26. Obtain the rectangular form of the equation $r^2 = \cos 2\theta$.
27. Graph the curve in Exercise 26, using polar coordinates. Note that if $\cos 2\theta$ is negative, r is not defined, and that if $\cos 2\theta$ is positive, there are two values of r, $\sqrt{\cos 2\theta}$ and $-\sqrt{\cos 2\theta}$.

■ ■

In Appendix B it is shown that the graph of $r = 1/(1 + e \cos \theta)$ is a parabola if $e = 1$, an ellipse if $0 \le e < 1$, and a hyperbola if $e > 1$. Exercises 28 to 30 concern such graphs.

28. Find an equation in rectangular coordinates for the curve $r = 1/(1 + \cos \theta)$.
29. (a) Graph
$$r = \frac{1}{(1 - \tfrac{1}{2} \cos \theta)}.$$
    (b) Find an equation in rectangular coordinates for the curve in (a).
30. (a) Graph
$$r = \frac{1}{(1 + 2 \cos \theta)}.$$
    (b) What angles do the asymptotes to the graph in (a) make with the positive x axis?
    (c) Find an equation in rectangular coordinates for the curve in (a).
31. (a) Graph
$$r = 3 + \cos \theta.$$
    (b) Find the point on the graph in (a) that has the maximum y coordinate.

10.8

Parametric equations

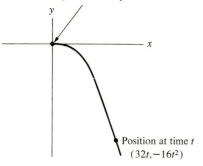

Ball is thrown horizontally to the right from this point at time $t = 0$

Position at time t
$(32t, -16t^2)$

If a ball is thrown horizontally with a speed of 32 feet per second, it falls in a curved path. Air resistance disregarded, its position t seconds later is given by $x = 32t$, $y = -16t^2$ relative to the coordinate system in the accompanying diagram. Here the curve is completely described, not by expressing y as a function of x, but by expressing both x and y as functions of a third variable t. The third variable is called a *parameter* (*para* meaning together, *meter* meaning measure). The equations $x = 32t$, $y = -16t^2$ are called *parametric equations* for the curve.

In this example it is easy to eliminate t and so find a direct relation between x and y:

$$t = \frac{x}{32},$$

hence

$$y = -16\left(\frac{x}{32}\right)^2 = -\frac{16}{(32)^2}x^2 = -\frac{1}{64}x^2.$$

The path of the falling ball is part of the curve $y = -\frac{1}{64}x^2$.

The next three examples present parametric equations in which the parameter is not time.

EXAMPLE 1   Let
$$\begin{cases} x = \cos 2\theta, \\ y = \sin \theta. \end{cases}$$
Determine a curve parametrically. Graph and describe this curve.

SOLUTION   Since x and y are the sine and cosine of angles, $|x|$ and $|y|$ never exceed 1. Moreover when θ increases by 2π, we obtain the same point again. Here is a table showing a few points on the graph.

θ	0	$\frac{\pi}{4}$	$\frac{\pi}{2}$	$\frac{3\pi}{4}$	π	$\frac{5\pi}{4}$	$\frac{3\pi}{2}$	$\frac{7\pi}{4}$	2π
x	1	0	-1	0	1	0	-1	0	1
y	0	$\frac{\sqrt{2}}{2}$	1	$\frac{\sqrt{2}}{2}$	0	$\frac{-\sqrt{2}}{2}$	-1	$\frac{-\sqrt{2}}{2}$	0

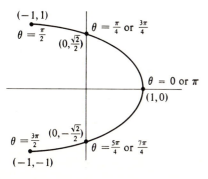

$(-1, 1)$
$\theta = \frac{\pi}{2}$
$\theta = \frac{\pi}{4}$ or $\frac{3\pi}{4}$
$(0, \frac{\sqrt{2}}{2})$
$\theta = 0$ or π
$(1, 0)$
$(0, -\frac{\sqrt{2}}{2})$
$\theta = \frac{3\pi}{2}$
$\theta = \frac{5\pi}{4}$ or $\frac{7\pi}{4}$
$(-1, -1)$

The last point duplicates the first. The first eight points suggest the shape of the graph, which has been drawn at the left. For θ in the interval $[0, 2\pi]$ the path runs over the graph twice. [Note that θ is *not* the polar angle of the point (x, y).]

The curve looks like part of a parabola. Fortunately, the equations for x and y are sufficiently simple that θ can be eliminated and a relation between x and y found:

$$x = \cos 2\theta = \cos^2 \theta - \sin^2 \theta$$
$$= 1 - 2\sin^2 \theta$$
$$= 1 - 2y^2.$$

Thus the path lies on the parabola

$$x = 1 - 2y^2.$$

But note that it is only a small part of the parabola and sweeps out this part infinitely often, like a pendulum. ●

In Example 1 it was easy to eliminate the parameter θ, thus obtaining a simple equation involving only x and y. In the next example, elimination of θ would lead to a complicated equation involving x and y. One great advantage of parametric equations is that they can provide a simple description of a curve, although it may be difficult or impossible to find an equation in x and y which describes the curve.

EXAMPLE 2   As a bicycle wheel of radius a rolls along, a tack stuck in its circumference traces out a curve called a *cycloid*, which consists of a sequence of arches, one arch for each revolution of the wheel.

Find the position of the tack as a function of the angle θ through which the wheel turns.

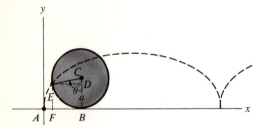

The rolling wheel has radius a

Note that because the wheel doesn't slip, arc $EB = \overline{AB}$, thus $\overline{AB} = a\theta$

When $\theta = 0$, the tack is on the ground at $x = 0, y = 0$

SOLUTION   The x coordinate of the tack, corresponding to θ, is

$$\overline{AF} = \overline{AB} - \overline{ED} = a\theta - a \sin \theta$$

and the y coordinate is

$$\overline{EF} = \overline{BC} - \overline{CD} = a - a \cos \theta.$$

Then the position of the tack, as a function of the parameter θ, is

$$\begin{cases} x = a\theta - a \sin \theta; \\ y = a - a \cos \theta. \end{cases}$$

In this case, eliminating θ would lead to a complicated relation between x and y. ●

Any curve can be described parametrically. For instance, consider the curve $y = e^x + x$. It is perfectly legal to introduce x itself as a parameter t, and write

$$\begin{cases} x = t \\ y = e^t + t. \end{cases}$$

This device may seem a bit artificial, but it will be useful in the next section in order to apply results for curves expressed by means of parametric equations to curves given in the form $y = f(x)$.

How can we find the slope of a curve which is described parametrically by the equations

$$x = g(t), \qquad y = h(t)?$$

The clumsy, perhaps impossible, way is to solve the equation $x = g(t)$ for t as a function of x and plug the result into the equation $y = h(t)$, thus expressing y explicitly in terms of x; then differentiate the result. Fortunately, there is a very easy way, which we will now describe.

Assume that the functions g and h are differentiable and that dx/dt is not 0. As mentioned in Sec. 5.10, t is a differentiable function of x, and

$$\frac{dt}{dx} = \frac{1}{dx/dt}.$$

Since y is a function of t, and t is a function of x, the chain rule says that

$$\frac{dy}{dx} = \frac{dy}{dt} \cdot \frac{dt}{dx}$$

How to Find the Slope of a Curve Given Parametrically

Consequently,

$$\frac{dy}{dx} = \frac{dy/dt}{dx/dt},$$

a formula which is as easy to apply as it is to remember.

EXAMPLE 3 At what angle does the arch of the cycloid shown in Example 2 meet the x axis at the origin?

SOLUTION The parametric equations of the cycloid are

$$x = a\theta - a \sin \theta, \qquad y = a - a \cos \theta.$$

Here θ plays the role of the parameter t. Then

$$\frac{dx}{d\theta} = a - a \cos \theta \qquad \text{and} \qquad \frac{dy}{d\theta} = a \sin \theta.$$

Consequently,

$$\frac{dy}{dx} = \frac{dy/d\theta}{dx/d\theta} = \frac{a \sin \theta}{a - a \cos \theta}$$

$$= \frac{\sin \theta}{1 - \cos \theta}.$$

When θ is near 0, (x, y) is near the origin. How does the slope, $\sin \theta/(1 - \cos \theta)$ behave as $\theta \to 0^+$? L'Hôpital's rule applies, and we have

$$\lim_{\theta \to 0^+} \frac{\sin \theta}{1 - \cos \theta} = \lim_{\theta \to 0^+} \frac{\cos \theta}{\sin \theta}$$

$$= \infty.$$

Thus the cycloid comes in vertically at the origin. ●

The Rotary Engine   The next two examples use parametric equations to describe the geometric principles of the rotary engine recognized by Felix Wankel in 1954. He found that it is possible for an equilateral triangle to revolve in a certain curve in such a way that its corners maintain contact with the curve and its center sweeps out a circle. These examples are not essential for understanding this section.

EXAMPLE 4   Let e and R be fixed positive numbers and consider the curve given parametrically by

$$x = e \cos 3\theta + R \cos \theta \qquad \text{and} \qquad y = e \sin 3\theta + R \sin \theta.$$

Show that an equilateral triangle can revolve in this curve while its center describes a circle or radius e.

SOLUTION   This diagram shows the typical point $P = (x, y)$ that corresponds to the

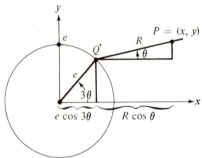

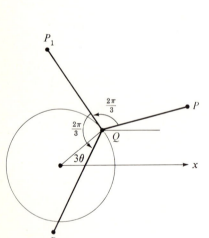

parameter value θ. As θ increases by $2\pi/3$ from any given angle, the point Q goes once around the circle of radius e and returns to its initial position. During this revolution of Q the point P moves to a point P_1 whose angle, instead of being θ, is $\theta + 2\pi/3$. Thus, if P is on the curve, so are the points P_1 and P_2 shown in the diagram in the margin.

Consequently, the equilateral triangle whose vertices are P, P_1, and P_2 revolves once around the curve, while its center Q goes three times around the circle of radius e. ●

What does the curve described in Example 4 look like? Wankel graphed it without knowing that mathematicians had met it long before in a different setting, described in Example 5, which provides a way of graphing the curve.

EXAMPLE 5   A circle of radius r rolls without slipping around a fixed circle of radius $2r$. Describe the path swept out by a point P located at a distance e from the center of the moving circle, $0 \le e \le r$.

SOLUTION   Place the rolling circle as shown in the first diagram, next page. Note that the center C of the rolling circle traces out a circle of radius $3r$. Let $R = 3r$.

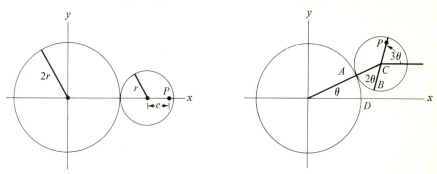

As the little circle rolls counterclockwise around the fixed circle without slipping, it traces out a path whose initial point is shown in the left-hand figure. The right-hand figure shows the typical point P on the path as the circle rolls around the larger circle.

Recall that in a triangle an exterior Angle is the Sum of the Two Opposite Interior Angles

Since the radius of the rolling circle is half that of the fixed circle, angle ACB is 2θ. Thus the angle that CP makes with the x axis is the sum of θ and 2θ, which is 3θ. Consequently, $P = (x, y)$ has coordinates given parametrically as

$$x = e \cos 3\theta + R \cos \theta \qquad \text{and} \qquad y = e \sin 3\theta + R \sin \theta.$$

Thus the curve swept out by P is precisely the curve Wankel studied.

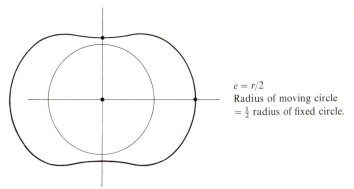

$e = r/2$
Radius of moving circle
$= \frac{1}{2}$ radius of fixed circle.

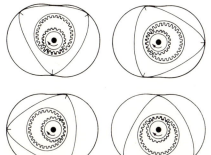

Long known to mathematicians as an *epitrochoid*, it appears typically as sketched above. ●

In order that the moving rotor in the rotary engine can turn the drive shaft, teeth are placed in it along a circle of radius $2e$ which engage teeth in the drive shaft, which has radius e. For each complete rotation of the rotor the drive shaft completes three rotations.

It was a Stuttgart professor, Othmar Baier, who showed that Wankel's curve was an epitrochoid. This insight was of practical importance, for it was of aid in simplifying the machining of the working surface of the motor.

Exercises

1. Consider the parametric equations $x = 2t + 1$, $y = t - 1$.
(a) Fill in this table:

t	-2	-1	0	1	2
x					
y					

(b) Plot the five points (x, y) obtained in (a).
(c) Graph the curve given by the parametric equations $x = 2t + 1$, $y = t - 1$.
(d) Eliminate t to find an equation for the graph involving only x and y.

2. Consider the parametric equations

$$x = t + 1, \qquad y = t^2.$$

(a) Fill in this table:

t	-2	-1	0	1	2
x					
y					

(b) Plot the five points (x, y) obtained in (a).
(c) Graph the curve.
(d) Find an equation in x and y that describes the curve.

3. Consider the parametric equations $x = t^2$, $y = t^2 + t$.
(a) Fill in this table:

t	-3	-2	-1	0	1	2	3
x							
y							

(b) Plot the seven points (x, y) obtained in (a).
(c) Graph the curve given by $x = t^2$, $y = t^2 + t$.
(d) Eliminate t and find an equation for the graph in terms of x and y.

4. (a) Letting $t = -1$, 0, and 1, find three points on the curve

$$\begin{cases} x = t^7 + t^2 + 1, \\ y = 2t^6 + 3t + 1. \end{cases}$$

(b) Can you eliminate t between the two equations?
(c) When t is large, what happens to y/x?
(d) What is the slope of the curve at the point corresponding to $t = 1$?

5. Consider the parametric equations $x = 2 \cos \theta$, $y = 3 \sin \theta$.
(a) Fill in this table, expressing the entries decimally.

θ	0	$\dfrac{\pi}{4}$	$\dfrac{\pi}{2}$	$\dfrac{3\pi}{4}$	π	$\dfrac{5\pi}{4}$	$\dfrac{3\pi}{2}$	$\dfrac{7\pi}{4}$	2π
x									
y									

(b) Plot the eight distinct points in (a).
(c) Graph the curve given by $x = 2 \cos \theta$, $y = 3 \sin \theta$.
(d) Using the identity

$$\cos^2 \theta + \sin^2 \theta = 1,$$

eliminate θ.
(e) Is θ the polar angle of $(2 \cos \theta, 3 \sin \theta)$?

6. Using the method mentioned after Example 2, describe each of these graphs in parametric form: (a) $y = x^3 - x^2$, x as parameter; (b) $x = e^y$, y as parameter; (c) $y = x \sin x$, x as parameter; (d) $x = \sin 3y$, y as parameter.

7. In Example 2, what is the value of θ when the point E is (a) at the top of the first arch? (b) at the right end of the first arch? (c) at the top of the second arch? (d) at the right end of the second arch?

8. A ball is thrown at an angle α and initial velocity v_0, as sketched below. It can be shown that if time is in seconds

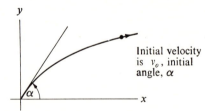

Initial velocity is v_0, initial angle, α

and distance in feet, then t seconds later the ball is at the point

$$\begin{cases} x = (v_0 \cos \alpha)t \\ y = (v_0 \sin \alpha)t - 16t^2. \end{cases}$$

Eliminate t. The resulting equation shows that the path is a parabola.

9. Eliminate t and plot

$$\begin{cases} x = e^t, \\ y = e^{2t}. \end{cases}$$

10. Instead of giving x and y in terms of t, it is sometimes convenient to give the polar coordinates r and θ in terms of *t*.   Eliminate *t* from $\begin{cases} r = 3t, \\ \theta = 6t, \end{cases}$

and plot the curve.

11. A curve is given parametrically by the equations

$$x = t^5 + \sin 2\pi t \qquad y = t + e^t.$$

(*a*) Find the slope of the curve at the point corresponding to *t* = 1.

(*b*) Try to solve for *y* explicitly in terms of *x*.

12. Find the area enclosed by one arch of the cycloid of Example 2 and the *x* axis.   *Hint:* The area is $\int_0^{2\pi a} y \, dx$, but make the substitution $x = a\theta - a \sin \theta$, expressing *y*, *dx*, and the limits of integration in terms of *θ*.

13. Consider the cardioid *r* = 1 + cos *θ*.

(*a*) Express the *x* and *y* coordinates of the typical point on the cardioid parametrically in terms of *θ*.

(*b*) What is the slope of the cardioid at (*x*, *y*) = (0, 1)?

(*c*) What happens to the slope of the cardioid near (*x*, *y*) = (0, 0) but above the *x* axis?

■

14. Consider the curve given parametrically by

$$x = t^2 + e^t, \qquad y = t + e^t \qquad \text{for } t \text{ in } [0, 1].$$

(*a*) Plot the points corresponding to *t* = 0, ½, and 1.

(*b*) Sketch the curve.

(*c*) Find the slope of the curve at the point (1, 1).

(*d*) Find the area of the region under the curve and above the interval [1, *e* + 1].

15. Let *a* and *b* be positive numbers.   Consider the curve given parametrically by the equations

$$x = a \cos t, \qquad y = b \sin t.$$

(*a*) Show that the curve is the ellipse

$$\frac{x^2}{a^2} + \frac{y^2}{b^2} = 1.$$

(*b*) The point (*x*, *y*) corresponding to the parameter value *t* has a polar angle *θ*.   For which values of *t* does *t* = *θ*?

(*c*) Find the area of the region bounded by the ellipse in (*a*) by making a substitution that replaces $4 \int_0^a y \, dx$ by an integral in which the variable is *t* and the range of integration is [0, *π*/2].

■ ■

16. Let *a* be a positive constant.   Consider the curve given parametrically by the equations $x = a \cos^3 t$, $y = a \sin^3 t$.

(*a*) Sketch the curve.

(*b*) Find the slope of the curve at the point corresponding to the parameter value *t*.

17. Consider a tangent line to the curve in Exercise 16 at a point *P* in the first quadrant.   Show that the length of the segment of that line intercepted by the coordinate axes is *a*.

This exercise is related to Examples 4 and 5, which concern the rotary engine.

18. In Example 5 a circle of radius *r* rolled around a circle of radius 2*r*.   Instead, consider the curve produced by a point *P* at a distance *e* from the center of the rolling circle, 0 ≤ *e* ≤ *r*, if the radius of the fixed circle is 3*r*.

(*a*) Find parametric equations of the curve produced.

(*b*) Sketch the curve.

(*c*) The curve in (*b*) is called a three-lobed epitrochoid.   Show that a square rotor can revolve in it, as the triangular rotor did in Example 4.   Engines with this design have been tried but are not as efficient as the standard rotary engine.

10.9

Arc length and speed on a curve

The path of some particle is given parametrically,

$$\begin{cases} x = g(t), \\ y = h(t). \end{cases}$$

(Think of *t* as time.)   A physicist might ask the following questions.   "How far does the particle travel from time *t* = *a* to time *t* = *b*?"   "What is the speed of the particle at time *t*?"

Consider the "distance traveled" question first.   The second will then be easy to answer, making use of the derivative. (In our reasoning we shall call the parameter t, and think of it as time, but the results apply to any parameter.)

HOW TO FIND ARC LENGTH   If a moving particle is at the point $(x, y) = (g(t), h(t))$ at time t, then we shall show that the length of its path from time $t = a$ to time $t = b$ is equal to

$$\int_a^b \sqrt{[g'(t)]^2 + [h'(t)]^2} \, dt,$$

or

$$\int_a^b \sqrt{\left(\frac{dx}{dt}\right)^2 + \left(\frac{dy}{dt}\right)^2} \, dt.$$

(Assume that g and h have continuous derivatives.)

We shall argue for the plausibility of the formula.

Partition the time interval $[a, b]$ and use this partition to inscribe a polygon in the curve of the moving particle.

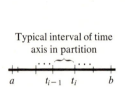

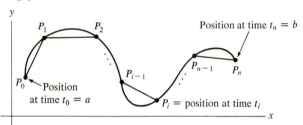

The length of such a polygon should approach the arc length as the mesh of the partition of $[a, b]$ shrinks toward 0 since the points P_i along the curve will get closer and closer together. The length of the typical straight-line segment $P_{i-1}P_i$, where $P_{i-1} = (g(t_{i-1}), h(t_{i-1}))$ and $P_i = (g(t_i), h(t_i))$, is

$$\sqrt{[g(t_i) - g(t_{i-1})]^2 + [h(t_i) - h(t_{i-1})]^2}$$

and so the length of the polygon is the sum

$$\sum_{i=1}^n \sqrt{[g(t_i) - g(t_{i-1})]^2 + [h(t_i) - h(t_{i-1})]^2}. \tag{1}$$

We shall relate this sum to sums of the type appearing in the definition of a definite integral over $[a, b]$.

By the law of the mean, there exist numbers T_i^* and T_i^{**}, both in the interval $[t_{i-1}, t_i]$, such that $g(t_i) - g(t_{i-1}) = g'(T_i^*)(t_i - t_{i-1})$ and $h(t_i) - h(t_{i-1}) = h'(T_i^{**})(t_i - t_{i-1})$. Thus the sum (1) can be rewritten as

$$\sum_{i=1}^n \sqrt{[g'(T_i^*)]^2 + [h'(T_i^{**})]^2}(t_i - t_{i-1}). \tag{2}$$

If T_i^{**} were equal to T_i^*, then this sum (2) would be an approximating sum used in defining

$$\int_a^b \sqrt{[g'(t)]^2 + [h'(t)]^2} \, dt.$$

To get around this difficulty, notice that, since h' is continuous, $h'(T_i^*)$ is near $h'(T_i^{**})$ when the mesh of the partition of $[a, b]$ is small. If the sum in (2) is a good approximation to the arc length, then presumably so is the sum

$$\sum_{i=1}^{n} \sqrt{[g'(T_i^*)]^2 + [h'(T_i^*)]^2}(t_i - t_{i-1}). \tag{3}$$

In other words, it is reasonable to expect that

$$\lim_{\text{mesh} \to 0} \sum_{i=1}^{n} \sqrt{[g'(T_i^*)]^2 + [h'(T_i^*)]^2}(t_i - t_{i-1}) = \text{arc length}.$$

But that limit is precisely the definition of the definite integral

$$\int_a^b \sqrt{[g'(t)]^2 + [h'(t)]^2}\, dt.$$

This shows why this definite integral should yield the arc length.

EXAMPLE 1 Find the distance s which the ball described at the beginning of Sec. 10.8 travels during the first b seconds.

SOLUTION Here $x = 32t$ and $y = -16t^2$. Thus $g'(t) = 32$ and $h'(t) = -32t$. Consequently,

$$s = \int_0^b \sqrt{(32)^2 + (-32t)^2}\, dt = 32 \int_0^b \sqrt{1 + t^2}\, dt,$$

a definite integral that can be evaluated with the aid of a table or the substitution $t = \tan \theta$; its value is $16b\sqrt{1 + b^2} + 16 \ln (b + \sqrt{1 + b^2})$. ●

EXAMPLE 2 Find the length of one arch of the cycloid in Example 2 of Sec. 10.8.

SOLUTION Here the parameter is θ, and we compute $dx/d\theta$ and $dy/d\theta$:

$$\frac{dx}{d\theta} = \frac{d(a\theta - a \sin \theta)}{d\theta} = a - a \cos \theta,$$

and

$$\frac{dy}{d\theta} = \frac{d(a - a \cos \theta)}{d\theta} = a \sin \theta.$$

To complete one arch, θ varies from 0 to 2π. By the formula for computing arc length, the length of one arch is $\int_0^{2\pi} \sqrt{(a - a \cos \theta)^2 + (a \sin \theta)^2}\, d\theta$. Thus

$$\text{Length of arc} = a \int_0^{2\pi} \sqrt{(1 - \cos \theta)^2 + (\sin \theta)^2}\, d\theta$$

$$= a \int_0^{2\pi} \sqrt{1 - 2 \cos \theta + (\cos^2 \theta + \sin^2 \theta)}\, d\theta$$

$$= a \int_0^{2\pi} \sqrt{2 - 2 \cos \theta}\, d\theta$$

$$= a\sqrt{2} \int_0^{2\pi} \sqrt{1 - \cos\theta} \; d\theta$$

$$= a\sqrt{2} \int_0^{2\pi} \sqrt{2} \sin\frac{\theta}{2} \; d\theta \qquad \text{trigonometry}$$

$$= 2a \int_0^{2\pi} \sin\frac{\theta}{2} \; d\theta \underset{\text{FTC}}{=} 2a\left(-2\cos\frac{\theta}{2}\right)\Big|_0^{2\pi}$$

$$= 2a[-2(-1) - (-2)(1)] = 8a.$$

While θ varies from 0 to 2π, the bicycle travels a distance $2\pi a \doteq 6.28a$, and the tack travels a distance $8a$. ●

From the formula for computing the length of a path given parametrically, it is easy to derive the formula for the length of a curve given in the form $y = f(x)$.

HOW TO FIND THE ARC LENGTH OF
$$y = f(x)$$

The length of the curve $y = f(x)$ for x in $[a, b]$ is equal to

$$\int_a^b \sqrt{1 + [f'(x)]^2} \; dx,$$

or

$$\int_a^b \sqrt{1 + \left(\frac{dy}{dx}\right)^2} \; dx.$$

(Assume that f has a continuous derivative.)

This formula follows from the one for the length of a parametrized path by the use of the parameterization

$$\begin{cases} x = t, \\ y = f(t). \end{cases}$$

Then
$$\frac{dx}{dt} = 1 \qquad \text{and} \qquad \frac{dy}{dt} = f'(t).$$

The formula
$$\int_a^b \sqrt{\left(\frac{dx}{dt}\right)^2 + \left(\frac{dy}{dt}\right)^2} \; dt$$

takes the form
$$\int_a^b \sqrt{1^2 + [f'(t)]^2} \; dt.$$

Since $x = t$, this is just
$$\int_a^b \sqrt{1 + [f'(x)]^2} \; dx.$$

EXAMPLE 3   Find the length of the curve $y = x^3$ for x in $[1, 2]$.

SOLUTION   In this case
$$\frac{dy}{dx} = 3x^2.$$

Thus the arc length is
$$\int_1^2 \sqrt{1 + (3x^2)^2} \; dx,$$

or
$$\int_1^2 \sqrt{1 + 9x^4}\, dx.$$

The fundamental theorem of calculus is useless in evaluating this definite integral, for the function $\sqrt{1 + 9x^4}$ does not have an elementary antiderivative. The definite integral could be estimated by an approximating sum based on a partition of the interval $[1, 2]$. ●

Incidentally, the length of the curve $y = x^a$, where a is an integer, usually *cannot* be computed with the aid of the fundamental theorem.   The only cases in which it can be computed by the fundamental theorem are $a = 0$ (the graph is a horizontal line), $a = 1$ (the graph is the line $y = x$), and $a = 2$ (the graph is the parabola $y = x^2$).   Exercise 21 treats this question.

In practice we are not interested as much in the length of the path as in the speed of the particle as it moves along the path.   The work done so far in this section easily helps us find this speed.

Consider a particle which at time t is at the point $(x, y) = (g(t), h(t))$. Choose a point B on the curve from which to measure distance along the curve. (Assume that the curve has an arc length.)   Let $s(t)$ denote the distance from B to $(g(t), h(t))$.   We shall always assume that B has been chosen in such a way that $s(t)$ is an *increasing* function of t.

DEFINITION   *Speed on a curved path.*   If ds/dt exists, it is called the *speed of the particle.*

Since $s(t)$ is assumed to be an increasing function, speed is not negative.

As early as the opening Sec. 1.1 we were able to treat the speed of a particle moving in a straight path.   Now it is possible to compute the speed of a particle moving on a curved path.

HOW TO FIND THE SPEED OF A PARTICLE MOVING ON A CURVED PATH

If a particle at time t is at the point $(x, y) = (g(t), h(t))$, where g and h are functions having continuous derivatives, then its speed at time t is equal to

$$\sqrt{[g'(t)]^2 + [h'(t)]^2}.$$

The argument is short.   Let $s(t)$ denote the arc length along the curve from some base point B to the particle at time t.

Now,
$$s(t) = \int_a^t \sqrt{[g'(T)]^2 + [h'(T)]^2}\, dT.$$

(The letter T is introduced, since t is already used to describe the interval of integration.)   Differentiation of this relation with respect to t (using the second fundamental theorem of calculus) yields

$$\frac{ds}{dt} = \sqrt{[g'(t)]^2 + [h'(t)]^2}.$$

EXAMPLE 4   Find the speed at time t of the ball described at the beginning of Sec. 10.8.

SOLUTION   At time t the ball is at the point

$$(x, y) = (32t, -16t^2).$$

Thus dx/dt, usually written $\dot{x}$, is 32, and $\dot{y}$ is $-32t$.   The speed of the ball is

$$\frac{ds}{dt} = \sqrt{\dot{x}^2 + \dot{y}^2} = \sqrt{32^2 + (-32t)^2}$$

$$= 32\sqrt{1 + t^2} \qquad \text{feet per second.} \quad \bullet$$

So far in this section curves have been described in rectangular coordinates. Next consider a curve given in polar coordinates by the equation $r = f(\theta)$.

HOW TO FIND THE ARC LENGTH OF
$r = f(\theta)$

The length of the curve $r = f(\theta)$ for θ in $[\alpha, \beta]$ is equal to

$$\int_\alpha^\beta \sqrt{[f(\theta)]^2 + [f'(\theta)]^2}\, d\theta$$

or

$$\int_\alpha^\beta \sqrt{r^2 + (r')^2}\, d\theta.$$

(Assume that f has a continuous derivative.)

This formula can be derived from that for the arc length of a parameterized curve in rectangular coordinates, as follows.   Find the rectangular coordinates of the point whose polar coordinates are

$$(r, \theta) = (f(\theta), \theta).$$

They are
$$\begin{cases} x = f(\theta)\cos\theta, \\ y = f(\theta)\sin\theta. \end{cases}$$

The curve is now given in rectangular form with parameter θ.   Thus its length is

$$\int_\alpha^\beta \sqrt{\left(\frac{dx}{d\theta}\right)^2 + \left(\frac{dy}{d\theta}\right)^2}\, d\theta.$$

Now,
$$\frac{dx}{d\theta} = f(\theta)(-\sin\theta) + f'(\theta)\cos\theta,$$

and
$$\frac{dy}{d\theta} = f(\theta)\cos\theta + f'(\theta)\sin\theta;$$

hence

$$\left(\frac{dx}{d\theta}\right)^2 + \left(\frac{dy}{d\theta}\right)^2 = [f(\theta)]^2\sin^2\theta - 2f(\theta)f'(\theta)\sin\theta\cos\theta + [f'(\theta)]^2\cos^2\theta$$

$$+ [f(\theta)]^2\cos^2\theta + 2f(\theta)f'(\theta)\sin\theta\cos\theta + [f'(\theta)]^2\sin^2\theta$$

which, by the identity $\sin^2\theta + \cos^2\theta = 1$, simplifies to $[f(\theta)]^2 + [f'(\theta)]^2$. This justifies the formula.

EXAMPLE 5   Find the length of the spiral $r = e^{-3\theta}$ for θ in $[0, 2\pi]$.

SOLUTION   First compute

$$r' = \frac{dr}{d\theta} = -3e^{-3\theta},$$

and then use the formula

$$\text{Arc length} = \int_{\alpha}^{\beta} \sqrt{r^2 + (r')^2}\, d\theta$$

$$= \int_{0}^{2\pi} \sqrt{(e^{-3\theta})^2 + (-3e^{-3\theta})^2}\, d\theta$$

$$= \int_{0}^{2\pi} \sqrt{e^{-6\theta} + 9e^{-6\theta}}\, d\theta$$

$$= \sqrt{10} \int_{0}^{2\pi} \sqrt{e^{-6\theta}}\, d\theta$$

$$= \sqrt{10} \int_{0}^{2\pi} e^{-3\theta}\, d\theta$$

$$= \sqrt{10}\, \frac{e^{-3\theta}}{-3}\Big|_{0}^{2\pi}$$

$$= \sqrt{10}\left(\frac{e^{-3 \cdot 2\pi}}{-3} - \frac{e^{-3 \cdot 0}}{-3}\right)$$

$$= \sqrt{10}\left(\frac{e^{-6\pi}}{-3} + \frac{1}{3}\right)$$

$$= \sqrt{10}\left(\frac{1}{3} - \frac{e^{-6\pi}}{3}\right). \quad \bullet$$

Memory Aids   Multiplying both sides of the equation

$$\left(\frac{ds}{dt}\right)^2 = \left(\frac{dx}{dt}\right)^2 + \left(\frac{dy}{dt}\right)^2$$

by $(dt)^2$ yields

$$(ds)^2 = (dx)^2 + (dy)^2.$$

$(ds)^2 = (dx)^2 + (dy)^2$

The right triangle in the margin serves as a reminder of the relation between the differentials dx, dy, and ds.

There is also a simple way to remember the formula for arc length in polar coordinates.   Since

$$s = \int_{\alpha}^{\beta} \sqrt{r^2 + (r')^2}\, d\theta,$$

the second fundamental theorem of calculus implies that

$$\frac{ds}{d\theta} = \sqrt{r^2 + (r')^2}, \quad \text{hence that} \quad \left(\frac{ds}{d\theta}\right)^2 = r^2 + \left(\frac{dr}{d\theta}\right)^2.$$

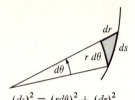

$$(ds)^2 = (rd\theta)^2 + (dr)^2$$

Multiplying by $(d\theta)^2$ yields

$$(ds)^2 = (r\,d\theta)^2 + (dr)^2.$$

This is easy to remember if you draw the "almost right" triangle which is shaded in the diagram at the left.

Exercises

1. Find the length of the curve $y = x^3/3 + x^{-1}/4$ from $x = 1$ to $x = 2$.

2. Find the length of the curve $y = 1/(2x^2) + x^4/16$ from $x = 2$ to $x = 3$.

3. Find the speed of the thrown ball referred to at the beginning of this section at times $t = 0$, $t = 1$, and $t = 2$.

4. At time t a particle is at the point

$$P = (\cos t + t \sin t, \sin t - t \cos t).$$

Find the distance it travels from time $t = 1$ to time $t = \pi$.

5. (a) How far does a bug travel from time $t = 1$ to time $t = 2$, if at time t it is at the point (t^2, t^3)?
   (b) How fast is it moving at time t?
   (c) Graph its path relative to an xy-coordinate system. Where is it at $t = 1$? At $t = 2$?
   (d) Eliminate t to find y as a function of x.

6. (a) At time t a bug is at $(\cos^3 t, \sin^3 t)$. Graph its path from time $t = 0$ to time $t = \pi/2$.
   (b) How far does it travel in that interval?
   (c) Show that it travels along the curve $x^{2/3} + y^{2/3} = 1$.

7. (a) Graph $y = (e^x + e^{-x})/2$.
   (b) Find the length of arc along the curve above $[0, b]$.

8. A particle is at $(t + \cos t, 2t - \sin t)$ at time t. Find its speed when $t = 1$.

9. A particle is at $(e^{2t} \cos t, t)$ at time t. Find its speed when $t = 0$.

10. Find the length of the curve $y = \frac{1}{2}x^2 - \frac{1}{4}\ln x$ from $x = 2$ to $x = 3$.

11. (a) Graph the spiral given in polar coordinates by $r = e^{\theta}$.
    (b) Find the length of the turn from $\theta = 0$ to $\theta = 2\pi$.
    (c) Find the length of the turn from $\theta = 2\pi$ to $\theta = 4\pi$.

12. (a) Graph the cardioid $r = 1 - \cos\theta$.
    (b) Find its perimeter.

13. Find the length of the curve given in polar coordinates by $r = 3\sin\theta$, for θ in $[0, \pi]$.

14. Find the length of the curve $r = \cos^2(\theta/2)$ for θ in $[0, \pi]$.

15. (a) Graph $y = x^3/3$.
    (b) Estimate its arc length from $(0, 0)$ to $(3, 9)$ by an inscribed polygon, whose vertices have x coordinates 0, 1, 2, 3. A table of square roots or a calculator will be useful.
    (c) Set up a definite integral for the arc length in question.
    (d) Estimate the definite integral in (c) by using a partition of $[0, 3]$ into three sections, each of length 1, and as sampling points X_i, the right endpoint of each section.
    (e) Proceed as in (d), but use left endpoints.

■

16. Assume that a curve is described in rectangular coordinates in the form $x = f(y)$. Show that

$$\text{Arc length} = \int_c^d \sqrt{1 + \left(\frac{dx}{dy}\right)^2}\, dy,$$

where y ranges in the interval $[c, d]$.

17. (See Exercise 16.) Consider the arc length of the curve $y = x^{2/3}$ for x in the interval $[1, 8]$.
    (a) Set up a definite integral for this arc length, using x as the parameter.
    (b) Set up a definite integral for this arc length, using y as the parameter.
    (c) Evaluate the easier of the integrals in (a) and (b).

18. At time t a particle has polar coordinates $r = g(t)$, $\theta = h(t)$. How fast is it moving?

19. Let $P = (x, y)$ depend on θ as shown in the accompanying diagram.

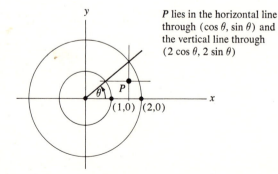

P lies in the horizontal line through $(\cos\theta, \sin\theta)$ and the vertical line through $(2\cos\theta, 2\sin\theta)$

(a) Sketch the curve that P sweeps out.

(b) Show that $P = (2 \cos \theta, \sin \theta)$.

(c) Set up a definite integral for the length of the curve described by P (do not evaluate it).

(d) Eliminate θ and show that P is on the curve

$$\frac{x^2}{4} + \frac{y^2}{1} = 1.$$

■ ■

20. Show that, if $\quad y = \dfrac{x^{m+1}}{m+1} + \dfrac{x^{1-m}}{4(m-1)},$

where m is any number other than 1 or -1, then the definite integral for the arc length of this curve can be computed with the aid of the fundamental theorem of

calculus. Consider only arcs corresponding to x in $[a, b]$, $0 < a < b$.

21. Consider the length of the curve $y = x^m$, where m is a rational number. Show that the fundamental theorem of calculus is of aid in computing this length if and only if $m = 1$ or if m is of the form $1 + 1/n$ for some integer n. *Hint:* The analog of Chebychev's theorem in Exercise 167 in Sec. 9.10 holds for $\int x^p (1 + x)^q \, dx$.

22. Consider the cardioid $r = 1 + \cos \theta$ for θ in $[0, \pi]$. We may consider r as a function of θ or as a function of s, arc length along the curve, measured, say from $(2, 0)$.

(a) Find the average of r with respect to θ.

(b) Find the average of r with respect to s. *Hint:* Express all quantities appearing in this average in terms of θ.

10.10

Area in polar coordinates

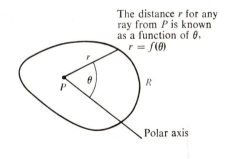

The distance r for any ray from P is known as a function of θ, $r = f(\theta)$

The area of the shaded region is $\frac{\theta}{2} r^2$

Section 7.4 showed how to compute the area of a region if the lengths of parallel cross sections are known. Sums based on estimating rectangles lead to the formula

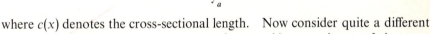

$$\text{Area} = \int_a^b c(x) \, dx,$$

where $c(x)$ denotes the cross-sectional length. Now consider quite a different situation, in which sectors, not rectangles, provide an estimate of the area.

Let R be a region in the plane and P a point inside it. Assume that the distance r from P to any point on the boundary of R is known as a function $r = f(\theta)$. (For convenience, assume that any ray from P meets the boundary of R just once.)

The cross sections made by the rays from P are *not* parallel. Instead, like spokes in a wheel, they all meet at the point P. It would be unnatural to use rectangles to estimate the area, but it is reasonable to use sectors of circles that have P as a common vertex.

Begin by recalling that in a circle of radius r a sector of central angle θ has area $(\theta/2) r^2$. This formula plays the same role now as the formula for the area of a rectangle did in Sec. 7.4.

Introduce a partition of $[0, 2\pi]$,

$$0 = \theta_0, \theta_1, \ldots, \theta_n = 2\pi,$$

and estimate the area of R bounded between the rays whose angles are

$$\theta_{i-1} \quad \text{and} \quad \theta_i.$$

Pick any angle θ_i^* in the interval $[\theta_{i-1}, \theta_i]$ and use a sector of a circle whose

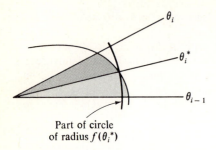

Part of circle
of radius $f(\theta_i^*)$

radius is $f(\theta_i^*)$ to approximate the area of the shaded region. The approximating sector has radius $f(\theta_i^*)$ and angle $\theta_i - \theta_{i-1}$; hence its area is

$$\frac{\theta_i - \theta_{i-1}}{2}[f(\theta_i^*)]^2,$$

or

$$\frac{[f(\theta_i^*)]^2}{2}(\theta_i - \theta_{i-1}).$$

The sum

$$\sum_{i=1}^{n} \frac{[f(\theta_i^*)]^2}{2}(\theta_i - \theta_{i-1})$$

is an estimate of the area of R.

As the mesh of the partition of $[0, 2\pi]$ approaches 0, these sums become better and better approximations to the area of R, and also approach the definite integral

$$\int_0^{2\pi} \frac{[f(\theta)]^2}{2}\, d\theta,$$

which, therefore, is the area of R.

To find the area of a region situated between two rays $\theta = \alpha$ and $\theta = \beta$, as in the diagram at the left, the same reasoning provides the following formula.

HOW TO FIND AREA IN POLAR COORDINATES

The area of the region bounded by the rays $\theta = \alpha$ and $\theta = \beta$ and by the curve $r = f(\theta)$ is

$$\int_\alpha^\beta \frac{[f(\theta)]^2}{2}\, d\theta$$

or

$$\int_\alpha^\beta \frac{r^2\, d\theta}{2}.$$

[Assume $f(\theta) \geq 0$.] It must be emphasized that no ray from the origin between α and β can cross the curve twice.

It may seem surprising to find $[f(\theta)]^2$, not just $f(\theta)$, in the integrand. But remember that area has the dimension "length times length." Since θ, given in radians, is dimensionless, being defined as "length of circular arc divided by length of radius," $d\theta$ is also dimensionless. Hence $f(\theta)\, d\theta$, having the dimension of length, not of area, could not be correct. But $\frac{1}{2}[f(\theta)]^2\, d\theta$, having the dimension of area (length times length), is plausible. For rectangular coordinates, in the expression $f(x)\, dx$, both $f(x)$ and dx have the dimension of length, one along the y axis, the other along the x axis; thus $f(x)\, dx$ has the dimension of area.

Memory device   This little sector at the left may call to mind the formula

$$\text{Area} = \int_\alpha^\beta \frac{r^2}{2}\, d\theta.$$

Area $= \frac{1}{2} \cdot r \cdot r\, d\theta$

EXAMPLE 1   Find the area of the region bounded by the curve

$$r = 3 + \cos\theta.$$

SOLUTION   By the formula just obtained, this area is

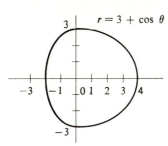

$r = 3 + \cos \theta$

$$\int_0^{2\pi} \frac{1}{2}(3 + \cos \theta)^2 \, d\theta = \frac{1}{2}\int_0^{2\pi}(9 + 6 \cos \theta + \cos^2 \theta) \, d\theta$$

$$= \frac{1}{2}\left(90 + 6 \sin \theta + \frac{\theta}{2} + \frac{\sin 2\theta}{4}\right)\Big|_0^{2\pi}$$

$$= \frac{1}{2}(19\pi) - \frac{1}{2}(0) = \frac{19\pi}{2}. \quad \bullet$$

Observe that any line through the origin intersects the region of Example 1 in a segment of length 6, since $(3 + \cos \theta) + [3 + \cos (\theta + \pi)] = 6$ for any θ. Also, any line through the center of a circle of radius 3 intersects the circle in a segment of length 6.   Thus two sets in the plane can have equal corresponding cross-sectional lengths through a fixed point and yet have different areas: the set in Example 1 has area $19\pi/2$, while the circle of radius 3 has area 9π. *Knowing the lengths of all the cross sections of a region through a given point is not enough to determine the area of the region!*

EXAMPLE 2   Find the area of the region inside one loop of the curve $r = \cos 4\theta$.

SOLUTION   First graph the curve.   Note that, when $4\theta = \pi/2, r = \cos \pi/2 = 0$.   Thus when $\theta = \pi/8, r = 0$.   This information appears in the following table.

θ	0	$\dfrac{\pi}{8}$	$\dfrac{\pi}{4}$	$\dfrac{3\pi}{8}$	$\dfrac{\pi}{2}$	$\dfrac{5\pi}{8}$	$\dfrac{3\pi}{4}$	$\dfrac{7\pi}{8}$	π	$\dfrac{9\pi}{8}$	$\dfrac{5\pi}{4}$	$\dfrac{11\pi}{8}$	$\dfrac{3\pi}{2}$	$\dfrac{13\pi}{8}$	$\dfrac{7\pi}{4}$	$\dfrac{15\pi}{8}$	2π
$r = \cos 4\theta$	1	0	-1	0	1	0	-1	0	1	0	-1	0	1	0	-1	0	1

It is then easy to graph the curve with the aid of the table.
The area of the loop which is bisected by the positive x axis is

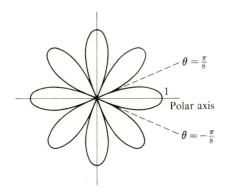

$\theta = \dfrac{\pi}{8}$

1
Polar axis

$\theta = -\dfrac{\pi}{8}$

$$\int_{-\pi/8}^{\pi/8} \frac{r^2}{2} \, d\theta = \int_{-\pi/8}^{\pi/8} \frac{\cos^2 4\theta}{2} \, d\theta$$

$$= \int_{-\pi/8}^{\pi/8} \frac{1 + \cos 8\theta}{4} \, d\theta$$

$$= \left[\frac{\theta}{4} + \frac{\sin 8\theta}{32}\right]_{-\pi/8}^{\pi/8}$$

$$= \frac{\pi}{32} + \frac{\sin \pi}{32} - \left[\frac{-\pi}{32} + \frac{\sin (-\pi)}{32}\right]$$

$$= \frac{\pi}{16}.$$

Incidentally, since the part of the loop above the x axis has the same

area as the part below,

$$\int_{-\pi/8}^{\pi/8} \frac{r^2}{2}\, d\theta = 2\int_{0}^{\pi/8} \frac{r^2}{2}\, d\theta.$$

This shortcut simplifies the arithmetic a little, reducing the chance for error when evaluating the integral. ●

Exercises

In Exercises 1 to 5, find the area of the region bounded by the indicated curve and rays.

1. $r = 2\theta$, $\alpha = 0$, $\beta = \dfrac{\pi}{2}$.

2. $r = \sqrt{\theta}$, $\alpha = 0$, $\beta = \pi$.

3. $r = \dfrac{1}{1+\theta}$, $\alpha = \dfrac{\pi}{4}$, $\beta = \dfrac{\pi}{2}$.

4. $r = \sqrt{\sin\theta}$, $\alpha = 0$, $\beta = \dfrac{\pi}{2}$.

5. $r = \tan\theta$, $\alpha = 0$, $\beta = \dfrac{\pi}{4}$.

6. (a) Graph the curve $r = 2\sin\theta$.
(b) Compute the area inside it.

7. (a) Graph the spiral $r = e^{\theta}$.
(b) Find the area within the first turn of the spiral, that is, for $0 \le \theta \le 2\pi$.

8. Find the area of the region inside the cardioid $r = 3 + 3\sin\theta$ and outside the circle $r = 3$.

9. (a) Graph the curve $r = \sqrt{\cos 2\theta}$. (Note that r is not defined for all θ.)
(b) Find the area inside one of its two loops.

In Exercises 10 to 13 find the area of the region described.

10. Inside one loop of the curve $r = \sin 3\theta$.
11. Inside one loop of the curve $r = \cos 2\theta$.
12. Inside one loop of the curve $r = 2\cos 2\theta$, but outside the circle $r = 1$.
13. Inside one loop of the curve $r = \sin 4\theta$.

■

14. (a) Show that the area of the triangle in the accompanying figure is $\int_{0}^{\beta} \frac{1}{2}\sec^2\theta\, d\theta$.

(b) From (a) and the fact that the area of a triangle is $\frac{1}{2}$(base)(height), show that $\tan\beta = \int_{0}^{\beta}\sec^2\theta\, d\theta$. With

the aid of this equation, obtain another proof that $D(\tan x) = \sec^2 x$.

15. Show that the area of the shaded crescent between the two circular arcs is equal to the area of square $ABCD$ (see the figure below). This type of result encouraged

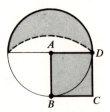

The outer arc has center A
The inner arc has center B

mathematicians from the time of the Greeks to try to find a method using only straightedge and compass for constructing a square whose area equals that of a given circle. This was proved impossible at the end of the nineteenth century.

■ ■

16. A point P in a region R bounded by a closed curve has the property that each chord through P cuts R into two regions of equal area. Must P bisect each chord through P? Explain. (Assume that each chord meets the curve at two points, namely at the ends of the chord.)

17. Let R be a region in the plane and P a point in R such that every line in the plane which passes through P intersects R in an interval of length at least a.
(a) Make a conjecture about the area of R.
(b) Prove your conjecture.

18. The function $r = f(\theta)$ describes, for θ in $[0, 2\pi]$, a curve in polar coordinates. Assume r' is continuous and $f(\theta) > 0$. Prove that the average of r as a function of arc length is at least as large as the quotient $2A/s$, where A is the area swept out by the radius and s is the arc length of the curve. When is the average equal to $2A/s$?

19. Prove that, if a region in the plane has the property that any two points in it are within a distance d of each other, then its area is at most $\pi d^2/4$. *Hint:* Use polar coordinates with the pole on the border of the region.

10.11

Area of a surface of revolution

This section develops a technique for computing the surface area of a solid of revolution, such as a sphere. The approach will be intuitive and will only justify the plausibility of defining the area of a surface of revolution as a certain definite integral.

Area of Surface of Cone

The cone

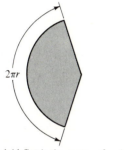

$2\pi r$

The cone laid flat is the sector of a circle

Begin by considering the area of a rather simple surface of revolution, the curved part of a cone whose base has radius r and whose slant height is l. If this cone is cut along a lateral edge and laid flat, it becomes a sector of a circle of radius l as illustrated at the left. Now, the area of a sector of radius l and angle θ (in radians) is $\frac{1}{2}l^2\theta$. Since $\theta = 2\pi r/l$, the area of this sector is $\frac{1}{2}l^2(2\pi r/l)$, which equals $\pi r l$. For this reason, define the lateral surface area of a cone as $\pi r l$. To remember this definition, keep in mind that the area of a sector is

$$\underbrace{\tfrac{1}{2}(2\pi r)}_{\text{Base}} \cdot \underbrace{(l)}_{\text{Height}} = \pi r l,$$

a formula similar to that for the area of a triangle.

Next, how shall we define the area of the surface formed when a line segment of length L is rotated about an axis at a distance r from its midpoint? This surface is a band, which may be considered to be the difference between two cones, often called a *frustum*. (See the diagram at the left.) Therefore define its area as the difference of the areas of the two cones whose dimensions are shown in the diagram. The area of the band is therefore

$$\pi r_2(l_1 + L) - \pi r_1 l_1 = \pi(r_2 l_1 - r_1 l_1 + r_2 L).$$

The inner cone has area $\pi r_1 l_1$, and the larger cone has area $\pi r_2 (l_1 + L)$

By similar triangles, $\quad \dfrac{L + l_1}{r_2} = \dfrac{l_1}{r_1}.$

Consequently,

$$r_1 L + r_1 l_1 = r_2 l_1 \qquad \text{or} \qquad r_2 l_1 - r_1 l_1 = r_1 L.$$

Thus the area of the band is

$$\pi(r_1 L + r_2 L) = \pi(r_1 + r_2)L = 2\pi \frac{r_1 + r_2}{2} L = 2\pi r L,$$

where $r = (r_1 + r_2)/2$.

The area of a surface formed by revolving a line segment has now been defined, and it can be used to justify the definition of the area swept out when we revolve a curve. The definition follows: Its justification is given afterward.

DEFINITION *Area of a surface of revolution.* Consider a curve given by the parametric equations $x = g(t)$, $y = h(t)$, where g and h have continuous derivatives and $h(t) \geq 0$. Let C be that portion of the curve corresponding to t in $[a, b]$. Then the area of the surface of revolution formed by revolving C about the

x axis is

$$\int_a^b 2\pi h(t)\sqrt{[g'(t)]^2 + [h'(t)]^2}\, dt$$

or, equivalently,

$$\int_a^b 2\pi y\sqrt{\left(\frac{dx}{dt}\right)^2 + \left(\frac{dy}{dt}\right)^2}\, dt.$$

To show that this is a reasonable definition, begin by partitioning the interval $[a, b]$ in the t axis, and consider a typical interval $[t_{i-1}, t_i]$. Probably the surface area of the solid of revolution corresponding to this small interval on the t axis is closely approximated by the area of the band obtained by revolving the line segment from (x_{i-1}, y_{i-1}) to (x_i, y_i) about the x axis.

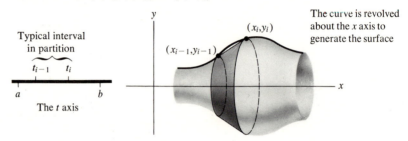

The approximating band has area

$$2\pi\, \frac{y_{i-1} + y_i}{2}\sqrt{(x_i - x_{i-1})^2 + (y_i - y_{i-1})^2}.$$

This area should be a good approximation to the area of the curved surface. This is not the typical summand of the type of summation leading to a definite integral over the t axis from a to b, since, in particular, $t_i - t_{i-1}$ does not appear. But the law of the mean can be used to introduce $t_i - t_{i-1}$ as it did in the preceding section.

By the law of the mean,

$$x_i - x_{i-1} = g'(T_i^*)(t_i - t_{i-1})$$

and

$$y_i - y_{i-1} = h'(T_i^{**})(t_i - t_{i-1}),$$

and by the intermediate-value theorem

$$\frac{y_{i-1} + y_i}{2} = h(T_i^{***}),$$

where T_i^*, T_i^{**}, T_i^{***} are all in $[t_{i-1}, t_i]$. So we now expect

$$2\pi h(T_i^{***})\sqrt{[g'(T_i^*)]^2 + [h'(T_i^{**})]^2}\,(t_i - t_{i-1})$$

to be a good approximation.

Since h and h' are continuous, $h(T_i^{***})$ is near $h(T_i^*)$, and $h'(T_i^{**})$ is near $h'(T_i^*)$, when the mesh of the partition of $[a, b]$ is small. Therefore *define the surface area to be*

$$\lim_{\text{mesh}\to 0}\ \sum_{i=1}^n 2\pi h(T_i^*)\sqrt{[g'(T_i^*)]^2 + [h'(T_i^*)]^2}\,(t_i - t_{i-1}).$$

That is,

$$\text{Surface area} = \int_a^b 2\pi h(t)\sqrt{[g'(t)]^2 + [h'(t)]^2}\ dt.$$

If a curve is given by $y = f(x)$, where f has a continuous derivative and $f(x) \geq 0$, it may be parameterized by the equations $x = t$ and $y = f(t)$. Then $dx/dt = dx/dx = 1$ and $dy/dt = dy/dx$. Hence the surface area swept out by rotating about the x axis that part of the curve above $[a, b]$ is

$$\text{Surface area} = \int_a^b 2\pi y \sqrt{1 + \left(\frac{dy}{dx}\right)^2}\ dx.$$

As the formulas are stated, they seem to refer to surfaces obtained by revolving a curve only about the x axis. In fact, they refer to revolution about any line. The factor y in the integrand,

$$2\pi y \sqrt{\left(\frac{dx}{dt}\right)^2 + \left(\frac{dy}{dt}\right)^2},$$

is the distance from the typical point on the curve to the axis of revolution. Call y simply R for *radius* to free ourselves from coordinate systems. (We use capital R to avoid confusion with polar coordinates.) Moreover, the expression

$$\sqrt{\left(\frac{dx}{dt}\right)^2 + \left(\frac{dy}{dt}\right)^2}\ dt$$

is simply ds, since

$$\frac{ds}{dt} = \sqrt{\left(\frac{dx}{dt}\right)^2 + \left(\frac{dy}{dt}\right)^2},$$

a formula resulting from our definition of arc length and the second fundamental theorem of calculus.

SHORT FORMULA FOR SURFACE AREA The simplest way to write the formula for surface area of revolution is then

$$\text{Surface area} = \int_a^b 2\pi R\ ds,$$

where the interval $[a, b]$ refers to the chosen parameter, whether it be x, t, or y.

$2\pi R$

ds

—Memory device

To remember this formula, think of unrolling a narrow circular band of width ds and radius R and pressing it flat.

EXAMPLE 1 Find the area of the surface obtained by revolving the part of the curve $y = \sqrt{x}$ that lies between $x = 2$ and $x = 6$ around the x axis.

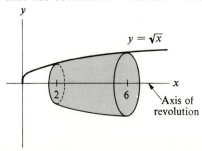

SOLUTION   The surface area is $$\int_a^b 2\pi R \; ds.$$

Since the function is given in terms of x, choose x as the parameter.   Then, by inspection of the diagram,

$$R = y = \sqrt{x}.$$

Next, introduce the differential dx:

$$ds = \frac{ds}{dx} \, dx = \sqrt{1 + \left(\frac{dy}{dx}\right)^2} \, dx.$$

Hence the surface area is

$$\int_2^6 2\pi\sqrt{x} \, \sqrt{1 + \left(\frac{dy}{dx}\right)^2} \, dx.$$

Next find dy/dx.   Since $y = \sqrt{x}$,

$$\frac{dy}{dx} = \frac{1}{2\sqrt{x}}.$$

Thus

$$\sqrt{1 + \left(\frac{dy}{dx}\right)^2} = \sqrt{1 + \left(\frac{1}{2\sqrt{x}}\right)^2} = \sqrt{1 + \frac{1}{4x}} = \sqrt{\frac{4x + 1}{4x}},$$

and

$$\text{Surface area} = \int_2^6 2\pi\sqrt{x} \, \sqrt{\frac{4x + 1}{4x}} \, dx.$$

Cancelling the $2\sqrt{x}$ in the numerator with $\sqrt{4x}$ in the denominator yields

$$\text{Surface area} = \int_2^6 \pi\sqrt{4x + 1} \; dx.$$

To evaluate this definite integral, use the substitution $u = 4x + 1$; hence $du = 4 \, dx$.   The new limits of integration are 9 and 25.   Thus

$$\text{Surface area} = \int_9^{25} \pi\sqrt{u} \, \frac{du}{4} = \frac{\pi}{4} \int_9^{25} \sqrt{u} \; du$$

$$= \frac{\pi}{4} \, \frac{2u^{3/2}}{3} \bigg|_9^{25}$$

$$= \frac{\pi}{6} [25^{3/2} - 9^{3/2}]$$

$$= \frac{\pi}{6} (125 - 27)$$

$$= \frac{\pi}{6} (98)$$

$$= \frac{49\pi}{3}. \quad \bullet$$

EXAMPLE 2   Find the surface area of the portion of a sphere enclosed between two parallel planes. Note that the sphere is the surface of a revolution of a semicircle revolved about the x axis.

SOLUTION   Let the radius of the sphere be a. For convenience, assume that the revolved semicircle has the equation $y = \sqrt{a^2 - x^2}$. Take the two planes to be perpendicular to the x axis and passing through $(d_1, 0)$, $(d_2, 0)$ with $-a \le d_1 < d_2 \le a$. The surface area of the sphere between the two planes is then equal to

$$\int_{d_1}^{d_2} 2\pi R \, ds,$$

where d_1 and d_2 refer to the range of x.
  Again use x as the parameter. Then

$$R = y = \sqrt{a^2 - x^2},$$

and

$$ds = \frac{ds}{dx} \, dx = \sqrt{1 + \left(\frac{dy}{dx}\right)^2} \, dx.$$

Thus the surface area is

$$\int_{d_1}^{d_2} 2\pi \sqrt{a^2 - x^2} \sqrt{1 + \left(\frac{dy}{dx}\right)^2} \, dx.$$

Now, $dy/dx = -x/\sqrt{a^2 - x^2}$. Hence

$$\sqrt{1 + \left(\frac{dy}{dx}\right)^2} = \sqrt{1 + \left(\frac{-x}{\sqrt{a^2 - x^2}}\right)^2} = \sqrt{1 + \frac{x^2}{a^2 - x^2}}$$

$$= \sqrt{\frac{(a^2 - x^2) + x^2}{a^2 - x^2}} = \frac{a}{\sqrt{a^2 - x^2}}.$$

The integral reduces to

$$\int_{d_1}^{d_2} 2\pi \sqrt{a^2 - x^2} \, \frac{a}{\sqrt{a^2 - x^2}} \, dx = \int_{d_1}^{d_2} 2\pi a \, dx$$

$$= 2\pi a \int_{d_1}^{d_2} 1 \, dx$$

$$= 2\pi a(d_2 - d_1).$$

  We obtain these interesting results: (1) The surface area in question depends only on the distance between the two planes—not on their location (as long as the two planes intersect the sphere); (2) the surface area is proportional to the distance between the two planes; (3) the surface area of the entire sphere is $4\pi a^2$, exactly four times the area of its equatorial cross section. (To see this, choose $d_2 = a$ and $d_1 = -a$.) ●

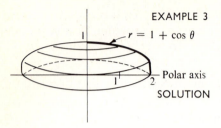

EXAMPLE 3   The portion of the curve $r = 1 + \cos\theta$ situated in the first quadrant of the xy plane is revolved around the y axis.   Set up a definite integral for the area of the surface formed.

SOLUTION   Begin with the formula   Surface area $= \displaystyle\int_a^b 2\pi R \, ds.$

Use the parameter θ, which goes from 0 to $\pi/2$.   A glance at the diagram shows that

$$R = x;$$

hence $$R = r \cos\theta = (1 + \cos\theta)\cos\theta.$$

To introduce $d\theta$, we write

$$ds = \frac{ds}{d\theta} \, d\theta$$

$$= \sqrt{r^2 + (r')^2} \, d\theta$$

$$= \sqrt{(1 + \cos\theta)^2 + (-\sin\theta)^2} \, d\theta$$

$$= \sqrt{1 + 2\cos\theta + \cos^2\theta + \sin^2\theta} \, d\theta$$

$$= \sqrt{2 + 2\cos\theta} \, d\theta.$$

Thus   Surface area $= \displaystyle\int_0^{\pi/2} 2\pi(1 + \cos\theta)\cos\theta\sqrt{2 + 2\cos\theta} \, d\theta.$

See Exercise 20(b) for the evaluation of this integral. ●

Exercises

In Exercises 1 to 5 set up definite integrals for the area of the indicated surface.   Use the suggested parameter.   Show radius R on a diagram.   Do *not* evaluate the definite integrals.

1. The curve $y = x^3$; x in $[1, 2]$; revolved about the x axis; parameter x.

2. The curve $y = x^3$; x in $[1, 2]$; revolved about the line $y = -1$; parameter x.

3. The curve $y = x^3$; x in $[1, 2]$; revolved about the y axis; parameter y.

4. The curve $y = x^3$; x in $[1, 2]$; revolved about the y axis; parameter x.

5. The curve $r = \sin 2\theta$; θ in $[0, \pi/2]$; revolved about the polar axis; parameter θ.

6. Consider the smallest tin can that contains a given sphere. (The height and diameter of the tin can equal the diameter of the sphere.)

   (a) Compare the volume of the sphere with the volume of the tin can. Archimedes, who obtained the solution about 2,200 years ago, considered it his greatest accomplishment.   Cicero wrote, about two centuries after Archimedes' death:

   I shall call up from the dust (the ancient equivalent of a blackboard) and his measuring-rod an obscure, insignificant person belonging to the same city (Syracuse), who lived many years after, Archimedes. When I was quaestor I tracked out his grave, which was unknown to the Syracusans (as they totally denied its existence), and found it enclosed all round and covered with brambles and thickets; for I remembered certain doggerel lines inscribed, as I had heard, upon his tomb, which stated that a sphere along with a cylinder has been set up on the top of his grave.

Accordingly, after taking a good look all round (for there are a great quantity of graves at the Agrigentine Gate), I noticed a small column rising a little above the bushes, on which there was the figure of a sphere and a cylinder. And so I at once said to the Syracusans (I had their leading men with me) that I believed it was the very thing of which I was in search. Slaves were sent in with sickles who cleared the ground of obstacles, and when a passage to the place was opened we approached the pedestal fronting us; the epigram was traceable with about half the lines legible, as the latter portion was worn away.

(Cicero, *Tusculan Disputations*, v. 23, translated by J. E. King, Loeb Classical Library, Harvard University, Cambridge, 1950.) Archimedes was killed by a Roman Soldier in 212 B.C. Cicero was quaestor in 75 B.C.

(b) Compare the surface area of the sphere with the area of the curved side of the can.

7. Find the area of the surface obtained by rotating that part of the curve $y = e^x$ that lies above $[0, 1]$ about the x axis.

8. Find the area of the surface formed by rotating one arch of the curve $y = \sin x$ about the x axis.

9. Find the area of the surface obtained when the cardioid $r = 1 + \cos \theta$ is rotated about the polar axis.

In Exercises 10 to 17 find the area of the surface formed by revolving the indicated curve about the indicated axis. Leave the answer as a definite integral, but indicate how it could be evaluated by the fundamental theorem of calculus.

10. $y = x^3$ for x in $[1, 2]$; about the x axis.
11. $y = 1/x$ for x in $[1, 2]$; about the x axis.
12. $y = x^2$ for x in $[1, 2]$; about the x axis.
13. $y = x^{4/3}$ for x in $[1, 8]$; about the y axis.
14. $y = x^{2/3}$ for x in $[1, 8]$; about the line $y = 1$.
15. $y = x^3/6 + 1/(2x)$ for x in $[1, 3]$; about the y axis.
16. $y = x^3/3 + 1/(4x)$ for x in $[1, 2]$; about the line $y = -1$.
17. The arc in Example 1; about the line $y = -1$.
18. One arch of the cycloid given parametrically by $x = \theta - \sin \theta$, $y = 1 - \cos \theta$ is revolved around the x axis. Find the area of the surface produced.
19. The curve given parametrically by $x = e^t \cos t$, $y = e^t \sin t$, t in $[0, \pi/2]$, is revolved around the x axis. Find the area of the surface produced.

20. (See Example 3.) Find the area of the surface formed by revolving the portion of the curve $r = 1 + \cos \theta$ about (a) the x axis, (b) the y axis. [In (b) the identity $1 + \cos \theta = 2 \cos^2 (\theta/2)$ may help.]

21. Show how the polar angle θ could have been used instead of x in Example 2.

■

22. The portion of the curve $x^{2/3} + y^{2/3} = 1$ situated in the first quadrant is revolved around the x axis. Find the area of the surface produced.

23. Two planes are perpendicular to the axis of the tin can in Exercise 6. Compare the areas that they intercept on the sphere and on the tin can. What does the comparison tell map makers?

24. Though the fundamental theorem of calculus is of no use in computing the perimeter of the ellipse $x^2/a^2 + y^2/b^2 = 1$, it is useful in computing the surface area of the "football" formed when the ellipse is rotated about one of its axes. Find that area. Does your answer give the correct formula for the surface area of a sphere of radius a, $4\pi a^2$?

25. The region bounded by $y = 1/x$ and the x axis and situated to the right of $x = 1$ is revolved around the x axis.
(a) Show that its volume is finite but its surface area is infinite.
(b) Does this mean that an infinite area can be painted by pouring a finite amount of paint into this solid?

26. Check that our definition of surface area as a definite integral agrees with our definition of the area of a cone.

■ ■

27. If the band formed by revolving a line segment is cut along the rotated segment and laid out in the plane, what shape will it be? *Warning:* It is generally *not* a rectangle. With this approach, compute its area.

28. Consider a solid of revolution. Its volume is approximated closely by the sum of the volumes of thin parallel circular slabs. Is the area of the surface of the solid approximated closely by the sum of the areas of the curved surfaces of the slabs?

10.12

Estimates of definite integrals

The definite integral $\int_0^1 \sqrt{1 - x^3} \, dx$ cannot be evaluated by the fundamental theorem of calculus, since $\sqrt{1 - x^3}$ does not have an elementary antiderivative. The method of partial fractions could be used to evaluate $\int_0^1 1/(1 + x^3 + x^4)\,dx$, but the procedure would be long and tedious. In the present section five ways of *estimating* a definite integral are presented. Their accuracy is examined in Sec. 15.4.

The definite integral $\int_a^b f(x)\,dx$, by definition, is a limit of sums of the form

$$\sum_{i=1}^{n} f(X_i)(x_i - x_{i-1}).$$

Any such sum consequently provides an estimate of $\int_a^b f(x)\,dx$. This observation is the basis of the first three of the five methods to be described in this section.

Left-Point, Right-Point, and Midpoint Rectangular Methods   Let n be a positive integer. Divide the interval $[a, b]$ into n sections of equal length, $h = (b - a)/n$, with

$$x_0 = a, \qquad x_1 = a + h, \qquad x_2 = a + 2h, \qquad \ldots, \qquad x_n = b.$$

A sum of the form

$$f(x_0)(x_1 - x_0) + f(x_1)(x_2 - x_1) + \cdots + f(x_{n-1})(x_n - x_{n-1}),$$

which is simply $h \sum_{i=1}^{n} f(x_{i-1}),$

is a *left-point estimate* of $\int_a^b f(x)\,dx$. (In the ith section the sampling point X_i is the left end of the section.)

If X_i is chosen to be x_i, the *right-point estimate* is formed,

$$h \sum_{i=1}^{n} f(x_i).$$

If X_i is chosen to be the midpoint of the ith section, $(x_{i-1} + x_i)/2$, the *midpoint estimate* is formed,

$$h \sum_{i=1}^{n} f\left(\frac{x_{i-1} + x_i}{2}\right).$$

These diagrams show the three estimates for $n = 4$. Note that, if f is a

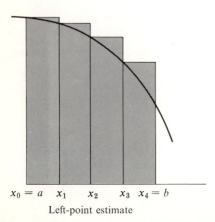

$x_0 = a \quad x_1 \quad x_2 \quad x_3 \quad x_4 = b$

Left-point estimate

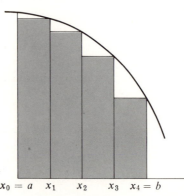

$x_0 = a \quad x_1 \quad x_2 \quad x_3 \quad x_4 = b$

Right-point estimate

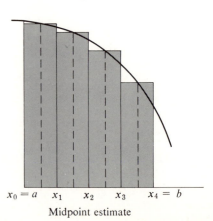

$x_0 = a \quad x_1 \quad x_2 \quad x_3 \quad x_4 = b$

Midpoint estimate

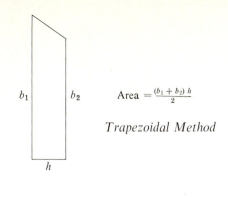

b_1 b_2 Area $= \frac{(b_1 + b_2) h}{2}$

Trapezoidal Method

h

decreasing function, the left-point method overestimates the integral, while the right-point method underestimates it.

The fourth method utilizes trapezoids instead of rectangles. Recall that the area of a trapezoid of height h and bases b_1 and b_2 is $(b_1 + b_2)h/2$.

Let n be a positive integer. Divide the interval $[a, b]$ into n sections of equal length $h = (b - a)/n$ with

$$x_0 = a, \qquad x_1 = a + h, \qquad x_2 = a + 2h, \qquad \ldots, \qquad x_n = b.$$

The sum

$$\frac{f(x_0) + f(x_1)}{2} \cdot h + \frac{f(x_1) + f(x_2)}{2} \cdot h + \cdots + \frac{f(x_{n-1}) + f(x_n)}{2} \cdot h$$

is the *trapezoidal estimate* of $\int_a^b f(x)\, dx$. It is usually written

$$\frac{h}{2}[f(x_0) + 2f(x_1) + 2f(x_2) + \cdots + 2f(x_{n-1}) + f(x_n)].$$

Note that $f(x_0)$ and $f(x_n)$ have coefficient 1, while all the other $f(x_i)$'s have coefficient 2. This is due to the double counting of the edges of the interior trapezoids.

The diagram in the margin illustrates the trapezoidal approximation for the case $n = 4$. Note that, if f is concave downward, the trapezoidal approximation underestimates $\int_a^b f(x)\, dx$. If f is a linear function, the trapezoidal method, of course, gives the integral exactly.

The fifth method is known as Simpson's or the prismoidal method. As Exercises 28 to 33 show, it gives the integral exactly if $f(x)$ is a polynomial of degree at most 3.

Divide the interval $[a, b]$ into $2n$ sections of equal length $h = (b - a)/(2n)$ with

$$x_0 = a, \qquad x_1 = a + h, \qquad x_2 = a + 2h, \qquad \ldots, \qquad x_{2n} = b.$$

Simpson's Method

$x_0 = a \quad x_1 \quad x_2 \quad x_3 \quad x_4 = b$

Note that $f(x_0)$ and $f(x_n)$ have coefficient 1, while the coefficients of the other $f(x_i)$'s alternate 4, 2, 4, 2, ..., 2, 4.

The sum

$$\frac{h}{3}[f(x_0) + 4f(x_1) + 2f(x_2) + 4f(x_3) + \cdots + 2f(x_{2n-2}) + 4f(x_{2n-1}) + f(x_{2n})]$$

is the *Simpson estimate* of $\int_a^b f(x)\, dx$.

Simpson's estimate is obtained by approximating the curve by parabolas. This is outlined in Exercises 28 to 33. Note that of the five methods it is the only one that requires that the interval be subdivided into an even number of sections.

The examples that follow apply the methods to estimate $\int_0^1 \sqrt{1 - x^3}\, dx$, in each case with a division of $[0, 1]$ into four sections.

EXAMPLE 1   Use the left-point method with $n = 4$ to estimate

$$\int_0^1 \sqrt{1 - x^3}\, dx.$$

SOLUTION The partition is $x_0 = 0$, $x_1 = \frac{1}{4}$, $x_2 = \frac{2}{4}$, $x_3 = \frac{3}{4}$, $x_4 = 1$, and $h = (1 - 0)/4 = \frac{1}{4}$. The left-point approximation is

A Calculator with a $\sqrt{x}$ Key or a Table of Square Roots Will Help Evaluate this Quantity.

$$\frac{1}{4}\left(\sqrt{1 - 0^3} + \sqrt{1 - (\tfrac{1}{4})^3} + \sqrt{1 - (\tfrac{2}{4})^3} + \sqrt{1 - (\tfrac{3}{4})^3}\right)$$

$$= \frac{1}{4}\left(\sqrt{\tfrac{64}{64}} + \sqrt{\tfrac{63}{64}} + \sqrt{\tfrac{56}{64}} + \sqrt{\tfrac{37}{64}}\right)$$

$$= \frac{1}{32}\left(\sqrt{64} + \sqrt{63} + \sqrt{56} + \sqrt{37}\right)$$

$$\doteq \frac{1}{32}(8.0000 + 7.9373 + 7.4833 + 6.0828)$$

$$\doteq 0.922$$

Since the function $\sqrt{1 - x^3}$ is decreasing for x in $[0, 1]$, this estimate is too large. ●

EXAMPLE 2 Use the right-point method with $n = 4$ to estimate

$$\int_0^1 \sqrt{1 - x^3}\, dx.$$

SOLUTION The partition of $[0, 1]$ is the same as that in Example 1. However, the function is now evaluated at the right end of each section. The approximation is

$$\frac{1}{4}\left(\sqrt{1 - (\tfrac{1}{4})^3} + \sqrt{1 - (\tfrac{2}{4})^3} + \sqrt{1 - (\tfrac{3}{4})^3} + \sqrt{1 - (1)^3}\right)$$

$$= \frac{1}{32}\left(\sqrt{63} + \sqrt{56} + \sqrt{37} + \sqrt{0}\right)$$

$$\doteq 0.672.$$

Since the function $\sqrt{1 - x^3}$ is decreasing for x in $[0, 1]$, this estimate is too small. ●

EXAMPLE 3 Use the midpoint method with $n = 4$ to estimate

$$\int_0^1 \sqrt{1 - x^3}\, dx.$$

SOLUTION The partition is the same as in Examples 1 and 2. The midpoints of the four sections are successively $\frac{1}{8}, \frac{3}{8}, \frac{5}{8}, \frac{7}{8}$. The midpoint approximation is therefore

$$\frac{1}{4}\left(\sqrt{1 - (\tfrac{1}{8})^3} + \sqrt{1 - (\tfrac{3}{8})^3} + \sqrt{1 - (\tfrac{5}{8})^3} + \sqrt{1 - (\tfrac{7}{8})^3}\right)$$

$$= \frac{1}{4}\left(\sqrt{\tfrac{511}{512}} + \sqrt{\tfrac{485}{512}} + \sqrt{\tfrac{387}{512}} + \sqrt{\tfrac{169}{512}}\right)$$

$$= \frac{\sqrt{2}}{128}\left(\sqrt{511} + \sqrt{485} + \sqrt{387} + \sqrt{169}\right)$$

$$\doteq \frac{\sqrt{2}}{128}(77.3003)$$

$$\doteq 0.854. ●$$

The estimate in Example 3 is too large; solving Exercise 24 will show why.

EXAMPLE 4   Use the trapezoidal method with $n = 4$ to estimate

$$\int_0^1 \sqrt{1 - x^3}\, dx.$$

SOLUTION   The partition employed in the first three examples is used here. The successive coefficients are 1, 2, 2, 2, 1, and the bookkeeping goes as follows.
The trapezoidal estimate is

$$\frac{h}{2}\left[f(0) + 2f\left(\frac{1}{4}\right) + 2f\left(\frac{2}{4}\right) + 2f\left(\frac{3}{4}\right) + f(1) \right].$$

Now $h/2 = \frac{1}{4}/2 = \frac{1}{8}$.   To compute the sum in brackets make a table.

x_i	$f(x_i)$	Coefficient	Summand	Decimal Form
0	$\sqrt{1 - 0^3}$	1	$1\sqrt{1}$	1.0000
$\frac{1}{4}$	$\sqrt{1 - \left(\frac{1}{4}\right)^3}$	2	$2\sqrt{\frac{63}{64}}$	1.9843
$\frac{2}{4}$	$\sqrt{1 - \left(\frac{2}{4}\right)^3}$	2	$2\sqrt{\frac{56}{64}}$	1.8708
$\frac{3}{4}$	$\sqrt{1 - \left(\frac{3}{4}\right)^3}$	2	$2\sqrt{\frac{37}{64}}$	1.5207
$\frac{4}{4}$	$\sqrt{1 - (1)^3}$	1	$1\sqrt{0}$	0.0000

The trapezoidal sum is therefore approximately

$$\frac{1}{8}(1.0000 + 1.9843 + 1.8708 + 1.5207 + 0.0000)$$
$$= \frac{1}{8}(6.3758)$$
$$\doteq 0.797. \ \bullet$$

The estimate in Example 4 is too small; solving Exercise 24 will show why.

EXAMPLE 5   Use Simpson's method with $2n = 4$ to estimate

$$\int_0^1 \sqrt{1 - x^3}\, dx.$$

SOLUTION   Again $h = \frac{1}{4}$.   Simpson's formula takes the form

$$\frac{1/4}{3}\left[f(0) + 4f\left(\frac{1}{4}\right) + 2f\left(\frac{2}{4}\right) + 4f\left(\frac{3}{4}\right) + f(1) \right].$$

A table helps put the computations in order.

x_i	$f(x_i)$	Coefficient	Summand	Decimal Form
0	$\sqrt{1 - 0^3}$	1	$1\sqrt{1}$	1.0000
$\frac{1}{4}$	$\sqrt{1 - \left(\frac{1}{4}\right)^3}$	4	$4\sqrt{\frac{63}{64}}$	3.9686
$\frac{2}{4}$	$\sqrt{1 - \left(\frac{2}{4}\right)^3}$	2	$2\sqrt{\frac{56}{64}}$	1.8708
$\frac{3}{4}$	$\sqrt{1 - \left(\frac{3}{4}\right)^3}$	4	$4\sqrt{\frac{37}{64}}$	3.0414
1	$\sqrt{1 - 1^3}$	1	$1\sqrt{0}$	0.0000

The approximation by Simpson's method is, consequently,

$$\frac{1/4}{3} (1.0000 + 3.9686 + 1.8708 + 3.0414 + 0.0000)$$

$$= \tfrac{1}{12}(9.8808)$$

$$\doteq 0.823. \ \bullet$$

This table compares the five estimates obtained to the value of $\int_0^1 \sqrt{1 - x^3}\, dx$, which, rounded to three decimal places, is 0.841.

Method	Estimate of $\int_0^1 \sqrt{1 - x^3}\, dx$
Left-point	0.922
Right-point	0.672
Midpoint	0.854
Trapezoidal	0.797
Simpson's	0.823

In this case the midpoint method gives the best approximation of the integral. However, generally Simpson's method is the most accurate. The following table describes general bounds on the errors of the five methods. In Section 15.4 it is shown how these bounds are obtained. The symbol M_k denotes the maximum value of $|f^{(k)}(x)|$ for x in $[a, b]$.

Error in estimating $\int_a^b f(x)\, dx$, using sections of length h

Method	Error Is at Most
Left-point	$\dfrac{(b - a)M_1 h}{2}$
Right-point	$\dfrac{(b - a)M_1 h}{2}$
Midpoint	$\dfrac{(b - a)M_2 h^2}{24}$
Trapezoidal	$\dfrac{(b - a)M_2 h^2}{12}$
Simpson's	$\dfrac{(b - a)M_4 h^4}{180}$

In the first two methods the size of $f^{(1)}(x)$ controls the error. Moreover, h appears to the first power. Thus cutting h in half tends to cut the error in half.

In the next two methods the error is controlled by the size of $f^{(2)}(x)$. In particular, if $f^{(2)}(x) = 0$ for all x in $[a, b]$, there is no error whatsoever. However, this would be clear on simple geometric considerations. If $f^{(2)}(x) = 0$ for all x in $[a, b]$, $f(x) = ax + b$ for some constants a and b. The graph of f is a straight line; in this case the rectangle of the midpoint method and the trapezoid have the same areas as the corresponding area under the curve.

Note the factor h^2 in the third and fourth methods. This indicates that cutting h in half will tend to cut the error to a fourth. Since $h^2 \to 0$ quickly as $h \to 0$, one would expect the midpoint and trapezoidal methods to be more accurate in general than the first two methods.

The bound on the error in the trapezoidal method turns out to be twice as large as the one for the midpoint method. Still, the trapezoidal method has an important advantage over the midpoint method: If you double n, the number of sections in the partition, the points of the original partition are also points of the finer partition. The values of f computed at the first stage can still be used at the second stage. However, in the midpoint method this is not the case; the midpoints at the first stage are not midpoints for the second stage.

The error in Simpson's method is controlled by the fourth derivative, $f^{(4)}(x)$, and h^4. If $f^{(4)}(x) = 0$ for all x in $[a, b]$, there is no error in Simpson's method. In other words, if $f(x)$ is a polynomial of degree at most 3, Simpson's method is exact. Moreover, as $h \to 0$, $h^4 \to 0$ very rapidly. This is why Simpson's method generally gives the best approximation of the five described.

Finding or estimating M_k for a particular function, especially when k is larger than 1, can be a difficult problem. Exercises 17 and 18 concern finding M_1 and M_2 for some simple cases.

Exercises

Exercises 1 to 3 concern integrals which are easy to evaluate and to compare with their estimates. In each case determine the exact errors when the integral is evaluated by each of the five methods using two sections $[h = (b - a)/2]$. The coefficients in Simpson's method are then 1, 4, 1; there are no 2s.

1. $\int_0^2 x^3 \, dx$ **2.** $\int_0^\pi \sin x \, dx$

3. $\int_1^2 \dfrac{dx}{x}$

The arithmetic in Exercises 4 to 7 can be simplified by use of tables in Appendix E or a calculator that has a $1/x$ key.

4. Estimate $\int_1^2 \dfrac{dx}{x}$ using the midpoint method with (a) $n = 1$
 (b) $n = 3$ (c) $n = 6$.

5. Estimate $\int_1^2 \dfrac{dx}{x}$ using the trapezoidal method with
 (a) $n = 1$ (b) $n = 3$ (c) $n = 6$.

6. Estimate $\int_1^2 \dfrac{dx}{x}$ using Simpson's method with (a) $2n = 2$
 (b) $2n = 4$ (c) $2n = 6$.

7. Compare the errors in using the estimates of Exercises 4(c), 5(c), and 6(c) with the general bounds in terms of M_2 and M_4 described at the end of the section.

Exercises 8 and 9 are easy to do with the aid of a calculator that has a $1/x$ key.

8. Estimate $\int_0^1 \dfrac{dx}{1 + x^4}$
 (a) by the midpoint method, $n = 4$;
 (b) by the trapezoidal method, $n = 4$;
 (c) by Simpson's method, $2n = 4$.

9. Estimate $\int_1^2 \dfrac{dx}{1 + x^4}$ by the same three approximations in Exercise 8.

Exercises 10 to 18 present integrals which cannot be evaluated by the fundamental theorem of calculus. A calculator with e^x, $\ln x$, and trigonometric keys will come in handy.

10. Estimate $\int_0^2 e^{-x^2} \, dx$
 (a) by the midpoint method, $n = 4$;
 (b) by the trapezoidal method, $n = 4$;
 (c) by Simpson's method, $2n = 4$.

11. Estimate $\int_1^5 \dfrac{e^x}{x} \, dx$

(a) by the midpoint method, $n = 4$;
(b) by the trapezoidal method, $n = 4$;
(c) by Simpson's method, $2n = 4$.

12. Estimate $\int_0^{\pi/4} \sqrt{1 + 2 \cos x} \, dx$ by Simpson's method with $2n = 6$.

13. Estimate $\int_1^3 (\ln x)/\sqrt{x} \, dx$
    (a) by the midpoint method, $n = 5$;
    (b) by the trapezoidal method, $n = 5$.

14. Estimate $\int_1^7 \sqrt[3]{1 + x^2} \, dx$ by the
    (a) right-point method, $n = 3$;
    (b) left-point method, $n = 3$;
    (c) midpoint method, $n = 3$.

15. Estimate $\int_0^1 \sin x^2 \, dx$ by the
    (a) right-point method, $n = 5$;
    (b) midpoint method, $n = 5$.

16. Estimate $\int_0^1 e^{-x^2} \, dx$ by the
    (a) right-point method, $n = 5$;
    (b) midpoint method, $n = 5$.

17. (a) Find the maximum value of $|f'(x)|$ over $[0, 1]$ if $f(x) = e^{-x^2}$.
    (b) Use (a) to put a bound on the error in Exercise 16(a).

18. (a) Find the maximum value of $|f''(x)|$ over $[0, 1]$ if $f(x) = e^{-x^2}$.
    (b) Use (a) to put a bound on the error in Exercise 16(b).

■

19. Show that, if $f(a) = f(b)$, the left-point, right-point and trapezoidal estimates for a given value of h are the same.

20. Show that for a given n the average of the left-point estimate and the right-point estimate equals the trapezoidal estimate.

21. Let T_n be the trapezoidal estimate and M_n be the midpoint estimate using the same n. Show that
$$\tfrac{2}{3}M_n + \tfrac{1}{3}T_n$$
equals Simpson's estimate with $2n$ sections.

22. For convenience, assume $f(x) \geq 0$. An estimate of $\int_a^b f(x) \, dx$ is made in the following manner. The interval $[a, b]$ is divided into n sections of equal length. For each i, $1 \leq i \leq n$, the tangent line to the curve $y = f(x)$ is constructed at the point
$$\left(\frac{x_{i-1} + x_i}{2}, f\left(\frac{x_{i-1} + x_i}{2} \right) \right).$$
The trapezoid bounded by this tangent line, the x axis, and the lines $x = x_{i-1}$ and $x = x_i$ is formed. Let its area

be A_i. The sum $\sum_{i=1}^n A_i$ is an estimate of $\int_a^b f(x) \, dx$. Show that it is the same as one of the five estimates described in this section.

This exercise is used in Exercise 24.

23. Let f be concave downward for x in $[a, b]$. Using diagrams and elementary geometry, show that the midpoint estimate of $\int_a^b f(x) \, dx$ is too large.

24. Let $f(x) = \sqrt{1 - x^3}$, as in the examples in this section. Show that f is concave downward for $0 < x < 1$. What does this imply about
    (a) the trapezoidal estimate of $\int_0^1 \sqrt{1 - x^3} \, dx$?
    (b) the midpoint estimate of $\int_0^1 \sqrt{1 - x^3} \, dx$?

The next three exercises present cases in which the bounds of maximum error are actually assumed.

25. Show that, if the trapezoidal method with $n = 1$ is used to estimate $\int_0^1 x^2 \, dx$, the error equals $(b - a)M_2 h^2/12$, where $a = 0$, $b = 1$, $h = 1$, and M_2 is the maximum value of $|D^2(x^2)|$ for x in $[0, 1]$.

26. Show that, if the midpoint method with $n = 1$ is used to estimate $\int_0^1 x^2 \, dx$, the error equals $(b - a)M_2 h^2/24$, where $a = 0$, $b = 1$, $h = 1$, and M_2 is as in Exercise 25.

27. Show that, if Simpson's method with $2n = 2$ is used to estimate $\int_0^1 x^4 \, dx$, the error equals $(b - a)M_4 h^4/180$, where $a = 0$, $b = 1$, $h = \frac{1}{2}$, and M_4 is the maximum value of $|D^4(x^4)|$ for x in $[0, 1]$.

■ ■

Exercises 28 to 33 describe the geometric motivation of Simpson's method.

28. Let $f(x) = Ax^2 + Bx + C$. Show that
$$\int_{-h}^h f(x) \, dx = \frac{h}{3} (f(-h) + 4f(0) + f(h)).$$

Hint: Just compute both sides.

29. Let f be a function. Show that there is a parabola

$y = Ax^2 + Bx + C$ that passes through the three points $(-h, f(-h))$, $(0, f(0))$, and $(h, f(h))$.

30. The equation in Exercise 28, which was known to the Greeks, is called the prismoidal formula. Use it to compute the volume of
(a) a sphere of radius a;
(b) a right circular cone of radius a and height h.

31. Let $f(x) = Ax^2 + Bx + C$. Show that

$$\int_{c-h}^{c+h} f(x)\, dx = \frac{h}{3}[f(c-h) + 4f(c) + f(c+h)].$$

Hint: Use the substitution $x = c + t$ to reduce this to Exercise 28. See the figure in the next column.

32. Exercises 29 and 31 are the basis of Simpson's method. First $[a, b]$ is divided into $2n$ sections, which are grouped into n pairs of adjacent sections. Over each pair the function is approximated by the parabola that passes through the three points of the graph with x coordinates equal to those that determine the two sections of the pair. Show that, when these n separate estimates are added, Simpson's formula results.

33. Since Simpson's method was designed to be accurate

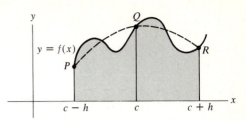

The dashed graph is a parabola, $y = Ax^2 + Bx + C$, through P, Q, R. The area of the region below the parabola is precisely $\frac{h}{3}[f(c-h) + 4f(c) + f(c+h)]$ and is an approximation of the area of the shaded region

when $f(x) = Ax^2 + Bx + C$, one would expect the error associated with it to involve $f^{(3)}(x)$. By a quirk of good fortune Simpson's method happens to be accurate even when $f(x)$ is a *cubic*, $Ax^3 + Bx^2 + Cx + D$. This suggests that the error involves $f^{(4)}(x)$, not $f^{(3)}(x)$.

(a) Show that, if $f(x) = x^3$,

$$\int_{-h}^{h} f(x)\, dx = \frac{h}{3}(f(-h) + 4f(0) + f(h)).$$

(b) Show that Simpson's method is accurate for cubics.

10.13

Summary

This chapter is a mixture, with four sections of analytic geometry and eight sections concerning various views of the definite integral.

As far back as Sec. 7.1 we knew that area $= \int_a^b c(x)\, dx$ and volume $= \int_a^b A(x)\, dx$. In Sec. 10.3 the techniques of Chaps. 8 and 9 were applied to compute some areas and volumes. The volume of a solid of revolution can be computed in another way—by shells—as shown in Sec. 10.4. This fact will be used in Chap. 16.

Section 10.5 introduced the notion of the average of a function, and Sec. 10.6, an improper integral.

Section 10.9, which showed that arc length can be expressed as an integral, is important primarily because it introduces the idea of speed along a curve. Two more geometric applications of integrals were given in Secs. 10.10 and 10.11, which treated area in polar coordinates and the area of a surface of revolution. Section 10.12 presented five ways of estimating a definite integral, of which the trapezoidal and Simpson's are the most frequently applied, since they generally provide the best estimates for a fixed amount of calculations.

Along the way, through examples and exercises, the integral was illustrated in other contexts, such as economics and the launch of a payload.

It is easy to lose one's way in such a potpourri. In the short run the chapter reinforces the idea of the definite integral and develops computational skill. In the long run it provides a foundation for concepts such as speed along a curve, average value of a continuous function, and integral from 0 to ∞, ideas which appear in many disciplines.

In the long run it is not skill in evaluating definite integrals that is so critical, but rather the ability to recognize that a certain concept can be expressed as a definite integral. Rare is the function whose graph has an arc length that can be computed by the fundamental theorem of calculus. And those few functions that do lead to elementary integrals tend to congregate in the exercise sets of calculus texts, perhaps misleading the reader to feel that the world is more subservient to man's tools than it is. So, recognizing when an integral is not elemen-

tary may save much time. If an antiderivative is messy or not elementary, then one should not hesitate to estimate the definite integral, especially in this age of computers and calculators. If an antiderivative is elementary, then its properties are easier to study. In the theory of differential equations, this could be quite helpful.

Though some of the reasoning leading to particular formulas was somewhat involved—for instance, the derivation for the formula for arc length used the law of the mean twice—the resulting integrands should seem reasonable and natural. This table summarizes the formulas and their respective memory aids.

Section	Concept	Memory Aid
10.3	$\text{Area} = \displaystyle\int_a^b c(x)\,dx$	
10.3	$\text{Volume} = \displaystyle\int_a^b A(x)\,dx$	
10.4	$\text{Volume} = \displaystyle\int_a^b 2\pi R(x)c(x)\,dx$	
10.9	$\text{Arc length} = \displaystyle\int_a^b \sqrt{\left(\dfrac{dx}{dt}\right)^2 + \left(\dfrac{dy}{dt}\right)^2}\,dt$ $= \displaystyle\int_a^b \sqrt{1 + \left(\dfrac{dy}{dx}\right)^2}\,dx$ $= \displaystyle\int_\alpha^\beta \sqrt{r^2 + (r')^2}\,d\theta$	

Section	Concept	Memory Aid
10.9	$\text{Speed} = \sqrt{\left(\dfrac{dx}{dt}\right)^2 + \left(\dfrac{dy}{dt}\right)^2}$	
10.10	$\text{Area} = \displaystyle\int_\alpha^\beta \dfrac{r^2}{2}\, d\theta$	 $\text{Area} = \tfrac{1}{2} r \cdot r\, d\theta$
10.11	$\text{Area} = \displaystyle\int_a^b 2\pi R\, ds$	

The slope of a curve given parametrically as $x = g(t)$, $y = h(t)$ is

$$\frac{dy}{dx} = \frac{dy/dt}{dx/dt} = \frac{\dot{y}}{\dot{x}}.$$

The area $\int_{x=a}^{x=b} y\, dx$ under a curve given parametrically is obtained by a substitution in which y, dx, and the limits of integration are all expressed in terms of the parameter t. Of course the parameter may be denoted by a different letter, such as θ, for instance.

Five different ways of estimating a definite integral $\int_a^b f(x)\, dx$ were described. Each depends on a choice of points where f is evaluated and a choice of coefficients or "weighting factors" for these values. This table summarizes these methods in brief.

Method	Choice of Points	Weights	Formula
Left-point	$x_0, x_1, \ldots, x_{n-1}, x_n$	$1, 1, \ldots, 1, 0$	$h \displaystyle\sum_{i=1}^n f(x_{i-1})$
Right-point	$x_0, x_1, \ldots, x_{n-1}, x_n$	$0, 1, \ldots, 1, 1$	$h \displaystyle\sum_{i=1}^n f(x_i)$
Midpoint	$X_1, \ldots, X_n$ (midpoints)	$1, 1, \ldots, 1$	$h \displaystyle\sum_{i=1}^n f(X_i)$
Trapezoidal	$x_0, x_1, \ldots, x_{n-1}, x_n$	$1, 2, 2, \ldots, 2, 2, 1$	$\dfrac{h}{2}[f(x_0) + 2f(x_1) + 2f(x_2) + \cdots + 2f(x_{n-1}) + f(x_n)]$
Simpson's (number of sections must be even)	$x_0, x_1, \ldots, x_{n-1}, x_n$	$1, 4, 2, 4, 2, \ldots, 2, 4, 1$	$\dfrac{h}{3}[f(x_0) + 4f(x_1) + 2f(x_2) + \cdots + 2f(x_{n-2}) + 4f(x_{n-1}) + f(x_n)]$

VOCABULARY AND SYMBOLS

$c(x)$, length of cross section of line perpendicular to interval of integration

$A(x)$, area of cross section by plane perpendicular to interval of integration

average of a function over an interval

improper integral (convergent and divergent)

$\int_a^\infty f(x)\,dx$ (unbounded interval of integration)

$$\int_{-\infty}^a f(x)\,dx \qquad \int_{-\infty}^\infty f(x)\,dx$$

$\int_a^b f(x)\,dx = \lim_{t\to a^+}\int_t^b f(x)\,dx$ (function unbounded near a)

arc length s

speed $= ds/dt$

R, $R(x)$, radius of the circle that a typical point sweeps out in solid or surface of revolution

left-point, right-point, midpoint, trapezoidal, and Simpson's methods

Guide quiz on Chap. 10

Set up the definite integrals required in Exercises 1 to 9. Do not evaluate them.

1. The area of the region above the parabola $y = x^2$ and below the line $y = 2x$, using (a) vertical cross sections, (b) horizontal cross sections.
2. The volume of the wedge cut from a right circular cylinder of height 5 inches and radius 3 inches by a plane that bisects one base and touches the other base at one point.
3. The arc length of the curve $y = \ln x$ from $x = 1$ to $x = 2$.
4. The arc length of the curve $r = \sin^3(\theta/3)$ corresponding to θ in $[0, \pi/2]$.
5. The area of the region within one loop of $r = \sin 2\theta$.
6. The area of the surface obtained by revolving the arc

$$\begin{cases} x = \cos^3 t, \\ y = \sin^3 t \end{cases} \quad t \text{ in } [0, \pi/2]$$

   about the x axis.

7. The area of the surface obtained by revolving the curve $r = 2\cos\theta - 2\cos 2\theta$, θ in $[0, \pi]$, about the polar axis.
8. The volume of the solid obtained by revolving the triangle whose vertices are $(2, 0)$, $(2, 1)$, and $(3, 2)$ about the x axis. (Use the shell technique.)
9. The area under the curve $x = \sin^2 t$, $y = \cos^3 t$, $0 \le t \le \pi/2$ and above the interval $[0, 1]$.
10. Find the slope of the curve in Exercise 9 when $t = \pi/6$.
11. (a) For which exponents a is $\int_1^\infty x^a\,dx$ convergent?
    (b) Show that

$$\int_0^\infty \frac{dx}{1 + x^2 + 2x^4}$$

   is convergent.
12. Set up the trapezoidal estimate, with $n = 6$, and the Simpson estimate, with $2n = 6$, of $\int_1^4 \sin x^2\,dx$. Do not evaluate them.

Review exercises for Chap. 10

Most of these exercises concern Chap. 10.

1. (a) Define "average value of a function."
   (b) Why is it a reasonable definition?
2. (a) Develop the formula for arc length.
   (b) What is the device for remembering the formula?
3. (a) Develop the formula for area in polar coordinates.
   (b) What is the device for remembering the formula?
4. (a) Develop the formula for the area of a surface of revolution.
   (b) What is the device for remembering the formula?

5. (a) Develop the shell formula for the volume of a solid of revolution.
   (b) What is the device for remembering the formula?
6. For what general class of functions $f(x)$ is $\int_a^b f(x)\,dx$ estimated without error by (a) the midpoint method? (b) the trapezoidal method? (c) Simpson's method?
7. Experimental data for values of a certain function f are known, as shown in this table.

x	1	1.5	2	2.5	3	3.5	4
$f(x)$	4	5	6	8	9	11	14

Use Simpson's method to estimate

$$\int_1^4 f(x)\, dx.$$

8. This table shows the temperature $f(t)$ as a function of time. (*a*) Use Simpson's method to estimate $\int_1^7 f(t)\, dt$, and then (*b*) use the result in (*a*) to estimate the average temperature.

Time	1	2	3	4	5	6	7
Temperature	81	75	80	83	78	70	60

9. When you cut h in half, what would you expect to happen to the error when you are using
(*a*) rectangles, right endpoints?
(*b*) midpoints?
(*c*) trapezoids?
(*d*) Simpson's formula?
(*e*) Why?

10. Estimate $\int_0^4 dx/(x^5 + 2x + 1)$ with $h = 1$ using
(*a*) the trapezoidal method;
(*b*) Simpson's formula.

11. Use the trapezoidal method with $h = 1$ to estimate $\int_1^5 \sqrt[3]{2 + x^2}\, dx$.

12. Consider a function f defined on $[1, 4]$ of which it is known that $M_1 = M_2 = M_3 = M_4 = 2$. (See the end of Sec. 10.12 for the definition of M_k.) How small must h be chosen in order that we are sure that the error in estimating $\int_1^4 f(x)\, dx$ is less than 0.001 if the estimate is made by
(*a*) the left-point method?
(*b*) the right-point method?
(*c*) the midpoint method?
(*d*) the trapezoidal method?
(*e*) Simpson's method?

13. Is $\displaystyle\int_0^\infty \frac{dx}{(x-1)^2}$ convergent or divergent?

14. Evaluate $\int_0^\infty e^{-x} \sin(2x + 3)\, dx$.

15. Define $G(x) = \int_0^\infty e^{-t} t^{x-1}\, dt$ for $x > 0$.
(*a*) Show that $G(x)$ is meaningful; that is, the integral is convergent.
(*b*) Show that $G(x + 1) = xG(x)$.
(*c*) Show that $G(1) = 1$.
(*d*) Deduce that $G(2) = 1$, $G(3) = 2$, $G(4) = 6$.

16. Is $\displaystyle\int_0^1 \frac{\ln x}{1 - x^2}\, dx$ convergent or divergent?

17. (*a*) Sketch $y = e^{-x}(1 + \sin x)$ for $x \ge 0$.
(*b*) The region beneath the curve in (*a*) and above the positive x axis is revolved around the y axis. Find the volume of the resulting solid.

18. Give an example of a region R in the xy plane such that the volume of the solid obtained by revolving R around the x axis can be computed by an elementary integral but the volume of the solid obtained by revolving R around the y axis cannot.

19. When calculating the surface-to-volume ratio of the rotary engine, engineers had to determine the arc length of a portion of the epitrochoid.
(*a*) Show that the length of a general arc of the epitrochoid given parametrically as in Example 5 of Sec. 10.8 is

$$\int_\alpha^\beta \sqrt{9e^2 + R^2 + 6eR \cos 2\theta}\, d\theta$$

(*b*) Show that the integral in (*a*) equals

$$\int_\alpha^\beta (3e + R) \sqrt{1 - k^2 \sin^2 \theta}\, d\theta,$$

where $k^2 = 12eR/(3e + R)^2$.
(*c*) Show that $k^2 \le 1$. Thus the integral in (*b*) is an elliptic integral, which is tabulated in many mathematical handbooks.

20. From the fact that $\int (e^x/x)\, dx$ is not elementary deduce that $\int e^x \ln x\, dx$ is not elementary.

21. In R. P. Feynman, *Lectures on Physics*, Addison-Wesley, Reading, Mass., 1963, appears this remark: "... The expression becomes

$$\frac{U}{V} = \frac{(kT)^4}{h^3 \pi^2 c^3} \int_0^\infty \frac{x^3\, dx}{e^x - 1}$$

The integral is just some number that we can get, approximately, by drawing a curve and taking the area by counting squares. It is roughly 6.5. The mathematician among us can show that the integral is exactly $\pi^4/15$."
Show at least that the integral is convergent. Consider both $x \to \infty$ and $x \to 0$.

22. Is $\int_0^\infty dx/\sqrt{x}\,\sqrt{x + 1}\,\sqrt{x + 2}$ convergent or divergent?

23. A drill of radius a inches bores a hole through the center of a sphere of radius b inches, $b > a$, leaving a ring whose height is 2 inches. Find the volume of the ring.

24. A barrel is made by rotating an ellipse and then cutting off equal caps, top and bottom. It is 3 feet high and 3 feet wide at its midsection. Its top and bottom have a diameter of 2 feet. What is its volume?

25. Consider the area of the triangle AOB and the area of the shaded region cut from the parabola $y = x^2$ by the horizontal line AB, as shown in the diagram on next page.
(*a*) What do you think happens to the ratio

$$\frac{\text{Area of triangle } AOB}{\text{Area of sector } AOB}$$

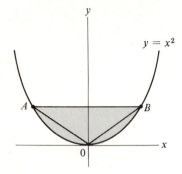

as A and B approaches O?

(b) Use calculus to find what happens to the ratio in (a) as A and B approach O.

26. Find the length of the curve $y = \ln x$ from $x = 1$ to $x = \sqrt{3}$. An integral table will save a lot of work.

27. By interpreting these improper integrals as expressions for the area of a certain region, show that

$$\int_0^\infty \frac{dx}{1 + x^2} = \int_0^1 \sqrt{\frac{1 - y}{y}}\, dy.$$

28. Define $G(a) = \int_0^\infty a/(1 + a^2 x^2)\, dx$. (a) Compute $G(0)$. (b) Compute $G(a)$ if a is negative. (c) Compute $G(a)$ if a is positive. (d) Graph G.

29. Prove that $\int_0^\infty (\sin x^2)/x\, dx = \frac{1}{2} \int_0^\infty (\sin x)/x\, dx$.

30. Evaluate the two integrals that are elementary (one is not).

(a) $\displaystyle \int \frac{\ln x}{\sqrt{x}}\, dx$ (b) $\displaystyle \int \sqrt{\ln x}\, dx$

(c) $\displaystyle \int \frac{\sqrt{\ln x}}{x}\, dx$.

31. Let a be a rational number. Consider the curve $y = x^a$ for x in the interval $[1, 2]$. Show that the area of the surface obtained by rotating this curve around the x axis can be evaluated by the fundamental theorem of calculus in the cases $a = 0, 1,$ or $1 + 2/n$, where n is any nonzero integer. These are the only rational a for which the pertinent integral is elementary. (See remark in Exercise 21 of Sec. 10.9.)

32. Is this computation correct?

$$\int_{-2}^1 \frac{dx}{2x + 1} = \frac{1}{2} \ln |2x + 1| \Big|_{-2}^1 = \frac{1}{2} \ln 3 - \frac{1}{2} \ln 3 = 0?$$

33. Is the total length of the curve $r = 1/(1 + \theta)$, $\theta \geq 0$, finite or infinite?

34. Is the total length of the curve $r = 1/(1 + \theta^2)$, $\theta \geq 0$, finite or infinite?

35. What is the length of the curve $r = e^{-\theta}$, $\theta \geq 0$?

36. At time $t \geq 1$ a plodding snail is at the point $x = 1/t$, $y = (1/t) \sin t$.
(a) Sketch the path of the snail.
(b) Find its speed at time t.
(c) If it were to continue its journey forever, would it travel a finite or an infinite distance?

37. Let a be a number, and let $x = g(t)$ and $y = h(t)$, $t \geq a$, describe a curve. Let $P = (g(a), h(a))$, and let $Q(t) = (g(t), h(t))$, $t > a$.
(a) Sketch the chord $PQ(t)$ and the tangent line $T(t)$ at $Q(t)$.
(b) What is the slope of $PQ(t)$? What is the slope of $T(t)$?
(c) Under what assumptions will the two slopes in (b) have the same limit as $t \to a^+$?

38. Let $r = e^\theta$, $0 \leq \theta \leq \pi/4$, describe a curve in polar coordinates. In parts (b) to (f) set up definite integrals for the quantities, and show that they could be evaluated by the fundamental theorem of calculus.
(a) Sketch the curve.
(b) The area of the region R below the curve and above the interval $[1, e^{\pi/4}\sqrt{2}/2]$ on the x axis.
(c) The volume of the solid obtained by revolving R, defined in (b), about the x axis.
(d) The volume of the solid obtained by revolving R about the y axis.
(e) The area of the surface obtained by rotating the curve in (a) around the x axis.
(f) The area of the surface obtained by rotating the curve in (a) around the y axis.
Rare is the curve for which the corresponding five integrals are all elementary.

39. Find the error in the following computations.

$$\int_{-1}^1 \frac{1}{x^2}\, dx = \frac{-1}{x} \Big|_{-1}^1 = \frac{-1}{1} - \frac{-1}{-1} = -2.$$

(The integrand is positive, yet the integral is negative.)

40. It can be proved that $\int_0^\infty (x^{n-1})/(1 + x)\, dx = \pi/\sin n\pi$ for $0 < n < 1$. Verify that this equation is correct for $n = \frac{1}{2}$.

41. Compute $\int_0^1 x^4 \ln x\, dx$.

42. (a) Let $G(a) = \int_0^\infty 1/[(1 + x^a)(1 + x^2)]\, dx$. Evaluate $G(0), G(1), G(2)$.
(b) Show, using the substitution $x = 1/y$, that

$$G(a) = \int_0^\infty \frac{x^a\, dx}{(1 + x^a)(1 + x^2)}.$$

(c) From (b), show that $G(a) = \pi/4$, independent of a.

43. Consider the curve $y = e^x$ for x in $[0, 1]$.

(a) Set up integrals for its arc length and for the areas of the surfaces obtained by rotating the curve around the x axis and also about the y axis.

(b) Two of the three integrals are elementary. Evaluate them.

44. Consider the curve $y = \sin x$ for x in $[0, \pi]$. Proceed as in Exercise 43. This time, however, only one of the three integrals is elementary. Evaluate it.

45. There are two values of a for which $\int \sqrt{1 + a \sin^2 \theta}\, d\theta$ is elementary. What are they?

46. From Exercise 45 deduce that there are two values of a for which

$$\int \frac{\sqrt{1 + ax^2}}{\sqrt{1 - x^2}}\, dx$$

is elementary.

47. There are three values of b for which $\int \sqrt{1 + b \cos \theta}\, d\theta$ is elementary. What are they?

48. From Exercise 47 deduce that there are three values of b for which

$$\int \frac{\sqrt{1 + bx}}{\sqrt{1 - x^2}}\, dx$$

is elementary.

49. Say that you wanted to estimate $\int_0^3 e^x\, dx$ (which of course equals $e^3 - 1$) in order to estimate e^3. What would be the bound on the error if you used $h = 0.1$ and

(a) the left-point method?
(b) the midpoint method?
(c) the trapezoidal method?
(d) Simpson's method?

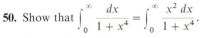

50. Show that $\displaystyle\int_0^\infty \frac{dx}{1 + x^4} = \int_0^\infty \frac{x^2\, dx}{1 + x^4}$.

   Hint: Let $x = 1/y$.

51. Is the area under the curve $y = (\ln x)/x^2$, above the x axis and to the right of the line $x = 1$, finite or infinite?

52. Is $\int_0^1 x \ln x\, dx$ convergent or divergent?

53. Graph $r = 3/(\cos \theta + 2 \sin \theta)$ after first finding the rectangular form of the equation.

54. For which values of k is the arc length of the curve $r = 1 + k \cos \theta$ computable by the fundamental theorem of calculus? Exercise 47 will come in handy.

55. Water flows out of a hole in the bottom of a cylindrical tank of radius r and height h at the rate of $\sqrt{y}$ cubic feet per second when the depth of the water is y feet. Initially the tank is full.

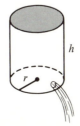

(a) How long will it take to become half full?
(b) How long will it take to empty?

56. A particle moves on a line in such a way that its time-average velocity over any interval of time $[a, b]$ is the same as its velocity at time $(a + b)/2$. Prove that the velocity $v(t)$ must be of the form $ct + d$ for appropriate constants c and d. *Hint:* Begin by differentiating the relation $\int_a^b v(t)\, dt = [v(a + b)/2](b - a)$ with respect to b and with respect to a.

57. A particle moves on a line in such a way that the time-average velocity over any interval of time $[a, b]$ is equal to the average of its velocities at the beginning and the end of the interval of time. Prove that the velocity $v(t)$ must be of the form $ct + d$ for appropriate constants c and d. *Hint:* Begin by differentiating the relation

$$\int_a^b v(t)\, dt = \frac{v(a) + v(b)}{2}(b - a)$$

with respect to a and with respect to b.

58. A disk of radius a is covered by a finite number of strips (perhaps overlapping). Prove that the sum of their widths is at least $2a$. (If the strips are parallel, the assertion is clearly true; do not assume that the strips are parallel.) A strip consists of the points between two parallel lines.

59. Show that the only positive integers x and y, $x \neq y$, such that $x^y = y^x$ are 2 and 4 (or 4 and 2).

In Exercise 60 a proof is outlined that the sum of the reciprocals of the squares of the positive integers is $\pi^2/6$. It may be the most elementary of the many proofs of this fact, the first two of which were discovered by Euler. This proof, due to D. P. Giesy, appeared in *Mathematics Magazine*, vol. 45, pp. 148–149, 1972.

60. Let $f_n(x) = \frac{1}{2} + \cos x + \cos 2x + \cdots + \cos nx$.

(a) Show that

$$f_n(x) = \frac{\sin \left[\dfrac{(2n + 1)x}{2}\right]}{2 \sin (x/2)} \qquad x \neq 2m\pi,\ m \text{ an integer}$$

for $n = 1$.

(b) Show that if the formula in (a) is valid for $n = k$, then it is valid for $n = k + 1$. Thus it holds for all n.

(c) Show that

$$E_n = \int_0^\pi x f_n(x)\, dx = \frac{\pi^2}{4} + \sum_{k=1}^{n} \frac{(-1)^k}{k^2} - \frac{1}{k^2}.$$

(*d*) Show that

$$\frac{1}{2} E_{2n-1} = \frac{\pi^2}{8} - \sum_{k=1}^{n} \frac{1}{(2k-1)^2}.$$

(*e*) Let

$$g(x) = D\left(\frac{x/2}{\sin(x/2)}\right).$$

Show that

$$E_{2n-1} = \frac{1}{4n-1}\left[2 + 2\int_{0}^{\pi} g(x) \cos\frac{(4n-1)x}{2} dx\right].$$

(*f*) Show that $0 < g(x) \leq \frac{1}{2}$ for $0 < x \leq \pi$.

(*g*) Show that $\lim_{n\to\infty} E_{2n-1} = 0$ and hence that

$$\lim_{n\to\infty} \sum_{k=1}^{n} \frac{1}{(2k-1)^2} = \frac{\pi^2}{8}.$$

(*h*) From (*g*) deduce that

$$\lim_{n\to\infty} \sum_{k=1}^{n} \frac{1}{k^2} = \frac{\pi^2}{6}.$$

61. Find a function $y = f(x)$ such that

$$\frac{dy}{dx} = \frac{1+y^2}{1+x}.$$

Hint: First get all terms involving y on one side of the equation and all those involving x on the other side.

ADDITIONAL APPLICATIONS OF THE DERIVATIVE

In Chapter 6 it was shown how to use the derivative and second derivative in graphing, finding maxima and minima, and determining the motion of an object. In the present chapter five additional topics involving the derivative are discussed. The opening section concerns the computation of a derivative when the function is described indirectly; the method is applied to maximum-minimum problems. Sec. 11.2 extends related rates to questions concerning the second derivative. In Sec. 11.3 the derivative is applied to estimate the solutions of an equation. In Sec. 11.4 the first and second derivatives are used to describe the "curviness" of a curve. In Sec. 11.5 it is shown how to find the angle between a line and a tangent line.

11.1

Implicit Differentiation

Sometimes a function $y = f(x)$ is given indirectly by an equation that relates x and y. For instance, consider the equation

$$x^2 + y^2 = 25. \tag{1}$$

This equation can be solved for y:

$$y^2 = 25 - x^2,$$
$$y = \sqrt{25 - x^2},$$

or
$$y = -\sqrt{25 - x^2}.$$

There are thus two continuous functions that satisfy (1).

The equation

$$x^2 + y^2 = 25$$

is said to describe the function $y = f(x)$ *implicitly*. The equations

$$y = \sqrt{25 - x^2} \quad \text{and} \quad y = -\sqrt{25 - x^2}$$

are said to describe the function $y = f(x)$ *explicitly*.

In this section it is shown first how to differentiate a function given implicitly without having to solve for the function and express it explicitly. An example will illustrate the method, which is simply to differentiate both sides of the equation that defines the function implicitly.   This procedure is called *implicit differentiation*.

EXAMPLE 1   Let $y = f(x)$ be the continuous function that satisfies the equation

$$x^2 + y^2 = 25$$

such that $y = 4$ when $x = 3$.   Find dy/dx when $x = 3$ and $y = 4$.

SOLUTION   Differentiating both sides of the equation

$$x^2 + y^2 = 25$$

with respect to x yields
$$\frac{d(x^2 + y^2)}{dx} = \frac{d(25)}{dx},$$

$$2x + 2y\frac{dy}{dx} = 0.$$

Hence
$$x + y\frac{dy}{dx} = 0.$$

In particular, when $x = 3$ and $y = 4$,

$$3 + 4\frac{dy}{dx} = 0,$$

and therefore
$$\frac{dy}{dx} = -\frac{3}{4}.$$

The problem could also be solved by differentiating $\sqrt{25 - x^2}$.   But the algebra involved is more complicated, since it is necessary to differentiate a square root. ●

In the next example implicit differentiation is the only way to find the derivative, for in this case there is no elementary formula giving y explicitly in terms of x.

EXAMPLE 2   Assume that the equation $2xy + \pi \sin y = 2\pi$

defines a function $y = f(x)$.   Find dy/dx when $x = 1$ and $y = \pi/2$.   (Note that $x = 1$ and $y = \pi/2$ satisfy the equation.)

SOLUTION   Implicit differentiation yields
$$\frac{d(2xy + \pi \sin y)}{dx} = \frac{d(2\pi)}{dx},$$

$$2\left(x\frac{dy}{dx} + y\frac{dx}{dx}\right) + \pi \cos y \frac{dy}{dx} = 0,$$

or
$$2x\frac{dy}{dx} + 2y + \pi \cos y \frac{dy}{dx} = 0.$$

For $x = 1$ and $y = \pi/2$ this last equation becomes

$$2 \cdot 1 \frac{dy}{dx} + 2\frac{\pi}{2} + \pi \cos \frac{\pi}{2} \frac{dy}{dx} = 0$$

or

$$2\frac{dy}{dx} + 2\frac{\pi}{2} = 0.$$

Hence

$$\frac{dy}{dx} = -\frac{\pi}{2}. \; \bullet$$

Logarithmic Differentiation

The next example presents a special case of implicit differentiation called *logarithmic differentiation*. This is a method for differentiating a function whose logarithm is simpler than the function itself.

EXAMPLE 3   Differentiate

$$y = \frac{\sqrt[3]{x}\sqrt{(1 + x^2)^3}}{x^{4/5}}.$$

SOLUTION   Rather than compute dy/dx directly, take logarithms of both sides of the equation first, obtaining

$$\ln y = \tfrac{1}{3} \ln x + \tfrac{3}{2} \ln (1 + x^2) - \tfrac{4}{5} \ln x.$$

Then differentiate this equation implicitly:

$$\frac{1}{y}\frac{dy}{dx} = \frac{1}{3x} + \frac{3 \cdot 2x}{2(1 + x^2)} - \frac{4}{5}\frac{1}{x}.$$

Solving for dy/dx yields

$$\frac{dy}{dx} = y\left(\frac{1}{3x} + \frac{3x}{1 + x^2} - \frac{4}{5x}\right)$$

$$= \frac{\sqrt[3]{x}\sqrt{(1 + x^2)^3}}{x^{4/5}}\left(\frac{1}{3x} + \frac{3x}{1 + x^2} - \frac{4}{5x}\right).$$

The reader is invited to find dy/dx directly from the explicit formula for y. Doing so will show the advantage of logarithmic differentiation. ●

In Example 4 of Sec. 6.8 we answered the question, "Of all the tin cans that enclose a volume of 100 cubic inches, which requires the least metal?" We found that the radius of the most economical can is $\sqrt[3]{50/\pi}$. From this and the fact that its volume is 100 cubic inches, its height could be calculated. In the next example implicit differentiation is used to answer the same question. Not only will the algebra be simpler than that in Example 4 of Sec. 6.8, but the answer will provide more information, since also the general shape—the proportion between height and radius—is revealed. Before reading the next example, it would be instructive to read over the solution in Sec. 6.8.

EXAMPLE 4   Of all the tin cans that enclose a volume of 100 cubic inches, which requires the least metal?

SOLUTION   The height h and radius r of any can of volume 100 cubic inches are related by the equation

$$\pi r^2 h = 100. \tag{2}$$

The surface area S of the can is

$$S = 2\pi r^2 + 2\pi rh. \tag{3}$$

Consider h, and hence S, as functions of r.   However, *it is not necessary to find these functions explicitly.*
  Differentiation of (2) and (3) with respect to r yields

$$\pi\left(r^2\frac{dh}{dr} + 2rh\right) = \frac{d(100)}{dr} = 0 \tag{4}$$

and

$$\frac{dS}{dr} = 4\pi r + 2\pi\left(r\frac{dh}{dr} + h\right). \tag{5}$$

Since when S is a minimum, $dS/dr = 0$, we have

$$0 = 4\pi r + 2\pi\left(r\frac{dh}{dr} + h\right). \tag{6}$$

Equations (4) and (6) yield, with a little algebra, a relation between h and r, as follows.
  Factoring πr out of (4) and 2π out of (6) shows that

$$\begin{cases} r\dfrac{dh}{dr} + 2h = 0; \\[2mm] 2r + r\dfrac{dh}{dr} + h = 0. \end{cases} \tag{7}$$

Elimination of dh/dr from (7) yields

$$2r + r\left(\frac{-2h}{r}\right) + h = 0,$$

which simplifies to $2r = h.$ (8)

Equation (8) asserts that the height of the most economical can is the same as its diameter.   Moreover, this is the ideal shape, no matter what the prescribed volume happens to be.   [Equation (4) follows from (2) merely because 100 is constant.]
  The specific dimensions of the most economical can are found by combining the equations

$$2r = h \tag{8}$$

and

$$\pi r^2 h = 100. \tag{2}$$

Elimination of h from these two equations shows that

$$\pi r^2 (2r) = 100,$$

or

$$r^3 = \frac{50}{\pi};$$

hence

$$r = \sqrt[3]{\frac{50}{\pi}},$$

and

$$h = 2r = 2\sqrt[3]{\frac{50}{\pi}}. \; \bullet$$

The general procedure illustrated by the preceding example consists of these steps:

General Procedure for Using Implicit Differentiation in an Applied Maximum Problem

1. Name the various quantities in the problem by letters, such as x, y, A, V.
2. Express the quantity to be maximized (or minimized) in terms of other letters, such as x and y.
3. Obtain an equation relating x and y. (This equation is called a constraint.)
4. Differentiate implicitly both the constraint and the expression to be maximized (or minimized), interpreting all the various quantities to be functions of x (or, perhaps, of y).
5. Use the equation obtained in step 4 to obtain an equation relating x and y at a maximum (or minimum).

An analogous procedure works for "minimum" problems.

Exercises

In Exercises 1 to 4 find dy/dx at the indicated values of x and y in two ways: explicitly (solving for y first) and implicitly.

1. $xy = 4$ at $(1, 4)$.
2. $x^2 - y^2 = 3$ at $(2, 1)$.
3. $x^2 y + xy^2 = 12$ at $(3, 1)$.
4. $x^2 + y^2 = 100$ at $(6, -8)$.

In Exercises 5 to 8 find dy/dx at the given point by implicit differentiation.

5. $\dfrac{2xy}{\pi} + \sin y = 2$ at $(1, \pi/2)$.

6. $xy + \ln y = 1$ at $(1, 1)$.
7. $2y^3 + 4xy + x^2 = 7$ at $(1, 1)$.
8. $e^y = \sin (x + y)$ at $(\pi/2, 0)$.

Use logarithmic differentiation to find the derivatives of the functions in Exercises 9 to 14.

9. $(x^3 - 1)^{3/2} (x^5)$

10. $\dfrac{(\sin 2x)^3 \sqrt{1 + x}}{(1 + x^2)^5}$

11. $(\sqrt{\cos x} \sqrt[3]{\ln x})^5$

12. x^{x^2}

13. $(1 + 3x)^x$

14. $y = \dfrac{\sqrt{(1 + 3x)^7}}{(1 + 4x)^8} e^{x^2}$

15. A function $y = f(x)$ satisfies the equation

$$e^{xy} + \sin x + y = 1.$$

Find $f'(0)$.

16. Solve Example 3 of Sec. 6.8 by the method illustrated in this section.

17. Solve Exercise 3 of Sec. 6.8 by the method of this section.

18. Solve Exercise 21 of Sec. 6.8 by the method of this section.
19. Solve Exercise 22 of Sec. 6.8 by the method of this section.
20. Solve Exercise 23 of Sec. 6.8 by the method of this section.
21. Solve Exercise 24 of Sec. 6.8 by the method of this section.
22. Solve Exercise 25 of Sec. 6.8 by the method of this section.
23. Solve Exercise 26 of Sec. 6.8 by the method of this section.
24. A rancher wishes to fence in a rectangular pasture 1 square mile in area, one side of which is along a road. The cost of fencing along the road is higher and equals 5 dollars a foot. The fencing for the other three sides costs 3 dollars a foot. What is the shape of the most economical pasture?
25. A rectangular box with square base is to contain a certain specified volume V. Find the proportion between height and side of the base so that the box has the least square area.

■

Exercise 26 shows how to find d^2y/dx^2 implicitly.
26. Let $y = f(x)$ satisfy the equation $x^3y + y^2 = 2$ and $y = 1$ when $x = 1$.
(a) Show that $x^3y' + 3x^2y + 2yy' = 0$.
(b) Differentiate the equation in (a) with respect to x to show that

$$x^3y'' + 6x^2y' + 6xy + 2yy'' + 2(y')^2 = 0.$$

(c) Solve the equation in (a) for y' in terms of x and y and substitute this result into the equation in (b). Then solve the resulting equation for y'' in terms of x and y.
(d) Determine y'' when $x = 1$ and $y = 1$.
27. Find y'' at $x = 1$, $y = 1$ for the function in Exercise 6.
28. Find y'' at $x = \pi/2$, $y = 0$ for the function in Exercise 8.

11.2

Related rates

Sometimes the rate at which one quantity is changing is known, and we wish to find the rate at which some related quantity is changing. Example 1 is typical of such problems and indicates a general method of attacking them.

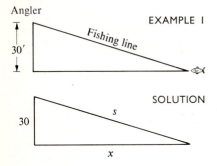

EXAMPLE I An angler has a fish at the end of his line, which is reeled in at 2 feet per second from a bridge 30 feet above the water. At what speed is the fish moving through the water when the amount of line out is 50 feet? 31 feet? Assume the fish is at the surface of the water.

SOLUTION Let s be the length of the line and x the horizontal distance of the fish from the bridge.
Since the line is reeled in at the rate of 2 feet per second,

$$\frac{ds}{dt} = -2.$$

The rate at which the fish moves through the water is given by the derivative

$$\frac{dx}{dt}.$$

The problem is to find dx/dt when $s = 50$ and 31.
The quantities x and s are related by the equation given by the pythagorean theorem:

$$x^2 + 30^2 = s^2.$$

Both x and s are functions of time t. Thus both sides of the equation may be differentiated with respect to t, yielding

$$\frac{d(x^2)}{dt} + \frac{d(30^2)}{dt} = \frac{d(s^2)}{dt},$$

or
$$2x\frac{dx}{dt} + 0 = 2s\frac{ds}{dt}.$$

Hence
$$x\frac{dx}{dt} = s\frac{ds}{dt}.$$

This last equation provides the tool for answering the questions.

Since
$$\frac{ds}{dt} = -2,$$

$$x\frac{dx}{dt} = s(-2).$$

Hence
$$\frac{dx}{dt} = \frac{-2s}{x}.$$

First of all, observe that dx/dt is negative. This simply means that x is decreasing as the fish is pulled in. The speed of the fish is $2s/x$, since speed is always taken to be positive.

Inspection of the right triangle whose hypotenuse has length s and whose legs have lengths 30 and x shows that

$$\frac{s}{x} > 1.$$

This implies that the fish is always moving faster than the line is being wound in. When the line is very long, the triangle is narrow and the ratio s/x is near 1.

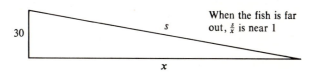

When the fish is far out, $\frac{s}{x}$ is near 1

Thus when the fish is far out it is approaching the bridge only slightly faster than the line is being wound in.

However, when the fish is almost under the bridge, x is small, and the ratio s/x is large. Then the fish is moving very quickly.

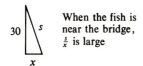

When the fish is near the bridge, $\frac{s}{x}$ is large

When $s = 50$, then
$$x^2 + 30^2 = 50^2,$$

from which it follows that
$$x = 40.$$

Thus, when 50 feet of line are out, the speed is

$$\frac{2s}{x} = \frac{2 \cdot 50}{40} = 2.5 \text{ feet per second.}$$

When $s = 31$,

$$x^2 + 30^2 = 31^2;$$

hence

$$x = \sqrt{31^2 - 30^2}$$
$$= \sqrt{961 - 900}$$
$$= \sqrt{61}.$$

Thus, when 31 feet of line are out, the fish is moving at the speed of

$$\frac{2s}{x} = \frac{2 \cdot 31}{\sqrt{61}} = \frac{62}{\sqrt{61}} = \frac{62\sqrt{61}}{61}$$

$$\doteq 7.9 \text{ feet per second.} \quad \bullet$$

General procedure for finding related rates   The method used in Example 1 applies to many related rate problems. This is the general procedure, broken into three steps:

1. Find an equation relating the varying quantities.
2. Differentiate both sides of the equation with respect to time.
3. Use the equation obtained in step 2 to determine the unknown rate from the given rates.

EXAMPLE 2   A woman on the ground is watching a jet through a telescope as it approaches at a speed of 10 miles per minute at an altitude of 7 miles. At what rate does the angle of the telescope change when the horizontal distance of the jet from the woman is 24 miles? When the jet is directly above the woman?

SOLUTION   To begin, sketch a diagram and label the parts that are of interest, as has been done in the accompanying diagram. Observe that

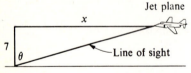

$$\frac{dx}{dt} = -10 \text{ miles per minute.}$$

The rate at which θ changes, $\dfrac{d\theta}{dt}$,

is to be found.

Step 1 consists of finding an equation relating θ and x. One such equation is

$$\tan \theta = \frac{x}{7}.$$

In step 2 this equation is differentiated with respect to time:

$$\frac{d(\tan \theta)}{dt} = \frac{d(x/7)}{dt};$$

hence, by the chain rule,

$$\sec^2 \theta \, \frac{d\theta}{dt} = \frac{1}{7} \frac{dx}{dt}.$$

Since $$\frac{dx}{dt} = -10,$$

it follows that $$\frac{d\theta}{dt} = \frac{1}{7}\frac{(-10)}{\sec^2\theta}$$

$$= \frac{-10}{7\sec^2\theta}$$

$$= -\frac{10}{7}\cos^2\theta \text{ radians per minute.}$$

The negative sign in the formula, $-(10/7)\cos^2\theta$, shows that θ is decreasing. When $x = 24$, let us find $d\theta/dt$. First it is necessary to compute $\cos\theta$ in the triangle in the accompanying diagram. Since

$$s^2 = 7^2 + 24^2$$
$$= 49 + 576$$
$$= 625,$$

it follows that $$s = 25.$$

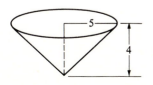

Thus $$\cos\theta = \frac{7}{25},$$

and $$\frac{d\theta}{dt} = -\frac{10}{7}\left(\frac{7}{25}\right)^2$$

$$= -\frac{70}{625}$$

$$\doteq -0.11 \text{ radian per minute}$$

$$\doteq -6° \text{ per minute}$$

When the plane is directly above the woman, $x = 0$, $\theta = 0$, $\cos\theta = 1$, and the formula

$$\frac{d\theta}{dt} = -\frac{10}{7}\cos^2\theta$$

shows that $$\frac{d\theta}{dt} = -\frac{10}{7}(1)^2$$

$$= -\frac{10}{7} \text{ radians per minute}$$

$$\doteq -81° \text{ per minute.}$$

The telescope moves much more rapidly when the plane is directly overhead. ●

The method described in Example 1 for determining unknown rates from known ones extends to finding an unknown acceleration. Just differentiate another time. Example 3 illustrates the procedure.

EXAMPLE 3   Water flows into a conical tank at the constant rate of 3 cubic meters per second. The radius of the cone is 5 meters and its height is 4 meters. Let

$h(t)$ represent the height of the water above the bottom of the cone at time t. Find dh/dt (the rate at which the water is rising in the tank) and d^2h/dt^2, when the tank is filled to a height of 2 meters.

SOLUTION   Let $V(t)$ be the volume of water in the tank at time t. The data imply that

$$\frac{dV}{dt} = 3,$$

and hence

$$\frac{d^2V}{dt^2} = 0.$$

To find dh/dt and d^2h/dt^2, first obtain an equation relating V and h.

When the tank is filled to the height h, the water forms a cone of height h and radius r.   By similar triangles,

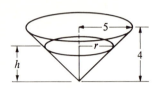

$$\frac{r}{h} = \frac{5}{4},$$

or

$$r = \tfrac{5}{4}h.$$

Thus

$$V = \tfrac{1}{3}\pi r^2 h = \tfrac{1}{3}\pi(\tfrac{5}{4}h)^2 h = \tfrac{25}{48}\pi h^3.$$

The equation relating V and h is

$$V = \frac{25\pi}{48} h^3. \tag{1}$$

From here on, the procedure is automatic: Just differentiate as often as needed. Differentiating once (using the chain rule) yields

$$\frac{dV}{dt} = \frac{25\pi}{48}\frac{d(h^3)}{dh}\frac{dh}{dt},$$

or

$$\frac{dV}{dt} = \frac{25\pi}{16} h^2 \frac{dh}{dt}. \tag{2}$$

Since $dV/dt = 3$ all the time, and $h = 2$ at the moment of interest, substitute these values into (2), obtaining

$$3 = \frac{25\pi}{16} 2^2 \frac{dh}{dt}.$$

Hence

$$\frac{dh}{dt} = \frac{12}{25\pi} \text{ feet per second}$$

when $h = 2$.

To find d^2h/dt^2, differentiate (2), obtaining

$$\frac{d^2V}{dt^2} = \frac{25\pi}{16}\left(h^2 \frac{d^2h}{dt^2} + 2h \frac{dh}{dt}\frac{dh}{dt}\right). \tag{3}$$

Since $d^2V/dt^2 = 0$ all the time, and, when $h = 2$, $dh/dt = 12/(25\pi)$, (3) implies

that

$$0 = \frac{25\pi}{16}\left[2^2 \frac{d^2h}{dt^2} + 2 \cdot 2\left(\frac{12}{25\pi}\right)^2\right]. \tag{4}$$

Solving Eq. (4) for d^2h/dt^2 shows that

$$\frac{d^2h}{dt^2} = \frac{-144}{625\pi^2} \text{ feet per second per second.}$$

Since d^2h/dt^2 is negative, the rate at which the water rises in the tank is slowing down. In general, the higher the water, the slower it rises. Even though V changes at a constant rate, h does not. ●

It may happen that $\dfrac{dx}{dt}$ and $\dfrac{dy}{dt}$

are given, and $\dfrac{dy}{dx}$

is to be found (assuming that y may be considered also a function of x). As in Sec. 10.8, by the chain rule,

$$\frac{dy}{dt} = \frac{dy}{dx}\frac{dx}{dt},$$

hence
$$\frac{dy}{dx} = \frac{dy/dt}{dx/dt}. \tag{5}$$

For convenience, use Newton's dot notation for derivatives, $\dot{x} = dx/dt$ and $\dot{y} = dy/dt$. Then (5) can be written simply as

$$\frac{dy}{dx} = \frac{\dot{y}}{\dot{x}}. \tag{6}$$

But how can d^2y/dx^2 be computed if, in addition, the second derivatives $\ddot{x}$ and $\ddot{y}$ are known? The next example obtains a general formula.

EXAMPLE 4   Given that x and y are functions of t, and that $\dot{x}$, $\ddot{x}$, $\dot{y}$, and $\ddot{y}$ are known, find d^2y/dx^2.

SOLUTION   Differentiation of (6) with respect to x yields

$$\frac{d^2y}{dx^2} = \frac{d(dy/dx)}{dx} = \frac{d(\dot{y}/\dot{x})}{dx}.$$

Now, the same reasoning that leads to Eq. (5) shows that

$$\frac{d(\dot{y}/\dot{x})}{dx} = \frac{d(\dot{y}/\dot{x})/dt}{dx/dt}. \tag{7}$$

The formula for the derivative of a quotient shows that

$$\frac{d(\dot{y}/\dot{x})}{dt} = \frac{\dot{x}\ddot{y} - \dot{y}\ddot{x}}{(\dot{x})^2}. \tag{8}$$

Combining (7) and (8) results in

$$\frac{d^2y}{dx^2} = \frac{\dot{x}\ddot{y} - \dot{y}\ddot{x}}{(\dot{x})^3},$$

a formula much less intuitively apparent than (6). ●

Rather than memorize the general formula for d^2y/dx^2 developed in Example 4, it is easier to go through the whole procedure in each case, as illustrated in Example 5.

EXAMPLE 5   Find d^2y/dx^2 for the cycloid $\begin{cases} x = a\theta - a\sin\theta, \\ y = a - a\cos\theta. \end{cases}$

SOLUTION   The parameter θ plays the role of t.   Now

$$\frac{dx}{d\theta} = a - a\cos\theta \qquad \text{and} \qquad \frac{dy}{d\theta} = a\sin\theta.$$

Thus

$$\frac{dy}{dx} = \frac{dy/d\theta}{dx/d\theta} = \frac{a\sin\theta}{a - a\cos\theta} = \frac{\sin\theta}{1 - \cos\theta},$$

and so

$$\frac{d^2y}{dx^2} = \frac{\dfrac{d(dy/dx)}{d\theta}}{dx/d\theta} = \frac{\dfrac{d[\sin\theta/(1 - \cos\theta)]}{d\theta}}{a - a\cos\theta}$$

$$= \frac{\left[\dfrac{(1 - \cos\theta)\cos\theta - \sin\theta\sin\theta}{(1 - \cos\theta)^2}\right]}{a(1 - \cos\theta)}$$

$$= \frac{\cos\theta - 1}{a(1 - \cos\theta)^3}$$

$$= -\frac{1}{a(1 - \cos\theta)^2}. \quad ●$$

Exercises

1. A 10-foot ladder is leaning against a wall. If a person pulls the base of the ladder away from the wall at the rate of 1 foot per second, how fast is the top going down the wall when the base of the ladder is (a) 6 feet from the wall? (b) 8 feet from the wall? (c) 9 feet from the wall?

2. A man is flying a kite at a height of 300 feet. A horizontal wind is blowing the kite away from the man. When 500 feet of string are out, the kite is pulling the string

out at a rate of 20 feet per second.   What is the wind velocity?

3. A beachcomber walks 2 miles per hour along the shore as a rotating light 3 miles offshore follows him.
   (a) Intuitively, what do you think happens to the rate at which the light rotates as the beachcomber walks further and further along the shore away from the lighthouse?
   (b) Letting x describe the position of the beachcomber

and θ the angle of the light, obtain an equation relating θ and x.

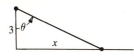

(c) With the aid of (b) show that $d\theta/dt = 6/(9 + x^2)$ (radians per hour).

(d) Does the formula in (c) agree with your guess in (a)?

4. A shrinking spherical balloon loses air at the rate of 1 cubic inch per second. At what rate is its radius changing when the radius is (a) 2 inches? (b) 1 inch? (The volume V of a sphere of radius r is $4\pi r^3/3$.)

5. Bulldozers are moving earth at the rate of 1,000 cubic yards per hour onto a conically shaped hill whose height remains equal to its radius. At what rate is the height of the hill increasing when the hill is (a) 20 yards high? (b) 100 yards high? (The volume of a cone of radius r and height h is $\pi r^2 h/3$.)

6. The lengths of the two legs of a right triangle depend on time. One, whose length is x, increases at the rate of 5 feet per second, while the other, of length y, decreases at the rate of 6 feet per second. At what rate is the hypotenuse changing when $x = 3$ and $y = 4$? Is the hypotenuse increasing or decreasing then?

7. Two sides of a triangle and their included angle are changing with respect to time. The angle increases at the rate of 1 radian per second; one side increases at the rate of 3 feet per second and the other side decreases at the rate of 2 feet per second. Find the rate at which the area is changing when the angle is $\pi/4$, the first side is 4 feet long, and the second side is 5 feet long. Is the area decreasing or increasing then?

In Exercises 8 to 10 find dy/dx and d^2y/dx^2, given that

8. $\dot{x} = 3$, $\dot{y} = 4$, $\ddot{x} = -2$, $\ddot{y} = 5$.

9. $\dot{x} = -1$, $\dot{y} = 1$, $\ddot{x} = 3$, $\ddot{y} = 2$.

10. $\dot{x} = 1$, $\dot{y} = 5$, $\ddot{x} = 0$, $\ddot{y} = 3$.

11. Find dy/dx and d^2y/dx^2 if $x = e^t + t$ and $y = \sin t + t^2$.

12. Find dy/dx and d^2y/dx^2 if $y = t^3 + t$ and $x = t^4 + t$.

13. What is the acceleration of the fish described in Example 1 when the length of line is (a) 300 feet? (b) 31 feet?

14. A particle moves on the parabola $y = x^2$ in such a way that $\dot{x} = 3$ throughout the journey. Find formulas for (a) $\dot{y}$ and (b) $\ddot{y}$.

15. Call one acute angle of a right triangle θ. The adjacent leg has length x and the opposite leg has length y.
(a) Obtain an equation relating x, y, and θ.
(b) Obtain an equation involving $\dot{x}$, $\dot{y}$, and $\dot{\theta}$ (and other variables).
(c) Obtain an equation involving $\ddot{x}$, $\ddot{y}$, and $\ddot{\theta}$ (and other variables).

16. Water is flowing into a hemispherical kettle of radius 5 feet at the constant rate of 1 cubic foot per minute.
(a) Find the rate at which the water is rising when its height above the bottom of the kettle is 3 feet, 4 feet, and 5 feet.
(b) If $h(t)$ is the depth in feet at time t, find $\ddot{h}$ when $h = 3$, 4, and 5.

17. In a certain board game a pen directed by two knobs is free to move over a piece of paper, which may be called the xy plane. One knob controls the y coordinate, and the other knob, the x coordinate. At a certain moment the x coordinate is being changed twice as fast as the y coordinate, but the y coordinate is accelerating twice as fast as the x coordinate. Determine dy/dx and d^2y/dx^2 at this moment if possible.

11.3

The second derivative and the curvature of a curve

The rate of change in y with respect to x measures the steepness of a curve. What is a reasonable measure of its "curviness" or curvature? A line, being perfectly straight, has no curvature. It seems reasonable that a line, therefore, should have curvature 0. Moreover, a large circle should have less curvature than a small circle. (The horizon of the earth, part of a circle of radius 4,000 miles, looks practically straight.)

If you walk around a small circle, your direction changes much more rapidly than when you walk around a large circle. To make this idea more precise, and to obtain a measure of the curvature of a circle, consider the diagram (next page) of a circle of radius a and a line tangent to it. Start at the bottom of the

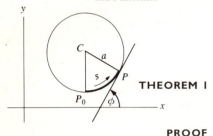

The tangent line turns as P moves counterclockwise and s increases

circle P_0 and walk counterclockwise. At a distance s along the curve from P_0 the direction is given by the angle ϕ from the positive x axis to the tangent line at P; the angle ϕ depends on s. Define *the curvature of a circle* to be the rate at which ϕ changes with respect to s, that is, $d\phi/ds$.

The next theorem shows that $d\phi/ds$ is small for a large circle and large for a small circle; in fact, it is simply the reciprocal of the radius.

THEOREM I For a circle of radius a, swept out counterclockwise, the curvature $d\phi/ds$ is constant and equals $1/a$, the reciprocal of the radius.

PROOF A little geometry shows that ϕ equals the angle $P_0 CP$ in the preceding diagram. By the definition of radian measure, the angle $P_0 CP$ has the measure s/a. Thus

$$\phi = \frac{s}{a} \quad \text{and} \quad \frac{d\phi}{ds} = \frac{1}{a},$$

and the theorem is proved. ●

Since $d\phi/ds$, the rate of change of direction with respect to arc length, gives a reasonable measure of curvature for a circle—the larger the circle, the less its curvature—it is common to use it as a measure of curvature for other curves.

Before defining curvature in general, we should discuss the arc length s and the angle ϕ in a little more detail. First of all, let us agree to measure arc length in such a way that it increases as we move along the curve away from the base point.

Second, consider the angle ϕ. There is ambiguity in the choice of ϕ, for if ϕ describes the angle so does $\phi + n\pi$ for any integer n. The particular choice is not of any importance; what does matter is that ϕ should vary continuously as we traverse the curve. Look back at the diagram which showed the typical tangent line to a circle of radius a. If we choose $\phi = 0$ for the (horizontal) tangent line at P_0, then our choice of ϕ for all other tangent lines to the circle, as we traverse the circle counterclockwise, is determined. The diagram in the margin shows that, as P goes once around the circle, ϕ increases by 2π.

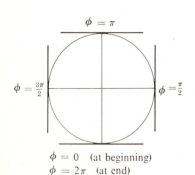

$\phi = \pi$

$\phi = \frac{3\pi}{2}$

$\phi = \frac{\pi}{2}$

$\phi = 0$ (at beginning)
$\phi = 2\pi$ (at end)

If we had chosen ϕ for the tangent line at P_0 initially to be, say, π, then as we traverse the circle once ϕ would increase to 3π. Since it is the *rate* at which ϕ changes that concerns us, not its actual value, this ambiguity, which may occur with any curve, does not affect the following definition.

DEFINITION *Curvature.* Assume that a curve is given parametrically, with the parameter of the typical point P being s, the distance along the curve from a fixed point P_0 to P. Let ϕ be the angle between the tangent line at P and the positive part of the x axis. The *curvature* at P is the derivative

$$\frac{d\phi}{ds}$$

(if this derivative exists).

Observe that a straight line has zero curvature everywhere, since ϕ is a con-

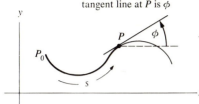

The arc length from P_0 to P is s, and the angle from the positive x axis to the tangent line at P is ϕ

stant.   Note also that in the diagram at the left the curvature is negative at the indicated point P, but positive at P_0.   If the curve is traversed in the opposite direction, however, the curvature is positive at P and negative at P_0.   (Why?) Inspection of the diagram shows that the curvature changes sign at the inflection point.   Therefore it is not surprising if the curvature, though defined as a derivative, is intimately connected with the second derivative d^2y/dx^2.   But more than the second derivative is involved, as the parabola $y = x^2$ shows (see the figure below).   Far from $(0, 0)$ the parabola $y = x^2$ has little curvature; near $(0, 0)$ it appears to bend more rapidly.   Thus the curvature is not constant; but the second derivative d^2y/dx^2 is constant (equaling 2 for all x).   As Theorem 2 shows, the curvature is determined by the *pair* of functions d^2y/dx^2 and dy/dx.

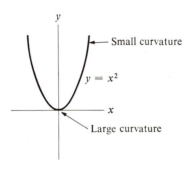

THEOREM 2   Assume that in the parameterization of the curve $y = f(x)$ with arc length s from a fixed point P_0 as the parameter, x increases as s increases.   Then

$$\text{Curvature} = \frac{d^2y/dx^2}{[1 + (dy/dx)^2]^{3/2}}.$$

PROOF   By the chain rule,

$$\frac{d\phi}{ds} = \frac{d\phi/dx}{ds/dx}.$$

As was shown in Sec. 10.9,

$$\frac{ds}{dx} = \left[1 + \left(\frac{dy}{dx}\right)^2\right]^{1/2}.$$

All that remains is to express $d\phi/dx$ in terms of dy/dx and d^2y/dx^2.

Note that

$$\tan\phi = \text{slope of tangent line}$$

$$= \frac{dy}{dx},$$

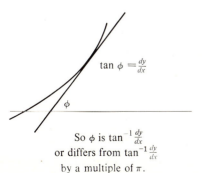

$\tan\phi = \frac{dy}{dx}$

ϕ

So ϕ is $\tan^{-1}\frac{dy}{dx}$ or differs from $\tan^{-1}\frac{dy}{dx}$ by a multiple of π.

or

$$\phi = \tan^{-1}\frac{dy}{dx} + n\pi,$$

for some fixed integer n.

Recall that

$$\frac{d(\tan^{-1} u)}{du} = \frac{1}{1 + u^2}.$$

Thus by the chain rule,

$$\frac{d\phi}{dx} = \frac{1}{1 + (dy/dx)^2}\frac{d(dy/dx)}{dx}$$

$$= \frac{d^2y/dx^2}{1 + (dy/dx)^2}.$$

Consequently, $\dfrac{d\phi}{ds} = \dfrac{d\phi/dx}{ds/dx} = \dfrac{d^2y/dx^2}{[1 + (dy/dx)^2]\sqrt{1 + (dy/dx)^2}}$,

and the theorem is proved. ●

EXAMPLE 1   Find the curvature at a typical point (x, y) on the curve $y = x^2$.

SOLUTION   In this case $dy/dx = 2x$ and $d^2y/dx^2 = 2$. Thus the curvature at (x, y) is $2/[1 + (2x)^2]^{3/2}$. Hence at $(0, 0)$ the curvature is 2, and the curve near the origin resembles a circle of radius $\frac{1}{2}$. As $|x|$ increases, the curvature approaches 0, and the curve gets straighter. ●

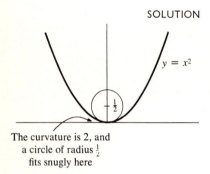

The curvature is 2, and
a circle of radius $\frac{1}{2}$
fits snugly here

Theorem 2 tells how to find the curvature if y is given as a function of x. If the curve is given parametrically, it is advisable to use the equations from the preceding section:

$$\frac{dy}{dx} = \frac{\dot{y}}{\dot{x}}$$

and

$$\frac{d^2y}{dx^2} = \frac{\dot{x}\ddot{y} - \dot{y}\ddot{x}}{(\dot{x})^3}.$$

Combining these two formulas with the formula from Theorem 2 provides a way of computing the curvature of a curve given parametrically.

THEOREM 3   If, as we move along the parametrized curve, $x = g(t)$, $y = f(t)$, to a point P, both x and the arc length s from a point P_0 increase as t increases, then

$$\text{Curvature} = \frac{\dot{x}\ddot{y} - \dot{y}\ddot{x}}{[(\dot{x})^2 + (\dot{y})^2]^{3/2}}.$$

To prove this, simply substitute in the formula of Theorem 2.

EXAMPLE 2   The cycloid determined by a wheel of radius 1 has the parametric equations

$$x = \theta - \sin\theta,$$

$$y = 1 - \cos\theta.$$

(See Sec. 10.8.)

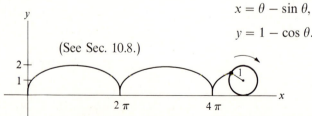

Find the curvature at a typical point on this curve.

SOLUTION   To do so, use Theorem 3, noting that θ plays the role of the parameter t.   First,

$$\frac{dx}{d\theta} = 1 - \cos\theta, \qquad \frac{dy}{d\theta} = \sin\theta,$$

$$\frac{d^2x}{d\theta^2} = \sin\theta, \qquad \frac{d^2y}{d\theta^2} = \cos\theta.$$

Then, by Theorem 3,

$$\text{Curvature} = \frac{(1 - \cos\theta)\cos\theta - (\sin\theta)\sin\theta}{[(1 - \cos\theta)^2 + (\sin\theta)^2]^{3/2}}$$

$$= \frac{\cos\theta - (\cos^2\theta + \sin^2\theta)}{[1 - 2\cos\theta + (\cos^2\theta + \sin^2\theta)]^{3/2}}$$

$$= \frac{\cos\theta - 1}{(2 - 2\cos\theta)^{3/2}}$$

$$= \frac{\cos\theta - 1}{2^{3/2}(1 - \cos\theta)^{3/2}}$$

$$= \frac{-1}{2^{3/2}} \frac{1 - \cos\theta}{(1 - \cos\theta)^{3/2}}$$

$$= \frac{-1}{2\sqrt{2}} \frac{1}{(1 - \cos\theta)^{1/2}}.$$

Since $y = 1 - \cos\theta$, the curvature is simply

$$-\frac{1}{2\sqrt{2}\sqrt{y}} = -\frac{1}{\sqrt{8y}}. \quad \bullet$$

The curvature of the cycloid in Example 2 is negative.   This means that $d\phi/ds$ is negative, or ϕ decreases as s increases.   This is plausible, for as s increases we move to the right on the arc, and the tangent line turns clockwise: ϕ decreases as a function of arc length s.   In general, the sign of the curvature depends on the base point from which arc length is measured and on the direction in which the curve is traversed.

It was noted in Example 1 that at $(0, 0)$ the parabola $y = x^2$ has curvature 2, the same as a circle of radius $\frac{1}{2}$.   In a sense, then, the parabola closely resembles a circle of radius $\frac{1}{2}$ in the vicinity of $(0, 0)$.   This observation suggests the following definition.

DEFINITION   *Radius of curvature.*   The radius of curvature of a curve at a point is the absolute value of the reciprocal of the curvature:

$$\text{Radius of curvature} = \left| \frac{1}{\text{curvature}} \right|.$$

As can be easily checked, the radius of curvature of a circle of radius a is, fortunately, a.

The cycloid in Example 2 has the radius of curvature at the point (x, y) equal to

$$\left| \frac{1}{-(1/\sqrt{8y})} \right| = \sqrt{8y}.$$

In particular, at the top of an arch, the radius of curvature is $\sqrt{8 \cdot 2} = \sqrt{16} = 4$.

Exercises

1. Find the radius of curvature of the curve in Example 1 at (a) (1, 1), (b) (2, 4).

In Exercises 2 to 6 find the radius of curvature of the indicated curve at the given point. Also sketch the graphs in Exercises 2 to 4.

2. $y = \cos x$ at $(0, 1)$.   3. $y = e^{-x}$ at $(1, 1/e)$.

4. $y = x^3$ at $(1, 1)$.
5. $\begin{cases} x = 2 \cos 3t, \\ y = 2 \sin 3t \text{ at } t = 0. \end{cases}$

6. $\begin{cases} x = 1 + t^2, \\ y = t^3 + t^4 \text{ at } t = 2. \end{cases}$

7. (a) Compute the curvature and radius of curvature for the curve $y = (e^x + e^{-x})/2$.
   (b) Show that the radius of curvature at (x, y) is y^2.
8. Find the radius of curvature along the curve $y = \sqrt{a^2 - x^2}$, where a is a constant.
9. Find the radius of curvature along the curve $y = \ln |\cos x|$.

In Exercises 10 to 12 find the curvature and radius of curvature.

10. $\begin{cases} x = e^{-t} \cos t \\ y = e^{-t} \sin t \end{cases}$
11. $\begin{cases} x = \cos^3 \theta \\ y = \sin^3 \theta \end{cases}$

12. $x + y + \ln x = 2y^5$
13. For what value of x is the radius of curvature of $y = e^x$ smallest?
14. For what value of x is the radius of curvature of $y = x^3$ smallest?
15. (a) Show that where a curve has its tangent parallel to the x axis its curvature is simply the second derivative d^2y/dx^2.
    (b) Show that the absolute value of the curvature is never larger than the absolute value of d^2y/dx^2.
16. An engineer lays out a railroad track as indicated in the diagram below. BC is part of a circle. AB and CD are straight and tangent to the circle. After the first train runs over this track, the engineer is fired. Why? How would you design a bend joining two straight sections?

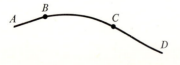

Exercises 17 to 19 are related.

17. Find the radius of curvature at a typical point on the curve
$$\begin{cases} x = a \cos \theta, \\ y = b \sin \theta. \end{cases}$$

18. (a) Show, by eliminating θ, that the curve in Exercise 17 is the ellipse
$$\frac{x^2}{a^2} + \frac{y^2}{b^2} = 1.$$
    (b) What is the radius of curvature of this ellipse at $(a, 0)$? at $(0, b)$?

19. An ellipse has a major diameter of length 6 and a minor diameter of length 4. Draw the circles that most closely approximate this ellipse at the four points which lie at the extremities of its diameters. (See Exercises 17 and 18.)

■

20. If on a curve $dy/dx = y^3$, express the curvature in terms of y.
21. In Theorem 2 we obtained a formula for curvature if the curve is given in rectangular form, $y = f(x)$. If the curve is given in polar form, $r = f(\theta)$, show that curvature equals $[r^2 + 2(r')^2 - rr'']/[r^2 + (r')^2]^{3/2}$. *Hint:* Consider the parametric representation of the curve as $x = r \cos \theta$, $y = r \sin \theta$, where $r = f(\theta)$.
22. Use the formula in Exercise 21 to show that the cardioid $r = 1 + \cos \theta$ has curvature $3\sqrt{2}/(4\sqrt{r})$ at (r, θ).
23. Use the formula in Exercise 21 to find the curvature of $r = a \cos \theta$.
24. Use the formula in Exercise 21 to find the curvature of $r = \cos 2\theta$.

■ ■

25. At the top of the cycloid in Example 2 the radius of curvature is twice the diameter of the rolling circle. What would you have guessed the radius of curvature to be at this point? Why is it not simply the diameter of the wheel, since the wheel at each moment is rotating about its point of contact with the ground?

Exercises 26 and 27 are related.

26. Let s denote arc length along a curve. Show that the curvature at a point is equal to $x'y'' - y'x''$ evaluated at that point, where differentiation is with respect to arc length s.

27. (See Exercise 26.) Show that (a) $(x')^2 + (y')^2 = 1$; (b) $x'x'' + y'y'' = 0$; (c) $x'y'' - y'x'' = y''[(x')^2 + (y')^2]/x' = y''/x'$.

28. Prove Theorem 3, using Theorem 2.

29. Prove Theorem 3 directly, without using Theorem 2.

11.4

Newton's method for solving an equation

Suppose that we wish to estimate a solution (or root) r of an equation $f(x) = 0$. If a first guess is, say x_1, then the accompanying diagram suggests that a better estimate of r may be x_2, the point at which the tangent line at $(x_1, f(x_1))$ crosses the x axis.

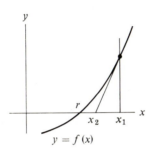

$y = f(x)$

Newton's Recursive Formula for Estimating a Root of $f(x) = 0$

To find x_2 explicitly, observe that the slope of the tangent line at $(x_1, f(x_1))$ is $f(x_1)/(x_1 - x_2)$ and is also $f'(x_1)$. Equating these two forms of the slope and solving for x_2 yields

$$x_2 = x_1 - \frac{f(x_1)}{f'(x_1)}, \tag{1}$$

which is meaningful if $f'(x_1)$ is not 0.

Formula (1), suggested by the diagram above, is the basis of Newton's method for estimating a root of an equation. Generally, this formula is applied several times, to increase accuracy.

EXAMPLE 1   Use Newton's method to estimate the square root of 3, that is, the positive root of the equation $x^2 - 3 = 0$.

SOLUTION   Here $f(x) = x^2 - 3$ and $f'(x) = 2x$. According to (1), if the first guess is x_1, then the next estimate x_2 should be

$$x_2 = x_1 - \frac{f(x_1)}{f'(x_1)} = x_1 - \frac{x_1^2 - 3}{2x_1} = \frac{x_1 + 3/x_1}{2}.$$

If $x_1 = 2$, say, then

$$x_2 = \frac{2 + \frac{3}{2}}{2} = 1.75.$$

For a better estimate of $\sqrt{3}$ repeat the process, using 1.75 instead of 2. Thus

$$x_3 = \frac{x_2 + 3/x_2}{2} = \frac{1.75 + 3/1.75}{2} \doteq 1.73214$$

is the third estimate, to five decimals. One more repetition of the process yields (to five decimals) $x_4 = 1.73205$, which is quite close to $\sqrt{3}$, whose decimal expansion begins 1.732051. ●

Since the recursive process represented by Newton's method is of practical use and is easily programmed on an electronic computer, it is important to know under what circumstances $|x_i - r|$ approaches 0 as $i \to \infty$. If the graph of f happens to be a nonhorizontal straight line, a quick sketch shows that no matter what choice of x_1 is made, the number x_2 is exactly the root r. In other words, if $f''(x)$ is identically 0, Newton's method is perfectly accurate. It is therefore reasonable to expect that the accuracy of Newton's method is influenced by $f''(x)$ [when $f''(x)$ is small, the method is probably more accurate]. On the other hand, if $f'(x_1)$ is near 0, the tangent line at $(x_1, f(x_1))$ is nearly horizontal, and may depart a great deal from the graph of f by the time it crosses the x axis. Hence $f'(x)$ also should influence the accuracy [when $f'(x)$ is large, the method is probably more accurate].

The following theorem shows that if $f''(x)$ is not too large nor $f'(x)$ too small, then $|x_i - r|$ does approach 0 as $i \to \infty$. Its proof is sketched in Exercise 22 of Sec. 15.1.

THEOREM Let r be a root of $f(x) = 0$ and x_i an estimate of r such that $f'(x_i)$ is not 0. Let

$$x_{i+1} = x_i - \frac{f(x_i)}{f'(x_i)}.$$

If f' and f'' are continuous and M is a number such that

$$\left| \frac{f''(x)}{f'(t)} \right| \le M$$

for all x and t in the interval from x_i to r, then

$$|x_{i+1} - r| \le \frac{M}{2}|x_i - r|^2. \tag{2}$$

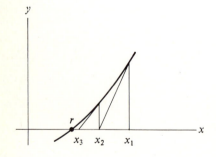

If $f''(x)$, $f'(x)$, and $f(x)$ are positive from $x = r$ to $x = x_i > r$, then, as the sketch at the left shows, $x_1 > x_2 > x_3 > \cdots > r$. This means that the successive estimates $x_2, x_3, \ldots$ are situated between the initial estimate x_1 and the root r. Hence the hypotheses of the theorem apply to x_i for all i.

In this case how swiftly does the decreasing sequence $x_1, x_2, x_3, \ldots$ approach r? Notice that if x_1 is close to r, say $|x_1 - r| \le 0.1$, then

$$|x_2 - r| \le \left(\frac{M}{2}\right)(0.1)^2 = \left(\frac{M}{2}\right)(0.01).$$

Thus, if M is not too large, x_2 is a much better approximation to r than x_1 is. For instance, if $M = 2$, then

$$|x_2 - r| \leq 0.01,$$

and

$$|x_3 - r| \leq \left(\frac{M}{2}\right)(x_2 - r)^2 \leq 0.0001,$$

and so on. Hence if x_1 is an accurate estimate of r to one decimal place, then x_2 is accurate to two decimal places, x_3 is accurate to four decimal places, and so on. The number of decimal places of accuracy tends to double at each step of the Newton recursion. For instance, the Newton recursion formula for $\sqrt{10}\ (\doteq 3.162278)$ is

$$x_{i+1} = \frac{x_i + 10/x_i}{2}.$$

The following table shows the results of the recursive process when the initial estimate x_1 is 3.

Step	Estimate	Correct Digits	Number of Correct Decimal Digits
1	$x_1 = 3$	3	0
2	$x_2 = 3.166667$	3.16	2
3	$x_3 = 3.162281$	3.1622	4

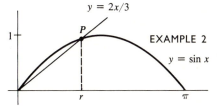

y = 2x/3

P

y = sin x

r

π

EXAMPLE 2   The line $y = 2x/3$ crosses the curve $y = \sin x$ at a point P, whose x coordinate r is between 0 and π, as shown in the accompanying diagram. The number r is a solution of the equation $2x/3 = \sin x$, since the graphs have equal y coordinates at $x = r$. Use Newton's method to approximate r.

SOLUTION   A glance at the graph suggests that r is approximately 1.5. To obtain a better estimate, note that r is a root of the equation

$$f(x) = \sin x - \frac{2x}{3} = 0.$$

Since $f'(x) = \cos x - \frac{2}{3}$, Newton's method provides this second estimate of r:

$$x_2 = 1.5 - \frac{f(1.5)}{f'(1.5)} = 1.5 - \frac{\sin 1.5 - 2(1.5)/3}{\cos 1.5 - \frac{2}{3}}.$$

Using a table of $\sin x$ and $\cos x$ in radians, we find that

$$x_2 \doteq 1.5 - \frac{0.997 - 1}{0.071 - 0.667} = 1.5 - \frac{0.003}{0.596} = 1.495.$$

Incidentally, to three decimals, $r = 1.496.$ ●

Exercises

1. Let a be a positive number. Show that the Newton recursion formula for estimating $\sqrt{a}$ is given by

$$x_{i+1} = \frac{x_i + a/x_i}{2}.$$

2. Use the formula of Exercise 1 to estimate $\sqrt{15}$. Choose $x_1 = 4$ and compute x_2 and x_3 to three decimals.

3. Use the formula of Exercise 1 to estimate $\sqrt{19}$. Choose $x_1 = 4$ and compute x_2 and x_3 to three decimals.

4. In estimating $\sqrt{3}$, an electronic computer began with $x_1 = 50$. What does Newton's method give for x_2, x_3, and x_4?

5. (a) Show that Newton's method gives this recursion formula for estimating $\sqrt[3]{7}$:

$$x_{i+1} = \frac{2}{3} x_i + \frac{7}{3x_i^2}.$$

   (b) Let $x_1 = 1$, and compute x_2 and x_3.
   (c) Let $x_1 = 2$, and compute x_2 and x_3.

6. Let $f(x) = x^4 + x - 19$.
   (a) Show that $f(2) < 0 < f(3)$ and that f must thus have a root r between 2 and 3.
   (b) Apply Newton's method, starting with $x_1 = 2$. Compute x_2 and x_3, and sketch the pertinent tangent lines on a graph.

7. Let $f(x) = x^5 + x - 1$.
   (a) Show that there is exactly one root of the equation $f(x) = 0$ in the interval $[0, 1]$. (Examine f'.)
   (b) Using $x_1 = \frac{1}{2}$ as a first estimate, apply Newton's method to find a second estimate x_2.

8. Let $f(x) = 2x^3 - x^2 - 2$.
   (a) Show that there is exactly one root of the equation $f(x) = 0$ in the interval $[1, 2]$.
   (b) Using $x_1 = \frac{3}{2}$ as a first estimate, apply Newton's method to find a second estimate x_2.

9. (a) Graph $y = e^x$ and $y = x + 2$ relative to the same axes.
   (b) With the aid of (a), estimate a root of $e^x - x - 2 = 0$.
   (c) Use Newton's method and a table of e^x to estimate the root to two-decimal accuracy.

■

Exercise 10 shows that care should be taken when using Newton's method.

10. Let $f(x) = 2x^3 - 4x + 1$.
    (a) Show that $f(1) < 0 < f(0)$ and that there must be a root r of $f(x) = 0$ in $[0, 1]$.
    (b) Take $x_1 = 1$, and apply Newton's method to obtain x_2 and x_3, estimates of r.
    (c) Graph f, and show what is happening in the sequence of estimates.

11. (a) Graph $y = \ln x$ and $y = \sin x$ relative to the same axes.
    (b) With the aid of the graphs in (a), estimate the x coordinate of the point such that $\sin x - \ln x = 0$.
    (c) Using the estimate in (b) as x_1, find another estimate x_2 by Newton's method.

12. Let $f(x) = x^2 + 1$.
    (a) Using Newton's method with $x_1 = 2$, compute x_2, x_3, x_4, and x_5 to two decimal places.
    (b) Using the graph of f, show geometrically what is happening in (a).
    (c) Using Newton's method with $x_1 = \sqrt{3}/3$, compute x_2 and x_3. What happens to x_n as $n \to \infty$?
    (d) Using Newton's method with $x_1 = (1 + \sqrt{2})/2$, examine x_2, x_3, and x_4.

13. (a) Graph $y = x \sin x$ for x in $[0, \pi]$.
    (b) Using the first and second derivatives, show that it has a unique relative maximum in the interval $[0, \pi]$.
    (c) Show that the maximum value of $x \sin x$ occurs when $x \cos x + \sin x = 0$.
    (d) Use Newton's method, with $x_1 = \pi/2$ to find an estimate x_2 for a root of the equation $x \cos x + \sin x = 0$.
    (e) Use Newton's method again, to find x_3.

■ ■

A calculator would be useful in the remaining two exercises.

14. (a) Graph $y = e^x$ and $y = \tan x$ relative to the same axes.
    (b) Show that the equation $e^x - \tan x = 0$ has a solution between 0 and $\pi/2$.
    (c) Choose x_1, an estimate of the solution in (b) on the basis of the graph in (a). Then determine x_2 by Newton's method.

15. (a) Show that the equation $3x + \sin x - e^x = 0$ has a root between 0 and 1.
    (b) Starting with $x_1 = 0.5$, compute x_2 and x_3, the estimates of the root in (a) by Newton's recursion.

11.5

The angle between a line and a tangent line

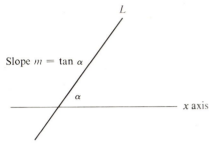

Slope $m = \tan \alpha$

In this section we deal with such questions as: "How do we find the angle between two intersecting curves?" "Why is a reflector parabolic?" "How can we find the angle between the radius arm to a point on a curve in polar coordinates and the tangent line at that point?" All the answers involve the angle between two lines, which we shall now discuss.

Consider a line L in the xy plane. It forms an angle of inclination α, $0 \le \alpha < \pi$, with the positive x axis. The slope m of L is $\tan \alpha$. (If $\alpha = \pi/2$, the slope is not defined.)

Consider two lines L and L', with angles of inclination α and α', and slopes m and m', respectively. There are two (supplementary) angles between the two lines. The following definition serves to distinguish one of these two angles as *the* angle between L and L'.

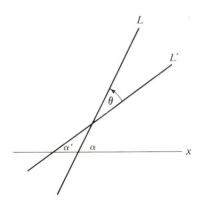

DEFINITION   *Angle between two lines.* Let L and L' be two lines, so named that L has the *larger* angle of inclination, $\alpha > \alpha'$. The angle θ between L and L' is defined to be

$$\theta = \alpha - \alpha'.$$

If L and L' are parallel, define θ to be 0.

These diagrams illustrate θ for some typical L and L'. In each case θ is the counterclockwise angle from L' to L. Note that $0 \le \theta < \pi$.

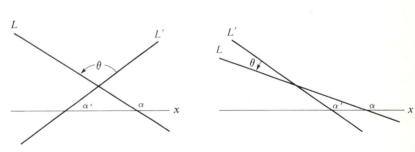

The tangent of θ is easily expressed in terms of the slopes m and m'. Using the trigonometric identity for $\tan (A - B)$, we obtain

$$\tan \theta = \tan (\alpha - \alpha')$$

$$= \frac{\tan \alpha - \tan \alpha'}{1 + \tan \alpha \tan \alpha'}$$

$$= \frac{m - m'}{1 + mm'}.$$

Formula for Tangent of Angle Between Two Lines in Terms of Their Slopes

Thus

$$\tan \theta = \frac{m - m'}{1 + mm'}$$

where m is the slope of the line with larger angle of inclination. If $mm' = -1$, then $\theta = \pi/2$; this corresponds to the fact that, as $mm' \to -1$, $|\tan \theta| \to \infty$.

The Angle between Two Curves

In the first example the formula just developed is applied to find the angle between two intersecting curves. The "angle between two curves" means "the angle between their tangent lines at the intersection."

EXAMPLE 1   The curves $y = x^2$ and $y = \sqrt{x}$ cross at the point (1, 1). Find the angle between them.

SOLUTION   A sketch shows that the tangent line to $y = x^2$ at (1, 1) has a larger inclination than the tangent line to $y = \sqrt{x}$ there. So let L denote the tangent line to $y = x^2$ at (1, 1) and let L' denote the tangent line to $y = \sqrt{x}$ there.
Since $D(x^2) = 2x$, L has slope $m = 2 \cdot 1 = 2$; since $D(\sqrt{x}) = 1/(2\sqrt{x})$, L' has slope $m' = \frac{1}{2}$. Thus

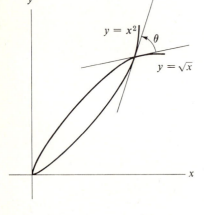

$$\tan \theta = \frac{2 - \frac{1}{2}}{1 + 2 \cdot \frac{1}{2}} = \frac{\frac{3}{2}}{2}$$

$$= \frac{3}{4}.$$

A trigonometric table or calculator then shows that $\theta \doteq 0.64$ radian. ●

In the next example it is shown why flashlight reflectors and microwave relay bowls should ideally be parabolic. The argument uses the fact that the angle of reflection of a light ray equals its angle of incidence.

Path of light reflected from a smooth surface.

EXAMPLE 2   Let P be any point on the parabola $y = \sqrt{x}$ and let F be the point $(\frac{1}{4}, 0)$. Show that the angle between the line FP and the tangent line to the parabola at P equals the angle between the x axis and the tangent line at P.

SOLUTION   The slope of the tangent line at a point on the parabola $y = \sqrt{x}$ is $1/(2\sqrt{x})$. The slope of the line FP is, by the two-point formula,

$$\frac{y - 0}{x - \frac{1}{4}} = \frac{y}{x - \frac{1}{4}}.$$

Let α be the angle between the line FP and the tangent line. Let β be the angle between FP and the line through P parallel to the x axis.

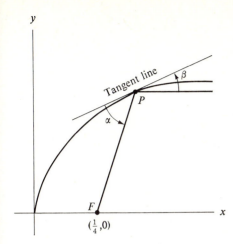

Then

$$\tan \alpha = \frac{\dfrac{y}{x - \frac{1}{4}} - \dfrac{1}{2\sqrt{x}}}{1 + \dfrac{y}{x - \frac{1}{4}} \dfrac{1}{2\sqrt{x}}}$$

and

$$\tan \beta = \text{slope of tangent line}$$

$$= \frac{1}{2\sqrt{x}}.$$

To show that $\alpha = \beta$, we show that $\tan \alpha = \tan \beta$, as follows:

$$\tan \alpha = \frac{\dfrac{y}{x - \frac{1}{4}} - \dfrac{1}{2\sqrt{x}}}{1 + \dfrac{y}{x - \frac{1}{4}} \dfrac{1}{2\sqrt{x}}}$$

$$= \frac{2\sqrt{x}\, y - x + \frac{1}{4}}{2(x - \frac{1}{4})\sqrt{x} + y}$$

$$= \frac{2\sqrt{x}\sqrt{x} - x + \frac{1}{4}}{2(x - \frac{1}{4})\sqrt{x} + \sqrt{x}} \qquad (x, y) \text{ lies on the parabola } y = \sqrt{x}.$$

$$= \frac{2x - x + \frac{1}{4}}{2x\sqrt{x} - (\sqrt{x}/2) + \sqrt{x}}$$

$$= \frac{x + \frac{1}{4}}{(2x + \frac{1}{2})\sqrt{x}} = \frac{1}{2\sqrt{x}} = \tan \beta. \;\bullet$$

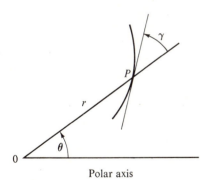

Polar axis

Similar reasoning shows that rays of light parallel to the x axis reflect off the parabola $y = \sqrt{2cx}$, where c is a positive constant, and pass through the point $(c/2, 0)$, which is called the *focus*. Appendix B3 offers a geometric definition of the focus.

Consider next a curve described in polar coordinates, $r = f(\theta)$. Let O be the pole and let P be a typical point on the curve. How can we find the angle between the line OP and the tangent to the curve at P? Denote this angle by the letter γ (gamma).

The memory device of Sec. 10.9, together with a little thinking in the style of the seventeenth-century founders of calculus will suggest a formula for $\tan \gamma$. Glance at the memory device shown in the margin. If $d\theta$ is small, the two sides of the sector are almost parallel, the curve locally looks like the tangent line, and the angle δ should be roughly γ. If that is so, it would be reasonable to hope that

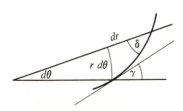

The Formula for tan γ

$$\tan \gamma = \frac{r\, d\theta}{dr} = \frac{r}{dr/d\theta} = \frac{r}{r'}.$$

The reasoning, suggestive but far from rigorous, leads to the correct formula, as is shown in the following theorem.

THEOREM Let γ be the angle between the line OP and the tangent line to a point P on the curve $r = f(\theta)$. Then

$$\tan \gamma = \frac{r}{dr/d\theta}.$$

PROOF We first find the slope of the tangent line at P. Since the curve can be written parametrically as

$$x = r \cos \theta \qquad y = r \sin \theta,$$

we have

$$\frac{dy}{dx} = \frac{dy/d\theta}{dx/d\theta} = \frac{r \cos \theta + r' \sin \theta}{-r \sin \theta + r' \cos \theta}$$

This is a formula for the slope of the tangent line at P. The slope of OP is simply $\tan \theta$.

For convenience assume that ϕ, the angle of inclination of the tangent line, is larger than θ, the angle of inclination of the line OP.

Consequently,

$$\tan \gamma = \tan (\phi - \theta)$$

$$= \frac{dy/dx - \tan \theta}{1 + (dy/dx) \tan \theta}$$

$$= \frac{\dfrac{r \cos \theta + r' \sin \theta}{-r \sin \theta + r' \cos \theta} - \dfrac{\sin \theta}{\cos \theta}}{1 + \dfrac{r \cos \theta + r' \sin \theta}{-r \sin \theta + r' \cos \theta} \dfrac{\sin \theta}{\cos \theta}}$$

$$= \frac{r}{r'} \qquad \text{algebra.}$$

Using some straightforward algebra and the identity $\cos^2 \theta + \sin^2 \theta = 1$, the reader may fill in the details. ●

Notice how much simpler the formula for $\tan \gamma$ is than that for $\tan \phi$. It is easier to work with γ; if ϕ is needed, express it in terms of γ and θ.

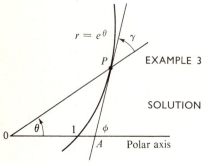

EXAMPLE 3 Let P be a point on the spiral $r = e^\theta$. Find γ and ϕ.

SOLUTION We have

$$\tan \gamma = \frac{r}{r'} = \frac{e^\theta}{(e^\theta)'} = \frac{e^\theta}{e^\theta} = 1.$$

The angle γ is constant, $\gamma = \pi/4$.

Inspection of triangle OAP shows that $\phi = \theta + \gamma = \theta + \pi/4$. ●

Exercises

1. What is the angle between a line with angle of inclination $\pi/4$ and a line with angle of inclination $3\pi/4$?

In Exercises 2 to 4 find the tangent of the angle between two lines with the given slopes and then the angle itself.

2. Slopes 2 and 3.

3. Slopes 2 and $-\frac{1}{2}$.

4. Slopes -2 and -3.

In Exercises 5 to 7 find the tangent of the angle between the two curves at the indicated point of intersection.

5. $y = \sin x$; $y = \cos x$ at $(\pi/4, \sqrt{2}/2)$.

6. $y = x^2$; $y = x^3$ at $(1, 1)$.

7. $y = e^x$; $y = e^{-x}$ at $(0, 1)$.

In Exercises 8 to 10 find γ and ϕ for the given curve and angle θ. Use a table or calculator to estimate γ.

8. $r = e^{\sqrt{3}\theta}$; $\theta = \pi/6$.

9. $r = 1 + \cos \theta$; $\theta = \pi/4$.

10. $r = \sin 2\theta$; $\theta = \pi/6$.

11. Fill in the missing algebra in the proof of the theorem of this section.

12. Show that, if $r = a \sin \theta$, then $\gamma = \theta$.

■

13. Show that, for the cardioid $r = 1 - \cos \theta$, $\gamma = \theta/2$.

14. (a) For the cardioid $r = 1 + \cos \theta$ find $\lim_{\theta \to \pi^-} \gamma$.

(b) Sketch $r = 1 + \cos \theta$, using the information obtained in (a).

15. If for the curve $r = f(\theta)$, γ always equals θ, what are all the possibilities for f?

16. If for the curve $r = f(\theta)$, γ is independent of θ, what are all the possibilities for f?

■ ■

17. Consider the curve $r = 1 + a \cos \theta$, where a is fixed, $0 \le a \le 1$.

(a) Relative to the same polar axis graph the curves corresponding to $a = 0, \frac{1}{4}, \frac{1}{2}, \frac{3}{4}, 1$.

(b) For $a = \frac{1}{4}$ the graph in (a) is convex, but not for $a = 1$. Show that for $\frac{1}{2} < a \le 1$ the curve is not convex.

This exercise explains why "whispering" rooms are elliptical.

18. The ellipse $x^2/a^2 + y^2/b^2 = 1$, $a > b > 0$, has foci at $F = (\sqrt{a^2 - b^2}, 0)$ and $F' = (-\sqrt{a^2 - b^2}, 0)$. Let P be any point on the ellipse, and let T be the tangent line

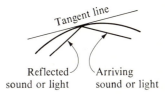

to the ellipse at P. Show that a sound wave starting at F, after bouncing off the ellipse at P, will pass through F'. (Assume that sound, like light, is reflected off P in such a way that the angle of reflection equals the angle of incidence.) The whisper of a person standing at one focus of an elliptical room can easily be heard at the other focus.

19. Four dogs are chasing each other counterclockwise at the same speed. Initially they are at the four vertices of a square of side a. As they chase each other, they approach the center of the square in a spiral path. How far does each dog travel?

(a) First find the equation of the path in polar coordinates; then find its arc length.

(b) Answer the question without calculus.

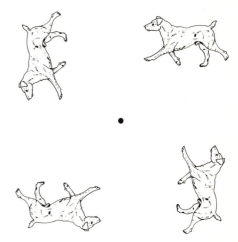

20. Let $r = f(\theta)$ describe a curve that does not pass through the origin O. Show that, at a point P on the curve closest to O, the line OP is perpendicular to the tangent line to the curve at P. (Assume that f is differentiable.)

11.6

The hyperbolic functions

Certain combinations of the exponential functions e^x and e^{-x} occur often enough in differential equations and engineering to be given names. This section defines these so-called *hyperbolic functions*, and obtains their basic properties. Since the letter x will be needed later for another purpose, we will use the letter t when writing the two preceding exponentials, namely e^t and e^{-t}.

DEFINITION   *The hyperbolic cosine.* Let t be a real number. The hyperbolic cosine of t, denoted $\cosh t$, is given by the formula

$$\cosh t = \frac{e^t + e^{-t}}{2}.$$

Note that

$$\cosh 0 = \frac{e^0 + e^{-0}}{2}$$

$$= \frac{1 + 1}{2}$$

$$= 1.$$

Also

$$\cosh(-t) = \frac{e^{(-t)} + e^{-(-t)}}{2}$$

$$= \frac{e^{-t} + e^t}{2}$$

$$= \cosh t.$$

Thus, $\cosh 0 = 1$ and $\cosh(-t) = \cosh t$, in marked similarity to the cosine function of trigonometry. However, the hyperbolic cosine is quite different from the ordinary cosine. In particular, as Example 1 shows, $\cosh t$ is always at least 1 and can be arbitrarily large.

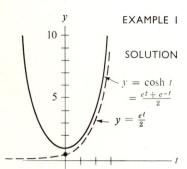

EXAMPLE I   Graph $f(t) = \cosh t$.

SOLUTION   First tabulate a few values of $f(t)$.

t	-3	-2	-1	0	1	2	3
$\dfrac{e^t + e^{-t}}{2}$	10.07	3.76	1.54	1	1.54	3.76	10.07

Note that, when t is large and positive, e^{-t} is small; thus $(e^t + e^{-t})/2$ is approximately $\frac{1}{2}e^t$. With the aid of the table and the observation, the graph is easily sketched. Incidentally, the curve sketched is called a *catenary* (from the Latin *catena*, meaning "chain"). A chain or rope, suspended from its ends, forms a curve that is part of a catenary. ●

The other hyperbolic function used in practice is defined as follows.

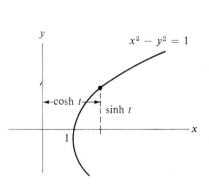

DEFINITION *The hyperbolic sine.* Let t be a real number. The hyperbolic sine of t, denoted sinh t, is given by the formula

$$\sinh t = \frac{e^t - e^{-t}}{2}.$$

It is a straightforward matter to check that $\sinh 0 = 0$ and $\sinh(-t) = -\sinh t$. The graph of $\sinh t$ is sketched in the margin.

Note the contrast between $\sinh t$ and $\sin t$. As t becomes large, the hyperbolic sine becomes large, $\lim_{t \to \infty} \sinh t = \infty$ and $\lim_{t \to -\infty} \sinh t = -\infty$. While the trigonometric functions are periodic, the hyperbolic functions are not.

The next example shows why the functions $(e^t + e^{-t})/2$ and $(e^t - e^{-t})/2$ are called hyperbolic.

EXAMPLE 2 Show that for any real number t the point

$$x = \cosh t, \qquad y = \sinh t$$

lies on the hyperbola $\qquad x^2 - y^2 = 1.$

SOLUTION Compute $\cosh^2 t - \sinh^2 t$ and see whether it equals 1. We have

$$\cosh^2 t - \sinh^2 t = \left(\frac{e^t + e^{-t}}{2}\right)^2 - \left(\frac{e^t - e^{-t}}{2}\right)^2$$

$$= \frac{e^{2t} + 2e^t e^{-t} + e^{-2t}}{4} - \frac{e^{2t} - 2e^t e^{-t} + e^{-2t}}{4}$$

$$= \frac{2 + 2}{4}$$

$$= 1.$$

Observe that, since $\cosh t \geq 1$, the point $(\cosh t, \sinh t)$ is on the right half of the hyperbola $x^2 - y^2 = 1$. ●

One reason that the function $\cosh t$ is important is that its inverse, denoted $\cosh^{-1} x$, is the function $\ln(x + \sqrt{x^2 - 1})$, which is an antiderivative of $1/\sqrt{x^2 - 1}$. The next example establishes this formula for $\cosh^{-1} x$.

EXAMPLE 3 For $t \geq 0$ the function $x = \cosh t$ is one-to-one. Find t in terms of x, and thus determine the inverse function $\cosh^{-1} x$.

SOLUTION Let $x = \dfrac{e^t + e^{-t}}{2}$.

Solve for t in terms of x, as follows:

$$e^t + e^{-t} = 2x;$$

$$e^t + \frac{1}{e^t} = 2x;$$

$$(e^t)^2 + 1 = 2xe^t;$$

$$(e^t)^2 - 2xe^t + 1 = 0.$$

The last equation is a quadratic in which the unknown is e^t. By the quadratic formula,

$$e^t = \frac{2x \pm \sqrt{(-2x)^2 - 4 \cdot 1 \cdot 1}}{2}$$

$$= \frac{2x \pm \sqrt{4x^2 - 4}}{2}$$

$$= x \pm \sqrt{x^2 - 1}.$$

Since $t \geq 0$, $e^t \geq 1$. As is shown in Exercise 10, $x - \sqrt{x^2 - 1}$ is less than 1 if $x > 1$. Thus

$$e^t = x + \sqrt{x^2 - 1},$$

and consequently,

$$t = \ln\left(x + \sqrt{x^2 - 1}\right).$$

In short,

$$\cosh^{-1} x = \ln\left(x + \sqrt{x^2 - 1}\right). \; \bullet$$

Other properties of the hyperbolic functions are presented in the exercises.

Exercises

1. Show that

(a) $\dfrac{d(\cosh t)}{dt} = \sinh t$ and (b) $\dfrac{d(\sinh t)}{dt} = \cosh t.$

2. Show that
(a) $\cosh(x + y) = \cosh x \cosh y + \sinh x \sinh y$;
(b) $\sinh(x + y) = \sinh x \cosh y + \cosh x \sinh y.$

3. Show that
(a) $\cosh(x - y) = \cosh x \cosh y - \sinh x \sinh y$;
(b) $\sinh(x - y) = \sinh x \cosh y - \cosh x \sinh y.$

4. Show that $\cosh^2 x + \sinh^2 x = \cosh 2x.$

5. Show that $\sinh 2x = 2 \sinh x \cosh x.$

6. Show that $2 \sinh^2(x/2) = \cosh x - 1.$

7. Show that $2 \cosh^2(x/2) = \cosh x + 1.$

8. Let $y = \cosh^{-1} x$. Show that, for $x > 1$,

$$\frac{dy}{dx} = \frac{1}{\sqrt{x^2 - 1}}$$

(a) by differentiating the equation $x = \cosh y$ with respect to x;
(b) by using the fact that $\cosh^{-1} x = \ln\left(x + \sqrt{x^2 - 1}\right).$

9. Show that $\sinh^{-1} x = \ln\left(x + \sqrt{x^2 + 1}\right).$

■

10. Show that, for $x > 1$, $x - \sqrt{x^2 - 1} < 1$. *Hint:* Multiply by $(x + \sqrt{x^2 - 1})/(x + \sqrt{x^2 - 1}).$

11. The *hyperbolic tangent* is defined as

$$\tanh x = \frac{\sinh x}{\cosh x}.$$

(a) Show that $\tanh x = \dfrac{e^x - e^{-x}}{e^x + e^{-x}}.$

(b) Show that $\tanh x$ is defined for all x.

(c) Show that $\tanh(-x) = -\tanh x$.

(d) Find $\lim_{x \to \infty} \tanh x$.

12. (See Exercise 11.) Show that

$$\tanh(x + y) = \frac{\tanh x + \tanh y}{1 + \tanh x \tanh y}.$$

13. The *hyperbolic secant* is defined as

$$\operatorname{sech} x = \frac{1}{\cosh x}.$$

Graph $y = \operatorname{sech} x$.

14. Show that $D(\tanh x) = \operatorname{sech}^2 x$.

15. Show that $D(\operatorname{sech} x) = \operatorname{sech} x \tanh x$.

16. The *hyperbolic cotangent* is defined as

$$\coth x = \frac{1}{\tanh x}.$$

Graph $y = \coth x$.

17. The *hyperbolic cosecant* is defined as

$$\operatorname{csch} x = \frac{1}{\sinh x}.$$

(a) What is the domain of $\operatorname{csch} x$?

(b) Graph $y = \operatorname{csch} x$.

18. Show that $D(\coth x) = -\operatorname{csch}^2 x$.

19. Show that $D(\operatorname{csch} x) = -\operatorname{csch} x \coth x$.

20. Show that

(a) $\operatorname{sech}^2 x + \tanh^2 x = 1$;

(b) $\operatorname{csch}^2 x - \coth^2 x = -1$.

21. Show that $D(\tanh^{-1} x) = 1/(1 - x^2)$.

■ ■

Exercises 22 to 24 use the calculator.

22. (a) How close is $\cosh x$ to $\frac{1}{2}e^x$ when $x = 3$?

(b) How close is $\tanh x$ to 1 when $x = 3$?

23. By methods in Chap. 15, it can be shown that

$$\cosh x = 1 + \frac{x^2}{2!} + \frac{x^4}{4!} + \frac{x^6}{6!} + \cdots.$$

(The more summands used, the better the approximation.)

(a) Using just four summands, estimate $\cosh 1$ and $\cosh 2$.

(b) How close are the estimates in (a) to the correct values?

24. By methods in Chap. 15, it can be shown that, for $0 \le x < 1$,

$$\tanh^{-1} x = x + \frac{x^3}{3} + \frac{x^5}{5} + \frac{x^7}{7} + \cdots.$$

(a) Using just four summands, estimate $\tanh^{-1} \frac{1}{2}$.

(b) How close is the estimate in (a) to the correct value?

25. Some integral tables contain the formulas

$$\int \frac{dx}{\sqrt{ax+b}\sqrt{cx+d}} = \frac{2}{\sqrt{-ac}} \tan^{-1} \sqrt{\frac{-c(ax+b)}{a(cx+d)}}$$

$$= \frac{2}{\sqrt{ac}} \tanh^{-1} \sqrt{\frac{c(ax+b)}{a(cx+d)}}.$$

The first formula is used if a and c have opposite signs, the second if they have the same signs. This exercise discusses the function $\tanh^{-1} x$ that appears in the second formula.

(a) Graph $y = \tanh x$. See Exercise 11 for the definition of $\tanh x$.

(b) Show that the function $\tanh x$ is one-to-one.

(c) Its inverse function is denoted $\tanh^{-1} x$. Show that

$$\tanh^{-1} x = \frac{1}{2} \ln \frac{x+1}{x-1}.$$

Thus $\tanh^{-1} x$ is expressible in terms of logarithms. Some calculators have keys for $\tanh x$ and $\tanh^{-1} x$.

11.7

Summary

In this chapter several topics involving the derivative were presented. In the opening section implicit differentiation was discussed and applied to maximizing a function of two variables where those variables are linked by another equation. Logarithmic differentiation was also discussed.

Section 11.2 concerned related rates. If several varying quantities are related by some equation, differentiate that equation to find the relationship between their rates of change. The resulting equation was differentiated to find relationships

between their accelerations (and rates of change.)

The curvature of a curve was expressed in terms of first and second derivatives in Sec. 11.3. Section 11.4 described Newton's recursive method for solving an equation. In Sec. 11.5 the angle between a line and a tangent line was examined, both in rectangular and polar coordinates. In particular the reflecting property of a parabola was established. The hyperbolic functions, which are certain combinations of e^x and e^{-x}, were the subject of Sec. 11.6.

VOCABULARY AND SYMBOLS

implicit function
implicit differentiation
constraint
logarithmic differentiation
related rates
curvature $d\phi/ds$
radius of curvature

Newton's method
angle between two lines
angle γ between radius and
   tangent line of $r = f(\theta)$
hyperbolic cosine   cosh t
hyperbolic sine   sinh t

$$\text{Curvature} = \frac{d\phi}{ds}; \text{radius of curvature} = 1 \left/ \left| \frac{d\phi}{ds} \right| \right.$$

$$\text{Curvature} = \frac{d^2y/dx^2}{[1 + (dy/dx)^2]^{3/2}} = \frac{\dot{x}\ddot{y} - \dot{y}\ddot{x}}{[(\dot{x})^2 + (\dot{y})^2]^{3/2}}$$

if curve is given parametrically.

$$\text{Newton's recursion: } x_2 = x_1 - \frac{f(x_1)}{f'(x_1)}.$$

If θ is the angle between two lines, then

KEY FACTS

If x and y are functions of t, then

$$\frac{dy}{dx} = \frac{\dot{y}}{\dot{x}} \quad \text{and} \quad \frac{d^2y}{dx^2} = \frac{\dfrac{d(dy/dx)}{dt}}{dx/dt}.$$

The base point from which arc length is measured on a curve is chosen so that s (arc length) increases as the curve is traversed. The angle of inclination ϕ of the tangent line is chosen to vary continuously with s.

$$\tan \theta = \frac{m - m'}{1 + mm'} \quad \begin{array}{l} m \text{ is slope of line with larger angle of} \\ \text{inclination} \end{array}$$

$$\tan \gamma = \frac{r}{r'} \quad \text{memory device:}$$

$$\cosh t = \frac{e^t + e^{-t}}{2}$$

$$\sinh t = \frac{e^t - e^{-t}}{2}$$

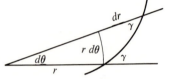

Guide quiz on Chap. 11

1. Let $y = f(x)$ be given as a solution for y of the equation
$$x^5y + y^3x^4 = 2 \quad \text{with} \quad y = 1 \quad \text{when} \quad x = 1.$$
Find dy/dx and d^2y/dx^2 when $(x, y) = (1, 1)$.

2. Differentiate
$$y = \left(\frac{\sec^3 x}{\sqrt[5]{1 + 2x}} \right)^{1/4},$$
using logarithmic differentiation.

3. A sealed container is to be built in the form of a cylinder with a hemispherical top. Find the proportion between h and r, shown in the diagram, that produces the most

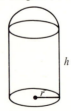

economical container. That is, maximize the volume, subject to using a given amount of metal.

4. The height h of a right circular cone is increasing at the rate of 0.3 foot per second, while the radius r is decreasing at the rate of 0.4 foot per second.
   (a) Find the rate at which the volume is changing when $h = 4$ and $r = 2$.
   (b) Does the volume change at a constant rate?

5. (a) Find the slope of the line that most closely resembles the curve $y = x^3$ near $(1, 1)$.
   (b) Sketch the line in (a).
   (c) Find the radius of the circle that most closely resembles the curve $y = x^3$ near $(1, 1)$.
   (d) Sketch the circle in (c).

6. Show that the maximum curvature of $y = x^3$, $x > 0$, occurs when $x = \sqrt[4]{\frac{1}{45}}$.

7. (a) Develop a recursion formula for estimating the positive root of the equation $x^4 = 7$.
   (b) Starting with the estimate $x_1 = 1.5$, find x_2.

8. Show that the parabola $y = \sqrt{2cx}$, where c is positive, has the property that a ray of light parallel to the x axis, reflecting off the parabola, passes through the point $(c/2, 0)$.

9. Let $r = 1 + \cos 2\theta$.

(a) Find γ when $\theta = \pi/4$. Give a decimal estimate of γ.

(b) Find the angle that the tangent line at the point on the curve corresponding to $\theta = \pi/4$ makes with the x axis. Give a decimal estimate of the angle.

10. (a) Show that when x is large and positive the curve $y = \sinh x$ is very close to the curve $y = \frac{1}{2}e^x$.

(b) Find the radius of curvature of $y = \cosh x$ at (0, 1).

Review exercises for Chap. 11

A few exercises review earlier chapters.

1. (a) Draw the diagram that suggests Newton's method for estimating a root of an equation.

(b) Obtain Newton's recursion by inspection of the diagram.

2. (a) Define curvature.

(b) Show that curvature equals $\dfrac{d^2y/dx^2}{[1 + (dy/dx)^2]^{3/2}}$.

3. Obtain the formula $\tan \gamma = r/r'$

(a) intuitively;

(b) with the aid of the parametric representation $x = r \cos \theta$, $y = r \sin \theta$.

4. (a) Why define curvature as $d\phi/ds$?

(b) Why can curvature be positive or negative?

(c) Can it be 0?

5. If x and y are related by the equation

$$x^2y + \sin x + \ln y = \pi^2,$$

find dy/dx and d^2y/dx^2 when $x = \pi$ and $y = 1$.

6. If $y = x^3$, find y, $\dot{y}$, $\ddot{y}$ when $x = 1$, $\dot{x} = 2$, and $\ddot{x} = 3$.

7. If $\begin{cases} x = \cos 2t, \\ y = \sin 3t, \end{cases}$

express dy/dx and d^2y/dx^2 in terms of t.

8. Find the radius of curvature of the curve $y = \ln x$ at $(e, 1)$.

9. One train departs at noon and travels east at 60 miles per hour. Another train departs an hour later and travels north at 80 miles per hour. (a) At what rate are they separating at 2 P.M.? (b) Is the rate at which they are separating increasing or decreasing at 2 P.M.?

10. A truck traveling 50 miles per hour to the north approaches an intersection. A car traveling 40 miles per hour to the east has passed through the intersection. When the truck is 1 mile from the intersection and the

car is 2 miles from the intersection, is the distance between them increasing or decreasing? At what rate?

In Exercises 11 and 12 find y' at the indicated values of x and y by implicit differentiation.

11. $y^7 + xy = 2$ at $x = 1$ and $y = 1$.

12. $\ln(x + y) + x + x^2 + y = 2$ at $x = 1$ and $y = 0$.

Use logarithmic differentiation to find y' in Exercises 13 and 14.

13. $y = (1 + x^2)^3 e^{x^2} \sin^5 3x$

14. $y = \dfrac{(\sqrt{x^3 + 1})^5}{(x^2 + 1)^4}$

15. A man 6 feet tall walks away from a lamp that is 20 feet high at the rate of 5 feet per second. At what rate is his shadow lengthening when he is (a) 10 feet from the lamp? (b) 100 feet from the lamp?

16. The length of a rectangle is increasing at the rate of 7 feet per second, and the width is decreasing at the rate of 3 feet per second. When the length is 12 feet and the width is 5 feet, find the rate of change of (a) the area, (b) the perimeter, (c) the diagonal.

17. A woman is walking on a bridge that is 20 feet above a river, as a boat passes directly under the center of the bridge (at a right angle to the bridge) at 10 feet per second. At that moment the woman is 50 feet from the center and approaching it at the rate of 5 feet per second. (a) At what rate is the distance between the boat and woman changing at that moment? (b) Is the rate at which they are approaching or separating increasing or is it decreasing?

18. A spherical raindrop evaporates at a rate proportional to its surface area. Show that the radius shrinks at a constant rate.

19. A couple is riding a Ferris wheel at high noon. The diameter of the wheel is 50 feet, and its speed is 0.1 revolution per second. What is the speed of their shadow on the ground when they are at a two-o'clock position? A one-o'clock position? Show that the shadow is moving its fastest when they are at the top, and its slowest when they are at the three-o'clock position.

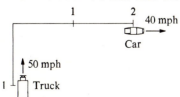

20. Sketch what the graph of f may look like near the point $(1, 3)$ if $f(1) = 3$, f' and f'' are continuous, and (a) $f'(1) = 0$, $f''(1) = 4$; (b) $f'(1) = 0$, $f''(1) = -2$; (c) $f'(1) = 0, f''(1) = 0$.

21. A function f has a derivative and second derivative for all x. If $f(1) = f(2) = f(3) = f(4)$, what can we conclude about (a) f'? (b) f''?

22. Let f possess a first and second derivative for all x. Let $g(x) = 1/f(x)$, where $f(x)$ is not 0. If $f(1) = 2$, $f'(1) = 3$, $f''(1) = 4$.   Find $g(1)$, $g'(1)$, and $g''(1)$.

23. If $d^2y/dx^2 = 6$ for all x, show that $y = 3x^2 + bx + c$ for some constants b and c.

24. Draw a small section of a graph of $y = f(x)$ if $f(x)$ is negative, $f'(x)$ is positive, and $f''(x)$ is negative.

25. Assume that x and y are functions of t. Obtain the formula for d^2y/dx^2 in terms of the derivatives $\dot{x}$, $\ddot{x}$, $\dot{y}$, $\ddot{y}$.

26. Let $f(x) = e^{x^2} \sin x$.   Find the critical numbers of f in the interval $[0, \pi]$.   Use Newton's method to estimate them if they are not easily determined.

27. Differentiate (for practice):

(a) $2^x \sec^{-1} 3x$ (b) $\sqrt{x} \cot^{-1} 4x$ (c) $e^{-x} \cos 5x$

(d) $\ln \left(\dfrac{\sqrt[5]{1 + 2x}}{\sqrt[4]{1 + 3x}} \right)^3$ (e) $\dfrac{(\sqrt[5]{1 + 2x})^3}{\sin^5 x}$ (f) x^{2x}.

28. Two of these three antiderivatives are elementary.   Find them.

(a) $\displaystyle\int \sqrt{e^x}\, dx$ (b) $\displaystyle\int \sqrt{\ln x}\, dx$ (c) $\displaystyle\int \frac{1}{x} \sqrt{\ln x}\, dx$.

■

29. "If $dy/dx = \dot{y}/\dot{x}$, why is not d^2y/dx^2 equal to $\ddot{y}/\ddot{x}$?" How would you answer this question?

30. A set R in the plane bounded by a curve is convex if, whenever P and Q are points in R, the line segment PQ also lies in R.   A curve is convex if it is the border of a convex set.

(a) Show why the average radius of curvature with respect to angle ϕ as you traverse a convex curve is length of curve/2π.

(b) Deduce from (a) that a convex curve of length L has a radius of curvature equal to $L/2\pi$ somewhere on the curve.

31. Prove that the average value of the curvature as a function of arc length s as you sweep out a convex curve in the counterclockwise direction is 2π/length of curve. (See Exercise 30 for the definition of convex curve.)

32. The flexure formula in the theory of beams asserts that the bending moment M required to bend a beam is proportional to the desired curvature, $M = k/R$, where k is a constant depending on the beam and R is the radius of curvature.   A beam is bent to form the parabola $y = x^2$. What is the ratio between the moments required at $(0, 0)$ and at $(2, 4)$?

33. Railroad curves are banked to reduce wear on the rails and flanges.   The greater the radius of curvature, the less the curve must be banked.   The best bank angle A satisfies the equation $\tan A = v^2/(32R)$, where v is speed in feet per second and R is radius of curvature in feet. A train travels in the elliptical track $x^2/1{,}000^2 + y^2/500^2 = 1$ (where x and y are measured in feet) at 60 miles per hour (equals 88 feet per second).   Find the best angle A at the points $(1{,}000, 0)$ and $(0, 500)$.

34. The larger the radius of curvature of a turn, the faster a given car can travel around that turn.   The radius of curvature required is proportional to the square of the maximum speed.   Or conversely, the maximum speed around a turn is proportional to the square root of the radius of curvature.   If a car moving on the path $y = x^3$ (x and y measured in miles) can go 30 miles per hour at $(1, 1)$ without sliding off, how fast can it go at $(2, 8)$?

■ ■

35. Read the generalized law of the mean, stated in Exercise 92 of Sec. 6.11.

(a) What does it say about the curve given parametrically by the equations $x = g(t)$, $y = f(t)$ (here t plays the role of the x in Exercise 92)?

(b) Let $h(t)$, for t in $[a, b]$, be defined as the vertical distance from $(g(t), f(t))$ to the line through $(g(a), f(a))$ and $(g(b), h(b))$, as in the proof of the law of the mean in Sec. 5.2.   Use the function h to prove the generalized law of the mean.

12

PARTIAL DERIVATIVES

The volume V of a cylindrical can of radius r and height h is given by the formula

$$V = \pi r^2 h.$$

The volume depends on two numbers, r and h. The function $\pi r^2 h$ is an example of a function of two variables, r and h. Almost all the work of the first 11 chapters concerns functions of one variable; Chap. 12 is devoted to functions of two variables, their graphs, their derivatives, and their maxima and minima. Since the graphs of such functions are usually surfaces in space (rather than curves in the plane), the first three sections concentrate on the geometry of space.

12.1

Rectangular coordinates in space

By means of an xy-coordinate system every point in a plane is described by two numbers. To describe points in space, three numbers are required. The simplest way to do this is by an xyz-coordinate system, which is constructed as follows.

Introduce three mutually perpendicular lines, called the x axis, the y axis, and the z axis. They are usually chosen in such a way that the positive portions of the x, y, and z axes match the thumb, index, and middle finger, respectively, of the right hand. (This "right-handed" coordinate system is shown below.)

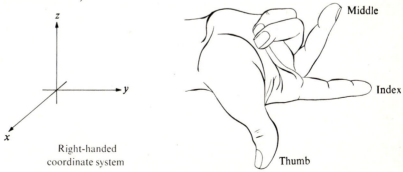

Right-handed coordinate system

Any point Q in space is now described by three numbers: First, two numbers

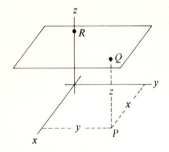

specify the x and y coordinates of the point P in the xy plane directly below (or above) Q; then the height of Q above (or below) the xy plane is recorded by the z coordinate of the point R where the plane through Q and parallel to the xy plane meets the z axis. The point Q is then denoted by (x, y, z). Note in the figure to the left that points on the xy plane have $z = 0$.

The four diagrams below display the points $(1, 0, 0)$, $(0, 1, 0)$, $(-1, 0, 0)$, and $(1, 2, 3)$. The rectangular box shown in the fourth of these diagrams helps to show the location of the point. It is also a convenient way of indicating spatial perspective.

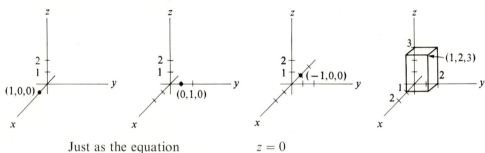

Just as the equation
$$z = 0$$
describes the xy plane, the equation
$$x = 0$$
describes the yz plane, and the equation
$$y = 0$$
describes the xz plane. These three planes are called the *coordinate planes*.

The distance between the points (x_1, y_1) and (x_2, y_2) in the xy plane is $\sqrt{(x_1 - x_2)^2 + (y_1 - y_2)^2}$, a formula based on the pythagorean theorem. The next theorem generalizes this formula to space.

THEOREM The distance between the points $Q_1 = (x_1, y_1, z_1)$ and $Q_2 = (x_2, y_2, z_2)$ is
$$\sqrt{(x_1 - x_2)^2 + (y_1 - y_2)^2 + (z_1 - z_2)^2}.$$

PROOF The diagram below shows the two points Q_1 and Q_2 and the box that they

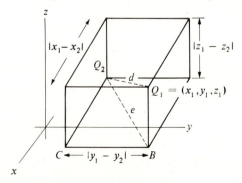

determine with its faces parallel to the three coordinate planes.   Note that the dimensions of the box are $|x_1 - x_2|$, $|y_1 - y_2|$, and $|z_1 - z_2|$.   Label two of the corners of the box B and C, as shown.   Let d be the distance between Q_1 and Q_2.   Let e be the distance between Q_2 and B.

From the right triangle $Q_1 B Q_2$,

$$d^2 = e^2 + |z_1 - z_2|^2.$$

From the right triangle $Q_2 CB$,

$$e^2 = |x_1 - x_2|^2 + |y_1 - y_2|^2.$$

Hence

$$d^2 = |x_1 - x_2|^2 + |y_1 - y_2|^2 + |z_1 - z_2|^2,$$

and

$$d = \sqrt{|x_1 - x_2|^2 + |y_1 - y_2|^2 + |z_1 - z_2|^2}$$
$$= \sqrt{(x_1 - x_2)^2 + (y_1 - y_2)^2 + (z_1 - z_2)^2}.$$

This completes the proof. ●

EXAMPLE 1   Find the distance between the points $(5, 4, 3)$ and $(2, 1, 1)$.

SOLUTION   The distance is

$$\sqrt{(5 - 2)^2 + (4 - 1)^2 + (3 - 1)^2} = \sqrt{3^2 + 3^2 + 2^2}$$
$$= \sqrt{22}. \ ●$$

EXAMPLE 2   Find the distance between $(2, -7, 3)$ and $(4, 1, -2)$.

SOLUTION   The distance is

$$\sqrt{(2 - 4)^2 + (-7 - 1)^2 + (3 - (-2))^2} = \sqrt{(-2)^2 + (-8)^2 + 5^2}$$
$$= \sqrt{2^2 + 8^2 + 5^2}$$
$$= \sqrt{93}. \ ●$$

EXAMPLE 3   Find the distance between $(8, 4, 1)$ and $(0, 0, 0)$.

SOLUTION   The distance is

$$\sqrt{(8 - 0)^2 + (4 - 0)^2 + (1 - 0)^2} = \sqrt{64 + 16 + 1}$$
$$= \sqrt{81}$$
$$= 9. \ ●$$

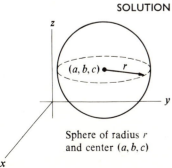

Sphere of radius r
and center (a, b, c)

The set of all points that are a fixed distance r from a given point (a, b, c) is a sphere of radius r and center (a, b, c).   To sketch this sphere, dot in the horizontal equator in perspective.   (See the figure to the left.)

A point (x, y, z) is on this sphere when the distance between it and (a, b, c)

is r, that is, when $\quad \sqrt{(x-a)^2 + (y-b)^2 + (z-c)^2} = r$,

or, equivalently, when

$$(x-a)^2 + (y-b)^2 + (z-c)^2 = r^2.$$

This last equation is an algebraic description of the sphere of radius r and center (a, b, c).

In practice, the origin $(0, 0, 0)$ of the xyz-coordinate system is usually placed at the center of the sphere. The equation of a sphere of radius r and center $(0, 0, 0)$ thus becomes

$$x^2 + y^2 + z^2 = r^2.$$

For instance, the equation

$$x^2 + y^2 + z^2 = 25$$

describes a sphere of radius 5 and center $(0, 0, 0)$. As another example, the equation

$$x^2 + y^2 + z^2 = 3$$

describes a sphere of radius $\sqrt{3}$ and center $(0, 0, 0)$.

Exercises

1. Plot the points $(0, 0, 0)$, $(2, 0, 0)$, $(0, 2, 0)$, $(0, 0, 2)$.

2. Plot the points $(1, 2, 0)$, $(-1, 0, 0)$, $(0, 3, 0)$, $(0, 0, -2)$.

3. Two opposite corners of a box whose faces are parallel to the coordinate planes are $(1, 4, 4)$ and $(5, 6, 7)$.

(a) Draw the box.

(b) What are the coordinates of the other six corners of the box?

(c) What is the volume of the box?

(d) What is the surface area of the box?

(e) What is the total length of the edges of the box?

4. How far apart are the points

(a) $(2, 0, 0)$ and $(-3, 0, 0)$?

(b) $(2, 1, 4)$ and $(-4, 7, -3)$?

(c) $(1, 1, 1)$ and $(3, 4, 7)$?

5. Find the distance from the origin $(0, 0, 0)$ to (a) $(2, 3, 6)$, (b) $(6, 6, 7)$, (c) $(1, 1, 1)$.

6. Show that, if $x^2 + y^2 \le 1$, then the point $(x, y, \sqrt{1 - x^2 - y^2})$ is a distance 1 from the origin $(0, 0, 0)$.

7. The three dimensions of a rectangular room are 10, 13, and 8 feet. What is the length of the longest piece of string that can be stretched straight in this room?

8. (a) Sketch the sphere whose center is $(1, 2, 3)$ and whose radius is 4.

(b) Is the point $(5, 2, 3)$ on this sphere?

(c) Is the point $(3, 0, 6)$ on this sphere?

9. Where does the sphere $x^2 + y^2 + z^2 = 49$ meet (a) the x axis? (b) the y axis? (c) the z axis?

10. What is the equation of the sphere (a) whose center is $(-3, 2, 1)$ and whose radius is 1? (b) whose center is $(1, 1, -1)$ and whose radius is $\sqrt{2}$?

11. Is the point $(4, 4, 4)$ inside, outside, or on the sphere $x^2 + y^2 + z^2 = 49$?

12. (a) Make a neat, clear sketch of the sphere $x^2 + y^2 + z^2 = 1$.

(b) Sketch its (great circle) intersections with the three coordinate planes.

12.2

Graphs of equations

The set of points (x, y, z) that satisfy some given equation in x, y, and z is called the *graph* of that equation. As shown in the preceding section, the

graph of
$$x^2 + y^2 + z^2 = 25$$

is the sphere of radius 5 and center $(0, 0, 0)$. Also it was observed that the graph of
$$z = 0$$

is the xy plane. (In this equation only the letter z appears. It is not necessary that all three letters appear in an equation.)

This section describes some of the graphs used later in this chapter and in following chapters.

EXAMPLE 1 Sketch the graph of the equation $z = 1$.

SOLUTION The graph consists of all points whose z coordinate is 1. This is a plane parallel to the plane $z = 0$ (the xy plane) and a distance 1 above it. The plane $z = 1$ is endless. The jagged border in the diagram to the left is one method of suggesting this. ●

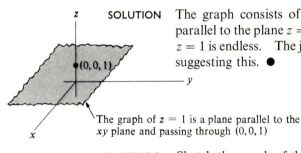

The graph of $z = 1$ is a plane parallel to the xy plane and passing through $(0, 0, 1)$

EXAMPLE 2 Sketch the graph of the equation $x = y$.

SOLUTION The point (x, y, z) is on this graph if $x = y$. There is no restriction on z. For instance, the points $(1, 1, 0)$, $(1, 1, 3)$, $(1, 1, -7)$ are all on this graph. For convenience, first sketch the part of the graph for which $z = 0$. In other words, consider the part of the graph that lies in the xy plane.

In the xy plane, the equation
$$x = y$$

describes a line, as shown in the diagram at the left. As observed above, there is no restriction on z. This means that if $(x, y, 0)$ is on the graph, so is (x, y, z) for any value of z. Thus the graph of the equation $x = y$ in space is the plane perpendicular to the xy plane, passing through the line $x = y$ in the xy plane. ●

The line $x = y$ in the xy plane is part of the graph of $x = y$ in space

Warning: The Graph of $x = y$ in Space is a Plane, Not a Line.

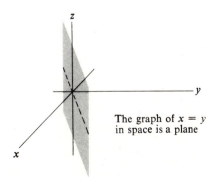

The graph of $x = y$ in space is a plane

EXAMPLE 3   Sketch the graph of $z = x + 2y$.

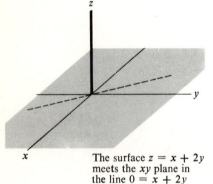

The choice $x = 1$, $y = 2$ produces the point $(1, 2, 5)$ on the surface

SOLUTION   The point (x, y, z) is on the graph when its z coordinate is related to its x and y coordinates by the equation $z = x + 2y$. In other words, the graph consists of all points of the form

$$(x, y, x + 2y),$$

where the numbers x and y can be chosen arbitrarily. For instance, the choice $x = 1$, $y = 2$ yields the point

$$(1, 2, 1 + 2 \cdot 2) = (1, 2, 5)$$

on the graph of

$$z = x + 2y.$$

In this manner, the choice of a point $(x, y, 0)$ in the xy plane produces a unique point $(x, y, z) = (x, y, x + 2y)$ on the graph of $z = x + 2y$. The graph of $z = x + 2y$ is therefore some kind of surface.

In order to sketch this surface, consider first its intersection with the three coordinate planes. First, its intersection with the xy plane consists of all points that satisfy the two equations

$$z = x + 2y \qquad \text{and} \qquad z = 0.$$

This intersection consists of all points $(x, y, 0)$ for which

$$0 = x + 2y,$$

or equivalently,

$$y = -\frac{x}{2}.$$

This is a line in the xy plane, shown dotted in the figure to the left.

The surface $z = x + 2y$ meets the xy plane in the line $0 = x + 2y$

The intersection of the surface $z = x + 2y$ and the xz plane consists of all points (x, y, z) for which

$$z = x + 2y \qquad \text{and} \qquad y = 0.$$

It is therefore the line

$$z = x$$

in the xz plane.

The surface $z = x + 2y$ meets the xz plane in the line $z = x$

In a similar manner, it can be shown that the surface $z = x + 2y$ meets the yz plane in the line $z = 2y$.

In Chap. 20 it is proved that the graph of any equation that has the form $z = Ax + By + C$, where A, B, and C are constants, is a plane. Each dotted line shown in the last three diagrams lies on this plane. Any two of these determine the plane and can be used to sketch it.

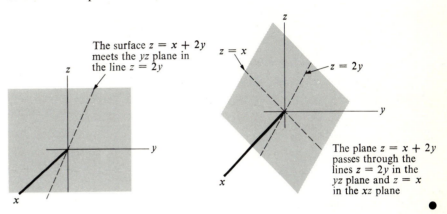

The surface $z = x + 2y$ meets the yz plane in the line $z = 2y$

The plane $z = x + 2y$ passes through the lines $z = 2y$ in the yz plane and $z = x$ in the xz plane

EXAMPLE 4   Sketch the graph of $y = x^2$.

SOLUTION   The graph of $y = x^2$ consists of all points (x, y, z) such that

$$y = x^2.$$

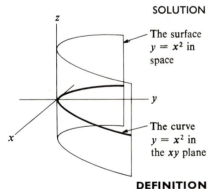

The surface $y = x^2$ in space

The curve $y = x^2$ in the xy plane

For instance $(3, 9, 0)$ is on the graph of $y = x^2$. So is $(3, 9, z)$ for any choice of z. The simplest way to graph the equation $y = x^2$ (in space) is to graph $y = x^2$ in the xy plane and then use the information that z is unrestricted. Since the graph of $y = x^2$ in the xy plane is a parabola, the graph of $y = x^2$ in space is a curved surface, as sketched in the diagram in the margin. ●

Example 4 is a special case of what is called a *cylinder*.

DEFINITION   *Cylinder.* Let R be a set in a plane. The set formed by all lines that are perpendicular to the given plane and that meet R is called the *cylinder* determined by R.

The surface $y = x^2$ in space, of Example 4, is a cylinder. The set R in this case is the parabola $y = x^2$ in the xy plane. The plane $z = 0$ is also a cylinder, for it consists of all lines perpendicular to the yz plane that pass through the y axis. If R is a circle, then the cylinder it determines is like the cylinder of everyday life, except that it is infinite and therefore has no base or top.

If an equation involves at most two of the letters x, y, and z, its graph will be a cylinder.

EXAMPLE 5   Sketch the graph of $$z^2 = \frac{x^2}{4} + \frac{y^2}{9}.$$

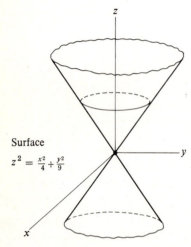

Surface
$z^2 = \frac{x^2}{4} + \frac{y^2}{9}$

SOLUTION If $z = 0$, the equation is

$$0 = \frac{x^2}{4} + \frac{y^2}{9},$$

which has only one solution, $(x, y) = (0, 0)$. So the intersection of the surface with the xy plane is just the point $(0, 0, 0)$.

When $z = 1$, the equation is

$$1 = \frac{x^2}{4} + \frac{y^2}{9},$$

which is an ellipse, sketched in the margin.

More generally, each plane parallel to the xy plane meets the surface in an ellipse (or just a point if the plane is the xy plane).

What is the intersection of the surface with the xz plane? Set $y = 0$ to find out. A point $(x, 0, z)$ in the intersection satisfies the equation

$$z^2 = \frac{x^2}{4}.$$

Now, the equation $z^2 = x^2/4$ is equivalent to the two equations

$$x = 2z \qquad \text{and} \qquad x = -2z.$$

So the intersection of the surface with the xz plane consists of two lines through the origin.

The graph of $z^2 = x^2/4 + y^2/9$ is a double cone with elliptical cross sections. ●

The table below summarizes this section.

Equation	Graph in Space
$x^2 + y^2 + z^2 = r^2$ (r a positive constant)	Sphere of radius r and center at the origin
$z = k$ (k constant)	A plane parallel to the xy plane
$x = y$	The plane through the line $x = y$ in the xy plane and perpendicular to that plane (it is a cylinder)
$z = Ax + By + C$ (A, B, C constant)	A plane
$y = x^2$	The cylinder determined by the parabola $y = x^2$ in the xy plane

Usually, the graph of an equation in space is a surface. This is an analog of the fact that the graph of an equation in the plane is usually a curve.

Exercises

In Exercises 1 to 12 sketch the graph in space of the given equation.

1. $y = 1$ **2.** $x = 3$ **3.** $y = x$ **4.** $z = y^2$
5. $y = x + 3$ **6.** $z = x + y$
7. $y = x^3$ **8.** $x^2 + y^2 + z^2 = 36$

9. $x + y = 1$
10. $x^2 + y^2 = 1$
11. $z = x + 2y + 1$
12. $z = x$

■

13. Sketch the intersection of the planes $y = x$ and $z = 2y$.
14. Sketch the surface $x^2 - y^2 + z^2 = 1$.
15. Sketch the surface $x^2 - y^2 - z^2 = 1$.
16. Sketch the intersection of the cylinder $x^2 + y^2 = 4$ and the cylinder (plane) $z = x$.
17. Sketch the intersection of the cylinders $x^2 + y^2 = 1$ and $x^2 + z^2 = 1$.

■ ■

18. (a) Sketch neatly the part of the surface $x^2 + y^2 = 1$ that lies above the xy plane and below the plane $z = x$.
    (b) Find its area.
19. (a) Sketch the triangle in the plane $y = x$ that lies above the xy plane, below the plane $z = y$, and to the left of the plane $y = 3$.
    (b) Find the coordinates of its vertices.
    (c) Find its area.

12.3

Functions and their graphs

Consider a region R in the plane, occupied by a very thin piece of metal, which may be idealized to have zero thickness. Imagine that a match is put under it. Then at each point P in R the metal has a temperature which may be denoted $f(P)$. This is an example of a function whose domain is a set in the plane. The function f assigns to each point P in R a number $f(P)$.

Perhaps the metal is not homogeneous. Its density, in grams per square centimeter, may vary from point to point. The function that assigns to each point P in R the density at P is another example of a function whose domain is a set in the plane.

The next example shows how to graph a function whose domain is a set in the plane.

EXAMPLE 1 Let f be the function that assigns to the point $P = (x, y)$ in the xy plane the number $f(P) = x + 2y$. For instance, $f(1, 2) = 5$. Graph this function.

SOLUTION To record the information that $f(1, 2) = 5$, draw the point $Q = (1, 2, 5)$ in space. To do this, first plot the point $P = (1, 2)$ in the xy plane, and then go up from it a distance 5. In brief, the z coordinate of Q is used to record the value of $f(P)$.

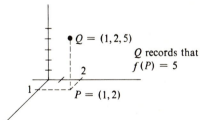

Q records that $f(P) = 5$

More generally, the information that $f(x, y) = x + 2y$ is recorded by plotting the point $Q = (x, y, x + 2y)$. The z coordinate of the point Q is equal to $x + 2y$. Thus Q lies on the plane

$$z = x + 2y.$$

As P wanders through the xy plane, the point Q sweeps out the surface $z = x + 2y$. In this case the graph of f is a plane, sketched in Example 3 of the preceding section. ●

Example 1 illustrates the following definition, which is similar to the definition of the graph of a function $y = f(x)$.

DEFINITION *Graph.* The *graph* of a function f whose domain is a set R in the xy plane consists of all points in space of the form $(x, y, f(x, y))$, for (x, y) in R. In other words, the graph of f consists of all points (x, y, z) that satisfy the equation

$$z = f(x, y)$$

and for which (x, y) is in R. Such a function is frequently called a *function of two variables.* Its graph is usually a surface.

It will be assumed that the function f is *continuous*, in the following sense. If (a, b) is in the domain of f and the point (x, y) in the domain of f approaches the point (a, b), then the number $f(x, y)$ approaches the number $f(a, b)$. Intuitively, nearby inputs produce nearby outputs.

The next two examples discuss the graphs of two functions that will be used later.

EXAMPLE 2 Let O be a fixed point in the plane and P an arbitrary point in the plane. Define $f(P)$ to be the square of the distance from P to O. Graph f.

SOLUTION Introduce a rectangular coordinate system in the plane, with origin at O; then f has the formula

$$f(x, y) = x^2 + y^2.$$

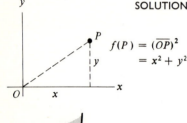
$f(P) = (\overline{OP})^2$
$= x^2 + y^2$

The graph of f is the surface $z = x^2 + y^2$.

Since $x^2 + y^2 \geq 0$, this surface lies above the xy plane, meeting the xy plane only at $(0, 0, 0)$. The further $P = (x, y)$ is from $(0, 0)$, the higher is the point $Q = (x, y, x^2 + y^2)$ on the surface.

In order to help visualize and sketch the graph of

$$z = x^2 + y^2,$$

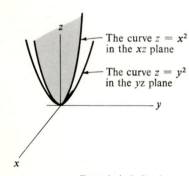

The curve $z = x^2$ in the xz plane

The curve $z = y^2$ in the yz plane

consider its intersections with the coordinate planes and planes parallel to these planes.

As already mentioned, the plane $z = 0$ meets the surface $z = x^2 + y^2$ only in one point, $(0, 0, 0)$.

The yz plane, that is, the plane $x = 0$, meets the surface $z = x^2 + y^2$ in the parabola

$$z = y^2 \qquad \text{in the } yz \text{ plane.}$$

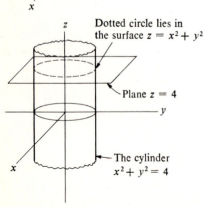

Dotted circle lies in the surface $z = x^2 + y^2$

Plane $z = 4$

The cylinder $x^2 + y^2 = 4$

Similarly, the plane $y = 0$ meets the surface in the parabola

$$z = x^2 \qquad \text{in the } xz \text{ plane.}$$

The two parabolas do not provide quite enough information to graph f. Consider next the intersection of a plane parallel to the xy plane with the surface $z = x^2 + y^2$. For instance, consider the plane $z = 4$. If a point (x, y, z) is on the surface $z = x^2 + y^2$ and the plane $z = 4$, it satisfies the equation

$$4 = x^2 + y^2;$$

hence it lies on the cylinder determined by the circle $4 = x^2 + y^2$ in the xy

plane.   Thus the intersection of

$$\text{the plane } z = 4$$

and $$\text{the surface } z = x^2 + y^2$$

is the same as the intersection of

$$\text{the plane } z = 4$$

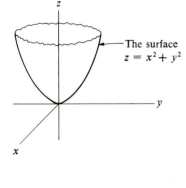

The surface
$z = x^2 + y^2$

and $$\text{the circular cylinder } x^2 + y^2 = 4.$$

This intersection is dotted in the third figure on page 480.   More generally, for any positive constant k, the intersection of the plane $z = k$ with the surface $z = x^2 + y^2$ is a circle of radius $\sqrt{k}$.   As k increases, so does the radius of this circle.   The surface $z = x^2 + y^2$ is built up of circles whose centers are on the z axis.   The easiest way to think of it is as the surface obtained by revolving the parabola $z = y^2$ around the z axis.   It has the shape of a headlight.   ●

Though the function in the next example differs from the function graphed in Example 2 only in the replacement of x^2 by $-x^2$, the graphs of the two functions are quite different.

EXAMPLE 3   Let f be the function defined by the formula

$$f(x, y) = y^2 - x^2.$$

Graph f, that is, the surface $z = y^2 - x^2.$

SOLUTION   The intersection of the surface with the xz plane is obtained by setting $y = 0$.   This intersection is the parabola

$$z = -x^2,$$

The Graph Described in this Example Will Be Referred to Later.   It Is Called a Saddle.

which lies below the xy plane.
The intersection of the surface with the yz plane is obtained by setting $x = 0$. This intersection is the parabola

$$z = y^2,$$

which lies above the xy plane.
Setting $z = 0$ determines the intersection of the surface with the xy plane. This intersection has the equation

$$0 = y^2 - x^2.$$

Since the right-hand side of this equation factors,

$$0 = (y - x)(y + x),$$

its graph in the xy plane consists of the two lines

$$0 = y - x \qquad \text{and} \qquad 0 = y + x,$$

that is, the lines $y = x$ and $y = -x.$

The surface is shaped like a saddle.   ●

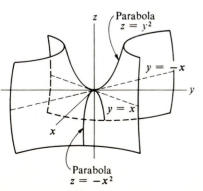

z Parabola
$z = y^2$

$y = -x$

y

$y = x$

x

Parabola
$z = -x^2$

As the examples suggest, when graphing $z = f(x, y)$, keep these facts in mind:

If $f(x, y)$ involves only x or only y, the graph is a cylinder.
If $f(x, y)$ is of the form $ax + by + c$, the graph is a plane.

To graph $z = f(x, y)$, determine the intersection of the graph with the co-ordinate planes and planes parallel to these planes.

Exercises

In Exercises 1 to 6 graph the function.
1. $f(x, y) = x + y$
2. $f(x, y) = x$
3. $f(x, y) = 3$ (a constant function)
4. $f(x, y) = x^2$
5. $f(x, y) = x^2 + 2y^2$
6. $f(x, y) = \dfrac{1}{x^2 + y^2}$

7. Consider the surface $z = xy$. [This is the graph of the function f given by $f(x, y) = xy$.]
   (a) Sketch the intersection of the surface and the xy plane.
   (b) Sketch the intersection of the surface and the xz plane.
   (c) Sketch the intersection of the surface and the yz plane.
   (d) Sketch the intersection of the surface and the plane $z = 1$.
   (e) Sketch the intersection of the surface and the plane $x = 2$.
   (f) Sketch the surface.
8. Sketch the intersection of the surface $z = y^2 - x^2$ (Example 3) with the planes (a) $z = 1$ and (b) $z = -1$.
9. Make a clay or soap model of the surface in Example 3.
10. Graph $z = \sqrt{x^2 + y^2}$.

11. (a) Show that the graph of $z = \sqrt{1 - x^2 - y^2}$ is part of the surface of a sphere.
    (b) Draw the graph of the function f defined by $f(x, y) = \sqrt{1 - x^2 - y^2}$.
    (c) For which points (x, y) is $f(x, y)$ defined?

    ■

12. The surface $z = 2x + y$ passes through the point $(1, 2, 4)$. Sketch the intersection of the surface with the plane (a) $x = 1$, (b) $y = 2$, (c) $y = 2x$. Note that all three planes pass through the point $(1, 2, 4)$.
13. The surface $z = xy$ passes through the point $(1, 1, 1)$. Sketch the intersection of the surface with the plane (a) $x = 1$, (b) $y = 1$, (c) $y = x$.
14. Sketch the intersection of the surface $z = xy$ and the plane $y = x$.
15. (a) Show that, if $f(x, y)$ can be written in the form $g(\sqrt{x^2 + y^2})$, then the graph of f is a surface of revolution about the z axis.
    (b) Graph $z = e^{-(x^2 + y^2)}$. (This function will appear in Chap. 13.)

12.4

Partial derivatives

Let f be a function of x and y. The graph of $z = f(x, y)$ is a surface. Consider a point (a, b) in the xy plane. The graph of $z = f(x, y)$ for (x, y) near (a, b) may look like the surface shown below. Let P be the point on the

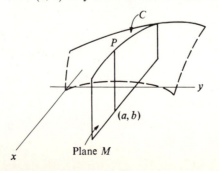

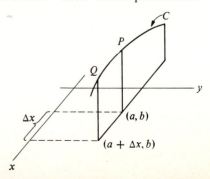

surface directly above (a, b). The plane M through (a, b) perpendicular to the y axis meets the surface in a curve, labeled C.

To find the slope of this curve in the plane M at the point P, consider, as in Sec. 4.1, a nearby point Q on the curve C directly above $(a + \Delta x, b)$. The limit of the slope of the line through P and Q, as Q approaches P, is

$$\lim_{\Delta x \to 0} \frac{f(a + \Delta x, b) - f(a, b)}{\Delta x}$$

This is the definition of the slope of the curve C at P, and suggests the following definition.

DEFINITION *Partial derivatives.* If the domain of f includes the region within some circle around the point (a, b), and if

$$\lim_{\Delta x \to 0} \frac{f(a + \Delta x, b) - f(a, b)}{\Delta x}$$

exists, this limit is called the *partial derivative of f with respect to x.* Similarly, if

$$\lim_{\Delta y \to 0} \frac{f(a, b + \Delta y) - f(a, b)}{\Delta y}$$

exists, it is called the *partial derivative of f with respect to y.*

The following notations are used for the partial derivative of f with respect to x:

$$f_x, \qquad \frac{\partial f}{\partial x}, \qquad z_x, \qquad \frac{\partial z}{\partial x}, \qquad f_1, \qquad \text{or} \qquad D_1 f,$$

and the following for the partial derivative of f with respect to y:

$$f_y, \qquad \frac{\partial f}{\partial y}, \qquad z_y, \qquad \frac{\partial z}{\partial y}, \qquad f_2, \qquad \text{or} \qquad D_2 f.$$

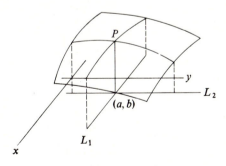

The curve above line L_1 in the figure has slope $f_x(a, b)$ at the point P. The curve above line L_2, which is parallel to the y axis, has slope $f_y(a, b)$ at P.

Note that on L_1 the y coordinate is constant while x is free to vary; on L_2 the x coordinate is constant while y is free to vary. (In earlier chapters, y was almost always the value of a function. Now it is part of the input.)

EXAMPLE I Find the partial derivative f_x at (a, b) if $f(x, y) = x^2 y$.

SOLUTION From the definition,

$$f_x(a, b) = \lim_{\Delta x \to 0} \frac{f(a + \Delta x, b) - f(a, b)}{\Delta x}$$

$$= \lim_{\Delta x \to 0} \frac{(a + \Delta x)^2 b - a^2 b}{\Delta x}$$

$$= \lim_{\Delta x \to 0} b \left[\frac{(a + \Delta x)^2 - a^2}{\Delta x} \right]$$

$$= \lim_{\Delta x \to 0} b \left[\frac{a^2 + 2a \, \Delta x + (\Delta x)^2 - a^2}{\Delta x} \right]$$

$$= \lim_{\Delta x \to 0} b(2a + \Delta x)$$

$$= 2ab. \ \bullet$$

As the computations in Example 1 suggest, to compute a partial derivative f_x, just treat y in the expression for f as a constant and differentiate with respect to x.

EXAMPLE 2 Compute f_x and f_y at (x, y), where f is given by the formula $x^2 y^3 + e^{x^2 y}$.

SOLUTION Treating y as a constant, we differentiate with respect to x:

$$f_x(x, y) = 2xy^3 + 2xye^{x^2 y}.$$

Similarly, treating x as a constant and differentiating with respect to y, we obtain

$$f_y(x, y) = 3x^2 y^2 + x^2 e^{x^2 y}. \ \bullet$$

EXAMPLE 3 Find the two partial derivatives of f at $(1, 3)$ if f is given by the formula

$$f(x, y) = 2x + 4y + x^2 + \ln(x^2 + y^2).$$

SOLUTION To find $\partial f/\partial x$, treat y as a constant and compute the partial derivative,

$$\frac{\partial f}{\partial x} = 2 + 0 + 2x + \frac{2x}{x^2 + y^2}.$$

At $(x, y) = (1, 3)$, this yields

$$\frac{\partial f}{\partial x}(1, 3) = 2 + 2 \cdot 1 + \frac{2 \cdot 1}{1^2 + 3^2} = 4.2.$$

To find $\partial f/\partial y$, differentiate with respect to y, treating x as a constant:

$$\frac{\partial f}{\partial y} = 0 + 4 + 0 + \frac{2y}{x^2 + y^2}.$$

At $(x, y) = (1, 3)$, this yields

$$\frac{\partial f}{\partial y}(1, 3) = 4 + \frac{2 \cdot 3}{1^2 + 3^2} = 4.6. \ \bullet$$

Just as there are derivatives of derivatives, so are there partial derivatives of partial derivatives. For instance, if

$$z = 2x + 5x^4 y^7,$$

then $$z_x = 2 + 20x^3y^7 \quad \text{and} \quad z_y = 35x^4y^6.$$

We may go on and compute partial derivatives of z_x and z_y:

$$(z_x)_x = 60x^2y^7 \qquad (z_x)_y = 140x^3y^6$$
$$(z_y)_x = 140x^3y^6 \qquad (z_y)_y = 210x^4y^5.$$

It is customary to denote $(z_x)_x$ simply by z_{xx}, $(z_x)_y$ by z_{xy}, and so on, omitting the parentheses. There are four possible partial derivatives of the second order,

$$z_{xx}, \qquad z_{xy}, \qquad z_{yx}, \qquad z_{yy}.$$

They are also denoted

$$f_{xx}, \qquad f_{xy}, \qquad f_{yx}, \qquad f_{yy}.$$

In the example just given, observe that z_{xy} equals z_{yx}. For most functions met in practice, the two "mixed partials" z_{xy} and z_{yx} are equal. In view of the importance of this remark we state it as a theorem. The proof, to be found in Sec. 23.1, may be read at this point.

THEOREM   If $z = f(x, y)$ has continuous partial derivatives z_x, z_y, z_{xy}, and z_{yx}, then

$$z_{xy} = z_{yx}.$$

The ∂ notation is also used for naming the four second-order partial derivatives. Thus

$$\frac{\partial(\partial f/\partial x)}{\partial y},$$

already denoted z_{xy}, is also denoted

$$\frac{\partial^2 f}{\partial y\,\partial x}.$$

(In the ∂ notation, differentiate from right to left in the order of the subscripts.) Thus

$$z_{xx} = \frac{\partial^2 f}{\partial x^2}, \qquad z_{xy} = \frac{\partial^2 f}{\partial y\,\partial x}, \qquad z_{yx} = \frac{\partial^2 f}{\partial x\,\partial y}, \qquad z_{yy} = \frac{\partial^2 f}{\partial y^2}.$$

EXAMPLE 4   Compute $\partial^2 z/\partial x^2$ and $\partial^2 z/\partial y\,\partial x$ for $z = y\cos xy$.

SOLUTION   First compute

$$\frac{\partial z}{\partial x} = -y^2\sin xy.$$

Then

$$\frac{\partial^2 z}{\partial x^2} = \frac{\partial(\partial z/\partial x)}{\partial x}$$

$$= \frac{\partial(-y^2\sin xy)}{\partial x}$$

$$= -y^3\cos xy$$

$$\text{and} \qquad \frac{\partial^2 z}{\partial y\,\partial x} = \frac{\partial(\partial z/\partial x)}{\partial y}$$

$$= \frac{\partial(-y^2 \sin xy)}{\partial y}$$

$$= -2y \sin xy - xy^2 \cos xy. \quad \bullet$$

Exercises

In Exercises 1 to 10 compute f_x and f_y for the given function.

1. $f(x, y) = x^3 y^4$ **2.** $f(x, y) = x^3 + y^3$

3. $f(x, y) = x/y$ **4.** $f(x, y) = x^2 \cos y$

5. $f(x, y) = \sin(xy^3)$ **6.** $f(x, y) = (1 + x^2)y$

7. $f(x, y) = x + 3y + \sqrt{x} + \tan^{-1} xy$

8. $f(x, y) = \sqrt{x^2 + y^2}$ **9.** $f(x, y) = e^{x/y}$

10. $f(x, y) = \sin^2(xy)$.

11. Compute (a) $D_1(x^2 y^4)$ and (b) $D_2(x^2 y^4)$.

12. Compute (a) $D_1[xy \cos(x + y)]$ and (b) $D_2[xy \cos(x + y)]$.

13. If $z = x/(x^2 + y^2)$, compute (a) z_x and (b) z_y.

14. Compute $D_1(\cos x \sin y)$.

In Exercises 15 to 19 compute f_{xx}, f_{yy}, f_{xy}, and f_{yx} and check that $f_{xy} = f_{yx}$.

15. $f(x, y) = 5x^2 - 3xy + 6y^2$ **16.** $f(x, y) = e^{x^2 y}$

17. $f(x, y) = x^4 y^7$ **18.** $f(x, y) = \ln(x^3 + y^2)$

19. $f(x, y) = 1/\sqrt{x^2 + y^2}$

20. A function $z = f(x, t)$, where z denotes temperature, x position, and t time, is said to satisfy the heat equation if

$$a^2 f_{xx} = f_t,$$

where a is a constant. Show that $f(x, t) = e^{-\pi^2 a^2 t} \sin \pi x$ satisfies the heat equation.

21. Let $T = f(x, y, z)$ be the temperature at the point (x, y, z) within a solid whose surface has a fixed temperature distribution. It can be shown that, if T does not vary with time, then $T_{xx} + T_{yy} + T_{zz} = 0$. Similarly, if $P(x, y, z)$ is the work done in moving a particle from a fixed base point in a gravitational field to the point (x, y, z), then $P_{xx} + P_{yy} + P_{zz} = 0$. The equation $f_{xx} + f_{yy} + f_{zz} = 0$ is called Laplace's equation (in three dimensions). Verify

that the functions $1/\sqrt{x^2 + y^2 + z^2}$, $x^2 - y^2 - z$, and $e^x \cos y + z$ satisfy Laplace's equation.

22. Compute (a) $D_1(e^{x\sqrt{y}})$ and (b) $D_2(e^{x\sqrt{y}})$.

23. Compute $D_1 f$ and $D_2 f$ if $f(x, y) = \int_x^y e^{t^2}\, dt$.

In Exercises 24 to 27 find the slope of the curve formed by the intersection of the given surface with a plane perpendicular to the y axis and passing through the given point.

24. $z = xy^2$ at $(1, 2)$.   **25.** $z = x/y$ at $(1, 1)$.

26. $z = \cos(x + 2y)$ at $(\pi/4, \pi/2)$.   **27.** $z = x^2 e^{xy}$ at $(1, 0)$.

∎

In Exercises 28 to 30 find the slope of the curve formed by the intersection of the given surface with a plane perpendicular to the x axis and passing through the given point.

28. $z = ye^{xy}$ at $(1, 1)$.   **29.** $z = e^{x/y}$ at $(0, 1)$.

30. $z = x^2 + \sqrt{x + 3y}$ at $(1, 5)$.

31. Compute (a) $\partial(x^y)/\partial x$ and (b) $\partial(x^y)/\partial y$.

∎ ∎

32. Is there a function f such that

$$f_x = e^x \cos y \quad \text{and} \quad f_y = e^x \sin y?$$

33. Let $f(x, y) = \int_0^1 \cos(x + 2y + t)\, dt$. Find f_x and f_y.

34. Let $f(x, y) = \int_0^y \sqrt{x + t}\, dt$. Find f_x and f_y.

35. Let $f(x, y) = \int_x^y g(t)\, dt$, where g is continuous. Find f_x and f_y.

12.5

The change Δf and the differential df

In the case of a function of one variable it was shown in Sec. 6.9 that the differential is a good approximation to the change in the function: The change

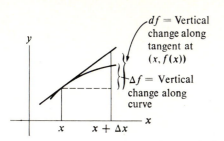

df = Vertical change along tangent at $(x, f(x))$

Δf = Vertical change along curve

$$\Delta f = f(x + \Delta x) - f(x)$$

is approximated by the differential

$$df = f'(x)\,\Delta x.$$

This section considers the analogous problem for a function of two variables,

$$z = f(x, y).$$

If Δx and Δy are small, what is a reasonable estimate of the difference

$$f(x + \Delta x,\ y + \Delta y) - f(x,\ y)?$$

It is to be expected that the estimate will involve the partial derivatives f_x and f_y evaluated at (x, y).

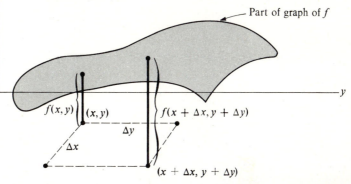

— Part of graph of f

$f(x, y)$ (x, y) $f(x + \Delta x, y + \Delta y)$

Δy

Δx

$(x + \Delta x,\ y + \Delta y)$

Consider a function f of two variables. Assume that f assigns to every point (x, y) in the plane (or in some plane region) a number, which is denoted $f(x, y)$ or z. If we go from point (x, y) to the point $(x + \Delta x,\ y + \Delta y)$, the value of the function changes by an amount which will be denoted Δf or Δz:

$$\Delta f = \Delta z = f(x + \Delta x,\ y + \Delta y) - f(x,\ y).$$

In the diagram shown below, Δf (if positive) is the length of the segment ST.

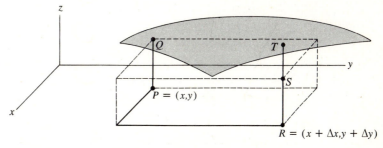

Q T

S

$P = (x, y)$

$R = (x + \Delta x, y + \Delta y)$

The diagram shows two points Q and T on the graph of f. For convenience, assume $f(x, y)$ and $f(x + \Delta x,\ y + \Delta y)$ are positive. Observe that

$$\overline{PQ} = f(x,\ y)$$

and

$$\overline{RT} = f(x + \Delta x,\ y + \Delta y).$$

Note that $\overline{RS} = \overline{PQ}$.   Thus

$$\Delta f = f(x + \Delta x, y + \Delta y) - f(x, y)$$
$$= \overline{RT} - \overline{PQ}$$
$$= \overline{RT} - \overline{RS}$$
$$= \overline{ST}.$$

The diagram suggests a way to estimate Δf.   Consider the path in the xy plane from $P = (x, y)$ to $R = (x + \Delta x, y + \Delta y)$ that passes through the point $C = (x + \Delta x, y)$ and consists of two line segments.   From P to C only the x coordinate changes; from C to R only the y coordinate changes.   The change in f on each of the two parts of the path can be estimated by a differential as discussed in Sec. 6.9, since from P to C only one variable, namely x, changes, and from C to R only one variable, namely y, changes.   The sum of these two estimates provides an estimate of the total change Δf.   (See the following diagram.)

Consider only that part of the graph of f which lies above the segments PC and CR, namely, the two curves QU and UT in the following diagram, in which the points Q, V, and S lie on one horizontal plane, and the line UW is horizontal.

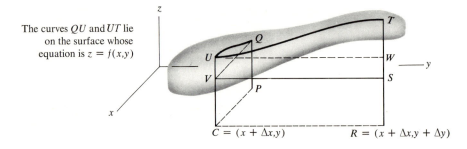

The curves QU and UT lie on the surface whose equation is $z = f(x,y)$

Inspection of the diagram, as before, shows that

$$\Delta f = \overline{ST}$$
$$= \overline{SW} + \overline{WT}$$
$$= \overline{VU} + \overline{WT}.$$

Now, by the law of the mean,

$$\overline{VU} = f(x + \Delta x, y) - f(x, y) = f_x(X, y)\, \Delta x$$

for some X between x and $x + \Delta x$ (since y is fixed and only x changes).   Also,

$$\overline{WT} = f(x + \Delta x, y + \Delta y) - f(x + \Delta x, y) = f_y(x + \Delta x, Y)\, \Delta y$$

for some Y between y and $y + \Delta y$ (since y changes but the x coordinate has the fixed value $x + \Delta x$).   Therefore

This Equation Is the Key to the Next Theorem.   Study It Carefully.

$$\Delta f = f_x(X, y)\, \Delta x + f_y(x + \Delta x, Y)\, \Delta y. \tag{1}$$

When Δx and Δy approach 0, both (X, y) and $(x + \Delta x, Y)$ approach (x, y). By the continuity of f_x and f_y,

$$f_x(X, y) = f_x(x, y) + \varepsilon_1 \quad \text{and} \quad f_y(x + \Delta x, Y) = f_y(x, y) + \varepsilon_2$$

where $\varepsilon_1 \to 0$ and $\varepsilon_2 \to 0$ as Δx and Δy approach 0. From (1) we obtain the following important result, which will be needed several times.

THEOREM   If f has continuous partial derivatives f_x and f_y, then Δf, defined as $f(x + \Delta x, y + \Delta y) - f(x, y)$, is of the form

$$\Delta f = f_x(x, y) \, \Delta x + f_y(x, y) \, \Delta y + \varepsilon_1 \, \Delta x + \varepsilon_2 \, \Delta y,$$

where $\varepsilon_1 \to 0$ and $\varepsilon_2 \to 0$ as Δx and Δy approach 0. ●

When Δx and Δy are small, the quantities $\varepsilon_1 \, \Delta x$ and $\varepsilon_2 \, \Delta y$, being the products of small quantities, are usually negligible when compared with $f_x(x, y) \, \Delta x$ and $f_y(x, y) \, \Delta y$ [if $f_x(x, y)$ and $f_y(x, y)$ are not 0]. For this reason, $f_x(x, y) \, \Delta x + f_y(x, y) \, \Delta y$ is often a good estimate of Δf when Δx and Δy are small. The similarity with the case of a function of one variable suggests the following definition.

DEFINITION   *Differential.* If f is a function of two variables, and if x, y, Δx, and Δy are numbers, the expression

$$f_x(x, y) \, \Delta x + f_y(x, y) \, \Delta y$$

is called the differential of f at x, y, Δx, and Δy. It is denoted df (or dz).

Observe that the differential df is a function of the four variables x, y, Δx, and Δy. However, the differential is usually applied only when Δx and Δy are small.

EXAMPLE 1   Let f be given by the formula $f(x, y) = x^2 y^3$. Compute Δf and df when $x = 2$, $y = 1$, $\Delta x = 0.1$, and $\Delta y = 0.05$.

SOLUTION   In this case $x + \Delta x = 2.1$ and $y + \Delta y = 1.05$. Thus

$$\Delta f = f(2.1, 1.05) - f(2, 1)$$
$$= (2.1)^2(1.05)^3 - 2^2 \cdot 1^3$$
$$= 5.10512625 - 4$$
$$= 1.10512625.$$

To compute $\qquad df = f_x(2, 1) \, \Delta x + f_y(2, 1) \, \Delta y,$

begin by finding the partial derivatives,

$$f_x(x, y) = 2xy^3 \quad \text{and} \quad f_y(x, y) = 3x^2 y^2.$$

Then
$$df = 2(2)(1)^3(0.1) + 3(2)^2(1)^2(0.05)$$
$$= 0.4 + 0.6$$
$$= 1.$$

In this case $\Delta f = 1.10512625$ and is tedious to compute, while $df = 1$, a good estimate of Δf, is easy to compute. ●

The next example is similar to Example 1 but has a practical context.

EXAMPLE 2   The dimensions of a cylindrical tin can change from radius 3 inches and height 4 inches to radius 2.9 inches and height 4.2 inches. Estimate the change in the volume.

SOLUTION   In this case the volume is a function of the two variables r and h, $V(r, h) = \pi r^2 h$. We wish to estimate
$$\Delta V = V(2.9, 4.2) - V(3, 4),$$

which can be thought of as
$$\Delta V = V(3 + (-0.1), 4 + (0.2)) - V(3, 4).$$

Rather than compute ΔV, use $dV = V_r(3, 4) \, \Delta r + V_h(3, 4) \, \Delta h$ as an estimate of ΔV.
    First of all, $\Delta r = -0.1$, $\Delta h = 0.2$, and $V_r = 2\pi r h$, $V_h = \pi r^2$. Thus
$$V_r(3, 4) = 24\pi \quad \text{and} \quad V_h(3, 4) = 9\pi.$$

Hence ΔV is approximately
$$(24\pi)(-0.1) + (9\pi)(0.2) = -0.6\pi.$$

A direct computation shows that $\Delta V = -0.678\pi$ (the minus sign indicates a decrease in volume). ●

EXAMPLE 3   A box of height y has a square base whose sides have length x. If we measure x with a possible error of 2 percent and y with a possible error of 3 percent, what is the maximum possible error that these errors may induce in an estimate of the volume $x^2 y$?

SOLUTION   That x may be off by 2 percent and y by 3 percent is recorded in these inequalities:
$$\frac{|\Delta x|}{x} \leq 0.02 \quad \text{and} \quad \frac{|\Delta y|}{y} \leq 0.03.$$

The volume $x^2 y$ is a function of x and y:
$$V = f(x, y) = x^2 y.$$

How large may

$$\frac{|\Delta f|}{f}$$

be?

Rather than use Δf, consider the more convenient quantity

$$df = f_x \, \Delta x + f_y \, \Delta y$$
$$= 2xy \, \Delta x + x^2 \, \Delta y.$$

Then
$$\frac{df}{f} = \frac{2xy \, \Delta x + x^2 \, \Delta y}{x^2 y}$$

$$= \frac{2xy \, \Delta x}{x^2 y} + \frac{x^2 \, \Delta y}{x^2 y}$$

$$= 2\frac{\Delta x}{x} + \frac{\Delta y}{y}.$$

Note that Exponents in the Function $x^2 y$ Show Up as Coefficients Here.

Consequently, $\quad \dfrac{|df|}{f} \le 2\dfrac{|\Delta x|}{x} + \dfrac{|\Delta y|}{y} \le 2(0.02) + 0.03 = 0.07.$

Since $|df|/f$ is a good estimate of $|\Delta f|/f$, the error in estimating the volume is presumably at most 7 percent. Of course, if we underestimate x and overestimate y, the error in estimating the volume may be much less, perhaps even zero. ●

In the case of a function of one variable the differential represents the change of y along a tangent line (see Sec. 6.9). It will now be shown that in the case of a function of two variables the differential represents the change of z along a certain plane. First this plane will be described.

Denote the curve lying in the graph of f and above the segment from (x, y) to $(x + \Delta x, y)$ by C_x (this is the curve QU in the figure on page 488). Similarly, denote the curve above the segment from (x, y) to $(x, y + \Delta y)$ by C_y. Let T_x be the tangent line to C_x at the point above (x, y); similarly, let T_y be the tangent line to C_y at the same point.

Preview of the Tangent Plane.

Then $f_x(x, y) \, \Delta x$ is the change in z as we move along T_x from the point above (x, y) to the point above $(x + \Delta x, y)$; similarly, $f_y(x, y) \, \Delta y$ is the change in z as we move from the point above (x, y) to the point above $(x, y + \Delta y)$. Lines T_x and T_y determine a plane (the *tangent plane* which will be discussed in Sec. 20.8).

The first diagram located below shows C_x, T_x, C_y, T_y and the tangent plane; the second diagram shows that $\overline{AB}$ is the change in z as we move along the tangent plane from the point above (x, y) to the point above $(x + \Delta x, y + \Delta y)$.

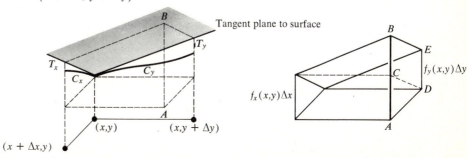

Note that $\overline{CB} = \overline{DE}$.   Then

$$\overline{AB} = \overline{AC} + \overline{CB}$$
$$= f_x(x, y)\,\Delta x + \overline{DE}$$
$$= f_x(x, y)\,\Delta x + f_y(x, y)\,\Delta y$$
$$= df.$$

This shows that the differential df equals $\overline{AB}$, the change along the tangent plane as we go from (x, y) to $(x + \Delta x, y + \Delta y)$.

Exercises

In Exercises 1 to 3 compute Δf and df for the indicated f, x, y, Δx, and Δy.

1. $f(x, y) = xy$, $x = 2$, $y = 3$, $\Delta x = 0.1$, $\Delta y = 0.2$.
2. $f(x, y) = 2x + 3y$, $x = 3$, $y = 4$, $\Delta x = 0.1$, $\Delta y = -0.3$.
3. $f(x, y) = x/y$, $x = 2$, $y = 1$, $\Delta x = 0.1$, $\Delta y = -0.2$.

In Exercises 4 to 6 compute df for the indicated f, x, y, Δx, and Δy.

4. $f(x, y) = x^3 y^4$, $x = 1$, $y = 2$, $\Delta x = 0.1$, $\Delta y = -0.1$.
5. $f(x, y) = \sqrt{x^2 + y^2}$, $x = 3$, $y = 4$, $\Delta x = 0.02$, $\Delta y = 0.03$.
6. $f(x, y) = \ln(x + 2y)$, $x = 2$, $y = 3$, $\Delta x = -0.1$, $\Delta y = 0.2$.

In Exercises 7 to 10, x is measured with a possible error of 3 percent and y with a possible error of 4 percent. Discuss the maximum possible percentage error in measuring the indicated quantity.

7. $x^3 y^2$ **8.** $x^5 y$
9. x^3/y^2 **10.** $x^m y^n$ (where m and n are constants)

In Exercises 11 to 14 use an appropriate differential to estimate the indicated difference.

11. $(3.1)^3(2.8)^2 - 3^3 3^2$
12. $\sqrt{(3.1)^2 + (4.2)^2} - \sqrt{3^2 + 4^2}$
13. $\ln[(1.1)^3 + (2.3)^3] - \ln 9$

14. $\dfrac{(1.1)^3}{(0.8)^2} - \dfrac{1^3}{1^2}$

■

15. Let $f(x, y) = Ax + By + C$, where A, B, and C are constants. Show that $df = \Delta f$ for all values of Δx and Δy.

16. (This exercise continues Example 2 and shows how dV compares with ΔV.) Fill in the following table [ΔV denotes $V(3 + \Delta r, 4 + \Delta h) - V(3, 4)$, while dV denotes $V_r(3, 4)\,\Delta r + V_h(3, 4)\,\Delta h$].

Δr	Δh	dV	ΔV	$\Delta V/dV$
-0.1	0.2	-0.6π	-0.678π	1.13
0.1	0.2			
0.01	0.003			
0.001	-0.001			

17. The theorem in this section was proved by using a path from (x, y) to $(x + \Delta x, y)$, and then to $(x + \Delta x, y + \Delta y)$. Prove the theorem by using the path that passes through $(x, y + \Delta y)$ instead of $(x + \Delta x, y)$.

■ ■

18. It is not necessary to draw any pictures to prove the theorem in this section. This is an outline of a purely algebraic approach:
(a) Prove that Δf equals $[f(x + \Delta x, y + \Delta y) - f(x + \Delta x, y)] + [f(x + \Delta x, y) - f(x, y)]$.
(b) Applying the law of the mean to each of the bracketed quantities in (a), prove the theorem.

19. Let u be a function of x, y, and z. (For instance, the volume of a rectangular box is the product of its three dimensions, xyz.)
(a) Obtain the formula for Δu analogous to that of the theorem for Δf. Use the technique of Exercise 18.
(b) From (a) show that $u_x\,\Delta x + u_y\,\Delta y + u_z\,\Delta z$ is a good approximation to Δu.
(c) Use the result in (a) to generalize the theorem to functions of three variables.

20. (a) Sketch the region in the plane $x + y = 1$ that lies above the xy plane, to the right of the xz plane, in front of the yz plane, and below the surface $z = x^2 + y^2$.
(b) Find the area of the region in (a).

12.6

The chain rules

The theorem in Sec. 12.5 is the basis for the chain rules for differentiating composite functions of more than one variable. Theorem 1 of this section concerns the case in which z is a function of x and y, and x and y are functions of one variable. Theorem 2 concerns the case in which z is a function of x and y, and x and y are, in turn, functions of two variables.

THEOREM I   (*Chain rule*). Let $z = f(x, y)$ have continuous partial derivatives f_x and f_y, and let $x = g(t)$ and $y = h(t)$ be differentiable functions of t. Then z is a composite function of t and

$$\frac{dz}{dt} = z_x \frac{dx}{dt} + z_y \frac{dy}{dt}.$$

PROOF   By definition,
$$\frac{dz}{dt} = \lim_{\Delta t \to 0} \frac{\Delta z}{\Delta t}.$$

Now, Δt induces changes Δx and Δy in x and y, respectively. According to the theorem in Sec. 12.5,

$$\Delta z = f_x(x, y)\, \Delta x + f_y(x, y)\, \Delta y + \varepsilon_1\, \Delta x + \varepsilon_2\, \Delta y,$$

where $\varepsilon_1 \to 0$ and $\varepsilon_2 \to 0$ as Δx and Δy approach 0. (Keep in mind that x and y are fixed.) Thus

$$\frac{\Delta z}{\Delta t} = f_x(x, y) \frac{\Delta x}{\Delta t} + f_y(x, y) \frac{\Delta y}{\Delta t} + \varepsilon_1 \frac{\Delta x}{\Delta t} + \varepsilon_2 \frac{\Delta y}{\Delta t},$$

and
$$\frac{dz}{dt} = \lim_{\Delta t \to 0} \frac{\Delta z}{\Delta t} = f_x(x, y) \frac{dx}{dt} + f_y(x, y) \frac{dy}{dt} + 0 \frac{dx}{dt} + 0 \frac{dy}{dt}.$$

The theorem is proved. ●

We illustrate Theorem 1 by examples.

EXAMPLE I   Let $z = x^2 y^3$, $x = 3t^2$, and $y = t/3$. Find dz/dt when $t = 1$.

SOLUTION   In order to apply Theorem 1, compute z_x, z_y, dx/dt, and dy/dt:

$$z_x = 2xy^3, \qquad z_y = 3x^2 y^2,$$

$$\frac{dx}{dt} = 6t, \qquad \frac{dy}{dt} = \frac{1}{3}.$$

By Theorem 1,
$$\frac{dz}{dt} = 2xy^3 \cdot 6t + 3x^2 y^2 \cdot \frac{1}{3}.$$

In particular, when $t = 1$, x is 3 and y is $\frac{1}{3}$. Therefore, when $t = 1$,

$$\frac{dz}{dt} = 2 \cdot 3\left(\frac{1}{3}\right)^3 6 + 3 \cdot 3^2 \left(\frac{1}{3}\right)^2 \frac{1}{3}$$

$$= \tfrac{36}{27} + \tfrac{27}{27}$$

$$= \tfrac{7}{3}.$$

The derivative dz/dt can also be found without using Theorem 1. To do this, express z explicitly in terms of t:

$$z = x^2 y^3$$

$$= (3t^2)^2 \left(\frac{t}{3}\right)^3$$

$$= \frac{t^7}{3}.$$

Then

$$\frac{dz}{dt} = \frac{7t^6}{3}.$$

When $t = 1$, this gives

$$\frac{dz}{dt} = \frac{7}{3},$$

in agreement with the first computation. ●

EXAMPLE 2 At time t, measured in minutes, a bug walking on the xy plane is at the point $(g(t), f(t))$, where distances are measured in feet. The temperature at (x, y) is e^{-x-2y} degrees. When the bug is at the point $(0, 0)$, it is moving east at 2 feet per minute $(dx/dt = 2)$ and north at 3 feet per minute $(dy/dt = 3)$. From the bug's point of view, how quickly is the temperature of the ground changing? That is, consider the temperature z to be a function of time, and find dz/dt.

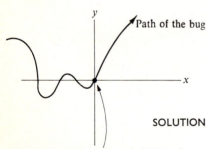

Path of the bug

As the bug passes through $(0,0)$ he is traveling east at 2 feet per minute and north at 3 feet per minute

SOLUTION In this case, $z = e^{-x-2y}$, and Theorem 1 says that

$$\frac{dz}{dt} = \frac{\partial z}{\partial x}\frac{dx}{dt} + \frac{\partial z}{\partial y}\frac{dy}{dt}.$$

Now,

$$\frac{\partial z}{\partial x} = -e^{-x-2y} \quad \text{and} \quad \frac{\partial z}{\partial y} = -2e^{-x-2y}.$$

At $(x, y) = (0, 0)$,

$$\frac{\partial z}{\partial x} = -1 \quad \text{and} \quad \frac{\partial z}{\partial y} = -2$$

$$\frac{dx}{dt} = 2 \quad \text{and} \quad \frac{dy}{dt} = 3.$$

Thus, as the bug passes through the origin it finds the temperature changing at the rate

$$\frac{dz}{dt} = (-1)(2) + (-2)(3)$$

$$= -8° \text{ per minute.}$$

The bug observes that the temperature is decreasing at the rate of $8°$ per minute.   Note that this rate depends not only on the temperature function e^{-x-2y}, but also upon the speed and direction of the bug.   ●

The next theorem is a generalization of Theorem 1 in that x and y are assumed to be functions of *two* variables t and u.

THEOREM 2   (*Chain rule*).   Let $z = f(x, y)$ have continuous partial derivatives f_x and f_y, and let x and y be differentiable functions of t and u.   Then z is indirectly a function of t and u, and

$$\frac{\partial z}{\partial t} = z_x \frac{\partial x}{\partial t} + z_y \frac{\partial y}{\partial t} \quad \text{and} \quad \frac{\partial z}{\partial u} = z_x \frac{\partial x}{\partial u} + z_y \frac{\partial y}{\partial u}.$$

PROOF   The proof, which is virtually the same as that for Theorem 1, is only sketched.   To examine

$$\frac{\partial z}{\partial t} = \lim_{\Delta t \to 0} \frac{\Delta z}{\Delta t},$$

hold u fixed and let t change by an amount Δt.   Then x changes by Δx, and y changes by Δy.   By the theorem in Sec. 12.5,

$$\Delta z = z_x \Delta x + z_y \Delta y + \varepsilon_1 \Delta x + \varepsilon_2 \Delta y, \tag{1}$$

where $\varepsilon_1 \to 0$ and $\varepsilon_2 \to 0$ as Δx and Δy approach 0.   Divide (1) by Δt and, recalling that

$$\lim_{\Delta t \to 0} \frac{\Delta x}{\Delta t} = \frac{\partial x}{\partial t} \quad \text{and} \quad \lim_{\Delta t \to 0} \frac{\Delta y}{\Delta t} = \frac{\partial y}{\partial t},$$

finish the proof in the same manner as for Theorem 1.   ●

EXAMPLE 3   Given that $z = e^{xy}$, $x = 3t + 2u$, and $y = 4t - 2u$, find

$$\frac{\partial z}{\partial t} \quad \text{and} \quad \frac{\partial z}{\partial u}.$$

SOLUTION   Begin by computing

$$z_x = ye^{xy}, \qquad z_y = xe^{xy},$$

$$\frac{\partial x}{\partial t} = 3, \qquad \frac{\partial y}{\partial t} = 4,$$

$$\frac{\partial x}{\partial u} = 2, \qquad \frac{\partial y}{\partial u} = -2.$$

By Theorem 2,
$$\frac{\partial z}{\partial t} = z_x \frac{\partial x}{\partial t} + z_y \frac{\partial y}{\partial t}$$

$$= ye^{xy} \cdot 3 + xe^{xy} \cdot 4$$

$$= (3y + 4x)e^{xy},$$

and
$$\frac{\partial z}{\partial u} = z_x \frac{\partial x}{\partial u} + z_y \frac{\partial y}{\partial u}$$

$$= ye^{xy} \cdot 2 + xe^{xy} \cdot (-2)$$

$$= (2y - 2x)e^{xy}.$$

It is also possible to compute $\partial z/\partial t$ and $\partial z/\partial u$ without using the chain rule. Write

$$z = e^{xy} = e^{(3t + 2u)(4t - 2u)}$$

and compute $\partial z/\partial t$ and $\partial z/\partial u$ directly. It will be instructive to do this and compare the solution with the one based on Theorem 2. ●

The chain rules stated in Theorems 1 and 2 are used in proving theorems, obtaining properties of a broad class of functions, and computing partial derivatives of particular functions. The next three examples and the proof of Theorem 3 will illustrate the first two uses of the chain rules.

EXAMPLE 4   Let $z = f(x, y)$, where $x = t - u$ and $y = -t + u$.   Show that

$$\frac{\partial z}{\partial t} + \frac{\partial z}{\partial u} = 0.$$

SOLUTION   First of all, direct computation shows that

$$\frac{\partial x}{\partial t} = 1, \qquad \frac{\partial x}{\partial u} = -1, \qquad \frac{\partial y}{\partial t} = -1, \qquad \frac{\partial y}{\partial u} = 1.$$

By Theorem 2,
$$\frac{\partial z}{\partial t} = f_x \frac{\partial x}{\partial t} + f_y \frac{\partial y}{\partial t}$$

$$= f_x \cdot 1 + f_y \cdot (-1)$$

$$= f_x - f_y.$$

Similarly,
$$\frac{\partial z}{\partial u} = f_x \frac{\partial x}{\partial u} + f_y \frac{\partial y}{\partial u}$$

$$= f_x \cdot (-1) + f_y \cdot (1)$$

$$= -f_x + f_y.$$

Thus
$$\frac{\partial z}{\partial t} + \frac{\partial z}{\partial u} = (f_x - f_y) + (-f_x + f_y) = 0.$$

Note that f can be any function with continuous partial derivatives; the equality above depends only on how x and y are defined in terms of u and t. ●

EXAMPLE 5   A function $y = g(x)$ is given implicitly as the solution of the equation $f(x, y) = 0$.   Show that

$$\frac{dy}{dx} = -\frac{f_x}{f_y}.$$

SOLUTION   Let $z = f(x, y)$.   Since $y = g(x)$ is a function of x, z is a function of x. Applying Theorem 1, with $t = x$, we have

$$\frac{dz}{dx} = f_x \frac{dx}{dx} + f_y \frac{dy}{dx}. \tag{2}$$

By the definition of g, $f(x, g(x)) = 0$.   Thus $dz/dx = 0$.   Moreover, $dx/dx = 1$. Thus (2) reduces to

$$0 = f_x + f_y \frac{dy}{dx}.$$

Solving this equation for dy/dx completes the proof. ●

EXAMPLE 6   Let $z = f(x, y)$ and $x = r \cos \theta$, $y = r \sin \theta$.   Thus z is a function of r and θ, $z = g(r, \theta)$.   Express g_{rr} in terms of partial derivatives of f.

SOLUTION   To begin, find g_r by the chain rule of Theorem 2:

$$g_r = f_x \frac{\partial x}{\partial r} + f_y \frac{\partial y}{\partial r},$$

or

$$g_r = f_x \cos \theta + f_y \sin \theta.$$

Having found g_r, differentiate again to obtain g_{rr}:

$$g_{rr} = \frac{\partial(g_r)}{\partial r} = f_x \frac{\partial(\cos \theta)}{\partial r} + \cos \theta \frac{\partial(f_x)}{\partial r} + f_y \frac{\partial(\sin \theta)}{\partial r} + \sin \theta \frac{\partial(f_y)}{\partial r}. \tag{3}$$

Now,

$$\frac{\partial(\cos \theta)}{\partial r} = 0 \quad \text{and} \quad \frac{\partial(\sin \theta)}{\partial r} = 0,$$

since θ is held constant during partial differentiation with respect to r.
To find $\partial(f_x)/\partial r$ and $\partial(f_y)/\partial r$ use the chain rule:

$$\frac{\partial(f_x)}{\partial r} = \frac{\partial(f_x)}{\partial x} \frac{\partial x}{\partial r} + \frac{\partial(f_x)}{\partial y} \frac{\partial y}{\partial r}$$

$$= f_{xx} \cos \theta + f_{xy} \sin \theta, \tag{4}$$

and

$$\frac{\partial(f_y)}{\partial r} = \frac{\partial(f_y)}{\partial x} \frac{\partial x}{\partial r} + \frac{\partial(f_y)}{\partial y} \frac{\partial y}{\partial r}$$

$$= f_{yx} \cos \theta + f_{yy} \sin \theta \tag{5}$$

Putting (3), (4), and (5) together shows that

$$g_{rr} = \cos\theta(f_{xx}\cos\theta + f_{xy}\sin\theta) + \sin\theta(f_{yx}\cos\theta + f_{yy}\sin\theta)$$

$$= f_{xx}\cos^2\theta + 2f_{xy}\cos\theta\sin\theta + f_{yy}\sin^2\theta. \; \bullet$$

While the discussion thus far has concerned only functions of two variables, it is but a slight generalization to consider functions of three or more variables. For instance, the number of air conditioners A sold in a summer is a function of (at least) four variables: temperature T, price p, number of new houses h, and total spendable income i; that is, $A = f(T, p, h, i)$. (Imagine f to be a continuous function that estimates the number actually sold.) In this case there are four partial derivatives; $\partial A/\partial T$ measures the influence of temperature, $\partial A/\partial p$ the influence of price, $\partial A/\partial h$ the influence of new construction, and $\partial A/\partial i$ the influence of spendable income on A. Each of these four partial derivatives indicates how A changes when one variable changes while the remaining three variables are held constant.

As another example, the temperature T in a solid depends on position (x, y, z) and time t; that is, $T = f(x, y, z, t)$. Then $\partial T/\partial t$ represents the rate of change in temperature as a function of time at a fixed point; $\partial T/\partial x$ is the rate of change in temperature along a line parallel to the x axis at a fixed time (with y, z, and t fixed); and so on.

Theorems 1 and 2 generalize to functions of more than two variables; the only change in the statement and proof is in the number of terms in the various sums, which depends on the number of variables under consideration.

The Remainder of this Section, which Treats a Special Topic, May Be Omitted.

A function f is called *homogeneous of degree n* if for all positive numbers k

$$f(kx, ky) = k^n f(x, y).$$

For instance, the function f given by the formula $f(x, y) = x^3 + y^3$ is homogeneous of degree 3, since

$$f(kx, ky) = (kx)^3 + (ky)^3 = k^3(x^3 + y^3) = k^3 f(x, y).$$

In economic theory functions that are homogeneous of degree 1 are important; such a function has the property that

$$f(kx, ky) = kf(x, y).$$

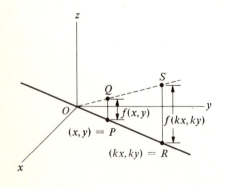

The diagram at the left records this information [for positive k, $f(x, y)$, and $f(kx, ky)$]. Note that

$$\frac{\overline{RS}}{\overline{PQ}} = \frac{f(kx, ky)}{f(x, y)} = \frac{kf(x, y)}{f(x, y)} = k,$$

and

$$\frac{\overline{OR}}{\overline{OP}} = \frac{\sqrt{(kx)^2 + (ky)^2}}{\sqrt{x^2 + y^2}} = \frac{k\sqrt{x^2 + y^2}}{\sqrt{x^2 + y^2}} = k.$$

Thus triangle OPQ is similar to triangle ORS, and the points O, Q, and S are collinear. This means that the part of the graph of f above (or below) the line OR through the origin is a *straight line.* The graph of a homogeneous function of degree 1 is composed of straight lines that pass through $(0, 0, 0)$.

For instance, suppose a firm that employs x workers and has a capital invest-ment of y dollars produces $f(x, y)$ units. Then, if production is proportional to input, the output of a firm employing kx workers and having an investment of ky dollars would be k times as large. That is,

$$f(kx, ky) = kf(x, y).$$

Homogeneous functions have an important property expressed in the follow-ing theorem.

THEOREM 3   (*Euler's theorem on homogeneous functions*).   If f is homogeneous of degree n, then

$$xf_1 + yf_2 = nf,$$

where f_1 denotes $\partial f / \partial x$ and f_2 denotes $\partial f / \partial y$.

PROOF   Consider $z = f(kx, ky)$ and $z = k^n f(x, y)$.   Both equations express z as a function of three variables, k, x, and y.   Differentiate both sides of the equation

$$f(kx, ky) = k^n f(x, y) \tag{6}$$

with respect to k, holding x and y fixed.   By Theorem 2, the derivative of the left side of (6) with respect to k is

$$f_1(kx, ky) \frac{\partial(kx)}{\partial k} + f_2(kx, ky) \frac{\partial(ky)}{\partial k},$$

which is $xf_1(kx, ky) + yf_2(kx, ky).$

On the other hand, the derivative of the right side of (6) with respect to k is simply

$$nk^{n-1}f(x, y).$$

Thus for all k, x, and y,

$$xf_1(kx, ky) + yf_2(kx, ky) = nk^{n-1}f(x, y). \tag{7}$$

Put $k = 1$ in (7) to obtain

$$xf_1(x, y) + yf_2(x, y) = nf(x, y).$$

This proves Euler's theorem.   ●

EXAMPLE 7   Check that Euler's theorem is true for the function f given by

$$f(x, y) = x^3 + y^3.$$

SOLUTION   (Note that f is homogeneous of degree 3.)   In this case

$$f_1(x, y) = 3x^2 \quad \text{and} \quad f_2(x, y) = 3y^2.$$

Is $xf_1(x, y) + yf_2(x, y)$ equal to $3f(x, y)$?

Now
$$xf_1(x, y) + yf_2(x, y) = x \cdot 3x^2 + y \cdot 3y^2$$
$$= 3x^3 + 3y^3$$
$$= 3(x^3 + y^3)$$
$$= 3f(x, y),$$

and Theorem 3 is verified. ●

In the example from economics, Euler's theorem asserts that

$$xf_1(x, y) + yf_2(x, y) = f(x, y), \qquad (8)$$

since f is homogeneous of degree 1. In this case f_1 is called the *marginal product of labor* and f_2 is called the *marginal product of capital*. Equation (8) asserts, "The number of workers times their marginal product, when added to the total capital times the marginal product of capital gives the total productivity of the enterprise." The term $xf_1(x, y)$ in (8) thus measures the contribution of labor, and the term $yf_2(x, y)$ measures the contribution of capita .

Exercises

In Exercises 1 to 3 check Theorem 1 by expressing z, dz/dt, z_x, z_y, dx/dt, and dy/dt explicitly in terms of t.

1. $z = x^2y^3$, $x = t^2$, $y = t^3$.
2. $z = xe^y$, $x = t$, $y = 1 + 3t$.
3. $z = x^2y^3 + y$, $x = \sin t$, $y = e^t$.
4. Find dz/dt if $z_x = 4$, $z_y = 3$, $\dot{x} = -2$, and $\dot{y} = 1$.
5. If z is a function of x and y, and x and y are functions of t and u, find $\partial z/\partial t$ and $\partial z/\partial u$ if $z_x = 3$, $z_y = 5$, $\partial x/\partial t = 2$, $\partial x/\partial u = -3$, $\partial y/\partial t = 5$, $\partial y/\partial u = 4$.

In Exercises 6 to 8 use Theorem 2.

6. Find $\partial z/\partial u$ if $z = \cos(x + 2y)$, $x = 2t + 3u$, and $y = 3t - 4u$.
7. (a) Find $\partial z/\partial r$ if $z = x^3 + y^3$, $x = r \cos \theta$, and $y = r \sin \theta$.
   (b) In the partial derivative $\partial z/\partial r$ which variable is held constant?
8. Find $\partial z/\partial \theta$ and $\partial z/\partial r$ if $z = 1/\sqrt{x^2 + y^2}$, $x = r \cos \theta$, and $y = r \sin \theta$.
9. If $z = x^3 + \cos xy + utx$, find $\partial z/\partial x$, $\partial z/\partial y$, $\partial z/\partial u$, $\partial z/\partial t$.
10. If $z = x_1/x_2 + e^{x_3 x_4} + \cos x_1$, find $\partial z/\partial x_1$ and $\partial z/\partial x_3$.
    Note: When z is a function of many variables, it is more convenient to name the variables $x_1, x_2, \ldots$ than to use different letters of the alphabet.
11. Let $z = f(x, y)$ and $x = u + v$, $y = u - v$. Show that $(\partial z/\partial x)^2 - (\partial z/\partial y)^2 = (\partial z/\partial u)(\partial z/\partial v)$.
12. Let u and v be differentiable functions of t and let f be continuous.

(a) Using Theorem 2, show that
$$\frac{d(\int_u^v f(x)\,dx)}{dt} = f(v)\frac{dv}{dt} - f(u)\frac{du}{dt}.$$
(b) Verify this for the special case $f(x) = \cos x$, $u = t$, $v = t^2$.

13. If $z = f(a + bt, c + dt)$, where a, b, c, d are constants, express dz/dt in terms of partial derivatives of f.
14. If $z = f(u^2 - v^2, v^2 - u^2)$, show that
$$u\frac{\partial z}{\partial v} + v\frac{\partial z}{\partial u} = 0.$$
15. This continues Example 6. Show that
$$g_{rr} + \frac{1}{r} g_r + \frac{1}{r^2} g_{\theta\theta} = f_{xx} + f_{yy},$$
a formula of use in physics.
16. Let $u = f(r)$ and $r = (x^2 + y^2 + z^2)^{1/2}$. Show that
$$u_{xx} + u_{yy} + u_{zz} = u_{rr} + \frac{2}{r} u_r.$$
17. Let z be a function of x and y and let $x = e^u$ and $y = e^v$. Show that
$$z_{uu} + z_{vv} = x^2 z_{xx} + y^2 z_{yy} + xz_x + yz_y.$$
18. Let $x = u \cos \theta - v \sin \theta$ and $y = u \sin \theta + v \cos \theta$. Let $f(x, y)$ be given and define $g(u, v) = f(u \cos \theta - v \sin \theta, u \sin \theta + v \cos \theta)$. Show that
$$f_x^2 + f_y^2 = g_u^2 + g_v^2.$$

19. Let $z = r^2 + s^2 + t^2$ and let $t = rsu$.
  (a) The symbol $\partial z/\partial t$ has two interpretations. What are they?
  (b) Evaluate $\partial z/\partial t$ in both cases in (a).
20. If $u = x^4 f(y/x, z/x)$, show that $xu_x + yu_y + zu_z = 4u$.
21. Let $z = \sin(x - 3y) + \cos(x - 3y)$. Show that

$$\frac{\partial^2 z}{\partial y^2} = 9 \frac{\partial^2 z}{\partial x^2}.$$

■

22. Let (r, θ) be polar coordinates for the point (x, y) given in rectangular coordinates.
  (a) From the relation $r = \sqrt{x^2 + y^2}$ show that $\partial r/\partial x = \cos\theta$.
  (b) From the relation $r = x/\cos\theta$ show that $\partial r/\partial x = 1/\cos\theta$.
  (c) Explain why (a) and (b) are not contradictory.
23. (a) Let $z = uv$, where u and v are differentiable functions of x. Use Theorem 1 to obtain the formula $(uv)' = uv' + vu'$.
  (b) Let $z = u + v$, where u and v are differentiable functions of x. Use Theorem 1 to obtain the formula $(u + v)' = u' + v'$.
24. Let $T = f(x, y, z)$, and let x, y, and z each be functions of t. (For instance, temperature T may depend on position in space, while x, y, and z describe the position of an astronaut in space at time t.) State the analog of Theorem 1 for dT/dt.

Theorem 1 is stated for a function of two variables. A similar result holds for functions of any finite number of variables.

For instance, if z is a function of the variables x_1, x_2, and x_3, and x_1, x_2, and x_3 are functions of t, then

$$\frac{dz}{dt} = \frac{\partial z}{\partial x_1}\frac{dx_1}{dt} + \frac{\partial z}{\partial x_2}\frac{dx_2}{dt} + \frac{\partial z}{\partial x_3}\frac{dx_3}{dt}.$$

Use this formula in Exercises 25 and 26.
25. At what rate is the volume of a rectangular box changing when its width is 3 feet and is increasing at the rate of 2 feet per second, its length is 8 feet and decreasing at the rate of 5 feet per second, and its height is 4 feet and increasing at the rate of 2 feet per second?
26. If T is the temperature at (x, y, z) in space, $T = f(x, y, z)$, and an astronaut is traveling in such a way that his x and y coordinates increase at the rate of 4 miles per second and his z coordinate decreases at the rate of 3 miles per second, compute dT/dt at a point where

$$\frac{\partial T}{\partial x} = 4, \qquad \frac{\partial T}{\partial y} = 7, \qquad \text{and} \qquad \frac{\partial T}{\partial z} = 9.$$

■ ■

27. (a) Is there a function $z = f(x, y)$ such that $z_x = 2xy$ and $z_y = 2xy$? If so, find it.
  (b) Is there a function $z = f(x, y)$ such that $z_x = 2xy$ and $z_y = x^2 + 3y^2$? If so, find it.
28. Show that each of these functions is homogeneous.
  (a) $f(x, y) = x^2(\ln x - \ln y)$;
  (b) $f(x, y) = 1/\sqrt{x^2 + y^2}$;
  (c) $f(x, y) = \sin(y/x)$.
29. Verify that Theorem 3 holds for each function in Exercise 28.

12.7

Critical points

Just as in the case of a function of one variable, calculus provides tools for finding a maximum (or minimum) of a function of two variables. In Chap. 6 the derivative helped to find a highest (or lowest) point on a *curve*. Now, partial derivatives help find a highest (or lowest) point on a *surface*.

The number M is called the *maximum* (or *global maximum*) of f over a set R in the plane if it is the largest value of $f(x, y)$ for (x, y) in R. A *relative maximum* of f occurs at a point (a, b) in R if there is a circle around (a, b) such that $f(a, b)$ is the maximum value of $f(x, y)$ for all points (x, y) within the circle. *Minimum* and *relative minimum* are defined similarly.

Let us look closely at the surface above a point (a, b) where a relative maximum of f occurs. Assume that f is defined for all points within some circle around (a, b). Let L_1 be the line $y = b$ in the xy plane; let L_2 be the line $x = a$ in the xy plane. (We are assuming, for convenience, that the values of f are positive.)

Let C_1 be the curve in the surface directly above the line L_1.   Let C_2 be the curve in the surface directly above the line L_2.   Let P be the point on the surface directly above (a, b).

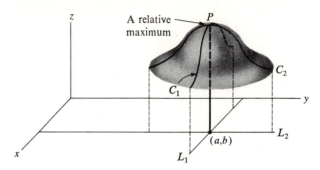

Since f has a relative maximum at (a, b), no point on the surface near P is higher than P.   Thus P is a highest point on the curve C_1 and on the curve C_2 (for points near P).   The study of functions of one variable showed that both these curves have horizontal tangents at P.   In other words, at (a, b) both partial derivatives of f must be 0

$$f_x(a, b) = 0, \quad \text{and} \quad f_y(a, b) = 0.$$

This conclusion is summarized in the following theorem.

THEOREM   Let f be defined on a domain that includes the point (a, b) and all points within some circle whose center is (a, b).   If f has a relative maximum (or relative minimum) at (a, b) and if f_x and f_y exist at (a, b), then both these partial derivatives are 0 at (a, b); that is,

$$f_x(a, b) = 0 = f_y(a, b). \quad \bullet$$

A function need not have a global maximum or a relative maximum.   For instance, the function $f(x, y) = x$ has no maximum value.   (Indeed, its graph is the plane $z = x$, which has no highest point.)   Nor does this function have any relative maximum.   After all, a relative maximum for a function corresponds to a point on the graph of f which is at least as high as all nearby points.

A point (a, b) where both partial derivatives, f_x and f_y, are 0, is clearly of importance.   The following definition is analogous to that of a critical point of a function of one variable.

DEFINITION   *Critical point.*   If $f_x(a, b) = 0$ and $f_y(a, b) = 0$, the point (a, b) is a critical point of the function $f(x, y)$.

EXAMPLE I   Discuss the maximum and minimum values of

$$f(x, y) = 6x^2 + 2y^2 - 24x + 36y + 2.$$

SOLUTION   When x and y have large absolute value, so does $f(x, y)$.   To see this, rewrite $f(x, y)$ in the form

$$f(x, y) = x^2\left(6 - \frac{24}{x}\right) + y^2\left(2 + \frac{36}{y}\right) + 2.$$

When $|x|$ and $|y|$ are large, $6 - 24/x$ is close to 6 and $2 + 36/y$ is close to 2. Hence $f(x, y)$ behaves like

$$6x^2 + 2y^2,$$

which is large when $|x|$ and $|y|$ are large.   Thus f has no maximum value.
  It can be proved that if the continuous function f whose domain is the xy plane has only large values when $|x|$ and $|y|$ are large, then f must have a minimum value.   Since this minimum is also a relative minimum, the theorem applies.   Therefore, to find where the minimum value occurs, look among the critical points.   Straightforward computations show that

$$\begin{cases} \dfrac{\partial f}{\partial x} = 12x - 24, \\[2mm] \dfrac{\partial f}{\partial y} = 4y + 36. \end{cases}$$

The only numbers x and y for which both partial derivatives are 0 are

$$x = 2 \qquad \text{and} \qquad y = -9.$$

Since there is only one point where both partial derivatives are 0, this point must be the place where the minimum occurs.   The minimum value of f is therefore

$$f(2, -9) = 6(2)^2 + 2(-9)^2 - 24(2) + 36(-9) + 2$$

$$= -184. \ \bullet$$

EXAMPLE 2   Examine the function f given by

$$f(x, y) = y^2 - x^2$$

for maximum and minimum values.

SOLUTION   First of all, $\qquad\qquad f(x, 0) = 0^2 - x^2 = -x^2.$

So when $|x|$ is large, $f(x, 0)$ is negative and of large absolute value; consequently f has no minimum value.
  Second,

$$f(0, y) = y^2 - 0^2 = y^2.$$

When $|y|$ is large, so is $f(0, y)$, and f has no maximum value either.
  Finally, let us see whether there are any relative maxima or minima.   Since

$$f_x(x, y) = \frac{\partial(y^2 - x^2)}{\partial x} = -2x$$

and
$$f_y(x, y) = \frac{\partial(y^2 - x^2)}{\partial y} = 2y,$$

only at
$$(x, y) = (0, 0)$$

are both the partial derivatives zero. Thus, if there is any relative maximum or minimum, it would have to be at $(0, 0)$. However, inspection of the graph of

$$z = y^2 - x^2$$

shows that $(0, 0)$ is neither a relative maximum nor a relative minimum. [Example 3 in Sec. 12.3 showed that in the vicinity of $(0, 0)$ the surface is saddle-shaped.]

The second partial derivatives, $z_{yy} = 2$ and $z_{xx} = -2$ at $(0, 0)$, also show what is happening. Considered only on the line $x = 0$, the function f has a local minimum; on the line $y = 0$ it has a local maximum. ●

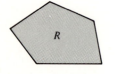

A continuous function on R (which includes the border) has a maximum value at some point in R

If R consists of the region bounded by a polygon including its border, and if f is continuous throughout R, then f has a maximum (and a minimum) value at some point in R. This is similar to Theorem 1 of Sec. 6.1, which concerns a continuous function defined on a closed interval $[a, b]$. Proofs of both results are to be found in any advanced calculus text. The theorem also holds if R is bounded by a curve. (R is also assumed to be "finite" in the sense that it lies within some circle.)

To find a maximum in this case can be rather involved. The procedure is similar to that for maximizing a function on a closed interval:

How to Maximize f on a Region with a Border

1. First find any points that are in R but not on the boundary of R where both f_x and f_y are 0. These are called *critical* points. (If there are no critical points, the maximum occurs on the boundary.)
2. If there are critical points, evaluate f at them. Also find the maximum of f on the boundary. (The next example shows how to do this.) The maximum of f on R is the largest value of f on the boundary and at critical points.

EXAMPLE 3   Let R be the triangle whose vertices are $(0, 0)$, $(2, 0)$, and $(1, 1)$. Let $f(x, y) = 2x + 3y$. What are the maximum and minimum values of f for points in R?

SOLUTION   The maximum or minimum may occur at a point in R that is not on the border. At such a point both f_x and f_y will be 0. Alternately, the maximum or minimum may occur on the boundary. Let us see which cases occur for this particular function.

Begin by computing f_x and f_y. Since $f(x, y) = 2x + 3y$,

$$f_x = 2 \quad \text{and} \quad f_y = 3$$

at all points. The partial derivatives are never 0, and therefore neither the maximum nor minimum of f for points in R can occur at a point not on the

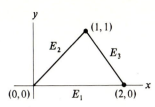

boundary. In other words, both the maximum and minimum must occur on the boundary.

It is necessary to consider the behavior of f on the boundary, which consists of three edges labeled E_1, E_2, and E_3 in the diagram at the left.

On E_1, y is 0. Therefore, if (x, y) is a point on E_1,

$$f(x, y) = f(x, 0) = 2x + 3 \cdot 0 = 2x.$$

Since x goes from 0 to 2 on E_1, the minimum of f on E_1 is 0 and the maximum on E_1 is 4.

On E_2, $y = x$. Thus, if (x, y) is on E_2,

$$f(x, y) = f(x, x) = 2x + 3x = 5x.$$

Since x goes from 0 to 1 on E_2, the minimum of f on E_2 is 0 and the maximum on E_2 is 5. On E_3, $y = 2 - x$. Thus, if (x, y) is on E_3,

$$f(x, y) = f(x, 2 - x) = 2x + 3(2 - x) = 6 - x.$$

Edge	Minimum of f on Edge	Maximum of f on Edge
E_1	0	4
E_2	0	5
E_3	4	5

Since x goes from 1 to 2 on E_3, the minimum of f on E_3 is 4 and the maximum is 5. The table shows that the maximum of f on the border is 5 and the minimum is 0.

The minimum value of f on R is 0, occurring at $(0, 0)$. The maximum value is 5, occurring at $(1, 1)$. ●

It would be natural to expect that, if at a critical point of f both second partial derivatives f_{xx} and f_{yy} are positive, then f would have a local minimum there. The next example destroys such a hope and is included now to prepare the reader to receive the message of the next section.

EXAMPLE 4 Let $z = f(x, y) = x^2 + 3xy + y^2$. Show that $(0, 0)$ is a critical point of f, $f_{xx}(0, 0) > 0$, and $f_{yy}(0, 0) > 0$, but $(0, 0)$ is *not* a local minimum of f.

SOLUTION Easy computations show that

$$z_x = 2x + 3y \qquad z_y = 3x + 2y$$

$$z_{xx} = 2 \qquad\qquad z_{yy} = 2.$$

At $(0, 0)$, then, $z_x = 0$, $z_y = 0$, $z_{xx} = 2$ and $z_{yy} = 2$.

All that remains is to show that there are points (x, y) near $(0, 0)$ where $f(x, y)$ is negative. To see that such points exist rewrite f:

$$f(x, y) = x^2 + 3xy + y^2$$

$$= (x + y)^2 + xy.$$

Thus, on the line $x + y = 0$ in the xy plane,

$$f(x, y) = 0^2 + xy$$

$$= -x^2.$$

For $(x, y) \neq (0, 0)$ on the line $x + y = 0$, f takes on negative values. Thus the critical point $(0, 0)$ is *not* a local minimum. It is neither a local maximum nor a local maximum, but a saddle point, like that illustrated in Example 3 of Sec. 12.3.

Note that for (x, y) on the x axis ($y = 0$) the function has the formula x^2; hence it has a local minimum at $(0, 0)$ when considered only on the x axis. Similarly, it has a local minimum when considered only on the y axis.

The reader might pause to sketch the saddle in this case. ●

Exercises

1. Consider $f(x, y) = 2x^2 + 4x + y^2 + 8y$.
  (a) Show that f has no maximum value.
  (b) Find the minimum value of f.
2. Let $z = -3x^2 - 5y^2 + 6x - 9y$.
  (a) Why does z have no minimum value?
  (b) Find the point (x, y) at which z is a maximum.
3. For a typical point $P = (x, y)$ in the (x, y) plane let $f(x, y)$ be the sum of the squares of the distances from P to the points $(0, 0)$, $(1, 1)$, $(2, 0)$. Find the point that minimizes this function.
4. Let $z = xy$.
  (a) Show that $(0, 0)$ is a critical point.
  (b) Does this mean that z necessarily has a maximum or a minimum at $(0, 0)$?
  (c) How does z behave on the line $x = 0$? on the line $y = 0$? on the line $y = -x$? on the line $y = x$?
  (d) Make a drawing or model of this surface, especially in the vicinity of $(0, 0)$.
5. Find the maximum value of $f(x, y) = xy$ for points in the triangular region whose vertices are $(0, 0)$, $(1, 0)$, and $(0, 1)$.
6. (a) Show that $z = x^2 - y^2 + 2xy + 2$ has no maximum and no minimum.
  (b) Find the minimum and maximum of z if we consider only (x, y) on the circle of radius 1 and center $(0, 0)$, that is, all (x, y) such that $x^2 + y^2 = 1$. *Hint*: To deal with this, use $x = \cos\theta$, $y = \sin\theta$.
  (c) Find the minimum and maximum of z if we consider all (x, y) in the disk of radius 1 and center $(0, 0)$, that is, all (x, y) such that $x^2 + y^2 \leq 1$.
7. Find the maximum value of $f(x, y) = 3x^2 - 4y^2 + 2xy$ for points (x, y) in the square region whose vertices are $(0, 0)$, $(0, 1)$, $(1, 0)$, and $(1, 1)$.

8. Maximize the function $-x + 3y + 6$ on the quadrilateral whose vertices are $(1, 1)$, $(4, 2)$, $(0, 3)$, $(5, 6)$.

■

Compare Exercises 9 to 12.
9. Show that $x^2 + 3xy + 4y^2$ has a local minimum at $(0, 0)$. *Hint*: Rewrite it as $(x + \frac{3}{2}y)^2 - 9y^2/4 + 4y^2$, that is, complete the square.
10. Show that $x^2 + 4xy + 4y^2$ has a local minimum at $(0, 0)$.
11. Show that $x^2 + 5xy + 4y^2$ does *not* have a local minimum at $(0, 0)$.
12. For which values of the constant k does $x^2 + kxy + 4y^2$ have a local minimum at $(0, 0)$?
13. Explain why Δf is approximately $f_x(a, b) \Delta x + f_y(a, b) \Delta y$. In particular, why is there a plus sign?
14. Let $z = f(x, y)$, $x = g(u, v)$, $y = h(u, v)$, thus, $z = m(u, v)$. Assume that $g_u = h_v$ and $g_v = -h_u$. Show that

$$m_{uu} + m_{vv} = (f_{xx} + f_{yy})(g_u{}^2 + g_v{}^2).$$

15. A fence perpendicular to the xy plane has as its base the line segment whose ends are $(1, 0, 0)$ and $(0, 1, 0)$. It is bounded on the top by the cylinder $z = y^2$.
  (a) Draw the fence.
  (b) Find its area.

■ ■

16. Let $f(x, y) = (y - x^2)(y - 2x^2)$.
  (a) Show that f has neither a local minimum nor a local maximum at $(0, 0)$.
  (b) Show that f has a local minimum at $(0, 0)$ when considered only on any fixed line through $(0, 0)$.

12.8

Local maximum or minimum of $f(x, y)$ and second-order partial derivatives

In the case of a function of one variable, $y = f(x)$, if $f'(a) = 0$ and $f''(a) > 0$, then $f(x)$ has a local minimum at a. In this section the analogous test for a function of two variables is presented.

Recall Example 4 of the preceding section, $f(x, y) = x^2 + 3xy + y^2$. Though at the critical point $(0, 0)$ both f_{xx} and f_{yy} are positive, $(0, 0)$ is not a local minimum for f. Contrast this with the following example, which sets the stage for the main result of this section.

EXAMPLE 1   Show that, if k is a constant, $k^2 < 4$, then the function $f(x, y) = x^2 + kxy + y^2$ has a local minimum at $(0, 0)$.

SOLUTION   Completing the square shows that

$$x^2 + kxy + y^2 = x^2 + kxy + \left(\frac{ky}{2}\right)^2 + y^2 - \left(\frac{ky}{2}\right)^2$$

$$= \left(x + \frac{ky}{2}\right)^2 + \left[1 - \left(\frac{k}{2}\right)^2\right]y^2.$$

Since $k^2 < 4$,

$$1 - \left(\frac{k}{2}\right)^2 > 0.$$

Thus, recalling that the square of any real number is nonnegative, we obtain the inequality

$$\left(x + \frac{ky}{2}\right)^2 + \left[1 - \left(\frac{k}{2}\right)^2\right]y^2 \geq 0.$$

Since $f(0, 0) = 0$, the function has a local minimum at $(0, 0)$; in fact, a global minimum. Incidentally, note that, if $f(x, y) = 0$, both y and $x + ky/2$ must be 0. So only at $(0, 0)$ does the function assume its minimum value. ●

Example 1 suggests that the problem of determining for which (fixed) values of A, B, and C the function

$$f(x, y) = Ax^2 + Bxy + Cy^2$$

has a local minimum at $(0, 0)$ is worth studying. When $y = 0$, the function reduces to Ax^2. Thus A must be nonnegative. Similarly, C must be nonnegative. But as Example 1 illustrates, these conditions are not enough. The key is provided by the next theorem (where the coefficient of xy is denoted $2B$ rather than B to simplify the proof). To avoid trivial cases, assume that neither A nor C is 0.

THEOREM 1   Let A, B, and C be fixed numbers such that

$$A > 0, \qquad C > 0, \qquad \text{and} \qquad B^2 < AC.$$

Then the function $\qquad f(x, y) = Ax^2 + 2Bxy + Cy^2$

has a minimum at $(0, 0)$.

PROOF   Since $f(0, 0) = 0$, it suffices to prove that, if $(x, y) \neq (0, 0)$, then $f(x, y) > 0$. This amounts to showing that

$$A(Ax^2 + 2Bxy + Cy^2) > 0$$

if $(x, y) \neq (0, 0)$, since A is positive.

The method of completing the square establishes this inequality as follows:

$$
\begin{aligned}
A(Ax^2 + 2Bxy + Cy^2) &= A^2x^2 + 2ABxy + ACy^2 \\
&= A^2x^2 + 2ABxy + B^2y^2 - B^2y^2 + ACy^2 \\
&= (Ax + By)^2 + (AC - B^2)y^2.
\end{aligned}
$$

It is easy to show that

$$(Ax + By)^2 + (AC - B^2)y^2 \tag{1}$$

is positive if x and y are not both 0, as follows.

First of all, if y is not 0, the second summand $(AC - B^2)y^2$ is positive, since it is the product of the positive number $AC - B^2$ and the square of a nonzero real number. Note that $(Ax + By)^2$, the square of a real number, is not negative.

If y is 0, (1) reduces to A^2x^2, which is 0 only when x is 0. Since the case $(x, y) = (0, 0)$ is excluded, the theorem is established. ●

EXAMPLE 2   Show that $\qquad f(x, y) = 2x^2 - xy + y^2$

has a minimum at $(0, 0)$.

SOLUTION   Apply Theorem 1. In this case $A = 2$, $B = -\frac{1}{2}$, and $C = 1$. The hypotheses of Theorem 1 apply, for $A > 0$, $C > 0$, and $B^2 < AC$. Thus $f(x, y) = 2x^2 - xy + y^2$ has a minimum at $(0, 0)$. (Compare this with Example 1.) ●

Before stating the test for a local minimum of a more general function, we express Theorem 1 in terms of partial derivatives.

Observe that, if

$$f(x, y) = Ax^2 + 2Bxy + Cy^2,$$

then $\qquad f_x(x, y) = 2Ax + 2By, \qquad f_y(x, y) = 2Bx + 2Cy,$

$$f_{xx}(x, y) = 2A, \qquad f_{xy}(x, y) = 2B, \qquad \text{and} \qquad f_{yy}(x, y) = 2C.$$

Thus $\qquad A = \dfrac{f_{xx}(0, 0)}{2}, \qquad B = \dfrac{f_{xy}(0, 0)}{2}, \qquad C = \dfrac{f_{yy}(0, 0)}{2}.$

The condition $B^2 < AC$ now reads

$$\left(\frac{f_{xy}(0, 0)}{2}\right)^2 < \frac{f_{xx}(0, 0)}{2}\frac{f_{yy}(0, 0)}{2},$$

or simply $\qquad [f_{xy}(0, 0)]^2 < f_{xx}(0, 0)f_{yy}(0, 0).$

Theorem 1 can therefore be expressed as follows.

THEOREM 2 Let $f(x, y) = Ax^2 + 2Bxy + Cy^2$, where A, B, and C are fixed numbers. If

$$f_{xx}(0, 0) > 0, \qquad f_{yy}(0, 0) > 0, \qquad \text{and} \qquad [f_{xy}(0, 0)]^2 < f_{xx}(0, 0)f_{yy}(0, 0),$$

then f has a minimum at $(0, 0)$. ●

It is Theorem 2 which generalizes to a test for a local minimum of more general functions. This test is stated in Theorem 3, whose proof is given in Sec. 15.6.

THEOREM 3 (*Test for a local minimum*). Let f be a function that has continuous partial derivatives f_x, f_y, f_{xx}, f_{yy}, and $f_{xy} = f_{yx}$ at and near (a, b). If

1. $f_x(a, b)$ and $f_y(a, b)$ are 0,
2. $f_{xx}(a, b)$ and $f_{yy}(a, b)$ are positive, and
3. $[f_{xy}(a, b)]^2 < f_{xx}(a, b)f_{yy}(a, b)$,

then f has a local minimum at (a, b). ●

A similar test for a local maximum is given in the next theorem.

THEOREM 4 (*Test for a local maximum*). Let f be a function that has continuous partial derivatives f_x, f_y, f_{xx}, f_{yy}, and $f_{xy} = f_{yx}$ at and near (a, b). If

1. $f_x(a, b)$ and $f_y(a, b)$ are 0,
2. $f_{xx}(a, b)$ and $f_{yy}(a, b)$ are negative, and
3. $[f_{xy}(a, b)]^2 < f_{xx}(a, b)f_{yy}(a, b)$,

then f has a local maximum at (a, b). ●

Notice that in both Theorems 3 and 4 condition (3) remains the same: The square of $f_{xy}(a, b)$ is less than the product of $f_{xx}(a, b)$ and $f_{yy}(a, b)$. However, condition (2) in Theorem 4 is opposite condition (2) in Theorem 3.

Exercise 26 shows in the case $f(x, y) = Ax^2 + 2Bxy + Cy^2$ that if $[f_{xy}(a, b)]^2 > f_{xx}(a, b)f_{yy}(a, b)$, then f has neither a local maximum nor a local minimum. If $[f_{xy}(a, b)]^2 = f_{xx}(a, b)f_{yy}(a, b)$, no conclusion can be drawn, as is illustrated in Exercise 19.

EXAMPLE 3 Determine whether

$$f(x, y) = 3x^2 + 5y^2 + 6x - 20y$$

has any relative maxima or minima.

SOLUTION   First compute f_x and f_y: $f_x(x, y) = 6x + 6$

and $f_y(x, y) = 10y - 20.$

Both of these vanish only when

$$6x + 6 = 0 \quad \text{and} \quad 10y - 20 = 0,$$

therefore only at the point $(-1, 2)$.

Now utilize the second-derivative test of Theorems 3 and 4 to help determine whether f does have a relative maximum or minimum at $(-1, 2)$.

Compute the second-order partial derivatives:

$$f_{xx}(x, y) = 6, \qquad f_{yy}(x, y) = 10, \qquad f_{xy}(x, y) = 0$$

[independently of the point (x, y)]. Since f_{xx} and f_{yy} are positive and $(f_{xy})^2$ is less than the product $f_{xx} f_{yy}$, it follows from Theorem 3 that f has a relative minimum at $(-1, 2)$. ●

The next example shows how partial derivatives are useful in solving applied maximum and minimum problems.

EXAMPLE 4   Of all rectangular boxes with volume 1 cubic foot, which has the smallest surface area?

SOLUTION   Let the dimensions of a box whose volume is 1 cubic foot be x, y, and z feet.

The volume V is given by $V = 1 = xyz,$ (2)

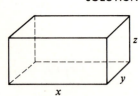

If $xyz = 1$ what is the minimal total area of the six faces?

and S, the surface area, is given by

$$S = 2xy + 2xz + 2yz.$$

By (2) $z = \dfrac{1}{xy}.$

Hence S can be expressed as a function of x and y:

$$S = 2xy + 2x\,\frac{1}{xy} + 2y\,\frac{1}{xy}$$

$$= 2\left(xy + \frac{1}{y} + \frac{1}{x}\right).$$

Observe that, when x and y are small or large, S is large. Thus there is no box of maximum surface area. But it is reasonable to expect that there may be a box of minimal surface area, and we shall assume that such a box exists.

To find it, determine those points (x, y) where $\partial S/\partial x$ and $\partial S/\partial y$ are both 0.

Now, $\dfrac{\partial S}{\partial x} = 2\left(y - \dfrac{1}{x^2}\right),$

and $\dfrac{\partial S}{\partial y} = 2\left(x - \dfrac{1}{y^2}\right).$

In order for both partial derivatives to be 0, these two equations must hold:

$$\begin{cases} 2\left(y - \dfrac{1}{x^2}\right) = 0, \\[2mm] 2\left(x - \dfrac{1}{y^2}\right) = 0. \end{cases}$$

Thus $$\begin{cases} x^2 y = 1 \end{cases} \tag{3}$$

and $$\begin{cases} xy^2 = 1. \end{cases} \tag{4}$$

From (3) it follows that $y = 1/x^2$. Replacing y by $1/x^2$ in (4) yields

$$x\left(\frac{1}{x^2}\right)^2 = 1,$$

or $$1 = x^3.$$

Hence $x = 1$. Thus $y = 1/1^2 = 1$. Finally, since $z = 1/xy$,

$$z = \frac{1}{(1)(1)} = 1.$$

Since $(x, y) = (1, 1)$ is the only critical point for the function $S = 2(xy + 1/y + 1/x)$ and we assumed that a box of minimal volume exists, it must be the cube of side 1. Still, it would be of interest to see what the test in Theorem 3 says about the point $(1, 1)$. Straightforward computation shows that, at $(1, 1)$,

$$\frac{\partial^2 S}{\partial x^2} = 4, \qquad \frac{\partial^2 S}{\partial y^2} = 4, \qquad \text{and} \qquad \frac{\partial^2 S}{\partial x \, \partial y} = 2.$$

Theorem 3 assures us that, since $\partial^2 S/\partial x^2$ and $\partial^2 S/\partial y^2$ are positive and $(\partial^2 S/\partial x \, \partial y)^2 < (\partial^2 S/\partial x^2)(\partial^2 S/\partial y^2)$, $(1, 1)$ is a local minimum. ●

Exercises

Using Theorems 3 and 4, determine all relative maxima or minima of the functions in Exercises 1 to 11.

1. $x^2 + 3xy + y^2$
2. $x^2 - xy + y^2$
3. $x^2 - y^2$
4. $x^2 + 2xy + 2y^2 + 4x$
5. $x^2 - 2xy + 2y^2 + 4x$
6. $2x^2 + 2xy + 5y^2 + 4x$
7. $x^4 + 8x^2 + y^2 - 4y$
8. $-4x^2 - xy - 3y^2$
9. $e^{-2x^2 - 2xy - 4y^2}$
10. $4/x + 2/y + xy$
11. $x^3 - y^3 + 3xy$

Let f be a function of x and y such that at (a, b) both f_x and f_y equal 0. In each of Exercises 12 to 17 values are specified for f_{xx}, f_{xy}, and f_{yy} at (a, b). Assume, as we do throughout this chapter, that all these partial derivatives are continuous. On the basis of the information decide whether (a) f has a relative maximum at (a, b); (b) f has a relative minimum at (a, b); (c) neither (a) nor (b) occurs; (d) there is inadequate information.

12. $f_{xy} = -3, f_{xx} = 2, f_{yy} = 4$.
13. $f_{xy} = 3, f_{xx} = 2, f_{yy} = 4$.
14. $f_{xy} = 2, f_{xx} = 3, f_{yy} = 4$.
15. $f_{xy} = -2, f_{xx} = -3, f_{yy} = -4$.
16. $f_{xy} = -2, f_{xx} = 3, f_{yy} = -4$.
17. $f_{xy} = 4, f_{xx} = 2, f_{yy} = 8$.

■

18. If at (x_0, y_0) $z_x = 0 = z_y$, $z_{xx} = 3$ and $z_{yy} = 12$, for what values of z_{xy} is it certain that z has a relative minimum at (x_0, y_0)?

19. This exercise shows that if $(z_{xy})^2 - z_{xx}z_{yy} = 0$, no conclusions of the type given in Theorems 3 and 4 can be drawn.
    (a) Let $z = x^2 + 2xy + y^2$. Show that at $(0, 0)$ z_{xx} and z_{yy} are positive, $(z_{xy})^2 = z_{xx}z_{yy}$, and z has a relative minimum.
    (b) Let $z = x^2 + 2xy + y^2 - x^4$. Show that at $(0, 0)$ z_{xx} and z_{yy} are positive, $(z_{xy})^2 = z_{xx}z_{yy}$, and z has neither a relative minimum nor a relative maximum.

20. Show that, of all rectangular boxes having a surface area of 6 square inches, the cube has maximal volume.

21. A rectangular box without a top is to have a volume of 1 cubic foot. Of all such boxes, what are the dimensions of the one with least surface area?

22. The material for the top and bottom of a rectangular box costs 3 cents per square foot, and that for the sides 2 cents per square foot. What is the least expensive box that has a volume of 1 cubic foot?

23. Let $U(x, y, z) = x^{1/2}y^{1/3}z^{1/6}$ be the "utility" or "desirability" to a given consumer of the amounts x, y, and z of three different commodities. Their prices are, respectively, 2 dollars, 1 dollar, and 5 dollars, and the consumer has 60 dollars to spend. How much of each product should he buy to maximize the utility?

24. A wire of length 1 is to be cut into three pieces which will be bent into a square, circle, and equilateral triangle. How should this be done to (a) minimize their total area? (b) maximize their total area?

25. Assume that the profits of two firms depend on the amount each produces. For the first firm let the profit be P_1 and its output Q_1; let P_2 and Q_2 be similarly defined for the second firm. Assume that

$$P_1 = 48Q_1 - 2Q_1{}^2 - 3Q_2{}^2 + 1,000$$

and $\qquad P_2 = 60Q_2 - 5Q_1{}^2 - 2Q_2{}^2 + 2,000.$

Thus the profit for each firm is dependent in part on the output of the other.
    (a) If each firm acts separately to maximize its profit, what will Q_1, Q_2, P_1, P_2, and $P_1 + P_2$ be?
    (b) If the firms agree to work together in order to maximize $P_1 + P_2$, what will Q_1, Q_2, P_1, P_2, and $P_1 + P_2$ be?

26. (a) Use expression (1) to prove the following analog of Theorem 1: Let A, B, and C be fixed numbers such that $\qquad A > 0, C > 0,$ and $B^2 > AC.$

Then there are numbers x and y, such that

$$Ax^2 + 2Bxy + Cy^2$$

is negative.
    (b) Deduce that if $B^2 > AC$, then $f(x, y) = Ax^2 + 2Bxy + Cy^2$ can have neither a local maximum nor a local minimum at $(0, 0)$.

■ ■

12.9

Summary

This chapter has developed for a function of two variables some analogs of properties of a function of one variable. The central result in this chapter is that

$$\Delta z = f_x \, \Delta x + f_y \, \Delta y + \varepsilon_1 \, \Delta x + \varepsilon_2 \, \Delta y,$$

where f_x and f_y are evaluated at a fixed point (a, b), and ε_1 and $\varepsilon_2 \to 0$ as Δx and $\Delta y \to 0$. This equation says that the change in z, caused by changes Δx and Δy at (a, b), is well approximated by

$$f_x(a, b) \, \Delta x + f_y(a, b) \, \Delta y, \tag{1}$$

which is the sum of differentials of two functions of a single variable x or y. The form of (1) is reflected in the chain rules developed in Sec. 12.6 and applied in the later sections. The second column of this table summarizes the chapter.

The Plane and $f(x)$	*Space and $f(x, y)$*
A point is determined by x and y $x = k$, a line perpendicular to x axis	A point is determined by x, y, and z $x = k$, a plane perpendicular to x axis; $y = k$, a plane perpendicular to y axis; $z = k$, a plane perpendicular to z axis

The Plane and $f(x)$	*Space and $f(x,y)$*

The distance between (x_1, y_1) and (x_2, y_2):

$$\sqrt{(x_1 - x_2)^2 + (y_1 - y_2)^2}$$

$(x - a)^2 + (y - b)^2 = r^2$ describes a circle of radius r, center (a, b)

The graph of $y = f(x)$ is usually a curve

The graph of $y = Ax + B$ is a line

The derivative f' measures the rate of change of f

Notation:
$$f', \frac{dy}{dx}, Df, y'$$

A continuous function on $[a, b]$ has a maximum

The maximum of a differentiable f on $[a, b]$ is the largest of the values of f at a, b, and at critical points

Higher derivatives: $d^2f/dx^2, \ldots$

$\Delta y = f'(x)\,\Delta x + \varepsilon\,\Delta x \;(\varepsilon \to 0 \text{ as } \Delta x \to 0)$

dy is defined as $f'(x)\,\Delta x$

dy is change along tangent line

Chain rule: $\dfrac{dy}{dx} = \dfrac{dy}{du}\dfrac{du}{dx}$

Second-derivative test for local maximum:

$$f' = 0, f'' < 0$$

Second-derivative test for local minimum:

$$f' = 0, f'' > 0$$

The distance between (x_1, y_1, z_1) and (x_2, y_2, z_2):

$$\sqrt{(x_1 - x_2)^2 + (y_1 - y_2)^2 + (z_1 - z_2)^2}$$

$(x - a)^2 + (y - b)^2 + (z - c)^2 = r^2$ describes a sphere of radius r, center (a, b, c)

The graph of $z = f(x, y)$ is usually a surface

The graph of $z = Ax + By + C$ is a plane

The partial derivatives f_x and f_y measure the rate of change of f in the x and y directions.

Notation:
$$f_x, \frac{\partial f}{\partial x}, z_x, \frac{\partial z}{\partial x}, f_1, D_1 f;$$

$$f_y, \frac{\partial f}{\partial y}, z_y, \frac{\partial z}{\partial y}, f_2, D_2 f$$

A continuous function on a region contained in some circle and bounded by a polygon or curve has a maximum

The maximum of f is the largest of the values of f at points on the border and at critical points

Higher partial derivatives: $\partial^2 f/\partial x^2, \partial^2 f/\partial y^2, \partial^2 f/\partial x\,\partial y, \partial^2 f/\partial y\,\partial x, \ldots$

$\Delta z = f_x\,\Delta x + f_y\,\Delta y + \varepsilon_1\,\Delta x + \varepsilon_2\,\Delta y\;(\varepsilon_1 \to 0, \varepsilon_2 \to 0$ as Δx and $\Delta y \to 0)$

dz is defined as $f_x\,\Delta x + f_y\,\Delta y$

dz is change along tangent plane

Chain rule: $\dfrac{dz}{dt} = \dfrac{\partial z}{\partial x}\dfrac{dx}{dt} + \dfrac{\partial z}{\partial y}\dfrac{dy}{dt}$ where x and y are

functions only of t)

Chain rule: $\dfrac{\partial z}{\partial t} = \dfrac{\partial z}{\partial x}\dfrac{\partial x}{\partial t} + \dfrac{\partial z}{\partial y}\dfrac{\partial y}{\partial t}$ (where x and y are

functions of t and u)

Second-partial-derivative test for local maximum:

$$f_x = 0, f_y = 0$$

$$f_{xx} < 0, f_{yy} < 0$$

$$(f_{xy})^2 < f_{xx} f_{yy}$$

Second-partial-derivative test for local minimum:

$$f_x = 0, f_y = 0$$

$$f_{xx} > 0, f_{yy} > 0$$

$$(f_{xy})^2 < f_{xx} f_{yy}$$

VOCABULARY

coordinate plane   function of two variables   chain rule   maximum, relative maximum   critical point
cylinder   partial derivative   differential   minimum, relative minimum

Guide quiz on Chap. 12

1. (a) Graph $z = 2x^2 + 2y^2$.
   (b) Graph $z^2 = x^2 + y^2$.
   (c) Graph $x^2 + y^2 + z^2 = 6$.
   (d) Graph $y = 2x$ (in space).
2. Compute:
   (a) $D_1(\sqrt{x^2 + y^2})$

   (b) $D_2\left[\ln\left(\dfrac{y^2}{3y + x}\right)\right]$

   (c) $\dfrac{\partial(ye^{x^3y})}{\partial x}$

   (d) $\dfrac{\partial^2(\cos(2x + 3y))}{\partial x\,\partial y}$

   (e) $\dfrac{\partial^2(3/xy)}{\partial y^2}$

   (f) $\dfrac{\partial^2(\sin^3 xy)}{\partial x^2}$.

3. Compute:

   (a) $\dfrac{d(\int_0^1 e^{xy}\,dx)}{dy}$

   (b) $\displaystyle\int_0^1 \dfrac{\partial(e^{xy})}{\partial y}\,dx$.   A theorem in Chap. 23 implies that the

   two are equal.
4. (a) Sketch the graph of a typical function $z = f(x, y)$ in the vicinity of the point $(a, b, f(a, b))$.
   (b) Sketch the tangent plane at $(a, b, f(a, b))$.

(c) Using the sketches in (a) and (b) indicate the line segments that correspond to dz and Δz.
5. Give examples of two functions f that have a critical point at $(0, 0)$, $f_{xx}(0, 0) > 0$, $f_{yy}(0, 0) > 0$, and
   (a) $(0, 0)$ is a local minimum;
   (b) $(0, 0)$ is not a local minimum.
6. A house in the form of a box is to hold 10,000 cubic feet. The glass walls admit heat at the rate of 5 units per minute per square foot, the roof at the rate of 3 units per minute per square foot, and the floor at the rate of 1 unit per minute per square foot. What should the shape of the house be in order to minimize the rate at which heat enters?
7. (a) Using a table of natural logarithms or a calculator, evaluate $(1.1)^2 \ln(1.2)$.
   (b) Using the differential of the function $f(x, y) = x^2 \ln y$, estimate $(1.1)^2 \ln(1.2)$, making use of $f(1, 1)$.
8. The kinetic energy of a particle of mass m and velocity v is given by $K = \frac{1}{2}mv^2$. If the maximum error in measuring m is 1 percent and in measuring v is 3 percent, estimate the maximum error in measuring K.
9. Let $u = f(x, y)$, $x = r \cos \theta$, and $y = r \sin \theta$. Show that

$$u_{rr} + \frac{1}{r} u_r + \frac{1}{r^2} u_{\theta\theta} = u_{xx} + u_{yy}.$$

10. For what values of the constant k does the function $x^2 + kxy + 9y^2$ have a relative minimum at $(0, 0)$?
11. Let f be a function of x and y. Assume that $f(2, 3) = 5$, $f_x(2, 3) = 4$, $f_y(2, 3) = -1$. Estimate $f(2.1, 2.8)$.
12. Prove one of the two chain rules for functions of two variables.

Review exercises for Chap. 12

1. (a) Show that $(0, 0)$ is a critical point of $f(x, y) = xy$.
   (b) Sketch enough of the graph of $z = xy$ to show that $(0, 0)$ provides neither a relative maximum nor a relative minimum.
2. On the graph of $z = f(x, y)$ sketch the curves whose slopes are given by f_x and f_y. In particular, draw the two tangent lines whose slopes are $f_x(a, b)$ and $f_y(a, b)$.

3. (a) Without using calculus, show that $f(x, y) = (x - y)^2$ has a minimum but no maximum. Where does the minimum occur?
   (b) Without using calculus, find the minimum value of $f(x, y) = (2x + 3y - 5)^2 + (x - y)^2$.
4. Using differentials, estimate:
   (a) $\sqrt{(3.01)^2 + (4.02)^2}$;

(b) $\sqrt{(3.04)^2 + (3.97)^2}$

5. Let $f_x(0, 0) = 0$ and $f_y(0, 0) = 0$. In each of these cases decide, if there is enough information, whether f at $(0, 0)$ has a relative maximum, relative minimum, or neither.
   (a) $f_{xx}(0, 0) = -1; f_{yy}(0, 0) = 2$.
   (b) $f_{xx}(0, 0) = 3; f_{yy}(0, 0) = 2; f_{xy}(0, 0) = -2.2$.
   (c) $f_{xx}(0, 0) = -3; f_{yy}(0, 0) = -5; f_{xy}(0, 0) = 4$.
   (d) $f_{xx}(0, 0) = 3; f_{yy}(0, 0) = 12; f_{xy}(0, 0) = -6$.

6. Let $u = f(x, y)$ and $v = g(x, y)$. Assume that
$$u_x = v_y \quad \text{and} \quad v_x = -u_y.$$
   Prove that $\quad u_{xx} + u_{yy} = 0$
   and $\quad v_{xx} + v_{yy} = 0$.

7. Let $z = u^3 v^5$, where $u = x + y$ and $v = x - y$.
   (a) Express z explicitly as a function of x and y and use this explicit expression to find z_x and z_y.
   (b) Find z_x and z_y by the chain rule. Do your answers agree? Which method is easier to use?

8. Let $z = e^{uv}$, where $u = y \sin x$ and $v = x + \cos y$.
   (a) Compute z_x and z_y by the chain rule.
   (b) Express z explicitly in terms of x and y and use this expression to compute z_x and z_y.
   (c) Does your answer to (a) agree with your answer to (b)?

9. Graph the planes
   (a) $x + y = 1$ (b) $x + y + z = 1$
   (c) $\dfrac{x}{2} + \dfrac{y}{3} + \dfrac{z}{4} = 1$.

10. Sketch the surface $\dfrac{x^2}{1} + \dfrac{y^2}{4} + \dfrac{z^2}{9} = 1$.

11. Sketch the cylinders
    (a) $y = x^2$ (b) $x^2 - y^2 = 1$
    (c) $z = x^3$.

12. The pressure P, volume V, and temperature T of a gas are related by the equation $(P + a/V^2)(V - b) = cT$, where a, b, c are constants. Thus any two of P, V, T determine the third.
    (a) Compute $\partial V/\partial T$, $\partial T/\partial P$, and $\partial P/\partial V$.
    (b) Show that the product of the three partial derivatives in (a) is -1.

13. Let $f(x, y) = x^2 + 2xy - y^2$.
    (a) Show that f considered only on the x axis has a local minimum at $(0, 0)$.
    (b) Show that f considered only on the y axis has a local maximum at $(0, 0)$.
    (c) For which values of m does f have a local minimum at $(0, 0)$ when considered only on the line $y = mx$?

14. Consider Euler's partial differential equation
$$az_{xx} + 2bz_{xy} + cz_{yy} = 0,$$

where $a, b,$ and c are constants and $b^2 \neq ac$. Show that
$$z = f(x + r_1 y) + g(x + r_2 y),$$
where r_1 and r_2 are the roots of $a + 2bx + cx^2 = 0$, is a solution of the differential equation. The functions f and g are differentiable.

15. Determine the minimum value: $f(x, y) = x^4 - x^2 y^2 + y^4$.

16. The maximum combined height and girth of packages that can be sent through the mails is 100 inches. Find the dimensions of a rectangular box of largest volume that can be sent through the mails.

∎

17. The volume V occupied by a mixture of several gases depends on the pressure p, the temperature T, and the amounts $x_1, x_2, \ldots, x_n$ of the several gases,
$$V = f(p, T, x_1, x_2, \ldots, x_n).$$
Assume that V is proportional to the amounts of the gases.
    (a) Show that $\sum_{i=1}^n x_i \, \partial V/\partial x_i = V$.
    (b) From (a) deduce that $\sum_{i=1}^n x_i \, \partial^2 V/\partial x_j \, \partial x_i = 0$ for each fixed j, with $j = 1, 2, \ldots, n$.

18. Let $y = f(x, t)$ describe the vertical displacement of a particle in a wave corresponding to the horizontal coordinate x at time t. It can be shown on physical grounds that f satisfies the wave equation
$$a^2 f_{xx} = f_{tt},$$
where a is a constant. Show that any function $f(x, t)$ of the form $g(x + at)$ satisfies the wave equation, where g is a function of a single variable that possesses first and second derivatives.

19. Let $z = x^2 y$, where $y = e^{3x} u$. Thus z may be considered a function of x and y or of x and u.
    (a) Compute $\partial z/\partial x$, considering z to be a function of x and y.
    (b) Compute $\partial z/\partial x$, considering z to be a function of x and u.
    The notation $(\partial z/\partial x)_y$ is used in (a) and $(\partial z/\partial x)_u$ in (b). The symbol $\partial z/\partial x$ is ambiguous in this context.

∎ ∎

20. Review the argument in Exercise 63 of Sec. 6.11 that shows that $\Delta y - dy$ is small when compared with Δx if Δx is small. Prove that $\Delta z - dz$ is small when compared with $\sqrt{(\Delta x)^2 + (\Delta y)^2}$ if Δx and Δy are small.

21. Tell what is wrong with this "proof" that if $z = f(x, y)$ and x and y are functions of t, then dz/dt is just $(\partial z/\partial x)(dx/dt)$: "Since dz/dt is $\lim_{\Delta t \to 0} \Delta z/\Delta t$, $dz/dt = \lim_{\Delta t \to 0} (\Delta z/\Delta x)(\Delta x/\Delta t)$, where Δx is the change in x

induced by the change in t. Thus $dz/dt = (\partial z/\partial x)(dx/dt)$."

22. Prove the converse of Euler's theorem: If $xf_1(x, y) + yf_2(x, y) = nf(x, y)$, then $f(kx, ky) = k^n f(x, y)$.

23. Let $z = f(u, v)$, where u and v are functions of x and y. Then, indirectly, $z = g(x, y)$. Show that, if $du = u_x\, dx + u_y\, dy$ and $dv = v_x\, dx + v_y\, dy$, then the two expressions for dz,

$$dz = z_u\, du + z_v\, dv \quad \text{and} \quad dz = z_x\, dx + z_y\, dy,$$

have equal values.

24. Let (x, y) be retangular coordinates in the plane, and (X, Y, Z) in space. Assume that F is a one-to-one correspondence between the plane and space, such that x and y depend continuously on X, Y, and Z and have continuous partial derivatives with respect to them. Similarly, assume that through the inverse function F^{-1}, X, Y, and Z are continuous functions of x and y and have continuous partial derivatives with respect to them. From this deduce that $2 = 3$. *Hint:* $2 = dx/dx + dy/dy$ and $3 = dX/dX + dY/dY + dZ/dZ$. Use the chain rule. Incidentally, there *is* a one-to-one correspondence between the plane and space, but it does not have the specified properties of continuity and differentiability.

25. Prove that, if f is homogeneous of degree n and has continuous second-order partial derivatives, then

$$x^2 f_{xx} + 2xy f_y + y^2 f_{yy} = n(n-1)f.$$

26. Let x, y, z be functions of u, v, w. Assume that u, v, w can be found as functions of x, y, z. Thus $\partial x/\partial u$ can be expressed as a function of x, y, z. The following argument shows that $\partial x/\partial u$ is *independent* of x, that is, $\partial(\partial x/\partial u)/\partial x = 0$. Argument:

$$\frac{\partial(\partial x/\partial u)}{\partial x} = \frac{\partial^2 x}{\partial x\, \partial u} = \frac{\partial^2 x}{\partial u\, \partial x} = \frac{\partial}{\partial u}\left(\frac{\partial x}{\partial x}\right)$$

$$= \frac{\partial}{\partial u}(1)$$

$$= 0.$$

Apply the result to the case

$$x = (2u)^{1/2}, \quad y = (2v)^{1/2}, \quad z = (2w)^{1/2}.$$

We have

$$\frac{\partial x}{\partial u} = \frac{1}{(2u)^{1/2}} = \frac{1}{x}.$$

Thus $1/x$ is independent of x. Where did the argument go wrong?

Exercises 27 to 34 are of importance in elementary thermodynamics.

27. (See Exercise 19.) Let $z = f(x, y)$, where $y = h(x, u)$. Then z is indirectly a function of x and u, $z = F(x, u)$.

Show that

$$\frac{\partial F}{\partial x} = \frac{\partial f}{\partial x} + \frac{\partial f}{\partial y}\frac{\partial h}{\partial x}.$$

Another notation for this result is used in thermodynamics:

$$\left(\frac{\partial z}{\partial x}\right)_u = \left(\frac{\partial z}{\partial x}\right)_y + \left(\frac{\partial z}{\partial y}\right)_x\left(\frac{\partial y}{\partial x}\right)_u.$$

Hint: Note that x itself is a function of x and u, $x = g(x, u) = x + 0u$. Then use a chain rule.

28. Let $w = f(x_1, x_2, x_3)$ be a function of three variables. The partial derivative of w with respect to x_1 (holding x_2 and x_3 constant) is denoted

$$\frac{\partial w}{\partial x_1} \quad \text{or} \quad \frac{\partial f}{\partial x_1} \quad \text{or } f_1.$$

The symbols $\partial w/\partial x_2$ and $\partial w/\partial x_3$ have similar definitions. Now assume that x_1, x_2, x_3 are functions of t and u:

$$x_1 = g(t, u), \quad x_2 = h(t, u), \quad x_3 = i(t, u).$$

Prove that

$$\frac{\partial w}{\partial t} = \frac{\partial w}{\partial x_1}\frac{\partial x_1}{\partial t} + \frac{\partial w}{\partial x_2}\frac{\partial x_2}{\partial t} + \frac{\partial w}{\partial x_3}\frac{\partial x_3}{\partial t}.$$

29. Consider $w = f(x_1, x_2, x_3)$ and $y = h(t, u)$ with the property that

$$f(t, h(t, u), u) = 0$$

for all t and u. Show that

$$\frac{\partial h}{\partial t} = -\frac{f_1}{f_2} \quad \text{and} \quad \frac{\partial h}{\partial u} = -\frac{f_3}{f_2}.$$

Hint: Consider $x_1 = t$, $x_2 = h(t, u)$, $x_3 = u$ and use a chain rule.

30. (See Exercise 29.) Show that if, instead,

$$f(h(t, u), t, u) = 0,$$

then

$$\frac{\partial h}{\partial t} = -\frac{f_2}{f_1} \quad \text{and} \quad \frac{\partial h}{\partial u} = -\frac{f_3}{f_1}.$$

31. (See Exercise 29.) Show that if, instead,

$$f(t, u, h(t, u)) = 0,$$

then

$$\frac{\partial h}{\partial t} = -\frac{f_1}{f_3} \quad \text{and} \quad \frac{\partial h}{\partial u} = -\frac{f_2}{f_3}.$$

32. The symbols P, T, V denote the pressure, temperature, and volume of a gas. They are not independent. There is a function f of three variables such that the equation

$$f(P, T, V) = 0$$

determines any one of the three variables P, T, V in terms of the other two.

Show that $$\left(\frac{\partial P}{\partial T}\right)_V = \frac{-(\partial V/\partial T)_P}{(\partial V/\partial P)_T}.$$

Hint: See Exercises 29 to 31.

33. Let E (internal energy) be a function of T and P. Recall that T is a function of V and P. Then E is indirectly a function of V and P. Show that

(a) $$\left(\frac{\partial E}{\partial V}\right)_P = \left(\frac{\partial E}{\partial T}\right)_V\left(\frac{\partial T}{\partial V}\right)_P + \left(\frac{\partial E}{\partial V}\right)_T;$$

(b) $$\left(\frac{\partial E}{\partial P}\right)_V = \left(\frac{\partial E}{\partial T}\right)_P\left(\frac{\partial T}{\partial P}\right)_V + \left(\frac{\partial E}{\partial P}\right)_T.$$

34. (a) Prove that $$\left(\frac{\partial P}{\partial T}\right)_V\left(\frac{\partial T}{\partial P}\right)_V = 1.$$

(b) Prove that $$\left(\frac{\partial P}{\partial T}\right)_V\left(\frac{\partial T}{\partial V}\right)_P\left(\frac{\partial V}{\partial P}\right)_T = -1.$$

35. Let $f_x = f_y$ throughout the xy plane. Show that there is a function $g(t)$ such that $f(x, y) = g(x + y)$.

13

DEFINITE INTEGRALS OVER PLANE REGIONS

In Chap. 7 the definite integral of a function f over an interval $[a, b]$ was discussed. We partitioned the interval into sections, chose a sampling point in each section, and then formed an approximating sum. The definite integral is the limit of these sums as the partitions are chosen finer and finer.

A similar procedure can be carried out for regions in the plane, such as triangles, circles, rectangles, etc. Such a region is partitioned into smaller regions, a sampling point selected in each of them, and an analogous sum formed. This leads to a definite integral over a plane region, a tool for computing volumes, centers of gravity, moments of inertia, etc. In Sec. 13.1 the definite integral over a plane region is defined; the sections that follow present two ways of computing it.

13.1

The definite integral of a function over a region in the plane

Two problems will introduce the definite integral of a function over a region in the plane.

A region in the plane shall be a set of points in the plane enclosed by a curve or polygon. When a region is partitioned into subsets, the smaller regions will also be bounded by curves or polygons.

PROBLEM 1   Estimate the volume of the solid S which we now describe. Above each point P in a 4-inch by 2-inch rectangle R erect a line segment whose length, in inches, is the square of the distance from P to the corner A. These segments form a solid S which looks like the figure on the right below. Note that the highest point of S is above the corner of R opposite A; there its height, by the pythagorean theorem, is $4^2 + 2^2 = 20$ inches.

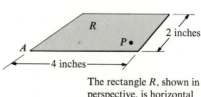

The rectangle R, shown in
perspective, is horizontal

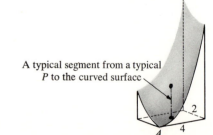

A typical segment from a typical
P to the curved surface

The set R is cut into smaller sets:

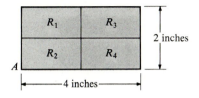

To begin, observe that the volume of S is certainly less than $4 \cdot 2 \cdot 20 = 160$ cubic inches, since S can be put into a box whose base has area $4 \cdot 2$ square inches and whose height is 20 inches.

In order to make more accurate estimates, we cut the rectangular base into smaller pieces. For convenience, cut it into four congruent rectangles R_1, R_2, R_3, and R_4, as in the accompanying drawing. To estimate the volume of S, estimate the volume of that portion of S above each of the rectangles R_1, R_2, R_3, and R_4, and add these estimates. To do this, select a point in each of the four rectangles, say the center of each, and above each rectangle form a box whose height is the height of S above the center of the corresponding rectangle.

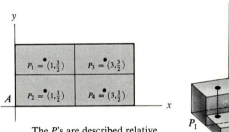

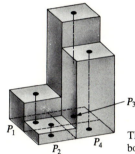

The P's are described relative to the indicated x and y axes

The area of the base of each box is 2 square inches

For instance, the height of the box above R_1 is $1^2 + (\frac{3}{2})^2 = \frac{13}{4}$ inches, and the height of the box above R_2 is $1^2 + (\frac{1}{2})^2 = \frac{5}{4}$ inches. Adding the volumes of the four boxes, we obtain the following estimate of the volume of S:

$$\frac{13}{4}2 + \frac{5}{4}2 + \frac{45}{4}2 + \frac{37}{4}2 = 50 \text{ cubic inches.}$$

This is only an estimate. With the same partition of R we could make other estimates by choosing other P's to determine the heights of approximating boxes.

In general, to estimate the volume of S, begin by partitioning R into smaller subsets R_1, R_2, ..., R_n and select a point P_1 in R_1, P_2 in R_2, ..., P_n in R_n.

A typical partition of R

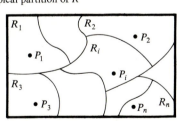

The cylinder has base R_i with area A_i, height $c(P_i)$, and volume $c(P_i)A_i$

Denote the height of S above a typical point P_i in R_i by $c(P_i)$, and the area of R_i by A_i. Then

$$c(P_1) \cdot A_1 + c(P_2) \cdot A_2 + \cdots + c(P_n) \cdot A_n$$

is an estimate of the volume of S by a sum of the volumes of n solids; a typical one may look like the figure shown above. Such a solid is called a cylinder.

PROBLEM 2   Estimate the mass of the rectangular sheet R described as follows. Its dimensions are 4 centimeters by 2 centimeters. The material is sparse near A and dense far from A. Indeed, assume that the density in the vicinity of any point P is equal numerically to the square of the distance from P to A (grams per square centimeter). Note that it is densest at the corner opposite A, where its density is $4^2 + 2^2 = 20$ grams per square centimeter.

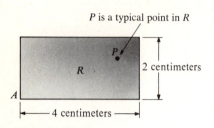

P is a typical point in R

To begin, observe that the total mass is certainly less than $4 \cdot 2 \cdot 20 = 160$ grams, since the area of R is $4 \cdot 2$ square centimeters and the maximum density is 20 grams per square centimeter.

In order to make more accurate estimates, cut the rectangle into smaller pieces, to be specific, into the four congruent rectangles shown in the diagram to the left. Estimate the total mass by estimating the mass in each of the rectangles R_1, R_2, R_3, and R_4. To do this, select a point, say the center, in each of the four rectangles and compute the density at each of these four points. The density at P_1 is $1^2 + (\frac{3}{2})^2 = \frac{13}{4}$ grams per square centimeter. As an estimate of the mass in R_1, we have $\frac{13}{4} \cdot 2$ grams, since the area of R_1 is 2 square centimeters. The sum of the estimates for each of the four rectangles,

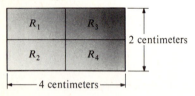

The region R is cut into smaller regions

$$\frac{13}{4}2 + \frac{5}{4}2 + \frac{45}{4}2 + \frac{37}{4}2 = 50 \text{ grams,}$$

is an estimate of the total mass in R.

This is only an estimate. Just as for the volume in Prob. 1 other estimates can be made in the same way. To do so, partition R into small subsets $R_1, R_2, \ldots,$ R_n and select a point P_1 in R_1, P_2 in R_2, $\ldots$, P_n in R_n. Denote the density at P_i by $f(P_i)$ and the area of R_i by A_i. Then

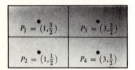

A point is chosen in each region of the partition

$$f(P_1) \cdot A_1 + f(P_2) \cdot A_2 + \cdots + f(P_n) \cdot A_n$$

is an estimate of the total mass.

Even though we have found neither the volume in Problem 1 nor the mass in Problem 2, it is clear that, if we know the answer to one, we have the answer to the other: The arithmetic for calculating any estimate for the volume is the same as that for an estimate of the mass.

The similarity of the sums formed for both problems to the sums met in Chap. 7 suggests that the idea of the definite integral can be generalized from intervals $[a, b]$ to regions in the plane. First, in order to speak of "fine" partitions on the plane, two definitions are required.

DEFINITION   *Diameter of a region.* Let S be a region in the plane bounded by a curve or polygon. The diameter of S is the largest distance between two points of S.

Note that the diameter of a square of side s is $s\sqrt{2}$, and the diameter of a circle whose radius is r is $2r$, its usual diameter.

DEFINITION   *Mesh of a partition in the plane.*   Let $R_1, R_2, \ldots, R_n$ be a partition of a region R in the plane into smaller regions.   The mesh of this partition is the largest of the diameters of the regions $R_1, R_2, \ldots, R_n$.

The mesh of the partitions used in both Problems 1 and 2 is $\sqrt{5}$.

Now the definite integral over a plane region can be defined.

DEFINITION   *Definite integral of a function f over a region R in the plane.*   Let f be a function that assigns to each point P in a region R in the plane a number $f(P)$.   Consider the typical sum

$$f(P_1)A_1 + f(P_2)A_2 + \cdots + f(P_n)A_n,$$

formed from a partition of R, where A_i is the area of R_i, and P_i is in R_i.   If these sums approach a certain number as the mesh of the partitions shrinks toward 0 (no matter how P_i is chosen in R_i), that number is called the definite integral of f over the set R and is written

$$\int_R f(P)\, dA.$$

That is, in shorthand,

$$\lim_{\text{mesh} \to 0} \sum_{i=1}^{n} f(P_i)A_i = \int_R f(P)\, dA.$$

It is illuminating to compare this definition with the definition of the definite integral over an interval.   Both are numbers that are approached by certain sums of products.   The sums are formed in a similar manner, as the table that follows shows.

Given:	For each subset in a partition compute	and select in each of the subsets	and take the limits of sums of the form
An interval and a function defined there	its length $x_i - x_{i-1}$	a point (described by its coordinate X_i)	$\sum_{i=1}^{n} f(X_i)(x_i - x_{i-1})$
A set in the plane and function defined there	its area A_i	a point P_i	$\sum_{i=1}^{n} f(P_i)A_i$

The definite integral is not defined as a sum formed in this table, but rather as the number approached by these sums when the mesh approaches 0.   The definite integral of f over R is sometimes called the *integral of f over R*, or the *integral of $f(P)$ over R*.

For example, the volume in Problem 1 and the mass in Problem 2 are both given by the definite integral

$$\int_R (x^2 + y^2)\, dA,$$

where R is the rectangle that has vertices $(0, 0)$, $(4, 0)$, $(0, 2)$, $(4, 2)$.

The two illustrations given of the definite integral over a plane region are quite important, and we emphasize them by stating them in full generality.

VOLUME OF A SOLID EXPRESSED AS AN INTEGRAL OVER A PLANE REGION R

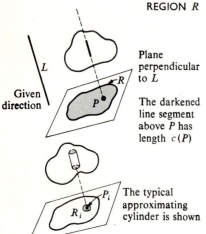

Plane perpendicular to L

The darkened line segment above P has length $c(P)$

The typical approximating cylinder is shown

Consider a solid set S and pick a line L in space. Assume that all lines parallel to L that meet S intersect S in a line segment or a point. Pick a plane perpendicular to L. Let R be the "shadow" or "projection" of S on that plane, that is, the set of all points where lines parallel to L that meet S intersect the plane. For each point P in R let $c(P)$ be the length of the intersection of the line through P parallel to L, with S. Partition R into smaller regions R_1, R_2, ..., R_n and pick a sampling point P_i in R_i for each $i = 1, 2, ..., n$. Let the area of R_i be A_i. Then approximate the volume of S above R_i by a cylinder of height $c(P_i)$ and base congruent to R_i. The volume of the ith cylinder is

$$c(P_i)A_i.$$

Hence

$$\sum_{i=1}^{n} c(P_i)A_i$$

is an estimate of the volume of S. Thus

$$\text{Volume of } S = \int_R c(P)\, dA.$$

MASS OF A FLAT REGION (LAMINA) EXPRESSED AS AN INTEGRAL OVER A PLANE REGION R

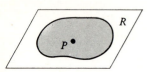

A plane distribution of matter. The density at P is $f(P)$ (grams per square centimeter)

Consider a plane distribution of mass through a region R. The density may vary throughout the region. Denote the density at P by $f(P)$ (in grams per square centimeter). To estimate the total mass in the region R, partition R into small regions R_1, R_2, ..., R_n and pick a sampling point P_i in R_i for each $i = 1, 2, ..., n$. Then the mass in R_i is approximately

$$f(P_i)A_i,$$

since density times area gives mass if the density is constant. Thus

$$\sum_{i=1}^{n} f(P_i)A_i$$

is an estimate of the mass in R. Consequently,

$$\text{Mass in } R = \int_R f(P)\, dA.$$

For engineers and physicists this is an important interpretation of the two-dimensional integral: The definite integral of density equals mass.

That the definite integral of density over an interval gives the total mass for matter distributed along a string was shown in Chap. 7. In Chap. 18 the analogous fact is shown to hold for matter distributed in space, once the definite integral over a region in space is defined.

To emphasize further the similarity between integrals over plane sets and integrals over intervals, we define the average of a function over a plane set.

DEFINITION *Average value.* The average value of f over the region R is

$$\frac{\int_R f(P)\, dA}{\text{area of } R}.$$

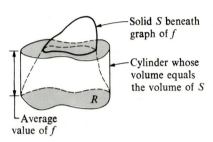

Solid S beneath graph of f

Cylinder whose volume equals the volume of S

R

Average value of f

If $f(P)$ is positive for all P in R, a simple geometric interpretation of the average of f over R can be given. Let S be the solid situated below the graph of f (a surface) and above the region R. The average value of f over R is the height of the cylinder whose base is R and whose volume is the same as the volume of S. This is the analog of the observation in Sec. 10.5 concerning the average of $f(x)$ over $[a, b]$.

The integral $\int_R f(P)\, dA$ is called a *two-dimensional integral* to distinguish it from $\int_a^b f(x)\, dx$, which, for contrast, is called a *one-dimensional integral*.

Exercises

1. In the estimates for the volume in Problem 1, the centers of the subrectangles were used as the P_i's. Make an estimate for the volume in Problem 1 by using the same partition but taking as the P_i's (a) the lower left corner of each R_i, (b) the upper right corner of each R_i. (c) What do (a) and (b) tell about the volume of the solid?

2. Estimate the mass in Problem 2 using a partition of R into eight congruent squares and taking as the P_i's (a) centers, (b) upper right corners, (c) lower left corners.

3. Let R be a set in the plane whose area is A. Let f be the function such that $f(P) = 5$ for every point P in R.
   (a) What can be said about any approximating sum $\sum_{i=1}^{n} f(P_i)A_i$ formed for this R and this f?
   (b) What is the numerical value of $\int_R f(P)\, dA$?

4. Let $f(P) = 1$ for every point P in the plane region R. Show that $\int_R f(P)\, dA$ is simply the area of R.

5. (a) Let f and R be as in Prob. 1 or 2. Use the estimate of $\int_R f(P)\, dA$ obtained in the text to estimate the average of f over R.
   (b) Using the information from Exercise 2, show that the average is between 4 and 10.

6. (a) Prove that, if the diameter of a set in the plane is d, then its area is less than d^2.
   (b) Which has the largest area: a circle of diameter d, a square of diameter d, or an equilateral triangle of diameter d?
   (c) How large an area do you think a set of diameter d can have?

(d) Can a circle of diameter d always be drawn to contain a given set of diameter d?

In Exercises 7 to 10 describe the general shape of the projection (shadow) of the given solid on the given plane.

7. A sphere; any plane.

8. A right circular cylinder; a plane parallel to the axis of the cylinder.

9. A right circular cone; a plane parallel to the axis of the cone.

10. A right circular cone; a plane parallel to the base.

11. Assume that, for all P in R, $m \le f(P) \le M$. Let A be the area of R. By examining approximating sums, show that

$$mA \le \int f(P)\, dA \le MA.$$

■

A calculator or tables would be of aid in Exercises 12 to 14.

12. (a) Let R be the rectangle with vertices $(0, 0)$, $(2, 0)$, $(2, 3)$, $(0, 3)$. Let $f(x, y) = \sqrt{x + y}$. Estimate $\int_R \sqrt{x + y}\, dA$ by partitioning R into six squares and choosing the sampling points to be their centers.
   (b) Use (a) to estimate the average value of f over R.

13. (a) Let R be the square with vertices $(0, 0)$, $(0.8, 0)$, $(0.8, 0.8)$, $(0, 0.8)$. Let $f(P) = f(x, y) = e^{xy}$. Estimate $\int_R e^{xy}\, dA$ by partitioning R into 16 squares and choosing the sampling points to be their centers.

(b) Use (a) to estimate the average value of $f(P)$ over R.

(c) Show that $0.64 \leq \int_R f(P)\, dA \leq 0.64 e^{0.64}$

14. (a) Let R be the triangle with vertices $(0, 0)$, $(4, 0)$, $(0, 4)$.

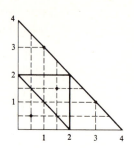

Let $f(x, y) = x^2 y$. Use the partition into four triangles and sampling points shown in the diagram to estimate $\int_R f(P)\, dA$.

(b) What is the maximum value of $f(x, y)$ in R?

(c) From (b) obtain an upper bound on $\int_R f(P)\, dA$.

■ ■

15. (a) Let R be the disk of radius 1. Let $f(P)$ be the distance from P to the center of the circle. By partitioning R with the aid of rays through the center and concentric circles, show that $\int_R f(P)\, dA = 2\pi/3$.

(b) Find the average value of f over R.

13.2

How to describe a plane region by coordinates

In Sec. 13.3 we shall evaluate definite integrals over plane regions by means of two integrals over intervals. The method requires a description of these regions in terms of a coordinate system. The present section is devoted to this aspect of analytic geometry. Examples illustrate the method first for rectangular coordinates, and then for polar coordinates.

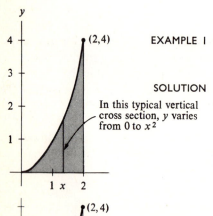

EXAMPLE 1   Let R be the region bounded by $y = x^2$, the x axis, and the line $x = 2$. Describe R in terms of cross sections parallel to the y axis.

SOLUTION   A glance at R shows that, for points (x, y) in R, x ranges from 0 to 2. To describe R completely, we shall describe the behavior of y for any x in the interval $[0, 2]$.

In this typical vertical cross section, y varies from 0 to x^2

Hold x fixed and consider only the cross section above the point $(x, 0)$. It extends from the x axis to the curve $y = x^2$. For any x, the y coordinate varies from 0 to x^2. This is a complete description of R by vertical cross sections, written in compact notation:

$$0 \leq x \leq 2, \qquad 0 \leq y \leq x^2. \; \bullet$$

EXAMPLE 2   Describe the region R of Example 1 by cross sections parallel to the x axis, that is, by horizontal cross sections.

SOLUTION   A glance at R shows that y varies from 0 to 4. For any y in the interval $[0, 4]$, x varies from a smallest value $x_1(y)$ to a largest value $x_2(y)$. Note that $x_2(y) = 2$ for each value of y in $[0, 4]$. To find $x_1(y)$, utilize the fact that the point $(x_1(y), y)$ is on the curve $y = x^2$,

$$x_1(y) = \sqrt{y}.$$

The description of R in terms of horizontal cross sections is

$$0 \leq y \leq 4, \qquad \sqrt{y} \leq x \leq 2. \quad \bullet$$

EXAMPLE 3   Describe the region R whose vertices are $(0, 0)$, $(6, 0)$, $(4, 2)$, and $(0, 2)$ by vertical cross sections, and then by horizontal cross sections.

SOLUTION   Clearly x varies between 0 and 6.   For any x in the interval $[0, 4]$, y ranges from 0 to 2 (independently of x).   For x in $[4, 6]$, y ranges from 0 to the value of y on the line through $(4, 2)$ and $(6, 0)$.   This line has the equation

$$y = 6 - x.$$

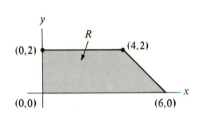

The description of R by vertical cross sections therefore requires two separate statements:

$$0 \leq x \leq 4, \qquad 0 \leq y \leq 2;$$

and $\qquad\qquad 4 \leq x \leq 6, \qquad 0 \leq y \leq 6 - x.$

Use of horizontal cross sections provides a simpler description.   First, y goes from 0 to 2.   For each y in $[0, 2]$, x goes from 0 to the value of x on the line

$$y = 6 - x.$$

Solving this equation for x yields $x = 6 - y$.

The description is much shorter:

$$0 \leq y \leq 2, \qquad 0 \leq x \leq 6 - y \quad \bullet$$

These three examples are typical.   First determine the range of one coordinate and then how the other coordinate varies for any fixed value of the first coordinate.

The method is the same when polar coordinates are used: First determine the range of θ, and then see how r varies for any fixed value of θ.

EXAMPLE 4   Let R be the circle of radius a and center at the pole of a polar coordinate system. Describe R in terms of cross sections by rays emanating from the pole.

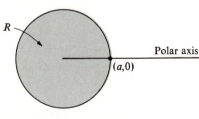

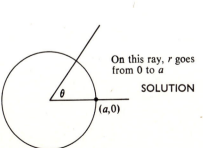

On this ray, r goes from 0 to a

SOLUTION   To sweep out R, θ goes from 0 to 2π.   Hold θ fixed and consider the behavior of r on the ray of angle θ.   Clearly r goes from 0 to a, independently of θ.   The complete description is

$$0 \leq \theta \leq 2\pi, \qquad 0 \leq r \leq a. \quad \bullet$$

EXAMPLE 5 Let R be the region between the circles $r = 2 \cos \theta$ and $r = 4 \cos \theta$. Describe R in terms of cross sections by rays from the pole.

SOLUTION To sweep out this region, use the rays from $\theta = -\pi/2$ to $\theta = \pi/2$. For each such θ, r varies from $2 \cos \theta$ to $4 \cos \theta$. The complete description is

$$-\frac{\pi}{2} \leq \theta \leq \frac{\pi}{2}, \qquad 2 \cos \theta \leq r \leq 4 \cos \theta. \quad \bullet$$

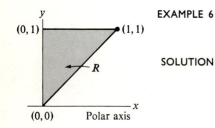

As Examples 4 and 5 suggest, polar coordinates provide simple descriptions for regions bounded by circles. The next example shows that polar coordinates may also provide simple descriptions of regions bounded by straight lines, especially if the lines pass through the origin.

EXAMPLE 6 Let R be the triangular region whose vertices, in rectangular coordinates, are $(0, 0)$, $(1, 1)$, and $(0, 1)$. Describe R in polar coordinates.

SOLUTION Inspection of R shows that θ varies from $\pi/4$ to $\pi/2$. For each θ, r goes from 0 until the point (r, θ) is on the line $y = 1$, that is, on the line $r \sin \theta = 1$. Thus the upper limit of r for each θ is $1/\sin \theta$. The description of R is:

$$\frac{\pi}{4} \leq \theta \leq \frac{\pi}{2}, \qquad 0 \leq r \leq \frac{1}{\sin \theta}. \quad \bullet$$

In practice, the region R often has the property that its intersection with a horizontal or vertical line or a ray from some pole is a line segment. This greatly simplifies the description of R if the appropriate coordinates are used. If vertical cross sections are used, the description will be of the form

$$a \leq x \leq b, \qquad y_1(x) \leq y \leq y_2(x),$$

where $y_1(x)$ and $y_2(x)$ are functions of x (perhaps constant). In the case of horizontal cross sections, the description will have the form

$$c \leq y \leq d, \qquad x_1(y) \leq x \leq x_2(y).$$

Cross sections by rays will lead to this type of description:

$$\alpha \leq \theta \leq \beta, \qquad r_1(\theta) \leq r \leq r_2(\theta).$$

Exercises

1. (a) Draw the triangular region whose vertices are $(0, 0)$, $(2, 1)$, and $(0, 1)$.
   (b) Describe it by vertical cross sections.
2. Describe the region in Exercise 1 by horizontal cross sections.
3. (a) Draw the region between $y = x$ and $y = x^2$.
   (b) Describe it by vertical cross sections.

4. Describe the region in Exercise 3 by horizontal cross sections.
5. Describe the circle in Example 4 by vertical cross sections (placing the xy-coordinate system in standard position).
6. Describe the region in Example 5 by (a) vertical cross sections, (b) horizontal cross sections.

In Exercises 7 to 12 draw the region described.

7. $0 \le x \le 2,\ 2x \le y \le 3x.$ 8. $0 \le x \le 1,\ x^2 \le y \le x.$

9. $1 \le y \le e,\ \ln y \le x \le 1.$ 10. $0 \le \theta \le \dfrac{\pi}{2},\ 0 \le r \le 2.$

11. $-\dfrac{\pi}{4} \le \theta \le \dfrac{\pi}{4},\ 1 \le r \le 2.$

12. $0 \le \theta \le 2\pi,\ 1 \le r \le 2 + \cos\theta.$

13. Describe the region in Exercise 7 by horizontal cross sections.

14. Describe the region in Exercise 8 by horizontal cross sections.

15. Describe the region in Exercise 9 by vertical cross sections.

16. (a) Draw the region R whose description is $-2 \le y \le 2$, $-\sqrt{4 - y^2} \le x \le \sqrt{4 - y^2}.$

    (b) Describe R by vertical cross sections.

    (c) Describe R by polar coordinates.

17. (a) Draw the region R bounded by the four lines $y = 1$, $y = 2$, $y = x$, $y = x/3$.

    (b) Describe R in terms of cross sections. (Choose the direction that is most convenient.)

13.3

Computing $\int_R f(P)\, dA$ **by introducing rectangular coordinates in** R

This section gives an intuitive development of formulas for the rapid computation of definite integrals over plane regions. Such questions as, "What properties of f and R ensure the existence of $\int_R f(P)\, dA$?" will not concern us. Just as the reasoning in Sec. 8.1 made the fundamental theorem of calculus plausible, the reasoning in this chapter will be more persuasive than rigorous. No proofs will be included. We are able to discuss the proof of the fundamental theorem of calculus because an interval $[a, b]$ offers less complication than a region R in the plane. It suffices to say that, if R is bounded by fairly simple curves and f is continuous, then the various formulas are valid.

We first develop a method for computing a definite integral over a rectangle. After applying this formula in Example 1, we make the slight modification needed to evaluate integrals over more general regions.

Consider a rectangular region R whose description by vertical cross sections is

$$a \le x \le b, \qquad c \le y \le d.$$

If $f(P) \ge 0$ for all P in R, then $\int_R f(P)\, dA$ is the volume V of the solid whose base is R and which has above P a linear cross section with the solid of height $f(P)$. Let $A(x)$ be the area of the cross section made by a plane perpendicular to the x axis and having abscissa x.

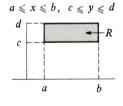

$a \le x \le b,\ c \le y \le d$

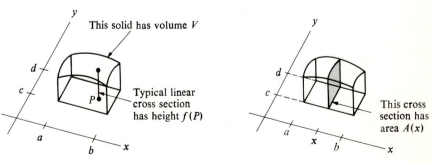

This solid has volume V

Typical linear cross section has height $f(P)$

This cross section has area $A(x)$

As was shown in Sec. 7.4, $$V = \int_a^b A(x)\,dx.$$

But the area $A(x)$ is itself expressible as a definite integral:

$$A(x) = \int_c^d f(x, y)\,dy.$$

Note that x is held fixed throughout this integration.   This reasoning provides a repeated integral whose value is $V = \int_R f(P)\,dA$, namely

$$\int_R f(P)\,dA = V$$

$$= \int_a^b A(x)\,dx$$

$$= \int_a^b \left[\int_c^d f(x, y)\,dy \right] dx.$$

Of course, cross sections by planes perpendicular to the y axis could be used. Then similar reasoning shows that

$$\int_R f(P)\,dA = \int_c^d \left[\int_a^b f(x, y)\,dx \right] dy.$$

The quantities $\int_a^b [\int_c^d f(x, y)\,dy]\,dx$ and $\int_c^d [\int_a^b f(x, y)\,dx]\,dy$ are called *repeated integrals* or *iterated integrals*.

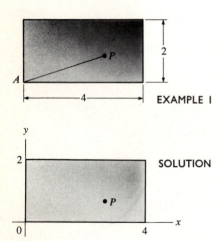

EXAMPLE I   Compute the definite integral $\int_R f(P)\,dA$, where R is the first rectangle shown in the margin, and the function f is defined by $f(P) = \overline{AP}^2$.   (This integral was estimated in Sec. 13.1.)

SOLUTION   Introduce xy coordinates in the convenient manner depicted in the second figure in the margin.   Then f has this description in rectangular coordinates:

$$f(x, y) = \overline{AP}^2 = x^2 + y^2.$$

To describe R, observe that x takes all values from 0 to 4 and that for each x the number y takes all values between 0 and 2.   Thus

$$\int_R f(P)\,dA = \int_0^4 \left[\int_0^2 (x^2 + y^2)\,dy \right] dx.$$

We must first compute $$\int_0^2 (x^2 + y^2)\,dy,$$

where x is fixed in $[0, 4]$.   [This is the cross-sectional area $A(x)$.]   To apply the fundamental theorem of calculus, first find a function $F(y)$ such that

$$\frac{dF}{dy} = x^2 + y^2.$$

Keep in mind that x is constant during this first integration.

$$F(y) = x^2 y + \frac{y^3}{3}$$

is such a function.   The appearance of x in its formula should not disturb us, since x is fixed for the time being.   In fact, the derivative dF/dy is actually a partial derivative, $\partial F/\partial y$.   By the fundamental theorem of calculus,

$$\int_0^2 (x^2 + y^2)\, dy = \left(x^2 y + \frac{y^3}{3} \right) \Bigg|_{y=0}^{y=2}$$

$$= \left[x^2(2) + \frac{2^3}{3} \right] - \left[x^2(0) + \frac{0^3}{3} \right]$$

$$= 2x^2 + \tfrac{8}{3}.$$

[The integral $2x^2 + \tfrac{8}{3}$ is the area $A(x)$ discussed earlier in this section.]

Now compute $$\int_0^4 (2x^2 + \tfrac{8}{3})\, dx.$$

By the fundamental theorem of calculus,

$$\int_0^4 \left(2x^2 + \frac{8}{3} \right) dx = \left(\frac{2x^3}{3} + \frac{8x}{3} \right) \Bigg|_0^4 = \frac{160}{3}.$$

Hence the two-dimensional definite integral has the value $\frac{160}{3}$.   The volume of the region in Problem 1 of Sec. 13.1 is $\frac{160}{3}$ cubic inches.   The mass in Problem 2 is $\frac{160}{3}$ grams. ●

The Repeated Integrals for $\int_R f(P)\, dA$ *in Rectangular Coordinates*   If R is not a rectangle, the repeated integral that equals $\int_R f(P)\, dA$ differs from that for $\int_R f(P)\, dA$, where R is a rectangle, only in the intervals of integration. If R has the description

$$a \le x \le b, \qquad y_1(x) \le y \le y_2(x),$$

by cross sections parallel to the y axis, then

$$\int_R f(P)\, dA = \int_a^b \left[\int_{y_1(x)}^{y_2(x)} f(x, y)\, dy \right] dx.$$

Similarly, if R has the description

$$c \le y \le d, \qquad x_1(y) \le x \le x_2(y),$$

by cross sections parallel to the x axis, then

$$\int_R f(P)\, dA = \int_c^d \left[\int_{x_1(y)}^{x_2(y)} f(x, y)\, dx \right] dy.$$

The intervals of integration are determined by R; the function f influences only the integrand.

The next example illustrates the method.

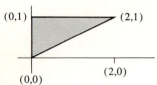

(0,1)   (2,1)

(2,0)

(0,0)

EXAMPLE 2   A triangular lamina is located as in the diagram at the left.   Its density at (x, y) is e^{y^2}.   Find its mass, that is,

$$\int_R f(P)\, dA,$$

where $f(x, y) = e^{y^2}$.

SOLUTION   The description of R by vertical cross sections is

$$0 \le x \le 2, \quad \frac{x}{2} \le y \le 1.$$

Hence

$$\int_R f(P)\, dA = \int_0^2 \left(\int_{x/2}^1 e^{y^2}\, dy \right) dx.$$

Unfortunately, the fundamental theorem of calculus is useless in computing

$$\int_{x/2}^1 e^{y^2}\, dy,$$

so try horizontal cross sections instead.

The description of R is now

$$0 \le y \le 1, \quad 0 \le x \le 2y.$$

Thus

$$\int_R f(P)\, dA = \int_0^1 \left(\int_0^{2y} e^{y^2} dx \right) dy.$$

The first integration $\int_0^{2y} e^{y^2}\, dx$ is easy, since y is fixed; the integrand is constant. Thus

$$\int_0^{2y} e^{y^2}\, dx = e^{y^2} \int_0^{2y} 1\, dx = e^{y^2} x \Big|_{x=0}^{x=2y}$$

$$= e^{y^2} 2y.$$

The second definite integral in the repeated integral is thus $\int_0^1 e^{y^2}\, 2y\, dy$, which luckily can be evaluated by the fundamental theorem of calculus, since $d(e^{y^2})/dy = e^{y^2} 2y$:

$$\int_0^1 e^{y^2} 2y\, dy = e^{y^2} \Big|_0^1 = e^{1^2} - e^{0^2} = e - 1.$$

The total mass is $e - 1$.   ●

Notice that computing a definite integral over R involves, first, a wise choice of an xy-coordinate system; second, a description of R and f relative to this coordinate system; and finally, the computation of two successive definite integrals over intervals.   The order of these integrations should be considered carefully since computation may be much simpler in one than in the other. This order is determined by the description of R by cross sections.   For instance, if the description is in the form

$$a \le x \le b, \quad y_1(x) \le y \le y_2(x),$$

then the repeated integral has the form

$$\int_a^b \left[\int_{y_1(x)}^{y_2(x)} f(x, y) \, dy \right] dx,$$

and the y integration is performed first.

EXAMPLE 3   Let R be the region in the plane bounded by $y = x^2$, $x = 2$, and $y = 0$. At each point $P = (x, y)$ erect a line segment of height $3xy$. What is the volume of the resulting solid? (This region R is described in Example 2 of Sec. 13.2.)

SOLUTION   The volume is $\int_R f(P) \, dA$, where $f(P)$ is the length of the line segment above P cut off by the solid. In this case, this integral is $\int_R 3xy \, dA$.

The base R is bounded by $y = x^2, x = 2$, and $y = 0$

$y = x^2$

If cross sections parallel to the y axis are used, then R is described by

$$0 \le x \le 2, \qquad 0 \le y \le x^2.$$

Thus

$$\int_R 3xy \, dA = \int_0^2 \left(\int_0^{x^2} 3xy \, dy \right) dx,$$

which is easy to compute. First with x fixed,

$$\int_0^{x^2} 3xy \, dy = \left(3x \frac{y^2}{2} \right) \Big|_{y=0}^{y=x^2}$$

$$= 3x \frac{(x^2)^2}{2} - 3x \frac{(0)^2}{2}$$

$$= \frac{3x^5}{2}.$$

Then

$$\int_0^2 \frac{3x^5}{2} \, dx = \frac{3x^6}{12} \Big|_0^2 = 16.$$

R can also be described in terms of cross sections parallel to the x axis:

$$0 \le y \le 4, \qquad \sqrt{y} \le x \le 2.$$

Then

$$\int_R 3xy \, dA = \int_0^4 \left(\int_{\sqrt{y}}^2 3xy \, dx \right) dy,$$

which, as the reader may verify, equals 16. ●

Exercises

1. Solve Example 3 using cross sections parallel to the x axis.

In Exercises 2 to 7 evaluate the repeated integral.

2. $\int_0^1 \left[\int_0^x (x + 2y) \, dy \right] dx$ **3.** $\int_1^2 \left(\int_x^{2x} dy \right) dx$

4. $\int_0^2 \left(\int_0^{x^2} xy^2 \, dy \right) dx$ **5.** $\int_1^2 \left(\int_0^{\sqrt{y}} yx^2 \, dx \right) dy$

6. $\int_1^2 \left(\int_0^y e^{x+y} \, dx \right) dy$ **7.** $\int_0^1 \left(\int_0^x y \sin \pi x \, dy \right) dx$

In Exercises 8 to 15 find the volume of the solid whose cross section by a line perpendicular to the xy plane has the given length, and whose projection on the xy plane is the given region R.

8. $x + 2y$; R is the triangle whose vertices are $(0, 0)$, $(1, 0)$, $(1, 1)$.

9. xy; R is bordered by $y = x^2$, the x axis, and the line $x = 2$.

10. $x + 3y$; R is bordered by $y = x^2$, the x axis, and the line $x = 2$.

11. $x^2 + 2y^2$; R is the rectangle whose vertices are $(0, 0)$, $(3, 0)$, $(3, 2)$, $(0, 2)$.

12. $x + y$; R is the triangle whose vertices are $(1, 1)$, $(2, 1)$, $(1, 2)$.

13. y; R is the portion of a circle of radius a, center at $(0, 0)$ above the x axis.

14. xy; R is bordered by $y = \ln x$, the x axis, the y axis, and the line $y = 1$.

15. xy; R is bordered by $y = \sin x$, the x axis, the line $x = \pi/2$ and the y axis.

16. Find the mass of a thin lamina occupying the square whose vertices are $(0, 0)$, $(0, 2)$, $(2, 2)$, $(2, 0)$ and whose density at (x, y) is $\sqrt{xy}$.

17. Find the mass of a thin lamina occupying the triangle whose vertices are $(0, 0)$, $(1, 0)$, $(1, 1)$ and whose density at (x, y) is $1/(1 + x^2)$.

18. Let $f(x, y) = y^2 e^{y^2}$ and let R be the triangle bounded by $y = a$, $y = x/2$, and $y = x$. Assume that a is positive.
(a) Set up two repeated integrals for $\int_R f(P)\, dA$.
(b) Evaluate the easier one.

19. Find the average value of the cross-sectional length in Exercise 8.

20. Find the average value of the cross-sectional length in Exercise 9.

21. Find the average density of the lamina in Exercise 17.

22. Let R be the set bounded by the curve $y = \sqrt{x}$ and the line $y = x$.   Let $f(x, y) = (\sin y)/y$ if $y \neq 0$ and $f(x, 0) = 1$. Compute $\int_R f(P)\, dA$.

23. Compute $\displaystyle\int_0^3 \left(\int_x^3 \frac{\sin y}{y}\, dy \right) dx$.

■ ■

24. Let $f(x, y) = e^{y^3}$.
(a) Devise a region R in the plane such that $\int_R f(P)\, dA$ can be evaluated with the aid of a repeated integral.
(b) Devise a region R in the plane such that $\int_R f(P)\, dA$ cannot be evaluated with the aid of a repeated integral, and describe the difficulty.

13.4

The center of gravity of a flat object (lamina)

In this section the definite integral over a plane set is used to deal with balancing lines and the center of gravity.   The results obtained would be difficult to derive if our only tool were the definite integral over intervals.   Moreover the argument shows how the *concept* of a definite integral over a plane region is useful not only for computations.

A small boy on one side of a seesaw (which we regard as weightless) can balance a bigger boy on the other side.   For example, these two boys balance.

Mass of 40 pounds   Mass of 90 pounds

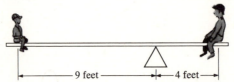

|← 9 feet →|← 4 feet →|

The small mass with the long lever arm balances the large mass with the small lever arm.   Each contributes the same tendency to turn—but in opposite directions.   To be more precise, introduce on the seesaw an x axis with its origin 0 at the fulcrum.   Define the moment about 0 of a mass m located at the point x on the x axis to be the product mx.   Then the bigger boy has a moment $(90)(4)$, while the smaller boy has a moment $(40)(-9)$.   The total moment of the lever-mass system is 0, and the boys balance.

If a mass m is located on a line at coordinate x, define its moment about the point having coordinate a as the product $m(x - a)$.

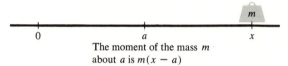

The moment of the mass m
about a is $m(x - a)$

Now consider several masses $m_1, m_2, \ldots, m_n$.   If mass m_i is located at x_i, with $i = 1, 2, \ldots, n$, then $\sum_{i=1}^{n} m_i(x_i - a)$ is the total moment of all the masses about the point a.   If a fulcrum is placed at a, then the seesaw rotates clockwise if the total moment is greater than 0, counterclockwise if it is less than 0, and is in equilibrium if the total moment is 0.

EXAMPLE I   Where should the fulcrum be placed so that these three masses will be in equilibrium?

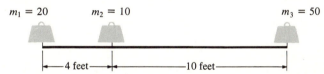

SOLUTION   Introduce an x axis with origin at mass m_1 and compute the moments about a typical fulcrum having coordinate a; then select a to make the total moment 0.

The total moment about a is

$$20(0 - a) + 10(4 - a) + 50(14 - a).$$

We seek a such that this expression is equal to 0 or, equivalently,

$$(20 \cdot 0) + (10 \cdot 4) + (50 \cdot 14) = a(20 + 10 + 50).$$

Hence
$$a = \frac{(20 \cdot 0) + (10 \cdot 4) + (50 \cdot 14)}{80} = 9.25.$$

This means the fulcrum is to the right of the midpoint, which was to be expected. ●

Matter of "0–thickness" occupies region
R. How should its moment about L be
defined?

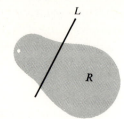

How should we define the moment about a line L in the plane of a distribution
of matter in the plane, a so-called *lamina*?

Assume that the lamina occupies the region R and has a density $f(P)$ at each
point P in R.

Introduce an x axis perpendicular to L in order to measure the "lever arm."

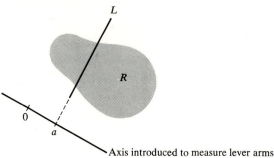

Axis introduced to measure lever arms

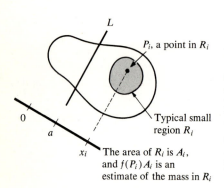

P_i, a point in R_i

Typical small
region R_i

x_i The area of R_i is A_i,
and $f(P_i)A_i$ is an
estimate of the mass in R_i

Assume that L passes through the point $x = a$ on this axis.   Imagine R partitioned into small regions R_1, R_2, ..., R_n.   What would be a reasonable
estimate of the moment about L contributed by the mass in R_i?   If we choose a
point P_i in R_i and denote the area of R_i by A_i, then the mass in R_i is approximately $f(P_i)A_i$.   The lever arm of P_i is $(x_i - a)$ if the line through P_i and parallel
to L meets the x axis at x_i.   The typical small region R_i contributes a moment
about L roughly equal to (lever arm) · (mass), or $(x_i - a)f(P_i)A_i$.   If finer
partitions are made, such estimates should be more accurate.   Therefore define
the moment of the lamina about L as

$$\int_R (x - a)f(P)\,dA.$$

If the total moment about L is 0, then L is called a *balancing line*.

THEOREM I   There is exactly one balancing line for any given lamina in each direction in
the plane of that lamina.

PROOF   Introduce an x axis perpendicular to the given direction.   We seek a such that
$\int_R (x - a)f(P)\,dA = 0$.   Now, just as in definite integrals over intervals,

$$\int_R (x - a)f(P)\,dA = \int_R xf(P)\,dA - a\int_R f(P)\,dA.$$

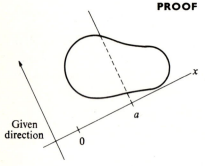

Given
direction

So a satisfies the equation

$$a\int_R f(P)\,dA = \int_R xf(P)\,dA.$$

This equation has a unique solution,

$$a = \frac{\int_R xf(P)\,dA}{\int_R f(P)\,dA} = \frac{\int_R xf(P)\,dA}{\text{mass of lamina}},$$

and the theorem is proved. ●

THEOREM 2 All the balancing lines for a given lamina pass through a single point.

PROOF Let L_1 and L_2 be two perpendicular balancing lines. Let L be any line through the intersection of L_1 and L_2. We shall prove that L is a balancing line. (That will be sufficient to establish Theorem 2, since Theorem 1 shows that there is only one balancing line in any direction.) For convenience, use L_1 and L_2 as x and y axes. Since L_1 and L_2 are balancing lines, by Theorem 1,

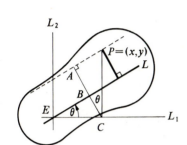

$$\int_R yf(P)\,dA = 0 \quad \text{and} \quad \int_R xf(P)\,dA = 0.$$

Now examine the lever arm of $P = (x, y)$ around L. Inspection of the second diagram in the margin shows that this lever arm is

$$\overline{AC} - \overline{BC} = y\cos\theta - x\sin\theta,$$

an expression which is positive or negative depending on the side of L on which the point P lies.

It is necessary to show that

$$\int_R (y\cos\theta - x\sin\theta)f(P)\,dA = 0. \tag{1}$$

Since L is fixed, $\cos\theta$ and $\sin\theta$ are constants; hence the definite integral in (1) equals

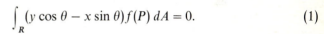

$$\cos\theta \int_R yf(P)\,dA - \sin\theta \int_R xf(P)\,dA.$$

$AC = y\cos\theta$ (by right triangle CAP)
$BC = x\sin\theta$ (by right triangle EBC)

Since $\int_R yf(P)\,dA = 0$ and $\int_R xf(P)\,dA = 0$, Eq. (1) follows, and the theorem is proved. ●

The point through which all the balancing lines pass is called the *center of gravity* of the lamina. Observe that two distinct balancing lines suffice to determine the center of gravity. Call the coordinates of the center of gravity $(\bar{x}, \bar{y})$. Then,

$$\bar{x} = \frac{\int_R xf(P)\,dA}{\text{mass of lamina}} \quad \text{and} \quad \bar{y} = \frac{\int_R yf(P)\,dA}{\text{mass of lamina}}.$$

In the particular case where the density is constant, $f(P) = k$ for all P in R, and then the formulas become much simpler. In this case the center of gravity is called the *centroid* of R, and its coordinates are given by these formulas:

$$\bar{x} = \frac{\int_R x\,dA}{\text{area of } R} \quad \text{and} \quad \bar{y} = \frac{\int_R y\,dA}{\text{area of } R}.$$

To show this, observe that

$$\bar{x} = \frac{\int_R xk\,dA}{\int_R k\,dA} = \frac{k\int_R x\,dA}{k\int_R 1\,dA} = \frac{\int_R x\,dA}{\text{area of } R}.$$

Similar reasoning obtains the formula for $\bar{y}$.

If the density is constant, $\bar{x}$ is the average value of x over the region R, and $\bar{y}$

is the average value of y.   The integral $\int_R x\, dA$ is also called the *moment of R about the y axis*, and the integral $\int_R y\, dA$ is called the *moment of R about the x axis*.

Keep in mind that an object balances on any line through its center of gravity (or centroid).   Conversely, its center of gravity (or centroid) lies on every balancing line.

In Example 2 we compute a centroid and in Example 3 a center of gravity.

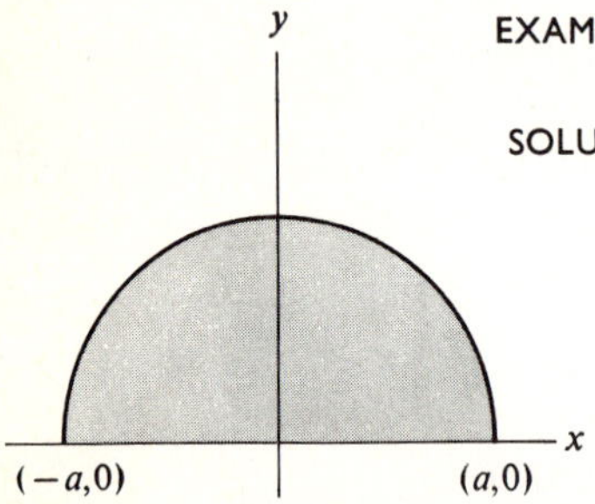

EXAMPLE 2   Find the centroid of a semicircle of radius a.

SOLUTION   To begin, introduce a rectangular-coordinate system, placing the origin at the center of the semicircle and the x axis along the diameter.

In this case, $\bar{x}$ is 0 since the y axis is a balancing line.   (By symmetry, the tendency of the semicircle to turn around the y axis is 0.)   The formula for $\bar{y}$ is

$$\bar{y} = \frac{\int_R y\, dA}{\text{area of } R}.$$

Now, the area of R is $\pi a^2/2$.   All that remains is to evaluate $\int_R y\, dA$ by means of a repeated integral.

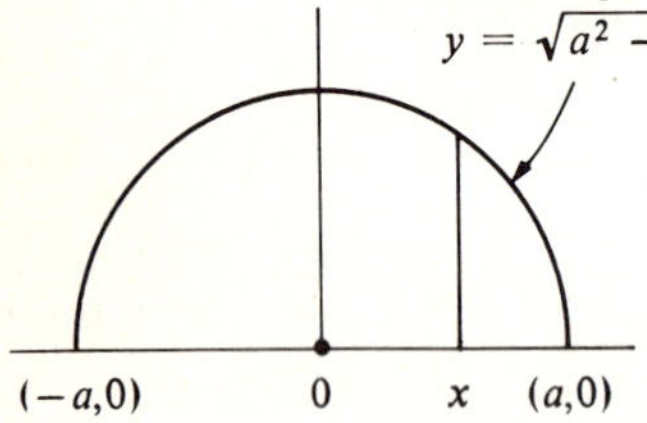

A typical cross section of R by a line parallel to the y axis is shown in the second diagram at the left.   The description of R by these cross sections is

$$-a \leq x \leq a, \qquad 0 \leq y \leq \sqrt{a^2 - x^2}.$$

Hence

$$\int_R y\, dA = \int_{-a}^{a} \left(\int_0^{\sqrt{a^2 - x^2}} y\, dy \right) dx.$$

The inner integral is easy to compute:

$$\int_0^{\sqrt{a^2 - x^2}} y\, dy = \frac{y^2}{2} \Big|_{y=0}^{y=\sqrt{a^2 - x^2}}$$

$$= \frac{(\sqrt{a^2 - x^2})^2}{2} - \frac{0^2}{2}$$

$$= \frac{a^2 - x^2}{2}.$$

The second integral is then

$$\int_{-a}^{a} \frac{a^2 - x^2}{2}\, dx = \left(\frac{a^2 x}{2} - \frac{x^3}{6} \right) \Big|_{-a}^{a}$$

$$= \left(\frac{a^2(a)}{2} - \frac{a^3}{6} \right) - \left(\frac{a^2(-a)}{2} - \frac{(-a)^3}{6} \right)$$

$$= \frac{2a^3}{3}.$$

Thus

$$\bar{y} = \frac{2a^3/3}{\pi a^2/2} = \frac{4}{3\pi} a \doteq 0.4244\, a.$$

This semicircular curve has the equation

$$y = \sqrt{a^2 - x^2}$$

The centroid is on the vertical radius, about 44 percent of the way from the center to the border. ●

EXAMPLE 3   A lamina occupies the triangle whose vertices are $(0, 0)$, $(1, 0)$, $(1, 2)$ and has density x^2y at the point (x, y).   Find its center of gravity.

SOLUTION   In this case it will be necessary to compute the three definite integrals:

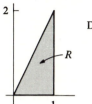

Density at (x,y) is x^2y

$$\text{Mass} = \int_R x^2y \, dA,$$

$$\int_R x \cdot \text{density} \cdot dA = \int_R x \cdot x^2y \, dA = \int_R x^3y \, dA,$$

and

$$\int_R y \cdot \text{density} \cdot dA = \int_R y \cdot x^2y \, dA = \int_R x^2y^2 \, dA.$$

To evaluate these integrals, describe R by, say, vertical cross sections,

$$0 \le x \le 1, \qquad 0 \le y \le 2x.$$

Then

$$\text{Mass} = \int_R x^2y \, dA = \int_0^1 \left(\int_0^{2x} x^2y \, dy \right) dx.$$

The first integration is fairly simple:

$$\int_0^{2x} x^2y \, dy = \frac{x^2y^2}{2} \Big|_{y=0}^{y=2x}$$

$$= \frac{x^2(2x)^2}{2} - \frac{x^20^2}{2}$$

$$= 2x^4.$$

The second integration gives

$$\int_0^1 2x^4 \, dx = \frac{2x^5}{5} \Big|_0^1 = \frac{2}{5}.$$

Thus

$$\text{Mass} = \int_R x^2y \, dA = \tfrac{2}{5}.$$

Next, compute $\int_R x^3y \, dA = \int_0^1 (\int_0^{2x} x^3y \, dy) \, dx$.   The inner integral (the one in parentheses) is

$$\int_0^{2x} x^3y \, dy = \frac{x^3y^2}{2} \Big|_{y=0}^{y=2x}$$

$$= \frac{x^3(2x)^2}{2} - \frac{x^30^2}{2}$$

$$= 2x^5.$$

The outer integral is then $\int_0^1 2x^5 \, dx = \dfrac{2x^6}{6} \Big|_0^1 = \dfrac{1}{3}.$

Thus

$$\int_R x^3 y\, dA = \tfrac{1}{3}.$$

Consequently

$$\bar{x} = \frac{\tfrac{1}{3}}{\tfrac{2}{5}} = \frac{1}{3}\frac{5}{2} = \frac{5}{6}.$$

Next, compute $\int_R x^2 y^2\, dA$. This equals $\int_0^1 \left(\int_0^{2x} x^2 y^2\, dy \right) dx$. The inner integral is

$$\int_0^{2x} x^2 y^2\, dy = \frac{x^2 y^3}{3}\Big|_{y=0}^{y=2x}$$

$$= \frac{8x^5}{3}.$$

The outer integral is

$$\int_0^1 \frac{8x^5}{3}\, dx = \frac{8x^6}{18}\Big|_0^1$$

$$= \tfrac{4}{9}.$$

Hence

$$\bar{y} = \frac{\tfrac{4}{9}}{\tfrac{2}{5}} = \frac{4}{9}\frac{5}{2} = \frac{10}{9}.$$

The center of gravity is therefore

$$(\bar{x}, \bar{y}) = (\tfrac{5}{6}, \tfrac{10}{9}). \quad \bullet$$

Exercises

In Exercises 1 to 5 find the centroid of the indicated region.
1. The triangle bounded by $y = x$, $x = 1$, and the x axis.
2. The triangle bounded by the two axes and the line $x/4 + y/3 = 1$.
3. The triangle bounded by $y = x$, $y = 2x$, and $x = 2$.
4. The region bounded by $y = x^2$, the x axis, and $x = 1$.
5. The region bounded by $y = x^2$ and the line $y = 4$.

In Exercises 6 to 11 find the center of gravity of the lamina occupying the indicated region and having the indicated density.
6. The rectangle whose vertices are $(0, 0)$, $(1, 0)$, $(1, 1)$, $(0, 1)$; density at (x, y) equal to x.
7. The triangle whose vertices are $(0, 0)$, $(1, 0)$, $(1, 2)$; density at $(x, y) = x + y$.
8. The triangle whose vertices are $(0, 0)$, $(1, 2)$, $(1, 3)$; density at (x, y) is xy.
9. The top half of the circle of radius 1, center $(0, 0)$; density at (x, y) is $x + y + 3$.
10. The region bounded by $y = x^2$, the x axis, and $x = 2$; density at (x, y) is xy.
11. The region bounded by $y = x^2$ and $y = x + 6$; situated to the right of the y axis; density at $(x, y) = 2x$.

12. Where should the fulcrum be located in order to have this seesaw in equilibrium?

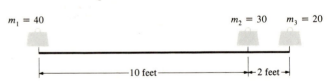

$m_1 = 40$ $m_2 = 30$ $m_3 = 20$

10 feet   2 feet

13. Does every balancing line through the centroid of a region cut that region into two regions of equal area?
14. Cut an irregular shape out of cardboard and find three balancing lines for it experimentally. Are they concurrent? Glue a small piece of cardboard onto it so that its density is not constant, and carry out the same experiment.

■ ■

15. A lamina of mass M and center of gravity $(\bar{x}, \bar{y})$ is cut into two objects, one of mass M_1 and center of gravity $(\bar{x}_1, \bar{y}_1)$, the other of mass M_2 and center of gravity $(\bar{x}_2, \bar{y}_2)$.

(a) Prove that the masses and centers of gravity of the two smaller pieces completely determine the center of gravity of the original lamina, i.e., express $(\bar{x}, \bar{y})$ in terms of $(\bar{x}_1, \bar{y}_1)$, $(\bar{x}_2, \bar{y}_2)$, M_1 and M_2.

(b) Why must the center of gravity of the original lamina lie on the line segment joining $(\bar{x}_1, \bar{y}_1)$ to $(\bar{x}_2, \bar{y}_2)$? *Hint:* Do (b) by considering moments about a wisely chosen line.

13.5

Computing ∫R f(P) dA **by introducing polar coordinates in** R

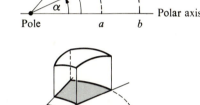

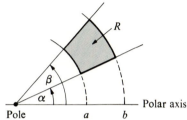

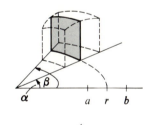

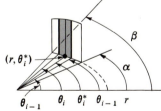

Section 13.3 showed how a repeated integral depending on rectangular coordinates can be used to evaluate $\int_R f(P)\,dA$. It could happen that f or R has a more convenient description in polar than in rectangular coordinates. Let us see what kind of repeated integral relative to polar coordinates has a value equal to $\int_R f(P)\,dA$.

Again assume that $f(P) \geq 0$ and interpret $\int_R f(P)\,dA$ as the volume of a solid. For convenience, first consider the case in which R has the simple description

$$\alpha \leq \theta \leq \beta, \qquad a \leq r \leq b.$$

R is bounded by two line segments and two arcs of concentric circles. The solid whose base is R and whose height above P is $f(P)$ may appear in perspective as in the diagram in the margin. Compute the volume of this solid by considering curved cross sections for fixed values of r, letting θ vary from α to β.

The area of the typical cross section shown at the left depends on r and will be denoted by $A(r)$. It seems reasonable that

$$\text{Volume} = \int_a^b A(r)\,dr.$$

(Think of straightening the curved sections, as in the shell technique of Sec. 10.4.) All that remains is to represent $A(r)$ itself as a definite integral over an interval.

Now, $A(r)$ is the area of a curved wall, corresponding to a fixed value of r. This area can be approximated by the sum of the areas of narrow curved rectangularlike strips of which a typical one is shown at the left. Partition the θ interval $[\alpha, \beta]$ by means of the angles

$$\alpha = \theta_0, \theta_1, \ldots, \theta_n = \beta.$$

Choose θ_i^* in the ith section $[\theta_{i-1}, \theta_i]$. The height of the typical narrow approximating strip is $f(r, \theta_i^*)$. Its base has length $r \cdot (\theta_i - \theta_{i-1})$.

Hence

$$\sum_{i=1}^{n} f(r, \theta_i^*)r \overbrace{(\theta_i - \theta_{i-1})}^{\text{Notice the } r}$$

is an estimate of the area $A(r)$. When the partition is chosen finer, the sums approach $A(r)$. Hence the definite integral

$$\int_\alpha^\beta f(r, \theta)r\,d\theta$$

equals $A(r)$.   Observe that the integrand is *not* just $f(r, \theta)$; an extra r appears in the integrand, because the typical narrow strip has a base whose length is $r(\theta_i - \theta_{i-1})$, and not simply $\theta_i - \theta_{i-1}$.

Putting these facts together yields

$$\int_R f(P) \, dA = \text{volume of solid}$$

$$= \int_a^b A(r) \, dr$$

$$= \int_a^b \left(\int_\alpha^\beta f(r, \theta) r \, d\theta \right) dr. \qquad \overset{\text{Remember the } r}{}$$

This repeated integral provides a way of evaluating definite integrals by polar coordinates, if the region has the simple description

$$a \le r \le b, \qquad \alpha \le \theta \le \beta.$$

If the region is more general, having the description

$$a \le r \le b, \qquad \theta_1(r) \le \theta \le \theta_2(r),$$

the formula becomes

$$\int_R f(P) \, dA = \int_a^b \left[\int_{\theta_1(r)}^{\theta_2(r)} f(r, \theta) r \, d\theta \right] dr.$$

The big difference between the repeated integral in polar coordinates and the one in rectangular coordinates is the presence of the extra factor r in the integrand.   It is also present in the repeated integral in the reverse order:

The More Useful Repeated Integral in Polar Coordinates. Remember the r in the Integrand.

$$\int_R f(P) \, dA = \int_\alpha^\beta \left[\int_{r_1(\theta)}^{r_2(\theta)} f(r, \theta) r \, dr \right] d\theta.$$

This is the more useful order whenever R is easily described in the order

$$\alpha \le \theta \le \beta, \qquad r_1(\theta) \le r \le r_2(\theta).$$

In fact, it is the order most commonly required.

Three examples will now illustrate the use of the repeated integral

$$\int_\alpha^\beta \left[\int_{r_1(\theta)}^{r_2(\theta)} f(r, \theta) r \, dr \right] d\theta.$$

EXAMPLE 1   Example 2 of Sec. 13.4 required the computation of

$$\int_R y \, dA,$$

where R is the semicircle pictured in the margin.   Evaluate $\int_R y \, dA$ by a repeated integral in polar coordinates.

SOLUTION   In polar coordinates, R has the simple description

$$0 \le \theta \le \pi, \qquad 0 \le r \le a.$$

Since $y = r \sin \theta$, the function has the formula

$$f(r, \theta) = r \sin \theta.$$

Notice this r, needed when using polar coordinates

Thus

$$\int_R y\, dA = \int_R r \sin \theta\, dA = \int_0^\pi \left(\int_0^a r \sin \theta \cdot r\, dr \right) d\theta.$$

The repeated integral is easy to compute. The integral in parentheses is

$$\int_0^a r \sin \theta\, r\, dr = \sin \theta \int_0^a r^2\, dr$$

$$= \sin \theta \left(\frac{r^3}{3} \right) \Big|_0^a$$

$$= \frac{a^3 \sin \theta}{3}.$$

The second integral is therefore

$$\int_0^\pi \frac{a^3 \sin \theta}{3}\, d\theta = \frac{a^3}{3} \int_0^\pi \sin \theta\, d\theta$$

$$= \frac{a^3}{3} (-\cos \theta) \Big|_0^\pi$$

$$= \frac{a^3}{3} [(-\cos \pi) - (-\cos 0)]$$

$$= \frac{a^3}{3} (1 + 1) = \frac{2a^3}{3}.$$

Thus

$$\int_R y\, dA = \frac{2a^3}{3}. \ \bullet$$

EXAMPLE 2   A solid has as its base the top half of the cardioid $r = 1 + \cos \theta$. Above the point (r, θ) its cross-sectional length is $y (= r \sin \theta)$. What is its volume?

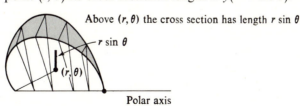

Above (r, θ) the cross section has length $r \sin \theta$

$r \sin \theta$

(r, θ)

Polar axis

SOLUTION   Recall that

$$\text{Volume} = \int_R c(P)\, dA,$$

where $c(P)$ is the length of the cross section by a line through P. In terms of polar coordinates,

$$c(r, \theta) = r \sin \theta.$$

R has the description $0 \le \theta \le \pi$, $0 \le r \le 1 + \cos \theta$.

Hence
$$\int_R c(P)\, dA = \int_0^\pi \left(\int_0^{1+\cos\theta} r\sin\theta \cdot r\, dr \right) d\theta.$$

Notice this r

The integral in parentheses (the inner integral) is:

$$\int_0^{1+\cos\theta} r^2 \sin\theta\, dr = \sin\theta \int_0^{1+\cos\theta} r^2\, dr = \sin\theta \left(\frac{r^3}{3}\right)\Big|_0^{1+\cos\theta}$$

$$= \sin\theta \left[\frac{(1+\cos\theta)^3}{3} - \frac{(0)^3}{3} \right]$$

$$= \frac{(1+\cos\theta)^3 \sin\theta}{3}.$$

The second integral is
$$\int_0^\pi \frac{(1+\cos\theta)^3 \sin\theta}{3}\, d\theta,$$

which can be evaluated by the substitution

$$u = 1 + \cos\theta,$$

$$du = -\sin\theta\, d\theta.$$

This leads to the integral
$$\int_2^0 \frac{u^3(-du)}{3} = \int_0^2 \frac{u^3\, du}{3}$$

$$= \frac{u^4}{12}\Big|_0^2$$

$$= \tfrac{16}{12}$$

$$= \tfrac{4}{3}.$$

The total volume is $\tfrac{4}{3}$. ●

EXAMPLE 3   A sphere of radius a has its center at the pole of a polar-coordinate system. Find the volume of the part of the sphere that lies above the plane region bounded by the curve $r = a\cos\theta$.

Pole

a

Polar axis

The circle
$r = a\cos\theta$

R: bounded by the
circle $r = a\cos\theta$

SOLUTION   Observe that the lateral surface of the solid in question is part of the surface of a right circular cylinder. Its top surface is part of the surface of a sphere, and its base is the region R. It is not a solid of revolution; its cross sections by parallel planes—no matter how we choose the direction—are difficult to compute. It will, however, be easy to compute the volume by an appropriate repeated integral in polar coordinates.

It is necessary to describe f and R in polar coordinates, where $f(P)$ is the length of a cross section of the solid made by a vertical line through P. R is described as follows: r goes from 0 to $a\cos\theta$ for each θ in $[-\pi/2, \pi/2]$,

$$-\frac{\pi}{2} \le \theta \le \frac{\pi}{2}, \qquad 0 \le r \le a\cos\theta.$$

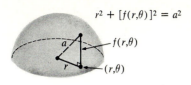

$$r^2 + [f(r,\theta)]^2 = a^2$$

To obtain $f(r,\ \theta)$, use the pythagorean theorem to deduce that $f(r,\ \theta) = \sqrt{a^2 - r^2}$. Thus

$$\text{Volume} = \int_R f(P)\, dA = \int_{-\pi/2}^{\pi/2} \left(\int_0^{a\cos\theta} \sqrt{a^2 - r^2}\, r\, dr \right) d\theta.$$

(Notice the factor r in the integrand.)   Exploiting symmetry, compute half the volume, keeping θ in $[0, \pi/2]$, and then double the result.

$$\int_0^{a\cos\theta} \sqrt{a^2 - r^2}\, r\, dr = \frac{-(a^2 - r^2)^{3/2}}{3} \Big|_0^{a\cos\theta}$$

$$= -\left[\frac{(a^2 - a^2\cos^2\theta)^{3/2}}{3} - \frac{(a^2)^{3/2}}{3} \right]$$

$$= \frac{a^3}{3} - \frac{(a^2 - a^2\cos^2\theta)^{3/2}}{3}$$

$$= \frac{a^3}{3} - \frac{a^3(1 - \cos^2\theta)^{3/2}}{3} = \frac{a^3}{3}(1 - \sin^3\theta).$$

(The trigonometric formula used above, $\sin\theta = \sqrt{1 - \cos^2\theta}$, is true when $0 \le \theta \le \pi/2$ but not when $-\pi/2 \le \theta < 0$.)

The second integration is then carried out:

$$\int_0^{\pi/2} \frac{a^3}{3}(1 - \sin^3\theta)\, d\theta = \int_0^{\pi/2} \left[\frac{a^3}{3} - \frac{a^3}{3}(1 - \cos^2\theta)\sin\theta \right] d\theta$$

$$= \left(\frac{a^3}{3}\theta + \frac{a^3}{3}\cos\theta - \frac{a^3}{3}\frac{\cos^3\theta}{3} \right) \Big|_0^{\pi/2}$$

$$= \frac{a^3}{3}\frac{\pi}{2} - \left(\frac{a^3}{3} - \frac{a^3}{9} \right) = a^3\left(\frac{\pi}{6} - \frac{1}{3} + \frac{1}{9} \right)$$

$$= a^3\left(\frac{3\pi - 6 + 2}{18} \right) = a^3\left(\frac{3\pi - 4}{18} \right).$$

The total volume is twice as large:

$$a^3\left(\frac{3\pi - 4}{9} \right). \quad \bullet$$

Having now worked with repeated integrals in rectangular coordinates and in polar coordinates, let us stop a moment to discuss the basic reason why the factor r must be introduced into the integrand in the case of polar coordinates.

A rectangle of dimensions dx and dy has area $dx\, dy$.   As a memory device let us say

The area is $dx\, dy$;
no extra coefficient is needed

$$dA = dx\, dy.$$

Now consider the area of a little region corresponding to the changes dr and $d\theta$ in polar coordinates.   Though two of its sides are curves, it closely

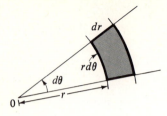

Shaded area is roughly $r\, dr\, d\theta$

resembles a rectangle of dimensions dr and $r\, d\theta$.   The corresponding memory device is

$$dA = dr \cdot r\, d\theta$$

or

$$dA = r\, dr\, d\theta.$$

Straightforward geometry shows that the shaded area in the figure is precisely $[r + (dr/2)]\, dr\, d\theta$.

Exercises

Evaluate the repeated integrals in Exercises 1 to 4.

1. $\displaystyle\int_0^\pi \left(\int_0^{\cos\theta} r \sin\theta\, dr \right) d\theta$

2. $\displaystyle\int_0^\pi \left(\int_0^3 r^2 \sin\theta\, dr \right) d\theta$

3. $\displaystyle\int_0^\pi \left(\int_0^{1+\cos\theta} r^3 \sin\theta\, dr \right) d\theta$

4. $\displaystyle\int_0^\pi \left(\int_0^{\sin\theta} r\, dr \right) d\theta$

In Exercises 5 to 8 set up (but do not evaluate) a repeated integral in polar coordinates for $\int_R f(P)\, dA$, for the given f and R.

5. $f(P) = $ distance from P to pole; R described by $0 \le \theta \le \pi/2$, $1 \le r \le 2 + \cos\theta$.

6. $f(P) = $ square of distance from P to pole; R described by $0 \le \theta \le \pi/6$, $0 \le r \le \sin 3\theta$.

7. $f(x, y) = y$; R described by $0 \le \theta \le \pi/2$, $0 \le r \le 1 + \sin\theta$.

8. $f(P) = $ square of distance from P to y axis; R described by $0 \le \theta \le \pi/4$, $0 \le r \le \sin 2\theta$.

The integrals $\int_0^{\pi/2} \cos^n\theta\, d\theta$ and $\int_0^{\pi/2} \sin^n\theta\, d\theta$ frequently arise when computing repeated integrals in polar coordinates.

A table of integrals lists the formulas

$$\int_0^{\pi/2} \cos^n\theta\, d\theta = \int_0^{\pi/2} \sin^n\theta\, d\theta = \frac{2 \cdot 4 \cdot 6 \cdots (n-1)}{1 \cdot 3 \cdot 5 \cdots (n)}$$

if n is an odd integer, and

$$\int_0^{\pi/2} \cos^n\theta\, d\theta = \int_0^{\pi/2} \sin^n\theta\, d\theta = \frac{1 \cdot 3 \cdot 5 \cdots (n-1)}{2 \cdot 4 \cdot 6 \cdots (n)}\, \frac{\pi}{2}$$

if n is an even integer.

Use the formulas to evaluate the integrals in Exercises 9 to 14.

9. $\displaystyle\int_0^{\pi/2} \sin^3\theta\, d\theta$

10. $\displaystyle\int_0^{\pi/2} \cos^2\theta\, d\theta$

11. $\displaystyle\int_0^{\pi/2} \sin^4\theta\, d\theta$

12. $\displaystyle\int_0^{\pi/2} \sin^5\theta\, d\theta$

13. $\displaystyle\int_0^{\pi/2} \cos^6\theta\, d\theta$

14. $\displaystyle\int_0^\pi \sin^7\theta\, d\theta$

15. (a) Show that $\int_0^\pi \cos^n\theta\, d\theta = 0$ when n is an odd integer.
(b) Evaluate $\int_0^\pi \cos^6\theta\, d\theta$ using the appropriate formula preceding Exercise 9.

16. The density of material at $P = (r, \theta)$ is r grams per square centimeter.   The region is bounded by the cardioid $r = 1 + \cos\theta$.   Find the total mass.

17. Find the average distance from points in the circle $r = 2 \sin\theta$ to the pole.

The integral $\int_R r^2\, dA$, denoted I_0, is called the *polar moment of inertia of R.* It is of importance in the study of bodies spinning about a line through the pole and perpendicular to the plane of R.   Find I_0 for the regions described in Exercises 18 to 21.

18. $-\pi/2 \le \theta \le \pi/2$, $0 \le r \le \cos\theta$.

19. $0 \le \theta \le \pi/2$, $0 \le r \le \sin^2\theta$.

20. $0 \le \theta \le 2\pi, 0 \le r \le 1 + \cos \theta$.
21. $0 \le \theta \le \pi/2, 0 \le r \le \sin 2\theta$.

The integral $\int_R y^2 \, dA$, denoted I_x, is called the *moment of inertia of R about the x axis*. It is of importance in the study of bodies spinning around the x axis. Find I_x for the regions described in Exercises 22 to 25.

22. The circle of radius a, center at the pole.
23. The circle of radius a, center at $(a, \pi/2)$ in polar coordinates.
24. $0 \le \theta \le \pi/2, 0 \le r \le \sqrt{\cos \theta}$.
25. $0 \le \theta \le \pi, 0 \le r \le \sqrt[4]{\sin \theta}$.
26. Evaluate the integral in Exercise 5.
27. Evaluate the integral in Exercise 6.
28. Evaluate the integral in Exercise 7.
29. Evaluate the integral in Exercise 8.

■

30. Find the volume of a sphere of radius a using a repeated integral in polar coordinates.
31. Find the moment of inertia I_x (defined before Exercise 22) of the semicircle of radius a shown below.

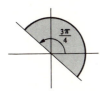

32. Describe f in rectangular and in polar coordinates if $f(P)$ is (a) the square of the distance from P to the polar axis (which is placed to coincide with the x axis); (b) the square of the distance from P to the origin.

33. Describe f in polar coordinates if (a) $f(x, y) = x$; (b) $f(x, y) = \sqrt{x^2 + y^2}$.
34. Let R be a disk of radius 3. Let $f(P)$ be the square of the distance from P to a certain fixed diameter of the disk. Compute $\int_R f(P) \, dA$.

35. Compute $\displaystyle\int_0^{\pi/4} \left(\int_{\sin \theta}^{\cos \theta} r \, dr \right) d\theta$.

36. Compute the mass of the lamina inside one loop of the curve $r = \sin 2\theta$ if the density at (r, θ) is r^2.

37. Compute $\displaystyle\int_0^{\pi/2} \left(\int_0^{\cos \theta} r \sin \theta \, dr \right) d\theta$.

38. (a) Compute $\displaystyle\int_0^{\pi/2} \left(\int_0^{\cos \theta} r^2 \sin \theta \, dr \right) d\theta$.

(b) Compute $\displaystyle\int_0^{\pi/2} \left(\int_0^{\cos \theta} r^3 \sin \theta \, dr \right) d\theta$.

■ ■

39. Using elementary geometry, show that in polar coordinates the area of the region corresponding to changes dr and $d\theta$ is precisely $(r + dr/2) \, dr \, d\theta$.
40. In Example 3 we computed half the volume and doubled the result. Evaluate the repeated integral

$$\int_{-\pi/2}^{\pi/2} \left(\int_0^{a \cos \theta} \sqrt{a^2 - r^2} \, r \, dr \right) d\theta$$

directly. The result should still be $a^3(3\pi - 4)/9$. *Caution:* Use trigonometric formulas with care.

13.6

Summary

Let R be a region in the plane and f a function that assigns to each point P in R a number. Then the definite integral of f over R is defined with the aid of partitions of R and sampling points as

$$\lim_{\text{mesh} \to 0} \sum_{i=1}^{n} f(P_i) A_i.$$

This number is denoted $\int_R f(P) \, dA$.

Some applications of the definite integral are reviewed in this table:

If $f(P)$ is	then $\int_R f(P)\, dA$ is
density at P	mass
length of cross section of solid by line through P perpendicular to R	volume of solid
$x - a$ where $P = (x, y)$	moment about line $x = a$
x where $P = (x, y)$	moment about y axis (note that the y axis is the line $x = 0$)
y where $P = (x, y)$	moment about x axis
y^2 where $P = (x, y)$	I_x, moment of inertia about x axis
r^2 where $P = (r, \theta)$	I_0, polar moment of inertia

The average value of f over R is defined as

$$\text{Average value} = \frac{\int_R f(P)\, dA}{\text{area of } R}.$$

One way to estimate $\int_R f(P)\, dA$ is to compute a particular approximating sum $\sum_{i=1}^{n} f(P_i)A_i$. This is not too unpleasant if a calculator is available.

Repeated integrals can be used to evaluate $\int_R f(P)\, dA$. It is necessary to use the fundamental theorem of calculus twice in succession. Except in the case of the simplest integrals (which are fortunately the ones most often applied), there is little chance of such good luck. The repeated integrals are

$$\int_a^b \left[\int_{y_1(x)}^{y_2(x)} f(x, y)\, dy \right] dx$$

and

$$\int_c^d \left[\int_{x_1(y)}^{x_2(y)} f(x, y)\, dx \right] dy$$

in rectangular coordinates, and

$$\int_\alpha^\beta \left[\int_{r_1(\theta)}^{r_2(\theta)} f(r, \theta) r\, dr \right] d\theta$$

in polar coordinates (the other order is seldom convenient). Remember the extra r in the integrand of the repeated integral in polar coordinates. It is present because $r\, dr\, d\theta$ (not $dr\, d\theta$) is the approximate area of the little region corresponding to changes of dr and $d\theta$ in the coordinates. These diagrams will serve as reminders.

Moments, balancing lines, center of gravity, and centroid were defined. A balancing line is a line about which the moment is zero. All balancing lines of a lamina intersect in one point, which is called the center of gravity.

The center of gravity $(\bar{x}, \bar{y})$ can be computed by these formulas:

$$\bar{x} = \frac{\int_R xf(P)\, dA}{\text{mass}}$$

$$\bar{y} = \frac{\int_R yf(P)\, dA}{\text{mass}},$$

where f is density.

If the density is constant, then the center of gravity is a purely geometric concept, also called the centroid.

The centroid $(\bar{x}, \bar{y})$ can be computed by the formulas:

$$\bar{x} = \frac{\int_R x\, dA}{\text{area}},$$

$$\bar{y} = \frac{\int_R y\, dA}{\text{area}}.$$

VOCABULARY AND SYMBOLS

two-dimensional integral $\int_R f(P)\, dA$
diameter of a plane region
repeated integral (rectangular, polar)
moment about line $x = a$,

$$\int_R (x - a)f(P)\, dA,$$

where f is density
balancing line
center of gravity $(\bar{x}, \bar{y})$
centroid $(\bar{x}, \bar{y})$

$dA = dx\, dy$ dy dx

dr $r\, d\theta$ $d\theta$ r 0

$dA = r\, dr\, d\theta$

Guide quiz on Chap. 13

1. (a) Describe in rectangular coordinates the region whose description in polar coordinates is

$$0 \le \theta \le \pi, \qquad 0 \le r \le a.$$

(b) Describe in polar coordinates the region whose description in rectangular coordinates is

$$1 \le x \le 2, \qquad \frac{x}{\sqrt{3}} \le y \le x.$$

2. Find the moment about the x axis of the triangle whose vertices are $(0, 0)$, $(2, 0)$, $(2, 2)$, using (a) a repeated integral in rectangular coordinates, (b) a repeated integral in polar coordinates.

3. (a) Find the average distance from points in a circle of radius a to the center.

(b) Why is the average larger than $a/2$? (Give an intuitive explanation.)

4. Transform this repeated integral to a repeated integral in polar coordinates, and evaluate the latter:

$$\int_0^a \left[\int_0^{\sqrt{a^2-x^2}} (x^2 + y^2)^{3/2} \, dy \right] dx.$$

5. (a) Find the moment of inertia of one loop of the curve $r = \sin 2\theta$ about the pole.

(b) Find the mass within the loop in (a) if the density at (r, θ) is r^2 grams per square centimeter.

(c) Find the volume of a solid whose base is the loop in (a) and whose cross section above (r, θ) has length r^2.

(d) The temperature at the point (r, θ) in the loop in (a) is r^2 degrees Celsius. What is the average temperature?

6. An agricultural sprinkler distributes water in a circle of radius 100 feet. By placing a few random cans in this circle, it is determined that the sprinkler supplies water at a depth of e^{-r} feet of water at a distance of r feet from the sprinkler in 1 hour. How much water does the sprinkler supply to the region within (a) 100 feet of the sprinkler? (b) 50 feet of the sprinkler?

Review exercises for Chap. 13

1. Describe the region between $y = x^2$ and $y = 4$ by (a) vertical cross sections, (b) horizontal cross sections.

2. Compute:

(a) $\displaystyle\int_0^1 x^2 y \, dy$ (b) $\displaystyle\int_0^1 x^2 y \, dx.$

3. Compute:

(a) $\displaystyle\int_1^{x^2} (x + y) \, dy$ (b) $\displaystyle\int_y^{y^2} (x + y) \, dx.$

4. Compute the easier of

(a) $\displaystyle\int_0^1 \sin (x^2 y) \, dy$ (b) $\displaystyle\int_0^1 \sin (x^2 y) \, dx.$

5. Describe the accompanying figure in terms of
(a) rectangular coordinates and vertical cross sections,
(b) rectangular coordinates and horizontal cross sections,
(c) polar coordinates.

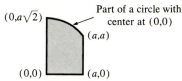

$(0, a\sqrt{2})$ Part of a circle with center at $(0,0)$

(a,a)

$(0,0)$ $(a,0)$

Translate the repeated integrals in Exercises 6 to 8 to repeated integrals in polar coordinates and in each case evaluate the latter.

6. $\displaystyle\int_0^1 \left(\int_0^x x^2 \, dy \right) dx$ 7. $\displaystyle\int_0^{1/\sqrt{2}} \left(\int_x^{\sqrt{1-x^2}} \sqrt{x^2 + y^2} \, dy \right) dx$

8. $\displaystyle\int_0^2 \left(\int_0^{\sqrt{2x-x^2}} x \, dy \right) dx$

9. Consider the repeated integral $\displaystyle\int_{\pi/4}^{\pi/2} \left(\int_0^a r^2 \sin \theta \, dr \right) d\theta.$

(a) Draw R and describe f such that $\int_R f(P) \, dA$ is represented by the given repeated integral.

(b) Choose a convenient repeated integral in rectangular coordinates equal to $\int_R f(P) \, dA.$

(c) Evaluate $\int_R f(P) \, dA$ by the simplest method.

Translate the repeated integrals in Exercises 10 to 12 to repeated integrals in rectangular coordinates and evaluate (choose the more convenient direction).

10. $\displaystyle\int_0^{\pi/4} \left(\int_0^a r^2 \cos \theta \, dr \right) d\theta$ 11. $\displaystyle\int_0^{\pi/4} \left(\int_0^a r^3 \, dr \right) d\theta$

12. $\displaystyle\int_{\pi/4}^{3\pi/4} \left(\int_0^a r^3 \, dr \right) d\theta$

13. Evaluate:

(a) $\displaystyle\int_0^{\pi/2} \cos^2 \theta \, d\theta$ (b) $\displaystyle\int_0^{\pi/2} \cos^3 \theta \, d\theta$

(c) $\int_0^\pi \cos^3 \theta \, d\theta$ (d) $\int_0^\pi \sin^3 \theta \, d\theta.$

14. Find the moment of a square of side a about (a) a side, (b) a diagonal.

15. Find the moment of inertia of a square of side a about (a) a side, (b) a diagonal.

16. Find the centroid of the region outside the circle $r = 1$ and inside the cardioid $r = 1 + \cos \theta$.

17. Find the moment of inertia of the region bounded by the curve $y = x^3$, $y = 8$, and the y axis about
(a) the x axis,   (b) the y axis.
(c) Find the polar moment I_0 about the origin.

In Exercises 18 to 20 compute $\int_R f(P) \, dA$ if $f(x, y) = xy$ and R are described in coordinates as

18. $0 \le x \le 2$, $x^3 \le y \le 2x^3$.

19. $0 \le x \le \pi/2$, $0 \le y \le \sin x$.

20. $0 \le \theta \le \pi/4$, $0 \le r \le 2 \sin \theta$.

21. Let R be the region bounded by the curve $y = \cos x$, the x axis, and the lines $x = \pi/2$ and $x = -\pi/2$.
(a) Find the moment of R about the x axis.
(b) Find the moment of inertia I_x.
(c) Find $(\bar{x}, \bar{y})$.

22. Find the centroid of the region bounded by $y = x^2$ and $y = \sqrt{x}$.

23. The depth of water provided by a water sprinkler is approximately 2^{-r} feet at a distance of r feet from the sprinkler. Find the total amount of water within a distance of a feet of the sprinkler.

Consider a distribution of matter in a plane region with density $\delta(P)$ at P. Its moment of inertia about any line in the plane is defined to be $\int_R x^2 \, \delta(P) \, dA$, where x is the distance from P to the line. (Think of $\delta(P) \, dA$ as density times small area or a small mass.) Usually the density is constant, $\delta(P) = M/A$, where M is the total mass, and A is the area of the region. This is the so-called homogeneous distribution. This concept is illustrated in Exercises 24 to 26.

24. Find the moment of inertia of a homogeneous rectangular plate of mass M whose sides are of lengths a and b about (a) a side of length a, (b) a side of length b, (c) a diagonal.

25. A homogeneous right triangular plate of mass M has legs of lengths a and b and hypotenuse of length c. Find its moment of inertia about (a) side a, (b) side b, (c) the hypotenuse.

26. Find the polar moment of inertia of a homogeneous square lamina of mass M and side a about its center, using (a) rectangular coordinates, (b) polar coordinates.

In Exercises 27 to 31 describe f and draw R, if $\int_R f(P) \, dA$ is equal to the given repeated integral.

27. $\int_0^1 \left(\int_2^3 x^2 \, dy \right) dx$

28. $\int_0^{\pi/4} \left(\int_1^{1+\cos\theta} r^2 \cos\theta \, dr \right) d\theta$

29. $\int_0^1 \left[\int_{\sqrt{1-y^2}}^1 (x^2 + y) \, dx \right] dy$

30. $\int_0^1 \left(\int_{x^3}^{x^2} dy \right) dx$

31. $\int_0^{\pi/2} \left(\int_1^2 dr \right) d\theta$

■

32. Find $\int_R xy \, dA$ where R is the region described as $0 \le \theta \le \pi/4$, $0 \le r \le \cos 2\theta$.

33. Let R be a triangle. Place an xy-coordinate system in such a way that its origin is at one vertex of the triangle and the x axis is parallel to the opposite side. Call the coordinates of the two other vertices (a, b) and (c, b). Show that $\bar{y} = 2b/3$.

34. (a) Using the result of Exercise 33, show that the centroid of a triangle is at the intersection of the three medians.
(b) How might one have guessed this result without using calculus?

35. Let R be a circle of radius a and center at $(0, 0)$.
(a) Without evaluating them, explain why the integrals $\int_R x^2 \, dA$ and $\int_R y^2 \, dA$ are equal.
(b) Without evaluating any of these integrals, show that $\int_R x^2 \, dA + \int_R y^2 \, dA = \int_R r^2 \, dA$.
(c) Evaluate $\int_R r^2 \, dA$ by using polar coordinates.
(d) Combining (a), (b), and (c), compute $\int_R x^2 \, dA$.

36. (a) Draw the region R whose description is

$$\frac{\sqrt{2}}{2} \le x \le 1, \qquad \sqrt{1 - x^2} \le y \le x.$$

(b) Describe R in polar coordinates.
(c) Transform the repeated integral

$$\int_{\sqrt{2}/2}^1 \left(\int_{\sqrt{1-x^2}}^x \frac{1}{\sqrt{x^2 + y^2}} \, dy \right) dx$$

into polar coordinates.
(d) Evaluate the repeated integral in polar coordinates.

37. Evaluate $\int_R \ln (x^2 + y^2) \, dA$ over the region in Exercise 36.

This exercise outlines a simple proof that $\int_0^\infty e^{-x^2} \, dx = \sqrt{\pi}/2$.

38. Let R_1, R_2, R_3 be the three regions indicated in the following diagram, and $f(P) = e^{-r^2}$, where r is the distance from

P to the origin. Hence $f(r, \theta) = e^{-r^2}$ and $f(x, y) = e^{-x^2-y^2}$.

| Quadrant of a circle | Square | Quadrant of a circle |

(Observe that R_1 is inside R_2, and R_2 is inside R_3.)

(a) Show that $\int_{R_1} f(P)\, dA = (\pi/4)(1 - e^{-a^2})$ and $\int_{R_3} f(P)\, dA = (\pi/4)(1 - e^{-2a^2})$.

(b) By considering $\int_{R_2} f(P)\, dA$ and the results in (a), show that

$$\frac{\pi}{4}(1 - e^{-a^2}) < \left(\int_0^a e^{-x^2}\, dx \right)^2 < \frac{\pi}{4}(1 - e^{-2a^2}).$$

(c) Show that $\int_0^\infty e^{-x^2}\, dx = \sqrt{\pi}/2$.

Once when lecturing to a class he [the physicist Lord Kelvin] used the word 'mathematician' and then interrupting himself asked his class: 'Do you know what a mathematician is?' Stepping to his blackboard he wrote upon it: $\int_{-\infty}^\infty e^{-x^2}\, dx = \sqrt{\pi}$. Then putting his finger on what he had written, he turned to his class and said, 'A mathematician is one to whom that is as obvious as that twice two makes four is to you.'
From S. P. Thompson, *Life of Lord Kelvin*, Macmillan, London, 1910.

The mathematician Littlewood wrote, "Many things are not accessible to intuition at all, the value of $\int_0^\infty e^{-x^2}\, dx$ for instance." From J. E. Littlewood, *Newton and the Attraction of the Sphere, Mathematical Gazette,* vol. 63, 1948.

39. Using the fact that $\int_0^\infty e^{-x^2}\, dx = \sqrt{\pi}/2$, show that

(a) $\displaystyle\int_0^\infty e^{-4x^2}\, dx = \frac{\sqrt{\pi}}{4}$; (b) $\displaystyle\int_0^\infty \frac{e^{-x}}{\sqrt{x}}\, dx = \sqrt{\pi}$;

(c) $\displaystyle\int_0^\infty x^2 e^{-x^2}\, dx = \frac{\sqrt{\pi}}{4}$; (d) $\displaystyle\int_0^\infty \sqrt{x}\, e^{-x}\, dx = \frac{\sqrt{\pi}}{2}$;

(e) $\displaystyle\int_0^1 \frac{dx}{\sqrt{\ln(1/x)}} = \sqrt{\pi}$; (f) $\displaystyle\int_0^1 \sqrt{\ln(1/x)}\, dx = \frac{\sqrt{\pi}}{2}$.

Transportation problems lead to integrals over plane sets, as Exercises 40 to 45 illustrate.

40. Show that the average travel distance from the center of a circle of area A to points in the circle is approxi-

mately $0.376\sqrt{A}$ (precisely $2\sqrt{A}/3\sqrt{\pi}$).

41. Show that the average travel distance from the center of a regular hexagon of area A to points in the hexagon is approximately $0.377\sqrt{A}$ {precisely $(\sqrt{2}\sqrt{A}/\sqrt{3}\sqrt[4]{3})[\frac{1}{3} + (\ln 3)/4]$}.

42. Show that the average travel distance from the center of a square of area A to points in the square is approximately $0.383\sqrt{A}$ {precisely $[(\sqrt{2} + \ln \tan 3\pi/8)\sqrt{A}]/6)$}.

43. Show that the average travel distance from the centroid of an equilateral triangle of area A to points in the triangle is approximately $0.404\sqrt{A}$ {precisely $(\sqrt{A}/9\sqrt[4]{3})[2\sqrt{3} + \ln \tan (5\pi/12)]$}.

In Exercises 40 to 43 the distance from the warehouse to a point is the ordinary straight-line distance. In cities the usual street pattern suggests that the "metropolitan" distance between the points (x_1, y_1) and (x_2, y_2) is closer to $|x_1 - x_2| + |y_1 - y_2|$.

44. Show that, if in Exercise 40 metropolitan distance is used, then the average is $8\sqrt{A}/3\pi^{3/2} \doteq 0.479\sqrt{A}$.

45. Show that, if in Exercise 42 metropolitan distance is used, then the average is $\sqrt{A}/2$. In most cities the metropolitan average tends to be about 25 percent larger than the direct-distance average.

This exercise presents a gravitational paradox.

46. Consider a mass distributed uniformly throughout the entire plane and a point mass a distance a from the plane. The gravitational attraction of the planar mass on the point mass is directed toward the plane and has magnitude

$$\int_R \frac{a}{(\sqrt{a^2 + r^2})^3}\, dA.$$

(The integral, which is taken over the entire plane, is improper in the sense that an integral over the x axis is improper. Treat it similarly.) In the integrand r refers to polar coordinates where the pole is the point in the plane closest to the point mass.

(a) Show that the integral has the value 2π.

(b) According to (a), the attractive force of the plane on the point mass is independent of the distance between the point mass and the plane. Does that make sense?

In this exercise a plane integral is used in the theory of the spread of epidemics.

47. In the theory of a spreading epidemic it is assumed that the probability that a contagious individual infects an individual D miles away depends only on D. Consider a population that is uniformly distributed in a circular city whose radius is 1 mile. Assume that the probability we mentioned is proportional to $2 - D$. For a fixed point Q let $f(P) = 2 -$ (distance from P to Q). Let R be the region occupied by the city.

(a) Why is $\int_R f(P)\, dA$ a reasonable measure of the exposure of a person residing at Q?

(b) Compute this definite integral when Q is the center of town and when Q is on the edge of town.

(c) In view of (b), which is the safer place?

48. Let L_1 be a line through the centroid of the plane region R, and L_2 a line parallel to L_1 and a distance d from L_1. Both lines are in the plane of R. Prove that the moment of inertia of R about L_2 is equal to the moment of inertia of R about L_1 plus $d^2 A$, where A is the area of R.

49. The moment of inertia of the region R about the point (x_0, y_0) is defined as

$$\int_R [(x - x_0)^2 + (y - y_0)^2]\, dA.$$

Let R be the circle of radius a and center at the origin of an xy-coordinate system.

(a) Find its moment of inertia about the point $(b, 0)$. *Hint:* Use polar coordinates. First use the law of cosines to show that $(x - b)^2 + (y - 0)^2 = r^2 + b^2 - 2rb \cos \theta$.

(b) Show that this moment of inertia is always larger than (area of R) $\cdot b^2$, which would be the moment of inertia if the entire region R were at the center of R.

50. (a) Show that the moment of inertia of R about (x_0, y_0) (defined in Exercise 49) is equal to the moment of inertia of R about the centroid of R plus $d^2 A$, where d is the distance from the centroid to (x_0, y_0).

(b) About what point (x_0, y_0) is the moment of inertia least?

This exercise involves estimating an integral by choosing points randomly.

51. A computing machine can be used to generate random numbers and thus random points in the plane which can be used to estimate definite integrals, as we now show. Say that a complicated region R lies in the square whose vertices are $(0,0)$, $(2,0)$, $(2, 2)$, and $(0, 2)$, and a complicated function f is defined in R. The machine generates 100 random points (x, y) in the square. Of these, 73 lie in R. The average value of f for these 73 points is 2.31.

(a) What is a reasonable estimate of the area of R?

(b) What is a reasonable estimate of the two-dimensional integral $\int_R f(P)\, dA$?

Techniques such as this one which utilize randomness are called *Monte Carlo methods*.

■ ■

52. Define $f(t)$ to be $\int_t^1 e^{x^2}\, dx$. Find the average value of f over the interval $[0, 1]$.

53. What is wrong with this reasoning? We shall obtain a repeated integral in polar coordinates equal to $\int_R f(P)\, dA$, where R has the description $\alpha \le \theta \le \beta$, $r_1(\theta) \le r \le r_2(\theta)$. Assuming that $f(P) \ge 0$, construct the solid whose base is R and whose height at P is $f(P)$. The cross section of this solid by a plane perpendicular to R and passing through the ray of angle θ is a plane section of area $A(\theta) = \int_{r_1(\theta)}^{r_2(\theta)} f(r, \theta)\, dr$. Since $V = \int_\alpha^\beta A(\theta)\, d\theta$,

$$\int_R f(P)\, dA = \int_\alpha^\beta \left(\int_{r_1(\theta)}^{r_2(\theta)} f(r, \theta)\, dr \right) d\theta.$$

This is a *wrong formula*, since r is not present in the integrand. Find the error.

The exercise illustrates the use of a planar integral to estimate fire hazards.

54. (See Exercise 50.) Where is the most effective place to station a helicopter for fighting fires in a large forest R? Assume that the helicopter travels at a constant speed in a straight line toward a fire, that it goes to the fire as soon as the fire starts, that the fire spreads in a circle whose radius is proportional to the duration of the fire, and that the forest is flat, with the trees uniformly distributed.

55. We outline another proof of Schwarz' inequality given in Exercise 12 of Sec. 10.5. Let R be the square $a \le x \le b$, $a \le y \le b$.

(a) Why is $\int_R [f(x)g(y) - f(y)g(x)]^2\, dA \ge 0$?

(b) From (a) deduce Schwarz' inequality.

56. Show that the moment about the y axis of a region R whose description in polar coordinates is "for each θ in $[\alpha, \beta]$, r goes from 0 to $f(\theta)$" is equal to

$$\int_\alpha^\beta [f(\theta)]^3 \frac{\cos \theta}{3}\, d\theta.$$

57. Let f be a continuous function defined for all θ in $[0, \pi]$, with $f(0) = f(\pi)$. Next assume that $\int_0^\pi f(\theta)\, d\theta = 0 = \int_0^\pi f(\theta) \cos \theta\, d\theta = \int_0^\pi f(\theta) \sin \theta\, d\theta$. Prove there are at least two numbers a and b such that $0 < a < b < \pi$ and $f(a) = 0$ and $f(b) = 0$. *Hint:* Why is there at least one a such that $f(a) = 0$? If there is only one root a, such that $0 < a < \pi$, then consider

$$\int_0^\pi f(x) \sin (x - a)\, dx.$$

58. (See Exercises 56 and 57.) Prove that the centroid of a region R bounded by a curve which has the equation $r = g(\theta)$ in polar coordinates (where g is continuous and the centroid is at the pole) bisects at least three chords.

59. Let f be a continuous nonconstant function such that $f(\theta + 2\pi) = f(\theta)$, and $f(0) \neq 0$. Show if $\int_0^{2\pi} f(\theta)\, d\theta = 0$, then there are at least two numbers a_1 and a_2 such that $0 < a_1 < a_2 < 2\pi, f(a_1) = 0 = f(a_2)$, and f changes sign at each of the roots.

60. (See Exercise 59.) Assume that f satisfies the hypotheses in Exercise 59 and that

$$\int_0^{2\pi} f(\theta) \sin \theta \, d\theta = 0 = \int_0^{2\pi} f(\theta) \cos \theta \, d\theta.$$

Show that there are at least four numbers $0 < a_1 < a_2 < a_3 < a_4 < 2\pi$ where $f(\theta) = 0$ and f changes sign. *Suggestions:* Why can there not be exactly three such numbers? If a_1 and a_2 are the only places between 0 and 2π where f changes sign, consider

$$\int_0^{2\pi} f(\theta) \sin \frac{\theta - a_1}{2} \sin \frac{\theta - a_2}{2} \, d\theta$$

and argue as in Exercise 57. Recall that $\sin A \sin B = \frac{1}{2}[\cos (A - B) - \cos (A + B)].$]

61. (See Exercises 57 and 60.) Let P be the centroid of a plane region R bounded by a smooth curve that does not pass through P. Show that there are at least four rays from P that cut the curve at a right angle. *Hint:* Let $r = g(\theta)$ describe the curve in polar coordinates with P as the pole, and apply Exercise 60 to the function $f(\theta) = g^2(\theta)g'(\theta)$. This implies that the lamina can be balanced on at least four points of its border. Experiments with models suggest that any homogeneous solid can be balanced on at least four points of its surface, but this has not been proved.

14

SERIES

Some functions, though not themselves polynomials, can be closely approximated by polynomials. As an example, in Sec. 14.2 it will be shown that for $|x| < 1$ and for large values of n, the polynomial

$$1 + x + x^2 + x^3 + \cdots + x^n$$

is a good approximation of the function

$$\frac{1}{1 - x}.$$

This particular fact has been used in high-speed computations to replace division by $1 - x$ with multiplication and addition, which are more convenient.

This chapter also obtains polynomials that approximate $\ln(1 + x)$, e^x, $\sin x$, $\cos x$, and other functions. Such polynomials can be used to estimate definite integrals. For instance, $\int_0^1 e^{x^2}\, dx$, which cannot be evaluated by the fundamental theorem of calculus, can be approximated by

$$\int_0^1 \left(1 + x^2 + \frac{x^4}{2!} + \frac{x^6}{3!} + \cdots + \frac{x^{2n}}{n!}\right) dx,$$

which can be evaluated easily.

14.1

Sequences

A *sequence* of real numbers,

$$a_1, a_2, a_3, \ldots, a_n, \ldots,$$

is a function that assigns to each positive integer n a number a_n. The number a_n is called the nth *term* of the sequence. For example, the sequence

$$\left(1 + \frac{1}{1}\right)^1, \left(1 + \frac{1}{2}\right)^2, \left(1 + \frac{1}{3}\right)^3, \ldots, \left(1 + \frac{1}{n}\right)^n, \ldots$$

of Sec. 3.2 was used to define the number e. In this case

$$a_n = \left(1 + \frac{1}{n}\right)^n.$$

Sometimes the notation $\{a_n\}$ is used as an abbreviation of the sequence a_1,

$a_2, \ldots, a_n, \ldots$. For instance, e is defined in terms of the sequence $\{(1 + 1/n)^n\}$. If, as n gets larger, a_n approaches a number L, then L is called the *limit* of the sequence. If the sequence $a_1, a_2, \ldots$ has a limit L, we write

$$\lim_{n \to \infty} a_n = L.$$

In Sec. 3.2 it was shown that the sequence $(1 + \frac{1}{1})^1, (1 + \frac{1}{2})^2, \ldots$ has a limit; this limit is denoted by e,

$$\lim_{n \to \infty} \left(1 + \frac{1}{n}\right)^n = e.$$

The next example introduces a simple but important sequence.

EXAMPLE 1   [*The sequence* $\{(\frac{1}{2})^n\}$]. A certain radioactive substance decays, losing half its mass in an hour. In the long run how much is left?

SOLUTION   If the initial mass is 1 gram, after an hour only $\frac{1}{2}$ gram remains. During the next hour, half of this amount is lost, and half of it remains. Thus after 2 hours

$$(\tfrac{1}{2})^2 = 0.25 \text{ gram}$$

remains. After 3 hours, $(\tfrac{1}{2})^3 = 0.125 \text{ gram}$

remains. In general, after n hours

$$(\tfrac{1}{2})^n \text{ gram}$$

remains.

When n is large, the amount remaining is very small. In other words, the sequence

$$\{(\tfrac{1}{2})^n\}$$

approaches 0 when n gets large. This is summarized by the equation

$$\lim_{n \to \infty} (\tfrac{1}{2})^n = 0. \ \bullet$$

An important fact about sequences to keep in mind at all times is that the terms of the sequence $\{a_n\}$ may perhaps never equal the value of their limit L but merely approach it arbitrarily closely.

Furthermore, not every sequence has a limit, as the next example illustrates.

EXAMPLE 2   *A sequence that does not have a limit.* Let $a_n = (-1)^n$ for $n = 1, 2, 3, \ldots$. What happens to a_n when n is large?

SOLUTION   The first four terms of the sequence are:

$$a_1 = (-1)^1 = -1$$
$$a_2 = (-1)^2 = \ \ \ 1$$
$$a_3 = (-1)^3 = -1$$
$$a_4 = (-1)^4 = \ \ \ 1.$$

The numbers in this sequence continue to alternate $-1, 1, -1, 1, \ldots.$ This sequence does not approach a single number. Therefore it does not have a limit. ●

DEFINITION   *Convergent and divergent sequences.* A sequence that has a limit is said to *converge* or to be *convergent*. A sequence that does not have a limit is said to *diverge* or to be *divergent*.

The sequence $\{(\tfrac{1}{2})^n\}$ of Example 1 converges to 0. The sequence $\{(-1)^n\}$ of Example 2 is divergent. There is no general procedure for deciding whether a sequence is convergent or divergent. However, there are methods for dealing with most sequences that arise in practice. The following theorem describes one such method.

THEOREM I   Let $\{a_n\}$ be an increasing sequence with the property that there is a number B such that $a_n \leq B$ for all $n.$ That is,

$$a_1 \leq a_2 \leq a_3 \leq a_4 \leq \cdots \leq a_n \leq \cdots,$$

and $a_n \leq B$ for all $n.$ Then the sequence $\{a_n\}$ is convergent, and a_n approaches a number L less than or equal to $B.$
  Similarly, if $\{a_n\}$ is a decreasing sequence and there is a number B such that $a_n \geq B$ for all $n,$ then the sequence $\{a_n\}$ is convergent, and its limit is greater than or equal to $B.$ ●

The proof, which depends on a fundamental property of the real numbers, called *completeness*, is omitted. (The "completeness" of the real numbers amounts to the fact that there are no holes in the x axis.) This diagram shows that Theorem 1 is at least plausible.

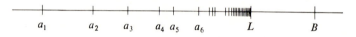

(The a_n's increase but, being less than B, approach some number L, and that number is not larger than B.)
  This theorem will be applied in later sections of the chapter.

EXAMPLE 3   A certain appliance depreciates in value over the years. In fact, at the end of any year it has only 80 percent of the value it had at the beginning of the year. What happens to its value in the long run if its value when new is 1?

SOLUTION   Let a_n be the value of the appliance at the end of the nth year. Thus $a_1 = 0.8$ and $a_2 = (0.8)(0.8) = 0.8^2 = 0.64.$ Similarly, $a_3 = 0.8^3.$ The question concerns the sequence $\{0.8^n\}.$
  This table lists a few values of 0.8^n, rounded off to four decimal places.

n	1	2	3	4	5	10	20
0.8^n	0.8	0.64	0.512	0.4096	0.3277	0.1074	0.0115

The entries in the table suggest that

$$\lim_{n \to \infty} 0.8^n = 0.$$

To verify this assertion, it is necessary to show that the decreasing sequence $\{0.8^n\}$ gets arbitrarily small. To indicate why it does, let us estimate how large n must be so that

$$0.8^n < 0.0001.$$

Taking logarithms to base 10 translates this inequality into the inequality

$$n \log_{10} 0.8 < -4,$$

Division of an Inequality by a Negative Number Reverses the Direction of the Inequality.

or

$$n > \frac{-4}{\log_{10} 0.8}.$$

Thus

$$n > \frac{-4}{-0.097} = 41.2.$$

For $n \geq 42$, the number 0.8^n is less than 0.0001. A similar argument shows that 0.8^n approaches 0 as closely as we please when n is large. Consequently,

$$\lim_{n \to \infty} 0.8^n = 0,$$

just as, in Example 1,

$$\lim_{n \to \infty} \left(\tfrac{1}{2}\right)^n = 0.$$

[The only difference is that $(0.8)^n$ approaches 0 more slowly than $\left(\tfrac{1}{2}\right)^n$ does.]
In the long run the appliance will be worth less than a nickel, then less than a penny, etc. ●

Reasoning similar to that used in Example 3 proves the next theorem.

THEOREM 2 If r is a number in the open interval $(-1, 1)$, then

$$\lim_{n \to \infty} r^n = 0. \quad ●$$

The next sequence to be considered involves the products

$$1 \cdot 2, \qquad 1 \cdot 2 \cdot 3, \qquad 1 \cdot 2 \cdot 3 \cdot 4, \qquad 1 \cdot 2 \cdot 3 \cdot 4 \cdot 5,$$

and so on. The product of the integers from 1 up to n is denoted $n!$ and is called *n factorial* (also, by definition, $0! = 1$). Therefore,

$$0! = 1,$$
$$1! = 1,$$
$$2! = 1 \cdot 2 = 2,$$
$$3! = 1 \cdot 2 \cdot 3 = 6,$$
$$4! = 1 \cdot 2 \cdot 3 \cdot 4 = 24,$$
$$5! = 1 \cdot 2 \cdot 3 \cdot 4 \cdot 5 = (4!)5 = 24 \cdot 5 = 120,$$

and

$$6! = 6(5!) = 720.$$

Note that $(n + 1)!$ is the product of $n!$ and $n + 1$.

In Example 4 a type of sequence is introduced that occurs later in the chapter in the study of $\sin x$, $\cos x$, and e^x.

EXAMPLE 4 Does the sequence defined by $$a_n = \frac{3^n}{n!}$$

converge or diverge?

SOLUTION The first terms of this sequence are computed (to two decimal places) with the aid of this table:

n	1	2	3	4	5	6	7	8
3^n	3	9	27	81	243	729	2,187	6,561
$n!$	1	2	6	24	120	720	5,040	40,320
$a_n = \dfrac{3^n}{n!}$	3.00	4.50	4.50	3.38	2.03	1.01	0.43	0.11

Though a_2 is larger than a_1, and a_3 is equal to a_2, from a_4 through a_8, as the table shows, the terms decrease.

The numerator 3^n becomes large as $n \to \infty$, influencing a_n to grow large. But the denominator $n!$ also becomes large as $n \to \infty$, influencing the quotient a_n to shrink toward 0. For $n = 1$ and 2 the first influence dominates, but then, as the table shows, the denominator $n!$ seems to grow faster than the numerator 3^n, forcing a_n toward 0.

To see why $$\frac{3^n}{n!} \to 0$$

as $n \to \infty$, consider, for instance, a_{10}. Express a_{10} as the product of 10 fractions:

$$a_{10} = \tfrac{3}{1}\tfrac{3}{2}\tfrac{3}{3}\tfrac{3}{4}\tfrac{3}{5}\tfrac{3}{6}\tfrac{3}{7}\tfrac{3}{8}\tfrac{3}{9}\tfrac{3}{10}.$$

The first three fractions are ≥ 1. But all the seven remaining fractions are $\leq \tfrac{3}{4}$. Thus

$$a_{10} < \tfrac{3}{1}\tfrac{3}{2}\tfrac{3}{3}(\tfrac{3}{4})^7.$$

Similarly, $$a_{100} < \tfrac{3}{1}\tfrac{3}{2}\tfrac{3}{3}(\tfrac{3}{4})^{97}.$$

By Theorem 2, $$\lim_{n \to \infty} (\tfrac{3}{4})^n = 0.$$

Thus $$\lim_{n \to \infty} a_n = 0.$$

This Limit Will Be Needed Several Times in Chaps. 14 and 15. Similar reasoning shows that, for any fixed number k,

$$\lim_{n \to \infty} \frac{k^n}{n!} = 0.$$

This means that the factorial function grows faster than any exponential k^n.

(In contrast, Sec. 6.3 shows that k^n, for $k > 1$, grows faster than n^a for any fixed a.) ●

Exercises

In Exercises 1 to 15 determine whether the sequence with the given value of a_n converges or diverges. If the sequence converges, give its limit.

1. $a_n = 0.3^n$

2. $a_n = 0.99^n$

3. $a_n = (-\frac{1}{2})^n$

4. $a_n = \dfrac{1}{n!}$

5. $a_n = \dfrac{20^n}{n!}$

6. $a_n = \dfrac{n+1}{n}$

7. $a_n = 2^n$

8. $a_n = \cos n\pi$

9. $a_n = \sin n\pi$

10. $a_n = \dfrac{n^2 - 1}{n^2}$

11. $a_n = \dfrac{10^n}{n!}$

12. $a_n = \dfrac{n^2}{n!}$

13. $a_n = (1.01)^n$

14. $a_n = \left(1 + \dfrac{2}{n}\right)^n$

15. $a_n = n \sin\left(\dfrac{1}{n}\right)$

∎

16. Let $a_n = 100^n/n!$.
 (a) Show that $a_1 < a_2 < \cdots < a_{99}$.
 (b) Show that $a_{99} = a_{100}$.
 (c) Show that $a_{100} > a_{101} > a_{102} > \cdots$.
17. Let $a_n = 200^n/n!$. What is the smallest n such that a_n is larger than the succeeding term a_{n+1}?

In Exercises 18 to 20 determine the given limits by first showing that each limit is a definite integral $\int_a^b f(x)\, dx$ for a suitable interval $[a, b]$ and function f.

18. $\displaystyle\lim_{n \to \infty} \sum_{i=1}^{n} \left(\dfrac{i}{n}\right)^2 \dfrac{1}{n}$

19. $\displaystyle\lim_{n \to \infty} \left[\dfrac{1}{n+1} + \dfrac{1}{n+2} + \cdots + \dfrac{1}{2n}\right]$

20. $\displaystyle\lim_{n \to \infty} \sum_{i=1}^{n} \dfrac{n}{n^2 + i^2}$

∎ ∎

21. Let $a_n = \dfrac{1}{1 \cdot 2} + \dfrac{1}{2 \cdot 3} + \cdots + \dfrac{1}{n(n+1)}$.
 (a) Compute a_n for $n = 1, 2, 3, 4$, in each case expressing a_n as a single fraction.
 (b) Using the identity
 $$\dfrac{1}{i(i+1)} = \dfrac{1}{i} - \dfrac{1}{i+1},$$
 find a short formula for a_n.
 (c) Show that $\displaystyle\lim_{n \to \infty} a_n = 1$.

22. Let $a_n = \dfrac{1}{2^2} + \dfrac{1}{3^2} + \cdots + \dfrac{1}{n^2}$ for $n \geq 2$. Show that $\displaystyle\lim_{n \to \infty} a_n$ exists and is ≤ 1. *Hint:* See Exercise 21.

23. Let $a_n = (1 - 1/2^2)(1 - 1/3^2) \cdots [1 - 1/(n+1)^2]$.
 (a) Compute a_1, a_2, a_3, a_4 as fractions and as decimals.
 (b) Find $\lim_{n \to \infty} a_n$.

In Exercises 24 to 28 use a calculator.

24. Let $a_n = 1 - \dfrac{1}{2} + \dfrac{1}{3} - \dfrac{1}{4} + \cdots + (-1)^{n-1} \dfrac{1}{n}$.
 (a) Compute a_n at least up to $n = 7$.
 (b) What do you think happens to a_n as $n \to \infty$?

25. Let $a_n = 1(\frac{1}{2}) + 2(\frac{1}{2})^2 + 3(\frac{1}{2})^3 + \cdots + n(\frac{1}{2})^n$.
 (a) Compute a_n at least up to $n = 7$.
 (b) What do you think happens to a_n as $n \to \infty$?

26. Let $a_n = \dfrac{1}{0!} + \dfrac{1}{1!} + \dfrac{1}{2!} + \dfrac{1}{3!} + \cdots + \dfrac{1}{n!}$, where $0! = 1 = 1!$ and $n! = 1 \cdot 2 \cdots \cdot n$.
 (a) Compute a_n at least up to $n = 6$.
 (b) What do you think happens to a_n as $n \to \infty$?

27. Let $a_n = 1 - \dfrac{1}{2} + \dfrac{1}{4} - \dfrac{1}{8} + \cdots + (-1)^n \dfrac{1}{2^n}$.
 (a) Compute a_n at least up to $n = 7$.
 (b) As $n \to \infty$, a_n approaches a famous number. What is it?

28. Let $a_n = \pi - \dfrac{\pi^3}{3!} + \dfrac{\pi^5}{5!} - \cdots + (-1)^{n-1} \dfrac{\pi^{2n-1}}{(2n-1)!}$.
 (a) Compute a_n at least up to $n = 5$.
 (b) As $n \to \infty$, a_n approaches a famous number. What is it?

29. How large must n be to satisfy the inequality $0.99^n < 0.00001$?

This exercise completes the proof of Theorem 2 in the style of Example 3.

30. Let r be a positive number less than 1. Let p be a positive number. Show that there is an integer n such that $r^n < p$. *Hint:* Use logarithms.

31. This exercise provides a neat but tricky proof of Theorem 2. Let r be a positive number less than 1.
 (*a*) Using Theorem 1, show that $\lim_{n \to \infty} r^n$ exists.
 (*b*) Call the limit in (*a*) L. Why is $rL = L$?
 (*c*) From (*b*) deduce that $L = 0$.

14.2

Series

Frequently a sequence is formed by summing terms of a given sequence. Example 1 illustrates this way of constructing a sequence.

EXAMPLE 1   Suppose that the government spends an extra billion dollars without raising an extra billion dollars in taxes. What is the total increase in spending in the United States due to this action? Assume that each business or individual spends 80 percent of any income or revenue and invests or saves the remaining 20 percent.

SOLUTION   The initial billion dollars leads directly to the spending of another 0.8 billion. In turn, 80 percent of this 0.8 billion is spent; that is, another 0.8^2 billion is spent. Thus, at the end of three transactions $1 + 0.8 + 0.8^2$ billion dollars are spent.

The total amount in billions of dollars spent at the end of n transactions is

$$S_n = 1 + 0.8 + 0.8^2 + \cdots + 0.8^{n-1}.$$

(S is short for *sum*.) Then

$$S_1 = 1$$
$$S_2 = 1 + 0.8 = 1.8$$
$$S_3 = 1 + 0.8 + 0.8^2 = 1 + 0.8 + 0.64 = 2.44$$
$$S_4 = 1 + 0.8 + 0.8^2 + 0.8^3 = 1 + 0.8 + 0.64 + 0.512 = 2.952$$
$$S_5 = 1 + 0.8 + 0.8^2 + 0.8^3 + 0.8^4$$
$$= 1 + 0.8 + 0.64 + 0.512 + 0.4096$$
$$= 3.3616.$$

The sequence $S_1, S_2, S_3, S_4, \ldots$ is obtained from the sequence $1, 0.8, 0.8^2, 0.8^3, \ldots$ by adding more and more terms. The long-range impact of the government's expenditure is determined by the behavior of $\{S_n\}$ when n is large.

Two influences affect the growth of S_n as n increases. On the one hand, the number of summands increases, causing S_n to get larger. On the other hand, the summands approach 0, so that S_n grows more and more slowly as n increases. Theorem 1 of this section shows that the sequence S_n converges and that its limit is 5: $$\lim_{n \to \infty} S_n = 5.$$

The total increase in spending due to the extra billion dollars spent by the government is 5 billion dollars. In economic theory the "multiplier" is said to be 5. ●

The rest of this section extends the ideas introduced in Example 1.

Let $a_1, a_2, a_3, \ldots, a_n, \ldots$ be a sequence. From this sequence a new sequence $S_1, S_2, S_3, \ldots, S_n, \ldots$ can be formed:

$$S_1 = a_1$$
$$S_2 = a_1 + a_2$$
$$S_3 = a_1 + a_2 + a_3$$
$$\cdots\cdots\cdots\cdots$$
$$S_n = a_1 + a_2 + a_3 + \cdots + a_n.$$

The sequence of sums $S_1, S_2, \ldots$ is called the *series* obtained from the sequence $a_1, a_2, \ldots$. Traditionally, though imprecisely, it is referred to as *the series whose nth term is* a_n. Common notations for the sequence $\{S_n\}$ are $\sum_{n=1}^{\infty} a_n$ and $a_1 + a_2 + a_3 + \cdots + a_n + \cdots$. The sum

$$S_n = a_1 + a_2 + \cdots + a_n$$

is called a *partial sum* or the *nth partial sum*. If the sequence of partial sums of a series converges to L, then L is called the *sum* of the series. Frequently the sum L of a series is written $a_1 + a_2 + \cdots + a_n + \cdots$. Remember, however, that we do not add an infinite number of numbers; we take the limit of finite sums.

Example 1 concerns the series whose nth term is 0.8^{n-1};

$$S_n = 1 + 0.8^1 + 0.8^2 + \cdots + 0.8^{n-1}.$$

It is a special case of a geometric series, which will now be defined.

DEFINITION *Geometric series.* Let a and r be real numbers. The series

$$a + ar + ar^2 + \cdots + ar^{n-1} + \cdots$$

is called the *geometric series with initial term a and ratio r.*

The series in Example 1 is a geometric series with initial term 1 and ratio 0.8.

THEOREM I If $-1 < r < 1$, the geometric series

$$a + ar + \cdots + ar^{n-1} + \cdots$$

converges to $a/(1 - r)$.

PROOF Let S_n be the sum of the first n terms:

$$S_n = a + ar + \cdots + ar^{n-1}.$$

Multiplication by r yields

$$rS_n = ar + \cdots + ar^{n-1} + ar^n.$$

Subtraction gives

$$S_n - rS_n = (a + ar + \cdots + ar^{n-1}) - (ar + \cdots + ar^{n-1} + ar^n),$$

or $(1 - r)S_n = a - ar^n$ due to cancellations.

Thus $$S_n = \frac{a - ar^n}{1 - r},$$

or $$S_n = \frac{a}{1 - r}(1 - r^n).$$

By Sec. 14.1, $$\lim_{n \to \infty} r^n = 0.$$

Thus $$\lim_{n \to \infty} S_n = \frac{a}{1 - r},$$

proving the theorem. ●

In particular, if $a = 1$ and $r = 0.8$, as in Example 1, the geometric series has the sum

$$\frac{1}{1 - 0.8} = \frac{1}{0.2} = 5.$$

Theorem 1 says nothing about geometric series in which the ratio r is ≥ 1 or ≤ -1. The next theorem, which concerns series in general, not just geometric series, will be of use in settling this case.

THEOREM 2   (*The nth-term test*).   If $\lim_{n \to \infty} a_n \neq 0$, then the series $a_1 + a_2 + \cdots + a_n + \cdots$ diverges.   (The same conclusion holds if $\{a_n\}$ has no limit.)

PROOF   Assume that the series $a_1 + a_2 + \cdots$ converges.   Since S_n is the sum $a_1 + a_2 + \cdots + a_n$, while S_{n-1} is the sum of the first $n - 1$ terms, $a_1 + a_2 + \cdots + a_{n-1}$, $S_n = S_{n-1} + a_n$, or

$$a_n = S_n - S_{n-1}.$$

Let $$S = \lim_{n \to \infty} S_n.$$

Then we also have $$S = \lim_{n \to \infty} S_{n-1},$$

since $S_{2-1}, S_{3-1}, S_{4-1}, \ldots$ run through the same numbers as $S_1, S_2, S_3, \ldots$.
Thus

$$\lim_{n \to \infty} a_n = \lim_{n \to \infty} (S_n - S_{n-1})$$

$$= \lim_{n \to \infty} S_n - \lim_{n \to \infty} S_{n-1}$$

$$= S - S$$
$$= 0.$$

This proves the theorem. ●

Theorem 2 implies that, if $a \neq 0$ and $r \geq 1$, the geometric series

$$a + ar + \cdots + ar^{n-1} + \cdots$$

diverges.   For instance, if $r = 1$,

$$\lim_{n \to \infty} ar^n = \lim_{n \to \infty} a1^n = a,$$

which is not 0.   If $r > 1$, then r^n gets arbitrarily large as n increases; hence $\lim_{n \to \infty} ar^n$ does not exist.   Similarly, if $r \leq -1$, $\lim_{n \to \infty} ar^n$ does not exist. The above results and Theorem 1 can be summarized by this statement: The geometric series

$$a + ar + ar^2 + \cdots + ar^{n-1} + \cdots,$$

for $a \neq 0$, converges if and only if $|r| < 1$.

Theorem 2 implies that the nth term of a convergent series approaches 0 as n gets large.   The next example shows that the converse is not true.   If the nth term of a series approaches 0 as n gets large, it does *not* necessarily follow that the series is convergent.

EXAMPLE 3   Show that the series

$$\frac{1}{\sqrt{1}} + \frac{1}{\sqrt{2}} + \frac{1}{\sqrt{3}} + \cdots + \frac{1}{\sqrt{n}} + \cdots$$

diverges.

SOLUTION   Consider

$$S_n = \frac{1}{\sqrt{1}} + \frac{1}{\sqrt{2}} + \cdots + \frac{1}{\sqrt{n}}.$$

Each of the n summands in S_n is $\geq 1/\sqrt{n}$.   Hence

$$S_n \geq \underbrace{\frac{1}{\sqrt{n}} + \frac{1}{\sqrt{n}} + \cdots + \frac{1}{\sqrt{n}}}_{n \text{ summands}} = \frac{n}{\sqrt{n}} = \sqrt{n}.$$

As n increases, $\sqrt{n}$ increases without bound.   Since $S_n \geq \sqrt{n}$,

$$\lim_{n \to \infty} S_n \text{ does not exist.}$$

In short, the series $\dfrac{1}{\sqrt{1}} + \dfrac{1}{\sqrt{2}} + \cdots + \dfrac{1}{\sqrt{n}} + \cdots$

diverges even though its nth term, $1/\sqrt{n}$, approaches 0. ●

In the next example the nth term approaches 0 much faster than $1/\sqrt{n}$ does.

Still the series diverges. The series in the next example is called the *harmonic series*. The argument that it diverges is due to the French mathematician Nicolas of Oresme, about the year 1360.

EXAMPLE 4 Show that the series
$$\frac{1}{1} + \frac{1}{2} + \cdots + \frac{1}{n} + \cdots$$
diverges.

SOLUTION Collect the summands in longer and longer groups in the manner indicated below. (The number of summands is a power of 2, doubling at each step.)

$$\tfrac{1}{1} + \tfrac{1}{2} + \tfrac{1}{3} + \tfrac{1}{4} + \tfrac{1}{5} + \tfrac{1}{6} + \tfrac{1}{7} + \tfrac{1}{8} + \tfrac{1}{9} + \tfrac{1}{10} + \cdots + \tfrac{1}{16} + \tfrac{1}{17} + \cdots$$

The sum of the terms in each group is at least $\frac{1}{2}$. For instance,

$$\tfrac{1}{5} + \tfrac{1}{6} + \tfrac{1}{7} + \tfrac{1}{8} > \tfrac{1}{8} + \tfrac{1}{8} + \tfrac{1}{8} + \tfrac{1}{8} = \tfrac{4}{8} = \tfrac{1}{2},$$

and
$$\tfrac{1}{9} + \tfrac{1}{10} + \cdots + \tfrac{1}{16} > \tfrac{1}{16} + \tfrac{1}{16} + \cdots + \tfrac{1}{16} = \tfrac{8}{16} = \tfrac{1}{2}.$$

Since the repeated addition of $\frac{1}{2}$s produces sums as large as we please, the series diverges. ●

An Important Moral If the series $a_1 + a_2 + \cdots + a_n + \cdots$ converges, it follows that $a_n \to 0$. However, if $a_n \to 0$, it *does not necessarily follow* that $a_1 + a_2 + \cdots + a_n + \cdots$ converges. Indeed, there is no general, practical rule for determining whether a series converges or diverges. Fortunately, a few rules suffice to decide on the convergence or divergence of the series most commonly needed; they will be presented in this chapter and the next.

It should be pointed out for the record that, if $a_1 + a_2 + \cdots + a_n + \cdots = L$ and c is a number, then the series $ca_1 + ca_2 + \cdots + ca_n + \cdots$ has the sum cL.

Exercises

In Exercises 1 to 4 find the sum of the given geometric series.

1. $1 + \frac{1}{2} + (\frac{1}{2})^2 + \cdots + (\frac{1}{2})^{n-1} + \cdots$
2. $1 - \frac{1}{3} + \frac{1}{9} - \frac{1}{27} + \cdots + (-\frac{1}{3})^{n-1} + \cdots$
3. $3/10 + 3/100 + 3/1000 + \cdots + 3/10^n + \cdots$
4. $0.99 + 0.99^2 + \cdots + 0.99^n + \cdots$

In Exercises 5 to 10 determine whether the given series converges or diverges.

5. $\dfrac{1}{10\sqrt{1}} + \dfrac{1}{10\sqrt{2}} + \dfrac{1}{10\sqrt{3}} + \dfrac{1}{10\sqrt{4}} + \cdots + \dfrac{1}{10\sqrt{n}} + \cdots$

6. $-5 + 5 - 5 + \cdots + (-1)^n 5 + \cdots$
7. $100 + 90 + 81 + \cdots + 100(0.9)^{n-1} + \cdots$
8. $2 + 2^2 + 2^3 + \cdots + 2^n + \cdots$

9. $\frac{3}{1} + \frac{3}{2} + \frac{3}{3} + \cdots + 3/n + \cdots$
10. $\frac{1}{10} + \frac{1}{20} + \frac{1}{30} + \cdots + 1/10n + \cdots$
11. If consumers were persuaded to spend 90 percent of their income instead of the 80 percent assumed in Example 1, what would be the value of the multiplier?
12. A certain rubber ball, when dropped on concrete, always rebounds 90 percent of the distance it falls.
    (a) If the ball is dropped from a height of 6 feet, how far does it travel during the first three descents and ascents?
    (b) How far does it travel before coming to rest?

    ∎

13. In Sec. 6.3 it was shown that $\lim_{x \to \infty} x/b^x = 0$ if $b > 1$.

Use this information to show that, if $|x| < 1$, then the nth term of the series

$$1 + 2x + 3x^2 + \cdots + (n + 1)x^n + \cdots$$

approaches 0 as $n \to \infty$.

14. (a) Use the proof of Theorem 1 to show that, if $x \neq 1$, then

$$1 + x + x^2 + \cdots + x^{n-1} = \frac{1 - x^n}{1 - x}.$$

(b) Differentiate the equation in (a) to show that

$$1 + 2x + 3x^2 + \cdots + (n - 1)x^{n-2} =$$
$$\frac{1 - x^n + nx^n - nx^{n-1}}{(1 - x)^2}$$

(c) Using Exercise 13, show that if $|x| < 1$, then

$$\lim_{n \to \infty} (1 + 2x + 3x^2 + \cdots + (n - 1)x^{n-2}) = \frac{1}{(1 - x)^2}.$$

■ ■

15. (See Exercise 12.) A falling object drops $16t^2$ feet during the first t seconds of fall. How long does the ball in Exercise 12 continue to bounce?

16. Write the decimal

$$0.5\overset{\frown}{2}\overset{\frown}{5}\overset{\frown}{2}\overset{\frown}{5}\overset{\frown}{2} \cdots \qquad \text{52s continuing}$$

as a geometric series, and find its sum by Theorem 1.

In Exercises 17 to 20 a short formula for estimating $n!$ is obtained.

17. Let f have the properties that, for $x \geq 1, f(x) \geq 0, f'(x) > 0$ and $f''(x) < 0$. Let a_n be the area of the region below the graph of $y = f(x)$ and above the line segment that joins $(n, f(n))$ with $(n + 1), f(n + 1))$.

(a) Draw a large-scale version of the accompanying

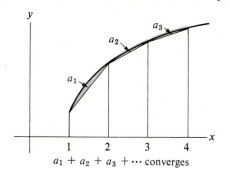

$a_1 + a_2 + a_3 + \cdots$ converges

diagram. The individual regions of areas $a_1, a_2, a_3,$ and a_4 should be clear, not too narrow.

(b) Using geometry, show that the sequence $a_1 + a_2 + a_3 + \cdots$ converges and has a sum no larger than the area of the triangle with vertices $(1, f(1)), (2, f(2)), (1, f(2))$.

18. Let $y = \ln x$.

(a) Using Exercise 17, show that, as $n \to \infty$,

$$\int_1^n \ln x \, dx - \left(\frac{\ln 1 + \ln 2}{2} + \frac{\ln 2 + \ln 3}{2} + \cdots \right.$$
$$\left. + \frac{\ln (n - 1) + \ln n}{2} \right)$$

has a limit; denote this limit by C.

(b) Show that (a) is equivalent to the assertion

$$\lim_{n \to \infty} (n \ln n - n + 1 - \ln n! + \ln \sqrt{n}) = C.$$

19. From Exercise 18(b) deduce that there is a constant k such that

$$\lim_{n \to \infty} \frac{n!}{k(n/e)^n \sqrt{n}} = 1.$$

20. Using Exercise 166 of Sec. 9.10 show that $k = \sqrt{2\pi}$. Thus a good estimate of $n!$ is provided by the formula

$$n! \sim \sqrt{2\pi n} \, (n/e)^n.$$

This is known as Stirling's formula.

A calculator will be of use in Exercises 21 and 22.

21. (a) Compute:

$$S_n = \frac{1}{\sqrt{1}} - \frac{1}{\sqrt{2}} + \frac{1}{\sqrt{3}} - \cdots + (-1)^{n-1} \frac{1}{\sqrt{n}}$$

for $n = 1, 2, 3, 4, 5, 6$.

(b) Do you think that $\lim_{n \to \infty} S_n$ exists?

22. (a) Compute:

$$S_n = 1 - \frac{(\pi/2)^2}{2!} + \frac{(\pi/2)^4}{4!} - \frac{(\pi/2)^6}{6!} + \cdots$$
$$+ (-1)^{n-1} \frac{(\pi/2)^{2n-2}}{(2n - 2)!}$$

at least for $n = 2, 3, 4, 5, 6$.

(b) What do you think happens to S_n as $n \to \infty$?

14.3

The alternating-series test

There is a general type of series for which the converse of the nth-term test is valid. This section discusses this type of series.

DEFINITION   *Alternating series.* If $p_1, p_2, \ldots, p_n, \ldots$ is a sequence of positive numbers,

then the series $p_1 - p_2 + p_3 - p_4 + \cdots + (-1)^{n+1} p_n + \cdots$

and the series $-p_1 + p_2 - p_3 + p_4 - \cdots + (-1)^n p_n + \cdots$

are called *alternating series.*

$$\text{For instance,}\quad 1 - \frac{1}{3} + \frac{1}{5} - \frac{1}{7} + \cdots + (-1)^{n+1}\frac{1}{2n-1} + \cdots$$

and $-1 + 1 - 1 + 1 - \cdots + (-1)^n + \cdots$

are alternating series.

By the nth-term test, the second series diverges. The main result of this section will show that the first series converges. Exercise 10 of Sec. 14.7 shows that its sum is $\pi/4$.

THEOREM   (*The alternating-series test*). If $p_1, p_2, \ldots, p_n, \ldots$ is a decreasing sequence of positive numbers such that $\lim_{n \to \infty} p_n = 0$, then the series whose nth term is $(-1)^{n+1} p_n$,

$$p_1 - p_2 + p_3 - \cdots + (-1)^{n+1} p_n + \cdots,$$

converges.

PROOF   The idea of the proof is easily conveyed by a specific case. For the sake of concreteness and simplicity, consider the series in which $p_n = 1/n$, that is, the series

$$1 - \frac{1}{2} + \frac{1}{3} - \frac{1}{4} + \cdots + (-1)^{n+1}\frac{1}{n} + \cdots.$$

Consider first the partial sums of an *even* number of terms, $S_2, S_4, S_6, \ldots$. For clarity, group the summands in pairs:

$$S_2 = \left(1 - \tfrac{1}{2}\right)$$
$$S_4 = \left(1 - \tfrac{1}{2}\right) + \left(\tfrac{1}{3} - \tfrac{1}{4}\right) = S_2 + \left(\tfrac{1}{3} - \tfrac{1}{4}\right)$$
$$S_6 = \left(1 - \tfrac{1}{2}\right) + \left(\tfrac{1}{3} - \tfrac{1}{4}\right) + \left(\tfrac{1}{5} - \tfrac{1}{6}\right) = S_4 + \left(\tfrac{1}{5} - \tfrac{1}{6}\right)$$

$$\cdots\cdots\cdots\cdots\cdots\cdots\cdots\cdots\cdots\cdots\cdots\cdots\cdots\cdots\cdots\cdots\cdots$$

Since $\tfrac{1}{3}$ is larger than $\tfrac{1}{4}$,

$$\tfrac{1}{3} - \tfrac{1}{4} \text{ is positive.}$$

Thus S_4, which equals $S_2 + \left(\tfrac{1}{3} - \tfrac{1}{4}\right)$,

is larger than S_2.

Similarly, $S_6 > S_4$.

More generally, then, $S_2 < S_4 < S_6 < S_8 < \cdots$.

The sequence S_{2n} is increasing.

Next it will be shown that S_{2n} is less than 1, the first term of the given sequence.

First of all $S_2 = 1 - \frac{1}{2} < 1.$

Next, consider S_4:

$$S_4 = 1 - \frac{1}{2} + \frac{1}{3} - \frac{1}{4}$$
$$= 1 - (\frac{1}{2} - \frac{1}{3}) - \frac{1}{4}$$
$$< 1 - (\frac{1}{2} - \frac{1}{3}).$$

Since $\frac{1}{2} - \frac{1}{3}$ is positive, this shows that

$$S_4 < 1.$$

Similarly,

$$S_6 = 1 - (\frac{1}{2} - \frac{1}{3}) - (\frac{1}{4} - \frac{1}{5}) - \frac{1}{6}$$
$$< 1 - (\frac{1}{2} - \frac{1}{3}) - (\frac{1}{4} - \frac{1}{5})$$
$$< 1.$$

In general then, $S_{2n} < 1$

for all n.

The sequence $S_2, S_4, S_6, \dots$

is therefore increasing and yet bounded by the number 1:

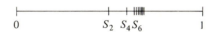

By Theorem 1 of Sec. 14.1, $\lim_{n \to \infty} S_{2n}$ exists.   Call this limit S.

All that remains to be shown is that the numbers

$$S_1, S_3, S_5, \dots$$

also converge to S.

Note that $S_3 = 1 - \frac{1}{2} + \frac{1}{3} = S_2 + \frac{1}{3}$

$$S_5 = 1 - \frac{1}{2} + \frac{1}{3} - \frac{1}{4} + \frac{1}{5} = S_4 + \frac{1}{5}.$$

In general, $S_{2n+1} = S_{2n} + \dfrac{1}{2n+1}.$

(The term $1/(2n+1)$ will be p_{2n+1} in the proof of the general case.)

Thus $$\lim_{n \to \infty} S_{2n+1} = \lim_{n \to \infty} \left(S_{2n} + \frac{1}{2n+1} \right)$$

$$= \lim_{n \to \infty} S_{2n} + \lim_{n \to \infty} \frac{1}{2n+1}$$

$$= S + 0$$

$$= S.$$

Since the partial sums $S_2, S_4, S_6, \dots$

and the partial sums $S_1, S_3, S_5, \dots$

both have the same limit S, it follows that

$$\lim_{n \to \infty} S_n = S.$$

Thus the sequence $1 - \frac{1}{2} + \frac{1}{3} - \frac{1}{4} + \frac{1}{5} \cdots$
converges.

A similar argument applies to any alternating series whose nth term approaches 0 and whose terms decrease in absolute value. ●

(In Sec. 14.7 it is shown that

$$1 - \frac{1}{2} + \frac{1}{3} - \frac{1}{4} + \cdots = \ln 2 \doteq 0.69.)$$

EXAMPLE 1 Estimate the sum S of the series

$$1 - \frac{1}{2} + \frac{1}{3} - \frac{1}{4} + \cdots.$$

SOLUTION These are the first five partial sums:

$$S_1 = 1 = 1.00$$
$$S_2 = 1 - \frac{1}{2} = 0.500$$
$$S_3 = 1 - \frac{1}{2} + \frac{1}{3} \doteq 0.500 + 0.333 = 0.833$$
$$S_4 = S_3 - \frac{1}{4} \doteq 0.833 - 0.250 = 0.583$$
$$S_5 = S_4 + \frac{1}{5} \doteq 0.583 + 0.200 = 0.783.$$

At the left is a graph of S_n as a function of n.   The sums $S_1, S_3, \ldots$ approach S from above.   The sums $S_2, S_4, \ldots$ approach S from below.   For instance,

$$S_4 < S < S_5$$

gives the information that $0.583 < S < 0.783.$

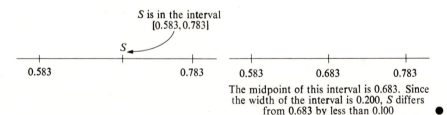

The midpoint of this interval is 0.683. Since the width of the interval is 0.200, S differs from 0.683 by less than 0.100   ●

As the first diagram in Example 1 suggests, any partial sum of a series satisfying the hypothesis of the alternating-series test differs from the sum of the series by less than the absolute value of the first omitted term.

EXAMPLE 2 Does the series

$$\frac{3}{1!} - \frac{3^2}{2!} + \frac{3^3}{3!} - \frac{3^4}{4!} + \frac{3^5}{5!} - \cdots + (-1)^{n+1} \frac{3^n}{n!} + \cdots$$

converge or diverge?

SOLUTION   This is an alternating series.   By Example 4 of Sec. 14.1, its nth term approaches 0.   Let us see whether the absolute value of the terms decreases in size, term by term.   These first few absolute values are:

$$\frac{3}{1!} = 3$$

$$\frac{3^2}{2!} = \frac{9}{2} = 4.5$$

$$\frac{3^3}{3!} = \frac{27}{6} = 4.5$$

$$\frac{3^4}{4!} = \frac{81}{24} = 3.75.$$

At first they increase.   However, the fourth term is less than the third.   It is easy to see that the rest of the terms decrease in size.   For instance,

$$\frac{3^5}{5!} = \frac{3}{5}\frac{3^4}{4!} < \frac{3^4}{4!}$$

and

$$\frac{3^6}{6!} = \frac{3}{6}\frac{3^5}{5!} < \frac{3^5}{5!}.$$

By the alternating-series test then, the series that begins

$$-\frac{3^4}{4!} + \frac{3^5}{5!} - \frac{3^6}{6!} + \cdots$$

converges.   Call its sum S.   If the first three terms

$$\frac{3}{1!} - \frac{3^2}{2!} + \frac{3^3}{3!}$$

are added on, the resulting series still converges and has the sum

$$\frac{3}{1!} - \frac{3^2}{2!} + \frac{3^3}{3!} + S.$$

It will be shown later that

$$\frac{3}{1!} - \frac{3^2}{2!} + \frac{3^3}{3!} - \frac{3^4}{4!} + \cdots = 1 - e^{-3}. \bullet$$

In the Alternating-Series Test the Absolute Values of the Terms Must Eventually Be Decreasing.

As Example 2 illustrates, the alternating-series test works as long as the terms decrease in size from some point on in the series.

It may seem that any alternating series whose nth term approaches 0 converges.   *This is not the case*, as is shown by this series:

$$\frac{2}{1} - \frac{1}{1} + \frac{2}{2} - \frac{1}{2} + \frac{2}{3} - \frac{1}{3} + \frac{2}{4} - \frac{1}{4} + \cdots + \frac{2}{n} - \frac{1}{n} + \cdots. \tag{1}$$

Let S_n be the sum of the first n terms of (1). Then

$$S_2 = \frac{2}{1} - \frac{1}{1} = \frac{1}{1},$$

$$S_4 = \left(\frac{2}{1} - \frac{1}{1}\right) + \left(\frac{2}{2} - \frac{1}{2}\right) = \frac{1}{1} + \frac{1}{2},$$

$$S_6 = \left(\frac{2}{1} - \frac{1}{1}\right) + \left(\frac{2}{2} - \frac{1}{2}\right) + \left(\frac{2}{3} - \frac{1}{3}\right) = \frac{1}{1} + \frac{1}{2} + \frac{1}{3},$$

and, more generally, $\quad S_{2n} = \dfrac{1}{1} + \dfrac{1}{2} + \dfrac{1}{3} + \cdots + \dfrac{1}{n}.$

Since S_{2n} gets arbitrarily large as $n \to \infty$ (the harmonic series of Example 4 of Sec. 14.2), the sequence (1) diverges.

The computations in Example 2 also illustrate another, more general, point. If you delete a finite number of terms from a series and what is left converges, then the series you started with converges. Another way to look at this is to note that a "front end," $a_1 + a_2 + \cdots + a_n$, does not influence convergence or divergence. It is rather a "tail end," $a_{n+1} + a_{n+2} + \cdots$ that matters.

Exercises

Which of the series in Exercises 1 to 6 converge? Diverge? Explain your answer.

1. $\dfrac{1}{2} - \dfrac{2}{3} + \dfrac{3}{4} - \dfrac{4}{5} + \cdots + (-1)^{n+1} \dfrac{n}{n+1} + \cdots$

2. $\dfrac{1}{1^2} - \dfrac{1}{2^2} + \dfrac{1}{3^2} - \dfrac{1}{4^2} + \cdots + (-1)^{n+1} \dfrac{1}{n^2} + \cdots$

3. $\dfrac{1}{\sqrt{1}} - \dfrac{1}{\sqrt{2}} + \dfrac{1}{\sqrt{3}} - \dfrac{1}{\sqrt{4}} + \cdots + (-1)^{n+1} \dfrac{1}{\sqrt{n}} + \cdots$

4. $\dfrac{5}{1!} - \dfrac{5^2}{2!} + \dfrac{5^3}{3!} - \dfrac{5^4}{4!} + \cdots + (-1)^{n+1} \dfrac{5^n}{n!} + \cdots$

5. $\dfrac{3}{\sqrt{1}} - \dfrac{2}{\sqrt{1}} + \dfrac{3}{\sqrt{2}} - \dfrac{2}{\sqrt{2}} + \dfrac{3}{\sqrt{3}} - \dfrac{2}{\sqrt{3}} + \cdots$

6. $\dfrac{1}{3} - \dfrac{2}{5} + \dfrac{3}{7} - \dfrac{4}{9} + \dfrac{5}{11} - \cdots + (-1)^{n+1} \dfrac{n}{2n+1} + \cdots$

7. Show that the sum of the series $1 - \frac{1}{3} + \frac{1}{5} - \frac{1}{7} + \frac{1}{9} - \frac{1}{11} + \cdots$ is between 0.72 and 0.84. (Use S_4 and S_5.)

8. (a) Using the method of Example 1, show that the sum of the series $1 - \frac{1}{2} + \frac{1}{4} - \frac{1}{8} + \cdots + (-1)^{n+1}1/2^n + \cdots$ is between 0.750 and 0.625.

(b) What is the sum of the series in (a)?

9. Show that the sums $S_1, S_3, S_5, \ldots$ in the proof of the theorem in this section decrease.

10. In Sec. 14.7 it will be shown that

$$e^{-1} = \frac{1}{2!} - \frac{1}{3!} + \frac{1}{4!} - \cdots + (-1)^{n+1} \frac{1}{(n+1)!} + \cdots.$$

(a) Compute S_3 and S_4 to three decimal places.
(b) Deduce that e^{-1} is between 0.366 and 0.375.

■

In Exercises 11 to 16 determine whether the series $a_1 + a_2 + \cdots + a_n + \cdots$ converges.

11. $a_n = \left(1 - \dfrac{1}{n}\right)^n$

12. $a_n = (-1)^n \dfrac{1}{n^2}$

13. $a_n = 1.01^{-n}$

14. $a_n = (-1.01)^n/n^3$

15. $a_n = (-0.99)^n$

16. $a_n = n^{-3}$

17. Let $a_n = (-1)^{n+1}[x^{2n-1}/(2n-1)!]$.
(a) Evaluate a_1, a_2, and a_3.
(b) Show that the series whose nth term is a_n converges for every value of x. (In Sec. 14.7 it will be shown that its sum is $\sin x$.)

18. Show that the series

$$1 - \frac{x^2}{2!} + \frac{x^4}{4!} - \frac{x^6}{6!} + \cdots$$

converges for every value of x. (Its sum is $\cos x$.)

19. Find a formula for the nth term of the series in Exercise 18.

■ ■

20. Prove the theorem of this section in generality, showing that $S_2 < S_4 < S_6 < \cdots < p_1$, that $\lim_{n \to \infty} S_{2n}$ exists, and that $\lim_{n \to \infty} S_{2n+1} = \lim_{n \to \infty} S_{2n}$.

21. Give an example of a series $a_1 + a_2 + a_3 + \cdots + a_n + \cdots$ such that it diverges but the series $a_1 - a_2 + a_3 - a_4 + \cdots + (-1)^{n+1}a_n + \cdots$ converges.

14.4

The integral test

This section develops a method for determining whether a series of a certain type converges or diverges. In particular it will show once again that the harmonic series

$$\frac{1}{1} + \frac{1}{2} + \frac{1}{3} + \frac{1}{4} + \cdots + \frac{1}{n} + \cdots$$

diverges, but that the series

$$\frac{1}{1^{1.01}} + \frac{1}{2^{1.01}} + \frac{1}{3^{1.01}} + \frac{1}{4^{1.01}} + \cdots + \frac{1}{n^{1.01}} + \cdots$$

converges.

The method is based on a comparison of sums with integrals, as expressed in Theorem 1.

THEOREM I Let f be a decreasing positive function. Then for any integer $n \geq 2$,

$$f(1) + f(2) + \cdots + f(n-1) \geq \int_1^n f(x)\,dx$$

and

$$f(2) + f(3) + \cdots + f(n) \leq \int_1^n f(x)\,dx.$$

PROOF To obtain the first inequality, compare the total area of the rectangles shown to the left with the area under the curve $y = f(x)$ and above $[1, n]$.

Since f is a decreasing function, the $n - 1$ rectangles indicated have a total area greater than or equal to the area under the graph of $y = f(x)$ and above $[1, n]$.

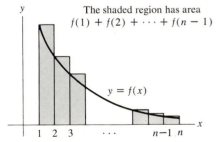

The shaded region has area
$$f(1) + f(2) + \cdots + f(n-1)$$
$y = f(x)$

Now, each rectangle has a width 1. The height of the rectangle above $[1, 2]$ is $f(1)$. Hence the area of this rectangle is $f(1) \cdot 1 = f(1)$. The rectangle above $[2, 3]$ has area $f(2)$. The total area of the $n - 1$ rectangles is therefore

$$f(1) + f(2) + \cdots + f(n-1).$$

Since the area of the region under the curve $y = f(x)$ and above $[1, n]$ is $\int_1^n f(x)\,dx$, the first inequality,

$$f(1) + f(2) + \cdots + f(n-1) \geq \int_1^n f(x)\,dx,$$

follows.

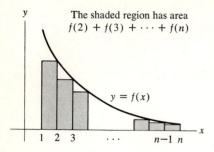

y

The shaded region has area
$f(2) + f(3) + \cdots + f(n)$

$y = f(x)$

1 2 3 $\quad\cdots\quad$ $n-1$ n $\qquad$ *x*

A slightly different diagram (shown at the left) leads to the second inequality. This time the rectangles are situated below the curve. For instance, the rectangle above $[1, 2]$ has height $f(2)$. Comparison of the total area of the $n-1$ rectangles with the area under the curve shows that

$$f(2) + f(3) + \cdots + f(n) \le \int_1^n f(x)\,dx.$$

This proves the theorem. ●

The next example shows how Theorem 1 may be used to estimate the sum of the first n terms of a series.

EXAMPLE I   Use Theorem 1 to estimate

$$\frac{1}{\sqrt{1}} + \frac{1}{\sqrt{2}} + \frac{1}{\sqrt{3}} + \cdots + \frac{1}{\sqrt{n}}.$$

SOLUTION   The function $f(x) = 1/\sqrt{x}$ (defined when x is positive) is positive and decreasing. By Theorem 1,

$$f(1) + f(2) + \cdots + f(n-1) \ge \int_1^n f(x)\,dx$$

or

$$\frac{1}{\sqrt{1}} + \frac{1}{\sqrt{2}} + \cdots + \frac{1}{\sqrt{n-1}} \ge \int_1^n \frac{1}{\sqrt{x}}\,dx.$$

Now, by the fundamental theorem,

$$\int_1^n \frac{1}{\sqrt{x}}\,dx = 2\sqrt{x}\,\Big|_1^n$$

$$= 2\sqrt{n} - 2.$$

Consequently

$$\frac{1}{\sqrt{1}} + \frac{1}{\sqrt{2}} + \cdots + \frac{1}{\sqrt{n-1}} \ge 2\sqrt{n} - 2,$$

and therefore

$$\frac{1}{\sqrt{1}} + \frac{1}{\sqrt{2}} + \cdots + \frac{1}{\sqrt{n}} \ge 2\sqrt{n} - 2 + \frac{1}{\sqrt{n}}. \tag{1}$$

The second inequality in Theorem 1 implies that

$$\frac{1}{\sqrt{2}} + \frac{1}{\sqrt{3}} + \cdots + \frac{1}{\sqrt{n}} \le \int_1^n \frac{1}{\sqrt{x}}\,dx = 2\sqrt{n} - 2.$$

Adding $1/\sqrt{1}(=1)$ to both sides of this inequality yields

$$\frac{1}{\sqrt{1}} + \frac{1}{\sqrt{2}} + \cdots + \frac{1}{\sqrt{n}} \le 2\sqrt{n} - 1. \tag{2}$$

Together, (1) and (2) put an upper bound and a lower bound on the sum of the first n terms:

$$2\sqrt{n} - 2 + \frac{1}{\sqrt{n}} \leq \frac{1}{\sqrt{1}} + \frac{1}{\sqrt{2}} + \cdots + \frac{1}{\sqrt{n}} \leq 2\sqrt{n} - 1. \tag{3}$$

For instance, these two inequalities, when $n = 1,000,000$, show that the sum of the first million terms is between

$$1,998.001 \qquad \text{and} \qquad 1,999,$$

as the reader may check by substitution in (3). ●

In Example 3 of Sec. 14.2 it was shown that the sum of the first n terms of the series

$$\frac{1}{\sqrt{1}} + \frac{1}{\sqrt{2}} + \frac{1}{\sqrt{3}} + \cdots$$

is larger than $\sqrt{n}$. Inequality (1) says more: The sum of the first n terms is greater than $2\sqrt{n} - 2 + 1/\sqrt{n}$. Inequality (1) shows that $\sum_{n=1}^{\infty} 1/\sqrt{n}$ diverges (as did the argument in Example 3 of Sec. 14.2). This suggests that Theorem 1 may be of use in testing whether certain series converge or diverge. A simple test is provided by the next theorem.

THEOREM 2   (*The integral test*).   Let f be a decreasing positive function for $x \geq 1$. Then the series

$$f(1) + f(2) + f(3) + \cdots \tag{4}$$

is convergent if the improper integral

$$\int_{1}^{\infty} f(x)\, dx$$

is convergent, and divergent if the improper integral is divergent.

PROOF   Assume that $\int_{1}^{\infty} f(x)\, dx$ is convergent and has the value B.   Then $\int_{1}^{n} f(x)\, dx \leq B$ and, by Theorem 1,

$$f(2) + f(3) + \cdots + f(n) \leq B.$$

Consequently $\qquad f(1) + f(2) + \cdots + f(n) \leq f(1) + B.$

Now, $f(1) + B$ is a fixed number, independent of n.   The sequence of partial sums

$$f(1), \; f(1) + f(2), \; f(1) + f(2) + f(3), \ldots$$

is increasing and always $\leq f(1) + B$.   By Theorem 1 of Sec. 14.1 the sequence has a limit [no larger than $f(1) + B$].   In other words, the series (4) converges [and has a sum no larger than $f(1) + B$].

On the other hand, if $\int_{1}^{\infty} f(x)\, dx$ is divergent, the definite integral

$$\int_{1}^{n} f(x)\, dx$$

gets arbitrarily large as $n \to \infty$. By the first inequality of Theorem 1, the sum

$$f(1) + f(2) + \cdots + f(n - 1)$$

also gets arbitrarily large as $n \to \infty$; hence (4) diverges.
This proves the theorem. ●

Before applying the theorem, we need a definition.

DEFINITION   *p series*. For a fixed positive number p, the series

$$\frac{1}{1^p} + \frac{1}{2^p} + \frac{1}{3^p} + \cdots$$

is called the *p series* (with exponent p). In particular, when $p = 1$, the *p series* is the harmonic series.

Note that Example 1 concerns the *p* series for the case $p = \frac{1}{2}$.

EXAMPLE 2   Use Theorem 2 (the integral test) to determine for which values of p the *p* series $\sum_{n=1}^{\infty} 1/n^p$ converges.

SOLUTION   The function $f(x) = 1/x^p = x^{-p}$ is decreasing for $x \geq 1$. Thus the integral test may be applied. Now if $p \neq 1$,

$$\int_1^{\infty} x^{-p}\, dx = \lim_{b \to \infty} \int_1^b x^{-p}\, dx$$

$$= \lim_{b \to \infty} \frac{x^{1-p}}{1-p}\Big|_1^b$$

$$= \lim_{b \to \infty} \left(\frac{b^{1-p}}{1-p} - \frac{1}{1-p} \right).$$

For $p > 1$ this limit is $-1/(1-p) = 1/(p-1)$ since

$$\lim_{b \to \infty} b^{1-p} = 0 \qquad p > 1.$$

Thus $\int_1^{\infty} x^{-p}\, dx$ is convergent.
For $p < 1$, b^{1-p} gets arbitrarily large as $b \to \infty$; hence $\int_1^{\infty} x^{-p}\, dx$ is divergent.

If $p = 1$,

$$\int_1^{\infty} x^{-p}\, dx = \int_1^{\infty} \frac{1}{x}\, dx$$

$$= \lim_{b \to \infty} \int_1^b \frac{1}{x}\, dx$$

$$= \lim_{b \to \infty} (\ln b - \ln 1).$$

Since $\ln b$ gets arbitrarily large as $b \to \infty$, $\int_1^{\infty} 1/x\, dx$ is divergent.

Consequently, the series $\dfrac{1}{1^p} + \dfrac{1}{2^p} + \dfrac{1}{3^p} + \cdots$

is divergent if $p \le 1$, and convergent if $p > 1$. ●

The idea behind the integral test also shows how integrals can be used to estimate the sum of a convergent series: Estimate the sum of some tail end of a series, and then add the sum of the corresponding front end.

EXAMPLE 3 By using integrals to estimate

$$\frac{1}{5^3} + \frac{1}{6^3} + \frac{1}{7^3} + \cdots,$$

estimate the sum of the convergent series

$$\frac{1}{1^3} + \frac{1}{2^3} + \frac{1}{3^3} + \frac{1}{4^3} + \frac{1}{5^3} + \cdots.$$

The Further Out in the Series You Choose the Tail End, the More Accurate Your Estimate Will Be.

SOLUTION The sum of the first four terms is

$$\tfrac{1}{1} + \tfrac{1}{8} + \tfrac{1}{27} + \tfrac{1}{64} \doteq 1.17766.$$

To control the size of the tail end, or remainder,

$$\frac{1}{5^3} + \frac{1}{6^3} + \frac{1}{7^3} + \cdots,$$

note that it is more than

$$\int_5^\infty \frac{dx}{x^3} = -\frac{1}{2x^2}\bigg|_5^\infty = \frac{1}{50} = 0.02.$$

and less than

$$\int_4^\infty \frac{dx}{x^3} = -\frac{1}{2x^2}\bigg|_4^\infty = \frac{1}{32} = 0.03125.$$

Hence $1.17766 + 0.02 < \displaystyle\sum_{n=1}^\infty \frac{1}{n^3} < 1.17766 + 0.03125,$

or $1.19766 < \displaystyle\sum_{n=1}^\infty \frac{1}{n^3} < 1.20891.$ ●

Exercises

In Exercises 1 to 9, determine whether the series converges or diverges.

1. $\displaystyle\sum_{n=1}^\infty \frac{1}{n^{1.01}}$

2. $\displaystyle\sum_{n=1}^\infty \frac{1}{n^{0.99}}$

3. $\displaystyle\sum_{n=1}^\infty \frac{n}{n^2 + 1}$

4. $\displaystyle\sum_{n=1}^\infty \frac{1}{n^2 + 1}$

5. $\displaystyle\sum_{n=2}^\infty \frac{1}{n \ln n}$

6. $\displaystyle\sum_{n=1}^\infty \frac{1}{n + 1{,}000}$

7. $\displaystyle\sum_{n=1}^{\infty} \frac{\ln n}{n}$

8. $\displaystyle\sum_{n=1}^{\infty} \frac{n^2}{n^3 + 100}$

9. $\displaystyle\sum_{n=1}^{\infty} \frac{n}{n^2 + 100}$

10. Show that

$$\ln(n+1) < \frac{1}{1} + \frac{1}{2} + \frac{1}{3} + \cdots + \frac{1}{n} < 1 + \ln n.$$

11. Show that the sum of the first million terms of the harmonic series is between 13.8 and 14.8.

12. (a) By comparing the sum with integrals, show that

$$\ln \tfrac{201}{100} < \tfrac{1}{100} + \tfrac{1}{101} + \tfrac{1}{102} + \cdots + \tfrac{1}{200} < \ln \tfrac{200}{99}.$$

(b) Show that $\displaystyle \lim_{n\to\infty} \sum_{i=n}^{2n} \frac{1}{i} = \ln 2.$

13. Show that

$$\frac{n+1}{(2n+1)n} < \frac{1}{n^2} + \frac{1}{(n+1)^2} + \cdots + \frac{1}{(2n)^2} < \frac{n+1}{2n(n-1)}.$$

■

14. For which values of x does the series $\sum_{n=1}^{\infty} x^n/n^2$ converge? Diverge?

15. (a) Use a picture to show that the sum of the first five terms of $\sum_{n=1}^{\infty} 1/n^2$ differs from the sum of the series by less than $\int_5^{\infty} 1/x^2\, dx$, but by more than $\int_6^{\infty} 1/x^2\, dx$.

(b) Use (a) to show that

$$1.63 < \sum_{n=1}^{\infty} \frac{1}{n^2} < 1.67.$$

16. Use elementary geometry to show that the area of that part of the endless staircase above the curve $y = 1/x$ is between $\frac{1}{2}$ and 1.

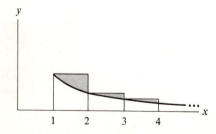

The total shaded area is known as Euler's constant γ, whose decimal representation begins 0.577. It is not known whether γ is rational or irrational.

17. Four of these integrals are elementary. Compute them.

(a) $\displaystyle\int x \sin x^2\, dx$ (b) $\displaystyle\int \frac{\sin x}{x^2}\, dx$ (c) $\displaystyle\int x^2 \sin x\, dx$

(d) $\displaystyle\int \frac{\sin x}{x}\, dx$ (e) $\displaystyle\int \frac{\ln x}{x}\, dx$ (f) $\displaystyle\int \frac{\ln x}{x^2}\, dx.$

18. Compute:

(a) $\dfrac{\partial(e^{-xt} \cos \pi t)}{\partial t}$

(b) $\dfrac{\partial(e^{-xt} \cos \pi t)}{\partial x}$

(c) $\dfrac{\partial(\csc xy^2)}{\partial x}$

(d) $\dfrac{\partial(\csc xy^2)}{\partial y}$

(e) $\dfrac{\partial(\sqrt{a + x^2})}{\partial x}$

(f) $\dfrac{\partial(\sqrt{a + x^2})}{\partial a}$

■ ■

19. Here is an argument that there is an infinite number of primes. Assume that there is only a finite number of primes, $p_1, p_2, \ldots, p_m$.

(a) Show then that

$$\frac{1}{1 - 1/p_1} \frac{1}{1 - 1/p_2} \cdots \frac{1}{1 - 1/p_m} = \sum_{n=1}^{\infty} \frac{1}{n}.$$

(b) From (a) obtain a contradiction.

20. Cards of unit length from a playing deck are piled on top of each other as shown in the diagram.

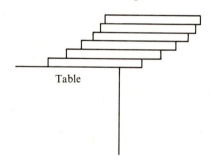

Table

(a) If there are n cards, show that it is possible to arrange them so that the total distance that the right edge of the top card extends beyond the table is $\sum_{i=1}^{n} (1/2i)$. (The length of a card is 1.)

(b) Show that it is possible to place three cards in such a pile that the right edge of the top card extends $\frac{11}{12}$ of a card beyond the table.

(c) If you have 52 cards, estimate how far beyond the table the top card can extend.

(d) If you have an unlimited supply of cards, how far beyond the table can you arrange to have the top card extend?

14.5

The comparison test and the ratio test

So far in this chapter four tests for convergence (or divergence) of a series have been presented. The first concerned a special type of series, a geometric series. The second, the nth-term test, asserts that the nth term of a convergent series must approach 0; so, if the nth term does not approach 0, the series must diverge. The third applies to alternating series for which the absolute value of the nth term decreases from some term on and approaches 0. The fourth, the integral test, applies to certain series of positive terms. In this section three further tests of a much more general character are developed. They are the ones most frequently applied.

THEOREM 1   (*The comparison test*).

1. If the series
$$c_1 + c_2 + \cdots + c_n + \cdots$$
converges, where each c_n is positive, and if
$$0 \le p_n \le c_n$$
for each n, then the series
$$p_1 + p_2 + \cdots + p_n + \cdots$$
converges.

2. If the series
$$c_1 + c_2 + \cdots + c_n + \cdots$$
diverges, where each c_n is positive, and if
$$p_n \ge c_n$$
for each n, then the series $p_1 + p_2 + \cdots + p_n + \cdots$ diverges.

PROOF   First let us establish case 1. Let the sum of the series $c_1 + c_2 + \cdots$ be C. Let S_n denote the partial sum $p_1 + p_2 + \cdots + p_n$. Then, for each n,
$$S_n = p_1 + p_2 + \cdots + p_n \le c_1 + c_2 + \cdots + c_n < C.$$
Since the p_n's are nonnegative,
$$S_1 \le S_2 \le \cdots \le S_n \le \cdots.$$
Since each S_n is less than C, Theorem 1 of Sec. 14.1 assures us that the sequence
$$S_1, S_2, \ldots, S_n, \ldots$$
converges to a number L (less than or equal to C). In other words, the series $p_1 + p_2 + \cdots$ converges (and its sum is less than or equal to the sum $c_1 + c_2 + \cdots$).

The second part of the theorem, the divergence test (case 2), follows immediately from the convergence test (case 1). For if the series $p_1 + p_2 + \cdots$ converged, so would the series $c_1 + c_2 + \cdots$, which is assumed to diverge. ●

To apply the comparison tests, compare a series to one known to converge (or diverge).

EXAMPLE 1   Show that the series $\dfrac{2}{3}\dfrac{1}{1^2} + \dfrac{3}{4}\dfrac{1}{2^2} + \cdots + \dfrac{n+1}{n+2}\dfrac{1}{n^2} + \cdots$

converges.

SOLUTION   The series resembles the series

$$\frac{1}{1^2} + \frac{1}{2^2} + \cdots + \frac{1}{n^2} + \cdots,$$

which was shown by the integral test to be convergent. Since the fraction $(n+1)/(n+2) < 1$,

$$\frac{n+1}{n+2}\frac{1}{n^2} < \frac{1}{n^2}.$$

Thus, by the comparison test for convergence, the series

$$\frac{2}{3}\frac{1}{1^2} + \frac{3}{4}\frac{1}{2^2} + \cdots + \frac{n+1}{n+2}\frac{1}{n^2} + \cdots$$

also converges. ●

The next test is similar to the comparison test. It is useful when two series of positive terms resemble each other a great deal, even though the terms of one series are not less than the terms of the other.

THEOREM 2   (*The limit-comparison test*).   Let

$$c_1 + c_2 + \cdots + c_n + \cdots$$

and $$p_1 + p_2 + \cdots + p_n + \cdots$$

be series of positive terms such that

$$\lim_{n\to\infty} \frac{p_n}{c_n}$$

exists and is not 0.   Then

1. If $c_1 + c_2 + \cdots + c_n + \cdots$ converges, so does $p_1 + p_2 + \cdots + p_n + \cdots$.
2. If $c_1 + c_2 + \cdots + c_n + \cdots$ diverges, so does $p_1 + p_2 + \cdots + p_n + \cdots$.

PROOF   Let $\lim_{n\to\infty} p_n/c_n = a$. We shall prove case 1. Since as $n \to \infty$, $p_n/c_n \to a$, there must be an integer N such that, for all $n \geq N$, p_n/c_n remains less than, say, $a + 1$. Thus

$$p_N < (a+1)c_N$$
$$p_{N+1} < (a+1)c_{N+1}$$
$$\cdots\cdots\cdots\cdots\cdots\cdots$$
$$p_n < (a+1)c_n \qquad n \geq N.$$

Now the series

$$(a + 1)c_N + (a + 1)c_{N+1} + \cdots + (a + 1)c_n + \cdots,$$

being $a + 1$ times the tail end of a convergent series, is itself convergent. By the comparison test

$$p_N + p_{N+1} + \cdots + p_n + \cdots$$

is convergent. Hence $p_1 + p_2 + \cdots + p_n + \cdots$ is convergent.

Case 2 follows from case 1, as in Theorem 1. ●

EXAMPLE 2 Show that

$$\sum_{n=1}^{\infty} \frac{(1 + 1/n)^n(1 + (-\frac{1}{2})^n)}{2^n}$$

converges.

SOLUTION Note that, as $n \to \infty$, $[1 + 1/n]^n \to e$ and $1 + (-\frac{1}{2})^n \to 1$. The major influence is the 2^n in the denominator. So use the limit-comparison theorem, with the convergent series $c_1 + c_2 + \cdots + c_n + \cdots = 1 + 1/2 + 1/4 + \cdots + 1/2^n + \cdots$, which the given series resembles.

$$\text{Then} \quad \lim_{n \to \infty} \frac{\dfrac{\left(1 + \dfrac{1}{n}\right)^n\left(1 + \left(-\dfrac{1}{2}\right)^n\right)}{2^n}}{\dfrac{1}{2^n}} = \lim_{n \to \infty} \left(1 + \frac{1}{n}\right)^n\left(1 + \left(-\frac{1}{2}\right)^n\right) = e \cdot 1 = e.$$

Since $\sum_{n=1}^{\infty} 2^{-n}$ is convergent, so is the given series. ●

The next test is suggested by the test for the convergence of a geometric series.

THEOREM 3 (*The ratio test*). Let $p_1 + p_2 + \cdots + p_n + \cdots$ be a series of positive terms.

1. If $\lim\limits_{n \to \infty} \dfrac{p_{n+1}}{p_n}$ exists and is less than 1, the series converges.

2. If $\lim\limits_{n \to \infty} \dfrac{p_{n+1}}{p_n}$ exists and is greater than 1, the series diverges.

PROOF To prove case 1, first let $\lim\limits_{n \to \infty} \dfrac{p_{n+1}}{p_n} = s < 1.$

Select a number r such that $s < r < 1.$

Then there is an integer N such that, for all $n \geq N$,

$$\frac{p_{n+1}}{p_n} < r;$$

hence $p_{n+1} < r p_n.$

Thus

$$p_{N+1} < rp_N;$$

$$p_{N+2} < rp_{N+1} < r(rp_N) = r^2 p_N$$

$$p_{N+3} < rp_{N+2} < r(r^2 p_N) = r^3 p_N$$

. . . .

Thus the terms of the series

$$p_N + p_{N+1} + p_{N+2} + \cdots$$

are less than the corresponding terms of the *geometric series*

$$p_N + rp_N + r^2 p_N + \cdots$$

(except for the first term p_N, which equals the first term of the geometric series). Since $r < 1$, the latter series converges. By the comparison test, $p_N + p_{N+1} + p_{N+2} + \cdots$ converges. Adding in the front end,

$$p_1 + p_2 + \cdots + p_{N-1},$$

still results in a convergent series.

The argument for case 2 is much shorter. If $\lim_{n \to \infty} p_{n+1}/p_n$ is greater than 1, then for all n from some point on p_{n+1} is larger than p_n. Thus the nth term of the series $p_1 + p_2 + \cdots$ cannot approach 0. By the nth-term test the series diverges. This completes the proof. ●

No mention has been made in Theorem 3 of the case $\lim_{n \to \infty} p_{n+1}/p_n = 1$. The reason for this omission is that anything can happen; the series may diverge or it may converge. (Exercise 16 illustrates these possibilities.) Also, $\lim_{n \to \infty} p_{n+1}/p_n$ may not exist. In that case, one must look to other tests to determine whether the series diverges or converges.

The ratio test is a natural one to try if the nth term of a series involves powers of a fixed number, as the next example shows.

EXAMPLE 3 Show that the series $p + 2p^2 + 3p^3 + \cdots + np^n + \cdots$

converges for any fixed number p, for which $0 < p < 1$.

SOLUTION Let a_n denote the nth term of the series. Then

$$a_n = np^n$$

and

$$a_{n+1} = (n+1)p^{n+1}.$$

The ratio between consecutive terms is

$$\frac{a_{n+1}}{a_n} = \frac{(n+1)p^{n+1}}{np^n}$$

$$= \frac{n+1}{n} p.$$

Thus
$$\lim_{n \to \infty} \frac{a_{n+1}}{a_n} = p$$

and the series converges. (In Example 1 of Sec. 14.7 it is shown that its sum is $p/(1 - p)^2$. In fact this holds whenever $|p| < 1$.) ●

EXAMPLE 4 Find for which positive values of x the series

$$\frac{x}{1!} + \frac{x^2}{2!} + \frac{x^3}{3!} + \cdots + \frac{x^n}{n!} + \cdots$$

converges and for which it diverges.

SOLUTION The nth term, a_n, is $x^n/n!$ Thus

$$a_{n+1} = \frac{x^{n+1}}{(n+1)!},$$

and, therefore,

$$\frac{a_{n+1}}{a_n} = \frac{\dfrac{x^{n+1}}{(n+1)!}}{\dfrac{x^n}{n!}}$$

$$= x \frac{n!}{(n+1)!}$$

$$= \frac{x}{n+1}.$$

Since x is fixed,
$$\lim_{n \to \infty} \frac{x}{n+1} = 0.$$

By the ratio test, the series converges for all positive x. By methods of the next section it can be shown that the series converges for all negative x also. In Sec. 14.7 it will be shown that its sum is $e^x - 1$. Note that, by the nth-term test, $x^n/n! \to 0$ as $n \to \infty$. ●

The next example uses the ratio test to establish divergence.

EXAMPLE 5 Show that the series $\quad \dfrac{2}{1} + \dfrac{2^2}{2} + \dfrac{2^3}{3} + \cdots + \dfrac{2^n}{n} + \cdots$

diverges.

SOLUTION In this case $a_n = 2^n/n$ and

$$\frac{a_{n+1}}{a_n} = \frac{\dfrac{2^{n+1}}{n+1}}{\dfrac{2^n}{n}}$$

$$= \frac{2^{n+1}}{n+1} \frac{n}{2^n}$$

$$= 2 \frac{n}{n+1}.$$

Thus $$\lim_{n \to \infty} \frac{a_{n+1}}{a_n} = 2,$$

which is larger than 1.   By the ratio test, the series diverges. ●

It is not really necessary to call on the powerful ratio test to establish the divergence of the series in Example 5.   As was shown in Sec. 6.3, its nth term gets arbitrarily large; by the nth-term test, the series diverges.

Exercises

Test for convergence or divergence.

1. $\displaystyle\sum_{n=1}^{\infty} \frac{1}{1 + n^2}$

2. $\displaystyle\sum_{n=1}^{\infty} \frac{n + 2}{(n + 1)\sqrt{n}}$

3. $\displaystyle\sum_{n=1}^{\infty} \frac{n^3}{2^n}$

4. $\displaystyle\sum_{n=1}^{\infty} \frac{\sin^2 n}{n^2}$

5. $\displaystyle\sum_{n=1}^{\infty} \frac{(n + 1)^2}{n!}$

6. $\displaystyle\sum_{n=1}^{\infty} \frac{5n + 1}{(n + 2)n^2}$

7. $\displaystyle\sum_{n=1}^{\infty} \frac{n!}{n^n}$

8. $\displaystyle\sum_{n=1}^{\infty} \frac{2^n + n}{3^n}$

9. $\displaystyle\sum_{n=1}^{\infty} \frac{4n + 1}{(2n + 3)n^2}$

10. $\displaystyle\sum_{n=1}^{\infty} \frac{n + 1}{(5n + 2)\sqrt{n}}$

11. $\displaystyle\sum_{n=1}^{\infty} \frac{1}{n^n}$

12. $\displaystyle\sum_{n=1}^{\infty} \frac{(1 + 1/n)^n}{n(n + 1)}$

13. $\displaystyle\sum_{n=1}^{\infty} \frac{(2n + 1)(2^n + 1)}{3^n + 1}$

14. $\displaystyle\sum_{n=1}^{\infty} \frac{\ln n}{n}$

15. $\displaystyle\sum_{n=1}^{\infty} \frac{1 + \cos n}{n^2}$

This exercise shows why the ratio test is useless when $\lim_{n \to \infty} p_{n+1}/p_n = 1$.

16. (a) Show that if $p_n = 1/n$, then $\sum_{n=1}^{\infty} p_n$ diverges and $\lim_{n \to \infty} p_{n+1}/p_n = 1$.

(b) Show that if $p_n = 1/n^2$, then $\sum_{n=1}^{\infty} p_n$ converges and

$\lim_{n \to \infty} p_{n+1}/p_n = 1$.

17. For which positive values of x does the series

$$\frac{x}{1} + \frac{x^2}{2} + \frac{x^3}{3} + \cdots + \frac{x^n}{n} + \cdots$$

converge?   Diverge?

18. For which positive values of x does the series

$$\frac{x}{\sqrt{1}} + \frac{x^2}{\sqrt{2}} + \frac{x^3}{\sqrt{3}} + \cdots + \frac{x^n}{\sqrt{n}} + \cdots$$

converge?   Diverge? ■

19. In Theorem 2 no mention is made of the case $\lim_{n \to \infty} p_n/c_n = 0$.
(a) Show that if $\sum_{n=1}^{\infty} c_n$ converges and $\lim_{n \to \infty} p_n/c_n = 0$, then $\sum_{n=1}^{\infty} p_n$ converges.
(b) Show that, if $\sum_{n=1}^{\infty} c_n$ diverges and $\lim_{n \to \infty} p_n/c_n = 0$, then $\sum_{n=1}^{\infty} p_n$ may diverge or it may converge.   (Give examples of the two possibilities.)

20. (a) Show that $\sum_{n=1}^{\infty} 1/(1 + 2^n)$ converges.
(b) Show that the sum of the series in (a) is between 0.64 and 0.77.   (Use the first three terms and control the sum of the rest of the series by comparing it to the sum of a geometric series.)

21. [See Exercise 20(b).]
(a) Show that $\sum_{n=1}^{\infty} (n + 1)/[(n + 2)2^n]$ converges.
(b) Show that the sum of the series in (a) is between

$$\tfrac{2}{3}\tfrac{1}{2} + \tfrac{3}{4}\tfrac{1}{4} + \tfrac{4}{5}\tfrac{1}{8} + \tfrac{5}{6}\tfrac{1}{8}$$

and

$$\tfrac{2}{3}\tfrac{1}{2} + \tfrac{3}{4}\tfrac{1}{4} + \tfrac{4}{5}\tfrac{1}{8} + \tfrac{1}{8}.$$

(c) From (b) deduce that the sum of the series in (a) is between 0.72 and 0.75. (The accuracy of this approach improves rapidly when larger front ends are used.)

■ ■ ■

22. Let $\{a_n\}$ be a sequence of positive terms. If $\sum_{n=1}^{\infty} a_n^2$ converges, does $\sum_{n=1}^{\infty} a_n/n$ necessarily converge?

23. (a) Show that, if $\{a_n\}$ is a sequence of positive terms and $\sum_{n=1}^{\infty} a_n$ converges, so does $\sum_{n=1}^{\infty} a_n^2$.
    (b) Give an example of a sequence $\{a_n\}$ such that $\sum_{n=1}^{\infty} a_n$

converges but $\sum_{n=1}^{\infty} a_n^2$ does not.

24. Determine whether $\sum_{n=1}^{\infty} (\ln n)/n^2$ converges or diverges.

25. The following result is used in the statistical theory of stochastic processes: Let $\{a_n\}$ and $\{c_n\}$ be two sequences of nonnegative numbers such that $\sum_{n=1}^{\infty} a_n c_n$ converges and $\lim_{n \to \infty} c_n = 0$. Prove that $\sum_{n=1}^{\infty} a_n c_n^2$ converges.

26. Determine for which x the series $\sum_{n=1}^{\infty} n^n x^n/n!$ converges or diverges. (Stirling's formula, which asserts that

$$\lim_{n \to \infty} n!/\sqrt{2\pi n}\,(n/e)^n = 1,$$

may be of use in part of the solution.)

14.6

Absolute convergence

Consider a series
$$a_1 + a_2 + \cdots + a_n + \cdots,$$
whose terms may be positive, negative, or zero. It is reasonable to expect it to behave at least as "nicely" as the series
$$|a_1| + |a_2| + \cdots + |a_n| + \cdots,$$
since by making all the terms positive we give the series more chance to diverge. The next theorem confirms this expectation.

THEOREM I (*Absolute convergence test*). If the series
$$|a_1| + |a_2| + \cdots + |a_n| + \cdots$$
converges, then so does the series
$$a_1 + a_2 + \cdots + a_n + \cdots.$$

PROOF Since the series $|a_1| + |a_2| + \cdots$ converges, so does the series $2|a_1| + 2|a_2| + \cdots$. (Its sum is $2\sum_{n=1}^{\infty} |a_n|$.)

Next, introduce the series whose nth term is
$$a_n + |a_n|.$$

Note that if a_n is negative, $a_n + |a_n| = 0$, while if a_n is nonnegative, $a_n + |a_n| = 2|a_n|$. Then, for all n,
$$0 \le a_n + |a_n| \le 2|a_n|.$$

By the comparison test $\sum_{n=1}^{\infty} (a_n + |a_n|)$ converges.

Let
$$\sum_{n=1}^{\infty} (a_n + |a_n|) = A \qquad \text{and} \qquad \sum_{n=1}^{\infty} |a_n| = B.$$

Now
$$\sum_{n=1}^{k} a_n = \sum_{n=1}^{k} (a_n + |a_n|) - \sum_{n=1}^{k} |a_n|.$$

Thus, as $k \to \infty$, $\sum_{n=1}^{k} a_n \to A - B$, and the theorem is proved. ●

The next example is a typical illustration of the way Theorem 1 is applied.

EXAMPLE 1   Show that
$$\frac{1}{1^2} + \frac{1}{2^2} - \frac{1}{3^2} + \frac{1}{4^2} + \frac{1}{5^2} - \frac{1}{6^2} + \cdots$$

(two positive terms alternating with one negative term) converges.

SOLUTION   The series whose nth term is the absolute value of the nth term of the given series is
$$\frac{1}{1^2} + \frac{1}{2^2} + \cdots + \frac{1}{n^2} + \cdots$$

In Sec. 14.4 this series was shown to converge (by the integral test). By the absolute convergence test the original series, with $+$'s and $-$'s, converges. ●

The alternating series $1 - \tfrac{1}{2} + \tfrac{1}{3} - \tfrac{1}{4} + \cdots$

converges, as shown in Sec. 14.3. However, when all the terms are replaced by their absolute values, the resulting series, the harmonic series, does not converge; that is,
$$1 + \tfrac{1}{2} + \tfrac{1}{3} + \tfrac{1}{4} + \cdots$$
diverges.

The following definitions are frequently used in describing these various cases of convergence or divergence.

DEFINITION   *Absolute convergence.*   A series $a_1 + a_2 + \cdots$ is said to *converge absolutely* if the series $|a_1| + |a_2| + \cdots$ converges.

Theorem 1 can be stated simply, "If a series converges absolutely, then it converges."

DEFINITION   *Conditional convergence.*   A series $a_1 + a_2 + \cdots$ is said to *converge conditionally* if it converges but does not converge absolutely.

The next example shows how the absolute convergence test can be combined with other tests to establish convergence of a series.

EXAMPLE 2   Show that $\dfrac{2}{1}\left(\dfrac{1}{2}\right) - \dfrac{3}{2}\left(\dfrac{1}{2}\right)^2 + \dfrac{4}{3}\left(\dfrac{1}{2}\right)^3 + \cdots + (-1)^{n+1}\dfrac{n+1}{n}\left(\dfrac{1}{2}\right)^n + \cdots$ (1)

converges.

SOLUTION   Consider the series of positive terms
$$\frac{2}{1}\left(\frac{1}{2}\right) + \frac{3}{2}\left(\frac{1}{2}\right)^2 + \frac{4}{3}\left(\frac{1}{2}\right)^3 + \cdots + \frac{n+1}{n}\left(\frac{1}{2}\right)^n + \cdots.$$

Its typical term is
$$a_n = \frac{n+1}{n}\left(\frac{1}{2}\right)^n.$$

The presence of the power $\left(\frac{1}{2}\right)^n$ suggests using the ratio test:

$$\frac{a_{n+1}}{a_n} = \frac{\dfrac{n+2}{n+1}\left(\dfrac{1}{2}\right)^{n+1}}{\dfrac{n+1}{n}\left(\dfrac{1}{2}\right)^{n}}$$

$$= \frac{n+2}{n+1}\frac{n}{n+1}\cdot\frac{1}{2}.$$

Thus
$$\lim_{n\to\infty}\frac{a_{n+1}}{a_n} = \frac{1}{2},$$

which is less than 1. Consequently, the given series (1), with positive and negative terms, converges absolutely. Thus it converges. ●

EXAMPLE 3   Examine the series

$$\frac{\cos x}{1^2} + \frac{\cos 2x}{2^2} + \frac{\cos 3x}{3^2} + \cdots + \frac{\cos nx}{n^2} + \cdots \qquad (2)$$

for convergence or divergence.

SOLUTION   The number x is fixed. The numbers $\cos nx$ may be positive, negative, or zero, in an irregular manner. However, for all n, $|\cos nx| \le 1$.
Recall that the series

$$\frac{1}{1^2} + \frac{1}{2^2} + \frac{1}{3^2} + \cdots + \frac{1}{n^2} + \cdots$$

converges, as shown in Sec. 14.4. Since $|\cos nx|/n^2 \le 1/n^2$, the series

$$\frac{|\cos x|}{1^2} + \frac{|\cos 2x|}{2^2} + \frac{|\cos 3x|}{3^2} + \cdots + \frac{|\cos nx|}{n^2} + \cdots$$

converges by the comparison test. Series (2) thus converges absolutely for all x. Hence it converges.
Incidentally, advanced calculus shows in the theory of Fourier series that for $0 \le x \le 2\pi$, series (2) has the sum $(3x^2 - 6\pi x + 2\pi^2)/12$. ●

A series of the form

$$a_0 + a_1 x + a_2 x^2 + \cdots + a_n x^n + \cdots$$

is called a *power series*. For instance, the power series

$$1 + x + x^2 + \cdots + x^n + \cdots,$$

$-1 < x < 1$, being a geometric series with initial term 1 and ratio x, has the sum $1/(1 - x)$.   Note that this series converges for $|x| < 1$.   The next theorem is concerned with the values of x for which a power series converges.

THEOREM 2   Let $a_0, a_1, a_2, \ldots$ be a sequence and c a nonzero number.   Assume that

$$a_0 + a_1 c + a_2 c^2 + \cdots$$

converges.   Then, if $|x| < |c|$,

$$a_0 + a_1 x + a_2 x^2 + \cdots$$

converges.   In fact it converges absolutely.

PROOF   Since $\sum_{n=0}^{\infty} a_n c^n$ converges, the nth term $a_n c^n$ approaches 0 as $n \to \infty$. Consequently, there is a fixed number M such that, for all $n > 0$,

$$|a_n c^n| \le M.$$

Now,

$$a_n x^n = a_n c^n \left(\frac{x}{c}\right)^n.$$

Hence

$$|a_n x^n| = |a_n c^n| \left|\frac{x}{c}\right|^n$$

$$\le M \left|\frac{x}{c}\right|^n.$$

The series

$$\sum_{n=0}^{\infty} M \left|\frac{x}{c}\right|^n$$

is a geometric series with first term M and ratio $|x/c| < 1$.   Hence it converges. Since $|a_n x^n| \le M |x/c|^n$, the series

$$\sum_{n=0}^{\infty} |a_n x^n|$$

converges.   Thus $\sum_{n=0}^{\infty} a_n x^n$ converges (in fact, absolutely).   ●

Theorem 2 implies that the set of numbers x such that $a_0 + a_1 x + a_2 x^2 + \cdots$ converges does not have gaps.   In other words, the set consists of one unbroken piece.   Either

1. $a_0 + a_1 x + a_2 x^2 + \cdots$ converges for all x, or
2. there is a number R such that $a_0 + a_1 x + a_2 x^2 + \cdots$ converges for all x such that $|x| < R$ but diverges when $|x| > R$.

In case 2 R is called the *radius of convergence* of the series.   In case 1 the radius of convergence is said to be infinite, $R = \infty$.

A power series $a_0 + a_1 x + a_2 x^2 + \cdots$ could converge only for $x = 0$.   (For example,

$$1 + 1x + 2^2 x^2 + 3^3 x^3 + \cdots + n^n x^n + \cdots$$

converges only when $x = 0$.   If $x \ne 0$, the nth term is

$$n^n x^n = (nx)^n,$$

which gets large when $n \to \infty$.) In this case its radius of convergence is said to be 0.

Note in Theorem 2 that it was not assumed that $\sum_{n=0}^{\infty} a_n c^n$ converges *absolutely*, even though it was shown that $\sum_{n=0}^{\infty} a_n x^n$ converges absolutely for $|x| < |c|$. The next example shows that $\sum_{n=0}^{\infty} a_n c^n$ might converge conditionally.

EXAMPLE 4   Use Theorem 2 to find all values of x for which

$$x - \frac{x^2}{2} + \frac{x^3}{3} - \frac{x^4}{4} + \cdots$$

converges.

SOLUTION   For $x = 1$ the alternating harmonic series

$$1 - \tfrac{1}{2} + \tfrac{1}{3} - \tfrac{1}{4} + \cdots$$

is obtained. This series converges, by the alternating-series test; that is, $x - x^2/2 + x^3/3 - x^4/4 + \cdots$ converges when $x = 1$. Hence $x - x^2/2 + x^3/3 - x^4/4 + \cdots$ converges for $-1 < x \leq 1$. The reader may establish by the nth-term test that it diverges when $|x| > 1$. A separate check shows that it diverges at -1. The radius of convergence R is 1.

A diagram records this information about $x - x^2/2 + x^3/3 - x^4/4 + \cdots$.

[Incidentally, as shown in Sec. 14.7, the sum of the series is $\ln(1 + x)$.] ●

If you replace x in $a_0 + a_1 x + a_2 x^2 + \cdots$ by $x - a$, you obtain $a_0 + a_1(x - a) + a_2(x - a)^2 + \cdots$, which is called a *power series* in $x - a$. It will have a radius of convergence R; that is, the series will converge absolutely for all x such that $|x - a| < R$. Again R may be infinite. Moreover, if it converges for $x = x_1$, it will converge absolutely for any x such that $|x - a| < |x_1 - a|$.

EXAMPLE 5   Find all values of x for which

$$(x - 1) - \frac{(x - 1)^2}{2} + \frac{(x - 1)^3}{3} - \frac{(x - 1)^4}{4} + \cdots \tag{3}$$

converges.

SOLUTION   Note that this is Example 4 with x replaced by $x - 1$. Thus $x - 1$ plays the role that x played in Example 4. Consequently, series (3) converges for

$$-1 < x - 1 \leq 1$$

and diverges for all other values of $x - 1$. Its radius of convergence is $R = 1$. The set of values where the series converges is an interval whose midpoint is $x = 1$. The convergence of (3) is recorded in this diagram

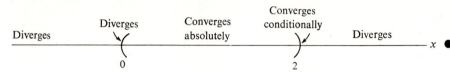

Exercises

In Exercises 1 to 6 determine whether the series diverges, converges conditionally, or converges absolutely.

1. $\displaystyle\sum_{n=1}^{\infty} \frac{(-1)^n}{\sqrt{n}}$

2. $\displaystyle\sum_{n=1}^{\infty} \frac{(-1)^n}{n^3}$

3. $\displaystyle\sum_{n=1}^{\infty} \frac{(-2)^n}{n!}$

4. $\displaystyle\sum_{n=1}^{\infty} \frac{(-1)^n(n+5)}{n^2}$

5. $\displaystyle\sum_{n=1}^{\infty} \frac{(-1)^n(n^2+n)}{3n^4+5n+2}$

6. $\displaystyle\sum_{n=1}^{\infty} \frac{(-9)^n}{10^n}$

In Exercises 7 to 15 determine the values of x where the power series diverges, converges conditionally, or converges absolutely. What is R in each case?

7. $\displaystyle\sum_{n=1}^{\infty} \frac{x^n}{n!}$

8. $\displaystyle\sum_{n=1}^{\infty} \frac{x^n}{2^n}$

9. $\displaystyle\sum_{n=1}^{\infty} \frac{x^n}{n2^n}$

10. $\displaystyle\sum_{n=1}^{\infty} \frac{(x-2)^n}{n!}$

11. $\displaystyle\sum_{n=1}^{\infty} \frac{(x+3)^n}{5^n}$

12. $\displaystyle\sum_{n=1}^{\infty} \frac{(x-1)^n}{n3^n}$

13. $\displaystyle\sum_{n=1}^{\infty} \frac{x^n}{n}$

14. $\displaystyle\sum_{n=1}^{\infty} \frac{x^n}{n^n}$

15. $\displaystyle\sum_{n=1}^{\infty} n(x+1)^n$

16. Assume that $\sum_{n=1}^{\infty} a_n x^n$ converges when $x = 9$ and diverges when $x = -12$. What, if anything, can be said about

   (a) convergence when $x = 7$?
   (b) absolute convergence when $x = -7$?
   (c) absolute convergence when $x = 9$?
   (d) convergence when $x = -9$?
   (e) divergence when $x = 10$?
   (f) divergence when $x = -15$?
   (g) divergence when $x = 15$?

17. If $\sum_{n=1}^{\infty} a_n 6^n$ converges, what can be said about the convergence of

   (a) $\displaystyle\sum_{n=1}^{\infty} a_n(-6)^n$?   (b) $\displaystyle\sum_{n=1}^{\infty} a_n 5^n$?   (c) $\displaystyle\sum_{n=1}^{\infty} a_n(-5)^n$?

18. Prove that, if the series $\sum_{n=0}^{\infty} a_n x^n$ converges whenever x is positive, it converges whenever x is negative.

19. If $\sum_{n=0}^{\infty} a_n(x-3)^n$ converges for $x = 7$, at what other values of x must the series necessarily converge?

   ■

20. Let $\{a_n\}$ be a sequence of nonzero terms such that
$$\lim_{n \to \infty} \left| \frac{a_{n+1}}{a_n} \right| = s < 1.$$
   Prove that $\sum_{n=1}^{\infty} a_n$ converges.

21. Let $\{a_n\}$ be a sequence of nonzero terms such that
$$\lim_{n \to \infty} \frac{a_{n+1}}{a_n} = -\frac{1}{2}.$$
   Prove that $\sum_{n=1}^{\infty} a_n$ converges.

   ■ ■

22. Let a be any real number that is *not* 0 or a positive integer, and form the series
$$1 + \frac{a}{1}x + \frac{a(a-1)}{1 \cdot 2}x^2 + \frac{a(a-1)(a-2)}{1 \cdot 2 \cdot 3}x^3 + \cdots.$$

   (a) Show that for $x \neq 0$ each term is not zero.
   (b) Show that for $a = -1$ and $|x| < 1$, the sum of the series is $(1 + x)^a$.
   (c) Show that the series converges absolutely (hence converges) whenever $|x| < 1$. In Chapter 15 it is shown that the series converges to $(1 + x)^a$ whenever $|x| < 1$.
   Remark: If a is 0 or a positive integer, the series has only a finite number of nonzero terms. Newton was the

first to study the series for other values of a. For any value of a, the series is known as the binomial series.

23. Show that if $\sum_{n=1}^{\infty} a_n 3^n$ converges, then so does $\sum_{n=1}^{\infty} na_n 2^n$.

14.7

Manipulating power series

The function $1/(1 - x)$, for $|x| < 1$, can be represented by a power series,

$$\frac{1}{1 - x} = 1 + x + x^2 + \cdots.$$

Similarly, as mentioned in Example 4 of the preceding section, for $|x| < 1$,

$$\ln (1 + x) = x - \frac{x^2}{2} + \frac{x^3}{3} - \cdots.$$

In this section power series for other common functions such as e^x, $\sin x$, and $\cos x$ are obtained, and rules are given for manipulating power series.

THEOREM 1 (*Differentiating a power series*). Assume that $R > 0$ and that $f(x) = a_0 + a_1 x + a_2 x^2 + a_3 x^3 + \cdots$ for $|x| < R$. Then f is differentiable, and

$$f'(x) = a_1 + 2a_2 x + 3a_3 x^2 + \cdots \qquad |x| < R. \quad \bullet$$

This theorem is *not* covered by the fact that the derivative of the sum of a *finite* number of functions is the sum of their derivatives. See Sec. 23.3 for a proof of Theorem 1.

EXAMPLE 1 Obtain a power series for the function $x/(1 - x)^2$.

SOLUTION Apply Theorem 1 to the function

$$\frac{1}{1 - x} = 1 + x + x^2 + \cdots \qquad |x| < 1,$$

obtaining, by differentiation,

$$\frac{1}{(1 - x)^2} = 1 + 2x + 3x^2 + \cdots \qquad |x| < 1.$$

Therefore the power series for the function $x/(1 - x)^2$ is

$$\frac{x}{(1 - x)^2} = x + 2x^2 + 3x^3 + \cdots \qquad |x| < 1. \quad \bullet$$

If a function $f(x)$ has as a power series representation $a_0 + a_1 x + a_2 x^2 + \cdots$, Theorem 1 enables us to find what the coefficients a_0, a_1, a_2, ... must be. The formula for a_n appears in Theorem 2. In this theorem $f^{(n)}$ denotes the nth derivative of f; $f^{(1)} = f'$, $f^{(0)}$ stands for f itself, and $0! = 1$.

THEOREM 2 (*Formula for* a_n). Assume that $R > 0$ and that $f(x) = a_0 + a_1 x + a_2 x^2 + \cdots + a_n x^n + \cdots$ for $|x| < R$. Then

$$a_n = \frac{f^{(n)}(0)}{n!}. \tag{1}$$

PROOF When $x = 0$, we obtain $f(0) = a_0 + a_1 \cdot 0 + a_2 0^2 + \cdots$. Hence

$$f(0) = a_0,$$

which agrees with (1). To obtain a_1, differentiate $f(x)$, obtaining

$$f^{(1)}(x) = a_1 + 2a_2 x + 3a_3 x^2 + \cdots + na_n x^{n-1} + \cdots. \tag{2}$$

Set $x = 0$ in (2), obtaining $\qquad f^{(1)}(0) = a_1$.

This establishes (1) for $n = 1$.
  To obtain a_2 differentiate (2), obtaining

$$f^{(2)}(x) = 2a_2 + 3 \cdot 2a_3 x + \cdots + n(n-1)a_n x^{n-2} + \cdots. \tag{3}$$

Letting $x = 0$, we obtain $\qquad f^{(2)}(0) = 2a_2$.

Hence $\qquad\qquad\qquad\qquad a_2 = \dfrac{f^{(2)}(0)}{2}$

and (1) is established for $n = 2$.
  Differentiate (3) to obtain a_3, as follows:

$$f^{(3)}(x) = 3 \cdot 2a_3 + 4 \cdot 3 \cdot 2a_4 x + \cdots + n(n-1)(n-2)a_n x^{n-3} + \cdots. \tag{4}$$

Set $x = 0$, and we have $\qquad f^{(3)}(0) = 3 \cdot 2a_3$,

or $\qquad\qquad\qquad\qquad a_3 = \dfrac{f^{(3)}(0)}{3!}$.

This establishes (1) for $n = 3$ and also shows why the factorial appears in the denominator of (1). The reader should differentiate (4) and verify (1) for $n = 4$. The argument applies for all n. ●

DEFINITION *Maclaurin series.* Assume that the function $f(x)$ can be represented by the power series

$$a_0 + a_1 x + a_2 x^2 + \cdots + a_n x^n + \cdots,$$

for all x in some interval including 0. The series is called the Maclaurin series for $f(x)$.

EXAMPLE 2 If e^x can be represented as a Maclaurin series for all x, what must that series be?

SOLUTION Let $e^x = a_0 + a_1 x + a_2 x^2 + \cdots + a_n x^n + \cdots$ for all x. By Theorem 2,

$$a_n = \frac{n\text{th derivative of } e^x \text{ at } x = 0}{n!}.$$

Now $(e^x)' = e^x$, $(e^x)'' = e^x$, and so on. All the higher derivatives of e^x are e^x again. At $x = 0$ they all have the value 1. Thus

$$a_n = \frac{1}{n!}$$

If e^x can be represented as a Maclaurin series, we must have

$$e^x = \frac{1}{0!} + \frac{x}{1!} + \frac{x^2}{2!} + \frac{x^3}{3!} + \frac{x^4}{4!} + \cdots + \frac{x^n}{n!} + \cdots,$$

or

$$e^x = 1 + x + \frac{x^2}{2} + \frac{x^3}{6} + \frac{x^4}{24} + \cdots. \ \bullet$$

In Sec. 15.2 it will be proved without the aid of Theorem 1 that $e^x = \sum_{n=0}^{\infty} x^n/n!$, for all x.

Calculations similar to those in Example 2 show that, *if* $\sin x$ is representable as a Maclaurin series, then

$$\sin x = x - \frac{x^3}{3!} + \frac{x^5}{5!} - \frac{x^7}{7!} + \cdots. \tag{5}$$

Differentiation of (5) then yields

$$\cos x = 1 - \frac{x^2}{2!} + \frac{x^4}{4!} - \frac{x^6}{6!} + \cdots. \tag{6}$$

The next theorem, which is proved in Sec. 23.2, justifies the term-by-term integration of power series.

THEOREM 3 (*Integrating a power series*). Assume that $R > 0$ and that

$$f(x) = a_0 + a_1 x + a_2 x^2 + \cdots + a_n x^n + \cdots \qquad \text{for} \quad |x| < R$$

Then f is continuous. Moreover,

$$a_0 x + \frac{a_1 x^2}{2} + \frac{a_2 x^3}{3} + \cdots + \frac{a_n x^{n+1}}{n+1} + \cdots$$

converges for $|x| < R$, and

$$\int_0^x f(t)\, dt = a_0 x + \frac{a_1 x^2}{2} + \frac{a_2 x^3}{3} + \cdots. \ \bullet$$

Note that the t is used to avoid writing $\int_0^x f(x)\, dx$, an expression in which x describes both the interval of integration $[0, x]$ and the independent variable of the function. To avoid double use of x with two different meanings, a different letter should be used in the integrand. The next Example demonstrates the power of Theorem 3.

EXAMPLE 3 Integrate the power series for $1/(1 + x)$ to obtain the Maclaurin series for $\ln (1 + x)$.

SOLUTION   Start with the series

$$\frac{1}{1-x} = 1 + x + x^2 + \cdots \qquad |x| < 1.$$

Replace x by $-x$, obtaining

$$\frac{1}{1+x} = 1 - x + x^2 - x^3 + x^4 - \cdots \qquad |x| < 1.$$

By Theorem 3,

$$\int_0^x \frac{dt}{1+t} = x - \frac{x^2}{2} + \frac{x^3}{3} - \frac{x^4}{4} + \cdots \qquad |x| < 1.$$

Now

$$\int_0^x \frac{dt}{1+t} = \ln(1+t)\Big|_0^x$$

$$= \ln(1+x) - \ln(1+0)$$

$$= \ln(1+x).$$

Thus

$$\ln(1+x) = x - \frac{x^2}{2} + \frac{x^3}{3} - \frac{x^4}{4} + \cdots \qquad |x| < 1.$$

Exercise 26 shows that this equation is also valid when $x = 1$, that is,

$$\ln 2 = 1 - \tfrac{1}{2} + \tfrac{1}{3} - \tfrac{1}{4} + \cdots. \ \bullet$$

In addition to differentiating and integrating power series, we may also add, subtract, multiply, and divide them just like polynomials. The following theorem states the rules for these operations. The first two are easy to establish; proofs of the latter two are reserved for an advanced calculus course.

THEOREM 4   (*The algebra of power series*). Assume that

$$f(x) = a_0 + a_1 x + a_2 x^2 + \cdots \qquad \text{and} \qquad g(x) = b_0 + b_1 x + b_2 x^2 + \cdots$$

for $|x| < R$. Then for $|x| < R$

1. $f(x) + g(x) = a_0 + b_0 + (a_1 + b_1)x + (a_2 + b_2)x^2 + \cdots;$
2. $f(x) - g(x) = a_0 - b_0 + (a_1 - b_1)x + (a_2 - b_2)x^2 + \cdots;$
3. $f(x)g(x) = a_0 b_0 + (a_0 b_1 + a_1 b_0)x + (a_0 b_2 + a_1 b_1 + a_2 b_0)x^2 + \cdots;$
4. $f(x)/g(x)$ is obtainable by long division, if $g(x) \neq 0$ for $|x| < R$. $\bullet$

Two examples will illustrate the usefulness of Theorem 4.

EXAMPLE 4   Find the first few terms of the Maclaurin series for $e^x/(1-x)$.

SOLUTION   By Theorem 4, we just multiply the series as we would polynomials:

$$e^x \frac{1}{1-x} = \left(1 + x + \frac{x^2}{2!} + \frac{x^3}{3!} + \cdots\right)\left(1 + x + x^2 + x^3 + \cdots\right)$$

$$= 1 + (1 + 1)x + \left(1 \cdot 1 + 1 \cdot 1 + \frac{1}{2!} \cdot 1\right)x^2$$

$$+ \left(1 \cdot 1 + 1 \cdot 1 + \frac{1}{2!} \cdot 1 + \frac{1}{3!} \cdot 1\right)x^3 + \cdots$$

$$= 1 + 2x + \tfrac{5}{2}x^2 + \tfrac{8}{3}x^3 + \cdots \qquad |x| < 1. \ \bullet$$

EXAMPLE 5 Find the first few terms of the Maclaurin series for $\cos x/e^x$.

SOLUTION Recall the Maclaurin series for e^x and for $\cos x$ [Eq. (6)] and arrange the long division as follows.

$$
1 + x + \frac{x^2}{2} + \frac{x^3}{6} + \frac{x^4}{24} + \cdots \overline{\left)1 + 0x - \dfrac{x^2}{2} + 0x^3 + \dfrac{x^4}{24} + 0 - \cdots\right.}
$$

with quotient $1 - x + 0x^2 + \tfrac{1}{3}x^3 - \tfrac{1}{6}x^4 + \cdots$

$$1 + x + \frac{x^2}{2} + \frac{x^3}{6} + \frac{x^4}{24} + \cdots$$

$$- x - x^2 - \frac{x^3}{6} + 0x^4 + \cdots$$

$$- x - x^2 - \frac{x^3}{2} - \frac{x^4}{6} - \cdots$$

$$\frac{x^3}{3} + \frac{x^4}{6} + \cdots$$

$$\frac{x^3}{3} + \frac{x^4}{3} + \cdots$$

$$- \frac{x^4}{6} + \cdots.$$

Thus the Maclaurin series for $\cos x/e^x$ begins

$$1 - x + \frac{x^3}{3} - \frac{x^4}{6} + \cdots. \ \bullet$$

Section 6.10 presented l'Hôpital's rule as a method for dealing with $\lim_{x \to a} f(x)/g(x)$ when both numerator and denominator $\to 0$ as $x \to a$. The next example shows that to calculate such limits, power series may on occasion be far more efficient.

EXAMPLE 6 Find

$$\lim_{x \to 0} \frac{\sin^2 x - x^2}{(e^{x^2} - 1)^2}.$$

SOLUTION Since

$$\sin x = x - \frac{x^3}{6} + \frac{x^5}{120} - \cdots,$$

the Maclaurin series for the numerator begins

$$\sin^2 x - x^2 = \left(x - \frac{x^3}{6} + \frac{x^5}{120} - \cdots\right)^2 - x^2$$

$$= \left(x^2 - \frac{x^4}{3} + \cdots\right) - x^2$$

$$= -\frac{x^4}{3} + \text{terms of degree more than 4.}$$

To develop a Maclaurin series for the denominator first replace x in the relation

$$e^x = 1 + x + \frac{x^2}{2} + \frac{x^3}{6} + \cdots$$

by x^2, obtaining

$$e^{x^2} = 1 + x^2 + \frac{x^4}{2} + \frac{x^6}{6} + \cdots.$$

Thus

$$(e^{x^2} - 1)^2 = \left(x^2 + \frac{x^4}{2} + \frac{x^6}{6} + \cdots\right)^2$$

$$= x^4 + \text{terms of higher degree.}$$

Hence

$$\frac{\sin^2 x - x^2}{(e^{x^2} - 1)^2} = \frac{-x^4/3 + \text{terms of degree more than 4}}{x^4 + \text{terms of degree more than 4}}$$

$$= \frac{x^4(-\frac{1}{3} + \text{terms of degree at least 1})}{x^4(1 + \text{terms of degree at least 1})}.$$

Cancel the x^4 in numerator and denominator, obtaining

$$\frac{\sin^2 x - x^2}{(e^{x^2} - 1)^2} = \frac{-\frac{1}{3} + \text{terms of degree at least 1}}{1 + \text{terms of degree at least 1}}.$$

Thus

$$\lim_{x \to 0} \frac{\sin^2 x - x^2}{(e^{x^2} - 1)^2} = \frac{-\frac{1}{3}}{1} = -\frac{1}{3}.$$

The reader may find the limit by l'Hôpital's rule, but the computations are messier. ●

The various theorems and methods of this section were stated for power series in x. But analogous theorems hold for power series in $x - a$. Such series may be differentiated and integrated inside the interval in which they converge. For instance, Theorem 2 generalizes to the following assertion.

THEOREM 5   (*Formula for a_n*).   Assume that $R > 0$ and that

$$f(x) = a_0 + a_1(x - a) + a_2(x - a)^2 + \cdots + a_n(x - a)^n + \cdots$$

for

$$|x - a| < R.$$

Then
$$a_n = \frac{f^{(n)}(a)}{n!}. \quad \bullet \tag{7}$$

The proof is similar to that of Theorem 2: Differentiate n times and replace x by a.

The next example illustrates the use of formula (7).

EXAMPLE 7 If $\sin x$ can be represented as a series in powers of $x - \pi/4$ for all x, find what the series must be.

SOLUTION In this case $f(x) = \sin x$ and $a = \pi/4$. In order to use (7), it is necessary to evaluate all the derivatives of $\sin x$ at $\pi/4$. This table records the computations.

n	$f^{(n)}(x)$	$f^{(n)}(\pi/4)$	$a_n = \dfrac{f^{(n)}(\pi/4)}{n!}$
0	$\sin x$	$\sqrt{2}/2$	$(\sqrt{2}/2)/0!$
1	$\cos x$	$\sqrt{2}/2$	$(\sqrt{2}/2)/1!$
2	$-\sin x$	$-\sqrt{2}/2$	$-(\sqrt{2}/2)/2!$
3	$-\cos x$	$-\sqrt{2}/2$	$-(\sqrt{2}/2)/3!$
4	$\sin x$	$\sqrt{2}/2$	$(\sqrt{2}/2)/4!$
5	$\cos x$	$\sqrt{2}/2$	$(\sqrt{2}/2)/5!$
...	....	....	....

The higher derivatives of $\sin x$ repeat in blocks of four:
$$\sin x, \qquad \cos x, \qquad -\sin x, \qquad -\cos x.$$

The series for $\sin x$ in powers of $x - \pi/4$ begins therefore

$$\sin x = \frac{\sqrt{2}}{2} + \sqrt{2}/2\left(x - \frac{\pi}{4}\right) - \frac{\sqrt{2}/2}{2!}\left(x - \frac{\pi}{4}\right)^2 - \frac{\sqrt{2}/2}{3!}\left(x - \frac{\pi}{4}\right)^3 + \frac{\sqrt{2}/2}{4!}\left(x - \frac{\pi}{4}\right)^4 + \cdots.$$

Two $+$'s continue to alternate with two $-$'s. $\bullet$

A series in powers of $x - a$ is also called a *series expansion around a*. In the special case when $a = 0$ the powers are simply x^n.

Exercises

Use the formula of Theorem 2 to obtain the first three nonzero terms of the Maclaurin series for the functions given in Exercises 1 to 6.

1. $\sin x$ **2.** $\cos x$ **3.** $\sqrt{1 + x}$
4. $\tan^{-1} x$ **5.** $\ln(1 + x)$ **6.** $\ln(1 - x)$
7. Obtain the series for $\sin x$ stated as (5).

8. Obtain the series (6) for $\cos x$ by use of the formula in Theorem 2.
9. Use series (5) to show that, for $0 \le x \le 1$,

$$x - \frac{x^3}{6} < \sin x < x - \frac{x^3}{6} + \frac{x^5}{120}.$$ (*continued*)

Consequently, if $x - x^3/6$ is used as an approximation of $\sin x$ for $0 \le x \le 1$, the error is less than $\frac{1}{120} < 0.01$.

10. (a) Show that $1/(1 + x^2) = 1 - x^2 + x^4 - x^6 + \cdots$ if $|x| < 1$.

(b) From (a) deduce that, for $|x| < 1$,

$$\tan^{-1} x = x - \frac{x^3}{3} + \frac{x^5}{5} - \frac{x^7}{7} + \cdots.$$

(c) It can be shown that the equation in (b) also holds when $x = 1$. Deduce that

$$\frac{\pi}{4} = 1 - \frac{1}{3} + \frac{1}{5} - \frac{1}{7} + \cdots.$$

In Exercises 11 to 13 obtain the first three nonzero terms in the Maclaurin series for the indicated function by algebraic operations with known series.

11. $e^x \sin x$ **12.** $\dfrac{x}{\cos x}$ **13.** $e^{-x} \cos \sqrt{x}$

In Exercises 14 to 16 use power series to determine the limits.

14. $\displaystyle\lim_{x \to 0} \frac{(1 - \cos \sqrt{x})^2}{1 - e^{x^2}}$ **15.** $\displaystyle\lim_{x \to 0} \frac{\sin^2 x^3}{(1 - \cos x^2)^3}$

16. $\displaystyle\lim_{x \to 0} \frac{1}{\sin x} - \frac{1}{\ln(1 + x)}$

In Exercises 17 to 19 use the formula $a_n = f^{(n)}(a)/n!$ to obtain the indicated series. Write out the first five nonzero terms.

17. Series for $\sin x$ in powers of $x - \pi/6$.

18. Series for $\cos x$ in powers of $x + \pi/4$.

19. Series for e^x in powers of $x - 1$.

■

20. Let $f(x) = a_0 + a_1 x + a_2 x^2 + \cdots$ for $|x| < R$.

(a) If only even powers appear, that is, $a_n = 0$ for all odd n, show that $f(-x) = f(x)$.

(b) If only odd powers appear, that is, $a_n = 0$ for all even n, show that $f(-x) = -f(x)$.

A function f such that $f(-x) = f(x)$ is called *even*; if $f(-x) = -f(x)$, f is called *odd*.

21. (a) Use the first five terms of the Maclaurin series for e^x to estimate $e = e^1$.

(b) Show that the error in (a) is less than the sum of the geometric progression

$$\frac{1}{5!} + \frac{1}{6 \cdot 5!} + \frac{1}{6^2 \cdot 5!} + \cdots.$$

(c) Deduce from (a) and (b) that

$$2.708 < e < 2.719.$$

22. (a) Noting that $\sqrt{e} = e^{1/2}$, use the first five terms of the

Maclaurin series for e^x to estimate $\sqrt{e}$.

(b) Estimate the error in (a) by comparing it to the sum of a geometric progression.

23. Obtain formula (7) in Theorem 5.

■ ■

24. Show that the function $\sqrt[3]{x}$ cannot be represented as a Maclaurin series with a nonzero radius of convergence.

This exercise involves a very flat function.

25. Let $f(x) = e^{-1/x^2}$ if $x \ne 0$, and let $f(0) = 0$.

(a) Show that f is continuous.

(b) Show that f is differentiable.

(c) Show that $f^{(1)}(0) = 0$ and $f^{(2)}(0) = 0$.

(d) Explain why $f^{(n)}(0) = 0$ for all $n \ge 0$.

(e) Show that $f(x)$ is not representable by a Maclaurin series with a nonzero radius of convergence.

26. This exercise shows that $\ln 2 = 1 - \frac{1}{2} + \frac{1}{3} - \frac{1}{4} + \frac{1}{5} - \cdots$.

(a) Sketch the curve $y = 1/x$.

(b) Using the diagram in (a), show that, if a_n is defined by the equation

$$1 + \frac{1}{2} + \cdots + \frac{1}{n} = \ln(n + 1) + a_n,$$

then $\lim_{n \to \infty} a_n$ exists.

(c) Show that

$$\lim_{n \to \infty} \left[\left(1 + \frac{1}{2} + \frac{1}{3} + \cdots + \frac{1}{2n} \right) - \left(1 + \frac{1}{2} + \cdots + \frac{1}{n} \right) \right] = \ln 2.$$

(d) From (c) deduce that

$$\ln 2 = 1 - \frac{1}{2} + \frac{1}{3} - \frac{1}{4} + \cdots + (-1)^{(n+1)} \frac{1}{n} + \cdots.$$

27. Prove parts (a) and (b) of Theorem 4.

In Exercises 28 to 30 use a calculator.

28. (a) Use the first 10 terms of the series $e^x = 1 + x + x^2/2 + \cdots$ to estimate $e = e^1$.

(b) Show that the error in the estimate in (a) is less than $11/(10 \cdot 10!) \doteq 0.0000003$.

29. (a) Use the first three nonzero terms of the series $\sin x = x - x^3/6 + \cdots$ to estimate $\sin \pi/5 \,(= \sin 36°)$.

(b) Show that the error in the estimate in (a) is less than $(\pi/5)^7/7! \doteq 0.000008$.

30. The integral $\int_0^1 e^{-x^2}\, dx$ cannot be evaluated by the fundamental theorem of calculus.

(a) Replacing x in the power series $e^x = 1 + x + x^2/2! + \cdots$ by $-x^2$, obtain the power series for e^{-x^2}.

(b) Show that

$$\int_0^1 e^{-x^2}\, dx = 1 - \frac{1}{3} + \frac{1}{5 \cdot 2!} - \frac{1}{7 \cdot 3!} + \cdots.$$

(c) Use the first six summands in the series in (b) to estimate $\int_0^1 e^{-x^2}\,dx$.

(d) Put a bound on the error in (c).

14.8

Summary

This chapter was devoted to sequences and their limits. In particular it concentrated on sequences formed by adding the terms of another sequence. Of special interest are the power series, for they can represent such important functions as $\sin x$, $\cos x$, and e^x. By using a partial sum of such a power series, the function values can be estimated. In this manner, such an intractable integral as $\int_0^1 e^{x^2}\,dx$ can easily be estimated to many decimal places.

There is no general method of deciding whether a series diverges or converges.

This chapter presented seven of the many known tests: the nth-term test, alternating-series test, integral test, comparison test, limit-comparison test, ratio test, and absolute convergence test.

Of particular importance is the representation of functions by a Maclaurin series,

$$f(x) = a_0 + a_1 x + a_2 x^2 + \cdots + a_n x^n + \cdots.$$

In such a case

$$a_n = \frac{f^{(n)}(0)}{n!}.$$

If

$$f(x) = a_0 + a_1(x - a) + a_2(x - a)^2 + \cdots + a_n(x - a)^n + \cdots,$$

then

$$a_n = \frac{f^{(n)}(a)}{n!}.$$

Associated with each power series is its radius of convergence R, which may be finite or infinite.

Various operations on power series—such as multiplication, division, differentiation, and integration—were discussed. Series were used to determine such limits as $\lim_{x \to a} f(x)/g(x)$ in case $f(x)$ and $g(x) \to 0$ as $x \to a$.

We assumed in Sec. 14.7 that such functions as e^x can be represented by a power series. In the next chapter this assumption will be justified.

KEY FACTS

$\lim_{n \to \infty} r^n = 0$ if $|r| < 1$.

$\lim_{n \to \infty} \dfrac{k^n}{n!} = 0$ for all k.

If $a_1 < a_2 < a_3 < \cdots$ and $a_n \leq B$ for all n, then $\lim_{n \to \infty} a_n$ exists and is $\leq B$. (This is the completeness property of the real numbers.)

$$a + ar + \cdots + ar^{n-1} = \frac{a(1 - r^n)}{1 - r} \qquad \text{for } r \neq 1.$$

$$a + ar + ar^2 + \cdots = \frac{a}{1 - r} \qquad \text{for } |r| < 1.$$

If $\lim_{n \to \infty} a_n \neq 0$, then $\sum_{n=1}^{\infty} a_n$ diverges.

If $p_1 > p_2 > p_3 > \cdots$ and $\lim_{n \to \infty} p_n = 0$, then $p_1 - p_2 + p_3 - \cdots$ converges. S_n differs from S by less than p_{n+1}.

If f is a decreasing positive function, then $\sum_{n=1}^{\infty} f(n)$ is convergent if and only if $\int_1^{\infty} f(x)\,dx$ is convergent. The series diverges if $\int_1^{\infty} f(x)\,dx$ diverges.

If $\sum_{n=1}^{\infty} c_n$ converges, $c_n > 0$, and if $0 \leq p_n \leq c_n$, then $\sum_{n=1}^{\infty} p_n$ converges.

If $\sum_{n=1}^{\infty} c_n$ diverges, $c_n > 0$, and if $p_n \geq c_n$, then $\sum_{n=1}^{\infty} p_n$ diverges.

If $\sum_{n=1}^{\infty} c_n$ converges, $c_n > 0$, and if $\lim_{n \to \infty} p_n/c_n$ exists, $p_n > 0$, then $\sum_{n=1}^{\infty} p_n$ converges. If $\sum_{n=1}^{\infty} c_n$ diverges, $c_n > 0$, and if $\lim_{n \to \infty} p_n/c_n$ exists and is not 0, then $\sum_{n=1}^{\infty} p_n$ diverges.

If $p_n > 0$ and if $\lim_{n \to \infty} p_{n+1}/p_n = r < 1$. then $\sum_{n=1}^{\infty} p_n$ converges. If $\lim_{n \to \infty} p_{n+1}/p_n = r > 1$, then the series diverges.

If $\sum_{n=1}^{\infty} |a_n|$ converges, then $\sum_{n=1}^{\infty} a_n$ converges.

If $\sum_{n=0}^{\infty} a_n x^n$ converges for $x = c$, then it converges absolutely for $|x| < c$.

If $\sum_{n=0}^{\infty} a_n(x - a)^n$ converges for $x = c$, then it converges absolutely for $|x - a| < |c - a|$.

A power series that converges for $|x| < R$ may be differentiated throughout the open interval $(-R, R)$ and integrated in any closed interval situated in $(-R, R)$.

If $f(x) = \sum_{n=0}^{\infty} a_n(x - a)^n$ and the series has a nonzero radius of convergence, then

$$a_n = \frac{f^{(n)}(a)}{n!};$$

that is,

$$f(x) = f(a) + \frac{f'(a)}{1!}(x - a) + \frac{f^{(2)}(a)}{2!}(x - a)^2$$

$$+ \cdots + \frac{f^{(n)}(a)}{n!}(x - a)^n + \cdots.$$

$$\sin x = x - \frac{x^3}{3!} + \frac{x^5}{5!} - \frac{x^7}{7!} + \cdots \qquad \text{for all } x.$$

$$\cos x = 1 - \frac{x^2}{2!} + \frac{x^4}{4!} - \frac{x^6}{6!} + \cdots \qquad \text{for all } x.$$

$$e^x = 1 + x + \frac{x^2}{2!} + \frac{x^3}{3!} + \cdots \qquad \text{for all } x.$$

$$\frac{1}{1 - x} = 1 + x + x^2 + \cdots \qquad \text{for } |x| < 1.$$

$$\ln(1 + x) = x - \frac{x^2}{2} + \frac{x^3}{3} - \cdots \qquad \text{for } -1 < x \le 1.$$

VOCABULARY AND SYMBOLS

sequence $\{a_n\}$
limit of sequence $\lim_{n \to \infty} a_n$
completeness
convergent (divergent) sequence
factorial $n! = 1 \cdot 2 \cdots n$.
geometric series
series $\sum_{n=1}^{\infty} a_n$
nth partial sum $S_n = a_1 + a_2 + \cdots + a_n$
sum of a series, S
convergent (divergent) series
nth-term test
alternating series

power series
integral test
p series
harmonic series
comparison test
limit-comparison test
ratio test
absolute convergence
conditional convergence
power series in x or $x - a$
radius of convergence
Maclaurin series

Guide quiz for Chap. 14

1. Give an example of a series which diverges, yet whose nth term approaches 0.
2. What happens to the sum of the p series as $p \to 1$ through values larger than 1?
3. Let $p_1 - p_2 + p_3 - \cdots$ satisfy the hypotheses of the alternating-series test. Explain why S_5 is larger than S, the sum of the series, and S is larger than S_6.
4. (a) Show that $\sum_{n=1}^{\infty} 1/n^3$ is convergent.
   (b) Show that its sum S is less than $(\frac{1}{1})^3 + (\frac{1}{2})^3 + (\frac{1}{3})^3 + (\frac{1}{4})^3 + \int_4^{\infty} 1/x^3 \, dx$.
   (c) Deduce that $1.17 < S < 1.21$. It is not known whether S is a rational multiple of π^3.
5. (a) Show that the series

$$1 - \tfrac{1}{4} + \tfrac{1}{27} - \tfrac{1}{256} + \cdots + (-1)^{n+1} 1/n^n + \cdots$$

   converges.
   (b) Show that the sum of the series is between 0.783 and 0.784.
6. Give an example of a conditionally convergent series.
7. Prove that absolute convergence of a series implies its convergence.

8. Does this alternating series converge or diverge?

$$1 - \tfrac{1}{2} + \tfrac{2}{3} - \tfrac{1}{3} + \tfrac{2}{4} - \tfrac{1}{4} + \tfrac{2}{5} - \tfrac{1}{5} + \tfrac{2}{6} - \tfrac{1}{6} + \cdots$$

9. By carrying out the necessary differentiation, obtain the formula

$$a_4 = \frac{f^{(4)}(a)}{4!}$$

   for the coefficient a_4 in $f(x) = \sum_{n=0}^{\infty} a_n(x - a)^n$.
10. Which of these series converge? diverge? Explain.

(a) $\dfrac{5}{1} + \dfrac{5^2}{2} + \dfrac{5^3}{3} + \cdots$

(b) $\dfrac{10}{1!} + \dfrac{10^2}{2!} + \dfrac{10^3}{3!} + \dfrac{10^4}{4!} + \cdots$

(c) $\tfrac{1}{1} - \tfrac{1}{2} + \tfrac{1}{3} - \tfrac{1}{4} + \cdots$

(d) $\tfrac{1}{1} + \tfrac{1}{2} + \tfrac{1}{3} + \tfrac{1}{4} + \cdots$

(e) $\tfrac{2}{6} + \tfrac{2}{7} + \tfrac{2}{8} + \tfrac{2}{9} + \cdots$

(f) $\tfrac{2}{3} + \tfrac{3}{4} + \tfrac{4}{5} + \cdots.$

11. (a) Estimate $e^{-1/2}$ using the sum of the first four terms of the series for e^x.
   (b) Discuss the possible error.
12. (a) Estimate $\int_{1/4}^{1} x \sin \sqrt{x}\, dx$, using three terms of the power series for $\sin \sqrt{x}$.
   (b) Discuss the possible error.
13. (a) Estimate $\ln 1.2$ using three terms of the power series for $\ln (1 + x)$.
   (b) Discuss the possible error.
14. What is the sum of each of these convergent series?

   (a) $1 - \dfrac{\pi^2}{2!} + \dfrac{\pi^4}{4!} - \dfrac{\pi^6}{6!} + \cdots$

   (b) $1 - \dfrac{1}{1!} + \dfrac{1}{2!} - \dfrac{1}{3!} + \dfrac{1}{4!} - \cdots$

   (c) $\dfrac{1}{3} - \dfrac{1}{3^2} + \dfrac{1}{3^3} - \cdots$

(d) $\frac{1}{3} - \frac{1}{2}(\frac{1}{3})^2 + \frac{1}{3}(\frac{1}{3})^3 - \frac{1}{4}(\frac{1}{3})^4 + \cdots$

(e) $\frac{1}{3} + \frac{1}{2}(\frac{1}{3})^2 + \frac{1}{3}(\frac{1}{3})^3 + \cdots$.

15. (a) Obtain the Maclaurin series for $\tan^{-1} x$ from that for $1/(1 + x^2)$.
   (b) Use the series in (a) to show that

$$\frac{\pi\sqrt{3}}{9} = 1 - \frac{3}{3} + \frac{3^2}{5} - \cdots.$$

16. Find the first two nonzero terms of the Maclaurin series for $\tan x$:
   (a) by dividing the Maclaurin series for $\sin x$ by the Maclaurin series for $\cos x$;
   (b) by using the formula $a_n = f^{(n)}(0)/n!$.
17. Find

$$\lim_{x \to 0} \frac{e^{-x} - \cos \sqrt{2x}}{e^{x^2} - 1}.$$

Review exercises for Chap. 14

In Exercises 1 to 21 determine whether the series converges or diverges. If the sum of the convergent series is easily determined, give its value.

1. $\displaystyle\sum_{n=1}^{\infty} e^{-n}$

2. $\displaystyle\sum_{n=1}^{\infty} \frac{5n^3 + 6n + 1}{n^5 + n^3 + 2}$

3. $\displaystyle\sum_{n=2}^{\infty} \frac{1}{n \ln n}$

4. $\displaystyle\sum_{n=1}^{\infty} \ln\left(\frac{n+1}{n}\right)$

5. $\displaystyle\sum_{n=1}^{\infty} \left(-\frac{3}{4}\right)^n$

6. $\displaystyle\sum_{n=1}^{\infty} \frac{2^{-n}}{n}$

7. $\displaystyle\sum_{n=1}^{\infty} (-1)^n \ln\left(\frac{n+1}{n}\right)$

8. $\displaystyle\sum_{n=0}^{\infty} (-1)^n \frac{\pi^{2n}}{(2n)!}$

9. $\displaystyle\sum_{n=1}^{\infty} \frac{(-2)^n}{n}$

10. $\displaystyle\sum_{n=0}^{\infty} \frac{10^n}{n!}$

11. $\displaystyle\sum_{n=1}^{\infty} n \sin \frac{1}{n}$

12. $\displaystyle\sum_{n=1}^{\infty} \frac{\ln n}{n}$

13. $\displaystyle\sum_{n=1}^{\infty} \frac{5n^2 - 3n + 1}{2n^3 + n^2 - 1}$

14. $\displaystyle\sum_{n=1}^{\infty} \frac{1}{n\sqrt{n}}$

15. $\displaystyle\sum_{n=1}^{\infty} \frac{n[1 - \cos (1/\sqrt{n})]}{1 - n \sin (1/n)}$

16. $\displaystyle\sum_{n=0}^{\infty} \frac{(-1)^n (\frac{1}{2})^n}{n!}$

17. $\displaystyle\sum_{n=1}^{\infty} \frac{n^3}{2^n}$

18. $\displaystyle\sum_{n=1}^{\infty} \sin \frac{1}{n}$

19. $\displaystyle\sum_{n=1}^{\infty} \frac{\cos^3 n}{n^2}$

20. $\displaystyle\sum_{n=1}^{\infty} \frac{1 + (-1)^n}{n^2}$

21. $\displaystyle\sum_{n=0}^{\infty} (-1)^n \frac{\pi^{2n+1}}{2^{2n+1}(2n + 1)!}$

In Exercises 22 to 29 determine for which x the series converges, diverges, converges absolutely, and converges conditionally. Give the radius of convergence in each case and the sum of the series if it is easily determined.

22. $\displaystyle\sum_{n=1}^{\infty} \frac{2^n x^n}{n}$

23. $\displaystyle\sum_{n=1}^{\infty} (-n)^n x^n$

24. $\displaystyle\sum_{n=1}^{\infty} n x^{n-1}$

25. $\displaystyle\sum_{n=1}^{\infty} \frac{x^n}{n}$

26. $\displaystyle\sum_{n=0}^{\infty} \frac{(x - 3)^n}{n!}$

27. $\displaystyle\sum_{n=1}^{\infty} \frac{3^n (x - \frac{2}{3})^n}{4^n}$

28. $\displaystyle\sum_{n=0}^{\infty} \frac{x^{2n}}{n!}$

29. $\displaystyle\sum_{n=1}^{\infty} \frac{n^5 + 2}{n^3 + 1} (x + 1)^n$

30. If $\sum_{n=1}^{\infty} a_n(x - 2)^n$ converges for $x = 7$ and diverges for $x = -3$, determine its radius of convergence.

31. Obtain the first three nonzero terms of the Maclaurin series for $\sin 2x$:
   (a) by replacing x by $2x$ in the Maclaurin series for $\sin x$;
   (b) by using the formula $a_n = f^{(n)}(0)/n!$;
   (c) by using the identity $\sin 2x = 2 \sin x \cos x$ and the Maclaurin series for $\sin x$ and $\cos x$.

32. Obtain the first three nonzero terms of the Maclaurin series for $\sin^2 x$:
   (a) by using the formula $a_n = f^{(n)}(0)/n!$;
   (b) by using the identity $\sin^2 x = (1 - \cos 2x)/2$ and the series for $\cos 2x$.

33. An engineer wishes to use a partial sum of the Maclaurin series for e^x to approximate e^x. How many terms of the series should he use to be sure that the error is less than
   (a) 0.01 for $|x| \le 1$?   (b) 0.001 for $|x| \le 1$?
   (c) 0.01 for $|x| \le 2$?   (d) 0.001 for $|x| \le 2$?

In Exercises 34 to 36 determine the limits, using power series. It might be instructive to solve the problems by l'Hôpital's rule also.

34. $\lim\limits_{x \to 0} \dfrac{\ln(1 + x^2) - \sin^2 x}{\tan x^2}$

35. $\lim\limits_{x \to 0} \dfrac{(e^{x^2} - 1)^2}{1 - x^2/2 - \cos x}$

36. $\lim\limits_{x \to 0} \dfrac{(1 - \cos x^2)^5}{(x - \sin x)^{20}}$

37. Estimate $\int_0^1 e^{-x^2}\, dx$ with error ≤ 0.001.
38. Estimate $\int_0^1 \sin x\, dx/x$ with error ≤ 0.001.
39. Estimate $\int_0^2 \sin x\, dx/x$ with error ≤ 0.001.
40. Explain why it is that, if $f(x) = \sum_{n=0}^{\infty} a_n x^n$ for $|x| < 1$, then $a_n = f^{(n)}(0)/n!$. Show details.
41. Explain why it is that, if $f(x) = \sum_{n=0}^{\infty} a_n(x - 3)^n$ for all x, then $a_n = f^{(n)}(3)/n!$. Show details.
42. Use the Maclaurin series for $\ln(1 + x)$ to estimate $\ln(1.5)$ with error less than 0.001.
43. Estimate $\int_0^{1/2} x \cos \sqrt{x}\, dx$ with error less than 0.001.

∎

44. In R. P. Feynman, *Lectures on Physics*, Addison-Wesley, Reading, Mass., 1963, appears this remark,

Thus the average energy is

$$\langle E \rangle = \frac{\hbar\omega(0 + x + 2x^2 + 3x^3 + \cdots)}{1 + x + x^2 + \cdots},$$

where $x = e^{-\hbar\omega/kT}$. Now the two sums which

appear here we shall leave for the reader to play with and have some fun with. When we are all finished summing and substituting for x in the sum, we should get—if we make no mistakes in the sum—

$$\langle E \rangle = \frac{\hbar\omega}{e^{\hbar\omega/kT} - 1}.$$

This, then, was the first quantum-mechanical formula ever known, or ever discussed, and it was the beautiful culmination of decades of puzzlement.

Have some fun.

45. Give an example of a Maclaurin series whose radius of convergence is 1 and which
   (a) converges at 1 and -1;
   (b) diverges at 1 and -1;
   (c) converges at 1 and diverges at -1.

46. Explain why absolute convergence implies convergence.

47. Explain why convergence of a Maclaurin series at $x = c$ implies its absolute convergence at any x for which $|x| < c$.

48. Let $\sum_{n=1}^{\infty} a_n$ be a series of positive terms. Assume that $\lim_{n \to \infty} a_{n+1}/a_n = r < 1$. Explain why the series converges.

49. Let $\sum_{n=1}^{\infty} a_n$ be a series of positive terms. Assume that there is a decreasing function f such that $f(n) = a_n$ and that $\int_1^{\infty} f(x)\, dx$ diverges. Show why the series $\sum_{n=1}^{\infty} a_n$ diverges. (A picture will suffice.)

50. Prove that, if $\sum_{n=1}^{\infty} a_n x^n$ equals 0 for all x in the interval (5.2, 5.3), then $a_n = 0$ for all n.

51. Prove that, if $\sum_{n=1}^{\infty} a_n x^n = \sum_{n=1}^{\infty} b_n x^n$ for all x in the interval (2, 2.01), then $a_n = b_n$ for all n.

52. (a) Show that, if $\sin x$ denotes the sine of an angle of x degrees, then

$$\frac{\pi}{180} x - \frac{[(\pi/180)x]^3}{3!}$$

   differs from $\sin x$ by less than 0.01 for x in [0, 57.3].
   (b) Using the value $\pi/180 \doteq 0.017$, estimate $\sin 10°$ and $\sin 20°$.

53. (a) Show that

$$\sum_{n=0}^{\infty} \frac{\cos(2n + 1)t}{(2n + 1)^2}$$

   converges for all t.
   (b) In the theory of Fourier series it is shown that, for $0 \le t < \pi$, the sum of the series in (a) is $(\pi^2 - 2\pi t)/8$.

   Deduce that $\dfrac{\pi^2}{8} = \dfrac{1}{1^2} + \dfrac{1}{3^2} + \dfrac{1}{5^2} + \cdots$.

54. (a) Show that

$$\sum_{n=1}^{\infty} \frac{\cos 2nt}{4n^2 - 1}$$

converges for all t.

(b) In the theory of Fourier series it is shown that, for $0 \le t < \pi$, the sum of the series in (a) is $\frac{1}{2} - \pi(\sin t)/4$. Deduce that

$$\frac{1}{4^2 - 1} - \frac{1}{8^2 - 1} + \frac{1}{12^2 - 1} - \cdots = \frac{1}{2} - \frac{\pi\sqrt{2}}{8}.$$

This exercise shows that rearranging the terms of a series is a delicate matter.

55. It was observed in Sec. 14.7 that

$$\ln 2 = 1 - \tfrac{1}{2} + \tfrac{1}{3} - \tfrac{1}{4} + \tfrac{1}{5} - \tfrac{1}{6} + \cdots;$$

hence

$$\tfrac{1}{2} \ln 2 = \quad \tfrac{1}{2} - \quad \tfrac{1}{4} + \quad \tfrac{1}{6} - \cdots.$$

(a) Add the two series to obtain

$$\tfrac{3}{2} \ln 2 = 1 + \tfrac{1}{3} - \tfrac{1}{2} + \tfrac{1}{5} + \tfrac{1}{7} - \tfrac{1}{4} + \cdots.$$

(b) Does it follow that

$$\ln 2 = \tfrac{3}{2} \ln 2 ?$$

It is easily proved in advanced calculus that, if a series $\sum_{n=1}^{\infty} a_n$ converges absolutely, then any series obtained by rearranging its terms converges and has the same sum as the original series. Also, if a series converges conditionally, it is always possible to rearrange its terms in such a way that the new series has a sum different from that of the original series.

56. Prove that if $\sum_{n=1}^{\infty} a_n^2$ and $\sum_{n=1}^{\infty} b_n^2$ converge, then so does $\sum_{n=1}^{\infty} a_n b_n$.

In this exercise the power series for $\ln(1 + x)$ will be obtained without borrowing from advanced calculus the result on integration of power series. It has the further advantage that it takes care of $x = 1$.

57. (a) Show that, for $t \ne -1$.

$$\frac{1}{1 + t} = 1 - t + \cdots + (-1)^{n-1}t^{n-1} + (-1)^n \frac{t^n}{1 + t}.$$

(b) Use the identity in (a) to show that, for $x > -1$,

$$\ln(1 + x) = x - \frac{x^2}{2} + \frac{x^3}{3} - \cdots + (-1)^{n-1} \frac{x^n}{n}$$

$$+ (-1)^n \int_0^x \frac{t^n}{1 + t}\, dt.$$

(c) Show that, if x is in $[0, 1]$, then $\int_0^x t^n/(1 + t)\, dt$ approaches 0 as $n \to \infty$. *Hint:* $1 + t \ge 1$.

(d) Show that, if $-1 < x \le 0$, then $\int_0^x t^n/(1 + t)\, dt$ approaches 0 as $n \to \infty$. *Hint:* $1 + t \ge 1 + x$.

(e) Conclude that, if $-1 < x \le 1$, then

$$\ln(1 + x) = x - \frac{x^2}{2} + \frac{x^3}{3} - \cdots + (-1)^{n-1} \frac{x^n}{n} + \cdots.$$

In this exercise the power series for $\tan^{-1} x$ will be obtained without using the result from advanced calculus concerning the integration of power series. Moreover it shows that $\tan^{-1} x = x - x^3/3 + x^5/5 - \cdots$, even when $|x| = 1$.

58. (a) Using the identity in Exercise 57(a), show that

$$\frac{1}{1 + t^2} = 1 - t^2 + \cdots + (-1)^{n-1}t^{2n-2} + (-1)^n \frac{t^{2n}}{1 + t^2}.$$

(b) From (a) deduce that

$$\tan^{-1} x = x - \frac{x^3}{3} + \frac{x^5}{5} - \cdots + (-1)^{n-1} \frac{x^{2n-1}}{2n - 1}$$

$$+ (-1)^n \int_0^x \frac{t^{2n}}{1 + t^2}\, dt.$$

(c) Show that, if $0 \le x \le 1$, $\int_0^x t^{2n}/(1 + t^2)\, dt \to 0$ as $n \to \infty$. *Hint:* Note that $1 + t^2 \ge 1$.

(d) From (c) deduce that, for $|x| \le 1$,

$$\tan^{-1} x = x - \frac{x^3}{3} + \frac{x^5}{5} - \frac{x^7}{7} + \cdots.$$

59. (a) Graph $f(x) = (\sin x)/x$, if $x > 0$, $f(0) = 1$.

(b) Show that, if n is an integer, $n \ge 1$, then

$$\int_{2n\pi}^{(2n+2)\pi} \frac{\sin x}{x}\, dx < \int_{2n\pi}^{(2n+1)\pi} \frac{\pi}{x(x + \pi)}\, dx$$

$$< \int_{2n\pi}^{(2n+1)\pi} \frac{\pi}{x^2}\, dx < \frac{1}{4n^2}.$$

(c) From (a) and (b), deduce that $\int_0^{\infty} f(x)\, dx$ is convergent.

60. What theorems in the text justify the assertion that

$$\lim_{x \to 0} (a_0 + a_1 x + a_2 x^2 + a_3 x^3 + \cdots) = a_0 ?$$

Assume the series has a nonzero radius of convergence.

61. In advanced mathematics a certain function $E(x)$ is *defined* as the sum $\sum_{n=0}^{\infty} x^n/n!$. Pretending that you have never heard of e or e^x, solve the following problems.

(a) Show that $E(0) = 1$.

(b) Show that $E'(x) = E(x)$.

(c) Show that $E(x)E(-x) = 1$. *Hint:* Differentiate $E(x)E(-x)$ and use (a) and (b).

(d) Deduce that $E(x + y)/E(x)$ is independent of x.

(e) Show from (d) that $E(x + y) = E(x)E(y)$, the basic law of exponents.

62. Let $P(x)$ be a polynomial of degree p and let $Q(x)$ be a polynomial of degree q. (Assume $Q(n) \neq 0$ for $n \geq 1$.) For which values of p and q will $\sum_{n=1}^{\infty} P(n)/Q(n)$ (a) converge? (b) diverge?

63. Show that if $a_n > 0$, $\lim_{n \to \infty} \sqrt[n]{a_n} = r < 1$, then the series $\sum_{n=1}^{\infty} a_n$ converges.

64. There is an insurance company that makes the following offer. If you pay it $520 annually at the beginning of each of the next 37 years, it will pay you at the end of those 37 years a lump sum of $35,880.47. (If you die before then, it will refund an amount which is roughly what you put in.) Assuming that you will live to collect, would you agree to the deal?
    (a) How much would you put in altogether?
    (b) If instead you put $520 into a bank that pays 4 percent compounded annually and left it and the interest earned to accumulate, how much would you have at the end of 37 years?
    (c) Like (b) but use 3.058 percent instead of 4 percent.
    (d) Like (b) but use 7 percent instead of 4 percent.

65. Let
$$a_n = \left(\frac{1}{n}\right)^2 \frac{1}{n} + \left(\frac{2}{n}\right)^2 \frac{1}{n} + \cdots + \left(\frac{k}{n}\right)^2 \frac{1}{n} + \cdots + \left(\frac{n}{n}\right)^2 \frac{1}{n}.$$
    (a) Compute a_1, a_2, a_3 and a_4 to three decimal places.
    (b) Find $\lim_{n \to \infty} a_n$ by interpreting a_n as an approximation of an appropriate definite integral.

66. Define $f(x)$ to be $\lim_{n \to \infty} (\sin x)^{2n}$ if the limit exists.
    (a) Compute $f(\pi/4)$.
    (b) Compute $f(\pi/2)$.
    (c) Compute $f(-\pi/2)$.
    (d) Why is the domain of f the entire x axis?
    (e) For which numbers a does $\lim_{x \to a} f(x)$ exist?
    (f) At which numbers a is f not continuous?

67. Let $\sum_{n=1}^{\infty} a_n x^n = A$ and $\sum_{n=1}^{\infty} b_n x^n = B$. Explain why $\sum_{n=1}^{\infty} (a_n + b_n)x^n = A + B$. Hint: Examine the typical partial sum $S_n = \sum_{k=1}^{n} (a_k + b_k)x^k$.

68. In College Is a Waste of Money, *Psychology Today*, May 1975, by Caroline Bird, the following argument appeared:

> A man who completed four years of college would expect to earn $199,000 more between the ages of 22 and 64 than a man who had only a high school diploma.
>
> If a 1972 Princeton-bound high school graduate had put the $34,181 that his four years of college would have cost him into a savings bank at 7.5 percent interest compounded daily, he would have had at age 64 a total of $1,129,200, or $528,200 more than the earnings of a male college graduate.

Assume that the figures $199,000, $34,181, 7.5 per cent and $1,129,200 are correct. Is the conclusion concerning $528,200 valid?

15

TAYLOR'S SERIES AND THE GROWTH OF A FUNCTION

In this chapter it will be shown how the higher derivatives of a function influence the way a function grows. With the aid of this knowledge, we will be able to measure the difference

$$f(x) - \left| f(a) + \frac{f^{(1)}(a)}{1!}(x - a) + \cdots + \frac{f^{(n)}(a)(x - a)^n}{n!} \right|.$$

It will then be easy to show that e^x, $\sin x$, and $\cos x$ can be represented by power series, as assumed in Chap. 14.

The same control on the growth of a function will also provide a simple way to determine a bound for the error in estimating a definite integral by such approximations as the trapezoidal and Simpson's method.

15.1

Higher derivatives and the growth of a function

In this section it is shown how information about the size of a higher derivative of a function f can be used to obtain information about the size of the function f itself.

The reasoning of most of the chapter rests on the following lemma, which tells what happens when you integrate an inequality.

LEMMA Let $f(x)$ and $g(x)$ be functions continuous at least in the open interval $(-b, b)$. Assume that

$$f(x) \le g(x) \qquad \text{for all } x \text{ in } (-b, b).$$

Then

$$\int_0^x f(t)\, dt \le \int_0^x g(t)\, dt \qquad \text{if } 0 < x < b,$$

and

$$\int_0^x f(t)\, dt \ge \int_0^x g(t)\, dt \qquad \text{if } -b < x < 0.$$

PROOF Let $h(x) = g(x) - f(x)$. Note that

$$h(x) \ge 0 \qquad \text{for } -b < x < b.$$

For $x > 0$, $\int_0^x h(t)\, dt$ is a definite integral. Since the integrand is nonnegative,

$$\int_0^x h(t)\, dt \geq 0.$$

Hence
$$\int_0^x [g(t) - f(t)]\, dt \geq 0,$$

or
$$\int_0^x g(t)\, dt - \int_0^x f(t)\, dt \geq 0,$$

or
$$\int_0^x f(t)\, dt \leq \int_0^x g(t)\, dt.$$

If $x < 0$, then $\int_0^x h(t)\, dt$ is defined as

$$\int_0^x h(t)\, dt = -\int_x^0 h(t)\, dt.$$

Thus
$$\int_0^x h(t)\, dt \leq 0.$$

Consequently
$$\int_0^x [g(t) - f(t)]\, dt \leq 0,$$

or
$$\int_0^x g(t)\, dt \leq \int_0^x f(t)\, dt.$$

This establishes the second inequality of the lemma. ●

The lemma says, "If you integrate an inequality involving two functions from 0 to a positive number, the inequality is *preserved*. If you integrate from 0 to a negative number, the inequality is *reversed*."

EXAMPLE I Let f be a function with continuous first, second, and third derivatives such that $f(0) = 0$, $f^{(1)}(0) = 0$, $f^{(2)}(0) = 0$, and

$$5 \leq f^{(3)}(x) \leq 7$$

for all x. Show that

$$\frac{5x^3}{3!} \leq f(x) \leq \frac{7x^3}{3!} \qquad \text{for } x > 0$$

and
$$\frac{5x^3}{3!} \geq f(x) \geq \frac{7x^3}{3!} \qquad \text{for } x < 0.$$

SOLUTION Consider the case $x > 0$ first. By the lemma,

$$\int_0^x 5\, dt \leq \int_0^x f^{(3)}(t)\, dt \leq \int_0^x 7\, dt.$$

By the fundamental theorem of calculus, therefore,

$$5x \leq f^{(2)}(x) - f^{(2)}(0) \leq 7x.$$

Since $f^{(2)}(0) = 0$,

$$5x \leq f^{(2)}(x) \leq 7x \qquad \text{for } x > 0. \tag{1}$$

Inequalities (1) control the size of $f^{(2)}(x)$. Next, integrate (1) to obtain information about $f^{(1)}(x)$.

By the lemma,

$$\int_0^x 5t\, dt \leq \int_0^x f^{(2)}(t)\, dt \leq \int_0^x 7t\, dt,$$

or

$$\frac{5x^2}{2} \leq f^{(1)}(x) - f^{(1)}(0) \leq \frac{7x^2}{2}$$

Since $f^{(1)}(0) = 0$,

$$\frac{5x^2}{2} \leq f^{(1)}(x) \leq \frac{7x^2}{2} \qquad \text{for } x > 0. \tag{2}$$

Thus the growth of $f^{(1)}(x)$ is controlled. Integrate once again, and $f(x)$ itself will appear in the middle of two inequalities. We have, by the lemma,

$$\int_0^x \frac{5t^2}{2}\, dt \leq \int_0^x f^{(1)}(t)\, dt \leq \int_0^x \frac{7t^2}{2}\, dt,$$

or

$$\frac{5x^3}{3!} \leq f(x) - f(0) \leq \frac{7x^3}{3!}$$

Since $f(0) = 0$,

$$\frac{5x^3}{3!} \leq f(x) \leq \frac{7x^3}{3!} \qquad \text{for } x > 0. \tag{3}$$

The first inequality of the example is established.

To establish the second inequality, which concerns $x < 0$, integrate three times as before, but keep in mind that in each integration the direction of inequality changes, so that, in this case, inequalities (1) have $\geq$ signs, inequalities (2) will be the same as before, but inequalities (3) have again $\geq$ signs, which completes the solution. ●

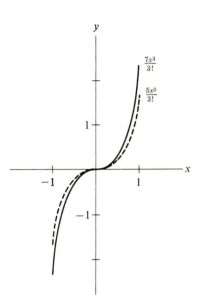

The graph of f lies between the graphs of $y = \frac{5x^3}{3!}$ and $y = \frac{7x^3}{3!}$

It may be illuminating to express Example 1 in terms of the graph of f. The assumptions that $f(0) = 0$, $f^{(1)}(0) = 0$, and $f^{(2)}(0) = 0$ mean that the graph passes through $(0, 0)$, the tangent line there is horizontal, and the slope is changing slowly $[f^{(2)}(0) = 0]$. In short, the graph is fairly flat near $(0, 0)$. From this information and the constraint on $f^{(3)}(x)$, it follows that the graph of f lies between the graphs of $y = 5x^3/3!$ and $y = 7x^3/3!$.

The method of repeatedly integrating inequalities establishes the following useful theorem.

THEOREM I   Let f be a function with continuous derivatives up to and including the nth order throughout some open interval $(-b, b)$. Assume that

$$f(0) = f^{(1)}(0) = f^{(2)}(0) = \cdots = f^{(n-1)}(0) = 0,$$

and that there exist numbers m and M such that

$$m \leq f^{(n)}(x) \leq M$$

for all x in $(-b, b)$.   Then

$$\frac{mx^n}{n!} \leq f(x) \leq \frac{Mx^n}{n!} \qquad \text{for } 0 < x < b. \tag{4}$$

Also, $$\frac{mx^n}{n!} \leq f(x) \leq \frac{Mx^n}{n!} \qquad \text{for even } n \text{ and } -b < x < 0 \tag{5}$$

and $$\frac{mx^n}{n!} \geq f(x) \geq \frac{Mx^n}{n!} \qquad \text{for odd } n \text{ and } -b < x < 0. \; \bullet \tag{6}$$

The conclusion of Theorem 1, expressed in inequalities (4), (5), and (6), is summarized in the assertion

$$f(x) \text{ lies between } \frac{mx^n}{n!} \quad \text{and} \quad \frac{Mx^n}{n!}.$$

The next result, which is an immediate consequence of Theorem 1, will be of use several times in the rest of the chapter.

THEOREM 2   Let f be a function with continuous derivatives up to and including the nth order throughout some open interval $(-b, b)$.   Assume that

$$f(0) = f^{(1)}(0) = f^{(2)}(0) = \cdots = f^{(n-1)}(0) = 0.$$

Let x be any number in the interval $(-b, b)$.   Then there exists a number X between 0 and x such that

$$f(x) = \frac{f^{(n)}(X)x^n}{n!}.$$

PROOF   For convenience, take the case $0 < x < b$.   Let M be the maximum value of $f^{(n)}(t)$ for t in $[0, x]$.   Let m be the minimum value of $f^{(n)}(t)$ for t in $[0, x]$. By Theorem 1,

$$\frac{mx^n}{n!} \leq f(x) \leq \frac{Mx^n}{n!}.$$

Thus $$f(x) = \frac{Cx^n}{n!}$$

for some number C for which $m \leq C \leq M$.   By the intermediate-value theorem of Sec. 6.1, there is at least one number X in the interval $[0, x]$ such that

$$f^{(n)}(X) = C.$$

Consequently $$f(x) = \frac{f^{(n)}(X)x^n}{n!},$$

which was to be proved.

A similar argument works for $-b < x < 0$. The case $x = 0$ is trivial, since

$$0 = f(0) = \frac{f^{(n)}(0)0^n}{n!} \cdot \quad \bullet$$

EXAMPLE 2   Let x be positive.   Show that there is a number X in $[0, x]$ such that

$$e^x = 1 + x + \frac{x^2}{2!} + \frac{x^3}{3!} + \frac{e^X x^4}{4!}.$$

SOLUTION   We will apply Theorem 2 to the function

$$f(t) = e^t - \left(1 + t + \frac{t^2}{2!} + \frac{t^3}{3!}\right).$$

Then

$$f(0) = e^0 - \left(1 + 0 + \frac{0^2}{2!} + \frac{0^3}{3!}\right) = 0.$$

Next

$$f^{(1)}(t) = e^t - \left(1 + t + \frac{t^2}{2!}\right);$$

thus

$$f^{(1)}(0) = e^0 - \left(1 + 0 + \frac{0^2}{2!}\right) = 0.$$

Next

$$f^{(2)}(t) = e^t - (1 + t);$$

hence

$$f^{(2)}(0) = e^0 - (1 + 0) = 0.$$

Next

$$f^{(3)}(t) = e^t - 1;$$

hence

$$f^{(3)}(0) = e^0 - 1 = 0.$$

Finally

$$f^{(4)}(t) = e^t.$$

Since $f(0) = 0$, $f^{(1)}(0) = 0$, $f^{(2)}(0) = 0$, $f^{(3)}(0) = 0$, and $f^{(4)}(t) = e^t$, it follows immediately from Theorem 2, with $n = 4$, that

$$f(x) = \frac{e^X x^4}{4!}$$

for some X in $[0, x]$.   By the definition of the function f, it follows that

$$e^x - \left(1 + x + \frac{x^2}{2!} + \frac{x^3}{3!}\right) = \frac{e^X x^4}{4!},$$

which solves the problem. $\bullet$

The lemma and Theorems 1 and 2 concern the behavior of functions in an interval around 0.   The following theorems are more general in that they concern the behavior of functions in the vicinity of any number a.   Their proofs, which are similar to those of Theorems 1 and 2, are outlined in Exercises 13 and 14.

THEOREM 3   Let f be a function with continuous derivatives up to and including the nth order throughout some open interval $(a - b, a + b)$.   Assume that

$$f(a) = f^{(1)}(a) = f^{(2)}(a) = \cdots = f^{(n-1)}(a) = 0,$$

and that there exist numbers m and M such that

$$m \le f^{(n)}(x) \le M$$

for all x in $(a - b, a + b)$.   Then

$$\frac{m(x - a)^n}{n!} \le f(x) \le \frac{M(x - a)^n}{n!} \qquad \text{for } a < x < a + b.$$

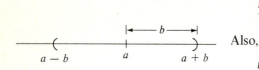

Also,

$$\frac{m(x - a)^n}{n!} \le f(x) \le \frac{M(x - a)^n}{n!} \qquad \text{for even } n \text{ and } a - b < x < a$$

and

$$\frac{m(x - a)^n}{n!} \ge f(x) \ge \frac{M(x - a)^n}{n!} \qquad \text{for odd } n \text{ and } a - b < x < a. \quad \bullet$$

THEOREM 4   Let f be a function with continuous derivatives up to and including the nth order throughout some open interval $(a - b, a + b)$.   Assume that

$$f(a) = f^{(1)}(a) = f^{(2)}(a) = \cdots = f^{(n-1)}(a) = 0.$$

Let x be any number in the interval $(a - b, a + b)$.   Then there exists a number X between a and x such that

$$f(x) = \frac{f^{(n)}(X)(x - a)^n}{n!}. \quad \bullet$$

In Sec. 6.9 the differential was used to approximate the change in a function, yielding the estimate

$$f(x) \doteq f(a) + f'(a)(x - a).$$

The next example shows that the error of this estimate depends on the second derivative of f.

EXAMPLE 3   Let f be a function with continuous first and second derivatives throughout some open interval that contains the number a.   Let x be a number in that interval.   Then

$$f(x) = f(a) + f'(a)(x - a) + \frac{f^{(2)}(X)(x - a)^2}{2}$$

for some X between a and x.

SOLUTION   Introduce the function

$$g(t) = f(t) - [f(a) + f'(a)(t - a)]. \tag{7}$$

Note that $\quad g(a) = f(a) - [f(a) + f'(a)(a - a)] = 0.$

Also, $\qquad\qquad\qquad g'(t) = f'(t) - f'(a);$ $\qquad\qquad$ (8)

hence $\qquad\qquad\qquad g'(a) = f'(a) - f'(a) = 0.$

Thus g satisfies the assumptions of Theorem 4 with $n = 2$. Consequently,

$$g(x) = \frac{g^{(2)}(X)(x - a)^2}{2!} \qquad\qquad (9)$$

for some X between a and x. But inspection of (8) shows that

$$g^{(2)}(X) = f^{(2)}(X). \qquad\qquad (10)$$

Combining (7), (9), and (10) shows that

$$g(x) = f(x) - [f(a) + f'(a)(x - a)] = \frac{f^{(2)}(X)(x - a)^2}{2!}$$

for some X between a and x. From this fact the solution follows immediately. ●

Exercises

In all the exercises assume that the functions have continuous derivatives through all orders and for all x.

1. If $f(0) = f^{(1)}(0) = f^{(2)}(0) = f^{(3)}(0) = 0$ and $0 \le f^{(4)}(x) \le 5$ for all x, what can be said about $f(2)$?

2. (a) If $f(0) = f^{(1)}(0) = f^{(2)}(0) = f^{(3)}(0) = 0$ and $2 \le f^{(4)}(x) \le 3$ for all x, show that

$$\frac{x^4}{12} \le f(x) \le \frac{x^4}{8}$$

for all x.

(b) Sketch the curves $y = x^4/12$ and $y = x^4/8$ and indicate where the graph of f lies.

3. (a) Assume that $f(1) = f^{(1)}(1) = 0$ and that $2 \le f^{(2)}(x) \le 3$ for all x. Show that

$$(x - 1)^2 \le f(x) \le \frac{3(x - 1)^2}{2}$$

for all x.

(b) Graph $y = (x - 1)^2$ and $y = 3(x - 1)^2/2$ and indicate where the graph of f lies.

4. (a) Assume that $f(1) = f^{(1)}(1) = f^{(2)}(1) = 0$ and that $1 \le f^{(3)}(x) \le 2$ for all x. Show that $f(x)$ is between $(x - 1)^3/6$ and $(x - 1)^3/3$.

(b) Graph the curves $y = (x - 1)^3/6$ and $y = (x - 1)^3/3$ and indicate where the graph of f lies.

5. Show that, if $f(0) = f^{(1)}(0) = f^{(2)}(0) = f^{(3)}(0) = 0$ and if $|f^{(4)}(x)| \le 8$ for all x, then

$$|f(x)| \le x^4/3$$

for all x.

6. (a) Let $\quad f(x) = e^x - \left(1 + x + \dfrac{x^2}{2!} + \dfrac{x^3}{3!} + \dfrac{x^4}{4!}\right).$

Show that $f^{(n)}(0) = 0$ for $n = 0, 1, 2, 3, 4$.

(b) Show that $f^{(5)}(x) = e^x$.

(c) Show that for any x there is a number X between 0 and x such that

$$e^x = 1 + x + \frac{x^2}{2!} + \frac{x^3}{3!} + \frac{x^4}{4!} + \frac{e^X x^5}{5!}.$$

(d) Deduce that, if $0 \le x \le 1$, then $1 + x + x^2/2! + x^3/3! + x^4/4!$ differs from e^x by less than $\frac{1}{40}$.

This exercise shows that

$$e^x \text{ equals } 1 + x + \frac{x^2}{2!} + \frac{x^3}{3!} + \cdots.$$

7. (a) Let x be fixed and let n be a positive integer. Show that there is a number X_n such that

$$e^x - \left(1 + x + \frac{x^2}{2!} + \cdots + \frac{x^n}{n!}\right) = \frac{e^{X_n} x^{n+1}}{(n + 1)!}$$

and that X_n is between 0 and x. (Note that X_n depends on n.)

(b) Deduce that, for positive x,

$$\lim_{n \to \infty} \left(1 + x + \frac{x^2}{2!} + \cdots + \frac{x^n}{n!}\right) = e^x.$$

(c) Deduce that, for negative x,

$$\lim_{n \to \infty} \left(1 + x + \frac{x^2}{2!} + \cdots + \frac{x^n}{n!}\right) = e^x.$$

The equations in (b) and (c) clearly are valid when $x = 0$. Thus for all x the function e^x is represented by the power series $\sum_{n=0}^{\infty} x^n/n!$.

8. (a) Show that

$$\sin x = x - \frac{x^3}{3!} + \frac{x^5}{5!} - \frac{\sin (X)x^6}{6!}$$

for some number X between 0 and x.

(b) Deduce that, for $0 \le x \le \pi/6$, the polynomial $x - x^3/6 + x^5/120$ differs from $\sin x$ by at most $\pi^6/(6^6 \cdot 720) \doteq 0.0000286$.

9. (a) Show that

$$\ln (1 + x) = x - \frac{x^2}{2} + \frac{x^3}{3(1 + X)^3}$$

for some number X between 0 and x.

(b) Show that, for $0 \le x \le \frac{1}{2}$, $1 - x + x^2/2$ differs from $\ln (1 + x)$ by at most $\frac{1}{24} \doteq 0.04$.

10. Let $E(h)$ be a function of h with $E(0) = 0 = E^{(1)}(0)$ and $|E^{(2)}(x)| \le 3$ for all x. What can be said about $|E(h)|$?

11. Let $g(x)$ be a function such that $g(0) = 2$, $g^{(1)}(0) = 0 = g^{(2)}(0)$, and $1 \le g^{(3)}(x) \le 2$ for all x. What can be said about $g(3)$?

12. Show that

$$\cos x = 1 - \frac{x^2}{2!} + \frac{\sin X \cdot x^3}{3!}$$

for some number X between 0 and x.

■

13. Prove Theorem 3 for $x > a$ as follows:

(a) Show that

$$m(x - a) \le f^{(n-1)}(x) \le M(x - a).$$

(b) Show that

$$\frac{m(x - a)^2}{2} \le f^{(n-2)}(x) \le \frac{M(x - a)^2}{2}.$$

(c) Complete the proof.

14. Prove Theorem 1 for negative x and (a) $n = 4$ (b) $n = 5$.

15. Sketch the graph of $y = (x - a)^n$ for (a) $n = 4$ (b) $n = 5$.

16. Prove Theorem 4.

17. Let f be a function and let

$$R(x) = f(x) - \left[f(0) + f^{(1)}(0)x + \frac{f^{(2)}(0)x^2}{2!} + \frac{f^{(3)}(0)x^3}{3!}\right].$$

(a) Show that $R^{(n)}(0) = 0$ for $n = 0, 1, 2, 3$.

(b) Show that $R^{(4)}(x) = f^{(4)}(x)$ for all x.

18. Let f be a function and a be a number. Let

$$R(x) = f(x) - \left[f(a) + f^{(1)}(a)(x - a)\right.$$
$$\left. + \frac{f^{(2)}(a)(x - a)^2}{2!} + \frac{f^{(3)}(a)(x - a)^3}{3!}\right].$$

(a) Show that $R^{(n)}(a) = 0$ for $n = 0, 1, 2, 3$.

(b) Show that $R^{(4)}(x) = f^{(4)}(x)$ for all x.

In Exercises 19 to 21 the power series for $\sin x$ and $\cos x$ are obtained.

19. Let $f(x)$ be a function with the property that $|f^{(n)}(x)| \le 20$ for all x and all n. (a) Show that

$$f(x) = \sum_{n=0}^{\infty} f^{(n)}(0)x^n/n!.$$

(b) Show that $\quad f(x) = \sum_{n=0}^{\infty} f^{(n)}(2)(x - 2)^n/n!.$

(c) Show that $\quad f(x) = \sum_{n=0}^{\infty} f^{(n)}(-2)(x + 2)^n/n!.$

20. (See Exercise 19.) Show that, for all x,

$$\sin x = x - \frac{x^3}{3!} + \frac{x^5}{5!} - \frac{x^7}{7!} + \cdots.$$

21. (See Exercise 19.) Show that, for all x,

$$\cos x = 1 - \frac{x^2}{2!} + \frac{x^4}{4!} - \frac{x^6}{6!} + \cdots.$$

■ ■

This exercise is concerned with Newton's method.

22. (a) Read the theorem in Sec. 11.4 concerning Newton's method for estimating a root r of an equation $f(x) = 0$. The notation there will be used in the remainder of the exercise.

(b) Show that

$$0 = f(r) = f(x_i) + f'(x_i)(r - x_i) + f''(X)\frac{(r - x_i)^2}{2}$$

for some X between x_i and r.

(c) From (b) deduce that

$$x_{i+1} - r = \frac{(r - x_i)^2}{2}\frac{f''(X)}{f'(x_i)}.$$

(d) From (c) deduce that

$$|x_{i+1} - r| \le \frac{M}{2} |x_i - r|^2,$$

which is the inequality claimed in Sec. 11.4.

23. If $f(0) = f^{(1)}(0) = f^{(2)}(0) = 1$ and $2 \le f^{(3)}(x) \le 4$ for all x, what can be said about $f(x)$ for all positive x?

15.2

Taylor's series

In Sec. 14.7 it was pointed out that a likely power series to represent a function $f(x)$ is

$$f(0) + f^{(1)}(0)x + \frac{f^{(2)}(0)}{2!} x^2 + \cdots + \frac{f^{(n)}(0)x^n}{n!} + \cdots. \tag{1}$$

To show that such a series actually represents $f(x)$ it is necessary to establish that, as $n \to \infty$,

$$f(x) - \left[f(0) + f^{(1)}(0)x + \frac{f^{(2)}(0)x^2}{2!} + \cdots + \frac{f^{(n)}(0)x^n}{n!} \right] \to 0.$$

For $f(x) = e^x$, $\cos x$, or $\sin x$ series (1) represents the function for all x. For $f(x) = \ln(1 + x)$, series (1) represents the function only between -1 and 1. For the function of Exercise 31 series (1) represents the function only at $x = 0$. In this section it will be shown that whether series (1) represents $f(x)$ depends on how quickly $f^{(n)}(x)$ grows as $n \to \infty$.

We will consider series in powers of $x - a$ in order to achieve full generality. That is, we will examine the difference between $f(x)$ and

$$f(a) + f^{(1)}(a)(x - a) + \frac{f^{(2)}(a)(x - a)^2}{2!} + \cdots + \frac{f^{(n)}(a)(x - a)^n}{n!}.$$

DEFINITION   *Taylor polynomial of degree n, $P_n(x; a)$.* If the function f has derivatives through order n at a, then the Taylor polynomial of degree n of f at a is

$$f(a) + f'(a)(x - a) + \frac{f^{(2)}(a)}{2!} (x - a)^2 + \cdots + \frac{f^{(n)}(a)}{n!} (x - a)^n.$$

This polynomial of degree n is denoted $P_n(x; a)$.

EXAMPLE 1   Find the Taylor polynomial $P_1(x; a)$ associated with the function f at a.

SOLUTION   By the definition of the Taylor polynomial of degree 1,

$$P_1(x; a) = f(a) + f'(a)(x - a).$$

Thus $P_1(x; a)$ is precisely the polynomial used in the preceding section to approximate $f(x)$ near a. It is the unique polynomial P of first degree such that $P(a) = f(a)$ and $P'(a) = f'(a)$. ●

EXAMPLE 2   Find the Taylor polynomial of degree 4 associated with e^x at $a = 0$.

SOLUTION   In this case $f(x) = e^x$. Repeated differentiation yields

$$f^{(1)}(x) = e^x, \qquad f^{(2)}(x) = e^x, \qquad f^{(3)}(x) = e^x, \qquad \text{and} \qquad f^{(4)}(x) = e^x.$$

At $x = 0$ all these derivatives have the value 1. The Taylor polynomial of degree 4 at 0 is therefore

$$P_4(x;0) = 1 + \frac{1}{1!}(x - 0) + \frac{1}{2!}(x - 0)^2 + \frac{1}{3!}(x - 0)^3 + \frac{1}{4!}(x - 0)^4,$$

or simply

$$1 + x + \frac{x^2}{2!} + \frac{x^3}{3!} + \frac{x^4}{4!}. \quad \bullet$$

The key property of the Taylor polynomial $P_n(x; a)$ is expressed in the following theorem. The index j is used to avoid using n for two purposes.

THEOREM I   Let $P_n(x; a)$ be the Taylor polynomial of degree n at a associated with the function f. Then, for $j = 0, 1, \ldots, n$, the jth derivative of $P_n(x; a)$ at a equals the jth derivative of $f(x)$ at a:

$$D^{(j)}(P_n(x; a)) = f^{(j)}(x) \qquad \text{at } x = a.$$

PROOF   Simply differentiate $P_n(x; a)$ repeatedly and plug in $x = a$. This table shows the argument at a glance.

	$D^{(j)}(P_n(x; a))$	$D^{(j)}(P_n(x; a))$ at $x = a$
$j = 0$	$f(a) + f^{(1)}(a)(x - a) + \dfrac{f^{(2)}(a)(x - a)^2}{2!} + \dfrac{f^{(3)}(a)(x - a)^3}{3!} + \cdots + \dfrac{f^{(n)}(a)(x - a)^n}{n!}$	$f(a) = f^{(0)}(a)$
$j = 1$	$f^{(1)}(a) + f^{(2)}(a)(x - a) + \dfrac{f^{(3)}(a)(x - a)^2}{2!} + \cdots + \dfrac{f^{(n)}(a)(x - a)^{n-1}}{(n-1)!}$	$f^{(1)}(a)$
$j = 2$	$f^{(2)}(a) + f^{(3)}(a)(x - a) + \cdots + \dfrac{f^{(n)}(a)(x - a)^{n-2}}{(n-2)!}$	$f^{(2)}(a)$
$j = n$	$f^{(n)}(a)$	$f^{(n)}(a)$

$\bullet$

Theorem 1 tells us that $P_n(x; a)$ is a polynomial of degree n whose derivatives up to order n coincide with those of $f(x)$ when $x = a$. [Actually $P_n(x; a)$ is the only polynomial of degree n with these properties.]

DEFINITION   *The remainder (error) $R_n(x; a)$.* Let f be a function and let $P_n(x; a)$ be the associated Taylor polynomial of degree n at a. The number $R_n(x; a)$ defined by the equation

$$f(x) = P_n(x; a) + R_n(x; a)$$

is called the *remainder* (or *error*) in using the Taylor polynomial $P_n(x; a)$ to approximate $f(x)$.

Our expectation is that, for well-behaved functions f,

$$R_n(x; a) \to 0$$

as $n \to \infty$. Theorem 2 will show that this hope is justified. This theorem is based on the following lemma, which restates Theorem 4 of the preceding section in a form convenient for our purposes; n is replaced by $n + 1$, and f is replaced by R.

LEMMA Let R be a function with continuous derivatives up to and including the $(n + 1)$st order throughout some open interval $(a - b, a + b)$. Assume that

$$R(a) = R^{(1)}(a) = R^{(2)}(a) = \cdots = R^{(n)}(a) = 0.$$

Let x be any number in the interval $(a - b, a + b)$. Then there exists a number X between a and x such that

$$R(x) = \frac{R^{(n+1)}(X)(x - a)^{n+1}}{(n + 1)!}$$

THEOREM 2 [*Derivative (Lagrange's) formula for* $R_n(x; a)$]. Let f be defined at least in $(a - b, a + b)$. Assume that the first $n + 1$ derivatives of f exist and are continuous in that interval. Then

$$R_n(x; a) = f^{(n+1)}(X) \frac{(x - a)^{n+1}}{(n + 1)!}$$

for some number X between a and x.

PROOF Let $R(x)$ denote the remainder $R_n(x; a)$:

$$R(x) = f(x) - P_n(x; a).$$

By Theorem 1,

$$R(a) = R^{(1)}(a) = R^{(2)}(a) = \cdots = R^{(n)}(a) = 0.$$

Also, since $P_n(x; a)$ is a polynomial of degree n, its $(n + 1)$st derivative is 0 for all x. Thus

$$R^{(n+1)}(x) = f^{(n+1)}(x).$$

By the lemma it follows immediately that

$$R(x) = \frac{f^{(n+1)}(X)(x - a)^{n+1}}{(n + 1)!}$$

for some number X between a and x. This concludes the proof. ●

Even though we may not know the precise value of X, and even though for fixed a and x the number X depends on n, still the information contained in Theorem 2 is quite useful, as the following examples illustrate.

EXAMPLE 3   Use Theorem 2 to show that for any positive number x,

$$e^x = 1 + x + \frac{x^2}{2!} + \frac{x^3}{3!} + \cdots.$$

SOLUTION   The Taylor polynomial of degree n associated with e^x at $a = 0$ is

$$P_n(x;0) = 1 + x + \frac{x^2}{2!} + \frac{x^3}{3!} + \frac{x^4}{4!} + \cdots + \frac{x^n}{n!}.$$

(See Example 2 for the case $n = 4$.)   The difference between e^x and this polynomial is denoted by $R_n(x;0)$.   By Theorem 2,

$$R_n(x;0) = \frac{f^{(n+1)}(X)(x-0)^{n+1}}{(n+1)!}$$

for some number X between 0 and x.   In this case $f(x) = e^x$, and therefore $f^{(n+1)}(x) = e^x$.   Consequently,

$$R_n(x;0) = \frac{e^X x^{n+1}}{(n+1)!}.$$

Since $1 \le e^X \le e^x$,

$$\frac{x^{n+1}}{(n+1)!} \le R_n(x;0) \le e^x \frac{x^{n+1}}{(n+1)!}.$$

As $n \to \infty$, $x^{n+1}/(n+1)! \to 0$, by Example 4 of Sec. 14.5.

Thus $$\lim_{n \to \infty} R_n(x;0) = 0,$$

and we conclude that

$$1 + x + \frac{x^2}{2!} + \frac{x^3}{3!} + \cdots + \frac{x^n}{n!} + \cdots = e^x.$$

The reader may show that the same conclusion holds for x negative.   ●

As Example 3 illustrates, the derivative form of $R_n(x;a)$ provides a way of showing that the power series defined by the Taylor polynomials of a function f actually converges to f.

DEFINITION   *Taylor's series and the Maclaurin series.*   Let f have derivatives of all orders at a.   The series

$$f(a) + f^{(1)}(a)(x-a) + \frac{f^{(2)}(a)(x-a)^2}{2!} + \cdots + \frac{f^{(n)}(a)(x-a)^n}{n!} + \cdots$$

is called the Taylor's series in powers of $x - a$ associated with the function f. If $a = 0$, the Taylor's series takes the simpler form

$$f(0) + f'(0)x + \frac{f^{(2)}(0)x^2}{2!} + \cdots + \frac{f^{(n)}(0)x^n}{n!} + \cdots$$

and is also called the Maclaurin series associated with f.

In Example 3 the Taylor's series for the function f was shown to converge to the given function. Exercise 31 describes a very pathological function (e^{-1/x^2}) whose Taylor series converges to 0 for all values of x, certainly not to the given function.

EXAMPLE 4   Find the Maclaurin series associated with the function $\sin x$. Does it converge to $\sin x$?

SOLUTION   The coefficient of x^n in the Maclaurin series associated with the function $\sin x$ is

$$\frac{\sin^{(n)} 0}{n!}$$

n	$\sin^{(n)} x$	$\sin^{(n)} 0$
0	$\sin x$	0
1	$\cos x$	1
2	$-\sin x$	0
3	$-\cos x$	-1
4	$\sin x$	0
5	$\cos x$	1
6	$-\sin x$	0

The first few derivatives of $\sin x$ are listed in the table in the margin, along with their values at 0: The pattern continues to repeat in blocks of four. Since $\sin^{(n)} 0 = 0$ when n is even, the Maclaurin series for $\sin x$ does not contain terms in which the power of x is even. (More precisely, there is no need to write them down.) The remaining terms are either

$$\frac{x^n}{n!}$$

or

$$\frac{-x^n}{n!},$$

and the signs alternate. The Maclaurin series is

$$x - \frac{x^3}{3!} + \frac{x^5}{5!} - \frac{x^7}{7!} + \cdots.$$

The assertion that this series converges to $\sin x$ is equivalent to the claim that $R_n(x; 0)$ approaches 0 as $n \to \infty$. By the derivative form for $R_n(x; 0)$, there is a number X in $(0, x)$ such that

$$R_n(x; 0) = \frac{\sin^{(n+1)} (X) x^{n+1}}{(n+1)!}.$$

(X depends on n and x.) Since the higher derivatives of $\sin x$ are either $\pm \cos x$ or $\pm \sin x$, it follows that $|\sin^{(n+1)} (X)| \le 1$, hence that

$$|R_n(x; 0)| \le \frac{|x|^{n+1}}{(n+1)!}.$$

By Example 4 of Sec. 14.5, $\quad \lim_{n \to \infty} \dfrac{|x|^{n+1}}{(n+1)!} = 0.$

Thus $\quad\quad\quad\quad\quad\quad \lim_{n \to \infty} R_n(x; 0) = 0.$

Therefore the Maclaurin series associated with $\sin x$ converges to $\sin x$ for all values of x. ●

In both examples a Taylor's series was found for $a = 0$.   In the next example a is not 0.

EXAMPLE 5   Find the Taylor's series in powers of $x - \pi/4$ associated with the function $\cos x$.

SOLUTION   In this case the derivatives are all to be evaluated at $\pi/4$.   The computations for $n \leq 6$ are displayed in this table:

n	$f^{(n)}(x) = \cos^{(n)} x$	$f^{(n)}\left(\dfrac{\pi}{4}\right)$	n	$f^{(n)}(x) = \cos^{(n)} x$	$f^{(n)}\left(\dfrac{\pi}{4}\right)$
0	$\cos x$	$\dfrac{\sqrt{2}}{2}$	4	$\cos x$	$\dfrac{\sqrt{2}}{2}$
1	$-\sin x$	$-\dfrac{\sqrt{2}}{2}$	5	$-\sin x$	$-\dfrac{\sqrt{2}}{2}$
2	$-\cos x$	$-\dfrac{\sqrt{2}}{2}$	6	$-\cos x$	$-\dfrac{\sqrt{2}}{2}$
3	$\sin x$	$\dfrac{\sqrt{2}}{2}$			

The Taylor's series at $a = \pi/4$ therefore begins

$$\frac{\sqrt{2}}{2} - \frac{\sqrt{2}}{2}\left(x - \frac{\pi}{4}\right) - \frac{\sqrt{2}}{2}\frac{(x - \pi/4)^2}{2!} + \frac{\sqrt{2}}{2}\frac{(x - \pi/4)^3}{3!} + \frac{\sqrt{2}}{2}\frac{(x - \pi/4)^4}{4!} - \cdots .$$

(Two pluses continue to alternate with two minuses.)   It can be shown with the aid of the derivative form for $R_n(x; \pi/4)$ that the series converges to $\cos x$ for all values of x. ●

Incidentally, Taylor, who discussed the relation between a function and the values of its higher derivatives, published his result in 1715 but did not apply it. Maclaurin in 1742, giving credit to Taylor, discussed power series around $a = 0$.

The final two examples show how Taylor's series can be used to provide very good estimates of definite integrals.

EXAMPLE 6   Estimate $\int_0^{1/2} e^{-x^2}\, dx$.

SOLUTION   Using the first four terms of the Maclaurin series for e^x and the derivative form of $R_3(x; 0)$, we have

$$e^t = 1 + t + \frac{t^2}{2!} + \frac{t^3}{3!} + \frac{e^T}{4!}t^4$$

for some number T between 0 and t (T depends on t).   When t is replaced with $-x^2$, the above equation becomes

$$e^{-x^2} = 1 - x^2 + \frac{x^4}{2!} - \frac{x^6}{3!} + \frac{e^T x^8}{4!}$$

for some T between 0 and $-x^2$. Thus

$$\int_0^{1/2} e^{-x^2}\, dx = \int_0^{1/2} \left(1 - x^2 + \frac{x^4}{2!} - \frac{x^6}{3!}\right) dx + \int_0^{1/2} \frac{e^T x^8}{4!}\, dx.$$

The error in using $\qquad \displaystyle\int_0^{1/2} \left(1 - x^2 + \frac{x^4}{2!} - \frac{x^6}{3!}\right) dx$

to estimate $\qquad\qquad \displaystyle\int_0^{1/2} e^{-x^2}\, dx$

is $\qquad\qquad\qquad\quad \displaystyle\int_0^{1/2} \frac{e^T x^8}{4!}\, dx.$

Since T is negative for each value of x,

$$0 \le \frac{e^T x^8}{4!} \le \frac{e^0 x^8}{4!} = \frac{x^8}{4!}.$$

Thus $\qquad \displaystyle\int_0^{1/2} \frac{e^T x^8}{4!}\, dx \le \int_0^{1/2} \frac{x^8}{4!}\, dx = \frac{x^9}{4! \cdot 9}\bigg|_0^{1/2} = \frac{1}{4! \cdot 9 \cdot 2^9}.$

Hence

$$\int_0^{1/2} \left(1 - x^2 + \frac{x^4}{2!} - \frac{x^6}{3!}\right) dx = \frac{1}{2} - \frac{1}{2^3 \cdot 3} + \frac{1}{2^5 \cdot 5 \cdot 2!} - \frac{1}{2^7 \cdot 7 \cdot 3!} \doteq 0.46127$$

is an estimate of $\int_0^{1/2} e^{-x^2}\, dx$, with an error less than

$$\frac{1}{4! \cdot 9 \cdot 2^9} = \frac{1}{110{,}592} < 0.00001. \quad \bullet$$

EXAMPLE 7   Estimate $\int_0^1 \sin x/x\, dx$ by using the first four nonzero terms of the **Maclaurin** series for $\sin x$.

SOLUTION $\qquad\qquad\qquad \sin x = x - \frac{x^3}{3!} + \frac{x^5}{5!} - \frac{x^7}{7!} + R_7(x; 0),$

where $\qquad\qquad R_7(x; 0) = \frac{\sin^{(8)}(X) x^8}{8!} = \frac{(\sin X) x^8}{8!}$

for some number X between 0 and x (X depends on x).   Thus

$$\frac{\sin x}{x} = 1 - \frac{x^2}{3!} + \frac{x^4}{5!} - \frac{x^6}{7!} + \frac{(\sin X) x^7}{8!}.$$

Hence $\qquad \displaystyle\int_0^1 \frac{\sin x}{x}\, dx = \int_0^1 \left(1 - \frac{x^2}{3!} + \frac{x^4}{5!} - \frac{x^6}{7!}\right) dx + \int_0^1 \frac{(\sin X) x^7}{8!}\, dx. \qquad (1)$

Now, since $0 \le \sin X \le 1$ for any number X,

$$0 \le \int_0^1 \frac{(\sin X)x^7}{8!}\, dx \le \int_0^1 \frac{1x^7}{8!}\, dx = \frac{1}{8 \cdot 8!} < 0.00001.$$

Thus the first definite integral in the right side of (1), which can easily be shown to equal approximately 0.94608, is an estimate of $\int_0^1 \sin x/x\, dx$ with an error of less than 0.00001. ●

Exercises

In Exercises 1 to 10 find $P_n(x; a)$, the Taylor polynomial of degree n associated with the given function and the given value of a.

1. e^{-x}, $n = 4$, $a = 0$.   2. e^{-x}, $n = 4$, $a = 1$.
3. x^4, $n = 4$, $a = 1$.   4. $\ln(1 + x)$, $n = 3$, $a = 0$.
5. $\sin x$, $n = 5$, $a = 0$.   6. $\sin x$, $n = 4$, $a = \pi/6$.
7. $\ln x$, $n = 3$, $a = 1$.   8. $\sqrt{1 + x}$, $n = 3$, $a = 0$.
9. $1/(1 + x)$, $n = 4$, $a = 0$.   10. $\tan^{-1} x$, $n = 2$, $a = 0$.

11. Expand $\sin x$ in powers of $x - \pi/2$. Write out the first three nonzero terms.

12. Expand $\cos x$ in powers of $x - \pi/3$. Write out the first four terms.

13. (a) Show that, if you use $1 - x^2/2 + x^4/24$ to estimate $\cos x$ for $|x| \le 1$, the error is less than $1/5! \le 0.009$. [Use the derivative form of $R_4(x; 0)$ to estimate $R_4(x; 0)$.]
    (b) Why is the error less than $1/6!$?

14. (a) Find the Maclaurin series associated with $\cos x$.
    (b) Prove that it converges to $\cos x$.

15. Show that $\tan^{-1} x = x - x^2 X/(1 + X^2)^2$ for some number X between 0 and x.

16. Prove that the Taylor's series in Example 5 converges to $\cos x$.

17. Find the Taylor polynomial $P_4(x; 0)$ associated with $\tan^{-1} x$.

In Exercises 18 to 21 use just enough terms of the Maclaurin series for e^x to make the indicated estimates. [Use the derivative form of $R_n(x; 0)$ to discuss the size of the error.]

18. e^{-1}, error less than 0.01.
19. $e^{2/3}$, error less than 0.01.
20. e^2, error less than 0.1.
21. e^{-2}, error less than 0.1.
22. This exercise concerns the MacLaurin series for $f(x) = \ln(1 + x)$.
    (a) Copy and fill in this table.

n	$f^{(n)}(x)$	$f^{(n)}(0)$	n	$f^{(n)}(x)$	$f^{(n)}(0)$
0			3		
1			4		
2			5		

(b) Write out the Maclaurin series for $\ln(1 + x)$.
(c) Prove that, if $|x| > 1$, the series in (b) does not converge.
(d) Prove that, if $-\frac{1}{2} < x < 1$, the series in (b) converges to $\ln(1 + x)$.

23. Show that $e^x = 1 + x + x^2/2! + \cdots$ when x is negative.

∎

The coefficient of x^j in the expansion of $(1 + x)^n$ for n a positive integer is

$$\frac{n!}{j!\,(n - j)!} = \frac{n(n - 1)\cdots(n - j + 1)}{1 \cdot 2 \cdots j}$$

and is called a binomial coefficient. It is denoted $\binom{n}{j}$ or C_j^n.

In this exercise the formula above is proved.

24. Let $f(x) = (1 + x)^n$, where n is a fixed positive integer.
    (a) Show that $f(x) = P_n(x; 0)$.
    (b) For $j = 0, 1, 2, \ldots, n$, show that

$$f^{(j)}(0) = n(n - 1)(n - 2) \cdots (n - j + 1) = \frac{n!}{(n - j)!}.$$

    (c) Deduce that

$$(1 + x)^n = 1 + \frac{n!}{1!\,(n - 1)!}\, x + \frac{n!}{2!\,(n - 2)!}\, x^2$$

$$+ \cdots + \frac{n!}{n!\,0!}\, x^n.$$

The coefficient of x^j is

$$\frac{n!}{j!\,(n-j)!}.$$

(d) From (c) obtain the following more common form of the binomial theorem:

$$(a+b)^n = a^n + \frac{n!}{1!\,(n-1)!}\,a^{n-1}b + \frac{n!}{2!\,(n-2)!}\,a^{n-2}b^2$$

$$+ \cdots + \frac{n!}{n!\,0!}\,b^n.$$

25. (a) Use Exercise 24(c) to show that $(1+x)^3 = 1 + 3x + 3x^2 + x^3$.

(b) Use Exercise 24(c) to obtain the first three terms of $(1+x)^{10}$.

(c) Use Exercise 24(c) to obtain the last three terms of $(1+x)^{10}$.

26. [See Exercise 24(c).] Show that the coefficient of x^3 in the expansion of $(1+x)^{100}$ is 161,700.

27. (a) Graph, relative to the same axes, the curves $y = 1 + 2x$ and $y = e^x$.

(b) Graphically estimate the x coordinate of the point off the y axis where they meet.

(c) Using $1 + x + x^2/2 + x^3/6$ as an approximation of e^x, estimate where the curves in (a) meet.

28. Let $f(x) = \cos x^2$.

(a) Compute $f(0)$, $f^{(1)}(0)$, $f^{(2)}(0)$, and $f^{(3)}(0)$.

(b) As the computations in (a) suggest, finding the Maclaurin series for $\cos x^2$ in the straightforward manner is difficult. Instead, obtain the series from that for $\cos x$ by replacing x by x^2.

29. What is the Maclaurin series for

(a) $\cos 2x$?

(b) $\cos \sqrt{x}$?

30. (a) From the Maclaurin series for $\cos x$ in powers of x, obtain the Maclaurin series for $\cos 2x$.

(b) Exploiting the identity $\sin^2 x = (1 - \cos 2x)/2$, obtain the Maclaurin series for $(\sin^2 x)/x^2$.

(c) Estimate $\int_0^1 (\sin x/x)^2\,dx$ using the first three nonzero terms of the series.

(d) Find a bound on the error entailed in the estimate in (c).

■ ■

31. Let $f(x) = e^{-1/x^2}$ if $x \neq 0$ and $f(0) = 0$.

(a) Graph $y = f(x)$ for x in $[-1, 1]$.

(b) Show that $f'(0) = 0$.

(c) Show that $f''(0) = 0$.

(d) Why is $f^{(n)}(0) = 0$ for all positive integers n?

(e) What is the Maclaurin series associated with f?

(f) Does the series in (e) represent f?

15.3

The differential equation of harmonic motion

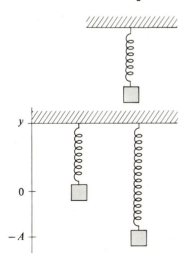

Consider a mass hanging motionless at the end of a spring, as in the accompanying diagram. It is at rest and will remain in that position. If we now pull it down and release it, the mass will bob up and down. Because of air resistance and friction within the spring, the oscillating motion will gradually die out. Disregarding these influences, assume that there is no damping effect. With this assumption, a formula will be obtained that completely describes the motion of the bobbing mass.

Introduce a y axis which has its origin at the rest position of the mass and its positive part above the origin. If the initial displacement of the mass is a distance A, then the y coordinate of the mass at time 0 is $-A$.

When we let go, the spring will pull the mass up. It will go above the rest position, and then be pulled down by gravity until the spring is stretched so far that the tension in the spring dominates and slows the mass down. The mass, pulled by the earth and pushed and pulled by the spring, bobs up and down.

On the basis of physical assumptions it can be shown that if $y = f(t)$ is the y

coordinate of the mass at time t, then

$$\frac{d^2y}{dt^2} \quad \text{is proportional to } y,$$

and the constant of proportionality is negative (if the constant were positive, the mass would continue to move in one direction).   That is, there is a number k such that

$$\frac{d^2y}{dt^2} = -k^2y. \tag{1}$$

(Note that k^2 is positive, whether or not k is; it will be assumed that k is positive also.)

DEFINITION   *Harmonic motion.*   Motion satisfying (1) is called *harmonic.*

The differential equation (1) involves the second derivative and the function. The differential equation of natural growth in Sec. 6.4,

$$\frac{dx}{dt} = kx,$$

related the derivative and the function.   Its solutions are exponential functions. What kind of solutions will (1) have?

Since the motion of the mass is periodic, it is to be expected that the possible solutions involve trigonometric functions.   In fact, the function

$$y = c_1 \cos kt + c_2 \sin kt$$

satisfies (1) for any choice of constants c_1 and c_2, as these straightforward computations show:

$$\frac{dy}{dt} = -kc_1 \sin kt + kc_2 \cos kt;$$

$$\frac{d^2y}{dt^2} = -k^2c_1 \cos kt - k^2c_2 \sin kt$$

$$= -k^2(c_1 \cos kt + c_2 \sin kt)$$

$$= -k^2y.$$

It will now be shown that the *only solutions of the differential equation* (1) *are functions of the form*

$$c_1 \cos kt + c_2 \sin kt.$$

THEOREM   Any solution $y = f(t)$ of the differential equation (1) is of the form

$$c_1 \cos kt + c_2 \sin kt$$

for suitable constants c_1 and c_2.

PROOF   The idea of the proof is to use the information that

$$\frac{d^2y}{dt^2} = -k^2y$$

to find the form of the Maclaurin series associated with $y = f(t)$, and then show that $R_n(t; 0) \to 0$ as $t \to \infty$. In order to simplify the presentation, the proof will be given for the case $k = 1$. The proof for other k is identical in spirit, but powers of k will appear in the formulas.

It is to be proved, then, that any solution $y = f(t)$ of the differential equation

$$\frac{d^2y}{dt^2} = -y \tag{2}$$

is of the form $$y = c_1 \cos t + c_2 \sin t.$$

In preparation for finding the Maclaurin series associated with the function f, examine the higher derivatives of f. From the assumption

$$f^{(2)} = -f$$

it follows that

$$f^{(3)} = D(f^{(2)}) = D(-f) = -f^{(1)}$$

$$f^{(4)} = D(f^{(3)}) = D(-f^{(1)}) = -f^{(2)} = -(-f) = f$$

$$f^{(5)} = D(f^{(4)}) = D(f) = f^{(1)}$$

$$f^{(6)} = D(f^{(5)}) = D(f^{(1)}) = f^{(2)} = -f$$

and so on.

When n is even, $f^{(n)}$ is either f or $-f$. When n is odd, $f^{(n)}$ is either $f^{(1)}$ or $-f^{(1)}$.

If we call $$f(0) = c_1 \quad \text{and} \quad f^{(1)}(0) = c_2,$$

then $$f^{(2)}(0) = -c_1, \quad f^{(4)}(0) = c_1, \quad f^{(6)}(0) = -c_1, \ldots,$$

and $$f^{(3)}(0) = -c_2, \quad f^{(5)}(0) = c_2, \quad f^{(7)}(0) = -c_2, \ldots,$$

the signs alternating.

The Taylor polynomial of degree $2n - 1$ in powers of t (for time) associated with f is therefore

$$c_1 + c_2 t - \frac{c_1 t^2}{2!} - \frac{c_2 t^3}{3!} + \frac{c_1 t^4}{4!} + \frac{c_2 t^5}{5!} + \cdots + (-1)^{n+1} \frac{c_1 t^{2n-2}}{(2n-2)!}$$

$$+ (-1)^{n+1} \frac{c_2 t^{2n-1}}{(2n-1)!}. \tag{3}$$

This equals

$$c_1 \left[1 - \frac{t^2}{2!} + \frac{t^4}{4!} - \cdots + (-1)^{n+1} \frac{t^{2n-2}}{(2n-2)!} \right]$$

$$+ c_2 \left[t - \frac{t^3}{3!} + \frac{t^5}{5!} - \cdots + (-1)^{n+1} \frac{t^{2n-1}}{(2n-1)!} \right]. \tag{4}$$

The two polynomials in brackets in (4) are precisely the Taylor polynomials for cos t and sin t, respectively.

All that remains is to show that the Taylor polynomial (3) for f actually approaches $f(t)$ as $n \to \infty$. That is, it must be shown that $R_{2n-1}(t) \to 0$ as $n \to \infty$, where $R_{2n-1}(t)$ is the difference between $f(t)$ and (3).

By the preceding section,

$$R_{2n-1}(t) = \frac{t^{2n}}{(2n)!} f^{(2n)}(T),$$

where T (which depends on n) is between 0 and t. Now, $f^{(2n)}$ is either f or $-f$. Since f is continuous, f is bounded on $[0, t]$. Thus there is a number B such that

$$|f^{(2n)}(T)| \leq B$$

for all n. Consequently

$$|R_{2n-1}(t)| \leq \frac{|t|^{2n}}{(2n)!} B.$$

Since

$$\lim_{n \to \infty} \frac{|t|^{2n}}{(2n)!} = 0,$$

it follows that

$$\lim_{n \to \infty} R_{2n-1}(t) = 0,$$

and thus the sum (3) approaches $f(t)$ as $n \to \infty$. This implies that

$$f(t) = c_1 \cos t + c_2 \sin t,$$

and the theorem is proved. ●

The proof of the theorem shows how Taylor's series can be used to solve differential equations. The method was introduced by Newton in the last quarter of the seventeenth century.

From the theorem it follows that the y coordinate of the bobbing mass at time t is

$$c_1 \cos kt + c_2 \sin kt.$$

Now, both the cosine and the sine functions are periodic with period 2π. Thus

$$\cos\left[k\left(t + \frac{2\pi}{k}\right)\right] = \cos(kt + 2\pi) = \cos kt,$$

and similarly,

$$\sin\left[k\left(t + \frac{2\pi}{k}\right)\right] = \sin kt.$$

This means that the motion of the spring is periodic with period $2\pi/k$. The mass continues to bob up and down, repeating its cycle in a time interval $2\pi/k$. It completes $k/2\pi$ cycles during one unit of time.

Exercises

Solve Exercises 1 to 6 by the method of this section.

1. Show that any solution of the equation

$$\frac{dy}{dt} = y$$

must be of the form ce^t for a suitable constant c.

2. Show that any solution of the equation

$$\frac{dy}{dt} = -y$$

must be of the form ce^{-t} for a suitable constant c.

3. Show that any solution of the equation

$$\frac{d^2y}{dt^2} = y$$

must be of the form $c_1 \cosh t + c_2 \sinh t$ for suitable constants c_1 and c_2.

4. Show that any solution of the equation

$$\frac{dy}{dt} = ky$$

must be of the form $y = Ae^{kt}$.

5. Show that any solution of the equation

$$\frac{d^2y}{dt^2} = k^2 y$$

must be of the form $c_1 \cosh kt + c_2 \sinh kt$.

6. Show that any solution of the equation

$$\frac{d^2y}{dt^2} = -k^2 y$$

must be of the form $c_1 \cos kt + c_2 \sin kt$.

7. If $y = c_1 \cos kt + c_2 \sin kt$ describes the motion of a particle, find (a) its position when $t = 0$ and (b) its velocity when $t = 0$.

8. (a) Show that $y = \sin (kt + k_0)$ is a solution of Eq. (1).
(b) By (a) and the theorem of this section, show that there are constants c_1 and c_2 such that

$$\sin (kt + k_0) = c_1 \cos kt + c_2 \sin kt.$$

Using a trigonometric identity, find c_1 and c_2.

9. Let c_1 and c_2 be numbers. Show that there is a number k_0 such that

$$c_1 = \sqrt{c_1{}^2 + c_2{}^2} \sin k_0$$

and

$$c_2 = \sqrt{c_1{}^2 + c_2{}^2} \cos k_0.$$

Hint: Draw the circle with center $(0, 0)$ passing through the point (c_2, c_1).

10. Using Exercise 9 and the theorem of this section, prove that any solution of the differential equation (1) is of the form

$$A \sin (kt + k_0)$$

for suitable constants A and k_0.

11. In view of Exercise 10, the motion of the bobbing mass is described by the equation

$$y = A \sin (kt + k_0).$$

What is the highest point that the mass reaches? The lowest?

■

12. A wheel rotates with angular velocity k, that is, $d\theta/dt = k$, where θ denotes the total angle the wheel has turned up through time t. A bug is on the wheel at a distance A from the center. Introduce an xy-coordinate system whose origin is at the center of the wheel.
(a) Show that the y coordinate of the bug is of the form $y = A \sin (kt + k_0)$ for a suitable constant k.
(b) Describe the motion of the shadow of the bug on the y axis cast by light parallel to the x axis.

13. Graph:
(a) $y = 3 \sin t$;
(b) $y = 3 \sin 2\pi t$.
(c) Find the period of the function in (b).

14. (a) Graph $y = 4 \sin 200\pi t$.
(b) What is the period of the function in (a)?
(c) What is the maximum value of y?
(d) How many cycles does the harmonic motion described in (a) complete in unit time?

15. Let $y = A \sin (kt + c)$ describe harmonic motion.
(a) Show that the particle reaches a highest coordinate $|A|$ and a lowest coordinate $-|A|$. ($|A|$ is called the *amplitude* of the motion.)
(b) Show that the time to complete a cycle is $2\pi/k$. ($2\pi/k$ is called the *period*.)
(c) Show that the number of cycles per unit time is $k/2\pi$.

■ ■

A differential equation of the form

$$\frac{d^2y}{dx^2} + a(x)\frac{dy}{dx} + b(x)y = g(x)$$

is called a *linear differential equation* of the second order. If $g(x)$ is 0 for all x, the equation is called *homogeneous*. If $a(x)$ and $b(x)$ are constant, then the equation is said to have *constant coefficients*. Such equations arise in the study of resonance, damping, oscillation, and vibration. The following exercises concern a few specific such equations.

16. (*a*) Find all numbers k such that e^{kx} is a solution of

$$\frac{d^2 y}{dx^2} - 2\frac{dy}{dx} - 3y = 0.$$

(*b*) Is there a nonzero k such that $\sin kx$ is a solution of the equation in (*a*)?

17. Show that, for any constants A and B, $Ae^{2x} + Be^{3x}$ is a solution of the differential equation

$$y'' - 5y' + 6y = 0.$$

18. (*a*) Find constants A and B such that $y = A\cos x + B\sin x$ is a solution of

$$y'' + 3y' + 2y = \sin x.$$

(*b*) Show that any function of the form

$$-\tfrac{3}{10}\cos x + \tfrac{1}{10}\sin x + k_1 e^{-x} + k_2 e^{-2x}$$

is a solution of the equation in (*a*).

15.4

The error in estimating the definite integral

Section 10.12 presented five ways of estimating a definite integral and formulas for bounds on their errors. Now we will obtain these formulas. The method rests on Sec. 15.1, where higher derivatives were shown to control the growth of a function.

We will consider only the error associated with the trapezoidal method. Exercises will outline similar arguments for the other methods.

THEOREM 1   (*Error in trapezoidal method*).   Let f be a function with a continuous second derivative throughout $[a, b]$. Let n be a positive integer. Divide the interval $[a, b]$ into n sections of equal length $h = (b - a)/n$, with $x_0 = a$, $x_1 = a + h$, $x_2 = a + 2h$, ..., $x_n = b$.   Then

$$\text{Error} = \int_a^b f(x)\,dx - \sum_{i=1}^{n} \frac{f(x_i) + f(x_{i-1})}{2} h = -\frac{f^{(2)}(X)(b - a)}{12} h^2$$

for some number X in $[a, b]$.

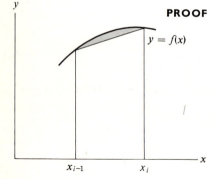

PROOF   Consider the typical section $[x_{i-1}, x_i]$ used in forming the estimate. We will examine the difference between

$$\int_{x_{i-1}}^{x_i} f(x)\,dx \qquad \text{and the estimate} \qquad \frac{f(x_i) + f(x_{i-1})}{2} h,$$

indicated by the shaded region in the diagram. Then we will add up the n differences to obtain the error.

For the sake of simplicity move the coordinate system along the x axis so that its origin is at the left endpoint of this typical interval. The right endpoint then has the coordinate h.

Let

$$E(h) = \int_0^h f(x)\,dx - \frac{f(0) + f(h)}{2} h,$$

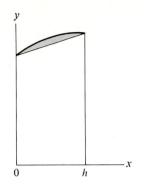

the error of the trapezoidal estimate associated with the typical small section. In order to estimate $E(h)$, introduce a function $E(t)$ defined for $0 \le t \le h$ by

$$E(t) = \int_0^t f(x)\,dx - \frac{f(0) + f(t)}{2}\,t. \tag{1}$$

Note that

$$E(0) = \int_0^0 f(x)\,dx - \frac{f(0) + f(0)}{2} \cdot 0 = 0.$$

We will obtain information about $E(h)$ by controlling the growth of the function $E(t)$ with the methods of Sec. 15.1.

Differentiation of (1) yields

$$E^{(1)}(t) = \left[\int_0^t f(x)\,dx \right]' - \frac{f(0) + f(t)}{2} - \frac{t[f(0) + f(t)]'}{2}$$

$$= f(t) - \frac{f(0)}{2} - \frac{f(t)}{2} - \frac{t f^{(1)}(t)}{2}$$

hence

$$E^{(1)}(t) = \frac{f(t)}{2} - \frac{f(0)}{2} - \frac{t f^{(1)}(t)}{2}. \tag{2}$$

Note that

$$E^{(1)}(0) = \frac{f(0)}{2} - \frac{f(0)}{2} - \frac{0\,f^{(1)}(0)}{2} = 0.$$

Differentiation of (2) yields

$$E^{(2)}(t) = \frac{f^{(1)}(t)}{2} - \frac{t f^{(2)}(t)}{2} - \frac{f^{(1)}(t)}{2}$$

or

$$E^{(2)}(t) = -\frac{t f^{(2)}(t)}{2}. \tag{3}$$

Let M_2 be the maximum value of $f^{(2)}(x)$ and let m_2 be the minimum value of $f^{(2)}(x)$ for x in $[a, b]$.

Then

$$m_2 \le f^{(2)}(t) \le M_2.$$

Keeping t in $[0, h]$, we have therefore, by (3),

$$-\frac{m_2 t}{2} \ge E^{(2)}(t) \ge -\frac{M_2 t}{2}. \tag{4}$$

Having Differentiated Twice, We Next Integrate Twice.

Integration of inequalities (4), together with the fact that $E^{(1)}(0) = 0$, gives

$$\frac{-m_2 t^2}{4} \ge E^{(1)}(t) \ge \frac{-M_2 t^2}{4}. \tag{5}$$

Integration of (5), together with the fact that $E(0) = 0$, gives

$$\frac{-m_2 t^3}{12} \ge E(t) \ge \frac{-M_2 t^3}{12}.$$

In particular, when $t = h$,

$$-\frac{m_2 h^3}{12} \geq E(h) \geq -\frac{M_2 h^3}{12}.$$

The error in the trapezoidal estimate, being the sum of the n individual errors, is bounded by the inequalities

$$-\frac{m_2 nh^3}{12} \geq \text{error} \geq -\frac{M_2 nh^3}{12}.$$

But $$nh = b - a.$$

Thus $$-\frac{m_2(b - a)h^2}{12} \geq \text{error} \geq -\frac{M_2(b - a)h^2}{12}.$$

Since $f^{(2)}(x)$ is continuous, the function

$$-\frac{f^{(2)}(x)(b - a)h^2}{12}$$

assumes every value between $-m_2(b - a)h^2/12$ and $-M_2(b - a)h^2/12$. Consequently, there is a number X in $[a, b]$ such that

$$\text{Error} = -\frac{f^{(2)}(X)(b - a)h^2}{12}.$$

This concludes the proof. ●

From Theorem 1 it follows immediately that, if $|f^{(2)}(x)| \leq M_2$, then the absolute value of the error in the trapezoidal method is at most

$$\frac{M_2(b - a)h^2}{12},$$

the bound given in Sec. 10.12.

Theorem 1 says much more. First of all, it shows that, if f is concave down $[f^{(2)}(x) < 0]$, then the error is positive; that is, the estimate is less than the integral. This is clear from a sketch of the graph of f if $f(x) \geq 0$. The trapezoids lie below the curve. More interesting, Theorem 1 under certain circumstances, will tell also *at least how large the error must be.* Example 1 illustrates this.

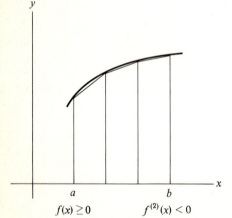

y

a b x

$f(x) \geq 0$ $f^{(2)}(x) < 0$

EXAMPLE 1 Let f be a function with a continuous second derivative throughout the interval $[a, b]$. Assume that $2 \leq f^{(2)}(x) \leq 3$ for all x in $[a, b]$. Find a lower and an upper bound on the error in the trapezoidal estimate when n trapezoids are used.

SOLUTION By Theorem 1,

$$\text{Error} = -\frac{f^{(2)}(X)(b - a)h^2}{12}$$

for some X in $[a, b]$.

Thus
$$-\frac{3(b-a)h^2}{12} \le \text{error} \le -\frac{2(b-a)h^2}{12},$$

or since $h = (b-a)/n$,

$$-\frac{3(b-a)^3}{12n^2} \le \text{error} \le -\frac{2(b-a)^3}{12n^2},$$

so that the absolute value of the error is at least

$$\frac{2(b-a)^3}{12n^2},$$

but at most
$$\frac{3(b-a)^3}{12n^2}.$$

(The trapezoidal approximation, since the function is concave upward, over-estimates the integral.) ●

EXAMPLE 2   If $\int_1^2 e^{x^2}\,dx$ is to be estimated by the trapezoidal method with an error at most 0.01, how large should n be?

SOLUTION   Let $f(x) = e^{x^2}$.   The error is equal to

$$-\frac{f^{(2)}(X)(b-a)h^2}{12} = -\frac{f^{(2)}(X)}{12n^2}$$

for some value of X in $[1, 2]$.

Now
$$f^{(1)}(x) = 2xe^{x^2}$$

and
$$f^{(2)}(x) = (4x^2 + 2)e^{x^2}.$$

Since $f^{(2)}(x)$ increases for $x \ge 0$, the largest value of $f^{(2)}(x)$ for x in $[1, 2]$ is $f^{(2)}(2) = 18e^4$.   Consequently,

$$|\text{error}| \le \frac{18e^4}{12n^2}$$

or
$$|\text{error}| \le \frac{3e^4}{2n^2}.$$

Now, $e^4 \doteq 55$.   To be assured that the error is less than 0.01, choose n such that

$$\frac{3 \cdot 55}{2n^2} \le 0.01$$

or
$$\frac{2n^2}{3 \cdot 55} \ge 100;$$

hence
$$n^2 \ge \frac{16{,}500}{2} = 8{,}250.$$

Consequently,
$$n \ge 91 \text{ suffices.} \quad ●$$

Exercises

The error in the left-point method is examined in this exercise.

1. (a) Review the left-point method of Sec. 10.12.
   (b) Show that one wishes to estimate

$$E(h) = \int_0^h f(x)\,dx - f(0)h.$$

   (c) Introduce the function

$$E(t) = \int_0^t f(x)\,dx - f(0)t.$$

   Show that $E(0) = 0$, $E^{(1)}(0) = 0$, and that

$$E^{(2)}(t) = f^{(1)}(t).$$

   (d) Let the maximum of $f^{(1)}(X)$ be M_1 and the minimum be m_1 for X in $[a, b]$. Deduce that

$$\frac{m_1 h^2}{2} \le E(h) \le \frac{M_1 h^2}{2}.$$

   (e) Show that the error of the left-point method equals

$$\frac{f^{(1)}(X)(b - a)h}{2}$$

   for some number X in $[a, b]$.
   (f) From (e) obtain the bound stated in Sec. 10.12.

The error in the midpoint method is discussed in the next exercise.

2. (a) Review the midpoint method of Sec. 10.12.
   (b) Show that one wishes to estimate

$$E(h) = \int_{-h/2}^{h/2} f(x)\,dx - f(0)h.$$

   (c) Introduce the function

$$E(t) = \int_{-t/2}^{t/2} f(x)\,dx - f(0)t.$$

   Show that $E(0) = 0$, $E^{(1)}(0) = 0$, and that

$$E^{(2)}(t) = \frac{1}{4}\left[f^{(1)}\left(\frac{t}{2}\right) - f^{(1)}\left(-\frac{t}{2}\right) \right].$$

   (d) Use the law of the mean to show that

$$E^{(2)}(t) = \frac{t}{4} f^{(2)}(T)$$

   for some number T between $-t/2$ and $t/2$.

(e) Let the maximum value of $f^{(2)}(x)$ be M_2, and the minimum value be m_2 for x in $[a, b]$. From (c) and (d) deduce that

$$\frac{m_2 h^3}{24} \le E(h) \le \frac{M_2 h^3}{24}.$$

   (f) Show that the error of the midpoint method equals

$$\frac{f^{(2)}(X)(b - a)h^2}{24}$$

   for some number X in $[a, b]$.
   (g) from (f) obtain the bound stated in Sec. 10.12.

   ∎

The next exercise concerns the error in Simpson's method.

3. Review Simpson's method in Sec. 10.12, in particular Exercise 32 of that section. Define

$$E(h) = \int_{-h}^h f(x)\,dx - \frac{h}{3}[f(-h) + 4f(0) + f(h)]$$

   and

$$E(t) = \int_{-t}^t f(x)\,dx - \frac{t}{3}[f(-t) + 4f(0) + f(t)].$$

   (a) Show that

$$E^{(1)}(t) = \frac{2}{3}[f(t) + f(-t)]$$

$$- \frac{4}{3}f(0) - \frac{t}{3}[f^{(1)}(t) - f^{(1)}(-t)].$$

   (b) Show that

$$E^{(2)}(t) = \frac{1}{3}[f^{(1)}(t) - f^{(1)}(-t)] - \frac{t}{3}[f^{(2)}(t) + f^{(2)}(-t)].$$

   (c) Show that

$$E^{(3)}(t) = \frac{-t}{3}[f^{(3)}(t) - f^{(3)}(-t)].$$

   (d) Show that

$$E^{(3)}(t) = \frac{-2t^2}{3} f^{(4)}(T) \quad \text{for some } T \text{ in } (-t, t).$$

   (e) Show that

$$E(0) = 0, \qquad E^{(1)}(0) = 0, \qquad \text{and } E^{(2)}(0) = 0.$$

   (f) Let the maximum of $f^{(4)}(x)$ be M_4 and the minimum be m_4 for x in $[a, b]$.

Deduce that $-\dfrac{M_4\,h^5}{90} \leq E(h) \leq -\dfrac{m_4\,h^5}{90}$.

(g) Recalling that $h = (b - a)/2n$, show that the error equals

$$-\frac{f^{(4)}(X)(b - a)h^4}{180}$$

for some number X in $[a, b]$.

(h) From (g) obtain the bound stated in Sec. 10.12.

■ ■

4. According to Exercise 2, the midpoint estimate for a function that is concave upward is too small. Show that this is true with the aid of a sketch of the graph of f.

This exercise concerns the error in the right-point method.

5. (a) Review the right-point method in Sec. 10.12.

(b) Show that one wishes to estimate

$$E(h) = \int_0^h f(x)\,dx - hf(h).$$

(c) Introduce the function

$$E(t) = \int_0^t f(x)\,dx - tf(t).$$

Show that

$$E(0) = 0 \qquad \text{and} \qquad E^{(1)}(t) = -tf^{(1)}(t).$$

(d) Let M_1 be the maximum of $f^{(1)}(x)$ and let m_1 be the minimum of $f^{(1)}(x)$ for x in $[a, b]$. Show that

$$-\frac{M_1 t^2}{2} \leq E(h) \leq -\frac{m_1 t^2}{2}.$$

(e) Show that there is a number X in $[a, b]$ such that

$$\text{Error} = -\frac{f^{(1)}(X)(b - a)h}{2}.$$

(f) From (e) obtain the bound stated in Sec. 10.12.

6. If $f^{(4)}(x)$ is positive for all x in $[a, b]$, does Simpson's approximation overestimate or underestimate $\int_a^b f(x)\,dx$?

7. Let $f(x) = e^{x^2}$. Find the maximum value of $\left|f^{(2)}(x)\right|$ for x in $[0, 1]$.

8. Let $f(x) = e^{x/2}$. Find the maximum value of $\left|f^{(2)}(x)\right|$ for x in $[0, 2]$.

9. At least how large is the error in estimating $\int_1^2 e^{x^2}\,dx$ with $h = 0.1$ in the case of

(a) the trapezoidal method?

(b) Simpson's method?

10. How large must $2n$ be taken if the error in using Simpson's method to estimate $\int_1^2 e^{x^2}\,dx$ is to be less than 0.01? Recall that $h = (b - a)/2n$.

15.5

The binomial theorem for any exponent

If n is a positive integer and x is any number, then

$$(1 + x)^n = 1 + nx + \frac{n(n - 1)}{1 \cdot 2}x^2 + \frac{n(n - 1)(n - 2)}{1 \cdot 2 \cdot 3}x^3 + \cdots + \frac{n(n - 1)\cdots 1}{1 \cdot 2 \cdots n}x^n, \quad (1)$$

as shown in Exercise 24 of Sec. 15.2. In fact, the right side of (1) is the Maclaurin series for $(1 + x)^n$; all powers from x^{n+1} on have coefficient 0. In this section we examine the Maclaurin series for $(1 + x)^r$, when r is not a positive integer or 0. It turns out that for $|x| < 1$ the function $(1 + x)^r$ is indeed represented by its associated Maclaurin series. However, the formula

$$R_n(x; 0) = \frac{f^{(n+1)}(X)x^{n+1}}{(n + 1)!}$$

is inadequate for showing that $\lim_{n \to \infty} R_n(x; 0) = 0$.

In this section the integral form for $R_n(x; 0)$ is developed. Exercise 13 uses it to show that, as $n \to \infty$, $R_n(x; 0) \to 0$. It is a generalization of the fundamental theorem of calculus, which asserts that

$$f(b) = f(a) + \int_a^b f^{(1)}(x) \, dx.$$

In order to emphasize this fact we will use the letter b instead of x in the statement of the theorem.

THEOREM I   (*Integral form of the remainder.*)   Assume that a function f has continuous derivatives up through order $n + 1$ in the interval $[a, b]$. Let $P_n(b; a)$ be the Taylor polynomial of degree n associated with f in powers of $b - a$. Let $R_n(b; a)$ be defined by

$$f(b) = P_n(b; a) + R_n(b; a).$$

Then $$R_n(b; a) = \frac{1}{n!} \int_a^b (b - x)^n f^{(n+1)}(x) \, dx.$$

PROOF   We give the proof for the case $n = 3$. The argument for any other n is similar.   We have

$$R_3(b; a) = f(b) - P_3(b; a)$$

$$= f(b) - \left[f(a) + f^{(1)}(a)(b - a) + \frac{f^{(2)}(a)(b - a)^2}{2!} + \frac{f^{(3)}(a)(b - a)^3}{3!} \right].$$

Now hold b fixed and let a vary.   So, for convenience, introduce the function

$$g(x) = f(b) - \left[f(x) + f^{(1)}(x)(b - x) + \frac{f^{(2)}(x)(b - x)^2}{2!} + \frac{f^{(3)}(x)(b - x)^3}{3!} \right] \quad (2)$$

Note that $g(x)$ has a very simple derivative:

$$g^{(1)}(x) = - \left[\overbrace{f^{(1)}(x) + f^{(1)}(x)(-1) + (b - x)f^{(2)}(x)} \right.$$

$$+ \frac{\overbrace{f^{(2)}(x)(-2)(b - x) + f^{(3)}(x)(b - x)^2}}{2!}$$

$$\left. + \frac{\overbrace{f^{(3)}(x)(-3)(b - x)^2 + f^{(4)}(x)(b - x)^3}}{3!} \right] \quad (3)$$

[The long braces show the derivatives of the individual terms in (2).]   After cancellation in (3), all that remains is

$$g^{(1)}(x) = - \frac{f^{(4)}(x)(b - x)^3}{3!}.$$

Observe also that

$$g(a) = R_3(b; a) \qquad \text{and that} \qquad g(b) = 0.$$

Thus
$$g(b) - g(a) = \int_a^b g^{(1)}(x)\, dx$$

$$= -\frac{1}{3!} \int_a^b f^{(4)}(x)(b - x)^3\, dx,$$

or
$$0 - R_3(b; a) = -\frac{1}{3!} \int_a^b f^{(4)}(x)(b - x)^3\, dx.$$

Thus the theorem is proved for $n = 3$. ●

In the usual notation for the remainder, $R_n(x; a)$, we have

$$R_n(x; a) = \frac{1}{n!} \int_a^x (x - t)^n f^{(n+1)}(t)\, dt.$$

EXAMPLE 1 Use the integral form of the remainder to show that e^x is represented by its Maclaurin series.

SOLUTION In this case $f(x) = e^x$ and $a = 0$. Note that $f^{(n+1)}(t) = e^t$ for all n. Thus

$$R_n(x; 0) = \frac{1}{n!} \int_0^x (x - t)^n e^t\, dt.$$

Consider $x > 0$. In this case, $e^t < e^x$ when t is in $[0, x]$. Thus

$$R_n(x; 0) \le \frac{1}{n!} \int_0^x (x - t)^n e^x\, dt$$

$$= \frac{e^x}{n!} \int_0^x (x - t)^n\, dt$$

$$= -\frac{e^x}{n!} \frac{(x - t)^{n+1}}{n + 1}\Bigg|_{t=0}^{t=x}$$

$$= \frac{e^x x^{n+1}}{(n + 1)!}.$$

Since $x^{n+1}/(n + 1)! \to 0$ as $n \to \infty$,

$$\lim_{n \to \infty} R_n(x; 0) = 0.$$

In other words,

$$e^x = 1 + x + \frac{x^2}{2!} + \frac{x^3}{3!} + \cdots + \frac{x^n}{n!} + \cdots \quad ●$$

Now let us consider the Maclaurin series for $f(x) = (1 + x)^r$, where r is *not* a positive integer or 0. The following table will help in computing $f^{(n)}(0)$.

n	$f^{(n)}(x)$	$f^{(n)}(0)$
0	$(1 + x)^r$	1
1	$r(1 + x)^{r-1}$	r
2	$r(r - 1)(1 + x)^{r-2}$	$r(r - 1)$
3	$r(r - 1)(r - 2)(1 + x)^{r-3}$	$r(r - 1)(r - 2)$
. . .	. . . .	. . . .
n	$r(r - 1) \cdots (r - n + 1)(1 + x)^{r-n}$	$r(r - 1)(r - 2) \cdots (r - n + 1)$
. . .	. . . .	. . . .

Consequently, the Maclaurin series associated with $(1 + x)^r$ is

$$1 + rx + \frac{r(r - 1)}{1 \cdot 2} x^2 + \frac{r(r - 1)(r - 2)}{1 \cdot 2 \cdot 3} x^3 + \cdots. \qquad (4)$$

Note that the series does not stop, for r is not a positive integer or 0. For which x does series (4) converge? If it does converge, does it represent $(1 + x)^r$?

Just to get a feeling for (4), consider the case $r = -1$. When $r = -1$, (4) becomes

$$1 + (-1)x + \frac{(-1)(-2)}{1 \cdot 2} x^2 + \frac{(-1)(-2)(-3)}{1 \cdot 2 \cdot 3} x^3 + \cdots$$

or

$$1 - x + x^2 - x^3 + \cdots.$$

This series converges for $|x| < 1$. Moreover it does represent the function $(1 + x)^r = (1 + x)^{-1}$, for it is a geometric series with first term 1 and ratio $-x$.

EXAMPLE 2   Show that series (4) converges when $|x| < 1$.

SOLUTION   For $x = 0$ the series clearly converges. So consider $0 < |x| < 1$. Let a_n be the term containing the power x^n. Then

$$a_n = \frac{r(r - 1)(r - 2) \cdots (r - n + 1)}{1 \cdot 2 \cdot 3 \cdots n} x^n,$$

and

$$a_{n+1} = \frac{r(r - 1)(r - 2) \cdots (r - n)}{1 \cdot 2 \cdot 3 \cdots (n + 1)} x^{n+1}.$$

Thus

$$\left| \frac{a_{n+1}}{a_n} \right| = \frac{\left| \dfrac{r(r - 1)(r - 2) \cdots (r - n)}{1 \cdot 2 \cdot 3 \cdots (n + 1)} x^{n+1} \right|}{\left| \dfrac{r(r - 1)(r - 2) \cdots (r - n + 1)}{1 \cdot 2 \cdot 3 \cdots n} x^n \right|}$$

$$= \left| \frac{r - n}{n + 1} x \right|.$$

Since r is fixed, $$\lim_{n \to \infty} \left| \frac{a_{n+1}}{a_n} \right| = |x|.$$

By the ratio test, series (4) converges.   Moreover, it diverges for $|x| > 1$. ●

Exercises 12 and 13 use the integral form of the remainder to show that the binomial series (4) actually converges to $(1 + x)^r$ for $|x| < 1$.   It may be valuable to discuss here in detail why Lagrange's form of $R_n(x; 0)$ is inadequate for showing that, for *all* x in $(-1, 1)$, $R_n(x; 0) \to 0$ as $n \to \infty$.

First of all, as Exercise 9 shows, Lagrange's form is adequate for $0 \le x < 1$: Exercise 10 shows that it is adequate for $-\frac{1}{2} < x < 0$.   But it runs into trouble for $x \le -\frac{1}{2}$.   To be specific, consider $x = -\frac{1}{2}$.   Lagrange's form of $R_n(x; 0)$ is

$$\frac{f^{(n+1)}(X)x^{n+1}}{(n+1)!} = \frac{r(r-1)\cdots(r-n)}{1 \cdot 2 \cdots (n+1)}(1+X)^{r-n-1}x^{n+1}$$

$$= \frac{r(r-1)\cdots(r-n)}{1 \cdot 2 \cdots (n+1)}(1+X)^r \left(\frac{x}{1+X}\right)^{n+1}. \tag{5}$$

Since $x = -\frac{1}{2}$, we have $$-\tfrac{1}{2} \le X \le 0;$$

recall that X depends on n.

The factor $(1 + X)^r$ gives no trouble since r is fixed and $\frac{1}{2} \le 1 + X \le 1$. $(1 + X)^r$ is bounded, independently of n.   However, $x/(1 + X)$ does give trouble.   There is no way to bound it below 1 uniformly for all n, since $x = -\frac{1}{2}$ and $1 + X$ may be arbitrarily close to $\frac{1}{2}$ or perhaps equal to $\frac{1}{2}$.

For $x = -0.4$, the Lagrange formula shows that $R_n(x; 0) \to 0$ as $n \to \infty$. For then,

$$X \ge -0.4.$$

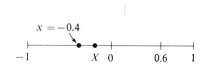

and therefore $$1 + X \ge 0.6.$$

Consequently $$\left| \frac{x}{1+X} \right| \le \frac{0.4}{0.6}$$

for all values of n.   By the reasoning in Example 2, with $x = 0.4/0.6$, and the nth-term test,

$$\lim_{n \to \infty} \left| \frac{r(r-1)\cdots(r-n)}{1 \cdot 2 \cdots (n+1)} \right| \left(\frac{0.4}{0.6}\right)^{n+1} = 0.$$

Since $|(1 + X)^r| \le 1^r$ for all n and $|x/(1 + X)| \le 0.4/0.6$, the product (5) approaches 0 as $n \to \infty$.

No mention has been made of the behavior of the Maclaurin series for $(1 + x)^r$ when $x = 1$ or -1.   When $x = 1$ the series diverges for $r \le -1$, but for $r > -1$ converges to $(1 + x)^r = 2^r$.   When $x = -1$, the series converges to $(1 + x)^r = 0$ when $r > 0$, but diverges when $r < 0$.   These assertions are usually justified in advanced calculus.

Exercises

1. Show that for $n = 0$ the integral form of $R^{(n)}(b; a)$ reduces to the fundamental theorem of calculus.
2. Prove the theorem concerning the integral form of $R^{(n)}(b; a)$ when $n = 2$.
3. Use the integral form of the remainder to show that the Maclaurin series associated with $\sin x$ converges to $\sin x$ for all x.
4. Use the integral form of the remainder to show that the Maclaurin series associated with $\ln(1 + x)$ converges to $\ln(1 + x)$ for all $|x| < 1$.
5. Write out the first five terms of the Maclaurin series for $\sqrt{1 + x} = (1 + x)^{1/2}$.
6. Write out the first five terms of the Maclaurin series for $1/\sqrt{1 + x} = (1 + x)^{-1/2}$.
7. (a) Write out the first five terms of the Maclaurin series for $1/(1 + x)^2 = (1 + x)^{-2}$.
   (b) What is the coefficient of x^n in the series in (a)?
8. (a) Use the five terms in Exercise 5 to estimate $\sqrt{1.5}$.
   (b) Discuss the error in the estimate in (a) with the aid of the integral form of the remainder.
9. Use the Lagrange form of $R^{(n)}(x; 0)$ to show that the binomial series associated with $(1 + x)^r$ converges to $(1 + x)^r$ for $0 < x < 1$.
10. Use the Lagrange form of $R^{(n)}(x; 0)$ to show that the binomial series associated with $(1 + x)^r$ converges to $(1 + x)^r$ for $-\frac{1}{2} < x < 0$.
11. Let f be a function with continuous first and second derivatives. Assume that $f(1) = 4, f'(1) = 0.3$, and $0.5 \le f^{(2)}(x) \le 0.6$ for x in $[1, 2]$. Use the integral form of the remainder to show that

$$4.55 \le f(2) \le 4.60.$$

■

In Exercises 12 and 13 it will be shown that the binomial series associated with $(1 + x)^r$ converges to $(1 + x)^r$ for $|x| < 1$.

12. (This exercise will be needed in Exercise 13.) For x a fixed number in $(-1, 1)$ define the function g by $g(t) = (x - t)/(1 + t)$.
   (a) Show that if $0 \le x < 1$ and $0 \le t \le x$, then $0 \le g(t) \le x$.
   (b) Show that if $-1 < x \le 0$ and $x \le t \le 0$, then $x \le g(t) \le 0$.
   (c) Combining (a) and (b), show that if $|x| < 1$ and t is between 0 and x, then

$$\left| \frac{x - t}{1 + t} \right| \le |x|.$$

13. This exercise outlines a proof of the binomial theorem for any exponent r, and $-1 < x < 1$. Begin by defining f by

$$f(x) = (1 + x)^r.$$

   (a) Show that the coefficient of x^n in the Maclaurin series associated with f is

$$\frac{r(r - 1)(r - 2) \cdots (r - n + 1)}{n!}.$$

   (b) Using the integral form of the remainder, show that

$$R_n(x; 0) = \frac{r(r - 1)(r - 2) \cdots (r - n)}{n!}$$
$$\times \int_0^x \left(\frac{x - t}{1 + t} \right)^n (1 + t)^{r-1} \, dt.$$

   (c) With the aid of Exercise 12, show that for $r > 1$,

$$|R_n(x; 0)| \le \frac{|r(r - 1)(r - 2) \cdots (r - n)|}{n!} |x|^{n+1} 2^{r-1}.$$

   (d) Show that, for $r > 1$, $R_n(x; 0) \to 0$ as $n \to \infty$.
   (e) Show that for $r < 1$, $R_n(x; 0) \to 0$ as $n \to \infty$.

14. Use the integral form of the remainder to show that the Maclaurin series associated with $\ln(1 + x)$ converges to $\ln(1 + x)$ for $x = 1$.

■ ■

15. (a) Write out the first four nonzero terms of the Maclaurin series associated with $\sqrt{1 - x^3}$. *Hint:* Use the binomial theorem.
   (b) Use (a) to estimate $\int_0^{1/2} \sqrt{1 - x^3} \, dx$.
16. Show that if r is a positive integer, then

$$(1 - x)^{-r-1} = \sum_{n=0}^{\infty} \binom{n + r}{r} x^n \qquad |x| < 1.$$

17. (a) From Exercise 16 deduce that

$$2^{r+1} = \sum_{n=0}^{\infty} \binom{n + r}{r} 2^{-n}.$$

   (b) Write out the first five terms of the sum on the right side of the equation in (a).

15.6

Taylor's series for $f(x, y)$

The higher partial derivatives of $z = f(x, y)$ were introduced in Sec. 12.4.   Recall that $\partial \left(\dfrac{\partial z}{\partial x} \right) \Big/ \partial y$ is denoted $\partial^2 z / \partial y \, \partial x$, $\partial^2 f / \partial y \, \partial x$, or z_{xy}.   There are four partial derivatives of the second order:

$$z_{xx}, \qquad z_{xy}, \qquad z_{yx}, \qquad z_{yy}.$$

If they are continuous, z_{xy} equals z_{yx}.   Each of these four partial derivatives may be differentiated with respect to x or with respect to y.   Thus there are eight possible partial derivatives of order 3.   For instance, two of them are

$$\frac{\partial(z_{xx})}{\partial x} \qquad \text{and} \qquad \frac{\partial(z_{xx})}{\partial y},$$

which will be denoted $\qquad z_{xxx} \qquad$ and $\qquad z_{xxy}$.

The eight are:

$$z_{xxx}, \qquad z_{xxy}, \qquad z_{xyx}, \qquad z_{xyy}, \qquad z_{yxx}, \qquad z_{yxy}, \qquad z_{yyx}, \qquad z_{yyy}.$$

If they are continuous, however, many of them are equal.   For instance,

$$z_{xxy} = z_{xyx}$$

for $\qquad\qquad z_{xxy} = (z_x)_{xy} \qquad$ and $\qquad z_{xyx} = (z_x)_{yx}$.

In general, the order of differentiation does not affect the result; all differentiations with respect to x may be done first, and afterward the differentiations with respect to y.   Thus

$$z_{xyy} = z_{yxy} = z_{yyx},$$

or in the ∂ notation, $\qquad \dfrac{\partial^3 z}{\partial y^2 \, \partial x} = \dfrac{\partial^3 z}{\partial y \, \partial x \, \partial y} = \dfrac{\partial^3 z}{\partial x \, \partial y^2}.$

Similar statements and notations hold for partial derivatives of higher orders.

Thus $\qquad\qquad z_{xyxyy} = z_{xxyyy} = \dfrac{\partial^5 z}{\partial y^3 \, \partial x^2}.$

EXAMPLE I   Compute the partial derivatives of $x^4 y^7$ up through order 3.

SOLUTION   To begin: $\qquad\qquad z_x = 4x^3 y^7 \qquad$ and $\qquad z_y = 7x^4 y^6$.

Then $z_{xx} = 12x^2 y^7$, $z_{xy} = z_{yx} = 28x^3 y^6$, $z_{yy} = 42x^4 y^5$.   The third-order partial derivatives are

$$z_{xxx} = 24xy^7;$$
$$z_{xxy} = z_{xyx} = z_{yxx} = 84x^2 y^6;$$

$$z_{xyy} = z_{yxy} = z_{yyx} = 168x^3y^5;$$
$$z_{yyy} = 210x^4y^4.$$

Note that on account of duplication there are in practice only four partial derivatives of order 3. Similarly, there are in practice only five different partial derivatives of order 4, and $n + 1$ different partial derivatives of order n. ●

Just as a function of a single variable $f(x)$ may be expressed as a power series in $x - a$, so may a function of two variables $f(x, y)$ be expressed as a series involving powers of $x - a$ and $y - b$. For instance, consider $f(x, y) = \cos(x + y)$, $a = 0$, and $b = 0$. Then

$$f(x, y) = \cos(x + y) = 1 - \frac{(x + y)^2}{2!} + \frac{(x + y)^4}{4!} - \cdots$$

$$= 1 - \frac{x^2 + 2xy + y^2}{2!} + \frac{x^4 + 4x^3y + 6x^2y^2 + 4xy^3 + y^4}{4!} - \cdots.$$

Usually the series for $f(x, y)$ can be obtained by manipulation of the series for a function of one variable. Theorem 1 of this section shows the general relation between the series for $f(x, y)$ and the partial derivatives of $f(x, y)$. It is primarily of theoretical importance and yields, for instance, the test for a local maximum or minimum in terms of second-order partial derivatives, expressed in Theorem 2.

Let f be a function of x and y that possesses partial derivatives of all orders. Let a, b, h, and k be fixed numbers. Define a function g as follows:

$$g(t) = f(a + th, b + tk).$$

The Taylor's series for f can be obtained from that for g by expressing the derivatives $g'(0)$, $g^{(2)}(0)$, $g^{(3)}(0)$, ... in terms of partial derivatives of f.

To compute $g'(t)$, observe that g is a composite function:

$$g(t) = f(x, y) \qquad \text{where } x = a + th, \ y = b + tk. \tag{1}$$

See the diagram on page 636. By the chain rule, Theorem 1 of Sec. 12.6,

$$g'(t) = \frac{\partial f}{\partial x}\frac{dx}{dt} + \frac{\partial f}{\partial y}\frac{dy}{dt}.$$

Now, by (1)

$$\frac{dx}{dt} = h \qquad \text{and} \qquad \frac{dy}{dt} = k.$$

Thus
$$g'(t) = f_x \cdot h + f_y \cdot k, \tag{2}$$

where f_x and f_y are evaluated at $(a + th, b + tk)$. In particular,

$$g'(0) = f_x(a, b)h + f_y(a, b)k. \tag{3}$$

Next, express $g^{(2)}(t)$ in terms of partial derivatives of the function f. To do so, differentiate (2) with respect to t:

$$g^{(2)}(t) = \frac{d[g'(t)]}{dt} = \frac{d(f_x h + f_y k)}{dt}$$

$$= \frac{\partial(f_x h + f_y k)}{\partial x}\frac{dx}{dt} + \frac{\partial(f_x h + f_y k)}{\partial y}\frac{dy}{dt}$$

$$= \frac{\partial(f_x h + f_y k)}{\partial x} h + \frac{\partial(f_x h + f_y k)}{\partial y} k$$

$$= (f_{xx} h + f_{yx} k)h + (f_{xy} h + f_{yy} k)k.$$

Hence
$$g^{(2)}(t) = f_{xx} h^2 + 2f_{xy} hk + f_{yy} k^2, \tag{4}$$

where all the partial derivatives are evaluated at $(a + th, b + tk)$. Thus

$$g^{(2)}(0) = f_{xx}(a, b)h^2 + 2f_{xy}(a, b)hk + f_{yy}(a, b)k^2. \tag{5}$$

Notice the similarity of the right side of (5) to the binomial expansion,

$$(c + d)^2 = c^2 + 2cd + d^2,$$

in the coefficients, the powers of h and k, and the subscripts. To make use of this similarity, introduce the expression

$$(h\, \partial_x + k\, \partial_y)^2 f,$$

where $(h\, \partial_x + k\, \partial_y)^2$ is treated formally like an algebraic product: for instance, $(\partial_x \partial_x)f$ is interpreted as f_{xx}.

Thus (5) may be written in this shorthand as

$$g^{(2)}(0) = (h\, \partial_x + k\, \partial_y)^2 f \bigg|_{(a,\, b)}. \tag{6}$$

Differentiating (4) with respect to t sufficiently often, and then setting $t = 0$, we can show similarly that

$$g^{(n)}(0) = (h\, \partial_x + k\, \partial_y)^n f \bigg|_{(a,\, b)} \tag{7}$$

for $n = 1, 2, 3, \ldots$.

THEOREM I (*Taylor's series for a function of two variables*). Let f have continuous partial derivatives of all orders up to and including $n + 1$, at and near the point (a, b). If $(x, y) = (a + h, b + k)$ is sufficiently near (a, b), then

$$f(x, y) = f(a, b) + (h\, \partial_x + k\, \partial_y)f \bigg|_{(a,\, b)} + \frac{(h\, \partial_x + k\, \partial_y)^2}{2} f \bigg|_{(a,\, b)}$$

$$+ \cdots + \frac{(h\, \partial_x + k\, \partial_y)^n}{n!} f \bigg|_{(a,\, b)} + \frac{(h\, \partial_x + k\, \partial_y)^{n+1}}{(n+1)!} f \bigg|_{(X,\, Y)}$$

where (X, Y) is some point on the line segment joining (a, b) and (x, y).

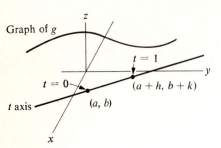

Graph of g

PROOF   This theorem follows from Theorem 2 of Sec. 15.2.   Introduce a function g, defined as follows:

$$g(t) = f(a + th, b + tk).$$

Observe that

$$g(0) = f(a, b) \tag{8}$$

and

$$g(1) = f(a + h, b + k) = f(x, y). \tag{9}$$

By Taylor's formula,

$$g(1) = g(0) + g'(0)(1) + \frac{g^{(2)}(0)}{2!} 1^2 + \cdots + \frac{g^{(n)}(0)}{n!} 1^n + \frac{g^{(n+1)}(T)}{(n+1)!} 1^{n+1} \tag{10}$$

for a suitable number T, $0 \le T \le 1$.   Combining (7) and (10) completes the proof. ●

As a consequence of the theorem, the coefficient of $h^r k^s$ in the Taylor's series for f is

$$\frac{1}{(r+s)!} \binom{r+s}{r} \frac{\partial^{r+s} f}{\partial x^r \, \partial y^s},$$

where the partial derivative is evaluated at (a, b), and $\binom{r+s}{r}$ denotes the binomial coefficient

$$\frac{(r+s)!}{r! \, s!}.$$

Observe that $(r + s)!$ in the numerator and denominator cancel.   Thus the coefficient of $h^r k^s$ is

$$\frac{1}{r! \, s!} \frac{\partial^{r+s} f}{\partial x^r \, \partial y^s}.$$

EXAMPLE 2   Use Theorem 1 of this section to express $f(x, y) = x^2 y$ in powers of $x - 1$ and $y - 2$.

SOLUTION   In this case $a = 1$, $b = 2$, $h = x - 1$, and $k = y - 2$.   To begin, compute the partial derivatives of f at $(1, 2)$.   We have

$$f_x = 2xy, \quad f_{xx} = 2y, \quad f_{xy} = 2x, \quad f_{xxy} = 2, \quad f_y = x^2, \quad f_{yy} = 0.$$

All higher partial derivatives of f are identically 0.   Thus $f(1, 2) = 2$, $f_x(1, 2) = (2)(1)(2) = 4$, and so on.   Therefore

$$f(x, y) = f(1 + h, 2 + k)$$

$$= f(1, 2) + [hf_x(1, 2) + kf_y(1, 2)]$$

$$+ \left[\frac{h^2 f_{xx}(1, 2) + 2hk f_{xy}(1, 2) + k^2 f_{yy}(1, 2)}{2!} \right]$$

$$+ \left[\frac{h^3 f_{xxx}(1, 2) + 3h^2 k f_{xxy}(1, 2) + 3hk^2 f_{xyy}(1, 2) + k^3 f_{yyy}(1, 2)}{3!} \right]$$

$$= 2 + 4h + k + \frac{4h^2 + 4hk + 0k^2}{2!} + \frac{6}{3!} h^2 k$$

or $\quad x^2 y = 2 + 4(x - 1) + (y - 2) + 2(x - 1)^2$

$$+ 2(x - 1)(y - 2) + (x - 1)^2(y - 2). \quad (11)$$

This can be checked by expanding the right side of the equation. ●

Theorem 1, with $n = 1$, provides the basis for the proof of the following test for a local minimum of $f(x, y)$, quoted from Sec. 12.8.

THEOREM 2   (*Test for a local minimum*).   Let f be a function that has continuous partial derivatives f_x, f_y, f_{xx}, f_{yy}, and $f_{xy} = f_{yx}$ at and near (a, b).   If

1. $f_x(a, b)$ and $f_y(a, b)$ are 0;
2. $f_{xx}(a, b)$ and $f_{yy}(a, b)$ are positive; and
3. $[f_{xy}(a, b)]^2 < f_{xx}(a, b) f_{yy}(a, b)$,

then f has a local minimum at (a, b).

PROOF   Theorem 1, with $n = 1$, asserts that

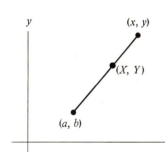

$$f(x, y) = f(a, b) + h f_x(a, b) + k f_y(a, b)$$
$$+ \tfrac{1}{2}[h^2 f_{xx}(X, Y) + 2hk f_{xy}(X, Y) + k^2 f_{yy}(X, Y)], \quad (12)$$

where (X, Y) is somewhere on the line segment joining (a, b) and $(x, y) = (a + h, b + k)$.

Now, $f_x(a, b) = 0$ and $f_y(a, b) = 0$.   Thus (12) reduces to

$$f(x, y) = f(a, b) + \tfrac{1}{2}[h^2 f_{xx}(X, Y) + 2hk f_{xy}(X, Y) + k^2 f_{yy}(X, Y)]. \quad (13)$$

Since f_{xx}, f_{yy}, and f_{xy} are continuous, when (x, y) is sufficiently close to (a, b), $f_{xx}(X, Y) > 0$, $f_{yy}(X, Y) > 0$, and $f_{xy}^2(X, Y) < f_{xx}(X, Y) f_{yy}(X, Y)$, from assumptions 2 and 3 in the statement of the theorem.   Now, let

$$A = f_{xx}(X, Y), \qquad B = f_{xy}(X, Y), \qquad \text{and} \qquad C = f_{yy}(X, Y).$$

We have $A > 0$, $B > 0$, and $B^2 < AC$.

In view of (13), all that remains is to show that

$$Ah^2 + 2Bhk + Ck^2 > 0. \quad (14)$$

The argument now follows that of Theorem 1 of Sec. 12.8 exactly.   Since A is positive, it suffices to show that $A(Ah^2 + 2Bhk + Ck^2)$ is positive.   This is accomplished by completing the square:

$$A^2 h^2 + 2ABhk + ACk^2 = (Ah + Bk)^2 + (AC - B^2)k^2.$$

The expression on the right, being the sum of a square and a positive multiple of a square, is never negative.   If it is 0, then

$$Ah + Bk = 0 \qquad \text{and} \qquad k = 0,$$

from which it follows that $h = 0$.   Thus, for $(h, k) \neq (0, 0)$ but sufficiently close to $(0, 0)$, (14) is established. Thus $f(x, y)$ has a local minimum at (a, b). ●

Theorem 4, Sec. 12.8, which concerns a local maximum, is proved similarly.

Exercises

In Exercises 1 to 4 compute all eight partial derivatives of the third order of the given function.

1. $x^5 y^7$ **2.** e^{2x+3y}

3. x/y **4.** $\sin(x^2 + y^3)$

5. Verify (11) by expanding the right side of the equation.

6. Using (11), compute the difference in the volumes of these two boxes: One has a square base of side 1 foot and height 2 feet; the other has a square base of side 1.1 feet and height 2.1 feet.

7. (a) Using partial derivatives, obtain the first four nonzero terms in the Taylor's series for e^{x+y^2} in powers of x and y.

    (b) Noticing that $e^{x+y^2} = e^x e^{y^2}$, and using a few terms of the Maclaurin series for e^x and for e^{y^2}, solve (a) again.

8. (a) Using partial derivatives, express $x^2 y^2$ as a polynomial in $x - 1$ and $y - 1$.

    (b) Verify your answer to (a) by expanding it.

9. The binomial theorem for $n = 4$ implies that $(x + y)^4 = x^4 + 4x^3 y + 6x^2 y^2 + 4xy^3 + y^4$. Expand $f(x, y) = (x + y)^4$ in a Taylor's series in x and y with the aid of partial derivatives, and check that the result is in agreement with the binomial theorem.

10. Verify that the expansion of $\sqrt{1 + x + y}$ begins $1 + x/2 + y/2 - \frac{1}{8}x^2 - \frac{1}{4}xy - \frac{1}{8}y^2 + \cdots$ (a) by using Theorem 1 of this section; (b) by using the first three terms of the expansion for $\sqrt{1 + t}$ and replacing t with $(x + y)$.

11. (a) Find the first three nonzero terms of the Taylor's series for $\sqrt[3]{e^x} + \sin y$ in powers of x and y.

    (b) With the aid of (a), estimate $\int_R f(P)\, dA$, where R is the triangle whose vertices are $(0, 0)$, $(1, 0)$, and $(1, 2)$ and where f is the function given in (a).

12. (a) Obtain the first three nonzero terms of a Taylor's series in powers of x and y for $f(x, y) = \sqrt{x^2 + y^2 + 1}$.

    (b) Use the result of (a) to estimate $\int_R f(P)\, dA$, where R is the square whose vertices are $(0, 0)$, $(0.1, 0)$, $(0, 0.1)$ and $(0.1, 0.1)$.

■

13. (a) Prove that, if all the partial derivatives of f are continuous, then

$$f_{xyxy} = f_{yxyx}.$$

    (b) Prove that each of the 16 partial derivatives of f of the fourth order is equal to one of these five:

$$\frac{\partial^4 f}{\partial x^4}, \quad \frac{\partial^4 f}{\partial y\, \partial x^3}, \quad \frac{\partial^4 f}{\partial y^2\, \partial x^2}, \quad \frac{\partial^4 f}{\partial y^3\, \partial x}, \quad \frac{\partial^4 f}{\partial y^4}.$$

14. What is the coefficient of $x^5 y^7$ in the Taylor's series expansion of the function f around the point $(a, b) = (0, 0)$?

15. Prove Theorem 4 of Sec. 12.8.

15.7

Summary

In this chapter the higher derivatives of a function were shown to give information about the function itself. In particular, if the first $n - 1$ derivatives of a function were all equal to 0 at a and the nth derivative bounded by some number M for x in some interval near a, then we were able to compare the value of the function at x to $M(x - a)^n/n!$. This idea (with n replaced by $n + 1$) was applied to the function

$$f(x) - \left(f(a) + f^{(1)}(a)(x - a) + \cdots + \frac{f^{(n)}(a)(x - a)^n}{n!} \right)$$

to show that, for some functions f:

$$f(x) = \sum_{n=0}^{\infty} \frac{f^{(n)}(a)}{n!}(x - a)^n.$$

The same idea was applied in Sec. 15.4 to obtain a bound for the error in the trapezoidal estimate of a definite integral, as well as the other estimates.

Power series were used in Sec. 15.3 to show that all solutions of the differential equation $d^2y/dt^2 = -k^2y$ must be of the form $c_1 \cos kt + c_2 \sin kt$. Definite integrals were estimated by the use of power series in Sec. 15.4.

In Sec. 15.5 the error $R_n(x; a)$ was expressed as an integral.

This in turn was used to obtain the power series for $(1 + x)^r$ when $|x| < 1$.

Functions of two variables were treated in Sec. 15.6. There the higher-order partial derivatives were used to obtain a power series for the function. As a particular application, the criterion in terms of second-order partial derivatives for a local minimum was derived.

KEY FACTS

If $f^{(j)}(a) = 0$, $0 \le j \le n - 1$, and $m \le f^{(n)}(x) \le M$, then $f(x)$ is between $m(x - a)^n/n!$ and $M(x - a)^n/n!$.
If $f^{(j)}(a) = 0$, $0 \le j \le n - 1$, then

$$f(x) = \frac{f^{(n)}(X)}{n!}(x - a)^n$$

for some X between a and x.
Associated with the function $f(x)$ and the number a is the Taylor's series

$$f(a) + f^{(1)}(a)(x - a) + \frac{f^{(2)}(a)(x - a)^2}{2!} + \cdots.$$

When $x = a$, it certainly represents $f(x)$, but it may not at other values of x. For instance, there is a very flat function $f(x)$ all of whose derivatives at 0 are 0, yet is 0 only at 0. Its Taylor's series at $a = 0$ is simply 0 and represents the function only at 0. (See Exercise 31 of Sec. 15.2.)

The Taylor polynomial $P_n(x; a)$ is defined as

$$\sum_{j=0}^{n} \frac{f^{(j)}(a)(x - a)^j}{j!}.$$

The remainder $R_n(x; a)$ is defined by the equation

$$f(x) = P_n(x; a) + R_n(x; a).$$

The derivative form of the remainder is

$$R_n(x; a) = \frac{f^{(n+1)}(X)(x - a)^{n+1}}{(n + 1)!}$$

for some X between a and x.

The integral form of the remainder is

$$R_n(x; a) = \frac{1}{n!} \int_a^x f^{(n+1)}(t)(x - t)^n \, dt.$$

The error in using the trapezoidal estimate of $\int_a^b f(x) \, dx$ equals

$$\frac{-f^{(2)}(X)(b - a)h^2}{12}$$

for some X in $[a, b]$.
The error in using Simpson's estimate of $\int_a^b f(x) \, dx$ equals

$$\frac{-f^{(4)}(X)(b - a)h^4}{180}$$

for some X in $[a, b]$.
The binomial theorem,

$$(1 + x)^r = 1 + rx + \frac{r(r - 1)}{2!}x^2 + \frac{r(r - 1)(r - 2)}{3!}x^3 + \cdots,$$

is valid for any positive integer r and all x in which case the series is just a polynomial of degree r. It is also valid for all r and $|x| < 1$.
Higher-order partial derivatives (if continuous) do not depend on the order of differentiation.

For a function of two variables

$$f(x, y) = f(a, b) + (h \, \partial_x + k \, \partial_y)f \bigg|_{(a, b)} + \frac{(h \, \partial_x + k \, \partial_y)^2}{2} f \bigg|_{(a, b)}$$

$$+ \cdots + \frac{(h \, \partial_x + k \, \partial_y)^n}{n!} f \bigg|_{(a, b)} + \frac{(h \, \partial_x + k \, \partial_y)^{n+1}}{(n + 1)!} f \bigg|_{(X, Y)}$$

where (X, Y) is some point on the line segment joining (a, b) and (x, y).

VOCABULARY AND SYMBOLS

Taylor polynomial $P_n(x; a)$
remainder $R_n(x; a)$
Lagrange or derivative form of $R_n(x; a)$
Taylor's series
Maclaurin series
harmonic motion
integral form of $R_n(x; a)$

higher-order partial derivatives

$$\frac{\partial^3 z}{\partial y x^2} = \frac{\partial}{\partial y}\left(\frac{\partial^2 z}{\partial x^2}\right) = z_{xxy}, \text{ etc.}$$

binomial theorem
binomial coefficient $\binom{n}{j}$ or

$$C_j^n = \frac{n!}{j!(n - j)!} = \frac{n(n - 1) \cdots (n - j + 1)}{1 \cdot 2 \cdot 3 \cdots j}$$

Guide quiz for Chap. 15

1. Assume that f is a function with continuous derivatives $f^{(1)}(x)$, $f^{(2)}(x)$, and $f^{(3)}(x)$. Assume that $f(1) = 0$, $f^{(1)}(1) = 0$, and $f^{(2)}(1) = 0$. Assume that, for all x, we have $m \leq f^{(3)}(x) \leq M$. Show in detail that,
   (a) for $x > 1$,
   $$\frac{m(x-1)^3}{3!} \leq f(x) \leq \frac{M(x-1)^3}{3!};$$
   (b) for $x < 1$,
   $$\frac{m(x-1)^3}{3!} \geq f(x) \geq \frac{M(x-1)^3}{3!};$$
   (c) if M is the maximum value and m is the minimum value of $f(x)$, then
   $$f(x) = \frac{f^{(3)}(X)(x-1)^3}{3!}$$
   for some X between 1 and x.
2. (a) What is $P_2(x; 0)$ for the function e^x?
   (b) Show in detail why there is an X between 0 and x such that
   $$e^x = P_2(x; 0) + \frac{e^X x^3}{3!}.$$
3. Estimate $\int_{1/2}^1 x \cos \sqrt{x} \, dx$ by using the first three nonzero

terms of the Maclaurin series for $\cos \sqrt{x}$.
4. Express $\sin x$ as a Taylor's series in powers of $(x - \pi/3)$.
5. (a) What is the differential equation of harmonic motion?
   (b) What is its most general solution?
6. The integral $\int_0^{\pi/4} \sin x^2 \, dx$ is to be estimated with an error of less than 0.01
   (a) by the trapezoidal method; show that division of the interval into at least five sections ensures the given accuracy;
   (b) by using the Maclaurin series of $\sin x^2$; show that, if the first two nonzero terms are used, the error in estimating the integral is less than 0.01.
7. Let f be a function such that $f(0, 0) = 4$, $f_x(0, 0) = 2$, $f_y(0, 0) = -1$, $f_{xx}(0, 0) = 3$, $f_{yy}(0, 0) = 2$, $f_{xy}(0, 0) = -1$, $f_{xxx}(0, 0) = 5$, $f_{xxy}(0, 0) = 7$, $f_{xyy}(0, 0) = -2$, $f_{yyy}(0, 0) = 3$. Write out the first 10 terms of the Taylor's series in powers of x and y associated with f.
8. State the integral form for the remainder $R_n(x; a)$.
9. (a) Write out the binomial series associated with the function $(1 + x)^{-3}$.
   (b) For which values of x does it represent the function $(1 + x)^{-3}$?

Review exercises for Chap. 15

1. Estimate $\ln 3 = \int_1^3 (1/x) \, dx$, with $h = \frac{1}{2}$,
   (a) by the trapezoidal formula;
   (b) by Simpson's formula.
2. Combine Exercise 1 with the error formula for
   (a) the trapezoidal estimate to show that $1.034 \leq \ln 3 \leq 1.114$.
   (b) Simpson's formula to show that $1.083 \leq \ln 3 \leq 1.099$.
3. (a) Use the first two nonzero terms of the Taylor's series for e^{x^3} to estimate
   $$\int_0^1 e^{x^3} \, dx.$$
   (b) Find a bound on the error.
4. The first four Taylor polynomials for e^x in powers of x are 1, $1 + x$, $1 + x + x^2/2$, and $1 + x + x^2/2 + x^3/6$.
   (a) Graph them and the function e^x relative to the same axes.
   (b) When $|x|$ is large, are any of these Taylor polynomials a good approximation of e^x? Explain.
5. Express $x^2 + x + 2$ as a polynomial in powers of $x - 5$.

6. Estimate $f(0.6)$ if $f(0) = 2$, $f'(0) = 0.5$, and $2 \leq f^{(2)}(x) \leq 3$ for x in $[0, 0.7]$.
7. Estimate $f(1.2)$ if $f(0) = 0$, $f'(0) = 0$, and we have $0.2 \leq f^{(2)}(x) \leq 0.4$ for x in $[0, 1.2]$.
8. In the expansion of $(1 + x)^{20}$ what is the coefficient of (a) x^4, (b) x^{16}?
9. (a) Considering expansions around $a = 0$, define the symbol $R_3(x; 0)$.
   (b) What is the derivative form of $R_3(x; 0)$?
   (c) What is the integral form of $R_3(x; 0)$?
10. (a) State the binomial theorem for $(1 + x)^n$, where n is a positive integer.
    (b) Prove it.
11. Evaluate:
    (a) $D^3(x^4)$ at $x = 0$ and at $x = 2$;
    (b) $D^4(x^4)$ at $x = 0$ and at $x = 2$;
    (c) $D^5(x^4)$ at $x = 0$ and at $x = 2$.
12. Compute: (a) $D^{100}(2e^x)$, (b) $D^{100}(e^{2x})$.
13. (a) Using the formula for the binomial coefficient $\binom{n}{j}$, compute $\binom{7}{j}$ for $j = 0, 1, \ldots, 7$.
    (b) Using (a), write out the expansion of $(1 + x)^7$.

14. (c) Using (a), write out the expansion of $(a + b)^7$.
Use a Maclaurin's series to estimate $\sin 25°$ with an error less than 0.01 (use $\pi/180 \doteq 0.017$).

15. Let $f(x) = \sqrt{1 + x}$. Find the Taylor polynomial of degree 3 for f at $a = 0$.

16. State the test for a local maximum of $f(x, y)$ in terms of first and second partial derivatives.

17. Show why $z_{xyxy} = z_{xyyx}$.

In Exercises 18 to 22 use a Taylor polynomial of degree 3 in powers of x to estimate the indicated quantities.

18. $(1.1)^7$ **19.** $e^{0.3}$

20. $\cos 10°$ **21.** $\sin 18°$

22. $1/e$

Using the derivative formula for $R_n(x; 0)$, put a bound on the estimate in the following exercises.

23. Exercise 18 **24.** Exercise 19

25. Exercise 20 **26.** Exercise 21

27. Exercise 22

28. (a) Write out the first five terms in the Taylor's series in powers of $x - \pi/3$ for $\sin x$.
(b) Use three terms of the series in (a) to estimate $\sin 65°$.

29. Write out the first three nonzero terms in the Taylor's series at $a = \pi$ for $\sin x$.

30. Show that if a polynomial is constant over some interval, then it is constant everywhere.

31. (a) Show that $\sqrt{1 + x} = 1 + x/2 - x^2/8 - (x^3/16) \times (1 + X)^{5/2}$ for some X between 0 and x.
(b) Use (a) to show that $1.21825 < \sqrt{1.5} < 1.21825 + \frac{1}{128}$.

32. (See Exercise 31.) Using $1 + x/2 - x^2/8$ as an approximation of $\sqrt{1 + x}$, estimate $\sqrt{2}, \sqrt{1.1}, \sqrt{1.01}$.

33. (a) Using derivatives of f, write $f(x) = 4x^3 + 6x^2 - 5x + 2$ in powers of $x - 2$ and verify your result by expanding it.
(b) Use the expansion in (a) to estimate $f(2.1)$ and $f(1.9)$.

34. (a) Repeat Exercise 33(a), but expand $f(x)$ in powers of $x + 2$.
(b) Use (a) to estimate $f(-1.9)$.

35. Show that $\cos \sqrt{x}$ can be expressed as a power series, and use the series to estimate $\cos \sqrt{2}$ with an error less than 0.01.

36. Compute $\sin 20°$, with error less than 0.001, using a Taylor's series for $\sin x$.

37. A certain function f has derivatives of all orders. Also, $f(1) = f(2) = f(3) = f(4)$. Prove that there is a number X in $(1, 4)$ such that $f^{(3)}(X) = 0$.

38. If it is known that $\frac{1}{3} \le f(0) \le \frac{1}{2}$, $\frac{1}{3} \le f'(0) \le \frac{1}{2}$, and $\frac{1}{3} \le f^{(2)}(x) \le \frac{1}{2}$ for all x in $[0, 2]$, what can be said about $f(2)$?

39. If $f(0) = 0$, $f'(0) = 0$, and $f^{(2)}(x) \ge 1$ for all x in $[0, 3]$, what can be said about $f(3)$?

40. If $f(2) = 0$, $f'(2) = 0$, and $f^{(2)}(x) \le 1$ for all x in $[2, 3]$, what can be said about $f(3)$?

41. (a) From the identity $1/(1 + x) = 1 - x + x^2 - x^3 + x^4/(1 + x)$ deduce that $1/(1 + x^3) = 1 - x^3 + x^6 - x^9 + x^{12}/(1 + x^3)$.
(b) Use (a) to estimate $\int_0^1 1/(1 + x^3) \, dx$.
(c) Estimate the error in (b).

42. (a) Using the Maclaurin series for $1/(1 + x)$, obtain the Maclaurin series for $1/(1 + x^4)$.
(b) Use the first three terms of the series in (a) to estimate

$$\int_0^1 \frac{dx}{1 + x^4}.$$

43. Use the first four terms of the Maclaurin series for e^{-x} to obtain an estimate of

$$\int_0^1 \frac{1 - e^{-x}}{x} \, dx.$$

44. Find a bound on the error in Exercise 43.

45. Using the first three nonzero terms of the Maclaurin series for $\sin x$, estimate

$$\int_0^1 \sqrt{x} \sin x \, dx.$$

46. Put a bound on the error in Exercise 45.

47. Since the function $\sin x^3$ does not have an elementary antiderivative, the fundamental theorem of calculus is not helpful in evaluating $\int_0^{1/2} \sin x^3 \, dx$.
(a) What is the Taylor's series for $\sin x^3$ in powers of x?
(b) Using x^3 as an approximation of $\sin x^3$, estimate $\int_0^{1/2} \sin x^3 \, dx$.
(c) Discuss the error in (b). *Hint:* See (a).

48. (a) Estimate $\int_0^{1/2} (\sin x/\sqrt{x}) \, dx$ using the first three nonzero terms of the Maclaurin series for $\sin x$.
(b) Find a bound on the error of the estimate made in (a).

49. Compute:
(a) $D^5(x^{20})$, (b) $D^3((x - 1)^{50})$,
(c) $D^{83}((x - \pi)^{83})$.

50. Compute:
(a) $D^3((x - 1)^5)$ at $x = 1$,
(b) $D^4((x - 1)^5)$ at $x = 1$,
(c) $D^5((x - 1)^5)$ at $x = 1$,
(d) $D^6((x - 1)^5)$ at $x = 1$.

51. Obtain a formula for the polynomial P such that $P(1) = 2$, $P'(1) = 3$, $P^{(2)}(1) = 1$, $P^{(3)}(1) = -1$, and $P^{(j)}(1) = 0$ for $j > 3$.

52. Obtain a formula for the polynomial P such that $P(0) = 0, P'(0) = -1, P^{(2)}(0) = 0, P^{(3)}(0) = 2, P^{(4)}(0) = \frac{1}{2}$, and $P^{(j)}(0) = 0$ for $j > 4$.

53. Obtain a formula for the polynomial P such that $P(-1) = 1, \quad P'(-1) = 1, \quad P^{(2)}(-1) = 1, \quad P^{(3)}(-1) = 1, P^{(4)}(-1) = 1$, and $P^{(j)}(-1) = 0$ for $j > 4$.

54. Show that in the expansion of $(1 + x)^n$ the coefficients of x^j and x^{n-j} are equal.

55. Write out fully the expansion of
(a) $(1 + x)^3$,
(b) $(1 + x)^4$,
(c) $(1 + x)^5$.

56. What is the coefficient of x in the expansion of
(a) $(1 + x)^3$?
(b) $(1 + x)^5$?
(c) $(1 + x)^{10}$?

57. What is the coefficient of
(a) x^4 in the expansion of $(1 + x)^{10}$?
(b) x^6 in the expansion of $(1 + x)^{10}$?

58. Show that, if a is a fixed number and n and j are positive integers, then

$$D^j((x - a)^n) = \begin{cases} 0 & \text{if } j > n; \\ n! & \text{if } j = n; \\ \dfrac{n!}{(n-j)!}(x - a)^{n-j} & \text{if } 1 \leq j \leq n - 1. \end{cases}$$

■

59. (a) Using the fact that $2^n = (1 + 1)^n$, show that

$$2^n > 1 + n + \frac{n(n-1)}{2} \qquad \text{for} \quad n > 2.$$

(b) Using (a), show that

$$\lim_{n \to \infty} \frac{n}{2^n} = 0.$$

60. Let f and g be polynomials. Assume that $f^{(n)}(1) = g^{(n)}(1)$ for $n = 0, 1, 2, 3, \ldots$. Must f and g be the same function?

61. The Taylor's series for a certain function f begins

$$f(x, y) = 3 + 2x + 5y + 6x^2 - xy + 5y^2 + \cdots.$$

From this information, deduce the values of $f, f_x, f_y, f_{xx}, f_{xy}$, and f_{yy} at $(0, 0)$.

62. (a) Using the expansion of $(1 + x)^r$, obtain the expansion of $1/\sqrt{1 - x^2}$. (Use $r = -\frac{1}{2}$ and replace x by $-x^2$.)
(b) Show that the coefficient of x^{2n} in the expansion of $1/\sqrt{1 - x^2}$ is

$$\frac{1 \cdot 3 \cdot 5 \cdot 7 \cdots (2n - 1)}{2^n n!}.$$

(c) Show that the coefficient in (b) equals

$$\frac{(2n)!}{(n!)^2 4^n}.$$

63. Find $\lim_{p \to \infty} (p^3 + p^2 + 2p + 1)^{1/3} - p$.

64. Let $c_0, c_1, \ldots, c_n$ be given numbers.
(a) Find a polynomial $P(x)$ of degree n such that

$$P^{(j)}(0) = c_j \qquad 0 \leq j \leq n.$$

(b) Show that there is only one polynomial that satisfies the conditions in (a).

65. The motion of a particle satisfies the differential equation $d^2y/dt^2 = -(100\pi)^2 y$. At time $t = 0$ its velocity is 300π, and its position is given by $y = 3$.
(a) Find its position at any time.
(b) What is its maximum displacement from 0?
(c) How many cycles does it complete in unit time?

66. Show that when $f(x, y)$ is expressed as a series in powers of $x - a$ and $y - b$, the coefficient of $(x - a)^r(y - b)^s$ is

$$\frac{1}{r! s!} \frac{\partial^{r+s} f}{\partial x^r \partial y^s},$$

where the partial derivative is evaluated at (a, b).

67. Let f be a function with derivatives $f^{(1)}, f^{(2)}, \ldots, f^{(n)}$. Let $g = f + f^{(1)} + \cdots + f^{(n-1)}$. Prove that g is a solution of the differential equation

$$y - \frac{dy}{dx} = f(x) - f^{(n)}(x).$$

68. Solve the equation $e^x = 2x + 1$ for the positive root, using a Taylor polynomial of degree 3 as an approximation of e^x.

69. Solve the equation $e^x = 2x + 1$ for the positive root by applying Newton's method to the function $f(x) = e^x - 2x - 1$. Browse through the table of e^x to make an initial estimate.

70. Use the Maclaurin series for $\ln(1 + x)$ to estimate
(a) $\ln(2)$,
(b) $\ln(\frac{3}{2})$,
(c) $\ln(\frac{5}{2})$.
(d) From (a) and (b) obtain an estimate of $\ln 3$.
(e) From (a) and (c) obtain an estimate of $\ln 5$.

71. Obtain the Taylor polynomial of degree 5 at $a = 0$ associated with $f(x) = \sin^{-1} x$.

72. Obtain the Maclaurin series for $\sin^{-1} x$ from that for

$$\frac{1}{\sqrt{1 - x^2}},$$

which is discussed in Exercise 62.

73. It is proved in trigonometry that $\cos^3 x = (\cos 3x + 3\cos x)/4$. Which is easier to compute, $D^{20}(\cos^3 x)$ or $D^{20}[(\cos 3x + 3\cos x)/4]$? Why? Compute the easier of the two.

74. Applying the binomial theorem to $(x + \Delta x)^n$, show that $D(x^n) = nx^{n-1}$. (It is possible to prove the binomial theorem without using derivatives, so this need not be a circular proof.)

75. As the Maclaurin series for $\sin x$ indicates, $x - x^3/6$ is a good approximation to $\sin x$ when x is small (and when radian measure of angle is used).
    (a) If θ measures an angle in degrees, show that $\pi\theta/180 - (\pi\theta/180)^3/6$ is an approximation of $\sin\theta$.
    (b) Use (a) to estimate $\sin 10°$, taking 0.017 for $\pi/180$.

76. Use the formula in Exercise 75 to estimate $\sin 20°$, and compare your result with $\sin 20°$ as listed in a table of sines.

77. Carry out this alternative proof that $\lim_{n\to\infty} (x^x/n!) = 0$.
    (a) Let $F(x) = (1 + x + x^2/2! + \cdots + x^n/n!)e^{-x}$ and show that $F'(x) = -e^{-x}(x^n/n!)$.
    (b) Deduce that for $x \geq 0$, we have $F(x) \leq 1$.
    (c) From (b) conclude that $1 + x + \cdots + x^n/n! < e^x$ if $x > 0$, and therefore that $x^n/n! \to 0$ as $n \to \infty$.

78. Prove that if $D^5(f) = 0$, for all x, then f is a polynomial of degree at most 4.

79. A certain function f has $f(0) = 3$, $f^{(1)}(0) = 2$, $f^{(2)}(0) = 5$, $f^{(3)}(0) = \frac{1}{2}$, and $f^{(j)}(0) = 0$ if $j > 3$. Give an explicit formula for $f(x)$.

80. The higher derivatives are occasionally useful in computing an antiderivative. Verify that, if f is a polynomial, then
$$\int e^x f(x)\, dx = e^x[f(x) - f^{(1)}(x) + f^{(2)}(x) - f^{(3)}(x) + \cdots$$
$$+ (-1)^n f^{(n)}(x) + \cdots].$$

Note that the number of terms on the right is finite, since f is a polynomial.

81. Use Exercise 80 to compute:
    (a) $\int e^x x^5\, dx$,   (b) $\int e^x x^8\, dx$.

82. (See Exercise 81.) Let f be a polynomial. Develop a formula for $\int e^{-x} f(x)\, dx$ involving the higher derivatives of f.

83. (a) Show that $D[F'(x)\sin(\pi x) - \pi F(x)\cos(\pi x)] = [F''(x) + \pi^2 F(x)]\sin(\pi x)$.
    (b) Now let f be a polynomial of degree $2n$ (or less) and let
$$F(x) = \pi^{2n} f(x) - \pi^{2n-2} f^{(2)}(x)$$
$$+ \pi^{2n-4} f^{(4)}(x) - \cdots + \pi^0 f^{(2n)}(x).$$

Show that $F''(x) + \pi^2 F(x) = \pi^{2n+2} f(x)$.

(c) Use (a) and (b) to obtain a formula for $\int f(x)\sin(\pi x)\, dx$, where f is a polynomial.
(d) Apply the formula in the case $f(x) = x^4$.

84. Write the (valid) equation $e^x - D(e^x) = 0$ as $(1 - D) \times (e^x) = 0$. This suggests that a polynomial in D can operate on a function. For instance, $(2 + 3D - 4D^2) \times f(x)$ will be short for $2f(x) + 3D(f(x)) - 4D^2(f(x))$. Verify that
    (a) $(1 + D)e^{-x} = 0$;   (b) $(1 + D^2)(\sin x) = 0$;
    (c) $(1 - D^4)(\sin x) = 0$.

85. A number r is algebraic if it is a root of a nonzero polynomial with rational coefficients. Show that this definition is equivalent to the following: A number r is algebraic if it is a root of some function f (not identically zero), such that $D^n(f) = 0$ for some n and $f^{(j)}(0)$ is rational for all j.

86. (See Exercises 84 and 85.) Call a number r quasi-algebraic if it is a root of some function f (not identically zero), such that $f^{(j)}(0)$ is rational for all j, and there is some polynomial in D with integer coefficients $P(D)$ (not identically zero) such that $[P(D)]f = 0$.
    (a) Show that if r is algebraic, it is quasialgebraic.
    (b) Show that π is quasialgebraic.

87. Justify this statement, found in a biological monograph: Expanding the equation $a \cdot \ln(x + p) + b \cdot \ln(y + q) = M$, we obtain
$$a\left(\ln p + \frac{x}{p} - \frac{x^2}{2p^2} + \frac{x^3}{3p^3} - \cdots\right)$$
$$+ b\left(\ln q + \frac{y}{q} - \frac{y^2}{2q^2} + \frac{y^3}{3q^3} + \cdots\right) = M.$$

88. Justify the second sentence in this statement, quoted from a biological monograph: Hence the probability of extinction $1 - y$ will be given by $1 - y = e^{-(1+k)y}$. If k is small, y is approximately equal to $2k$.

89. At 1 P.M. a departing car is 10 miles away on a straight highway, and has a speed of 20 miles per hour. Its acceleration during the next half hour is kept within the range of 20 miles per hour per hour and -20 miles per hour per hour. Discuss the possible locations of the car at the end of this half-hour interval.

90. (a) If $f^1(0) = 2$ and $f^{(1)}(3) = 7$, what can be concluded about the value of $f^{(2)}(x)$ for some x?
    (b) If $f(0) = 0$, $f(2) = 3$, and $f(4) = 6$, what can be concluded about the value of $f^{(2)}(x)$ for some x?

91. If $f(0) = 0$, $f'(0) = 1$, and $f(1) = 5$, show that there is a number X in $[0, 1]$ such that $f^{(2)}(X) = 8$.

92. Estimate $\int_0^{1.2} (\sin x)/x\, dx$, using $h = 0.2$ and
    (a) the trapezoidal method;
    (b) Simpson's formula.

(c) Estimate $\int_0^{1.2} (\sin x)/x \, dx$ using the first three nonzero terms of the Taylor's series for $\sin x$.

93. Find:
   (a) $D^{100}(\tan^{-1} x)$ at $x = 0$;
   (b) $D^{101}(\tan^{-1} x)$ at $x = 0$.

94. Find:
   (a) $D^{99}(e^{x^3})$ at $x = 0$;
   (b) $D^{100}(e^{x^3})$ at $x = 0$;
   (c) $D^{101}(e^{x^3})$ at $x = 0$.

95. Prove that $\sin x$ cannot be written as a polynomial $P(x)$, even if we demand that $\sin x = P(x)$ only throughout a small interval $[a, b]$.

96. Define $f(x)$ to be x^6 if $x \geq 0$ and to be x^4 if $x \leq 0$ (see the accompanying graph).

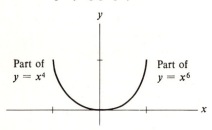

Part of $y = x^4$ Part of $y = x^6$

   (a) For which x is $f^{(4)}(x)$ defined?
   (b) Does $\lim_{x \to 0} f^{(4)}(x)$ exist?
   (c) Is $f^{(4)}(x)$ continuous at $x = 0$?
   (d) Does $\lim_{x \to 0} f^{(5)}(x)$ exist?
   (e) Is $f^{(5)}(x)$ defined when $x = 0$?

97. Use the binomial series for $(1 + 1/n)^n$, where n is a positive integer, to show that
   (a) $(1 + 1/n)^n$ increases as n increases;
   (b) $\left(1 + \dfrac{1}{n}\right)^n \leq \dfrac{1}{0!} + \dfrac{1}{1!} + \dfrac{1}{2!} + \cdots + \dfrac{1}{n!}$;
   (c) $(1 + 1/n)^n < 3$.

98. Assume that $e^x = a_0 + a_1 x + a_2 x^2 + \cdots + a_n x^n + \cdots$. Set $x = 0$ and get $e^0 = a_0$. Thus $a_0 = 1$. Differentiation yields $e^x = a_1 + 2a_2 x + 3a_3 x^2 + \cdots + na_n x^{n-1} + \cdots$; set $x = 0$, and get $a_1 = 1$. Differentiate again and obtain $a_2 = \frac{1}{2}$. Continuing, obtain $a_n = 1/n!$. Hence $e^x = \sum_{n=0}^{\infty} (1/n!)x^n$. The conclusion of the reasoning is correct, but there is a hole in the argument. What is it?

In Exercises 99 to 102 a proof that π^2 is irrational is outlined.

99. Define $f(x) = x^n(1 - x)^n/n!$ for any fixed positive integer n. Show that
   (a) $f(x) = f(1 - x)$;
   (b) $f'(x) = -f'(1 - x)$;

(c) $f^{(2)}(x) = f^{(2)}(1 - x)$;
(d) $f^{(j)}(x) = (-1)^j f^{(j)}(1 - x)$.

100. Let f be the function defined in Exercise 99.
   (a) Show that $f^{(j)}(0) = 0$ for $j < n$ and that $f^{(j)}(0)$ is an integer for $j \geq n$.
   (b) Combining (a) with Exercise 99, show that $f^{(j)}(1)$ is an integer.

101. From the assumption that $\pi^2 = a/b$, where a and b are positive integers, a contradiction will be obtained in this exercise and Exercise 102 by showing that $\pi a^n \int_0^1 f(x) \sin(\pi x) \, dx$ is a positive integer for any n [recall that $f(x)$ depends on n], but that for large n it is less than 1.
   (a) Show that if x is in $[0, 1]$, then $0 \leq f(x) \leq 1/n!$.
   (b) Show that $0 < \pi a^n \int_0^1 f(x) \sin(\pi x) \, dx < \pi a^n/n!$.
   (c) Show that $\pi a^n/n!$ is less than 1 if n is large.

102. (See Exercises 83 and 101.) Now it will be shown that $\pi a^n \int_0^1 f(x) \sin(\pi x) \, dx$ is an integer. This will contradict Exercise 101(b) and (c). Let F be defined as in Exercise 83, and let $G(x) = b^n F(x)$.
   (a) Show that $G(0)$ and $G(1)$ are integers. (See Exercise 100.)
   (b) Show that

$$D[G'(x) \sin(\pi x) - \pi G(x) \cos(\pi x)]$$
$$= b^n \pi^{2n+2} f(x) \sin(\pi x) = \pi^2 a^n f(x) \sin(\pi x).$$

   (c) Show that

$$\pi a^n \int_0^1 f(x) \sin(\pi x) \, dx = \left[\frac{G'(x) \sin(\pi x)}{\pi} \right.$$
$$\left. - G(x) \cos(\pi x) \right]_0^1 = G(1) + G(0).$$

   (d) From (c) and Exercise 101, conclude that π^2 is irrational.
   This proof, based on a method of I. Niven, is due to J. D. Dixon, π Is Not Algebraic, of Degree One or Two, *American Mathematical Monthly*, vol. 69, p. 636, 1962.

103. Which is the stronger statement: "π is irrational" or "π^2 is irrational"? Explain.

104. Complex numbers are of the form $a + bi$, where a and b are real, and $i^2 = -1$. Using complex numbers, we can exhibit the intimate relation between the exponential and trigonometric functions, a relation that cannot be expressed within the real numbers. The arguments are made rigorous in the theory of complex variables.
   (a) If z is a complex number, define e^z to be $1 + z + z^2/2! + z^3/3! + \cdots$.

Show that $\qquad e^{ix} = \cos x + i \sin x$

and $\qquad e^{-ix} = \cos x - i \sin x$

for any real number x.

(b) From (a) deduce that

$$\cos x = \frac{e^{ix} + e^{-ix}}{2}$$

and $\qquad \sin x = \dfrac{e^{ix} - e^{-ix}}{2i}.$

Trigonometry can therefore be based on the exponential function. Contrast the formulas in (b) with those for the hyperbolic cosine and sine.

105. Let $F_k(x) = \sum_{n=0}^{k} x^n/n!$. Prove that $F_k(x) = 0$ has exactly one real root if k is odd, and no real roots if k is even.

106. Let f be a function having continuous $f^{(1)}$, $f^{(2)}$, and $f^{(3)}$ for all x. Assume that $\lim_{x \to \infty} f(x) = 1$, and $\lim_{x \to \infty} f^{(3)}(x) = 0$. Prove that $\lim_{x \to \infty} f^{(1)}(x) = 0 = \lim_{x \to \infty} f^{(2)}(x)$. *Hint:* Express $f(a + 1)$ and $f(a - 1)$ in terms of derivatives of f at a.

107. Let f be defined for all x and have a continuous $f^{(1)}$ and $f^{(2)}$. Prove that if $|f(x)| \le 1$ and $|f^{(2)}(x)| \le 1$ for all x in $[0, 2]$, then $|f^{(1)}(x)| \le 2$ for all x in $[0, 2]$. *Hint:* Express both $f(0)$ and $f(2)$ in terms of derivatives of f at x.

108. Assume that f, $f^{(1)}$, and $f^{(2)}$ are continuous and that $f^{(2)}(a) \ne 0$. By the law of the mean, $f(a + h) = f(a) + hf'(a + \theta h)$ for some θ in $[0, 1]$.
(a) When h is small, why is θ unique?
(b) Prove that $\theta \to \frac{1}{2}$ as $h \to 0$.

109. Though $f'(a)$ is the limit of $[f(a + \Delta x) - f(a)]/\Delta x$, there is a better way to estimate $f'(a)$ than by that quotient. Assume that f, $f^{(1)}$, $f^{(2)}$, and $f^{(3)}$ are continuous. Show that

(a) $\qquad \dfrac{f(a + \Delta x) - f(a)}{\Delta x} = f'(a) + \dfrac{f^{(2)}(X_1)}{2} \Delta x$

for some X_1 between a and $a + \Delta x$, and

(b) $\qquad \dfrac{f(a + \Delta x) - f(a - \Delta x)}{2\Delta x} = f'(a) + \dfrac{[f^{(3)}(X_2)]}{6}(\Delta x)^2,$

where X_2 is in $[a - \Delta x, a + \Delta x]$. [Since the error in using the quotient in (b) involves $(\Delta x)^2$, while the error in using the standard quotient involves Δx, the quotient in (b) is more accurate when Δx is small.] Test this observation on the function $y = x^3$ at $a = 2$.

110. (a) Using the inequality $e^x > 1 + x$ for $x > 0$, prove this theorem: If $u_1, u_2, u_3, \ldots$ is a sequence of positive

numbers such that the sequence of sums u_1, $u_1 + u_2$, $u_1 + u_2 + u_3$, $\ldots$ has a limit, then the sequence of products $(1 + u_1)$, $(1 + u_1)(1 + u_2)$, $(1 + u_1) \times (1 + u_2)(1 + u_3)$, $\ldots$ also has a limit.
(b) Prove the converse.

111. Consider $\int_0^b xe^{-x}\, dx$, when b is a small positive number. Since e^{-x} is then close to $1 - x$, the definite integral behaves like $\int_0^b (x - x^2)\, dx = b^2/2 - b^3/3$, hence approximately like $b^2/2$. On the other hand, $\int_0^b xe^{-x}\, dx = 1 - e^{-b}(1 + b)$, and since e^{-b} is approximately $1 - b$, we have $1 - e^{-b}(1 + b)$ approximately equal to $1 - (1 - b)(1 + b) = b^2$. Hence $\int_0^b xe^{-x}\, dx$ behaves like b^2. Which is correct, $b^2/2$ or b^2? Find the error.

112. Prove that e is not rational. *Hint:* Start with the assumption that $e = m/n$, and show that

$$n!\left[e - \left(1 + 1 + \frac{1}{2!} + \cdots + \frac{1}{n!}\right)\right]$$

is both an integer and *not* an integer.

This exercise is used in Exercise 114.

113. Let Δf denote $f(x + 1) - f(x)$ and Df denote, as usual, the derivative of f. Just as we speak of $D^n f$, we can define $\Delta^n f$ by repeated applications of Δ to f. For instance,

$$\Delta^2 f = (\Delta f)(x + 1) - (\Delta f)(x)$$
$$= f(x + 2) - f(x + 1) - [f(x + 1) - f(x)].$$

(a) Show that $D\,\Delta f = \Delta Df$.
(b) Show that $(\Delta f)x = (Df)X_1$ for some number X_1, $x < X_1 < x + 1$.
(c) Show that for any positive integer k there is a number X_k, $x < X_k < x + k$, such that

$$(\Delta^k f)x = (D^k f)X_k.$$

(d) Let r be a positive number that is not an integer. Show that

$$\Delta^k x^r = r(r - 1) \cdots (r - k + 1)X_k^{r-k}$$

for some X_k, $x < X_k < x + k$.

114. Let r be a positive number such that n^r is an integer for all positive integers n. Show that r is a positive integer. The preceding exercise may be of aid.

Incidentally, it is known that, if p^r is an integer for three distinct primes p, then r is an integer. It is not known whether the assumption that p^r is an integer for two distinct primes forces r to be an integer.

16

THE MOMENT OF A FUNCTION

This chapter is devoted primarily to definite integrals that have the form $\int_a^b (x - k) f(x)\, dx$, where f is a function and k is a number. This type of integral appears in such varied topics as the volume of a solid of revolution, the work done in emptying a tank, the force against a dam, the centroid of a plane region, and the center of gravity of a rod.

16.1

Work

The work required to raise a weight of W pounds a distance D feet is defined to be $W \times D$ foot-pounds. An elevator that lifts a 150-pound person 100 feet thus accomplishes

150 × 100 foot-pounds,

or   15,000 foot-pounds of work.

When all parts of an object are lifted the same distance, the work is simply the product of two numbers. We now pose a problem in which different parts of an object are lifted different distances.

A water tank is built in the form of a cylinder of height h feet and base R, a plane region (not necessarily a circle). It is full of water (weighing 62.4 pounds per cubic foot). How much work is accomplished in pumping all the water out of the outlet situated a given distance above the tank?

Water at the bottom of the tank must be pumped farther than water at the top: The lower the water is in the tank, the more it has to be raised.

To treat this problem mathematically, introduce a vertical x axis with positive part below the origin. (In each particular problem choose the origin in such a way as to obtain a convenient description of R in terms of coordinates.) The

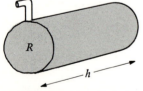

The water in a cylinder of base R and height h is to be pumped out

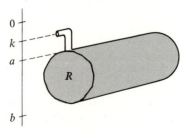

region R extends over the interval $[a, b]$, and the x coordinate of the outlet is k. Observe that water whose x coordinate is x is lifted a distance

$$x - k \text{ feet.}$$

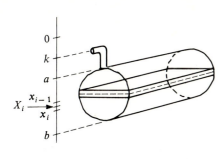

To find out how much work is accomplished in emptying this tank, consider a thin horizontal layer of water (all the water in the layer is raised about the same distance). The layer consists of all water whose x coordinate is between x_{i-1} and x_i. Let us imagine that all of this layer is at X_i.

Approximate it by a rectangular layer as shown in the diagram below. The

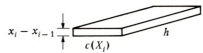

volume of the rectangular layer is approximately

$$h \cdot c(X_i)(x_i - x_{i-1}) \qquad \text{cubic feet,}$$

where $c(x)$ is the length of the cross section of R at the x coordinate x.

The work needed to raise this layer a distance $X_i - k$ is approximately

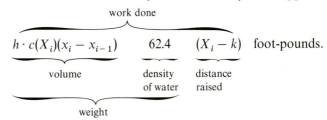

This is only an estimate, since not all the water in the layer is raised exactly the same distance. Thus the work is approximately

$$62.4h(X_i - k)c(X_i)(x_i - x_{i-1}) \quad \text{foot-pounds.}$$

Likewise, if we partition the interval $[a, b]$ with numbers $x_0 = a$, x_1, x_2, $\ldots$, $x_n = b$, where $x_0 < x_1 < \cdots < x_n$, then

$$62.4h \sum_{i=1}^{n} (X_i - k)c(X_i)(x_i - x_{i-1}) \quad \text{foot-pounds}$$

is an estimate of the total work accomplished in pumping out the tank.

As the layers are chosen thinner, these sums should provide more accurate estimates of the total work. For this reason a physicist defines the work required to empty the tank as

$$62.4h \int_a^b (x - k)c(x)\, dx \qquad \text{foot-pounds.} \tag{1}$$

This formula applies to any cylindrical tank.

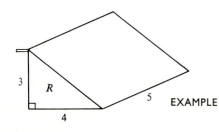

EXAMPLE How much work is required to empty the tank shown in the margin whose base R is a triangle and whose height is 5 feet?

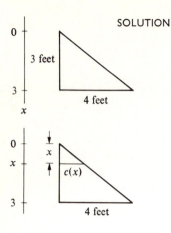

SOLUTION Choose the origin of the x axis to be level with the top vertex of R in order to make the formula for $c(x)$, the cross-sectional length, as simple as possible. The coordinate of the outlet is then 0. To find $c(x)$, use similar triangles. Inspection of the second diagram at the left shows that

$$\frac{c(x)}{4} = \frac{x}{3},$$

or

$$c(x) = \frac{4x}{3}.$$

Hence (1) shows that the work is

$$(62.4)(5) \int_0^3 (x - 0)\left(\frac{4x}{3}\right) dx \qquad \text{foot-pounds}$$

$$= (62.4)(5)(\tfrac{4}{3}) \int_0^3 x^2 \, dx \qquad \text{foot-pounds}$$

$$= (62.4)(5)(\tfrac{4}{3}) \left.\frac{x^3}{3}\right|_0^3 \qquad \text{foot-pounds}$$

$$= (62.4)(5)(\tfrac{4}{3})(9) \qquad \text{foot-pounds}$$

$$= (62.4)\, 60 \text{ foot pounds.} \bullet$$

A more general definition of work will be developed in Chap. 21.

Exercises

In Exercises 1 to 6, find the work required to pump the water out of the given tank.

1.

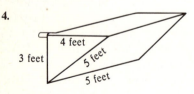

Tank with triangular base is full of water

2.

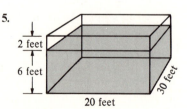

Tank is full of water

3.

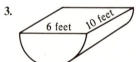

Tank with semicircular base is full of water

4.

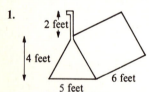

Tank is full of water

5.

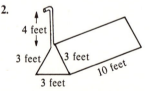

Surface of water is 2 feet below top of tank

6.

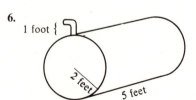

Tank with circular base is full of water

7. Find the work required to pump the water out of the cylindrical tank with an isosceles trapezoidal base, pictured below, through the outlet.

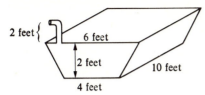

In Exercises 8 to 11, compute $\int_a^b (x - k)c(x)\, dx$ for the given data:

8. $a = 1$, $b = 3$, $k = 0$, and $c(x) = 1/x$.

9. $a = 1$, $b = 2$, $k = 1$, and $c(x) = \sqrt{x}$.

10. $a = 0$, $b = 1$, $k = -1$, and $c(x) = e^x$.

11. $a = 0$, $b = \pi$, $k = 0$, and $c(x) = \sin x$.

■

12. A tank, not necessarily cylindrical, is filled with water. An x axis is introduced with the positive part below the origin. The coordinate of the outlet is k. The plane perpendicular to the x axis, and passing through the point on the x axis whose coordinate is x, meets the water in a

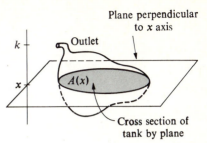

plane of cross-sectional area $A(x)$. Show that the work required to empty the tank is $(62.4) \int_a^b (x - k)A(x)\, dx$ foot-pounds if all dimensions are in feet, and $[a, b]$ is the interval on the axis corresponding to the tank.

13. Use the formula in Exercise 12 to compute the work required to pump the water in a hemispherical tank of radius 3 feet over its side. (The equator of the tank is horizontal and is the top of the tank.) See the figure above.

14. Use the formula in Exercise 12 to compute the work required to pump the water in a conical tank over its top. The axis of the tank is vertical and its top is a circle of radius 3 feet; its height is 5 feet.

16.2

Force against a dam

The deeper you swim, the greater is the pressure of the water. This pressure may be computed precisely, as follows: Imagine a flat piece of material of area A square feet placed horizontally in the water at a depth of h feet. It supports a column of water, whose volume is $A \cdot h$ cubic feet. The weight of this water is $Ah62.4$ pounds. This is the *force* of the water against the piece of material. The pressure is defined to be *force per unit area*, in this case

$$\text{Pressure} = \frac{Ah62.4}{A} = 62.4h \qquad \text{pounds per square foot.}$$

The pressure at a depth of h feet is therefore $62.4h$ pounds per square foot.

As the depth h increases, so does the pressure. Moreover, the pressure is the same in all directions (a submerged swimmer does not avoid the pressure against his eardrums by turning his head).

Let us see how to compute the force against a submerged vertical surface R. Because the pressure varies over R, a definite integral will be needed. To see what it is, construct a sum which is a good estimate of the total force. Begin by introducing a vertical x axis whose origin is arbitrary and whose positive part is directed downward. Call the x coordinate of the surface of the water k. Define a, b, and the cross-section function for R as usual, then partition the interval $[a, b]$. Consider the force against a typical narrow strip of R from

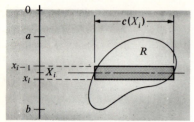

The horizontal dimension of the rectangle is $c(X_i)$

depth x_{i-1} to depth x_i approximated by a rectangle as in the diagram in the margin. The depth of all points in this strip is close to, say, $X_i - k$. The force against this typical strip is therefore approximately

$$62.4(X_i - k)c(X_i)(x_i - x_{i-1}) \qquad \text{pounds.}$$

Hence the sum

$$\sum_{i=1}^{n} 62.4(X_i - k)c(X_i)(x_i - x_{i-1})$$

is a good estimate of the force against R, when the mesh is small. For this reason the physicist defines the total force against R to be

$$62.4 \int_a^b (x - k)c(x)\, dx \qquad \text{pounds.} \tag{1}$$

EXAMPLE What is the force of the water against the isosceles triangular plate pictured in the diagram in the margin? Its top vertex is 3 feet below the surface of the water, and its horizontal base is 5 feet below the surface.

SOLUTION In order to get a simple description of the triangle, introduce the x axis in such a way that its origin is at the top vertex of the triangle. By similar triangles

$$\frac{c(x)}{x} = \frac{6}{2},$$

or

$$c(x) = 3x.$$

By (1), the force against the triangle is

$$62.4 \int_0^2 (x - (-3))(3x)\, dx = 62.4 \int_0^2 (x + 3)3x\, dx$$

$$= 62.4 \int_0^2 (3x^2 + 9x)\, dx$$

Now,

$$\int_0^2 (3x^2 + 9x)\, dx = \left(x^3 + \frac{9x^2}{2} \right)\Big|_0^2 = 26.$$

Hence the total force is $(62.4)(26)$ pounds. ●

Exercises

In Exercises 1 to 5 find the total force of the water against the region shown.

1.
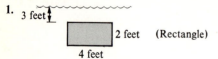

2.
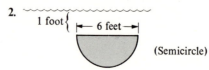

3.
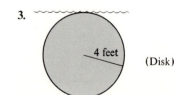

4.

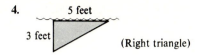

(Right triangle)

5. 1 foot {

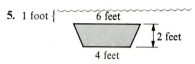

(Isosceles trapezoid)

In Exercises 6 to 8, compute $\int_a^b (x - k)c(x)\, dx$ for the following values.

6. $a = 1$, $b = 3$, $k = 0$, $c(x) = 1/x^2$.
7. $a = 0$, $b = 1$, $k = 1$, $c(x) = \sqrt{1 - x^2}$.
8. $a = 1$, $b = 3$, $k = 0$, $c(x) = e^{-x^2}$.
9. Who puts a greater pressure on the ground, a 5-ton elephant each of whose feet is a circle 8 inches in diameter or a 100-pound woman balancing on two stilts with circular cross section each of which has a diameter of $\frac{1}{2}$ inch?

■

10. Find the force of the water against the rectangle, shown below, inclined at an angle of 30° to the vertical and whose top base lies on the water surface. Use a definite integral (*a*) in which the interval of integration is vertical; (*b*) in which the interval of integration is inclined at 30° to the vertical; (*c*) in which the interval of integration is horizontal. In each case draw a neat picture that shows the interval [*a*, *b*] of integration and compute the integrand carefully.

The uppermost edge of the rectangle touches the surface of the water

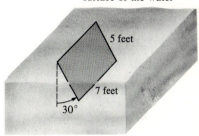

16.3

The moment of a function

In Sec. 16.1 the work required to empty a cylindrical tank is expressed as the definite integral

$$62.4h \int_a^b (x - k)c(x)\, dx \qquad \text{foot-pounds.}$$

If the origin of the *x* axis is at the outlet, this integral becomes simply

$$62.4h \int_a^b xc(x)\, dx.$$

In Sec. 16.2 the force against a submerged plane region is expressed as the definite integral

$$62.4 \int_a^b (x - k)c(x)\, dx \qquad \text{pounds.}$$

If the origin of the *x* axis is at the surface of the water, this integral becomes simply

$$62.4 \int_a^b xc(x)\, dx.$$

The integral $\qquad \int_a^b xc(x)\, dx$

also appeared in computing the volume of a solid of revolution. In Sec. 10.4 it was shown that the volume of the solid formed by revolving a region R about the y axis is

$$2\pi \int_a^b xc(x)\,dx.$$

$\int_a^b xc(x)\,dx$ is clearly important; it is given a name in the following definition.

DEFINITION *Moment of a function.* Let f be a function. The integral

$$\int_a^b xf(x)\,dx$$

is called the *moment* of the function f over the interval $[a, b]$.

Thus the work in emptying a tank, the force against a submerged plane region, and the volume of a solid of revolution all involve the moment of the cross-section function of the region R. The moment of a function is used also in statistics. For this reason the definition is phrased to cover this case also, and not just cross-section functions.

EXAMPLE I Compute the moment of the function x^2 over the interval $[2, 4]$.

SOLUTION By definition, this moment is $\displaystyle\int_2^4 x \cdot x^2\,dx,$

which is easily computed:

$$\int_2^4 x^3\,dx = \frac{x^4}{4}\bigg|_2^4 = \frac{4^4}{4} - \frac{2^4}{4} = 60. \ \bullet$$

The moment of the cross-section function of a region R,

$$\int_a^b xc(x)\,dx,$$

is also related to the centroid of the region.

Recall that the x coordinate of the centroid of the plane region R is given by the formula

$$\bar{x} = \frac{\int_R x\,dA}{\text{area of } R}$$

(see Sec. 13.4). Evaluating $\int_R x\,dA$ by a repeated integral in rectangular coordinates yields

$$\int_R x\,dA = \int_a^b \int_{y_1(x)}^{y_2(x)} x\,dy\,dx,$$

where R is described by

$$a \le x \le b, \qquad y_1(x) \le y \le y_2(x).$$

The inner integral,

$$\int_{y_1(x)}^{y_2(x)} x \, dy,$$

equals

$$x \int_{y_1(x)}^{y_2(x)} 1 \, dy.$$

This expression is easily evaluated:

$$x \int_{y_1(x)}^{y_2(x)} 1 \, dy = x[y_1(x) - y_2(x)] = xc(x).$$

Consequently,

$$\bar{x} = \frac{\int_a^b xc(x) \, dx}{\text{area of } R}. \tag{1}$$

Equation (1) is important; it implies that if $\bar{x}$ and the area of R are known, then the moment of the cross-section function can be found:

$$\int_a^b xc(x) \, dx = \bar{x} \cdot \text{area of } R. \tag{2}$$

As the next three theorems show, (2) is quite useful.

THEOREM I   The work required to pump the water out of a cylindrical tank of the type described in Sec. 16.1 is the product

$$W \cdot D \qquad \text{in foot-pounds,}$$

where W is the total weight of water in the tank and D is the distance the water at the centroid of R is lifted.

PROOF   Place the x axis with its positive part directed downward, and with its origin at the outlet.   The total work is then

$$62.4h \int_a^b xc(x) \, dx \qquad \text{foot-pounds.}$$

By (2) this equals

$$62.4h\bar{x} \cdot \text{area of } R \qquad \text{foot-pounds.}$$

But $62.4h \cdot \text{area of } R$

is the weight W of water in the tank.   Hence

$$\text{Work} = W \cdot \bar{x}.$$

Since $\bar{x}$ is the distance from the centroid to the outlet, this proves the theorem. ●

EXAMPLE 2   Use Theorem 1 to solve the example in Sec. 16.1.

SOLUTION   As shown in Exercise 2 of Sec. 13.4, the centroid of the triangle is 2 feet below the top vertex, which is at the level of the outlet.   The total weight of the water is

$$62.4 \frac{(3)(4)(5)}{2} \text{ pounds.}$$

Hence the work is

$$(62.4) \frac{(3)(4)(5)}{2} (2) = 62.4 \cdot 60 \text{ foot-pounds, as before.} \ \bullet$$

As Example 2 shows, the work is the same as if all the water were located at the depth of the centroid.   The centroid acts as an average point.

THEOREM 2   The force of water against a submerged vertical region R is the product

$$p \cdot \text{area of } R,$$

where p is the pressure at the centroid of R. $\bullet$

The proof is similar to that of Theorem 1 and is left to the reader.

EXAMPLE 3   Use Theorem 2 to find the force against a submerged circular region, if its radius is 3 feet and its center is at a depth of 5 feet.

SOLUTION   The centroid of a circle is its center.   The pressure at the centroid is therefore $(62.4)(5)$ pounds per square foot.   By Theorem 2, the force against the circle is

$$\underbrace{(62.4)(5)}_{\substack{\text{pressure at}\\\text{centroid}}} \cdot \underbrace{\pi \, 3^2}_{\substack{\text{area of}\\R}} \quad \text{pounds,}$$

which is $45\pi(62.4)$ pounds. $\bullet$

The relation between the centroid of R and the solid of revolution obtained by revolving R about a line is expressed in the next theorem, due to the fourth-century Greek mathematician Pappus.

THEOREM 3   (*Pappus' theorem*).   Let R be a region in the plane and L a line in the plane that either does not meet R or else just meets R at its border.   Then the volume of the solid formed by revolving R about L is equal to the product:

$$(\text{Distance centroid of } R \text{ is revolved}) \cdot (\text{area of } R).$$

PROOF   Introduce an xy-coordinate system in such a way that L is the y axis and R lies to the right of it.   The volume of the solid of revolution is

$$2\pi \int_a^b x c(x) \, dx.$$

By (2),

$$\int_a^b x c(x) \, dx = \bar{x} \cdot \text{area of } R.$$

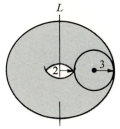

Hence the volume is $2\pi\bar{x} \cdot$ area of R.

Since $2\pi\bar{x}$ is the distance the centroid of R is revolved, this proves the theorem. ●

EXAMPLE 4 Use Pappus' theorem to find the volume of the "doughnut" formed by revolving a circle of radius 3 inches about a line 5 inches from the center.

SOLUTION In this case the area of R is $\pi 3^2$. The centroid of a circle is its center. Hence the distance it is revolved is $2\pi(5)$. The volume of the doughnut is

$$2\pi(5)\pi 3^2 = 90\pi^2 \text{ cubic inches.} \quad ●$$

Pappus' theorem can be used to find the centroid of a region R if the volume of the solid of revolution obtained by revolving R is known.

EXAMPLE 5 Find the centroid of a semicircle R of radius a.

SOLUTION By symmetry, the centroid lies somewhere on the radius that is perpendicular to the diameter of the semicircle. Let $\bar{x}$ be its distance from the diameter. When R is revolved about its diameter, it produces a sphere of radius a. The volume of a sphere of radius a is $4\pi a^3/3$.

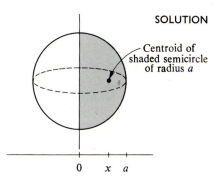

Centroid of shaded semicircle of radius a

By Pappus' theorem

$$2\pi\bar{x} \cdot \text{area of } R = \text{volume of sphere of radius } a,$$

or

$$2\pi\bar{x}\,\frac{\pi a^2}{2} = \frac{4}{3}\pi a^3.$$

Consequently

$$\pi^2 a^2 \bar{x} = \frac{4}{3}\pi a^3$$

and thus

$$\bar{x} = \frac{4\pi a^3}{3\pi^2 a^2} = \frac{4a}{3\pi}. \quad ●$$

As the last four examples show, the relation between the centroid of R and the moment of R, $\int_a^b xc(x)\,dx$, can be useful.

Exercises

In Exercises 1 to 5 compute the moment of the given function over the given interval.

1. x^{-2}; $[1, 2]$.

2. $\sqrt{1 + x^2}$; $[0, 1]$.

3. $\sin x^2$; $[0, \sqrt{\pi}]$.

4. $\dfrac{1}{1 + x^2}$; $[-1, 1]$.

5. $\dfrac{1}{1 + x^2}$; $[0, 1]$.

Use Theorem 1 to solve the following exercises.

6. Exercise 1, Sec. 16.1. **7.** Exercise 2, Sec. 16.1.

8. Exercise 3, Sec. 16.1. **9.** Exercise 4, Sec. 16.1.
10. Exercise 5, Sec. 16.1. **11.** Exercise 6, Sec. 16.1.

Use Theorem 2 to solve the following exercises.

12. The example in Sec. 16.2.
13. Exercise 1, Sec. 16.2. **14.** Exercise 2, Sec. 16.2.
15. Exercise 3, Sec. 16.2. **16.** Exercise 4, Sec. 16.2.

In Exercises 17 to 20 use Pappus' theorem to find the volume of the solid of revolution formed by revolving the given region about the y axis.

17. The triangle whose vertices are $(1, 0)$, $(3, -1)$ $(3, 2)$.

18. The semicircle whose center is at $(4, 0)$, whose radius is 3, and which lies to the right of the line $x = 4$.
19. The semicircle whose center is at $(4, 0)$, whose radius is 3, and which lies to the left of the line $x = 4$.
20. The rectangle with vertices $(4, 0)$, $(4, 2)$, $(7, 0)$, and $(7, 2)$.
21. Use Pappus' theorem to find the centroid of a right triangle whose vertices are $(0, 0)$, $(a, 0)$ $(0, b)$, where $a > 0$, $b > 0$.
    (a) Revolve it around the y axis to obtain $\bar{x}$.
    (b) Revolve it around the x axis to obtain $\bar{y}$.

    ■

22. Consider the region bounded by $y = e^{x^2}$, $x = 2$, and the axes.
    (a) Set up definite integrals for its area and the moment of the cross-section function.
    (b) Compute the integral in (a) that can be evaluated by the fundamental theorem of calculus.

In Exercises 23 to 25, use the formula

$$\bar{x} = \frac{\int_a^b xc(x)\, dx}{\text{area of } R}$$

to find $\bar{x}$ for the indicated region.
23. The region bounded by $y = x^2$, the x axis, and the line $x = 2$.
24. The region bounded by $y = \sin x$, $y = 0$, $x = 0$, and $x = \pi$.
25. The region bounded by $y = e^x$, $x = -1$, $x = 1$, and $y = 0$.
26. The integral $\int_a^b (x - k)f(x)\, dx$ is called the moment of f about k, over the interval $[a, b]$. Show that, if $f(x) = c(x)$, the cross-section function of a region R, then the moment of f about k is $(\bar{x} - k)A$, where $\bar{x}$ is the x coordinate of the centroid of R, and A is the area of R.
27. (See Exercise 26.) Show that the moment of the cross-section function about $\bar{x}$ is 0.

28. Using Pappus' theorem, find the volume of the solid formed by revolving the triangle whose vertices are $(5, 0)$, $(6, 1)$, $(6, -1)$ about (a) the y axis, (b) the line $x = 2$, (c) the line $y = 1$.
29. Prove Theorem 2.
30. Prove that Theorem 2 holds for any submerged flat surface, vertical or not.

Exercises 31 to 36 are related.
31. Consider a curve of length L in the xy plane. Assume that its mass is distributed over it uniformly. (Think of the curve as a wire.)
    (a) Show that it has a *center of gravity*, or *centroid*, that is, a point such that the curve balances on every line through that point.
    (b) Show that the y coordinate of the center of gravity is

    $$\bar{y} = \frac{\int_0^L y\, ds}{L}$$

    where s is arc length on the curve.
32. Use the formula in Exercise 31 to compute the y coordinate of the centroid of the semicircle $y = \sqrt{a^2 - x^2}$.
    (a) Use rectangular coordinates.
    (b) Use polar coordinates.
    (c) Explain without calculus why $\bar{y}$ is greater than $a/2$.
33. Prove the following analog of Theorem 3, Pappus' theorem: The area of the surface of revolution obtained by revolving about a line a curve situated on one side of the line is equal to the length of the curve times the distance the centroid of the curve is revolved. (The centroid of a curve is defined in Exercise 31.)
34. Using the theorem stated in Exercise 33, compute the surface area of the "doughnut" (torus) of Example 3.
35. Using the theorem stated in Exercise 33, compute the lateral surface area of a cone.
36. Using the theorem stated in Exercise 33, find the centroid of a semicircular arc.

16.4

Summary

Section 16.1 showed that the work required to pump out a cylindrical tank filled with water is

$$62.4h \int_a^b (x - k)c(x)\, dx \qquad \text{foot-pounds},$$

where k is the x coordinate of the outlet, $c(x)$ is the cross-sectional length corresponding to the x coordinate x, h is the length of the tank, and 62.4 is the weight of 1 cubic foot of water.

If the origin of the vertical x axis used is chosen to be at the level of the outlet, the work is simply

$$62.4h \int_a^b xc(x)\, dx \qquad \text{foot-pounds}.$$

In both cases the positive part of the x axis is downward.

Section 16.2 showed that the force against a submerged vertical flat region is $62.4 \int_a^b (x-k)c(x)\, dx$ pounds,

where k is the x coordinate of the surface of the water. If the origin of the vertical x axis is chosen to be at the level of the water's surface, the force is simply

$$62.4 \int_a^b xc(x)\, dx \qquad \text{pounds.}$$

Section 16.3 opened with the observation that $\int_a^b xc(x)\, dx$ appears in the formula for the volume of a solid of revolution. If the origin of the x axis is on the axis of revolution, then

$$\text{Volume} = 2\pi \int_a^b xc(x)\, dx.$$

For any function f, the moment of f over $[a, b]$ is defined to be

$$\int_a^b xf(x)\, dx,$$

if that integral exists.   The moment of the cross-section function $\int_a^b xc(x)\, dx$ appears in the formulas for the work accomplished in pumping water from a cylindrical tank, the force against a submerged flat object, and the volume of a solid of revolution. Moreover, it also appears in the formula for $\bar{x}$, the x coordinate of the centroid of a plane region:

$$\bar{x} = \frac{\int_a^b xc(x)\, dx}{\text{area of region}}.$$

This relation between $\bar{x}$ and the moment $\int_a^b xc(x)\, dx$ provides a shortcut for computing the work, or force, or volume discussed above:

> The work done pumping is the same as if all the water were located at the centroid of the end of the cylindrical tank.
> The force of the water against the submerged flat object is the same as if the water pressure were constant, and equal to its value at the depth of the centroid.
> The volume of a solid of revolution is the product of the distance the centroid of the region moves and the area of the region (Pappus' theorem).

Engineering students are warned that the shortcuts in the first two cases apply only to homogeneous fluids. If the density of the fluid varies, say, with depth, these shortcuts do not work.

There are other moments of a function, discussed in the review exercises of this chapter. The second moment $\int_a^b x^2 f(x)\, dx$ is used extensively in physics, engineering, and statistics.

VOCABULARY AND SYMBOLS

work	moment of a function $\int_a^b xf(x)\, dx$
pressure	Pappus' theorem
force exerted by fluid	

Guide quiz on Chap. 16

1. Use Pappus' theorem to find $\bar{x}$, the x coordinate of the centroid of the region in the left-hand figure below.

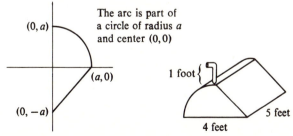

The arc is part of a circle of radius a and center $(0,0)$

$(0, a)$ $(a, 0)$ $(0, -a)$

2. Find the work required to pump water out of the tank shown in the right-hand figure above whose end is shaped like the figure in Exercise 1.

3. Find the force of the water against the region shaped like the figure in Exercise 1 and shown below.

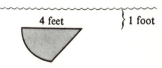

4 feet   1 foot

4. Solve either Exercise 2 or Exercise 3 without making use of the centroid.

5. Compute the moment of (a) $\cos x^2$ over $[0, 1]$; (b) $\cos x$ over $[0, \pi/2]$.

6. Develop the general formula for the work required to

empty a tank, placing the origin of the x axis at the level of the outlet.

7. Develop the general formula for the force against a

submerged vertical object, placing the origin of the x axis at the level of the surface of the water.

8. Prove Pappus' theorem.

Review exercises for Chap. 16

The exercises are supplementary, exploring various moments of a function. The nth moment M_n of a function f over the interval $[a, b]$ is defined to be $\int_a^b x^n f(x)\, dx$. In this chapter we studied the case $n = 1$. In physics the second moment $M_2 = \int_a^b x^2 f(x)\, dx$ is also of great importance. In statistics one sees, in addition, M_3 and M_4. The following exercises discuss these "higher" moments, M_2, M_3, ...

1. Compute M_0, M_1, and M_2 for $f(x) = \sqrt{x}$ and $[a, b] = [0, 1]$.

2. Compute M_0, M_1, and M_2 for $f(x) = e^x$ and $[a, b] = [-1, 1]$.

3. Show that, if $f(x)$ is the cross-section function of a plane region, then $\bar{x} = M_1/M_0$.

Exercises 4 to 7 are related.

4. The kinetic energy of an object of mass m, all of whose parts are moving with velocity v, is defined as $\frac{1}{2}mv^2$. This represents the work required to bring it from a state of rest to that velocity. Consider a rod spinning 100 times per second about one of its ends. Assume that the material of which it is composed has density $f(x)$ grams per centimeter at a distance of x centimeters from the fixed end. Let the length of the rod be b. Show that the kinetic energy of the rod is

$$\tfrac{1}{2}(200\pi)^2 \int_0^b x^2 f(x)\, dx \qquad \text{ergs.}$$

The second moment $M_2 = \int_0^b x^2 f(x)\, dx$ is called the *inertia* of the rod.

5. (See Exercise 4.) M_1/M_0 gives the center of gravity of the rod. M_2/M_1 also has a physical interpretation. If we suspend the rod from its left end and start it oscillating, it will do so at a certain frequency. This frequency is the same as that of a pendulum which consists of a point mass at the end of a "weightless" string of length M_2/M_1. Moreover, if we hold the stick at its left end and use it as a bat, then we should strike the ball at a point a distance M_2/M_1 from our hands. This point is called the *center of percussion*. Prove that the center of percussion lies to the right of the center of gravity. The Schwarz inequality will be needed.

6. A flat piece of steel of mass m and covering a region R in the plane is spun around the y axis 100 times per second. The region lies between $x = a$ and $x = b$ and has cross section $c(x)$ at x, and an area A. Explain in detail why its kinetic energy should be defined as $(\frac{1}{2})(200\pi)^2(m/A) \int_a^b x^2 c(x)\, dx$. This is a physical application of the second moment of the cross-section function.

7. A homogeneous steel triangle of mass m has vertices $(0, 0)$, $(2, 0)$, $(0, 3)$. Compute its kinetic energy if it is spun with an angular velocity w about (a) the x axis, (b) the y axis. Why would you expect the answer to (b) to be smaller than the answer to (a)?

8. Find a polynomial of degree 2 whose moments M_0, M_1, and M_2 over $[0, 1]$ are $M_0 = 1$, $M_1 = 1$, $M_2 = 1$.

■

9. The moment of inertia of a rod situated on the x axis about the point whose coordinate is c is defined as $\int_a^b (x - c)^2 f(x)\, dx$, where f is its density function. Assume now that the coordinate system is placed in such a way that the center of gravity of the rod is at $x = 0$.

(a) Prove that the moment of inertia of the rod about c is $M_2 + c^2 m$, where M_2 is the moment of inertia about the center of gravity and m is the mass of the rod.

(b) In view of (a), prove that it is easiest to spin the rod about its center of gravity. By virtue of (a), the center of gravity could be defined as "the point relative to which the moment of inertia is least."

10. Explain why the tendency of the water in a dam to rotate a submerged gate about a horizontal hinge on the surface of the water is proportional to $\int_0^b y^2 c(y)\, dy$, where $c(y)$ is the length of the horizontal cross section of the gate at depth y and b is the height of the gate. Note that this is a second moment.

■ ■

11. Let f be a continuous function defined on $[a, b]$.

(a) Prove that if $\int_a^b f(x)\,dx = 0$, then f has at least one root in $[a, b]$.

(b) Prove that if $\int_a^b f(x)\,dx = 0 = \int_a^b xf(x)\,dx$, then f has at least two roots in $[a, b]$.

(c) Prove that if $\int_a^b f(x)\,dx = 0 = \int_a^b xf(x)\,dx = \int_a^b x^2 f(x)\,dx$, then f has at least three roots in $[a, b]$.

(d) Generalize.

12. (See Exercise 11.) Prove that, if f and g are polynomials such that the nth moment of f and g over $[a, b]$ are equal for all integers $n \geq 0$, then $f = g$. Thus we may say that "a polynomial is known by its moments."

13. (a) Define for $n > 0$, $f(n) = \int_0^\infty x^{n-1}e^{-x}\,dx$, the $(n-1)$st moment of the function e^{-x} over the positive part of the x axis. Show that $f(1) = 1$, $f(2) = 1$, and $f(n+1) = nf(n)$.

(b) Prove that for a positive integer n, $f(n) = (n-1)!$.

(c) Prove that $f(\frac{1}{2}) = \sqrt{\pi}$. (Use the fact that $\int_0^\infty e^{-x^2}\,dx = \sqrt{\pi}/2$.)

14. Prove that the kinetic energy of a homogeneous fluid is greater than if all its particles were moving at their average speed. (The kinetic energy of a mass M moving at the velocity v is defined as $Mv^2/2$. In the proof you may need Schwarz's inequality, which holds in all dimensions.)

17

MATHEMATICAL MODELS

In Sec. 6.4 the differential equation

$$\frac{dy}{dt} = ky$$

was used to analyze growth and decay. This equation is an example of a mathematical model of a phenomenon in the real world. Such a model may be simply a differential equation, as in the study of growth and decay, or may be given by axioms (assumptions) phrased in the language of set theory, algebra, or other branches of mathematics.

In this chapter a model of random traffic is developed. Though expressed initially in terms of automobile traffic, this model can be applied to the study of customer arrivals at a checkout counter, machine breakdowns, and telephone calls. The general notion of mathematical modeling is explored in the concluding section.

17.1

The basic ideas of probability

When we toss a die, any one of the numbers 1, 2, 3, 4, 5, or 6 can turn up. If the die is not "loaded," each of the six possibilities is equally likely to occur. That is, on the average, in one-sixth of the tosses a 1 will appear, in one-sixth of the tosses a 2 will appear, and so on. When the die is thrown many times, the fraction of times any particular number appears should be approximately one-sixth. We state briefly: The probability of tossing a 1 is $\frac{1}{6}$.

The probability of an event occurring when there are n equally likely possible outcomes is $1/n$. Thus the probability of obtaining heads when tossing a fair coin is $\frac{1}{2}$.

When a die is thrown many times, a 1 or 2 tends to appear in about

$$\tfrac{1}{6} + \tfrac{1}{6} = \tfrac{2}{6}$$

of the tosses. Thus, the probability of obtaining a 1 or a 2 is $\frac{2}{6}$. Similarly, the probability of obtaining a 3, 4, 5, or 6 is $\frac{4}{6}$. This illustrates the following principle: The probability of one of several mutually exclusive events occurring is the sum of their respective probabilities. (Events are called *mutually exclusive* if only one of them can occur at a time. For instance,

The probability of at least one of several mutually exclusive events

when a die is thrown, the two events "obtaining a 1" and "obtaining a 2" are mutually exclusive.)

When we toss two dice at a time, the score can be anywhere from 2 $(= 1 + 1)$ to 12 $(= 6 + 6)$.   Let us compute the probability that the score is 5.

For convenience, imagine that one of the dice is red and the other is white. The red die can show a 1, 2, 3, 4, 5, or 6; so can the white die.   There are now $6 \cdot 6 = 36$ equally likely outcomes.   In how many of them is the score 5?   A total of 5 can be obtained in the ways listed in the table in the margin.

Red Die	1	2	3	4
White Die	4	3	2	1

Since in 4 out of the 36 cases the score is 5, the probability of scoring a 5 is $\frac{4}{36}$.

Note that the probability of obtaining a 2 with the red die together with a 3 with the white die is $\frac{1}{36}$, which is $\frac{1}{6} \cdot \frac{1}{6}$.   This illustrates a second fundamental

The probability of all of several independent events

principle in probability: The probability of all of several independent events occurring is the product of their respective probabilities.   (Two or more events are called *independent* if the probability that any one of them will occur is not influenced by the occurrence of the other events.   For instance, obtaining a 2 with the red die is independent of obtaining a 3 with the white die.)   The next example applies this principle.

EXAMPLE 1   A person tosses a die repeatedly until a 3 appears.   What is the probability that a 3 will come up for the first time on the fourth toss?

SOLUTION   For the 3 to occur for the first time on the fourth toss, four independent events must happen:

On the first toss she must *not* get a 3.
On the second toss she must *not* get a 3.
On the third toss she must *not* get a 3.
On the fourth toss she must get a 3.

Since the probability of not getting a 3 is $\frac{5}{6}$, and the probability of getting a 3 is $\frac{1}{6}$, the probability of all four listed events occurring is the product

$$\frac{5}{6} \cdot \frac{5}{6} \cdot \frac{5}{6} \cdot \frac{1}{6} = \left(\frac{5}{6}\right)^3 \frac{1}{6}. \quad \bullet$$

Throwing a die illustrates another fundamental principle.   The probability of getting a 5 is $\frac{1}{6}$.   The probability of *not* getting a 5 is $\frac{5}{6}$.   Observe that the two probabilities add up to 1,

$$\frac{1}{6} + \frac{5}{6} = \frac{6}{6} = 1.$$

The probability that an event does not occur

The general principle is: If the probability that an event will occur is p, then the probability that it will not occur is $1 - p$.

Pursuing further the problem raised in Example 1, we raise the question, "What is the average number of times a person must toss a die until she gets a 3?"   She may toss a 3 on the very first try, or she may have to take 1,000 tries, or more.   Example 2 records the results of an experimental approach to this problem.   Example 3 obtains a theoretical solution.

Before starting Example 2, let us recall how the average of several numbers is computed, where by *average* we refer to what is sometimes called the *arithmetic mean*. For instance, how is the average of the 10 numbers

$$5, 6, 5, 7, 9, 5, 8, 6, 9, 5$$

computed? The numbers are added, and the sum is divided by 10:

$$\text{Average} = \frac{5 + 6 + 5 + 7 + 9 + 5 + 8 + 6 + 9 + 5}{10}.$$

To simplify the computation, first collect equal summands in the numerator. (For instance, 5 appears four times.) Thus

$$\text{Average} = \frac{4 \cdot 5 + 2 \cdot 6 + 1 \cdot 7 + 1 \cdot 8 + 2 \cdot 9}{10},$$

a shorter expression. Carrying out the arithmetic shows that

$$\text{Average} = \frac{20 + 12 + 7 + 8 + 18}{10}$$

$$= \tfrac{65}{10}$$

$$= 6.5.$$

This type of shortcut will be used in Example 2.

EXAMPLE 2   A person tosses a die until a 3 appears. He carries out this experiment 100 times. The results are recorded in the table below. For instance, it shows that in 15 cases he threw a 3 on the first try, while in 14 cases he obtained a 3 for the first time on his second toss.

Required Tosses	1	2	3	4	5	6	7	8	9	10	11	12	13	14	15	16
Number of Times	15	14	11	12	9	6	5	6	4	4	3	2	3	1	3	2

What is the average number of tosses?

SOLUTION   The average number of tosses required is the total number of tosses divided by 100, the number of repetitions of the experiment, or the quotient of

$$1 \cdot 15 + 2 \cdot 14 + 3 \cdot 11 + 4 \cdot 12 + 5 \cdot 9 + 6 \cdot 6 + 7 \cdot 5 + 8 \cdot 6 + 9 \cdot 4 + 10 \cdot 4 +$$
$$11 \cdot 3 + 12 \cdot 2 + 13 \cdot 3 + 14 \cdot 1 + 15 \cdot 3 + 16 \cdot 2$$

divided by 100, which is $\tfrac{551}{100} = 5.51.$

The experimental average is 5.51 tosses to reach a 3. ●

If the experiment described in Example 2 is performed again, the average need not be 5.51, though it would very likely be near it. In the next example the average is computed on a theoretical basis, without performing any experiments.

EXAMPLE 3 Find the average number of tosses of a die required to reach a 3.

SOLUTION Notice, first of all, that the experimental result of Example 2 can be rewritten as

$$1(\tfrac{15}{100}) + 2(\tfrac{14}{100}) + 3(\tfrac{11}{100}) + 4(\tfrac{12}{100}) + 5(\tfrac{9}{100}) + 6(\tfrac{6}{100}) + 7(\tfrac{5}{100})$$
$$+ 8(\tfrac{6}{100}) + 9(\tfrac{4}{100}) + 10(\tfrac{4}{100}) + 11(\tfrac{3}{100}) + 12(\tfrac{2}{100})$$
$$+ 13(\tfrac{3}{100}) + 14(\tfrac{1}{100}) + 15(\tfrac{3}{100}) + 16(\tfrac{2}{100}). \qquad (1)$$

The first fraction in (1), $\tfrac{15}{100}$, is the fraction of the time that only one toss was required. Theoretically, this fraction would be $\tfrac{1}{6}$. The second fraction in (1), $\tfrac{14}{100}$, is the fraction of the time that precisely two tosses were required. Theoretically, this fraction is $\tfrac{5}{6} \cdot \tfrac{1}{6}$. And so on, for the other fractions in (1). Thus the theoretical analog of (1) is

$$1(\tfrac{1}{6}) + 2(\tfrac{5}{6})\tfrac{1}{6} + 3(\tfrac{5}{6})^2 \tfrac{1}{6} + 4(\tfrac{5}{6})^3 \tfrac{1}{6} + \cdots + n(\tfrac{5}{6})^{n-1}\tfrac{1}{6} + \cdots, \qquad (2)$$

the sum of an infinite series.

The series is of the form

$$\tfrac{1}{6}(1 + 2x + 3x^2 + \cdots + nx^{n-1} + \cdots),$$

where $x = \tfrac{5}{6}$. To sum it, recall that

$$\frac{1}{1-x} = 1 + x + x^2 + \cdots + x^n + \cdots \qquad (3)$$

According to Sec. 14.7, differentiation of (3) yields

$$\frac{1}{(1-x)^2} = 1 + 2x + 3x^2 + \cdots + nx^{n-1} + \cdots.$$

Thus the sum of series (2) is $\tfrac{1}{6}[1/(1 - \tfrac{5}{6})^2] = 6$. The theoretical average is six tosses. ●

Example 3 suggests the following definition.

DEFINITION *Expected value (average value) of n.* If the probability that the integer value n occurs is $p(n)$, then the expected (average) value of n is

$$1 \cdot p(1) + 2 \cdot p(2) + 3 \cdot p(3) + \cdots + np(n) + \cdots$$

(assuming that this series converges).

As Example 3 shows, the expected number of tosses required to obtain a 3 on a die is 6.

EXAMPLE 4 Find the average value of the number showing when one die is tossed.

SOLUTION The numbers 1, 2, 3, 4, 5, and 6 each occur with probability $\tfrac{1}{6}$. The numbers 7, 8, 9, ... never occur, hence occur with probability 0. Thus

$$p(n) = \begin{cases} \frac{1}{6} & \text{for } n = 1, 2, 3, 4, 5, 6 \\ 0 & \text{for } n = 7, 8, 9, \ldots. \end{cases}$$

The average value, defined as

$$1 \cdot p(1) + 2 \cdot p(2) + 3 \cdot p(3) + \cdots + n \cdot p(n) + \cdots,$$

is simply

$$1(\tfrac{1}{6}) + 2(\tfrac{1}{6}) + 3(\tfrac{1}{6}) + 4(\tfrac{1}{6}) + 5(\tfrac{1}{6}) + 6(\tfrac{1}{6}) \text{ ,}$$

or

$$\tfrac{21}{6} = 3.5. \; \bullet$$

Exercises

1. Compute the probability of tossing a 4 with one die.
2. Compute the probability of tossing a total of 7 with two dice.
3. Compute the probability of tossing a total of 7 or 11 with two dice.
4. Compute the probability of tossing heads twice with two throws of a coin.
5. Compute the probability of tossing one head and one tail in two throws of a coin.
6. Compute the probability of tossing a 5 with one die for the first time on the seventh throw.
7. Compute the probability of tossing heads with a coin for the first time on the
   (a) first throw;
   (b) second throw;
   (c) third throw;
   (d) nth throw.
8. (a) Toss a coin until it turns up heads and record the number of throws required.
   (b) Do (a) 20 times and record the average number of throws.
9. Compute the theoretical expected number of tosses of a coin required in order for heads to turn up.
10. (a) Toss a die until it turns up 1 and record the number of throws required.

(b) Do (a) 20 times and record the average number of throws.
(c) What is the theoretical average (expected value)?

■

11. (a) Toss a die until it turns up 1 or 6 and record the number of throws required.
    (b) Do (a) 20 times and record the average number of throws.
    (c) What is the theoretical average (expected value)?
12. Let $p(n) = 3^n e^{-3}/n!$ for $n = 0, 1, 2, \ldots$.
    (a) Show that $\sum_{n=0}^{\infty} p(n) = 1$.
    (b) Show that the expected value of n is 3.
13. Let $p(n) = \frac{1}{10}$ for $n = 0, 1, 2, 3, 4, 5, 6, 7, 8,$ and 9, and 0 otherwise. Compute the expected value of n.
14. Let $p(n) = (\tfrac{1}{2})^n$ for $n = 1, 2, 3, \ldots$.
    (a) Show that $\sum_{n=1}^{\infty} p(n) = 1$.
    (b) Show that the expected value of n is 2.

■ ■

15. What is the probability, when you throw a coin three times, of getting exactly two heads?
16. What is the probability, when you throw a die three times, of getting a 1 exactly twice?

17.2

Probability distributions

When a die is thrown, the result will be an integer, either 1, 2, 3, 4, 5, or 6. But in many types of experiments involving chance, the result may be any real number, not merely an integer. As an example, the gap (measured in feet) between cars in a single lane of traffic can be any nonnegative real number.

For instance, a helicopter photograph of a single lane of traffic provided the following data concerning 100 gaps:

Length of gap less than or equal to	100 feet	200 feet	300 feet	400 feet	500 feet	600 feet	700 feet
Number of such gaps observed	32	56	76	88	95	99	100

Let $F(x)$ denote the fraction of gaps that are less than or equal to x feet. For instance

$$F(200) = \tfrac{56}{100} = 0.56.$$

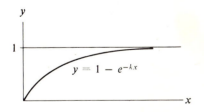

On the basis of the data shown in the table above, we may compute and graph F, obtaining the results shown in the figure at the left. The heavy dots record the data in the table and the assumption that $F(0) = 0$, that is, no cars are bumper-to-bumper. Note that $F(x) = 1$ for $x \geq 700$.

The form of this table is used quite often to record data. Thus rather than including an entry "numbers of gaps between 100 and 200 feet," we obtain this number by subtracting the number of gaps ≤ 100 feet from the number of gaps ≤ 200 feet. Such a table can be made for any variable that takes on only nonnegative values. The resulting function F will resemble the function F derived from the helicopter photograph in several respects. For instance if,

$$x_1 > x_2,$$

then
$$F(x_1) \geq F(x_2),$$

since the probability that x is at most x_1 is at least as great as the probability that x is at most x_2. Also
$$0 \leq F(x) \leq 1,$$

since $F(x)$ is a probability.

DEFINITION *Cumulative probability distribution.* A cumulative probability distribution (over the nonnegative real numbers) is any function F such that $F(0) = 0$, $\lim_{x \to \infty} F(x) = 1$, and the graph of F rises as we move to the right. (Often such a function is called a *probability distribution* or simply a *distribution*.)

The graph shown at the beginning of this section records a cumulative probability distribution. But probability distributions also may arise theoretically, as in Sec. 17.3 where the probability distribution $F(x) = 1 - e^{-kx}$ $(k > 0)$ is used to describe the gaps in traffic. The graph of this particular probability distribution does not reach the line $y = 1$ but does approach it.

EXAMPLE 1 Check that the function F given by the formula

$$F(x) = 1 - e^{-kx},$$

with $k > 0$, is a probability distribution.

SOLUTION   First of all $F(0) = 1 - e^{-k(0)} = 1 - e^0 = 1 - 1 = 0.$

Second, we have to show that $\lim_{x \to \infty} F(x) = 1$:

$$\lim_{x \to \infty} F(x) = \lim_{x \to \infty} (1 - e^{-kx}) = 1 - 0 = 1.$$

Third, it must be shown that, if $x_1 > x_2$, then $F(x_1) \geq F(x_2)$. To do this, show that F' is nonnegative:

$$F'(x) = D(1 - e^{-kx}) = ke^{-kx}.$$

Since k is positive, so is $F'(x)$; hence F is an increasing function. ●

Consider a typical probability distribution F. What is the probability that x is between t and $t + \Delta t$, where $\Delta t > 0$? If Δt is small, this probability is near 0. In any case, it is simply

$$F(t + \Delta t) - F(t). \tag{1}$$

This expression is reminiscent of the quotient used to define the derivative of F. Therefore rewrite (1) as

$$\frac{F(t + \Delta t) - F(t)}{\Delta t} \Delta t. \tag{2}$$

When Δt is small, (2) is approximately

$$F'(t) \cdot \Delta t. \tag{3}$$

Thus the probability that x is in the narrow interval $[t, \ t + \Delta t]$ is approximately $F'(t) \Delta t$.

Now consider the average gap in the traffic. The following computation motivates the definition that will be given for *average value* of a variable that takes on only nonnegative values. Of the 100 gaps recorded in the table at the beginning of this section, 32 were between 0 and 100 feet. For simplicity, say that these were all 0 feet. (Of course it would be more reasonable to use 50 feet, but since our purpose is theoretical, let us keep the arithmetic simple.) Similarly, there were $F(200) - F(100) = 56 - 32 = 24$ gaps between 100 and 200 feet. Assume that these were all 100 feet. And so on, for the other cases. Then the average gap is approximately

$$\frac{0 \cdot 32 + 100 \cdot 24 + 200 \cdot 20 + 300 \cdot 12 + 400 \cdot 7 + 500 \cdot 4 + 600 \cdot 1}{100}$$

$$= \frac{15{,}400}{100} = 154 \text{ feet.} \tag{4}$$

To relate this average to the function F, rewrite (4) as

$$0\left(\tfrac{32}{100}\right) + 100\left(\tfrac{24}{100}\right) + 200\left(\tfrac{20}{100}\right) + 300\left(\tfrac{12}{100}\right) + 400\left(\tfrac{7}{100}\right) + 500\left(\tfrac{4}{100}\right) + 600\left(\tfrac{1}{100}\right). \tag{5}$$

Now, $\tfrac{32}{100}$ is an experimental estimate of the probability that the gap is between 0 and 100. By (3), the theoretical estimate of this quantity is $F'(0)100$. Similarly, $\tfrac{24}{100}$ is approximately $F'(100)100$, and so on.

Thus (5) is approximately

$$0F'(0)100 + 100F'(100)100 + 200F'(200)100 + 300F'(300)100 + 400F'(400)100$$
$$+ 500F'(500)100 + 600F'(600)100,$$

which is an approximating sum, based on a partition of $[0, 700]$ into seven sections, for the definite integral

$$\int_0^{700} xF'(x)\,dx. \tag{6}$$

This suggests how to define the average value of x when x is described by the probability distribution F.

DEFINITION   *Expected value (average value).* If the probability that a variable is in the interval $[0, x]$ is $F(x)$, then the average value of x is the improper integral

$$\int_0^\infty xF'(x)\,dx$$

(if it converges).

EXAMPLE 2   Compute the expected value of x for the probability distribution $F(x) = 1 - e^{-kx}$, with $k > 0$.

SOLUTION   Since $F'(x) = ke^{-kx}$, by definition,

$$\text{Expected value} = \int_0^\infty xke^{-kx}\,dx.$$

Now,

$$\int_0^b xke^{-kx}\,dx = \frac{1}{k}e^{-kx}(-kx - 1)\Big|_0^b$$

$$= \left[\frac{1}{k}e^{-kb}(-kb - 1)\right] - \left\{\frac{e^{-k0}}{k}[-k(0) - 1]\right\}$$

$$= \frac{1}{k}[1 - e^{-kb}(kb + 1)].$$

As shown in Sec. 6.3,

$$\lim_{x\to\infty} xe^{-x} = 0.$$

Thus

$$\lim_{b\to\infty} e^{-kb}(kb + 1) = 0$$

and

$$\int_0^\infty xke^{-kx}\,dx = \frac{1}{k}(1 - 0) = \frac{1}{k}.$$

Consequently the average gap is $1/k$. ●

Since the average gap in the traffic, as observed from the helicopter, was 154 feet, Example 2 suggests that the value for k that should be used in the formula $F(x) = 1 - e^{-kx}$ to best approximate the traffic is $k = \frac{1}{154}$.

Exercises

1. Show that $F(x) = (2/\pi) \tan^{-1} x$ is a probability distribution.

2. Let $F(x) = 1 - [1/(x + 1)^2]$ for $x \geq 0$.
   (a) Show that F is a probability distribution.
   (b) Find its expected value.

3. Show that $F(x) = 1 - e^{-x^2}$ describes a probability distribution.

4. The following data were compiled for 100 transistor radios.

Time interval, hours	0–10	10–20	20–30	30–40	40–50	50–60
Number that failed during the interval	53	24	11	6	4	2

   Let $F(x)$ be the fraction of the 100 radios that failed within their first x hours.
   (a) Find $F(60)$.
   (b) Find $F(50)$, $F(40)$, $F(30)$, $F(20)$, $F(10)$.
   (c) Assuming that $F(0) = 0$, graph F.
   (d) What is a reasonable estimate of the life expectancy of a transistor radio?

5. What is the expected value of the distribution F given by $F(x) = 1 - e^{-0.2x}$?

6. Let $F(x) = x$ for x in $[0, 1]$ and $F(x) = 1$ if $x \geq 1$.
   (a) Show that F is a distribution.
   (b) Find its expected value.

7. Let $F(x) = \sqrt{x}$ for x in $[0, 1]$ and $F(x) = 1$ if $x \geq 1$.
   (a) Show that F is a distribution.
   (b) Find its expected value.

8. The distribution function for gaps in sparse traffic may be of the form $F(x) = 1 - e^{-kx}$ for a suitable positive number k. If the average gap observed from a helicopter is 200 feet, what k should be chosen? Explain.

9. Assume of a distribution F that for $x \geq a$, $F(x) = 1$. Prove that the expected value
$$\int_0^\infty xF'(x)\, dx \quad \text{equals}$$
$$\int_0^\infty [1 - F(x)]\, dx = \int_0^a [1 - F(x)]\, dx.$$

   Hint: Use integration by parts.

10. Use the result in Exercise 9 and the table below to estimate the expected life-span of males in rural India.

■

11. In reliability theory the Weibull distribution, defined as $F(x) = 1 - e^{-(x^a)/b}$, where a and b are appropriate constants, is used. If $a = 1$, it reduces to the exponential distribution. In a study of ball bearings it was found that 10 percent did not perform for more than 20,000,000 revolutions and that 50 percent did not perform for more than 80,000,000 revolutions. Let $F(x)$ denote the proportion that fail before x million revolutions.
    (a) Translate the data to assertions about F.
    (b) It has also been observed that for ball bearings F takes the form of a Weibull distribution. Show that a satisfies the equation $4^a = \ln 0.5/\ln 0.9$.

■ ■

12. Let F be a probability distribution. Assuming that $\lim_{t \to \infty} t[1 - F(t)] = 0$, prove that the expected value is $\int_0^\infty [1 - F(t)]\, dt$. *Hint:* Use integration by parts.

13. The design of transmitting satellites, rockets, and other complex mechanisms makes use of redundant components to obtain longer life.
    (a) Assume that the distribution function F for a component is given by $F(t) = 1 - e^{-kt}$. Show that the distribution function G for a device that fails only when two such components fail is given by $G(t) = (1 - e^{-kt})^2$.
    (b) By Example 2 the expected life of a component is $1/k$. Show that the expected life of the device described in (a) is $\frac{3}{2}(1/k)$.

14. Prove that, if F is a probability distribution and $\int_0^\infty xF'(x)\, dx$ converges, then it necessarily follows that $\lim_{x \to \infty} x[1 - F(x)] = 0$.

15. A machine that has survived a hours has a *mean residual life* defined as the expected number of hours remaining before it breaks down. A person a years old has a life expectancy, the expected number of years of life remaining at age a.
    (a) Explain why these quantities are measured by $\int_a^\infty (t - a)F'(t)\, dt/[1 - F(a)]$.
    (b) Show that $\int_a^b (t - a)F'(t)\, dt = (a - b)[1 - F(b)] + \int_a^b [1 - F(t)]\, dt$.

Age, years		0	1	5	15	25	35	45	55	65	75	85	95
Number surviving to the age		1,000	857	738	699	675	647	610	536	385	179	26	1

17.3

The exponential (Poisson) model of random traffic

We shall construct a mathematical model for automobile traffic on a one-lane road.   First of all, disregard the lengths of the cars and treat each car as a point. It is therefore possible theoretically to have an arbitrarily large number of cars in a finite interval.   Moreover, assume that all cars travel at the same speed (say, the speed limit) and that each car enters the traffic flow independently of the other cars.   (These assumptions are more realistic for sparse than for dense traffic.)

To construct the model, introduce the functions P_0, P_1, P_2, ..., P_n, ..., where $P_n(x)$ shall be the probability that an interval of length x contains exactly n cars (independently of the location of the interval).   Thus $P_0(x)$ is the probability that an interval of length x is empty.   Assume that

$$P_0(x) + P_1(x) + \cdots + P_n(x) + \cdots = 1 \qquad \text{for any } x.$$

Also assume that $P_0(0) = 1$ (read as "the probability is 1 that a given point contains no cars").

For this model we make the following two major assumptions:

ASSUMPTION 1   The probability that exactly one car is in any fixed short section of the road is approximately proportional to the length of the section.   That is, there is some positive number k such that

$$\lim_{\Delta x \to 0} \frac{P_1(\Delta x)}{\Delta x} = k.$$

ASSUMPTION 2   The probability that there is more than one car in any fixed short section of the road is negligible, even when compared with the length of the section.   That is,

$$\lim_{\Delta x \to 0} \frac{P_2(\Delta x) + P_3(\Delta x) + P_4(\Delta x) + \cdots}{\Delta x} = 0. \tag{1}$$

The reader should pause and convince himself that these assumptions are reasonable.   Think them over now, for soon some very delicate reasoning will be based on them.

Assumptions 1 and 2 will now be put into more useful form.   Let

$$\varepsilon = \frac{P_1(\Delta x)}{\Delta x} - k, \tag{2}$$

where ε depends on Δx.   Assumption 1 asserts that $\lim_{\Delta x \to 0} \varepsilon = 0$.   Thus, solving for $P_1(\Delta x)$ shows that assumption 1 can be rephrased as

$$P_1(\Delta x) = k \, \Delta x + \varepsilon \, \Delta x, \tag{3}$$

where $\varepsilon \to 0$ as $\Delta x \to 0$.

Since $P_0(\Delta x) + P_1(\Delta x) + \cdots + P_n(\Delta x) + \cdots = 1$, assumption 2 may be expressed as

$$\lim_{\Delta x \to 0} \frac{1 - P_0(\Delta x) - P_1(\Delta x)}{\Delta x} = 0. \tag{4}$$

In view of assumption 1, (4) is equivalent to

$$\lim_{\Delta x \to 0} \frac{1 - P_0(\Delta x)}{\Delta x} = k. \tag{5}$$

In the same manner in which we obtained (3), it can be shown that

$$1 - P_0(\Delta x) = k \, \Delta x + \delta \Delta x,$$

where $\delta \to 0$ as $\Delta x \to 0$. Thus

$$P_0(\Delta x) = 1 - k \, \Delta x - \delta \Delta x, \tag{6}$$

where $\delta \to 0$ as $\Delta x \to 0$. On the basis of assumptions 1 and 2, as expressed in (3) and (6), it is possible to obtain an explicit formula for each P_n.

First, it will be shown that $P_0(x) = e^{-kx}$. To begin, observe in the figure at the left that a section of length $x + \Delta x$ is vacant if its left-hand part, of length x, is vacant and its right-hand part, of length Δx, is also vacant. Since the cars move independently of each other, the probability of the whole interval of length $x + \Delta x$ being empty is the product of the probabilities that the two smaller intervals of lengths x and Δx are both empty. Thus

$$P_0(x + \Delta x) = P_0(x)P_0(\Delta x). \tag{7}$$

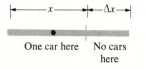

No cars in a section of length $x + \Delta x$

In view of (6), (7) can be written as

$$P_0(x + \Delta x) = P_0(x)(1 - k \, \Delta x - \delta \Delta x). \tag{8}$$

A little algebra transforms (8) to

$$\frac{P_0(x + \Delta x) - P_0(x)}{\Delta x} = -(k + \delta)P_0(x). \tag{9}$$

Taking limits on both sides of (9) as $\Delta x \to 0$, we obtain

$$P_0'(x) = -kP_0(x). \tag{10}$$

From (10) it follows that there is a constant A such that $P_0(x) = Ae^{-kx}$ (see Sec. 6.4). Since $1 = P_0(0) = Ae^{-k0} = A$, it follows that $A = 1$; hence

The formula for $P_0(x)$

$$P_0(x) = e^{-kx}. \tag{11}$$

This explicit formula for P_0 is plausible; e^{-kx} is a decreasing function of x, so that the larger an interval is, the less chance there is that it is empty.

Next we show that $P_1(x) = kxe^{-kx}$. To do so, examine $P_1(x + \Delta x)$ and relate it to $P_0(x)$, $P_0(\Delta x)$, $P_1(x)$, and $P_1(\Delta x)$, with the goal of finding an equation involving the derivative of P_1. Again, imagine an interval of length $x + \Delta x$ cut

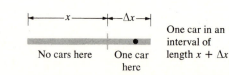

One car in an interval of length $x + \Delta x$

into two intervals, the left-hand subinterval of length x and the right-hand subinterval of length Δx. Then there is precisely one car in the whole interval if either there is exactly one car in the left-hand subinterval and none in the right-hand subinterval or there is none in the left-hand subinterval and exactly one in the right-hand subinterval.

Thus
$$P_1(x + \Delta x) = P_1(x)P_0(\Delta x) + P_0(x)P_1(\Delta x). \tag{12}$$

In view of (3) and (6), Eq. (12) may be written

$$P_1(x + \Delta x) = P_1(x)(1 - k\,\Delta x - \delta\Delta x) + P_0(x)(k\,\Delta x + \varepsilon\,\Delta x).$$

A little algebra changes this to

$$\frac{P_1(x + \Delta x) - P_1(x)}{\Delta x} = -(k + \delta)P_1(x) + (k + \varepsilon)P_0(x). \tag{13}$$

Letting $\Delta x \to 0$ in (13) and remembering that $\delta \to 0$ and $\varepsilon \to 0$ as $\Delta x \to 0$, we obtain
$$P_1'(x) = -kP_1(x) + kP_0(x).$$

Since $P_0(x) = e^{-kx}$, it follows that

$$P_1'(x) = -kP_1(x) + ke^{-kx}. \tag{14}$$

From (14) we shall obtain an explicit formula for $P_1(x)$. Since $P_0(x)$ involves e^{-kx} and so does (14), it is reasonable to guess that $P_1(x)$ involves e^{-kx}. Therefore express $P_1(x)$ as $g(x)e^{-kx}$ and determine the form of $g(x)$. (Since $P_1(x) = [P_1(x)e^{kx}]e^{-kx}$, $g(x)$ exists.) According to (14),

$$[g(x)e^{-kx}]' = -kg(x)e^{-kx} + ke^{-kx};$$

hence
$$g(x)(-ke^{-kx}) + g'(x)e^{-kx} = -kg(x)e^{-kx} + ke^{-kx},$$

from which it follows that $g'(x) = k$. Hence $g(x) = kx + c_1$, where c_1 is some constant. Thus

$$P_1(x) = (kx + c_1)e^{-kx}.$$

Since $P_1(0) = 0$, we have $P_1(0) = (k0 + c_1)e^{-k0} = c_1$, and hence $c_1 = 0$. All this is summarized in the formula

The formula for $P_1(x)$
$$P_1(x) = kxe^{-kx}, \tag{15}$$

and P_1 is completely determined.

To obtain P_2, argue as in the case for P_1. Instead of (12) start with

$$P_2(x + \Delta x) = P_2(x)P_0(\Delta x) + P_1(x)P_1(\Delta x) + P_0(x)P_2(\Delta x), \tag{16}$$

an equation which records the three ways in which two cars in a section of length $x + \Delta x$ can be situated in a section of length x and a section of length Δx:

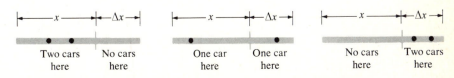

| Two cars here | No cars here | One car here | One car here | No cars here | Two cars here |

Making use of (3) and (6), we can rewrite (16) as

$$P_2(x + \Delta x) = P_2(x)(1 - k\,\Delta x - \delta\Delta x) + P_1(x)(k\,\Delta x + \varepsilon\,\Delta x) + P_0(x)P_2(\Delta x).$$

Hence

$$\frac{P_2(x + \Delta x) - P_2(x)}{\Delta x} = -kP_2(x) - \delta P_2(x) + kP_1(x) + \varepsilon P_1(x) + P_0(x)\frac{P_2(\Delta x)}{\Delta x}.$$

(17)

Recalling (1), we see that $\lim_{\Delta x \to 0} P_2(\Delta x)/\Delta x = 0$. Letting $\Delta x \to 0$ in (17) yields

$$P_2'(x) = -kP_2(x) + kP_1(x).$$ (18)

Now, $P_1(x) = kxe^{-kx}$. Write $P_2(x) = h(x)e^{-kx}$. Then (18) becomes

$$-kh(x)e^{-kx} + h'(x)e^{-kx} = -kh(x)e^{-kx} + k^2xe^{-kx}.$$

This yields $h'(x) = k^2x$. Thus

$$h(x) = \frac{k^2x^2}{2} + c_2.$$

To determine c_2, use the information that $P_2(0) = 0$. Since $P_2(x) = h(x)e^{-kx} = [(k^2x^2)/2 + c_2]e^{-kx}$, $P_2(0) = c_2$; thus $c_2 = 0$. Therefore P_2 is completely determined:

The formula for $P_2(x)$

$$P_2(x) = \frac{k^2x^2}{2}e^{-kx}.$$ (19)

Similar reasoning carried out inductively shows that

The Poisson Formulas

$$P_n(x) = \frac{(kx)^ne^{-kx}}{n!}.$$ (20)

Equation (20) presents the formulas on which the rest of the analysis will be based. Note that these formulas refer to a road section of any length, though the assumptions 1 and 2 refer only to short sections. What has enabled us to go from the "microscopic" to the "macroscopic" is the additional assumption that the traffic in any one section is independent of the traffic in any other section. The formulas (20) are known as the *Poisson formulas.*

Before these formulas can be applied to practical problems, it is necessary to know how to determine the constant k experimentally. The next example shows how to do this.

EXAMPLE I Consider a section of length x. There may be no cars in it, or one car, or two cars, or more. What is the average number to be expected in this section?

SOLUTION By definition of expected value given in Sec. 17.1, this average is

$$0P_0(x) + 1P_1(x) + 2P_2(x) + \cdots + nP_n(x) + \cdots.$$

(Note that a term corresponding to $n = 0$ is present, but does not contribute to the sum.)

Next compute this sum:

$$\sum_{n=0}^{\infty} nP_n(x) = \sum_{n=1}^{\infty} nP_n(x)$$

$$= \sum_{n=1}^{\infty} n\frac{(kx)^n e^{-kx}}{n!}$$

$$= \sum_{n=1}^{\infty} \frac{(kx)^n e^{-kx}}{(n-1)!}$$

$$= kxe^{-kx} \sum_{n=1}^{\infty} \frac{(kx)^{n-1}}{(n-1)!}$$

$$= kxe^{-kx}\left[\frac{(kx)^0}{0!} + \frac{(kx)^1}{1!} + \frac{(kx)^2}{2!} + \frac{(kx)^3}{3!} + \cdots\right]$$

$$= kxe^{-kx}\left[1 + kx + \frac{(kx)^2}{2!} + \frac{(kx)^3}{3!} + \cdots\right]$$

$$= kxe^{-kx}e^{kx}$$

$$= kx.$$

The meaning of k   The expected number of cars in a section is proportional to the length of the section. This shows that the k appearing in assumption 1 is a measure of traffic density, the number of cars per unit length of road. ●

As Example 1 shows, to determine k in practice, experimentally find the average number of cars per unit length of road. The same principle can be used in other types of traffic, as long as the random events satisfy assumptions 1 and 2, as the next two examples show.

EXAMPLE 2   In 100 days a salesperson sells 50 sets of encyclopedias. What is the probability that on a given day she sells (*a*)   no sets? (*b*)   one set? (*c*)   two sets?

SOLUTION   Let $P_n(x)$ denote the probability of the salesperson selling exactly n encyclopedias in x days (x need not be an integer). It seems reasonable to assume that the probability of selling an encyclopedia during a short time interval is small and roughly proportional to its duration. That is,

$$\lim_{\Delta x \to 0} \frac{P_1(\Delta x)}{\Delta x} = k$$

for some constant k. It is also reasonable to assume that the likelihood of selling two or more encyclopedias during a very short interval is negligible. (This is assumption 2 in a Poisson distribution.) Next, assume that the sales occur independently of each other. In this case all the assumptions of a Poisson model are satisfied. Therefore $P_0(x) = e^{-kx}$ and, more generally,

$$P_n(x) = \frac{(kx)^n e^{-kx}}{n!}.$$

To find k, use Example 1, where it was shown that

$$k = \text{density of traffic.}$$

But in this case

$$k = \text{average number of encyclopedias per unit time}$$

$$= \tfrac{50}{100}$$

$$= 0.5.$$

It is now a simple matter to solve the three parts of the problem.

1. The probability of selling no sets during one day is

$$P_0(1) = e^{-(0.5)1}$$

$$= e^{-0.5}$$

$$\doteq 0.607.$$

2. The probability of selling one set during one day is

$$P_1(1) = \frac{[0.5)(1)]^1 e^{-0.5}}{1!}$$

$$= (0.5)e^{-0.5}$$

$$\doteq 0.303.$$

3. The probability of selling two sets in one day is

$$P_2(1) = \frac{[(0.5)(1)]^2 e^{-0.5}}{2}$$

$$= \frac{(0.5)^2 e^{-0.5}}{2}$$

$$\doteq (0.125)(0.607)$$

$$\doteq 0.076.$$

Thus it is to be expected that on 61 of those 100 days she sold no encyclopedias, on 30 days she sold exactly one, and on 8 exactly two. ●

EXAMPLE 3   *Traffic at a checkout counter.*   Customers arrive at a checkout counter at the rate of 15 per hour.   What is the probability that exactly five customers will arrive in any given 20-minute period?

SOLUTION   Assume that the probability of exactly one customer arriving in a short interval of time is roughly proportional to the duration of that interval, and that there is only a negligible probability that more than one customer may arrive in a brief interval of time.   Therefore assumptions 1 and 2 hold, if "length of section" is replaced by "length of time."   Without further ado, it follows that the probability of exactly n customers arriving in a period of x minutes is given by the formula $$P_n(x) = \frac{(kx)^n e^{-kx}}{n!}.$$

Moreover, the "customer density" is one per 4 minutes; hence $k = \frac{1}{4}$, and thus the probability that exactly five customers arrive during a 20-minute period, $P_5(20)$, is

$$\left(\frac{1}{4} 20\right)^5 \frac{e^{-(1/4)(20)}}{5!} = \frac{5^5 e^{-5}}{120} \doteq 0.18. \quad \bullet$$

EXAMPLE 4 *Airport traffic.* Planes arrive randomly at an airport at the rate of one per 2 minutes. What is the probability that more than three planes arrive in a 1-minute interval?

SOLUTION Let $P_n(t)$ be the probability that exactly n planes arrive in a time interval of t minutes. Because of the random nature of the arrivals, we conclude that $P_n(t) = (kt)^n e^{-kt}/n!$, where $k = \frac{1}{2}$. The probability that more than three planes arrive in a 1-minute interval is $P_4(1) + P_5(1) + \cdots = 1 - P_0(1) - P_1(1) - P_2(1) - P_3(1)$. Since $P_n(1) = e^{-1/2}(\frac{1}{2})^n/n!$, the probability in question is

$$1 - e^{-1/2} - e^{-1/2}\left(\frac{1}{2}\right) - \frac{e^{-1/2}(\frac{1}{2})^2}{2!} - \frac{e^{-1/2}(\frac{1}{2})^3}{3!},$$

which equals

$$1 - e^{-1/2}\left[1 + \frac{1}{2} + \frac{(\frac{1}{2})^2}{2!} + \frac{(\frac{1}{2})^3}{3!}\right] \doteq 1 - 0.99818 = 0.00182.$$

Thus the odds are about 1 in 600 that more than three planes arrive in a given 1-minute period. $\bullet$

Exercises

1. Find k in the Poisson distribution if the average number of cars per foot of road is 0.002.

2. Find k in the Poisson distribution if the average number of cars in
  (a) 1,000 feet of road is 5;
  (b) 1,000 feet of road is 20.

3. If a mile of road contains, on the average, two cars, find (and express to two decimal places) the probability that
  (a) a mile of road has no cars;
  (b) a mile of road has exactly one car;
  (c) a mile of road has exactly two cars;
  (d) a mile of road has more than two cars.

4. If the average number of cars in 1,000 feet of road is 10, what is the probability that
  (a) a section of length 100 feet has no cars;
  (b) a section of length 100 feet has exactly one car;
  (c) a section of length 100 feet has exactly two cars?

5. In a large, continually operating factory there are, on the average, two accidents per hour. Let $P_n(x)$ denote the probability that there are exactly n accidents in a time interval of length x hours.
  (a) Why is it reasonable to assume that there is a constant k such that $P_0(x)$, $P_1(x)$, ... satisfy assumptions 1 and 2?
  (b) If these assumptions are satisfied, show that $P_n(x) = (kx)^n e^{-kx}/n!$.
  (c) Why must k equal 2?
  (d) Compute $P_0(1)$, $P_1(1)$, $P_2(1)$, $P_3(1)$, and $P_4(1)$.

6. A brief rain left, on the average, two raindrops per square inch on a large flat surface. Let $P_n(x)$ be the probability that exactly n drops fell on a region of area x square inches. Explain carefully why you would expect $P_n(x) = (2x)^n e^{-2x}/n!$.

7. A typesetter makes an average of one mistake per page. Let $P_n(x)$ be the probability that a section of x pages (x need not be an integer) has exactly n errors.
(a) Why would you expect $P_n(x) = x^n e^{-x}/n!$?
(b) Approximately how many pages would be error-free in a 300-page book?

8. A cloud chamber registers an average of four cosmic rays per second.
(a) What is the probability that no cosmic rays are registered in a period of 6 seconds?
(b) What is the probability that exactly two are registered in a period of 4 seconds?

9. Telephone calls between 9 and 10 A.M. arrive at a rate of three per minute. What is the probability that during that time none arrive in a period of
(a) $\frac{1}{2}$ minute;
(b) 1 minute;
(c) 3 minutes?

10. (a) Look at 25 consecutive pages of an illustrated dictionary and record the number of pages that have 0, 1, 2, ... illustrations.
(b) What is the average number of illustrations per page?
(c) Let $P_n(x)$ be the probability that x pages have exactly n illustrations. With the data in (a), estimate $P_n(1)$ for $n = 0, 1, \ldots, 6$.
(d) Compare the result in (c) with $P_n(x) = (kx)^n e^{-kx}/n!$, where k is given in (b) of this exercise.

11. A town has an average of two automobile accidents per day. What is the probability that there will be more than two accidents in a given day?

12. Let $P_0(x)$ denote the probability that there are no cars in a section of road of length x.
(a) Why would you expect that $P_0(a + b) = P_0(a) \cdot P_0(b)$ for any a and b?
(b) Verify that $P_0(x) = e^{-kx}$ satisfies the equation in (a).

■

13. Write x^2 in the form $g(x)e^{-kx}$.
14. Show that $P_3(x) = (kx)^3 e^{-kx}/3!$.
15. $P_n(x)$ denotes the probability that there are n cars in a section of road of length x.

(a) Why would you expect $P_3(a + b) = P_0(a)P_3(b) + P_1(a)P_2(b) + P_2(a)P_1(b) + P_3(a)P_0(b)$?
(b) Do the formulas $P_n(x) = (kx)^n e^{-kx}/n!$ satisfy the equation in (a)?

16. (a) Why would you expect $\lim_{n \to \infty} P_n(x) = 0$?
(b) Using the formula $P_n(x) = (kx)^n e^{-kx}/n!$, show that the limit is 0.

17. (a) Why would you expect $\lim_{x \to \infty} P_n(x) = 0$?
(b) Using the formula $P_n(x) = (kx)^n e^{-kx}/n!$, show that the limit is 0.

18. Describe the behavior of $P_n(x)$, with $n > 0$, for small x and for large x, (a) using only your intuition; (b) using the formula $P_n(x) = (kx)^n e^{-kx}/n!$.

19. From the fact that $P_n(x) = (kx)^n e^{-kx}/n!$, obtain the Maclaurin series for e^x.

20. We obtained $P_0(x) = e^{-kx}$ and $P_1(x) = kxe^{-kx}$. Verify that $\lim_{\Delta x \to 0} P_1(\Delta x)/\Delta x = k$, and $\lim_{\Delta x \to 0} [1 - P_0(\Delta x)]/\Delta x = 1 - k$. Hence show that

$$\lim_{\Delta x \to 0} [P_2(\Delta x) + P_3(\Delta x) + \cdots]/\Delta x = 0,$$

and that assumptions 1 and 2 are indeed satisfied.

21. What length of road is most likely to contain exactly one car? That is, what x maximizes $P_1(x)$? Express the answer in terms of k.

22. What length of road is most likely to contain three cars?

23. Obtain assumption 1 from Eq. (3), Eq. (6) from assumption 2, assumption 2 from Eq. (6).

24. A person walking on the side of a road counted 500 beer cans in 5,000 feet. What is the probability that a section of 10 feet has (a) no cans, (b) exactly one can, (c) exactly two cans?

■ ■

25. (a) Obtain the formula $P_n(x) = (kx)^n e^{-kx}/n!$ for $n = 4$ and $n = 5$.
(b) Using mathematical induction, obtain the formula for $P_n(x)$ for all n.

26. (See Exercise 12.) Assuming that f is a continuous function and that $f(x + y) = f(x)f(y)$ for all x and y, show that $f(x) = a^x$ for some fixed number a.

17.4

Summary

In this chapter the Poisson distribution, a mathematical model for random traffic is developed. After introducing some basic concepts in probability, it was shown how definite integrals, improper integrals, derivatives, and infinite series can be used to draw conclusions from this model. The review exercises at the end of this section present more illustrations of mathematical modeling.

KEY FACTS

To find the probability that one of several mutually exclusive events occurs, *add* their individual probabilities.

To find the probability that all of several independent events occur, *multiply* their individual probabilities.

If the probability that an experiment results in the integer n is $p(n)$, then the expected or average value of n is defined as $\sum_{n=0}^{\infty} np(n)$. [For example, $p(n)$ may be the probability of throwing the sum n with two dice, or it may be the probability that there are n cars in a certain section of road.]

If the probability that an experiment results in a nonnegative number $\leq x$ is $F(x)$, then the expected or average value of x is defined as $\int_0^{\infty} xF'(x)\, dx$.

The Poisson distribution is defined by two assumptions:

1. The probability that an event occurs in a very small interval (of a line, of time, etc.) is approximately k times the length of that interval.
2. The probability that more than one event occurs in a very small interval is negligible.

From this it is deduced that

$$P_n(x) = \frac{(kx)^n e^{-kx}}{n!},$$

where $P_n(x)$ is the probability that in an interval of length x exactly n events occur. k turns out to be the average number of events per unit length (of distance, of time, etc.).

The Poisson distribution is descriptive of such random (and independent) events as telephone calls, arrivals at a checkout counter, and automobiles moving without congestion.

VOCABULARY AND SYMBOLS

expected (average) value
$\sum_{n=0}^{\infty} np(n)$

cumulative probability distribution F (also probability distribution or distribution)

expected (average) value
$\int_0^{\infty} xF'(x)\, dx$

Poisson model of random traffic

$P_n(x)$ probability of exactly n events in an interval of length x in the Poisson model

k average number of events in unit length in the Poisson model

Guide quiz on Chap. 17

1. (a) What is the probability of throwing both 1s ("snake eyes") with two dice?
   (b) On the average, how many times will you have to toss two dice in order to get both 1s?
2. (a) What is the probability of throwing a total of 4 when tossing two dice?
   (b) What is the probability that you throw a total of 4 for the first time on the third toss?
3. Find the expected value of the distribution $1 - e^{-5x}$, beginning with the definition of expected value.
4. There are on the average two fires a day in a certain town. What is the probability that there are
   (a) exactly two fires in a day?
   (b) more than two fires in a day?
5. There are on the average 30 raisins in a loaf of bread. If it is sliced into 10 pieces, what is the probability that a given slice has (a) no raisins, (b) exactly one raisin, (c) exactly two raisins, (d) exactly three raisins, (e) more than three raisins?
6. What is the probability of obtaining one head and one tail when you throw two coins?
7. (a) What is the probability of throwing a total of 5 with two dice?
   (b) What is the probability of not throwing a total of 5?
   (c) What is the average number of throws required to throw a total of 5?
8. (a) Define *probability distribution*.
   (b) If F is a probability distribution, show why its expected value is defined to be $\int_0^{\infty} xF'(x)\, dx$.
9. State the two assumptions of the Poisson model.
10. Give three different examples of random events that presumably satisfy assumptions 1 and 2 for a Poisson model.

Other models

The preceding guide quiz serves as a review of the Poisson distribution. The discussion and exercises here will develop the notion of mathematical models a little more fully.

The procedure of mathematical modeling may be described by this diagram:

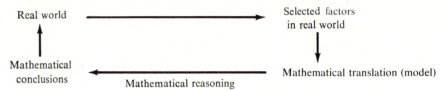

The model (or translation) may be a differential equation or a larger mathematical structure (like euclidean geometry) described by axioms, which records the assumptions.

In the first step, based on experiments in the real world and on intuition, certain factors may be singled out as relevant. Next, relations between the factors are translated into mathematics. Then, by purely mathematical reasoning, conclusions are drawn about the mathematical model. These are then interpreted back into the real world. A good model should yield more information about the real world than was put into it. The value of a mathematical model is the light it sheds on the real world.

As we saw, the Poisson model tells a good deal about the real world, far more than we had in mind when framing its two basic assumptions. In fact, this model has even been used to analyze delay at an intersection. Calculations indicate that the delay increases much more quickly than the traffic density, which is related to the size of the population. A 4-fold increase in density can produce a 72-fold increase in delay. It may be that other nuisances grow disproportionately as the population grows.

Newton, whose inverse square law of gravitation is a model for the motions of all the planets and comets, observed in his *Principia* (Univ. of Calif. Press, Berkeley, p. 550):

> ...we do not know in what manner the ancients explained ... how the planets came to be retained ... into regular revolutions in curvilinear orbits. Probably it was to give some sort of satisfaction to this difficulty that solid orbs had been introduced.
>
> The later philosophers pretend to account for it either by the action of certain vortices, as Kepler and Descartes; or by some other principle of impulse or attraction, as Borelli, Hooke, and others of our nation.
>
> But our purpose is only to trace out the quantity and properties of this force from the phenomena, and to apply what we discover in some simple cases as principles, by which, in a mathematical way, we may estimate the effects thereof in more involved cases ...
>
> We said, *in a mathematical way*, to avoid all questions about the nature or quality of this force, which we would not be understood to determine by any hypothesis.

In modern times mathematical modeling is required in many areas, where rules of thumb give no clue and experiments are impossible or too expensive to carry out. Mathematical models have been made to help answer such questions as, "If the tax on a gallon of gasoline is raised x cents, how much less gasoline will be consumed?" "What will the population of the nation be x years from now?" "How much traffic can a freeway with x lanes carry?" "How can we control a mixture of inflation and recession?"

A few simple mathematical models included in this text are listed in this table

Model	Reference
Natural growth and decay	Sec. 6.4
Inhibited growth	Sec. 6.4, Exercise 28
Ideal lot	Sec. 6.8, Example 5
Traffic in tunnel	Sec. 6.8, Exercise 33
Present value of future income	Sec. 10.6, Example 6
Economic production	Sec. 12.6
Warehouse location	Sec. 13.6, Exercises 40 to 45
Epidemics	Sec. 13.6, Exercise 47
Location of fire station	Sec. 13.6, Exercise 54
Multiplier effect	Sec. 14.2, Example 1

Review exercises for Chap. 17

In this exercise a simple model is developed to explain the periodic depressions and recoveries of the economy.

1. Let $y(t)$ be the total economic activity at time t, measured in dollars of goods and services purchased.

(a) What is the meaning of dy/dt?

(b) What is the meaning of d^2y/dt^2?

(c) What is the meaning of the equation

$$\frac{d^2y}{dt^2} = -k(y - A),$$

where k and A are positive constants?

(d) From (c) deduce that

$$\frac{d^2(y - A)}{dt^2} = -k(y - A).$$

(e) Recalling harmonic motion, show that the function $y(t)$ is periodic.

In this exercise a simple model is developed to describe the behavior of an epidemic as a function of time.

2. Consider a population of size n. At any time t let there be $x(t)$ susceptible individuals, $y(t)$ infected individuals, and $z(t)$ who are immune, by isolation, vaccination, or recovery. Let $t = 0$ represent the initial time, and assume that $x(0)$ is approximately n and that $z(0) = 0$. What is the meaning of these differential equations in which a and b are positive constants?

(a) $\dfrac{dx}{dt} = -axy,$ (b) $\dfrac{dy}{dt} = axy - by,$

(c) $\dfrac{dz}{dt} = by.$

(d) Show that, if $x(0) < b/a$, the number of infected individuals decreases as a function of time. This means the epidemic is not serious.

(e) The quotient b/a, denoted ρ, is the *critical threshold*. Show that $dx/dz = -x/\rho$.

(f) Show that $x(t) = x(0)e^{-z/\rho}$.

(g) Show that $dz/dt \doteq b[n - z - x_0(1 - z/\rho + z^2/2\rho^2)]$.

Public health officials usually report daily or weekly the number of recoveries; this corresponds to dz/dt.

(h) Show, on the basis of (g) and the assumption that $x(0) \doteq n$, that as $dz/dt \to 0$, $z \to 2\rho(1 - \rho/n)$.

(i) As mentioned in (d), if $n < \rho$, there is no serious

epidemic. If $n > \rho$, let $n = \rho + v$. Show that, if v is small, then as $dz/dt \to 0$, $z \to 2v$.

(j) Show that, if v is small, $x(t) \to n - 2v$ as $dz/dt \to 0$. Thus $x(t) \to n - 2v = \rho - v$, which is below the threshold ρ [by as much as $x(0)$ was above the threshold]. According to (d), the epidemic runs its course and is no longer serious. This is known as the *threshold theorem*.

This exercise describes a learning model.

3. A learning experiment was set up in such a way that a dog could avoid an electric shock by jumping over a hurdle within 10 seconds after being placed in a compartment. Each dog tested gradually learned to avoid the shock, though none knew how to avoid it at the beginning of the experiment. We describe a theoretical model of the learning process: Let p be the probability that the dog avoids the shock at a certain trial. If he avoids the shock, then the probability that he will avoid it on the next trial is $0.80p + 0.20$. If he is shocked, then the probability that he will avoid the shock on the next trial is $0.92p + 0.08$.

(a) What is p at the beginning of the experiment?

(b) What is p when the dog fully understands how to avoid the shock?

(c) Let us use p as a measure of the dog's understanding of the experiment. Does the dog's understanding increase at each trial? Is the dog's understanding increased more by his being shocked or by his jumping in time?

(d) Show that p is at least 0.08 at the second trial and at least 0.1536 at the third trial.

(e) Show that p approaches 1 as the trials continue.

The behavior of 30 statistical dogs (stat-dogs) run off randomly on a computer in accordance with this theory is in remarkably close agreement with that of 30 real dogs.

This exercise illustrates a problem involving international politics.

4. Let x and y denote the arms budgets of two rival nations. L. F. Richardson, a pioneer in the application of mathematics to political science, made these two assumptions about the rate at which the budget changes with respect to time:

$$\frac{dx}{dt} = k_1 y - k_2 x + k_3 \qquad \frac{dy}{dt} = c_1 x - c_2 y + c_3$$

where k_1, k_2, k_3, c_1, c_2, and c_3 are constants (k_1, k_2, c_1, and c_2 are positive).

(a) Which term represents "fatigue"?

(b) Which term represents the "threat of the rival's actions"?

(c) Which term represents the "general attitude" of one nation about the other?

These equations have been used to analyze Russian-American relations, in particular to decide whether Russia is motivated by ideology or nationalism.

18

DEFINITE INTEGRALS OVER SOLID REGIONS

In Chapter 7 the definite integral over an interval was introduced, and in Chap. 13, the definite integral over a plane region. In this chapter definite integrals over solid regions are considered. These arise, for instance, in the study of fluid flow, gravity, and rotation.

In Section 18.1 the definite integral of a function over a solid region is defined. In the remaining sections in this chapter repeated integrals are developed for computing it. As always, keep in mind that only for the simplest integrands and regions can the repeated integrals be computed by the fundamental theorem of calculus. This is not a tragedy, since with the aid of computing machines accurate approximating sums can be quickly computed. Moreover, the primary use of these definite integrals over solid regions is conceptual: to help provide an adequate vocabulary for the application of mathematics to the physical world.

Throughout the chapter the terms "spherical ball" and "ball" refer to the solid sphere.

18.1

The definite integral of a function over a region in three-dimensional space

The notion of a definite integral over an interval in the line, or over a plane region, generalizes easily to integrals over solids located in space. (These solids will be assumed to be bounded by smooth surfaces or planes.) Rather than plunge directly into the definition, let us first illustrate the idea with a problem.

PROBLEM   A cube of side 4 inches is made of a material of varying density. Near one corner A it is very light; at the opposite corner it is very dense. In fact, the density $f(P)$ at any point P in the cube is the square of the distance from A to P (ounces per cubic inch). How do we estimate the mass of the cube?

SOLUTION   We can proceed exactly as in the case of the string of Sec. 7.1 and the rectangular plate of Sec. 13.1. First, partition the cube into regions $R_1, R_2, \ldots, R_n$; then compute the density at a selected point P_i in each R_i and form the sum

$$f(P_1)V_1 + f(P_2)V_2 + \cdots + f(P_n)V_n,$$

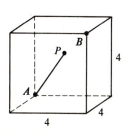

The density at P is the square of the distance AP

681

where V_i is the volume of R_i. As the R_i's become smaller, we obtain more reliable estimates of the total mass of the cube.

Observe first of all that the maximum density is the square of the length of the longest diagonal. This density is $\overline{AB}^2 = 4^2 + 4^2 + 4^2 = 48$ ounces per cubic inch. Since the total volume is $4 \cdot 4 \cdot 4 = 64$ cubic inches, the total mass is less than $48 \cdot 64 = 3{,}072$ ounces.

The arithmetic in evaluating even the simplest approximating sum is tedious. It may be of value, though, to go through the agony of computing one such sum. The following is a sample.

Partition the cube into four 2- by 2- by 4-inch boxes, as shown and labeled in the diagram at the left. This table displays the computation:

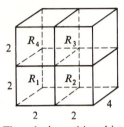

The cube is partitioned into 4 boxes, R_1, R_2, R_3, R_4

Region	Volume	Minimum Density	Maximum Density	Mass Is Between
R_1	16	0	$2^2 + 2^2 + 4^2 = 24$	0 and 384 ounces
R_2	16	$2^2 = 4$	$4^2 + 4^2 + 2^2 = 36$	64 and 576 ounces
R_3	16	$2^2 + 2^2 = 8$	$4^2 + 4^2 + 4^2 = 48$	128 and 768 ounces
R_4	16	$2^2 = 4$	$2^2 + 4^2 + 4^2 = 36$	64 and 576 ounces

Thus the mass of the cube is between

$$0 + 64 + 128 + 64 = 256 \text{ ounces}$$

and
$$384 + 576 + 768 + 576 = 2{,}304 \text{ ounces.}$$

This is much more information than the fact that the mass is less than 3,072 ounces. ●

If the cube is cut into smaller regions, perhaps sixty-four 1- by 1- by 1-inch cubes, more accurate estimates can be made. The important idea is that the procedure for making an approximation is practically the same that led to the sums

$$\sum_{i=1}^{n} f(X_i)(x_i - x_{i-1})$$

of Chap. 7 and to the sums $$\sum_{i=1}^{n} f(P_i)A_i$$

of Chap. 13.

Two definitions are needed before defining the definite integral of a function over a region R in space.

DEFINITION *Diameter of a region in space.* Let S be a set of points in space bounded by some surface or polyhedron. The *diameter* of S is the largest distance between two points of S.

For instance, the diameter of a cube of side s is $s\sqrt{3}$, the length of its longest diagonal. The diameter of a sphere is its customary diameter (twice the radius).

DEFINITION   *Mesh of a partition in space.*   Let R_1, R_2, ..., R_n be a partition of a region R in space.   The *mesh* of this partition is the largest of the diameters of the regions R_1, R_2, ..., R_n.

The functions of interest in this chapter will have some region R in space in their domain.   A function f will assign to each point P in R a number, denoted $f(P)$.   For the sake of concreteness, think of $f(P)$ as the density at P or temperature at P.

The graph of a function of one variable, $y = f(x)$, is a curve in the xy plane, the set of points $(x, f(x))$.   The graph of a function of two variables, $z = f(x, y)$, is a surface in space, the set of points $(x, y, f(x, y))$.   The graph of a function of three variables (that is, a function defined on a region in space) is a set in four-dimensional space, the set of the points $(x, y, z, f(x, y, z))$.   Since our eyes and intuition are accustomed to a three-dimensional world, this graph is of little use.   For this reason, it is best to think of a function defined on a solid region as, say, describing the varying density of a distribution of matter.

DEFINITION   *The definite integral of a function f over a set R in space.*   Let f be a function that assigns to each point P of a region R in space a number $f(P)$.   Consider the sum

$$f(P_1)V_1 + f(P_2)V_2 + \cdots + f(P_n)V_n,$$

formed from a partition R_1, R_2, ..., R_n of R, where V_i is the volume of R_i, and P_i is in R_i.   If these sums approach a certain number as the mesh of the partition shrinks toward 0 (no matter how P_i is chosen in R_i), we call that certain number the definite integral of f over the region R.   The definite integral of f over R is denoted

$$\int_R f(P)\, dV.$$

If $f(P)$ is thought of as the density at P of some solid matter, the definite integral can be interpreted as the total mass of the solid.

EXAMPLE   If $f(P) = 1$ for each point P in a solid region R, compute $\int_R f(P)\, dV$.

SOLUTION   Each approximating sum

$$\sum_{i=1}^{n} f(P_i)V_i$$

has the value

$$\sum_{i=1}^{n} 1 \cdot V_i = V_1 + V_2 + \cdots + V_n$$

$$= \text{volume of } R.$$

Hence

$$\int_R f(P)\, dV = \text{volume of } R,$$

a fact that will be useful later in this chapter for computing volumes.   ●

Average of a Function   The average value of a function f defined on a region R in space is defined as

$$\frac{\int_R f(P)\, dV}{\text{volume of } R}.$$

This is the analog of the definition of the average of a function over an interval (Sec. 10.5) or the average of a function over a plane region (Sec. 13.1). If f describes the density of matter in R, then the average value of f is the density of a *homogeneous* solid occupying R and having the same total mass as the given solid. [For if the number

$$\frac{\int_R f(P)\, dV}{\text{volume of } R}$$

is multiplied by   volume of R,

the result is $\displaystyle\int_R f(P)\, dV,$

which is the total mass.]

The average value and the mass are important applications of the definite integral over a solid, as they are for definite integrals over an interval or planar region. The total gravitational attraction of the sun on the earth or of the earth on a satellite and the centers of gravity of physical bodies will provide further applications.

Exercises

1. Find upper and lower estimates for the mass of the cube in this section by partitioning it into eight cubes.
2. Using the same partition as in the text, estimate the mass of the cube, but select as the P_i's the center of each of the four rectangular boxes.
3. If R is a ball of radius r and $f(P) = 5$ for each point in R, compute $\int_R f(P)\, dV$ by examining approximating sums. Assume that the ball has volume $\frac{4}{3}\pi r^3$.
4. How would you define the average distance from points of a certain set in space to a fixed point F?
5. Estimate the mass of the cube described in the problem by cutting it into eight congruent cubes and using their centers as the P_i's.
6. If R is a three-dimensional set, and $f(P)$ is never more than 8 for all P in R,
   (a) What can we say about the maximum possible value of $\int_R f(P)\, dV$? *Hint:* Consider approximating sums.
   (b) What can we say about the average of f over R?
7. What is the mesh of the partition of the cube used in the text?
8. What is the mesh of the partition used in Exercise 1?

9. A point Q is 5 units away from the center of a ball R of radius 3. For any point P in R define $f(P)$ to be the reciprocal of the distance from P to Q. The definite integral $\int_R f(P)\, dV$ is of great importance in gravitational theory. Show that it is between $9\pi/2$ and 18π.
10. The work done in lifting a weight of w pounds a vertical distance of x feet is wx foot-pounds. Imagine that through geological activity a mountain is formed consisting of material originally at sea level. Let the density of the material near point P in the mountain be $g(P)$ pounds per cubic foot and the height of P be $h(P)$. What definite integral represents the total work expended in forming the mountain? This type of problem is important in the geological theory of mountain formation.

■ ■

11. What can be said about the volume of a solid if its diameter is 10?

18.2

Describing solid regions with rectangular coordinates

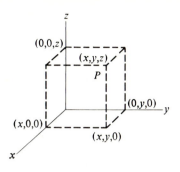

Rectangular coordinates in space are a natural extension of rectangular coordinates in the plane. A point P is described by three numbers (x, y, z), where z records the distance from P to the xy plane (positive if P is above the xy plane, negative if P is below).

The set of all points $P = (x, y, z)$ for which $x = 2$ is a plane parallel to the yz plane. The diagram below shows the plane and three points in it.

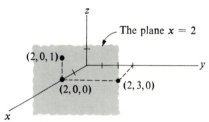

Rectangular coordinates are especially suitable for describing planes and solids bounded by planes. (In Sec. 19.4 it is shown that the equation $Ax + By + Cz + D = 0$ describes a plane.) In preparation for the repeated integrals that will be used to evaluate integrals over three-dimensional sets, we illustrate how to describe solid regions with rectangular coordinates in space.

EXAMPLE I   Describe in terms of x, y, and z the rectangular box shown in this diagram:

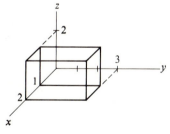

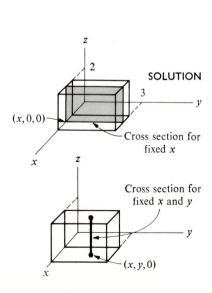

SOLUTION   First of all, x may be any number between 1 and 2. For each x, the cross section of the box made by a plane through $(x, 0, 0)$ and parallel to the yz plane is a rectangle. On this typical rectangle y varies from 0 to 3, independently of x.

Each pair x and y that satisfies the above conditions, $1 \le x \le 2$ and $0 \le y \le 3$, determines a cross section of the box by the line parallel to the z axis passing through $(x, y, 0)$. On the cross section illustrated at the left below, which is a line segment, z varies from 0 to 2, independently of x and y.

The description of the box is simply:

$$1 \le x \le 2, \qquad 0 \le y \le 3, \qquad 0 \le z \le 2.$$

This description is to be read from left to right. ●

EXAMPLE 2 Describe the cross sections of the tetrahedron bounded by the planes $x = 0$, $y = 0$, $z = 0$, and $x + y + z = 1$.

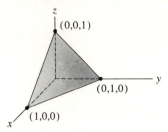

SOLUTION For any x in $[0, 1]$, the cross section of the tetrahedron made by the plane parallel to the yz plane and passing through $(x, 0, 0)$ is a triangle, such as the one shaded in the accompanying diagram.

On this typical triangle, y varies from 0 up to the value for y that satisfies the equation $x + y = 1$, that is, up to $y = 1 - x$.

Finally, for each x and y given above, z varies from 0 up to the value of z that satisfies the equation $x + y + z = 1$, that is, up to $z = 1 - x - y$. This is a description of the behavior of z on the line parallel to the z axis and passing through $(x, y, 0)$. The tetrahedron is described by these equations

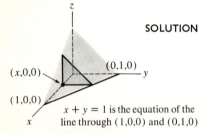

$x + y = 1$ is the equation of the line through $(1,0,0)$ and $(0,1,0)$

$$0 \le x \le 1, \qquad 0 \le y \le 1 - x, \qquad 0 \le z \le 1 - x - y. \quad \bullet$$

In the next example the description of the solid is obtained by first holding x and y fixed and letting z vary.

EXAMPLE 3 Describe in rectangular coordinates the ball of radius 4 whose center is at the origin.

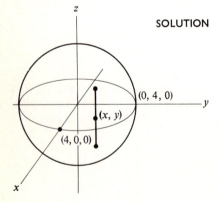

SOLUTION Hold (x, y) fixed in the xy plane and consider the way z varies on the line parallel to the z axis that passes through the point $(x, y, 0)$. Since the sphere that bounds the ball has the equation

$$x^2 + y^2 + z^2 = 16,$$

for each appropriate (x, y), z varies from

$$-\sqrt{16 - x^2 - y^2} \qquad \text{to} \qquad \sqrt{16 - x^2 - y^2}.$$

This describes the line segment shown in the diagram.

Next describe the possible values of x and y. Since (x, y) ranges over a disk of radius 4 and center $(0, 0)$ in the xy plane,

$$-4 \le x \le 4, \qquad -\sqrt{16 - x^2} \le y \le \sqrt{16 - x^2}.$$

The ball, therefore, has the description

$$-4 \le x \le 4, \qquad -\sqrt{16 - x^2} \le y \le \sqrt{16 - x^2}, \qquad -\sqrt{16 - x^2 - y^2} \le z \le \sqrt{16 - x^2 - y^2}. \quad \bullet$$

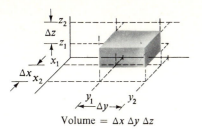

Volume $= \Delta x \, \Delta y \, \Delta z$

Consider all points (x, y, z) that satisfy the conditions

$$x_1 \le x \le x_2, \qquad y_1 \le y \le y_2, \qquad z_1 \le z \le z_2,$$

where $x_1, x_2, y_1, y_2, z_1, z_2$ are fixed numbers. As in Example 1, this set is a box, shown in the diagram at the left.

The volume of this box is the product of the lengths of three perpendicular edges:

$$(x_2 - x_1)(y_2 - y_1)(z_2 - z_1).$$

Consequently, the solid consisting of all points whose x coordinate is between x and $x + \Delta x$, whose y coordinate is between y and $y + \Delta y$, and whose z coordinate is between z and $z + \Delta z$, is a rectangular box whose volume is precisely

$$\Delta x \, \Delta y \, \Delta z.$$

This fact will be needed when setting up repeated integrals in rectangular coordinates.

Exercises

1. Describe the tetrahedron in Example 2 in rectangular coordinates by taking, first, cross sections for fixed z, then letting y vary for each z, and then letting x vary for each pair of y and z.
2. Describe in rectangular-coordinate cross sections the tetrahedron bounded by the three coordinate planes and the plane $(x/2) + (y/3) + (z/4) = 1$.
3. Describe the solid region below the plane $z = x + y + 2$ and above the circle in the xy plane of radius 1 and center $(0, 0)$. Draw the cross sections for fixed x and for fixed x and y.
4. (a) Describe in rectangular coordinates the cylinder of radius a and height h shown below.

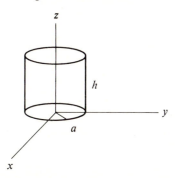

(b) Draw the cross sections for fixed x and for fixed x and y.
5. (a) Describe in rectangular coordinates the solid region below the plane $z = 3x + 4y$ and above the triangle

in the xy plane whose vertices are $(0, 0, 0)$, $(2, 0, 0)$, and $(0, 1, 0)$.
(b) Draw the cross sections for fixed x and for fixed x and y.
6. Describe in rectangular coordinates the right circular cone of radius r and height h if its axis is on the positive z axis and its vertex is at the origin. Draw the cross sections for fixed x and for fixed x and y.
7. Describe in rectangular coordinates the spherical ball of radius a and center $(0, 0, 0)$.
8. Describe in rectangular coordinates the solid bounded by the ellipsoid

$$\frac{x^2}{a^2} + \frac{y^2}{b^2} + \frac{z^2}{c^2} = 1,$$

where a, b, and c are positive.
9. Describe in rectangular coordinates the ball of radius 4 whose center is at $(1, 2, 3)$.
10. Describe in rectangular coordinates the cone of radius a and height h shown below.

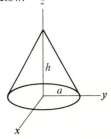

In Exercises 11 to 15 draw the solid described.

11. $1 \le x \le 3, 0 \le y \le 2, 0 \le z \le x$.
12. $0 \le x \le 1, 0 \le y \le 1, 1 \le z \le 1 + x + y$.
13. $0 \le x \le 1, 0 \le y \le x^2, 0 \le z \le 1$.
14. $0 \le x \le 1, x^2 \le y \le x, 0 \le z \le x + y$.
15. $-1 \le x \le 1, -1 \le y \le 1, 0 \le z \le \sqrt{4 - x^2 - y^2}$.

■

16. Draw the intersection of the solid cylinders bounded by $x^2 + z^2 = a^2$ and $y^2 + z^2 = a^2$.
17. Draw the typical cross section, if the solid of Exercise 16 is intersected by a plane parallel to the xy plane, but less than a units from that plane.

18.3

Describing solid regions with cylindrical or spherical coordinates

Cylindrical Coordinates

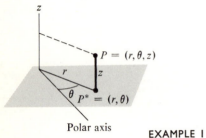

Cylindrical coordinates combine polar coordinates in the plane with the z of rectangular coordinates in space. Each point P in space receives the name (r, θ, z), as in the accompanying diagram. We are free to choose the direction of the polar axis; usually it will coincide with the x axis of an (x, y, z) system. Note that (r, θ, z) is directly above (or below) $P^* = (r, \theta)$ in the $r\theta$ plane. Since the set of all points $P = (r, \theta, z)$ for which $r = k$, some constant, is a circular cylinder, this coordinate system is especially convenient for describing such cylinders.

EXAMPLE 1   Describe a solid cylinder of radius a and height h in cylindrical coordinates. Assume that the axis of the cylinder is on the positive z axis, and the lower base has its center at the pole.

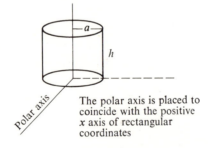

The polar axis is placed to coincide with the positive x axis of rectangular coordinates

SOLUTION   First of all, θ varies from 0 to 2π. If we hold θ fixed and consider only positive r, the cross section we obtain is a rectangle perpendicular to the $r\theta$ plane. Next, examine the behavior of r and z on the cross section pictured. First of

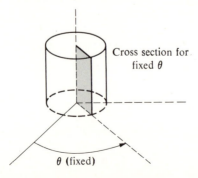

Cross section for fixed θ

θ (fixed)

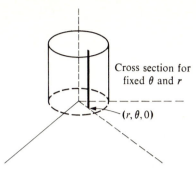

Cross section for fixed θ and r

$(r, \theta, 0)$

all, r goes from 0 to a independently of θ. Thus far it has been shown that

$$0 \le \theta \le 2\pi, \qquad 0 \le r \le a,$$

which is just a description of the shadow of the cylinder on the $r\theta$ plane cast by light parallel to the z axis.

Finally, hold r and θ fixed, and determine the behavior of z. The cross section for fixed r and θ is a line segment. On this line segment z varies from 0 to h. Hence the cylinder has this description:

$$0 \le \theta \le 2\pi, \qquad 0 \le r \le a, \qquad 0 \le z \le h. \ \bullet$$

EXAMPLE 2 Describe in cylindrical coordinates the region in space formed by the intersection of a solid cylinder of radius 3 with a ball of radius 5 whose center is on the axis of the cylinder. Locate the cylindrical coordinate system as shown in this diagram:

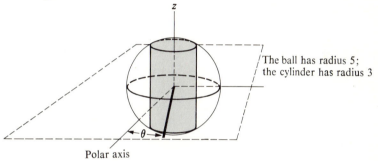

The ball has radius 5; the cylinder has radius 3

Polar axis

SOLUTION First, describe the surface of the ball in cylindrical coordinates. To do this, note that the point $P = (r, \theta, z)$ is a distance $\sqrt{r^2 + z^2}$ from the origin O. For, by the pythagorean theorem,

$$r^2 + z^2 = \overline{OP}^2.$$

Now consider the description of the solid. First of all, θ varies from 0 to 2π and r from 0 to 3, bounds determined by the cylinder. For fixed θ and r, the cross section of the solid is a line segment determined by the sphere that bounds the ball, as shown in the diagram in the margin. Now, since the sphere has radius 5, for any point (r, θ, z) on it,

$$r^2 + z^2 = 25,$$

or

$$z = \pm\sqrt{25 - r^2}.$$

Thus, on the line segment determined by fixed r and θ, z varies from $-\sqrt{25 - r^2}$ to $\sqrt{25 - r^2}$.

The solid has this description:

$$0 \le \theta \le 2\pi, \qquad 0 \le r \le 3, \qquad -\sqrt{25 - r^2} \le z \le \sqrt{25 - r^2}. \ \bullet$$

The set of all points (r, θ, z) such that

$$r_1 \le r \le r_2, \qquad \theta_1 \le \theta \le \theta_2, \qquad z_1 \le z \le z_2$$

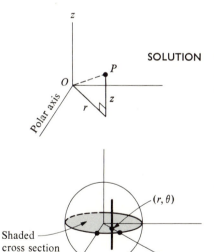

(r, θ)

Shaded cross section lies in the $r\theta$ plane

θ

is a solid with four flat surfaces and two curved surfaces, shown in this diagram.

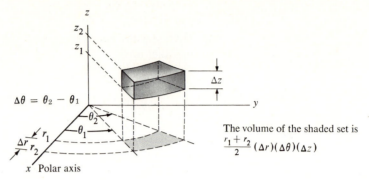

The volume of the shaded set is
$$\frac{r_1 + r_2}{2} (\Delta r)(\Delta\theta)(\Delta z)$$

The volume of this solid is the product of its height $z_2 - z_1$ by the area of its base. As mentioned at the end of Sec. 13.5, the area of the base is precisely

$$\frac{r_1 + r_2}{2} \Delta r \, \Delta\theta.$$

The volume is therefore $\dfrac{r_1 + r_2}{2} \Delta r \, \Delta\theta \, \Delta z.$

Consequently, the solid consisting of all points (r, θ, z) whose r coordinate is between r and $r + \Delta r$, whose θ coordinate is between θ and $\theta + \Delta\theta$, and whose z coordinate is between z and $z + \Delta z$, has volume equal to

$$\frac{r + (r + \Delta r)}{2} \Delta r \, \Delta\theta \, \Delta z.$$

If Δr, $\Delta\theta$, and Δz are small, the volume is then approximately

$$r \, \Delta r \, \Delta\theta \, \Delta z.$$

Note the extra factor r. It will appear whenever cylindrical coordinates are used to evaluate integrals over three-dimensional sets.

Spherical Coordinates   Spherical coordinates provide compact descriptions of cones and spheres. A point P is described by three numbers (ρ, ϕ, θ), as in the figure shown below. (The Greek letter ρ is pronounced "roe.") Observe that the sphere $x^2 + y^2 + z^2 = 5^2$ now has the simpler equation $\rho = 5$. Note also that the set of points for which ϕ is fixed is the surface of a cone whose axis is the z axis.

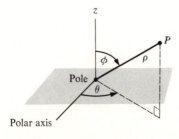

θ is the same as in cylindrical coordinates; ϕ is the indicated angle from the z axis $(0 \leqslant \phi \leqslant \pi)$; and ρ is the distance from P to the pole

The set of points for which θ is prescribed is a half plane bordered by the z axis. Thus the set of points for which both ϕ and θ are fixed is a ray emanating from the origin, namely, the intersection of the surface of a cone and a half plane whose edge is the axis of the cone.

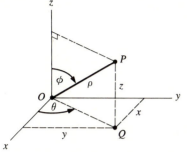

$z = OP \cos \phi = \rho \cos \phi$
$y = OQ \sin \theta = \rho \sin \phi \sin \theta$
$x = OQ \cos \theta = \rho \sin \phi \cos \theta$

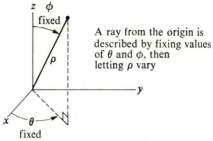

A ray from the origin is described by fixing values of θ and ϕ, then letting ρ vary

The rectangular coordinates of (ρ, ϕ, θ) can be found by inspection of the diagram in the margin.

EXAMPLE 3   Find the equation of the plane $z = 4$ in spherical coordinates.

SOLUTION   Since $$z = \rho \cos \phi$$

describes z in terms of spherical coordinates, the equation of the plane $z = 4$ is

$$\rho \cos \phi = 4,$$

or $$\rho = \frac{4}{\cos \phi},$$

or $$\rho = 4 \sec \phi. \bullet$$

EXAMPLE 4   Describe in spherical coordinates the solid cone of height 4 shown in the accompanying diagram.

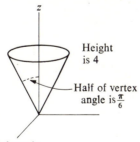

Height is 4

Half of vertex angle is $\frac{\pi}{6}$

Polar axis

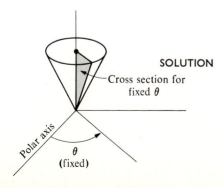

Cross section for fixed θ

Polar axis

θ (fixed)

SOLUTION   It is usually most convenient to examine first how θ varies. In this case θ goes from 0 to 2π. For each fixed θ the cross section is a triangle.

On this typical triangle ϕ varies from 0 to $\pi/6$. For each fixed θ and ϕ the cross section is a segment on a ray emanating from the origin, as will be noted in the top figure on page 692. On this cross section ρ varies from 0 to its value where the ray meets the plane $z = 4$. By Example 3, this plane has the

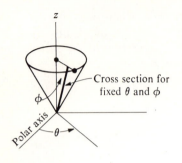

Cross section for fixed θ and ϕ

Polar axis

equation $\qquad\qquad\qquad \rho = 4 \sec \phi.$

Hence for fixed θ and ϕ, ρ varies from 0 to $4 \sec \phi$.

This is the description of the cone:

$$0 \le \theta \le 2\pi, \qquad 0 \le \phi \le \frac{\pi}{6}, \qquad 0 \le \rho \le 4 \sec \phi. \ \bullet$$

Notice the simplicity of the description of θ and ϕ for the cone in Example 4. Clearly, spherical coordinates are convenient for describing cones. They are also fine for a ball whose center is at the origin. To be specific,

$$0 \le \theta \le 2\pi, \qquad 0 \le \phi \le \pi, \qquad 0 \le \rho \le a$$

is the description of a ball of radius a whose center is at the origin.

The set of all points (ρ, ϕ, θ) such that

$$\rho_1 \le \rho \le \rho_2, \qquad \phi_1 \le \phi \le \phi_2, \qquad \theta_1 \le \theta \le \theta_2$$

is a solid R with two flat surfaces and four curved surfaces. We shall estimate its volume when $\rho_2 - \rho_1, \phi_2 - \phi_1,$ and $\theta_2 - \theta_1$ are small. This estimate will be needed when setting up repeated integrals in spherical coordinates.

R is bordered by the six surfaces shown below: by spheres of radius ρ_1 and ρ_2; by cones of half-vertex angles ϕ_1 and ϕ_2; by half planes of polar angles θ_1 and θ_2.

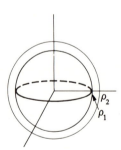

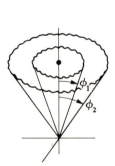

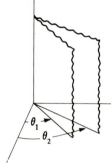

The little solid R looks like this:

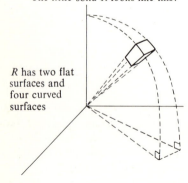

R has two flat surfaces and four curved surfaces

The little solid R appears as shown at the left. Lable the eight corners as has been done in the accompanying figure. $ABCD$ and $EFGH$ are spherical. $BCGF$ and $ADHE$ are conical. $ABFE$ and $DCGH$ are flat.

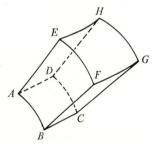

Since the small solid R resembles a rectangular box, its volume is approximated by the product $\overline{AE} \cdot \widehat{AB} \cdot \widehat{AD}$.

First of all, $\overline{AE}$ is just the difference in the radii of the two spheres:

$$\overline{AE} = \rho_2 - \rho_1.$$

Next, AB is an arc of a circle of radius ρ_1 and subtends an angle $\phi_2 - \phi_1$.

Thus

$$\widehat{AB} = \rho_1(\phi_2 - \phi_1).$$

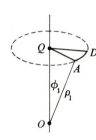

$\widehat{AD}$ subtends an angle $\theta_2 - \theta_1$

Finally, consider AD. It is an arc of a circle that is perpendicular to the z axis, as depicted at the left. AD subtends an angle $\theta_2 - \theta_1$. The radius of the dashed circle is computed from right triangle AOQ:

Radius of dashed circle $= \overline{QA} = \rho_1 \sin \phi_1$.

Thus arc AD has length $\rho_1 \sin \phi_1 (\theta_2 - \theta_1)$.

The volume of R is, therefore, approximately

$$\overline{AE} \cdot \widehat{AB} \cdot \widehat{AD} = (\rho_2 - \rho_1)\rho_1(\phi_2 - \phi_1)\rho_1 \sin \phi_1(\theta_2 - \theta_1)$$
$$= \rho_1{}^2 \sin \phi_1(\rho_2 - \rho_1)(\phi_2 - \phi_1)(\theta_2 - \theta_1).$$

Consequently, the solid consisting of all points (ρ, ϕ, θ) whose ρ coordinate is between ρ and $\rho + \Delta\rho$, whose ϕ coordinate is between ϕ and $\phi + \Delta\phi$, and whose θ coordinate is between θ and $\theta + \Delta\theta$, has a volume approximately

$$\rho^2 \sin \phi \, \Delta\rho \, \Delta\phi \, \Delta\theta. \tag{1}$$

Notice the factor $\rho^2 \sin \phi$. It will be needed in forming repeated integrals in spherical coordinates. Note that in (1) ρ and $\Delta\rho$ have the dimension of length, whereas $\Delta\phi$ and $\Delta\theta$ are dimensionless (radian measure is defined as the quotient of two lengths). Thus the dimension of (1) is *length cubed*, as it should be if it is to measure volume.

Exercises

Exercises 1 to 18 concern cylindrical coordinates. In Exercises 1 to 6 sketch a graph of all points (r, θ, z) such that

1. $r = 1$.

2. $\theta = \pi/4$ (consider only nonnegative r).

3. $z = 1$.

4. $r = z$.

5. $r^2 + z^2 = 9$.

6. $r = 2 \cos \theta$.

In Exercises 7 and 8 draw the cross section corresponding to fixed θ of the region described. (Restrict r to nonnegative values.)

7. The region R consists of all points within a distance a of the origin of the $r\theta z$-coordinate system.

8. The region R consists of all points of the solid of Exercise 7 that are within the cylinder whose equation is $r = a \cos \theta$, where a is a fixed positive number.

9. Describe the solid in Exercise 7 in cylindrical coordinates.

10. Describe the solid in Exercise 8 in cylindrical coordinates.

11. (a) What are the cylindrical coordinates of the point $P = (x, y, z)$?

   (b) What are the rectangular coordinates of the point $P = (r, \theta, z)$?

12. Describe in cylindrical coordinates the solid cone shown in the diagram below.

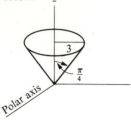

13. Give the equation in cylindrical coordinates of
   (a) the xy plane,
   (b) the plane $x = y$,
   (c) the z axis.

In Exercises 14 to 18 draw the solid described.

14. $0 \leq \theta \leq \pi/2, 1 \leq z \leq 2, 1 \leq r \leq z$.

15. $0 \leq \theta \leq 2\pi, 0 \leq r \leq 1, r \leq z \leq 1$.

16. $0 \leq \theta \leq \pi/2, 0 \leq r \leq \cos \theta, 1 \leq z \leq 2$.

17. $0 \leq \theta \leq \pi/2, 0 \leq r \leq 1, 0 \leq z \leq r \cos \theta$.

18. $0 \leq \theta \leq 2\pi, 0 \leq r \leq \cos \theta, r \leq z \leq \sqrt{4 - r^2}$.

Exercises 19 to 40 concern spherical coordinates. In Exercises 19 to 24 sketch the graph of all points (ρ, ϕ, θ) satisfying the given equation.

19. $\rho = 2$ 20. $\phi = \pi/6$

21. $\theta = \pi/2$ 22. $\phi = \pi/2$

23. $\phi = 0$ 24. $\phi = \pi$

25. What are the cylindrical coordinates of the point (ρ, ϕ, θ)?

26. Sketch the set of all points (ρ, ϕ, θ) such that $\phi = \pi/2$ and $\theta = \pi/2$.

In Exercises 27 to 31 describe in spherical coordinates the region R.

27. R is the ball of radius a, center at the origin.

28. R is the top half of the ball of radius a, center at the origin, above the xy plane.

29. R is the ice cream cone-shaped intersection of a solid cone and a ball shown in the figure below.

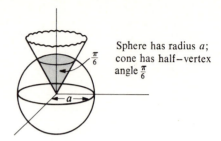

Sphere has radius a; cone has half–vertex angle $\frac{\pi}{6}$

30. R is the region between two spheres, both with center at the origin, of radii a and b, $a < b$.

31. R is the region in the cone shown in Exercise 29 below the plane $z = 3a/5$.

In Exercises 32 to 35 sketch the region described.

32. $0 \leq \theta \leq \pi/2, 0 \leq \phi \leq \pi/2, 0 \leq \rho \leq 1$.

33. $0 \leq \theta \leq 2\pi, \pi/2 \leq \phi \leq \pi, 1 \leq \rho \leq 2$.

34. $0 \leq \theta \leq \pi, 0 \leq \phi \leq \pi/4, 0 \leq \rho \leq \sec \phi$.

35. $0 \leq \theta \leq \pi/2, 0 \leq \phi \leq \pi/4, 1 \leq \rho \leq 2 \sec \phi$.

36. Fill in the blanks and explain with the aid of a sketch:
   Rectangular coordinates describe a point by specifying three planes on which it lies. Spherical coordinates describe a point by specifying _____, _____, and _____ on which it lies.
   Cylindrical coordinates describe a point by specifying _____, _____, and _____ on which it lies.

37. Find the spherical coordinates of the point whose cylindrical coordinates are $(r, \theta, z), r > 0$.

38. Sketch the solid whose description is $0 \leq \theta \leq \pi/2, \pi/4 \leq \phi \leq \pi/2, 1 \leq \rho \leq 2$.

39. Find the spherical coordinates of the point whose rectangular coordinates are (x, y, z).

40. Find the equation in spherical coordinates of the plane
   (a) $x = 2$,
   (b) $2x + 3y + 4z = 1$.

18.4

Computing $\int_R f(P)\, dV$ with rectangular coordinates

In Chap. 13 repeated integrals were used to evaluate definite integrals over two-dimensional sets. Similarly, appropriate repeated integrals can be used to evaluate integrals over three-dimensional sets. In this section we use repeated integrals in rectangular coordinates; in the next, repeated integrals in cylindrical and spherical coordinates.

Before computing any integrals, $\int_R f(P)\,dV$, we list some of their important three-dimensional applications:

$f(P)$	Significance of $\int_R f(P)\,dV$
1	$\int_R 1\,dV = $ volume of R
Density at P, usually denoted δ	$\int_R \delta\,dV = $ mass in R
$s^2\delta$, where δ is density at P and s is distance from P to some fixed line L (usually δ is 1).	$\int_R s^2\delta\,dV = $ moment of inertia of the mass in R about the line L. (Important in the study of a rotating body.)
$x\delta$, where δ is density at P and x is the x coordinate of P (usually δ is 1).	$\int_R x\delta\,dV = $ moment of the mass in R relative to yz plane. (Important in computing center of gravity of a body. See formula after table.)
δ/q, where δ is density at P and q is the distance from P to some fixed point A (usually δ is 1.)	$\int_R \dfrac{\delta}{q}\,dV = $ work done by the gravitational pull of the mass in R when moving a unit mass from "infinity" to the point A. The negative of this quantity is called the "potential."

Note: If δ is constant, then $\delta = M/V$, where M is the mass and V is the volume of the solid.

The center of gravity of the mass in R has the coordinates

$$\bar{x} = \frac{\int_R x\delta\,dV}{\text{mass in } R}, \qquad \bar{y} = \frac{\int_R y\delta\,dV}{\text{mass in } R}, \qquad \bar{z} = \frac{\int_R z\delta\,dV}{\text{mass in } R},$$

formulas similar to those in Sec. 13.4.

If $\delta = 1$, the body is homogeneous, and the formulas for the center of gravity become

$$\bar{x} = \frac{\int_R x\,dV}{\text{volume of } R}, \qquad \bar{y} = \frac{\int_R y\,dV}{\text{volume of } R}, \qquad \bar{z} = \frac{\int_R z\,dV}{\text{volume of } R}.$$

The center of gravity in this case is called the *centroid*.

The repeated integral in rectangular coordinates for evaluating $\int_R f(P)\,dV$ is similar to that for evaluating $\int_R f(P)\,dA$. It involves three, instead of two, integrations. The limits of integration are determined by the description of R in rectangular coordinates. If R has the description

$$a \le x \le b, \qquad y_1(x) \le y \le y_2(x), \qquad z_1(x, y) \le z \le z_2(x, y),$$

then
$$\int_R f(P)\, dV = \int_a^b \left[\int_{y_1(x)}^{y_2(x)} \left(\int_{z_1(x,y)}^{z_2(x,y)} f(P)\, dz \right) dy \right] dx.$$

Some examples illustrate how this formula is applied. In Exercise 15 an argument for its plausibility is presented.

EXAMPLE 1 Compute $\int_R z\, dV$, where R is the tetrahedron in Example 2 of Sec. 18.2.

SOLUTION The description of the tetrahedron is

$$0 \le x \le 1, \qquad 0 \le y \le 1 - x, \qquad 0 \le z \le 1 - x - y.$$

Hence
$$\int_R z\, dV = \int_0^1 \int_0^{1-x} \int_0^{1-x-y} z\, dz\, dy\, dx.$$

Compute the inner integral first, treating x and y as constants. By the fundamental theorem,

$$\int_0^{1-x-y} z\, dz = \frac{z^2}{2} \Big|_{z=0}^{z=1-x-y} = \frac{(1-x-y)^2}{2}.$$

The next integration, where x is fixed, is

$$\int_0^{1-x} \frac{(1-x-y)^2}{2}\, dy = -\frac{(1-x-y)^3}{6} \Big|_{y=0}^{y=1-x}$$

$$= -\frac{0^3}{6} + \frac{(1-x)^3}{6}$$

$$= \frac{(1-x)^3}{6}.$$

The third integration is

$$\int_0^1 \frac{(1-x)^3}{6}\, dx = -\frac{(1-x)^4}{24} \Big|_0^1$$

$$= -\frac{0^4}{24} + \frac{1^4}{24}$$

$$= \tfrac{1}{24}. \quad \bullet$$

With the aid of Example 1 it is easy to find $\bar{z}$, the z coordinate of the center of gravity of the tetrahedron. To be specific,

$$\bar{z} = \frac{\int_R z\, dV}{\text{volume of } R}$$

$$= \frac{\tfrac{1}{24}}{\text{volume of } R}.$$

The volume of a tetrahedron is $\frac{1}{3} \cdot$ height $\cdot$ area of base $= \frac{1}{3} \cdot 1 \cdot \frac{1}{2} = \frac{1}{6}$.

Thus
$$\bar{z} = \frac{\frac{1}{24}}{\frac{1}{6}} = \frac{1}{4}.$$

The center of gravity lies on a plane parallel to the xy plane and three-quarters of the way from the vertex $(0, 0, 1)$ to the xy plane.

EXAMPLE 2   Compute the moment of inertia of a homogeneous cube of side a and mass M about an edge.

SOLUTION   For a convenient description of the cube place the origin of a rectangular coordinate system at a corner. We shall compute the moment of inertia about the edge lying on the x axis.

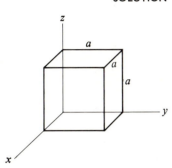

The density at any point P in the cube is M/a^3, since the volume of the cube is a^3. The square of the distance from $P = (x, y, z)$ to the x axis is $y^2 + z^2$. The definition of moment of inertia in the table provides this integral:

$$\int_R (y^2 + z^2)\frac{M}{a^3}\, dV.$$

To evaluate the integral, it is necessary to describe the cube R in rectangular coordinates:

$$0 \le x \le a, \qquad 0 \le y \le a, \qquad 0 \le z \le a.$$

Thus
$$\int_R (y^2 + z^2)\frac{M}{a^3}\, dV = \int_0^a \int_0^a \int_0^a (y^2 + z^2)\frac{M}{a^3}\, dz\, dy\, dx.$$

The three integrations can be carried out with the aid of the fundamental theorem of calculus.

The first integration is:

$$\int_0^a (y^2 + z^2)\frac{M}{a^3}\, dz = \left(y^2 z + \frac{z^3}{3}\right)\frac{M}{a^3}\Bigg|_{z=0}^{z=a}$$

$$= \left(y^2 a + \frac{a^3}{3}\right)\frac{M}{a^3}$$

$$= \frac{M}{a^2}\left(y^2 + \frac{a^2}{3}\right).$$

The second is:
$$\int_0^a \frac{M}{a^2}\left(y^2 + \frac{a^2}{3}\right) dy = \frac{M}{a^2}\left(\frac{y^3}{3} + \frac{a^2 y}{3}\right)\Bigg|_{y=0}^{y=a}$$

$$= \frac{M}{a^2}\left(\frac{a^3}{3} + \frac{a^3}{3}\right)$$

$$= \frac{2Ma}{3}.$$

Finally,
$$\int_0^a \frac{2Ma}{3}\, dx = \frac{2Ma^2}{3}.$$

The moment of inertia about an edge is $2Ma^2/3$.   (This is as though all the mass were concentrated at a point a distance $\sqrt{2/3}a$ from the edge.) ●

Observe that to evaluate a three-dimensional integral by repeated integrations, the fundamental theorem must be used three times.   Only for the simplest integrands will we be able to evaluate the integral.   Fortunately, this covers most applications in engineering and physics.   Moreover, the primary importance of $\int_R f(P)\, dV$ is conceptual, as a means of defining physical quantities and stating physical assumptions.

Exercises

1. (a) Compute $\int_R y\, dV$ for the tetrahedron of Example 1.
   (b) Find $\bar{y}$ for the tetrahedron.
2. Find the volume of the tetrahedron in Example 1, evaluating the integral $\int_R 1\, dV$ by a repeated integral.
3. Find the moment of inertia of the tetrahedron in Example 1 about the z axis, if its mass is M and it is homogeneous.
4. Evaluate the repeated integral

$$\int_0^1 \int_{x^3}^{x^2} \int_0^{x+y} z\, dz\, dy\, dx.$$

5. A rectangular solid box has mass M and sides of length a, b, and c.   Find its moment of inertia about an edge of length a.
6. Compute $\int_R z\, dV$, where R is the region above the rectangle whose vertices are $(0, 0, 0)$, $(2, 0, 0)$, $(2, 3, 0)$, $(0, 3, 0)$ and below the plane $z = x + 2y$.
7. Without using a repeated integral, compute $\int_R x\, dV$, where R is a spherical ball whose center is $(0, 0, 0)$ and whose radius is a.
8. Find the average value of the square of the distance from a corner of a cube of side a to points in the cube.
9. Compute $\int_R xy\, dV$ for the tetrahedron of Example 1.
10. A solid consists of all points below the surface $z = xy$ that are above the triangle whose vertices are $(0, 0, 0)$, $(1, 0, 0)$, and $(0, 2, 0)$.   If the density at (x, y, z) is $x + y$, find the total mass.
11. A right solid circular cone has altitude h, radius a, constant density, and mass M.
    (a) Why is its moment of inertia about its axis less than Ma^2?
    (b) Show that its moment of inertia about its axis is $3Ma^2/10$.

    ■

12. Compute $\bar{z}$ for the top half of the ball of radius a situated above the xy plane and whose center is at the origin.

13. A rectangular homogeneous box of mass M has dimensions a, b, and c.   Show that the moment of inertia of the box around a line through its center and parallel to the side of length a is $M(b^2 + c^2)/12$.
14. A pyramid of mass M has a square base of side a and height h, as shown in the accompanying diagram.   Show

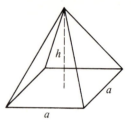

that its moment of inertia about the line through its top vertex and perpendicular to its base is $Ma^2/10$.

    ■ ■

15. In Section 13.3 an intuitive argument was presented for the equality

$$\int_R f(P)\, dA = \int_a^b \left(\int_{y_1(x)}^{y_2(x)} f\, dy \right) dx.$$

Here is an intuitive argument for the equality

$$\int_R f(P)\, dV = \int_{x_1}^{x_2} \left[\int_{y_1(x)}^{y_2(x)} \left(\int_{z_1(x,\,y)}^{z_2(x,\,y)} f\, dz \right) dy \right] dx.$$

To start, interpret $f(P)$ as "density."
(a) Let $R(x)$ be the plane cross section consisting of all points in R whose x coordinate is x.   Show that the average density in $R(x)$ is

$$\frac{\displaystyle\int_{y_1(x)}^{y_2(x)} \left(\int_{z_1(x,\,y)}^{z_2(x,\,y)} f\, dz \right) dy}{\text{area of } R(x)}.$$

(b) Show that the mass of R between the plane sections $R(x)$ and $R(x + \Delta x)$ is approximately

$$\int_{y_1(x)}^{y_2(x)} \left(\int_{z_1(x, y)}^{z_2(x, y)} f\, dz \right) dy\, \Delta x.$$

(c) From (b) obtain a repeated integral in rectangular coordinates for $\int_R f(P)\, dV$.

18.5

Computing $\int_R f(P)\, dV$ with cylindrical or spherical coordinates

Repeated integrals in polar coordinates in Sec. 13.5 involve the introduction of a factor r in the integrand. This must also be done when using cylindrical coordinates (r, θ, z), for the volume of the little solid obtained by letting each coordinate change a small amount is, as was shown in Sec. 18.3, approximately

$$r\, \Delta r\, \Delta\theta\, \Delta z,$$

where Δr, $\Delta\theta$, Δz are the changes in the coordinates.

When setting up repeated integrals in spherical coordinates, it is necessary to insert the extra factor

$$\rho^2 \sin \phi$$

in the integrand. The reason for this is that the little solid obtained by letting each coordinate change a small amount is approximately

$$\rho^2 \sin \phi\, \Delta\rho\, \Delta\phi\, \Delta\theta,$$

where $\Delta\rho$, $\Delta\phi$, and $\Delta\theta$ are the changes in the coordinates, as was shown in Sec. 18.3.

The limits of integration are determined by the description of the region R relative to the coordinate system. A few examples will show how to set up and compute these repeated integrals; we will not stop to justify the technique.

EXAMPLE 1   Find the volume of a ball R of radius a using cylindrical coordinates.

SOLUTION   Place the origin of a cylindrical coordinate system at the center of the ball, as in the figure at the left.

The volume of the ball is $\int_R 1\, dV$. The description of R in cylindrical coordinates is

$$0 \le \theta \le 2\pi, \qquad 0 \le r \le a, \qquad -\sqrt{a^2 - r^2} \le z \le \sqrt{a^2 - r^2}.$$

The repeated integral for the volume is thus

$$\int_R 1\, dV = \int_0^{2\pi} \left(\int_0^a \left(\int_{-\sqrt{a^2 - r^2}}^{\sqrt{a^2 - r^2}} 1 \cdot r\, dz \right) dr \right) d\theta.$$

(Note the r in the integrand.)

A ball of radius a

z

Polar axis

Evaluation of the first integral, where r and θ are fixed, yields:

$$\int_{-\sqrt{a^2-r^2}}^{\sqrt{a^2-r^2}} r\,dz = rz \Big|_{z=-\sqrt{a^2-r^2}}^{z=\sqrt{a^2-r^2}}$$

$$= 2r\sqrt{a^2-r^2}.$$

Evaluation of the second integral, where θ is fixed, yields:

$$\int_0^a 2r\sqrt{a^2-r^2}\,dr = \frac{-2(a^2-r^2)^{3/2}}{3}\Big|_{r=0}^{r=a}$$

$$= \frac{2a^3}{3}.$$

Evaluation of the third integral yields:

$$\int_0^{2\pi} \frac{2a^3}{3}\,d\theta = \frac{2a^3}{3}\,2\pi = \frac{4\pi a^3}{3}. \;\bullet$$

EXAMPLE 2 Find $\bar{z}$, the z coordinate of the center of gravity of a solid homogeneous hemisphere R of radius a whose base is on the xy plane.

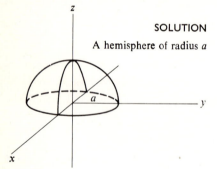

A hemisphere of radius a

SOLUTION Use spherical coordinates. By definition,

$$\bar{z} = \frac{\int_R z\,dV}{\text{volume of hemisphere}}.$$

The volume is $2\pi a^3/3$. All that remains is to compute $\int_R z\,dV$.

Placing the origin of the spherical coordinate system at the center of the hemisphere, we have this description of R:

$$0 \le \theta \le 2\pi, \qquad 0 \le \phi \le \pi/2, \qquad 0 \le \rho \le a.$$

Before we set up a repeated integral for $\int_R z\,dV$, it is necessary to express z in spherical coordinates also,

$$z = \rho \cos \phi.$$

Notice the
extra $\rho^2 \sin \phi$

Then

$$\int_R z\,dV = \int_0^{2\pi} \left(\int_0^{\pi/2} \left(\int_0^a \rho \cos \phi \overbrace{\rho^2 \sin \phi\,d\rho} \right) d\phi \right) d\theta.$$

Evaluation of the first integral, where ϕ and θ are fixed, yields:

$$\int_0^a \rho^3 \cos \phi \sin \phi\,d\rho = \frac{\rho^4 \cos \phi \sin \phi}{4}\Big|_{\rho=0}^{\rho=a}$$

$$= \frac{a^4 \cos \phi \sin \phi}{4}.$$

Evaluation of the second integral, where θ is fixed, yields:

$$\int_0^{\pi/2} \frac{a^4 \cos\phi \sin\phi \, d\phi}{4} = \frac{a^4 \sin^2\phi}{8} \Big|_{\phi=0}^{\phi=\pi/2}$$

$$= \frac{a^4}{8}.$$

Evaulation of the third integral yields:

$$\int_0^{2\pi} \frac{a^4}{8} \, d\theta = \frac{\pi a^4}{4}.$$

Hence $\int_R z \, dV = \pi a^4/4$, and

$$\bar{z} = \frac{\pi a^4/4}{2\pi a^3/3} = \frac{3a}{8}. \quad \bullet$$

The next example is of importance in the theory of gravitational attraction. Students of the physical sciences will see later that it implies that a homogeneous ball attracts a particle (or satellite) as if all the mass of the ball were at its center.

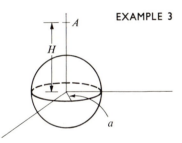

EXAMPLE 3   Let R be a homogeneous ball of mass M and radius a. Let A be a point at a distance H from the center of the ball, $H > a$. Compute the potential

$$-\int_R \frac{\delta}{q} \, dV,$$

where δ is density, and q is the distance from a point P in R to A. (See the diagrams in the margin.)

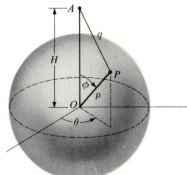

SOLUTION   First, express q in terms of spherical coordinates. To do so, choose a spherical coordinate system whose origin is at the center of the sphere and such that the ϕ coordinate of A is 0.

Let $P = (\rho, \phi, \theta)$ be a typical point in the sphere. Applying the law of cosines to triangle AOP, we find that

$$q^2 = H^2 + \rho^2 - 2\rho H \cos\phi.$$

Hence

$$q = \sqrt{H^2 + \rho^2 - 2\rho H \cos\phi}.$$

Since the ball is homogeneous,

$$\delta = \frac{M}{\frac{4}{3}\pi a^3} = \frac{3M}{4\pi a^3}.$$

Hence

$$\int_R \frac{\delta}{q} \, dV = \int_R \frac{3M}{4\pi a^3 q} \, dV,$$

or

$$\int_R \frac{\delta}{q} \, dV = \frac{3M}{4\pi a^3} \int_R \frac{1}{q} \, dV. \qquad (1)$$

Now evaluate
$$\int_R \frac{1}{q} \, dV$$

by a repeated integral in spherical coordinates,

$$\int_R \frac{1}{q} \, dV = \int_0^{2\pi} \left(\int_0^a \left(\int_0^\pi \frac{\rho^2 \sin \phi}{\sqrt{H^2 + \rho^2 - 2\rho H \cos \phi}} \, d\phi \right) d\rho \right) d\theta.$$

(Integrate with respect to ϕ first, rather than ρ, because it is easier to do so in this case.)

Evaluation of the first integral, where ρ and θ are constants, is accomplished with the aid of the fundamental theorem:

$$\int_0^\pi \frac{\rho^2 \sin \phi}{\sqrt{H^2 + \rho^2 - 2\rho H \cos \phi}} \, d\phi = \frac{\rho \sqrt{H^2 + \rho^2 - 2\rho H \cos \phi}}{H} \Bigg|_{\phi=0}^{\phi=\pi}$$

$$= \frac{\rho}{H} \left(\sqrt{H^2 + \rho^2 + 2\rho H} - \sqrt{H^2 + \rho^2 - 2\rho H} \right).$$

Now, $\sqrt{H^2 + \rho^2 + 2\rho H} = H + \rho$. Since $\rho \le a < H$, $H - \rho$ is positive and
$$\sqrt{H^2 + \rho^2 - 2\rho H} = H - \rho.$$

Thus the first integral equals

$$\frac{\rho}{H} (H + \rho - (H - \rho)) = \frac{2\rho^2}{H}.$$

Evaluation of the second integral yields:

$$\int_0^a \frac{2\rho^2}{H} \, d\rho = \frac{2a^3}{3H}.$$

Evaluation of the third integral yields:

$$\int_0^{2\pi} \frac{2a^3}{3H} \, d\theta = \frac{4\pi a^3}{3H}.$$

Hence
$$\int_R \frac{1}{q} \, dV = \frac{4\pi a^3}{3H}.$$

By (1),   Potential $= -\int_R \frac{\delta}{q} \, dV = \frac{-3M}{4\pi a^3} \frac{4\pi a^3}{3H} = \frac{-M}{H}.$

Though the computations are long, the answer is short. It says that the potential is just the negative of the mass of the ball times the reciprocal of the distance from A to the center. In other words, the potential is the same as if all the mass of the ball were concentrated at its center. ●

Exercises

1. Compute the volume of a ball of radius a using spherical coordinates.

2. Find the moment of inertia of a ball of radius a and mass M about a diameter, using cylindrical coordinates.

3. Solve Exercise 2 using spherical coordinates.
4. Solve Exercise 11 of the preceding section using cylindrical coordinates.
5. Solve Exercise 11 of the preceding section using spherical coordinates.
6. A right circular cylinder has mass M, radius a, and height h. Find its moment of inertia about its axis.
7. Find the moment of inertia of the cylinder in Exercise 6 about a line L on its surface and parallel to the z axis.
   (a) Use cylindrical coordinates, the z axis coinciding with L.
   (b) Use cylindrical coordinates, the z axis coinciding with the axis of the cylinder.
8. Solve Example 2 using cylindrical coordinates.
9. A solid of mass M consists of that part of a ball of radius a that lies within a cone of half-vertex-angle $\phi = \pi/6$, the vertex being at the center of the ball. Set up repeated integrals for $\int_R z \, dV$ in all three coordinate systems and evaluate the simplest.
10. Show by using a repeated integral that the volume of the little solid estimated in Sec. 18.3 is precisely
$$\frac{\rho_2{}^3 - \rho_1{}^3}{3} (\theta_2 - \theta_1)(\cos \phi_1 - \cos \phi_2).$$
11. Find the average distance from the center of a ball of radius a to other points of the ball by setting up appropriate repeated integrals in the three types of coordinate systems and evaluating the two easiest.
12. Find the average length of the shadow on the xy plane of all line segments whose length is at most a and one of whose ends is the origin. Assume that the light is parallel to the z axis.
13. Find the moment of inertia of a solid hemisphere of radius a and mass M about a diameter in its circular base,
    (a) using cylindrical coordinates;
    (b) using spherical coordinates;

(c) using the result of Exercise 2.
14. A certain ball of radius a is *not* homogeneous. However, its density at P depends only on the distance from P to the center of the ball. Call $\delta(P) = \delta(\rho, \phi, \theta) = g(\rho)$. Using a repeated integral, show that the mass of the ball is $4\pi \int_0^a g(\rho)\rho^2 \, d\rho$.

∎

15. Show that $\int_R (x^3 + y^3 + z^3) \, dV = 0$, where R is a ball whose center is the origin of a rectangular coordinate system. (Do not use a repeated integral.)
16. Let R be a solid ball of radius a with center at the origin of the coordinate system.
    (a) Explain why $\int_R x^2 \, dV = \frac{1}{3} \int_R (x^2 + y^2 + z^2) \, dV$.
    (b) Evaluate the second integral by spherical coordinates.
    (c) Use (b) to find $\int_R x^2 \, dV$.
17. Combining the fact that the volume of R equals $\int_R 1 \, dV$ with a repeated integral in rectangular coordinates,
    (a) obtain the formula $V = \int_a^b A(x) \, dx$;
    (b) obtain the formula for volume as a definite integral over a plane set.
18. Combining the fact that the volume of R equals $\int_R 1 \, dV$ with a repeated integral in cylindrical coordinates, obtain the shell technique of Sec. 10.4.

∎ ∎

19. When the dimensions of the solid in Exercise 10 are small, does the formula in that exercise become, approximately, the formula $\rho^2 \sin \phi \, \Delta\rho \, \Delta\phi \, \Delta\theta$?
20. Using the method of Example 3, find the average value of q for all points P in the ball. Note that it is *not* the same as if the entire ball were placed at its center.
21. Show that the result of Example 3 holds if the density $\delta(P)$ depends only on ρ, the distance to the center. (This is the case with the planet Earth, which is not homogeneous.) Call $\delta(\rho, \phi, \theta) = g(\rho)$. See Exercise 14 and Example 3.

18.6

Summary

This chapter began with a definition of the definite integral over a solid region. If R is such a region, perhaps a cube, a cone, a cylinder, a sphere, etc., and f is a function that assigns to each point P a number $f(P)$, then $\int_R f(P) \, dV$ is defined as

$$\lim_{\text{mesh} \to 0} \sum_{i=1}^n f(P_i)V_i,$$

where the sums arise from partitions of R. In the case of the usual functions and regions met in practice, this limit does exist.

The table in Sec. 18.4 lists the most important applications of the definite integral over a solid region.

Repeated integrals, one for each type of coordinate system, were introduced to evaluate the definite integral over a solid region R. When cylindrical coordinates are used, a factor r

must be inserted in the integrand. A factor $\rho^2 \sin \phi$ is to be included in the repeated integral with spherical coordinates.

VOCABULARY AND SYMBOLS

definite integral of a function over a solid region $\int_R f(P)\, dV$
partition, diameter, mesh

average value of a function over a solid region
rectangular coordinates (x, y, z)

cylindrical coordinates (r, θ, z)
spherical coordinates (ρ, ϕ, θ)

repeated integral
moment of inertia
center of gravity $(\bar{x}, \bar{y}, \bar{z})$
centroid

Guide quiz on Chap. 18

1. A solid circular cylinder of radius a and height h is composed of a uniform material of mass M. Show that its moment of inertia about a line perpendicular to the axis and midway between the two ends of the cylinder is

$$\frac{Ma^2}{4} + \frac{Mh^2}{12}.$$

2. Find $\bar{z}$ of the solid region bounded by two concentric spherical shells of radii a and b, $a < b$ centered at the origin and by the xy plane, as shown below.

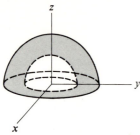

3. (a) What is meant by the symbol $\int_R f(P)\, dV$?
   (b) Using the definition in (a), show that, if $2 \le f(P) \le 3$ for all P in R, then $2 \cdot$ volume of $R \le \int_R f(P)\, dV \le 3 \cdot$ volume of R.
4. (a) Find the cylindrical coordinates of the point whose rectangular coordinates are $(3, 4, \pi/2)$.
   (b) Find the rectangular coordinates of the point whose spherical coordinates are $(3, 1, \pi/2)$.
   (c) Find the spherical coordinates of the point whose cylindrical coordinates are $(2, \pi/4, 2)$.
5. Draw the cross section of a ball of radius 1, whose center is at the origin of the coordinate system, determined by
   (a) $x = \frac{1}{2}$;
   (b) $\phi = \pi/3$;
   (c) $\rho = \frac{1}{2}$;
   (d) $\theta = \pi/2$;
   (e) $z = -\frac{1}{2}$.
6. (a) What extra factor must be introduced when setting up a repeated integral in cylindrical or in spherical coordinates?
   (b) What reason was suggested?
7. (a) Draw the little solid region corresponding to changes $\Delta\rho$, $\Delta\phi$, and $\Delta\theta$ in the spherical coordinates.
   (b) Show why its volume is approximately

   $$\rho^2 \sin \phi \; \Delta\rho \; \Delta\phi \; \Delta\theta.$$

8. A solid right circular cylinder has radius a and height h. Find the average over R of the function f, where $f(P)$ is the square of the distance from the center of the cylinder to P.
   (a) Set up repeated integrals in at least two of the three coordinate systems.
   (b) Evaluate the easier repeated integral in (a).
9. (a) Evaluate the repeated integral

   $$\int_0^1 \left(\int_0^1 \left(\int_0^x y e^{x^2}\, dz \right) dy \right) dx.$$

   (b) Draw the region R described by the range of integrations in (a).
10. A solid homogeneous right circular cone of radius a and height h has a mass M.
    (a) What is meant by its "moment of inertia about a line through its vertex and parallel to its base"?
    (b) Set up repeated integrals in all three coordinate systems for the moment of inertia in (a).
    (c) Evaluate at least one of the repeated integrals in (b).
11. What is the equation of:
    (a) the plane $z = 3$ in spherical coordinates;
    (b) the cylindrical surface $r = 2$ in rectangular coordinates;
    (c) the cylindrical surface $r = 2$ in spherical coordinates;
    (d) the spherical surface $\rho = 3$ in cylindrical coordinates?
12. What is the average of the square of the distance from points within a cube of side a to a fixed corner of the cube?

Review exercises for Chap. 18

A few exercises concern earlier chapters.

1. See Example 3 of Sec. 18.5. Show that the average of the reciprocal of the distance from a fixed point A outside a ball to points in the ball is equal to the reciprocal of the distance from A to the center of the ball.

2. A homogeneous right circular cylindrical shell has inner radius r, outer radius R, and height h. Its mass is M. Show that its moment of inertia
   (a) about its axis is $M(R^2 + r^2)/2$;
   (b) about a line through its center of gravity and perpendicular to its axis is $M(R^2 + r^2 + h^2/3)/4$.

3. A homogeneous solid of mass M occupies the space between two concentric spheres of radii r and R, $r < R$. Show that its moment of inertia around a diameter is $2M(R^5 - r^5)/[5(R^3 - r^3)]$.

4. Let R be the region bounded by a circle of radius a. Let A be a point in the plane of the circle at a distance $H > a$ from the center of the circle. Define a function f by setting $f(P) =$ square of distance from P to A. Show that the average value of f over R is $H^2 + a^2/2$.

5. The axes of two right circular cylinders of radius a meet at right angles. Show that the surface area of one cylinder that lies inside the other is $8a^2$. Note that it is eight times the area of the shaded surface.

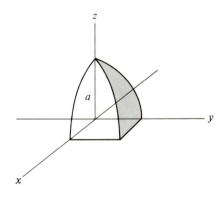

6. Let R be the ball of radius a. For any point P in the ball other than the center of the ball, define $f(P)$ to be the reciprocal of the distance from P to the origin. The average value of f over R involves an improper integral, since the function blows up near the origin. Does this improper integral converge or diverge? What is the average value of f over R?

7. The gravitational attraction between a homogeneous ball of radius s and a point mass, as shown in the top figure, next column, involves evaluation of the integral

$$\int_S \frac{\cos \alpha}{x^2} \, dV.$$

Show that its value is $\dfrac{4\pi s^3}{3H^2}$.

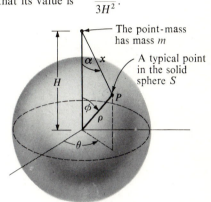

The radius of the sphere is s

8. A right rectangular pyramid has a base of dimensions a by b and height h. Its mass is M. Show that the moment of inertia of the pyramid around the line that is perpendicular to the base and passes through its top vertex is $M(a^2 + b^2)/20$.

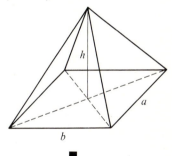

■

9. A doughnut (torus) is formed by spinning a circle of radius a in a plane around a line L in that plane that is a distance $b > a$ from the center of the circle. Its mass is M. Show that the moment of inertia of the doughnut around the line L is $M(b^2 + 3a^2/4)$.

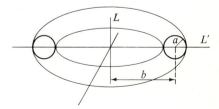

10. Show that the moment of inertia of the doughnut of Exercise 9 around the line L that passes through the center of the doughnut and is perpendicular to L is $M(b^2/2 + 5a^2/8)$.

11. A solid right circular cone of radius a and height h has mass M. Show that the moment of inertia of the cone around a line through its center of gravity and parallel to its base is $3M(a^2 + h^2/4)/20$.

■ ■

12. Let $z = g(y)$ be a decreasing function of y such that $g(1) = 0$. Let R be the solid of revolution formed by rotating about the z axis the region in the yz plane bounded by $y = 0$, $z = 0$, and $z = g(y)$. Using appropriate repeated integrals in cylindrical coordinates, show that $\int_R z \, dV = \int_0^1 \pi y[g(y)]^2 \, dy$ or $\int_0^{g(0)} \pi[g^{-1}(z)]^2 z \, dz$.

13. (See Exercise 12.)
    (a) Show that the z coordinate of the centroid of the solid described in Exercise 12 is $\int_0^1 \{x[g(x)]^2/2\} \, dx / \int_0^1 xg(x) \, dx$, while the z coordinate of the centroid of the plane region that was revolved is

    $$\frac{\int_0^1 \{[g(x)]^2/2\} \, dx}{\int_0^1 g(x) \, dx}.$$

    (b) By considering

    $$\int_0^1 \left(\int_0^1 g(x)g(y)(x - y)[g(x) - g(y)] \, dx \right) dy.$$

show that the centroid of the solid of revolution is below that of the plane region. *Hint:* Why is the repeated integral less than or equal to 0?

14. The center of gravity of a solid of mass M is located at $(0, 0, 0)$. Let its moment of inertia about the x axis be I.
    (a) Find the moment of inertia of the solid about a line parallel to the x axis and a distance k from it.
    (b) About which line parallel to the x axis is the moment of inertia of the solid least?

15. Let F be a fixed point in a solid of mass M. Show that, for all choices of three mutually perpendicular lines that meet at F, the sum of the moments of inertia of the solid about the lines is the same.

16. In R. P. Feynman, *Lectures on Physics*, Addison-Wesley, Reading, Mass., 1963, this statement appears in Sec. 15.8:

> An approximate formula to express the increase of mass, for the case when the velocity is small, can be found by expanding $m_0/\sqrt{1 - v^2/c^2} = m_0(1 - v^2/c^2)^{-1/2}$ in a power series, using the binomial theorem. We get
>
> $$m_0\left(1 - \frac{v^2}{c^2}\right)^{-1/2} = m_0\left(1 + \frac{1}{2}\frac{v^2}{c^2} + \frac{3}{8}\frac{v^4}{c^4} + \cdots\right).$$
>
> We can see clearly from the formula that the series converges rapidly when v is small and the terms after the first two or three are negligible.

Check the expansion and justify the statement.

19

ALGEBRAIC OPERATIONS ON VECTORS

In this chapter we discuss the algebra of vectors in the plane and in space. Section 19.6 develops enough of the theory of determinants to permit a simple presentation of certain vector concepts. Only in Sec. 19.5, where vectors help generalize the notion of partial derivatives, does calculus enter the picture.

19.1

The algebra of vectors

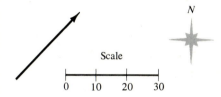

Scale

An adequate description of the wind indicates both its speed and its direction. One way to describe a wind of 30 miles per hour from the southwest is to draw an arrow aimed in the direction in which the wind blows, scaled so that its length represents a magnitude of 30.

Relative to this same scale, here are some more wind arrows:

Wind from the
south at 10 miles
per hour

Wind from the
northeast at 20 miles
per hour

No wind: the
arrow of length 0

Similarly the flow of water on the surface of a stream is best indicated by a few sample arrows:

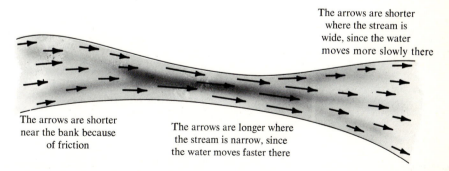

The arrows are shorter
where the stream is
wide, since the water
moves more slowly there

The arrows are shorter
near the bank because
of friction

The arrows are longer where
the stream is narrow, since
the water moves faster there

Of course, associated with *each* point on the surface is an arrow representing the velocity of the water at that point.

An arrow describes the magnitude and direction of the force applied to the rock

Similarly when we pull a heavy rock by a rope, the force we exert has both a magnitude and a direction.   The magnitude describes how hard we pull; the direction of our pull is along the rope.

The mathematical word for our arrow is *vector*.   A vector is determined by specifying two points P and Q, one of which is the "tail" of the vector and one the "head."   For this reason a vector is formally defined as an ordered pair of points P and Q, and is sometimes denoted $\overrightarrow{PQ}$ (P is the tail and Q is the head).

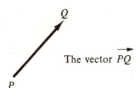

The vector $\overrightarrow{PQ}$

If $P = Q$, then $\overrightarrow{PQ}$ is called the *zero vector*, denoted by **0**, a boldface 0. The location of a vector is of no importance; two vectors with the same direction and length are considered the same.   Thus all the arrows in the diagram in the margin represent the same vector.

In print, boldface letters, such as **A**, **B**, **F**, **R**, and **V**, are used to denote vectors.   In handwriting the symbols $\mathbb{A}$ and $\vec{A}$ are used.   The length of **A** is denoted by $|\mathbf{A}|$.   The length of **A** is also called the *magnitude* of **A**.

Lowercase letters, such as a, b, x, y, z will be used to name numbers. Numbers will also be called *scalars* to contrast them with vectors.   Thus **A** is a vector but $|\mathbf{A}|$ is a scalar.   It is important to distinguish between the number 0 and the zero vector **0**.

For the moment consider only vectors **A** situated in the xy plane which we will call *plane vectors*.

The numbers x and y describe the vector **A**

If the origin of a rectangular coordinate system is at the tail of **A**, then the head of **A** has coordinates (x, y).   The numbers x and y are called the *scalar components* of **A** relative to the coordinate system.   The two scalar components determine **A**, which lies along the diagonal of a rectangle whose sides have lengths $|x|$ and $|y|$.   For instance, the four vectors shown in the last figure in the margin all have x component 3 and y component -4.

A vector whose head is at $(13, 18)$ and whose tail is at $(10, 22)$ also has scalar components 3 and -4, as a sketch will show.   A vector with scalar components x and y will be denoted $\overrightarrow{(x, y)}$ to distinguish it from the point (x, y).   The vector $\overrightarrow{(0, 0)}$ of length 0 is the zero vector and is denoted by **0**. Observe that $\overrightarrow{(x_1, y_1)} = \overrightarrow{(x_2, y_2)}$ if and only if $x_1 = x_2$ and $y_1 = y_2$.

An arbitrary vector **A** in *space* has *three* scalar components.   Place the origin of a rectangular xyz-coordinate system at the tail of **A**.   Then the head of **A** has coordinates (x, y, z).   The numbers x, y, and z are the *scalar components* of **A** relative to the coordinate system.   Example 1 suggests a way of indicating the scalar components graphically.

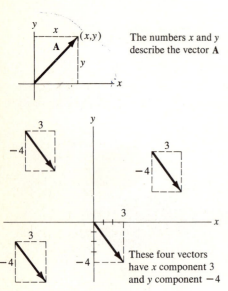

These four vectors have x component 3 and y component -4

EXAMPLE 1   Sketch the vector $\mathbf{A} = \overrightarrow{PQ}$ if

$$P = (0, 0, 0) \quad \text{and} \quad Q = (3, -4, 2).$$

SOLUTION Plot the points P and Q and indicate with dashed lines the "box" they determine. The scalar components correspond (with appropriate signs) to the three dimensions of the box. The vector $\overrightarrow{PQ}$ is also denoted $(3, -4, 2)$. Its scalar components are 3, -4, and 2. ●

Our first theorem shows how to compute the length of a vector in terms of its scalar components.

THEOREM I $|(x, y, z)| = \sqrt{x^2 + y^2 + z^2}$. That is, the magnitude of a vector is the square root of the sum of the squares of its scalar components.

PROOF This is an immediate consequence of the distance formula of Sec. 12.1. That x, y, or z might be negative causes no trouble, since they are all squared. ●

In the case of a plane vector $\mathbf{A} = (x, y)$, its length is simply $\sqrt{x^2 + y^2}$. Thus $|(3, -4)| = \sqrt{3^2 + (-4)^2} = 5$. The vector $(3, -4, 2)$ has magnitude $\sqrt{3^2 + (-4)^2 + 2^2} = \sqrt{29}$.

DEFINITION Any vector whose length is 1 is called a *unit vector*.

The vector $(-1, 0, 0)$ is a unit vector, since $\sqrt{(-1)^2 + 0^2 + 0^2} = \sqrt{1} = 1$. Similarly, for any angle θ, the plane vector $(\cos \theta, \sin \theta)$ is a unit vector, since

$$\sqrt{(\cos \theta)^2 + (\sin \theta)^2} = \sqrt{1} = 1.$$

The vector $(\frac{3}{13}, \frac{4}{13}, \frac{-12}{13})$ is a unit vector, for its magnitude is

$$\sqrt{\left(\frac{3}{13}\right)^2 + \left(\frac{4}{13}\right)^2 + \left(\frac{-12}{13}\right)^2} = \sqrt{\frac{9 + 16 + 144}{13^2}} = \frac{\sqrt{169}}{13} = 1.$$

The sum of two vectors $\mathbf{A}$ and $\mathbf{B}$ is defined as follows. Place $\mathbf{B}$ in such a way that its tail is at the head of $\mathbf{A}$. Then the vector sum $\mathbf{A} + \mathbf{B}$ goes from the tail of $\mathbf{A}$ to the head of $\mathbf{B}$. Observe that $\mathbf{B} + \mathbf{A} = \mathbf{A} + \mathbf{B}$, since both sums lie on the diagonal of a parallelogram.

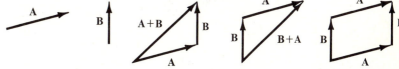

For example, if $\mathbf{W}$ is a wind vector (describing the motion of the air relative to the earth) and $\mathbf{A}$ is a vector describing the motion of an airplane relative to the air, then $\mathbf{W} + \mathbf{A}$ is the vector describing the motion of the airplane relative to the earth.

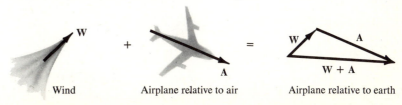

Wind   Airplane relative to air   Airplane relative to earth

The concept of the sum of two vectors is also important in the study of force. If $\mathbf{F}_1$ and $\mathbf{F}_2$ describe the forces in two ropes lifting a heavy rock, then a single rope with the force $\mathbf{F}_1 + \mathbf{F}_2$ pulling from the same point has the same effect on the rock.

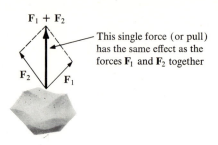

This single force (or pull) has the same effect as the forces $\mathbf{F}_1$ and $\mathbf{F}_2$ together

EXAMPLE 2 Find the sum of the vectors $\mathbf{A} = (\overrightarrow{2, 3})$ and $\mathbf{B} = (\overrightarrow{4, 1})$.

SOLUTION Place the tail of $\mathbf{B}$ at the head of $\mathbf{A}$. Inspection of the diagram in the margin shows that $\mathbf{A} + \mathbf{B}$ has the scalar components 6 and 4; that is,

$$\mathbf{A} + \mathbf{B} = (\overrightarrow{6, 4}). \; \bullet$$

Subtraction of vectors will be defined the way it is for numbers. If x and y are numbers, $x - y$ is that number that must be added to y to obtain x. [That is, $y + (x - y) = x$.] This suggests the following definition.

DEFINITION *Subtraction of vectors.* Let $\mathbf{A}$ and $\mathbf{B}$ be vectors. The vector $\mathbf{V}$ such that $\mathbf{B} + \mathbf{V}$ equals $\mathbf{A}$ is called the *difference* of $\mathbf{A}$ and $\mathbf{B}$ and is denoted $\mathbf{A} - \mathbf{B}$.

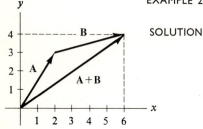

Thus $\mathbf{B} + (\mathbf{A} - \mathbf{B}) = \mathbf{A}.$

To illustrate the definition, let $\mathbf{A}$ and $\mathbf{B}$ be given as in the first diagram at the left. $\mathbf{A} - \mathbf{B}$ is the dotted vector from the head of $\mathbf{B}$ to the head of $\mathbf{A}$.

The *negative* of the vector $\mathbf{A}$ is defined as the vector having the same magnitude as $\mathbf{A}$ but the opposite direction. It is denoted $-\mathbf{A}$. If $\mathbf{A} = \overrightarrow{PQ}$, then $-\mathbf{A} = \overrightarrow{QP}$.

Observe that $\mathbf{A} + (-\mathbf{A}) = \mathbf{0}$, just as with scalars. More generally, subtracting a vector gives the same result as adding its negative; that is, $\mathbf{A} - \mathbf{B} = \mathbf{A} + (-\mathbf{B})$.

The definitions of addition and subtraction and the negative of a vector are geometric. The next theorem tells how to compute the sum and difference of vectors and the negative of a vector in terms of scalar components.

THEOREM 2

$$\overrightarrow{(x_1, y_1, z_1)} + \overrightarrow{(x_2, y_2, z_2)} = \overrightarrow{(x_1 + x_2, y_1 + y_2, z_1 + z_2)}$$

$$\overrightarrow{(x_1, y_1, z_1)} - \overrightarrow{(x_2, y_2, z_2)} = \overrightarrow{(x_1 - x_2, y_1 - y_2, z_1 - z_2)}$$

$$-\overrightarrow{(x, y, z)} = \overrightarrow{(-x, -y, -z)}.$$

Similar equations hold for vectors in the plane.

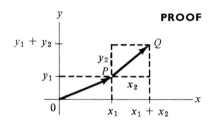

PROOF In order to have simple diagrams, consider the case of plane vectors. To justify the first assertion, let $\overrightarrow{OP} = (x_1, y_1)$ and $\overrightarrow{PQ} = (x_2, y_2)$. What is the scalar component of $\overrightarrow{OP} + \overrightarrow{PQ}$ along, say, the x axis? The accompanying sketch, in which x_1 and x_2 are positive, shows that $\overrightarrow{OQ} = \overrightarrow{OP} + \overrightarrow{PQ}$ has scalar components $x_1 + x_2$ and $y_1 + y_2$.

If some of the components are negative, the sketch needs to be modified slightly.

The second assertion follows from the first. For, what must x and y be if

$$\overrightarrow{(x_2, y_2)} + \overrightarrow{(x, y)} = \overrightarrow{(x_1, y_1)}?$$

We have, by the first assertion,

$$\overrightarrow{(x_2 + x, y_2 + y)} = \overrightarrow{(x_1, y_1)}.$$

Hence $\qquad x_2 + x = x_1 \qquad$ and $\qquad y_2 + y = y_1.$

Thus $\qquad x = x_1 - x_2 \qquad$ and $\qquad y = y_1 - y_2.$

Consequently, $\qquad \overrightarrow{(x_1, y_1)} - \overrightarrow{(x_2, y_2)} = \overrightarrow{(x_1 - x_2, y_1 - y_2)}.$

To justify the third assertion, add the two vectors $\overrightarrow{(x, y)}$ and $\overrightarrow{(-x, -y)}$, obtaining

$$\overrightarrow{(0, 0)}.$$

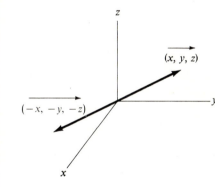

The sketch in the margin shows a typical vector $\overrightarrow{(x, y, z)}$ and its negative $\overrightarrow{(-x, -y, -z)}$, placed with their tails at the origin. ●

Theorem 2 is illustrated by these computations:

$$\overrightarrow{(1, 4)} + \overrightarrow{(2, 5)} = \overrightarrow{(3, 9)}$$

$$\overrightarrow{(2, 5, 7)} + \overrightarrow{(-1, 3, 4)} = \overrightarrow{(1, 8, 11)};$$

$$\overrightarrow{(3, -6, 8)} - \overrightarrow{(-2, 4, 5)} = \overrightarrow{(5, -10, 3)};$$

$$-\overrightarrow{(5, -2, 0)} = \overrightarrow{(-5, 2, 0)}.$$

For convenience we list the terms defined in this section: vector, zero vector, length (magnitude) of a vector, scalar, scalar components, unit vector, sum of vectors, difference of vectors, negative of a vector.

Exercises

Exercises 1 to 10 concern vectors in the xy plane.

1. Draw the vector $\overrightarrow{(2, 3)}$, placing its tail at
   (a) $(0, 0)$,   (b) $(-1, 2)$,   (c) $(1, 1)$.

2. Find $|\mathbf{A}|$ if $\mathbf{A}$ is
   (a) $\overrightarrow{(4, 0)}$,   (b) $\overrightarrow{(5, 12)}$,   (c) $\overrightarrow{(-12, 5)}$,
   (d) $\overrightarrow{(8, -6)}$.

3. Find $\mathbf{A} + \mathbf{B}$ geometrically and express the result in the form (x, y) if
(a) $\mathbf{A} = (2, 1)$, $\mathbf{B} = (1, 4)$;
(b) $\mathbf{A} = (2, 2)$, $\mathbf{B} = (1, -1)$.

4. Give an example of plane vectors $\mathbf{A}$ and $\mathbf{B}$ such that
(a) $|\mathbf{A} + \mathbf{B}|$ is not equal to $|\mathbf{A}| + |\mathbf{B}|$.
(b) $|\mathbf{A} + \mathbf{B}| = |\mathbf{A}| + |\mathbf{B}|$.

5. Find $\mathbf{A} - \mathbf{B}$ geometrically and express the result in the form (x, y) if
(a) $\mathbf{A} = (4, 3)$, $\mathbf{B} = (2, 0)$;
(b) $\mathbf{A} = (3, 4)$, $\mathbf{B} = (5, 1)$;
(c) $\mathbf{A} = (1, 1)$, $\mathbf{B} = (-2, 4)$.

6. Show pictorially that $(2, 3) + (-1, 2) = (1, 5)$.

7. Write $\mathbf{A}$ in the form (x, y) if
(a) its tail is at $(1, 3)$ and its head at $(3, 6)$;
(b) its tail is at $(2, 7)$ and its head at $(2, 4)$;
(c) its tail is at $(2, 4)$ and its head at $(2, 7)$;
(d) its tail is at $(5, 3)$ and its head at $(-1, -6)$.

8. Find the scalar components of $\mathbf{A}$ if
(a) $|\mathbf{A}| = 10$, and $\mathbf{A}$ points to the northwest;
(b) $|\mathbf{A}| = 6$, and $\mathbf{A}$ points to the south;
(c) $|\mathbf{A}| = 9$, and $\mathbf{A}$ points to the southeast;
(d) $|\mathbf{A}| = 5$, and $\mathbf{A}$ points to the east.
(North is indicated by the positive y axis.)

9. Sketch a diagram to show that $\mathbf{A} + (\mathbf{B} + \mathbf{C}) = (\mathbf{A} + \mathbf{B}) + \mathbf{C}$.

10. Show with a diagram that $\mathbf{A} + (-\mathbf{B}) = \mathbf{A} - \mathbf{B}$.

In Exercises 11 to 14 the vectors are in space and have the form (x, y, z).

11. Compute $\mathbf{A} - \mathbf{B}$ if
(a) $\mathbf{A} = (2, 3, 4)$ and $\mathbf{B} = (1, 5, 0)$;
(b) $\mathbf{A} = (3, 4, 2)$ and $\mathbf{B} = (0, 0, 0)$;
(c) $\mathbf{A} = (0, 3, 2)$ and $\mathbf{B} = (2, 4, -1)$.

12. Find $|\mathbf{A}|$ if $\mathbf{A}$ is
(a) $\left(-\frac{1}{2}, \frac{1}{2}, \sqrt{2}/2\right)$,
(b) $(-3, 4, 12)$,
(c) $\left(\dfrac{2}{\sqrt{38}}, \dfrac{-3}{\sqrt{38}}, \dfrac{5}{\sqrt{38}}\right)$.

13. Find $|\mathbf{A} + \mathbf{B}|$ if
(a) $\mathbf{A} = (1, 2, 0)$ and $\mathbf{B} = (2, 3, 5)$;
(b) $\mathbf{A} = (3, -2, 1)$ and $\mathbf{B} = (-4, 3, 2)$.

14. Determine $|(\cos \theta, \sin \theta, 1)|$, where θ is arbitrary.

■

15. The wind is 30 miles per hour to the northeast. An airplane is traveling 100 miles per hour relative to the wind, and the vector from the tail of the plane to its front tip points to the southeast:

30
100

(a) What is the speed of the plane relative to the ground?
(b) What is the direction of the flight relative to the ground?

16. Let a, b, and c be scalars, not all 0. Show that
$$\left(\frac{a}{\sqrt{a^2 + b^2 + c^2}}, \frac{b}{\sqrt{a^2 + b^2 + c^2}}, \frac{c}{\sqrt{a^2 + b^2 + c^2}}\right)$$
is a unit vector.

17. If $\mathbf{A}$ has tail P and head Q, and $\mathbf{B}$ has tail Q and head R, find
(a) the tail and head of $\mathbf{A} + \mathbf{B}$;
(b) the tail and head of $-\mathbf{A}$;
(c) the tail and head of $-\mathbf{B}$.

■ ■

18. What does the inequality $|\mathbf{A} + \mathbf{B}| \leq |\mathbf{A}| + |\mathbf{B}|$ say about a triangle, two of whose sides are $\mathbf{A}$ and $\mathbf{B}$?

19. From Exercise 18 deduce that, for any real numbers $x_1, y_1, z_1, x_2, y_2, z_2$,
$$x_1 x_2 + y_1 y_2 + z_1 z_2 \leq \sqrt{x_1^2 + y_1^2 + z_1^2}\,\sqrt{x_2^2 + y_2^2 + z_2^2}.$$

20. (a) When does equality hold in Exercise 18?
(b) When does equality hold in Exercise 19?

21. Prove by using scalar components that, for space vectors $\mathbf{A}$, $\mathbf{B}$, and $\mathbf{C}$, $\mathbf{A} + (\mathbf{B} + \mathbf{C}) = (\mathbf{A} + \mathbf{B}) + \mathbf{C}$.

22. (a) What is the sum of the five vectors shown below?
(b) Sketch the pentagon corresponding to the sum $\mathbf{A} + \mathbf{C} + \mathbf{D} + \mathbf{E} + \mathbf{B}$.

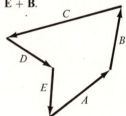
C
B
D
E
A

23. Show that for any angles α and β, the vector
$$(\cos \alpha \cos \beta, \cos \alpha \sin \beta, \sin \alpha)$$
has length 1.

19.2

The product of a scalar and a vector

The algebra of vectors described in Sec. 19.1 closely resembles the algebra of scalars. The present section introduces a type of multiplication that does not resemble anything in ordinary algebra. It concerns the product of a scalar and a vector, and the result will be a vector. Another way to say this is, "Scalars will operate on vectors."

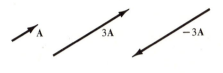

Scalars will operate on vectors by magnifying or shrinking them. Thus 3**A** shall mean a vector three times as long as **A** having the same direction as **A**. Also, (-3)**A** shall be $-(3$**A**$)$, a vector three times as long as **A** but in the opposite direction.

More generally, the product of a scalar c and a vector **A** is defined as follows. Note that c need not be an integer.

DEFINITION   *The product of a scalar and a vector.*   If c is a scalar and **A** a vector, the product c**A** is the vector whose length is $|c|$ times the length of **A** and whose direction is the same as that of **A** if c is positive, and opposite that of **A** if c is negative.

Observe that 0**A** has length 0 and thus is the zero vector **0** to which no direction is assigned. The vector c**A** is called a *scalar multiple* of the vector **A**.

EXAMPLE I   If **A** is the vector shown in the margin, sketch (-1)**A**, $\frac{1}{2}$**A**, (-2)**A**.

SOLUTION (-1)**A** has the same length as **A** but the opposite direction. Thus (-1)**A** $= -$**A**. $\frac{1}{2}$**A** is half as long as **A** and has the same direction as **A**. (-2)**A** is twice as long as **A** but its direction is opposite that of **A**.

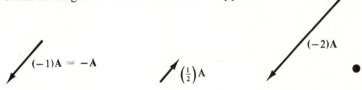

The next theorem shows how to express the scalar components of c**A** in terms of c and the scalar components of **A**.

THEOREM I $$c\overrightarrow{(x,\ y)} = \overrightarrow{(cx,\ cy)},$$
and $$c\overrightarrow{(x,\ y,\ z)} = \overrightarrow{(cx,\ cy,\ cz)}.$$

PROOF   For the sake of simplicity, we will carry through the argument for plane vectors, showing that $c\overrightarrow{(x,\ y)} = \overrightarrow{(cx,\ cy)}$. The proof for vectors in space is similar, but the diagrams are messier.

First it will be shown that the length of $\overrightarrow{(cx,\ cy)}$ is $|c|$ times the length of $\overrightarrow{(x,\ y)}$. To do this, compute the length of $\overrightarrow{(cx,\ cy)}$;

$$|\overrightarrow{(cx,\ cy)}| = \sqrt{(cx)^2 + (cy)^2} = \sqrt{c^2(x^2 + y^2)} = \sqrt{c^2}\sqrt{x^2 + y^2}.$$

Now, $\sqrt{x^2 + y^2}$ is the length of $\overrightarrow{(x, y)}$.

Furthermore, $\sqrt{c^2} = |c|$,

where c is positive or negative. [For instance, $\sqrt{(-3)^2} = \sqrt{9} = 3 = |-3|$.]
  Thus the length of $\overrightarrow{(cx, cy)}$ is $|c|$ times the length of $\overrightarrow{(x, y)}$.
  Next it will be shown that the direction of $\overrightarrow{(cx, cy)}$ is the same as that of $c\overrightarrow{(x, y)}$.

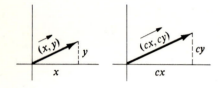

  Consider the case $c > 0$.   Then $\overrightarrow{(x, y)}$ and $\overrightarrow{(cx, cy)}$ lie on corresponding sides of similar triangles.   Hence they are parallel and point in the same direction. Thus if c is positive, $c\overrightarrow{(x, y)}$ and $\overrightarrow{(cx, cy)}$ have the same direction.   Since $c\overrightarrow{(x, y)}$ and $\overrightarrow{(cx, cy)}$ have the same length and the same direction, they are the same vector.
  If c is negative, similar reasoning shows that $c\overrightarrow{(x, y)} = \overrightarrow{(cx, cy)}$.   ●

By Theorem 1,
$$4\overrightarrow{(2, 5)} = \overrightarrow{(8, 20)};$$
$$\tfrac{1}{2}\overrightarrow{(4, 2, -5)} = \overrightarrow{(2, 1, -2.5)};$$
$$0.03\overrightarrow{(2, -1, 4)} = \overrightarrow{(0.06, -0.03, 0.12)};$$
$$-1\overrightarrow{(2, 3, 4)} = \overrightarrow{(-2, -3, -4)}.$$

DEFINITION *The division of a vector by a scalar.* Let **A** be a vector and let c be a scalar other than 0.   Then
$$\frac{\mathbf{A}}{c}$$
is defined as
$$\left(\frac{1}{c}\right)\mathbf{A}.$$

EXAMPLE 2   Sketch the vector **A**/2 for a typical vector **A**.

SOLUTION   **A**/2 is defined to be $\tfrac{1}{2}$**A**, which is a vector half as long as **A** and in the same direction as **A**.

EXAMPLE 3   Compare the vectors **A** and **A**/0.1.

SOLUTION   **A**/0.1 = (1/0.1)**A** = 10**A**, a vector 10 times as long as **A** and in the same direction as **A**. ●

EXAMPLE 4   Compute $\overrightarrow{(4, 7, 6)}/3$.

SOLUTION
$$\frac{\overrightarrow{(4, 7, 6)}}{3} = \frac{1}{3}\overrightarrow{(4, 7, 6)} = \overrightarrow{\left(\frac{4}{3}, \frac{7}{3}, 2\right)}. ●$$

Example 4 generalizes to the formula

$$\frac{\overrightarrow{(x,\, y,\, z)}}{c} = \overrightarrow{\left(\frac{x}{c},\, \frac{y}{c},\, \frac{z}{c}\right)} \qquad \text{if } c \neq 0$$

The next theorem will be used often to produce a unit vector with a prescribed direction.

THEOREM 2 For any vector $\mathbf{A}$ not equal to $\mathbf{0}$, the vector

$$\frac{\mathbf{A}}{|\mathbf{A}|}$$

is a unit vector (and has the same direction as $\mathbf{A}$).

PROOF Since $|\mathbf{A}|$ is a positive number, $\mathbf{A}/|\mathbf{A}|$ has the same direction as $\mathbf{A}$. What is the length of $\mathbf{A}/|\mathbf{A}|$? Since $\mathbf{A}/|\mathbf{A}|$ is defined as

$$\frac{1}{|\mathbf{A}|}\mathbf{A},$$

its length is $\qquad \dfrac{1}{|\mathbf{A}|}$ times the length of $\mathbf{A}$,

that is, $\qquad \dfrac{1}{|\mathbf{A}|}\,|\mathbf{A}|,$

which is 1.

Thus $\mathbf{A}/|\mathbf{A}|$ is a unit vector.   This concludes the proof. ●

EXAMPLE 5 Compute $\mathbf{A}/|\mathbf{A}|$ when $\mathbf{A} = \overrightarrow{(4,\, -5,\, 20)}$.

SOLUTION We have

$$|\mathbf{A}| = \sqrt{4^2 + (-5)^2 + (20)^2} = \sqrt{16 + 25 + 400} = \sqrt{441} = 21.$$

Thus $\qquad \dfrac{\mathbf{A}}{|\mathbf{A}|} = \overrightarrow{\left(\dfrac{4}{21},\, \dfrac{-5}{21},\, \dfrac{20}{21}\right)}.$ ●

The next definition introduces and names three special vectors which will be used often.

DEFINITION *The basic unit vectors.* The vectors

$$\mathbf{i} = \overrightarrow{(1,\, 0,\, 0)},$$
$$\mathbf{j} = \overrightarrow{(0,\, 1,\, 0)},$$
$$\text{and} \qquad \mathbf{k} = \overrightarrow{(0,\, 0,\, 1)}$$

are called the *basic unit vectors.*

EXAMPLE 6 Express the vector $3\mathbf{i} + 4\mathbf{j} + 5\mathbf{k}$ in the form $\overrightarrow{(x,\, y,\, z)}$.

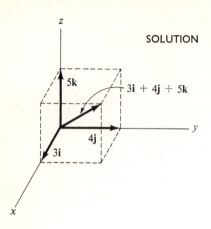

SOLUTION

$$3\mathbf{i} + 4\mathbf{j} + 5\mathbf{k} = 3\overrightarrow{(1, 0, 0)} + 4\overrightarrow{(0, 1, 0)} + 5\overrightarrow{(0, 0, 1)}$$
$$= \overrightarrow{(3, 0, 0)} + \overrightarrow{(0, 4, 0)} + \overrightarrow{(0, 0, 5)}$$
$$= \overrightarrow{(3, 4, 5)}.$$

The diagram in the margin shows the relation between $\mathbf{i}$, $\mathbf{j}$, and $\mathbf{k}$ and the vector $3\mathbf{i} + 4\mathbf{j} + 5\mathbf{k}$. ●

As Example 6 suggests,

$$x\mathbf{i} + y\mathbf{j} + z\mathbf{k} = \overrightarrow{(x, y, z)}.$$

Generally the notation $x\mathbf{i} + y\mathbf{j} + z\mathbf{k}$ is preferable to $\overrightarrow{(x, y, z)}$. First of all, it is more geometric. Second, when x, y, and z are messy expressions, the notation $x\mathbf{i} + y\mathbf{j} + z\mathbf{k}$ is easier to read.

When representing a plane vector $\mathbf{A} = \overrightarrow{(x, y)}$, we do not need the vector $\mathbf{k}$; we have simply

$$\mathbf{A} = x\mathbf{i} + y\mathbf{j},$$

since the coefficient of $\mathbf{k}$ is 0.

While $\mathbf{i}$, $\mathbf{j}$, and $\mathbf{k}$ will be the three mutually perpendicular unit vectors most often referred to, occasionally other such triplets will provide a convenient reference.

Let $\mathbf{u}_1$, $\mathbf{u}_2$, and $\mathbf{u}_3$ be three unit vectors that are perpendicular to each

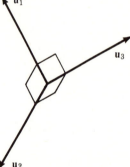

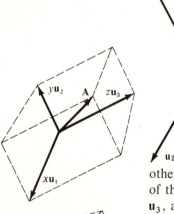

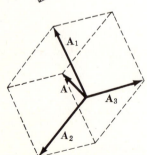

other. For any scalars x, y, and z, the vector $\mathbf{A} = x\mathbf{u}_1 + y\mathbf{u}_2 + z\mathbf{u}_3$ is the sum of three vectors, one parallel to $\mathbf{u}_1$, one parallel to $\mathbf{u}_2$, and one parallel to $\mathbf{u}_3$, as shown in the margin. (If x, say, is negative, then $x\mathbf{u}_1$ points in the direction opposite that of $\mathbf{u}_1$.) Note that since $\mathbf{u}_1$ is a unit vector, $x\mathbf{u}_1$ has length $|x|$. The length of $x\mathbf{u}_1 + y\mathbf{u}_2 + z\mathbf{u}_3$ is $\sqrt{x^2 + y^2 + z^2}$.

The procedure can be reversed. It is possible to start with a vector $\mathbf{A}$ and the three unit vectors $\mathbf{u}_1$, $\mathbf{u}_2$, and $\mathbf{u}_3$ (which are perpendicular to each other) and express $\mathbf{A}$ as the sum of a vector $\mathbf{A}_1$ parallel to $\mathbf{u}_1$, a vector $\mathbf{A}_2$ parallel to $\mathbf{u}_2$, and a vector $\mathbf{A}_3$ parallel to $\mathbf{u}_3$. Draw the box whose diagonal is $\mathbf{A}$ and whose edges are parallel to $\mathbf{u}_1$, $\mathbf{u}_2$, and $\mathbf{u}_3$. Thus, $\mathbf{A}_1 = x\mathbf{u}_1$, $\mathbf{A}_2 = y\mathbf{u}_2$, and $\mathbf{A}_3 = z\mathbf{u}_3$ for some scalars x, y, and z.

In the case of a plane vector $\mathbf{A}$ and a pair of perpendicular unit vectors

$\mathbf{u}_1$ and $\mathbf{u}_2$, the diagram is much easier to sketch.

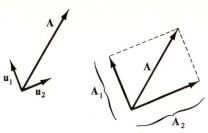

DEFINITION   *Vector and scalar components.*   Let $\mathbf{u}_1$, $\mathbf{u}_2$, and $\mathbf{u}_3$ be perpendicular unit vectors.   Let $\mathbf{A}$ be an arbitrary vector.   Then $\mathbf{A}$ is expressible in the form

$$\mathbf{A} = \mathbf{A}_1 + \mathbf{A}_2 + \mathbf{A}_3,$$

where $\mathbf{A}_1$ is parallel to $\mathbf{u}_1$,

$\qquad\quad\mathbf{A}_2$ is parallel to $\mathbf{u}_2$,

and $\quad\ \mathbf{A}_3$ is parallel to $\mathbf{u}_3$.

$\mathbf{A}_1$ is the *vector component* of $\mathbf{A}$ in the direction of $\mathbf{u}_1$. $\mathbf{A}_2$ is the *vector component* of $\mathbf{A}$ in the direction of $\mathbf{u}_2$. $\mathbf{A}_3$ is the *vector component* of $\mathbf{A}$ in the direction of $\mathbf{u}_3$.

Furthermore, if $\mathbf{A}_1 = x\mathbf{u}_1$, $\mathbf{A}_2 = y\mathbf{u}_2$, and $\mathbf{A}_3 = z\mathbf{u}_3$, the number x is called the *scalar component* of $\mathbf{A}$ in the direction of $\mathbf{u}_1$; y is the *scalar component* of $\mathbf{A}$ in the direction of $\mathbf{u}_2$; z is the *scalar component* of $\mathbf{A}$ in the direction of $\mathbf{u}_3$.   A scalar component may be negative.   By the scalar component of a vector $\mathbf{A}$ along the vector $\mathbf{R}$ is meant the scalar component of $\mathbf{A}$ in the direction of the unit vector $\mathbf{R}/|\mathbf{R}|$.

Scalar components, being numbers, are frequently easier to work with than vectors.   The next chapter will use scalar components extensively, for instance in examining the forces acting on a particle moving in a curved path.

EXAMPLE 7   Find the vector components of $\mathbf{A} = \overrightarrow{(3, -4)}$ along $\mathbf{u}_1 = \overrightarrow{(1, 0)}$ and $\mathbf{u}_2 = \overrightarrow{(0, 1)}$.

SOLUTION   The accompanying diagram displays $\mathbf{A}$, $\mathbf{u}_1$, and $\mathbf{u}_2$.   Inspection of the diagram shows that

$$\mathbf{A} = 3\mathbf{u}_1 - 4\mathbf{u}_2.$$

Thus the vector component of $\mathbf{A}$ along $\mathbf{u}_1$ is $3\mathbf{u}_1$ and along $\mathbf{u}_2$ is $-4\mathbf{u}_2$.

The scalar component of $\mathbf{A}$ along $\mathbf{u}_1$ is 3 and along $\mathbf{u}_2$ is -4. ●

EXAMPLE 8   Find the vector component of $\mathbf{A} = \overrightarrow{(3, -4, 5)}$ along

$$\mathbf{u}_1 = \overrightarrow{(-1, 0, 0)}, \ \mathbf{u}_2 = \overrightarrow{(0, 1, 0)}, \text{ and } \mathbf{u}_3 = \overrightarrow{(0, 0, 1)}.$$

SOLUTION $$\mathbf{A} = -3\mathbf{u}_1 - 4\mathbf{u}_2 + 5\mathbf{u}_3.$$

Thus the vector components are $-3\mathbf{u}_1$, $-4\mathbf{u}_2$, and $5\mathbf{u}_3$. The scalar components of $\mathbf{A}$ along $\mathbf{u}_1$, $\mathbf{u}_2$, and $\mathbf{u}_3$ are, respectively, -3, -4, and 5. ●

The vectors $\mathbf{i}$, $\mathbf{j}$, $\mathbf{k}$ are sometimes denoted $\mathbf{e}_1$, $\mathbf{e}_2$, $\mathbf{e}_3$. The three scalar components of $\mathbf{A}$ are sometimes denoted A_x, A_y, A_z.

Thus

$$\mathbf{A} = A_x \mathbf{i} + A_y \mathbf{j} + A_z \mathbf{k}.$$

The scalar components of $\mathbf{A}$ are also denoted a_1, a_2, a_3; then we have

$$\mathbf{A} = a_1 \mathbf{i} + a_2 \mathbf{j} + a_3 \mathbf{k} \quad \text{and} \quad \mathbf{A} = \overrightarrow{(a_1, a_2, a_3)}.$$

The next section shows how to find the scalar and vector components of any vector $\mathbf{A}$ along any unit vector $\mathbf{u}$.

Exercises

1. Compute and sketch $c\mathbf{A}$ if $\mathbf{A} = 2\mathbf{i} + 3\mathbf{j} + \mathbf{k}$ and c is

   (a) 2   (b) -2 (c) $\frac{1}{2}$ (d) $-\frac{1}{2}$.

2. Express each of the following vectors in the form $c(2\mathbf{i} + 3\mathbf{j} + 4\mathbf{k})$ for suitable c:

   (a) $\overrightarrow{(4, 6, 8)}$
   (b) $-2\mathbf{i} - 3\mathbf{j} - 4\mathbf{k}$
   (c) $\mathbf{0}$
   (d) $\frac{2}{11}\mathbf{i} + \frac{3}{11}\mathbf{j} + \frac{4}{11}\mathbf{k}$.

3. If $\mathbf{u}$ is a unit vector, what is the length of $-3\mathbf{u}$?

4. Find $|5(2\mathbf{i} + 4\mathbf{j} - \mathbf{k})|$.

5. What is the vector component of $-4\mathbf{i} + 5\mathbf{j} + 0\mathbf{k}$ along

   (a) $\mathbf{i}$?   (b) $\mathbf{j}$?   (c) $\mathbf{k}$?

6. What is the scalar component of $-4\mathbf{i} + 5\mathbf{j} + 0\mathbf{k}$ along

   (a) $\mathbf{i}$?   (b) $\mathbf{j}$?   (c) $\mathbf{k}$?

7. Sketch an appropriate diagram and find the vector component of $2\mathbf{i} + 2\mathbf{j} + 3\mathbf{k}$ along

   (a) $\mathbf{i}$ (b) $-\mathbf{i}$ (c) $3\mathbf{i}$ (d) $\dfrac{\mathbf{i}}{2} + \dfrac{\mathbf{j}}{2} + 0\mathbf{k}$.

8. Find the unit vector that has the same direction as $\mathbf{i} + 2\mathbf{j} + 3\mathbf{k}$.

9. If the vector component of $\mathbf{A}$ along the unit vector $\mathbf{u}$ is $4\mathbf{u}$ and the vector component of $\mathbf{B}$ along $\mathbf{u}$ is $-3\mathbf{u}$, what is

   (a) the scalar component of $\mathbf{A}$ along $\mathbf{u}$?
   (b) the vector component of $\mathbf{A} + \mathbf{B}$ along $\mathbf{u}$?

10. Let $\mathbf{u}_1$ and $\mathbf{u}_2$ be the unit vectors in the xy plane shown below.

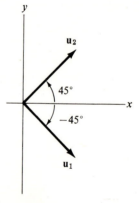

Express in the form $x\mathbf{u}_1 + y\mathbf{u}_2$ the vectors
(a) $\mathbf{i}$ (b) $\mathbf{j}$ (c) $3\mathbf{i} - 4\mathbf{j}$.

11. Let $\mathbf{u}$ be the unit vector that makes an angle $\pi/3$ with the x axis, as shown in the diagram.

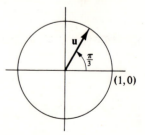

(a) Express $\mathbf{u}$ in the form $x\mathbf{i} + y\mathbf{j}$.
(b) Draw the vector components of $\mathbf{u}$ along $\mathbf{i}$ and $\mathbf{j}$.

■

12. Draw a unit vector tangent to the curve $y = \sin x$ at $(0, 0)$. What are its vector components along $\mathbf{i}$ and $\mathbf{j}$?

13. Draw a unit vector tangent to the curve $y = x^3$ at $(1, 1)$. What are its scalar components along $\mathbf{i}$ and $\mathbf{j}$?

14. What angle does the vector $2\mathbf{i} + 3\mathbf{j}$ make with the positive x axis?

15. (a) Show that the vectors

$$\mathbf{u}_1 = \frac{1}{2}\mathbf{i} + \frac{\sqrt{3}}{2}\mathbf{j} \quad \text{and} \quad \mathbf{u}_2 = \frac{\sqrt{3}}{2}\mathbf{i} - \frac{1}{2}\mathbf{j}$$

are perpendicular unit vectors. *Hint:* What angles do they make with the x axis?

(b) Find scalars x and y such that $\mathbf{i} = x\mathbf{u}_1 + y\mathbf{u}_2$.

16. (a) Show that

$$\mathbf{u}_1 = \frac{\sqrt{2}}{2}\mathbf{i} + \frac{\sqrt{2}}{2}\mathbf{j} \quad \text{and} \quad \mathbf{u}_2 = \frac{-\sqrt{2}}{2}\mathbf{i} + \frac{\sqrt{2}}{2}\mathbf{j}$$

are perpendicular unit vectors. *Hint:* Draw them.

(b) Express $\mathbf{i}$ in the form $x\mathbf{u}_1 + y\mathbf{u}_2$. *Hint:* Draw $\mathbf{i}$, $\mathbf{u}_1$, $\mathbf{u}_2$.

(c) Express $\mathbf{j}$ in the form $x\mathbf{u}_1 + y\mathbf{u}_2$.

(d) Express $-2\mathbf{i} + 3\mathbf{j}$ in the form $x\mathbf{u}_1 + y\mathbf{u}_2$.

■ ■

17. Let $\mathbf{A}$, $\mathbf{B}$, and $\mathbf{C}$ be three vectors in the plane. Show, with the aid of a sketch, that there are always x, y, and z, not all 0, such that

$$x\mathbf{A} + y\mathbf{B} + z\mathbf{C} = \mathbf{0}.$$

This exercise gives a vector formula for the midpoint.

18. Let P and Q be points in the xy plane whose origin is O. Let M be the midpoint of PQ. Show that

$$\overrightarrow{OM} = \overrightarrow{OP} + \tfrac{1}{2}(\overrightarrow{PQ})$$
$$= \overrightarrow{OP} + \tfrac{1}{2}(\overrightarrow{OQ} - \overrightarrow{OP})$$
$$= \tfrac{1}{2}(\overrightarrow{OP} + \overrightarrow{OQ}).$$

19. (See Exercise 18.) Prove, using vectors, that the line segment joining the midpoints of two sides of a triangle is parallel to the third side and half as long.

20. (See Exercise 19.) The midpoints of a quadrilateral in space are joined to form another quadrilateral. Prove that this second quadrilateral is a parallelogram.

21. Draw two plane vectors $\mathbf{A}$ and $\mathbf{B}$ with their tails at the origin. Sketch the vectors $t\mathbf{A} + (1 - t)\mathbf{B}$ with their tails at the origin for

(a) $t = 2$; (b) $t = -2$; (c) $t = 0$; (d) $t = 1$; (e) $t = \tfrac{1}{2}$; (f) $t = \tfrac{1}{3}$. (g) On what familiar geometric object do the heads of the six vectors sketched seem to lie?

22. Consider two vectors $\mathbf{A}$ and $\mathbf{B}$, with their tails at the origin.

(a) Why is the vector from the origin to any point on the line through their heads of the form $\mathbf{B} + t(\mathbf{A} - \mathbf{B})$ for some number t?

(b) Use (a) to discuss Exercise 21 (g).

23. (See Exercise 18). Let P, Q, and R be vertices of a triangle in the xy plane, and let O be the origin of the xy plane. Let S be a point two-thirds of the way from one of the three vertices to the midpoint of the opposite side of the triangle.

(a) Prove that

$$\overrightarrow{OS} = \tfrac{1}{3}(\overrightarrow{OP} + \overrightarrow{OQ} + \overrightarrow{OR}).$$

(b) From (a) deduce that the three medians of a triangle intersect at a point.

24. Let $\mathbf{A}$, $\mathbf{B}$, and $\mathbf{C}$ be three vectors in space with the property that they do not lie on one plane. That is, when located so that their tails are at the origin, they are not contained in a plane.

(a) Show with the aid of sketches that any vector $\mathbf{D}$ is of the form $x\mathbf{A} + y\mathbf{B} + z\mathbf{C}$ for suitable scalars x, y, z.

(b) For a given vector are the scalars in (a) unique?

19.3

The dot product of two vectors

Consider a rock being pulled along the level ground by a rope inclined at a fixed angle to the ground. Let the force applied to the rock be represented by the vector $\mathbf{F}$. The force $\mathbf{F}$ can be expressed as the sum of a vertical force $\mathbf{F}_2$ and a horizontal force $\mathbf{F}_1$.

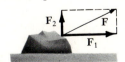

We may replace $\mathbf{F}$ with $\mathbf{F}_1$ and $\mathbf{F}_2$, which together have the same effect on the rock as $\mathbf{F}$

How much work is done by the force **F** in moving the rock along the ground? The physicist defines the work accomplished by a constant force **F** (whatever direction it may have) in moving a particle on a straight path from the tail to the head of the vector **R** as the following product:

magnitude of vector component **F** in the direction of **R** times the length of **R**.

Thus the force F_2 shown in the preceding diagram accomplishes no work in moving the rock; only the force F_1 does so. The work that **F** accomplishes is the work done by F_1 in overcoming friction, and hence equals the product

$$|F_1| \text{ (distance the rock moves).}$$

More generally, let a force represented by the vector **F** move an object along a straight line from the tail to the head of a vector **R** as in this diagram.

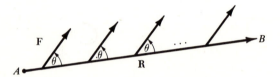

A constant force **F** moves an object from A to B, perhaps against gravity, air resistance, friction, etc.

The work accomplished is defined to be

$$\underbrace{|F| \cos \theta}_{\substack{\text{Magnitude} \\ \text{of force in} \\ \text{direction} \\ \text{object is} \\ \text{moved}}} \cdot \underbrace{|R|}_{\substack{\text{Distance} \\ \text{the object} \\ \text{is moved}}}$$

where θ is the angle between **R** and **F**. The angle θ can be anywhere in $[0, \pi]$.

This important physical concept illustrates the *dot product* of two vectors, which will be introduced after the following definition.

DEFINITION *Angle between two nonzero vectors.* Let **A** and **B** be two nonparallel and nonzero vectors. They determine a triangle and an angle θ, shown in the diagram at the left. The *angle between* **A** *and* **B** is θ. Note that

$$0 < \theta < \pi.$$

If **A** and **B** are parallel, the angle between them is 0 (if they have the same direction) or π (if they have opposite directions). The angle between **0** and another vector is not defined.

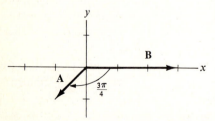

The angle between **i** and **j** is $\pi/2$. The angle between $A = -i - j$ and $B = 3i$ is $3\pi/4$, as the accompanying diagram shows. The angle between **k** and $-k$ is π; the angle between $2i + 2j + 2k$ and $5i + 5j + 5k$ is 0. The dot product of vectors can now be defined.

DEFINITION   *Dot product.*   Let **A** and **B** be two nonzero vectors.   Their dot product is the number

$$|\mathbf{A}|\,|\mathbf{B}|\cos\theta,$$

where θ is the angle between **A** and **B**.   If **A** or **B** is **0**, their dot product is 0.   The dot product is denoted $\mathbf{A}\cdot\mathbf{B}$.   It is a scalar and is also called the *scalar* product of **A** and **B**.

EXAMPLE 1   Compute the dot product $\mathbf{A}\cdot\mathbf{B}$ if $\mathbf{A}=3\mathbf{i}+3\mathbf{j}$ and $\mathbf{B}=5\mathbf{i}$.

SOLUTION   Inspection of the diagram shows that θ, the angle between **A** and **B**, is $\pi/4$.
Also,

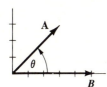

$$|\mathbf{A}|=\sqrt{3^2+3^2}=\sqrt{18},$$

and

$$|\mathbf{B}|=\sqrt{5^2+0^2}=5.$$

Thus

$$\mathbf{A}\cdot\mathbf{B}=|\mathbf{A}|\,|\mathbf{B}|\cos\theta$$

$$=\sqrt{18}\cdot 5\,\frac{\sqrt{2}}{2}$$

$$=15. \;\bullet$$

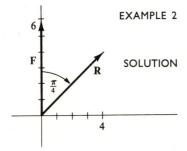

EXAMPLE 2   How much work is accomplished by the force $\mathbf{F}=6\mathbf{j}$, in moving an object from the tail to the head of $\mathbf{R}=4\mathbf{i}+4\mathbf{j}$?

SOLUTION   The work is defined as the dot product

$$\mathbf{F}\cdot\mathbf{R}=|\mathbf{F}|\,|\mathbf{R}|\cos\theta.$$

The diagram in the margin shows that $\theta=\pi/4$.   Hence

$$\text{Work}=\sqrt{0^2+6^2}\sqrt{4^2+4^2}\cos\frac{\pi}{4}$$

$$=6\sqrt{32}\,\frac{\sqrt{2}}{2}$$

$$=\frac{6\sqrt{64}}{2}$$

$$=24.$$

If force is measured in pounds and distance in feet, the work is 24 foot-pounds.   ●

Observe that $\mathbf{A}\cdot\mathbf{A}=|\mathbf{A}|^2$, a fact that will be used often.

The Meaning of $\mathbf{A}\cdot\mathbf{B}=0$ Note also that, if **A** is perpendicular to **B**, then $\mathbf{A}\cdot\mathbf{B}=0$.   Thus the dot product of two perpendicular vectors is 0.   This fact is an important property of the dot product.   Moreover, its converse is valid: If $\mathbf{A}\cdot\mathbf{B}=0$ and neither **A** nor **B** is **0**, then the cosine of the angle between **A** and **B** is 0.

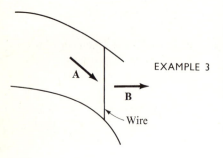

Thus **A** and **B** are perpendicular. *Consequently, the vanishing of the dot product is a test for perpendicularity.*

The next example, which will be refined in Chaps. 21 and 22, gives another physical interpretation of the dot product.

EXAMPLE 3 Let **A** represent the direction and rate of flow of water in a stream as it passes an imaginary fixed wire. Let **B** represent a vector on the surface of the water and perpendicular to the wire. Let $|\mathbf{B}|$ = length of wire and $|\mathbf{A}|$ = speed of water. Show that $\mathbf{A} \cdot \mathbf{B}$ is proportional to the rate at which water passes the wire.

SOLUTION If **A** is parallel to the wire, no water passes over the wire. If **A** is perpendicular to the wire, hence parallel to **B**, water passes over the wire. If the angle θ between **A** and **B** is, say, $\pi/4$, water passes over the wire but not as quickly as when **A** is parallel to **B**. The rate at which water passes over the wire depends on the speed of the water, the direction in which the water moves, and the length of the wire. The simplest measure of this rate is the product

$$\left(\begin{array}{c}\text{Scalar component of } \mathbf{A}\\ \text{in direction of } \mathbf{B}\end{array}\right)\left(\text{length of wire}\right).$$

If θ denotes the angle between **A** and **B**, then this product is

$$(|\mathbf{A}| \cos \theta)|\mathbf{B}|,$$

the dot product of **A** and **B**, $\mathbf{A} \cdot \mathbf{B}$. ●

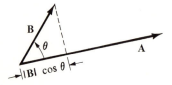

The Dot Product and Scalar Components

Note that the factor $|\mathbf{B}| \cos \theta$ in the definition of the dot product, $\mathbf{A} \cdot \mathbf{B} = |\mathbf{A}||\mathbf{B}| \cos \theta$, is the scalar component of **B** in the direction along **A**. Often $|\mathbf{B}| \cos \theta$ is called the *projection* of **B** on **A**. If $0 \le \theta < \pi/2$, this scalar component is positive, as in the diagram at the left. If $\pi/2 < \theta \le \pi$, this scalar component is negative. In any case,

$$\mathbf{A} \cdot \mathbf{B} = |\mathbf{A}| \,(\text{scalar component of } \mathbf{B} \text{ along } \mathbf{A})$$

and, similarly,

$$\mathbf{A} \cdot \mathbf{B} = |\mathbf{B}| \,(\text{scalar component of } \mathbf{A} \text{ along } \mathbf{B}).$$

The next theorem shows how to compute $\mathbf{A} \cdot \mathbf{B}$ in terms of the components of **A** and **B**.

THEOREM I (*Formula for dot product in scalar components*).
If $\mathbf{A} = x_1\mathbf{i} + y_1\mathbf{j}$ and $\mathbf{B} = x_2\mathbf{i} + y_2\mathbf{j}$, then

$$\mathbf{A} \cdot \mathbf{B} = x_1 x_2 + y_1 y_2.$$

If $\mathbf{A} = x_1\mathbf{i} + y_1\mathbf{j} + z_1\mathbf{k}$ and $\mathbf{B} = x_2\mathbf{i} + y_2\mathbf{j} + z_2\mathbf{k}$, then

$$\mathbf{A} \cdot \mathbf{B} = x_1 x_2 + y_1 y_2 + z_1 z_2.$$

PROOF We will prove the theorem for plane vectors only, since the diagrams are easier and the reasoning is the same as for vectors in space.

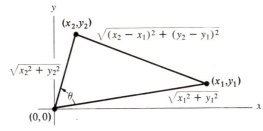

The angle between the vectors is θ

If either $\mathbf{A}$ or $\mathbf{B}$ is $\mathbf{0}$, a simple computation verifies the theorem. In the remainder of the proof, assume that neither is $\mathbf{0}$. For convenience, put the tails of $\mathbf{A}$ and $\mathbf{B}$ at the origin, as in the diagram in the margin. Now, by the definition of the dot product,

$$\mathbf{A} \cdot \mathbf{B} = \sqrt{x_1^2 + y_1^2}\sqrt{x_2^2 + y_2^2}\cos\theta. \tag{1}$$

To express $\cos\theta$ in terms of x_1, y_1, x_2, and y_2, apply the law of cosines to the triangle whose vertices are $(0, 0)$, (x_1, y_1), and (x_2, y_2). The law of

cosines ($c^2 = a^2 + b^2 - 2ab\cos\theta$, where a, b, c are the lengths of the sides of a triangle, and c is opposite angle θ) asserts in this case that

$$[\sqrt{(x_2 - x_1)^2 + (y_2 - y_1)^2}]^2 = (\sqrt{x_1^2 + y_1^2})^2 + (\sqrt{x_2^2 + y_2^2})^2$$
$$- 2\sqrt{x_1^2 + y_1^2}\sqrt{x_2^2 + y_2^2}\cos\theta.$$

Squaring and cancelling yields

$$-2x_1x_2 - 2y_1y_2 = -2\sqrt{x_2^2 + y_2^2}\sqrt{x_1^2 + y_1^2}\cos\theta.$$

Dividing this equation by -2, we obtain

$$x_1x_2 + y_1y_2 = \sqrt{x_1^2 + y_1^2}\sqrt{x_2^2 + y_2^2}\cos\theta. \tag{2}$$

Comparison of (1) and (2) establishes the theorem. ●

EXAMPLE 4 In Example 1 the dot product of $3\mathbf{i} + 3\mathbf{j}$ and $5\mathbf{i}$ was computed geometrically. Compute it now with the aid of Theorem 1.

SOLUTION $$(3\mathbf{i} + 3\mathbf{j}) \cdot 5\mathbf{i} = (3)(5) + (3)(0) = 15. \quad ●$$

EXAMPLE 5 Are the vectors $3\mathbf{i} + 7\mathbf{j}$ and $9\mathbf{i} - 4\mathbf{j}$ perpendicular? (Sketch them before reading the solution; do they seem perpendicular?)

SOLUTION Recall that two nonzero vectors are perpendicular if their dot product is 0. So compute the dot product:

$$(3\mathbf{i} + 7\mathbf{j}) \cdot (9\mathbf{i} - 4\mathbf{j}) = (3)(9) + (7)(-4)$$
$$= 27 - 28 = -1.$$

Since the dot product is *not* 0, the vectors are not perpendicular. ●

EXAMPLE 6 Find the angle θ between the vectors $\mathbf{A} = 2\mathbf{i} - \mathbf{j} + 3\mathbf{k}$ and $\mathbf{B} = \mathbf{i} + \mathbf{j} + 2\mathbf{k}$.

SOLUTION Compute $\mathbf{A} \cdot \mathbf{B}$ in two ways: first by its geometric definition, and then in terms of scalar components.

$$\mathbf{A} \cdot \mathbf{B} = |\mathbf{A}| |\mathbf{B}| \cos \theta$$
$$= \sqrt{2^2 + (-1)^2 + 3^2} \sqrt{1^2 + 1^2 + 2^2} \cos \theta$$
$$= \sqrt{14} \sqrt{6} \cos \theta$$
$$= \sqrt{84} \cos \theta.$$

Also, by Theorem 1,

$$\mathbf{A} \cdot \mathbf{B} = (2)(1) + (-1)(1) + (3)(2)$$
$$= 2 - 1 + 6$$
$$= 7.$$

Comparing the two versions of $\mathbf{A} \cdot \mathbf{B}$ shows that

$$\sqrt{84} \cos \theta = 7,$$

or
$$\cos \theta = \frac{7}{\sqrt{84}} = \frac{7\sqrt{84}}{84} = \frac{\sqrt{84}}{12} = \frac{\sqrt{21}}{6} \doteq 0.764.$$

A calculator or trigonometric table shows that θ is about $40.2°$, or 0.702 radian. ●

How to Find the Angle between Two Vectors   The technique illustrated in Example 6 is a very easy way to find the angle between two line segments if their endpoints are given or can be expressed simply in rectangular coordinates. Exercise 19, for instance, concerns the angle between diagonals in a cube. The angle between two lines can be determined by the formula

$$\cos \theta = \frac{\mathbf{A} \cdot \mathbf{B}}{|\mathbf{A}| |\mathbf{B}|}.$$

In economics the dot product is used as an algebraic convenience, without any regard for its geometric significance. For instance, a shopper buys 20 pounds of potatoes at 5 cents a pound and 10 pounds of oranges at 12 cents a pound. The vector $20\mathbf{i} + 10\mathbf{j}$ records how much she bought of each item, while the vector $5\mathbf{i} + 12\mathbf{j}$ records the corresponding prices. The dot product

$$(20\mathbf{i} + 10\mathbf{j}) \cdot (5\mathbf{i} + 12\mathbf{j}) = (20)(5) + (10)(12) = 220 \text{ cents}$$

records the total cost of the purchase.

The next theorem is useful in finding scalar components of a vector.

THEOREM 2 *Scalar component along a unit vector.* Let $\mathbf{A}$ be a vector and let $\mathbf{u}$ be a unit vector. Then the scalar component of $\mathbf{A}$ in the direction $\mathbf{u}$ is the dot product

$$\mathbf{A} \cdot \mathbf{u}.$$

PROOF By definition of the dot product of $\mathbf{A}$ and $\mathbf{u}$,

$$\mathbf{A} \cdot \mathbf{u} = |\mathbf{A}||\mathbf{u}| \cos \theta,$$

where θ is the angle between $\mathbf{u}$ and $\mathbf{A}$. Since $\mathbf{u}$ is a unit vector, $|\mathbf{u}| = 1$. Thus

$$\mathbf{A} \cdot \mathbf{u} = |\mathbf{A}| \cos \theta,$$

which is the scalar component of $\mathbf{A}$ in the direction $\mathbf{u}$. ●

EXAMPLE 7   Find the scalar component of $\mathbf{A} = 2\mathbf{i} + 6\mathbf{j}$ in the direction of the unit vector

$$\mathbf{u} = \frac{\sqrt{2}}{2}\mathbf{i} - \frac{\sqrt{2}}{2}\mathbf{j}.$$

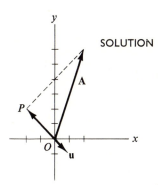

SOLUTION   By Theorem 2, this scalar component is $\mathbf{A} \cdot \mathbf{u}$. By Theorem 1,

$$\mathbf{A} \cdot \mathbf{u} = (2\mathbf{i} + 6\mathbf{j}) \cdot \left(\frac{\sqrt{2}}{2}\mathbf{i} - \frac{\sqrt{2}}{2}\mathbf{j}\right) = \frac{2\sqrt{2}}{2} - \frac{6\sqrt{2}}{2} = -2\sqrt{2}.$$

Hence the scalar component of $\mathbf{A}$ along $\mathbf{u}$ is $-2\sqrt{2}$. (The negative sign shows that the angle between $\mathbf{A}$ and $\mathbf{u}$ is greater than $\pi/2$.) The vector component of $\mathbf{A}$ along $\mathbf{u}$ is then $-2\sqrt{2}\,\mathbf{u}$.

A diagram shows that these results are plausible: $\overrightarrow{OP}$ is the vector component of $\mathbf{A}$ in the direction of $\mathbf{u}$. Its length is $2\sqrt{2}$. ●

Exercises

In Exercises 1 to 6 compute $\mathbf{A} \cdot \mathbf{B}$.

1. $\mathbf{A}$ has length 3, $\mathbf{B}$ has length 4, and the angle between $\mathbf{A}$ and $\mathbf{B}$ is $\pi/4$.
2. $\mathbf{A}$ has length 2, $\mathbf{B}$ has length 3, and the angle between $\mathbf{A}$ and $\mathbf{B}$ is $3\pi/4$.
3. $\mathbf{A}$ has length 5, $\mathbf{B}$ has length $\frac{1}{2}$, and the angle between $\mathbf{A}$ and $\mathbf{B}$ is $\pi/2$.
4. $\mathbf{A}$ is the zero vector $\mathbf{0}$, and $\mathbf{B}$ has length 5.
5. $\mathbf{A} = 2\mathbf{i} - 3\mathbf{j} + 5\mathbf{k}$, and $\mathbf{B} = \mathbf{i} - \mathbf{j} - \mathbf{k}$.
6. $\mathbf{A} = \overrightarrow{PQ}$, and $\mathbf{B} = \overrightarrow{PR}$, where $P = (1, 0, 2)$, $Q = (1, 1, -1)$, $R = (2, 3, 5)$.
7. (a) Draw the vectors $7\mathbf{i} + 12\mathbf{j}$ and $9\mathbf{i} - 5\mathbf{j}$.
   (b) Do they seem to be perpendicular?
   (c) Determine whether they are perpendicular by examining their dot product.
8. (a) Draw the vectors $\mathbf{i} + 2\mathbf{j} + 3\mathbf{k}$ and $\mathbf{i} + \mathbf{j} - \mathbf{k}$.
   (b) Do they seem to be perpendicular?
   (c) Determine whether they are perpendicular by examining their dot product.
9. (a) Estimate the angle between $\mathbf{A} = 3\mathbf{i} + 4\mathbf{j}$ and $\mathbf{B} = 5\mathbf{i} + 12\mathbf{j}$ by drawing them.
   (b) Find the angle between $\mathbf{A}$ and $\mathbf{B}$.
10. Prove Theorem 1 for the two vectors in space, $\mathbf{A} = x_1\mathbf{i} + y_1\mathbf{j} + z_1\mathbf{k}$ and $\mathbf{B} = x_2\mathbf{i} + y_2\mathbf{j} + z_2\mathbf{k}$.

11. Use Theorem 1 to prove that $x\mathbf{A} \cdot y\mathbf{B} = xy(\mathbf{A} \cdot \mathbf{B})$.
12. Use Theorem 1 to prove that
    (a) $\mathbf{A} \cdot (\mathbf{B} + \mathbf{C}) = \mathbf{A} \cdot \mathbf{B} + \mathbf{A} \cdot \mathbf{C}$;
    (b) $(x\mathbf{A}) \cdot \mathbf{B} = \mathbf{A} \cdot (x\mathbf{B}) = x(\mathbf{A} \cdot \mathbf{B})$.
13. Draw two vectors $\mathbf{A}$ and $\mathbf{B}$ such that
    (a) $\mathbf{A} \cdot \mathbf{B} < 0$;
    (b) $\mathbf{A} \cdot \mathbf{B} > 0$;
    (c) $\mathbf{A} \cdot \mathbf{B} = 0$.
14. Prove that for any vector $\mathbf{A}$, $\mathbf{A} \cdot \mathbf{A} = |\mathbf{A}|^2$, using
    (a) the definition of the dot product;
    (b) Theorem 1.

■

Exercises 15 to 19 refer to the diagram of the cube shown below.

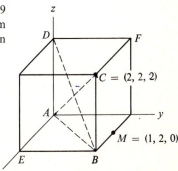

15. Find the cosine of the angle between $\overrightarrow{AC}$ and $\overrightarrow{BD}$.
16. Find the cosine of the angle between $\overrightarrow{AF}$ and $\overrightarrow{BD}$.
17. Find the cosine of the angle between $\overrightarrow{AC}$ and $\overrightarrow{AM}$.
18. Find the cosine of the angle between $\overrightarrow{MD}$ and $\overrightarrow{MF}$.
19. Find the cosine of the angle between $\overrightarrow{EF}$ and $\overrightarrow{BD}$.
20. Show that if **B** is a nonzero vector, then the scalar component of the vector **A** in the direction of **B** is

$$\frac{\mathbf{A} \cdot \mathbf{B}}{|\mathbf{B}|}.$$

In Exercises 21 to 23 find the scalar component of **A** in the direction parallel to **B** (use Exercise 20).

21. $\mathbf{A} = 3\mathbf{i} + 4\mathbf{j}, \mathbf{B} = 2\mathbf{i} - \mathbf{j}$.
22. $\mathbf{A} = \mathbf{i} + 2\mathbf{j} + 3\mathbf{k}, \mathbf{B} = \mathbf{j} + \mathbf{k}$.
23. $\mathbf{A} = 2\mathbf{i} - 3\mathbf{j} + 4\mathbf{k}, \mathbf{B} = -3\mathbf{k}$.

■ ■

Exercises 24 to 26 are related.
24. Let **A** be a vector and let **u** be a unit vector. Prove that $\mathbf{A} - (\mathbf{A} \cdot \mathbf{u})\mathbf{u}$ is perpendicular to **u**
  (a) by computing the dot product of $\mathbf{A} - (\mathbf{A} \cdot \mathbf{u})\mathbf{u}$ and **u**;
  (b) by drawing a picture of **A**, **u**, and $\mathbf{A} - (\mathbf{A} \cdot \mathbf{u})\mathbf{u}$.
25. (See Exercise 24.) Let **A** and **B** be nonzero vectors. Show that the vector

$$\mathbf{C} = \mathbf{A} - \frac{(\mathbf{A} \cdot \mathbf{B})\mathbf{B}}{|\mathbf{B}|^2}$$

is perpendicular to **B**.
26. Let **A** and **B** be nonzero vectors.
  (a) Using a sketch, explain why there is a scalar x and a vector **C** perpendicular to **B** such that

$$\mathbf{A} = x\mathbf{B} + \mathbf{C}.$$

  (b) Give a formula for **C** and x in terms of **A** and **B**.
27. Let $P = (1, 2, 3)$, $Q = (2, -1, 4)$, and $R = (1, 3, 5)$. Find the distance from P to the line determined by Q and R.
  Hint: Construct a vector perpendicular to $\overrightarrow{QR}$ whose

length equals the distance in question. Exercise 25 may be of aid.
28. Find a nonzero vector **A** that is perpendicular to both vectors $2\mathbf{i} - 3\mathbf{j} + 4\mathbf{k}$ and $\mathbf{i} + 2\mathbf{j} + 3\mathbf{k}$.
29. A firm sells x chairs at C dollars per chair and y desks at D dollars per desk. It costs the firm c dollars to make a chair and d dollars to make a desk. What is the economic interpretation of
  (a) Cx?
  (b) $(x\mathbf{i} + y\mathbf{j}) \cdot (C\mathbf{i} + D\mathbf{j})$?
  (c) $(x\mathbf{i} + y\mathbf{j}) \cdot (c\mathbf{i} + d\mathbf{j})$?
  (d) $(x\mathbf{i} + y\mathbf{j}) \cdot (C\mathbf{i} + D\mathbf{j}) > (x\mathbf{i} + y\mathbf{j}) \cdot (c\mathbf{i} + d\mathbf{j})$?

Exercise 30 is applied in Exercise 31.
30. Let $\mathbf{u}_1$, $\mathbf{u}_2$ and $\mathbf{u}_3$ be perpendicular unit vectors, and let **A** be any vector. Show that

$$\mathbf{A} = (\mathbf{A} \cdot \mathbf{u}_1)\mathbf{u}_1 + (\mathbf{A} \cdot \mathbf{u}_2)\mathbf{u}_2 + (\mathbf{A} \cdot \mathbf{u}_3)\mathbf{u}_3$$

by finding the dot product of both sides with $\mathbf{u}_1$, $\mathbf{u}_2$, and $\mathbf{u}_3$.
31. Let $\mathbf{u}_1 = \dfrac{\sqrt{2}}{2}\mathbf{i} - \dfrac{\sqrt{2}}{2}\mathbf{j}$, $\mathbf{u}_2 = \dfrac{\sqrt{2}}{2}\mathbf{i} + \dfrac{\sqrt{2}}{2}\mathbf{j}$ and $\mathbf{u}_3 = \mathbf{k}$.
  Let $\mathbf{A} = 2\mathbf{i} + 3\mathbf{j} - 4\mathbf{k}$. Use Exercise 30 to find x, y, and z such that $\mathbf{A} = x\mathbf{u}_1 + y\mathbf{u}_2 + z\mathbf{u}_3$.
32. By considering the dot product of the two unit vectors $\mathbf{u}_1 = \cos\theta_1\,\mathbf{i} + \sin\theta_1\,\mathbf{j}$ and $\mathbf{u}_2 = \cos\theta_2\,\mathbf{i} + \sin\theta_2\,\mathbf{j}$, prove that

$$\cos(\theta_1 - \theta_2) = \cos\theta_1 \cos\theta_2 + \sin\theta_1 \sin\theta_2.$$

33. Prove that, for any vectors **A** and **B**, $(\mathbf{A} + \mathbf{B}) \cdot (\mathbf{A} - \mathbf{B}) = |\mathbf{A}|^2 - |\mathbf{B}|^2$.
34. $OPQR$ is a parallelogram. Let $\mathbf{A} = \overrightarrow{OP}$ and $\mathbf{B} = \overrightarrow{OR}$.
  (a) Show that one diagonal of the parallelogram, $\overrightarrow{OQ}$, is $\mathbf{A} + \mathbf{B}$, while the other, $\overrightarrow{RP}$, is $\mathbf{A} - \mathbf{B}$.
  (b) Prove that the sides of the given parallelogram are all of equal length if and only if the diagonals are perpendicular. (Use Exercise 33.)

19.4

Lines and planes

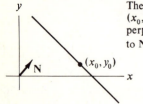

The line through (x_0, y_0) and perpendicular to N

The equations of lines and planes are easy to develop with the aid of vectors, especially with the dot product test for perpendicularity. This section develops the basic geometry of lines and planes, in particular, formulas for determining the distance between a point and a line or plane.

Let $\mathbf{N} = A\mathbf{i} + B\mathbf{j}$ be a nonzero vector and (x_0, y_0) be a point in the xy plane. There is a unique line through (x_0, y_0) that is perpendicular to **N**, as shown in the diagram on the bottom of page 726. **N** is called a *normal* to the line. The next theorem provides an algebraic criterion for determining whether the point (x, y) lies on this line.

THEOREM I   An equation of the line (in the xy plane) passing through (x_0, y_0) and perpendicular to the nonzero vector $\mathbf{N} = A\mathbf{i} + B\mathbf{j}$ is given by

$$A(x - x_0) + B(y - y_0) = 0.$$

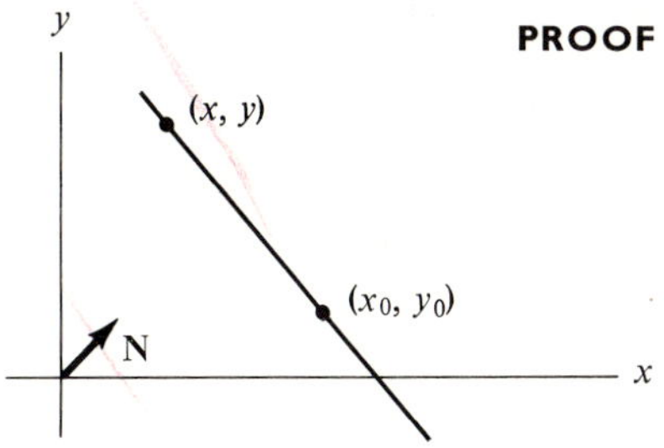

PROOF   Let (x, y) be a point on the line perpendicular to **N**. Then the vector $(x - x_0)\mathbf{i} + (y - y_0)\mathbf{j}$ is perpendicular to **N**. Hence

$$0 = [(x - x_0)\mathbf{i} + (y - y_0)\mathbf{j}] \cdot \mathbf{N}$$
$$= A(x - x_0) + B(y - y_0).$$

Conversely, it must be shown that if $A(x - x_0) + B(y - y_0) = 0$, then (x, y) is on the line through (x_0, y_0) perpendicular to **N**. The number $A(x - x_0) + B(y - y_0)$ is the scalar product of **N** and $(x - x_0)\mathbf{i} + (y - y_0)\mathbf{j}$. If this scalar product is zero, the two vectors **N** and $(x - x_0)\mathbf{i} + (y - y_0)\mathbf{j}$ are perpendicular. Thus (x, y) lies on the line through (x_0, y_0) perpendicular to **N**. ●

EXAMPLE I   Find an equation of the line through $(2, -7)$ and perpendicular to the vector $4\mathbf{i} + \mathbf{j}$.

SOLUTION   By Theorem 1 an equation is

$$4(x - 2) + 1(y + 7) = 0,$$

which, when multiplied out, is

$$4x + y = 1. \ ●$$

As Theorem 1 and Example 1 show, to find a vector perpendicular to a given line $Ax + By + C = 0$, read off the coefficients of x and y in order, A and B, and form the vector $A\mathbf{i} + B\mathbf{j}$. It will be perpendicular to the line. The constant term C plays no role in determining the direction of the line or of a vector perpendicular to it.

THEOREM 2   The distance from the origin to the line L whose equation is

$$Ax + By + C = 0$$

is

$$\frac{|C|}{\sqrt{A^2 + B^2}}.$$

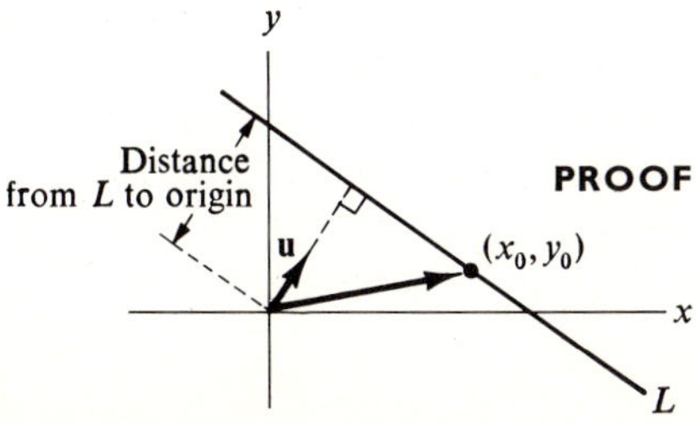

PROOF   Let (x_0, y_0) be a point on the line L. Let **u** be a unit vector perpendicular to L. Inspection of the right triangle pictured in the diagram at the left shows that the distance from the line $Ax + By + C = 0$ to the origin is the absolute value of the scalar component of the vector $x_0\mathbf{i} + y_0\mathbf{j}$ in the direction $\mathbf{u}$. Thus, by Theorem 2 of Sec. 19.3,

$$\left| \mathbf{u} \cdot (x_0\mathbf{i} + y_0\mathbf{j}) \right| = \text{distance from } L \text{ to the origin.}$$

Since $A\mathbf{i} + B\mathbf{j}$ is perpendicular to L,

$$\mathbf{u} = \frac{A\mathbf{i} + B\mathbf{j}}{\sqrt{A^2 + B^2}}$$

is a unit normal to L. Thus

$$\text{Distance from } L \text{ to origin} = \left| \frac{A\mathbf{i} + B\mathbf{j}}{\sqrt{A^2 + B^2}} \cdot (x_0\mathbf{i} + y_0\mathbf{j}) \right|$$

$$= \frac{|Ax_0 + By_0|}{\sqrt{A^2 + B^2}}. \tag{1}$$

But, since (x_0, y_0) lies on the line L,

$$Ax_0 + By_0 + C = 0$$

or

$$Ax_0 + By_0 = -C.$$

By (1), then,

$$\text{Distance from } L \text{ to origin} = \left| \frac{-C}{\sqrt{A^2 + B^2}} \right|$$

$$= \frac{|C|}{\sqrt{A^2 + B^2}}$$

This concludes the proof. ●

EXAMPLE 2 Find the distance from the line $4x + y = 1$ to the origin.

SOLUTION The line has the equation

$$4x + y - 1 = 0.$$

In this case $A = 4$, $B = 1$, and $C = -1$. By Theorem 2, the distance from the line to the origin is

$$\frac{|-1|}{\sqrt{4^2 + 1^2}} = \frac{1}{\sqrt{17}}. \; ●$$

Theorem 2 generalizes to give a formula for finding the distance from any point in the xy plane to a line in the plane.

THEOREM 3 The distance from the point (x_1, y_1) to the line L whose equation is

$$Ax + By + C = 0$$

is

$$\frac{|Ax_1 + By_1 + C|}{\sqrt{A^2 + B^2}}. \; ●$$

The proof, which is similar to that of Theorem 2, is omitted.   Note how easy it is to find the mentioned distance: Just plug the coordinates of the point into the equation and divide by $\sqrt{A^2 + B^2}$.

EXAMPLE 3   Find the distance from the point $(3, 7)$ to the line $2x - 4y + 5 = 0$.

SOLUTION   By Theorem 3, the distance is

$$\frac{|2 \cdot 3 - 4 \cdot 7 + 5|}{\sqrt{2^2 + 4^2}} = \frac{|6 - 28 + 5|}{\sqrt{20}}$$

$$= \frac{|-17|}{\sqrt{20}}$$

$$= \frac{17}{\sqrt{20}}$$

$$= \frac{17\sqrt{5}}{10}. \quad \bullet$$

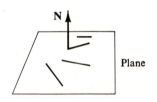

N is perpendicular to the plane.
(four lines in the plane are shown)

Analogs of Theorems 1, 2 and 3 hold for planes located in space.   First of all, the notion of a vector being perpendicular to a plane must be defined.   A vector $\mathbf{N}$ is said to be *perpendicular* to a plane if $\mathbf{N}$ is perpendicular to every line situated in the plane.

The next theorem is analogous to Theorem 1; its proof, which is practically the same as that of Theorem 1, is omitted.

THEOREM 4   An equation of the plane passing through (x_0, y_0, z_0) and perpendicular to the nonzero vector $A\mathbf{i} + B\mathbf{j} + C\mathbf{k}$ is given by

$$A(x - x_0) + B(y - y_0) + C(z - z_0) = 0.$$

EXAMPLE 4   Find an equation of the plane through $(1, -2, 4)$ that is perpendicular to the vector $5\mathbf{i} + 3\mathbf{j} + 6\mathbf{k}$.

SOLUTION   By Theorem 4, an equation of the plane is

$$5(x - 1) + 3(y - (-2)) + 6(z - 4) = 0,$$

which simplifies to

$$5x + 3y + 6z = 23. \quad \bullet$$

In Chap. 11 it was asserted that the graph of $z = Ax + By + C$ is a plane. The next example shows why this is the case.

EXAMPLE 5   Show that the graph of

$$z = 2x + 3y + 4$$

is a plane.   Find a vector perpendicular to it.

SOLUTION   The equation is equivalent to

$$2x + 3y - z = -4.$$

Choose (x_0, y_0, z_0) to be some fixed point on the graph of

$$2x + 3y - z = -4.$$

Then

$$2x_0 + 3y_0 - z_0 = -4.$$

If (x, y, z) is any point on the graph, then

$$2x + 3y - z = -4. \tag{2}$$

Subtracting

$$2x_0 + 3y_0 - z_0 = -4$$

from (2) yields the equation

$$2(x - x_0) + 3(y - y_0) - 1(z - z_0) = 0. \tag{3}$$

Equation (3) is the description of the plane through (x_0, y_0, z_0) and perpendicular to the vector $2\mathbf{i} + 3\mathbf{j} - \mathbf{k}$. ●

Similar reasoning shows that the graph of

$$Ax + By + Cz + D = 0$$

is a plane and that the vector $A\mathbf{i} + B\mathbf{j} + C\mathbf{k}$ is perpendicular to it. The next theorem is the natural companion in space of Theorem 2.

THEOREM 5   The distance from the origin to the plane

$$Ax + By + Cz + D = 0$$

is

$$\frac{|D|}{\sqrt{A^2 + B^2 + C^2}}.$$

The proof is practically the same as that of Theorem 2. The diagram that shows the critical right angle is harder to draw. It appears as shown in the margin.

Theorem 5 is a special case of the next theorem.

THEOREM 6   The distance from the point (x_1, y_1, z_1) to the plane

$$Ax + By + Cz + D = 0$$

is

$$\frac{|Ax_1 + By_1 + Cz_1 + D|}{\sqrt{A^2 + B^2 + C^2}}.$$

EXAMPLE 6   How far is the point $(2, 1, 5)$ from the plane $x - 3y + 4z + 8 = 0$?

SOLUTION   By Theorem 6 the desired distance is

$$\frac{|1 \cdot 2 - 3 \cdot 1 + 4 \cdot 5 + 8|}{\sqrt{1^2 + (-3)^2 + 4^2}} = \frac{|2 - 3 + 20 + 8|}{\sqrt{26}}$$

$$= \frac{27}{\sqrt{26}}$$

$$= \frac{27\sqrt{26}}{26}. \quad \bullet$$

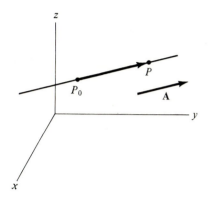

So far in this section we have been concerned with lines in the xy plane and planes in space. Vectors also provide a neat way to treat the geometry of lines in space, as will now be shown.

Consider the line L through the point $P_0 = (x_0, y_0, z_0)$ and parallel to the vector $\mathbf{A} = a_1\mathbf{i} + a_2\mathbf{j} + a_3\mathbf{k}$. A point $P = (x, y, z)$ is on this line if and only if

the vector $\overrightarrow{P_0 P}$ is parallel to $\mathbf{A}$.

One way to express that $\overrightarrow{P_0 P}$ is parallel to $\mathbf{A}$ is to assert that there is a scalar t such that

$$\overrightarrow{P_0 P} = t\mathbf{A},$$

that is, $\quad (x - x_0)\mathbf{i} + (y - y_0)\mathbf{j} + (z - z_0)\mathbf{k} = ta_1\mathbf{i} + ta_2\mathbf{j} + ta_3\mathbf{k}.$

In short,

$$x - x_0 = ta_1, \qquad y - y_0 = ta_2, \qquad z - z_0 = ta_3.$$

Consequently, we have these *parametric equations* for the line through (x_0, y_0, z_0) parallel to $\mathbf{A} = a_1\mathbf{i} + a_2\mathbf{j} + a_3\mathbf{k}$:

Parametric Equations of a Line in Space

$$\begin{cases} x = x_0 + a_1 t \\ y = y_0 + a_2 t \\ z = z_0 + a_3 t \end{cases}$$

EXAMPLE 7 Write parametric equations for the line through $(1, 2, 2)$ and parallel to the vector $3\mathbf{i} - \mathbf{j} + 5\mathbf{k}$. Does the point $(10, -1, 16)$ lie on the line?

SOLUTION Parametric equations of the line are given by

$$\begin{cases} x = 1 + 3t \\ y = 2 - t \\ z = 2 + 5t \end{cases}$$

Does $(10, -1, 16)$ lie on this line? To find out, determine if there is a number t such that these three equations are *simultaneously* satisfied:

$$10 = 1 + 3t$$
$$-1 = 2 - t$$
$$16 = 2 + 5t$$

The first equation, $10 = 1 + 3t$, has the solution $t = 3$. This value, 3, does satisfy the second equation, since $-1 = 2 - 3$. But it does *not* satisfy the third equation, since $16 \neq 2 + 5 \cdot 3$. Hence $(10, -1, 16)$ is *not* on the line. ●

DEFINITION *Direction numbers of a line.* If the vector $\mathbf{A} = a_1\mathbf{i} + a_2\mathbf{j} + a_3\mathbf{k}$ is parallel to the line L, then the numbers a_1, a_2, and a_3 are called *direction numbers* of L.

The next definition is closely related to the preceding one.

The direction of a vector in the plane is described by a single angle, the angle it makes with the positive x axis. The direction of a vector in space involves three angles, two of which almost determine the third.

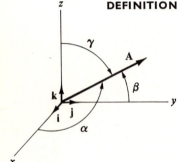

DEFINITION *Direction angles of a vector.* Let $\mathbf{A}$ be a nonzero vector in space. The angle between

$$\mathbf{A} \text{ and } \mathbf{i} \text{ is denoted } \alpha,$$

$$\mathbf{A} \text{ and } \mathbf{j} \text{ is denoted } \beta,$$

$$\mathbf{A} \text{ and } \mathbf{k} \text{ is denoted } \gamma.$$

The angles α, β, γ are called the direction angles of $\mathbf{A}$.

The three direction angles are not independent of each other, as is shown by the next theorem, which concerns the *direction cosines*, $\cos \alpha$, $\cos \beta$, $\cos \gamma$.

THEOREM 7 If α, β, and γ are the direction angles of the vector $\mathbf{A}$, then

$$\cos^2 \alpha + \cos^2 \beta + \cos^2 \gamma = 1.$$

PROOF It is no loss of generality to assume that $\mathbf{A}$ is a unit vector,

$$\mathbf{A} = x\mathbf{i} + y\mathbf{j} + z\mathbf{k},$$

where
$$x^2 + y^2 + z^2 = 1.$$

Now,
$$\mathbf{A} \cdot \mathbf{i} = |\mathbf{A}||\mathbf{i}| \cos \alpha$$
$$= 1 \cdot 1 \cdot \cos \alpha$$
$$= \cos \alpha.$$

But $\mathbf{A} \cdot \mathbf{i} = (x\mathbf{i} + y\mathbf{j} + z\mathbf{k}) \cdot (\mathbf{i} + 0\mathbf{j} + 0\mathbf{k}) = (x)(1) + (y)(0) + (z)(0) = x.$

Thus
$$\cos \alpha = x.$$

Similarly,
$$\cos \beta = y,$$

and
$$\cos \gamma = z.$$

Hence
$$\cos^2 \alpha + \cos^2 \beta + \cos^2 \gamma = x^2 + y^2 + z^2 = 1. ●$$

EXAMPLE 8   The vector **A** makes angles of 60° with the x and y axes.   What angle does it make with the z axis?

SOLUTION   Here $\alpha = 60°$ and $\beta = 60°$; hence

$$\cos \alpha = \tfrac{1}{2}, \qquad \text{and} \qquad \cos \beta = \tfrac{1}{2}.$$

Since

$$\cos^2 \alpha + \cos^2 \beta + \cos^2 \gamma = 1,$$

it follows that

$$\left(\tfrac{1}{2}\right)^2 + \left(\tfrac{1}{2}\right)^2 + \cos^2 \gamma = 1,$$

or

$$\cos^2 \gamma = \tfrac{1}{2}.$$

Thus

$$\cos \gamma = \frac{\sqrt{2}}{2} \qquad \text{or} \qquad \cos \gamma = -\frac{\sqrt{2}}{2}.$$

Hence

$$\gamma = 45° \qquad \text{or} \qquad 135°.$$

The diagrams in the margin show the two possibilities for the direction of **A**. ●

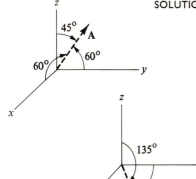

Exercises

In Exercises 1 to 3 find an equation of the line through the given point and perpendicular to the given vector.

1. $(2, 3)$, $4\mathbf{i} + 5\mathbf{j}$.   2. $(1, 0)$, $2\mathbf{i} - \mathbf{j}$.
3. $(4, 5)$, $2\mathbf{i} + 3\mathbf{j}$.

In Exercises 4 to 7 find a vector perpendicular to the given line.

4. $2x - 3y + 8 = 0$ 5. $2x - 3y = 7$
6. $2(x - 1) + 5(y + 2) = 0$ 7. $y = 3x + 7$
8. Does the line through $(2, 3)$ and perpendicular to the vector $4\mathbf{i} + \mathbf{j}$ pass through the point $(3, -6)$?
9. Use Theorem 2 to find the distance from the origin to the line
   (a) $3x + 4y - 10 = 0$,   (b) $5x - 12y + 2 = 0$.

In Exercises 10 to 12 find the distance from the given point to the given plane.

10. The point $(0, 0, 0)$ to the plane $2x - 4y + 3z + 2 = 0$.
11. The point $(1, 2, 3)$ to the plane $x + 2y - 3z + 5 = 0$.
12. The point $(2, 2, -1)$ to the plane that passes through $(1, 4, 3)$ and has a normal $2\mathbf{i} - 7\mathbf{j} + 2\mathbf{k}$.
13. Prove Theorem 3.
14. Prove Theorem 4.
15. Prove Theorem 5.
16. Give parametric equations for the line
    (a) through $(1, 2, -3)$ with direction numbers 2, 5, and -8;
    (b) through $(2, 4, 5)$ and $(3, 6, 4)$.

17. Are the three points $(1, 2, -3)$, $(1, 6, 2)$, and $(7, 14, 11)$ on a single line?

In Exercises 18 to 22 find the direction cosines of the vector.

18. $\mathbf{i}$ 19. $2\mathbf{i} - 3\mathbf{j} + 4\mathbf{k}$
20. $2\mathbf{i} + \mathbf{j} + 6\mathbf{k}$ 21. $-\mathbf{k}$
22. $-3\mathbf{k}$
23. (a) Find the direction cosines of $\mathbf{A} = 2\mathbf{i} - 3\mathbf{j} + 6\mathbf{k}$.
    (b) Find the direction angles of the vector in (a).
24. A vector makes angles of 70° and 80° with two of the axes. What possible angles can it make with the third axis?

■

25. Find the distance between the origin and the plane:
    (a) $x/2 + y/3 + z/4 = 1$,   (b) $z = 3x + 4y + 5$,
    (c) $x + 2y - 3z = 0$.
26. Where does the line through $(1, 2, 4)$ and $(2, 1, -1)$ meet the plane $x + 2y + 5z = 0$?
27. Develop a general formula for determining the distance from the point $P_1 = (x_1, y_1, z_1)$ to the line through the point $P_0 = (x_0, y_0, z_0)$ and parallel to the vector $\mathbf{A} = a_1\mathbf{i} + a_2\mathbf{j} + a_3\mathbf{k}$. The formula should be expressed in terms of the vectors $\overrightarrow{P_0 P_1}$ and $\mathbf{A}$.
28. How far is the point $(1, 2, -1)$ from the line through $(1, 3, 5)$ and $(2, 1, -3)$?
    (a) Solve by calculus, minimizing a certain function.

(b) Solve by vectors.

29. Let θ be the angle between the nonzero vectors $\mathbf{A} = a_1\mathbf{i} + a_2\mathbf{j} + a_3\mathbf{k}$ and $\mathbf{B} = b_1\mathbf{i} + b_2\mathbf{j} + b_3\mathbf{k}$. Show that

$$\sin^2 \theta = \frac{(a_2 b_3 - a_3 b_2)^2 + (a_1 b_3 - a_3 b_1)^2 + (a_1 b_2 - a_2 b_1)^2}{|\mathbf{A}|^2 |\mathbf{B}|^2}.$$

Hint: Begin with $\sin^2 \theta = 1 - \cos^2 \theta$.

30. Let $P_0 = (x_0, y_0, z_0)$ and let $\mathbf{A} = a_1\mathbf{i} + a_2\mathbf{j} + a_3\mathbf{k}$.
(a) Show that $P = (x, y, z)$ is on the line L through P_0 parallel to $\mathbf{A}$ if and only if

$$\frac{x - x_0}{a_1} = \frac{y - y_0}{a_2} = \frac{z - z_0}{a_3}.$$

(b) Use the equations in (a) to give the equations of three planes on which L lies.

31. Does the line through $(0, 0, 0)$ and $(2, 1, 1)$ meet the line through $(3, 5, 6)$ and $(13, 3, 4)$?

■ ■

Exercises 32 to 34 are related.

32. Let $\mathbf{A} = a_1\mathbf{i} + a_2\mathbf{j} + a_3\mathbf{k}$ and $\mathbf{B} = b_1\mathbf{i} + b_2\mathbf{j} + b_3\mathbf{k}$ be nonparallel nonzero vectors. Find a vector $\mathbf{C} = c_1\mathbf{i} + c_2\mathbf{j} + c_3\mathbf{k} \neq \mathbf{0}$ that is perpendicular to both $\mathbf{A}$ and $\mathbf{B}$.

33. Find a formula for the distance from the point $P_0 = (x_0, y_0, z_0)$ to the plane through the points $P_1 = (x_1, y_1, z_1)$, $P_2 = (x_2, y_2, z_2)$, and $P_3 = (x_3, y_3, z_3)$. Assume, of course, that P_1, P_2, and P_3 are not colinear.

34. Find a formula for the distance between the line through $Q_1 = (x_1, y_1, z_1)$ and $Q_2 = (x_2, y_2, z_2)$ and the line through $Q_3 = (x_3, y_3, z_3)$ and $Q_4 = (x_4, y_4, z_4)$.

19.5

Directional derivative and the gradient

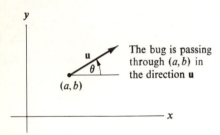

The bug is passing through (a, b) in the direction $\mathbf{u}$

Imagine that $f(x, y)$ is the temperature at the point (x, y) of a thin layer of metal occupying the xy plane. A bug is passing through the point (a, b) in the direction given by the unit vector $\mathbf{u}$, which makes an angle θ with the positive x axis. The bug observes that the temperature changes as it moves. The rate at which it changes per unit distance that the bug travels depends not only on the point (a, b) but on the direction $\mathbf{u}$ in which the bug is moving.

If, for example, $\mathbf{u} = \mathbf{i}[= (1, 0)]$, the bug is walking east, and the rate of change in the temperature is given by the partial derivative $f_x(a, b)$. If the bug is going in the opposite direction (west), and $\mathbf{u} = -\mathbf{i}$, the rate of change is $-f_x(a, b)$. (For instance, if the temperature increases as the bug walks east, then it decreases as it walks west.) Similarly, when $\mathbf{u} = \mathbf{j} = (0, 1)$, the rate of change is simply $f_y(a, b)$.

This section shows how to find the rate of change in the temperature in any direction $\mathbf{u}$ and introduces an important concept, the gradient.

To begin, let (a, b) be a point in the xy plane and consider a line in the plane through (a, b). On this line introduce a coordinate system with the same scale as the x or y axis: Call the line the t axis and place $t = 0$ at (a, b). Let θ be the angle from the positive x axis to the positive t axis. Then, if we consider $z = f(x, y)$ only at points (x, y) on the t axis, z is a (composite) function of t; that is, $z = g(t)$. The graph of g is a curve situated in the graph of f.

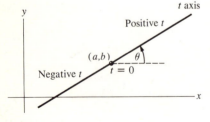

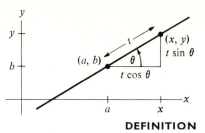

When $t = 0$, $(x, y) = (a, b)$. More generally, the point on the t axis having coordinate t is

$$(x, y) = (a + t \cos \theta, b + t \sin \theta),$$

as inspection of the right triangle in the figure at the left shows.

DEFINITION *Directional derivative of* $f(x, y)$. The derivative of f at (a, b) in the direction θ is $g'(t)$, where g is defined by $g(t) = f(a + t \cos \theta, b + t \sin \theta)$. A directional derivative at (a, b) is a number that depends on θ, as well as on a and b.

There are two ways to think of the directional derivative: either as the slope of a certain curve or as the rate at which $f(x, y)$ changes as you move from the point (a, b) in a fixed direction in the xy plane. It is the latter interpretation which easily generalizes at the end of this section to the directional derivative of functions $f(x, y, z)$.

THEOREM I If $f(x, y)$ has continuous partial derivatives f_x and f_y, then the directional derivative of f at (a, b) in the direction θ is

$$f_x(a, b) \cos \theta + f_y(a, b) \sin \theta.$$

PROOF The directional derivative of f at (a, b) in the direction θ is the derivative of the function

$$g(t) = f(a + t \cos \theta, b + t \sin \theta),$$

when $t = 0$.

Now g is a composite function

$$g(t) = f(x, y) \qquad \text{where} \begin{cases} x = a + t \cos \theta \\ y = b + t \sin \theta. \end{cases}$$

By a chain rule,

$$g'(t) = f_x \frac{dx}{dt} + f_y \frac{dy}{dt}.$$

Now,

$$\frac{dx}{dt} = \cos \theta \qquad \text{and} \qquad \frac{dy}{dt} = \sin \theta.$$

Thus

$$g'(0) = f_x(a, b) \cos \theta + f_y(a, b) \sin \theta,$$

and the theorem is proved. ●

When $\theta = 0$, the formula given by Theorem 1 becomes

$$f_x(a, b) \cos 0 + f_y(a, b) \sin 0 = f_x(a, b)(1) + f_y(a, b)(0) = f_x(a, b).$$

This agrees with the earlier observation that when the bug goes east the rate of change in the temperature is $f_x(a, b)$.

When $\theta = \pi$, Theorem 1 asserts that the directional derivative is

$$f_x(a, b) \cos \pi + f_y(a, b) \sin \pi = f_x(a, b)(-1) + f_y(a, b)(0) = -f_x(a, b).$$

This corresponds to the bug walking west and agrees with the observation that the rate of change in the temperature is $-f_x(a, b)$.

When $\theta = \pi/2$, Theorem 1 asserts that the directional derivative is

$$f_x(a, b) \cos \frac{\pi}{2} + f_y(a, b) \sin \frac{\pi}{2} = f_x(a, b)(0) + f_y(a, b)(1) = f_y(a, b),$$

which also agrees with an earlier observation.

EXAMPLE 1   Compute the derivative of $f(x, y) = x^2 y^3$ at $(1, 2)$ in the direction given by the angle $\pi/3$.   Interpret the result if f describes a temperature distribution.

SOLUTION   First of all, $f_x = 2xy^3$ and $f_y = 3x^2 y^2$.

Hence $f_x(1, 2) = 16$ and $f_y(1, 2) = 12$.

Second, $\cos \dfrac{\pi}{3} = \dfrac{1}{2}$ and $\sin \dfrac{\pi}{3} = \dfrac{\sqrt{3}}{2}$.

Thus the derivative of f in the direction given by $\theta = \pi/3$ is

$$16\left(\frac{1}{2}\right) + 12\left(\frac{\sqrt{3}}{2}\right) = 8 + 6\sqrt{3}.$$

If $x^2 y^3$ is the temperature in degrees at the point (x, y), where x and y are measured in centimeters, the rate at which the temperature changes at $(1, 2)$, in the direction given by $\theta = \pi/3$, is $(8 + 6\sqrt{3})$ degrees per centimeter.   ●

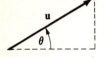

The scalar components of **u** are $\cos \theta$ and $\sin \theta$

Let $\mathbf{u} = \cos \theta \mathbf{i} + \sin \theta \mathbf{j}$ be the unit vector in the direction of angle θ, as pictured at the left.   The derivative of $f(x, y)$ in the direction corresponding to θ is denoted

$$D_{\mathbf{u}} f.$$

For example, $D_{\mathbf{i}} f = f_x$,

$$D_{-\mathbf{i}} f = -f_x,$$
$$D_{\mathbf{j}} f = f_y.$$

In Example 1 $D_{\mathbf{u}} f$ was computed for $f(x, y) = x^2 y^3$ and

$$\mathbf{u} = \cos \frac{\pi}{3} \mathbf{i} + \sin \frac{\pi}{3} \mathbf{j} = \frac{1}{2} \mathbf{i} + \frac{\sqrt{3}}{2} \mathbf{j}.$$

Theorem 1 asserts that, if $\mathbf{u} = \cos \theta \mathbf{i} + \sin \theta \mathbf{j}$, then

$$D_{\mathbf{u}} f = f_x(a, b) \cos \theta + f_y(a, b) \sin \theta. \tag{1}$$

Formula (1) bears a close resemblance to the formula for the dot product.   To exploit this similarity, it is useful to introduce the vector whose scalar components are $f_x(a, b)$ and $f_y(a, b)$.

DEFINITION   *The gradient of f(x, y).*   The vector $f_x(a, b)\mathbf{i} + f_y(a, b)\mathbf{j}$ is the *gradient* of f at (a, b) and is denoted ∇f.   (It is also called del f, because of the upside-down delta ∇.)

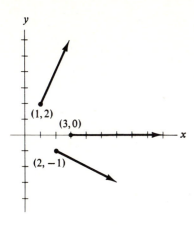

The del symbol is in boldface to emphasize that the gradient of f is a vector.

For instance, let $f(x, y) = x^2 + y^2$. We compute and draw $\mathbf{\nabla} f$ at a few points, listed in the following table.

(x, y)	$f_x = 2x$	$f_y = 2y$	$\mathbf{\nabla} f$
$(1, 2)$	2	4	$2\mathbf{i} + 4\mathbf{j}$
$(3, 0)$	6	0	$6\mathbf{i}$
$(2, -1)$	4	-2	$4\mathbf{i} - 2\mathbf{j}$

The diagram at the left shows $\mathbf{\nabla} f$ in each case, with the tail of $\mathbf{\nabla} f$ placed at the point where $\mathbf{\nabla} f$ is computed.

In vector notation, Theorem 1 reads as follows:

THEOREM I   (*Rephrased*).   If $z = f(x, y)$ has continuous partial derivatives f_x and f_y, then

$$D_{\mathbf{u}} f = \mathbf{\nabla} f \cdot \mathbf{u}.$$

The gradient is introduced not merely to simplify the statement of Theorem 1. Its importance is made clear in the next theorem.

THEOREM 2   (*Significance of* $\mathbf{\nabla} f$).   Let $z = f(x, y)$ have continuous partial derivatives f_x and f_y.   Let (a, b) be a point in the plane where $\mathbf{\nabla} f$ is not $\mathbf{0}$.   Then

the magnitude of $\mathbf{\nabla} f$ at (a, b) is the largest directional derivative of f at (a, b);

the direction of $\mathbf{\nabla} f$ is the direction in which the directional derivative at (a, b) has its largest value.

PROOF   By Theorem 1 (rephrased), if $\mathbf{u}$ is a unit vector, then at (a, b)

$$D_{\mathbf{u}} f = \mathbf{\nabla} f \cdot \mathbf{u}.$$

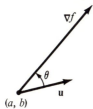

By the definition of the dot product,

$$\mathbf{\nabla} f \cdot \mathbf{u} = |\mathbf{\nabla} f| |\mathbf{u}| \cos \theta,$$

where θ is the angle between $\mathbf{\nabla} f$ and $\mathbf{u}$.   Since $|\mathbf{u}| = 1$,

$$D_{\mathbf{u}} f = |\mathbf{\nabla} f| \cos \theta. \tag{2}$$

The largest value of $\cos \theta$, for $0 \le \theta \le \pi$, occurs when $\theta = 0$; that is, $\cos \theta = 1$. Thus, by (2), the largest directional derivative of $f(x, y)$ at (a, b) occurs when the direction is that of $\mathbf{\nabla} f$ at (a, b).   For that choice of $\mathbf{u}$, $D_{\mathbf{u}} f = |\mathbf{\nabla} f|$.   This proves the theorem. ●

What does Theorem 2 tell the wandering bug about the flat piece of metal? If it is at the point (a, b) and wishes to get warmer as quickly as possible, it should compute the gradient of the temperature function, and then go in the direction indicated by the gradient.   If, instead, it wishes to cool off as quickly as possible, it should go in the direction opposite the gradient.

EXAMPLE 2   Let $f(x, y) = x^2y^3$. What is the largest directional derivative of f at $(2, 3)$? In what direction does this maximum directional derivative occur?

SOLUTION   At the point (x, y) $\nabla f = 2xy^3\mathbf{i} + 3x^2y^2\mathbf{j}$.

Thus, at $(2, 3)$ $\nabla f = 108\mathbf{i} + 108\mathbf{j}$,

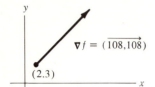

$\nabla f = \overline{(108,108)}$

(2.3)

which is sketched at the left (not to scale). Note that its angle θ is $\pi/4$. The maximal directional derivative of x^2y^3 at $(2, 3)$ is $|\nabla f| = 108\sqrt{2}$. This is achieved at the angle $\theta = \pi/4$, relative to the x axis, that is, for

$$\mathbf{u} = \cos\frac{\pi}{4}\mathbf{i} + \sin\frac{\pi}{4}\mathbf{j} = \frac{\sqrt{2}}{2}\mathbf{i} + \frac{\sqrt{2}}{2}\mathbf{j}. \quad \bullet$$

Incidentally, if $f(x, y)$ denotes the temperature at (x, y), the gradient ∇f helps to indicate the direction in which heat flows. It tends to flow "toward the coldest," which boils down to the mathematical assertion, "Heat tends to flow in the direction of $-\nabla f$."

The gradient and directional derivative have been interpreted in terms of a temperature distribution in the plane and a wandering bug. It is also of interest to interpret these concepts in terms of a hiker on the surface of a mountain.

Consider a mountain above the xy plane. The altitude of the point on the surface above the point (x, y) will be denoted by $f(x, y)$. The directional derivative

$$D_\mathbf{u}f$$

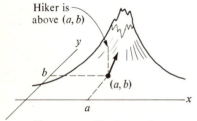

Hiker is above (a, b)

b

(a, b)

a

The vector ∇f in the xy plane points in the direction of steepest ascent

indicates the rate at which altitude changes per unit change in *horizontal* distance. The gradient ∇f at (a, b) points in the direction the hiker should choose if he wishes to climb in the direction of steepest ascent. The magnitude of ∇f tells the hiker the steepest slope available to him.

The notions of directional derivative and gradient can be generalized with little effort to functions of three (or more) variables. However, the "slope of a curve" interpretation of the directional derivative no longer applies. It is easiest to interpret the directional derivative of $f(x, y, z)$ in a particular direction in space as indicating the rate of change in the function in a certain direction in space. A useful interpretation is how fast the temperature changes in a certain direction. The formal definition will now be given.

DEFINITION   *Directional derivative of $f(x, y, z)$.* The derivative of f at (a, b, c) in the direction of the unit vector $\mathbf{u} = \cos\alpha\mathbf{i} + \cos\beta\mathbf{j} + \cos\gamma\mathbf{k}$ is $g'(t)$, where g is defined by

$$g(t) = f(a + t\cos\alpha, b + t\cos\beta, c + t\cos\gamma).$$

It is denoted $D_\mathbf{u}f$.

Note that t is simply the measure of length along the line through (a, b, c) with direction cosines $\cos\alpha$, $\cos\beta$, and $\cos\gamma$. So $D_\mathbf{u}f$ is just the derivative df/dt along the t axis.

The following theorem is proved just like Theorem 1.

THEOREM 3   If $f(x, y, z)$ has continuous partial derivatives f_x, f_y, and f_z, then the directional derivative of f at (a, b, c) in the direction of the unit vector $\cos \alpha \mathbf{i} + \cos \beta \mathbf{j} + \cos \gamma \mathbf{k}$ is

$$f_x(a, b, c) \cos \alpha + f_y(a, b, c) \cos \beta + f_z(a, b, c) \cos \gamma.$$

DEFINITION   *The gradient of* $f(x, y, z)$.   The vector

$$f_x(a, b, c)\mathbf{i} + f_y(a, b, c)\mathbf{j} + f_z(a, b, c)\mathbf{k}$$

is the *gradient* of f at (a, b, c) and is denoted ∇f.

Theorem 3 thus asserts that the derivative of $f(x, y, z)$ in the direction of the unit vector $\mathbf{u}$ is simply

$$D_{\mathbf{u}} f = \nabla f \cdot \mathbf{u}.$$

Just as in the case of a function of two variables, ∇f, evaluated at (a, b, c), points in the direction $\mathbf{u}$ that produces the largest directional derivative at (a, b, c).   Moreover $|\nabla f|$ is that largest directional derivative.   The proof is practically identical with that of Theorem 2.

EXAMPLE 3   The temperature at the point (x, y, z) in a solid piece of metal is given by the formula $f(x, y, z) = e^{2x + y + 3z}$ degrees.   In what direction at the point $(0, 0, 0)$ does the temperature increase most rapidly?

SOLUTION   First compute

$$\nabla f = 2e^{2x + y + 3z}\mathbf{i} + e^{2x + y + 3z}\mathbf{j} + 3e^{2x + y + 3z}\mathbf{k}.$$

At $(0, 0, 0)$ $\nabla f = 2\mathbf{i} + \mathbf{j} + 3\mathbf{k}.$

Consequently, the direction of most rapid increase in temperature is that given by the vector $2\mathbf{i} + \mathbf{j} + 3\mathbf{k}$.   The rate of increase is then

$$|2\mathbf{i} + \mathbf{j} + 3\mathbf{k}| = \sqrt{14} \text{ degrees per unit length.}$$

If the line through $(0, 0, 0)$ parallel to $2\mathbf{i} + \mathbf{j} + 3\mathbf{k}$ is given on a coordinate system so that it becomes the t axis, with $t = 0$ at the origin, then $df/dt = \sqrt{14}$ at 0.   ●

Exercises

In Exercises 1 to 5 compute the directional derivative of $x^4 y^5$ at $(1, 1)$ in the indicated direction.

1. $\theta = \pi/2$ 2. $\theta = 0$
3. $\theta = 3\pi/4$ 4. $\theta = \pi$
5. $\theta = 3\pi/2$

In Exercises 6 to 9 compute and draw ∇f at the indicated point for the given function.

6. $f(x, y) = 3x + 4y$ at $(2, 5)$.
7. $f(x, y) = 1/\sqrt{x^2 + y^2}$ at $(3, 4)$.
8. $f(x, y) = e^x y$ at $(0, 1)$.

9. $f(x, y) = e^x \cos y$ at $(1, \pi)$.

10. Find the directional derivative of e^{x+2y} at $(0, 0)$ in the direction of the vector $\mathbf{A} = 2\mathbf{i} + 3\mathbf{j}$. *Warning:* $\mathbf{A}$ is not a unit vector.

11. Let $f(x, y, z) = 2x + 3y + z$.
    (a) Compute ∇f at $(0, 0, 0)$ and at $(1, 1, 1)$.
    (b) Draw ∇f for the two points in (a), in each case putting its tail at the point.

12. Let $f(x, y, z) = x^2 + y^2 + z^2$.
    (a) Compute ∇f at $(2, 0, 0)$, $(0, 2, 0)$, and $(0, 0, 2)$.
    (b) Draw ∇f for the three points in (a), in each case putting its tail at the point.

13. Find the directional derivative of $f(x, y, z) = x^3 y^2 z$ at $(1, 1, 1)$ in the direction of (a) $\mathbf{i}$, (b) $\mathbf{j}$, (c) $\mathbf{k}$, (d) $-\mathbf{i}$, (e) $\mathbf{i} + \mathbf{j} + \mathbf{k}$. *Warning:* In (e) the given vector does not have unit length.

14. If $f_x(a, b) = 2$ and $f_y(a, b) = 3$, in what direction should a directional derivative at (a, b) be computed in order that it is
    (a) 0?
    (b) as large as possible?
    (c) as small as possible?

15. Assume that ∇f at (a, b) is not $\mathbf{0}$. Show that there are two unit vectors $\mathbf{u}_1$ and $\mathbf{u}_2$ such that the directional derivatives of f at (a, b) in the directions of $\mathbf{u}_1$ and $\mathbf{u}_2$ are 0.

16. (a) If $f_x(a, b, c) = 2$, $f_y(a, b, c) = 3$, and $f_z(a, b, c) = 1$, find three different unit vectors $\mathbf{u}$ such that $D_{\mathbf{u}} f$ at (a, b, c) is 0.
    (b) How many unit vectors $\mathbf{u}$ are there such that $D_{\mathbf{u}} f$ at (a, b, c) is 0?

∎

17. Let $f(x, y) = xy$. Prove that, if (a, b) is an arbitrary point on the curve $xy = 5$, then ∇f computed at (a, b) is perpendicular to the tangent line to that curve at (a, b).

18. Let $f(x, y) = x^2 + y^2$. Prove that, if (a, b) is an arbitrary point on the curve $x^2 + y^2 = 9$, then ∇f computed at (a, b) is perpendicular to the tangent line to that curve at (a, b).

19. If at (a, b) $D_{\mathbf{u}} f = 3$, find $D_{-\mathbf{u}} f$.

20. Let $f(x, y, z) =$ temperature at (x, y, z). Let $P = (a, b, c)$ and Q be a point very near (a, b, c). Show that $\nabla f \cdot \overrightarrow{PQ}$ is a good estimate of the change in temperature from point P to point Q.

21. Let $f(x, y) = 1/\sqrt{x^2 + y^2}$; the function f is defined everywhere except at $(0, 0)$. (This function is the potential in a gravitational field due to a point-mass.) Let $\mathbf{R} = (\vec{x, y})$.
    (a) Show that $\nabla f = -\mathbf{R}/|\mathbf{R}|^3$.
    (b) Show that $|\nabla f| = 1/|\mathbf{R}|^2$.
    (The gradient is closely related to the gravitational force of attraction.)

22. What happens to ∇f when f has a local maximum? What happens to $D_{\mathbf{u}} f$ there? Explain. (Assume that f is defined in the entire plane and has continuous partial derivatives.)

23. If $f(P)$ is the electric potential at the point P, then the electric field $\mathbf{E}$ at P is given by $-\nabla f$. Calculate $\mathbf{E}$ if $f(x, y) = \sin \alpha x \cos \beta y$, where α and β are constants.

∎ ∎

24. Let f have continuous partial derivatives f_x, f_y, f_{xy}, and f_{yx} (hence $f_{xy} = f_{yx}$). Let $\mathbf{u}_1$ and $\mathbf{u}_2$ be two unit vectors. Prove that $D_{\mathbf{u}_2} D_{\mathbf{u}_1} f = D_{\mathbf{u}_1} D_{\mathbf{u}_2} f$.

25. Show that the maximum of $D_{\mathbf{u}} f$ at (a, b) is $\sqrt{f_x^2 + f_y^2}$, where f_x and f_y are evaluated at (a, b).

26. Prove the first part of Theorem 2 without the aid of vectors. That is, prove that the maximum value of $g(\theta) = f_x(a, b) \cos \theta + f_y(a, b) \sin \theta$, is $\sqrt{f_x^2(a, b) + f_y^2(a, b)}$.

19.6

Determinants of orders 2 and 3

This section develops a small part of the extensive field of linear algebra. It should be kept in mind that the few definitions and theorems we present have broad generalizations and applications that only a full course could adequately explore.

DEFINITION *Matrix.* An array of four numbers a_1, a_2, b_1, b_2 forming a square is called a *matrix of order 2*, and is denoted

$$\begin{pmatrix} a_1 & a_2 \\ b_1 & b_2 \end{pmatrix}.$$

Similarly, an array of nine numbers arranged in a square is called a *matrix of order 3*,

$$\begin{pmatrix} a_1 & a_2 & a_3 \\ b_1 & b_2 & b_3 \\ c_1 & c_2 & c_3 \end{pmatrix}.$$

For instance, $\begin{pmatrix} 2 & 3 \\ 4 & -1 \end{pmatrix}$ and $\begin{pmatrix} 1 & 5 & 0 \\ 0 & 2 & 3 \\ 8 & 4 & -1 \end{pmatrix}$

are matrices of orders 2 and 3, respectively.

The set of entries of a matrix in a line parallel to the bottom of the page is called a *row* of the matrix. The set of entries in a line parallel to the margin is called a *column*. A matrix of order 3 thus consists of three rows; it may also be thought of as being composed of three columns. For convenience, number the rows from top to bottom and the columns from left to right:

$$\begin{pmatrix} \text{row 1} \\ \text{row 2} \\ \text{row 3} \end{pmatrix} \quad \text{or} \quad \begin{pmatrix} \text{c} & \text{c} & \text{c} \\ \text{o} & \text{o} & \text{o} \\ \text{l} & \text{l} & \text{l} \\ \text{u} & \text{u} & \text{u} \\ \text{m} & \text{m} & \text{m} \\ \text{n} & \text{n} & \text{n} \\ 1 & 2 & 3 \end{pmatrix}.$$

In the matrix $\begin{pmatrix} 2 & 3 \\ 4 & -1 \end{pmatrix}$

the first row is (2, 3), and the second column is $\begin{pmatrix} 3 \\ -1 \end{pmatrix}$.

Associated with each matrix is an important number, called its *determinant*.

DEFINITION   *Determinant of a second-order matrix.*   The determinant of the matrix

$$\begin{pmatrix} a_1 & a_2 \\ b_1 & b_2 \end{pmatrix}$$

is the number $a_1 b_2 - a_2 b_1.$

It is denoted $\begin{vmatrix} a_1 & a_2 \\ b_1 & b_2 \end{vmatrix}.$

EXAMPLE 1   Compute the determinants:

(a) $\begin{vmatrix} 2 & 3 \\ 1 & 4 \end{vmatrix}$ (b) $\begin{vmatrix} 0 & -5 \\ 2 & 8 \end{vmatrix}.$

SOLUTION   (a) $\begin{vmatrix} 2 & 3 \\ 1 & 4 \end{vmatrix} = 2 \cdot 4 - 3 \cdot 1 = 8 - 3 = 5.$

(b) $\begin{vmatrix} 0 & -5 \\ 2 & 8 \end{vmatrix} = 0 \cdot 8 - (-5) \cdot 2 = 0 + 10 = 10.$ ●

The determinant of a third-order matrix may be defined with the aid of second-order matrices, as follows.

DEFINITION   *Determinant of a third-order matrix.*   The determinant of the matrix

$$\begin{pmatrix} a_1 & a_2 & a_3 \\ b_1 & b_2 & b_3 \\ c_1 & c_2 & c_3 \end{pmatrix}$$

is the number $a_1 \begin{vmatrix} b_2 & b_3 \\ c_2 & c_3 \end{vmatrix} - a_2 \begin{vmatrix} b_1 & b_3 \\ c_1 & c_3 \end{vmatrix} + a_3 \begin{vmatrix} b_1 & b_2 \\ c_1 & c_2 \end{vmatrix}.$

It is denoted $\begin{vmatrix} a_1 & a_2 & a_3 \\ b_1 & b_2 & b_3 \\ c_1 & c_2 & c_3 \end{vmatrix}.$

EXAMPLE 2   Compute $\begin{vmatrix} 3 & 2 & -4 \\ 1 & 5 & 0 \\ 2 & -7 & 3 \end{vmatrix}.$

SOLUTION   By definition this determinant equals

$$3 \begin{vmatrix} 5 & 0 \\ -7 & 3 \end{vmatrix} - 2 \begin{vmatrix} 1 & 0 \\ 2 & 3 \end{vmatrix} + (-4) \begin{vmatrix} 1 & 5 \\ 2 & -7 \end{vmatrix}.$$

Now $\begin{vmatrix} 5 & 0 \\ -7 & 3 \end{vmatrix} = 15,$ $\begin{vmatrix} 1 & 0 \\ 2 & 3 \end{vmatrix} = 3,$ and $\begin{vmatrix} 1 & 5 \\ 2 & -7 \end{vmatrix} = -17.$

Hence the third-order determinant equals

$$3 \cdot 15 - 2 \cdot 3 + (-4)(-17) = 45 - 6 + 68 = 107. ●$$

Note that in the definition of a third-order determinant a_2 has a negative sign.   Also observe that the second-order matrix associated with a_1 is obtained by blotting out of the third-order matrix the row and column on which a_1 lies.   A similar procedure works for a_2 and a_3 (but remember the minus sign that goes with a_2).

$$\begin{pmatrix} \cancel{a_1} & \cancel{a_2} & \cancel{a_3} \\ \cancel{b_1} & b_2 & b_3 \\ \cancel{c_1} & c_2 & c_3 \end{pmatrix}$$

We now obtain some useful theorems about determinants that are true of matrices of either order 2 or 3.   However, the proofs, which are straightforward computations, will be given in only one of the two cases.

THEOREM 1   If two rows (or two columns) of a matrix are identical, then the determinant of the matrix is 0.

PROOF Let us show that when two rows are identical, the matrix has determinant 0; for instance,

$$\begin{vmatrix} a_1 & a_2 & a_3 \\ a_1 & a_2 & a_3 \\ c_1 & c_2 & c_3 \end{vmatrix} = 0.$$

This determinant equals

$$a_1 \begin{vmatrix} a_2 & a_3 \\ c_2 & c_3 \end{vmatrix} - a_2 \begin{vmatrix} a_1 & a_3 \\ c_1 & c_3 \end{vmatrix} + a_3 \begin{vmatrix} a_1 & a_2 \\ c_1 & c_2 \end{vmatrix}$$

$$= a_1(a_2 c_3 - a_3 c_2) - a_2(a_1 c_3 - a_3 c_1) + a_3(a_1 c_2 - a_2 c_1)$$

$$= a_1 a_2 c_3 - a_1 a_3 c_2 - a_2 a_1 c_3 + a_2 a_3 c_1 + a_3 a_1 c_2 - a_3 a_2 c_1$$

$$= 0 \qquad \text{the terms cancel in pairs.}$$

The other cases may be proved similarly. ●

THEOREM 2 If two rows (or two columns) of a matrix are switched with each other, the determinant of the resulting matrix is the determinant of the original matrix with its sign changed.

PROOF Take the case, for instance, in which the second and third columns of a third-order matrix are switched. We will prove that

$$\begin{vmatrix} a_1 & a_3 & a_2 \\ b_1 & b_3 & b_2 \\ c_1 & c_3 & c_2 \end{vmatrix} = - \begin{vmatrix} a_1 & a_2 & a_3 \\ b_1 & b_2 & b_3 \\ c_1 & c_2 & c_3 \end{vmatrix} \qquad (1)$$

Simply calculate both determinants and compare the results:

$$\begin{vmatrix} a_1 & a_3 & a_2 \\ b_1 & b_3 & b_2 \\ c_1 & c_3 & c_2 \end{vmatrix} = a_1 \begin{vmatrix} b_3 & b_2 \\ c_3 & c_2 \end{vmatrix} - a_3 \begin{vmatrix} b_1 & b_2 \\ c_1 & c_2 \end{vmatrix} + a_2 \begin{vmatrix} b_1 & b_3 \\ c_1 & c_3 \end{vmatrix} \qquad (2)$$

While

$$\begin{vmatrix} a_1 & a_2 & a_3 \\ b_1 & b_2 & b_3 \\ c_1 & c_2 & c_3 \end{vmatrix} = a_1 \begin{vmatrix} b_2 & b_3 \\ c_2 & c_3 \end{vmatrix} - a_2 \begin{vmatrix} b_1 & b_3 \\ c_1 & c_3 \end{vmatrix} + a_3 \begin{vmatrix} b_1 & b_2 \\ c_1 & c_2 \end{vmatrix}. \qquad (3)$$

Since

$$\begin{vmatrix} b_3 & b_2 \\ c_3 & c_2 \end{vmatrix} = b_3 c_2 - b_2 c_3 \qquad \text{and} \qquad \begin{vmatrix} b_2 & b_3 \\ c_2 & c_3 \end{vmatrix} = b_2 c_3 - b_3 c_2,$$

direct comparison of the three summands in (2) and in (3) establishes (1). The other cases are proved similarly. ●

Determinants are used to describe the solution of linear equations. Indeed This is probably their oldest application. The next two theorems describe this use. The proofs, which are straightforward calculations, are left to the reader.

THEOREM 3   Assume that
$$\begin{vmatrix} a_1 & a_2 \\ b_1 & b_2 \end{vmatrix} \neq 0.$$

Then the simultaneous equations

$$a_1 x + a_2 y = k_1$$
$$b_1 x + b_2 y = k_2$$

have a unique solution (x, y) given by

$$x = \frac{\begin{vmatrix} k_1 & a_2 \\ k_2 & b_2 \end{vmatrix}}{\begin{vmatrix} a_1 & a_2 \\ b_1 & b_2 \end{vmatrix}} \quad \text{and} \quad y = \frac{\begin{vmatrix} a_1 & k_1 \\ b_1 & k_2 \end{vmatrix}}{\begin{vmatrix} a_1 & a_2 \\ b_1 & b_2 \end{vmatrix}}.$$

Theorem 3 tells us how to find where two nonparallel lines in the xy plane intersect.   Notice that the two denominators (for x and for y) are the same, being the determinant of the matrix formed by the coefficients.   The matrices in the numerators are obtained from that matrix by replacing an appropriate column by the column

$$\begin{pmatrix} k_1 \\ k_2 \end{pmatrix}.$$

EXAMPLE 3   Use determinants to find where the lines

$$2x - y = 5$$
and
$$3x + 4y = -1$$

intersect.

SOLUTION   First check that the determinant of the coefficient matrix is not 0:

$$\begin{vmatrix} 2 & -1 \\ 3 & 4 \end{vmatrix} = 2 \cdot 4 - (-1)(3) = 8 + 3 = 11.$$

Thus
$$x = \frac{\begin{vmatrix} 5 & -1 \\ -1 & 4 \end{vmatrix}}{11} \quad \text{and} \quad y = \frac{\begin{vmatrix} 2 & 5 \\ 3 & -1 \end{vmatrix}}{11}.$$

Straightforward calculations show that

$$x = \frac{19}{11} \quad \text{and} \quad y = \frac{-17}{11}.$$

The skeptical reader may check this solution by plugging it into both the original equations, $2x - y = 5$ and $3x + 4y = -1$. ●

The next theorem is the companion for third-order matrices of Theorem 3.

THEOREM 4 Assume that
$$\begin{vmatrix} a_1 & a_2 & a_3 \\ b_1 & b_2 & b_3 \\ c_1 & c_2 & c_3 \end{vmatrix} \neq 0.$$

Then the simultaneous equations

$$a_1 x + a_2 y + a_3 z = k_1$$
$$b_1 x + b_2 y + b_3 z = k_2$$
$$c_1 x + c_2 y + c_3 z = k_3$$

have a unique solution (x, y, z) given by

$$x = \frac{\begin{vmatrix} k_1 & a_2 & a_3 \\ k_2 & b_2 & b_3 \\ k_3 & c_2 & c_3 \end{vmatrix}}{\begin{vmatrix} a_1 & a_2 & a_3 \\ b_1 & b_2 & b_3 \\ c_1 & c_2 & c_3 \end{vmatrix}} \qquad y = \frac{\begin{vmatrix} a_1 & k_1 & a_3 \\ b_1 & k_2 & b_3 \\ c_1 & k_3 & c_3 \end{vmatrix}}{\begin{vmatrix} a_1 & a_2 & a_3 \\ b_1 & b_2 & b_3 \\ c_1 & c_2 & c_3 \end{vmatrix}} \qquad z = \frac{\begin{vmatrix} a_1 & a_2 & k_1 \\ b_1 & b_2 & k_2 \\ c_1 & c_2 & k_3 \end{vmatrix}}{\begin{vmatrix} a_1 & a_2 & a_3 \\ b_1 & b_2 & b_3 \\ c_1 & c_2 & c_3 \end{vmatrix}}.$$

Note that Theorem 4 tells how to find the point where three planes meet.

EXAMPLE 4 Find the point of intersection of the three planes

$$3x + 2y - 4z = 6$$
$$1x + 5y + 0z = -3$$
$$2x - 7y + 3z = 4.$$

SOLUTION The matrix of coefficients already appeared in Example 2, where its determinant was shown to be 107, which is not 0. Thus

$$x = \frac{\begin{vmatrix} 6 & 2 & -4 \\ -3 & 5 & 0 \\ 4 & -7 & 3 \end{vmatrix}}{107} \qquad y = \frac{\begin{vmatrix} 3 & 6 & -4 \\ 1 & -3 & 0 \\ 2 & 4 & 3 \end{vmatrix}}{107} \qquad z = \frac{\begin{vmatrix} 3 & 2 & 6 \\ 1 & 5 & -3 \\ 2 & -7 & 4 \end{vmatrix}}{107}.$$

Straightforward calculations show that

$$x = \frac{104}{107} \qquad y = \frac{-85}{107} \qquad z = -\frac{125}{107}. \; \bullet$$

The theory introduced in this brief section goes much further. For any positive integer n there are matrices and determinants of the nth order. The theorems of this section generalize to all orders. It is possible to define a determinant of order n without referring to matrices or determinants of lower orders. In a linear algebra or vector space course such a definition is needed in order to establish such properties of determinants as indicated in Theorems 1 to 4.

Exercises

In Exercises 1 to 6 evaluate the determinants.   Use a shortcut if there is one.

1. $\begin{vmatrix} 3 & 4 \\ 7 & 2 \end{vmatrix}$ **2.** $\begin{vmatrix} 3 & 3 \\ 7 & 7 \end{vmatrix}$ **3.** $\begin{vmatrix} 4 & 2 & 0 \\ 5 & 6 & -1 \\ 1 & -1 & 2 \end{vmatrix}$

4. $\begin{vmatrix} 1 & 3 & 1 \\ 2 & 1 & 2 \\ 4 & 5 & 4 \end{vmatrix}$ **5.** $\begin{vmatrix} 2 & 1 & 4 \\ 2 & 1 & 4 \\ 3 & 5 & 7 \end{vmatrix}$ **6.** $\begin{vmatrix} 0 & 0 & 0 \\ 1 & 5 & 9 \\ 3 & -1 & 2 \end{vmatrix}$

7. Prove that if the first and third columns of a third-order matrix are identical, then the determinant of the matrix is 0.

8. Show that Theorem 2 implies Theorem 1.

9. Prove that if the first and third rows of a third-order matrix are switched, the determinant of the resulting matrix is the determinant of the original matrix with its sign changed.

10. Prove that

$$\begin{vmatrix} ka_1 & ka_2 \\ b_1 & b_2 \end{vmatrix} = k \begin{vmatrix} a_1 & a_2 \\ b_1 & b_2 \end{vmatrix}.$$

This illustrates a general theorem: If the entries in a single row (or single column) are all multiplied by the number k, the determinant of the resulting matrix is k times the determinant of the original matrix.

11. Prove that

$$\begin{vmatrix} a_1 & a_2 & a_3 \\ b_1 & b_2 & b_3 \\ c_1 & c_2 & c_3 \end{vmatrix} = \begin{vmatrix} a_1 & b_1 & c_1 \\ a_2 & b_2 & c_2 \\ a_3 & b_3 & c_3 \end{vmatrix}.$$

This says that if you spin a matrix around the diagonal that stretches from the top left corner to the bottom right, the determinant of the resulting matrix is the same as that of the original matrix.

12. (a) Use determinants to find x and y that satisfy both equations simultaneously:

$$2x - 3y = 6$$
$$3x + 5y = 15.$$

(b) Check your solution to (a) by substituting it in the two equations.

(c) Graph the two lines in (a) and show their intersection.

13. (a) Use determinants to find the point common to these three planes:

$$x - y + z = 2$$
$$2x + y - 3z = 0$$
$$x + 3y + 4z = 1.$$

(b) Check that your solution satisfies the equations in (a).

■

14. If the matrix of coefficients of the equations

$$a_1 x + a_2 y = k_1,$$
$$b_1 x + b_2 y = k_2 \tag{4}$$

has determinant 0, Theorem 3 gives no information.

(a) Give an example where Eqs. (4) have no solution.

(b) Give an example where Eqs. (4) have an infinite number of solutions.

15. Sketch three planes whose common intersection is (a) a single point, (b) a line, (c) no points at all.

16. (a) Show that the planes

$$x + 2y + 3z = 1$$
$$2x + 4y + 6z = 3$$
$$x - y + z = 4$$

have no points in common.   *Hint*: Look carefully at the first two equations.

(b) What must the determinant of the matrix of coefficients be?

■ ■

17. (a) Show that the planes

$$x + 2y + 3z = 1$$
$$x + 3y + 5z = 2$$
$$2x + 5y + 8z = 100$$

have no points in common.

(b) What must the determinant of the matrix of coefficients be?

18. Prove that

$$\begin{vmatrix} a_1 + kb_1 & a_2 + kb_2 & a_3 + kb_3 \\ b_1 & b_2 & b_3 \\ c_1 & c_2 & c_3 \end{vmatrix} = \begin{vmatrix} a_1 & a_2 & a_3 \\ b_1 & b_2 & b_3 \\ c_1 & c_2 & c_3 \end{vmatrix}.$$

This illustrates the general theorem: If you multiply a row by a scalar and add the result to a different row, the resulting matrix has the same determinant as the original one.   A corresponding theorem holds for columns.

19. Two plane vectors $\mathbf{A} = a_1 \mathbf{i} + a_2 \mathbf{j}$ and $\mathbf{B} = b_1 \mathbf{i} + b_2 \mathbf{j}$ are located in the first quadrant in the relative positions shown in the diagram.

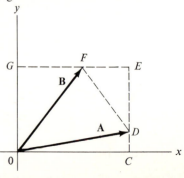

(a) Express in terms of a_1, a_2, b_1, b_2 the areas of the rectangle $OCEG$ and the triangles OCD, DEF, and OFG.

(b) Deduce that the area of triangle ODF is

$$\frac{1}{2}\begin{vmatrix} a_1 & a_2 \\ b_1 & b_2 \end{vmatrix}.$$

From (b) it follows that $\begin{vmatrix} a_1 & a_2 \\ b_1 & b_2 \end{vmatrix}$ may be interpreted

as the area of the parallelogram three of whose vertices are O, D, and F. This illustrates a general result valid for any plane vectors $\mathbf{A} = a_1\mathbf{i} + a_2\mathbf{j}$ and $\mathbf{B} = b_1\mathbf{i} + b_2\mathbf{j}$:

The determinant $\begin{vmatrix} a_1 & a_2 \\ b_1 & b_2 \end{vmatrix}$

is either $+$ or $-$ the area of the parallelogram determined by $\mathbf{A}$ and $\mathbf{B}$.

19.7

The cross product of two vectors in space

It is frequently necessary in applications of vectors in space to construct a nonzero vector perpendicular to two given vectors $\mathbf{A}$ and $\mathbf{B}$ in space. This section provides a formula for finding such a vector.

If $\mathbf{A}$ and $\mathbf{B}$ are not parallel and are drawn with their tails at a single point, they determine a plane, as in the left figure shown below. Any vector $\mathbf{C}$

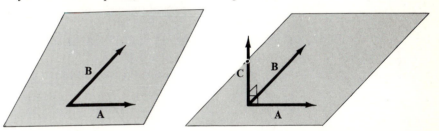

perpendicular to this plane is perpendicular to both $\mathbf{A}$ and $\mathbf{B}$. There are many such vectors, all parallel to each other and having various lengths. For instance, any vector perpendicular to both $\mathbf{i}$ and $\mathbf{j}$ is of the form $c\mathbf{k}$, where c is an arbitrary scalar, positive or negative.

In the following definition vectors appear as entries in a matrix. Though such a matrix and its determinant do not fit into the general definition in the preceding section, the notation is so useful and unambiguous that we shall not hesitate to employ it.

DEFINITION   *Cross product (vector product).*   Let

$$\mathbf{A} = a_1\mathbf{i} + a_2\mathbf{j} + a_3\mathbf{k} \quad \text{and} \quad \mathbf{B} = b_1\mathbf{i} + b_2\mathbf{j} + b_3\mathbf{k}.$$

The vector $\begin{vmatrix} \mathbf{i} & \mathbf{j} & \mathbf{k} \\ a_1 & a_2 & a_3 \\ b_1 & b_2 & b_3 \end{vmatrix}$

is called the cross product (or vector product) of $\mathbf{A}$ and $\mathbf{B}$. It is denoted

$$\mathbf{A} \times \mathbf{B}.$$

$$\begin{pmatrix} \mathbf{i} & \mathbf{j} & \mathbf{k} \\ a_1 & a_2 & a_3 \\ b_1 & b_2 & b_3 \end{pmatrix} \qquad \begin{pmatrix} \mathbf{i} & \mathbf{j} & \mathbf{k} \\ a_1 & a_2 & a_3 \\ b_1 & b_2 & b_3 \end{pmatrix} \qquad \begin{pmatrix} \mathbf{i} & \mathbf{j} & \mathbf{k} \\ a_1 & a_2 & a_3 \\ b_1 & b_2 & b_3 \end{pmatrix}$$

Delete the two lines through **i**. The determinant of the remaining square is the coefficient of **i** in **A** × **B**.

Delete the two lines through **j**. The negative of the determinant of the remaining square is the coefficient of **j** in **A** × **B**.

Delete the two lines through **k**. The determinant of the remaining square is the coefficient of **k** in **A** × **B**.

EXAMPLE 1   Compute **A** × **B** if $\mathbf{A} = 2\mathbf{i} - \mathbf{j} + 3\mathbf{k}$ and $\mathbf{B} = 3\mathbf{i} + 4\mathbf{j} + \mathbf{k}$.

SOLUTION   By definition,

$$\mathbf{A} \times \mathbf{B} = \begin{vmatrix} \mathbf{i} & \mathbf{j} & \mathbf{k} \\ 2 & -1 & 3 \\ 3 & 4 & 1 \end{vmatrix} = \mathbf{i} \begin{vmatrix} -1 & 3 \\ 4 & 1 \end{vmatrix} - \mathbf{j} \begin{vmatrix} 2 & 3 \\ 3 & 1 \end{vmatrix} + \mathbf{k} \begin{vmatrix} 2 & -1 \\ 3 & 4 \end{vmatrix}$$

$$= -13\mathbf{i} + 7\mathbf{j} + 11\mathbf{k}. \quad \bullet$$

Note that **A** × **B** is a vector, while **A** · **B** is a scalar. The most important property of the cross product is expressed in the following theorem.

THEOREM 1   **A** × **B** is a vector perpendicular to both **A** and **B**.

PROOF   Let us show, for instance, that $\mathbf{A} \cdot (\mathbf{A} \times \mathbf{B})$ is 0. Now, $\mathbf{A} = a_1\mathbf{i} + a_2\mathbf{j} + a_3\mathbf{k}$,

and
$$\mathbf{A} \times \mathbf{B} = \mathbf{i} \begin{vmatrix} a_2 & a_3 \\ b_2 & b_3 \end{vmatrix} - \mathbf{j} \begin{vmatrix} a_1 & a_3 \\ b_1 & b_3 \end{vmatrix} + \mathbf{k} \begin{vmatrix} a_1 & a_2 \\ b_1 & b_2 \end{vmatrix}.$$

Thus
$$\mathbf{A} \cdot (\mathbf{A} \times \mathbf{B}) = a_1 \begin{vmatrix} a_2 & a_3 \\ b_2 & b_3 \end{vmatrix} - a_2 \begin{vmatrix} a_1 & a_3 \\ b_1 & b_3 \end{vmatrix} + a_3 \begin{vmatrix} a_1 & a_2 \\ b_1 & b_2 \end{vmatrix},$$

which is precisely the definition of the determinant

$$\begin{vmatrix} a_1 & a_2 & a_3 \\ a_1 & a_2 & a_3 \\ b_1 & b_2 & b_3 \end{vmatrix}.$$

Since two rows are identical, the determinant equals 0. Thus $\mathbf{A} \cdot (\mathbf{A} \times \mathbf{B}) = 0$. A similar argument shows that $\mathbf{B} \cdot (\mathbf{A} \times \mathbf{B}) = 0$. This completes the proof. ●

EXAMPLE 2   Verify Theorem 1 in the case of the vectors **A**, **B**, and **A** × **B** of Example 1.

SOLUTION   In the case of Example 1,

$$\mathbf{A} \cdot (\mathbf{A} \times \mathbf{B}) = (2\mathbf{i} - \mathbf{j} + 3\mathbf{k}) \cdot (-13\mathbf{i} + 7\mathbf{j} + 11\mathbf{k})$$
$$= -26 - 7 + 33$$
$$= 0.$$

$$\mathbf{B} \cdot (\mathbf{A} \times \mathbf{B}) = (3\mathbf{i} + 4\mathbf{j} + \mathbf{k}) \cdot (-13\mathbf{i} + 7\mathbf{j} + 11\mathbf{k})$$
$$= -39 + 28 + 11$$
$$= 0.$$

These results verify the theorem in this case. ●

EXAMPLE 3   Compute $\mathbf{i} \times \mathbf{j}$ and $\mathbf{j} \times \mathbf{i}$.

SOLUTION

$$\mathbf{i} \times \mathbf{j} = \begin{vmatrix} \mathbf{i} & \mathbf{j} & \mathbf{k} \\ 1 & 0 & 0 \\ 0 & 1 & 0 \end{vmatrix} = 0\mathbf{i} - 0\mathbf{j} + \mathbf{k} = \mathbf{k}.$$

$$\mathbf{j} \times \mathbf{i} = \begin{vmatrix} \mathbf{i} & \mathbf{j} & \mathbf{k} \\ 0 & 1 & 0 \\ 1 & 0 & 0 \end{vmatrix} = 0\mathbf{i} - 0\mathbf{j} - \mathbf{k} = -\mathbf{k}.$$

Thus $\mathbf{i} \times \mathbf{j}$ is the negative of $\mathbf{j} \times \mathbf{i}$. ●

The Order of the Factors in the Vector Product is Critical.

Example 3 shows that $\mathbf{A} \times \mathbf{B}$ may be different from $\mathbf{B} \times \mathbf{A}$. Indeed it is easy to show that for all vectors $\mathbf{A}$ and $\mathbf{B}$

$$\mathbf{B} \times \mathbf{A} = -(\mathbf{A} \times \mathbf{B}).$$

This property corresponds to the fact that, when two rows of a matrix are interchanged, the determinant changes sign.

Another surprising property of the operation $\times$ is that, for all vectors $\mathbf{A}$,

$$\mathbf{A} \times \mathbf{A} = \mathbf{0}.$$

This corresponds to the fact that, if two rows of a matrix are identical, then its determinant is 0. More generally, if $\mathbf{A}$ and $\mathbf{B}$ are parallel,

$$\mathbf{A} \times \mathbf{B} = \mathbf{0}.$$

After these shocks it may be comforting to know that

$$\mathbf{A} \times (\mathbf{B} + \mathbf{C}) = \mathbf{A} \times \mathbf{B} + \mathbf{A} \times \mathbf{C},$$

which is reminiscent of the arithmetic of numbers. This distributive law is easy to establish by a straightforward computation.

The definition of $\mathbf{A} \times \mathbf{B}$ is purely algebraic. For many applications it is important to have a *completely geometric* description of $\mathbf{A} \times \mathbf{B}$, one that expresses the direction and magnitude of $\mathbf{A} \times \mathbf{B}$ in terms of those of $\mathbf{A}$ and $\mathbf{B}$.

We know already that $\mathbf{A} \times \mathbf{B}$ is perpendicular to both $\mathbf{A}$ and $\mathbf{B}$. However, there are two directions in which $\mathbf{A} \times \mathbf{B}$ may point. Which is the correct one? The clue is given by the case

$$\mathbf{i} \times \mathbf{j} = \mathbf{k}.$$

The vectors $\mathbf{i}$, $\mathbf{j}$, and $\mathbf{k}$ match, in order, the thumb, index finger, and middle finger of the *right hand* (just as the positive x, y, and z axes do).

This phenomenon always holds: $\mathbf{A}$, $\mathbf{B}$, and $\mathbf{A} \times \mathbf{B}$ always match the thumb,

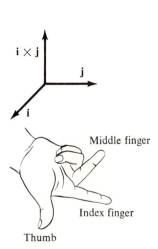

$\mathbf{i} \times \mathbf{j}$

$\mathbf{j}$

$\mathbf{i}$

Middle finger

Index finger

Thumb

index finger, and middle finger of the *right hand*. (The reader might pause and check this for **j**, **i**, and **j** × **i**.)

The direction of **A** × **B** is completely determined. But what is the length of **A** × **B**? It is given by the following theorem.

THEOREM 2   The magnitude of **A** × **B** is equal to the area of the parallelogram spanned by **A** and **B**.

PROOF   Let θ be the angle between **A** and **B**. Then the area of the parallelogram spanned by **A** and **B** is

$$|\mathbf{A}||\mathbf{B}| \sin \theta.$$

To avoid square roots, consider the square of the area of the parallelogram. We wish to show that

$$|\mathbf{A}|^2|\mathbf{B}|^2 \sin^2 \theta = |\mathbf{A} \times \mathbf{B}|^2;$$

The parallelogram has area $|\mathbf{A}||\mathbf{B}| \sin \theta$

that is, $$|\mathbf{A}|^2|\mathbf{B}|^2 \sin^2 \theta = \begin{vmatrix} a_2 & a_3 \\ b_2 & b_3 \end{vmatrix}^2 + \begin{vmatrix} a_1 & a_3 \\ b_1 & b_3 \end{vmatrix}^2 + \begin{vmatrix} a_1 & a_2 \\ b_1 & b_2 \end{vmatrix}^2. \tag{1}$$

Now, $$\sin^2 \theta = 1 - \cos^2 \theta$$
$$= 1 - \frac{(\mathbf{A} \cdot \mathbf{B})^2}{|\mathbf{A}|^2|\mathbf{B}|^2}$$
$$= \frac{|\mathbf{A}|^2|\mathbf{B}|^2 - (\mathbf{A} \cdot \mathbf{B})^2}{|\mathbf{A}|^2|\mathbf{B}|^2}.$$

Thus
$$\sin^2 \theta = \frac{(a_1{}^2 + a_2{}^2 + a_3{}^2)(b_1{}^2 + b_2{}^2 + b_3{}^2) - (a_1 b_1 + a_2 b_2 + a_3 b_3)^2}{|\mathbf{A}|^2|\mathbf{B}|^2} \tag{2}$$

Comparison of (1) and (2) with the aid of elementary algebra completes the proof. The details are left to the reader. ●

A × B *Described Geometrically*   In short, **A** × **B** *is that vector perpendicular to both* **A** *and* **B**, *whose direction is obtained by the right-hand rule and whose length is the area of the parallelogram spanned by* **A** *and* **B**. Of course, if **A** or **B** is **0**, or if **A** is parallel to **B**, **A** × **B** is the vector **0**.

EXAMPLE 4   A parallelogram in the plane has the vertices $(0, 0)$, (a_1, a_2), (b_1, b_2), and $(a_1 + b_1, a_2 + b_2)$. Find its area.

SOLUTION   The parallelogram is spanned by the vectors

$$\mathbf{A} = a_1\mathbf{i} + a_2\mathbf{j} + 0\mathbf{k}$$

and $$\mathbf{B} = b_1\mathbf{i} + b_2\mathbf{j} + 0\mathbf{k}.$$

Consequently, its area is the magnitude of the vector

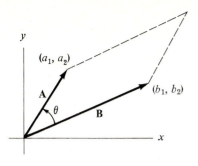

$$\begin{vmatrix} \mathbf{i} & \mathbf{j} & \mathbf{k} \\ a_1 & a_2 & 0 \\ b_1 & b_2 & 0 \end{vmatrix} = 0\mathbf{i} - 0\mathbf{j} + \mathbf{k} \begin{vmatrix} a_1 & a_2 \\ b_1 & b_2 \end{vmatrix}.$$

Thus the area is
$$\pm \begin{vmatrix} a_1 & a_2 \\ b_1 & b_2 \end{vmatrix}. \quad \bullet$$

Example 4 provides a general result, namely that the absolute value of a second-order determinant may be interpreted as the area of a certain parallelogram. The next example is typical of the geometric applications of the cross product.

EXAMPLE 5 Find a vector normal to the plane determined by the three points

$$P = (1, 3, 2) \qquad Q = (4, -1, 1) \qquad R = (3, 0, 2).$$

Then find an equation of the plane.

SOLUTION The vectors $\overrightarrow{PQ}$ and $\overrightarrow{PR}$ lie in the plane. The vector $\overrightarrow{PQ} \times \overrightarrow{PR}$, being perpendicular to both $\overrightarrow{PQ}$ and $\overrightarrow{PR}$, is a normal to the plane. Now,

$$\overrightarrow{PQ} = 3\mathbf{i} - 4\mathbf{j} - \mathbf{k},$$

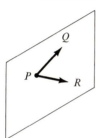

and
$$\overrightarrow{PR} = 2\mathbf{i} - 3\mathbf{j} + 0\mathbf{k}.$$

Thus
$$\mathbf{N} = \begin{vmatrix} \mathbf{i} & \mathbf{j} & \mathbf{k} \\ 3 & -4 & -1 \\ 2 & -3 & 0 \end{vmatrix} = -3\mathbf{i} - 2\mathbf{j} - \mathbf{k}$$

is a normal to the plane.

An equation of the plane can easily be obtained by using $\mathbf{N}$ and any of the three points P, Q, or R. Using $P = (1, 3, 2)$, we obtain the equation

$$-3(x - 1) - 2(y - 3) - (z - 2) = 0,$$

which reduces to
$$-3x + 3 - 2y + 6 - z + 2 = 0$$

or
$$3x + 2y + z - 11 = 0.$$

The reader may check that each of the three points lies in this plane. $\bullet$

The cross product is used in physics to express various concepts. We cite three examples.

Consider a particle of mass m moving in a straight line with constant speed. Let $\mathbf{V}$ be the vector whose length is the speed of the particle and which points in the direction of motion. The vector $m\mathbf{V}$ is called the *momentum* of the particle.

If O is a fixed point and P is the location of the particle at a certain instant, the vector

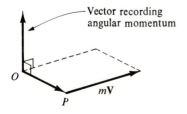

Vector recording angular momentum

$$\overrightarrow{OP} \times m\mathbf{V}$$

is called the *angular momentum* of the particle about O.

If **F** is the force applied at P, then

$$\overrightarrow{OP} \times \mathbf{F}$$

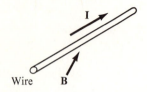

Wire **B**

represents the *torque* or turning tendency of the force.

If the vector **I** represents the current in a wire (it has a direction and a magnitude), and **B** represents a magnetic field, then the force per unit length on the wire is the vector

$$\mathbf{I} \times \mathbf{B}.$$

Exercises

In Exercises 1 to 3 compute and sketch $\mathbf{A} \times \mathbf{B}$.

1. $\mathbf{A} = \mathbf{k}, \mathbf{B} = \mathbf{j}$.
2. $\mathbf{A} = \mathbf{i} + \mathbf{j}, \mathbf{B} = \mathbf{i} - \mathbf{j}$.
3. $\mathbf{A} = \mathbf{i} + \mathbf{j} + \mathbf{k}, \mathbf{B} = \mathbf{i} + \mathbf{j}$.

In Exercises 4 to 6 draw $\mathbf{A}$, $\mathbf{B}$, and $\mathbf{A} \times \mathbf{B}$, and check that they obey the right-hand rule.

4. $\mathbf{A} = \mathbf{j}, \mathbf{B} = \mathbf{j} + \mathbf{k}$.
5. $\mathbf{A} = \mathbf{i} + \mathbf{j}, \mathbf{B} = \mathbf{j} + \mathbf{k}$.
6. $\mathbf{A} = \mathbf{k}, \mathbf{B} = \mathbf{i}$.
7. Find the area of a parallelogram three of whose vertices are $(0, 0, 0)$, $(1, 5, 4)$, and $(2, -1, 3)$.
8. Find the area of a parallelogram three of whose vertices are $(1, 2, -1)$, $(2, 1, 4)$, and $(3, 5, 2)$.
9. Find an equation of the plane determined by the points $(0, 0, 0)$, $(4, 1, 2)$, and $(2, 5, 0)$.
10. Find an equation of the plane determined by the points $(1, -1, 2)$, $(2, 1, 3)$, $(3, 3, 5)$.
11. How far is the point $(3, 1, 5)$ from the plane determined by the points $(1, 3, 3)$, $(2, -1, 7)$, $(1, 2, 4)$?
12. In the proof of Theorem 1 it was shown that $\mathbf{A}$ is perpendicular to $\mathbf{A} \times \mathbf{B}$. Show that $\mathbf{B}$ is perpendicular to $\mathbf{A} \times \mathbf{B}$.
13. Prove that $\mathbf{A} \times (\mathbf{B} + \mathbf{C}) = \mathbf{A} \times \mathbf{B} + \mathbf{A} \times \mathbf{C}$.
14. Complete the algebra in the proof of Theorem 2 that shows that Eq. (1) holds.
15. Prove that $\mathbf{B} \times \mathbf{A} = -(\mathbf{A} \times \mathbf{B})$ in two ways:
    (a) using the algebraic definition of the cross product;
    (b) using the geometric description of the cross product.
16. Show that, if $\mathbf{B} = c\mathbf{A}$, then $\mathbf{A} \times \mathbf{B} = \mathbf{0}$:
    (a) using the algebraic definition of the cross product;
    (b) using the geometric description of the cross product.
17. We needed the fact that the area of a parallelogram of adjacent sides of lengths a and b and included angle θ is $ab \sin \theta$. Use elementary geometry to show that this formula is correct.

■

18. Let $\mathbf{A}$ and $\mathbf{B}$ be nonparallel vectors and let $\mathbf{C}$ be a vector.
    (a) Using a sketch, show that there are scalars x and y such that

    $$\mathbf{C} \times (\mathbf{A} \times \mathbf{B}) = x\mathbf{A} + y\mathbf{B}.$$

    (b) Why is $x(\mathbf{C} \cdot \mathbf{A}) + y(\mathbf{C} \cdot \mathbf{B}) = 0$?
    (c) From (a) and (b) deduce that there is a scalar z such that

    $$\mathbf{C} \times (\mathbf{A} \times \mathbf{B}) = z[(\mathbf{C} \cdot \mathbf{B})\mathbf{A} - (\mathbf{C} \cdot \mathbf{A})\mathbf{B}].$$

19. Using the formulas for dot products and cross products in terms of components, show that

    $$\mathbf{C} \times (\mathbf{A} \times \mathbf{B}) = (\mathbf{C} \cdot \mathbf{B})\mathbf{A} - (\mathbf{C} \cdot \mathbf{A})\mathbf{B}.$$

20. The planes $2x + 3y + 5z = 8$ and $x - 2y + 4z = 9$ meet in a line. Find a vector parallel to that line.
21. Show by an example that $\mathbf{A} \times (\mathbf{B} \times \mathbf{C})$ need not equal $(\mathbf{A} \times \mathbf{B}) \times \mathbf{C}$.
22. Prove that $\mathbf{A} \cdot (\mathbf{B} \times \mathbf{C}) = (\mathbf{A} \times \mathbf{B}) \cdot \mathbf{C}$.
23. Prove that $|\mathbf{A} \times \mathbf{B}|^2 = |\mathbf{A}|^2|\mathbf{B}|^2 - (\mathbf{A} \cdot \mathbf{B})^2$.
24. Prove that $(\mathbf{A} + \mathbf{B}) \times (\mathbf{A} - \mathbf{B}) = 2(\mathbf{B} \times \mathbf{A})$.

■ ■

Exercises 25 to 28 are related.

25. Three vectors $\mathbf{A}$, $\mathbf{B}$, and $\mathbf{C}$ determine (span) a parallelepiped, as shown below.

Show that the volume of the parallelepiped is the absolute value of $\mathbf{A} \cdot (\mathbf{B} \times \mathbf{C})$.

This exercise provides a geometric interpretation of a third-order determinant.

26. (See Exercise 25.) Show that, if $\mathbf{A} = a_1\mathbf{i} + a_2\mathbf{j} + a_3\mathbf{k}$, $\mathbf{B} = b_1\mathbf{i} + b_2\mathbf{j} + b_3\mathbf{k}$, and $\mathbf{C} = c_1\mathbf{i} + c_2\mathbf{j} + c_3\mathbf{k}$, then the parallelepiped spanned by $\mathbf{A}$, $\mathbf{B}$, and $\mathbf{C}$ has a volume equal to the absolute value of the determinant

$$\begin{vmatrix} a_1 & a_2 & a_3 \\ b_1 & b_2 & b_3 \\ c_1 & c_2 & c_3 \end{vmatrix}.$$

27. (See Exercise 26.)
   (a) Show geometrically that these numbers all have the same absolute value: $(\mathbf{A} \times \mathbf{B}) \cdot \mathbf{C}$; $\mathbf{A} \cdot (\mathbf{B} \times \mathbf{C})$; $(\mathbf{A} \times \mathbf{C}) \cdot \mathbf{B}$.
   (b) Which of the three numbers in (a) are equal?

28. Show that the points $(0, 0, 0)$, (x_1, y_1, z_1), (x_2, y_2, z_2), (x_3, y_3, z_3) lie on a plane if and only if

$$\begin{vmatrix} x_1 & y_1 & z_1 \\ x_2 & y_2 & z_2 \\ x_3 & y_3 & z_3 \end{vmatrix} = 0.$$

19.8

Summary

This chapter is devoted to vectors. A vector may be thought of intuitively as an arrow, or formally as an ordered pair (or triple) of numbers.   Vectors, which record both a direction and a magnitude, are used in the study of force, velocity, etc.

The following table summarizes the basic concepts concerning vectors in the plane. For vectors in space the only difference is that the formulas involve one more component.

Concept	Reminder of Definition	Expression in Components if $\mathbf{A} = a_1\mathbf{i} + a_2\mathbf{j}$ and $\mathbf{B} = b_1\mathbf{i} + b_2\mathbf{j}$
Vector $\mathbf{A}$	Direction and magnitude	$a_1\mathbf{i} + a_2\mathbf{j}$
$|\mathbf{A}|$	Length or magnitude of $\mathbf{A}$	$\sqrt{a_1{}^2 + a_2{}^2}$
$-\mathbf{A}$		$-a_1\mathbf{i} - a_2\mathbf{j}$
$\mathbf{A} + \mathbf{B}$		$(a_1 + b_1)\mathbf{i} + (a_2 + b_2)\mathbf{j}$
$\mathbf{A} - \mathbf{B}$		$(a_1 - b_1)\mathbf{i} + (a_2 - b_2)\mathbf{j}$
$c\mathbf{A}$		$ca_1\mathbf{i} + ca_2\mathbf{j}$
$\mathbf{A} \cdot \mathbf{B}$	$|\mathbf{A}||\mathbf{B}|$ (cosine of angle between $\mathbf{A}$ and $\mathbf{B}$)	$a_1b_1 + a_2b_2$

Similar definitions and formulas hold for

$$\mathbf{A} = a_1\mathbf{i} + a_2\mathbf{j} + a_3\mathbf{k} \quad \text{and} \quad \mathbf{B} = b_1\mathbf{i} + b_2\mathbf{j} + b_3\mathbf{k}.$$

Matrices and determinants were introduced, primarily for the definition of $\mathbf{A} \times \mathbf{B}$. If $\mathbf{A} = a_1\mathbf{i} + a_2\mathbf{j} + a_3\mathbf{k}$ and $\mathbf{B} = b_1\mathbf{i} + b_2\mathbf{j} + b_3\mathbf{k}$, the vector $\mathbf{A} \times \mathbf{B}$ was defined as the determinant

$$\begin{vmatrix} \mathbf{i} & \mathbf{j} & \mathbf{k} \\ a_1 & a_2 & a_3 \\ b_1 & b_2 & b_3 \end{vmatrix}.$$

For a unit vector $\mathbf{u}$ the directional derivative $D_{\mathbf{u}}(f)$ was defined and related to the gradient vector ∇f by the formula

$$D_{\mathbf{u}} f = \nabla f \cdot \mathbf{u}.$$

VOCABULARY AND SYMBOLS

vector $\mathbf{A}$, $\mathbf{B}$, $(\overrightarrow{x, y})$, $(\overrightarrow{x, y, z})$
sum of vectors
scalar
product of scalar and vector
$\quad c\mathbf{A}$
unit vector
basic unit vectors $\mathbf{i}$, $\mathbf{j}$, $\mathbf{k}$
scalar component
vector component

dot product (scalar product) $\mathbf{A} \cdot \mathbf{B}$
parametric equations of a line
direction cosines
direction angles α, β, γ
directional derivatives $D_{\mathbf{u}} f$
gradient ∇f
matrix
determinant
cross product (vector product) of vectors in space $\mathbf{A} \times \mathbf{B}$

KEY FACTS

If $\mathbf{A}$ is not $\mathbf{0}$, then $\mathbf{A}/|\mathbf{A}|$ is a unit vector.
  If $\mathbf{A} \cdot \mathbf{B} = 0$, then $\mathbf{A} = \mathbf{0}$, $\mathbf{B} = \mathbf{0}$, or $\mathbf{A}$ is perpendicular to $\mathbf{B}$.
  The cosine of the angle between $\mathbf{A}$ and $\mathbf{B}$ is

$$\frac{\mathbf{A} \cdot \mathbf{B}}{|\mathbf{A}||\mathbf{B}|}.$$

An equation of the line through (x_0, y_0) and perpendicular to the vector $A\mathbf{i} + B\mathbf{j}$ is

$$A(x - x_0) + B(y - y_0) = 0.$$

The vector $A\mathbf{i} + B\mathbf{j}$ is perpendicular to the line

$$Ax + By + C = 0.$$

The distance from the point (x_1, y_1) to the line

$$Ax + By + C = 0$$

is

$$\frac{|Ax_1 + By_1 + C|}{\sqrt{A^2 + B^2}}.$$

An equation of the plane through (x_0, y_0, z_0) and perpendicular to the vector $A\mathbf{i} + B\mathbf{j} + C\mathbf{k}$ is

$$A(x - x_0) + B(y - y_0) + C(z - z_0) = 0.$$

The vector $A\mathbf{i} + B\mathbf{j} + C\mathbf{k}$ is perpendicular to the plane

$$Ax + By + Cz + D = 0.$$

The distance from the point (x_1, y_1, z_1) to the plane

$$Ax + By + Cz + D = 0$$

is

$$\frac{|Ax_1 + By_1 + Cz_1 + D|}{\sqrt{A^2 + B^2 + C^2}}.$$

The direction angles α, β, γ of a vector are the angles it makes with $\mathbf{i}$, $\mathbf{j}$, and $\mathbf{k}$. Cos α, cos β, and cos γ are the direction angles of the vector (or of a line parallel to the vector). They are related by the equation

$$\cos^2 \alpha + \cos^2 \beta + \cos^2 \gamma = 1.$$

The line through (x_0, y_0, z_0) parallel to the vector $\mathbf{A} = a_1\mathbf{i} + a_2\mathbf{j} + a_3\mathbf{k}$ is given parametrically as

$$x = x_0 + a_1 t$$
$$y = y_0 + a_2 t$$
$$z = z_0 + a_3 t.$$

Also, the line has the description

$$\frac{x - x_0}{a_1} = \frac{y - y_0}{a_2} = \frac{z - z_0}{a_3}.$$

The gradient of a function was first defined algebraically, relative to an xyz-coordinate system. Thus, if f is defined in a region in space at the point (a, b, c),

$$\nabla f = f_x(a, b, c)\mathbf{i} + f_y(a, b, c)\mathbf{j} + f_z(a, b, c)\mathbf{k}.$$

It was then shown that ∇f can be described without reference to a coordinate system: It is the vector that points in the direction in which f increases most rapidly at (a, b, c); its magnitude is the largest directional derivative of f at (a, b, c).

A similar definition holds for the gradient of $f(x, y)$ at the point (a, b).

An arrangement $\begin{pmatrix} a_1 & a_2 \\ b_1 & b_2 \end{pmatrix}$

is a matrix of order 2. A determinant of order 2 is defined as

$$\begin{vmatrix} a_1 & a_2 \\ b_1 & b_2 \end{vmatrix} = a_1 b_2 - a_2 b_1.$$

A determinant of order 3 is defined as

$$\begin{vmatrix} a_1 & a_2 & a_3 \\ b_1 & b_2 & b_3 \\ c_1 & c_2 & c_3 \end{vmatrix} = a_1 \begin{vmatrix} b_2 & b_3 \\ c_2 & c_3 \end{vmatrix} - a_2 \begin{vmatrix} b_1 & b_3 \\ c_1 & c_3 \end{vmatrix} + a_3 \begin{vmatrix} b_1 & b_2 \\ c_1 & c_2 \end{vmatrix}.$$

If two rows (or columns) of a matrix are the same, its determinant is 0. If two rows (or columns) are interchanged, the determinant changes sign.

Determinants provide formulas for the solution of simultaneous linear equations. In particular they may be used to find where two lines or three planes meet.

Like the gradient, the cross product was defined algebraically, and afterward described geometrically. The vector $\mathbf{A} \times \mathbf{B}$ is perpendicular to both $\mathbf{A}$ and $\mathbf{B}$, pointing in the direction that makes $\mathbf{A}$, $\mathbf{B}$, $\mathbf{A} \times \mathbf{B}$ right-handed. The length of $\mathbf{A} \times \mathbf{B}$ is the area of the parallelogram spanned by $\mathbf{A}$ and $\mathbf{B}$.

The cross product has various unusual properties:

$$\mathbf{A} \times \mathbf{A} = \mathbf{0}.$$

$$\mathbf{B} \times \mathbf{A} = -\mathbf{A} \times \mathbf{B}.$$

$\mathbf{A} \times (\mathbf{B} \times \mathbf{C})$ usually is *not* equal to $(\mathbf{A} \times \mathbf{B}) \times \mathbf{C}$. In fact,

$$\mathbf{A} \times (\mathbf{B} \times \mathbf{C}) = (\mathbf{A} \cdot \mathbf{C})\mathbf{B} - (\mathbf{A} \cdot \mathbf{B})\mathbf{C}.$$

It can be shown that

$$(\mathbf{A} \times \mathbf{B}) \times \mathbf{C} = -[(\mathbf{C} \cdot \mathbf{B})\mathbf{A} - (\mathbf{C} \cdot \mathbf{A})\mathbf{B}].$$

$$\mathbf{A} \cdot (\mathbf{B} \times \mathbf{C}) = \pm \text{volume of parallelepiped}$$
spanned by $\mathbf{A}$, $\mathbf{B}$, and $\mathbf{C}$

$$= \begin{vmatrix} a_1 & a_2 & a_3 \\ b_1 & b_2 & b_3 \\ c_1 & c_2 & c_3 \end{vmatrix}.$$

(See Exercises 25 and 26, pages 752–753.)

Guide quiz on Chap. 19

1. Given $\mathbf{A} = \mathbf{i} + 2\mathbf{j} - \mathbf{k}$ and $\mathbf{B} = 2\mathbf{i} - \mathbf{j} + 3\mathbf{k}$, find

(a) $\mathbf{A} \cdot \mathbf{B}$; (b) $|\mathbf{A}|$;
(c) a unit vector in the direction of $\mathbf{A}$;
(d) the scalar component of $\mathbf{B}$ along $\mathbf{A}$;
(e) the vector component of $\mathbf{B}$ along $\mathbf{A}$;
(f) the scalar component of $\mathbf{A}$ along $\mathbf{B}$;
(g) the cosine of the angle between $\mathbf{A}$ and $\mathbf{B}$;
(h) the angle between $\mathbf{A}$ and $\mathbf{B}$;
(i) $\mathbf{A} \times \mathbf{B}$; (j) $\mathbf{B} \times \mathbf{A}$;
(k) a unit vector perpendicular to both $\mathbf{A}$ and $\mathbf{B}$;
(l) the area of the parallelogram spanned by $\mathbf{A}$ and $\mathbf{B}$.

2. Draw the necessary diagrams and explain why the distance from the point (x_1, y_1) to the line $Ax + By + C = 0$ is

$$\frac{|Ax_1 + By_1 + C|}{\sqrt{A^2 + B^2}}.$$

3. Find the direction cosines of a vector normal to the plane $x - 2y + 2z = 16$.

4. Prove that

$$\begin{vmatrix} a_1 & a_2 & a_3 \\ b_1 & b_2 & b_3 \\ c_1 + b_1 & c_2 + b_2 & c_3 + b_3 \end{vmatrix} = \begin{vmatrix} a_1 & a_2 & a_3 \\ b_1 & b_2 & b_3 \\ c_2 & c_2 & c_3 \end{vmatrix}.$$

5. (a) Find the volume of the parallelepiped shown below.

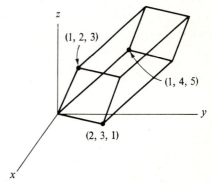

(b) Justify the general formula you used in (a).

6. Let $f(x, y) = y - x^2$. Show that ∇f, evaluated at $(1, 1)$, is perpendicular to the curve $y = x^2$ at $(1, 1)$.

7. Let $\mathbf{A}$ and $\mathbf{B}$ be two nonparallel nonzero vectors in space.
(a) Explain why there are scalars x and y such that

$$\mathbf{A} \times (\mathbf{A} \times \mathbf{B}) = x\mathbf{A} + y\mathbf{B}.$$

(b) Show that $\mathbf{A} \times (\mathbf{A} \times \mathbf{B})$ is not $\mathbf{0}$.
(c) Show that $(\mathbf{A} \times \mathbf{A}) \times \mathbf{B} = \mathbf{0}$.

8. Where does the line through $(1, 2, 1)$ and $(3, 1, 1)$ meet the plane determined by the three points $(2, -1, 1)$, $(5, 2, 3)$, and $(4, 1, 3)$?

9. Let $f(x, y, z) = x^2 y z^3$.
   (a) Find the largest directional derivative of f at $(1, 1, 1)$.
   (b) Find the direction cosines of the vector in the direction that produces the largest directional derivative at $(1, 1, 1)$.
   (c) Find a unit vector **u** such that $D_\mathbf{u} f = 0$ at $(1, 1, 1)$.

Review exercises for Chap. 19

1. (a) Define $\mathbf{A} \cdot \mathbf{B}$, where **A** and **B** are vectors in space.
   (b) What is the formula for $\mathbf{A} \cdot \mathbf{B}$ in terms of the components of **A** and **B**?
   (c) Prove that if $\mathbf{A} \cdot \mathbf{B} = 0$ and neither **A** nor **B** is the zero vector, then **A** is perpendicular to **B**.

2. What is the scalar component of $\mathbf{A} = -2\mathbf{i} + 3\mathbf{j}$ along the vector
   (a) **i**?
   (b) **j**?
   (c) $0.6\mathbf{i} + 0.8\mathbf{j}$?
   (d) $4\mathbf{i} - 5\mathbf{j}$?

3. What is the vector component of $\mathbf{A} = -2\mathbf{i} + 3\mathbf{j}$ along the vector
   (a) **i**?
   (b) **j**?
   (c) $0.6\mathbf{i} + 0.8\mathbf{j}$?
   (d) $4\mathbf{i} - 5\mathbf{j}$?

4. (a) Give an application of the dot product in physics and in economics.
   (b) Give an application of the cross product in physics.

5. Prove that $(a_1\mathbf{i} + a_2\mathbf{j} + a_3\mathbf{k}) \cdot (b_1\mathbf{i} + b_2\mathbf{j} + b_3\mathbf{k}) = a_1 b_1 + a_2 b_2 + a_3 b_3$.

6. Let $f(x, y) = e^{-xy^2}$.
   (a) Compute and draw ∇f at $(1, 2)$.
   (b) What is the largest directional derivative of f at $(1, 2)$?
   (c) Draw the unit vector **u** such that $D_\mathbf{u} f$ is the maximum directional derivative at $(1, 2)$.
   (d) Draw the two unit vectors **u** such that $D_\mathbf{u} f = 0$ at $(1, 2)$.

7. (a) Define ∇f.
   (b) Describe ∇f geometrically.

8. (a) Define $\mathbf{A} \times \mathbf{B}$.
   (b) Describe $\mathbf{A} \times \mathbf{B}$ geometrically.

9. Find the point on the plane $2x - y + 3z + 12 = 0$ that is nearest the origin. Use vectors, not calculus.

10. Let $f(x, y) = x^2 + 3y^2$.
    (a) Compute and draw ∇f at $(1, 2)$.
    (b) Show that ∇f is perpendicular to the tangent line to the curve $x^2 + 3y^2 = 13$ at $(1, 2)$.
    (c) Find $D_\mathbf{u} f$ at $(1, 2)$ if **u** has the angle $3\pi/4$.

11. Consider a function $f(x, y, z)$.
    (a) Define $D_\mathbf{u} f$ at (a, b, c), where **u** is a unit vector.
    (b) Show that $D_\mathbf{u} f = \nabla f \cdot \mathbf{u}$.

12. Use determinants to find x, y, and z such that
$$x(3\mathbf{i} + \mathbf{j} + 2\mathbf{k}) + y(\mathbf{i} - \mathbf{j} + 2\mathbf{k})$$
$$+ z(2\mathbf{i} + 2\mathbf{j} + \mathbf{k}) = 3\mathbf{i} + 3\mathbf{j} + 3\mathbf{k}.$$

13. The determinant of a matrix in which a column has only 0s is 0. Prove this for the matrix
$$\begin{pmatrix} a_1 & 0 & a_3 \\ b_1 & 0 & b_3 \\ c_1 & 0 & c_3 \end{pmatrix}.$$

14. Show that
$$\begin{vmatrix} 0 & a & b \\ -a & 0 & c \\ -b & -c & 0 \end{vmatrix} = 0$$
for all numbers a, b, c.

15. Find the length of a vector from $(1, 3, 3)$ to the plane $x - 4y + 5z + 4 = 0$ that makes an angle of $45°$ with that plane.

16. The accompanying diagram shows a pyramid with a square base.

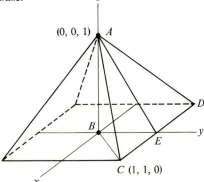

Find the cosine of the angle between
   (a) $\overrightarrow{CA}$ and $\overrightarrow{CB}$;
   (b) $\overrightarrow{EA}$ and $\overrightarrow{EB}$;
   (c) $\overrightarrow{AD}$ and $\overrightarrow{AC}$.

17. The planes $2x + 5y + z = 10$ and $3x - y + 4z = 11$ meet in a line. For this line find
   (a) the direction numbers;

(b) direction cosines;
(c) direction angles;
(d) a point on the line.

18. Prove that the numbers

$$x = \frac{\begin{vmatrix} k_1 & a_2 \\ k_2 & b_2 \end{vmatrix}}{\begin{vmatrix} a_1 & a_2 \\ b_1 & b_2 \end{vmatrix}} \quad \text{and} \quad y = \frac{\begin{vmatrix} a_1 & k_1 \\ b_1 & k_2 \end{vmatrix}}{\begin{vmatrix} a_1 & a_2 \\ b_1 & b_2 \end{vmatrix}}$$

satisfy both equations $a_1 x + a_2 y = k_1$ and $b_1 x + b_2 y = k_2$. Assume that $a_1 b_2 - a_2 b_1$ is not 0.

19. Find parametric equations of the line through $(1, 1, 2)$ that is parallel to the planes $x + 2y + 3z = 0$ and $2x - y + 3z + 4 = 0$.

20. Does the plane through $(1, 1, -1)$, perpendicular to $2\mathbf{i} + 4\mathbf{j} + 5\mathbf{k}$, pass through the point $(4, 5, -7)$?

21. Is the line through $(1, 4, 7)$ and $(5, 10, 15)$ perpendicular to the plane $2x + 3y + 4z = 17$?

22. Two planes that interesect in a line determine a *dihedral angle* θ, as shown below.

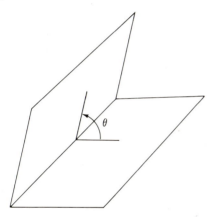

Find the dihedral angle between the planes $x - 3y + 4z = 10$ and $2x + y + z = 11$.

23. Find the point on the plane

$$\frac{x}{2} + \frac{y}{3} + \frac{z}{4} = 1$$

(a) nearest the origin;
(b) nearest the point $(1, 2, 3)$.

24. Let $\mathbf{A} = 2\mathbf{i} + 3\mathbf{j} + 4\mathbf{k}$. Find:
(a) the scalar component of $\mathbf{A}$ in the direction $\mathbf{i}$;
(b) the vector component of $\mathbf{A}$ in the direction $\mathbf{i}$;
(c) the scalar component of $\mathbf{A}$ in the direction $-\mathbf{i}$;
(d) the vector component of $\mathbf{A}$ in the direction $-\mathbf{i}$;
(e) the projection of $\mathbf{A}$ on $\mathbf{i} + \mathbf{j}$.

25. How would you determine whether the points (x_1, y_1)

and (x_2, y_2) are on the same side or on opposite sides of the line $Ax + By + C = 0$?

26. How would you determine whether the points (x_1, y_1, z_1) and (x_2, y_2, z_2) are on the same side or on opposite sides of the plane $Ax + By + Cz + D = 0$?

27. How far apart are the planes parallel to the plane $Ax + By + Cz + D = 0$ that pass through the points (x_1, y_1, z_1) and (x_2, y_2, z_2)?

■

28. A parallelepiped is spanned by the three vectors $\mathbf{A} = \mathbf{i} + 2\mathbf{j} + 3\mathbf{k}$, $\mathbf{B} = 2\mathbf{i} + \mathbf{j} + \mathbf{k}$, $\mathbf{C} = 3\mathbf{i} + 3\mathbf{j} + \mathbf{k}$.
(a) Find the volume of the parallelepiped.
(b) Find the area of the face spanned by $\mathbf{A}$ and $\mathbf{C}$.
(c) Find the angle between $\mathbf{A}$ and the face spanned by $\mathbf{B}$ and $\mathbf{C}$.

29. Find the angle between the line through $(0, 0, 0)$ and $(1, 1, 1)$ and the plane through $(1, 2, 3)$, $(4, 1, 5)$, and $(2, 0, 6)$.

30. Let $\mathbf{u}_1$ and $\mathbf{u}_2$ be perpendicular unit vectors in space. Find $(\mathbf{u}_1 \times \mathbf{u}_2) \times \mathbf{u}_1$ and $\mathbf{u}_1 \times (\mathbf{u}_1 \times \mathbf{u}_2)$.

31. Consider a company that produces x_1 washing machines, x_2 refrigerators, and x_3 dishwashers. The profit from each washing machine is p_1 dollars, from each refrigerator p_2 dollars, and from each dishwasher p_3 dollars. (The p's need not be positive.)

The three-dimensional vector $\mathbf{V} = (\overrightarrow{x_1, x_2, x_3})$ is called the *production vector*, and $\mathbf{P} = (\overrightarrow{p_1, p_2, p_3})$ is the *profit vector*.
(a) Show that the total profit is $\mathbf{P} \cdot \mathbf{V}$.
(b) Assume that the company also manufactures x_4 freezers at a profit of p_4 dollars on each, and x_5 electric brooms at a profit of p_5 dollars on each. How would you define its production and profit vectors then?

32. Express in terms of cross products and dot products the fact that the points P_4 and P_5 are situated on the same side of the plane through P_1, P_2, and P_3.

33. Let L_1 and L_2 be two lines in the xy plane through $(0, 0)$ neither being the y axis, one of slope m_1 and one of slope m_2.
(a) Show that the point $(1, m_1)$ is on L_1 and that $\mathbf{i} + m_1\mathbf{j}$ is a vector parallel to L_1.
(b) Show that the point $(1, m_2)$ is on L_2 and that $\mathbf{i} + m_2\mathbf{j}$ is a vector parallel to L_2.
(c) Using (a) and (b), prove that L_1 is perpendicular to L_2 if and only if $m_1 m_2 = -1$.

34. Express in terms of cross products and dot products the fact that the line through P_1 and P_2 is parallel to the plane through P_3, P_4, and P_5.

35. Produce as simply as possible a vector in the xy plane that is perpendicular to the vector $2\mathbf{i} - 3\mathbf{j} + \mathbf{k}$.

36. Let P_1, P_2, P_3, and P_4 be four points in space. Outline a general procedure for finding the distance between the line through P_1 and P_2 and the line through P_3 and P_4. Use vectors, not calculus. Assume that the lines do not intersect and are not parallel.

37. (See Exercise 36.) Find the distance between the line through the points $(0, 0, 0)$ and $(1, 1, 1)$ and the line through the points $(3, 1, 2)$ and $(1, 4, 4)$.

38. (a) Make a sketch to show that for any vector $\mathbf{V}$ in the plane there are scalars x and y such that
$$\mathbf{V} = x(2\mathbf{i} + 3\mathbf{j}) + y(\mathbf{i} + 5\mathbf{j}).$$

(b) Use determinants to find x and y such that
$$\mathbf{i} + 4\mathbf{j} = x(2\mathbf{i} + 3\mathbf{j}) + y(\mathbf{i} + 5\mathbf{j}).$$

Exercises 39 to 42 can be solved with the aid of the identities
$$\mathbf{D} \times \mathbf{E} = -\mathbf{E} \times \mathbf{D} \quad \text{and} \quad \mathbf{D} \times (\mathbf{E} \times \mathbf{F}) = (\mathbf{D} \cdot \mathbf{F})\mathbf{E} - (\mathbf{D} \cdot \mathbf{E})\mathbf{F}.$$
Keep in mind also that $\mathbf{D} \cdot (\mathbf{E} \times \mathbf{F}) = (\mathbf{D} \times \mathbf{E}) \cdot \mathbf{F}$.

39. Show that
$$(\mathbf{A} \times \mathbf{B}) \times \mathbf{C} = (\mathbf{A} \cdot \mathbf{C})\mathbf{B} - (\mathbf{B} \cdot \mathbf{C})\mathbf{A}.$$

40. Show that
$$(\mathbf{A} \times \mathbf{B}) \times (\mathbf{C} \times \mathbf{D}) = (\mathbf{A} \times \mathbf{B} \cdot \mathbf{D})\mathbf{C} - (\mathbf{A} \times \mathbf{B} \cdot \mathbf{C})\mathbf{D}.$$

41. Show that
$$(\mathbf{A} \times \mathbf{B}) \cdot (\mathbf{C} \times \mathbf{D}) = (\mathbf{B} \cdot \mathbf{D})(\mathbf{A} \cdot \mathbf{C}) - (\mathbf{B} \cdot \mathbf{C})(\mathbf{A} \cdot \mathbf{D}).$$

42. Show that
$$\mathbf{A} \times (\mathbf{B} \times \mathbf{C}) + \mathbf{B} \times (\mathbf{C} \times \mathbf{A}) + \mathbf{C} \times (\mathbf{A} \times \mathbf{B}) = \mathbf{0}.$$

43. Though defined at first algebraically, the gradient turned out to be an intrinsic geometric property of the function, for its direction and magnitude are expressible in terms of the function, without reference to any particular coordinate system.

Consider, then, a function $f(P)$, where P runs over a portion of the plane. If f is given in rectangular coordinates, then
$$\nabla f = f_x \mathbf{i} + f_y \mathbf{j}.$$
What if f is described in *polar coordinates*, $f(r, \theta)$? How would ∇f be expressed? The natural unit vectors to use, instead of $\mathbf{i}$ and $\mathbf{j}$, are $\mathbf{u}_r$ and $\mathbf{u}_\theta$ shown in the diagram.

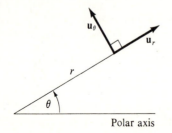

Polar axis

(a) If $\nabla f = A\mathbf{u}_r + B\mathbf{u}_\theta$ for scalars A and B, show that $A = D_{\mathbf{u}_r}(f)$ and $B = D_{\mathbf{u}_\theta}(f)$.

(b) Show that $D_{\mathbf{u}_r}(f) = \dfrac{df}{dr}$.

(c) Why would $D_{\mathbf{u}_\theta}(f) = \dfrac{1}{r}\dfrac{df}{d\theta}$? $\left(\text{Not } \dfrac{df}{d\theta}\ !!\right)$
Give a persuasive argument, not necessarily a rigorous proof.

(d) From (a), (b), and (c) deduce that
$$\nabla f = \frac{df}{dr}\mathbf{u}_r + \frac{1}{r}\frac{df}{d\theta}\mathbf{u}_\theta.$$

44. (See Exercise 43.)
Let $f(P)$ be the reciprocal of distance from P in the plane to the origin.

(a) In rectangular coordinates f has the formula $f(x, y) = 1/\sqrt{x^2 + y^2}$. Calculate ∇f using rectangular coordinates.

(b) In polar coordinates f has the formula $f(r, \theta) = 1/r$. Calculate ∇f using polar coordinates.

(c) Sketch ∇f as calculated in (a) and in (b). Show that the results are the same.

45. (See Exercise 43.) Consider a spherical coordinate system ρ, ϕ, θ. At a given point let $\mathbf{u}_\rho$ be the unit vector pointing in the direction of increasing ρ. Let $\mathbf{u}_\phi$ be the unit vector pointing in the direction of increasing ϕ. Let $\mathbf{u}_\theta$ be the unit vector pointing in the direction of increasing θ.

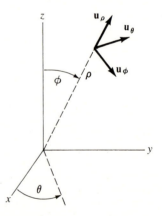

Let f be a function defined on space. Show why it is reasonable that
$$\nabla f = \frac{\partial f}{\partial \rho}\mathbf{u}_\rho + \frac{1}{\rho}\frac{\partial f}{\partial \phi}\mathbf{u}_\phi + \frac{1}{\rho \sin \phi}\frac{\partial f}{\partial \theta}\mathbf{u}_\theta.$$

■ ■

46. Do there exist three points P, Q, and R in the plane such that their rectangular coordinates are all rational and angle $\angle PQR$ is $60°$?

47. Let $Q = (1, 0, 0)$ and $R = (-1, 0, 0)$. Consider all points $P = (x, y, z)$ such that $\angle QPR = \pi/3$. Show that all such points form a sphere (the surface of a ball).

48. In Sec. 13.4 it was shown that all balancing lines of a flat object pass through a single point. This exercise obtains the analogous result for three-dimensional solids, and thus shows that a material solid has a center of gravity which is independent of the particular choice of a coordinate system.

Let R be a solid region with density at P equal to $\delta(P)$. Assume that, for a certain choice of an xyz coordinate system,

$$\int_R x\delta(P)\, dV = 0,$$

$$\int_R y\delta(P)\, dV = 0,$$

and

$$\int_R z\delta(P)\, dV = 0.$$

(See Sec. 18.4, where the moment about a plane was defined.) Let $Ax + By + Cz = 0$ be an arbitrary plane through the origin. For any point P let $f(P)$ be the distance to the arbitrary plane, choosing $f(P)$ to be positive if P is on one selected side of the plane, and negative if P is on the other side. Prove that

$$\int_R f(P)\delta(P)\, dV = 0.$$

Why does this show that a material solid has a center of gravity?

49. In Sec. 18.4 the moment of inertia of a material solid about a line was defined as the integral

$$\int_R [s(P)]^2 \delta(P)\, dV,$$

where $\delta(P)$ is the density at P and $s(P)$ is the distance from P to the line.

Now consider an xyz-coordinate system and assume that the moments of inertia of the solid about the three axes are known. Assume also that the integrals

$$\int_R xy\, \delta(P)\, dV, \quad \int_R xz\, \delta(P)\, dV, \quad \text{and} \int_R yz\, \delta(P)\, dV$$

are known. Show that the moment of inertia about any line through $(0, 0, 0)$ is determined. In fact, express it in terms of the three integrals displayed, the moments of inertia about the three axes, and the direction cosines of the line.

50. (a) Consider any triangle in the plane and a point P_0 inside the triangle. Let $f(P)$ denote the sum of the (positive) perpendicular distances from P to the three lines of the triangle. Show that there is a line segment whose ends lie on the border of the triangle such that $f(P) = f(P_0)$ for all points P on the segment.

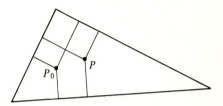

(b) Does this generalize to all polygons in the plane?

51. Consider any tetrahedron and a point P_0 inside the tetrahedron. Let $f(P)$ denote the sum of the (positive) distances from P to the four planes of the tetrahedron. Show that there is a plane through P_0 such that, if P is on the plane and inside the tetrahedron, then $f(P) = f(P_0)$.

Exercises 52 to 55 are related.

52. In a more abstract approach than the one followed in this book, a plane vector is defined as an ordered pair of real numbers (x, y). Taking this purely algebraic approach, define

(a) the vector $\mathbf{0}$;

(b) the product of a scalar and a vector;

(c) the sum of two vectors;

(d) the magnitude of a vector;

(e) the dot product.

Just as it is possible to define a plane vector as an ordered pair of real numbers (x, y), or a space vector as an ordered triple of real numbers (x, y, z), it is possible to define vectors algebraically in higher dimensions. A four-dimensional vector can be defined to be an ordered quadruple of numbers $\overrightarrow{(x_1, x_2, x_3, x_4)}$. The *dot product* of two vectors $\overrightarrow{(x_1, x_2, x_3, x_4)}$ and $\overrightarrow{(y_1, y_2, y_3, y_4)}$ is defined as the sum $x_1 y_1 + x_2 y_2 + x_3 y_3 + x_4 y_4$. The *magnitude* of a vector $\overrightarrow{(x_1, x_2, x_3, x_4)}$ is defined as $\sqrt{x_1^2 + x_2^2 + x_3^2 + x_4^2}$. This can be done in any number of dimensions.

53. (a) If the sequence $\overrightarrow{(x_1, x_2, x_3, \ldots, x_n, \ldots)}$ is considered an *infinite-dimensional vector*, how would its magnitude be defined?

(b) What is the magnitude of the vector

$$\overrightarrow{(1/2, 1/4, 1/8, 1/16, \ldots, 1/2^n, \ldots)}?$$

54. The output of a firm that manufactures x_1 washing machines, x_2 refrigerators, x_3 dishwashers, x_4 stoves, and

x_5 clothes dryers is recorded by the five-dimensional production vector $\mathbf{P} = \overrightarrow{(x_1, x_2, x_3, x_4, x_5)}$. Similarly, the cost vector $\mathbf{C} = \overrightarrow{(y_1, y_2, y_3, y_4, y_5)}$ records the cost of producing each item; for instance, each refrigerator costs the firm y_2 dollars.

(a) What is the economic significance of $\mathbf{P} \cdot \mathbf{C} = \overrightarrow{(20, 0, 7, 9, 15)} \cdot \overrightarrow{(50, 70, 30, 20, 10)}$?

(b) If the firm doubles its production of all items in (a), what is its new production vector?

55. Let P_1 be the profit from selling a washing machine, and P_2, P_3, P_4, and P_5 be defined analogously for the firm of Exercise 54. (Some of the P's may be negative.) What does it mean to the firm to have $\overrightarrow{(P_1, P_2, P_3, P_4, P_5)}$ "perpendicular" to $\overrightarrow{(x_1, x_2, x_3, x_4, x_5)}$?

20

THE DERIVATIVE OF A VECTOR FUNCTION

The motion of a particle traveling along a straight line is most easily described with the aid of a coordinate system on the line. But the motion of a particle along a curved path, as well as the forces acting on that particle, are most easily described with vectors. For example, the attraction of the earth on an astronaut's capsule is represented by a vector directed toward the center of the earth.

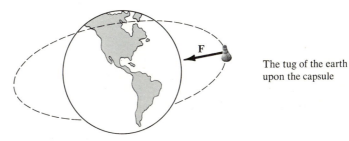

The tug of the earth upon the capsule

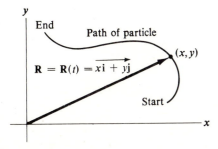

When the astronaut fires this engine, he experiences the shove **F** owing to the escaping gas pushing against the capsule

The astronaut influences his or her flight through small rockets. Their thrust is described by a vector. When you spin an object in a circle at the end of a rope, it is subject to two forces: one is toward your hand, and the other is the pull of gravity straight down.

This chapter uses vectors to investigate the motion of a particle whose direction and speed may both change with time, because of some external forces.

20.1

The derivative of a vector function

Consider an object moving in a plane. It might be a mass at the end of a rope, a ball, a satellite, a comet, a raindrop, or an astronaut's capsule. Call this object a "particle" and assume that all its mass is located at a single point. Denote the position of the particle at time t relative to an xy-coordinate system by (x, y). We shall describe its position with the position vector $\mathbf{R}$, whose tail is at $(0, 0)$ and whose head is at (x, y). Thus $\mathbf{R} = x\mathbf{i} + y\mathbf{j}$, where x and y depend on time t. Therefore $\mathbf{R}$ depends on t, and may be written as $\mathbf{R} = \mathbf{G}(t)$.

If the particle is moving on a curve in space, its position at time t is described by a position vector $\mathbf{R} = \mathbf{G}(t) = x\mathbf{i} + y\mathbf{j} + z\mathbf{k}$, where x, y, and z are functions of time.

This brings us to an important definition, which calls attention to a new type of function.

DEFINITION *Vector function.* A function whose inputs are scalars and whose outputs are vectors is called a vector function. It is denoted by a boldface letter, such as $\mathbf{F}$ or $\mathbf{G}$.

Usually the scalar input may be thought of as time, and $\mathbf{G}(t)$ as the position vector at time t. The scalar input might sometimes be arc length along the curve, and $\mathbf{G}(s)$ would be the position vector when the particle has swept out a distance s along the curve.

EXAMPLE 1 Sketch the path of a particle that has the position vector

$$\mathbf{G}(t) = 2t\mathbf{i} + 4t^2\mathbf{j}$$

at time t.

SOLUTION At time t the particle is at the point (x, y) where

$$x = 2t \quad \text{and} \quad y = 4t^2.$$

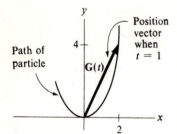

The path is given parametrically by these equations. Elimination of t shows that $y = x^2$. The path is a parabola, sketched in the diagram at the left, in which $\mathbf{G}(1)$ is also depicted. ●

In the case of motion on a horizontal line the derivative of position with respect to time is sufficient to describe the motion of the particle. If the derivative is positive, the particle is moving to the right. If the derivative is negative, the particle is moving to the left. The speed is simply the absolute value of the derivative. But the study of motion in the plane depends on the concept of the derivative of a vector function. First, a definition.

DEFINITION *Limit of a vector function.* Let $\mathbf{H}$ be a vector function. Let a be a real number and $\mathbf{A}$ a vector. We say that

$$\lim_{t \to a} \mathbf{H}(t) = \mathbf{A}$$

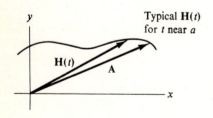

if

$$\lim_{t \to a} |\mathbf{H}(t) - \mathbf{A}| = 0.$$

In other words, if $\mathbf{H}(t)$ and $\mathbf{A}$ are drawn with their tails at the origin, the head of $\mathbf{H}(t)$ gets close to the head of $\mathbf{A}$ as $t \to a$.

EXAMPLE 2 Let $\mathbf{H}(t) = t^2\mathbf{i} + 2t\mathbf{j}$. Use the above definition to show that

$$\lim_{t \to 1} \mathbf{H}(t) = \mathbf{i} + 2\mathbf{j}.$$

SOLUTION First compute

$$|\mathbf{H}(t) - (\mathbf{i} + 2\mathbf{j})|,$$

which is

$$|(t^2\mathbf{i} + 2t\mathbf{j}) - (\mathbf{i} + 2\mathbf{j})| = |(t^2 - 1)\mathbf{i} + (2t - 2)\mathbf{j}|$$
$$= \sqrt{(t^2 - 1)^2 + (2t - 2)^2}.$$

Since

$$\lim_{t \to 1} \sqrt{(t^2 - 1)^2 + (2t - 2)^2} = 0,$$

it follows that

$$\lim_{t \to 1} \mathbf{H}(t) = \mathbf{i} + 2\mathbf{j}. \; \bullet$$

Note in Example 2 that

$$\lim_{t \to 1} \mathbf{H}(t) = \left(\lim_{t \to 1} t^2\right)\mathbf{i} + \left(\lim_{t \to 1} 2t\right)\mathbf{j}.$$

This suggests a practical method for finding the limit of a vector function: If its scalar components are known, just find the limits of these components. This observation is justified by the following theorem.

THEOREM I Suppose a single vector function $\mathbf{H}$ is described by two scalar functions g and h,

$$\mathbf{H}(t) = g(t)\mathbf{i} + h(t)\mathbf{j}.$$

Then

$$\lim_{t \to a} \mathbf{H}(t) = A\mathbf{i} + B\mathbf{j},$$

if and only if

$$\lim_{t \to a} g(t) = A \quad \text{and} \quad \lim_{t \to a} h(t) = B.$$

PROOF If $\lim_{t \to a} g(t) = A$ and $\lim_{t \to b} g(t) = B$,

then

$$\lim_{t \to a} |\mathbf{H}(t) - (A\mathbf{i} + B\mathbf{j})| = \lim_{t \to a} \sqrt{(g(t) - A)^2 + (h(t) - B)^2}$$
$$= 0.$$

Thus

$$\lim_{t \to a} \mathbf{H}(t) = A\mathbf{i} + B\mathbf{j}.$$

Conversely, if $\lim_{t \to a} \mathbf{H}(t) = A\mathbf{i} + B\mathbf{j}$, then

$$\lim_{t \to a} \sqrt{(g(t) - A)^2 + (h(t) - B)^2} = 0,$$

so that $\lim_{t \to a} g(t) = A$ and $\lim_{t \to a} h(t) = B$. $\bullet$

Theorem 1 says that, to find the limit of a vector function, find the limit of each of its component scalar functions. Though Theorem 1 concerns only functions whose values are *plane vectors*, it extends to functions whose values are vectors in space, that is, vectors that have three scalar components. The proof is essentially unchanged.

EXAMPLE 3 Find $\lim_{t \to 2} \mathbf{H}(t)$, where $\mathbf{H}(t) = t^2\mathbf{i} + t^3\mathbf{j} + t^4\mathbf{k}$.

SOLUTION $\quad \lim_{t \to 2} \mathbf{H}(t) = \left(\lim_{t \to 2} t^2 \right)\mathbf{i} + \left(\lim_{t \to 2} t^3 \right)\mathbf{j} + \left(\lim_{t \to 2} t^4 \right)\mathbf{k} = 4\mathbf{i} + 8\mathbf{j} + 16\mathbf{k}.$ ●

Now it is possible to define the derivative of a vector function. The definition will be modeled after that of the derivative of a scalar function,

$$f'(x) = \lim_{\Delta x \to 0} \frac{f(x + \Delta x) - f(x)}{\Delta x}.$$

After the definition is given, its geometric meaning will be examined.

DEFINITION $\quad$ *Derivative of a vector function.* Let **G** be a vector function. The limit

$$\lim_{\Delta t \to 0} \frac{\mathbf{G}(t + \Delta t) - \mathbf{G}(t)}{\Delta t} \qquad (1)$$

(if it exists) is called the derivative of **G** at *t*. It is denoted **G**′(*t*).

What does the quotient in (1) mean geometrically? The numerator, which may be called

$$\Delta \mathbf{G},$$

is a vector. The denominator Δt is a scalar. The quotient

$$\frac{\Delta \mathbf{G}}{\Delta t}$$

is a vector.

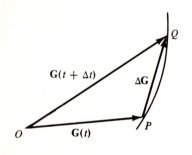

If the derivative exists, then when Δt is small, $\Delta \mathbf{G}$ is a short vector. It is the vector from the head of $\mathbf{G}(t)$ to the head of $\mathbf{G}(t + \Delta t)$. That is, if $\mathbf{G}(t) = \overrightarrow{OP}$ and $\mathbf{G}(t + \Delta t) = \overrightarrow{OQ}$, then $\Delta \mathbf{G} = \overrightarrow{PQ}$. The vector $\Delta \mathbf{G}$ lies on a chord of the path; when Δt is small, the direction of $\Delta \mathbf{G}$ is close to that of the tangent line to the curve at *P*. Since Δt is a scalar (for simplicity, consider it to be positive), the vector

$$\frac{\Delta \mathbf{G}}{\Delta t}$$

is parallel to the vector $\Delta \mathbf{G}$ and points in the same direction as $\Delta \mathbf{G}$. Therefore the vector

$$\frac{\Delta \mathbf{G}}{\Delta t},$$

when Δt is small and positive, would presumably point almost along a tangent line at *P*.

Thus $\mathbf{G}'(t)$ is presumably parallel to the tangent line at *P*. The direction of $\mathbf{G}'(t)$ is the direction in which the particle is moving as it passes through *P*. These observations suggest the following definition.

DEFINITION $\quad$ *Tangent vector to a curve.* Let $\mathbf{R} = \mathbf{G}(t)$ describe a curve in the plane or in space. The vector $\mathbf{G}'(t)$ is called a *tangent vector* to the curve.

For any parameterization of the curve, $\mathbf{G}'(t)$ is a tangent vector. Its length may depend on the particular parameterization.

It is easy to compute $\mathbf{G}'(t)$ if the scalar components of $\mathbf{G}(t)$ are known. For if $\mathbf{G}(t) = g(t)\mathbf{i} + h(t)\mathbf{j}$, then

$$\frac{\Delta \mathbf{G}}{\Delta t} = \frac{\Delta g \mathbf{i} + \Delta h \mathbf{j}}{\Delta t}$$

$$= \frac{\Delta g}{\Delta t}\mathbf{i} + \frac{\Delta h}{\Delta t}\mathbf{j}.$$

By Theorem 1, $\qquad \mathbf{G}'(t) = \lim_{\Delta t \to 0} \frac{\Delta \mathbf{G}}{\Delta t} = \lim_{\Delta t \to 0} \frac{\Delta g}{\Delta t}\mathbf{i} + \lim_{\Delta t \to 0} \frac{\Delta h}{\Delta t}\mathbf{j}$

$$= g'(t)\mathbf{i} + h'(t)\mathbf{j}.$$

In short, to differentiate a vector function just differentiate each of its scalar component functions. The following theorem states this formally in the case in which $\mathbf{G}(t)$ is a plane vector. The principle applies just as well to the more general case when $\mathbf{G}(t)$ is a space vector.

THEOREM 2 Let $\mathbf{G}(t) = g(t)\mathbf{i} + h(t)\mathbf{j}$. If both g and h have a derivative at $t = a$, then $\mathbf{G}$ has a derivative at $t = a$, and

$$\mathbf{G}'(a) = g'(a)\mathbf{i} + h'(a)\mathbf{j}.$$

The proof, which is left to the reader, follows almost immediately from Theorem 1.

Consider the case $\mathbf{G}(t) = g(t)\mathbf{i} + h(t)\mathbf{j}$, where $\mathbf{G}'(t) = g'(t)\mathbf{i} + h'(t)\mathbf{j}$ is a tangent vector to a plane curve. Hence the slope of the curve at the point in question is $h'(t)/g'(t)$. This formula agrees with the formula in Sec. 10.8 for finding the slope of a curve given parametrically:

$$\frac{dy}{dx} = \frac{dy/dt}{dx/dt}.$$

The vector approach to the definition of a tangent line is more general, for it also applies to curves in space. The next example illustrates this for a particular curve.

EXAMPLE 4 At time t a particle has the position vector

$$R = \mathbf{G}(t) = 3 \cos 2\pi t \mathbf{i} + 3 \sin 2\pi t \mathbf{j} + 5t \mathbf{k}.$$

Describe its path and sketch a typical tangent vector $\mathbf{G}'(t)$.

SOLUTION At time t the particle is at the point

$$\begin{cases} x = 3 \cos 2\pi t \\ y = 3 \sin 2\pi t \\ z = 5t. \end{cases}$$

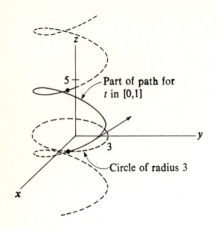

5 — Part of path for
t in [0,1]

3

Circle of radius 3

Notice that $x^2 + y^2 = (3 \cos 2\pi t)^2 + (3 \sin 2\pi t)^2 = 9$. Thus the point is always on the cylinder

$$x^2 + y^2 = 9 \qquad z \text{ is unrestricted.}$$

Moreover, as t increases, $z = 5t$ increases.

The path is thus the spiral spring sketched in the margin. When t increases by 1, the angle $2\pi t$ increases by 2π, and the particle goes once around the spiral. This type of corkscrew path is called a *helix*.

At time t

$$\mathbf{G}'(t) = -6\pi \sin 2\pi t \mathbf{i} + 6\pi \cos 2\pi t \mathbf{j} + 5\mathbf{k}. \quad \bullet$$

The length of the vector $\mathbf{G}'(t) = g'(t)\mathbf{i} + h'(t)\mathbf{j}$ is

$$\sqrt{[g'(t)]^2 + [h'(t)]^2}$$

or, equivalently,

$$\sqrt{\left(\frac{dx}{dt}\right)^2 + \left(\frac{dy}{dt}\right)^2}.$$

This is the formula obtained in Sec. 10.9 for the speed of a particle moving on a plane curve. Now, a similar assertion holds for curves in space. It can be shown that the arc length s swept out on a space curve during the time interval $[a, b]$ is

$$\int_a^b \sqrt{\left(\frac{dx}{dt}\right)^2 + \left(\frac{dy}{dt}\right)^2 + \left(\frac{dz}{dt}\right)^2} \, dt.$$

Consequently, the speed of the particle at time t is

$$\sqrt{\left(\frac{dx}{dt}\right)^2 + \left(\frac{dy}{dt}\right)^2 + \left(\frac{dz}{dt}\right)^2},$$

which a physicist prefers to write in the dot notation of Newton as

$$\sqrt{(\dot{x})^2 + (\dot{y})^2 + (\dot{z})^2}.$$

If $\mathbf{G}(t) = x(t)\mathbf{i} + y(t)\mathbf{j} + z(t)\mathbf{k}$, then $\mathbf{G}'(t) = \sqrt{(\dot{x})^2 + (\dot{y})^2 + (\dot{z})^2} = \text{speed}$. Thus the magnitude of the vector $\mathbf{G}'(t)$ may be interpreted as the *speed* of the particle:

$$|\mathbf{G}'(t)| = \text{speed of the particle at time } t.$$

For this reason it is customary to call $\mathbf{G}'(t)$ the *velocity vector*. It points in the direction the particle is moving at a given instant, and its magnitude is the speed of the particle. Frequently $\mathbf{G}'(t)$ will be denoted $\mathbf{V}$.

EXAMPLE 5   Find the speed at time t of the particle described in Example 4.

SOLUTION

$$\begin{aligned} \text{Speed} = |\mathbf{G}'(t)| &= \sqrt{(-6\pi \sin 2\pi t)^2 + (6\pi \cos 2\pi t)^2 + 5^2} \\ &= \sqrt{36(\sin^2 2\pi t + \cos^2 2\pi t) + 25} \\ &= \sqrt{61}. \end{aligned}$$

The particle travels at a constant speed along its helical path.   In t units of time it travels the distance $\sqrt{61}\,t$.

Note that the velocity vector is not constant; its direction is always changing. However, its length in this example remains constant, for the speed is constant.   ●

EXAMPLE 6   Sketch the path of a particle whose position vector at time $t \geq 0$ is

$$\mathbf{G}(t) = \cos t^2\mathbf{i} + \sin t^2\mathbf{j}.$$

Find its speed at time t.

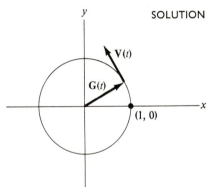

SOLUTION   Note that $$|\mathbf{G}(t)| = \sqrt{\cos^2 t^2 + \sin^2 t^2} = 1.$$

So the path of the particle is the circle of radius 1 and center $(0, 0)$.   The speed of the particle is

$$|\mathbf{V}(t)| = |\mathbf{G}'(t)| = |-2t \sin t^2\mathbf{i} + 2t \cos t^2\mathbf{j}|$$
$$= \sqrt{(-2t \sin t^2)^2 + (2t \cos t^2)^2}$$
$$= 2t\sqrt{\sin^2 t^2 + \cos^2 t^2}$$
$$= 2t.$$

The particle travels faster and faster around a circle of radius 1.   ●

Exercises

1. At time t the position vector of a thrown ball is $\mathbf{R}(t) = 32t\mathbf{i} - 16t^2\mathbf{j}$.
   (a) Draw $\mathbf{R}(0)$, $\mathbf{R}(1)$, and $\mathbf{R}(2)$.
   (b) Sketch the path.
   (c) Compute and draw $\mathbf{V}(0)$, $\mathbf{V}(1)$, and $\mathbf{V}(2)$.   In each case place the tail of the vector at the head of the corresponding position vector.
2. At time $t \geq 0$ a particle is at the point $x = 2t$, $y = 4t^2$.
   (a) What is the position vector $\mathbf{R}(t)$ at time t?
   (b) Sketch the path.
   (c) How fast is the particle moving when $t = 1$?
   (d) Draw $\mathbf{V}(1)$ with its tail at the head of $\mathbf{R}(1)$.
3. Let $\mathbf{G}(t) = t^2\mathbf{i} + t^3\mathbf{j}$.
   (a) Sketch the vector $\Delta\mathbf{G} = \mathbf{G}(1.1) - \mathbf{G}(1)$.
   (b) Sketch the vector $\Delta\mathbf{G}/\Delta t$, where $\Delta\mathbf{G}$ is given in (a), and $\Delta t = 0.1$.
   (c) Sketch $\mathbf{G}'(1)$.
   (d) Find $|\Delta\mathbf{G}/\Delta t - \mathbf{G}'(1)|$, where $\Delta\mathbf{G}/\Delta t$ is given in (b).
4. At time t the position vector of a particle is

   $$\mathbf{G}(t) = 2 \cos 4\pi t\mathbf{i} + 2 \sin 4\pi t\mathbf{j} + t\mathbf{k}.$$

   (a) Sketch its path.

   (b) Find its speed.
   (c) Find a unit tangent vector to the path at time t.
5. At time t the position vector of a particle is

   $$\mathbf{G}(t) = t \cos 2\pi t\mathbf{i} + t \sin 2\pi t\mathbf{j} + t\mathbf{k}.$$

   (a) Show that the particle lies on the surface of a cone whose equation in cylindrical coordinates is $z = r$.
   (b) Sketch the path of the particle.
6. A particle at time $t = 0$ is at the point (x_0, y_0, z_0).   It moves on a line through that point in the direction of the unit vector $\mathbf{u} = \cos \alpha\mathbf{i} + \cos \beta\mathbf{j} + \cos \gamma\mathbf{k}$.   It travels at the constant speed of 3 feet per second.
   (a) Give a formula for its position vector $\mathbf{R} = \mathbf{G}(t)$.
   (b) Find its velocity vector $\mathbf{V} = \mathbf{G}'(t)$.
7. A particle moves in a circular orbit of radius a.   At time t its position vector is

   $$\mathbf{R}(t) = a \cos 2\pi t\mathbf{i} + a \sin 2\pi t\mathbf{j}.$$

   (a) Draw its position vector when $t = 0$ and when $t = \frac{1}{4}$.
   (b) Draw its velocity vector when $t = 0$ and when $t = \frac{1}{4}$.
   (c) Using the dot product, prove that its velocity vector is always perpendicular to its position vector.

8. A combination of rockets, ropes, and air resistance influences a particle in such a way that its coordinates are $x = t$ and $y = t^{-1}$ at time $t \geq 1$.
   (a) Draw the path of the particle.
   (b) Draw $\mathbf{R}(1)$, $\mathbf{R}(2)$, and $\mathbf{R}(3)$.
   (c) Draw $\mathbf{V}(1)$, $\mathbf{V}(2)$, $\mathbf{V}(3)$.
   (d) As time goes on, what happens to dx/dt, dy/dt, $|\mathbf{V}|$, and $\mathbf{V}$?

9. An electron travels clockwise in a circle of radius 100 feet 200 times a second.   At time $t = 0$ it is at $(100, 0)$.
   (a) Compute $\mathbf{R}(t)$ and $\mathbf{V}(t)$.
   (b) Draw $\mathbf{R}(0)$, $\mathbf{R}(\frac{1}{800})$, $\mathbf{V}(0)$, $\mathbf{V}(\frac{1}{800})$.
   (c) How do $|\mathbf{R}(t)|$ and $|\mathbf{V}(t)|$ behave as time goes on?

10. At time t a particle is at $(3t^2, 6t^2)$.
   (a) Show that the particle moves on the line $y = 2x$.
   (b) Draw $\mathbf{R}(0)$, $\mathbf{R}(1)$, $\mathbf{R}(2)$ and $\mathbf{V}(0)$, $\mathbf{V}(1)$, $\mathbf{V}(2)$.

11. At time t a particle is at $(\cos t^2, \sin t^2)$.
   (a) Show that it moves on the circle $x^2 + y^2 = 1$.
   (b) Compute $\mathbf{R}(t)$ and $\mathbf{V}(t)$.
   (c) How does $|\mathbf{V}(t)|$ behave for large t?   What does this say about the particle?

12. At time t a particle is at $(4t, 16t^2)$.
   (a) Show that the particle moves on the curve $y = x^2$.
   (b) Draw $\mathbf{R}(t)$ and $\mathbf{V}(t)$ for $t = 0$, $\frac{1}{4}$, $\frac{1}{2}$.
   (c) What happens to $|\mathbf{V}(t)|$ and the direction of $\mathbf{V}(t)$ for large t?

■

13. A ball is thrown up at an initial speed of 200 feet per second and at an angle of $60°$ from the horizontal.   If we disregard air resistance, then at time t it is at $(100t, 100\sqrt{3}\, t - 16t^2)$, as long as it is in flight.   Compute and draw $\mathbf{R}(t)$ and $\mathbf{V}(t)$ (a) when $t = 0$; (b) when the ball reaches its maximum height; (c) when the ball strikes the ground.

14. Instead of time t, use arc length s along the path as a parameter, $\mathbf{R} = \mathbf{G}(s)$.
   (a) [Recall from Sec. 10.9 that for any parameterization of a curve $(ds/dt)^2 = (dx/dt)^2 + (dy/dt)^2$.]   Show that $d\mathbf{R}/ds$ is a unit vector.
   (b) Draw $\Delta\mathbf{R}$ and Δs.   Why is it reasonable that $|\Delta\mathbf{R}/\Delta s|$ is near 1 when Δs is small?

■ ■

15. A rock is thrown up at an angle θ from the horizontal and at a speed v_0.   Show that

$$\mathbf{R}(t) = \overrightarrow{((v_0 \cos \theta)t, (v_0 \sin \theta)t - 16t^2)}.$$

[At time $t = 0$, the rock is at $(0, 0)$; the x axis is horizontal.]

16. (See Exercise 15.)   The moment a ball is dropped straight down, you shoot an arrow directly at it.   Assume that there is no air resistance.   Show that if you shoot the arrow fast enough it will hit the ball.
   (a) Solve with the aid of the formulas in Exercise 18.
   (b) Solve with a maximum of intuition and a minimum of formula.
   (c) Must the acceleration be constant for the reasoning in (b) to be valid?   Explain.

17. (a) Show that the horizontal distance traveled by the rock in Exercise 15 is the same whether the angle is θ or its complement $(\pi/2) - \theta$.
   (b) What value of θ yields the maximum range?

18. Prove that, if the velocity vector $\mathbf{V}(t)$ remains perpendicular to the position vector $\mathbf{R}(t)$, then the path lies on a sphere with center at the origin.   *Hint:* Show that the moving particle remains at a constant distance from the origin.

19. (a) Show that the paths $\mathbf{G}(t) = t\mathbf{i} + t^2\mathbf{j} + t^3\mathbf{k}$ and $\mathbf{H}(t) = t^2\mathbf{i} + t^3\mathbf{j} + t^4\mathbf{k}$ intersect when $t = 1$.
   (b) At what angle do they intersect?

20.2

Properties of the derivative of a vector function

The theorems to follow concern some useful properties of the derivative of a vector function.

We now have two types of functions, vector functions and numerical functions.   A vector function $\mathbf{G}$ assigns to a number t a vector $\mathbf{G}(t)$.   A numerical function f assigns to a number t a number $f(t)$; for emphasis we shall call such a function a *scalar function*.   The function that assigns to the number t the vector $f(t)\mathbf{G}(t)$ is a vector function.   Thus the product of a scalar function and a vector function is a vector function. It is denoted $f\mathbf{G}$. The notation $\mathbf{G}f$, where the scalar appears second, is seldom used. The first theorem concerns the derivative of $f\mathbf{G}$. Note its similarity to the theorem about the derivative of the product of two scalar functions in Sec. 5.5.

THEOREM I   If f is a scalar function and if $\mathbf{G}$ is a vector function, the derivative of the vector function $f\mathbf{G}$ is

$$f\mathbf{G}' + f'\mathbf{G}.$$

PROOF   The proof is almost the same as that which appears in Sec. 5.5.   First work with $\Delta(f\mathbf{G})$, as follows:

$$\begin{aligned}
\Delta(f\mathbf{G}) &= f(t + \Delta t)\mathbf{G}(t + \Delta t) - f(t)\mathbf{G}(t) \\
&= [f(t) + \Delta f][\mathbf{G}(t) + \Delta \mathbf{G}] - f(t)\mathbf{G}(t) \\
&= f(t)\mathbf{G}(t) + f(t)\,\Delta\mathbf{G} + \Delta f\,\mathbf{G}(t) + \Delta f\,\Delta\mathbf{G} - f(t)\mathbf{G}(t) \\
&= f(t)\,\Delta\mathbf{G} + \Delta f\,\mathbf{G}(t) + \Delta f\,\Delta\mathbf{G}.
\end{aligned}$$

Thus

$$\lim_{\Delta t \to 0} \frac{\Delta(f\mathbf{G})}{\Delta t} = \lim_{\Delta t \to 0}\left[f(t)\frac{\Delta\mathbf{G}}{\Delta t} + \frac{\Delta f}{\Delta t}\mathbf{G}(t) + \Delta f\,\frac{\Delta\mathbf{G}}{\Delta t}\right]$$

$$= f(t)\mathbf{G}'(t) + f'(t)\mathbf{G}(t) + 0\mathbf{G}'(t).$$

This proves the theorem. ●

EXAMPLE I   Find the derivative of the vector function

$$t^2(\cos t\mathbf{i} + \sin t\mathbf{j}).$$

SOLUTION   By Theorem 1, the derivative is

$$\begin{aligned}
t^2(\cos t\mathbf{i} &+ \sin t\mathbf{j})' + (t^2)'(\cos t\mathbf{i} + \sin t\mathbf{j}) \\
&= t^2(-\sin t\mathbf{i} + \cos t\mathbf{j}) + 2t(\cos t\mathbf{i} + \sin t\mathbf{j}) \\
&= (-t^2 \sin t + 2t \cos t)\mathbf{i} + (t^2 \cos t + 2t \sin t)\mathbf{j}.
\end{aligned}$$

Actually, Theorem 1 is not needed for this particular example. We could have written the function as

$$t^2 \cos t\mathbf{i} + t^2 \sin t\mathbf{j}$$

and differentiated each scalar component as in the preceding section. ●

From two vector functions $\mathbf{G}$ and $\mathbf{H}$ we can obtain a scalar function $\mathbf{G} \cdot \mathbf{H}$ by defining the value of $\mathbf{G} \cdot \mathbf{H}$ at t to be the dot product

$$\mathbf{G}(t) \cdot \mathbf{H}(t).$$

THEOREM 2   If $\mathbf{G}$ and $\mathbf{H}$ are vector functions, then

$$(\mathbf{G} \cdot \mathbf{H})' = \mathbf{G} \cdot \mathbf{H}' + \mathbf{H} \cdot \mathbf{G}'.$$

The proof of Theorem 2 is outlined in Exercise 3.

EXAMPLE 2   Use Theorem 2 to show that if the magnitude of $\mathbf{G}(t)$ is independent of t, then $\mathbf{G}'(t)$ is perpendicular to $\mathbf{G}(t)$.

SOLUTION   Theorem 2, with $\mathbf{H}(t) = \mathbf{G}(t)$, says that

$$(\mathbf{G} \cdot \mathbf{G})' = \mathbf{G} \cdot \mathbf{G}' + \mathbf{G} \cdot \mathbf{G}' \tag{1}$$
$$= 2\mathbf{G} \cdot \mathbf{G}'.$$

Now $\mathbf{G}(t) \cdot \mathbf{G}(t) = |\mathbf{G}(t)|^2$. Call the length $|\mathbf{G}(t)|$ simply r. Then (1) asserts that

$$\frac{d(r^2)}{dt} = 2\mathbf{G} \cdot \mathbf{G}';$$

hence

$$2r\frac{dr}{dt} = 2\mathbf{G} \cdot \mathbf{G}'.$$

Thus

$$r\frac{dr}{dt} = \mathbf{G} \cdot \mathbf{G}'. \tag{2}$$

If r is constant, (2) implies that $\mathbf{G}(t) \cdot \mathbf{G}'(t)$ is 0. Thus, if neither $\mathbf{G}(t)$ nor $\mathbf{G}'(t)$ is $\mathbf{0}$, they are perpendicular nonzero vectors. ●

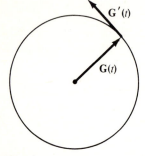

If $|\mathbf{G}(t)|$ is constant, then $\mathbf{G}'(t)$ is perpendicular to $\mathbf{G}(t)$.

The result obtained in Example 2 also follows from high school geometry. If $|\mathbf{G}(t)|$ is constant, the particle moves on a sphere (or circle). $\mathbf{G}'(t)$ is tangent to the sphere and therefore perpendicular to $\mathbf{G}(t)$, since any ray drawn from the center of a sphere is perpendicular to the tangent at the point where the ray meets the sphere. This result deserves emphasis, since it will be needed several times.

Let $\mathbf{G}$ be a vector function of the scalar s, $\mathbf{R} = \mathbf{G}(s)$, and let s in turn be a function of the scalar t, $s = f(t)$. Then we may consider $\mathbf{R}$ to be a function of t

$$\mathbf{R} = \mathbf{G}(f(t)).$$

(Think of s as arc length along a curve and t as time.) Denote this (composite) function $\mathbf{G} \circ f$.

THEOREM 3   *Chain rule.* If $\mathbf{G}$ is a vector function and f is a scalar function, then $\mathbf{G} \circ f$ is a vector function, and its derivative at t is

$$\mathbf{G}'(f(t))f'(t).$$

In other words, if $\mathbf{R} = \mathbf{G}(s)$ and $s = f(t)$,

then

$$\frac{d\mathbf{R}}{dt} = \frac{d\mathbf{R}}{ds}\frac{ds}{dt}.$$

The proof of Theorem 3 is outlined in Exercise 12.

EXAMPLE 3   Let $\mathbf{R} = s^2\mathbf{i} + s^3\mathbf{j}$ where $s = e^{2t}$. Compute $d\mathbf{R}/dt$.

SOLUTION

$$\frac{d\mathbf{R}}{ds} = 2s\mathbf{i} + 3s^2\mathbf{j} \quad \text{and} \quad \frac{ds}{dt} = 2e^{2t}.$$

Thus

$$\frac{d\mathbf{R}}{dt} = (2s\mathbf{i} + 3s^2\mathbf{j})2e^{2t},$$

which is usually written with the scalar coefficient in front:

$$\frac{d\mathbf{R}}{dt} = 2e^{2t}(2s\mathbf{i} + 3s^2\mathbf{j}) = 2e^{2t}(2e^{2t}\mathbf{i} + 3e^{4t}\mathbf{j}). \; \bullet$$

Exercises

1. Let $\mathbf{G}(t) = \cos 2t\mathbf{i} + \sin 2t\mathbf{j}$.
   (a) Show that $\mathbf{G}(t)$ is a unit vector.
   (b) Is $\mathbf{G}'(t)$ a unit vector?
2. Theorem 1 can be proved with the aid of scalar components. Write $\mathbf{G}(t) = g(t)\mathbf{i} + h(t)\mathbf{j}$. Then express $f(t)\mathbf{G}(t)$ in components and differentiate.
3. This outlines a proof of Theorem 2.
   (a) Express $\Delta(\mathbf{G} \cdot \mathbf{H})$ as simply as possible with the aid of the equations $\mathbf{G}(t + \Delta t) = \mathbf{G}(t) + \Delta\mathbf{G}$ and $\mathbf{H}(t + \Delta t) = \mathbf{H}(t) + \Delta\mathbf{H}$.
   (b) Using (a), find $\lim\limits_{\Delta t \to 0} \dfrac{\Delta(\mathbf{G} \cdot \mathbf{H})}{\Delta t}$.
4. A particle at time $t \geq 0$ is at the point (t, t^2, t).
   (a) Plot its path.
   (b) Find its speed at time t.
   (c) Show that its path lies above a parabola in the xy plane.
   (d) Show that its path lies in a plane.
5. A particle moves in a path such that $\mathbf{R}(t) = e^t \cos t\mathbf{i} + e^t \sin t\mathbf{j}$. Show that the angle between its position vector and velocity vector has the constant value $\pi/4$.
6. (a) If $\mathbf{G}$ and $\mathbf{H}$ are vector functions, one obtains a new vector function $\mathbf{G} \times \mathbf{H}$ by defining $(\mathbf{G} \times \mathbf{H})(t)$ to be $\mathbf{G}(t) \times \mathbf{H}(t)$. Prove that

   $$(\mathbf{G} \times \mathbf{H})' = \mathbf{G} \times \mathbf{H}' + \mathbf{G}' \times \mathbf{H}$$

   (b) Would it be correct to write, as in the case of a derivative of the product of scalar functions,

$(\mathbf{G} \times \mathbf{H})'$ equals $\mathbf{G} \times \mathbf{H}' + \mathbf{H} \times \mathbf{G}'$?

7. If the velocity vector $\mathbf{V}(t)$ is constant, show that the path lies on a straight line.

■

8. A particle moves on the path $\mathbf{R} = \mathbf{G}(t)$, which does not pass through the origin. Prove that, if P is a point on the path closest to the origin O, then the position vector $\overrightarrow{OP}$ is perpendicular to the velocity vector at P.
9. (a) If $\mathbf{G}(t)$ has a constant direction, does $\mathbf{G}'(t)$ necessarily also?
   (b) If $\mathbf{G}(t)$ has a constant magnitude, does $\mathbf{G}'(t)$ necessarily also?
10. Let $\mathbf{G}$ be a vector function such that $\mathbf{G}(t)$ is never $\mathbf{0}$. Show that $\mathbf{G}$ can be written in the form $f\mathbf{H}$, where $f(t) \geq 0$ and $|\mathbf{H}(t)| = 1$ for all t.
11. Let $\mathbf{G}$ be a vector function such that $\mathbf{G}(t)$ is never $\mathbf{0}$. Let $r(t) = |\mathbf{G}(t)|$.

   (a) Show that $\left|\dfrac{\mathbf{G}(t)}{r(t)}\right|$ is 1.

   (b) By Example 2, $\quad \left(\dfrac{\mathbf{G}}{r}\right)' \cdot \dfrac{\mathbf{G}}{r} = 0.$

   Deduce that $r\mathbf{G} \cdot \mathbf{G}' = r'\mathbf{G} \cdot \mathbf{G}$.
   (c) Verify (b) by expressing $\mathbf{G}$ and $\mathbf{G}'$ in components.
12. To prove Theorem 3 write $\mathbf{G}(s) = g(s)\mathbf{i} + h(s)\mathbf{j}$ and work with components. Carry out the details.

20.3

The acceleration vector

If $\mathbf{R} = \mathbf{G}(t)$ is the position vector at time t, then $\mathbf{V} = \mathbf{G}'(t)$ is the velocity vector at time t. The definition of the acceleration vector is motivated by the definition of acceleration in the case of a particle moving on a line.

DEFINITION *Acceleration vector.* The derivative of the velocity vector is called the *acceleration vector* and is denoted **A**:

$$\mathbf{A} = \frac{d\mathbf{V}}{dt}.$$

EXAMPLE I Let $\mathbf{G}(t) = 32t\mathbf{i} - 16t^2\mathbf{j}$ be the position of a thrown ball at time t. Compute $\mathbf{V}(t)$ and $\mathbf{A}(t)$:

SOLUTION
$$\mathbf{V}(t) = 32\mathbf{i} - 32t\mathbf{j}$$
and
$$\mathbf{A}(t) = -32\mathbf{j}.$$

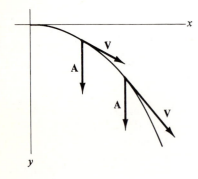

In this case the acceleration vector is constant in direction and length. It points directly downward, as does the vector that represents the force of gravity. The diagram that is located to the left shows **V** and **A** at two points on the path. ●

As Example 1 shows, it is a simple matter to compute the velocity vector and the acceleration vector **A** when the components of the position vector **R**(t) are given. If

$$\mathbf{R}(t) = x(t)\mathbf{i} + y(t)\mathbf{j} + z(t)\mathbf{k},$$

then
$$\mathbf{V}(t) = \frac{dx}{dt}\mathbf{i} + \frac{dy}{dt}\mathbf{j} + \frac{dz}{dt}\mathbf{k},$$

and
$$\mathbf{A}(t) = \frac{d^2x}{dt^2}\mathbf{i} + \frac{d^2y}{dt^2}\mathbf{j} + \frac{d^2z}{dt^2}\mathbf{k}.$$

The dot notation for derivatives ($\dot{x} = dx/dt$, $\ddot{x} = d^2x/dt^2$) provides these simpler formulas: If

$$\mathbf{R}(t) = x(t)\mathbf{i} + y(t)\mathbf{j} + z(t)\mathbf{k},$$

then
$$\mathbf{V}(t) = \dot{x}\mathbf{i} + \dot{y}\mathbf{j} + \dot{z}\mathbf{k},$$

and
$$\mathbf{A}(t) = \ddot{x}\mathbf{i} + \ddot{y}\mathbf{j} + \ddot{z}\mathbf{k}.$$

If no forces act on a moving particle, **V** is constant in direction and magnitude; hence $\mathbf{A} = \mathbf{0}$. That is, if the vector **F**, representing the forces, is **0**, then $\mathbf{A} = \mathbf{0}$. Newton's second law asserts universally that **F**, **A**, and the mass m of the particle are related by the vector equation:

$$\mathbf{F} = m\mathbf{A}.$$

This little equation says several things: (1) The direction of the acceleration vector **A** is the same as the direction of **F**. (2) A force **F** applied to a heavy mass produces a shorter acceleration vector **A** than the same force applied to a light mass. (3) For a given mass, the magnitude of **A** is proportional to the magnitude of **F**.

We may always think of the acceleration vector as representing the effect of a force on the particle, since the acceleration vector **A** and the force vector **F** point in the same direction.   If the mass of the particle is 1, then **F** = **A**.

Consider now a particle moving in a circular orbit at constant speed v. It may be, perhaps, a heavy mass at the end of a rope or a satellite in a circular orbit around the earth.   The following theorem describes the acceleration vector associated with this motion.

THEOREM   If a particle moves in a circular path of radius r at a constant speed v, its acceleration vector is directed toward the center of the circle and has magnitude

$$\frac{v^2}{r}.$$

PROOF   Introduce an xy-coordinate system such that $(0, 0)$ is at the center of the orbit and the particle is at $(r, 0)$ at time 0.   Let $\mathbf{R} = \mathbf{R}(t)$ be the position vector of the particle at time t.   Assume that the particle travels counterclockwise; then θ is positive, as shown in the diagram at the left.

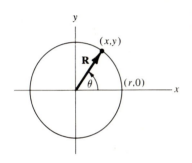

Since the particle moves at the constant speed v, it sweeps out an arc length vt up to time t.   By definition of radian measure,

$$\theta = \frac{vt}{r},$$

where θ is the angle of **R** at time t.   Thus

$$\mathbf{R} = x\mathbf{i} + y\mathbf{j}$$

$$= r\cos\theta\,\mathbf{i} + r\sin\theta\,\mathbf{j}$$

$$= r\cos\frac{vt}{r}\,\mathbf{i} + r\sin\frac{vt}{r}\,\mathbf{j}.$$

Now **V** and **A** can be computed explicitly.   Remembering that r is constant, we have then

$$\mathbf{V} = \frac{d\mathbf{R}}{dt}$$

$$= \frac{-rv}{r}\sin\frac{vt}{r}\,\mathbf{i} + \frac{rv}{r}\cos\frac{vt}{r}\,\mathbf{j}$$

$$= -v\sin\frac{vt}{r}\,\mathbf{i} + v\cos\frac{vt}{r}\,\mathbf{j}.$$

Hence

$$\mathbf{A} = \frac{d\mathbf{V}}{dt}$$

$$= \frac{-v^2}{r}\cos\frac{vt}{r}\,\mathbf{i} - \frac{v^2}{r}\sin\frac{vt}{r}\,\mathbf{j}$$

$$= \frac{v^2}{r}\left(-\cos\theta\,\mathbf{i} - \sin\theta\,\mathbf{j}\right).$$

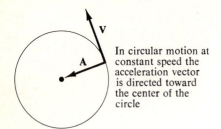

In circular motion at constant speed the acceleration vector is directed toward the center of the circle

From this last equation we can read off the direction and length of **A**. First of all,

$$-\cos\theta\mathbf{i} - \sin\theta\mathbf{j}$$

is a unit vector pointing in the direction opposite that of

$$\mathbf{R} = r\cos\theta\mathbf{i} + r\sin\theta\mathbf{j}.$$

Hence **A** points toward the center of the circle. Since **A** is v^2/r times a unit vector, its magnitude is v^2/r. This concludes the proof. ●

This theorem was discovered in 1657 by Huygens while developing a theory of clock mechanisms. Anyone who has spun a pail of water at the end of a rope should find the theorem plausible. First of all, to hold the pail in its orbit one must pull on the rope. Thus the force of the rope on the pail is directed toward the center of its circular orbit. Second, the faster one spins the pail (keeping the rope at fixed length), the harder one must pull on the rope. Hence the appearance of v^2 in the numerator is reasonable. Third, if the same speed is maintained but the radius of the circle is decreased, more force is required.

With the aid of the theorem it is a simple matter to determine the speed that a satellite requires in order to achieve a circular orbit around the earth. Disregard the resistance of the air and assume that the launch is made *horizontally* from a high tower. The satellite would then be swung around the earth like a pail at the end of a rope. Instead of the tension on a rope, the force of gravity pulls the satellite in toward the earth from a linear path. Now, if a particle moves in a circle of radius 4,000 miles with velocity v miles per second, it has an acceleration toward the center of the earth of $v^2/4{,}000$ miles per second per second. This acceleration must coincide with the acceleration of 32 feet per second per second $\doteq 0.006$ mile per second per second which gravity imparts to any object at the surface of the earth. Thus $0.006 \doteq v^2/4{,}000$, and

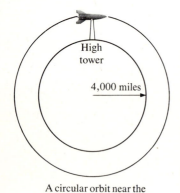

High tower

4,000 miles

A circular orbit near the surface of the earth

$$\text{Orbital velocity} = v = \sqrt{(4{,}000)(0.006)}$$

$$= \sqrt{24}$$

$$\doteq 4.90 \text{ miles per second.}$$

The velocity necessary to maintain an object in orbit at the surface of the earth is 4.90 miles per second, which is less than the escape velocity, 6.92 miles per second, that was computed in Sec. 6.6.

Exercises

1. Let $\mathbf{R}(t) = t^2\mathbf{i} + t^3\mathbf{j}$. Compute and sketch $\mathbf{R}(1)$, $\mathbf{V}(1)$, and $\mathbf{A}(1)$.

2. Let

$$\mathbf{R}(t) = \cos\left(\frac{\pi e^t}{2}\right)\mathbf{i} + \sin\left(\frac{\pi e^t}{2}\right)\mathbf{j}.$$

Compute and sketch $\mathbf{R}(0)$, $\mathbf{V}(0)$, and $\mathbf{A}(0)$.

3. Let $\mathbf{R}(t) = 4t\mathbf{i} - 16t^2\mathbf{j}$ describe a falling ball. Compute and draw $\mathbf{R}$, $\mathbf{V}$, and $\mathbf{A}$ for $t = 0$, $t = 1$, and $t = 2$.

4. Let $\mathbf{R}(t) = 10\cos 2\pi t\mathbf{i} + 10\sin 2\pi t\mathbf{j}$ denote the position vector of a particle at time t.
(a) What is the shape of the path this particle follows?
(b) Compute $\mathbf{V}$ and $\mathbf{A}$.

(c) Draw **R**, **V**, and **A** for $t = \frac{1}{4}$.

(d) Show that **A** is always in the direction opposite that of **R**.

5. Let $\mathbf{R}(t) = t\mathbf{i} + (1/t)\mathbf{j}$.

(a) Sketch the path of the particle for $t > 0$.

(b) Compute and sketch $\mathbf{R}(1)$, $\mathbf{V}(1)$, $\mathbf{A}(1)$.

(c) What happens to $\mathbf{V}(t)$ and $\mathbf{A}(t)$ when t is large?

(d) What happens to the speed when t is large?

6. At time t a particle has the position vector $\mathbf{R}(t) = (t + \cos t)\mathbf{i} + (t - \sin t)\mathbf{j}$.

(a) Show that **A** has constant magnitude.

(b) Sketch the path corresponding to t in $[0, 4\pi]$.

(c) Sketch **R**, **V**, and **A** for $t = 0$, $\pi/2$, π, $3\pi/2$, and 2π.

7. If $\mathbf{R}(t)$ is a unit vector for all t, is $\mathbf{R}'(t)$ necessarily also a unit vector? Explain your answer.

8. (a) Prove that if the horizontal component of **V** is constant, then **A** is vertical.

(b) What does (a) say physically, in view of Newton's law?

9. (a) If **V** has constant direction, does **A** have constant direction? Explain.

(b) If **V** has constant magnitude, does **A** have constant magnitude? Explain.

10. The momentum of a particle of mass m and velocity vector **V** is the vector $m\mathbf{V}$. Newton stated his second law in the form $\mathbf{F} = (m\mathbf{V})'$.

(a) Prove that $\mathbf{F} = m\mathbf{A} + m'\mathbf{V}$.

(b) Deduce that if m is constant, then $\mathbf{F} = m\mathbf{A}$. In relativity theory m is not necessarily constant.

11. Prove that if **A** is always perpendicular to **V** on a certain path, the speed of the particle is constant.

12. (a) Prove that if the speed of a particle is constant, **A** is perpendicular to **V**.

(b) Does this make sense physically?

13. Let the position vector **R** be a function of arc length s on a curve. Prove that $d\mathbf{R}/ds$ is a *unit* vector tangent to the curve.

■

14. (a) Prove that if a particle moves in a circular orbit [centered at $(0, 0)$], then $\mathbf{R} \cdot \mathbf{A} + \mathbf{V} \cdot \mathbf{V} = 0$.

(b) From (a) deduce that $\mathbf{A} \cdot \mathbf{R}$ is ≤ 0.

(c) What does (b) say about the direction of the force vector **F**?

15. A particle moves in the circular orbit $\mathbf{R} = \cos t^2 \mathbf{i} + \sin t^2 \mathbf{j}$.

(a) Compute **V** and **A**.

(b) Verify that $\mathbf{R} \cdot \mathbf{A} \leq 0$.

16. The acceleration that gravity imparts to an object decreases with the square of the distance of the object from the center of the earth.

(a) Show that if an object is r miles from the center of the earth, it has an acceleration of $(0.006)(4,000/r)^2$ miles per second per second.

(b) With the aid of (a), find the velocity of a satellite in orbit at an altitude of 1,000 miles.

17. How long would it take a satellite to orbit the earth just above the earth's surface? One hundred miles above the earth's surface? One thousand miles above the surface?

18. What altitude must an orbiting satellite have in order to stay directly above a fixed spot on the equator?

19. A certain satellite in circular orbit goes around the earth once every 92 minutes. How high is it above the earth?

20. A boy is spinning a pail of water at the end of a rope.

(a) If he doubles the speed of the pail, how many times as hard must he pull on the rope?

(b) If instead he doubles the length of the rope, but keeps the speed of the pail the same, will he have to pull more or less? How much?

21. Give an example of a position vector function $\mathbf{R} = \mathbf{G}(t)$ such that the speed v is constant, yet **A** is not **0**.

22. (a) If a curve in the plane is given parametrically, x and y being functions of t, how are $\dot{x}$, $\dot{y}$, and dy/dx related? Express d^2y/dx^2 in terms of $\dot{x}$, $\dot{y}$, $\ddot{x}$, and $\ddot{y}$.

(b) If at a certain instant $\mathbf{V} = 2\mathbf{i} + 3\mathbf{j}$ and $\mathbf{A} = \mathbf{i} + 4\mathbf{j}$, find dy/dx, d^2y/dx^2, and the radius of curvature at that instant.

■ ■

23. Show that the orbital velocity at the surface of the earth is $1/\sqrt{2}$ times the escape velocity.

24. A particle in space moves under the influence of a force that is always directed toward a fixed point. (For instance, the particle may be moving under the influence of the gravitational field of the sun.) Prove that its path lies in a plane:

(a) Let **R**, **V**, and $\mathbf{V}' = \mathbf{A}$ be the position, velocity, and acceleration vectors. Show that $\mathbf{A}(t) = f(t)\mathbf{R}(t)$, where f is a scalar function.

(b) Assume that if **G** and **H** are vector functions, then

$$(\mathbf{G} \times \mathbf{H})' = \mathbf{G}' \times \mathbf{H} + \mathbf{G} \times \mathbf{H}'.$$

Show that

$$(\mathbf{R} \times \mathbf{V})' = \mathbf{0}.$$

(c) From (b) it follows that $\mathbf{R} \times \mathbf{V}$ is a constant vector **C**. Show that, if **C** is not **0**, then the particle travels in a plane to which **C** is perpendicular.

(d) If **C** in part (c) is **0**, what is the path of the particle?

20.4

The unit vectors T and N

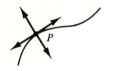

There are two unit vectors tangent to the curve at P and two unit vectors perpendicular to the curve at P

In the study of motion along a curve in the plane, two unit vectors defined at a point P on the curve are of special importance. One is tangent to the curve at P. The other is perpendicular to the curve at P. There are two possible choices for each, as the diagram at the left shows.

This section is devoted to making a precise mathematical choice of one of the two unit tangent vectors and of one of the two unit vectors perpendicular to the curve at P. Their physical interpretation will also be given in this section; in the next section they will be used to examine the acceleration vector.

Consider a particle whose position vector at time t is $\mathbf{R}(t)$. Then $\mathbf{R}'(t)$ is its velocity vector $\mathbf{V}(t)$. If $\mathbf{V}(t)$ is not the zero vector,

$$\frac{\mathbf{V}(t)}{|\mathbf{V}(t)|}$$

is a unit vector that points in the same direction as $\mathbf{V}(t)$. Denote $|\mathbf{V}(t)|$, the speed of the particle at time t, by v. The vector

$$\frac{\mathbf{V}(t)}{v}$$

is thus a unit vector tangent to the curve and pointing in the direction in which the particle is moving.

DEFINITION *Unit tangent vector* $\mathbf{T}$. The vector

$$\frac{\mathbf{V}(t)}{v}$$

is denoted $\mathbf{T}$ or $\mathbf{T}(t)$ and is called the *unit tangent vector*. [It is defined whenever $\mathbf{V}(t)$ is not $\mathbf{0}$.]

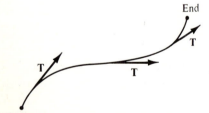

We have thus picked out one of the two tangent vectors, and the choice records the direction of motion. As the particle traverses the path, $\mathbf{T}$ remains of unit length, but changes in direction.

EXAMPLE I Let $\mathbf{R}(t) = 3 \cos 2t\mathbf{i} + 3 \sin 2t\mathbf{j}$. Find $\mathbf{T}(t)$.

SOLUTION By differentiation, $\mathbf{V}(t) = -6 \sin 2t\mathbf{i} + 6 \cos 2t\mathbf{j}$. Thus

$$\begin{aligned} v &= |\mathbf{V}(t)| \\ &= \sqrt{(-6 \sin 2t)^2 + (6 \cos 2t)^2} \\ &= 6. \end{aligned}$$

Hence

$$\mathbf{T}(t) = \frac{\mathbf{V}(t)}{6} = -\sin 2t\mathbf{i} + \cos 2t\mathbf{j}.$$

The curve being swept out is a circle of radius 3.   The direction of $\mathbf{T}(t)$ shows that the particle moves counterclockwise. ●

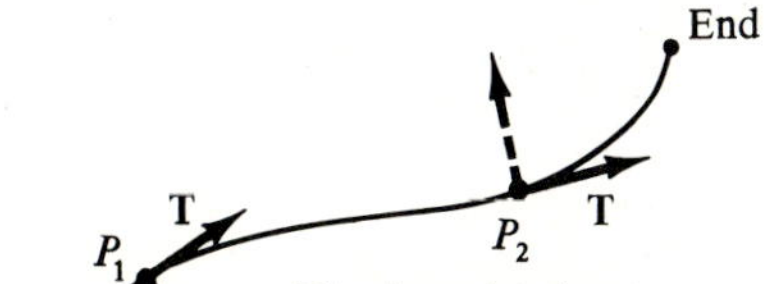

A unit vector perpendicular to $\mathbf{T}$ will also be needed.   There are two choices.   The one we shall take indicates the direction in which the particle is veering—to the right or to the left.   For instance, at P_1 in the accompanying diagram the particle is veering to the right, as if making a right turn; at P_2 it is veering to the left.

This choice of a vector perpendicular to $\mathbf{T}$ is intuitive and physical.   The computations in the next section require that a more formal mathematical definition for this vector be given.

End

T

P_1 **T**   **T** P_2

The dotted unit vectors indicate the direction of turning

Start

DEFINITION   *Principal unit normal vector* **N**.   If $\mathbf{T}'(t)$ is not $\mathbf{0}$, the vector

$$\frac{\mathbf{T}'(t)}{|\mathbf{T}'(t)|}$$

is called the *principal unit normal* vector and is denoted **N**.

The word "normal" frequently is used instead of "perpendicular" in the study of vectors.

Let us show that

$$\mathbf{N} = \frac{\mathbf{T}'(t)}{|\mathbf{T}'(t)|}$$

is a unit vector, is perpendicular to $\mathbf{T}(t)$, and points in the direction in which the particle is turning.

First of all, for any nonzero vector **B**,

$$\frac{\mathbf{B}}{|\mathbf{B}|}$$

is a unit vector.   Thus **N** is a unit vector.

Second, since $\mathbf{T}(t)$ is of constant length one, then by Example 2 of Sec. 20.2, $\mathbf{T}'(t)$ is perpendicular to $\mathbf{T}(t)$.   Since **N** is just a scalar multiple of $\mathbf{T}'(t)$, **N** is perpendicular to $\mathbf{T}(t)$.

Third, we show that $\mathbf{T}'$ (hence **N**) points in the direction in which **T** is turning as t increases.   To check that this is so, go back to the definition of $\mathbf{T}'(t)$:

$$\mathbf{T}'(t) = \lim_{\Delta t \to 0} \frac{\mathbf{T}(t + \Delta t) - \mathbf{T}(t)}{\Delta t}$$

$$= \lim_{\Delta t \to 0} \frac{\Delta \mathbf{T}}{\Delta t}.$$

When Δt is small, the direction of

$$\frac{\Delta \mathbf{T}}{\Delta t}$$

is approximately that of $\mathbf{T}'(t)$.   Moreover, when Δt is positive,

$$\frac{\Delta \mathbf{T}}{\Delta t} \quad \text{and} \quad \Delta \mathbf{T}$$

have the same direction.   Let us sketch $\Delta \mathbf{T}$ for $\Delta t > 0$.   When both $\mathbf{T}(t + \Delta t)$ and $\mathbf{T}(t)$ are placed with their tails at the origin, $\Delta \mathbf{T}$ is the vector from the head of $\mathbf{T}(t)$ to the head of $\mathbf{T}(t + \Delta t)$.   As the accompanying diagram shows,

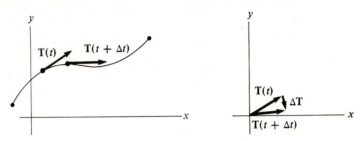

$\Delta \mathbf{T}$, hence $\mathbf{T}'(t)$, points in the direction in which $\mathbf{T}$ is turning (to the right in this case).

EXAMPLE 2   Compute $\mathbf{N}(t)$ for the curve described in Example 1, namely for

$$\mathbf{R}(t) = 3 \cos 2t \mathbf{i} + 3 \sin 2t \mathbf{j}.$$

SOLUTION   As was shown in Example 1,

$$\mathbf{T}(t) = -\sin 2t \mathbf{i} + \cos 2t \mathbf{j}.$$

Thus
$$\mathbf{T}'(t) = -2 \cos 2t \mathbf{i} - 2 \sin 2t \mathbf{j},$$

and
$$|\mathbf{T}'(t)| = \sqrt{(-2 \cos 2t)^2 + (-2 \sin 2t)^2}$$
$$= 2.$$

Consequently,
$$\mathbf{N}(t) = \frac{\mathbf{T}'(t)}{|\mathbf{T}'(t)|}$$
$$= \frac{-2 \cos 2t \mathbf{i} - 2 \sin 2t \mathbf{j}}{2}$$
$$= -\cos 2t \mathbf{i} - \sin 2t \mathbf{j}.$$

Observe that in this example
$$\mathbf{N}(t) = -\tfrac{1}{3}\mathbf{R}(t).$$

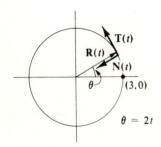

Thus $\mathbf{N}(t)$, pointing toward the center of the circular path, is, as was expected, perpendicular to the curve.   It also indicates the direction in which $\mathbf{T}$ is turning: It is always veering toward the center of the circle. ●

The symbol t denotes the parameter, which we have generally interpreted as time.   But arc length s along the path can also be chosen as a parameter. Let s denote the arc length as measured from some base point B on the curve. If P is a point on the curve, the position vector $\overrightarrow{OP}$ depends on this arc

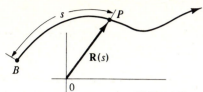

length s and may be denoted $\overrightarrow{OP} = \mathbf{R}(s)$.

With this special choice of parameterization, the formula for a unit tangent vector becomes very simple.

THEOREM 1   Let s denote arc length along a curve and let $\mathbf{R}(s)$ be the position vector corresponding to the arc length s.   Then

$$\frac{d\mathbf{R}}{ds} = \mathbf{T},$$

a unit tangent vector.

PROOF   By the definition of **T**, $\mathbf{T} = \dfrac{d\mathbf{R}/ds}{|d\mathbf{R}/ds|}$.

It suffices to show that

$$\left|\frac{d\mathbf{R}}{ds}\right| = 1.$$

Since $\mathbf{R} = x(s)\mathbf{i} + y(s)\mathbf{j}$,

$$\frac{d\mathbf{R}}{ds} = \frac{dx}{ds}\mathbf{i} + \frac{dy}{ds}\mathbf{j}.$$

Hence

$$\left|\frac{d\mathbf{R}}{ds}\right| = \sqrt{\left(\frac{dx}{ds}\right)^2 + \left(\frac{dy}{ds}\right)^2}$$

$$= \sqrt{\left(\frac{ds}{ds}\right)^2} \qquad \text{by Sec. 10.9}$$

$$= 1. \;\bullet$$

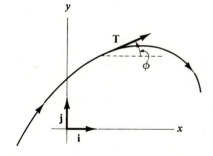

There is another natural geometric parameter associated with a plane curve, namely the angle that **T** makes with the vector **i**.   This is the angle ϕ in the diagram at the left (and used in Sec. 11.3 for the study of curvature).   The next theorem is concerned with $d\mathbf{T}/d\phi$.

THEOREM 2   Let $\mathbf{R}(t)$ be the position vector on a curve at time t.   Let ϕ denote the angle that

$$\mathbf{T}(t) = \frac{\mathbf{V}(t)}{|\mathbf{V}(t)|}$$

makes with the vector **i**.   Then

$$\left|\frac{d\mathbf{T}}{d\phi}\right| = 1.$$

PROOF   Since **T** is a unit vector and makes an angle ϕ with **i**, the vector components of **T** along **i** and **j** are $\cos \phi\,\mathbf{i}$ and $\sin \phi\,\mathbf{j}$.

Thus $\mathbf{T} = \cos \phi\,\mathbf{i} + \sin \phi\,\mathbf{j}.$

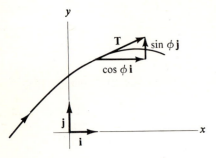

Hence

$$\frac{d\mathbf{T}}{d\phi} = -\sin \phi \mathbf{i} + \cos \phi \mathbf{j}$$

and

$$\left| \frac{d\mathbf{T}}{d\phi} \right| = \sqrt{(-\sin \phi)^2 + (\cos \phi)^2}$$

$$= \sqrt{1}$$

$$= 1. \ \bullet$$

The conclusion of Theorem 2 is similar to that of Theorem 1. In fact, Theorem 2 is a consequence of Theorem 1, as discussed in Exercise 14.

Exercises

1. True or false: At a given point P on a curve, $\mathbf{T}$ depends on the direction in which the curve is being swept out.
2. True or false: At a given point P on a curve, $\mathbf{N}$ depends on the direction in which the curve is being swept out.
3. Let $\mathbf{R}(t) = t\mathbf{i} + t^2\mathbf{j}$ describe the path of a particle.
   (a) Graph the path.
   (b) Compute $\mathbf{R}(1)$, $\mathbf{V}(1)$, $\mathbf{T}(1)$, and $\mathbf{N}(1)$.
   (c) Draw the vectors in (b), placing the tails of the last three at the head of $\mathbf{R}(1)$.
4. Let $\mathbf{R}(t) = \cos 2\pi t\mathbf{i} - \sin 2\pi t\mathbf{j}$.
   (a) Sketch the path corresponding to $0 \le t \le 1$.
   (b) Compute $\mathbf{R}(0)$, $\mathbf{V}(0)$, $\mathbf{T}(0)$, and $\mathbf{N}(0)$.
   (c) Draw the four vectors in (b).
   (d) Check that $\mathbf{T}(0)$ points in the direction of motion when the particle is at $\mathbf{R}(0)$.
5. Let $\mathbf{R}(t) = t^2\mathbf{i} + t^3\mathbf{j}$, for $0 \le t \le 2$.
   (a) Sketch the path.
   (b) Compute $\mathbf{R}(1)$, $\mathbf{V}(1)$, $\mathbf{A}(1)$ (the acceleration vector), $\mathbf{T}(1)$, and $\mathbf{N}(1)$.
   (c) Draw the five vectors in (b).
   (d) What is the speed of the particle when $t = 1$?
   (e) What is the scalar component of $\mathbf{A}(1)$ along $\mathbf{T}(1)$?
   (f) What is the scalar component of $\mathbf{A}(1)$ along $\mathbf{N}(1)$?
6. (Distance is measured in feet and time in seconds.) A particle has the position vector $\mathbf{R}(t)$ at time t. Its speed when $t = 1$ is 3 feet per second. Find $\mathbf{V}(1) \cdot \mathbf{T}(1)$.
7. If $\mathbf{V}$ is the velocity vector and $\mathbf{N}$ is the principal normal vector, what can be said about the magnitude of $\mathbf{V} \cdot \mathbf{N}$? $\mathbf{V} \cdot \mathbf{T}$?

Exercises 8 to 11 are related.

8. Let $\mathbf{A}$ be any vector and let $\mathbf{u}_1$ and $\mathbf{u}_2$ be two unit vectors that are perpendicular to each other.
   (a) Explain why there are scalars x and y such that

$$\mathbf{A} = x\mathbf{u}_1 + y\mathbf{u}_2.$$

   (b) Show that if $\mathbf{A} = x\mathbf{u}_1 + y\mathbf{u}_2$, then

$$\mathbf{A} \cdot \mathbf{u}_1 = x \quad \text{and} \quad \mathbf{A} \cdot \mathbf{u}_2 = y.$$

   (c) Show that for any vector $\mathbf{A}$,

$$\mathbf{A} = (\mathbf{A} \cdot \mathbf{u}_1)\mathbf{u}_1 + (\mathbf{A} \cdot \mathbf{u}_2)\mathbf{u}_2.$$

9. Let $\mathbf{u}_1$ and $\mathbf{u}_2$ be perpendicular unit vectors. What is the scalar component of $3\mathbf{u}_1 + 4\mathbf{u}_2$ along $\mathbf{u}_1$?
10. Let $\mathbf{R}(t) = 32t\mathbf{i} - 16t^2\mathbf{j}$ be the position vector of a thrown ball.
   (a) Sketch the path.
   (b) Compute and sketch $\mathbf{T}(1)$ and $\mathbf{N}(1)$.
   (c) Find the scalar components of $\mathbf{V}(1)$ along $\mathbf{T}(1)$ and along $\mathbf{N}(1)$.
   (d) Find the scalar components of $\mathbf{A}(1)$ along $\mathbf{T}(1)$ and $\mathbf{N}(1)$.
11. (This continues Exercise 10.)
   (a) Express $\mathbf{V}(1)$ in the form $x\mathbf{i} + y\mathbf{j}$ for suitable scalars x and y.
   (b) Express $\mathbf{V}(1)$ in the form $x\mathbf{T}(1) + y\mathbf{N}(1)$.
   (c) Express $\mathbf{A}(1)$ in the form $x\mathbf{T}(1) + y\mathbf{N}(1)$. [See Exercise 8(b).]
12. (a) Does $\mathbf{V}$ depend on the speed of the particle?
   (b) Does $\mathbf{T}$ depend on the speed of the particle?
13. The interesting part of Theorem 1 is that $d\mathbf{R}/ds$ has length 1.
   (a) Make a sketch showing $\mathbf{R}(s + \Delta s)$, $\mathbf{R}(s)$, $\Delta \mathbf{R}$, and Δs.
   (b) Why is it plausible that

$$\lim_{\Delta s \to 0} \left| \frac{\Delta \mathbf{R}}{\Delta s} \right| = 1?$$

■

14. A plane curve is given by the function $\mathbf{R}(t)$. Then $\mathbf{T}(t)$ is defined as $\mathbf{V}(t)/|\mathbf{V}(t)|$. Now parameterize the curve by the angle ϕ that was used in Theorem 2. Define $\mathbf{T}(\phi) = \mathbf{V}(\phi)/|\mathbf{V}(\phi)|$, where $\mathbf{V}(\phi)$ is now the derivative of the position vector with respect to ϕ.
  (a) Show by a sketch that $\mathbf{T}(t)$ need not equal $\mathbf{T}(\phi)$. In fact, sketch an example in which $\mathbf{T}(\phi) = -\mathbf{T}(t)$.
  (b) As ϕ varies, plot $\mathbf{T}(\phi)$ with its tail at the origin. In other words, consider $\mathbf{T}(\phi)$ the position vector of a new curve. What is that curve?
  (c) Why may ϕ be interpreted as arc length on the new curve?
  (d) Deduce from Theorem 1 that

$$\left|\frac{d\mathbf{T}(\phi)}{d\phi}\right| = 1,$$

  as Theorem 2 asserts.

15. Let $\mathbf{R}(t) = \cos t^2 \mathbf{i} + \sin t^2 \mathbf{j}$. Show that $\mathbf{T}$ is not defined when $t = 0$.

16. Let $\mathbf{R}(t) = t\mathbf{i} + t^3\mathbf{j}$.
  (a) Compute $\mathbf{T}(t)$.
  (b) Show that $\mathbf{N}$ is not defined when $t = 0$.
  (c) Sketch the path. What property of the path causes $\mathbf{N}$ not to be defined when $t = 0$?

17. Show that there is a scalar c, not necessarily constant, such that

$$\mathbf{N}' = c\mathbf{T}.$$

18. Prove that

$$\mathbf{T}' = \frac{v\mathbf{A} - (dv/dt)\mathbf{V}}{v^2}.$$

19. Prove that if a particle travels at a constant speed, then

$$\mathbf{T}' = \frac{\mathbf{A}}{v}.$$

20.5

The scalar components of the acceleration vector along T and N

Let $\mathbf{R}(t) = x(t)\mathbf{i} + y(t)\mathbf{j}$ be the position vector at time t. The acceleration vector is then

$$\mathbf{A}(t) = \frac{d^2x}{dt^2}\mathbf{i} + \frac{d^2y}{dt^2}\mathbf{j}.$$

The scalar components of $\mathbf{A}(t)$ along $\mathbf{i}$ and $\mathbf{j}$ are simply

$$\frac{d^2x}{dt^2} \quad \text{and} \quad \frac{d^2y}{dt^2},$$

the second derivatives of each of the two scalar components.

In the study of a particle moving along a plane curve it is of use to find the scalar components of $\mathbf{A}(t)$ along $\mathbf{T}$ and $\mathbf{N}$. These scalars will be denoted A_T and A_N, respectively. In other words,

$$\mathbf{A}(t) = A_T\mathbf{T} + A_N\mathbf{N}.$$

A_T is called the *tangential component* and A_N the *normal component* of the acceleration vector.

EXAMPLE 1 Find A_T and A_N if a particle moves at a constant speed v in a circular orbit of radius r.

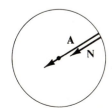

SOLUTION As the theorem in Sec. 20.3 shows, $\mathbf{A}(t)$ is directed toward the center of the circle and has length

$$\frac{v^2}{r}.$$

$|\mathbf{N}| = 1$

$|\mathbf{A}| = \dfrac{v^2}{r}$

The principal normal vector **N** points in the same direction.   Thus

$$\mathbf{A} = \frac{v^2}{r}\,\mathbf{N}.$$

The tangential component of **A** is 0, since **A** is perpendicular to **T**.   For emphasis we may write

$$\mathbf{A} = 0\mathbf{T} + \frac{v^2}{r}\,\mathbf{N}.$$

Thus $A_T = 0$ and $A_N = \dfrac{v^2}{r}.$ ●

Example 1 determines A_T and A_N for circular motion at constant speed. The goal of this section is to obtain a formula for A_T and A_N for *motion on any curve* and with *speed not necessarily constant*.   The key will be the following little lemma.   Before reading it, review the definition of radius of curvature from Sec. 11.3.

LEMMA   Let s represent arc length along the path of a moving particle.   Let $v = ds/dt$ be the speed (which is always assumed to be positive).   Then

$$\frac{d\mathbf{T}}{dt} = \frac{v}{r}\,\mathbf{N},$$

where r is the radius of curvature.

PROOF   By the definition of **N** given in Sec. 20.4,

$$\frac{d\mathbf{T}}{dt} = \left|\frac{d\mathbf{T}}{dt}\right|\mathbf{N}. \tag{1}$$

To compute
$$\left|\frac{d\mathbf{T}}{dt}\right|,$$

use the chain rule from Sec. 20.2,

$$\frac{d\mathbf{T}}{dt} = \frac{d\mathbf{T}}{d\phi}\frac{d\phi}{dt}, \tag{2}$$

where ϕ is the angle between the tangent line and the positive x axis. Theorem 2 of Sec. 20.2 shows that

$$\left|\frac{d\mathbf{T}}{d\phi}\right| = 1.$$

Thus (2) implies that

$$\left|\frac{d\mathbf{T}}{dt}\right| = \left|\frac{d\phi}{dt}\right|. \tag{3}$$

By the chain rule for scalar functions,

$$\frac{d\phi}{dt} = \frac{d\phi}{ds}\frac{ds}{dt}.$$

Hence
$$\left|\frac{d\phi}{dt}\right| = \left|\frac{d\phi}{ds}\right|\left|\frac{ds}{dt}\right|. \tag{4}$$

By the definition of radius of curvature r in Sec. 11.3,

$$\left|\frac{d\phi}{ds}\right| = \frac{1}{r}. \tag{5}$$

Since
$$\frac{ds}{dt} = v,$$

it follows from Eqs. (3) to (5) that

$$\left|\frac{d\mathbf{T}}{dt}\right| = \left|\frac{d\phi}{dt}\right|$$

$$= \left|\frac{d\phi}{ds}\right|\left|\frac{ds}{dt}\right|$$

$$= \frac{1}{r}v$$

$$= \frac{v}{r}.$$

Thus (1) becomes

$$\frac{d\mathbf{T}}{dt} = \frac{v}{r}\mathbf{N},$$

and the lemma is proved. ●

The lemma is the basis of the proof of the following theorem, which provides formulas for A_T and A_N for a parameterized plane curve. For simplicity the theorem is stated in terms of a moving particle.

THEOREM   If a particle moves in a plane curve, and arc length s is measured in such a way that $v = ds/dt$ is positive, then

$$\mathbf{A} = \frac{d^2s}{dt^2}\mathbf{T} + \frac{v^2}{r}\mathbf{N},$$

where r is the radius of curvature. In other words,

$$A_T = \frac{d^2s}{dt^2} \quad \text{and} \quad A_N = \frac{v^2}{r}.$$

PROOF   By definition of **T**,

$$\frac{\mathbf{V}}{v} = \mathbf{T};$$

hence $$V = v\mathbf{T}.$$

By Theorem 1 of Sec. 20.2,

$$\mathbf{V}' = v\mathbf{T}' + v'\mathbf{T}. \tag{6}$$

The lemma of this section asserts that

$$\mathbf{T}' = \frac{v}{r}\mathbf{N}. \tag{7}$$

Consequently, $$\mathbf{A} = \mathbf{V}' = v\left(\frac{v}{r}\mathbf{N}\right) + v'\mathbf{T} \tag{8}$$

$$= \frac{v^2}{r}\mathbf{N} + v'\mathbf{T}$$

$$= \frac{d^2s}{dt^2}\mathbf{T} + \frac{v^2}{r}\mathbf{N}.$$

This proves the theorem. ●

Observe that A_T is determined by the "scalar" acceleration, the rate at which the speed is changing.   On the other hand, A_N depends only on the speed and the radius of curvature; it is the normal component of the acceleration of a particle moving in a circle of radius r with constant speed v.   Note that the scalar component of $\mathbf{A}$ along $\mathbf{N}$, being v^2/r, is positive.   (See Example 1.)

The fact that A_N is positive is reasonable, since $\mathbf{A}$ points in the same direction as the external force $\mathbf{F}$; $\mathbf{F}$ causes the particle to veer, and $\mathbf{N}$ records the direction in which the particle veers. A_T can be positive or negative, depending on whether the particle is speeding up or slowing down.

The craft is veering to the right and is accelerating

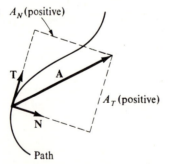

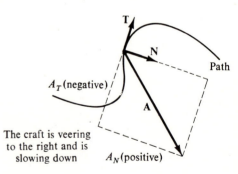

The craft is veering to the right and is slowing down

EXAMPLE 2   At time t a particle is at $(t^2, \frac{1}{2}t^4)$ on the curve $2y = x^2$.   Determine $\mathbf{R}$, $\mathbf{V}$, $\mathbf{A}$, A_N, and A_T when $t = 1$.

SOLUTION   First, since $\mathbf{R} = t^2\mathbf{i} + \frac{1}{2}t^4\mathbf{j}$, we have $\mathbf{V} = 2t\mathbf{i} + 2t^3\mathbf{j}$.   Thus,

$$\mathbf{A} = 2\mathbf{i} + 6t^2\mathbf{j}.$$

From this it follows that $|\mathbf{A}| = \sqrt{4 + 36t^4}$.

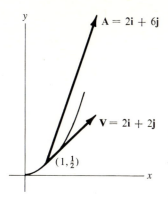

Let us sketch **A**, **V**, and **R** at $t = 1$. At that time,

$$\mathbf{A} = 2\mathbf{i} + 6\mathbf{j}, \qquad \mathbf{V} = 2\mathbf{i} + 2\mathbf{j}, \qquad \text{and} \qquad \mathbf{R} = \mathbf{i} + \tfrac{1}{2}\mathbf{j}.$$

A_T is simply dv/dt, where $v = |\mathbf{V}| = \sqrt{4t^2 + 4t^6}$. Thus

$$A_T = \frac{4t + 12t^5}{\sqrt{4t^2 + 4t^6}}.$$

Rather than compute A_N by the formula $A_N = v^2/r$, which involves computation of the radius of curvature, use the pythagorean relation

$$A_N{}^2 + A_T{}^2 = |\mathbf{A}|^2.$$

At $t = 1$, for instance, this becomes

$$A_N{}^2 + \left(\frac{16}{\sqrt{8}}\right)^2 = (\sqrt{2^2 + 6^2})^2 = 40.$$

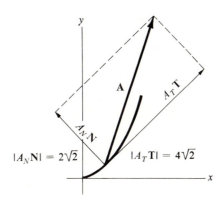

Hence $A_N{}^2 = 40 - 32 = 8$, and $A_N = 2\sqrt{2}$. The diagram above shows **A**, $A_T\mathbf{T}$, and $A_N\mathbf{N}$ at $t = 1$. ●

Exercises

In Exercises 1 to 3 find A_N and A_T, given that at a certain value of t:

1. $\dfrac{dv}{dt} = 3,\ v = 4,\ r = 5.$

2. $\dfrac{d^2s}{dt^2} = -2,\ v = 3,\ \text{curvature} = -\tfrac{1}{3}.$

3. $\dfrac{d^2s}{dt^2} = 0,\ v = 1,\ r = 3.$

In Exercises 4 to 6 note that the path is a circle. Find A_T and A_N in each case when $t = 1$, if

4. $\mathbf{R}(t) = 5 \cos 3\pi t \mathbf{i} + 5 \sin 3\pi t \mathbf{j}.$

5. $\mathbf{R}(t) = 5 \cos \pi t^2 \mathbf{i} + 5 \sin \pi t^2 \mathbf{j}.$

6. $\mathbf{R}(t) = \cos t \mathbf{i} + \sin t \mathbf{j}.$

7. Show that
(a) $\mathbf{A} \cdot \mathbf{T} = d^2s/dt^2$;
(b) $\mathbf{A} \cdot \mathbf{N} = v^2/r.$

8. At a certain moment $\mathbf{R} = \mathbf{i} + \mathbf{j}$, $\mathbf{V} = 3\mathbf{i} + 4\mathbf{j}$, $\mathbf{A} = 3\mathbf{i} - 3\mathbf{j}$.
(a) Draw **R**, **V**, and **A**.
(b) Is the particle speeding up or slowing down? Explain.
(c) Estimate A_T and A_N graphically.
(d) Compute A_T and A_N.

9. Repeat Exercise 8 for $\mathbf{R} = 0$, $\mathbf{V} = -5\mathbf{i} + 12\mathbf{j}$, $\mathbf{A} = 3\mathbf{i} + \mathbf{j}$.

10. Let **A**, the acceleration vector, be expressed in the form

$$\mathbf{A} = A_x\mathbf{i} + A_y\mathbf{j},$$

and
$$A = A_T \mathbf{T} + A_N \mathbf{N}.$$

(A_x and A_y do *not* denote partial derivatives.)   Show that
$$A_x{}^2 + A_y{}^2 = A_T{}^2 + A_N{}^2.$$

11. At a certain moment $v = 3$, $dv/dt = -2$, and $\mathbf{A} = 2\mathbf{i} + 3\mathbf{j}$. Find r, the radius of curvature, using the equation in Exercise 10 that links A_x, A_y, A_T, and A_N.

12. What can be said about $\mathbf{A} \cdot \mathbf{V}$ if the particle is
    (a) slowing down?
    (b) speeding up?

13. Prove that if a particle travels with constant speed, then $\mathbf{A} \cdot \mathbf{V} = 0$, using
    (a) Example 2 of Sec. 20.2.
    (b) the theorem in this section.

14. Let $\mathbf{R}$ describe the journey in Example 2 of this section.
    (a) Compute and draw $\mathbf{A}$ for $t = 1/\sqrt{6}$.
    (b) From your drawing estimate A_T and A_N at $t = 1/\sqrt{6}$.
    (c) Compute A_T and A_N at $t = 1/\sqrt{6}$.
    (d) Compare (b) and (c) in decimal form.

15. At time t a particle is at $(\cos t + t \sin t, \sin t - t \cos t)$.
    (a) Show that $|\mathbf{V}| = t$ and $|\mathbf{A}| = \sqrt{1 + t^2}$.
    (b) Show that $A_T = 1$ and $A_N = t$.
    (c) Show that r, the radius of curvature, equals t.

16. Let $\mathbf{R}(t) = e^t \mathbf{i} + e^{2t} \mathbf{j}$.
    (a) Compute $\mathbf{V}(t)$ and $\mathbf{A}(t)$ and express them in terms of $\mathbf{i}$ and $\mathbf{j}$.
    (b) Compute $v(t)$ and dv/dt.
    (c) Find A_T.
    (d) Find A_N.

■

17. At a certain moment, a particle headed northeast is speeding up and veering to the left.   Taking north as the positive y axis and east as the positive x axis, draw (a) $\mathbf{T}$, (b) $\mathbf{N}$, (c) enough typical $\mathbf{A}$'s to indicate the directions possible for $\mathbf{A}$.

18. Repeat Exercise 17 for a particle headed north, slowing down, and veering to the left.

19. At a certain instant $\mathbf{V} = 2\mathbf{i} + 3\mathbf{j}$ and $\mathbf{A} = 3\mathbf{i} - 4\mathbf{j}$.
    (a) Draw $\mathbf{V}$, $\mathbf{A}$, $\mathbf{T}$, and $\mathbf{N}$.
    (b) Find v and d^2s/dt^2.
    (c) Find the radius of curvature.

20. At a certain instant $\mathbf{V} = 5\mathbf{i} + 12\mathbf{j}$ and $\mathbf{A} = \mathbf{i} + 2\mathbf{j}$.
    (a) Draw $\mathbf{V}$ and $\mathbf{A}$.
    (b) On the basis of the sketch for (a) estimate A_T and A_N.
    (c) Compute A_T and A_N.

■ ■

21. From the theorem in this section deduce this formula for obtaining r, the radius of curvature:
$$\frac{v^4}{r^2} = \left(\frac{d^2x}{dt^2}\right)^2 + \left(\frac{d^2y}{dt^2}\right)^2 - \left(\frac{d^2s}{dt^2}\right)^2.$$

22. According to Sec. 11.3, the radius of curvature of the curve $y = f(x)$ is given by the formula
$$r = \frac{[1 + (dy/dx)^2]^{3/2}}{|d^2y/dx^2|}$$

Obtain this formula from the result in Exercise 21.

20.6

Level curves and level surfaces

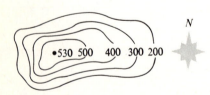

N

A U.S. Geodetic Survey map indicates height by contour lines.   For instance, one curve may show all the points at the 300-foot altitude.   This is the 300-foot contour.   The map to the left depicts a 530-foot hill by using four contour lines at intervals of 100 feet.   A hiker walking along that part of the hill which corresponds to a contour line neither rises nor descends.

Let $f(x, y)$ be the altitude of the hill corresponding to the point (x, y) on the map.   Then the 300-foot contour consists of those points where $f(x, y) = 300$. On the 300-foot contour the altitude function f is constant: Its value is 300.

The same idea carries over to other functions of x and y.   If f is a function of x and y, and if c is a fixed number, the set of points where
$$f(x, y) = c$$

is called a *level curve* for f.

EXAMPLE I   Discuss the level curves of the function $f(x, y) = x^2 + y^2$.

SOLUTION   The level curves of f are circles,

$$x^2 + y^2 = c.$$

The level curve

$$x^2 + y^2 = 0$$

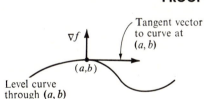

Three of the level curves of the function $z = x^2 + y^2$

$x^2 + y^2 = (7.2)^2$

$x^2 + y^2 = 36$

$x^2 + y^2 = 25$

$(0, 5)$

$(0, -6)$

7.2

At any point on this circle the function $z = x^2 + y^2$ has the value 36

consists of just the origin $(0, 0)$.   The level curve

$$x^2 + y^2 = -1$$

is empty, since $x^2 + y^2$ cannot be negative.   ●

Level curves go by many names, as this table shows:

Function f	Name of Level Curve
Altitude of land	Contour line
Air pressure	Isobar
Temperature	Isotherm
Utility (in economic theory)	Indifference curve
Gravitational potential	Equipotential curve

The next theorem presents an important relation between a level curve of $z = f(x, y)$ and the gradient vector of f.

THEOREM I   The gradient ∇f at (a, b) is perpendicular to the level curve of $z = f(x, y)$ passing through (a, b).

PROOF   The idea of the proof is to show that ∇f evaluated at (a, b) is perpendicular to a tangent vector to the level curve at (a, b).

Assume that the level curve that passes through the point (a, b) can be given parametrically as

$$\mathbf{R} = \mathbf{G}(t) = g(t)\mathbf{i} + h(t)\mathbf{j}.$$

∇f

Tangent vector to curve at (a, b)

(a,b)

Level curve through (a, b)

Then

$$\mathbf{V} = \mathbf{G}'(t) = g'(t)\mathbf{i} + h'(t)\mathbf{j}$$

is a tangent vector to the level curve at $(a, b) = \mathbf{G}'(t_0)$, say.

Let $\mathbf{G}'(t_0)$ be the tangent vector at (a, b) and let $\nabla f = f_x(a, b)\mathbf{i} + f_y(a, b)\mathbf{j}$ be the gradient at (a, b).   Then, ∇f is perpendicular to $\mathbf{G}'(t_0)$ if

$$\nabla f \cdot \mathbf{G}'(t_0) = 0.$$

In short, all we have to show is that

$$f_x(a, b)g'(t_0) + f_y(a, b)h'(t_0) = 0.$$

By Theorem 1 in Sec. 12.6 (a chain rule) the sum on the left side is just

$$\frac{dz}{dt},$$

evaluated at $t = t_0$.   But the function $z = f(x, y)$, considered on one of its level curves, is constant. Thus $$\frac{dz}{dt} = 0,$$

and consequently $$\mathbf{V}f \cdot \mathbf{G}'(t_0) = 0.$$

This shows that $\mathbf{V}f$ is perpendicular to $\mathbf{G}'(t_0)$ and establishes the theorem. ●

EXAMPLE 2   Give an example of a vector perpendicular to the curve $xy = 2$ at the point $(1, 2)$.

SOLUTION   Let $f(x, y) = xy$. Then the curve $xy = 2$ is a level curve of f, namely $f(x, y) = 2$. Thus the gradient of f, evaluated at $(1, 2)$, is a vector perpendicular to the curve $xy = 2$. Now,

$$\mathbf{V}f = y\mathbf{i} + x\mathbf{j}.$$

At $(1, 2)$ the gradient is $$\mathbf{V}f = 2\mathbf{i} + 1\mathbf{j}.$$

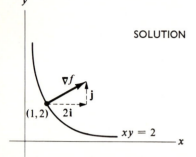

The accompanying diagram exhibits $\mathbf{V}f$ and the level curve. ●

Theorem 1 describes the second main use of the gradient: It provides a vector perpendicular to a level curve. The first use was given in Sec. 19.5: The gradient points in the direction in which f changes most rapdily. The two results together assert that f changes most rapidly in the direction perpendicular to a level curve.

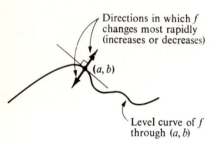

Directions in which f changes most rapidly (increases or decreases)

(a, b)

Level curve of f through (a, b)

If $f(x, y)$ represents the altitude of the land above the point (x, y), Theorem 1 tells a climber that the direction of steepest ascent (or descent) is perpendicular to the contour line at which he is located.

If $f(x, y)$ denotes air pressure at (x, y), the level curves of f are called *isobars* (and are shown on the daily weather map). Air tends to move from high pressure to low pressure, and $\mathbf{V}f$ points approximately in the direction of the wind vector. However, since influences other than just air pressure determine the wind, the wind vector is usually not perpendicular to the isobar.

EXAMPLE 3   Find a vector perpendicular to the curve

$$x^3y + 2x^2y + y^3 = 14$$

at the point $(1, 2)$.

SOLUTION   This curve is a level curve of the function f given by

$$f(x, y) = x^3y + 2x^2y + y^3.$$

The gradient of f is

$$\mathbf{V}f = (3x^2y + 4xy)\mathbf{i} + (x^3 + 2x^2 + 3y^2)\mathbf{j}.$$

At $(x, y) = (1, 2)$,

$$\mathbf{V}f = [3(1)^2(2) + 4(1)(2)]\mathbf{i} + [(1)^3 + 2(1)^2 + 3(2)^2]\mathbf{j}$$
$$= 14\mathbf{i} + 15\mathbf{j}.$$

Thus $14\mathbf{i} + 15\mathbf{j}$ is a vector perpendicular to the given curve at the point $(1, 2)$. ●

For later use we single out a result obtained in the course of proving Theorem 1.

THEOREM 2   Let f be a function of x and y: $z = f(x, y)$. Also let $x = g(t)$, $y = h(t)$ define parametrically the path of a particle. The position vector at time t is

$$\mathbf{G}(t) = g(t)\mathbf{i} + h(t)\mathbf{j}.$$

Then $z = f(g(t), h(t))$ is a composite function of t (and describes the behavior of z on the path of the particle), and its derivative is given by

$$\frac{dz}{dt} = \nabla f \cdot \mathbf{G}'(t).$$

For instance, let $z = f(x, y)$ be the temperature at the point (x, y) and let $\mathbf{G}$ describe the journey of a bug in the xy plane. Then the rate of change in the temperature as observed by the bug is the dot product of the temperature gradient ∇f and the velocity vector $\mathbf{V} = \mathbf{G}'$.

Theorem 2 generalizes from the plane to space, that is, to a function $u = f(x, y, z)$ and a curve in space given by a position function $\mathbf{G}(t) = x(t)\mathbf{i} + y(t)\mathbf{j} + z(t)\mathbf{k}$. The proof again is simply a translation of a chain rule into vector notation. In fact, the main ideas of this section all generalize from the plane to space. The concept of a level curve must be replaced by that of a level surface, which will now be defined.

Let f be a function defined in some region in space. Let

$$u = f(x, y, z).$$

If c is a fixed number, the set of points where

$$f(x, y, z) = c$$

is called a *level surface* for f.

EXAMPLE 4   Consider the function f defined by $f(x, y, z) = x^2 + y^2 + z^2$. Sketch the level surface of f that passes through the point $(2, 6, 3)$.

SOLUTION   Since $\qquad f(2, 6, 3) = 2^2 + 6^2 + 3^2 = 4 + 36 + 9 = 49,$

the level surface through $(2, 6, 3)$ has the equation

$$x^2 + y^2 + z^2 = 49.$$

This equation describes a sphere of radius 7 and center $(0, 0, 0)$. Compare this with Example 1. ●

The level surface of $x^2 + y^2 + z^2$ through $(2, 6, 3)$ is the sphere $x^2 + y^2 + z^2 = 49$

The next example suggests that Theorem 1 generalizes to functions of three variables, their gradients, and their level surfaces.

EXAMPLE 5 Let $f(x, y, z) = Ax + By + Cz$, where A, B, and C are constant, and at least one of them is not 0. Show that ∇f at (a, b, c) is perpendicular to the level surface of f through (a, b, c).

SOLUTION A level surface of f has an equation of the form

$$Ax + By + Cz = k,$$

and hence is a plane. The partial derivatives of f are

$$f_x = A, \qquad f_y = B, \qquad \text{and} \qquad f_z = C.$$

Thus $\nabla f = A\mathbf{i} + B\mathbf{j} + C\mathbf{k}$. As was shown in Sec. 19.4 the vector $A\mathbf{i} + B\mathbf{j} + C\mathbf{k}$ is perpendicular to the plane $Ax + By + Cz = k$. In this case ∇f is independent of the point where it is computed. ●

$\nabla f = A\mathbf{i} + B\mathbf{j} + C\mathbf{k}$ is perpendicular to the level surface of f

Plane $Ax + By + Cz = k$ is a level surface of $f(x, y, z) = Ax + By + Cz$

Before proceeding, it is necessary to state what is meant by a vector being perpendicular to a surface.

DEFINITION *Normal vector to a surface.* A vector is perpendicular to a surface at the point (a, b, c) on this surface if the vector is perpendicular to any curve on the surface through the point (a, b, c). [A vector is perpendicular to a curve at a point (a, b, c) on the curve if the vector is perpendicular to a tangent vector to the curve at (a, b, c).] Such a vector is called a *normal vector*.

Now Theorem 3, which generalizes Example 5, can be stated.

THEOREM 3 The gradient ∇f at (a, b, c) is a normal to the level surface of f passing through (a, b, c).

PROOF The proof is so close to that of Theorem 1 that we will only sketch it.

Let $\mathbf{G}(t) = g(t)\mathbf{i} + h(t)\mathbf{j} + i(t)\mathbf{k}$ be the parametrization of a curve in the level surface of f that passes through the point (a, b, c). Let $\mathbf{G}'(t_0)$ be the tangent vector to the curve at (a, b, c) and let $\nabla f = f_x(a, b, c)\mathbf{i} + f_y(a, b, c)\mathbf{j} + f_z(a, b, c)\mathbf{k}$ be the gradient at (a, b, c). We wish to show that

$$\nabla f \cdot \mathbf{G}'(t_0) = 0;$$

that is, $\qquad f_x(a, b, c)g'(t_0) + f_y(a, b, c)h'(t_0) + f_z(a, b, c)i'(t_0) = 0.$

If we let $u = f(x, y, z)$, then, as in Theorem 2, the sum on the left side is just

$$\frac{du}{dt} \qquad \text{at } t = t_0.$$

But the function $u = f(x, y, z)$, considered on one of its level surfaces, is constant. Thus

$$\frac{du}{dt} = 0.$$

This proves the theorem. ●

The sphere is the level surface of f through $(2, 6, 3)$

In the case of the function f of Example 4,

$$f(x, y, z) = x^2 + y^2 + z^2,$$

the gradient ∇f evaluated at $(2, 6, 3)$ is

$$4\mathbf{i} + 12\mathbf{j} + 6\mathbf{k}.$$

According to Theorem 3, this vector is perpendicular to the sphere of Example 4 at the point $(2, 6, 3)$.

This is reasonable, for notice that ∇f is twice the position vector from the origin to the point $(2, 6, 3)$:

$$\nabla f = 2(2\mathbf{i} + 6\mathbf{j} + 3\mathbf{k}).$$

Since any line through the center of a sphere is perpendicular to the sphere at the two points where it cuts the sphere, it follows that $2\mathbf{i} + 3\mathbf{j} + 4\mathbf{k}$, and hence ∇f, is perpendicular to the sphere.

Theorem 3 suggests the following definition.

DEFINITION   *Tangent plane to a surface.*   Consider a surface that is a level surface of a function $u = f(x, y, z)$.   Let (a, b, c) be a point on this surface where ∇f is not $\mathbf{0}$.   The tangent plane to the surface at the point (a, b, c) is that plane through (a, b, c) that is perpendicular to the vector ∇f evaluated at (a, b, c).

EXAMPLE 6   Find the tangent plane at $(2, 6, 3)$ to the sphere $x^2 + y^2 + z^2 = 49$.

SOLUTION   The sphere $x^2 + y^2 + z^2 = 49$ is a level surface of the function

$$f(x, y, z) = x^2 + y^2 + z^2.$$

The gradient of f is given by the formula

$$\nabla f = 2x\mathbf{i} + 2y\mathbf{j} + 2z\mathbf{k}.$$

At $(2, 6, 3)$, ∇f equals $\qquad 4\mathbf{i} + 12\mathbf{j} + 6\mathbf{k}.$

The equation of the tangent plane at $(2, 6, 3)$ is therefore

$$4(x - 2) + 12(y - 6) + 6(x - 3) = 0.$$

(See Sec. 19.4 for the method of finding the equation of a plane.) ●

The next theorem provides a way to find vectors that are perpendicular to a surface whose equation is given in the form $z = f(x, y)$.

THEOREM 4   Let f be a function of two variables, and let (a, b, c) be a point on the surface

$$z = f(x, y).$$

Then the vector $\qquad f_x\mathbf{i} + f_y\mathbf{j} - \mathbf{k},$

where f_x and f_y are evaluated at (a, b), is perpendicular to the surface at (a, b, c).

PROOF   The proof consists of introducing a function g of three variables such that the given surface is a level surface of g.   Application of Theorem 3 to this function g will prove the theorem.

Define $g(x, y, z)$ to be $f(x, y) - z$, and consider the particular level surface of g,

$$g(x, y, z) = 0.$$

This surface also has the description

$$f(x, y) - z = 0,$$

or, equivalently, $z = f(x, y).$

This shows that the given surface $z = f(x, y)$ is the same as a level surface of g. Now, Theorem 3 applied to g asserts that

$$g_x(a, b, c)\mathbf{i} + g_y(a, b, c)\mathbf{j} + g_z(a, b, c)\mathbf{k}$$

is a normal to the surface $g(x, y, z) = 0.$

But, since $g(x, y, z) = f(x, y) - z,$

it follows that

$$g_x(a, b, c) = f_x(a, b), \qquad g_y(a, b, c) = f_y(a, b) \qquad \text{and} \qquad g_z(a, b, c) = -1.$$

Hence $f_x(a, b)\mathbf{i} + f_y(a, b)\mathbf{j} - \mathbf{k}$

is perpendicular to the surface $z = f(x, y)$

at (a, b, c). ●

EXAMPLE 7   Find a vector that is perpendicular to the surface $z = y^2 - x^2$ at the point $(1, 2, 3)$.

SOLUTION   Apply Theorem 4 to $f(x, y) = y^2 - x^2$.   Since $f_x(x, y) = -2x$ and $f_y(x, y) = 2y$, the vector

$$-2x\mathbf{i} + 2y\mathbf{j} - \mathbf{k}$$

is perpendicular to the surface at (x, y, z).   In particular, at $(1, 2, 3)$, the vector

$$-2\mathbf{i} + 4\mathbf{j} - \mathbf{k}$$

is perpendicular to the surface.

Incidentally, this surface, which looks like a saddle near the origin, was graphed in Sec. 12.3.   The surface and the normal vector $-2\mathbf{i} + 4\mathbf{j} - \mathbf{k}$ are shown in the diagram at the left. ●

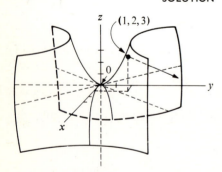

EXAMPLE 8   Use Theorem 4 to show that the tangent lines T_x and T_y discussed near the end of Sec. 12.5 lie in the tangent plane.

SOLUTION   Let us show, for instance, that T_x lies in the tangent plane at (x, y, z).   Inspection of the next to last diagram in Sec. 12.5 shows that the vector

$$\mathbf{i} + 0\mathbf{j} + f_x\,\mathbf{k}$$

is parallel to T_x. To show that T_x lies in the tangent plane, it suffices to check that the dot product

$$(\mathbf{i} + 0\mathbf{j} + f_x\mathbf{k}) \cdot (f_x\mathbf{i} + f_y\mathbf{j} - \mathbf{k})$$

equals 0. This dot product is

$$f_x + 0 - f_x,$$

which does equal 0. Similarly T_y is in the tangent plane. Hence the plane determined by T_x and T_y is indeed the tangent plane as defined in the present section. ●

Exercises

In Exercises 1 to 4 draw the level curve of the given function passing through the given point.

1. $x^2 + y^2$; (5, 12)
2. $3x + 4y$; (2, 1)
3. xy; (1, −1)
4. $y - x^2$; (0, 0)

In Exercises 5 to 8 compute ∇f at the given point and sketch it, where f and the point are those specified in:

5. Exercise 1
6. Exercise 2
7. Exercise 3
8. Exercise 4
9. Let $f(x, y) = xy$.
   (a) Draw ∇f at (1, 1), (1, 2), and (2, 3), each time placing the tail of ∇f at the point where ∇f is evaluated.
   (b) Draw the level curves $xy = 1$, $xy = 2$, $xy = 6$, which pass through the respective points in (a).
10. Let $f(x, y, z) = 2x + 3y + 4z$.
    (a) Where does the level surface

$$f(x, y, z) = 12$$

meet each of the three axes?
    (b) Sketch the level surface in (a).
    (c) Sketch the level surface

$$f(x, y, z) = 0.$$

11. Let $f(x, y, z) = x^2 + y^2 + z^2$. Sketch the level surfaces
    (a) $f(x, y, z) = 1$ and
    (b) $f(x, y, z) = 4$.
12. Let $f(x, y, z) = x^2/4 + y^2/1 + z^2/9$.
    (a) Where does the level surface

$$f(x, y, z) = 1$$

meet each of the three axes?

(b) Sketch the surface in (a).

In Exercises 13 to 15 find a vector perpendicular to the surface at the given point.

13. $2x^2 + 3y^2 + 4z^2 = 9$ at (1, 1, 1).
14. $2x - 3y + 4z = 0$ at (0, 0, 0).
15. $x^3yz = 6$ at (1, 2, 3).
16. Find an equation of the tangent plane to the surface $2x^2 + 3y^2 + 4z^2 = 9$ at (1, 1, 1).
17. (a) Show that (1, 1, 2) is on the surface whose equation is $z = x^2 + y^2$.
    (b) Sketch the surface.
    (c) Find a vector perpendicular to the surface at (1, 1, 2).
18. Find a vector perpendicular to the surface $z = x^2y$ at (1, 1, 1).

■

19. (a) Draw three level curves of the function f defined by $f(x, y) = xy$. Include the curve through (1, 1) as one of them.
    (b) Draw three level curves of the function g defined by $g(x, y) = x^2 - y^2$. Include the level curve through (1, 1) as one of them.
    (c) Prove that each level curve of f intersects each level curve of g at a right angle.
    (d) If we think of f as air pressure, how may we interpret the level curves of g?
20. (a) Draw a level curve for the function $2x^2 + y^2$.
    (b) Draw a level curve for the function y^2/x.
    (c) Prove that any level curve of $2x^2 + y^2$ crosses any level curve of y^2/x at a right angle.

The angle between two surfaces that pass through the point (a, b, c) is defined as the angle between the two lines through (a, b, c) that are perpendicular to the two surfaces at (a, b, c). This angle may be taken to be acute.

21. (a) Show that the point $(1, 1, 2)$ lies on the surfaces $xyz = 2$ and $x^3yz^2 = 4$.

(b) Find the cosine of the angle between the surfaces in (a) at the point $(1, 1, 2)$.

22. (a) Show that the point $(1, 2, 3)$ lies on the plane

$$2x + 3y - z = 5$$

and the sphere

$$x^2 + y^2 + z^2 = 14.$$

(b) Find the angle between them at the point $(1, 2, 3)$.

23. (a) Show that the surfaces $z = x^2y^3$ and $z = 2xy$ pass through the point $(2, 1, 4)$.

(b) At what angle do they cross at that point?

■ ■

24. Two surfaces $f(x, y, z) = 0$ and $g(x, y, z) = 0$ both pass through the point (a, b, c). Their intersection is a curve. How would you find a tangent vector to that curve at (a, b, c)?

20.7

Surface integrals

The main object of this section is to introduce the notion of the integral over a surface. This definition is very similar to that given in Sec. 13.1 for the integral of a function over a plane region, that is, over a flat surface.

We shall not be finicky about fine points. The subject of surface area is fraught with far more difficulties than the area of a plane region (a flat surface). We shall assume that the surfaces we deal with are smooth, or composed of a finite number of smooth pieces. A sphere or the surface of a cube would be typical.

DEFINITION   *Definite integral of a function f over a surface S.* Let f be a function that assigns to each point P in a surface S a number $f(P)$. Consider the typical sum

$$f(P_1)S_1 + f(P_2)S_2 + \cdots + f(P_n)S_n,$$

formed from a partition of S, where S_i is the area of the ith region in the partition, and P_i is a point in the ith region. If these sums approach a certain number as the mesh of the partitions shrinks toward 0 (no matter how P_i is chosen in the ith region), the number is called the definite integral of f over S, and is written

$$\int_S f(P)\, dS.$$

The notions of partition and mesh, used in the preceding definition, being like those in Sec. 13.1, will be omitted.

If $f(P)$ is the density of matter at P, then $\int_S f(P)\, dS$ is the total mass of the surface. If $f(P) = 1$ for all P, then $\int_S f(P)\, dS$ is the area of S. Though the primary interest in the surface integral is conceptual, we will stop to see how to compute it.

One method of computing a surface integral is to express it as an integral over a region in the plane. This method is based on the following geometric lemma.

LEMMA Let U be a region in a plane that is inclined at the angle $\gamma \neq \pi/2$ to the xy plane. Let V be the set in the xy plane directly below (or above) U. Then

$$\text{Area of } U = |\sec \gamma| \text{ area of } V.$$

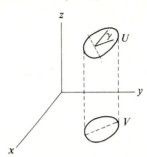

PROOF Introduce a t axis parallel both to the xy plane and to the plane of U. Let $c(t)$ be the cross-sectional length of U for a given t. Then the corresponding cross-sectional length for V is

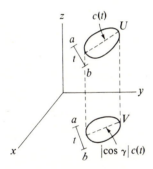

$$|\cos \gamma| c(t).$$

Thus

$$\text{Area of } V = \int_a^b |\cos \gamma| c(t)\, dt$$

$$= |\cos \gamma| \int_a^b c(t)\, dt$$

$$= |\cos \gamma| \text{ area of } U.$$

The lemma follows immediately. ●

EXAMPLE I Find the area of that portion, U, of the plane $x + 2y + 3z = 12$ that lies inside the cylinder $x^2 + y^2 = 9$.

SOLUTION The region V, being a circle of radius 3, has area 9π. To find the area of U, the portion of the plane above V, it is necessary to find $\sec \gamma$, where γ is the angle between the xy plane and the plane $x + 2y + 3z = 12$. Now, the angle between two planes is the same as the angle between their normals.

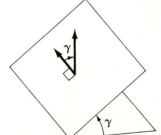

The vector

$$\mathbf{i} + 2\mathbf{j} + 3\mathbf{k}$$

is a normal to the tilted plane, and the vector $\mathbf{k}$ is a normal to the xy plane.

$$\cos \gamma = \frac{(\mathbf{i} + 2\mathbf{j} + 3\mathbf{k}) \cdot \mathbf{k}}{\sqrt{1^2 + 2^2 + 3^2}\sqrt{1^2}}$$

$$= \frac{3}{\sqrt{1^2 + 2^2 + 3^2}} = \frac{3}{\sqrt{14}}.$$

Hence $$\sec \gamma = \frac{\sqrt{14}}{3}.$$

By the lemma, the area of the tilted flat region U is thus

$$|\sec \gamma| \cdot \text{area of } V = \frac{\sqrt{14}}{3} \cdot 9\pi = 3\sqrt{14}\,\pi. \ \bullet$$

The lemma is the key to the following theorem.

THEOREM   Let S be a surface and let R be a region in the xy plane with the property that for each point Q in R the line through Q parallel to the z axis meets S in exactly one point P.   Let f be a function defined on S.   Define a function g on R by

$$g(Q) = f(P).$$

Think of This as Saying That $dA = |\cos \gamma|\, dS$ *or* $dS = |\sec \gamma|\, dA.$ Then $$\int_S f(P)\, dS = \int_R g(Q)|\sec \gamma|\, dA.$$

In this equation γ denotes the angle between the tangent plane to the surface at P and the xy plane.

PROOF   Imagine a typical set U_i in a partition of S.   Let its area be S_i.   Choose a point P_i in this set.   Let V_i be the "shadow," or the corresponding set in R. Let the area of V_i be A_i.   Let Q_i be the point located in R that corresponds to P_i.

    Now, if U_i is small, it looks very much like part of a plane (just as a small portion of a curve looks much like a line).   Hence it is reasonable to assume that, if S is smooth, then

Area of U_i is approximately $|\sec \gamma_i| \cdot$ area of V_i,

where γ_i is the angle that is located between the tangent plane at P_i and the xy plane.

Thus $$S_i \doteq |\sec \gamma_i|\, A_i,$$

and $$f(P_i)S_i \doteq f(P_i)|\sec \gamma_i|\, A_i$$
$$\doteq g(Q_i)|\sec \gamma_i|\, A_i.$$

Adding and taking limits as the mesh $\to 0$, shows that

$$\int_S f(P)\, dS = \int_R g(Q)|\sec \gamma|\, dA. \ \bullet$$

    The following example will show how the above theorem is applied. Note that the factor $|\sec \gamma|$ permits the evaluation of a surface integral by a plane integral.

V_i has area A_i

EXAMPLE 2   Let S be the top half of the sphere $x^2 + y^2 + z^2 = a^2$. Let $f(P)$ be the distance from point P on the sphere to the xy plane. Evaluate $\int_S f(P)\, dS$.

SOLUTION   In this case $f(P) = f(x, y, z) = z$. We wish to evaluate

$$\int_S z\, dS.$$

By the theorem,

$$\int_S z\, dS = \int_R z\,|\sec \gamma|\, dA,$$

where R is the shaded disk in the diagram.

To find γ at $P = (x, y, z)$ recall that a radius vector $\overrightarrow{OP}$ is normal to the sphere. Thus γ is the angle between the vectors $x\mathbf{i} + y\mathbf{j} + z\mathbf{k}$ and $\mathbf{k}$. Consequently,

$$\cos \gamma = \frac{(x\mathbf{i} + y\mathbf{j} + z\mathbf{k}) \cdot \mathbf{k}}{\sqrt{x^2 + y^2 + z^2}\sqrt{1^2}}$$

$$= \frac{z}{a}.$$

From this it follows that $\sec \gamma = \dfrac{a}{z}.$

$\left|\dfrac{a}{z}\right| = \dfrac{a}{z}$ *Since z Is Positive*   Thus

$$\int_R z\,|\sec \gamma|\, dA = \int_R z \cdot \frac{a}{z}\, dA = \int_R a\, dA = a \int_R dA = a \cdot \pi a^2.$$

In short, $\displaystyle\int_S z\, dS = \pi a^3.$ ●

Incidentally, the z coordinate $\bar{z}$ of the center of gravity of the hemisphere (a surface) in Example 2 is defined as $\int_S z\, dS/\text{area of } S$. Thus it is

$$\frac{\pi a^3}{2\pi a^2} = \frac{a}{2}.$$

In Example 2 the integrand in the planar integral was constant. The next example is more typical; it will be used in Example 5 of Sec. 21.3.

EXAMPLE 3   Let S be the portion of the surface $z = 9 - x^2 - y^2$ above the xy plane. Let

$$f(x, y, z) = \frac{2x^2 + 2y^2 + z}{\sqrt{4x^2 + 4y^2 + 1}}.$$

Compute $\displaystyle\int_S f(P)\, dS.$

SOLUTION   The region S, as well as its shadow R on the xy plane, is shown in the margin. In order to transform $\int_S f(P)\, dS$ to an integral over R we must compute $|\sec \gamma|$. To find a normal vector to the surface $z = 9 - x^2 - y^2$, first write the

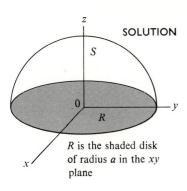

R is the shaded disk of radius a in the xy plane

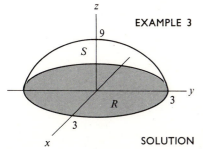

equation in the form $$x^2 + y^2 + z = 9,$$

showing that it is a level surface of the function $x^2 + y^2 + z$. The gradient of this function is normal to the surface; thus

$$2x\mathbf{i} + 2y\mathbf{j} + \mathbf{k}$$

is normal to the surface. Since γ is the angle between this normal and $\mathbf{k}$,

$$\cos \gamma = \frac{(2x\mathbf{i} + 2y\mathbf{j} + \mathbf{k}) \cdot \mathbf{k}}{\sqrt{4x^2 + 4y^2 + 1}\sqrt{1^2}}$$

$$= \frac{1}{\sqrt{4x^2 + 4y^2 + 1}}$$

and $$\sec \gamma = \sqrt{4x^2 + 4y^2 + 1}.$$

Thus

$$\int_S \frac{2x^2 + 2y^2 + z}{\sqrt{4x^2 + 4y^2 + 1}}\, dS = \int_R \frac{2x^2 + 2y^2 + z}{\sqrt{4x^2 + 4y^2 + 1}} \cdot \sqrt{4x^2 + 4y^2 + 1}\, dA$$

$$= \int_R (2x^2 + 2y^2 + z)\, dA.$$

The integrand, $2x^2 + 2y^2 + z$, which appears to be a function of x, y, and z, is actually a function only of x and y for recall that $z = 9 - x^2 - y^2$. Consequently, the integral to be evaluated is

$$\int_R [2x^2 + 2y^2 + (9 - x^2 - y^2)]\, dA = \int_R (9 + x^2 + y^2)\, dA.$$

This final integral is typical of the integrals met in Chap. 13. Since R is a disk and $x^2 + y^2$ appears in the integrand, it is natural to evaluate the integral using polar coordinates in R. We have then

$$\int_R (9 + x^2 + y^2)\, dA = \int_0^{2\pi} \left[\int_0^3 (9 + r^2)r\, dr \right] d\theta.$$

$$= \int_0^{2\pi} \left[\int_0^3 (9r + r^3)\, dr \right] d\theta.$$

It is easy to check that

$$\int_0^3 (9r + r^3)\, dr = \frac{243}{4},$$

and then that

$$\int_0^{2\pi} \frac{243}{4}\, d\theta = \frac{243\pi}{2}.$$

Thus $$\int_S \frac{2x^2 + 2y^2 + z}{\sqrt{4x^2 + 4y^2 + 1}}\, dS = \frac{243\pi}{2}. \quad \bullet$$

How to Find the Area of a Surface   Section 10.11 developed a formula for finding the area of a surface of revolution. The surface in the next example, being part of a saddle, is not a surface of revolution. Its area will be computed by exploiting the fact that $\int_S 1 \, dS = $ area of S. Of course, the resulting integrals may not always be expressible in terms of elementary functions.

EXAMPLE 4   Find the area of that portion of the surface $z = xy$ that is inside the cylinder $x^2 + y^2 = a^2$.

SOLUTION   Let S be the surface described. Then, area of $S = \int_S 1 \, dS$. Let R be the *The Reader May Sketch the Saddle* projection of S on the xy plane, that is, the disk of radius a and center $(0, 0)$.
$z = xy$ and See What the Surface
Looks Like.   Then,

$$\int_S 1 \, dS = \int_R 1 |\sec \gamma| \, dA = \int_R |\sec \gamma| \, dA.$$

So we must next compute $\sec \gamma$. Writing the equation of the surface in the form

$$z - xy = 0,$$

we obtain, by using the gradient, the normal vector

$$-y\mathbf{i} - x\mathbf{j} + \mathbf{k}.$$

Then

$$\cos \gamma = \frac{(-y\mathbf{i} - x\mathbf{j} + \mathbf{k}) \cdot \mathbf{k}}{\sqrt{y^2 + x^2 + 1}\sqrt{1^2}} = \frac{1}{\sqrt{y^2 + x^2 + 1}}.$$

Hence

$$\text{Area of } S = \int_R \sqrt{y^2 + x^2 + 1} \, dA.$$

Switching to polar coordinates in R, we obtain

$$\int_R \sqrt{y^2 + x^2 + 1} \, dA = \int_0^{2\pi} \left(\int_0^a \sqrt{r^2 + 1} \, r \, dr \right) d\theta.$$

First of all,

$$\int_0^a \sqrt{r^2 + 1} \, r \, dr = \frac{(r^2 + 1)^{3/2}}{3} \Big|_0^a$$

$$= \frac{(a^2 + 1)^{3/2} - 1^{3/2}}{3}.$$

Then

$$\int_0^{2\pi} \frac{(a^2 + 1)^{3/2} - 1}{3} \, d\theta = \frac{2\pi}{3}[(a^2 + 1)^{3/2} - 1],$$

which, therefore, is the area of that part of the saddle within the cylinder of radius a:

$$2\pi\left(\frac{(a^2 + 1)^{3/2} - 1}{3}\right). \quad \bullet$$

The principle illustrated by Example 4 is: to find the area of the surface S, evaluate $\int_S 1 \, dS$.

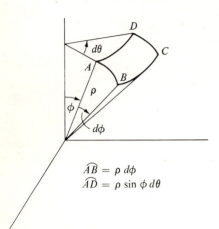

$\widehat{AB} = \rho \, d\phi$
$\widehat{AD} = \rho \sin \phi \, d\theta$

If S happens to be a sphere or part of a sphere, it is sometimes possible to evaluate an integral over it with the aid of spherical coordinates. There may be no need to use the theorem of this section.

If the center of a spherical coordinate system (ρ, ϕ, θ) is at the center of a sphere, then ρ is constant on the sphere. As the diagram suggests, the area of the small region on the sphere corresponding to slight changes $d\phi$ and $d\theta$ is approximately

$$(\rho \, d\phi)(\rho \sin \phi \, d\theta) = \rho^2 \sin \phi \, d\phi \, d\theta.$$

See Sec. 18.3 for a similar argument, where ρ was not constant. Thus we may write

$$dS = \rho^2 \sin \phi \, d\phi \, d\theta$$

and evaluate

$$\int_S f(P) \, dS$$

in terms of a repeated integral in ϕ and θ. The next example illustrates this technique. Keep in mind that a sphere is the surface bounding a ball.

EXAMPLE 5   Let S be the top half of the sphere $x^2 + y^2 + z^2 = a^2$. Evaluate

$$\int_S z^3 \, dS.$$

SOLUTION   In spherical coordinates $z = \rho \cos \phi$. Since the sphere has radius a, $\rho = a$. Thus

$$\int_S z^3 \, dS = \int_S (a \cos \phi)^3 \, dS$$

$$= \int_0^{2\pi} \left[\int_0^{\pi/2} (a \cos \phi)^3 a^2 \sin \phi \, d\phi \right] d\theta.$$

Now,
$$\int_0^{\pi/2} (a \cos \phi)^3 a^2 \sin \phi \, d\phi = a^5 \int_0^{\pi/2} \cos^3 \phi \sin \phi \, d\phi$$

$$= a^5 \frac{(-\cos^4 \phi)}{4} \Big|_0^{\pi/2}$$

$$= \frac{a^5}{4} [-0 - (-1)]$$

$$= \frac{a^5}{4},$$

so that
$$\int_S z^3 \, dS = \int_0^{2\pi} \frac{a^5}{4} \, d\theta$$

$$= \frac{\pi a^5}{2}. \quad \bullet$$

Exercises

1. Find the area of that portion of the parabolic cylinder $z = \frac{1}{2}x^2$ between the three planes $y = 0$, $y = x$, and $x = 2$.

2. Find the area of the part of the spherical surface $x^2 + y^2 + z^2 = 1$ that lies within the vertical cylinder erected on the circle $r = \cos \theta$ and above the xy plane.

3. Show that, if the surface S has the equation $F(x, y, z) = 0$, then

$$|\sec \gamma| = \sqrt{F_x{}^2 + F_y{}^2 + F_z{}^2}/|F_z|.$$

4. Show that if the surface S is given by the equation $z = f(x, y)$, then

$$\sec \gamma = \sqrt{f_x{}^2 + f_y{}^2 + 1}.$$

5. Let S be the triangle whose vertices are $(1, 0, 0)$, $(0, 2, 0)$, and $(0, 0, 3)$. Let $f(x, y, z) = 3x + 2y + 2z$. Evaluate $\int_S f(P)\, dS$.

6. Let S be the triangle whose vertices are $(1, 0, 0)$, $(0, 1, 0)$, and $(0, 0, 1)$. Let $f(x, y, z) = (1/\sqrt{3})(-3x + z - 1)$. Evaluate $\int_S f(P)\, dS$.

7. Let S be the portion of the surface $z = 9 - x^2 - 2y^2$ that lies above the xy plane. Let $f(x, y, z) = (2x^2 + 4y^2 + 2z)/\sqrt{1 + 4x^2 + 16y^2}$. Evaluate $\int_S f(P)\, dS$.

8. A sphere of radius $2a$ has its center at the origin of a rectangular-coordinate system. A circular cylinder of radius a has its axis parallel to the z axis and passes through the z axis. Find the area of that part of the sphere that lies within the cylinder.

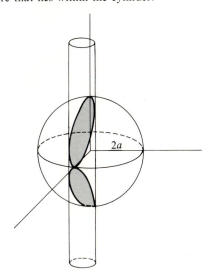

9. Find the area of the cylinder $x^2 + z^2 = 4$ that is above the square whose vertices are $(1, 1, 0)$, $(1, -1, 0)$, $(-1, 1, 0)$, $(-1, -1, 0)$ by

(a) elementary geometry;
(b) integration.

10. Find the area of the cone $z^2 = x^2 + y^2$ that lies above one loop of the curve $r = \sqrt{\cos 2\theta}$.

11. Evaluate the integral in Example 2 by introducing spherical coordinates.

■

In Exercises 12 and 13, let S be a sphere of radius a with center at the origin of a rectangular-coordinate system.

12. Evaluate each of these integrals with a minimum amount of labor.

(a) $\int_S x\, dS$ (b) $\int_S x^3\, dS$ (c) $\int_S \dfrac{2x + 4y^5}{\sqrt{2 + x^2 + 3y^2}}\, dS$.

13. (a) Why is $\int_S x^2\, dS = \int_S y^2\, dS$?
(b) Evaluate $\int_S (x^2 + y^2 + z^2)\, dS$ with a minimum amount of labor.
(c) In view of (a) and (b), evaluate $\int_S x^2\, dS$.
(d) Evaluate $\int_S (2x^2 + 3y^2)\, dS$.

14. An electric field radiates power at the rate of $k \sin^2 \phi/\rho^2$ units per square meter at the point $P = (\rho, \phi, \theta)$. Find the total power radiated to the sphere of radius a, that has the equation $\rho = a$.

15. Find the moment of inertia of the surface of a ball of radius a around a diameter.

16. Find the moment of inertia of the triangle whose vertices are $(a, 0, 0)$, $(0, b, 0)$, $(0, 0, c)$ around the z axis. Take a, b, and c to be positive.

The average of a function $f(P)$ over a surface S is defined as

$$\frac{\int f(P)\, dS}{\text{area of } S}.$$

This concept is illustrated in Exercises 17 and 18.

17. Let S be a sphere of radius a. Let A be a point a distance $R > a$ from the center of S. For P in S let $f(P)$ be $1/q$, where q is the distance from P to A. Show that the average of $f(P)$ over S is $1/R$.

18. The data are the same as in Exercise 17, but $R < a$. Show that in this case the average of $1/q$ is $1/a$. (The average does *not* depend on R.)

■ ■

19. In Example 4 the area of S should be greater than the area of R. Check that if $a > 0$,

$$\frac{2\pi[(a^2 + 1)^{3/2} - 1]}{3} > \pi a^2.$$

20.8

Lagrange multipliers

The goal of this section is to give an intuitive geometric explanation of the method of Lagrange multipliers, which is of use in maximizing functions subject to several constraints and is an important tool in economics and advanced mechanics.

In Sec. 6.8 we discussed the problem of finding the minimal surface area of a right circular can of volume 100. The problem was to minimize

$$2\pi r^2 + 2\pi rh, \tag{1}$$

subject to the constraint that

$$\pi r^2 h = 100. \tag{2}$$

The solution began by using Eq. (2) to eliminate h. Then, the expression $2\pi r^2 + 2\pi rh$ was written as a function of r alone.

In Sec. 11.1 the problem was solved with the aid of implicit differentiation. Though Eq. (2) was not solved to give h explicitly as a function of r, it was clear that h could be considered a function of r. Thus the derivative, dh/dr, made sense.

In both solutions the variables r and h assumed quite different roles. We singled out one of them, r, to be the independent variable, and the other, h, to be the dependent variable. In the method of Lagrange multipliers *all the variables are treated the same*. None is distinguished from the others. Variables that play similar roles in the assumptions will play similar roles in the details of the solution. Furthermore, the method of Lagrange multipliers generalizes easily to several variables and several constraints.

First let us illustrate the method of Lagrange multipliers by using it to solve the can problem just cited. The explanation of why it works will follow.

The first step is to form a certain function L of the variables r, h, and λ.

$$L(r, h, \lambda) = 2\pi r^2 + 2\pi rh - \lambda(\pi r^2 h - 100);$$

that is, $\quad L(r, h, \lambda) = $ (function to be minimized) $- \lambda$(constraint).

(λ is called a Lagrange multiplier.) Then compute the partial derivatives

$$\frac{\partial L}{\partial r}, \quad \frac{\partial L}{\partial h}, \quad \text{and} \quad \frac{\partial L}{\partial \lambda}$$

and find where they are all simultaneously 0:

$$0 = \frac{\partial L}{\partial r} = 4\pi r + 2\pi h - 2\pi \lambda rh$$

$$0 = \frac{\partial L}{\partial h} = 2\pi r - \lambda \pi r^2$$

$$0 = \frac{\partial L}{\partial \lambda} = -(\pi r^2 h - 100).$$

This gives three equations in three unknowns, r, h, and λ:

$$0 = 4\pi r + 2\pi h - 2\pi \lambda rh \tag{3}$$

$$0 = 2\pi r - \lambda \pi r^2 \tag{4}$$

$$0 = \pi r^2 h - 100. \tag{5}$$

Note that (5) is just the given constraint (2).

Since $r = 0$ does not satisfy (5), we may divide (4) by πr, obtaining

$$0 = 2 - \lambda r;$$

hence

$$\lambda r = 2. \tag{6}$$

Combining (6) with (3) yields

$$0 = 4\pi r + 2\pi h - (2\pi)(2)h,$$

or

$$0 = 4\pi r - 2\pi h;$$

hence

$$2r = h. \tag{7}$$

Equation (7) already shows that the can of smallest surface area has its diameter equal to its height. This illustrates the method of Lagrange multipliers.

To find r and h explicitly, combine (5) and (7), obtaining

$$0 = \pi r^2 (2r) - 100.$$

Hence

$$2\pi r^3 = 100,$$

or

$$r^3 = \frac{50}{\pi}.$$

Thus

$$r = \left(\frac{50}{\pi}\right)^{1/3}.$$

It is then possible to solve for λ. In economic theory the value of λ is called a *shadow price*.

Let us see why Lagrange's method works. Consider the following problem:

Maximize or minimize $u = f(x, y)$, given the constraint $g(x, y) = 0$.

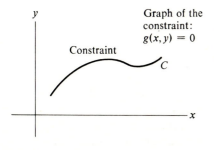

Graph of the constraint: $g(x, y) = 0$

The graph of $g(x, y) = 0$ is in general a curve C. Assume that f, *considered only on points of* C, takes a maximum (or minimum) value at the point P_0. Let C be parameterized by the vector function $\mathbf{G}$: $\mathbf{G}(t) = x(t)\mathbf{i} + y(t)\mathbf{j}$. Let $\mathbf{G}(t_0) = \overrightarrow{OP_0}$. Then u is a function of t:

$$u = f(x(t), y(t)),$$

and by Theorem 2 of Sec. 20.6,

$$\frac{du}{dt} = \nabla f \cdot \mathbf{G}'(t_0).$$

Since f, considered only on C, has a maximum at $\mathbf{G}(t_0)$,

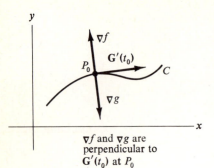

$\mathbf{\nabla}f$ and $\mathbf{\nabla}g$ are
perpendicular to
$\mathbf{G}'(t_0)$ at P_0

Thus $\mathbf{\nabla}f$ and $\mathbf{\nabla}g$
are parallel

$$\frac{du}{dt} = 0$$

at t_0.   Thus

$$\mathbf{\nabla}f \cdot \mathbf{G}'(t_0) = 0.$$

This means that

$$\mathbf{\nabla}f \text{ is perpendicular to } \mathbf{G}'(t_0) \text{ at } P_0. \qquad (8)$$

But

$$\mathbf{\nabla}g, \text{ evaluated at } P_0, \text{ is also perpendicular to } \mathbf{G}'(t_0),$$

since the gradient $\mathbf{\nabla}g$ is perpendicular to the level curve $g(x, y) = 0$.   By (8),

$$\mathbf{\nabla}f \text{ is parallel to } \mathbf{\nabla}g.$$

In other words, there is a scalar λ such that

$$\mathbf{\nabla}f = \lambda\mathbf{\nabla}g.$$

Thus at a maximum (or minimum) of f, subject to the constraint $g(x, y) = 0$, there is a scalar λ such that

$$f_x\mathbf{i} + f_y\mathbf{j} = \lambda(g_x\mathbf{i} + g_y\mathbf{j})$$

or, equivalently,

$$f_x = \lambda g_x$$

and

$$f_y = \lambda g_y.$$

Consequently, at such a maximum (or minimum), occurring at $P_0 = (x_0, y_0)$, these three conditions hold: There is a scalar λ such that

$$\begin{cases} f_x(x_0, y_0) = \lambda g_x(x_0, y_0), \\ f_y(x_0, y_0) = \lambda g_y(x_0, y_0), \end{cases} \qquad (9)$$

and

$$g(x_0, y_0) = 0. \qquad (10)$$

Conditions (9) and (10) provide three equations for three unknowns x_0, y_0, and λ.

Conditions (9) and (10) have a simple description in terms of the function L, defined as follows:

$$L(x, y, \lambda) = f(x, y) - \lambda g(x, y).$$

Conditions (9) are equivalent to

$$\frac{\partial L}{\partial x} = 0 \quad \text{and} \quad \frac{\partial L}{\partial y} = 0.$$

Condition (10) is equivalent to

$$\frac{\partial L}{\partial \lambda} = 0,$$

since this partial is $-g(x, y)$.

EXAMPLE 1   Maximize the function $x^2 y$ for points (x, y) on the circle

$$x^2 + y^2 = 1.$$

SOLUTION First, put the constraint in the form

$$x^2 + y^2 - 1 = 0.$$

The function L in this case is given by

$$L(x, y, \lambda) = x^2y - \lambda(x^2 + y^2 - 1).$$

The three partial derivatives, which are set equal to 0, are

$$\frac{\partial L}{\partial x} = 2xy - 2\lambda x = 0$$

$$\frac{\partial L}{\partial y} = x^2 - 2\lambda y = 0$$

$$\frac{\partial L}{\partial \lambda} = -(x^2 + y^2 - 1) = 0.$$

Since at a maximum of x^2y the number x is not 0, $2x$ can be canceled from the first of the three equations.   These equations then simplify to

$$y - \lambda = 0 \tag{11}$$

$$x^2 - 2\lambda y = 0 \tag{12}$$

$$x^2 + y^2 = 1. \tag{13}$$

By (11) and (12),

$$x^2 = 2y^2. \tag{14}$$

By (13) and (14),

$$2y^2 + y^2 = 1;$$

hence

$$y^2 = \tfrac{1}{3}.$$

Thus,

$$y = \frac{\sqrt{3}}{3} \quad \text{or} \quad y = -\frac{\sqrt{3}}{3}.$$

By (14)

$$x = \sqrt{2}\,y \quad \text{or} \quad x = -\sqrt{2}\,y.$$

There are only four points to be considered on the circle:

$$\left(\frac{\sqrt{6}}{3}, \frac{\sqrt{3}}{3}\right), \quad \left(\frac{-\sqrt{6}}{3}, \frac{\sqrt{3}}{3}\right), \quad \left(\frac{-\sqrt{6}}{3}, \frac{-\sqrt{3}}{3}\right), \quad \left(\frac{\sqrt{6}}{3}, \frac{-\sqrt{3}}{3}\right).$$

At the first and second points x^2y is positive, while at the third and fourth x^2y is negative.   The first two points provide the maximum of x^2y on the circle $x^2 + y^2 = 1$, namely

$$\left(\frac{\sqrt{6}}{3}\right)^2 \frac{\sqrt{3}}{3} = \frac{2\sqrt{3}}{9}.$$

The third and fourth points provide the minimum value of x^2y, namely

$$\frac{-2\sqrt{3}}{9}. \quad \bullet$$

The same method applies to finding a maximum or minimum of $f(x, y, z)$ subject to the constraint $g(x, y, z) = 0$. In this case, form $L(x, y, z) = f(x, y, z) - \lambda g(x, y, z)$ and set the four partial derivatives L_x, L_y, L_z, and L_λ equal to 0.

Lagrange multipliers can also be used to maximize $f(x, y, z)$ subject to more than one constraint; for instance, the constraints may be

$$g(x, y, z) = 0 \quad \text{and} \quad h(x, y, z) = 0. \tag{15}$$

The two surfaces (15) in general meet in a curve C. Assume that C is parameterized by the function $\mathbf{G}$. Then at a maximum (or minimum) of f at a point $P_0 = (x_0, y_0, z_0)$ on C,

$$\nabla f \cdot \mathbf{G}'(t) = 0.$$

Thus ∇f, evaluated at P_0, is perpendicular to $\mathbf{G}'(t_0)$. But ∇g and ∇h, being normal vectors at P_0 to the level surfaces $g(x, y, z) = 0$ and $h(x, y, z) = 0$, respectively, are both perpendicular to $\mathbf{G}'(t_0)$. Thus

$$\nabla f, \qquad \nabla g, \qquad \text{and} \qquad \nabla h$$

are all perpendicular to $\mathbf{G}'(t_0)$ at (x_0, y_0, z_0). Consequently, ∇f lies in the plane determined by the vectors ∇g and ∇h (which we assume are not parallel). Hence there are scalars λ and μ such that

$$\nabla f = \lambda \nabla g + \mu \nabla h.$$

This equation asserts that at P_0 there are scalars λ and μ such that

$$
\begin{aligned}
f_x &= \lambda g_x + \mu h_x \\
f_y &= \lambda g_y + \mu h_y \\
f_z &= \lambda g_z + \mu h_z .
\end{aligned}
\tag{16}
$$

Now (15) and (16) together are equivalent to the brief assertion that the five partial derivatives of the function L defined by

$$L(x, y, z, \lambda, \mu) = f(x, y, z) - \lambda g(x, y, z) - \mu h(x, y, z) \tag{17}$$

are 0. Hence to maximize (or minimize) f subject to the constraints $g(x, y, z) = 0$ and $h(x, y, z) = 0$, form the function L in (17) and proceed as in the example in this section.

A rigorous development of the material in this section belongs in an advanced calculus course. If a maximum occurs at an endpoint of the curves in question, or if the two surfaces do not meet in a curve, the method does not apply. We will content ourselves by illustrating the method with an example in which there are two constraints.

EXAMPLE 2   Minimize $x^2 + y^2 + z^2$ subject to the constraints $x + 2y + 3z = 6$ and $x + 3y + 9z = 9$.

SOLUTION   There are three variables and two constraints. Before forming the Lagrange

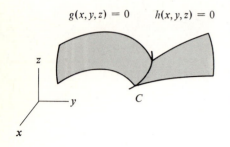

$g(x,y,z) = 0 \qquad h(x,y,z) = 0$

C

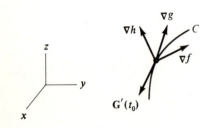

∇g

∇h

C

∇f

$\mathbf{G}'(t_0)$

function L write the constraints with 0 on one side of the equation:

$$x + 2y + 3z - 6 = 0$$
$$x + 3y + 9z - 9 = 0.$$

The Lagrange function is

$$L(x, y, z, \lambda, \mu) = x^2 + y^2 + z^2 - \lambda(x + 2y + 3z - 6) - \mu(x + 3y + 9z - 9).$$

Setting the five partial derivatives of L equal to 0 gives:

$$\begin{aligned} L_x &= 0 & 2x - \lambda - \mu &= 0 \\ L_y &= 0 & 2y - 2\lambda - 3\mu &= 0 \\ L_z &= 0 & 2z - 3\lambda - 9\mu &= 0 \end{aligned} \tag{18}$$

$$\begin{aligned} L_\lambda &= 0 & x + 2y + 3z - 6 &= 0 \\ L_\mu &= 0 & x + 3y + 9z - 9 &= 0. \end{aligned} \tag{19}$$

There are many ways to solve these five equations.   Let us use the first three to express x, y, and z in terms of λ and μ.   Then, after substituting the results in the last two equations, we will find λ and μ.

By Eqs. (18)

$$x = \frac{\lambda + \mu}{2}, \qquad y = \frac{2\lambda + 3\mu}{2}, \qquad z = \frac{3\lambda + 9\mu}{2}.$$

Equations (19) then become

$$\frac{\lambda + \mu}{2} + \frac{2(2\lambda + 3\mu)}{2} + \frac{3(3\lambda + 9\mu)}{2} - 6 = 0$$

and

$$\frac{\lambda + \mu}{2} + \frac{3(2\lambda + 3\mu)}{2} + \frac{9(3\lambda + 9\mu)}{2} - 9 = 0,$$

which a little algebra changes to

and

$$\begin{aligned} 14\lambda + 34\mu &= 12, \\ 34\lambda + 91\mu &= 18. \end{aligned} \tag{20}$$

By Sec. 19.6,

$$\lambda = \frac{\begin{vmatrix} 12 & 34 \\ 18 & 91 \end{vmatrix}}{\begin{vmatrix} 14 & 34 \\ 34 & 91 \end{vmatrix}} \qquad \text{and} \qquad \mu = \frac{\begin{vmatrix} 14 & 12 \\ 34 & 18 \end{vmatrix}}{\begin{vmatrix} 14 & 34 \\ 34 & 91 \end{vmatrix}};$$

hence

$$\lambda = \frac{240}{59} \qquad \text{and} \qquad \mu = -\frac{78}{59}.$$

Thus

$$x = \frac{\lambda + \mu}{2} = \frac{81}{59}$$

$$y = \frac{2\lambda + 3\mu}{2} = \frac{123}{59}$$

$$z = \frac{3\lambda + 9\mu}{2} = \frac{9}{59}.$$

The minimum of $x^2 + y^2 + z^2$ is thus

$$\left(\frac{81}{59}\right)^2 + \left(\frac{123}{59}\right)^2 + \left(\frac{9}{59}\right)^2 = \frac{21{,}771}{3{,}481} = \frac{369}{59}. \;\bullet$$

Each of the two constraints in Example 2 describes a plane. Thus the two constraints together describe a *line*. The function $x^2 + y^2 + z^2$ is the square of the distance from (x, y, z) to the origin. So in Example 2 the problem was essentially, "How far is the origin from a certain line?" (It could be solved by vector algebra.) When viewed in this perspective, the problem certainly has a solution; that is, there is clearly a minimum.

In Example 2 there were three variables x, y, z and two constraints. There may in some cases be many variables, $x_1, x_2, \ldots, x_n$, and many constraints. Write each constraint in the form "something $= 0$" and introduce Lagrange multipliers $\lambda_1, \lambda_2, \ldots, \lambda_m$, one for each constraint.

Exercises

In Exercises 1 to 17 use the method of Lagrange multipliers to solve the indicated problem.

1. Maximize xy subject to the constraint $x^2 + y^2 = 4$.

2. Minimize $x^2 + y^2$ for points on the line $2x + 3y = 6$.

3. Exercise 3 of Sec. 6.8.

4. Exercise 22 of Sec. 6.8.

5. Exercise 24 of Sec. 6.8.

6. Exercise 25 of Sec. 6.8.

7. Find the point on the plane $x + 2y + 3z = 6$ that is closest to the origin.

8. Find the dimensions of the box of largest volume whose surface area is to be 6 square inches.

9. Maximize $x^2 y^2 z^2$ subject to the constraint

$$x^2 + y^2 + z^2 = 1.$$

10. Minimize $x^2 + y^2 + z^2$ on the line common to the two planes $x + 2y + 3z = 0$ and $2x + 3y + z = 4$.

11. The plane $2y + 4z - 5 = 0$ meets the cone $z^2 = 4(x^2 + y^2)$ in a curve. Find the point on this curve nearest the origin.

12. Maximize $x^3 + y^3 + 2z^3$ on the intersection of the two spherical surfaces $x^2 + y^2 + z^2 = 4$ and $(x - 3)^2 + y^2 + z^2 = 4$.

■

13. (a) Maximize $x_1 x_2 \cdots x_n$ subject to the constraint

$\sum_{i=1}^n x_i = 1$ and all $x_i \geq 0$.

(b) Deduce that, for nonnegative numbers $a_1, a_2, \ldots, a_n$, $\sqrt[n]{a_1 a_2 \cdots a_n} \leq (a_1 + a_2 + \cdots + a_n)/n$. (The *geometric mean* is less than or equal to the *arithmetic mean*.)

14. (a) Maximize $\sum_{i=1}^n x_i y_i$ subject to the constraints $\sum_{i=1}^n x_i^2 = 1$ and $\sum_{i=1}^n y_i^2 = 1$.

(b) Deduce that, for any numbers $a_1, a_2, \ldots, a_n$ and $b_1, b_2, \ldots, b_n$, $\sum_{i=1}^n a_i b_i \leq (\sum_{i=1}^n a_i^2)^{1/2}(\sum_{i=1}^n b_i^2)^{1/2}$.

(c) How would you justify the inequality in (b), for $n = 3$, by vectors?

15. Maximize $x + 2y + 3z$ subject to the constraints $x^2 + y^2 + z^2 = 1$ and $x + y + z = 0$.

16. Let $a_1, a_2, \ldots, a_n$ be fixed nonzero numbers. Maximize $\sum_{i=1}^n a_i x_i$ subject to $\sum_{i=1}^n x_i^2 = 1$.

■ ■

17. Let p and q be positive numbers such that $1/p + 1/q = 1$. Obtain Hölder's inequality for nonnegative numbers a_i and b_i:

$$\sum_{i=1}^n a_i b_i \leq \left(\sum_{i=1}^n a_i{}^p\right)^{1/p}\left(\sum_{i=1}^n b_i{}^q\right)^{1/q}.$$

Hint: Recall Exercise 14.

18. Solve Example 2 by vector algebra.

19. The center of gravity of n points on the curve $y = \ln x$ lies below the curve. (Why?) From this fact deduce the inequality in Exercise 13(b).

20. Explain why at a maximum of $f(x, y, z)$ subject to the constraint $g(x, y, z) = 0$, ∇f is parallel to ∇g. (Assume that the surface $g(x, y, z) = 0$ is smooth and that the maximum of f does not occur on the border of this surface.)

21. A consumer has a budget of B dollars and may purchase n different items. The price of the ith item is p_i dollars. When the consumer buys x_i units of the ith item, the total cost is $\sum_{i=1}^{n} p_i x_i$. Assume that $\sum_{i=1}^{n} p_i x_i = B$ and that the consumer wishes to maximize his pleasure or utility $u(x_1, x_2, \ldots, x_n)$.

(a) Show that, when $x_1, \ldots, x_n$ are chosen to maximize utility, then

$$\frac{\partial u / \partial x_i}{p_i} = \frac{\partial u / \partial x_j}{p_j} \qquad i \neq j.$$

(b) Explain the result in (a) using just economic intuition. *Hint*: Consider a slight change in x_i and x_j, with the other x_k's held fixed.

20.9

Summary

In this chapter we defined and applied the derivative of a vector function $\mathbf{G}$, which assigns to a scalar t a vector $\mathbf{G}(t)$. $\mathbf{G}(t)$ may be thought of as the position vector in the plane or in space of a moving particle at time t.

The derivative of $\mathbf{G}$ is defined as

$$\lim_{t \to 0} \frac{\Delta \mathbf{G}}{\Delta t}.$$

The mathematician denotes it $\mathbf{G}'$, but a physicist denotes it $\mathbf{V}$, for its length is the speed of the moving particle, and its direction is that in which the particle is moving at a given instant.

The derivative of a vector function has many useful properties that resemble those of the ordinary derivative. For instance

$$(\mathbf{G} \cdot \mathbf{H})' = \mathbf{G} \cdot \mathbf{H}' + \mathbf{H} \cdot \mathbf{G}'.$$

In particular, if $\mathbf{G} = \mathbf{H}$, then

$$(\mathbf{G} \cdot \mathbf{G})' = 2\mathbf{G} \cdot \mathbf{G}'.$$

From this it follows that if $|\mathbf{G}|$ is constant, then $\mathbf{G}'(t)$ is perpendicular to $\mathbf{G}(t)$ for all t.

The acceleration vector is defined as the second derivative of $\mathbf{G}$,

$$\mathbf{A} = (\mathbf{G}')' = \mathbf{V}'.$$

If $\mathbf{G}$ is given in components relative to the basic unit vectors $\mathbf{i}$ and $\mathbf{j}$,

$$\mathbf{G}(t) = g(t)\mathbf{i} + h(t)\mathbf{j}.$$

Then

$$\mathbf{A}(t) = \frac{d^2 g}{dt^2} \mathbf{i} + \frac{d^2 h}{dt^2} \mathbf{j}.$$

In the study of a moving particle two other perpendicular unit vectors are of interest: $\mathbf{T}$, pointing straight ahead, and $\mathbf{N}$, pointing in the direction in which the particle is turning. $\mathbf{T}$ and $\mathbf{N}$ depend on time. It turns out that

$$\mathbf{A} = \frac{d^2 s}{dt^2} \mathbf{T} + \frac{v^2}{r} \mathbf{N}.$$

The gradient ∇f, defined in the preceding chapter, appears again in the present chapter. When evaluated at (a, b), it provides a vector perpendicular to the level curve of f that passes through (a, b). (The same holds for a function of three variables: ∇f is perpendicular to the level surface of f.) The tangent plane to a surface is defined in terms of ∇f.

The gradient is of use in rephrasing the chain rule

$$\frac{du}{dt} = \frac{\partial f}{\partial x} \frac{dx}{dt} + \frac{\partial f}{\partial y} \frac{dy}{dt}.$$

In vector notation this rule reads

$$\frac{du}{dt} = \nabla f \cdot \mathbf{G}'(t).$$

The same formula holds in three dimensions. This rephrasing in terms of the dot product is useful in showing that ∇f is perpendicular to $\mathbf{G}'$ if $\mathbf{G}$ parameterizes a curve on which f is constant.

The integral over a surface $\int_S f(P)\, dS$ was defined like an integral over a plane region R. It was shown how to evaluate such an integral by replacing dS by $|\sec \gamma|\, dA$ and integrating over a plane region. In particular,

$$\text{Area of } S = \int_R |\sec \gamma|\, dA.$$

The final topic, Lagrange multipliers, concerned a method of finding a maximum or minimum of a function of several variables subject to one or more constraints.

Vectors are used for two purposes in this and the preceding chapter. First, they are a bookkeeping device. For instance, the two equations

$$\frac{dx}{dt} = 3 \quad \text{and} \quad \frac{dy}{dt} = 4$$

are summarized in a single vector equation

$$\mathbf{V} = 3\mathbf{i} + 4\mathbf{j}.$$

Second, they provide a language in which the symbols closely correspond to intuitive concepts. For instance, instead of saying, "The x coordinate changes at the rate of 3 feet per second and the y coordinate changes at the rate of 4 feet per second," one may say simply, "The velocity vector $\mathbf{V}$ is $3\mathbf{i} + 4\mathbf{j}$," and draw it as shown below. The longer $\mathbf{V}$ is, the faster the particle is moving; the direction of $\mathbf{V}$ is the direction in which the particle moves.

Throughout the chapter we have switched back and forth between mathematical and physical terminology. This little glossary records some of the concepts in both languages.

Mathematical Formulation	Physical Interpretation
Vector function $\mathbf{G}$	Parameterized path of moving particle
Derivative of $\mathbf{G}$	Velocity vector $\mathbf{V}$
Second derivative of $\mathbf{G}$	Acceleration vector $\mathbf{A}$
$\mathbf{T} = \dfrac{\mathbf{V}}{|\mathbf{V}|}$	Unit vector pointing straight ahead
$\mathbf{N} = \dfrac{d\mathbf{T}/dt}{|d\mathbf{T}/dt|}$	Unit vector perpendicular to path and pointing in the direction the path is veering
$\mathbf{A} = \dfrac{d^2s}{dt^2}\mathbf{T} + \dfrac{v^2}{r}\mathbf{N}$	Acceleration vector is sum of two vectors: One is the acceleration if the particle were moving in a straight line; the other is the acceleration if the particle were moving at a constant speed on a circle whose radius is the radius of curvature

VOCABULARY AND SYMBOLS

vector function $\mathbf{G}$, $\mathbf{R}$	velocity vector $\mathbf{G}'$, $\mathbf{V}$	scalar components of $\mathbf{A}$, A_T and	principal normal
limit of a vector function	acceleration vector $\mathbf{G}''$, $\mathbf{A}$	A_N, along $\mathbf{T}$ and $\mathbf{N}$	tangent plane
position vector $\mathbf{R}$	unit tangent vector $\mathbf{T}$	level curve $f(x, y) = c$	surface integral
derivative of $\mathbf{G}$, $\mathbf{G}'$	unit normal vector $\mathbf{N}$	level surface $f(x, y, z) = c$	Lagrange multipliers

Guide quiz on Chap. 20

1. Let $\mathbf{R} = \mathbf{G}(t) = t\mathbf{i} + t^2\mathbf{j} + \mathbf{k}$ parameterize a curve.
   (a) Sketch the curve.
   (b) Calculate and sketch $\Delta\mathbf{R} = \mathbf{G}(1.1) - \mathbf{G}(1)$.
   (c) Calculate and sketch $\Delta\mathbf{R}/0.1$.
2. Explain why the gradient of $f(x, y, z)$ is perpendicular to the level surface $f(x, y, z) = c$.
3. Find a unit vector perpendicular to
   (a) the surface $z = x^2 + y^2$ at $(1, 1, 2)$;
   (b) the surface $x + y^2 + z^3 = 3$ at $(1, 1, 1)$;

   (c) the sphere $x^2 + y^2 + z^2 = 9$ at (x, y, z).
4. The temperature T at the point (x, y, z) is $x^2 + y^2 + z^2$. As a bird flies on the path $t\mathbf{i} + t^2\mathbf{j} + t^3\mathbf{k}$, it notices that the air is getting hotter. Find the rate of change in the temperature along the bird's flight at the point $(1, 1, 1)$ (a) with respect to time; (b) with respect to distance.
5. The surfaces $2x^2 + 3y^2 + z^2 = 6$ and $x^3 + y^3 + z^3 = 3$ pass through the point $(1, 1, 1)$. At what angle do they cross there?

6. Find an equation of the
   (a) tangent line to $x^5y + x^3y^2 = 2$ at $(1, 1)$;
   (b) tangent plane to $xy + x^3z + xy^3 = 3$ at $(1, 1, 1)$.

7. (a) If a plane curve is given in the form $\mathbf{R} = \mathbf{G}(t)$, how would you find a vector parallel to the curve? Perpendicular to the curve?
   (b) If a plane curve is given in the form $f(x, y) = 0$, how would you find a vector perpendicular to the curve?
   (c) If a curve is given as the intersection of two surfaces, how could you find a vector perpendicular to the curve? Parallel to the curve?
   (d) If a space curve is given in the form $\mathbf{R} = \mathbf{G}(t)$, how would you find a vector perpendicular to the curve?

8. At a certain time a particle moving on the curve $y = f(x)$ has
$$\mathbf{V} = 2\mathbf{i} + 3\mathbf{j} \text{ and } \mathbf{A} = \mathbf{i} - \mathbf{j}.$$
   (a) Is the particle speeding up or slowing down?
   (b) Estimate A_T and A_N graphically.
   (c) Compute A_T and A_N.
   (d) Find the radius of curvature.

(e) Find dy/dx.
(f) Find d^2y/dx^2.

9. (a) Sketch the part of the cylinder $x^2 + z^2 = a^2$ $(a > 0)$ that lies above the xy plane and between the planes $y = 0$ and $y = x$.
   (b) Find the area of the surface in (a).

10. Consider the curve given as the intersection of the surfaces $g(x, y, z) = 0$ and $h(x, y, z) = 0$. The function $u = f(x, y, z)$, considered only on this curve, has a maximum at (a, b, c). Why might we expect there to be, in general, scalars λ and μ such that
$$\mathbf{V}f = \lambda\mathbf{V}g + \mu\mathbf{V}h,$$
where the gradients are evaluated at (a, b, c)? (Include a sketch of the curve and the gradients.)

11. Find the point on the surface $z = x^2 + 2y^2$ closest to the plane $x + y - z = 1$, as follows:
   (a) Sketch the surface and the plane.
   (b) Prove that they do not meet.
   (c) Use the method of Lagrange multipliers to find the point on the surface closest to the plane.

Review exercises for Chap. 20

1. (a) Find a vector normal to the surface $z = x^2 + y^3$ at $(1, 1, 2)$.
   (b) Find a vector parallel to the surface $z = x^2 + y^3$ at $(1, 1, 2)$.

2. A triangle in the plane $z = x + y$ is directly above the triangle in the xy plane whose vertices are $(1, 2)$, $(3, 4)$, and $(2, 5)$. Find the area of (a) the triangle in the xy plane, (b) the triangle in the plane $z = x + y$.

3. An astronaut traveling in the path $\mathbf{G}(t) = t\mathbf{i} + t^2\mathbf{j} + t^3\mathbf{k}$ shuts off her motor when at the point $(1, 1, 1)$.
   (a) Show that she passes through the point $(3, 5, 7)$.
   (b) How near does she get to the point $(5, 8, 9)$?

4. Let S be that portion of the surface $z = x^3 - y^3$ that lies above the triangle in the xy plane that has vertices $(0, 0)$, $(1, 0)$, and $(1, 1)$. Let S^* be that portion of the surface $z = x^3 + y^3$ that lies above the same triangle.
   (a) Sketch S and S^*.
   (b) Show that they have equal areas.

5. Let
$$\mathbf{G}(t) = \frac{e^t + e^{-t}}{2}\mathbf{i} + \frac{e^t - e^{-t}}{2}\mathbf{j}.$$
   (a) Show that the particle moves on the hyperbola $x^2 - y^2 = 1$.

(b) Find A_T and A_N when $t = 1$.

6. (a) Draw the level surface of $f(x, y, z) = 2x + 3y + 4z$ that passes through $(1, 0, 0)$.
   (b) Compute $\mathbf{V}f$ at $(1, 0, 0)$.
   (c) Find two points on the level surface mentioned in (a).
   (d) Using these two points, construct a vector $\mathbf{B}$ lying on the surface.
   (e) Compute $\mathbf{B} \cdot \mathbf{V}f$.
   (f) Is the answer to (e) reasonable? Why?

7. At time $t = 0$ the velocity vector of a certain particle is $0\mathbf{i} + 1\mathbf{j}$. If the acceleration vector $\mathbf{A}(t)$ is $6t\mathbf{i} + e^t\mathbf{j}$, find
   (a) $\mathbf{V}$ and $\mathbf{A}$ when $t = 1$;
   (b) v, A_T, A_N, and r (the radius of curvature) when $t = 1$.

8. At what angle does the curve $6x^2 + 7xy + 3y^2 = 16$ cross the curve $x^2 + 3xy = 4$ at $(1, 1)$?

9. (a) Show that when $t = 1$ the particles on the paths
$$\mathbf{G}(t) = t^2\mathbf{i} + 6t\mathbf{j} + \mathbf{k} \text{ and } \mathbf{H}(t) = t^3\mathbf{i} + (t + 5)\mathbf{j} + \mathbf{k}$$
   collide.
   (b) At what angle do they collide?

10. Let $\mathbf{R}$ be the position vector of a moving particle and let $r = |\mathbf{R}|$.
   (a) Show that $r\, dr/dt = \mathbf{R} \cdot d\mathbf{R}/dt$.
   (b) Using the definition of $\mathbf{R}'$ as $\lim_{\Delta t \to 0} \Delta\mathbf{R}/\Delta t$, show

why $\mathbf{R}'$ is tangent to the curve whose position vector at time t is $\mathbf{R}(t)$.

11. Why is ∇f at (a, b) perpendicular to the level curve of f through (a, b)?

12. At a certain moment a particle moving on a curve has $\mathbf{V} = 2\mathbf{i} + 3\mathbf{j}$ and $\mathbf{A} = -\mathbf{i} + 2\mathbf{j}$.
 (*a*) What is the speed of the particle?
 (*b*) Find $\mathbf{T}$ and $\mathbf{N}$, expressing each in the form $x\mathbf{i} + y\mathbf{j}$, for suitable scalars x and y.
 (*c*) Draw $\mathbf{V}$, $\mathbf{A}$, $\mathbf{T}$, and $\mathbf{N}$.
 (*d*) Compute $\mathbf{A} \cdot \mathbf{N}$.
 (*e*) From (*a*) and (*d*) obtain the radius of curvature.
 (*f*) Compute $\mathbf{A} \cdot \mathbf{T}$.
 (*g*) Is the particle speeding up or slowing down? Compute d^2s/dt^2.

13. (*a*) Explain why
$$\frac{d\mathbf{R}}{ds}$$
is a unit vector (*s* denotes arc length).
 (*b*) From (*a*) deduce *as a special case* that
$$\frac{d\mathbf{T}}{d\phi}$$
is a unit vector (ϕ denotes the angle between the unit tangent vector $\mathbf{T}$ and the basic vector $\mathbf{i}$).

14. Assume that the surfaces $f(x, y, z) = 0$ and $g(x, y, z) = 0$ meet in a curve and that the point (a, b, c) is on this curve. Explain why
$$\nabla f \times \nabla g$$
is a tangent vector to this curve at (a, b, c). [∇f and ∇g are evaluated at (a, b, c).]

15. (See Exercise 14.) Find a tangent vector to the curve of intersection of the surfaces $x^2 + 2y^2 + 3z^2 = 36$ and $2x^2 - y^2 + z^2 = 7$ at the point $(1, 2, 3)$.

16. (*a*) Sketch the level curve of the function $5x^2 + 3y^2$ that passes through $(1, 1)$.
 (*b*) Do the same for the function y^5/x^3.
 (*c*) Show that the two curves cross at a right angle at $(1, 1)$.
 (*d*) Prove that each level curve of $5x^2 + 3y^2$ crosses each level curve of y^5/x^3 at a right angle.

17. (*a*) Show that the curves
$$x^3y + xy^5 = 2$$
and $\qquad x^4y + xy^6 = 2$
both pass through $(1, 1)$.
 (*b*) At what angle do they cross at $(1, 1)$?

In Exercises 18 to 20 let $z = f(x, y)$ and $\mathbf{G}(t) = x\mathbf{i} + y\mathbf{j}$. Find dz/dt when $t = t_0$ and $\mathbf{G}(t_0) = x_0\mathbf{i} + y_0\mathbf{j}$ if:

18. ∇f at (x_0, y_0) is $3\mathbf{i} + 4\mathbf{j}$ and $\mathbf{G}'(t_0) = 2\mathbf{i} - 3\mathbf{j}$.

19. ∇f at (x_0, y_0) is perpendicular to $\mathbf{G}'(t_0)$.

20. ∇f at (x_0, y_0) has length 5, the speed $|\mathbf{G}'(t_0)| = 4$, and the angle between ∇f and $\mathbf{G}'(t_0)$ is $\pi/4$.

21. Let C be a curve in the xy plane whose equation is $y = f(x)$ for x in $[a, b]$. Let $\gamma(x, y)$ be the angle between a line perpendicular to the curve at (x, y) and the y axis. Show that the length of C is $\int_a^b |\sec \gamma|\, dx$.

22. Let S be the portion of the surface $z = x^2 + 3y^2$ below the surface $z = 1 - x^2 - y^2$.
 (*a*) Set up a plane integral for the area of S.
 (*b*) Set up repeated integrals in rectangular coordinates for the integral in (*a*) and in polar coordinates, but do not evaluate them.

23. At time t the position vector of a particle is
$$\mathbf{G}(t) = t\mathbf{i} + 3t\mathbf{j} + 4t\mathbf{k}.$$
 (*a*) Show that any point in the path lies in the plane $3x = y$ and also in the plane $4y = 3z$.
 (*b*) Sketch the planes in (*a*) and indicate their intersection, which is the path of the particle.
 (*c*) Find the velocity vector at time t.
 (*d*) Find the speed at time t.

24. Let $\mathbf{G}(t) = e^t \cos t\,\mathbf{i} + e^t \sin t\,\mathbf{j} + e^t\mathbf{k}$.
 (*a*) Find the speed at time t.
 (*b*) Find the distance traveled during the time interval $[0, 1]$.

25. A particle moves on a straight line at a constant speed of 4 feet per second in the direction of the unit vector
$$\frac{\mathbf{i}}{3} + \frac{2\mathbf{j}}{3} + \frac{2\mathbf{k}}{3}$$
At time $t = 0$ the particle is at the point $(2, 5, 7)$. Find its position vector at time t.

26. A particle moves along a straight line at a constant speed. At time $t = 0$ it is at the point $(1, 2, 1)$, and at time $t = 2$ it is at $(4, 1, 5)$.
 (*a*) Find its speed.
 (*b*) Find the direction cosines of the line.
 (*c*) Find the position vector of the particle at time t.

■

27. Let $P_0 = (a, b, c)$ be a point not on the smooth surface whose equation is $f(x, y, z) = 0$. Let P_1 be a nearest point to P_0 on the surface. Prove that the vector $\overrightarrow{P_0 P_1}$ is a normal to the surface at P_1. (Use Lagrange multipliers.)

28. Let $P_0 = (a, b, c)$ be a point not on the smooth curve given as the intersection of the surfaces $f(x, y, z) = 0$ and $g(x, y, z) = 0$. Let P_1 be a nearest point to P_0 on the curve. Prove that $\overrightarrow{P_0 P_1}$ is perpendicular to the curve at P_1. (Use Lagrange multipliers.)

29. A particle travels in such a way that its velocity vector is always perpendicular to its position vector. Show that its path lies on a sphere (the surface of a ball).

30. Let $G(s)$, where s denotes arc length, parameterize a curve in space. Let $u = f(x, y, z)$ be a function of x, y, and z. Then u may be considered a composite function of s. Show that
(a) $du/ds = \nabla f \cdot G'(s)$, where ∇f is evaluated at $G(s)$.
(b) $du/ds = D_T(u)$, where T is a unit tangent vector to the curve at $G(s)$.

31. Let S be the triangle with vertices $(1, 1, 1)$, $(2, 3, 4)$, and $(3, 4, 5)$.
(a) Using vectors, find the area of S.
(b) Using the formula, area of $S = \int_R |\sec \gamma|\, dA$, find the area of S.

In the study of planetary motion the acceleration vector is expressed in terms of a unit vector u_r directed away from the sun, and a unit vector u_θ perpendicular to u_r, as in the accompanying diagram.

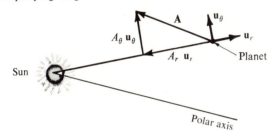

Exercises 32 to 39 illustrate their use.

32. Show that (a) $u_r = \cos \theta\, i + \sin \theta\, j$;
(b) $u_\theta = -\sin \theta\, i + \cos \theta\, j$.
33. Show that (a) $d(u_r)/d\theta = u_\theta$; (b) $d(u_\theta)/d\theta = -u_r$.
34. Show that (a) $d(u_r)/dt = u_\theta\, d\theta/dt$; (b) $d(u_\theta)/dt = -u_r\, dr/dt$.
35. Let R be the position vector at time t. From the fact that $R = ru_r$, deduce that $V = r\dot\theta u_\theta + \dot r u_r$.
36. Explain why the scalar components, $r\dot\theta$ and $\dot r$, in Exercise 34 are intuitively plausible.

37. By differentiating V, show that
$$A = (\ddot r - r\dot\theta^2)u_r + (r\ddot\theta + 2\dot r\dot\theta)u_\theta.$$
38. Explain why the terms $\ddot r$, $-r\dot\theta^2$, and $r\ddot\theta$, in Exercise 36, are plausible. The term $2\dot r\dot\theta$ is related to what physicists call the *Coriolis force*.
39. (a) Show that
$$A_\theta = \frac{1}{r}\frac{d(r^2\dot\theta)}{dt}.$$
(b) Show that A is directed toward the sun only if $r^2\dot\theta$ is constant.
(c) From (b) deduce Kepler's second law of planetary motion: The line joining a planet to the sun sweeps out equal areas in equal times. Kepler discovered this law in 1602, on the basis of extensive observations, and published it in 1609.

■ ■

40. Let $G(t)$ parameterize a curve in space. Let $T(t)$ be the unit tangent vector at time t. At a given point P_0 on the curve the curve seems to be almost planar. Making a sketch of $T(t)$, ΔT, and $d(T)/dt$, explain why $T(t) \times T'(t)$ is a plausible candidate for a vector normal to the plane in which the curve, locally, seems to be situated.

41. Let $z = f(y)$ be positive for $a \le y \le b$. The area of the surface formed by rotating this curve around the y axis is $2\pi \int_a^b f(y)\sqrt{1 + (df/dy)^2}\, dy$. Derive this formula from the general formula, area of $S = \int_R |\sec \gamma|\, dA$, as follows:
(a) Show that the surface of revolution has the equation $z^2 + x^2 = [f(y)]^2$.
(b) Show that $xi - f(y)\, df/dy\, j + zk$ is a normal to the surface at (x, y, z).
(c) Let R be the region in the xy plane bounded by the interval $a \le y \le b$ on the y axis, the curve $x = f(y)$, and the lines $y = a$ and $y = b$. Using (b), evaluate $\int_R |\sec \gamma|\, dA$.

21 INTEGRALS OF SCALAR AND VECTOR FIELDS

This chapter is concerned with vector fields, scalar fields, and various integrals associated with them. The emphasis is primarily on definition and notation. The following chapter, which continues in the same direction, develops interrelationships among the integrals introduced here.

Students who go on to study electromagnetism or fluid dynamics will see vector analysis used as a language whose compact expressions may represent concepts which are quite complex. Concentrate on this aspect; computational skill, while important, is of secondary consideration.

Throughout this chapter it will be assumed that all functions have derivatives or partial derivatives of all orders.

21.1

Vector and scalar fields

Imagine a loop of wire C held firmly in place on the surface of a stream. At some points of C water is entering the region bounded by C, and at some points of C water is leaving the region. In this diagram the arrows indicate the velocity at various points on the stream, and the curve is the wire C. How can the net amount entering or leaving be computed? Presumably some kind of integration is required. This chapter will develop methods for answering questions such as this.

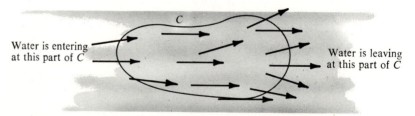

Water is entering at this part of C

Water is leaving at this part of C

The flow of water in a stream provides an example of a *vector field*, which will now be defined.

DEFINITION *Vector field.* A function that assigns a vector to each point in some region in the plane (or space) is called a *vector field*. It will usually be denoted **F**.

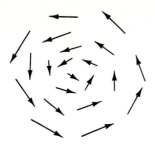

For example, the function **F** that assigns to each point on the surface of the stream the velocity vector of the water at that point is a vector field.

The daily weather map displays a few of the (vector) values of the vector field that assigns to each point on the surface of a portion of the earth (considered to be flat) the wind vector at that point. The diagram at the left shows a few of the wind vectors of the vector field associated with a hurricane. Near the eye of the hurricane the wind vectors are shorter: The air is relatively calm.

DEFINITION *Scalar field.* A function that assigns a number to each point in some region in the plane (or in space) is called a *scalar field.* It will usually be denoted *f.*

The function that assigns the temperature at a point is a scalar function; so is the function that describes the density at a point.

A vector field **F** in the plane is described by two scalar fields, the scalar components of **F**:

$$\mathbf{F}(x, y) = P(x, y)\mathbf{i} + Q(x, y)\mathbf{j}.$$

Both *P* and *Q* are scalar fields.

Let *f* be a scalar field. The vector field that assigns to each point (x, y) the gradient of *f* at (x, y), ∇f, is called the *gradient field* associated with *f*. A similar construction can be carried out on $f(x, y, z)$, again producing a vector field ∇f.

EXAMPLE I Compute and sketch the gradient field associated with the scalar field $f(x, y) = 1/\sqrt{x^2 + y^2}$. (These two fields are of importance in gravitational and electromagnetic theory.)

SOLUTION
$$\nabla f = f_x \mathbf{i} + f_y \mathbf{j}$$

$$= \frac{-x}{(\sqrt{x^2 + y^2})^3}\mathbf{i} + \frac{-y}{(\sqrt{x^2 + y^2})^3}\mathbf{j}.$$

Let
$$\mathbf{R} = x\mathbf{i} + y\mathbf{j}$$

be the position vector of the point (x, y).

Then
$$|\mathbf{R}| = \sqrt{x^2 + y^2}.$$

Thus
$$\nabla f = -\frac{\mathbf{R}}{|\mathbf{R}|^3}$$

and is therefore pointed toward the origin. Moreover,

$$|\nabla f| = \left| -\frac{\mathbf{R}}{|\mathbf{R}|^3} \right| = \frac{|\mathbf{R}|}{|\mathbf{R}|^3} = \frac{1}{|\mathbf{R}|^2}.$$

When **R** is short, ∇f is long, and when **R** is long, ∇f is short. In physics, the gradient field

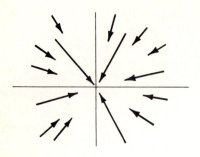

$$\frac{-\mathbf{R}}{|\mathbf{R}|^3}$$

corresponds to a force of attraction. The accompanying sketch shows a few of the values of this vector field. (However, not every vector field is the gradient of some scalar field, as Exercise 20 shows.) ●

In Example 1 we obtained a vector field from a scalar field by "taking the gradient." Also of importance in mathematics and physics is the following procedure, which obtains a scalar field from a vector field.

DEFINITION *Divergence of a vector field.* Let

$$\mathbf{F}(x, y) = P(x, y)\mathbf{i} + Q(x, y)\mathbf{j}$$

be a vector field in the plane. The scalar field

$$\frac{\partial P}{\partial x} + \frac{\partial Q}{\partial y}$$

is called the *divergence of* **F**. Similarly, if

$$\mathbf{F}(x, y, z) = P(x, y, z)\mathbf{i} + Q(x, y, z)\mathbf{j} + R(x, y, z)\mathbf{k},$$

a vector field in space, then the scalar field

$$\frac{\partial P}{\partial x} + \frac{\partial Q}{\partial y} + \frac{\partial R}{\partial z}$$

is called the *divergence of* **F**.

The next chapter will show why this particular function is called *divergence*. It turns out that if **F** describes the velocity of a fluid, then the divergence of **F** at a point describes the tendency of that fluid to accumulate or disperse near that point.

EXAMPLE 2 Compute the divergence of the vector field

$$5x^2y\mathbf{i} + xy\mathbf{j} + x^2z\mathbf{k}.$$

SOLUTION By definition, the divergence of the vector field $5x^2y\mathbf{i} + xy\mathbf{j} + x^2z\mathbf{k}$ is

$$\frac{\partial(5x^2y)}{\partial x} + \frac{\partial(xy)}{\partial y} + \frac{\partial(x^2z)}{\partial z}.$$

A short computation shows that this sum reduces to

$$10xy + x + x^2. ●$$

NOTATION For convenience introduce the formal "vector"

$$\nabla = \frac{\partial}{\partial x}\mathbf{i} + \frac{\partial}{\partial y}\mathbf{j}.$$

If $$\mathbf{F} = P\mathbf{i} + Q\mathbf{j},$$

compute $\mathbf{V} \cdot \mathbf{F}$ as a dot product of two ordinary vectors:

$$\mathbf{V} \cdot \mathbf{F} = \frac{\partial}{\partial x} P + \frac{\partial}{\partial y} Q.$$

Interpret this to mean $\dfrac{\partial P}{\partial x} + \dfrac{\partial Q}{\partial y}.$

This explains the customary notation for the divergence of $\mathbf{F}$,

$$\mathbf{V} \cdot \mathbf{F}.$$

Similarly, to provide a short notation for the divergence of a vector field in space, introduce the formal "vector"

$$\frac{\partial}{\partial x}\mathbf{i} + \frac{\partial}{\partial y}\mathbf{j} + \frac{\partial}{\partial z}\mathbf{k}.$$

Then if $$\mathbf{F} = P\mathbf{i} + Q\mathbf{j} + R\mathbf{k},$$

$\mathbf{V} \cdot \mathbf{F}$ is a shorthand for the divergence

$$\frac{\partial P}{\partial x} + \frac{\partial Q}{\partial y} + \frac{\partial R}{\partial z}.$$

The divergence of $\mathbf{F}$ is also written div $\mathbf{F}$.

EXAMPLE 3 Compute the divergence $\mathbf{V} \cdot \mathbf{F}$ of the vector field

$$\mathbf{F}(x, y, z) = x^2 y\mathbf{i} + e^{xy}\mathbf{j} + z^2\mathbf{k}.$$

SOLUTION The divergence of $\mathbf{F}$ is

$$\mathbf{V} \cdot \mathbf{F} = \frac{\partial(x^2 y)}{\partial x} + \frac{\partial(e^{xy})}{\partial y} + \frac{\partial(z^2)}{\partial z}$$

$$= 2xy + xe^{xy} + 2z. \;\bullet$$

Note how much information is packed into the compact notation $\mathbf{V} \cdot \mathbf{F}$, read as "del dot $\mathbf{F}$."   When you read $\mathbf{V} \cdot \mathbf{F}$, think "$\mathbf{F}$ is a vector field in the plane or space.   It is described by scalar fields P, Q, and, if in space, R:

$$\mathbf{F} = P\mathbf{i} + Q\mathbf{j} + R\mathbf{k}.$$

Then $\mathbf{V} \cdot \mathbf{F}$ is the sum of three partial derivatives,

$$\frac{\partial P}{\partial x} + \frac{\partial Q}{\partial y} + \frac{\partial R}{\partial z},"$$

In case $\mathbf{F}$ is a vector field in space, there is a way of deriving from it another vector field, called the *curl* of $\mathbf{F}$, which is important in physics.   The next chapter will show why it is called curl.

DEFINITION   *Curl of a vector field.*   Let $\mathbf{F} = P\mathbf{i} + Q\mathbf{j} + R\mathbf{k}$ be a vector field in space.   The function that assigns to each point the vector

$$\begin{vmatrix} \mathbf{i} & \mathbf{j} & \mathbf{k} \\ \dfrac{\partial}{\partial x} & \dfrac{\partial}{\partial y} & \dfrac{\partial}{\partial z} \\ P & Q & R \end{vmatrix}$$

is called the curl of $\mathbf{F}$.   It is denoted

$$\mathbf{curl\ F} \qquad or \qquad \mathbf{\nabla \times F}.$$

The latter is read, "del cross $\mathbf{F}$."
   The above formal determinant is shorthand for

$$\left(\frac{\partial R}{\partial y} - \frac{\partial Q}{\partial z}\right)\mathbf{i} - \left(\frac{\partial R}{\partial x} - \frac{\partial P}{\partial z}\right)\mathbf{j} + \left(\frac{\partial Q}{\partial x} - \frac{\partial P}{\partial y}\right)\mathbf{k}.$$

The similarity to the cross product of vectors suggests the useful notation $\mathbf{\nabla \times F}$.

EXAMPLE 4   Compute the curl of $\mathbf{F} = xyz\mathbf{i} + x^2\mathbf{j} - xy\mathbf{k}$.

SOLUTION   The curl of $\mathbf{F}$ is defined as

$$\begin{vmatrix} \mathbf{i} & \mathbf{j} & \mathbf{k} \\ \dfrac{\partial}{\partial x} & \dfrac{\partial}{\partial y} & \dfrac{\partial}{\partial z} \\ xyz & x^2 & -xy \end{vmatrix},$$

which is short for

$$\left[\frac{\partial(-xy)}{\partial y} - \frac{\partial(x^3)}{\partial z}\right]\mathbf{i} - \left[\frac{\partial(-xy)}{\partial x} - \frac{\partial(xyz)}{\partial z}\right]\mathbf{j} + \left[\frac{\partial(x^2)}{\partial x} - \frac{\partial(xyz)}{\partial y}\right]\mathbf{k}$$

$$= (-x - 0)\mathbf{i} - (-y - xy)\mathbf{j} + (2x - xz)\mathbf{k}$$

$$= -x\mathbf{i} + (y + xy)\mathbf{j} + (2x - xz)\mathbf{k}. \ \bullet$$

   There is another scalar field of general importance in engineering, physics, and mathematics.   Say that you start with a scalar field f and form the gradient field ∇f.   Then you may take the divergence of this vector field, obtaining a scalar field.   This scalar field is called the *Laplacian* of f.   If f is a function of two variables, $\nabla f = f_x\mathbf{i} + f_y\mathbf{j}$, and $\nabla \cdot \nabla f$ equals

$$\left(\frac{\partial}{\partial x}\mathbf{i} + \frac{\partial}{\partial y}\mathbf{j}\right) \cdot (f_x\mathbf{i} + f_y\mathbf{j}) = \frac{\partial f_x}{\partial x} + \frac{\partial f_y}{\partial y} = f_{xx} + f_{yy}.$$

The Laplacian   Thus the Laplacian of f is simply $f_{xx} + f_{yy}$.   In the case of a function $f(x, y, z)$, the Laplacian equals $f_{xx} + f_{yy} + f_{zz}$.

The symbols

$$\mathbf{V} \cdot \mathbf{V}f \quad \text{and} \quad \mathbf{V}^2 f$$

are also standard notations for the Laplacian.

EXAMPLE 5 Compute the Laplacian of $f(x, y) = x^3 - 3xy^2$.

SOLUTION First compute f_x and f_y:

$$f_x = 3x^2 - 3y^2 \quad \text{and} \quad f_y = -6xy.$$

Then $f_{xx} = 6x$ and $f_{yy} = -6x.$

Consequently, $\mathbf{V}^2 f = f_{xx} + f_{yy} = 6x + (-6x) = 0.$

The Laplacian in this special case has the constant value 0. ●

A function f whose Laplacian is identically 0 is called *harmonic*. Harmonic functions are important in the study of electricity and temperature distributions.

Exercises

In Exercises 1 to 3 compute $\mathbf{V} \cdot \mathbf{F} = \text{div } \mathbf{F}$.

1. $\mathbf{F} = 2x\mathbf{i} + 3y\mathbf{j} + 4z\mathbf{k}.$
2. $\mathbf{F} = y^2\mathbf{i} - z^2\mathbf{j} + x^2\mathbf{k}.$
3. $\mathbf{F} = \sin xy\mathbf{i} + \cos xy^2\mathbf{j}.$
4. If f is a scalar field and $\mathbf{F}$ is a vector field, which type of field is $f\mathbf{F}$?
5. If $\mathbf{F}$ and $\mathbf{G}$ are vector fields, which type of field is the function $\mathbf{F} \cdot \mathbf{G}$?
6. Which of the following are scalar fields? vector fields?
   (*a*) **curl F**   (*b*) $|\mathbf{F}|$ (*c*) $\mathbf{F} \cdot \mathbf{F}$ (*d*) div **F**
   (*e*) $\mathbf{V} \cdot \mathbf{F}$ (*f*) $\mathbf{F} \times \mathbf{i}$ (*g*) $\mathbf{V} \times \mathbf{F}.$

In Exercises 7 to 9 compute **curl** $\mathbf{F} = \mathbf{V} \times \mathbf{F}$.

7. $\mathbf{F} = x^3\mathbf{i} - y^3\mathbf{j} + z^2\mathbf{k}.$
8. $\mathbf{F} = yz\mathbf{i} + z^2x\mathbf{j} + xyz\mathbf{k}$
9. $\mathbf{F} = y^2z\mathbf{i} + (x + z)\mathbf{j} + \mathbf{k}$
10. By a straightforward computation verify that the curl of the gradient of f is **0**, that is,

$$\mathbf{V} \times (\mathbf{V}f) = \mathbf{0}.$$

11. By a straightforward computation, verify that the divergence of the curl of **F** is 0, that is,

$$\mathbf{V} \cdot (\mathbf{V} \times \mathbf{F}) = 0,$$

where $\mathbf{F} = P\mathbf{i} + Q\mathbf{j} + R\mathbf{k}$. Thus div (**curl F**) = 0.

12. The vector field $\mathbf{F}(x, y, z) = x\mathbf{i} + y\mathbf{j} + z\mathbf{k}$ is usually denoted **R**. It represents the position vector of the point

(x, y, z). The same notation is used for $\mathbf{F}(x, y) = x\mathbf{i} + y\mathbf{j}$. Show that (*a*) in space $\mathbf{V} \cdot \mathbf{R} = 3$; (*b*) in the plane $\mathbf{V} \cdot \mathbf{R} = 2.$

13. (*a*) Prove that the divergence of a constant vector field is 0.
    (*b*) Give an example of a vector field **F** that is not constant, yet $\mathbf{V} \cdot \mathbf{F} = 0.$

14. (*a*) If f is a scalar field and **F** is a vector field, prove that

$$\mathbf{V} \cdot (f\mathbf{F}) = f\mathbf{V} \cdot \mathbf{F} + \mathbf{V}f \cdot \mathbf{F}.$$

   (*b*) Express the equation in (*a*) in a sentence, using such terms as "divergence," "gradient," and "dot product."

15. If f and g are scalar fields, so is fg, their product. Prove that the gradient of fg equals f times the gradient of g plus g times the gradient of f, that is,

$$\mathbf{V}(fg) = f\mathbf{V}g + g\mathbf{V}f.$$

16. (*a*) Show that $f(x, y) = \ln (x^2 + y^2)$ is harmonic.
    (*b*) Show that $f(x, y, z) = \ln (x^2 + y^2 + z^2)$ is not harmonic.

■

17. Letting $\mathbf{F} = P\mathbf{i} + Q\mathbf{j} + R\mathbf{k}$, prove that

$$\mathbf{V} \times f\mathbf{F} = f\mathbf{V} \times \mathbf{F} + \mathbf{V}f \times \mathbf{F}.$$

18. Letting $\mathbf{F} = P\mathbf{i} + Q\mathbf{j} + R\mathbf{k}$, prove that

$$\text{div } \mathbf{F} \times \mathbf{G} = \mathbf{G} \cdot \text{curl } \mathbf{F} - \mathbf{F} \cdot \text{curl } \mathbf{G}.$$

19. For scalar fields f and g show that

$$\text{div}\,(\nabla f \times \nabla g) = 0.$$

■ ■

20. (a) Show that, if $\mathbf{F} = P\mathbf{i} + Q\mathbf{j}$ is a vector field and equals

∇f for some scalar field, then

$$\frac{\partial P}{\partial y} = \frac{\partial Q}{\partial x}.$$

(b) Show that the vector field $x^2 y\mathbf{i} + x^2 y^3\mathbf{j}$ is not of the form ∇f for any scalar field f.

21.2

Line integrals of scalar or vector fields

Consider a curve C in the plane (or in space). Assume that C is parameterized by $\mathbf{R} = \mathbf{G}(t)$, defined for all t in the interval $[a, b]$; think of t as time. Let f be a scalar function in the plane (or in space) defined at least on every point of the curve C. This section defines the definite integral of f over the curve C, and illustrates its use.

To begin, let $s(t)$ denote arc length along the curve, measured from some base point B on the curve. Then consider a typical partition of the interval $[a, b]$,

$$t_0 = a, \; t_1, \; \ldots, \; t_n = b.$$

Let $\Delta s_i = s(t_i) - s(t_{i-1})$. This is positive since we assume that s is increasing. Let P_i be a sample point on the curve corresponding to some instant chosen in the ith time interval.

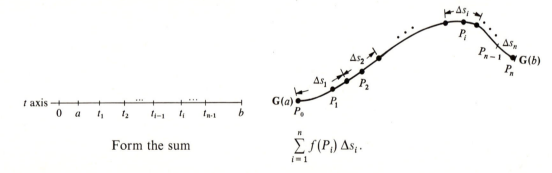

Form the sum

$$\sum_{i=1}^{n} f(P_i)\,\Delta s_i.$$

DEFINITION   *Line integral of a scalar field f.* The limit of sums of the form

$$\sum_{i=1}^{n} f(P_i)\,\Delta s_i,$$

as the mesh of the partition of the interval $[a, b]$ approaches 0, is the line integral of f over the curve given parametrically as $\mathbf{R} = \mathbf{G}(t)$. If the curve is called C, the line integral is denoted

$$\int_C f(P)\,ds.$$

In a moment a line integral will be used to describe the work accomplished by a varying force pushing a particle along a curve.   But first it is advisable to see how a line integral can be calculated.

EXAMPLE 1   Let C be the curve given by the position vector $\mathbf{R} = \mathbf{G}(t) = \cos 2t\,\mathbf{i} + \sin 2t\,\mathbf{j}$ for t in $[0, \pi/2]$, and let $f(P) = f(x, y) = x^2y$.   Compute $\int_C f(P)\,ds$.

SOLUTION   The curve C is given parametrically by $x = \cos 2t$ and $y = \sin 2t$.   Since $x^2 + y^2 = \cos^2 2t + \sin^2 2t = 1$, this curve is part of the circle $x^2 + y^2 = 1$. Since $\mathbf{R}(0) = \cos 0\,\mathbf{i} + \sin 0\,\mathbf{j} = \mathbf{i}$ and $\mathbf{R}(\pi/2) = \cos \pi\,\mathbf{i} + \sin \pi\,\mathbf{j} = -\mathbf{i}$, the curve C is the (top) semicircle from $(1, 0)$ to $(-1, 0)$.

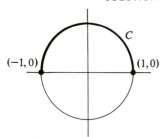

Then
$$\int_C f(P)\,ds = \int_C x^2y\,ds.$$

Now,
$$ds = \frac{ds}{dt}\,dt$$

$$= \sqrt{\left(\frac{dx}{dt}\right)^2 + \left(\frac{dy}{dt}\right)^2}\,dt.$$

Since
$$x = \cos 2t \qquad \text{and} \qquad y = \sin 2t,$$

it follows that
$$\frac{dx}{dt} = -2\sin 2t \qquad \text{and} \qquad \frac{dy}{dt} = 2\cos 2t.$$

Hence
$$\int_C x^2y\,ds = \int_C x^2y\sqrt{(-2\sin 2t)^2 + (2\cos 2t)^2}\,dt$$

$$= \int_0^{\pi/2} \cos^2 2t \sin 2t\sqrt{4}\,dt$$

$$= 2\int_0^{\pi/2} \cos^2 2t \sin 2t\,dt$$

$$= \frac{2(-\cos^3 2t)}{6}\bigg|_0^{\pi/2}$$

$$= -\tfrac{1}{3}(-1)^3 - (-\tfrac{1}{3}\cdot 1^3)$$

$$= \tfrac{2}{3}. \; \bullet$$

The next example introduces an important application of a line integral.

EXAMPLE 2   A particle is pushed along the curve C (in the plane or in space) by a varying force $\mathbf{F}$ (which may be a mixture of gravity, wind, friction, ropes, etc.).   How much work is done?

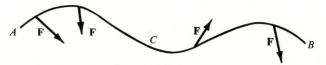

An object moves from A to B on curve C

SOLUTION In case the force **F** remains constant in magnitude and direction and pushes a particle in a straight path from the tail to the head of a vector **R**, the work is defined (in Sec. 19.3) as the dot product

$$\mathbf{F} \cdot \mathbf{R}.$$

This is the product of the scalar component of **F** in the direction of **R** and the distance the particle moves.

   To find the work accomplished by the varying force **F** pushing a particle along a curve, a line integral is required. First, break up the path into short sections of length Δs_i. Select a point P_i in the ith section, evaluate the force **F** at P_i, and find the unit tangent vector, $\mathbf{T}(P_i)$, at P_i.

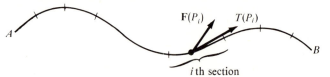

Then a reasonable estimate of the work required to push the particle over the ith section is the product

$$\underbrace{\mathbf{F}(P_i) \cdot \mathbf{T}(P_i)}_{\substack{\text{Scalar component} \\ \text{of } \mathbf{F} \text{ in direction} \\ \text{of motion}}} \quad \underbrace{\Delta s_i}_{\substack{\text{Distance the} \\ \text{particle is} \\ \text{pushed}}}$$

Thus, when all Δs_i are small, the sum

$$\sum_{i=1}^{n} \mathbf{F}(P_i) \cdot \mathbf{T}(P_i) \, \Delta s_i$$

is an estimate of the total work. Taking limits, as all Δs_i are chosen smaller, shows that the line integral

$$\int_C \mathbf{F}(P) \cdot \mathbf{T}(P) \, ds$$

represents the total work. [Note that $\mathbf{F}(P) \cdot \mathbf{T}(P)$ is a scalar function $f(P)$, defined on the curve C.]

In short,

$$\int_C \mathbf{F} \cdot \mathbf{T} \, ds = \text{work.} \quad \bullet$$

   The next example applies the formula developed in Example 2.

EXAMPLE 3 The force $\mathbf{F} = x\mathbf{i} + y^2\mathbf{j}$ acts on a particle that moves on the curve $\mathbf{R}(t) = 2t\mathbf{i} + 3t\mathbf{j}$ from time $t = 1$ to time $t = 2$. How much work is accomplished?

SOLUTION The work equals

$$\int_C \mathbf{F} \cdot \mathbf{T} \, ds = \int_1^2 \mathbf{F} \cdot \mathbf{T} \frac{ds}{dt} \, dt.$$

Next express **F**, **T**, and ds/dt in terms of t:

$$\mathbf{F}(x, y) = x\mathbf{i} + y^2\mathbf{j}$$
$$= 2t\mathbf{i} + (3t)^2\mathbf{j}$$
$$= 2t\mathbf{i} + 9t^2\mathbf{j}.$$

Recall that
$$\mathbf{T} = \frac{\mathbf{V}}{|\mathbf{V}|},$$

where **V** is the velocity vector,

$$\mathbf{V} = \frac{dx}{dt}\mathbf{i} + \frac{dy}{dt}\mathbf{j}$$
$$= 2\mathbf{i} + 3\mathbf{j}.$$

Thus
$$\mathbf{T} = \frac{2\mathbf{i} + 3\mathbf{j}}{\sqrt{2^2 + 3^2}}$$
$$= \frac{2\mathbf{i} + 3\mathbf{j}}{\sqrt{13}}.$$

The quantity ds/dt is the speed, which is the magnitude of the velocity vector **V**;

hence
$$\frac{ds}{dt} = \sqrt{13}.$$

Thus
$$\int_1^2 \mathbf{F} \cdot \mathbf{T}\,\frac{ds}{dt}\,dt = \int_1^2 (2t\mathbf{i} + 9t^2\mathbf{j}) \cdot \left(\frac{2\mathbf{i} + 3\mathbf{j}}{\sqrt{13}}\right)\sqrt{13}\,dt$$
$$= \int_1^2 (4t + 27t^2)\,dt$$
$$= 2t^2 + 9t^3 \Big|_1^2$$
$$= 8 + 72 - (2 + 9)$$
$$= 69. \;\bullet$$

In Example 3 an integral in terms of arc length s was replaced by an integral in terms of t. The next example illustrates a useful shortcut, which expresses the original integral in terms of x and y; t does not appear at all.

EXAMPLE 4   Work Example 3, expressing the integral in terms of x, y, dx, and dy.

SOLUTION   Since $\mathbf{F} = x\mathbf{i} + y^2\mathbf{j}$ and $\mathbf{T} = dx/ds\,\mathbf{i} + dy/ds\,\mathbf{j}$,

$$\mathbf{F} \cdot \mathbf{T} = x\frac{dx}{ds} + y^2\frac{dy}{ds}.$$

Hence
$$\int_C \mathbf{F} \cdot \mathbf{T} \, ds = \int_C \left(x \frac{dx}{ds} + y^2 \frac{dy}{ds} \right) ds$$

$$= \int_C (x \, dx + y^2 \, dy)$$

$$= \int_C x \, dx + \int_C y^2 \, dy.$$

To find the limits on x and y, recall that the curve is given as

$$\mathbf{R}(t) = 2t\mathbf{i} + 3t\mathbf{j}$$

for t in $[1, 2]$. When $t = 1$, we have $x = 2$ and $y = 3$; when $t = 2$, $x = 4$ and $y = 6$. Hence

$$\int_C x \, dx + \int_C y^2 \, dy = \int_2^4 x \, dx + \int_3^6 y^2 \, dy$$

$$= \frac{x^2}{2} \Big|_2^4 + \frac{y^3}{3} \Big|_3^6$$

$$= 6 + 63$$

$$= 69. \; \bullet$$

As Example 2 showed, $\int_C \mathbf{F} \cdot \mathbf{T} \, ds$ is of importance in physics. It is the integral of the tangential component of the vector field $\mathbf{F}$ along the curve. The following definition provides a shorter description.

DEFINITION   *Line integral of a vector field* $\mathbf{F}$.   The line integral

$$\int_C \mathbf{F} \cdot \mathbf{T} \, ds$$

is called the *line integral* of the vector field $\mathbf{F}$.

To evaluate a line integral of a vector field it is usually necessary to express the vector function in terms of appropriate scalar functions.

If $\mathbf{F} = P\mathbf{i} + Q\mathbf{j}$, where P and Q are functions of x and y, and since

$$\mathbf{T} = \frac{dx}{ds}\mathbf{i} + \frac{dy}{ds}\mathbf{j},$$

it follows that
$$\mathbf{F} \cdot \mathbf{T} = P\frac{dx}{ds} + Q\frac{dy}{ds}.$$

Hence
$$\int_C \mathbf{F} \cdot \mathbf{T} \, ds = \int_C \left(P\frac{dx}{ds} + Q\frac{dy}{ds} \right) ds$$

$$= \int_C P \, dx + Q \, dy.$$

This differential notation presents the integral in a computable form. If $\mathbf{F} = P\mathbf{i} + Q\mathbf{j} + R\mathbf{k}$ and C is a curve in space, the corresponding formulation is

$$\int_C \mathbf{F} \cdot \mathbf{T} \, ds = \int_C P \, dx + Q \, dy + R \, dz.$$

The next example illustrates the use of this notation. Note the resemblance to Example 4 in that the integration is expressed in terms of x and y.

EXAMPLE 5   Let $\mathbf{F}(x, \, y) = y\mathbf{i} + x\mathbf{j}$. Compute $\int_C \mathbf{F} \cdot \mathbf{T} \, ds$ on each of the following two curves from $(0, \, 0)$ to $(1, \, 2)$:

(a) The path along the parabola $y = 2x^2$.
(b) The path which goes in a straight line from $(0, \, 0)$ to $(1, \, 0)$ and then in a straight line from $(1, \, 0)$ to $(1, \, 2)$. Note that this path has the same beginning and end as the path in (a).

SOLUTION   (a) We compute the integral over the parabolic path. It is

$$\int_C \mathbf{F} \cdot \mathbf{T} \, ds = \int_C y \, dx + x \, dy.$$

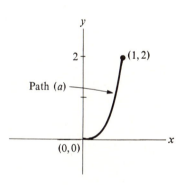

On this path $y = 2x^2$, hence $dy = 4x \, dx$. Thus

$$\int_C y \, dx + x \, dy = \int_0^1 2x^2 \, dx + x(4x \, dx)$$

$$= \int_0^1 6x^2 \, dx$$

$$= 2.$$

Therefore, on the parabolic path from $(0, \, 0)$ to $(1, \, 2)$, the line integral of $\mathbf{F}$ is 2.

(b) The path from $(0, \, 0)$ to $(1, \, 2)$ is composed of two line segments. The integral $\int_C y \, dx + x \, dy$ will be broken into two separate integrals, one over the horizontal segment C_1 that goes from $(0, \, 0)$ to $(1, \, 0)$ and the other over the vertical segment C_2 that goes from $(1, \, 0)$ to $(1, \, 2)$.

On C_1, $y = 0$, hence $dy = 0$. Thus

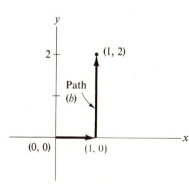

$$\int_{C_1} y \, dx + x \, dy = \int_0^1 0 \, dx = 0.$$

On the vertical segment C_2, $x = 1$, hence $dx = 0$. Thus

$$\int_{C_2} y \, dx + x \, dy = \int_0^2 1 \, dy = \int_0^2 dy = 2. \ \bullet$$

The answers in both (a) and (b) are 2. Section 22.2 will show that if $\mathbf{F}$ is given by the formula $\mathbf{F}(x, \, y) = y\mathbf{i} + x\mathbf{j}$, then the integral of $\mathbf{F} \cdot \mathbf{T}$ over any curve C depends only on the beginning and end of C.

Another notation for the line integral of **F** is based on introducing the differential of the position vector **R**. The differential of **R** is defined as the formal "vector"

$$dR = dx\mathbf{i} + dy\mathbf{j}.$$

Observe that

$$\mathbf{F} \cdot d\mathbf{R} = (P\mathbf{i} + Q\mathbf{j})(dx\mathbf{i} + dy\mathbf{j})$$

$$= P\, dx + Q\, dy.$$

Hence $\int_C \mathbf{F} \cdot \mathbf{T}\, ds$ is now expressible as

$$\int_C \mathbf{F} \cdot d\mathbf{R}.$$

Also note that

$$\mathbf{T}\, ds = \left(\frac{dx}{ds}\mathbf{i} + \frac{dy}{ds}\mathbf{j}\right) ds = dx\mathbf{i} + dy\mathbf{j} = d\mathbf{R}.$$

So the line integral $\int_C \mathbf{F} \cdot \mathbf{T}\, ds$ is also denoted by the symbol

$$\int_C \mathbf{F} \cdot (\mathbf{T}\, ds).$$

Often physicists denote this integral by

$$\int_C \mathbf{F} \cdot d\mathbf{s}.$$

The following table records these various notations for the line integral of the vector field **F**.

Notation	Idea
$\int_C (\mathbf{F} \cdot \mathbf{T})\, ds$	Integral of scalar component of **F** along **T**
$\int_C \mathbf{F} \cdot d\mathbf{R}$	Think of scalar component of **F** along curve times length of small displacement
$\int_C \mathbf{F} \cdot (\mathbf{T}\, ds)$	**T** ds is just another notation for $d\mathbf{R}$
$\int_C P\, dx + Q\, dy$	$\int_C \left(P\dfrac{dx}{ds} + Q\dfrac{dy}{ds}\right) ds$ or $\int_a^b \left(P\dfrac{dx}{dt} + Q\dfrac{dy}{dt}\right) dt$
$\int_C \mathbf{F} \cdot d\mathbf{s}$	Think of the curve as locally straight and $d\mathbf{s}$ as $d\mathbf{R}$

All these notations represent the work accomplished by the force $\mathbf{F} = P\mathbf{i} + Q\mathbf{j}$. (This was done in the plane; the same ideas carry over to curves and vector fields in space.) Courses that apply vector analysis may use all these notations.

The concluding example illustrates the calculation of $\int_C \mathbf{F} \cdot \mathbf{T}\, ds$ in the case that C is a curve in space. It will be referred to in Sec. 22.6.

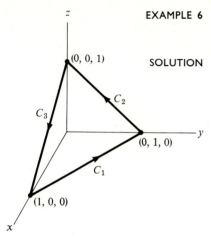

Note That, on C_1, x Goes from 1 down to 0.

EXAMPLE 6 Let $\mathbf{F} = y\mathbf{i} + xz\mathbf{j} + x^2\mathbf{k}$. Let C be the triangular curve that starts at $(1, 0, 0)$, goes to $(0, 1, 0)$, then to $(0, 0, 1)$, and then back to $(1, 0, 0)$. Compute $\int_C \mathbf{F} \cdot \mathbf{T} \, ds$.

SOLUTION Express the integral over C as the sum of integrals over the three straight paths, C_1, C_2, and C_3, that compose C. In each case $\mathbf{F} \cdot \mathbf{T} \, ds = y \, dx + xz \, dy + x^2 \, dz$.

On C_1, $z = 0$, hence $dz = 0$. Thus

$$\int_{C_1} y \, dx + xz \, dy + x^2 \, dz = \int_{C_1} y \, dx.$$

On the line C_1 in the xy plane x and y are related by the equation $x + y = 1$, hence $y = 1 - x$. Thus

$$\int_{C_1} y \, dx = \int_1^0 (1 - x) \, dx$$

$$= \int_0^1 (x - 1) \, dx$$

$$= -\tfrac{1}{2}.$$

On C_2, $x = 0$, hence $dx = 0$. Thus

$$\int_{C_2} y \, dx + xz \, dy + x^2 \, dz = \int_{C_2} 0 \, dy + 0 \, dz = 0.$$

On C_3, $y = 0$, hence $dy = 0$. Thus

$$\int_{C_3} y \, dx + xz \, dy + x^2 \, dz = \int_{C_3} x^2 \, dz$$

$$= \int_1^0 x^2 \, dz$$

$$= \int_1^0 (1 - z)^2 \, dz$$

$$= -\int_0^1 (1 - z)^2 \, dz$$

$$= -\tfrac{1}{3}.$$

Thus

$$\int_C \mathbf{F} \cdot \mathbf{T} \, ds = -\tfrac{1}{2} + 0 + \left(-\tfrac{1}{3}\right) = -\tfrac{5}{6}. \ \bullet$$

Exercises

In Exercises 1 to 4, C is the curve given parametrically by $\mathbf{R} = \mathbf{G}(t) = t^2\mathbf{i} + t^3\mathbf{j}$. In each case evaluate the line integral where t goes from 0 to 1.

1. $\int_C \mathbf{F} \cdot \mathbf{T} ds$, where $\mathbf{F}(x, y) = x\mathbf{i} + 3y\mathbf{j}$.

2. $\int_C \mathbf{F} \cdot d\mathbf{R}$, where $\mathbf{F}(x, y) = xy\mathbf{i} - x\mathbf{j}$.

3. $\int_C \mathbf{F} \cdot (\mathbf{T} \, ds)$, where $\mathbf{F}(x, y) = y^2\mathbf{i} + x^2\mathbf{j}$.

4. $\int_C P \, dx + Q \, dy$, where $P = xy^2$ and $Q = x + y$.

In Exercises 5 to 7 compute the work accomplished by the force

F in moving a particle along the curve parameterized by the function **G**.

5. $F(x, y) = x^2 i - xy j$;
 $G(t) = t i + (1 - t) j$, t in $[1, 2]$.

6. $F(x, y) = \dfrac{x}{\sqrt{x^2 + y^2}} i + \dfrac{y}{\sqrt{x^2 + y^2}} j$;
 $G(t) = t i + \sin t j$, t in $[0, \pi]$.

7. $F(x, y) = -x i - y j$;
 $G(t) = t i + t^2 j$, t in $[0, 2]$.

8. Compute $\int_C x^3 y\, dx$ if C goes from $(0, 1)$ to $(1, 0)$ along (a) the line $x + y = 1$, with the parameterization $x = t$, $y = 1 - t$; (b) the circle $x^2 + y^2 = 1$, with the parameterization $x = \sin t$, $y = \cos t$.

9. Compute:

$$\int_C \frac{-y\, dx}{x^2 + y^2} + \frac{x\, dy}{x^2 + y^2},$$

where C goes from $(1, 0)$ to $(1, 1)$ along (a) the straight line $x = 1$, parameterized as $x = 1$, $y = t$; (b) the circular path parameterized as $x = \cos 2\pi t$, $y = \sin 2\pi t$, t in $[0, \frac{1}{4}]$ and then followed by the path $x = t$, $y = 1$, t in $[0, 1]$.

10. (a) Let $x = f(t)$, $y = g(t)$ be any parameterization of a curve from $(0, 0)$ to $(1, 2)$. Show that $\int_C y\, dx + x\, dy$ equals 2.
 (b) Show that $\int_C y\, dx + x\, dy$ depends only on the endpoints of any curve C.

11. Let $F(x, y, z) = 2x i + 3yz j + y^2 k$. Let C be the straight path from $(1, 0, 0)$ to $(1, 1, 1)$. Evaluate $\int_C F \cdot T\, ds$.

In this exercise the three integrals are equal. This is not the case for all **F**.

12. Let $F(x, y, z) = 2xyz i + x^2 z j + x^2 y k$. Evaluate $\int_C F \cdot T\, ds$ on each of these three paths from $(0, 0, 0)$ to $(1, 1, 1)$.
 (a) C_1, the straight path from $(0, 0, 0)$ to $(1, 1, 1)$;
 (b) C_2, the polygonal path from $(0, 0, 0)$ to $(1, 0, 0)$, then to $(1, 1, 0)$, and then to $(1, 1, 1)$;
 (c) C_3, the polygonal path from $(0, 0, 0)$ to $(0, 1, 0)$ and then directly to $(1, 1, 1)$.

13. Let $F = xz i + x^2 j + xy k$. Let C be the path around the square whose vertices are $(0, 0, 1)$, $(1, 0, 1)$, $(1, 1, 1)$, and $(0, 1, 1)$. The path starts at $(0, 0, 1)$, sweeps out the vertices in the indicated order, and returns to $(0, 0, 1)$. Evaluate $\int_C F \cdot T\, ds$. (See diagram to the right.)

14. Let $f(x, y, z)$ be a scalar function and C a curve situated on a level surface of f. Evaluate $\int_C \nabla f \cdot T\, ds$.

■

15. Compute $\int_C xy\, dx + x^2\, dy$ if C goes from $(0, 0)$ to $(1, 1)$ on (a) the line $y = x$, parameterized as $x = t$, $y = t$; (b) the line $y = x$, parameterized as $x = t^2$, $y = t^2$; (c) the parabola $y = x^2$, parameterized as $y = t^2$, $x = t$; (d) the polygonal path from $(0, 0)$ to $(0, 1)$ to $(1, 1)$, parameterized conveniently.

16. The gravitational force **F** of the earth, located at the origin $(0, 0)$ of a rectangular coordinate system, on a certain particle at the point (x, y) is

$$\frac{-x i}{(\sqrt{x^2 + y^2})^3} + \frac{-y j}{(\sqrt{x^2 + y^2})^3}.$$

Compute the total work done by **F** if the particle goes from $(2, 0)$ to $(0, 1)$ along (a) the ellipse $x = 2 \cos t$, $y = \sin t$; (b) the line $x = 2 - 2t$, $y = t$.

■ ■

17. Let $R = G(t)$ parameterize a curve C that begins at the point A and ends at the point B.
 (a) Show that, if $F(x, y)$ is the unit tangent vector $T(x, y)$ for (x, y) on C, then

$$\int_C F \cdot T\, ds$$

is the length of the curve C.
 (b) Show that, if $F(x, y) = x i + y j$, then

$$\int_C F \cdot dR$$

equals $\qquad \frac{1}{2}(|\overrightarrow{OB}|^2 - |\overrightarrow{OA}|^2)$.

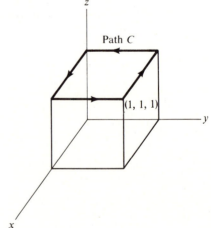

21.3

The integral of the normal component of a vector field

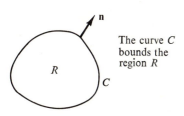

The curve C bounds the region R

The preceding section was concerned with the integral of the tangential component of a vector field over any curve, $\int_C \mathbf{F} \cdot \mathbf{T}\, ds$, which can be thought of as the work accomplished by $\mathbf{F}$. This section is concerned with a different type of integral of a vector field that can be defined over a curve that bounds a surface or over a surface that bounds a solid.

Let C be a curve enclosing some convex region R in the plane. At each point P on C let $\mathbf{n}$ be the unit vector perpendicular to the curve and pointing away from R. (This is the opposite of the principal normal.)

The curve C is given parametrically by the vector function $\mathbf{G}(t)$ for t in $[a, b]$. Since the curve forms a closed loop, its start or initial point $\mathbf{G}(a)$ coincides with its end or terminal point $\mathbf{G}(b)$,

$$\mathbf{G}(a) = \mathbf{G}(b).$$

Any curve whose initial and terminal points are the same is called a *closed curve*.

Imagine that a closed curve C is placed on top of a stream of liquid and kept fixed. Let $\mathbf{V}$ be the velocity vector of the fluid at each point on the surface. Let f be the density of the liquid. Let $\mathbf{F} = f\mathbf{V}$. This vector function records the density of the fluid, its speed, and its direction. It is called the *flux*. Note that $\mathbf{F}$ and $\mathbf{V}$ point in the same direction. Both the density f and the velocity $\mathbf{V}$ may vary with the point in the stream and with time.

We raise the question: At what rate is fluid escaping or entering the region R?

Since the fluid is escaping or entering R only along its boundary, it suffices to consider the total loss or gain past C. Where $\mathbf{V}$ is tangent to C, fluid neither enters nor leaves. Where $\mathbf{V}$ is not tangent to C, fluid is either entering or leaving across C, as indicated in the diagram below.

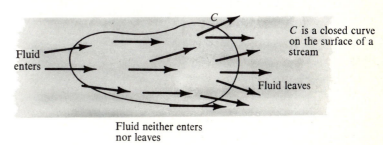

C is a closed curve on the surface of a stream

Fluid enters

Fluid leaves

Fluid neither enters nor leaves

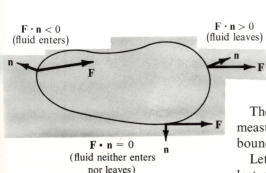

$\mathbf{F} \cdot \mathbf{n} < 0$
(fluid enters)

$\mathbf{F} \cdot \mathbf{n} > 0$
(fluid leaves)

$\mathbf{F} \cdot \mathbf{n} = 0$
(fluid neither enters nor leaves)

The critical component of $\mathbf{F}$ (or $\mathbf{V}$) is that along $\mathbf{n}$. Indeed, $\mathbf{F} \cdot \mathbf{n}$ is a measure of how much fluid is leaving or entering near a certain point on the boundary.

Let us now compute the total net loss of fluid across C. The amount of fluid lost crossing a short (nearly straight) section of the curve, of length Δs, is presumably proportional to Δs and is approximately

$$(\mathbf{F} \cdot \mathbf{n})\,\Delta s,$$

where $\mathbf{F} \cdot \mathbf{n}$ is evaluated at some point in the section.
Hence the line integral

$$\int_C \mathbf{F} \cdot \mathbf{n}\,ds$$

represents the rate at which fluid is leaving the region R.
Frequently, if C is a closed curve, the line integral

$$\int_C f(P)\,ds$$

is denoted

$$\oint f(P)\,ds.$$

Thus

$$\oint \mathbf{F} \cdot \mathbf{n}\,ds$$

represents the total rate at which fluid is leaving the region enclosed by C.

EXAMPLE 1   Let the flux of a fluid be given by the vector

$$\mathbf{F} = (x + 2)\mathbf{i}.$$

Let R be the disk of radius 1 and center $(0, 0)$.   Is fluid tending to leave or to enter R? In other words, is $\oint \mathbf{F} \cdot \mathbf{n}\,ds$ positive or is it negative?

SOLUTION   Observe that the flux $\mathbf{F}$ depends on x, not on y, and increases as x increases. Since the flux is larger where the fluid is escaping than where it is entering, the line integral

$$\oint \mathbf{F} \cdot \mathbf{n}\,ds$$

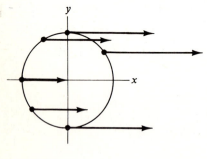

around the boundary should be positive.   Let us compute it and see whether it is.

To parameterize the circle of radius 1, let

$$\mathbf{R} = \mathbf{G}(\theta) = \cos\theta\,\mathbf{i} + \sin\theta\,\mathbf{j}, \text{ for } \theta \text{ in } [0, 2\pi].$$

Here the polar angle θ plays the role of parameter t.   Then

$$\mathbf{n}(\theta) = \cos\theta\,\mathbf{i} + \sin\theta\,\mathbf{j},$$

for the unit normal vector $\mathbf{n}$ is precisely $\mathbf{R}$ in the case of the unit circle.   Also,

$$\mathbf{F}(x, y) = (x + 2)\mathbf{i} = (\cos\theta + 2)\mathbf{i}$$

for (x, y) on C.   Thus the total flow out of R is

$$\oint \mathbf{F} \cdot \mathbf{n}\,ds = \int_0^{2\pi} (\cos\theta + 2)\mathbf{i} \cdot (\cos\theta\,\mathbf{i} + \sin\theta\,\mathbf{j})\,d\theta,$$

since $ds = d\theta$ on a *unit* circle. Thus

$$\oint \mathbf{F} \cdot \mathbf{n} \, ds = \int_0^{2\pi} (\cos \theta + 2) \cos \theta \, d\theta$$

$$= \int_0^{2\pi} (\cos^2 \theta + 2 \cos \theta) \, d\theta.$$

This definite integral can be evaluated by the fundamental theorem. However, a couple of observations provide a shortcut

Note that

$$\int_0^{2\pi} 2 \cos \theta \, d\theta = 0,$$

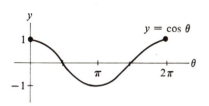

since the integral of the cosine function from π to 2π is the negative of the integral of the cosine function from 0 to π. This is shown in the graph at the left.

To evaluate $\int_0^{2\pi} \cos^2 \theta \, d\theta$, recall the shortcut after Example 1 in Sec. 9.8, and obtain $\int_0^{2\pi} \cos^2 \theta \, d\theta = \pi$.

Thus

$$\int_0^{2\pi} (\cos^2 \theta + 2 \cos \theta) \, d\theta = \pi + 0 = \pi.$$

The total flux is π, which is, as expected, positive. ●

In order to compute $\mathbf{F} \cdot \mathbf{n}$, where $\mathbf{F} = P\mathbf{i} + Q\mathbf{j}$, it will be useful to have a formula for $\mathbf{n}$, the exterior normal, comparable to the formula

$$\mathbf{T} = \frac{dx}{ds} \mathbf{i} + \frac{dy}{ds} \mathbf{j}$$

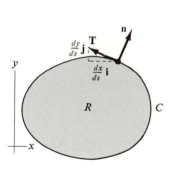

for the unit tangent vector.

To obtain such a formula, assume that C is swept out counterclockwise. A typical $\mathbf{T}$ and $\mathbf{n}$ are shown in the diagram to the left. As the diagram below shows, the exterior normal $\mathbf{n}$ has its x component equal to the y

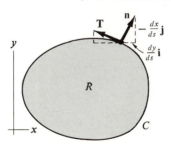

component of $\mathbf{T}$, and its y component equal to the negative of the x component of $\mathbf{T}$.

Thus

$$\mathbf{n} = \frac{dy}{ds} \mathbf{i} - \frac{dx}{ds} \mathbf{j}.$$

Let us check that $\mathbf{T}$ and $\mathbf{n}$ are perpendicular, by showing that $\mathbf{T} \cdot \mathbf{n} = 0$:

$$\mathbf{T} \cdot \mathbf{n} = \left(\frac{dx}{ds}\mathbf{i} + \frac{dy}{ds}\mathbf{j}\right) \cdot \left(\frac{dy}{ds}\mathbf{i} - \frac{dx}{ds}\mathbf{j}\right)$$

$$= \frac{dx}{ds}\frac{dy}{ds} - \frac{dy}{ds}\frac{dx}{ds}$$

$$= 0.$$

Thus if
$$\mathbf{F} = P\mathbf{i} + Q\mathbf{j},$$

then
$$\oint \mathbf{F} \cdot \mathbf{n} \, ds = \oint (P\mathbf{i} + Q\mathbf{j}) \cdot \left(\frac{dy}{ds}\mathbf{i} - \frac{dx}{ds}\mathbf{j}\right) ds$$

$$= \oint \left[P\left(\frac{dy}{ds}\right) + Q\left(-\frac{dx}{ds}\right) \right] ds$$

$$= \oint P \, dy - Q \, dx$$

or
$$= \oint - Q \, dx + P \, dy.$$

EXAMPLE 2 Compute
$$\oint \mathbf{F} \cdot \mathbf{n} \, ds,$$

where $\mathbf{F} = P\mathbf{i} + Q\mathbf{j} = x\mathbf{i} + 2y\mathbf{j}$ and the curve encloses the ellipse

$$\frac{x^2}{2^2} + \frac{y^2}{3^2} = 1.$$

SOLUTION First parameterize the ellipse, as in Sec. 10.8:

$$x = 2 \cos t, \qquad y = 3 \sin t;$$

that is,
$$\mathbf{G}(t) = 2 \cos t\mathbf{i} + 3 \sin t\mathbf{j}, \ t \text{ in } [0, 2\pi].$$

In this case $P = x$ and $Q = 2y$, and

$$\oint \mathbf{F} \cdot \mathbf{n} \, ds = \oint P \, dy - Q \, dx$$

$$= \int_{t=0}^{t=2\pi} x \, dy - 2y \, dx$$

$$= \int_0^{2\pi} (2 \cos t)(3 \cos t \, dt) - 2(3 \sin t)(-2 \sin t \, dt)$$

$$= \int_0^{2\pi} (6 \cos^2 t + 12 \sin^2 t) \, dt$$

$$= 18\pi. \ \bullet$$

EXAMPLE 3 Let $\mathbf{F} = x\mathbf{i} + xy\mathbf{j}$. Let C be the closed curve shown in the diagram in the

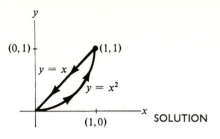

(0, 1)

(1, 1)

$y = x$

$y = x^2$

(1, 0)

SOLUTION

margin. The path is counterclockwise, begins at $(0, 0)$, goes along the parabola $y = x^2$ to $(1, 1)$, and then returns to $(0, 0)$ on the line $y = x$. Compute

$$\oint_C \mathbf{F} \cdot \mathbf{n} \, ds.$$

$$\oint_C \mathbf{F} \cdot \mathbf{n} \, ds = \oint_C (x\mathbf{i} + xy\mathbf{j}) \cdot \left(\frac{dy}{ds} \mathbf{i} - \frac{dx}{ds} \mathbf{j} \right) ds$$

$$= \oint_C x \, dy - xy \, dx.$$

The integral must be broken into two integrals, one along the parabola P from $(0, 0)$ to $(1, 1)$, the other along the line segment L from $(1, 1)$ to $(0, 0)$. On the parabola,

$$y = x^2 \qquad \text{and} \qquad dy = 2x \, dx.$$

Therefore the integral on the parabola

$$\int_P x \, dy - xy \, dx$$

reduces to

$$\int_0^1 x(2x \, dx) - x^3 \, dx = \int_0^1 (2x^2 - x^3) \, dx$$

$$= \left(\frac{2x^3}{3} - \frac{x^4}{4} \right)\Big|_0^1$$

$$= \tfrac{2}{3} - \tfrac{1}{4}$$

$$= \tfrac{5}{12}.$$

On the line $y = x$, use x as a parameter, varying from 1 down to 0. On the line from $(1, 1)$ to $(0, 0)$,

$$y = x \qquad \text{and} \qquad dy = dx.$$

The integral over the straight part of the path is

$$\int_L x \, dy - xy \, dx = \int_1^0 x \, dx - x^2 \, dx$$

$$= \int_0^1 (-x + x^2) \, dx$$

$$= -\tfrac{1}{6}.$$

Thus,

$$\oint \mathbf{F} \cdot \mathbf{n} \, ds = \tfrac{5}{12} - \tfrac{1}{6} = \tfrac{1}{4}. \quad \bullet$$

The flow of a fluid past a plane curve suggested the investigation of $\oint \mathbf{F} \cdot \mathbf{n} \, ds$ for a curve surrounding a region. But we can just as well examine the flow of a fluid past a surface S that is the boundary of some solid region R such as a cube or ball.

S is the surface of the solid R

S

R

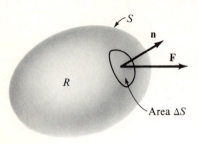

Consider such a surface S placed in a stream. Since it is imaginary—not material, fluid can flow through the surface. Let $\mathbf{F}$ be the flux. At what rate is fluid escaping or entering the region R? The reasoning is practically the same as in the case of a curve bounding a plane region.

First of all, at each point P let $\mathbf{n}$ be the unit vector perpendicular to the surface and pointing away from R. The amount of fluid lost crossing a small (nearly flat) portion of the surface, of area ΔS, is proportional to ΔS and is approximately

$$\mathbf{F} \cdot \mathbf{n}\, \Delta S,$$

where $\mathbf{F} \cdot \mathbf{n}$ is evaluated at some point in the small portion of the surface. Hence the surface integral

$$\int_S \mathbf{F} \cdot \mathbf{n}\, dS$$

represents the rate at which fluid is leaving the region R.

The next example evaluates such a surface integral.

EXAMPLE 4 Let $\mathbf{F} = x\mathbf{i} + y\mathbf{j} + z\mathbf{k}$ and let S be the sphere $x^2 + y^2 + z^2 = 9$. Evaluate $\int_S \mathbf{F} \cdot \mathbf{n}\, dS$.

SOLUTION Since a vector from the center of a sphere to a point P on the sphere is a normal to the sphere at P, the vector

$$x\mathbf{i} + y\mathbf{j} + z\mathbf{k}$$

is normal to the sphere at (x, y, z). It points away from the ball that the sphere bounds; but, having magnitude 3, it is not a unit normal. A unit normal is

$$\mathbf{n} = \frac{x}{3}\mathbf{i} + \frac{y}{3}\mathbf{j} + \frac{z}{3}\mathbf{k}.$$

Thus
$$\int_S \mathbf{F} \cdot \mathbf{n}\, dS = \int_S (x\mathbf{i} + y\mathbf{j} + z\mathbf{k}) \cdot \left(\frac{x}{3}\mathbf{i} + \frac{y}{3}\mathbf{j} + \frac{z}{3}\mathbf{k}\right) dS$$

$$= \int_S \left(\frac{x^2}{3} + \frac{y^2}{3} + \frac{z^2}{3}\right) dS.$$

On the sphere, $x^2 + y^2 + z^2 = 9$. Thus

$$\int_S \mathbf{F} \cdot \mathbf{n}\, dS = \int_S \frac{9}{3}\, dS = 3 \cdot \text{area of sphere}$$

$$= 3 \cdot 4\pi 3^2$$

$$= 108\pi. \quad \bullet$$

The integral in Example 4 was easy to evaluate because of the convenient relation between $\mathbf{F}$ and the sphere. The next example requires a little more work.

EXAMPLE 5 Let $\mathbf{F} = x\mathbf{i} + y\mathbf{j} + z\mathbf{k}$. Let S be the boundary of the solid which lies between the surface $z = 9 - x^2 - y^2$ and the xy plane. Evaluate $\int_S \mathbf{F} \cdot \mathbf{n}\, dS$.

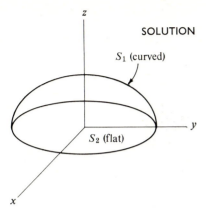

SOLUTION The surface S, shown in the margin, has a curved top and a flat base. Call the curved part S_1 and the flat part S_2.

To evaluate $\int_S \mathbf{F} \cdot \mathbf{n}\, dS$ we will evaluate $\int_{S_1} \mathbf{F} \cdot \mathbf{n}\, dS$ and $\int_{S_2} \mathbf{F} \cdot \mathbf{n}\, dS$ separately and add the results.

First consider $\int_{S_1} \mathbf{F} \cdot \mathbf{n}\, dS$. To find a normal to S_1, write its equation in the form

$$z + x^2 + y^2 = 9.$$

The gradient of $z + x^2 + y^2$ provides a normal, which is

$$2x\mathbf{i} + 2y\mathbf{j} + \mathbf{k}.$$

It points to the outside of S, as a quick sketch will indicate. The *unit* exterior normal is thus

$$\mathbf{n} = \frac{2x\mathbf{i} + 2y\mathbf{j} + \mathbf{k}}{\sqrt{4x^2 + 4y^2 + 1}}.$$

Hence
$$\int_{S_1} \mathbf{F} \cdot \mathbf{n}\, dS = \int_{S_1} (x\mathbf{i} + y\mathbf{j} + z\mathbf{k}) \cdot \frac{2x\mathbf{i} + 2y\mathbf{j} + \mathbf{k}}{\sqrt{4x^2 + 4y^2 + 1}}\, dS$$

$$= \int_{S_1} \frac{2x^2 + 2y^2 + z}{\sqrt{4x^2 + 4y^2 + 1}}\, dS.$$

By good fortune, this integral was evaluated in Example 3 of Sec. 20.7 (by replacing it by an integral over a plane region). From that example, we have

$$\int_{S_1} \mathbf{F} \cdot \mathbf{n}\, dS = \frac{243\pi}{2}.$$

The second integral, $\int_{S_2} \mathbf{F} \cdot \mathbf{n}\, dS$, over the flat base, is much easier. The unit exterior normal is just $\mathbf{n} = -\mathbf{k}$. Then

$$\mathbf{F} \cdot \mathbf{n} = (x\mathbf{i} + y\mathbf{j} + z\mathbf{k}) \cdot (-\mathbf{k}) = -z.$$

But at the base, $z = 0$. Hence

$$\int_{S_2} \mathbf{F} \cdot \mathbf{n}\, dS = \int_{S_2} 0\, dS = 0.$$

All told,
$$\int_S \mathbf{F} \cdot \mathbf{n}\, dS = \frac{243\pi}{2} + 0 = \frac{243\pi}{2}. \;\bullet$$

Incidentally, the integrals $\oint \mathbf{F} \cdot \mathbf{n}\, ds$ and $\int_S \mathbf{F} \cdot \mathbf{n}\, dS$ are also significant in the theory of heat flow. In that case $\mathbf{F}$ represents the flow of heat in a material body. The integral represents the net gain or loss of heat across the boundary.

Exercises

In Exercises 1 to 3 compute $\oint \mathbf{F} \cdot \mathbf{T} \, ds$ and $\oint \mathbf{F} \cdot \mathbf{n} \, ds$.

1. $\mathbf{F}(x, y) = 2x\mathbf{i}$ and the curve is the circle;
   $\mathbf{G}(\theta) = 3 \cos \theta \mathbf{i} + 3 \sin \theta \mathbf{j}$, θ in $[0, 2\pi]$.

2. $\mathbf{F}(x, y) = x\mathbf{i} + y\mathbf{j}$ and the curve is the ellipse;
   $\mathbf{G}(t) = a \cos t \mathbf{i} + b \sin t \mathbf{j}$, t in $[0, 2\pi]$.

3. $\mathbf{F}(x, y) = 3x\mathbf{i} + 4y\mathbf{j}$ and the curve is the circle;
   $\mathbf{G}(\theta) = a \cos \theta \mathbf{i} + a \sin \theta \mathbf{j}$, θ in $[0, 2\pi]$.

4. Compute $\oint_C - y \, dx + x \, dy$, where C is the ellipse

$$\begin{cases} x = a \cos t, \\ y = b \sin t \end{cases} \quad 0 \le t \le 2\pi.$$

(In the next chapter it will be shown that $\oint - y \, dx + x \, dy$, where the curve is swept out counterclockwise, is always twice the area of the region bounded by the curve.)

5. Let the flux vector $\mathbf{F}$ of fluid at the point (x, y) be $(x + 1)^2 \mathbf{i} + y\mathbf{j}$. Let C be the unit circle described parametrically as $x = \cos t$, $y = \sin t$, for t in $[0, 2\pi]$.
   (a) Draw $\mathbf{F}$ at eight convenient equally spaced points on the circle.
   (b) Is fluid tending to leave or enter the region bounded by C; that is, is the net outward flow positive or negative? [Answer on the basis of your diagram in (a).]
   (c) Compute the net outward flow with the aid of a line integral.

6. Let C be the closed curve $x = 2 \cos t$, $y = 3 \sin t$, for t in $[0, 2\pi]$.
   (a) Graph C.
   (b) Compute $\oint_C x^2 \, dx + (y + 1) \, dy$.
   (c) Devise a work problem whose answer is the integral in (b).
   (d) Devise a fluid-flow problem whose answer is the integral in (b).

7. Let $\mathbf{F}$ be the flux vector in fluid flow. Why do you think $\oint_C \mathbf{F} \cdot d\mathbf{R}$ is called the *circulation* around C?

In Exercises 8 to 11 check that

$$\int_R \mathbf{V} \cdot \mathbf{F} \, dA = \int_C \mathbf{F} \cdot \mathbf{n} \, ds,$$

where C is the boundary of R. (This equation describes a major theorem of the next chapter.)

8. $\mathbf{F}$ is the vector field of Example 2 and R is the elliptical region bounded by the curve of that example.

9. $\mathbf{F}$ is the vector field of Example 3 and R is the region bounded by the closed curve of that example.

10. $\mathbf{F}$ is the vector field $2xy\mathbf{i} - x^2 y\mathbf{j}$ and R is the triangular region whose vertices are $(0, 0)$, $(1, 0)$, $(1, 1)$.

11. $\mathbf{F}$ is the vector field $x^2\mathbf{i} + y^2\mathbf{j}$ and R is the region between $y = x^2$ and $y = x^3$.

In Exercises 12 to 15 check that $\int_R \mathbf{V} \cdot \mathbf{F} \, dV = \int_S \mathbf{F} \cdot \mathbf{n} \, dS$, where S is the boundary of R.

12. $\mathbf{F} = 3x\mathbf{i} + 2y\mathbf{j} + 6z\mathbf{k}$, and R is the tetrahedron with the four vertices $(0, 0, 0)$, $(1, 0, 0)$, $(0, 2, 0)$, $(0, 0, 3)$. (Exercise 5 in Sec. 20.7 is related to this one.)

13. $\mathbf{F} = x\mathbf{i} + y\mathbf{j} + z\mathbf{k}$, and R is bounded by the surface $z = 9 - x^2 - 2y^2$ and the xy plane. (Exercise 7 in Sec. 20.7 is related to this one.)

14. $\mathbf{F} = x^2\mathbf{i} + y\mathbf{j} + 3\mathbf{k}$, and R is the cube spanned by the basic unit vectors $\mathbf{i}$, $\mathbf{j}$, and $\mathbf{k}$.

15. $\mathbf{F} = 3z\mathbf{i} + xz\mathbf{j} + z\mathbf{k}$, and R is the region bounded by the three coordinate planes, the plane $z = 2$, and the plane $x + y = 1$.

■

16. Let $\mathbf{F} = xy^2\mathbf{i} + y\mathbf{j} + x\mathbf{k}$. Let S be the surface of the tetrahedron whose vertices are $(0, 0, 0)$, $(1, 0, 0)$, $(0, 2, 0)$, and $(0, 0, 3)$. Compute:

$$\int_S (\mathbf{curl} \, \mathbf{F}) \cdot \mathbf{n} \, dS.$$

According to a general result in the next chapter, the integral is equal to 0.

17. Verify that the integral $\oint_C - y \, dx + x \, dy$ is twice the area of the region enclosed by C, when C is (a) the square path from $(a, 0)$ to $(0, a)$ to $(-a, 0)$ to $(0, -a)$ and back to $(a, 0)$; (b) the triangular path from $(0, 0)$ to $(a, 0)$ to $(0, b)$ and back to $(0, 0)$. Assume that a and b are positive.

18. (a) Compute $\oint_C - y \, dx + x \, dy$ over the curve bordering the region bounded by the curve $y = x^3$ and the lines $x = a$ and $y = 0$. Assume that a is positive.
    (b) Compute the area of the region enclosed by the path.

19. Compute $\oint_C - y \, dx + x \, dy$, where C sweeps out the boundary of the rectangle whose vertices are $(0, 0)$, $(a, 0)$, (a, b), $(0, b)$. Assume that a and b are positive.

20. (a) Let C be any closed curve. Show that $\oint_C dx = 0$ and $\oint_C dy = 0$.
    (b) Show that $\oint_C \cos \phi \, ds = 0$ and $\oint_C \sin \phi \, ds = 0$, where ϕ is the angle between $\mathbf{T}$ and $\mathbf{i}$.
    (c) Show that $\oint_C \mathbf{i} \cdot d\mathbf{R} = 0$ and $\oint_C \mathbf{j} \cdot d\mathbf{R} = 0$.

21. Let $\mathbf{F}(x, y) = x\mathbf{i} + y\mathbf{j}$, usually denoted $\mathbf{R}$. Compute $\oint \mathbf{R} \cdot \mathbf{n} \, ds$ around the triangular curve going from $(0, 0)$ to $(a, 0)$, then to $(0, b)$, and then back to $(0, 0)$. (Here a and b are positive numbers.)

22. The convex plane region R is occupied by a sheet of metal. By various heating and cooling devices the temperature along the border is kept fixed (independent of time). Assume that the temperature in R eventually stabilizes also. The steady-state temperature at point P is denoted $T(P)$.
   (a) Why is it plausible that heat in the metal at P tends to flow in the direction of $-\nabla T$?
   (b) Why is it plausible that the rate at which heat moves through P is proportional to $|-\nabla T|$?
   (c) If C is any curve in R that surrounds some region, why is $\oint_C -\nabla T \cdot \mathbf{n} \, ds$ a plausible measure of the heat lost across C?
   (d) Why is $\oint_C \nabla T \cdot \mathbf{n} \, ds = 0$ for any such curve C?

   ■ ■

23. Let $\mathbf{R} = \mathbf{G}(t)$ be a curve C that forms the boundary of a convex region S. Assume that the origin of the coordinate system lies inside S. At each point on the curve C let $\mathbf{n}$ be a unit vector perpendicular to the curve and pointing away from S. By examining approximating sums, show

that
$$\int_C \mathbf{R} \cdot \mathbf{n} \, ds$$

is twice the area of S.

24. The gravitational force of the earth on a satellite located at the point (x, y, z) in space is
$$\frac{-(x\mathbf{i} + y\mathbf{j} + z\mathbf{k})}{(\sqrt{x^2 + y^2 + z^2})^3}.$$

Show that the total work done by gravity during one orbit of the satellite is 0.

25. Let S be a surface that encloses a solid region R. Assume that the origin O is in R. Let $\mathbf{F}(P)$ be the position vector $\overrightarrow{OP}$. Using approximating sums and a sketch, show that
$$\int_S \mathbf{F}(P) \cdot \mathbf{n} \, dS = 3 \cdot \text{volume of } R.$$

26. For the function $\mathbf{F}$ of Exercise 25 verify that
$$\int_S \mathbf{F} \cdot \mathbf{n} \, dS = \int_R \nabla \cdot \mathbf{F} \, dV.$$

21.4
Conservative vector fields

As mentioned in Sec. 21.1, one way of constructing a vector field $\mathbf{F}$ is to begin with a scalar field f and compute its gradient ∇f. This may be done in the plane or in space. In the present section we examine the line integral of a gradient field, a special case of the line integral of a vector field, namely the line integral
$$\int_C \nabla f \cdot \mathbf{T} \, ds.$$

The most important property of a gradient field ∇f is expressed in Theorem 1. It asserts that
$$\int_C \nabla f \cdot \mathbf{T} \, ds$$

depends only on the endpoints of the curve C, and not on the particular route between them. In particular, when f is the gravitational potential, then ∇f is the gravitational force, and Theorem 1 says that the work accomplished by gravity when a mass moves from one point to another is independent of the particular path swept out by the particle.

THEOREM I Let f be a scalar function and let C be a curve going from the point A to the point B. Then
$$\int_C \nabla f \cdot \mathbf{T} \, ds = f(B) - f(A).$$

PROOF For simplicity we take only the planar case. Let C be given by the parameterization $\mathbf{R} = \mathbf{G}(t)$ for t in $[a, b]$. First of all,

$$\nabla f = f_x \mathbf{i} + f_y \mathbf{j}.$$

Let $\mathbf{G}(t) = x(t)\mathbf{i} + y(t)\mathbf{j}$. Then

$$\int_C \nabla f \cdot \mathbf{T} \, ds = \int_C f_x \, dx + f_y \, dy$$

$$= \int_a^b \left(f_x \frac{dx}{dt} + f_y \frac{dy}{dt} \right) dt.$$

The integrand

$$f_x \frac{dx}{dt} + f_y \frac{dy}{dt}$$

is reminiscent of the first chain rule in Sec. 12.6. To be specific, introduce the function H defined by the formula

$$H(t) = f(x(t), y(t)).$$

Then, by Theorem 1 of Sec. 12.6,

$$\frac{dH}{dt} = f_x(x(t), y(t)) \frac{dx}{dt} + f_y(x(t), y(t)) \frac{dy}{dt}.$$

Thus

$$\int_a^b \left(f_x \frac{dx}{dt} + f_y \frac{dy}{dt} \right) dt = \int_a^b \frac{dH}{dt} \, dt = H(b) - H(a)$$

by the fundamental theorem of calculus.

But

$$H(b) = f(x(b), y(b)) = f(B)$$

and similarly

$$H(a) = f(A).$$

Consequently

$$\int_C \nabla f \cdot \mathbf{T} \, ds = f(B) - f(A),$$

and the theorem is proved. ●

Theorem 1 may also be written in the form

$$\int_C \nabla f \cdot d\mathbf{R} = f(B) - f(A).$$

Expressed in this form, it resembles the fundamental theorem of calculus,

$$\int_a^b \frac{df}{dx} \, dx = f(b) - f(a),$$

and extends it from closed intervals on a straight line to curves.

EXAMPLE 1 Let $f(x, y) = xy$ and let C be a curve starting at $(2, 3)$ and ending at $(4, 5)$. Compute $\int_C \nabla f \cdot d\mathbf{R}$.

SOLUTION $\nabla f = y\mathbf{i} + x\mathbf{j}$. According to Theorem 1, then,

$$\int_C \nabla f \cdot d\mathbf{R} = \int_C (y\mathbf{i} + x\mathbf{j}) \cdot d\mathbf{R} = f(4, 5) - f(2, 3)$$

$$= 4 \cdot 5 - 2 \cdot 3$$

$$= 14.$$

Since

$$\int_C (y\mathbf{i} + x\mathbf{j}) \cdot d\mathbf{R} = \int_C (y\mathbf{i} + x\mathbf{j}) \cdot (dx\mathbf{i} + dy\mathbf{j})$$

$$= \int_C y\,dx + x\,dy,$$

this shows that

$$\int_C y\,dx + x\,dy = 14$$

for any curve joining $(2, 3)$ to $(4, 5)$. ●

When the curve C is closed, Theorem 1 takes the following form.

THEOREM 2 If f is a scalar function, then the integral of the vector field ∇f around any closed curve is 0,

$$\oint \nabla f \cdot \mathbf{T}\,ds = 0.$$

PROOF This is a consequence of Theorem 1. For in the case of a closed path, the initial point A coincides with the terminal point B. Hence $f(B) - f(A) = 0$. This proves the theorem. ●

In the theory of gravitational attraction, the scalar function f defined by

$$f(x, y, z) = \frac{-1}{\sqrt{x^2 + y^2 + z^2}}$$

is of great importance. It is defined everywhere except at the origin and is called a *potential function*. The negative of its gradient is called the *force field*. A quick calculation shows that

$$-\nabla f = -\frac{x\mathbf{i} + y\mathbf{j} + z\mathbf{k}}{(\sqrt{x^2 + y^2 + z^2})^3}$$

$$= -\frac{\mathbf{R}}{|\mathbf{R}|^3},$$

where $\mathbf{R}$ is the position vector $x\mathbf{i} + y\mathbf{j} + z\mathbf{k}$. Since

$$-\nabla f = -\frac{\mathbf{R}}{|\mathbf{R}|^3},$$

$-\nabla f$ is pointed toward the origin, and its magnitude is inversely proportional to the square of the distance from the origin to the point (x, y, z), for

$$\left| \frac{-\mathbf{R}}{|\mathbf{R}|^3} \right| = \frac{|\mathbf{R}|}{|\mathbf{R}|^3}$$

$$= \frac{1}{|\mathbf{R}|^2}.$$

Note: The minus signs introduced above in the definition of potential and force are purely for the convenience of physicists.

Theorem 1 tells the physicist that the work accomplished by the gravitational field when moving a particle from one point to another is independent of the path. Theorem 2 says that the work accomplished by gravity on a satellite during one orbit is 0 (thus the satellite may remain in orbit indefinitely).

Now, instead of a gradient field $\mathbf{V}f$, consider any vector field $\mathbf{F}$. The field $\mathbf{F}$ may or may not have the property that $\int_C \mathbf{F} \cdot \mathbf{T} \, ds$ depends only on the endpoints of C. If it does have that property, it is called *conservative*, according to the following definition.

DEFINITION   *Conservative vector field.*   A vector field $\mathbf{F}$ is called *conservative* if, whenever C_1 and C_2 are two curves with the same endpoints,

$$\int_{C_1} \mathbf{F} \cdot \mathbf{T} \, ds = \int_{C_2} \mathbf{F} \cdot \mathbf{T} \, ds.$$

Theorem 1 asserts that any field $\mathbf{F}$ of the form $\mathbf{V}f$ for some scalar function f is conservative.

An equivalent definition of a conservative vector field $\mathbf{F}$ is that for any closed curve $\oint \mathbf{F} \cdot \mathbf{T} \, ds = 0$. (The reader may easily check this.)

The question may come to mind, "If $\mathbf{F}$ is conservative, is it necessarily the gradient of some scalar function?" The answer is "yes." This is the substance of the next theorem.

THEOREM 3   If $\mathbf{F}$ is conservative, then there is a scalar function f such that $\mathbf{F} = \mathbf{V}f$.

PROOF   For convenience we take $\mathbf{F}$ to be planar and defined throughout the plane. Hence $\mathbf{F} = P(x, y)\mathbf{i} + Q(x, y)\mathbf{j}$. Define a scalar function $f(x, y)$ as follows.

Let (x, y) be a point in the plane. Select a curve C that starts at $(0, 0)$ and ends at (x, y). Define $f(x, y)$ to be $\int_C \mathbf{F} \cdot \mathbf{T} \, ds$. Since $\mathbf{F}$ is conservative, the number $f(x, y)$ depends only on the point (x, y) and not on the choice of C.

All that remains is to show that $f_x = P$ and that $f_y = Q$. We will go through the details for the first case, $f_x = P$. The reasoning for the other is similar.

Let (x_0, y_0) be a fixed point in the plane and consider the quotient appearing in the definition of $f_x(x_0, y_0)$, namely,

$$\frac{f(x_0 + h, y_0) - f(x_0, y_0)}{h}.$$

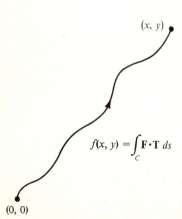

(x, y)

$$f(x, y) = \int_C \mathbf{F} \cdot \mathbf{T} \, ds$$

$(0, 0)$

Consider the numerator,

$$f(x_0 + h, y_0) - f(x_0, y_0).$$

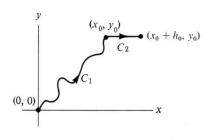

Let C_1 be a curve from $(0, 0)$ to (x_0, y_0) and let C_2 be the straight path from (x_0, y_0) to $(x_0 + h, y_0)$. Let C be the curve from $(0, 0)$ to $(x_0 + h, y_0)$ formed by taking C_1 first and then continuing on C_2.

Then

$$f(x_0, y_0) = \int_{C_1} \mathbf{F} \cdot \mathbf{T} \, ds$$

and

$$f(x_0 + h, y_0) = \int_C \mathbf{F} \cdot \mathbf{T} \, ds = \int_{C_1} \mathbf{F} \cdot \mathbf{T} \, ds + \int_{C_2} \mathbf{F} \cdot \mathbf{T} \, ds.$$

Thus

$$\frac{f(x_0 + h, y_0) - f(x_0, y_0)}{h} = \frac{\int_{C_2} \mathbf{F} \cdot \mathbf{T} \, ds}{h} = \frac{\int_{C_2} P(x, y) \, dx + Q(x, y) \, dy}{h}.$$

But on C_2, y is constant, $y = y_0$, hence $dy = 0$.

Thus

$$\int_{C_2} P(x, y) \, dx = \int_{x_0}^{x_0 + h} P(x, y_0) \, dx.$$

Now,

$$\lim_{h \to 0} \frac{\int_{x_0}^{x_0 + h} P(x, y_0) \, dx}{h}$$

is the definition of the derivative of $\int_{x_0}^t P(x, y_0) \, dx$ with respect to t at the value x_0. The second fundamental theorem of calculus asserts that this derivative is the value of the integrand at the upper limit of integration, which is x_0.

$$\frac{d[\int_a^t f(x) \, dx]}{dt} = f(t).$$

Consequently,

$$f_x(x_0, y_0) = P(x_0, y_0),$$

as was to be shown.

In a similar manner it can be shown that

$$f_y(x_0, y_0) = Q(x_0, y_0). \quad \bullet$$

For a vector field $\mathbf{F}$ defined throughout the plane (or space) the following three properties are therefore equivalent:

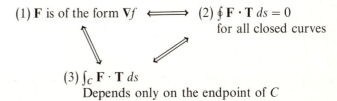

(1) $\mathbf{F}$ is of the form ∇f $\Longleftrightarrow$ (2) $\oint \mathbf{F} \cdot \mathbf{T} \, ds = 0$
for all closed curves

(3) $\int_C \mathbf{F} \cdot \mathbf{T} \, ds$
Depends only on the endpoint of C

An arrow $\Rightarrow$ means "implies." Any one of the three properties, (1), (2), or (3), describes a conservative field. In our approach (3) was used as the defining property.   However, both of the other two are employed as definitions of "conservative."

For the sake of simplicity we have assumed that **F** is defined throughout the plane (or space).   Actually, such a stringent assumption is not required in the argument.   Let R be the domain in which $\mathbf{F}(x, y)$ is defined.   Two properties of R are critical in the proof of Theorem 3: first, that there is a point P_0 in R such that any point P in R can be reached by a curve that begins at P_0 and lies in R; second, if P is a point in R, then there is a disk of nonzero radius and center at P that lies completely in R.   For instance, R might be all the points inside some bounding loop, or it might be all of the plane except for a finite number of points.   Similar assertions hold for $\mathbf{F}(x, y, z)$ defined in space.

Exercises

1. Let C be the curve $\mathbf{G}(t) = (t^4 + t)\mathbf{i} + t^5\mathbf{j}$ for t in $[0, 1]$. Evaluate $\int_C \nabla f \cdot d\mathbf{R}$ if
   (a) $f(x, y) = 3x^2 - 4y^2$;
   (b) $f(x, y) = \cos 2\pi xy$;
   (c) $f(x, y) = e^{x^3 y}$.
2. Let C be the curve $\mathbf{G}(t) = \cos t\mathbf{i} + \sin t\mathbf{j} + (t - \pi)^2\mathbf{k}$ for t in $[0, \pi]$.   Evaluate $\int_C \mathbf{F} \cdot \mathbf{T} \, ds$, where $\mathbf{F}$ is the gradient of
   (a) $x^2 yz$,
   (b) $x^5 + y^5 + z^5$,
   (c) $e^{xz} + \tan xyz$.
3. (a) Show that if $\mathbf{F} = P\mathbf{i} + Q\mathbf{j}$ is a gradient field, then

   $$P_y = Q_x.$$

   (b) Show that $\mathbf{F}(x, y) = x^2 y\mathbf{i} - xy^2\mathbf{j}$ is *not* a gradient field.

In Exercises 4 to 6 use the criterion in Exercise 3 to show that the given vector field is not conservative.
4. $\mathbf{F} = e^x \cos y\mathbf{i} + e^x \sin y\mathbf{j}$
5. $\mathbf{F} = \ln y\mathbf{i} + (y/x)\mathbf{j}$
6. $\mathbf{F} = x^2\mathbf{i} + xy\mathbf{j}$

In Exercises 7 to 9 verify that
7. $\mathbf{F} = e^{-x} \cos y\mathbf{i} + e^{-x} \sin y\mathbf{j}$ is the gradient of $-e^{-x} \cos y$.
8. $\mathbf{F} = \ln y\mathbf{i} + (x/y)\mathbf{j}$ is the gradient of $x \ln y$.
9. $\mathbf{F} = y^2\mathbf{i} + 2xy\mathbf{j}$ is the gradient of $y^2 x$.
10. Consider the vector field

   $$\mathbf{F} = \frac{-y}{x^2 + y^2}\mathbf{i} + \frac{x}{x^2 + y^2}\mathbf{j},$$

   defined everywhere except the origin.

   (a) Verify that $\dfrac{\partial P}{\partial y} = \dfrac{\partial Q}{\partial x}.$

   (b) Show that $\oint_C \mathbf{F} \cdot \mathbf{T} \, ds = 2\pi$, when C is the circle of radius 1 and center $(0, 0)$.
   (c) Show that $\mathbf{F}$ is not a gradient.
11. Let $\mathbf{F}(x, y, z)$ be conservative.   Show that **curl F** $= \mathbf{0}$.

In Exercises 12 to 14 use the criterion in Exercise 11 to show that the given $\mathbf{F}$ is *not* conservative.
12. $\mathbf{F} = x^2\mathbf{i} - xz\mathbf{j} + y^2\mathbf{k}$
13. $\mathbf{F} = zx\mathbf{i} + y^2\mathbf{j} + z^5\mathbf{k}$
14. $\mathbf{F} = zy\mathbf{i} + 3\mathbf{j} + 5\mathbf{k}$
15. There is a fairly simple function $f(x, y, z)$ such that

   $$\nabla f = 2xyz\mathbf{i} + x^2 z\mathbf{j} + x^2 y\mathbf{k}.$$

   Find one such f.

■

16. Explain why these two properties of a vector field $\mathbf{F}$ are equivalent: The integral of the tangential component of $\mathbf{F}$ around any closed curve is 0 ($\oint \mathbf{F} \cdot \mathbf{T} \, ds = 0$); the integral of the tangential component of $\mathbf{F}$ over any curve depends only on the endpoints of the curve ($\int_C \mathbf{F} \cdot \mathbf{T} \, ds$ depends only on the endpoints of C).
17. (a) Show that $\mathbf{F} = 2xy^2\mathbf{i} + 2x^2 y\mathbf{j}$ is conservative by exhibiting a scalar function f such that $\mathbf{F} = \nabla f$.
    (b) Show that $\mathbf{F} = 2xy^2\mathbf{i} + x^2 y\mathbf{j}$ is *not* conservative by exhibiting a closed curve over which $\oint \mathbf{F} \cdot \mathbf{T} \, ds$ is not 0.
18. Complete the proof of Theorem 2, showing that

   $$f_y(x_0, y_0) = Q(x_0, y_0).$$

19. Let $\mathbf{F}(x, y) = x^2\mathbf{i} + y^2\mathbf{j}$.
    (a) Show that $\mathbf{F}$ is conservative by showing that the integral $\oint \mathbf{F} \cdot \mathbf{T} \, ds$ is always 0.

(b) Use the construction employed in the proof of Theorem 3 to obtain a function f such that $\mathbf{F} = \mathbf{V}f$.

■ ■

20. Let f and g be scalar functions defined throughout space such that

$$\mathbf{V}f = \mathbf{V}g.$$

(a) Show that f need not equal g.
(b) How are f and g related? Explain.

Exercises 21 and 22 are related.

21. By a *special* curve in the plane we shall mean a polygonal path whose segments are parallel to either the x axis or the y axis. Let $\mathbf{F}(x, y)$ have the property that $\int_C \mathbf{F} \cdot \mathbf{T}\, ds$ depends only on the endpoints of C for all *special* curves. Deduce that $\mathbf{F}$ is conservative.

22. By a *rectangular* curve in the plane we shall mean a closed curve that surrounds a rectangle.

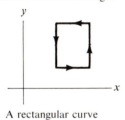

A rectangular curve

Let $\mathbf{F}$ be a vector field in the plane with the property that $\oint \mathbf{F} \cdot \mathbf{T}\, ds = 0$ whenever the curve is rectangular and has diameter at most 1. Deduce that $\oint \mathbf{F} \cdot \mathbf{T}\, ds = 0$ for any rectangular path. *Hint:* Partition a large rectangle into small rectangles and draw some paths.

23. This section, which concerns $\int_C \mathbf{V}f \cdot \mathbf{T}\, ds$, does not mention $\oint \mathbf{V}f \cdot \mathbf{n}\, ds$ (where $\mathbf{n}$ is the exterior normal to a curve bounding a convex region). Show that

$$\oint \mathbf{V}f \cdot \mathbf{n}\, ds = \oint D_{\mathbf{n}} f\, ds,$$

where $D_{\mathbf{n}} f$ denotes the derivative of f in the direction $\mathbf{n}$. This is surely not as interesting as Theorem 1.

24. The electric field $\mathbf{E}$ at any point (x, y) due to a point charge q at $(0, 0)$ is equal to

$$\mathbf{E} = \frac{q\mathbf{U}}{4\pi\varepsilon r^2},$$

where r is the distance from the charge to the point (x, y), and $\mathbf{U}$ is the unit vector directed from the charge to the point (x, y). Evaluate the work done by the field when a particle is moved from $(1, 0)$ to $(2, 0)$ along (a) the x axis; (b) the rectangular path from $(1, 0)$ to $(1, \frac{1}{2})$ to $(2, \frac{1}{2})$ to $(2, 0)$. Is there a difference between the work done in (a) and that done in (b)?

21.5

Summary

A vector field $\mathbf{F}$ assigns to a point in the plane (or in space) a vector. A numerical function f defined on part of or all the plane (or space) is called a scalar field to distinguish it from a vector field $\mathbf{F}$. If f is a scalar field, $\mathbf{V}f$ provides a vector field. From the vector field $\mathbf{F}$ a scalar field can be obtained by computing the "divergence" $\mathbf{V} \cdot \mathbf{F}$, denoted also div $\mathbf{F}$. Also, if $\mathbf{F}$ is defined in space, we may derive from it another vector field, **curl** $\mathbf{F} = \mathbf{V} \times \mathbf{F}$.

The integral of a scalar field f over a curve is defined as the limit of sums of the form

$$\sum_{i=1}^{n} f(P_i)\, \Delta s_i,$$

and is denoted $\int_C f(P)\, ds$. The integral of a vector field $\mathbf{F}$ over a curve is defined to be the line integral

$$\int_C \mathbf{F} \cdot \mathbf{T}\, ds.$$

(Note that $\mathbf{F} \cdot \mathbf{T}$ is a scalar function defined on the curve C.) Thus the integral of a vector field is the integral of the tangential component of $\mathbf{F}$. Another notation for this integral is

$$\int_C \mathbf{F} \cdot d\mathbf{R}.$$

If $\mathbf{F} = P\mathbf{i} + Q\mathbf{j}$, the line integral of $\mathbf{F}$ is also denoted

$$\int_C P\, dx + Q\, dy.$$

If $\mathbf{F}$ is force, the integral $\int_C \mathbf{F} \cdot \mathbf{T}\, ds$ is work.

When C is the curve bounding a region R in the plane, and $\mathbf{n}$ is the exterior unit normal, the integral

$$\oint_C \mathbf{F} \cdot \mathbf{n}\, ds$$

is of interest in fluid flow. In this case $\mathbf{F}$ denotes the flux of the fluid, defined as the product of the density (which is a scalar) and the velocity vector. The integral $\oint_C \mathbf{F} \cdot \mathbf{n}\, ds$ records the net rate at which fluid leaves R.

Similarly, if $\mathbf{F}$ denotes the flux of a fluid in a solid region R bounded by a surface S, the surface integral

$$\int_S \mathbf{F} \cdot \mathbf{n}\, dS$$

represents the net flow out of R or, equivalently, across S. Here $\mathbf{n}$ is the unit exterior normal to the surface.

A vector field $\mathbf{F}$ is called conservative if $\int_C \mathbf{F} \cdot \mathbf{T} \, ds$ depends only on the endpoints of the path C. It follows immediately that F is conservative if and only if $\oint \mathbf{F} \cdot \mathbf{T} \, ds = 0$ for all closed curves. It was also shown that $\mathbf{F}$ is conservative if and only if $\mathbf{F}$ is the gradient of a scalar field, $\mathbf{F} = \nabla f$. In that case, when the curve C goes from point A to point B, $\int_C \mathbf{F} \cdot \mathbf{T} \, ds = f(B) - f(A)$.

In the next chapter it will be shown that, if $\mathbf{F}$ is defined on all of space and $\mathbf{curl}\ \mathbf{F} = \mathbf{0}$, then $\mathbf{F}$ is conservative. It will follow that, if $P_y = Q_x$, then $\mathbf{F}(x, y) = P\mathbf{i} + Q\mathbf{j}$ is conservative.

VOCABULARY AND SYMBOLS

scalar field f
vector field $\mathbf{F}$
gradient of scalar field ∇f
divergence of a vector field $\nabla \cdot \mathbf{F}$, div $\mathbf{F}$
curl of a vector field $\nabla \times \mathbf{F}$, **curl F**
line integral of a scalar field $\int_C f \, ds$

line integral of a vector field $\int_C \mathbf{F} \cdot \mathbf{T} \, ds$
flux
exterior normal $\mathbf{n}$
closed curve
integral over a closed curve $\oint$ or $\oint_C$
conservative

Guide quiz on Chap. 21

1. Let $\mathbf{F}(x, y) = x^2 y \mathbf{i} + 4x^3 \mathbf{j}$.   Let C be the counterclockwise curve bordering the rectangle whose vertices are $(0, 0)$, $(1, 0)$, $(1, 2)$, $(0, 2)$.
   (a) Compute $\oint_C \mathbf{F} \cdot \mathbf{T} \, ds$.
   (b) Compute $\oint_C \mathbf{F} \cdot \mathbf{n} \, ds$, where $\mathbf{n}$ denotes the exterior normal.

2. Find the exterior normal to the ellipse

$$\mathbf{G}(t) = a \cos t \mathbf{i} + b \sin t \mathbf{j} \qquad 0 \le t \le 2\pi.$$

   Note: t is *not* angle, as it is in the next problem.

3. Let R be the disk bordered by the circle C:

$$\mathbf{G}(t) = a \cos t \mathbf{i} + a \sin t \mathbf{j} \qquad 0 \le t \le 2\pi.$$

   Let $\mathbf{F}(x, y) = x\mathbf{i} + y\mathbf{j}$.   Verify that

$$\int_R \nabla \cdot \mathbf{F} \, dA = \oint_C \mathbf{F} \cdot \mathbf{n} \, ds$$

   by computing both sides of the equation.

4. Show that
   (a) $\mathbf{F}(x, y) = ye^{xy}\mathbf{i} + xe^{xy}\mathbf{j}$ is conservative.
   (b) $\mathbf{F}(x, y, z) = e^y\mathbf{i} + e^x\mathbf{j} + z^5\mathbf{k}$ is not conservative by exhibiting a circular closed path over which $\oint \mathbf{F} \cdot \mathbf{T} \, ds$ is not 0.

5. Let $f = x^3 y^2 z^4$ be a scalar function defined for points in space.
   (a) Compute ∇f.

   (b) Compute the divergence of ∇f.
   (c) Compute the curl of ∇f.

6. Let $\mathbf{F} = P\mathbf{i} + Q\mathbf{j}$ be a vector field.
   (a) Explain why $\int_C P \, dx + Q \, dy$ represents the work accomplished by the force $\mathbf{F}$ in moving a particle along C.
   (b) Explain why $\int_C - Q \, dx + P \, dy$ represents the net loss of fluid past the closed curve C bounding a convex region if the flux is $\mathbf{F}$.

7. Using a diagram, show that, if a closed planar curve C is swept out counterclockwise, then

$$\frac{dy}{ds}\mathbf{i} - \frac{dx}{ds}\mathbf{j}$$

   is a unit exterior normal.

8. Compute $\oint_C \mathbf{F} \cdot \mathbf{T} \, ds$, where $\mathbf{F}(x, y, z) = xy\mathbf{i} + 3z\mathbf{j} + y\mathbf{k}$, and C is the polygonal curve that starts at $(0, 0, 0)$, goes to $(1, 1, 1)$, then to $(0, 1, 1)$, and then back to $(0, 0, 0)$.

9. Verify that

$$\int_R \nabla \cdot \mathbf{F} \, dV = \int_S \mathbf{F} \cdot \mathbf{n} \, dS,$$

   where R is the box bordered by the three coordinate planes and the three planes $x = a$, $y = b$, and $z = c$ (a, b, and c are positive), S is the surface of R, and $\mathbf{F}(x, y, z) = x^2\mathbf{i} + y^2\mathbf{j} + z^2\mathbf{k}$.

Review exercises for Chap. 21

A few exercises concern earlier chapters.

1. Find the work done by the force $F(x, y, z) = x\mathbf{i} - z\mathbf{j} + 2y\mathbf{k}$ along the parabola in the plane $z = 2$ described by the equation $y = 2x^2$, from $(0, 0, 2)$ to $(1, 2, 2)$.

2. Evaluate the following integrals in the clockwise sense [as viewed from the origin $(0, 0, 0)$] around the given curve C.
   (a) $\int_C z\, dx + x\, dy + y\, dz$; C: the boundary of the triangle with vertices $(2, 0, 0)$, $(0, 2, 0)$, $(0, 0, 1)$.
   (b) $\int_C \sin z\, dx - \cos x\, dy + \sin y\, dz$; C: the boundary of the rectangle, $0 \le x \le \pi$, $0 \le y \le 1$, $z = 2$.

3. Let R be the position vector of a particle at time t. Let s denote arc length.
   (a) Is $d\mathbf{R}/dt$ a scalar multiple of $d\mathbf{R}/ds$?
   (b) Is $d^2\mathbf{R}/dt^2$ a scalar multiple of $d^2\mathbf{R}/ds^2$?

4. Let R be the region bordered by $y = 0$, $x = 2$, and $y = x^2$. Let C be the counterclockwise curve bordering R. Let $\mathbf{F} = x^2\mathbf{i} + 3y\mathbf{j}$. Verify that

$$\int_R \nabla \cdot \mathbf{F}\, dA = \oint_C \mathbf{F} \cdot \mathbf{n}\, ds$$

   by evaluating both sides.

5. A moving particle achieves its maximum speed at the instant $t = 3$. Does it follow that its acceleration vector $\mathbf{A}$ is $\mathbf{0}$ when $t = 3$?

6. Let $\mathbf{F}(x, y, z) = x^3\mathbf{i} + y^3\mathbf{j} + z^3\mathbf{k}$. Let S be the sphere $x^2 + y^2 + z^2 = 1$, and R be the ball $x^2 + y^2 + z^2 \le 1$. By computing both sides verify that

$$\int_R \operatorname{div} \mathbf{F}\, dV = \int_S \mathbf{F} \cdot \mathbf{n}\, dS.$$

7. Which of these vector fields in the plane are conservative? not conservative? Explain.
   (a) $\mathbf{F}(x, y) = 3\mathbf{i} + 4\mathbf{j}$;
   (b) $\mathbf{F}(x, y) = y \sin xy\, \mathbf{i} + x \cos xy\mathbf{j}$;
   (c) $\mathbf{F}(x, y) = x\mathbf{i} + y\mathbf{j}$.

8. Let $\mathbf{F}(x, y, z) = ze^{xz}\mathbf{i} + xe^{xz}\mathbf{k}$.
   (a) Exhibit a function f such that $\mathbf{F} = \nabla f$.
   (b) Evaluate $\int_C \mathbf{F} \cdot \mathbf{T}\, ds$ for the curve given parametrically

   as $\quad \mathbf{R}(t) = t^2\mathbf{i} + \cos t\mathbf{j} + e^t\mathbf{k} \qquad 0 \le t \le 2\pi$.

9. Let $\mathbf{F}$ be a vector function defined at all points of a curve C and suppose $|\mathbf{F}| \le M$ on C, where M is some positive number. Show that $|\int_C \mathbf{F} \cdot \mathbf{T}\, ds| \le Ml$, where l is the length of C.

10. Let $\mathbf{F}(x, y, z) = 4xz\mathbf{i} - y^2\mathbf{j} + yz\mathbf{k}$. Let S be the surface of the cube bounded by the three coordinate planes and

the planes $x = 1$, $y = 1$, $z = 1$. Let R be the cube bounded by S. Verify that

$$\int_R \operatorname{div} \mathbf{F}\, dV = \int_S \mathbf{F} \cdot \mathbf{n}\, dS.$$

11. Let $\mathbf{F}(x, y, z) = 3xy\mathbf{i} + 4\mathbf{j} + z\mathbf{k}$. Let C be the closed polygonal curve that starts at $(0, 0, 0)$, goes to $(1, 0, 1)$, then to $(1, 1, 1)$, then to $(0, 1, 0)$, and back to $(0, 0, 0)$.
   (a) Evaluate $\int_C \mathbf{F} \cdot \mathbf{T}\, ds$.
   (b) Is $\mathbf{F}$ conservative?

12. Let $\mathbf{A} = \mathbf{i} - 4\mathbf{j} + 3\mathbf{k}$ and $\mathbf{B} = \mathbf{i} + \mathbf{j} - \mathbf{k}$. Write $\mathbf{A}$ in the form $\mathbf{C} + \mathbf{D}$, where $\mathbf{C}$ is parallel to $\mathbf{B}$ and $\mathbf{D}$ is perpendicular to $\mathbf{B}$.

13. Let C be the helical spiral $\mathbf{R}(t) = 2 \cos t\mathbf{i} + 2 \sin t\mathbf{j} + t\mathbf{k}$, $0 \le t \le 2\pi$. Let $\mathbf{F}(x, y, z) = 2xy\mathbf{i} + x^2\mathbf{j} + (1 - \sin z)\mathbf{k}$. Evaluate $\int_C \mathbf{F} \cdot d\mathbf{R}$.

14. Let C be the curve from $(0, 0, 0)$ to $(0, 4, 0)$ that lies in the xy plane and on the parabola $y^2 - 4y = x$. Let $\mathbf{F}(x, y, z) = x^2\mathbf{i} + y^2\mathbf{j} + xyz\mathbf{k}$. Evaluate $\int_C \mathbf{F} \cdot \mathbf{T}\, ds$.

15. (a) Show that the lines

$$\frac{x - 1}{2} = 3 - y = 2z \qquad \text{and} \qquad \frac{x - 1}{3} = \frac{y - 3}{4} = \frac{z}{5}$$

   intersect.
   (b) Find the angle between them.

16. The flux of a gas is given by the formula $\mathbf{F}(x, y, z) = -x\mathbf{i} + y\mathbf{j} + z\mathbf{k}$. Let R be the ball of radius 3 and center $(0, 0, 0)$. Is the amount of gas in R tending to increase or to decrease?

■

17. Let R be the solid cylinder consisting of all points (x, y, z) such that $x^2 + y^2 \le a^2$ $(a > 0)$ and $0 \le z \le h$ $(h > 0)$. Let S be the surface of R. Let $\mathbf{F}(x, y, z) = x\mathbf{i} + y\mathbf{j} + z\mathbf{k}$. Verify that

$$\int_R \nabla \cdot \mathbf{F}\, dV = \int_S \mathbf{F} \cdot \mathbf{n}\, ds.$$

Comparison of Exercises 18 and 19 presents an instructive contrast.

18. Let D consist of those points (x, y, z) not on the z axis. Let

$$\mathbf{F}(x, y, z) = \frac{x\mathbf{i}}{x^2 + y^2} + \frac{y\mathbf{j}}{x^2 + y^2} + 0\mathbf{k}$$

   for (x, y, z) in D. Let $f(x, y, z) = \ln(x^2 + y^2)$.
   (a) Show that div $\mathbf{F} = 0$.
   (b) Show that **curl** $\mathbf{F} = \mathbf{0}$.

(c) Show that $\mathbf{F} = \nabla f$.

(d) Is $\mathbf{F}$ conservative?

19. Let D be as in Exercise 18. Let

$$\mathbf{F}(x, y, z) = \frac{-y\mathbf{i}}{x^2 + y^2} + \frac{x\mathbf{j}}{x^2 + y^2}j + 0\mathbf{k}.$$

(a) Show that div $\mathbf{F} = 0$.

(b) Show that **curl** $\mathbf{F} = \mathbf{0}$.

(c) Let $f(x, y, z) = \tan^{-1}(y/x)$. Is $\mathbf{F} = \nabla f$?

(d) Evaluate $\int_C \mathbf{F} \cdot \mathbf{T}\, ds$, where C is the circle $x^2 + y^2 = 1$ in the xy plane.

(e) Show that $\mathbf{F}$ is not conservative.

(f) Does (e) contradict (c)? Explain.

20. Let $f(x, y, z, t)$ be the density of a fluid or gas at the point (x, y, z) at time t. Let $R(t)$ be a solid region which depends on time, and let $S(t)$ be the surface that bounds it. Let $\mathbf{W} = \mathbf{F}(x, y, z, t)$ be the velocity vector of the point (x, y, z) on $S(t)$ at time t. The equation

$$\frac{d}{dt} \int_{R(t)} f(P)\, dV = \int_{R(t)} \frac{\partial f}{\partial t}\, dV - \int_{S(t)} f(P)(\mathbf{W} \cdot \mathbf{n})\, dS$$

is known as the *general transport theorem* in fluid mechanics.

(a) Interpret $\displaystyle\int_{R(t)} f(P)\, dV$ physically.

(b) Interpret $\displaystyle\frac{d}{dt} \int_{R(t)} f(P)\, dV$ physically.

(c) Interpret $\dfrac{\partial f}{\partial t}$ physically.

(d) Interpret $\displaystyle\int_{R(t)} \frac{\partial f}{\partial t}\, dV$ physically.

(e) Interpret $\displaystyle\int_{S(t)} f(P)(\mathbf{W} \cdot \mathbf{n})\, dS$ physically.

(f) Is the general transport theorem plausible?

■ ■

21. If $\int_R (\nabla f \cdot \nabla f)\, dV = 0$, where R is a ball, what can we conclude about f? Explain.

This exercise is an introduction to Exercise 23.

22. A motorboat goes back and forth on a straight measured mile at a constant speed V relative to the water. Show that the time required for a round trip is always less when there is no current than when there is a constant current **W**. (Assume that the component of **W** along the route is less than V and greater than 0. If the component along the route were larger than V, the boat could not make a round trip.)

23. (See Exercise 22.) An aircraft traveling at constant air speed V traverses a closed horizontal curve marked on the ground. Show that the time required for one complete trip is always less when there is no wind than when there is a constant wind **W**. (Assume that $W = |\mathbf{W}|$ is less than V, so that the plane never meets an insuperable head wind.)

24. Let C be a closed curve of length L surrounding a convex region R in the plane. For each direction ϕ, $0 \leq \phi \leq 2\pi$, let $l(\phi)$ be the distance between the two parallel tangent lines to C corresponding to the angles ϕ and $\phi + \pi$. Prove that the average value of $l(\phi)$ is equal to L/π. (Note that for a circle of circumference L this is easy to check.) *Suggestion*: Select the origin of a coordinate system within R. Let $\mathbf{R}(\phi)$ be the position vector to the point of tangency of the tangent line having the angle ϕ. Let $\mathbf{R}'$ denote $d\mathbf{R}/d\phi$.

(a) Show that $\int_0^{2\pi} (\mathbf{R} \cdot \mathbf{T})'\, d\phi = 0$.

(b) From (a) show that $\int_0^{2\pi} \mathbf{R}' \cdot \mathbf{T}\, d\phi = \int_0^{2\pi} \mathbf{R} \cdot \mathbf{n}\, d\phi$, where **n** is the exterior normal.

(c) Show that $\int_0^{2\pi} \mathbf{R}' \cdot \mathbf{T}\, d\phi = L$.

(d) Show that $\int_0^{2\pi} \mathbf{R} \cdot \mathbf{n}\, d\phi = \int_0^{\pi} l(\phi)\, d\phi$.

This shows that the average shadow in all directions of a plane curve determines its perimeter.

22

GREEN'S THEOREM AND ITS GENERALIZATIONS

The fundamental theorem of calculus relates the integral of a function over an interval $[a, b]$ to the values of another function at the ends of the interval:

$$\int_a^b F'(x)\,dx = F(b) - F(a).$$

The first theorem of this chapter, Green's theorem, is similar in spirit. It relates the integral of one function over a region R in the plane to an integral of another function over its boundary curve C. It states that, if $\mathbf{F}$ is a vector field, then

$$\int_R \nabla \cdot \mathbf{F}\,dA = \oint_C \mathbf{F} \cdot \mathbf{n}\,ds.$$

Expressed in words, it is read, "The integral over R of the divergence of a vector field $\mathbf{F}$ equals the integral of the normal component of $\mathbf{F}$ around the boundary of R."

Just as there is an intuitive basis for the fundamental theorem of calculus, so is there for Green's theorem. Section 22.1 provides this basis by consideration of fluid flowing on the region R. The following section proves Green's theorem.

The analogy in space of Green's theorem asserts that

$$\int_R \nabla \cdot \mathbf{F}\,dV = \int_S \mathbf{F} \cdot \mathbf{n}\,dS,$$

where S is the surface bounding the solid region R. This theorem, known as the divergence theorem or Gauss' theorem, is discussed in Sec. 22.5.

Another generalization of Green's theorem concerns the curve in space bounding a portion of a surface. This theorem, known as Stokes' theorem, is discussed in Sec. 22.6.

The reader should keep one objective uppermost: to understand what Green's theorem, the divergence theorem, and Stokes' theorem say. The student who goes on to fluid mechanics or electromagnetism will see these theorems used conceptually in developing the basic theory of these fields. Computation will be of importance, primarily to illustrate concepts.

Engineering students, especially, should keep these words in mind, quoted from the preface of Stephen Whitaker's upper-division text, *Introduction to Fluid Mechanics*, "Vector notation is used freely throughout the text, not because it leads to elegance or rigor but simply because fundamental concepts are best expressed in a form which attempts to connect them with reality."

22.1

Statement and physical interpretation of Green's theorem

The main theorem of this chapter, Green's theorem, informally runs:
  Let R be a convex region in the plane, and C its boundary. Let $\mathbf{F}$ be a vector field on the plane. Then

$$\int_R \nabla \cdot \mathbf{F} \, dA = \oint_C \mathbf{F} \cdot \mathbf{n} \, ds,$$

where $\mathbf{n}$ is the unit exterior normal. Expressed in less condensed notation, it reads as follows:

GREEN'S THEOREM   Let R be a convex region in the plane and let C be its boundary (swept out counterclockwise).   Assume that the functions P and Q have continuous partial derivatives throughout R.   Then

$$\int_R (P_x + Q_y) \, dA = \oint_C P \, dy - Q \, dx. \quad \bullet$$

This restatement is based on the definition of divergence:

$$\nabla \cdot \mathbf{F} = \left(\frac{\partial}{\partial x} \mathbf{i} + \frac{\partial}{\partial y} \mathbf{j} \right) \cdot (P\mathbf{i} + Q\mathbf{j}) = P_x + Q_y,$$

and the definition of $\mathbf{F} \cdot \mathbf{n}$:

$$\mathbf{F} \cdot \mathbf{n} = (P\mathbf{i} + Q\mathbf{j}) \cdot \left(\frac{dy}{ds} \mathbf{i} - \frac{dx}{ds} \mathbf{j} \right)$$

$$= P \frac{dy}{ds} - Q \frac{dx}{ds}.$$

The rest of this section motivates Green's theorem by considering fluid flow. The physical interpretation of

$$\oint_C \mathbf{F} \cdot \mathbf{n} \, ds = \oint_C P \, dy - Q \, dx$$

was already given in Sec. 21.3. It is the rate of net loss of fluid across C.
  Recall that physical situation in detail. Liquid on the surface of a stream has a velocity vector $\mathbf{V}$ and density f. Both $\mathbf{V}$ and f may depend on the point on the surface and on time. (If the liquid is, like water, homogeneous, then f, considered as a function of the point in the stream, is constant.) The vector function $\mathbf{F} = f\mathbf{V}$, the flux, tells the speed, direction, and rate at which the liquid is passing a point. Also assume that there is no evaporation from the stream

and no rain or fog. Thus the liquid leaves or enters a region R on the surface only at the boundary of that region.

Now, the net loss of fluid past C is the same as the net loss from the region R bounded by C, since the only way that fluid enters or leaves R is across C. Let us calculate the net loss from the viewpoint of R rather than of C. Consider the net loss from a typical small rectangle of dimensions Δx and Δy in R. Let its lower-left-hand corner A have the coordinates (a, b), and let the other vertices be B, C, D, as in the diagram.

What is the Net Loss of Fluid from This Small Rectangle?

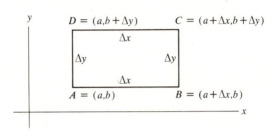

The loss from this small rectangle is precisely

$$\int_A^B P\,dy - Q\,dx + \int_B^C P\,dy - Q\,dx + \int_C^D P\,dy - Q\,dx + \int_D^A P\,dy - Q\,dx,$$

where the integrals are over straight paths. Let us estimate this sum when Δx and Δy are small.

Consider first the integrals over the parallel edges AB and CD. They are

$$\int_A^B P\,dy - Q\,dx = \int_A^B - Q\,dx$$

and

$$\int_C^D P\,dy - Q\,dx = \int_C^D - Q\,dx,$$

since in both cases y is constant and $dy\ [=dy/dt\ dt]$ is 0. Now

$$\int_A^B - Q\,dx = \int_a^{a+\Delta x} - Q(x, b)\,dx,$$

and

$$\int_C^D - Q\,dx = \int_D^C Q\,dx = \int_a^{a+\Delta x} Q(x, b + \Delta y)\,dx.$$

Hence $\displaystyle \int_A^B - Q\,dx + \int_C^D - Q\,dx = \int_a^{a+\Delta x} [Q(x, b + \Delta y) - Q(x, b)]\,dx.$ (1)

By the law of the mean,

$$Q(x, b + \Delta y) - Q(x, b) = Q_y(x, Y)\,\Delta y$$

for some Y, with $b < Y < b + \Delta y$.

From the continuity of Q_y it follows that $Q_y(x, Y)$ is approximately equal to $Q_y(a, b)$. Hence the integral on the right side of (1) is approximately

$$\int_{a}^{a+\Delta x} Q_y(a, b) \, \Delta y \, dx = Q_y(a, b) \, \Delta y \, \Delta x.$$

Thus

$$\int_{A}^{B} P \, dy - Q \, dx + \int_{C}^{D} P \, dy - Q \, dx$$

is approximately

$$Q_y(a, b) \, \Delta y \, \Delta x.$$

Similarly

$$\int_{B}^{C} P \, dy - Q \, dx + \int_{D}^{A} P \, dy - Q \, dx$$

is approximately

$$P_x(a, b) \, \Delta x \, \Delta y.$$

Thus the flow out of a typical small rectangle, whose area is $\Delta A = \Delta x \, \Delta y$,

is approximately

$$[P_x(a, b) + Q_y(a, b)] \, \Delta A.$$

Since the net flow out of R is the sum of the net flows out of all such small rectangles, we may expect that

$$\text{Net flow out of } R = \int_{R} (P_x + Q_y) \, dA.$$

Since

$$\text{Net flow out of } R = \text{net flow past } C,$$

it seems reasonable that

$$\int_{R} (P_x + Q_y) \, dA = \oint_{C} P \, dy - Q \, dx.$$

A Physical Interpretation of $\mathbf{V} \cdot \mathbf{F}$ This is the fluid-flow interpretation of Green's theorem. It also provides a physical interpretation of $\mathbf{V} \cdot \mathbf{F} = P_x + Q_y$. As we saw in the fluid-flow reasoning,

$$\mathbf{V} \cdot \mathbf{F} = P_x + Q_y$$

is a local measure of net loss—or "divergence"—of the fluid.

Actually, Green's theorem holds for any region bounded by a smooth curve or polygonal path. Most regions for which it is needed are convex.

Two examples are included as an extra check on Green's theorem for the skeptical reader.

EXAMPLE I   Let $\mathbf{F}(x, y) = x\mathbf{i} + 2y\mathbf{j}$ and let C be the ellipse

$$\frac{x^2}{2^2} + \frac{y^2}{3^2} = 1.$$

Show that

$$\int_{R} \mathbf{V} \cdot \mathbf{F} \, dA = \oint_{C} \mathbf{F} \cdot \mathbf{n} \, ds,$$

where R is the region inside the ellipse.

SOLUTION   First compute the divergence

$$\mathbf{V} \cdot \mathbf{F} = \frac{\partial(x)}{\partial x} + \frac{\partial(2y)}{\partial y} = 1 + 2 = 3.$$

Thus
$$\int_R \mathbf{V} \cdot \mathbf{F} \, dA = \int_R 3 \, dA$$
$$= 3 \cdot \text{area of } R$$
$$= 3 \cdot 6\pi$$
$$= 18\pi.$$

(The area inside the ellipse $x^2/a^2 + y^2/b^2 = 1$ is πab. See Exercise 15 in Sec. 10.8.)

The integral $\oint_C \mathbf{F} \cdot \mathbf{n} \, ds$ was already found in Example 3 of Sec. 21.3 to be 18π. Thus Green's theorem is verified in this case. ●

EXAMPLE 2 Let $\mathbf{F}(x, y) = x\mathbf{i} + xy\mathbf{j}$. Let C be the curve shown in the margin and R the region inside it. Check Green's theorem in this case.

SOLUTION By Example 2 in Sec. 21.3, $\oint \mathbf{F} \cdot \mathbf{n} \, ds = \frac{1}{4}.$

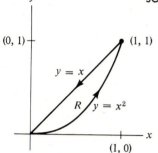

To evaluate the integral of the divergence over R, first note that
$$\mathbf{V} \cdot \mathbf{F} = \frac{\partial(x)}{\partial x} + \frac{\partial(xy)}{\partial y} = 1 + x.$$

Thus
$$\int_R \mathbf{V} \cdot \mathbf{F} \, dA = \int_0^1 \left[\int_{x^2}^x (1 + x) \, dy \right] dx.$$

The first integration gives
$$\int_{x^2}^x (1 + x) \, dy = (1 + x) \int_{x^2}^x dy$$
$$= x - x^3,$$

and the second,
$$\int_0^1 (x - x^3) \, dx = \frac{1}{4}.$$

Thus
$$\int_R \mathbf{V} \cdot \mathbf{F} \, dA = \frac{1}{4} = \oint \mathbf{F} \cdot \mathbf{n} \, ds. \ ●$$

Exercises

In Exercises 1 to 3 compute $\int_R \mathbf{V} \cdot \mathbf{F} \, dA$ and $\oint_C \mathbf{F} \cdot \mathbf{n} \, ds$ and verify Green's theorem.

1. $\mathbf{F} = 3x\mathbf{i} + 2y\mathbf{j}$, and R is the disk of radius 1 with center $(0, 0)$.

2. $\mathbf{F} = 5y^3\mathbf{i} - 6x^2\mathbf{j}$, and R is the disk of radius 2 with center $(0, 0)$.

3. $\mathbf{F} = xy\mathbf{i}$, and R is the region of Example 2.

4. State Green's theorem in words, without using any mathematical symbols.

5. Using Green's theorem with $\mathbf{F} = x\mathbf{i} + y\mathbf{j}$, show that
$$\oint_C x \, dy - y \, dx$$
is twice the area of the region enclosed by the curve C.

6. Let $\mathbf{F} = P\mathbf{i} + Q\mathbf{j}$ satisfy the condition:
$$P_x = -Q_y.$$
Show, by Green's theorem, that

$$\oint_C \mathbf{F} \cdot \mathbf{n} \, ds = 0$$

for any curve C bounding a convex region.

7. Let f be a scalar function. Using Green's theorem, show that

$$\int_R \left(\frac{\partial^2 f}{\partial x^2} + \frac{\partial^2 f}{\partial y^2} \right) dA = \oint_C \frac{\partial f}{\partial x} \, dy - \frac{\partial f}{\partial y} \, dx,$$

where R is a convex region and C its boundary taken counterclockwise.

8. Let f be a scalar function. Using Green's theorem, prove that

$$\int_R \mathbf{V} \cdot (\mathbf{V}f) \, dA = \oint_C f_x \, dy - f_y \, dx,$$

where C is the boundary of R taken counterclockwise.

9. Let R be a convex region and let C be its boundary. Let $\mathbf{F}$ be a vector field with the property that, at any point P on C, $\mathbf{F}(P)$ is tangent to the curve C. Prove that

$$\int_R \mathbf{V} \cdot \mathbf{F} \, dA = 0.$$

■

Exercises 10 to 12 are related.

10. Let f be a continuous scalar field defined throughout the plane. Assume that for any convex region R,

$$\int_R f(P) \, dA = 0.$$

Explain why $f(P)$ is 0 for all points P.

11. Let $\mathbf{F}$ be a vector field with the property that, for any closed curve C,

$$\oint_C \mathbf{F} \cdot \mathbf{n} \, ds = 0.$$

Show that the divergence of $\mathbf{F}$ is 0 everywhere. *Hint:* See Exercise 10.

12. Let g be a scalar field with the property that

$$\oint_C \mathbf{V}g \cdot \mathbf{n} \, ds = 0$$

for all closed curves. Prove that

$$g_{xx} + g_{yy} = 0.$$

Hint: See Exercise 11.

13. Let $\mathbf{F}$ and $\mathbf{G}$ be two vector fields with the property that $\mathbf{F}(P) = \mathbf{G}(P)$ for all points P on the curve C, which is

the boundary of a convex region R. Prove that

$$\int_R \mathbf{V} \cdot \mathbf{F} \, dA = \int_R \mathbf{V} \cdot \mathbf{G} \, dA.$$

■ ■

14. Show that, as stated in the discussion of Green's theorem,

$$\int_B^C P \, dy - Q \, dx + \int_D^A P \, dy - Q \, dx$$

is approximately

$$P_x(a, b) \, \Delta x \, \Delta y.$$

15. Let $T(x, y)$ be the temperature at the point (x, y) in a sheet of metal. Assume that the temperature is maintained independently of time. Now, at (x, y) heat tends to flow in the direction of $-\mathbf{V}T$, at a rate proportional to the magnitude of $\mathbf{V}T$. Assuming that the net loss of heat over any closed curve is 0, show that T satisfies Laplace's partial differential equation, $T_{xx} + T_{yy} = 0$. Exercise 12 may be of aid.

16. Let $\mathbf{F}(x, y) = Q^*(x, y)\mathbf{i} - P^*(x, y)\mathbf{j}$, where Q^* and P^* are scalar functions.

(a) Deduce from Green's theorem that

$$\int_R (Q_x^* - P_y^*) \, dA = \oint_C P^* \, dx + Q^* \, dy,$$

where C is the curve, taken counterclockwise, that bounds R.

(b) Deduce that, for $\mathbf{F}(x, y) = P(x, y)\mathbf{i} + Q(x, y)\mathbf{j}$,

$$\int_R (Q_x - P_y) \, dA = \oint_C \mathbf{F} \cdot \mathbf{T} \, ds.$$

17. Let $\mathbf{F}(x, y) = P(x, y)\mathbf{i} + Q(x, y)\mathbf{j}$ for all points (x, y) in the plane. Assume that $P_y = Q_x$. This exercise will show that there is a scalar function f such that $\mathbf{F} = \mathbf{V}f$.

(a) Show that $\oint_C \mathbf{F} \cdot \mathbf{T} \, ds = 0$ for any closed curve C. *Hint:* See Exercise 16.

(b) Show on the basis of (a) that $\int_C \mathbf{F} \cdot \mathbf{T} \, ds$ depends only on the initial and terminal points of C for any curve C in the plane.

(c) Define $f(x, y)$ as the value of $\int_C \mathbf{F} \cdot \mathbf{T} \, ds$, where C is any curve whose initial point is $(0, 0)$ and whose terminal point is (x, y). By examining

$$\frac{f(x + \Delta x, y) - f(x, y)}{\Delta x}$$

show that $f_x = P$. Similarly, show that $f_y = Q$.

22.2

Proof of Green's theorem

The proof of Green's theorem is perhaps simpler than the fluid-flow argument of Sec. 22.1. It may be instructive to have both the "pure" and the "applied" versions of the theorem. This section provides the "pure" viewpoint. However, the proof in this section may be omitted without affecting the rest of the development of this chapter. In any case, Theorem 2 should be studied, for it will be needed later.

THEOREM I   (*Green's theorem*).   Let R be a convex region in the plane, and C its boundary (swept out counterclockwise). Let P and Q have continuous partial derivatives throughout R. Then

$$\int_R (P_x + Q_y)\, dA = \oint_C P\, dy - Q\, dx.$$

PROOF   It will simplify the proof to assume that each of the two tangent lines parallel to the y axis meets C at only one point.

It will be proved that

$$\int_R Q_y\, dA = \oint_C - Q\, dx. \tag{1}$$

A similar proof will show that

$$\int_R P_x\, dA = \oint_C P\, dy.$$

Green's theorem follows immediately from these two equations.

Let the region R have the following description:

$$a \le x \le b, \qquad y_1(x) \le y \le y_2(x).$$

[Note by the assumption on C that $y_1(a) = y_2(a)$ and $y_1(b) = y_2(b)$.]   Then

$$\int_R Q_y\, dA = \int_a^b \left[\int_{y_1(x)}^{y_2(x)} \frac{\partial Q}{\partial y}\, dy \right] dx.$$

By the fundamental theorem of calculus,

$$\int_{y_1(x)}^{y_2(x)} \frac{\partial Q}{\partial y}\, dy = Q(x, y_2(x)) - Q(x, y_1(x)).$$

Hence
$$\int_R Q_y \, dA = \int_a^b [Q(x, y_2(x)) - Q(x, y_1(x))] \, dx. \tag{2}$$

Now consider the right side of (1),

$$\oint_C - Q \, dx.$$

Break the closed path C into two successive paths, one along the bottom part of R, described by $y = y_1(x)$, the other along the top part of R, described by $y = y_2(x)$. Denote the bottom path C_1 and the top path C_2.

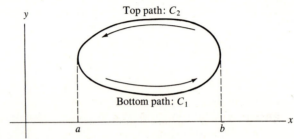

Then
$$\oint_C - Q \, dx = \int_{C_1} - Q \, dx + \int_{C_2} - Q \, dx. \tag{3}$$

But
$$\int_{C_1} - Q \, dx = \int_{C_1} - Q(x, y_1(x)) \, dx$$

$$= \int_a^b - Q(x, y_1(x)) \, dx,$$

and
$$\int_{C_2} - Q \, dx = \int_{C_2} - Q(x, y_2(x)) \, dx$$

$$= \int_b^a - Q(x, y_2(x)) \, dx$$

$$= \int_a^b Q(x, y_2(x)) \, dx.$$

Thus by (3),

$$\oint_C - Q \, dx = \int_a^b - Q(x, y_1(x)) \, dx + \int_a^b Q(x, y_2(x)) \, dx$$

$$= \int_a^b [Q(x, y_2(x)) - Q(x, y_1(x))] \, dx.$$

This is precisely (2) and concludes the proof. (A similar argument shows that

$$\int_R P_x \, dA = \oint_C P \, dy.) \; \bullet$$

Green's theorem holds for regions that are not convex; we will assume it in its full generality.

An important consequence of Green's theorem is the following theorem.

THEOREM 2 Let R be a plane region with boundary C, swept out counterclockwise. Then

$$\text{Area of } R = \frac{1}{2}\oint_C - y\, dx + x\, dy.$$

PROOF Consider Green's theorem in the special case that

$$P = x \qquad \text{and} \qquad Q = y.$$

The assertion that

$$\int_R \left(\frac{\partial P}{\partial x} + \frac{\partial Q}{\partial y}\right) dA = \oint_C P\, dy - Q\, dx$$

then becomes

$$\int_R (1 + 1)\, dA = \oint_C x\, dy - y\, dx.$$

Thus

$$2(\text{Area of } R) = \oint_C - y\, dx + x\, dy,$$

and the theorem follows on division by 2. ●

Exercises

1. By computations, verify Green's theorem when R is the disk of radius 1 and center $(0, 0)$, and
   (a) $P = x^2$, $Q = y^2$;
   (b) $P = y^2$, $Q = x^2$.

2. Prove that $\oint_C (a\, dx + b\, dy) = 0$, where a and b are constants
   (a) by Green's theorem;
   (b) without Green's theorem.

3. The moment of inertia about the z axis of a region R in the xy plane is defined as the integral $\int_R (x^2 + y^2)\, dA$. Show that this equals $\frac{1}{3}\oint_C (-y^3\, dx + x^3\, dy)$, where C bounds R.

4. Use Theorem 2 to find the area of the region bounded by the line $y = x$ and the curve
   $$\begin{cases} x = t^6 + t^4 \\ y = t^3 + t \end{cases}$$
   for t in $[0, 1]$.

5. A curve is given parametrically by $x = t(1 - t^2)$, $y = t^2(1 - t^3)$, for t in $[0, 1]$.
   (a) Sketch the points corresponding to $t = 0$, 0.2, 0.4, 0.6, 0.8, and 1.0, and use them to sketch the curve.
   (b) Let R be the region enclosed by the curve. What

difficulty arises when you try to compute the area of R by a definite integral involving vertical or horizontal cross sections?
   (c) Use Theorem 2 to find the area of R.

6. Repeat Exercise 5 for $x = \sin \pi t$ and $y = t - t^2$. In (a), let $t = 0$, $\frac{1}{4}$, $\frac{1}{2}$, $\frac{3}{4}$, 1.

7. (a) Using Green's theorem, prove that $\oint_C x\, dy = A = -\oint_C y\, dx$, where A is the area of the region bounded by a simple closed curve C.
   (b) Why is (a) a generalization of Theorem 2?

8. Work Exercise 7 without using Green's theorem, when C bounds a region of the type discussed in Green's theorem. That is, compute $\oint_C x\, dy$ and $\oint_C y\, dx$, using the definition of curve integrals.

9. (a) Let C be the circle of radius 1, with center at $(0, 0)$. Use Green's theorem to compute $\oint_C (x^2 - y^3)\, dx + (y^2 + x^3)\, dy$. *Hint:* To evaluate $\int_R (P_x + Q_y)\, dA$, use a repeated integral in polar coordinates.
   (b) Compute the line integral in (a) directly.

   ∎

10. Let R be the region swept out by the position vector

$$G(t) = \frac{e^t + e^{-t}}{2}\, \mathbf{i} + \frac{e^t - e^{-t}}{2}\, \mathbf{j},$$

for t in $[0, a]$, $a > 0$.

(a) Show that R is the region bounded by the x axis, the right half of the hyperbola $x^2 - y^2 = 1$, and the line $y = (e^a - e^{-a})x/(e^a + e^{-a})$.

(b) Using Theorem 2, show that the area of R is $a/2$.

11. Complete the proof of Green's theorem by showing that

$$\int_R P_x \, dA = \oint_C P \, dy.$$

■ ■

12. Use Theorem 2 to obtain the formula for area in polar coordinates,

$$\text{Area} = \frac{1}{2}\int_\alpha^\beta r^2 \, d\theta.$$

13. Show how Green's theorem for polygonal regions R would follow from Green's theorem for the special case in which R is triangular.

Exercises 14 and 15 are related.

14. A triangle has vertices $A = (a_1, a_2)$, $B = (b_1, b_2)$, and $C = (c_1, c_2)$ such that the triangular closed path $ABCA$ is counterclockwise.

(a) Using Theorem 2, find the area of the triangle. *Hint:* On each edge, use x or y as the parameter.

(b) Check your formula in a simple case.

15. (See Exercise 14.)

(a) How can you decide *without the use of a picture*, but with the knowledge of the coordinates of A, B, and C, whether the closed path around the triangle A, B, C in the order A, B, C, A is counterclockwise or clockwise?

(b) Test your criterion on three (noncollinear) points of your choice.

16. Let $\mathbf{F}(x, y)$ be a vector field in the plane with the property that, for all closed curves of length at most 1, $\oint \mathbf{F} \cdot \mathbf{n}\, ds = 0$. Deduce that for any closed curve C that bounds some region R, $\oint_C \mathbf{F} \cdot \mathbf{n}\, ds = 0$,

(a) using Green's theorem.

(b) not using Green's theorem.

22.3

Functions from the plane to the plane

This section introduces a new type of function. Before reading its description, recall the types of functions used in earlier chapters, as summarized in this table.

Symbol for Function	What the Function Does
$y = f(x)$	Assigns to a number x a number y
$z = f(x, y)$	Assigns to a point (x, y) in the plane a number z
$u = f(x, y, z)$	Assigns to a point (x, y, z) in space a number u
$\mathbf{R} = \mathbf{G}(t)$	Assigns to a number t (time) a position vector $\mathbf{R}$
$\mathbf{F}(x, y) = P\mathbf{i} + Q\mathbf{j}$	Assigns to a point in the plane a vector $P\mathbf{i} + Q\mathbf{j}$
$\mathbf{F}(x, y, z) = P\mathbf{i} + Q\mathbf{j} + R\mathbf{k}$	Assigns to a point in space a vector $P\mathbf{i} + Q\mathbf{j} + R\mathbf{k}$

Now we introduce another type of function. This type assigns to a point in the plane a point in the plane. It is called a *mapping*, and will be denoted F.

It will be convenient to denote the "input" point (u, v) and the "output" point (x, y). Thus

$$F(u, v) = (x, y).$$

Both x and y depend on u and v. So we may write

$$x = f(u, v) \qquad \text{and} \qquad y = g(u, v).$$

Thus $$F(u, v) = (f(u, v), g(u, v)),$$

and F is shorthand for a pair, f and g, of scalar functions of two variables. Any mapping from the uv plane to the xy plane is described by a pair of such functions, $x = f(u, v)$ and $y = g(u, v)$.

EXAMPLE 1   Let F be the mapping that assigns to the point (u, v) the point $(2u, 3v)$. Describe this mapping geometrically.

SOLUTION   In this case $$x = 2u \quad \text{and} \quad y = 3v.$$

The table below records the effect of the mapping on four typical points in the uv plane.

(u, v)	$(0, 0)$	$(1, 0)$	$(1, 1)$	$(0, 1)$
$(2u, 3v)$	$(0, 0)$	$(2, 0)$	$(2, 3)$	$(0, 3)$

In the notation $F(u, v) = (x, y)$, these data read:

$$F(0, 0) = (2(0), 3(0)) = (0, 0);$$
$$F(1, 0) = (2(1), 3(0)) = (2, 0);$$
$$F(1, 1) = (2(1), 3(1)) = (2, 3);$$
$$F(0, 1) = (2(0), 3(1)) = (0, 3).$$

Note that the first coordinate of $(x, y) = F(u, v)$ is $x = 2u$, twice the first coordinate of (u, v). Thus the mapping magnifies horizontally by a factor of 2. Similarly, it stretches vertically by a factor of 3. (This causes a sixfold magnification of areas.) If P is a point in the square R whose vertices are

$$(0, 0), \quad (1, 0), \quad (1, 1), \quad (0, 1),$$

then the image of P is the point in the rectangle S whose vertices are

$$(0, 0), \quad (2, 0), \quad (2, 3), \quad (0, 3).$$

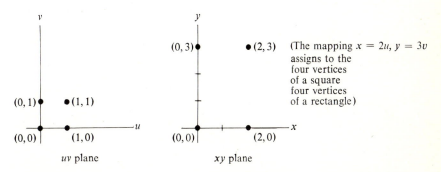

(The mapping $x = 2u$, $y = 3v$ assigns to the four vertices of a square four vertices of a rectangle)

Think of (u, v) as a point on a slide, and $(2u, 3v)$ as its image on the screen. Then the mapping F projects the square R on the slide onto a rectangle on the screen.

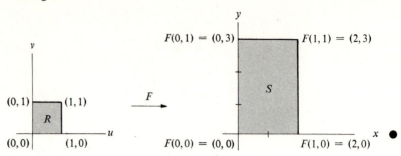

EXAMPLE 2   Let $F = (f, g)$ be the same as in Example 1. Let R be the set of points (u, v) in the uv plane (the slide) such that

$$u^2 + v^2 \leq 1.$$

In words, R is the disk of radius 1, center $(0, 0)$. What is the image of R on the xy plane (the screen)?

SOLUTION   Since $x = 2u$ and $y = 3v$, it follows that

$$u = \frac{x}{2} \quad \text{and} \quad v = \frac{y}{3}.$$

Thus the inequality

$$u^2 + v^2 \leq 1$$

implies that

$$\left(\frac{x}{2}\right)^2 + \left(\frac{y}{3}\right)^2 \leq 1;$$

that is,

$$\frac{x^2}{4} + \frac{y^2}{9} \leq 1.$$

Consequently, the image of the disk of radius 1 consists of all points inside the ellipse

$$\frac{x^2}{4} + \frac{y^2}{9} = 1.$$

The image of the disk R is the ellipse S.

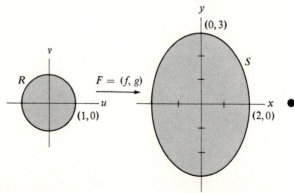

Pay special attention to the method used in Example 2 for finding the image of the curve $u^2 + v^2 = 1$. Solve for u and v in terms of x and y and substitute the results into the equation that relates u and v. This provides an equation linking x and y.

Examples 1 and 2 suggest why a function $F(u, v) = (f(u, v), g(u, v))$ is called a *mapping*. Think of the uv plane as the land and the xy plane as a map of this land. The point $(x, y) = (f(u, v), g(u, v))$ is the point on the map that corresponds to the point (u, v). (Of course, some functions may produce quite distorted maps.)

EXAMPLE 3   Let $(x, y) = (uv, v)$. If $P = (u, v)$ lies on the line $v = u$, where does $F(P) = (uv, v)$ lie?

SOLUTION   In this case $x = uv$ and $y = v.$

To find what the mapping does to the line $u = v$, find an equation linking x and y if (x, y) is on the image of the line $u = v$.

Observe that $v = y$ and that

$$u = \frac{x}{v} = \frac{x}{y}.$$

Substitution in the equation $u = v$ yields

$$\frac{x}{y} = y,$$

or $x = y^2.$

Thus as $P = (u, v)$ wanders about the line $u = v$, the image point (uv, v) wanders about the parabola $x = y^2$.

In order to make this fact more concrete, let us compute $F(P)$ for a few points on the line $v = u$. For instance,

$$\text{If } P = (2, 2), \qquad F(P) = (2 \cdot 2, 2) = (4, 2).$$

The table in the margin records the results of several such computations. The diagrams show these data (the dotted line joins the images).

P	*Image of P*
$P_1 = (-2, -2)$	$(4, -2)$
$P_2 = (-1, -1)$	$(1, -1)$
$P_3 = (0, 0)$	$(0, 0)$
$P_4 = (1, 1)$	$(1, 1)$
$P_5 = (2, 2)$	$(4, 2)$
$P_6 = (3, 3)$	$(9, 3)$

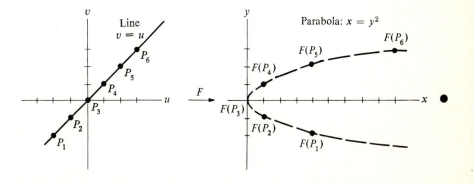

It can be proved that the mapping

$$(f(u, v), g(u, v)) = (au + bv, cu + dv),$$

for constants a, b, c, d such that $ad - bc \neq 0$, takes any straight line into a straight line. (See Exercise 15.) This fact may be assumed when doing the exercises.

The next section obtains a formula for the magnifying effect of a mapping. The magnification may vary from point to point. The function in Examples 1 and 2, which magnifies all areas by a factor of 6, is unusually simple, and its magnification was found by high school geometry.

Exercises

1. Let $(f(u, v), g(u, v)) = (2u, 3v)$.
   (a) Fill in this table.

(u, v)	$(x, y) = (2u, 3v)$
$(0, 0)$	
$(2, 0)$	
$(1, 1)$	

   (b) Plot the three points in the left column of the table in (a) in the uv plane.
   (c) Plot the three points entered in the right column of the table in (a) in the xy plane.
   (d) Draw the triangular region R whose vertices are listed in (a) in the uv plane.
   (e) Draw S, the image of R in the xy plane.
   (f) Compute the areas of R and S, and show that S is six times as large as R.
   (g) Show that R is a right triangle, but S is not.

2. Let $(f(u, v), g(u, v)) = (2u, 3v)$, as in Examples 1 and 2. Show that, if (u, v) is on the line $5u + 4v + 7 = 0$, then the image of (u, v) is on the line

$$\frac{5x}{2} + \frac{4y}{3} + 7 = 0.$$

3. Let $(f(u, v), g(u, v)) = (2u, 3v)$. Show that, if (u, v) is on the line $au + bv + c = 0$, then the image (x, y) is on the line $3ax + 2by + 6c = 0$.

4. Let $(f(u, v), g(u, v)) = (uv, v)$, as in Example 3.
   (a) Select four points P_1, P_2, P_3, P_4 on the hyperbola $uv = 1$.
   (b) Compute and plot the images of P_1, P_2, P_3, P_4 in the xy plane.
   (c) As (u, v) wanders over the hyperbola $uv = 1$, what curve does $(x, y) = (uv, v)$ sweep out in the xy plane?

5. Let $(f(u, v), g(u, v)) = (uv, v)$ as in Example 3.
   (a) As (u, v) sweeps out the line $u = 0$, what curve does $(x, y) = (uv, v)$ sweep out in the xy plane?

   (b) As (u, v) sweeps out the line $u + v = 1$, what curve does $(x, y) = (uv, v)$ sweep out in the xy plane?

6. Let $(f(u, v), g(u, v)) = (1/u, 1/v)$. [The mapping is defined whenever (u, v) is not on the u axis $(v = 0)$ or the v axis $(u = 0)$.]
   (a) As (u, v) sweeps out the line $v = u$, what curve does $(x, y) = (1/u, 1/v)$ sweep out in the xy plane?
   (b) As (u, v) sweeps out the line $u + v = 1$, what curve does $(x, y) = (1/u, 1/v)$ sweep out in the xy plane?

7. Let $(f(u, v), g(u, v)) = (\sqrt{u}, v)$ for $u \geq 0$ and $v \geq 0$.
   (a) Fill in this table.

(u, v)	$(x, y) = (\sqrt{u}, v)$
$(0, 0)$	
$(1, 1)$	
$(2, 2)$	
$(3, 3)$	
$(4, 4)$	

   (b) Plot the five points in the uv plane given in (a) and the five corresponding points in the xy plane.
   (c) As (u, v) moves along the line $v = u$, what curve does $(x, y) = (\sqrt{u}, v)$ trace out in the xy plane?

8. Let $(f(u, v), g(u, v)) = (\sqrt{u}, v)$ as in Exercise 7. Let R, in the uv plane, be bounded by the lines $v = u$, $u + v = 1$, and $u = 0$.
   (a) Draw R.
   (b) Find S, the image of R in the xy plane.

9. Let $(f(u, v), g(u, v)) = (2u + 3v, u + v)$.
   (a) Fill in this table.

(u, v)	$(x, y) = (2u + 3v, u + v)$
$(1, 0)$	
$(0, 1)$	
$(-1, 0)$	
$(0, -1)$	

(b) As (u, v) runs through the circle $u^2 + v^2 = 1$ counterclockwise, $(x, y) = (2u + 3v, u + v)$ sweeps out a curve C in the xy plane. Is C swept out clockwise or counterclockwise?

10. Let $(x, y) = (2u + 3v, u + v)$. Find (u, v) if $(x, y) = (7, 2)$.

11. Let $(x, y) = (u + v, u - 2v)$.
   (a) Fill in this table.

(u, v)	$(x, y) = (u + v, u - 2v)$
$(-1, 0)$	
$(0, 1)$	
$(1, 2)$	
$(2, 3)$	

   (b) Plot the four points in the left column of (a) in the uv plane.
   (c) Plot the four points computed in (a) in the xy plane.
   (d) Prove that if (u, v) lies on the line $v = u + 1$, then $(x, y) = (u + v, u - 2v)$ lies on the line $x + 2y + 3 = 0$. *Hint*: Solve for u and v in terms of x and y.

12. Let $F(u, v) = (2u + 3v, u + v)$. Find (u, v) if $F(u, v) = (x, y)$. (Express the answer in terms of x and y.)

■

13. Let $(f(u, v), g(u, v)) = (u \cos v, u \sin v)$. Let R be the rectangle in the uv plane described as follows:

$$1 \le u \le 2, \qquad 0 \le v \le 2\pi.$$

Let S be the region in the xy plane corresponding to R. Sketch S.

14. Let θ be a fixed number and let $(f(u, v), g(u, v)) = (u \cos \theta - v \sin \theta, u \sin \theta + v \cos \theta)$. Show that for any points P_1 and P_2 in the uv plane, the distance between $F(P_1)$ and $F(P_2)$ equals the distance between P_1 and P_2.

■ ■

15. Let a, b, c, and d be constants such that $ad - bc$ is not 0. Let $F(u, v) = (f(u, v), g(u, v)) = (au + bv, cu + dv)$.
   (a) Prove that the image of a straight line in the uv plane under F is a straight line in the xy plane.
   (b) Prove that the image of a conic section in the uv plane under F is a conic section in the xy plane.

16. Let F, from the uv plane to the xy plane, have the property that it "preserves lengths," that is, for all pairs of points P and Q in the uv plane the distance between $F(P)$ and $F(Q)$ is the same as the distance from P to Q.
   (a) Prove that if P_1, P_2, and P_3 lie on a line in the uv plane, then $F(P_1)$, $F(P_2)$, and $F(P_3)$ lie on a line in the xy plane.
   (b) Prove that if P_1, P_2, and P_3 are points in the uv plane, then the angle $\angle F(P_1)F(P_2)F(P_3)$ equals the angle $\angle P_1 P_2 P_3$.

22.4
Magnification in the plane: the Jacobian

In the discussion of the linear slide and linear screen in Sec. 3.1 it was shown that magnification at a point is represented by a derivative. This section generalizes the concept of magnification to mappings of plane sets. But consider for a moment more the linear slide projected onto the linear screen. Let the coordinate on the slide be u and the coordinate of its image on the screen be x; $x = f(u)$. The magnification at u is the derivative f' evaluated at u. For instance, if $x = 2u$, then $dx/du = 2$ for all u. There is a magnification by a factor of 2 at all points. If $x = -2u$, the magnification is -2 at all points; the negative value records the interchange of left and right.

Now consider the analogous magnification for the more common mapping from a plane slide to a plane screen. The slide is the uv plane; the screen is the xy plane. Let the mapping F be given by the functions

$$x = f(u, v) \text{ and } y = g(u, v).$$

This is the question: If a small region in the uv plane has area ΔA,

The image of a short section of length Δu has length Δx approximately $f'(u) \Delta u$.

approximately what is the area of its image in the xy plane?

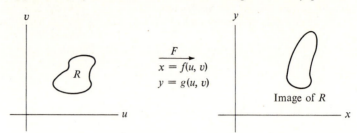

In the case of the mapping in Example 1 of the preceding section, the answer is easy.   There

$$x = 2u \text{ and } y = 3v.$$

It was observed that this mapping stretches horizontally by a factor of 2 and vertically by a factor of 3.   Thus this mapping magnifies the areas of *all* regions by a factor of 6, whether they be small or large.

But in a more complicated mapping the magnification may vary from point to point.   Theorem 1 is the basis for Theorem 2, which will show that the mapping

$$x = f(u, v) \qquad y = g(u, v)$$

has a "local" magnification given by the formula

$$f_u g_v - f_v g_u,$$

which is the determinant

$$\begin{vmatrix} f_u & f_v \\ g_u & g_v \end{vmatrix}.$$

The reader may check that it gives the right result for the mapping $x = 2u$, $y = 3v$.

THEOREM I (*The area of an image*).   Let R be a region in the uv plane bounded by the curve C_R.   Let S be a region in the xy plane bounded by the curve C_S. Let F be a one-to-one mapping from the uv plane to the xy plane, given by the formulas

$$x = f(u, v), \qquad y = g(u, v).$$

Assume that S is the image of R of this mapping.

Then
$$\text{Area of } S = \int_R \begin{vmatrix} f_u & f_v \\ g_u & g_v \end{vmatrix} dA$$

if $\begin{vmatrix} f_u & f_v \\ g_u & g_v \end{vmatrix}$ is nonnegative everywhere in R. $\left(\text{If } \begin{vmatrix} f_u & f_v \\ g_u & g_v \end{vmatrix} \text{ is nonpositive every-}\right.$

where in R, then the area of S equals

$$\left. -\int_R \begin{vmatrix} f_u & f_v \\ g_u & g_v \end{vmatrix} dA. \right)$$

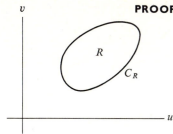

v

R

C_R

u

y

S is the image of R

S

C_S

(Counterclockwise)

x

PROOF If C_S is counterclockwise,

$$\text{Area of } S = \tfrac{1}{2} \oint_{C_S} - y \, dx + x \, dy, \tag{1}$$

by Theorem 2 of Sec. 22.2.

We shall express $\oint_{C_S} - y \, dx + x \, dy$ as a line integral over C_R. The same parameter t will be used for both C_R and C_S.

Let the curve C_R be counterclockwise and parameterized by $u = p(t)$, $v = q(t)$ for t in $[a, b]$. Then a parameterization of C_S is

$$x = f(p(t), q(t)) \qquad y = g(p(t), q(t))$$

for t in $[a, b]$. (After the proof is completed, it will be shown why the parameterization of C_S is also counterclockwise, a fact that will be needed in order to apply Theorem 2 of Sec. 22.2.) Thus, by a chain rule,

$$\frac{dx}{dt} = f_u(p(t), q(t)) \frac{dp}{dt} + f_v(p(t), q(t)) \frac{dq}{dt}$$

and

$$\frac{dy}{dt} = g_u(p(t), q(t)) \frac{dp}{dt} + g_v(p(t), q(t)) \frac{dq}{dt}.$$

Writing $(p(t), q(t))$ as (u, v), we therefore have

$$dx = f_u(u, v) \, du + f_v(u, v) \, dv$$

and

$$dy = g_u(u, v) \, du + g_v(u, v) \, dv.$$

Consequently,

$$\oint_{C_S} - y \, dx + x \, dy = \int_{C_R} [-g(f_u \, du + f_v \, dv) + f(g_u \, du + g_v \, dv)]. \tag{2}$$

Note that the line integral over C_S is now expressed as a line integral over C_R. All that remains is to express the integral over C_R as an integral over R. This is where Green's theorem comes in.

After the terms are collected, (2) becomes

$$\int_{C_R} (-g f_u + f g_u) \, du + (f g_v - g f_v) \, dv. \tag{3}$$

Now, Green's theorem (written with u and v replacing x and y) asserts that

$$\int_{C_R} P \, dv - Q \, du = \int_R (P_u + Q_v) \, dA.$$

Inspection of (3) shows that we are interested in the case when

$$-Q = -g f_u + f g_u \qquad \text{and} \qquad P = f g_v - g f_v;$$

hence

$$Q = g f_u - f g_u \qquad \text{and} \qquad P = f g_v - g f_v.$$

By Green's theorem, (3) is equal to

$$\int_R \left[\frac{\partial(f g_v - g f_v)}{\partial u} + \frac{\partial(g f_u - f g_u)}{\partial v} \right] dA. \tag{4}$$

Computing the partial derivatives in (4) and exploiting the identities $f_{uv} = f_{vu}$ and $g_{uv} = g_{vu}$ transforms (4) to

$$\int_R [(fg_{vu} + g_v f_u - gf_{vu} - f_v g_u) + (gf_{uv} + f_u g_v - fg_{uv} - g_u f_v)] \, dA$$

$$= \int_R (g_v f_u - f_v g_u + f_u g_v - g_u f_v) \, dA$$

$$= 2 \int_R (f_u g_v - f_v g_u) \, dA. \tag{5}$$

Comparison of (1), (2), (3), (4), and (5)—which represent the gradual transformation of an integral over C_S to an integral over R—shows that

$$\text{Area of } S = \int_R \begin{vmatrix} f_u & f_v \\ g_u & g_v \end{vmatrix} \, dA$$

and the theorem is proved. ●

Remark: Why is the parameterization of C_S counterclockwise? Recall that we have assumed that $f_u g_v - f_v g_u$ is nonnegative. Thus $\int_R (f_u g_v - f_v g_u) \, dA$ is positive, as the area of S must be. If C_S were *clockwise*, then $\frac{1}{2} \int_{C_S} - y \, dx + x \, dy$ would be the *negative* of the area of S, and hence a negative number. But our computations showed that $\frac{1}{2} \int_{C_S} - y \, dx + x \, dy$ equals $\frac{1}{2} \int_R (f_u g_v - f_v g_u) \, dA$, a positive number. Hence it is the assumption that $f_u g_v - f_v g_u$ is nonnegative that assures us that the mapping takes the counterclockwise C_R into a counterclockwise C_S. Just as the sign of dx/du records the preservation or interchange of right and left on a line, so does the sign of $f_u g_v - f_v g_u$ record the preservation or interchange of counterclockwise and clockwise in the plane.

Theorem 1 will be used in Example 1 to compute the area of a region. The mapping will be a special case of the general form

$$x = au + bv \qquad y = cu + dv,$$

where a, b, c, d, are constants, with $ad - bc \neq 0$. Any such mapping takes lines into lines.

EXAMPLE 1 Let R be the triangle in the uv plane with vertices $(0, 0)$, $(1, 0)$ and $(0, 1)$. Let F be the mapping,

$$x = 2u - 3v \qquad y = 5u + 7v.$$

Find the area of the image of R under this mapping.

SOLUTION Let S be the image of R under the mapping. According to Theorem 1,

$$\text{Area of } S = \int_R (f_u g_v - f_v g_u) \, dA,$$

where $f(u, v) = 2u - 3v$ and $g(u, v) = 5u + 7v.$

(Assume that $f_u g_v - f_v g_u$ is positive.)

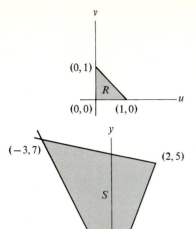

The image of R

In this case $\qquad f_u = 2, \qquad f_v = -3, \qquad g_u = 5, \qquad g_v = 7.$

Hence $\qquad\qquad$ Area of $S = \displaystyle\int_R [2\cdot 7 - (-3)5]\, dA$

$$= \int_R (14 + 15)\, dA$$

$$= \int_R 29\, dA$$

$$= 29 \cdot \text{area of } R.$$

The area of R is $\frac{1}{2} \cdot$ base $\cdot$ height $= \frac{1}{2}$. Hence the area of S is $\frac{29}{2}$.

Incidentally, since the mapping takes lines into lines, S is a triangle. It is easy to check that the mapping sends the three vertices of R to the points $(0, 0)$, $(2, 5)$, $(-3, 7)$ in the xy plane. R and S are shown in the diagrams in the margin. ●

DEFINITION *Jacobian.* Let $x = f(u, v)$, $y = g(u, v)$ describe a mapping from the uv plane to the xy plane. The function

$$f_u g_v - f_v g_u = \begin{vmatrix} f_u & f_v \\ g_u & g_v \end{vmatrix}$$

is called the *Jacobian* of the mapping. It is also denoted

$$\frac{\partial(f, g)}{\partial(u, v)} \qquad \text{or} \qquad \frac{\partial(x, y)}{\partial(u, v)}.$$

The Jacobian was introduced in 1841 by Jacobi.

The Jacobian in Example 1 is constant, having the value 29 at all points (u, v). In Example 2 the Jacobian is not constant.

EXAMPLE 2 Let $x = uv$, $y = v$ be the mapping described in Example 3 of the preceding section. Let R be the triangle in the uv plane whose vertices are $(0, 0)$, $(1, 1)$, and $(0, 1)$. Let S be the image of R under this mapping. Find the area of S.

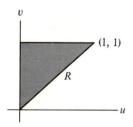

SOLUTION As shown in Example 3 of the preceding section, the mapping sends $(0, 0)$ in the uv plane to $(0, 0)$ in the xy plane, $(1, 1)$ in the uv plane to $(1, 1)$ in the xy plane, and the line through $(0, 0)$ and $(1, 1)$ to the parabola $x = y^2$. The line $v = 1$, which is the top edge of triangle R, goes to the line $y = 1$. Moreover, the line $u = 0$ goes to the line $x = 0$, since $x = uv$. The diagrams in the margin show R and S.

First let us compute the Jacobian and determine its sign.

Since $\qquad\qquad f(u, v) = uv \qquad$ and $\qquad g(u, v) = v,$

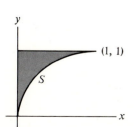

it follows that

$$f_u = v, \qquad f_v = u, \qquad g_u = 0, \qquad \text{and } g_v = 1.$$

Thus
$$\frac{\partial(f, g)}{\partial(u, v)} = \begin{vmatrix} v & u \\ 0 & 1 \end{vmatrix} = v.$$

Throughout R, $v \geq 0$, hence the Jacobian is nonnegative throughout R.

Thus
$$\text{Area of } S = \int_R \frac{\partial(f, g)}{\partial(u, v)} \, dA$$

$$= \int_R v \, dA$$

$$= \int_0^1 \left(\int_u^1 v \, dv \right) du$$

$$= \int_0^1 \left(\frac{1}{2} - \frac{u^2}{2} \right) du$$

$$= \tfrac{1}{3}.$$

Incidentally, because S is such a simple region, it is possible to calculate its area directly ($\int_0^1 y^2 \, dy = \frac{1}{3}$) and verify the result. ●

It is now possible to calculate the local magnification of the mapping $x = f(u, v)$, $y = g(u, v)$.

THEOREM 2   (*The Jacobian as local magnification*).   Let $P_0 = (u_0, v_0)$ be a point where the Jacobian of the mapping $x = f(u, v)$, $y = g(u, v)$ is not 0.   Let the value of the Jacobian at (u_0, v_0) be J_0.   Then the area of the image S of a small region R around P_0 is approximately

$$|J_0| \cdot \text{area of } R,$$

or, more precisely,
$$\lim_{\text{diameter } R \to 0} \frac{\text{Area of } S}{\text{Area of } R} = |J_0|.$$

PROOF   Take the case in which J_0 is positive.   Since the Jacobian is continuous, if R has sufficiently small diameter, it remains positive throughout R, in fact, as close to J_0 as we please.   For instance, if R is sufficiently small, the Jacobian remains, say, between $0.9J_0$ and $1.1J_0$.   Thus, since

$$\text{Area of } S = \int_R \frac{\partial(f, g)}{\partial(u, v)} \, dA,$$

we have
$$0.9 \, J_0 \text{ area of } R \leq \text{area of } S \leq 1.1 \, J_0 \text{ area of } R$$

for R of sufficiently small diameter.   In other words, for small R,

$$0.9 \, J_0 \leq \frac{\text{area of } S}{\text{area of } R} \leq 1.1 \, J_0.$$

Of course we could have used, instead of 0.9 and 1.1, two numbers even closer to 1, such as 0.999 and 1.001.   Thus we see that, as the diameter of R

approaches 0, $\dfrac{\text{Area of } S}{\text{Area of } R} \to J_0.$ ●

Just as the derivative of the function $x = f(u)$ records its local magnification, the Jacobian records the local magnification of the mapping $x = f(u, v)$, $y = g(u, v)$. When the Jacobian is positive at (u_0, v_0), the mapping takes a small counterclockwise curve around (u_0, v_0) into a counterclockwise curve— that is, preserves orientation. When the Jacobian is negative, it reverses orientation. The magnification in both cases is given by the absolute value of the Jacobian.

The Jacobian of the mapping in Example 2 is v at the point (u, v). Thus, for instance, it magnifies the area of a small region around $(2, 5)$ approximately fivefold. The area of a small region around $(2, -5)$ is also magnified approximately fivefold. But the image of a small counterclockwise curve around $(2, -5)$ is swept out clockwise.

Not only is the Jacobian of a mapping useful in expressing the area of a set S but it also can be used to transform an integral over S to an integral over R. The next theorem describes this in detail. The integrand in the theorem is given the uncustomary label h, because f and g have already been utilized to describe a mapping. The typical point in S is labeled Q—not P—because P denotes the typical point in R.

THEOREM 3 Let F be a mapping from the uv plane to the xy plane, $x = f(u, v)$, $y = g(u, v)$. Let S be the image in the xy plane of the set R in the uv plane. Let h be a numerical function defined on S. Then

$$\int_S h(Q)\, dS = \int_R h(F(P)) \left| \frac{\partial(f, g)}{\partial(u, v)} \right| dA.$$

The proof of this theorem is usually given in advanced calculus. However, it should appear plausible, since the area of a small region in the xy plane is approximately the absolute value of the Jacobian times the area of a corresponding small region in the uv plane. That is, dS is replaced by $|\text{Jacobian}| \cdot dA$.

EXAMPLE 3 Let S be the region in Example 2. Let h be the function $h(x, y) = xy$. Use Theorem 3 to compute $\int_S h(Q)\, dS$.

SOLUTION Let R be the triangular region described in Example 2. Let F be the mapping,

$$x = uv \qquad \text{and} \qquad y = v,$$

of that example. Recall that its Jacobian is v. According to Theorem 3,

$$\int_S h(Q)\, dS = \int_S xy\, dS = \int_R xyv\, dA$$

$$= \int_R uv \cdot v \cdot v\, dA$$

$$= \int_R uv^3 \, dA$$

$$= \int_0^1 \left(\int_u^1 uv^3 \, dv \right) du$$

$$= \int_0^1 \left(\frac{u}{4} - \frac{u^5}{4} \right) du$$

$$= \tfrac{1}{12}. \quad \bullet$$

Theorem 3 provides a new explanation of why the r appears in the integrand when you use polar coordinates. It turns out that this r is the Jacobian of a certain mapping. The explanation goes as follows.

Let $x = u \cos v$ and $y = u \sin v$ describe a mapping from the uv plane to the xy plane. In order to make it one-to-one keep $u > 0$ and $0 \le v < 2\pi$. Let R be the region in the uv plane described by the conditions

$$r_1 \le u \le r_2 \qquad \text{and} \qquad \alpha \le v \le \beta.$$

Let S be the image of R in the xy plane.

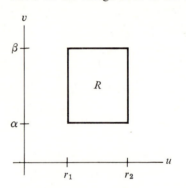 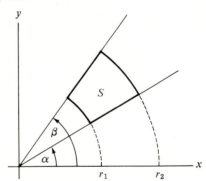

That the region S appears as in the diagram can be seen if you think of (u, v) as polar coordinates (r, θ) in the xy plane. After all,

$$x^2 + y^2 = (u \cos v)^2 + (u \sin v)^2 = u^2,$$

hence

$$u = \sqrt{x^2 + y^2}.$$

Also,

$$\frac{y}{x} = \frac{u \sin v}{u \cos v} = \tan v.$$

So (u, v) are in fact polar coordinates of the point (x, y).

The Jacobian of the mapping is

$$\begin{vmatrix} \dfrac{\partial x}{\partial u} & \dfrac{\partial x}{\partial v} \\[2mm] \dfrac{\partial y}{\partial u} & \dfrac{\partial y}{\partial v} \end{vmatrix} = \begin{vmatrix} \cos v & -u \sin v \\ \sin v & u \cos v \end{vmatrix}$$

$$= u \cos^2 v + u \sin^2 v$$
$$= u.$$

Theorem 3 then asserts that

$$\int_S f(Q)\, dS = \int_R f(u, v)u\, dA.$$

Note the appearance of u in the integrand of the integral over R.

Then
$$\int_R f(u, v)u\, dA = \int_\alpha^\beta \left[\int_{r_1}^{r_2} f(u, v)u\, du \right] dv.$$

But since u and v coincide with polar coordinates r and θ in the xy plane, we may replace the letters u and v by r and θ instead, obtaining

$$\int_S f(x, y)\, dS = \int_\alpha^\beta \left[\int_{r_1}^{r_2} f(r, \theta)r\, dr \right] d\theta$$

The r appearing in the integrand is a Jacobian.

It is shown in advanced calculus that, whenever you change from rectangular coordinates to another coordinate system, the compensating factor that must be inserted in the integrand is a Jacobian. In the case of spherical coordinates (ρ, ϕ, θ), we have

$$x = \rho \sin \phi \cos \theta, \qquad y = \rho \sin \phi \sin \theta, \qquad z = \rho \cos \phi.$$

The Jacobian is the third-order determinant:

$$\begin{vmatrix} \dfrac{\partial x}{\partial \rho} & \dfrac{\partial x}{\partial \phi} & \dfrac{\partial x}{\partial \theta} \\[2mm] \dfrac{\partial y}{\partial \rho} & \dfrac{\partial y}{\partial \phi} & \dfrac{\partial y}{\partial \theta} \\[2mm] \dfrac{\partial z}{\partial \rho} & \dfrac{\partial z}{\partial \phi} & \dfrac{\partial z}{\partial \theta} \end{vmatrix}.$$

Straightforward computations show that it equals $\rho^2 \sin \phi$, the expression met in Secs. 18.3 and 18.5.

Exercises

In Exercises 1 to 4 compute the Jacobian of the given mapping at the given points.

1. $x = 2u$, $y = -3v$ at (a) (4, 5) and (b) (1, 7).
2. $x = 3v$, $y = 2u$ at (a) (1, 2) and (b) (2, 3).
3. $x = u^2 v$, $y = v/u$ at (a) (1, 2) and (b) (−1, 3).
4. $x = 1/u$, $y = 1/v$ at (a) (1, 2) and (b) (3, 2).
5. Let a, b, c, d be constants such that $ad - bc \neq 0$. Let

$$x = au + bv, \qquad y = cu + dv.$$

Show that the Jacobian of the mapping is $ad - bc$ at all points.

6. Let the Jacobian of a mapping be 3 at (2, 4). Let R be a small region around (2, 4) of area 0.05. Approximately how large is the image of R under the mapping?
7. Show that the mapping

$$x = u \cos \theta - v \sin \theta, \quad y = u \sin \theta + v \cos$$

where θ is any constant, preserves area.

preserves area if the area of the image of R is equal to the area of R for all regions of the type we consider.

8. (See Exercise 7.)
   (a) Show that the mapping defined by $x = u - v^2 - 2u^2v - u^4$, $y = v + u^2$ preserves area.
   (b) Sketch the image of the square whose vertices are $(0, 0)$, $(1, 0)$, $(1, 1)$, $(0, 1)$.

9. Let $x = u - v$, $y = 2u + v$. Let R be the triangular region in the uv plane that has the vertices $(0, 0)$, $(1, 0)$, and $(0, 1)$. Let C_R be its border, and let S be the image of R and C_S the image of C_R.
   (a) Draw the images of the three vertices.
   (b) As you sweep out C_R counterclockwise, in what direction does the mapping sweep out C_s?
   (c) Compute the Jacobian. Are your answers in (b) and (c) compatible?

10. Verify (a) that the mapping $x = -u$, $y = v$ has a negative Jacobian, and (b) that, as P runs counterclockwise around the unit circle whose center is at $(0, 0)$, then the image of P runs clockwise.

11. Let $x = u^2$, $y = 2v$ be a mapping and let R be the triangle bordered by $u = v$, $v = 0$, and $u = 1$. Let S be the image of R.
    (a) Draw S, which has one curved side.
    (b) Compute the area of S with the aid of the theorem in this section.
    (c) Compute the area of S directly.

12. Let $x = u^2 - v^2$, $y = 2uv$ be a mapping and let R be the square whose vertices are $(1, 0)$, $(2, 0)$, $(2, 1)$, and $(1, 1)$.
    (a) Show that, when $u = 1$, the image of (u, v) lies on the curve $x = 1 - (y/2)^2$.
    (b) Show that, when $u = 2$, the image of (u, v) lies on the curve $x = 4 - (y/4)^2$.
    (c) Show that the image of the line $v = 0$ is the positive x axis.
    (d) Show that the image of the line $v = 1$ is the curve $x = (y/2)^2 - 1$.
    (e) Draw S, the image of R. (It has three curved sides and one straight side.)
    (f) Find the area of S.

13. Consider only positive u, v, x, y. Define a mapping by setting $x = u^{1/3}v^{2/3}$, $y = u^{2/3}v^{1/3}$.
    (a) Show that $x^2 = vy$ and $y^2 = ux$.
    (b) Let R in the uv plane be the rectangle bordered by the lines $u = 1$, $u = 2$, $v = 3$, and $v = 4$. Let S be the image in the xy plane of R. Show that S is bordered by the four parabolas $x^2 = y$, $x^2 = 2y$, $y^2 = 3x$, and $y^2 = 4x$.
    (c) Draw S.
    (d) Compute the area of S by integrating the Jacobian over R.

14. Let the mapping, R, and S, be the same as in Exercise 11. Use the Jacobian to compute $\int_S xy \, dA$.

15. Consider only positive u and $0 \le v \le \pi/2$.
    (a) Show that, if $x = u \cos v$ and $y = u \sin v$, then $u = \sqrt{x^2 + y^2}$ and $v = \tan^{-1}(y/x)$.
    (b) Show that $\partial(u, v)/\partial(x, y) = 1/\sqrt{x^2 + y^2}$.
    (c) Show that $\partial(u, v)/\partial(x, y)$ is the reciprocal of $\partial(x, y)/\partial(u, v)$.
    (d) Why is (c) to be expected? (Think optically.)

■

16. Let $x = 2u + v$, $y = 3u + 2v$ be a mapping from the uv plane to the xy plane.
    (a) Show that the mapping does not change areas (though, as (b) will show, it changes shapes).
    (b) Sketch the image S of the circle $u^2 + v^2 = 1$.
    (c) Let R be the region in the uv plane bounded by the circle $u^2 + v^2 = 1$. Let S be its image in the xy plane. Evaluate $\int_S x^2 \, dA$.

Exercises 17 to 19 concern the mapping $x = u + v$, $y = uv$ for nonnegative u and v.

17. (a) Show that the image of a horizontal or a vertical line in the uv plane is a line in the xy plane.
    (b) Show that the image of the line $u = v$ is part of the parabola $y = x^2/4$.

18. (a) Where is the Jacobian $\partial(x, y)/\partial(u, v)$ positive? negative?
    (b) The triangular path in the uv plane that sweeps out the points $P_1 = (1, 2)$, $P_2 = (2, 3)$, $P_3 = (1, 3)$, $P_1 = (1, 2)$, in that order, is counterclockwise. Sketch the path in the xy plane swept out by the image of this path. Is it clockwise or counterclockwise?
    (c) The triangular path in the uv plane that sweeps out the points $P_1 = (2, 1)$, $P_2 = (3, 2)$, $P_3 = (2, 2)$, $P_1 = (2, 1)$, in that order, is counterclockwise. Sketch the path in the xy plane swept out by the image of this path. Is it clockwise or counterclockwise?

19. Let R be the region in the uv plane between the parabola $v = u^2$ and the line $v = u$. Find the area of its image in the xy plane.

20. Consider the mapping $x = 2u + v$, $y = u + 2v$.
    (a) Show that the image of the square R with vertices $(0, 0)$, $(1, 0)$, $(1, 1)$, $(0, 1)$ is a parallelogram S.
    (b) Use (a) to help evaluate $\int_S x^2 \, dA$.

21. (a) Evaluate the third-order Jacobian defined at the end of the section. The result should be $\rho^2 \sin \phi$.
    (b) Carry out the analogous computations for cylindrical coordinates (r, θ, z). The result should be r.

■ ■

22. Prove Theorem 1 of this section when the Jacobian is negative at all points in R.

22.5

The divergence theorem

The divergence theorem, also called Gauss' theorem, is the analog in space of Green's theorem in the plane. Instead of a region in the plane and its bounding curve, the divergence theorem concerns a region in space and its bounding surface.

THE DIVERGENCE THEOREM

Let R be a convex region in space, and S its surface. Let $\mathbf{F}$ be a vector field in space.

Then

$$\int_R \nabla \cdot \mathbf{F} \, dV = \int_S \mathbf{F} \cdot \mathbf{n} \, dS,$$

where $\mathbf{n}$ is the unit exterior normal.

As with Green's theorem in Sec. 22.1, consideration of fluid flow makes the divergence theorem plausible. First of all, as mentioned in Sec. 21.3,

$$\int_S \mathbf{F} \cdot \mathbf{n} \, dS$$

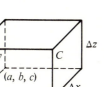

represents the net flow out of R across S. Now consider the total loss of fluid from a typical small box in R, as we considered the loss of fluid from a typical small rectangle in Sec. 22.1.

Let $\mathbf{F}(x, y, z) = P\mathbf{i} + Q\mathbf{j} + R\mathbf{k}$, and estimate the loss of fluid from the little box shown in the margin. One corner is (a, b, c); the opposite corner is $(a + \Delta x, b + \Delta y, c + \Delta z)$. The volume of the box is $\Delta x \, \Delta y \, \Delta z$.

Let us estimate the net flow out across the two planes parallel to the yz plane. These planes have the equations $x = a$ and $x = a + \Delta x$, respectively. For the plane $x = a + \Delta x$ the outer unit normal is $\mathbf{i}$. The net loss across this surface is thus approximately

$$[\mathbf{F}(a + \Delta x, b, c) \cdot \mathbf{i}] \cdot \text{area of face } ABCD$$

$$= [P(a + \Delta x, b, c)\mathbf{i} + Q(a + \Delta x, b, c)\mathbf{j} + R(a + \Delta x, b, c)\mathbf{k}] \cdot \mathbf{i} \, \Delta y \, \Delta z$$

$$= P(a + \Delta x, b, c) \, \Delta y \, \Delta z.$$

For the rectangular face corresponding to $x = a$ the outer unit normal is $-\mathbf{i}$. By similar reasoning the net flow past that rectangle is approximately

$$-P(a, b, c) \, \Delta y \, \Delta z.$$

The net flow across the two faces is thus approximately

$$[P(a + \Delta x, b, c) - P(a, b, c)] \, \Delta y \, \Delta z,$$

which is approximately

$$[P_x(a, b, c)\, \Delta x]\, \Delta y\, \Delta z = P_x(a, b, c)\, \Delta x\, \Delta y\, \Delta z.$$

Similar reasoning for the other four faces suggests that the loss of fluid out of the little box is approximately

$$[P_x(a, b, c) + Q_y(a, b, c) + R_z(a, b, c)]\, \Delta x\, \Delta y\, \Delta z.$$

Since the volume of the box is $\Delta x\, \Delta y\, \Delta z$, this suggests that the net loss of fluid out of R is the integral

$$\int_R (P_x + Q_y + R_z)\, dV,$$

which is

$$\int_R \mathbf{V} \cdot \mathbf{F}\, dV.$$

Hence

$$\int_S \mathbf{F} \cdot \mathbf{n}\, dS = \text{net loss} = \int_R \mathbf{V} \cdot \mathbf{F}\, dV,$$

as the divergence theorem asserts.

EXAMPLE 1   Let $\mathbf{F}(x, y, z) = x\mathbf{i} + 2y\mathbf{j} + 3z\mathbf{k}$.   Let S be the surface of the sphere with center $(1, 1, 1)$ and radius 5.   Find $\int_S \mathbf{F} \cdot \mathbf{n}\, dS$.

SOLUTION   Let R be the ball bounded by S.   Then, by the divergence theorem,

$$\int_S \mathbf{F} \cdot \mathbf{n}\, dS = \int_R \mathbf{V} \cdot \mathbf{F}\, dV$$

$$= \int_R \left[\frac{\partial(x)}{\partial x} + \frac{\partial(2y)}{\partial y} + \frac{\partial(3z)}{\partial z} \right] dV$$

$$= \int_R 6\, dV$$

$$= 6 \cdot \text{volume of } R$$

$$= 6 \cdot \frac{4\pi 5^3}{3}$$

$$= 1000\pi. \ \bullet$$

The next example represents an important application of the divergence theorem.

EXAMPLE 2   On the surface S of a solid piece of metal a fixed temperature distribution is maintained (by a mixture of ice cubes, heating coils, etc.).   The temperature on the surface does not vary with time, but it may vary from point to point. Assume that the temperature distribution inside the metal reaches a steady state—independent of time.   Let $T(x, y, z)$ be the temperature at (x, y, z). Show that T satisfies the partial differential equation

$$T_{xx} + T_{yy} + T_{zz} = 0.$$

SOLUTION   At each point in the metal ∇T points in the direction of most rapid increase in temperature. Thus heat flows in the direction of $-\nabla T$. Moreover, let us assume that the rate of flow (as a function of time) is proportional to the magnitude of the vector $-\nabla T$.

Now consider a fixed point P_0 in the metal and a small region R_0 around P_0. Let S_0 be the surface of R_0.

Since the temperature distribution in the metal is at a steady state, heat cannot accumulate or disappear from R_0. Thus

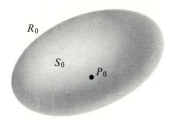

R_0 is a small solid region.
S_0 is its surface.
P_0 is inside R_0.

$$-\int_{S_0} \nabla T \cdot \mathbf{n}\, dS = 0.$$

By the divergence theorem,

$$\int_{R_0} \nabla \cdot \nabla T\, dV = 0.$$

Now

$$\nabla \cdot \nabla T = \nabla(T_x \mathbf{i} + T_y \mathbf{j} + T_z \mathbf{k})$$

$$= \frac{\partial(T_x)}{\partial x} + \frac{\partial(T_y)}{\partial y} + \frac{\partial(T_z)}{\partial z}$$

$$= T_{xx} + T_{yy} + T_{zz}.$$

Hence for any region R_0 around P_0, however small,

$$\int_{R_0} (T_{xx} + T_{yy} + T_{zz})\, dV = 0.$$

Thus $T_{xx} + T_{yy} + T_{zz}$, evaluated at P_0, is 0. ●

The proof of the divergence theorem, similar to that of Green's theorem in Sec. 22.2, is sketched in Exercise 22.

Exercises

1. Let $\mathbf{F}(x, y, z) = x\mathbf{i} + y\mathbf{j} + z\mathbf{k}$. Let R be the ball of radius a and center $(0, 0, 0)$. Let S be its surface. Verify the divergence theorem in this case.

2. Review Example 5 of Sec. 21.3 in which $\int_S \mathbf{F} \cdot \mathbf{n}\, dS$ was evaluated. Compute $\int_R \nabla \cdot \mathbf{F}\, dV$ and verify the divergence theorem.

3. Prove that, if $\mathbf{F}$ is a constant vector field and S is the surface of a convex solid, then

$$\int_S \mathbf{F} \cdot \mathbf{n}\, dS = 0.$$

4. Let $\mathbf{F} = 2x\mathbf{i} + 3y\mathbf{j} + (5z + 6x)\mathbf{k}$, and let $\mathbf{G} = (3x + 4z^2)\mathbf{i} + (2y + 5x)\mathbf{j} + 5z\mathbf{k}$. Prove that

$$\int_S \mathbf{F} \cdot \mathbf{n}\, dA = \int_S \mathbf{G} \cdot \mathbf{n}\, dS,$$

where S bounds a convex region.

5. If the length of $\mathbf{F}(P)$ is at most 5 for all points P on the surface S, what can be said about

$$\int \nabla \cdot \mathbf{F}\, dV?$$

6. Let $\mathbf{F}(x, y, z) = x^3\mathbf{i} + y^3\mathbf{j} + z^3\mathbf{k}$. Let S be the sphere of radius a and center $(0, 0, 0)$. Use the divergence theorem to help evaluate $\int_S \mathbf{F} \cdot \mathbf{n}\, dS$.

7. Explain these two steps, paraphrased from a fluid-mechanics text.

(a) Since $\int_R \frac{\partial f}{\partial t}\, dV + \int_S (f\mathbf{V} \cdot \mathbf{n})\, dS = 0$, it follows that

$$\int_R \left[\frac{\partial f}{\partial t} + \nabla \cdot f\mathbf{V}\right] dV = 0.$$

(b) Since R can be chosen arbitrarily small and the inte-

grand
$$\frac{\partial f}{\partial t} + \mathbf{V} \cdot f\mathbf{V}$$

is continuous, it follows that the integrand must be 0 everywhere. This provides the *fundamental continuity equation*

$$\frac{\partial f}{\partial t} + \mathbf{V} \cdot f\mathbf{V} = 0.$$

8. Explain why the fluid flow past the two surfaces of the box described in the text parallel to the xz plane is approximately $Q_y \, \Delta x \, \Delta y \, \Delta z$.

9. Show that $\mathbf{V} \cdot \mathbf{F}$, evaluated at P_0, equals

$$\lim_{a \to 0} \frac{\int_S \mathbf{F} \cdot \mathbf{n} \, dS}{\text{volume of } R},$$

where R is the sphere of radius a and center P_0.

10. Let $\mathbf{R}(x, y, z) = x\mathbf{i} + y\mathbf{j} + z\mathbf{k}$. Denote $|\mathbf{R}|$ by r. Let S be a smooth surface bounding a convex solid, and assume the origin is exterior to S. Using the divergence theorem, prove that

$$\int_S \frac{\mathbf{R} \cdot \mathbf{n}}{r^3} \, dS = 0.$$

 ■

11. Let $g(P)$ be a continuous function in space with the property that, for all solid regions R, $\int_R g(P) \, dV = 0$. Deduce that $g(P) = 0$ for all P.

12. Let $\mathbf{F}(x, y, z) = \mathbf{R}/|\mathbf{R}|^3$, where $\mathbf{R} = x\mathbf{i} + y\mathbf{j} + z\mathbf{k}$.
    (a) Show that $\mathbf{V} \cdot \mathbf{F} = 0$.
    (b) Show that $\int_S \mathbf{F} \cdot \mathbf{n} \, dS \neq 0$, where S is the sphere of radius 1 and center $(0, 0, 0)$.
    (c) Do (a) and (b) violate the divergence theorem?

 ■ ■

13. (See Exercise 6.) Find $\int_S x^4 \, dS$, where S is the spherical surface described in Exercise 6. *Suggestions*: (a) Why is $\int_S x^4 \, dS = \int_S z^4 \, dS = \frac{1}{3} \int_S (x^4 + y^4 + z^4) \, dS$? (b) Evaluate $\int_R (x^2 + y^2 + z^2) \, dV$, where R is the solid sphere of Exercise 6. (c) From (a), (b), and the results of Exercise 6 find $\int_S x^4 \, dS$.

14. Let f and g be scalar functions defined throughout space. Let S be the boundary of the convex solid R. Prove that

$$\int_S f(\mathbf{V}g \cdot \mathbf{n}) \, dS = \int_R (f\mathbf{V}^2 g + \mathbf{V}f \cdot \mathbf{V}g) \, dV,$$

where $\mathbf{V}^2 f$ is short for $\mathbf{V} \cdot (\mathbf{V}f)$ (which equals $f_{xx} + f_{yy} + f_{zz}$). ($\mathbf{V}^2$ is read "del squared.")

15. (See Exercise 14.) Prove that if $\mathbf{V}^2 f$ and $\mathbf{V}^2 g$ are both 0 everywhere, then

$$\int_S f(\mathbf{V}g \cdot \mathbf{n}) \, dS = \int_S g(\mathbf{V}f \cdot \mathbf{n}) \, dS.$$

16. In Example 2 it was assumed that the temperature distribution had reached a steady state. If that assumption is removed, what is the physical significance of $T_{xx} + T_{yy} + T_{zz}$? That is, what does it measure?

17. A continuous vector field $\mathbf{F}$ has the property that at each point P at a distance 1 from the origin O, $\mathbf{F}(P)$ is perpendicular to the vector $\overrightarrow{OP}$. Show that the divergence of $\mathbf{F}$ must be 0 at at least one point.

18. We proved the part of the divergence theorem corresponding to the component $P\mathbf{i}$ (because the diagram for this case is easiest to draw and visualize). Making the necessary sketches, carry out the proof for the component $R\mathbf{k}$.

Exercises 19 to 21 concern a solid region whose surface consists of two pieces, one interior to the other.

19. Let R be the solid region bounded by two separate surfaces S_1 and S_2, as indicated in the diagram.

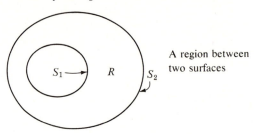

A region between two surfaces

The outer surface is S_2; the inner surface is S_1. Why is it reasonable that, for a vector field defined in R,

$$\int_R \mathbf{V} \cdot \mathbf{F} \, dV = \int_{S_2} \mathbf{F} \cdot \mathbf{n} \, dS + \int_{S_1} \mathbf{F} \cdot \mathbf{n} \, dS?$$

In both surface integrals the unit normal $\mathbf{n}$ points away from R.

20. This continues Exercise 19. Assume now that $\mathbf{V} \cdot \mathbf{F} = 0$ throughout R. Deduce that

$$\int_{S_2} \mathbf{F} \cdot \mathbf{n} \, dS = - \int_{S_1} \mathbf{F} \cdot \mathbf{n} \, dS.$$

21. Let $\mathbf{F} = \mathbf{R}/|\mathbf{R}|^3$, as in Exercise 12. Let S_2 be a surface that encloses the origin. Letting S_1 be a small sphere with center at the origin, show that

$$\int_{S_2} \frac{\mathbf{R} \cdot \mathbf{n}}{|\mathbf{R}|^3} \, dS = 4\pi.$$

This result is important in the study of an electrostatic field.

22. This exercise outlines a proof of the divergence theorem. (a) Review the proof of Green's theorem in Sec. 22.2.

(b) Let $\mathbf{F} = P\mathbf{i} + Q\mathbf{j} + R\mathbf{k}$, where P, Q, and R are functions of x, y, and z. Let the surface S bound the solid region R. (Though the letter R thus denotes both a function and a region, the context will indicate which meaning is intended.) Show that the divergence theorem is equivalent to the three equations

$$\int_R P_x \, dV = \int_S P(\mathbf{i} \cdot \mathbf{n}) \, dS, \qquad \int_R Q_y \, dV = \int_S Q(\mathbf{j} \cdot \mathbf{n}) \, dS,$$

and

$$\int_R R_z \, dV = \int_S R(\mathbf{k} \cdot \mathbf{n}) \, dS.$$

(c) Since the diagram corresponding to the third equation in (b) is easiest to sketch, establish that equation. Divide S into an upper surface S_2 and a lower surface S_1. Then reason as in Sec. 22.2, expressing the various integrals as integrals over a region R^* in the xy plane.

22.6
Stokes' theorem

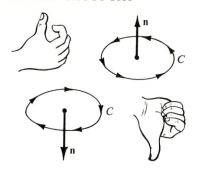

Another generalization of Green's theorem is Stokes' theorem, which concerns a vector field in space and a portion of some surface together with its bounding curve.

Let S be a surface, either curved or flat, with boundary curve C. Assume that S is part of the surface of a convex body. Let $\mathbf{n}$ denote the unit exterior normal to the surface. Since C does not lie in the xy plane, it makes no sense to speak of C being swept out counterclockwise. Instead, choose the direction in which to sweep out C in such a way that $\mathbf{n}$ matches the thumb of the right hand when the fingers of the right hand match the orientation of C. Now it is possible to state Stokes' theorem.

THEOREM I

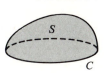

Stokes' theorem. Let S be part of the surface of a convex region in space and let C be its boundary curve. At each point of S let $\mathbf{n}$ be the unit exterior normal to S. Let C be oriented by the right-hand rule. Let $\mathbf{F}$ be a vector field in space. Then

$$\int_S (\mathbf{curl\ F}) \cdot \mathbf{n} \, dA = \oint_C \mathbf{F} \cdot \mathbf{T} \, ds.$$

Stokes' theorem relates the *tangential component* of a vector field along a closed curve to the *normal component* of the curl over a surface.

A mathematical proof of Stokes' theorem is outlined in Exercises 19 to 21. However, the proof, which obtains Stokes' theorem from Green's theorem, instead of illuminating Stokes' theorem, mainly enhances our appreciation of Green's theorem. For this reason we will focus on interpreting and applying Stokes' theorem.

First of all, let us verify Stokes' theorem in a specific example.

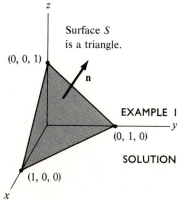

Surface S is a triangle.

EXAMPLE I

Let $\mathbf{F}(x, y, z) = y\mathbf{i} + xz\mathbf{j} + x^2\mathbf{k}$. Let S be the triangle with vertices $(1, 0, 0)$, $(0, 1, 0)$, and $(0, 0, 1)$. Verify Stokes' theorem in this case.

SOLUTION

Let $\mathbf{n}$ be the unit normal to S on the side opposite the origin. Let C be the curve that sweeps out the vertices in the order $(1, 0, 0)$, $(0, 1, 0)$, $(0, 0, 1)$, and then back to $(1, 0, 0)$.

By Example 6 in Sec. 21.2,

$$\oint_C \mathbf{F} \cdot \mathbf{T} \, ds = -\tfrac{5}{6}. \tag{1}$$

To compute

$$\int_S \mathbf{curl\ F} \cdot \mathbf{n} \, dS,$$

we must compute $\mathbf{n}$ and $\mathbf{curl\ F}$.

Since S is on the plane $x + y + z = 1$, a unit normal is

$$\mathbf{n} = \frac{\mathbf{i} + \mathbf{j} + \mathbf{k}}{\sqrt{3}}.$$

(Note that it points to the side away from the origin, as desired.)

Next

$$\mathbf{curl\ F} = \begin{vmatrix} \mathbf{i} & \mathbf{j} & \mathbf{k} \\ \dfrac{\partial}{\partial x} & \dfrac{\partial}{\partial y} & \dfrac{\partial}{\partial z} \\ y & xz & x^2 \end{vmatrix}$$

$$= \mathbf{i}(-x) - \mathbf{j}(2x) + \mathbf{k}(z - 1).$$

Thus

$$\mathbf{curl\ F} \cdot \mathbf{n} = (-x\mathbf{i} - 2x\mathbf{j} + (z - 1)\mathbf{k}) \cdot \frac{(\mathbf{i} + \mathbf{j} + \mathbf{k})}{\sqrt{3}}$$

$$= \frac{-x - 2x + z - 1}{\sqrt{3}}$$

$$= \frac{-3x + z - 1}{\sqrt{3}}.$$

So the integral to be computed is

$$\int_S \mathbf{curl\ F} \cdot \mathbf{n} \, dS = \int_S \frac{-3x + z - 1}{\sqrt{3}} \, dS.$$

To evaluate the latter integral, replace it by an integral over the triangular region R in the xy plane bounded by the lines $x = 0$, $y = 0$, and $x + y = 1$. By the theorem in Sec. 20.7,

$$\int_S \frac{-3x + z - 1}{\sqrt{3}} \, dS = \int_R \frac{-3x + z - 1}{\sqrt{3}} \sqrt{3} \, dA$$

$$= \int_R (-3x + z - 1) \, dA. \tag{2}$$

Recall that z is related to x and y by the equation

$$x + y + z = 1,$$

since S is on the plane with that equation. Hence

$$z = 1 - x - y,$$

and (2) becomes

$$\int_R [-3x + (1 - x - y) - 1]\, dA = \int_R (-4x - y)\, dA$$

$$= \int_0^1 \left[\int_0^{1-x} (-4x - y)\, dy \right] dx$$

$$= \int_0^1 \left(\frac{7x^2}{2} - 3x - \frac{1}{2} \right) dx$$

$$= -\tfrac{5}{6}.$$

This agrees with the result (1), and confirms Stokes' theorem. ●

Example 1 may at least bring Stokes' theorem down to earth. The next example provides a different perspective by showing that Stokes' theorem is a generalization of Green's theorem.

EXAMPLE 2 Show that Green's theorem is a special case of Stokes' theorem.

SOLUTION Let R be a region in the xy plane, bounded by the counterclockwise curve C. At each point of R use as a unit normal the vector $\mathbf{k}$. Note that it is compatible with the right-hand rule. Let $\mathbf{F}(x, y, z) = -Q(x, y)\mathbf{i} + P(x, y)\mathbf{j} + 0\mathbf{k}$. Then

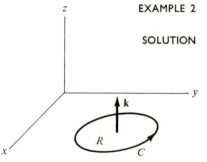

$$\mathbf{curl\ F} = \begin{vmatrix} \mathbf{i} & \mathbf{j} & \mathbf{k} \\ \dfrac{\partial}{\partial x} & \dfrac{\partial}{\partial y} & \dfrac{\partial}{\partial z} \\ -Q(x, y) & P(x, y) & 0 \end{vmatrix} = \mathbf{i} \cdot 0 - \mathbf{j} \cdot 0 + \mathbf{k}(P_x + Q_y)$$

$$= (P_x + Q_y)\mathbf{k}.$$

Stokes' theorem asserts that

$$\int_R \mathbf{curl\ F} \cdot \mathbf{k}\, dA = \oint_C \mathbf{F} \cdot \mathbf{T}\, ds,$$

which becomes

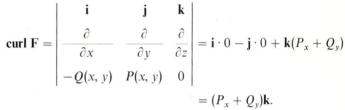

$$\int_R (P_x + Q_y)\mathbf{k} \cdot \mathbf{k}\, dA = \oint_C -Q\, dx + P\, dy$$

or

$$\int_R (P_x + Q_y)\, dA = \oint_C P\, dy - Q\, dx.$$

This is precisely Green's theorem, as stated in Sec. 22.1. ●

What does Stokes' theorem say in terms of the flux $\mathbf{F}$ of a fluid? In particular, what is the physical meaning of **curl F**? Let us answer both questions together.

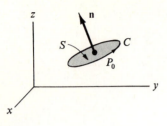

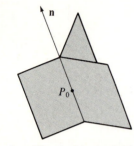

A paddle wheel in the fluid.

Consider a fixed point P_0 in space. Imagine a *small* circular disk S with center P_0. Let C be the boundary of S oriented in such a way that C and $\mathbf{n}$ fit the right-hand rule.

Now examine the two sides of the equation

$$\int_S (\mathbf{curl\ F}) \cdot \mathbf{n} \, dS = \oint_C \mathbf{F} \cdot \mathbf{T} \, ds. \tag{3}$$

The right side of (3) measures the tendency of the fluid to move along C (rather than, say, perpendicular to it). Thus $\oint_C \mathbf{F} \cdot \mathbf{T} \, ds$ might be thought of as the "circulation" or "whirling tendency" of the fluid along C. For each placement of the small disk S at P_0—or, equivalently—each choice of unit normal vector $\mathbf{n}$, $\oint_C \mathbf{F} \cdot \mathbf{T} \, ds$ measures a corresponding circulation. It records the tendency of a paddle wheel at P_0 with axis along $\mathbf{n}$ to rotate. So much, for the moment, for $\oint_C \mathbf{F} \cdot \mathbf{T} \, ds$, which is the right side of (3).

Now consider the left side of (3). If S is small, the integrand is almost constant and the integral is approximately

$$(\mathbf{curl\ F})_{P_0} \cdot \mathbf{n} \cdot \text{area of } S, \tag{4}$$

where $(\mathbf{curl\ F})_{P_0}$ denotes the curl at P_0.

Keeping the center of S at P_0, vary the vector $\mathbf{n}$. For which choice of $\mathbf{n}$ will (4) be largest? Answer: for that $\mathbf{n}$ which has the same direction as the fixed vector $(\mathbf{curl\ F})_{P_0}$. With that choice of $\mathbf{n}$, (4) becomes

$$\left| (\mathbf{curl\ F})_{P_0} \right| \text{ area of } S.$$

The Physical Interpretation of Curl

Thus a paddle wheel placed in the fluid at P_0 *rotates most quickly when its axis is in the direction of* **curl F** at P_0. The *magnitude* of **curl F** is a measure of *how fast* the paddle wheel can rotate when placed at P_0. Thus **curl F** records the direction and magnitude of maximum circulation at a given point.

Now imagine a surface S with boundary curve C. Assume the paddle wheel interpretation of curl (suggested by Stokes' theorem). Let us go in the reverse direction and show that it, in turn, may suggest Stokes' theorem.

Consider a partition of the surface S into small pieces with areas $S_1, S_2, \ldots, S_n$. Let P_i be a point in S_i for $i = 1, 2, \ldots, n$. Let $(\mathbf{curl\ F})_i$ be the curl of $\mathbf{F}$ at P_i; let $\mathbf{n}_i$ be the unit normal at P_i. Then

$$\sum_{i=1}^{n} (\mathbf{curl\ F})_i \cdot \mathbf{n}_i \, s_i \tag{5}$$

is an approximating sum for

$$\int_S (\mathbf{curl\ F}) \cdot \mathbf{n} \, dS.$$

But

$$(\mathbf{curl\ F})_i \cdot \mathbf{n}_i \, S_i,$$

because of our "paddle wheel" assumption, seems to be a likely estimate of the circulation of $\mathbf{F}$ around the border of S_i, which is

$$\int_{C_i} \mathbf{F} \cdot \mathbf{T} \, ds, \tag{6}$$

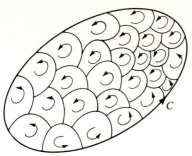

Typical small curve is C_i

where C_i is the boundary curve of S_i. Thus the sum (5) is approximately the sum of the line integrals (6) for $i = 1, 2, \ldots, n$. These line integrals are indicated schematically in the diagram in the margin. Where the little paths C_i overlap, the directions of integration are opposite and the corresponding portions of the integrals cancel.

Thus
$$\sum_{i=1}^{n} \int_{C_i} \mathbf{F} \cdot \mathbf{T} \, ds = \int_{C} \mathbf{F} \cdot \mathbf{T} \, ds. \tag{7}$$

Looking back at (5), (6), and (7), and letting all the S_i get small, we might expect the equation

$$\int_{S} (\mathbf{curl} \ \mathbf{F}) \cdot \mathbf{n} \, dS = \oint_{C} \mathbf{F} \cdot \mathbf{T} \, ds,$$

which is Stokes' theorem.

No attempt has been made to prove Stokes' theorem. We have merely explored some ideas suggested by paddle wheels immersed in a moving fluid. Now let us apply Stokes' theorem.

One of the major applications of Stokes' theorem is in the realm of *conservative* vector fields. Recall that a conservative vector field, as defined in Sec. 21.4, is a field $\mathbf{F}$ such that $\int_C \mathbf{F} \cdot \mathbf{T} \, ds$ depends only on the endpoints of the curve C (or, what is essentially the same, $\oint \mathbf{F} \cdot \mathbf{T} \, ds = 0$ for closed curves). Equivalently, $\mathbf{F}$ is a gradient field $\mathbf{F} = \nabla f$ for some scalar function f.

Now, a straightforward computation shows that

$$\nabla \times \nabla f = \mathbf{0}.$$

Thus the curl of a conservative field $\mathbf{F}$ is $\mathbf{0}$. If, conversely, $\mathbf{F}$ is a vector field whose curl is $\mathbf{0}$, does it follow that $\mathbf{F}$ is conservative? Not necessarily, as Exercise 19 in Sec. 21.5 shows. The next theorem, however, shows that, if $\mathbf{F}$ is defined throughout space, and $\mathbf{curl} \ \mathbf{F} = \mathbf{0}$, then $\mathbf{F}$ is conservative.

THEOREM 2 Let $\mathbf{F}$ be a vector field defined throughout space. Assume that $\mathbf{curl} \ \mathbf{F} = \mathbf{0}$. Then $\mathbf{F}$ is conservative.

PROOF Let C be any closed path in space that bounds some surface S that is part of the surface of a convex region. Then

$$\oint_{C} \mathbf{F} \cdot \mathbf{T} \, ds = \int_{S} (\mathbf{curl} \ \mathbf{F}) \cdot \mathbf{n} \, dS,$$

by Stokes' theorem. Since $\mathbf{curl} \ \mathbf{F} = \mathbf{0}$, it follows that $\oint_C \mathbf{F} \cdot \mathbf{T} \, ds = 0$. (By advanced mathematics, it can then be shown that $\oint_C \mathbf{F} \cdot \mathbf{T} \, ds = 0$ for any smooth curve.) Thus $\mathbf{F}$ is conservative. ●

EXAMPLE 3 Show that $\mathbf{F} = (y^3 + z^2)\mathbf{i} + 3xy^2\mathbf{j} + 2xz\mathbf{k}$ is conservative.

SOLUTION Note first that $\mathbf{F}$ is defined throughout space. According to Theorem 2, all that is necessary is to check whether $\mathbf{curl} \ \mathbf{F} = \mathbf{0}$. Now,

$$\text{curl } \mathbf{F} = \begin{vmatrix} \mathbf{i} & \mathbf{j} & \mathbf{k} \\ \dfrac{\partial}{\partial x} & \dfrac{\partial}{\partial y} & \dfrac{\partial}{\partial z} \\ y^3 + z^2 & 3xy^2 & 2xz \end{vmatrix}$$

$$= \mathbf{i}(0) - \mathbf{j}(2z - 2z) + \mathbf{k}(3y^2 - 3y^2)$$

$$= \mathbf{0}.$$

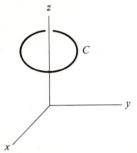

Thus **F** is conservative. ●

   In Theorem 2 it is assumed that the domain of **F** is all of space.   Actually, the theorem holds if it is assumed that the domain of **F** is a convex body or, more generally, a region R in which any closed curve can be gradually shrunk to a point while staying in R.   The set of points in space exclusive of the z axis does *not* have this property.   The closed curve C shown in the diagram, which loops around the z axis, cannot be shrunk to a point while staying away from the z axis.   That is what permits the vector field in Exercise 19 of Sec. 21.5 not to be conservative.

Exercises

1. State Stokes' theorem in words, not in mathematical symbols.
2. Let S be the top half of the sphere of radius 1 and with center $(0, 0, 0)$.   Let C be the unit circle in the xy plane with center $(0, 0, 0)$, swept out counterclockwise.   Let $\mathbf{F}(x, y, z) = y^2 x \mathbf{i} + y^3 \mathbf{j} + y^2 z \mathbf{k}$.   Verify Stokes' theorem in this case.

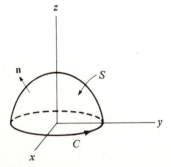

3. Verify Stokes' theorem for the special case when **F** has the form ∇f, that is, is a gradient field.
4. Let $\mathbf{F} = P\mathbf{i} + Q\mathbf{j} + R\mathbf{k}$.   A convex region has a flat base situated on the xy plane.   Let S be the part of its surface other than the base B.   Show that

$$\int_S (\text{curl } \mathbf{F}) \cdot \mathbf{n} \, dS = \int_B (Q_x - P_y) \, dA.$$

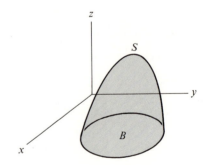

5. Using Stokes' theorem, evaluate $\int_S (\nabla \times \mathbf{F}) \cdot \mathbf{n} \, dS$, where $\mathbf{F} = (x^2 + y - 4)\mathbf{i} + 3xy\mathbf{j} + (2xz + z^2)\mathbf{k}$, and S is the portion of the surface $z = 4 - (x^2 + y^2)$ above the xy plane.
6. Use Stokes' theorem to evaluate $\oint_C \mathbf{F} \cdot \mathbf{T} \, ds$ around the circle given as the intersection of the cylinder $x^2 + y^2 - 2y = 2$ and the plane $z = -1$, where $\mathbf{F} = y\mathbf{i} + x\mathbf{j} + xyz^2\mathbf{k}$.
7. Evaluate as simply as possible $\int_S \mathbf{F} \cdot \mathbf{n} \, dS$, where S and **F** are the following: $\mathbf{F}(x, y, z) = x\mathbf{i} - y\mathbf{j}$, and S is the surface of the cube bounded by the three coordinate planes and the planes $x = 1$, $y = 1$, $z = 1$, exclusive of the surface in the plane $x = 1$.
8. Why cannot $\mathbf{F} = x\mathbf{i} + y\mathbf{j} + z\mathbf{k}$ be the curl of a vector field **G**?
9. Is $\mathbf{F} = (y + 2x + z)\mathbf{i} + x\mathbf{j} + (1 + x + y)\mathbf{k}$ conservative?

10. Let $\mathbf{F}(x, y) = P(x, y)\mathbf{i} + Q(x, y)\mathbf{j}$ be a vector field in the plane such that $P_y = Q_x$. Show that it is conservative. *Hint*: Consider the vector field in space, $\mathbf{G}(x, y, z) = P(x, y)\mathbf{i} + Q(x, y)\mathbf{j} + 0\mathbf{k}$.

11. Let $\mathbf{F}$ be a vector field in space with the property that, for any points A and B and any curve C from A to B, the line integral

$$\int_C \mathbf{F} \cdot \mathbf{T}\, ds$$

depends only on the endpoints A and B. Show that

$$\mathbf{V} \times \mathbf{F} = \mathbf{0}$$

at all points.

12. Let $\mathbf{F}$ be a vector field throughout space such that $\mathbf{F}(P)$ is perpendicular to the curve C at each point P on C, the boundary of a surface S. What can one conclude about

$$\int_S (\text{curl } \mathbf{F}) \cdot \mathbf{n}\, dS?$$

13. Let S be the half of the sphere $x^2 + y^2 + z^2 = 1$ for which $z \geq 0$. Let $\mathbf{F}(x, y, z)$ be the vector function $y\mathbf{i} + x\mathbf{j} + z\mathbf{k}$. Using Stokes' theorem, evaluate

$$\int_S (\mathbf{V} \times \mathbf{F}) \cdot \mathbf{n}\, dS.$$

14. Let $\mathbf{F}(x, y, z) = P(x, y)\mathbf{i} + Q(x, y)\mathbf{j} + 0\mathbf{k}$ for some scalar functions P and Q.
    (a) Show that each vector $\mathbf{F}(x, y, z)$ is parallel to the xy plane.
    (b) Show that **curl** $\mathbf{F}$ is parallel to the z axis.
    (c) Interpret (b) in terms of a paddle wheel.

15. Let $\mathbf{F}(x, y, z) = x\mathbf{j}$. The diagram below shows a few values of $\mathbf{F}$ near $(0, 0, 0)$.

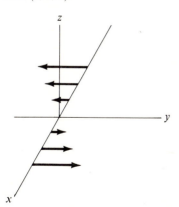

    (a) The diagram shows $\mathbf{F}(x, y, z)$ for a few cases when $z = 0$. Sketch $\mathbf{F}(x, y, z)$ for a few cases when $z = 0.1$.
    (b) Using your "fluid" intuition, decide in what direction the axis of a paddle wheel at $(0, 0, 0)$ should be

placed in order that the paddle wheel rotates most quickly.
    (c) Is your opinion in (b) compatible with **curl** $\mathbf{F}$ at the point $(0, 0, 0)$?

∎

16. Let S be the surface of a convex region R. Consider a very small loop C on S and let S_1 be the large part of S. Let $\mathbf{F}$ be a vector field in space.

    (a) Show that

$$\int_{S_1} (\mathbf{V} \times \mathbf{F}) \cdot \mathbf{n}\, dS = \int_C \mathbf{F} \cdot \mathbf{n}\, ds.$$

    (b) Letting C get arbitrarily small, deduce that

$$\int_S (\mathbf{V} \times \mathbf{F}) \cdot \mathbf{n}\, dS = 0.$$

    (c) Deduce that $\int_R \mathbf{V} \cdot (\mathbf{V} \times \mathbf{F})\, dV = 0$.
    (d) Deduce that the divergence of the curl is $\mathbf{0}$.

17. Make a diagram to show that the validity of Stokes' theorem for surfaces of small diameter implies its validity for surfaces of any size.

18. Let $\mathbf{F}(x, y, z)$ be a vector field defined throughout space with the property that $|\mathbf{F}(x, y, z)| \leq (\sqrt{x^2 + y^2 + z^2})^{-3}$ if (x, y, z) is not the origin. Let $g(r) = \int_R \mathbf{V} \cdot \mathbf{F}\, dV$, where R is the ball of radius r and center $(0, 0, 0)$. Show that $\lim_{r \to \infty} g(r) = 0$.

∎ ∎

Exercises 19 to 21 outline a proof of Stokes' theorem in the case that S is part of a surface $z = f(x, y)$.

19. Let $\mathbf{F} = P(x, y, z)\mathbf{i} + Q(x, y, z)\mathbf{j} + R(x, y, z)\mathbf{k}$ be a vector field in space. Let S be a surface on $z = f(x, y)$ and C its bounding curve. Let $\mathbf{n}$ be the unit normal to S that has *positive* z component. Orient C in accord with the right-hand rule.

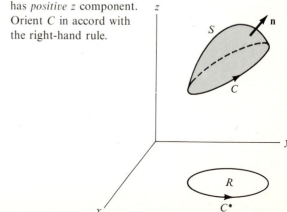

Let R be the projection (shadow) of S in the xy plane and let C^* be the projection of C, with a counterclockwise orientation. Show that

$$\int_S (\mathbf{V} \times \mathbf{F})\, dS = \oint_C \mathbf{F} \cdot \mathbf{n}\, dS,$$

by expressing the first integral as an integral over R and the second integral as an integral over C^*. Then use Green's theorem. The details are sketched in the rest of this exercise and in Exercises 20 and 21.

(a) Show that

$$\mathbf{n} = \frac{-f_x \mathbf{i} - f_y \mathbf{j} + \mathbf{k}}{\sqrt{1 + f_x^2 + f_y^2}}.$$

(b) Show that

$$\int_S (\mathbf{V} \times \mathbf{F}) \cdot \mathbf{n}\, dS$$

$$= \int_R [(Q_z - R_y)f_x + (R_x - P_z)f_y + (Q_x - P_y)]\, dA.$$

20. This continues Exercise 19.

Show that

$$\oint_C \mathbf{F} \cdot \mathbf{T}\, ds = \int_C (P\, dx + Q\, dy + R\, dz)$$

$$= \int_{C^*} [P(x, y, f(x, y))\, dx + Q(x, y, f(x, y))\, dy$$

$$+ R(x, y, f(x, y))(f_x\, dx + f_y\, dy)]$$

$$= \int_{C^*} [P(x, y, f(x, y)) + f_x R(x, y, f(x, y))]\, dx$$

$$+ [Q(x, y, f(x, y)) + f_y R(x, y, f(x, y))]\, dy.$$

21. Exercises 19 and 20 express the integrals in Stokes' theorem as integrals over a plane region R and a plane curve C^*. Use Green's theorem to complete the proof of Stokes' theorem. Be careful in computing partial derivatives. For instance,

$$\frac{\partial P(x, y, f(x, y))}{\partial x} = P_x + f_x P_z$$

by a chain rule.

22.7

Summary

This chapter concerned three theorems and their applications. These three theorems are described in this table.

Statement		Condition
Green's theorem:	$\int_R \mathbf{V} \cdot \mathbf{F}\, dA = \oint_C \mathbf{F} \cdot \mathbf{n}\, ds$	C bounds plane region R
	$\int_R (P_x + Q_y)\, dA = \oint_C P\, dy - Q\, dx$	Counterclockwise C
	$\int_R (Q_x - P_y)\, dA = \int_C P\, dx + Q\, dy$	Counterclockwise C
	$\int_R (Q_x - P_y)\, dA = \int_C \mathbf{F} \cdot \mathbf{T}\, ds$	Counterclockwise C, $\mathbf{F} = P\mathbf{i} + Q\mathbf{j}$
Divergence theorem:	$\int_R \mathbf{V} \cdot \mathbf{F}\, dV = \int_S \mathbf{F} \cdot \mathbf{n}\, dS$	S bounds solid region R, $\mathbf{n}$ exterior normal
Stokes' theorem:	$\int_S (\mathbf{curl}\ \mathbf{F}) \cdot \mathbf{n}\, dS = \int_C \mathbf{F} \cdot \mathbf{T}\, ds$	S part of surface of convex region, right-hand rule

KEY FACTS

Area of plane region R is $\frac{1}{2}\int_C -y\, dx + x\, dy$ (C counterclockwise). The local magnification of the mapping $x = f(u, v)$, $y = g(u, v)$ is equal to the absolute value of $\begin{vmatrix} f_u & f_v \\ g_u & g_v \end{vmatrix}$.

Mappings of the form $x = au + bv$, $y = cu + dv$ have Jacobian $ad - bc$ at all points. If $ad - bc \neq 0$ they are one-to-one and take lines to lines and conic sections to conic sections.

When a mapping takes a region R in the uv plane to a region S in the xy plane, an integral over S can be expressed as an integral over R:

$$\int_S h(x, y) \, dA = \int_R h(f(u, v), g(u, v)) \left| \frac{\partial(x, y)}{\partial(u, v)} \right| dA.$$

The vector field $\mathbf{V} \times \mathbf{F}$ may be interpreted as describing the direction in which a paddle wheel should be placed to rotate most swiftly.

If $\mathbf{F}$ is defined on all of space and **curl** $\mathbf{F} = \mathbf{0}$ everywhere, then $\mathbf{F}$ is conservative.

This table summarizes a good deal of integral calculus, as illustrated by definite integrals over plane regions.

f	$\int_R f(P) \, dA$
1	Area of R
Density	Mass in R
Divergence $\mathbf{V} \cdot \mathbf{F}$	Fluid flow out of R
Jacobian of a mapping from R to S (if positive)	Area of S

VOCABULARY AND SYMBOLS

exterior normal $\mathbf{n}$
Green's theorem
divergence theorem
Stokes' theorem

mapping $F = (f, g)$, $x = f(u, v)$, $y = g(u, v)$

Jacobian, $\dfrac{\partial(f, g)}{\partial(u, v)}$, $\dfrac{\partial(x, y)}{\partial(u, v)}$.

Guide quiz on Chap. 22

1. (a) State Green's theorem.
(b) What is its physical interpretation?
2. Why is $(dy/ds)\mathbf{i} - (dx/ds)\mathbf{j}$ (a) a unit vector? (b) an *exterior* normal? (State the assumptions.)
3. (a) Define the Jacobian.
(b) What information does the Jacobian carry?
4. (a) State the divergence theorem.
(b) What is its physical interpretation?
5. (a) State Stokes' theorem.
(b) What is its physical interpretation?
6. Let $\mathbf{F} = xy^2\mathbf{i} + y^2z\mathbf{j} + y^3\mathbf{k}$. Let S be the top half of the spherical surface $x^2 + y^2 + z^2 = 4$. Find

$$\int_S \mathbf{V} \times \mathbf{F} \cdot \mathbf{n} \, dS.$$

7. Let F be the mapping given by the formula $x = 2u + v$, $y = 3u - 2v$.

(a) Fill in this table.

(u, v)	(x, y)
$(1, 0)$	
$(0, 1)$,	
$(-1, 0)$	
$(0, -1)$	

(b) With the aid of (a) sketch the image in the xy plane of the circle $u^2 + v^2 = 1$.
(c) Let R be the disk bounded by the circle in (b). Find the area of S, the image of R in the xy plane.
(d) Evaluate $\int_S x^2 \, dA$.

8. Show that Green's theorem is a special case of Stokes' theorem.

Review exercises for Chap. 22

1. Check Green's theorem for $\mathbf{F} = (x - 5y)\mathbf{i} + xy\mathbf{j}$ and R the region bounded by $y = x^2$ and $y = \sqrt{x}$.
2. [R is a convex set in the plane and C its boundary (taken counterclockwise). $\mathbf{F}$ is a vector field in the plane.] Complete these equations.

(a) $\oint_C \mathbf{F} \cdot \mathbf{n} \, ds = \int_R \underline{\hspace{1cm}}$;

(b) $\oint_C \mathbf{F} \cdot \mathbf{T} \, ds = \int_R \underline{\hspace{1cm}}$.

3. Complete these statements, in words:
(*a*) The integral over_____of the normal component of the curl of a vector field equals the integral of _____over_____.
(*b*) The integral over_____of the normal component of a vector field in space equals the integral of_____over_____.

4. Verify Stokes' theorem in the case in which the vector field **F** is $(ax + by + cz)\mathbf{i}$ and S is the triangle with vertices $(0, 0, 0)$, $(1, 1, 1)$, and $(2, 3, 4)$.

5. Let $\mathbf{F}(x, y, z) = xz\mathbf{i} + x^2\mathbf{j} + xy\mathbf{k}$. Let S be all of the surface of the cube shown below *other than the top*. Verify Stokes' theorem in this case.

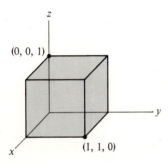

6. Let $\mathbf{F}(x, y) = xy\mathbf{i} + y\mathbf{j}$. Let R be the region between the parabola $y = x^2$ and the line $y = 2x$. Verify Green's theorem in this case.

7. Let S be the square in the xy plane with vertices $(0, 0)$, $(1, 0)$, $(1, 1)$ $(0, 1)$. Let $x = (u + v)/2, y = (u - v)/2$ describe a mapping from the uv plane to the xy plane.
(*a*) Sketch the region R in the uv plane whose image is S.
(*b*) Use (*a*) to evaluate
$$\int_R (u + v) \, dA$$

8. Consider the mapping $x = au + bv$, $y = cu + dv$, when $ad - bc \neq 0$.
(*a*) Show that f is one-to-one.
(*b*) Solve for u and v in terms of x and y, thus obtaining a mapping from the xy plane to the uv plane.
(*c*) Show that
$$\frac{\partial(x, y)}{\partial(u, v)} \cdot \frac{\partial(u, v)}{\partial(x, y)} = 1.$$
(*d*) Why is (*c*) to be expected?

9. Let $\mathbf{F}(x, y) = \mathbf{R}/|\mathbf{R}|^n$, where $\mathbf{R} = x\mathbf{i} + y\mathbf{j}$. For which values of n, if any, is $\nabla \cdot \mathbf{F} = 0$?

10. Let $\mathbf{F}(x, y, z) = \mathbf{R}/|\mathbf{R}|^n$, where $\mathbf{R} = x\mathbf{i} + y\mathbf{j} + z\mathbf{k}$. For which values of n, if any, is $\nabla \cdot \mathbf{F} = 0$?

Exercises 11 to 14 are a unit.

11. Let R be a plane region bounded by two curves C_1 and C_2, as in the diagram. Let $\mathbf{F}(x, y) = P\mathbf{i} + Q\mathbf{j}$ be a vector field defined on R. Why is it plausible that
$$\int_R \nabla \cdot \mathbf{F} \, dA = \int_{C_2} \mathbf{F} \cdot \mathbf{n} \, ds + \int_{C_1} \mathbf{F} \cdot \mathbf{n} \, ds,$$
where in both integrals **n** indicates the unit normal that points away from R?

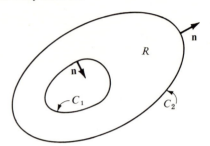

12. (*a*) Let $\mathbf{F}(x, y) = \mathbf{R}/|\mathbf{R}|^2$, where $\mathbf{R}(x, y) = x\mathbf{i} + y\mathbf{j}$. Show that $\nabla \cdot \mathbf{F} = 0$.
(*b*) Show that, for any curve C in the plane that encloses the origin $(0, 0)$,
$$\int_C \mathbf{F} \cdot \mathbf{n} \, ds = 2\pi.$$

13. Let $\mathbf{F}(x, y) = \mathbf{R}/|\mathbf{R}|^3$, where $\mathbf{R}(x, y) = x\mathbf{i} + y\mathbf{j}$. Consider any curve C in the plane that encloses the origin. Is $\int_C \mathbf{F} \cdot \mathbf{n} \, ds$ independent of the choice of C?

14. Let $\mathbf{F}(x, y)$ be a vector field defined everywhere except at the origin. Assume that the value of $\int_C \mathbf{F} \cdot \mathbf{n} \, ds$ for any curve that encloses the origin and bounds a region is the same (not necessarily 0, however). Must the divergence of **F** be 0?

15. In the assertion $\int_R \nabla \cdot \mathbf{F} \, dA = \int_C \mathbf{F} \cdot \mathbf{n} \, ds$, which is Green's theorem, there is no mention of orientation of C. In the statement,
$$\int_R (P_x + Q_y) \, dA = \oint_C P \, dy - Q \, dx,$$
which is also Green's theorem, the curve must be given an orientation. Why? For reference see Sec. 22.1.

16. Let $\mathbf{F} = x\mathbf{i} + y\mathbf{j} + z\mathbf{k}$. Let S be the sphere of radius 3 and center $(4, 5, 6)$. Evaluate $\int_S \mathbf{F} \cdot \mathbf{n} \, dS$.

17. Let $\mathbf{F}(x, y, z) = \mathbf{R}/|\mathbf{R}|^3$, where $\mathbf{R}(x, y, z) = x\mathbf{i} + y\mathbf{j} + z\mathbf{k}$.
(*a*) Let S be a surface that bounds a solid. Assume that the origin is *not* in the solid. Show that $\int_S \mathbf{F} \cdot \mathbf{n} \, dS = 0$.
(*b*) If the origin is inside the solid, what is $\int_S \mathbf{F} \cdot \mathbf{n} \, dS$? *Hint*: See Exercises 11 to 13.

18. The following is a paraphrase of a discussion about electric fields in a physics book. Explain the mathematical steps:

The potential ϕ varies as $1/r$, where r is the distance from the origin, and $|\nabla\phi|$ as $1/r^2$. Thus the value of

$$\int_S \phi\nabla\phi \cdot \mathbf{n}\, dS$$

over a sphere of large radius is almost 0. Thus if R is a ball of large radius

$$\int_R \nabla \cdot (\phi\nabla\phi)\, dV$$

is small and may be disregarded.

■

19. Let F be a one-to-one mapping from R onto all of S in the xy plane. Then x and y are functions of u and v, and u and v are functions of x and y, that is, $(u, v) = F^{-1}(x, y)$.
   (a) Using a chain rule, show that

$$1 = x_u u_x + x_v v_x \qquad 0 = x_u u_y + x_v v_y$$
$$1 = y_u u_y + y_v v_y \qquad 0 = y_u u_x + y_v v_x.$$

   (b) With the aid of (a), show that the Jacobians of F and F^{-1} are reciprocals of each other.
   (c) State (b) in optical terms. Is (b) to be expected?
20. Let F be a mapping from the uv plane to the xy plane, and let G be a mapping from the xy plane to the st plane. Then $G \circ F$ is a mapping from the uv plane to the st plane.
   (a) Why would you expect the Jacobian of $G \circ F$ to be the product of the Jacobians of F and G?
   (b) Prove the theorem suggested in (a).
   (c) What theorem about functions of a single variable does (a) generalize?

Throughout the last two chapters only integrals of scalar functions were considered. But integrals of vector functions are sometimes of use and are easily defined. If $\mathbf{F} = P\mathbf{i} + Q\mathbf{j} + R\mathbf{k}$ is defined throughout the solid R, define $\int_R \mathbf{F}(x, y, z)\, dV$ to be $(\int_R P\, dV)\mathbf{i} + (\int_R Q\, dV)\mathbf{j} + (\int_R R\, dV)\mathbf{k}$. In short, to integrate a vector function, integrate each of its scalar components. This concept is the subject of Exercises 21 to 27.

21. Let $\mathbf{F}(x, y, z) = f(x, y, z)(x\mathbf{i} + y\mathbf{j} + z\mathbf{k})$, where $f(x, y, z)$ denotes the density of matter at point (x, y, z) in the solid that occupies the region R. What is the physical significance of the equation

$$\int_R \mathbf{F}(x, y, z)\, dV = \mathbf{0}?$$

22. Let $\mathbf{F} = P\mathbf{i} + Q\mathbf{j} + R\mathbf{k}$ be a vector field and R a solid region. Let $\mathbf{c}$ be a fixed vector. Show that $\int_R \mathbf{c} \cdot \mathbf{F}\, dV =$

$\mathbf{c} \cdot \int_R \mathbf{F}\, dV.$ A similar theorem holds for surface integrals.
23. Let $\mathbf{A}$ and $\mathbf{B}$ be vectors such that for all vectors $\mathbf{c}$, $\mathbf{c} \cdot \mathbf{A} = \mathbf{c} \cdot \mathbf{B}$. Show that $\mathbf{A} = \mathbf{B}$.
24. Let $\mathbf{c}$ be any fixed vector. Let R be a solid region and S the surface that bounds it. Let f be a scalar function on R.
   (a) Show that $\int_R \nabla \cdot f\mathbf{c}\, dV = \int_S f\mathbf{c} \cdot \mathbf{n}\, dS$.
   (b) Show that $\int_R \nabla \cdot f\mathbf{c}\, dV = \int_R \mathbf{c} \cdot \nabla f\, dV$.
   (c) Show that $\mathbf{c} \cdot \int_R \nabla f\, dV = \mathbf{c} \cdot \int_S f\mathbf{n}\, dS$.
   (d) Deduce that $\int_R \nabla f\, dV = \int_S f\mathbf{n}\, dS$.
25. Let S be a surface bounding a solid region R. Let $\mathbf{c}$ be a fixed vector.
   (a) Show that $\int_S \mathbf{c} \cdot \mathbf{n}\, dS = 0$.
   (b) Deduce that $\int_S \mathbf{n}\, dS = \mathbf{0}$.
26. (See Exercise 25.) On each of the four faces of a tetrahedron a vector is constructed that is perpendicular to the face, points outward, and has magnitude equal to the area of the face. Show that the sum of the four vectors is $\mathbf{0}$.
27. If f is a scalar field, $\nabla^2 f = \nabla \cdot \nabla f = f_{xx} + f_{yy} + f_{zz}$. For a vector field $\mathbf{F} = P\mathbf{i} + Q\mathbf{j} + R\mathbf{k}$, define $\nabla^2 \mathbf{F}$ to be

$$(\nabla^2 P)\mathbf{i} + (\nabla^2 Q)\mathbf{j} + (\nabla^2 R)\mathbf{k}.$$

   (a) Show that $\nabla \times (\nabla \times \mathbf{F}) = \nabla(\nabla \cdot \mathbf{F}) - \nabla^2 \mathbf{F}$. This shows that the curl of the curl need not be $\mathbf{0}$.
   (b) Show that

$$\nabla \times \nabla \times (\nabla \times \mathbf{F}) = -\nabla^2(\nabla \times \mathbf{F}).$$

28. Consider an object submerged in water. It occupies a region R with surface S. The force $\mathbf{F}$ of the water against the object at each point is perpendicular to S. Assume that the surface of the water is the xy plane. Thus the z coordinate of points in the water is negative. The force against a small patch of S of area ΔS and depth $|z|$ is approximately $z\mathbf{n}\,\Delta S$, where $\mathbf{n}$ is the unit external normal. The z component of this force is $cz\mathbf{k} \cdot \mathbf{n}\,\Delta S$, where c is the density of water.
   (a) What integral over S represents the vertical component of the total force of the water against the submerged object?
   (b) Show that the vertical component of the force against the object is equal to the weight of the water displaced by the object.
   (c) Show that the x and y components of the total force against the object are 0.
29. Let $x = a_1 u + a_2 v + a_3 w$

$$y = b_1 u + b_2 v + b_3 w$$
$$z = c_1 u + c_2 v + c_3 w$$

describe a mapping from uvw space to xyz space. Assume

that the determinant

$$\begin{vmatrix} a_1 & a_2 & a_3 \\ b_1 & b_2 & b_3 \\ c_1 & c_2 & c_3 \end{vmatrix}$$

is *not* 0.

(*a*) Solve for u, v, w in terms of x, y, z. Thus the original mapping is one-to-one.

(*b*) Show that the original mapping takes a plane to a plane.

(*c*) Deduce that the image of a cube is a parallelepiped.

(*d*) Let R be the cube of volume 1 shown below.

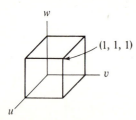

Find the volume of its image (a parallelepiped) in xyz space.

30. Assume that the origin O of the xyz-coordinate system is inside R. Assume also that for all points P on S, the position vector $\overrightarrow{OP}$ makes an angle of at most $\pi/6$ radians with the exterior normal **n**. Prove that

$$\int_S r\, dA \le 2\sqrt{3V},$$

where $r = |\overrightarrow{OP}|$ and V is the volume of R.

■ ■

31. That local magnification of the mapping $x = f(u, v)$, $y = g(u, v)$ is given by the Jacobian was shown in Sec. 22.4 with the aid of Green's theorem. Here is a simple plausibility argument for the same result. Consider a little rectangle R in the uv plane with vertices (a, b), $(a + \Delta u, b)$, $(a + \Delta u, b + \Delta v)$, $(a, b + \Delta v)$, as shown below.

Let S be its image in the xy plane. Let (A, B) be the image of (a, b).

(*a*) Define

$$dx = f_u\, \Delta u + f_v\, \Delta v$$
$$dy = g_u\, \Delta u + g_v\, \Delta v,$$

where the partials are evaluated at (a, b). Why might we expect the area of S to be approximately the area of the parallelogram whose vertices are (A, B), $(A + dx, B)$, $(A + dx, B + dy)$, $(A, B + dy)$?

(*b*) Show that the area of the parallelogram in the xy plane mentioned in (*a*) is equal to

$$|f_u g_v - f_v g_u| \cdot \text{area of } R.$$

32. Let $x = f(u, v, w)$, $y = g(u, v, w)$, $z = h(u, v, w)$ be a mapping from uvw space to xyz space. Reasoning as in Exercise 31, show that the local magnification of the mapping is very likely equal to the absolute value of the determinant

$$\begin{vmatrix} f_u & f_v & f_w \\ g_u & g_v & g_w \\ h_u & h_v & h_w \end{vmatrix}.$$

Exercises 33 and 34 concern harmonic functions. A function f is harmonic if $\nabla^2 f = 0$, that is, $f_{xx} + f_{yy} + f_{zz} = 0$.

33. Let R be a solid and let S be its surface. Let f and g be two scalar functions defined on R.

(*a*) Show that

$$\int_R \nabla(f \,\nabla g)\, dV = \int_R (f\nabla^2 g + \nabla f \cdot \nabla g)\, dV$$
$$= \int_S (f\nabla g) \cdot \mathbf{n}\, dS.$$

(*b*) Show that

$$\int_R \nabla(g\nabla f)\, dV = \int_R (g\,\nabla^2 f + \nabla g \cdot \nabla f)\, dV$$
$$= \int_S (g\nabla f) \cdot \mathbf{n}\, dS.$$

(*c*) Deduce that

$$\int_R (f\nabla^2 g - g\nabla^2 f)\, dV = \int_S [(f\nabla g - g\nabla f) \cdot \mathbf{n}]\, dS.$$

This result, also, is called Green's theorem.

34. Assume that f and g are harmonic on R and equal on S. This exercise will show that they must be equal on R. Let $h = f - g$.

(*a*) Use Exercise 33(*a*) to show that

$$\int_R h\,\nabla^2 h\, dV + \int_R \nabla h \cdot \nabla h\, dV = \int_S (h\,\nabla h) \cdot \mathbf{n}\, dS.$$

(*b*) Deduce that $\nabla h = \mathbf{0}$ throughout R.

(*c*) Deduce that h is constant throughout R.

(*d*) Show that $f = g$ throughout R.

■ ■

35. In an ordinary map an area on the map is proportional to the area depicted. But some maps for airline pilots have a larger scale near airports. It has been suggested that maps should be distorted so that they represent accurately various quantities. Thus a map of population would show the cities as large and sparsely populated states as small. In a map of rainfall the Olympic peninsula of Washington would be large, while Arizona would be small.

Consider a square region R in the uv plane, furnished with a positive function $Q(u, v)$. [Perhaps $Q(u, v)$ is the number of inches of rain per year at (u, v).]
(a) Why might we seek a mapping $x = f(u, v)$, $y = g(u, v)$ whose Jacobian equals $Q(u, v)$?
(b) Let the vertices of the square region in the uv plane be $(0, 0)$ $(1, 0)$, $(1, 1)$, $(0, 1)$. Define

$$x = f(u, v) = \int_0^u \left[\int_0^1 Q(u_1, v_1) \, dv_1 \right] du_1$$

and $\qquad y = g(u, v) = \dfrac{\int_0^v Q(u, v_1) \, dv_1}{\int_0^1 Q(u, v_1) \, dv_1}.$

Show that

$$f_u(u, v) = \int_0^1 Q(u, v_1) \, dv_1$$

and that $f_v(u, v) = 0$.
(c) Show that

$$g_v(u, v) = \frac{Q(u, v)}{\int_0^1 Q(u, v_1) \, dv_1}.$$

There is no need to compute g_u.
(d) Show that $\partial(x, y)/\partial(u, v) = Q(u, v)$, as desired.
(e) What is the image of R?

36. Consider a function that assigns to each point in the uv plane a point (x, y, z) in space. There are consequently three scalar functions, $x = f(u, v)$, $y = g(u, v)$, $z = h(u, v)$. Let $\mathbf{R}(u, v)$ be the position vector $x\mathbf{i} + y\mathbf{j} + z\mathbf{k}$ corresponding to the point (u, v). Assume that all the functions f, g, h are "nice" (in particular, have continuous partial derivatives). Then as (u, v) varies in a region R in the plane, (x, y, z) sweeps out a smooth surface S.

If v is held fixed, $\mathbf{R}(u, v)$ parameterizes a curve on S. Then $\partial\mathbf{R}/\partial u$ is a tangent vector to that curve. Similarly, $\partial\mathbf{R}/\partial v$ is a tangent vector to another curve on S.
(a) How would you obtain a normal vector to S?
(b) Consider a very small rectangle in the uv plane with vertices (a, b), $(a + \Delta u, b)$, $(a, b + \Delta v)$, $(a + \Delta u, b + \Delta v)$. Estimate the area of the image in S of this rectangle.
(c) Express the area of S as an integral over R.

37. Let $F = P\mathbf{i} + Q\mathbf{j} + R\mathbf{z}$ be a vector field whose divergence is 0 throughout space. This exercise will construct a vector field $\mathbf{G}$ such that $\mathbf{F} = \operatorname{curl} \mathbf{G}$. In fact, $\mathbf{G}$ will be constructed to be of the form $u\mathbf{i} + v\mathbf{j} + 0\mathbf{k}$.
(a) Show that we are seeking functions u and v such that

$$P = -v_z, \qquad Q = u_z, \qquad R = v_x - u_y.$$

(b) Define $v(x, y, z) = -\int_0^z P(x, y, z_1) \, dz_1$. Show that $P = -V_z$.
(c) Let $A(x, y)$ be a function of x and y to be determined later. Define $u(x, y, z) = \int_0^z Q(x, y, z_1) \, dz_1 + A(x, y)$. Show that $Q = u_z$.
(d) Show that $v_x - u_y =$

$$-\int_0^z [P_x(x, y, z_1) + Q_y(x, y, z_1)] \, dz_1 - A_y(x, y).$$

You will need the theorem of Sec. 23.2.
(e) Recalling that $\mathbf{\nabla} \cdot \mathbf{F} = 0$, show that

$$v_x - u_y = \int_0^z R_z(x, y, z_1) \, dz_1 - A_y(x, y)$$
$$= R(x, y, z) - R(0, y, z) - A_y(x, y).$$

(f) Define $A(x, y)$ to be $-\int_0^y R(x, y_1, 0) \, dy_1$. Show that

$$R = v_x - u_y.$$

In Theorem 3 of Sec. 22.6 it was shown that, if $\operatorname{curl} \mathbf{F} = \mathbf{0}$, then $\mathbf{F}$ is a gradient of some function f. Exercises 38 and 39 obtain this result without the use of Stokes' theorem.

38. Read the statement of the theorem in Sec. 23.2.
(a) Let $\mathbf{F}(x, y) = P(x, y)\mathbf{i} + Q(x, y)\mathbf{j} + 0\mathbf{k}$. Show that the equation $\operatorname{curl} \mathbf{F} = \mathbf{0}$ is equivalent to $P_y = Q_x$.
(b) Assume that $P_y = Q_x$. Define a scalar function $f(x, y)$ by

$$f(x, y) = \int_0^x P(t, y) \, dt + \int_0^y Q(0, u) \, du.$$

Show that $f_x = P$.
(c) Show that $f_y = Q$. You will need the theorem of Sec. 23.2.
(d) Show that $\mathbf{\nabla} f = \mathbf{F}$.

39. (a) Let $\mathbf{F}(x, y, z) = P(x, y, z)\mathbf{i} + Q(x, y, z)\mathbf{j} + R(x, y, z)\mathbf{k}$. Show that the equation $\operatorname{curl} \mathbf{F} = \mathbf{0}$ is equivalent to the equations $R_y = Q_z$, $R_x = P_z$, and $Q_x = P_y$.
(b) Assume that $\operatorname{curl} \mathbf{F} = \mathbf{0}$. Define a scalar function f by

$$f(x, y, z) =$$
$$\int_0^x P(t, y, z) \, dt + \int_0^y Q(0, u, z) \, du + \int_0^z R(0, 0, v) \, dv.$$

Show that $\mathbf{\nabla} f = \mathbf{F}$.

THE INTERCHANGE OF LIMITS

This chapter provides proofs of a few theorems whose validity was assumed in earlier chapters. In Sec. 23.1 the equality of the mixed partial derivatives f_{xy} and f_{yx} is examined; it can be read after Sec. 12.4. The next section is concerned with the derivative of $\int_a^b f(x, y)\, dx$ with respect to y, a concept used in a few exercises in Chap. 22. Term-by-term differentiation and integration of power series is the subject of Sec. 23.3; it can be read after Sec. 14.7. Section 23.3, a gentle introduction to advanced calculus, puts the first three sections in common perspective.

23.1

The equality of f_{xy} and f_{yx}

For most common functions $f(x, y)$,

$$\frac{\partial}{\partial y}\left(\frac{\partial f}{\partial x}\right) \qquad \text{equals} \qquad \frac{\partial}{\partial x}\left(\frac{\partial f}{\partial y}\right);$$

that is, the order in which we compute partial derivatives does not affect the result. This assertion is justified in the following theorem.

THEOREM Let f be a function defined on the xy plane. If f_{xy} and f_{yx} exist and are continuous at all points, then they are equal.

PROOF To keep the proof uncluttered, it will be shown that $f_{xy}(0, 0)$ equals $f_{yx}(0, 0)$. The identical argument holds for any point (a, b).

To begin, consider the definition of $f_{xy}(0, 0)$:

$$f_{xy}(0, 0) = \frac{\partial(f_x)}{\partial y}\bigg|_{\text{at } (0,0)} = \lim_{k \to 0} \frac{f_x(0, k) - f_x(0, 0)}{k}.$$

But, by definition of the partial derivative f_x,

$$f_x(0, k) = \lim_{h \to 0} \frac{f(h, k) - f(0, k)}{h} \qquad \text{and} \qquad f_x(0, 0) = \lim_{h \to 0} \frac{f(h, 0) - f(0, 0)}{h}.$$

Thus $f_{xy}(0, 0) = \lim_{k \to 0} \dfrac{f_x(0, k) - f_x(0, 0)}{k}$

$$= \lim_{k \to 0} \frac{\lim_{h \to 0} \dfrac{f(h, k) - f(0, k)}{h} - \lim_{h \to 0} \dfrac{f(h, 0) - f(0, 0)}{h}}{k}$$

$$= \lim_{k \to 0} \left\{ \lim_{h \to 0} \frac{[f(h, k) - f(0, k)] - [f(h, 0) - f(0, 0)]}{hk} \right\}. \tag{1}$$

Let us focus our attention on the numerator in (1):

$$\text{Numerator} = [f(h, k) - f(0, k)] - [f(h, 0) - f(0, 0)]. \tag{2}$$

Note that the second bracketed expression is obtained from the first bracketed expression by replacing k by 0. Define for fixed h, a function

$$u(y) = f(h, y) - f(0, y). \tag{3}$$

Then (2) takes the simple form

$$u(k) - u(0). \tag{4}$$

By the law of the mean (see Sec. 6.2),

$$u(k) - u(0) = u'(K)k \tag{5}$$

for some K between 0 and k. But by the definition of the function u, given in (3),

$$u'(K) = f_y(h, K) - f_y(0, K). \tag{6}$$

Thus, by the law of the mean, applied to the function $f_y(x, K)$, for fixed K,

$$u'(K) = f_{yx}(H, K)h \tag{7}$$

for some H between 0 and h.

Thus (2) becomes

$$\text{Numerator} = f_{yx}(H, K)hk \tag{8}$$

for some point (H, K) in the rectangle with vertices $(0, 0)$, $(h, 0)$, (h, k), and $(0, k)$.

Now rewrite (2) as

$$\text{Numerator} = [f(h, k) - f(h, 0)] - [f(0, k) - f(0, 0)]. \tag{9}$$

Note that the second bracketed expression is obtained from the first bracketed expression by replacing h with 0. Proceeding as before, with the aid of the law of the mean, we can show that

$$\text{Numerator} = f_{xy}(H^*, K^*)kh \tag{10}$$

for some point (H^*, K^*) in the same rectangle as mentioned. Comparison of (8) and (10) shows that

$$f_{yx}(H, K) = f_{xy}(H^*, K^*). \tag{11}$$

Now, let $(h, k) \to (0, 0)$, $f_{yx}(H, K) \to f_{yx}(0, 0)$, and $f_{xy}(H^*, K^*) \to f_{xy}(0, 0)$ by

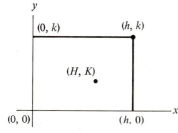

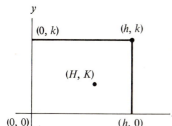

the continuity of the partial derivatives f_{xy} and f_{yx}.   Consequently,

$$f_{yx}(0, 0) = f_{xy}(0, 0),$$

as asserted. ●

In advanced calculus a stronger result than Theorem 1 is obtained: The assumption that f_{xy} is continuous is not needed.   Example 3 in Sec. 23.4 presents a function whose two mixed partial derivatives f_{xy} and f_{yx} are *not* equal at (0, 0).   In the theory of real variables it is proved that, if a function possesses continuous first partial derivatives and both mixed partials throughout a disk, then the latter are equal at at least one point in the disk.

Exercises

1. Obtain Eq. (10).
2. Prove the theorem without looking at the text.

3. Prove the theorem at an arbitrary point (a, b).

23.2

The derivative of $\int_a^b f(x, y)\, dx$ **with respect to** y

The integral

$$\int_a^b f(x, y)\, dx$$

depends on y.   Let

$$F(y) = \int_a^b f(x, y)\, dx.$$

It makes sense to speak of the derivative of F with respect to y.   It turns out that for most common functions f,

$$\frac{dF}{dy} = \int_a^b \frac{\partial f}{\partial y}\, dx.$$

That is,

$$\frac{d[\int_a^b f(x, y)\, dx]}{dy} = \int_a^b \frac{\partial f}{\partial y}\, dx.$$

Generally it is safe to differentiate the integral by differentiating the integrand.
The reader may check this assertion for the function $f(x, y) = x^3 + xy^2$ before going through the proof of the theorem.

THEOREM   Let f be defined on the xy plane, and assume that f and f_y are continuous. Assume also that f_{yy} is defined on the xy plane and that it is bounded on each rectangle.   [That is, if R is a rectangle, then there is a number M, depending on R, such that $|f_{yy}(x, y)| \le M$ for all (x, y) in R.]   Let F be defined by

$$F(y) = \int_a^b f(x, y)\, dx.$$

Then F is differentiable, and

$$\frac{dF}{dy} = \int_a^b \frac{\partial f}{\partial y}\,dx.$$

PROOF   To show that

$$\lim_{h \to 0} \frac{F(y + h) - F(y)}{h} = \int_a^b \frac{\partial f}{\partial y}\,dx,$$

consider, for a fixed y, the difference

$$\frac{F(y + h) - F(y)}{h} - \int_a^b f_y(x, y)\,dx, \tag{1}$$

which, by the definition of F, equals

$$\frac{\int_a^b f(x, y + h)\,dx - \int_a^b f(x, y)\,dx}{h} - \int_a^b f_y(x, y)\,dx. \tag{2}$$

Now, (2) equals

$$\int_a^b \left[\frac{f(x, y + h) - f(x, y)}{h} - f_y(x, y) \right] dx. \tag{3}$$

To show that (1) approaches 0 as $h \to 0$, it suffices to show that the integrand in (3) is small when h is small.   It may be assumed now that $|h| \le 1$.
First of all, the expression

$$\frac{f(x, y + h) - f(x, y)}{h}$$

in the integrand in (3) equals

$$\frac{h f_y(x, y + H)}{h} = f_y(x, y + H)$$

for some number H between 0 and h, by the law of the mean.   (H depends on x, y, and h.)
Thus the integrand in (3) equals

$$f_y(x, y + H) - f_y(x, y). \tag{4}$$

By the law of the mean, (4) equals

$$H f_{yy}(x, y + H^*) \tag{5}$$

for some number H^* between 0 and H.
Since $|H^*| \le |H| \le |h| \le 1$, the point $(x, y + H^*)$ lies somewhere in the rectangle whose vertices are

$$(a, y - 1), \quad (a, y + 1), \quad (b, y - 1), \quad (b, y + 1).$$

By assumption, $|f_{yy}| \le M$ on this rectangle.   By (4) and (5), the integrand in (3) has absolute value at most

$$|H| M \le |h| M.$$

Thus the absolute value of (3) is at most

$$|h| M(b - a),$$

which approaches 0 as $h \to 0$, since M and $b - a$ are fixed numbers.   This proves the theorem. ●

The assumption made in the theorem that f_{yy} is bounded in each rectangle is satisfied if f_{yy} is continuous.   So the theorem does cover the cases commonly encountered.   In advanced calculus the theorem is proved without any assumption on f_{yy}.   (See for instance, R. C. Buck, *Advanced Calculus*, 2d ed., p. 120, McGraw-Hill, New York, 1965.)

Exercises

1. Verify the theorem for $f(x, y) = x^3 y^4$.

2. Verify the theorem for $f(x, y) = \cos xy$.

3. For what value of y does the function

$$F(y) = \int_0^{\pi/2} (y - \cos x)^2 \, dx$$

have a minimum?   (Use the theorem of this section.)

4. Let $G(u, v, w) = \int_u^v f(w, x) \, dx$.   Find (a) $\partial G/\partial v$, (b) $\partial G/\partial u$, (c) $\partial G/\partial w$.

5. Let $G(u) = \int_0^u f(u, x) \, dx$.   Find $\partial G/\partial u$.

6. Let $G(u, v) = \int_0^u e^{-vx^2} \, dx$.   Find (a) $\partial G/\partial u$, (b) $\partial G/\partial v$.

7. Exercise 59 in Sec. 14.8 shows that $\int_0^\infty [(\sin x)/x] \, dx$ is convergent.   Fill in the details and find all the gaps or assumptions in the following argument that $\int_0^\infty [(\sin x)/$ $x] \, dx = \pi/2$: Set $F(y) = \int_0^\infty e^{-yx}[(\sin x)/x] \, dx$.   We are interested in $F(0)$.   Differentiating under the integral sign, we have $F'(y) = -\int_0^\infty e^{-yx} \sin x \, dx$, which, by the fundamental theorem of calculus, equals $-1/(1 + y^2)$.   Hence $F(y) = C - \tan^{-1} y$ for some constant C.   To find C, observe that $0 = \lim_{y \to \infty} F(y) = C - \pi/2$.   Thus $C = \pi/2$, and we have $F(y) = (\pi/2) - \tan^{-1} y$.   Hence $F(0) = \pi/2$.

8. Let $F(y) = \int_0^1 [(x^y - 1)/\ln x] \, dx$ for $y \geq 0$.
  (a) Assuming that one may differentiate F by differentiating under the integral sign, show that $dF/dy = 1/(1 + y)$.
  (b) From (a) deduce that $F(y) = \ln (1 + y) + C$.
  (c) Show that the constant C in part (b) is 0 by examining the case $y = 0$.

23.3

Differentiating and integrating power series

In Sec. 14.7 it was stated that if a function is represented as a power series, then the derivative (or integral) of the function is obtained by termwise differentiation or integration of the power series.   This section justifies that assertion.   We begin with two observations, which, for ease of reference, will be stated as lemmas.

LEMMA I   If $\sum_{n=1}^\infty c_n$ converges, then there is a number M such that, for all n,

$$|c_n| \leq M.$$

PROOF   Since the series converges, its nth term, c_n, approaches 0 as $n \to \infty$.   So there is an integer N such that

$$|c_n| \leq 1$$

for all $n \geq N$. (The choice of 1 is arbitrary; any positive number would work as well.) Let M be the largest of the finite list of numbers

$$|c_1|, |c_2|, \ldots, |c_{N-1}|, 1.$$

This M meets the demand of the lemma. ●

The next lemma depends on the inequality

$$|a_1 + a_2 + \cdots + a_n| \leq |a_1| + |a_2| + \cdots + |a_n|,$$

valid for all numbers $a_1, a_2, \ldots, a_n$. See its proof in Exercise 1.

LEMMA 2   If $f(x)$ is continuous and $b > a$, then

$$\left| \int_a^b f(x) \, dx \right| \leq \int_a^b |f(x)| \, dx.$$

PROOF   Consider any approximating sum

$$\sum_{i=1}^n f(X_i)(x_i - x_{i-1})$$

for $\int_a^b f(x) \, dx$. We have

$$\left| \sum_{i=1}^n f(X_i)(x_i - x_{i-1}) \right| \leq \sum_{i=1}^n |f(X_i)(x_i - x_{i-1})|$$
$$= \sum_{i=1}^n |f(X_i)| \, |x_i - x_{i-1}|$$
$$= \sum_{i=1}^n |f(X_i)|(x_i - x_{i-1}),$$

which is an approximation of $\int_a^b |f(x)| \, dx$. Taking limits as mesh $\to 0$, establishes the lemma. ●

The first theorem asserts that the series obtained from a power series by termwise differentiation converges. It does not say to what it converges.

THEOREM I   If $\sum_{n=0}^\infty a_n x^n$ converges for $|x| < R$, then so does the series

$$\sum_{n=1}^\infty n a_n x^{n-1}.$$

In fact, it converges absolutely.

PROOF   Let $|x| < R$. For the case $x = 0$ the theorem follows immediately. Hence, assume $x \neq 0$. It will be shown that $\sum_{n=1}^\infty n a_n x^{n-1}$ converges. Select a number c such that

$$|x| < c < R.$$

By Lemma 1, there is a number M such that

$$|a_n c^n| \leq M$$

for all n.   Now,

$$na_n x^{n-1} = n \frac{a_n}{c} \frac{x^{n-1}}{c^{n-1}} c^n.$$

Thus

$$|na_n x^{n-1}| = \frac{|a_n c^n|}{|c|} \left| n \left(\frac{x}{c}\right)^{n-1} \right| \leq \frac{M}{|c|} \left| n \left(\frac{x}{c}\right)^{n-1} \right|$$

Since $|x/c| < 1$, the series

$$\frac{M}{|c|} \sum_{n=1}^{\infty} n \left| \frac{x}{c} \right|^{n-1}$$

converges by the ratio test.   By the comparison test,

$$\sum_{n=1}^{\infty} |na_n x^{n-1}|$$

converges.   The absolute convergence test then implies the convergence of

$$\sum_{n=1}^{\infty} na_n x^{n-1}. \; \bullet$$

THEOREM 2   (*Continuity of* $\sum_{n=0}^{\infty} a_n x^n$).   Let

$$f(x) = \sum_{n=0}^{\infty} a_n x^n$$

converge for $|x| < R$.   Then $f(x)$ is continuous for $|x| < R$.

PROOF   Let c be a number in the open interval $(-R, R)$.   We wish to prove that as $x \to c$, $f(x) \to f(c)$.   To do so, it is necessary to examine $|f(x) - f(c)|$. We have

$$|f(x) - f(c)| = \left| \sum_{=0}^{\infty} a_n(x^n - c^n) \right| \leq \sum_{n=0}^{\infty} |a_n| |x^n - c^n|.$$

By the law of the mean,

$$|x^n - c^n| = |nX_n^{n-1}(x - c)|$$

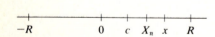

for some X_n between c and x.

Pick a fixed number k, such that $k < R$, $k > |x|$, and $k > |c|$.   Then for each n,

$$|X_n| < k.$$

Thus

$$|f(x) - f(c)| \leq \sum_{n=1}^{\infty} |na_n X_n^{n-1}(x - c)| \leq |x - c| \sum_{n=1}^{\infty} |na_n k^{n-1}|. \qquad (1)$$

Since $\sum_{n=1}^{\infty} |na_n k^{n-1}|$ is just some fixed number (by Theorem 1), the right side of (1) approaches 0 as $x \to c$.   Thus as $x \to c$,

$$|f(x) - f(c)| \to 0. \; \bullet$$

THEOREM 3 (*Termwise integration of a power series*).   Let $\sum_{n=0}^{\infty} a_n x^n$ converge for $|x| < R$. Let $f(x) = \sum_{n=0}^{\infty} a_n x^n$.

Then for $|t| < R$,

$$\int_0^t f(x)\,dx = \sum_{n=0}^{\infty} \frac{a_n t^{n+1}}{n+1}.$$

PROOF

It Follows from the Proof that the Series $\sum_{n=0}^{\infty} a_n t^{n+1}/(n+1)$ Converges. This Could Be Established Directly.

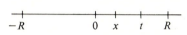

Since f is continuous, $\int_0^t f(x)\,dx$ exists.   All that remains is to show that

$$\left| \int_0^t f(x)\,dx - \sum_{n=0}^{m} \frac{a_n t^{n+1}}{n+1} \right| \tag{2}$$

approaches 0 as $m \to \infty$.   We consider the case $t > 0$; the proof of the case $t < 0$ is similar.

Now,

$$\int_0^t f(x)\,dx - \sum_{n=0}^{m} \frac{a_n t^{n+1}}{n+1} = \int_0^t f(x)\,dx - \int_0^t \left(\sum_{n=0}^{m} a_n x^n \right) dx$$

$$= \int_0^t \left(\sum_{n=0}^{\infty} a_n x^n - \sum_{n=0}^{m} a_n x^n \right) dx$$

$$= \int_0^t \left(\sum_{n=m+1}^{\infty} a_n x^n \right) dx.$$

Thus by Lemma 2,

$$\left| \int_0^t f(x)\,dx - \sum_{n=0}^{m} \frac{a_n t^{n+1}}{n+1} \right| \le \int_0^t \left| \sum_{n=m+1}^{\infty} a_n x^n \right| dx. \tag{3}$$

Now

$$\left| \sum_{n=m+1}^{\infty} a_n x^n \right| \le \sum_{n=m+1}^{\infty} |a_n x^n| \le \sum_{n=m+1}^{\infty} |a_n t^n|.$$

Since the series $\sum_{n=0}^{\infty} |a_n t^n|$ converges, the expression $\sum_{n=m+1}^{\infty} |a_n t^n|$ approaches 0 as $m \to \infty$.   Consequently, the integrand in the right side of (3) is small when m is large.   Thus, as $m \to \infty$, the right side of (3) approaches 0. Therefore, the left side of (3) approaches 0 as $m \to \infty$.   That is, (2) approaches 0 as $m \to \infty$.   This concludes the proof. ●

THEOREM 4 (*Termwise differentiation of a power series*).   Let $\sum_{n=0}^{\infty} a_n x^n$ converge for $|x| < R$.   Let $f(x) = \sum_{n=0}^{\infty} a_n x^n$.   Then for $|x| < R$,

$$\sum_{n=1}^{\infty} n a_n x^{n-1}$$

converges, and its sum is $f^{(1)}(x)$.

PROOF

We already know from Theorem 1 that $\sum_{n=1}^{\infty} n a_n x^{n-1}$ converges for $|x| < R$. Denote its sum by $g(x)$.   By Theorem 2, $g(x)$ is continuous.   All that remains

is to show that
$$f^{(1)}(x) = g(x).$$

Since we can integrate a power series term by term, it follows that for $|t| < R$,

$$\int_0^t g(x)\, dx = \sum_{n=1}^{\infty} a_n t^n = \sum_{n=0}^{\infty} a_n t^n - a_0$$

Thus
$$\int_0^t g(x)\, dx = f(t) - a_0.$$

Differentiation of both sides of this equation yields

$$g(t) = f^{(1)}(t),$$

and the theorem is proved. ●

Exercises

1. (a) Prove that $|a + b| \le |a| + |b|$. *Hint:* $|x| = \sqrt{x^2}$.
   (b) Use (a) to prove that

   $$|a_1 + a_2 + a_3| \le |a_1| + |a_2| + |a_3|.$$

   (c) From (a) and (b) deduce that $|a_1 + a_2 + a_3 + a_4| \le |a_1| + |a_2| + |a_3| + |a_4|$. Similarly, mathematical induction establishes $|\sum_{n=1}^{k} a_n| \le \sum_{n=1}^{k} |a_n|$.

2. Let $P(x)$ and $Q(x)$ be polynomials. Assume that there is no positive integer n such that $Q(n) = 0$. Show that, if $\sum_{n=1}^{\infty} a_n x^n$ converges for $|x| < R$, so does

   $$\sum_{n=1}^{\infty} a_n P(n) x^n / Q(n).$$

3. Show that the series $\sum_{n=0}^{\infty} a_n x^n$,

$$\sum_{n=1}^{\infty} n a_n x^{n-1}, \qquad \text{and} \qquad \sum_{n=0}^{\infty} a_n x^{n+1}/(n+1)$$

have the same radius of convergence.

4. Prove that if $f(x) = \sum_{n=0}^{\infty} a_n x^n$, $|x| < R$, then $f^{(2)}(x) = \sum_{n=2}^{\infty} n(n-1) a_n x^{n-2}$, $|x| < R$.

5. (a) Generalize Theorem 1 to series in powers of $x - a$.
   (b) Prove your generalization.

6. (a) Generalize Theorem 2 to series in powers of $x - a$.
   (b) Prove your generalization.

7. (a) Generalize Theorem 3 to series in powers of $x - a$.
   (b) Prove your generalization.

8. (a) Generalize Theorem 4 to series in powers of $x - a$.
   (b) Prove your generalization.

23.4

The interchange of limits

Though the various topics discussed in this chapter are independent, they all illustrate a certain type of problem which students who go on to advanced calculus will study, namely, the interchange of limits. To see what this means, let us take a new look at the theorem in Sec. 23.1, which concerns the equality of the mixed partials. By (1) in Sec. 23.1,

$$f_{xy}(0, 0) = \lim_{k \to 0} \left[\lim_{h \to 0} \frac{f(h, k) - f(0, k) - f(h, 0) + f(0, 0)}{hk} \right].$$

Similarly, from the definition of f_{yx}, it can be shown that

$$f_{yx}(0, 0) = \lim_{h \to 0} \left[\lim_{k \to 0} \frac{f(h, k) - f(0, k) - f(h, 0) + f(0, 0)}{hk} \right].$$

Note that the two quotients are identical, but that f_{xy} involves

$$\lim_{k \to 0} \left(\lim_{h \to 0} \right),$$

while f_{yx} involves

$$\lim_{h \to 0} \left(\lim_{k \to 0} \right).$$

It is tempting to claim that the order of taking limits should not matter. But it does.   (Recall that f_{xy} does not always equal f_{yx}.)   This instance raises the general question, "When can one interchange limits?"   Example 1 presents a simple case in which the order of taking the limits *does* matter.

EXAMPLE 1   Let $f(x, y) = x^y$ for $x > 0$ and $y > 0$.   Evaluate the two "repeated limits"

$$\lim_{y \to 0^+} \left(\lim_{x \to 0^+} x^y \right)$$

and

$$\lim_{x \to 0^+} \left(\lim_{y \to 0^+} x^y \right).$$

SOLUTION

$$\lim_{y \to 0^+} \left(\lim_{x \to 0^+} x^y \right) = \lim_{y \to 0^+} 0 = 0.$$

On the other hand,

$$\lim_{x \to 0^+} \left(\lim_{y \to 0^+} x^y \right) = \lim_{x \to 0^+} 1 = 1.$$

This shows that *the order of taking limits may affect the result.*   Moreover, it suggests why the symbol 0^0 is not given any meaning.   ●

The next example also illustrates the effect of switching the order of taking limits.   In Example 3 it becomes the basis of an illustration that shows the mixed partial derivatives f_{xy} and f_{yx}, are not always equal.

EXAMPLE 2   Let

$$g(x, y) = \begin{cases} \dfrac{x^2 - y^2}{x^2 + y^2} & \text{if } (x, y) \neq (0, 0); \\ 0 & \text{if } (x, y) = (0, 0). \end{cases}$$

Show that

$$\lim_{x \to 0} \left[\lim_{y \to 0} g(x, y) \right] \neq \lim_{y \to 0} \left[\lim_{x \to 0} g(x, y) \right].$$

SOLUTION

$$\lim_{y \to 0} g(x, y) = \lim_{y \to 0} \frac{x^2 - y^2}{x^2 + y^2}$$

$$= \frac{x^2}{x^2}$$

$$= 1.$$

Thus
$$\lim_{x \to 0} \left[\lim_{y \to 0} g(x, y) \right] = \lim_{x \to 0} 1 = 1.$$

On the other hand,
$$\lim_{x \to 0} g(x, y) = \lim_{x \to 0} \frac{x^2 - y^2}{x^2 + y^2}$$

$$= \frac{-y^2}{y^2},$$

$$= -1.$$

Thus
$$\lim_{y \to 0} \left[\lim_{x \to 0} g(x, y) \right] = \lim (-1) = -1. \quad \bullet$$

EXAMPLE 3   Let $f(x, y) = xyg(x, y)$, where g is given in Example 2.   Show that

$$f_{xy}(0, 0) \neq f_{yx}(0, 0).$$

SOLUTION   That f_x, f_y, f_{xy}, f_{yx} exist at all points is left for the reader (see Exercise 8). Note that $f(x, y) = 0$ whenever x or y is 0.
By (1) in Sec. 23.1,

$$f_{xy}(0, 0) = \lim_{k \to 0} \left\{ \lim_{h \to 0} \frac{[f(h, k) - f(0, k)] - [f(h, 0) - f(0, 0)]}{hk} \right\}$$

$$= \lim_{k \to 0} \left[\lim_{h \to 0} \frac{f(h, k)}{hk} \right]$$

$$= \lim_{k \to 0} \left[\lim_{h \to 0} g(h, k) \right] = -1.$$

From Example 2, similarly,

$$f_{yx}(0, 0) = \lim_{h \to 0} \left[\lim_{k \to 0} g(h, k) \right] = 1.$$

Therefore, $f_{xy}(0, 0) \neq f_{yx}(0, 0)$. $\bullet$

The theorem in Sec. 23.2, which asserts that in general

$$\frac{d[\int_a^b f(x, y) \, dx]}{dy} = \int_a^b f_y(x, y) \, dx,$$

also concerns the validity of switching limits.   After all, both the derivative and the definite integral are defined as limits.

The theorems in Sec. 23.3 about differentiating and integrating power series are basically theorems about the interchange of limits.   For instance, Theorem 3 on termwise integration of series can be rephrased as follows.

THEOREM I   (*Termwise integration of series*).   Let $f(x) = a_0 + a_1 x + a_2 x^2 + \cdots$ for x in $(-R, R)$.   Let a and b be numbers in $(-R, R)$.   For each positive integer n

let

$$f_n(x) = a_0 + a_1 x + \cdots + a_n x^n.$$

Then

$$\int_a^b \lim_{n \to \infty} f_n(x)\, dx = \lim_{n \to \infty} \int_a^b f_n(x)\, dx. \; \bullet$$

The reader should pause to check that this really is just a restatement of the termwise integration theorem.

Theorem 1 may seem to be just a special case of a universal theorem about all sequences of functions $f_1(x), f_2(x), f_3(x), \ldots$ that approach a function $f(x)$. The next example shows that this is *not* the case and should increase our gratitude for Theorem 1.

EXAMPLE 4 For each positive integer n define the function f_n as follows: The graph of f_n consists of an isosceles triangle whose base is the interval $[0, 1/n]$ and whose height is n, together with the portion of the x axis outside $[0, 1/n]$.
Does $\lim_{n \to \infty} \int_0^1 f_n(x)\, dx = \int_0^1 \lim_{n \to \infty} f_n(x)\, dx$?

SOLUTION Note that, as n increases, the triangular part of the graph of f_n gets narrower and higher but its area remains fixed:

$$\text{Area of triangle} = \frac{1}{2} \cdot \frac{1}{n} \cdot n = \frac{1}{2}.$$

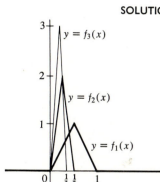

Thus for all n

$$\int_0^1 f_n(x)\, dx = \tfrac{1}{2}.$$

Note also for each fixed x that

$$\lim_{n \to \infty} f_n(x) = 0$$

since, for n large enough, $f_n(x) = 0$.

Thus

$$\lim_{n \to \infty} \int_0^1 f_n(x)\, dx = \tfrac{1}{2}$$

but

$$\int_0^1 \lim_{n \to \infty} f_n(x)\, dx = 0.$$

It is not always permissible to move "lim" past an integral sign.

Since an integral is defined as a limit, this example concerns an interchange of limits. $\bullet$

In advanced calculus a fairly general theorem, which includes Theorem 1, is established. The assumption is added that the graph of f_n looks a lot like the graph of f when n is large. More precisely, it is assumed that f_n and f are continuous on $[a, b]$ and that

$$\lim_{n \to \infty} (\text{maximum} \, | f(x) - f_n(x) | \, \text{for } x \text{ in } [a, b]) = 0.$$

Then it turns out that

$$\int_a^b \lim_{n\to\infty} f_n(x)\,dx = \lim_{n\to\infty} \int_a^b f_n(x)\,dx.$$

The exercises present more examples of the interchange of limits. The main point of the examples and the exercises is that the interchange of limits is a risky business. Fortunately there are theorems that imply that sometimes the order of taking limits has no effect on the outcome.

Exercises

1. Let $f(x, y) = 1$ if $y \geq x$ and let $f(x, y) = 0$ if $y < x$.
   (a) Shade in the part of the plane where $f(x, y) \geq 1$.
   (b) Show that $\lim_{x\to\infty} [\lim_{y\to\infty} f(x, y)] = 1$.
   (c) Show that $\lim_{y\to\infty} [\lim_{x\to\infty} f(x, y)] = 0$.

2. Show that $\lim_{x\to 0} [\lim_{n\to\infty} nx/(1 + nx)] = 1$, while
   $$\lim_{n\to\infty} [\lim_{x\to 0} nx/(1 + nx)] = 0. \quad \text{(Assume } x > 0.)$$

3. Let $f_n(x) = nx/(1 + n^2 x^4)$. Show that $\int_0^\infty \lim_{n\to\infty} f_n(x)\,dx$
   $= 0$, but $\lim_{n\to\infty} \int_0^\infty f_n(x)\,dx = \pi/4$.

4. Show that $\lim_{x\to 0} [\lim_{y\to 0} x^2/(x^2 + y^2)]$ is not equal to $\lim_{y\to 0} [\lim_{x\to 0} x^2/(x^2 + y^2)]$.

5. Let $f_n(x) = n\pi \sin(n\pi x)$ if $0 \leq x \leq 1/n$, and 0 otherwise.
   (a) Graph f_1, f_2, and f_3.
   (b) Show that $\lim_{n\to\infty} \int_0^1 f_n(x)\,dx = 2$, but that
   $\int_0^1 \lim_{n\to\infty} f_n(x)\,dx = 0$.

6. Compare
   $$\lim_{x\to\infty} \left(\lim_{y\to\infty} \frac{x^2}{x^2 + y^2 + 1} \right)$$
   and
   $$\lim_{y\to\infty} \left(\lim_{x\to\infty} \frac{x^2}{x^2 + y^2 + 1} \right).$$

7. Let $f_n(x) = (1/n) \sin nx$ for all x and all positive integers n.

   Show that $\displaystyle \lim_{n\to\infty} \left\{ \lim_{h\to 0} \frac{f_n(h) - f_n(0)}{h} \right\} = 1,$

   while $\displaystyle \lim_{h\to 0} \left\{ \lim_{n\to\infty} \frac{f_n(h) - f_n(0)}{h} \right\} = 0.$

8. Show that f_{xy} and f_{yx} exist at all points in the plane, where f is the pathological function in Example 3.

9. (a) Show that Theorem 2 of Sec. 23.3 concerns the interchange of limits. *Hint:* Define $f_n(x)$ to be
   $$a_0 + a_1 x + \cdots + a_n x^n.$$
   (b) Use Exercise 7 to show that the interchange of limits in part (a) is not valid for any sequence of functions

$f_1(x), f_2(x), \ldots, f_n(x), \ldots$ that approach a function $f(x)$.

◼

10. Show that the theorem of Sec. 23.2 is related to the assertion that
    $$\lim_{n\to\infty} \left\{ \lim_{k\to 0} \sum_{i=1}^n \frac{f(x_i, y + k) - f(x_i, y)}{kn} (b - a) \right\}$$
    equals
    $$\lim_{k\to 0} \left\{ \lim_{n\to\infty} \sum_{i=1}^n \frac{f(x_i, y + k) - f(x_i, y)}{kn} (b - a) \right\},$$
    where
    $$x_0 = a, \ x_1 = a + \frac{b - a}{n}, \ldots, x_i = a + \frac{i(b - a)}{n}, \ldots, x_n = b.$$

11. Let $f_n(x)$ be defined for each x in $[0, 1]$ and each positive integer n. Assume that f_n is continuous for each n and that $\lim_{n\to\infty} f_n(x)$ exists for each x in $[0, 1]$. Call this limit $f(x)$,
    $$\lim_{n\to\infty} f_n(x) = f(x).$$
    (a) Show that the statement, "f is continuous at a," is equivalent to the equation
    $$\lim_{n\to\infty} \left\{ \lim_{x\to a} f_n(x) \right\} = \lim_{x\to a} \left\{ \lim_{n\to\infty} f_n(x) \right\}.$$
    (b) In particular, let $f_n(x) = x^n$. Show that in this case the two repeated limits in (a) are not necessarily equal.

12. Let $f_n(x) = x^{2n}/(1 + x^{2n})$.
    (a) Let $f(x) = \lim_{n\to\infty} f_n(x)$. Graph f_4 and f.
    (b) For which a is
    $$\lim_{n\to\infty} [\lim_{x\to a} f_n(x)] = \lim_{x\to a} [\lim_{n\to\infty} f_n(x)]?$$

13. Show that

$$\int_0^\infty \left[\int_0^1 (2xy - x^2y^2)e^{-xy}\,dx\right]dy = 1,$$

but

$$\int_0^1 \left[\int_0^\infty (2xy - x^2y^2)e^{-xy}\,dy\right]dx = 0.$$

14. Show that l'Hôpital's rule in the zero-over-zero case concerns the equality of these two limits:

$$\lim_{\Delta t \to 0}\left\{\lim_{t \to a}\frac{f(t + \Delta t) - f(t)}{g(t + \Delta t) - g(t)}\right\}$$

and

$$\lim_{t \to a}\left\{\lim_{\Delta t \to 0}\frac{f(t + \Delta t) - f(t)}{g(t + \Delta t) - g(t)}\right\}.$$

■ ■

15. Define f by setting $f(x) = \lim_{n\to\infty}\{\lim_{m\to\infty}[\cos(n!\,\pi x)]^{2m}\}$. Recalling that π is irrational, prove that $f(x) = 1$ if x is rational and $f(x) = 0$ if x is irrational. (Thus the function f that is 1 at the rationals and 0 at the irrationals is the limit of a sequence of functions that are themselves limits of sequences of continuous functions. It can be proved that f is *not* the limit of a sequence of continuous functions.)

16. Devise an example of a function f such that $\sum_{m=1}^\infty [\sum_{n=1}^\infty f(m, n)] = 0$, but $\sum_{n=1}^\infty [\sum_{m=1}^\infty f(m, n)] = 1$. *Suggestion:* Place $f(m, n)$ on the point (m, n). Let us call the points (m, n) for fixed m, *column m* and for fixed n, *row n*. First fill in row 1 in such a way that $\sum_{m=1}^\infty f(m, 1) = \frac{1}{2}$. Then choose $f(1, 2)$ to be the negative of $f(1, 1)$ and all remaining $f(1, n) = 0$; hence $\sum_{n=1}^\infty f(1, n) = 0$. Fill in row 2 in such a way that its sum is $\frac{1}{4}$, and then column 2 in such a way that its sum is 0. For instance, we may begin as follows:

$\cdot 0$	$\cdot 0$					
$\cdot 0$	$\cdot 0$					
$\cdot 0$	$\cdot 0$					
$\cdot 0$	$\cdot 0$					
$\cdot 0$	$\cdot -\frac{3}{8}$	$\cdots\cdots\cdots\cdots\cdots\cdots$				
$\cdot -\frac{1}{4}$	$\cdot\frac{1}{4}$	$\cdot\frac{1}{8}$	$\cdot\frac{1}{16}$	$\cdot\frac{1}{32}$	$\cdots$	Row 2
$\cdot\frac{1}{4}$	$\cdot\frac{1}{8}$	$\cdot\frac{1}{16}$	$\cdot\frac{1}{32}$	$\cdot\frac{1}{64}$	$\cdots$	Row 1

Exercises 17 to 23 outline a proof due to Euler that

$$\frac{1}{1^2} + \frac{1}{2^2} + \frac{1}{3^2} + \cdots = \frac{\pi^2}{6}.$$

17. Show that if

$$\frac{1}{1^2} + \frac{1}{3^2} + \frac{1}{5^2} + \frac{1}{7^2} + \cdots + \frac{1}{(2n-1)^2} + \cdots = \frac{\pi^2}{8},$$

then

$$\frac{1}{1^2} + \frac{1}{2^2} + \frac{1}{3^2} + \cdots + \frac{1}{n^2} + \cdots = \frac{\pi^2}{6}.$$

18. Show that $\displaystyle\int_0^1 \frac{\sin^{-1}x}{\sqrt{1 - x^2}}\,dx = \frac{\pi^2}{8}.$

19. Use the binomial series of Sec. 15.5 to show that, if $0 \le t < 1$, then

$$\frac{1}{\sqrt{1 - t^2}} = 1 + \frac{1}{2}t^2 + \frac{1 \cdot 3}{2 \cdot 4}t^4 + \frac{1 \cdot 3 \cdot 5}{2 \cdot 4 \cdot 6}t^6 + \cdots.$$

20. (See Exercise 19.) Show that

$$\sin^{-1}x = x + \frac{1}{2}\frac{x^3}{3} + \frac{1 \cdot 3}{2 \cdot 4}\frac{x^5}{5} + \frac{1 \cdot 3 \cdot 5}{2 \cdot 4 \cdot 6}\frac{x^7}{7} + \cdots$$

for $0 \le x < 1$. This equation is also valid when $x = 1$.

21. Show that

$$\int_0^1 \frac{x^{2n+1}}{\sqrt{1 - x^2}}\,dx = \int_0^{\pi/2} \sin^{2n+1}\theta\,d\theta.$$

22. Assuming that it is safe to integrate term by term, even in the case of an improper integral, show that

$$\int_0^1 \frac{\sin^{-1}x}{\sqrt{1 - x^2}}\,dx = \frac{1}{1^2} + \frac{1}{3^2} + \frac{1}{5^2} + \frac{1}{7^2} + \cdots$$

$$+ \frac{1}{(2n-1)^2} + \cdots.$$

23. Deduce that

$$\sum_{n=1}^\infty n^{-2} = \frac{\pi^2}{6}.$$

23.5

Summary

This chapter established a few basic assumptions made earlier in the text and gave a glimpse into advanced calculus. The basic theme is expressed in the question, "When can we interchange the order of limits and not affect the result?"

Section 23.1 concerned the equation

$$f_{xy}(a, b) = f_{yx}(a, b),$$

which is valid wherever both the mixed partial derivatives are continuous.

Section 23.2 established that

$$\frac{d}{dy} \int_a^b f(x, y)\, dx = \int_a^b f_y(x, y)\, dx,$$

under conditions satisfied by most common functions.

In Sec. 23.3 differentiation and integration of power series were explored.

All these results were shown in Sec. 23.4 to be related to the general problem of the interchange of limits. There the perils of such an interchange were emphasized.

APPENDIX A
THE REAL NUMBERS

This appendix describes those properties of the real number system which are used in the text. While it is possible to construct the real numbers from the positive integers 1, 2, 3, 4, ..., a description of the procedure would require a small book. We shall assume that the set of real numbers exists and content ourselves with a summary of its important attributes.

A.1

The properties of addition and multiplication (the field axioms)

Let S be the set of real numbers. On S are defined two operations, addition, denoted $+$, and multiplication, denoted $\cdot$, which satisfy the following axioms.

A1. For each a and b in S, $a + b$ is in S.

A2. For each a and b in S, $a + b = b + a$.

A3. For each a, b, and c in S, $a + (b + c) = (a + b) + c$.

A4. There is an element in S, denoted 0, such that $0 + a = a$ for all a in S.

A5. For each a and b in S, there is a unique element c in S such that $a + c = b$.

M1. For each a and b in S, $a \cdot b$ is in S.

M2. For each a and b in S, $a \cdot b = b \cdot a$.

M3. For each a, b, and c in S, $a \cdot (b \cdot c) = (a \cdot b) \cdot c$.

M4. There is an element in S, denoted 1, such that $1 \cdot a = a$ for all a in S.

M5. For each element a in S (other than 0) and each element b in S, there is a unique element c in S such that $a \cdot c = b$.

D. For each a, b, and c in S,
$$a \cdot (b + c) = a \cdot b + a \cdot c.$$

The first four axioms for addition and for multiplication are analogous. The second axiom (A2 or M2) is the *commutative law*; the third axiom is the *associative law*. The element c, whose existence is assumed by A5, is usually denoted $b - a$ and is called *the difference of a and b*. In particular, $0 - a$ is denoted $-a$ and is called *the opposite* or *the additive inverse of a*. The element c, whose existence is assumed by M5, is usually denoted b/a and is called *the quotient of b by a*. In particular, $1/a$ is called *the reciprocal of a* or *the multiplicative inverse of a*. Every element has an additive inverse; every element except 0 has a multiplicative inverse. The distributive axiom D distinguishes addition from multiplication: We do *not* have a companion axiom relating $a + (b \cdot c)$ to $a + b$ and $a + c$.

A.2

The ordering axioms

There is a relation between real numbers, denoted " $>$ " and read as "greater than," that satisfies the following axioms.

O1. If a is not 0, then $a > 0$ or $-a > 0$, but not both.
O2. If $a > b$ and $b > c$, then $a > c$.
O3. If $a > b$ and $c > 0$, then $ca > cb$ and, for any c, $c + a > c + b$.

If $a > 0$, then a is called *positive*. If $-a > 0$, then a is called *negative*.

If we think of the real numbers as describing points on a number line, then "$a > b$" means that a is to the right of b; "a is positive" means that a is to the right of 0; and "a is negative" means that a is to the left of 0.

In the figure $a > b$, a is positive, and b is negative

Axiom O3 asserts that multiplication by a positive number preserves an inequality. Multiplication by a negative number reverses it: $4 > 3$, but $(-5)(3) > (-5)(4)$. Analogous remarks hold for division by a positive or by a negative number.

The *absolute value* of a is a if a is positive, and $-a$ if a is negative. The absolute value of 0 is 0. In terms of the number line, the absolute value of a is the distance from a to 0. The absolute value of a is denoted $|a|$. Thus $|3| = 3 = |-3|$.

The symbol $a \geq b$ shall mean $a > b$ or $a = b$. Thus $4 \geq 3$ and $3 \geq 3$. The symbol $a < b$ shall mean $b > a$. The symbol $a \leq b$ shall mean $b \geq a$. Thus $3 < 4$, $3 \leq 4$, $3 \leq 3$.

The absolute value has two important properties:

$$|ab| = |a||b| \qquad \text{and} \qquad |a + b| \leq |a| + |b|.$$

For instance,

$$|(-3) \cdot 7| = |-3||7| \qquad \text{and} \qquad |(-3) + 7| \leq |-3| + |7|,$$

as the reader may verify.

A.3

Rational and irrational numbers

Within the real numbers certain types of numbers are singled out. The *positive integers*, 1, 2, 3, ...; the *integers* ..., -3, -2, -1, 0, 1, 2, 3, ...; and the *rational numbers* a/b, where a and b are integers and b ≠ 0.   On the number line the rational numbers correspond to the points obtainable by dividing the interval between integers into equal divisions.   For instance, $\frac{7}{5}$ corresponds to the point between 1 and 2 shown in this diagram.

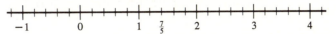

Note that any integer n is a rational number: $n = n/1$.

   Not every real number is rational.   The decimal representation of a rational number always *repeats*.   From some point on, the digits repeat in blocks.   For instance,

$$\tfrac{23}{14} = 1.6\overline{428571}\ \overline{428571}\ \overline{428571}\ \overline{428571}\ \cdots.$$

Conversely, any *repeating* decimal represents a rational number.   For instance, consider the number

$$x = 0.5\overline{31}\overline{31}\overline{31}\ \cdots.$$

Then
$$100x = 53.1\overline{31}\overline{31}\overline{31}\ \cdots,$$

and subtraction of the former from the latter yields

$$100x - x = 52.6,$$

or $99x = 52.6.$

Thus
$$x = \frac{52.6}{99} = \frac{526}{990}.$$

Hence $0.5\overline{31}\overline{31}\ \cdots$ is expressible as the quotient of two integers and is a rational number.   The same type of argument may be applied to any repeating decimal.

   A real number that is not rational is called *irrational*.   Irrational numbers (as well as rational numbers) are plentiful.   Any nonrepeating decimal, such as

$$0.12122122212222\ \cdots \tag{1}$$

(in which the number 1 alternates with lengthening blocks of 2s) is irrational. The ratio between the circumference and diameter of a circle is irrational.   So are $\sqrt{2}, \sqrt{3}, \sqrt{5}, \sqrt{6},$ and $\sqrt[3]{2}$.

EXAMPLE   How may irrational numbers are there between 0.12 and 0.13?

SOLUTION   One such number is given above in (1).   We may construct an infinite set of irrational numbers between 0.12 and 0.13 by, let us say, changing one 1 or 2 from the third decimal place on in (1) to a 3.   Thus

$$0.12322122212222\ \cdots$$
$$0.12132122212222\ \cdots$$
$$0.12123122212222\ \cdots$$
$$\cdots\cdots\cdots\cdots\cdots$$

are irrational numbers between 0.12 and 0.13.   Thus there is an infinite set of irrational numbers between 0.12 and 0.13.   Similarly, by constructing *repeating* decimals that begin $0.12 \cdots$, it can be shown that between 0.12 and 0.13 there is an infinite set of rational numbers. ●

The same reasoning shows that both the rational numbers and the irrational numbers are distributed abundantly on the number line: *Between any two real numbers there are an infinity of rational numbers and an infinity of irrational numbers.*

The sum, difference, product, and quotient of two rational numbers are also rational numbers. For instance,

$$\frac{3}{4} + \frac{5}{7} = \frac{3 \cdot 7 + 5 \cdot 4}{28} = \frac{41}{28}; \qquad \frac{3}{4} - \frac{5}{7} = \frac{3 \cdot 7 - 5 \cdot 4}{28} = \frac{1}{28};$$

$$\frac{3}{4} \cdot \frac{5}{7} = \frac{15}{28}; \qquad \frac{\frac{3}{4}}{\frac{5}{7}} = \frac{3}{4} \cdot \frac{7}{5} = \frac{21}{20}.$$

The set of rational numbers satisfies the field axioms and the order axioms. The set of irrational numbers satisfies the order axioms but not all the field axioms. For instance, the sum of two irrational numbers is not necessarily irrational: $(\sqrt{2}) + (-\sqrt{2}) = 0$.

Exercises

1. (a) If a and b are both negative, or else both positive, what is the relation between $|a + b|$, $|a|$, and $|b|$? Explain and illustrate by examples.
   (b) Prove that, if $|a - b| < c$, then $|a| < |b| + c$. *Hint:* Consider the cases a, b, $a - b$ positive or negative.
2. Find the decimal expansion of the following rational numbers, and show the repeating block in each case.
   (a) $\frac{4}{13}$, (b) $\frac{3}{7}$, (c) $\frac{5}{8}$, (d) $\frac{2}{17}$.
3. Which of these numbers are rational? Which are irrational?
   (a) -8,

(b) $5\sqrt{2}/2$,
(c) π,
(d) $-3/(-7)$,
(e) $\sqrt{4}$,
(f) 5.238.

4. Construct at least four irrational numbers and at least four rational numbers between 3.17 and 3.18.
5. Find integers m and n such that
   (a) $m/n = 6.2\overline{457}457457 \cdots$;
   (b) $m/n = 20.3\overline{65}6565 \cdots$.
6. Is there a smallest positive number?

A.4

Completeness of the real numbers

We now come to the property of the real numbers which is most important in the calculus. In order to state it precisely, two definitions are needed.

DEFINITION   *Upper bound.* Let X be a set of real numbers. The number u is an *upper bound* for the set X if $u \geq x$ for all x in X.

For instance, if $X = \{1, 2, 3\}$, then $u = 17$, $u = 6.2$, and $u = 3$ are some of the upper bounds for X. The set $X = \{1, 2, 3, 4, \ldots\}$ has no upper bound. The set $X = \{\frac{1}{2}, \frac{2}{3}, \frac{3}{4}, \frac{4}{5}, \frac{5}{6}, \ldots\}$ has $u = 1{,}000$, $u = 15$, and $u = 1$ as upper bounds. In fact, any number $u \geq 1$ is an upper bound of this set.

DEFINITION *Least upper bound.* A number is the *least upper bound* of a set X if it is an upper bound of X and is less than or equal to every upper bound of X.

For instance, the set $X = \{\frac{1}{2}, \frac{2}{3}, \frac{3}{4}, \frac{4}{5}, \ldots\}$ has the least upper bound 1. So has the set $Y = \{1, \frac{1}{2}, \frac{2}{3}, \frac{3}{4}, \frac{4}{5}, \ldots\}$.

We are now ready to state the most useful property of the real number system.

THE COMPLETENESS AXIOM *Any set X of real numbers that has an upper bound has a least upper bound.*

The set of rational numbers, which satisfies the field axioms and the order axioms, does *not* satisfy the completeness axiom, as the next example illustrates.

EXAMPLE Show that the set $X = \{1, 1.7, 1.73, 1.732, \ldots\}$, the set of successive (rational) decimal approximations to $\sqrt{3} = 1.732051\cdots$, does not have a rational least upper bound.

SOLUTION Since $2^2 > 3$, 2 is larger than any number in X, hence X has a rational upper bound. But there is no smallest *rational* upper bound for X. Specifically, if r is rational and is an upper bound of X, then $r > \sqrt{3}$. Between $\sqrt{3}$ and r, select any rational number r^*. Then r^* is also an upper bound of X but is less than r. Hence no rational number r can serve as a least upper bound for the set X. ●

Another version of the completeness axiom refers to sequences rather than sets. It is as follows:

Let $a_1, a_2, a_3, \ldots, a_n, \ldots$ *be a sequence of real numbers such that* $a_1 \leq a_2 \leq a_3 \leq \cdots \leq a_n \leq \cdots$. *Then if there is a number B such that $a_n \leq B$ for all n, there is a number L such that a_n approaches L as closely as we please.*

Exercises

1. Can a set of irrational numbers have a rational least upper bound? Explain.

2. (a) Show that
$$\frac{1}{1\cdot 2} + \frac{1}{2\cdot 3} + \frac{1}{3\cdot 4} + \cdots + \frac{1}{n(n+1)} = 1 - \frac{1}{n+1}.$$

   Hint: Use the identity
   $$\frac{1}{i} - \frac{1}{i+1} = \frac{1}{i(i+1)}.$$

(b) Show that 1 is an upper bound of the set
$$X = \left\{ \frac{1}{1\cdot 2}, \frac{1}{1\cdot 2} + \frac{1}{2\cdot 3}, \frac{1}{1\cdot 2} + \frac{1}{2\cdot 3} + \frac{1}{3\cdot 4}, \cdots \right\}.$$

(c) What is the least upper bound of X?

(d) Using (b) or (c), find an upper bound for the set
$$Y = \left\{ \frac{1}{2\cdot 2}, \frac{1}{2\cdot 2} + \frac{1}{3\cdot 3}, \frac{1}{2\cdot 2} + \frac{1}{3\cdot 3} + \frac{1}{4\cdot 4}, \cdots \right\}.$$

It can be shown that the least upper bound of Y is $(\pi^2 - 6)/6$.

3. (a) What is the least upper bound of the set of real numbers x such that $x \le 10$?

   (b) What is the least upper bound of the set of real numbers x such that $x < 10$?

4. (a) What is the least upper bound of the set of all negative numbers?

   (b) Is 15 an upper bound for the set of all negative numbers?

   (c) Is -2 an upper bound for the set of all negative numbers?

APPENDIX B
ANALYTIC GEOMETRY

The following 15 sections in the text cover the analytic geometry needed in calculus.

In most courses these sections will be sufficient. The four sections in this appendix are of a supplemental nature.

1. Analytic geometry and the distance formula. This reviews the way in which coordinates are introduced in the plane and distance is computed in terms of coordinates.

2. Equations of a line. While Sec. 2.2 in the text develops the equation of a line in terms of its slope and y intercept, this section presents other formulas occasionally of use in examples and exercises.

3. Conic sections. This section defines the conic sections and obtains their equations in rectangular coordinates.

4. Conic sections in polar coordinates.

B.1

Analytic geometry and the distance formula

Analytic geometry introduces numbers into geometry by a coordinate system. First introduce a number scale on a line. In this manner each point P in the

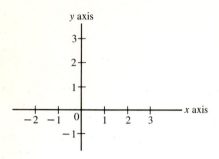

Point P is described by the number x

line is identified with a real number x. For instance, the point P shown in the diagram at the left is identified with $x = 4.5$.

Then select two lines in the plane, perpendicular to each other, and each furnished with a number scale. Place the 0 of each line at the intersection of the two lines and make their number scales equal. The horizontal line is called the x *axis*; the vertical line is called the y *axis*. Their intersection is called the *origin*.

The basis of analytic geometry in the plane is the observation that each point P in the plane can be described, with the aid of the perpendicular lines introduced in the accompanying figure, by two real numbers. Indeed the vertical line through a point P cuts the x axis at a point corresponding to some number x, and the horizontal line through P cuts the y axis at a point corresponding to some number y. We call x the x *coordinate* or *abscissa* of P and y the y *coordinate* or *ordinate* of P. We write $P = (x, y)$; x and y are the *rectangular coordinates* of P.

EXAMPLE 1   Plot the points $(0, 0)$, $(0, 3)$, $(2, 2)$, $(6, 2)$, and $(-5.5, 0)$.

SOLUTION   Observe that a point $(x, 0)$ lies on the x axis and a point $(0, y)$ lies on the y axis. If y is positive, the point (x, y) lies above the x axis. If y is negative, the point (x, y) lies below the x axis. If x is positive, the point (x, y) lies to the right of the y axis. If x is negative, the point (x, y) lies to the left of the y axis.

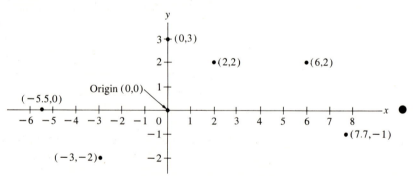

Next consider the question of finding the distance between two points if their coordinates are known. Example 2 presents a specific case, and is followed by the general formula.

EXAMPLE 2   Find the distance between $P_1 = (1, 2)$ and $P_2 = (7, 4)$.

SOLUTION   First plot the two points. Then sketch the right triangle whose legs lie on the horizontal line through P_1 and on the vertical line through P_2. Inspection of the top diagram, next page shows the horizontal side of the triangle has length

$$7 - 1 = 6$$

and the vertical side of the triangle has length

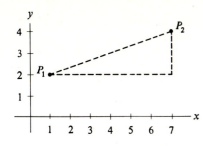

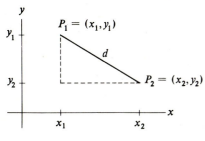

$$4 - 2 = 2.$$

Let d denote the distance between P_1 and P_2. By the pythagorean theorem,

$$d^2 = 2^2 + 6^2$$

$$= 4 + 36$$

$$= 40.$$

Thus
$$d = \sqrt{40} \text{ or } 2\sqrt{10}. \quad \bullet$$

The method used in the example can be applied to any two points in the plane. To find the distance d between $P_1 = (x_1, y_1)$ and $P_2 = (x_2, y_2)$, introduce a right triangle whose hypotenuse has the ends P_1 and P_2 and whose other sides are parallel to the axes, as in the accompanying diagram. The horizontal side of the triangle has length $x_1 - x_2$ or $-(x_1 - x_2)$, depending on whether $x_1 \geq x_2$ or $x_1 \leq x_2$. In either case the square of the length of the horizontal side is

$$(x_1 - x_2)^2.$$

Similarly, the vertical side has length $y_1 - y_2$ or $-(y_1 - y_2)$. The pythagorean theorem implies that

$$d^2 = (x_1 - x_2)^2 + (y_1 - y_2)^2.$$

Thus the distance between the points (x_1, y_1) and (x_2, y_2) is

$$d = \sqrt{(x_1 - x_2)^2 + (y_1 - y_2)^2}.$$

So far, this section has shown that the algebraic analog of a point P is a pair of numbers (x, y), and that the algebraic analog of distance is the formula $\sqrt{(x_1 - x_2)^2 + (y_1 - y_2)^2}$. Next, the algebraic description of a circle will be given. An example demonstrates the idea.

EXAMPLE 3 What algebraic condition must the numbers x and y satisfy if the point $P = (x, y)$ is on the circle of radius 3 and center $(1, 2)$?

SOLUTION The distance between (x, y) and $(1, 2)$ must be 3. The distance formula translates this condition into the equation

$$\sqrt{(x - 1)^2 + (y - 2)^2} = 3.$$

To remove the radical sign and simplify the equation, square both sides of the equation, obtaining

$$(x - 1)^2 + (y - 2)^2 = 9.$$

This is the equation of a circle of radius 3 and center $(1, 2)$. A point (x, y) is on that circle if and only if it satisfies this last equation. $\bullet$

The circle of radius 3 and center $(1, 2)$

Reasoning similar to that in Example 3 shows that the equation of the circle whose center is (h, k) and whose radius is r is

$$(x - h)^2 + (y - k)^2 = r^2.$$

Exercises

1. (a) Plot the points $(5, 8)$ and $(-7, 3)$.
   (b) Using your sketch, determine the distance between them.
   (c) Use the distance formula to find the distance between them.

2. Find the distance between
   (a) $(0, 0)$ and $(-5, 12)$;
   (b) $(1, -3)$ and $(4, 1)$;
   (c) $(5, -11)$ and $(-19, 14)$.

3. (a) Find the equation of the circle of radius 5 whose center is $(1, -1)$.
   (b) Sketch the circle.
   (c) Does the point $(4.5, 2.5)$ lie on this circle?   [Decide on the basis of your sketch and on the basis of (a).]

4. (a) Find the equation of the circle whose center is $(0, 0)$ and whose radius is 3.
   (b) Sketch the circle whose equation is $x^2 + y^2 = 25$.
   (c) Sketch the circle whose equation is $x^2 + y^2 = 5$.

5. (a) Find the equation of the circle whose center is $(3, 0)$ and whose radius is 3.
   (b) Show that the equation in (a) is equivalent to the equation

$$x^2 - 6x + y^2 = 0.$$

6. (a) What is the equation of the circle whose center is $(-3, -4)$ and whose radius is 5?
   (b) Sketch the circle whose equation is

$$(x + 2)^2 + (y - 1)^2 = (\sqrt{7})^2.$$

7. (a) Show that if A, B, and C are fixed numbers, then

$$x^2 + y^2 + Ax + By + C$$

   is equal to

$$\left(x + \frac{A}{2}\right)^2 + \left(y + \frac{B}{2}\right)^2 + C - \frac{A^2}{4} - \frac{B^2}{4}.$$

   (b) Show that if $A^2 + B^2 > 4C$, then the equation

$$x^2 + y^2 + Ax + By + C = 0$$

   describes a circle whose center is at $(-A/2, -B/2)$.
   (c) What is the radius of the circle in (b)?

8. (a) Plot the points $(1, 2)$, $(10, 3)$, and $(9, 6)$.
   (b) On the basis of your sketch, which point do you think is closer to $(1, 2)$, the point $(10, 3)$ or the point $(9, 6)$?
   (c) Use the distance formula to decide which of the points $(10, 3)$ and $(9, 6)$ is closer to $(1, 2)$.

9. Is the triangle whose vertices are $(7, 2)$, $(0, 0)$, and $(2, 7)$ equilateral?

10. Show that the triangle whose vertices are $(0, 0)$, $(5, 3)$, and $(2, 8)$ is isosceles.

B.2

Equations of a line

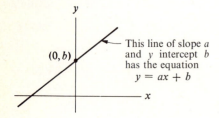

This line of slope a and y intercept b has the equation $y = ax + b$

$(0, b)$

This section, which assumes Sec. 2.3, develops various forms of equations for a line.

Section 2.3 shows that the line whose slope is a and whose y intercept is b has the equation

$$y = ax + b.$$

(Frequently in analytic geometry the letter m is used for slope.   The line of slope m and y intercept b has the equation $y = mx + b$.)

However, the equation of a line must often be determined from other information.   For example, the coordinates of two points P_1 and P_2 may be given, and the equation of the line through them may be needed.   Example 1 shows how to find the equation in this type of problem.

EXAMPLE 1 Find an equation of the line through the points $P_1 = (1, 2)$ and $P_2 = (5, 4)$.

SOLUTION Let $P = (x, y)$ be a point on the line through $(1, 2)$ and $(5, 4)$. Then the slopes of the segments PP_1 and P_2P_1 are equal; that is,

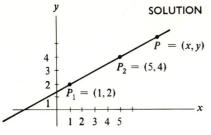

$$\frac{y - 2}{x - 1} = \frac{4 - 2}{5 - 1},$$

or

$$\frac{y - 2}{x - 1} = \frac{2}{4} = \frac{1}{2}.$$

Thus

$$2(y - 2) = 1(x - 1),$$

or

$$2y - 4 = x - 1,$$

or

$$x - 2y + 3 = 0. \quad \bullet$$

Two-Point Formula   The method illustrated by Example 1 provides the following *two-point formula for the equation of a line*. The line through the points $P_1 = (x_1, y_1)$ and $P_2 = (x_2, y_2)$, where x_1 is not equal to x_2, has the equation

$$\frac{y - y_1}{x - x_1} = \frac{y_2 - y_1}{x_2 - x_1}.$$

Intercept Formula   When P_1 and P_2 are on the x and y axes, there is an especially simple equation of the line, known as the *intercept* form: Let a and b be nonzero numbers. The equation of the line through $(a, 0)$ and $(0, b)$ is

$$\frac{x}{a} + \frac{y}{b} = 1.$$

The number a is called the x intercept. The number b is called the y intercept.

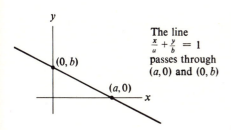

The line $\frac{x}{a} + \frac{y}{b} = 1$ passes through $(a, 0)$ and $(0, b)$

To obtain this formula, apply the two-point formula in the case $P_1 = (a, 0)$ and $P_2 = (0, b)$, obtaining the equation

$$\frac{y - 0}{x - a} = \frac{b - 0}{0 - a},$$

or

$$\frac{y}{x - a} = -\frac{b}{a},$$

or

$$ay = -b(x - a).$$

This equation is easily transformed into

$$bx + ay = ab.$$

Division by the product ab then yields

$$\frac{x}{a} + \frac{y}{b} = 1.$$

EXAMPLE 2 Find an equation of the line through the points $(2, 0)$ and $(0, 3)$.

SOLUTION   The line has x intercept $a = 2$ and y intercept $b = 3$. Its equation is thus

$$\frac{x}{2} + \frac{y}{3} = 1. \bullet$$

An equation of the form
$$Ax + By + C = 0,$$

where both A and B are not 0, always describes a line. The next example illustrates this in a specific case.

EXAMPLE 3   Show that the equation $-5x + 2y - 3 = 0$

describes a line. Find its slope and y intercept.

SOLUTION   Solve the equation $-5x + 2y - 3 = 0$

for y, obtaining $$y = \frac{5x + 3}{2},$$

or $$y = \tfrac{5}{2}x + \tfrac{3}{2}.$$

This last equation is the slope-intercept form. The slope is $\tfrac{5}{2}$ and the y intercept is $\tfrac{3}{2}$. $\bullet$

As long as B is not 0, the same method shows that

$$Ax + By + C = 0$$

is the equation of a line of slope $-A/B$. If $B = 0$, the equation is simply

$$Ax + C = 0$$

or $$x = -\frac{C}{A}.$$

This is the equation of a vertical line.

Exercises

1. Find the slope and y intercept of these lines:
   (a) $y = -2x + 3$; (b) $y = x + 1$; (c) $y = -x + 5$;
   (d) $y = 6x$.
2. Find an equation of the line through
   (a) $(-3, 1)$ and $(5, 3)$; (b) $(4, 0)$ and $(0, -2)$.
3. Where does the line $\dfrac{x}{4} + \dfrac{y}{3} = 1$
   meet (a) the x axis, (b) the y axis?
4. Where does the line $\dfrac{x}{2} - \dfrac{y}{3} = 1$ meet

(a) the x axis, (b) the y axis?
5. (a) Find an equation of the line through $(4, 1)$ and $(6, 0)$.
   (b) Where does the line in (a) meet the y axis?
6. Do the points $(1, 2)$, $(5, 3)$, and $(18, 6)$ lie on a line?
7. (a) Find an equation of the line passing through $(2, 5)$ and having slope $-3/2$.
   (b) Sketch the line.
   (c) Does the point $(4, 2)$ lie on the line?
8. Find the x and y intercepts of the line
   $$2x - 3y - 6 = 0$$

by rewriting the equation in the form

$$\frac{x}{a} + \frac{y}{b} = 1.$$

9. Find three points on the line

$$2x - 3y + 5 = 0.$$

10. Find the slope of the line

$$2x + 4y + 7 = 0.$$

B.3

Conic sections

The intersection of a plane and the surface of a double cone is called a *conic section*. If the plane cuts off a bounded curve, that curve is called an *ellipse*. (In particular a circle is an ellipse.)

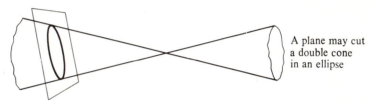

A plane may cut a double cone in an ellipse

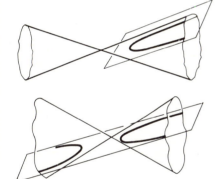

If the plane is parallel to the edge of the double cone, the intersection is called a *parabola*. In the cases of the ellipse and the parabola, the plane generally meets just one of the two cones.

If the plane meets both parts of the cone and is not parallel to an edge, the intersection is called a *hyperbola*. The hyperbola consists of two separate pieces. It can be proved that these two pieces are congruent, and that they are *not* congruent to parabolas.

For the sake of simplicity we shall use a definition of the conic sections that depends only on the geometry of the plane. It is shown in geometry courses that the two approaches yield the same curves.

DEFINITION   *Ellipse.* Let F and F' be points in the plane and let a be a fixed positive number such that $2a$ is greater than the distance between F and F'. A point P in the plane is on the *ellipse* determined by F, F', and $2a$ if and only if the sum of the distances from P to F and from P to F' equals $2a$. Points F and F' are the *foci* of the ellipse.

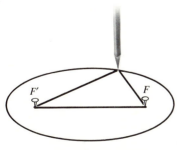

The string has length $2a$, greater than the distance between the tacks

To construct an ellipse, place two tacks in a piece of paper, tie a string of length $2a$ to them, and trace out a curve with a pencil held against the string, keeping the string taut by means of the pencil point. The foci are at the tacks. (Note that when $F = F'$ the ellipse is a circle of radius a.)

The equation of a circle whose center is at $(0, 0)$ and whose radius is a is

$$x^2 + y^2 = a^2,$$

or

$$\frac{x^2}{a^2} + \frac{y^2}{a^2} = 1. \tag{1}$$

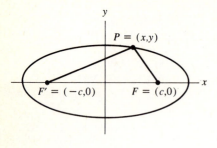

Let us generalize this result by determining the equation of an ellipse.   To make the equation as simple as possible, introduce the x and y axes in such a way that the x axis contains the foci and the origin is midway between them.   Thus $F = (c, 0)$ and $F' = (-c, 0)$, where $c \geq 0$ and $2c < 2a$, and thus $c < a$.

Now translate into algebra this assertion: The sum of the distances from $P = (x, y)$ to $F = (c, 0)$ and from P to $F' = (-c, 0)$ equals $2a$.   By the distance formula, the distance from P to F is

$$\sqrt{(x - c)^2 + (y - 0)^2}.$$

Similarly, the distance from P to F' is

$$\sqrt{(x + c)^2 + (y - 0)^2}.$$

Thus the point (x, y) is on the ellipse if and only if

$$\sqrt{(x - c)^2 + y^2} + \sqrt{(x + c)^2 + y^2} = 2a.$$

A few algebraic steps will transform this equation into an equation without square roots.

First, write the equation as

$$\sqrt{(x + c)^2 + y^2} = 2a - \sqrt{(x - c)^2 + y^2}.$$

Then square both sides, obtaining

$$(x + c)^2 + y^2 = 4a^2 - 4a\sqrt{(x - c)^2 + y^2} + (x - c)^2 + y^2.$$

Expanding yields

$$x^2 + 2cx + c^2 + y^2 = 4a^2 - 4a\sqrt{(x - c)^2 + y^2} + x^2 - 2cx + c^2 + y^2,$$

which a few cancellations reduce to

$$2cx = 4a^2 - 4a\sqrt{(x - c)^2 + y^2} - 2cx$$

or

$$4cx - 4a^2 = -4a\sqrt{(x - c)^2 + y^2}.$$

This equation can be changed to

$$a^2 - cx = a\sqrt{(x - c)^2 + y^2}.$$

Squaring gets rid of the square root:

$$a^4 - 2a^2cx + c^2x^2 = a^2(x^2 - 2cx + c^2 + y^2),$$

or

$$a^4 - 2a^2cx + c^2x^2 = a^2x^2 - 2a^2cx + a^2c^2 + a^2y^2,$$

or

$$a^4 + c^2x^2 = a^2x^2 + a^2c^2 + a^2y^2.$$

This equation can be transformed to

$$(a^2 - c^2)x^2 + a^2y^2 = a^2(a^2 - c^2).$$

Dividing both sides by $a^2(a^2 - c^2)$ results in the equation

$$\frac{x^2}{a^2} + \frac{y^2}{a^2 - c^2} = 1. \tag{2}$$

Since $a^2 - c^2 > 0$, there is a number b such that

$$b^2 = a^2 - c^2 \qquad b > 0,$$

and thus (2) takes the shorter form

$$\frac{x^2}{a^2} + \frac{y^2}{b^2} = 1. \tag{3}$$

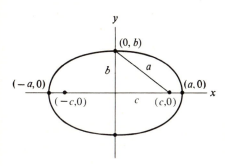

[Note that (3) generalizes (1), the equation for a circle.]

Setting $y = 0$ in (3), we obtain $x = a$ or $-a$; if we set $x = 0$ in (3), we obtain $y = b$ or $-b$. Thus the four "extreme" points of the ellipse have coordinates $(a, 0), (-a, 0), (0, b)$, and $(0, -b)$, as shown in the diagram at the left. Observe that the distance from F or F' to $(0, b)$ is a, half the length of string. The right triangle in the diagram is a reminder of the fact that $b^2 = a^2 - c^2$. Keep in mind that in the above ellipse a is larger than b. The semimajor axis is said to have length a; the semiminor axis has length b.

EXAMPLE 1 Discuss the foci and "length of string" of the ellipse whose equation is

$$\frac{x^2}{25} + \frac{y^2}{9} = 1.$$

SOLUTION Since the larger denominator is with the x^2, the foci lie on the x axis. In this case $a = 5$ and $b = 3$. The length of string is $2a$ or 10. The foci are at a distance

$$\sqrt{a^2 - b^2} = \sqrt{25 - 9} = 4$$

from the origin. ●

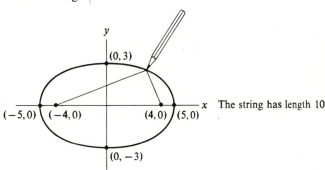

The string has length 10

EXAMPLE 2 Discuss the foci and "length of string" of the ellipse whose equation is

$$\frac{x^2}{9} + \frac{y^2}{25} = 1.$$

SOLUTION This is similar to Example 1. The only difference is that the roles of x and y are interchanged. The foci are at $(0, 4)$ and $(0, -4)$, and the ellipse is longer in the y direction than in the x direction. ●

The definition of the hyperbola is similar to that of the ellipse.

DEFINITION   *Hyperbola.*   Let F and F' be points in the plane and let a be a fixed positive number such that $2a$ is less than the distance between F and F'.   A point P in the plane is on the *hyperbola* determined by F, F', and $2a$ if and only if the difference between the distances from P to F and from P to F' equals $2a$ (or $-2a$). Points F and F' are the *foci* of the hyperbola.

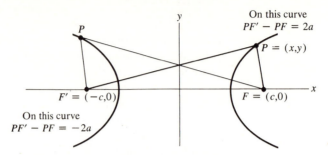

A hyperbola consists of two separate curves.   On one curve $\overline{PF'} - \overline{PF} = 2a$; on the other, $\overline{PF'} - \overline{PF} = -2a$.   (For simplicity, denote the distance between points P and Q as $\overline{PQ}$.)   If the distance $\overline{FF'}$ is $2c$, then $2a < 2c$; hence $a < c$. Again place the axes in such a way that $F = (c, 0)$ and $F' = (-c, 0)$.   Let $P = (x, y)$ be a typical point on the hyperbola.   Then x and y satisfy the equation

$$\sqrt{(x - c)^2 + y^2} - \sqrt{(x + c)^2 + y^2} = \pm 2a. \tag{4}$$

Some algebra similar to that used in simplifying the equation of the ellipse transforms (4) into

$$\frac{x^2}{a^2} + \frac{y^2}{a^2 - c^2} = 1. \tag{5}$$

But now $a^2 - c^2$ is *negative* and can be expressed as $-b^2$ for some number $b > 0$.   Hence the hyperbola has the equation

$$\frac{x^2}{a^2} - \frac{y^2}{b^2} = 1. \tag{6}$$

(If the foci are on the y axis, the equation is

$$\frac{y^2}{a^2} - \frac{x^2}{b^2} = 1.)$$

In both cases, $c^2 = a^2 + b^2$.

EXAMPLE 3   Sketch the hyperbola $\dfrac{x^2}{9} - \dfrac{y^2}{16} = 1.$

SOLUTION   Since the minus sign is with the y^2, the foci are on the x axis.   In this case $a^2 = 9$ and $b^2 = 16$.   Observe that the hyperbola meets the x axis at $(3, 0)$ and $(-3, 0)$.   The hyperbola does not meet the y axis.   The distance from the origin to a focus, c, is determined by the equation

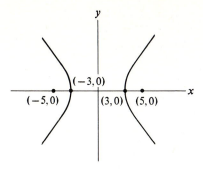

$$c^2 = 9 + 16 = 25;$$

hence
$$c = 5.$$

The hyperbola
$$\frac{x^2}{9} - \frac{y^2}{16} = 1$$

is shown in the accompanying diagram. ●

The definition of a parabola involves the distance to a point and the distance to a line.

DEFINITION   *Parabola.*   Let L be a line in the plane and let F be a point in the plane but not on the line.   A point P in the plane is on the *parabola* determined by F and L if and only if the distance from P to F equals the distance from P to the line L. Point F is the *focus* of the parabola; line L is its *directrix*.

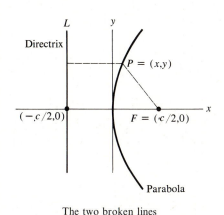

The two broken lines have equal lengths

To obtain an algebraic equation for a parabola, denote the distance from point F to line L by c, and introduce axes in such a way that $F = (c/2, 0)$ and L has the equation $x = -c/2$. The distance from P to F is $\sqrt{(x - c/2)^2 + (y - 0)^2}$.   Now, if $P = (x, y)$ is on the parabola, x is clearly not negative.   The distance from P to the line L is therefore $x + c/2$.   Thus the equation of the parabola is

$$\sqrt{\left(x - \frac{c}{2}\right)^2 + y^2} = x + \frac{c}{2}. \tag{7}$$

Squaring and simplifying reduces (7) to
$$y^2 = 2cx \qquad c > 0, \tag{8}$$

which is the equation of a parabola in "standard position."

If the focus is at $(0, c/2)$ and the directrix is the line $y = -c/2$, the parabola has the equation $x^2 = 2cy$.

EXAMPLE 4   Sketch the parabola $y = x^2$, showing its focus and directrix.

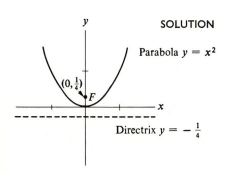

SOLUTION   The equation $y = x^2$ is equivalent to $x^2 = y$, which is of the form

$$x^2 = 2cy,$$

where $c = \frac{1}{2}$.
The focus is on the y axis at $(0, \frac{1}{4})$.   The directrix is the line $y = -\frac{1}{4}$. ●

It can be proved that any ellipse, hyperbola, or parabola, no matter where it lies relative to the axes, has an equation of the form

$$Ax^2 + Bxy + Cy^2 + Dx + Ey + F = 0, \tag{9}$$

where A, B, C, D, E, and F are appropriate constants.   Moreover, when $B^2 - 4AC$ is negative, (9) describes an ellipse; when $B^2 - 4AC$ is positive, (9) describes a hyperbola; and when $B^2 - 4AC = 0$, (9) describes a parabola. Thus the algebraic equivalent of a conic section is a second-degree equation in x and y.

This table summarizes the algebraic equivalents of the geometric ideas examined so far in this appendix.

Geometric Concept	*Algebraic Equivalent*
Point in the plane	An ordered pair of numbers (x, y)
Distance between points P_1 and P_2 in the plane	$\sqrt{(x_2 - x_1)^2 + (y_2 - y_1)^2}$
Circle with center at $(0, 0)$ and radius r	$x^2 + y^2 = r^2$
Line	First-degree equation $Ax + By + C = 0$
Conic section	Second-degree equation $Ax^2 + Bxy + Cy^2 + Dx + Ey + F = 0$
Ellipse in standard position	$\dfrac{x^2}{a^2} + \dfrac{y^2}{b^2} = 1 \quad a \geq b \quad$ (foci on x axis)
Hyperbola in standard position	$\dfrac{x^2}{a^2} - \dfrac{y^2}{b^2} = 1 \quad$ (foci on x axis)
Parabola in standard position	$y^2 = 2cx \quad$ (focus on x axis) $x^2 = 2cy \quad$ (focus on y axis) (c = distance from F to directrix)

Exercises

1. Sketch the ellipse $\dfrac{x^2}{49} + \dfrac{y^2}{25} = 1$ and its foci.

2. Sketch the ellipse $\dfrac{x^2}{4} + \dfrac{y^2}{36} = 1$ and its foci.

3. How would you inscribe an elliptical garden in a rectangle whose dimensions are 8 by 10 feet?

4. Sketch the hyperbola
$$x^2 - y^2 = 1$$
and its foci.

5. Sketch the hyperbola
$$y^2 - x^2 = 1$$
and its foci.

6. Sketch the parabola $y = 6x^2$, its focus, and its directrix.

7. Sketch the parabola $x = -6y^2$, its focus, and its directrix.

8. (a) Using the definition of the hyperbola, show that the hyperbola that has its foci at $(\sqrt{2}, \sqrt{2})$, $(-\sqrt{2}, -\sqrt{2})$, and $2a = 2\sqrt{2}$ has the equation $xy = 1$.
   (b) Graph $xy = 1$ and show the foci.

9. Obtain (5) from (4).

10. Obtain (8) from (7).

11. In the definition of the hyperbola it was assumed that $2a$ is less than the distance between the foci. Show that, if $2a$ were greater than the distance between the foci, the hyperbola would have no points.

12. A plane intersects the surface of a right circular cylinder in a curve. Prove that this curve is an ellipse, as defined in terms of foci and sum of distances. *Hint:* Consider the two spheres inscribed in the cylinder and tangent to the plane, the spheres being on opposite sides of the plane. Let $2a$ denote the distance between the equators of the spheres perpendicular to the axis of the cylinder, and let F and F' be the points at which they touch the plane.

13. (This exercise provides an aid in sketching a hyperbola. Incidentally it implies that a hyperbola is not made up

of two parabolas.) Let a be a fixed positive number. It can be shown that, when the positive number x is large (in comparison with a), then $\sqrt{x^2 - a^2}$ differs from x by only a small amount. This will be of use in (b).
(a) Show that, if $x^2/a^2 - y^2/b^2 = 1$, then y is equal to $\pm (b/a)\sqrt{x^2 - a^2}$.
(b) Show that, for large x the point (x, y) on the hyper-

bola in (a) is near the line $y = bx/a$ or the line $y = -bx/a$.
(c) Use (b) to help graph the hyperbola $x^2 - y^2 = 1$. In view of (b), far from the origin a hyperbola is closely approximated by two straight lines. This is not the case with a parabola.

B.4

Conic sections in polar coordinates

This section depends on Sec. 10.7.

For the study of the conic sections in terms of polar coordinates, it is convenient to use definitions that depend on the ratios of distances, rather than on their sums or differences. (Note that the definition of the parabola involves essentially the ratio of two distances being equal to 1.)

Consider the ellipse whose foci are at $F = (c, 0)$ and $F' = (-c, 0)$ and whose "length of string" is $2a$. In the algebraic treatment of the ellipse in the preceding section, the equation

$$a^2 - cx = a\sqrt{(x - c)^2 + y^2}$$

appears.

Some algebraic manipulations of this equation will show that this ellipse can be defined in terms of the focus F, a line, and a fixed ratio of the distances from P on the ellipse to F and from P to the line.

First observe that this equation asserts that

$$a^2 - cx = a\overline{PF}$$

or, equivalently,

$$\overline{PF} = a - \frac{c}{a}x.$$

Eccentricity of an Ellipse Denote the quotient c/a, which is less than 1, by e. The number e is called the *eccentricity* of the ellipse (when $e = 0$, the ellipse is a circle). Thus

$$\overline{PF} = a - ex = e\left(\frac{a}{e} - x\right),$$

an equation which is meaningful if the ellipse is not a circle. Now, $(a/e) - x$ is the distance from P to the vertical line through $(a/e, 0)$, a line which will be denoted L. Letting $Q = (a/e, y)$, the point on L and on the horizontal line through P, we have

$$\overline{PF} = e\overline{PQ}.$$

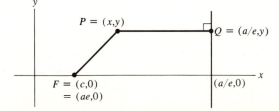

$\overline{PF} = ePQ$

L is the vertical line through $(a/e, 0)$

In other words, *the ratio* $\overline{PF}/\overline{PQ}$ *has a constant value*, less than 1. Thus the ellipse, like the parabola, can be defined in terms of a point F and a line L.

EXAMPLE 1   Find the eccentricity and draw the line L for the ellipse

$$\frac{x^2}{25} + \frac{y^2}{9} = 1.$$

SOLUTION   In this ellipse $a = 5$ and $b = 3$. Thus $c = \sqrt{a^2 - b^2} = \sqrt{25 - 9} = 4$. Consequently

$$e = \frac{c}{a} = \frac{4}{5}.$$

The line L has the equation

$$x = \frac{a}{e}$$

or

$$x = \frac{5}{\frac{4}{5}};$$

hence

$$x = \tfrac{25}{4} = 6.25.$$

Note that for each point P on the ellipse

$$\frac{\overline{PF}}{\overline{PQ}} = \frac{4}{5}. \quad \bullet$$

The hyperbola can be treated in a similar manner. The main difference is that the eccentricity of a hyperbola, again defined as c/a, is greater than 1. With this background, we now describe the approach to the conic sections in terms of the ratios of certain distances.

DEFINITION   *Conic section.* Let L be a line in the plane, and let F be a point in the plane but not on the line. Let e be a positive number. A point P in the plane is on the conic section determined by F, L, and e if and only if

$$\frac{\text{Distance from } P \text{ to } F}{\text{Distance from } P \text{ to } L} = e.$$

When $e = 1$, the conic section is a parabola (this is the definition of the parabola used in the preceding section). When $e < 1$, it is an ellipse. When $e > 1$, it is a hyperbola. The point F is called a *focus*; the line L is called the *directrix*.

To obtain the simplest description of the conic sections in polar coordinates, we place the pole at the focus F. Let the polar axis make an angle B with a line perpendicular to the directrix. The following diagram shows a typical point $P = (r, \theta)$ on the conic section, as well as the point Q, on the directrix, nearest P. Let the distance from F to the directrix be p.

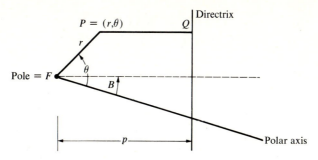

Then $\overline{PF}/\overline{PQ} = e$. But $\overline{PF} = r$ and $\overline{PQ} = p - r \cos(\theta - B)$. Thus

$$\frac{r}{p - r \cos(\theta - B)} = e. \tag{1}$$

Solving (1) for r yields *the equation of a conic section in polar coordinates*,

$$r = \frac{ep}{1 + e \cos(\theta - B)}. \tag{2}$$

EXAMPLE 2 Show that the graph of the equation

$$r = \frac{8}{5 + 6 \cos \theta}$$

is a conic section.

SOLUTION This can be put in the form (2) by dividing numerator and denominator by 5:

$$r = \frac{\frac{8}{5}}{1 + \frac{6}{5} \cos \theta} = \frac{\left(\frac{6}{5}\right)\left(\frac{8}{6}\right)}{1 + \frac{6}{5} \cos \theta}.$$

Hence the graph is a conic section for which $p = \frac{8}{6}$ and $e = \frac{6}{5}$. It is a hyperbola, since $e > 1$. ●

Exercises

1. (a) Sketch the four points on the graph of $r = 10/(3 + 2 \cos \theta)$ corresponding to $\theta = 0, \pi/2, \pi, 3\pi/2$.
   (b) Using (2), show that the curve in (a) is an ellipse.
2. Obtain (2) from (1).
3. Find the eccentricity of these conics:
   (a) $r = 5/(3 + 4 \cos \theta)$;
   (b) $r = 5/(4 + 3 \cos \theta)$;
   (c) $r = 5/(3 + 3 \cos \theta)$;
   (d) $r = 5/(3 - 4 \cos \theta)$.
4. (a) Show that $r = 8/[1 - (1/2) \cos \theta]$ is the equation of an ellipse. *Hint*: Set $B = \pi$ in (2).
   (b) Graph the ellipse and its foci.
   (c) Find a, where $2a$ is the fixed sum of the distances

from points on the ellipse to the foci.
5. In rectangular coordinates the focus of a certain parabola is at $(-1, 0)$ and its directrix is the line $x = 1$.
   (a) Show that its equation is $y^2 = -4x$.
   (b) Find the equation of the parabola relative to a polar coordinate system whose pole is at F and whose polar axis contains the positive x axis.
   (c) Find the equation of the parabola relative to a polar coordinate system whose pole is at the origin of the rectangular coordinate system and whose polar axis coincides with the positive x axis.
6. Assume that a rectangular coordinate system is placed in such a way that an ellipse has the equation $x^2/a^2 +$

$y^2/b^2 = 1$.   Place a polar coordinate system so that the polar axis coincides with the positive x axis (the pole thus being at the center of the ellipse, *not at a focus*). Show that the polar equation of the ellipse is (the relatively complicated)

$$r^2 = \frac{a^2 b^2}{b^2 \cos^2 \theta + a^2 \sin^2 \theta}.$$

Hint: Recall that $x = r \cos \theta$ and $y = r \sin \theta$.

7. (a) Show that $r = 3 \cos \theta + 4 \sin \theta$ is the equation of a circle.

(b) Show that $r = 1/(3 \cos \theta + 4 \sin \theta)$ is the equation of a line.

APPENDIX C
THEORY OF LIMITS

This appendix provides rigorous definitions of limits and proves several assertions about limits used repeatedly in the text. In particular it is shown that any polynomial is continuous. The theory of limits is developed more extensively in advanced calculus.

C.I

Precise definitions of limits

Three types of limits were introduced in the text: the limit of a sequence, denoted

$$\lim_{n \to \infty} a_n;$$

the limit of a function of a real variable, denoted

$$\lim_{x \to a} f(x);$$

and the limit of approximating sums,

$$\lim_{\text{mesh} \to 0} \sum_{i=1}^{n} f(X_i)(x_i - x_{i-1}),$$

used to define the definite integral.

Let us recall the definition given for each of these three limits and state the corresponding rigorous definition. The rigorous definitions are needed in advanced calculus for the study of limit problems more difficult than those included in this text.

Section 14.1 presents the following definition of the limit of a sequence $a_1, a_2, a_3, \ldots, a_n, \ldots$: If as n gets large a_n approaches a number L, then L is called the limit of the sequence.

Examine this definition carefully. It contains the phrase "n gets large." What does this mean? What is meant by a "large" number? What is meant by "gets large"? A similar criticism may be leveled at "approaches." Yet the definition does convey the idea of a limit. In order to describe this essential idea more precisely, rephrase the intuitive definition as follows: If, when n is chosen large enough, a_n is as close as we please to the number L, then L is called the limit of the sequence.

This definition avoids the word "approaches" but introduces "as close as we please," a phrase that is still somewhat vague. This definition suggests the following rigorous definition. (The Greek letter ε, appearing in the definition, is *epsilon*.)

DEFINITION *The limit of a sequence.* The number L is the *limit of the sequence* $\{a_n\}$, if for each positive number ε (however small) there is a positive integer N, depending on ε, such that a_n differs from L by less than ε, whenever n is larger than N.

Read Slowly! Study the   Read the definition aloud several times. Compare it with the two informal
Definition Carefully   definitions which precede it. Note that such terms as "large," "approaches," and "as close as we please" do not appear in the final definition. Of course, the phrase "however small" now appears, but it can be deleted without any loss of meaning. It is included parenthetically only for emphasis.

Think of ε as the challenge and N as the reply. Generally it is to be expected that the smaller that ε is, the larger N will have to be. The following example illustrates how this definition is applied.

EXAMPLE 1 Use the rigorous definition of the limit of a sequence to show that

$$\lim_{n \to \infty} \frac{1}{n} = 0.$$

SOLUTION In this case $L = 0$. The difference between a_n and L is

$$\left| \frac{1}{n} - 0 \right| = \frac{1}{n}.$$

The inequality
$$\frac{1}{n} < \varepsilon$$

is equivalent to
$$1 < n\varepsilon,$$

or
$$n > \frac{1}{\varepsilon}.$$

So let N be an integer greater than $1/\varepsilon$. ($1/\varepsilon$ might not be an integer.) Then if n is larger than N, it follows that

$$\left| \frac{1}{n} - 0 \right| = \frac{1}{n} < \varepsilon,$$

since
$$\frac{1}{n} < \frac{1}{N} < \varepsilon. \;\bullet$$

Next consider the definition of the limit of a function of a real variable given in Sec. 4.3: Let f be a function and a some fixed number. If, as x approaches a through numbers in the domain of f, $f(x)$ approaches a specific number L, then L is called the limit of $f(x)$ as x approaches a.

This definition is subject to the criticism that a number is locked in place on the number line and cannot "approach" another number.   The rigorous definition avoids this by using the "challenge and reply" concept introduced in the rigorous definition for the limit of a sequence.   The challenge is still ε, but the reply will be δ (the Greek letter *delta*).

DEFINITION   *Limit of a function of a real variable.*   The number L is *the limit of* $f(x)$ *at a* if, for any positive number ε (however small), there is a positive number δ, depending on ε, such that $f(x)$ differs from L by less than ε when x in the domain of f differs from a by less than δ (except, perhaps, when $x = a$).

Presumably, the smaller ε is, the smaller δ will have to be.   (Read the definition aloud several times.)   If f is thought of as the projection from a slide to a screen, then the definition reads informally as follows: The number L on the screen is the limit of $f(x)$ at a on the slide, if the following condition is met. For each small open interval around L of the form

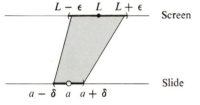

$$(L - \varepsilon, L + \varepsilon)$$

there must be an open interval on the slide,

$$(a - \delta, a + \delta)$$

which, with the possible exception of a, is projected entirely into the interval

$$(L - \varepsilon, L + \varepsilon).$$

Example 2 applies this definition in a simple case.

EXAMPLE 2   Prove that

$$\lim_{x \to 3} 2x = 6.$$

SOLUTION   Suppose a positive number ε is given.   A positive number δ must be found such that when

$$|x - 3| \text{ is less than } \delta, \quad \text{and} \quad x \neq 3,$$

it follows that

$$|2x - 6| \text{ is less than } \varepsilon.$$

Now, the inequality

$$|2x - 6| < \varepsilon$$

is equivalent to

$$2|x - 3| < \varepsilon,$$

or

$$|x - 3| < \frac{\varepsilon}{2}.$$

Thus choose $\delta = \varepsilon/2$.   Then, whenever

$$|x - 3| \text{ is less than } \delta,$$

it follows that

$$|2x - 6| \text{ is less than } \varepsilon.$$

This means that

$$\lim_{x \to 3} 2x = 6. \quad \bullet$$

The informal definition of the definite integral $\int_a^b f(x)\,dx$ is to be found in Sec. 7.4. The rigorous definition, modeled after the rigorous definitions of $\lim_{n \to \infty} a_n$ and $\lim_{x \to a} f(x)$, runs as follows.

DEFINITION   *The definite integral of f over* $[a, b]$.   The number L is the definite integral of f over $[a, b]$ if the following condition is met. For any positive number ε(however small) there must exist a positive number δ, depending on ε, such that any sum

$$\sum_{i=1}^{n} f(X_i)(x_i - x_{i-1})$$

formed with any partition of $[a, b]$ of mesh less than δ (no matter how X_i is chosen in $[x_{i-1}, x_i]$) differs from L by less than ε.

EXAMPLE 3   In Sec. 7.2 it was shown that any approximating sum for the function x^2 over the interval $[0, 3]$ differs from 9 by less than nine times the mesh of the partition. Use this fact to prove that $\int_0^3 x^2\,dx$ exists.

SOLUTION   Let ε be a positive number.   Let $\delta = \varepsilon/9$.   Any approximating sum whose mesh is $< \delta$ differs from 9 by less than $9\delta = 9\varepsilon/9 = \varepsilon$. Thus the condition of the ε, δ-definition is satisfied. ●

Exercises

1. Prove that $\lim\limits_{n \to \infty} \dfrac{1}{2n} = 0.$

2. Prove that $\lim\limits_{n \to \infty} \dfrac{1}{n^2} = 0.$

3. Prove that $\lim\limits_{n \to \infty} \dfrac{n}{n+1} = 1.$

4. Prove that $\lim\limits_{n \to \infty} (-1)^n$ does not exist.   (Show that for $\varepsilon = 1/2$ no N exists.)

5. Prove that $\lim\limits_{x \to 2} 3x + 1 = 7.$

6. Prove that $\lim\limits_{x \to 0} x^2 = 0.$

7. Prove that $\lim\limits_{x \to 1} x^2 = 1.$

8. Prove that $\lim\limits_{n \to \infty} 2^{-n} = 0.$

9. Show that *neither* of these proposed definitions for the limit of a sequence is acceptable.   (In each case give an example of a sequence that would have a limit according to the definition, but in fact does not have a limit.)

(a) If, for every positive integer N there is a positive number ε such that, when n is larger than N, a_n differs from L by less than ε, then L is the limit of the sequence $\{a_n\}$.

(b) If there is a positive integer N and a positive number ε such that a_n differs from L by less than ε whenever $n > N$, then L is the limit of the sequence $\{a_n\}$.

10. The following is a precise definition of the assertion, "$f(x)$ approaches infinity as x gets large."   Let $f(x)$ be defined for all x greater than or equal to some number a.   If for each number b there exists a number N such that $f(x) \geq b$ for all $x \geq N$, we say, "$f(x)$ approaches infinity as x gets large," and write

$$\lim_{x \to \infty} f(x) = \infty.$$

(a) Prove that $\lim\limits_{x \to \infty} x^2 = \infty.$

(b) Is $\lim\limits_{x \to \infty} x^2 \sin x = \infty$?

C.2

Proofs of some theorems about limits

With the aid of the definitions given in the preceding section, it is possible to prove such assertions as the following: If $\lim_{x \to a} f(x) = A$ and $\lim_{x \to a} g(x) = B$, then it follows that $\lim_{x \to a} [f(x) \cdot g(x)] = A \cdot B$, $\lim_{x \to a} [f(x) + g(x)] = A + B$, and (if $B \neq 0$) $\lim_{x \to a} f(x)/g(x) = A/B$. Their proofs are presented in Theorems 1 and 3 and in Exercises 4 to 7. For convenience assume that the functions f and g are defined for all x.

THEOREM I   If $\lim_{x \to a} f(x) = A$ and $\lim_{x \to a} g(x) = B$, then $\lim_{x \to a} [f(x) + g(x)] = A + B$.

PROOF   It must be shown that given any $\varepsilon > 0$, no matter how small, there exists a number $\delta > 0$ depending on ε, such that

$$|[f(x) + g(x)] - (A + B)| < \varepsilon, \tag{1}$$

whenever $|x - a| < \delta$ and $x \neq a$.

Rewrite $[f(x) + g(x)] - (A + B)$ as $[f(x) - A] + [g(x) - B]$, which is a sum of two quantities known to be small when x is near a. Since the absolute value of the sum of two numbers is not larger than the sum of their absolute values,

$$|[f(x) - A] + [g(x) - B]| \le |f(x) - A| + |g(x) - B|. \tag{2}$$

Since $\lim_{x \to a} f(x) = A$, there is a positive number δ_1 such that

$$|f(x) - A| < \frac{\varepsilon}{2},$$

when $|x - a| < \delta_1$, for $x \neq a$. (Why we pick $\varepsilon/2$ rather than ε will be clear in a moment.) Similarly there is a positive number δ_2 such that

$$|g(x) - B| < \frac{\varepsilon}{2}$$

when $|x - a| < \delta_2$, for $x \neq a$.

Now let δ be the smaller of δ_1 and δ_2. For any x (not equal to a) such that $|x - a| < \delta$ we have both

$$|x - a| < \delta_1 \quad \text{and} \quad |x - a| < \delta_2$$

and therefore, simultaneously,

$$|f(x) - A| < \frac{\varepsilon}{2} \quad \text{and} \quad |g(x) - B| < \frac{\varepsilon}{2}. \tag{3}$$

Combining (2) and (3), we conclude that when $|x - a| < \delta$, for $x \neq a$,

$$|[f(x) + g(x)] - (A + B)| < \frac{\varepsilon}{2} + \frac{\varepsilon}{2} = \varepsilon.$$

Thus for each $\varepsilon > 0$ there exists a suitable $\delta > 0$, depending, of course, on ε, f, and g. This ends the proof. ●

Theorem 1 is the basis of the proof of the next theorem.

THEOREM 2 The sum of two functions that are continuous at a is itself continuous at a.

PROOF Let f and g be continuous at a. Let h be their sum; that is, $h(x) = f(x) + g(x)$. We wish to show that h is continuous at a.

In view of the definition of continuity given in Sec. 4.7, it must be shown that $h(a)$ is defined and that $\lim_{x \to a} h(x) = h(a)$.

Since f and g are defined at a, so is h, and $h(a) = f(a) + g(a)$. All that remains is to show that $\lim_{x \to a} [f(x) + g(x)] = f(a) + g(a)$. But Theorem 1 assures us that $\lim_{x \to a} [f(x) + g(x)] = \lim_{x \to a} f(x) + \lim_{x \to a} g(x)$. Since f and g are continuous at a, $\lim_{x \to a} f(x) = f(a)$ and $\lim_{x \to a} g(x) = g(a)$, This concludes the proof. ●

THEOREM 3 If $\lim_{x \to a} f(x) = A$ and $\lim_{x \to a} g(x) = B$, then $\lim_{x \to a} f(x)g(x) = AB$.

Plan of proof: We know that $|f(x) - A|$ and $|g(x) - B|$ are small when x is near a. We wish to conclude that $|f(x)g(x) - AB|$ is small when x is near a. The algebraic identity

$$f(x)g(x) - AB = f(x)[g(x) - B] + B[f(x) - A] \qquad (4)$$

will be of use. From (4) and properties of the absolute value, it follows that

$$|f(x)g(x) - AB| \leq |f(x)||g(x) - B| + |B||f(x) - A|. \qquad (5)$$

Now $|B|$ is fixed, and $|f(x) - A|$ and $|g(x) - B|$ are small where x is near a. The real problem is to control $|f(x)|$. Watch carefully the way in which $|f(x)|$ is treated in the proof.

PROOF Consider the case for $B \neq 0$. Let $\varepsilon > 0$ be given. We wish to show that there is a number $\delta > 0$ such that $|f(x)g(x) - AB| < \varepsilon$ when $|x - a| < \delta$, for $x \neq a$. Observe that

$$|f(x)g(x) - AB| = |f(x)[g(x) - B] + B[f(x) - A]| \qquad (6)$$
$$\leq |f(x)||g(x) - B| + |B||f(x) - A|.$$

Since $\lim_{x \to a} f(x) = A$, there is a number $\delta > 0$ such that

$$|f(x) - A| < \frac{\varepsilon}{2|B|},$$

when $|x - a| < \delta_1$, for $x \neq a$. Thus the second summand in (6) is less than

$$\frac{|B|\varepsilon}{2|B|} = \frac{\varepsilon}{2}.$$

For x such that $|x - a| < \delta_1$, $|f(x)|$ does not become arbitrarily large, since $|f(x) - A| < \varepsilon/(2|B|)$. Indeed for such x,

$$|f(x)| < |A| + \frac{\varepsilon}{2|B|}.$$

Calling $|A| + \varepsilon/(2|B|)$ "C" we have $|f(x)| < C$ when $|x - a| < \delta_1$, for $x \neq a$. (This controls the size of $|f(x)|$.)

Since $\lim_{x \to a} g(x) = B$, there is a $\delta_2 > 0$ such that $|g(x) - B| < \varepsilon/(2C)$ when $|x - a| < \delta_2$, for $x \neq a$.

Now let δ be the smaller of δ_1 and δ_2. When $|x - a| < \delta$, both $|x - a| < \delta_1$ and $|x - a| < \delta_2$ hold; hence $|f(x) - A| < \varepsilon/(2|B|)$, $|f(x)| < C$, and $|g(x) - B| < \varepsilon/(2C)$. Inspection of (6) then shows that for x such that $|x - a| < \delta$, and for $x \neq a$,

$$|f(x)g(x) - AB| < C\frac{\varepsilon}{2C} + |B|\frac{\varepsilon}{2|B|} = \frac{\varepsilon}{2} + \frac{\varepsilon}{2} = \varepsilon.$$

The proof is completed. (The case $B = 0$ is easier and is left to the reader.) ●

THEOREM 4 The product of two functions that are continuous at a is itself continuous at a.

The proof is similar to that of Theorem 2, but depends on Theorem 3 instead of Theorem 1.

THEOREM 5 The function f, such that $f(x) = x$, is continuous everywhere. So are the functions x^2, x^3, x^4, ….

PROOF Since $f(a) = a$, it must be shown that $|f(x) - a|$ is small whenever $|x - a|$ is sufficiently small. More precisely, for $\varepsilon > 0$ we wish to exhibit $\delta > 0$ such that $|f(x) - a| < \varepsilon$ whenever $|x - a| < \delta$. But $f(x) = x$; hence $|f(x) - a|$ is simply $|x - a|$. Let $\delta = \varepsilon$. Thus when $|x - a| < \delta$, it follows that $|f(x) - a| < \varepsilon$. This shows that the function x is continuous.

Since the function x^2 is the product of the function x and the function x, Theorem 4 implies that x^2 is continuous. Similarly, x^3 is continuous, for $x^3 = x^2 x$. Similarly, x^4 and x^5 can be shown to be continuous. Mathematical induction establishes the continuity of x^n for all positive integers n. ●

THEOREM 6 Any constant function is continuous everywhere.

PROOF Let $f(x) = c$ for all x. For any $\varepsilon > 0$, choose $\delta = 1{,}776$ a perfectly fine positive number. Now $|f(x) - f(a)| = |c - c| = 0 < \varepsilon$ for any x, and hence for x such that $|x - a| < 1{,}776$. Thus f is continuous at any number a. ●

THEOREM 7 Any polynomial is continuous everywhere.

PROOF We illustrate the idea of the proof by showing that $6x^2 - 5x + 1$ is continuous everywhere.

By Theorem 6, the constant functions $f(x) = 6$, $f(x) = -5$, and $f(x) = 1$ are continuous everywhere.   By Theorem 5, the functions x and x^2 are continuous everywhere.   By Theorem 4, the functions $6x^2$ and $(-5)x$ are continuous.   By Theorem 2, the function $[6x^2 + (-5)x]$ is continuous.   Again, by Theorem 2, the function $[6x^2 + (-5)x] + 1$ is continuous.   Thus $6x^2 - 5x + 1$ is continuous.   The same argument applies to any polynomial. ●

Exercises

1. Prove that the function $2x$ is continuous at $x = 3$, using only the definition of continuity.
2. Using the definition of continuity, prove that $3x + 8$ is continuous at $x = 2$.
3. Using the definition of continuity, prove that x^2 is continuous at $x = 3$.
4. Prove that if f is continuous at a, and g is continuous at $f(a)$, then the composite function $g \circ f$ is continuous at a.
5. Prove that $1/x$ is continuous at any $a \neq 0$.
6. (a) Using Exercises 4 and 5, prove that $1/f(x)$ is continuous at any a such that $f(a) \neq 0$.
   (b) Using (a) and Theorem 4, deduce that the quotient of two continuous functions, $g(x)/f(x)$, is continuous at any a such that $f(a) \neq 0$.
   (c) Prove that $(x^3 + 1)/(x^2 + 1)$ is continuous everywhere.
7. Prove Theorem 3 for the case $B = 0$.

8. (a) Find $\delta > 0$ such that $|\sqrt{x} - \sqrt{9}| < 0.04$ when $|x - 9| < \delta$.
   (b) Prove that $\sqrt{x}$ is continuous at any $a \geq 0$.   *Hint:* Make use of the identity $\sqrt{b} - \sqrt{a} = (b - a)/(\sqrt{b} + \sqrt{a})$.
9. Find $\delta > 0$ such that
   (a) $|x^2 - 4| < 1$ when $|x - 2| < \delta$;
   (b) $|x^2 - 4| < 0.1$ when $|x - 2| < \delta$;
   (c) $|x^2 - 4| < 0.01$ when $|x - 2| < \delta$;
   (d) $|x^2 - 4| < t$ when $|x - 2| < \delta$;
   (e) $|x^2 - 4| < \varepsilon$ when $|x - 2| < \delta$.   *Hint:* Note that $|x^2 - 4| = |x - 2||x + 2|$, and be sure to control $|x + 2|$.
10. Prove that the difference of two continuous functions is continuous.

APPENDIX D
PARTIAL FRACTIONS

This appendix examines an algebraic result used in Sec. 9.7 for finding antiderivatives of rational functions, but is not required by the presentation in that section. It is intended only to provide some motivation for the partial-fractions representation.

D.1

Partial-fraction representation of rational numbers

We begin with a few definitions. An integer D *divides* an integer A if there is an integer Q such that $A = DQ$. The integer D is a *divisor* of A. An integer A, *greater than* 1, is *prime* if, whenever it is expressed as the product BC of integers B and C, at least one of B and C has absolute value 1. Thus 2, 3, 5, 7, 11, and 13 are primes, but 1, 4, 6, 8, 9, 10, and 12 are not primes.

A little arithmetic will show that

$$\frac{19}{24} = \frac{0}{2} + \frac{1}{2^2} + \frac{-1}{2^3} + \frac{2}{3}. \tag{1}$$

Note in (1) that the denominators are powers of prime numbers, that these prime powers divide the denominator in the left member of (1), and that the numerators in the right member of (1) have small absolute value. This illustrates the following theorem concerning rational numbers, that is, quotients of integers. (We shall not prove Theorem 1 or any of the theorems in this appendix.)

THEOREM I *Partial fractions for rational numbers.* Let a and b be integers such that $|a| < |b|$, and b is not 1 or -1. Then the rational number a/b can be represented as a sum of rational numbers of the form N/p^e, where p is prime, p^e divides b, and $|N| < p$.

The representation described in Theorem 1 is a *partial-fraction representation* of a/b; the summands N/p^e are the *partial fractions* in the representation.

If $a/b = 19/24$, then the prime powers that divide b are 2, 2^2, 2^3, and 3. Equation (1) illustrates Theorem 1. Note that in each partial fraction on the right side of (1), the absolute value of the numerator is less than the prime appearing in its denominator.

EXAMPLE 1   Discuss the representation of $\frac{37}{10}$.

SOLUTION   Since 37 is not less than 10, Theorem 1 makes no statement about $a/b = \frac{37}{10}$. In this case we may write $\frac{37}{10}$ as $3 + \frac{7}{10}$ and apply Theorem 1 to $\frac{7}{10}$. According to Theorem 1, there are integers N_1 and N_2 such that

$$\frac{7}{10} = \frac{N_1}{2} + \frac{N_2}{5} \quad \text{and} \quad |N_1| < 2, \quad |N_2| < 5.$$

A little experimentation shows that $N_1 = 1$ and $N_2 = 1$ do the job:

$$\frac{37}{10} = 3 + \frac{1}{2} + \frac{1}{5}. \quad \bullet$$

Exercises

1. Verify that these representations satisfy the conditions of Theorem 1:
   (a) $\frac{18}{35} = \frac{4}{5} + \frac{-2}{7} = \frac{5}{7} + \frac{-1}{5}$;
   (b) $\frac{7}{27} = \frac{2}{9} + \frac{1}{27}$.

2. Find partial-fraction representations for (a) $\frac{1}{6}$, (b) $\frac{4}{15}$,
   (c) $\frac{1}{36}$, (d) $\frac{1}{15}$, (e) $\frac{7}{9}$, (f) $\frac{19}{27}$.

3. Prove that, if a is an integer and $0 \le a < 16$, then $a/16 = N_1/2 + N_2/4 + N_3/8 + N_4/16$, where each N_i is 0 or 1. Do not use Theorem 1. *Hint:* Consider the base-2 representation of a.

D.2

Partial-fraction representation of rational functions

An expression of the form $a_0 x^n + a_1 x^{n-1} + \cdots + a_n$, where the a_i's are real numbers, is a *polynomial*. If $a_0 \ne 0$, then n is the *degree* of the polynomial. A *rational function* is the quotient of two polynomials. An *irreducible* (or *prime*) *polynomial* is a polynomial of degree at least 1 that is not the product of polynomials of lower degree. The polynomial D *divides* the polynomial A if there is a polynomial Q such that $A = DQ$. The polynomial D is a *divisor* of A.

The polynomial $1x^4 + 1$, which is usually written $x^4 + 1$, is a polynomial of degree four. It is *not* prime, since

$$x^4 + 1 = (x^2 + \sqrt{2}x + 1)(x^2 - \sqrt{2}x + 1)$$

as can be checked by multiplication.

Any first-degree polynomial $ax + b$ is prime. A second-degree polynomial $ax^2 + bx + c$ is prime if and only if $b^2 - 4ac < 0$. It can be proved that *no polynomial whose degree is larger than 2 is a prime.*

For instance, the polynomial $6x + 8$ is prime, since its degree is 1. The polynomial $x^2 + \sqrt{2}x + 1$ is prime, since $(\sqrt{2})^2 - 4(1)(1) = 2 - 4 < 0$. The polynomial $x^3 + 1$ is *not* prime, since its degree is greater than 2; indeed we have $x^3 + 1 = (x + 1)(x^2 - x + 1)$.

The following theorem is frequently helpful in finding first-degree divisors of a polynomial.

THEOREM 1 (*Factor theorem*).   Let c be a real number.   The polynomial $x - c$ divides the polynomial f if and only if $f(c) = 0$.

For instance, let $f(x) = x^3 + 1$.   Since $f(-1) = (-1)^3 + 1 = 0$, the factor theorem tells us that $x + 1$ divides $x^3 + 1$.   Similarly, since 2 is a root of the equation $4x^3 - 6x^2 - x - 6 = 0$, we know that $x - 2$ divides $4x^3 - 6x^2 - x - 6$.

Just as every integer greater than 1 is a prime number or the product of prime numbers, every polynomial of degree at least 1 is either a prime polynomial or the product of prime polynomials.

EXAMPLE 2   Express $x^4 - 2x^3 + 2x^2 - 2x + 1$ as the product of prime polynomials.

SOLUTION   Since 1 is a root of the equation $x^4 - 2x^3 + 2x^2 - 2x + 1 = 0$, we know that $x - 1$ is a divisor of $x^4 - 2x^3 + 2x^2 - 2x + 1$.   To determine the quotient, we carry out the division:

$$
\begin{array}{r}
x^3 - x^2 + x - 1 \\
x - 1 \overline{\smash{)}\ x^4 - 2x^3 + 2x^2 - 2x + 1} \\
\underline{x^4 - x^3} \\
- x^3 + 2x^2 \\
\underline{- x^3 + x^2} \\
x^2 - 2x \\
\underline{x^2 - x} \\
- x + 1 \\
\underline{- x + 1} \\
0
\end{array}
$$

Hence $x^4 - 2x^3 + 2x^2 - 2x + 1 = (x - 1)(x^3 - x^2 + x - 1)$.   But 1 is a root of $x^3 - x^2 + x - 1 = 0$; hence $x - 1$ is a divisor of $x^3 - x^2 + x - 1$.   Long division shows that $x^3 - x^2 + x - 1 = (x - 1)(x^2 + 1)$.   Since $x^2 + 1$ is prime, the representation of $x^4 - 2x^3 + 2x^2 - 2x + 1$ as the product of prime polynomials is

$$x^4 - 2x^3 + 2x^2 - 2x + 1 = (x - 1)(x - 1)(x^2 + 1). \quad \bullet$$

The prime polynomials are analogs of prime numbers; the next theorem shows that degree is the analog of absolute value.

THEOREM 2   (*Partial fractions for rational functions*).   Let $A(x)$ and $B(x)$ be polynomials such that the degree of $A(x)$ is less than the degree of $B(x)$.   Then the rational function $A(x)/B(x)$ can be represented as a sum of rational functions of the form $N(x)/[P(x)]^e$, where $P(x)$ is a prime polynomial, $[P(x)]^e$ divides $B(x)$, and the degree of $N(x)$ is less than the degree of $P(x)$ [or possibly all the coefficients of $N(x)$ are 0].

The representation in Theorem 3 is always essentially unique for given $P(x)$'s.   (The representations for rational numbers in Theorem 1 is not necessarily unique; for instance, $\frac{13}{35} = \frac{4}{5} + \left(-\frac{3}{7}\right) = -\frac{1}{5} + \frac{4}{7}$.)   For instance,

$$\frac{2}{1 - x^2} = \frac{1}{1 - x} + \frac{1}{1 + x}$$

$$\frac{x^2 + 4x - 10}{(x - 1)^3} = \frac{1}{x - 1} + \frac{6}{(x - 1)^2} - \frac{5}{(x - 1)^3}$$

$$\frac{1}{x^4 + 1} = \frac{(\sqrt{2}/4)x + \frac{1}{2}}{x^2 + \sqrt{2}x + 1} + \frac{(-\sqrt{2}/4)x + \frac{1}{2}}{x^2 - \sqrt{2}x + 1}.$$

Since any prime polynomial has degree 1 or 2, Theorem 3 may be expressed as follows.

THEOREM 3   (*Rephrased: Partial fractions for rational functions*).   Let $A(x)$ and $B(x)$ be polynomials such that the degree of $A(x)$ is less than the degree of $B(x)$.   Then the rational function $A(x)/B(x)$ can be represented as the sum of rational functions of the form

$$\frac{k_i}{(ax + b)^i} \qquad \text{and} \qquad \frac{c_j x + d_j}{(ax^2 + bx + c)^j},$$

where $ax^2 + bx + c$ is prime, $(ax^2 + bx + c)^j$ and $(ax + b)^i$ divide $B(x)$, and k_i, c_j, and d_j are real numbers.

It must be kept in mind that if $(ax + b)^n$ divides $B(x)$, then so do $(ax + b)^{n-1}$, $(ax + b)^{n-2}$, ..., and $ax + b$.

EXAMPLE 3   Express $\dfrac{4x - 17}{x^2 - x - 6}$ as the sum of partial fractions.

SOLUTION   According to the rephrased Theorem 3, there are real numbers a and b such that

$$\frac{4x - 17}{x^2 - x - 6} = \frac{a}{x - 3} + \frac{b}{x + 2}. \tag{1}$$

To find a and b, we multiply both sides of (2) by $x^2 - x - 6$, obtaining

$$4x - 17 = a(x + 2) + b(x - 3)$$

or

$$4x - 17 = (a + b)x + (2a - 3b). \tag{2}$$

Comparing the constant term and coefficients of x on both sides of (2), we have

$$4 = a + b \qquad \text{and} \qquad -17 = 2a - 3b,$$

simultaneous equations for a and b.   Solving these equations for a and b, we find that $a = -1$ and $b = 5$.   Thus

$$\frac{4x - 17}{x^2 - x - 6} = \frac{-1}{x - 3} + \frac{5}{x + 2}. \ \bullet$$

EXAMPLE 4   Represent

$$\frac{6x^2 + 2x + 1}{x^3 - 3x + 2}$$

as the sum of partial fractions.

SOLUTION   Express $f(x) = x^3 - 3x + 2$ as a product of prime polynomials. Since $f(1) = 0$, we know that $x - 1$ is a divisor of $x^3 - 3x + 2$. Long division shows that $x^3 - 3x + 2 = (x - 1)(x^2 + x - 2)$. But $x^2 + x - 2$ is *not* prime, for the quadratic formula shows that the equation $x^2 + x - 2 = 0$ has the roots 1 and -2. Thus $x^2 + x - 2 = (x - 1)(x + 2)$, and $x^3 - 3x + 2 = (x - 1)^2(x + 2)$. According to Theorem 3, then,

$$\frac{6x^2 + 2x + 1}{x^3 - 3x + 2} = \frac{a}{x - 1} + \frac{b}{(x - 1)^2} + \frac{c}{x + 2}$$

for suitable real numbers a, b, and c. Multiplying both sides of this equation by $x^3 - 3x + 2$, we obtain

$$6x^2 + 2x + 1 = a(x - 1)(x + 2) + b(x + 2) + c(x - 1)^2. \tag{3}$$

Rather than expand the right side of (3) and compare coefficients, as in Example 3, let us use a shortcut.

Since (3) holds for all x, it holds for $x = 1$, $x = -2$, and $x = 0$. (We choose 1 because it is a root of $x - 1 = 0$, and -2 because it is a root of $x + 2 = 0$; we choose 0 because it is easy to work with. We need three values to obtain three equations for the three unknowns a, b, c.) We find a, b, c as follows. Substitution of 1, -2, and 0 into (4) yields

$$9 = 3b \qquad \text{from } x = 1$$
$$21 = 9c \qquad \text{from } x = -2$$
$$1 = -2a + 2b + c \qquad \text{from } x = 0.$$

We quickly obtain $b = 3$ and $c = \frac{7}{3}$. Last, find a by means of the third of the above equations:

$$1 = -2a + 2 \cdot 3 + \tfrac{7}{3},$$

and $a = \frac{11}{3}$. Thus

$$\frac{6x^2 + 2x + 1}{x^3 - 3x + 2} = \frac{\frac{11}{3}}{x - 1} + \frac{3}{(x - 1)^2} + \frac{\frac{7}{3}}{x + 2}. \; \bullet$$

EXAMPLE 5   Represent

$$\frac{(3x^2 - 3x - 4)}{(x^2 - 1)}$$

as the sum of partial fractions.

SOLUTION   Just as Theorem 1 says nothing about $\frac{37}{10}$ (see Example 1), Theorem 3 says nothing about the partial-fraction representation of $(3x^2 - 3x - 4)/(x^2 - 1)$, since the degree of the numerator is *not* less than the degree of the denominator. Just as in Example 1, carry out a division first:

$$
\begin{array}{r}
3 \\
x^2 - 1 \overline{\smash{\big)}\ 3x^2 - 3x - 4} \\
\underline{3x^2 - 3} \\
-3x - 1
\end{array}
$$

which shows that

$$\frac{3x^2 - 3x - 4}{x^2 - 1} = 3 + \frac{-3x - 1}{x^2 - 1}.$$

Theorem 3 does hold for $(-3x - 1)/(x^2 - 1)$, and it is not difficult to show that

$$\frac{-3x - 1}{x^2 - 1} = \frac{-2}{x - 1} + \frac{-1}{x + 1}.$$

Hence

$$\frac{3x^2 - 3x - 4}{x^2 - 1} = 3 + \frac{-2}{x - 1} + \frac{-1}{x + 1}. \quad \bullet$$

Exercises

1. Factor into prime polynomials and check by multiplying:
   (a) $2x^3 + 7x^2 + 4x - 4$ *Hint:* Use Theorem 2;
   (b) $x^2 - 3x + 10$;
   (c) $2x^2 + x - 7$.

2. Factor into prime polynomials and check:
   (a) $x^3 - 8$,
   (b) $x^3 - 4$.

3. (a) Verify that $x^4 + x^2 + 1 = (x^2 + x + 1)(x^2 - x + 1)$.
   (b) Express $(6x + 1)/(x^4 + x^2 + 1)$ in partial fractions.

4. How many constants would we have to determine in the partial-fraction decomposition of
   (a) $\dfrac{A(x)}{(x^2 - 3x + 2)^{10}}$;
   (b) $\dfrac{A(x)}{(x^2 + 2x + 3)^{10}}$, where, in each case, the degree of $A(x)$ is less than 20?
   Explain.

5. Solve for a and b in Example 3 by substituting two convenient values for x in the equation
   $$4x - 17 = a(x + 2) + b(x - 3).$$

6. Find the partial-fraction decomposition of each of the following rational functions, and check your answers:
   (a) $\dfrac{2x^2 + 1}{(x - 2)^3}$,
   (b) $\dfrac{x}{(x + 1)^2}$,
   (c) $\dfrac{x^3 - 5x^2 + 9x + 1}{(x^2 + 1)(x - 3)^2}$.

7. Express $1/(4x^3 - 6x^2 - x - 6)$ as a sum of partial fractions. *Hint:* Use Theorem 2.

8. Express as a sum of partial fractions:
   (a) $x^2/(x^2 + 3x + 1)^2$;
   (b) $x/(x - 1)^2(x + 2)^2$.

9. Express as the sum of a polynomial and partial fractions
   (a) $(x^5 + 1)/(x^3 + 1)$;
   (b) $x^4/(x^2 + 1)^2$.

10. Let $ax^2 + bx + c$ be a polynomial of degree 2.
    (a) Prove that, if $b^2 - 4ac \geq 0$, then the polynomial *is not* prime.
    (b) Prove that, if $b^2 - 4ac < 0$, then the polynomial *is* prime.

11. (a) Using a graph, suggest why any polynomial of odd degree has at least one real root.
    (b) With the aid of (a), show that any polynomial whose degree is odd and greater than two *is not* prime.

12. Prove Theorem 3 for the special case $B(x) = x^{1,000}$.

13. Prove this frequently encountered special case of Theorem 3.
    THEOREM: If c and d are two unequal numbers, then there exist numbers a and b such that
    $$\frac{1}{(x + c)(x + d)} = \frac{a}{x + c} + \frac{b}{x + d}.$$

APPENDIX E
TABLES

E.I

Exponential function

x	e^x Value	e^{-x} Value	x	e^x Value	e^{-x} Value	x	e^x Value	e^{-x} Value	x	e^x Value	e^{-x} Value
0.00	1.0000	1.00000	0.35	1.4191	.70469	0.70	2.0138	.49659	1.60	4.9530	.20190
0.01	1.0101	0.99005	0.36	1.4333	.69768	0.71	2.0340	.49164	1.70	5.4739	.18268
0.02	1.0202	.98020	0.37	1.4477	.69073	0.72	2.0544	.48675	1.80	6.0496	.16530
0.03	1.0305	.97045	0.38	1.4623	.68386	0.73	2.0751	.48191	1.90	6.6859	.14957
0.04	1.0408	.96079	0.39	1.4770	.67706	0.74	2.0959	.47711	2.00	7.3891	.13534
0.05	1.0513	.95123	0.40	1.4918	.67032	0.75	2.1170	.47237	3.00	20.086	.04979
0.06	1.0618	.94176	0.41	1.5068	.66365	0.76	2.1383	.46767	4.00	54.598	.01832
0.07	1.0725	.93239	0.42	1.5220	.65705	0.77	2.1598	.46301	5.00	148.41	.00674
0.08	1.0833	.92312	0.43	1.5373	.65051	0.78	2.1815	.45841	6.00	403.43	.00248
0.09	1.0942	.91393	0.44	1.5527	.64404	0.79	2.2034	.45384	7.00	1096.6	.00091
0.10	1.1052	.90484	0.45	1.5683	.63763	0.80	2.2255	.44933	8.00	2981.0	.00034
0.11	1.1163	.89583	0.46	1.5841	.63128	0.81	2.2479	.44486	9.00	8103.1	.00012
0.12	1.1275	.88692	0.47	1.6000	.62500	0.82	2.2705	.44043	10.00	22026.5	.00005
0.13	1.1388	.87809	0.48	1.6161	.61878	0.83	2.2933	.43605			
0.14	1.1503	.86936	0.49	1.6323	.61263	0.84	2.3164	.43171			
0.15	1.1618	.86071	0.50	1.6487	.60653	0.85	2.3396	.42741			
0.16	1.1735	.85214	0.51	1.6653	.60050	0.86	2.3632	.42316			
0.17	1.1853	.84366	0.52	1.6820	.59452	0.87	2.3869	.41895			
0.18	1.1972	.83527	0.53	1.6989	.58860	0.88	2.4109	.41478			
0.19	1.2092	.82696	0.54	1.7160	.58275	0.89	2.4351	.41066			
0.20	1.2214	.81873	0.55	1.7333	.57695	0.90	2.4596	.40657			
0.21	1.2337	.81058	0.56	1.7507	.57121	0.91	2.4843	.40252			
0.22	1.2461	.80252	0.57	1.7683	.56553	0.92	2.5093	.39852			
0.23	1.2586	.79453	0.58	1.7860	.55990	0.93	2.5345	.39455			
0.24	1.2712	.78663	0.59	1.8040	.55433	0.94	2.5600	.39063			
0.25	1.2840	.77880	0.60	1.8221	.54881	0.95	2.5857	.38674			
0.26	1.2969	.77105	0.61	1.8404	.54335	0.96	2.6117	.38298			
0.27	1.3100	.76338	0.62	1.8589	.53794	0.97	2.6379	.37908			
0.28	1.3231	.75578	0.63	1.8776	.53259	0.98	2.6645	.37531			
0.29	1.3364	.74826	0.64	1.8965	.52729	0.99	2.6912	.37158			
0.30	1.3499	.74082	0.65	1.9155	.52205	1.00	2.7183	.36788			
0.31	1.3634	.73345	0.66	1.9348	.51685	1.20	3.3201	.30119			
0.32	1.3771	.72615	0.67	1.9542	.51171	1.30	3.6693	.27253			
0.33	1.3910	.71892	0.68	1.9739	.50662	1.40	4.0552	.24660			
0.34	1.4049	.71177	0.69	1.9937	.50158	1.50	4.4817	.22313			

Excerpted from Burington's *Handbook of Mathematical Tables and Formulas*, McGraw-Hill, New York, 1973.

E.2

Natural logarithms (base e)

x	$\ln x$	x	$\ln x$	x	$\ln x$	x	$\ln x$
0.01	−4.60517	0.50	−0.69315	1.00	0.00000	1.5	0.4 0547
.02	−3.91202	.51	.67334	1.01	.00995	1.6	7000
.03	.50656	.52	.65393	1.02	.01980	1.7	0.5 3063
.04	.21888	.53	.63488	1.03	.02956	1.8	8779
		.54	.61619	1.04	.03922	1.9	0.6 4185
.05	−2.99573	.55	.59784	1.05	.04879	2.0	9315
.06	.81341	.56	.57982	1.06	.05827	2.1	0.7 4194
.07	.65926	.57	.56212	1.07	.06766	2.2	8846
.08	.52573	.58	.54473	1.08	.07696	2.3	0.8 3291
.09	.40795	.59	.52763	1.09	.08618	2.4	7547
0.10	−2.30259	0.60	−0.51083	1.10	.09531	2.5	0.9 1629
.11	.20727	.61	.49430	1.11	.10436	2.6	5551
.12	.12026	.62	.47804	1.12	.11333	2.7	9325
.13	.04022	.63	.46204	1.13	.12222	2.8	1.0 2962
.14	−1.96611	.64	.44629	1.14	.13103	2.9	6471
.15	.89712	.65	.43078	1.15	.13976	3.0	9861
.16	.83258	.66	.41552	1.16	.14842		
.17	.77196	.67	.40048	1.17	.15700	4.0	1.3863
.18	.71480	.68	.38566	1.18	.16551		
.19	.66073	.69	.37106	1.19	.17395	5.0	1.6044
0.20	−1.60944	0.70	−0.35667	1.20	.18232	6.0	1.7918
.21	.56065	.71	.34249	1.21	.19062		
.22	.51413	.72	.32850	1.22	.19885	7.0	1.9459
.23	.46968	.73	.31471	1.23	.20701		
.24	.42712	.74	.30111	1.24	.21511	8.0	2.0794
.25	.38629	.75	.28768	1.25	.22314	9.0	2.1972
.26	.34707	.76	.27444	1.26	.23111		
.27	.30933	.77	.26136	1.27	.23902	10.0	2.3026
.28	.27297	.78	.24846	1.28	.24686		
.29	.23787	.79	.23572	1.29	.25464	20.0	2.9957
0.30	−1.20397	0.80	−0.22314	1.30	.26236	30.0	3.4012
.31	.17118	.81	.21072	1.31	.27003		
.32	.13943	.82	.19845	1.32	.27763	40.0	3.6889
.33	.10866	.83	.18633	1.33	.28518		
.34	.07881	.84	.17435	1.34	.29267	50.0	3.9120
.35	−1.04982	.85	−0.16252	1.35	.30010	60.0	4.0943
.36	.02165	.86	.15032	1.36	.30748		
.37	−0.99425	.87	.13926	1.37	.31481	70.0	4.2485
.38	.96758	.88	.12783	1.38	.32208		
.39	.94161	.89	.11653	1.39	.32930	80.0	4.3820
0.40	−0.91629	0.90	−0.10536	1.40	.33647	90.0	4.4998
.41	.89160	.91	.09431	1.41	.34359		
.42	.86750	.92	.08338	1.42	.35066	100.0	4.6052
.43	.84397	.93	.07257	1.43	.35767		
.44	.82098	.94	.06188	1.44	.36464		
.45	.79851	.95	.05129	1.45	.37156		
.46	.77653	.96	.04082	1.46	.37844		
.47	.75502	.97	.03046	1.47	.38526		
.48	.73397	.98	.02020	1.48	.39204		
.49	.71335	.99	.01005	1.49	.39878		

Warning: the integer to the left of the decimal point is not shown in each line.

E.3

Common logarithms (base 10)[1]

	0	1	2	3	4	5	6	7	8	9	N	0	1	2	3	4	5	6	7	8	9
10	0000	0043	0086	0128	0170	0212	0253	0294	0334	0374	55	7404	7412	7419	7427	7435	7443	7451	7459	7466	7474
11	0414	0453	0492	0531	0569	0607	0645	0682	0719	0755	56	7482	7490	7497	7505	7513	7520	7528	7536	7543	7551
12	0792	0828	0864	0899	0934	0969	1004	1038	1072	1106	57	7559	7566	7574	7582	7589	7597	7604	7612	7619	7627
13	1139	1173	1206	1239	1271	1303	1335	1367	1399	1430	58	7634	7642	7649	7657	7664	7672	7679	7686	7694	7701
14	1461	1492	1523	1553	1584	1614	1644	1673	1703	1732	59	7709	7716	7723	7731	7738	7745	7752	7760	7767	7774
15	1761	1790	1818	1847	1875	1903	1931	1959	1987	2014	60	7782	7789	7796	7803	7810	7818	7825	7832	7839	7846
16	2041	2068	2095	2122	2148	2175	2201	2227	2253	2279	61	7853	7860	7868	7875	7882	7889	7896	7903	7910	7917
17	2304	2330	2355	2380	2405	2430	2455	2480	2504	2529	62	7924	7931	7938	7945	7952	7959	7966	7973	7980	7987
18	2553	2577	2601	2625	2648	2672	2695	2718	2742	2765	63	7993	8000	8007	8014	8021	8028	8035	8041	8048	8055
19	2788	2810	2833	2856	2878	2900	2923	2945	2967	2989	64	8062	8069	8075	8082	8089	8096	8102	8109	8116	8122
20	3010	3032	3054	3075	3096	3118	3139	3160	3181	3201	65	8129	8136	8142	8149	8156	8162	8169	8176	8182	8189
21	3222	3243	3263	3284	3304	3324	3345	3365	3385	3404	66	8195	8202	8209	8215	8222	8228	8235	8241	8248	8254
22	3424	3444	3464	3483	3502	3522	3541	3560	3579	3598	67	8261	8267	8274	8280	8287	8293	8299	8306	8312	8319
23	3617	3636	3655	3674	3692	3711	3729	3747	3766	3784	68	8325	8331	8338	8344	8351	8357	8363	8370	8376	8382
24	3802	3820	3838	3856	3874	3892	3909	3927	3945	3962	69	8388	8395	8401	8407	8414	8420	8426	8432	8439	8445
25	3979	3997	4014	4031	4048	4065	4082	4099	4116	4133	70	8451	8457	8463	8470	8476	8482	8488	8494	8500	8506
26	4150	4166	4183	4200	4216	4232	4249	4265	4281	4298	71	8513	8519	8525	8531	8537	8543	8549	8555	8561	8567
27	4314	4330	4346	4362	4378	4393	4409	4425	4440	4456	72	8573	8579	8585	8591	8597	8603	8609	8615	8621	8627
28	4472	4487	4502	4518	4533	4548	4564	4579	4594	4609	73	8633	8639	8645	8651	8657	8663	8669	8675	8681	8686
29	4624	4639	4654	4669	4683	4698	4713	4728	4742	4757	74	8692	8698	8704	8710	8716	8722	8727	8733	8739	8745
30	4771	4786	4800	4814	4829	4843	4857	4871	4886	4900	75	8751	8756	8762	8768	8774	8779	8785	8791	8797	8802
31	4914	4928	4942	4955	4969	4983	4997	5011	5024	5038	76	8808	8814	8820	8825	8831	8837	8842	8848	8854	8859
32	5051	5065	5079	5092	5105	5119	5132	5145	5159	5172	77	8865	8871	8876	8882	8887	8893	8899	8904	8910	8915
33	5185	5198	5211	5224	5237	5250	5263	5276	5289	5302	78	8921	8927	8932	8938	8943	8949	8954	8960	8965	8971
34	5315	5328	5340	5353	5366	5378	5391	5403	5416	5428	79	8976	8982	8987	8993	8998	9004	9009	9015	9020	9025
35	5441	5453	5465	5478	5490	5502	5514	5527	5539	5551	80	9031	9036	9042	9047	9053	9058	9063	9069	9074	9079
36	5563	5575	5587	5599	5611	5623	5635	5647	5658	5670	81	9085	9090	9096	9101	9106	9112	9117	9122	9128	9133
37	5682	5694	5705	5717	5729	5740	5752	5763	5775	5786	82	9138	9143	9149	9154	9159	9165	9170	9175	9180	9186
38	5798	5809	5821	5832	5843	5855	5866	5877	5888	5899	83	9191	9196	9201	9206	9212	9217	9222	9227	9232	9238
39	5911	5922	5933	5944	5955	5966	5977	5988	5999	6010	84	9243	9248	9253	9258	9263	9269	9274	9279	9284	9289
40	6021	6031	6042	6053	6064	6075	6085	6096	6107	6117	85	9294	9299	9304	9309	9315	9320	9325	9330	9335	9340
41	6128	6138	6149	6160	6170	6180	6191	6201	6212	6222	86	9345	9350	9355	9360	9365	9370	9375	9380	9385	9390
42	6232	6243	6253	6263	6274	6284	6294	6304	6314	6325	87	9395	9400	9405	9410	9415	9420	9425	9430	9435	9440
43	6335	6345	6355	6365	6375	6385	6395	6405	6415	6425	88	9445	9450	9455	9460	9465	9469	9474	9479	9484	9489
44	6435	6444	6454	6464	6474	6484	6493	6503	6513	6522	89	9494	9499	9504	9509	9513	9518	9523	9528	9533	9538
45	6532	6542	6551	6561	6571	6580	6590	6599	6609	6618	90	9542	9547	9552	9557	9562	9566	9571	9576	9581	9586
46	6628	6637	6646	6656	6665	6675	6684	6693	6702	6712	91	9590	9595	9600	9605	9609	9614	9619	9624	9628	9633
47	6721	6730	6739	6749	6758	6767	6776	6785	6794	6803	92	9638	9643	9647	9652	9657	9661	9666	9671	9675	9680
48	6812	6821	6830	6839	6848	6857	6866	6875	6884	6893	93	9685	9689	9694	9699	9703	9708	9713	9717	9722	9727
49	6902	6911	6920	6928	6937	6946	6955	6964	6972	6981	94	9731	9736	9741	9745	9750	9754	9759	9763	9768	9773
50	6990	6998	7007	7016	7024	7033	7042	7050	7059	7067	95	9777	9782	9786	9791	9795	9800	9805	9809	9814	9818
51	7076	7084	7093	7101	7110	7118	7126	7135	7143	7152	96	9823	9827	9832	9836	9841	9845	9850	9854	9859	9863
52	7160	7168	7177	7185	7193	7202	7210	7218	7226	7235	97	9868	9872	9877	9881	9886	9890	9894	9899	9903	9908
53	7243	7251	7259	7267	7275	7284	7292	7300	7308	7316	98	9912	9917	9921	9926	9930	9934	9939	9943	9948	9952
54	7324	7332	7340	7348	7356	7364	7372	7380	7388	7396	99	9956	9961	9965	9969	9974	9978	9983	9987	9991	9996
N	0	1	2	3	4	5	6	7	8	9	N	0	1	2	3	4	5	6	7	8	9

[1] This table lists the logarithm to four decimal places to the base 10 of numbers in the range from 1 to 9.99. For instance, $\log 3.57 = 0.5527$. (First find 35 under N.) Logarithms of numbers outside this range can be calculated as shown in these two examples:

$$\log_{10} 357 = \log_{10} (10^2)(3.57) = \log_{10} 10^2 + \log_{10} 3.57 = 2 + 0.5527 = 2.5527$$

$$\log_{10} 0.357 = \log_{10} (10^{-1})(3.57) = \log_{10} 10^{-1} + \log_{10} 3.57 = (-1) + 0.5527 = -0.4473$$

E.4

Squares, cubes, square roots, and cube roots

n	n^2	n^3	$\sqrt{n}$	$\sqrt[3]{n}$	n	n^2	n^3	$\sqrt{n}$	$\sqrt[3]{n}$
1	1	1	1.000 000	1.000 000	50	2 500	125 000	7.071 068	3.684 031
2	4	8	1.414 214	1.259 921	51	2 601	132 651	7.141 428	3.708 430
3	9	27	1.732 051	1.442 250	52	2 704	140 608	7.211 103	3.732 511
4	16	64	2.000 000	1.587 401	53	2 809	148 877	7.280 110	3.756 286
					54	2 916	157 464	7.348 469	3.779 763
5	25	125	2.236 068	1.709 976	55	3 025	166 375	7.416 198	3.802 952
6	36	216	2.449 490	1.817 121	56	3 136	175 616	7.483 315	3.825 862
7	49	343	2.645 751	1.912 931	57	3 249	185 193	7.549 834	3.848 501
8	64	512	2.828 427	2.000 000	58	3 364	195 112	7.615 773	3.870 877
9	81	729	3.000 000	2.080 048	59	3 481	205 379	7.681 146	3.892 996
10	100	1 000	3.162 278	2.154 435	60	3 600	216 000	7.745 967	3.914 868
11	121	1 331	3.316 625	2.223 980	61	3 721	226 981	7.810 250	3.936 497
12	144	1 728	3.464 102	2.289 428	62	3 844	238 328	7.874 008	3.957 892
13	169	2 197	3.605 551	2.351 335	63	3 969	250 047	7.937 254	3.979 057
14	196	2 744	3.741 657	2.410 142	64	4 096	262 144	8.000 000	4.000 000
15	225	3 375	3.872 983	2.466 212	65	4 225	274 625	8.062 258	4.020 726
16	256	4 096	4.000 000	2.519 842	66	4 356	287 496	8.124 038	4.041 240
17	289	4 913	4.123 106	2.571 282	67	4 489	300 763	8.185 353	4.061 548
18	324	5 832	4.242 641	2.620 741	68	4 624	314 432	8.246 211	4.081 655
19	361	6 859	4.358 899	2.668 402	69	4 761	328 509	8.306 624	4.101 566
20	400	8 000	4.472 136	2.714 418	70	4 900	343 000	8.366 600	4.121 285
21	441	9 261	4.582 576	2.758 924	71	5 041	357 911	8.426 150	4.140 818
22	484	10 648	4.690 416	2.802 039	72	5 184	373 248	8.485 281	4.160 168
23	529	12 167,	4.795 832	2.843 867	73	5 329	389 017	8.544 004	4.179 339
24	576	13 824	4.898 979	2.884 499	74	5 476	405 224	8.602 325	4.198 336
25	625	15 625	5.000 000	2.924 018	75	5 625	421 875	8.660 254	4.217 163
26	676	17 576	5.099 020	2.962 496	76	5 776	438 976	8.717 798	4.235 824
27	729	19 683	5.196 152	3.000 000	77	5 929	456 533	8.774 964	4.254 321
28	784	21 952	5.291 503	3.036 589	78	6 084	474 552	8.831 761	4.272 659
29	841	24 389	5.385 165	3.072 317	79	6 241	493 039	8.888 194	4.290 840
30	900	27 000	5.477 226	3.107 233	80	6 400	512 000	8.944 272	4.308 869
31	961	29 791	5.567 764	3.141 381	81	6 561	531 441	9.000 000	4.326 749
32	1 024	32 768	5.656 854	3.174 802	82	6 724	551 368	9.055 385	4.344 481
33	1 089	35 937	5.744 563	3.207 534	83	6 889	571 787	9.110 434	4.362 071
34	1 156	39 304	5.830 952	3.239 612	84	7 056	592 704	9.165 151	4.379 519
35	1 225	42 875	5.916 080	3.271 066	85	7 225	614 125	9.219 544	4.396 830
36	1 296	46 656	6.000 000	3.301 927	86	7 396	636 056	9.273 618	4.414 005
37	1 369	50 653	6.082 763	3.332 222	87	7 569	658 503	9.327 379	4.431 048
38	1 444	54 872	6.164 414	3.361 975	88	7 744	681 472	9.380 832	4.447 960
39	1 521	59 319	6.244 998	3.391 211	89	7 921	704 969	9.433 981	4.464 745
40	1 600	64 000	6.324 555	3.419 952	90	8 100	729 000	9.486 833	4.481 405
41	1 681	68 921	6.403 124	3.448 217	91	8 281	753 571	9.539 392	4.497 941
42	1 764	74 088	6.480 741	3.476 027	92	8 464	778 688	9.591 663	4.514 357
43	1 849	79 507	6.557 439	3.503 398	93	8 649	804 357	9.643 651	4.530 655
44	1 936	85 184	6.633 250	3.530 348	94	8 836	830 584	9.695 360	4.546 836
45	2 025	91 125	6.708 204	3.556 893	95	9 025	857 375	9.746 974	4.562 903
46	2 116	97 336	6.782 330	3.583 048	96	9 216	884 736	9.797 959	4.578 857
47	2 209	103 823	6.855 655	3.608 826	97	9 409	912 673	9.848 858	4.594 701
48	2 304	110 592	6.928 203	3.634 241	98	9 604	941 192	9.899 495	4.610 436
49	2 401	117 649	7.000 000	3.659 306	99	9 801	970 299	9.949 874	4.626 065

E.5

Reciprocals of numbers

n	$1,000/n$	n	$1,000/n$
1	1,000.000	50	20.000 00
2	500.000 0	51	19.607 84
3	333.333 3	52	19.230 77
4	250.000 0	53	18.867 92
		54	18.518 52
5	200.000 0		
6	166.666 7	55	18.181 82
7	142.857 1	56	17.857 14
8	125.000 0	57	17.543 86
9	111.111 1	58	17.241 38
		59	16.949 15
10	100.000 0		
11	90.909 09	60	16.666 67
12	83.333 33	61	16.393 44
13	76.923 08	62	16.129 03
14	71.428 57	63	15.873 02
		64	15.625 00
15	66.666 67		
16	62.500 00	65	15.384 62
17	58.823 53	66	15.151 52
18	55.555 56	67	14.925 37
19	52.631 58	68	14.705 88
		69	14.492 75
20	50.000 00		
21	47.619 05	70	14.285 71
22	45.454 55	71	14.084 51
23	43.478 26	72	13.888 89
24	41.666 67	73	13.698 63
		74	13.513 51
25	40.000 00		
26	38.461 54	75	13.333 33
27	37.037 04	76	13.157 89
28	35.714 29	77	12.987 01
29	34.482 76	78	12.820 51
		79	12.658 23
30	33.333 33		
31	32.258 06	80	12.500 00
32	31.250 00	81	12.345 68
33	30.303 03	82	12.195 12
34	29.411 76	83	12.048 19
		84	11.904 76
35	28.571 43		
36	27.777 78	85	11.764 71
37	27.027 03	86	11.627 91
38	26.315 79	87	11.494 25
39	25.641 03	88	11.363 64
		89	11.235 96
40	25.000 00		
41	24.390 24	90	11.111 11
42	23.809 52	91	10.989 01
43	23.255 81	92	10.869 57
44	22.727 27	93	10.752 69
		94	10.638 30
45	22.222 22		
46	21.739 13	95	10.526 32
47	21.276 60	96	10.416 67
48	20.833 33	97	10.309 28
49	20.408 16	98	10.204 08
		99	10.101 01

E.6

Trigonometric functions (degrees)

Deg.	Rad	Sin	Cos	Tan	Cot	Sec	Csc		
0	0.0000	0.0000	1.0000	0.0000	......	1.0000	......	1.5708	90
1	0.0175	0.0175	0.9998	0.0175	57.290	1.0002	57.299	1.5533	89
2	0.0349	0.0349	0.9994	0.0349	28.636	1.0006	28.654	1.5359	88
3	0.0524	0.0523	0.9986	0.0524	19.081	1.0014	19.107	1.5184	87
4	0.0698	0.0698	0.9976	0.0699	14.301	1.0024	14.336	1.5010	86
5	0.0873	0.0872	0.9962	0.0875	11.430	1.0038	11.474	1.4835	85
6	0.1047	0.1045	0.9945	0.1051	9.5144	1.0055	9.5668	1.4661	84
7	0.1222	0.1219	0.9925	0.1228	8.1443	1.0075	8.2055	1.4486	83
8	0.1396	0.1392	0.9903	0.1405	7.1154	1.0098	7.1853	1.4312	82
9	0.1571	0.1564	0.9877	0.1584	6.3138	1.0125	6.3925	1.4137	81
10	0.1745	0.1736	0.9848	0.1763	5.6713	1.0154	5.7588	1.3963	80
11	0.1920	0.1908	0.9816	0.1944	5.1446	1.0187	5.2408	1.3788	79
12	0.2094	0.2079	0.9781	0.2126	4.7046	1.0223	4.8097	1.3614	78
13	0.2269	0.2250	0.9744	0.2309	4.3315	1.0263	4.4454	1.3439	77
14	0.2443	0.2419	0.9703	0.2493	4.0108	1.0306	4.1336	1.3265	76
15	0.2618	0.2588	0.9659	0.2679	3.7321	1.0353	3.8637	1.3090	75
16	0.2793	0.2756	0.9613	0.2867	3.4874	1.0403	3.6280	1.2915	74
17	0.2967	0.2924	0.9563	0.3057	3.2709	1.0457	3.4203	1.2741	73
18	0.3142	0.3030	0.9511	0.3249	3.0777	1.0515	3.2361	1.2566	72
19	0.3316	0.3256	0.9455	0.3443	2.9042	1.0576	3.0716	1.2392	71
20	0.3491	0.3420	0.9397	0.3640	2.7475	1.0642	2.9238	1.2217	70
21	0.3665	0.3584	0.9336	0.3839	2.6051	1.0711	2.7904	1.2043	69
22	0.3840	0.3746	0.9272	0.4040	2.4751	1.0785	2.6695	1.1868	68
23	0.4014	0.3907	0.9205	0.4245	2.3559	1.0864	2.5593	1.1694	67
24	0.4189	0.4067	0.9135	0.4452	2.2460	1.0946	2.4586	1.1519	66
25	0.4363	0.4226	0.9063	0.4663	2.1445	1.1034	2.3662	1.1345	65
26	0.4538	0.4384	0.8988	0.4877	2.0503	1.1126	2.2812	1.1170	64
27	0.4712	0.4540	0.8910	0.5095	1.9626	1.1223	2.2027	1.0996	63
28	0.4887	0.4695	0.8829	0.5317	1.8807	1.1326	2.1301	1.0821	62
29	0.5061	0.4848	0.8746	0.5543	1.8040	1.1434	2.0627	1.0647	61
30	0.5236	0.5000	0.8660	0.5774	1.7321	1.1547	2.0000	1.0472	60
31	0.5411	0.5150	0.8572	0.6009	1.6643	1.1666	1.9416	1.0297	59
32	0.5585	0.5299	0.8480	0.6249	1.6003	1.1792	1.8871	1.0123	58
33	0.5760	0.5446	0.8387	0.6494	1.5399	1.1924	1.8361	0.9948	57
34	0.5934	0.5592	0.8290	0.6745	1.4826	1.2062	1.7883	0.9774	56
35	0.6109	0.5736	0.8192	0.7002	1.4281	1.2208	1.7434	0.9599	55
36	0.6283	0.5878	0.8090	0.7265	1.3764	1.2361	1.7013	0.9425	54
37	0.6458	0.6018	0.7986	0.7536	1.3270	1.2521	1.6616	0.9250	53
38	0.6632	0.6157	0.7880	0.7813	1.2799	1.2690	1.6243	0.9076	52
39	0.6807	0.6293	0.7771	0.8098	1.2349	1.2868	1.5890	0.8901	51
40	0.6981	0.6428	0.7660	0.8391	1.1918	1.3054	1.5557	0.8727	50
41	0.7156	0.6561	0.7547	0.8693	1.1504	1.3250	1.5243	0.8552	49
42	0.7330	0.6691	0.7431	0.9004	1.1106	1.3456	1.4945	0.8378	48
43	0.7505	0.6820	0.7314	0.9325	1.0724	1.3673	1.4663	0.8302	47
44	0.7679	0.6947	0.7193	0.9657	1.0355	1.3902	1.4396	0.8029	46
45	0.7854	0.7071	0.7071	1.0000	1.0000	1.4142	1.4142	0.7854	45
		Cos	Sin	Cot	Tan	Csc	Sec	Rad	Deg.

For degrees indicated in the left-hand column use the column headings at the top. For degrees indicated in the right-hand column use the column headings at the bottom.

E.7

Trigonometric functions in radian measure

Rad	Sin	Tan	Cot	Cos	Rad	Sin	Tan	Cot	Cos
.00	.0000	.0000	......	1.0000	.50	.4794	.5463	1.830	.8776
.01	.0100	.0100	99.997	1.0000	.51	.4882	.5594	1.788	.8727
.02	.0200	.0200	49.993	.9998	.52	.4969	.5726	1.747	.8678
.03	.0300	.0300	33.323	.9996	.53	.5055	.5859	1.707	.8628
.04	.0400	.0400	24.987	.9992	.54	.5141	.5994	1.668	.8577
.05	.0500	.0500	19.983	.9988	.55	.5227	.6131	1.631	.8525
.06	.0600	.0601	16.647	.9982	.56	.5312	.6269	1.595	.8473
.07	.0699	.0701	14.262	.9976	.57	.5396	.6410	1.560	.8419
.08	.0799	.0802	12.473	.9968	.58	.5480	.6552	1.526	.8365
.09	.0899	.0902	11.081	.9960	.59	.5564	.6696	1.494	.8309
.10	.0998	.1003	9.967	.9950	.60	.5646	.6841	1.462	.8253
.11	.1098	.1104	9.054	.9940	.61	.5729	.6989	1.431	.8196
.12	.1197	.1206	8.293	.9928	.62	.5810	.7139	1.401	.8139
.13	.1296	.1307	7.649	.9916	.63	.5891	.7291	1.372	.8080
.14	.1395	.1409	7.096	.9902	.64	.5972	.7445	1.343	.8021
.15	.1494	.1511	6.617	.9888	.65	.6052	.7602	1.315	.7961
.16	.1593	.1614	6.197	.9872	.66	.6131	.7761	1.288	.7900
.17	.1692	.1717	5.826	.9856	.67	.6210	.7923	1.262	.7838
.18	.1790	.1820	5.495	.9838	.68	.6288	.8087	1.237	.7776
.19	.1889	.1923	5.200	.9820	.69	.6365	.8253	1.212	.7712
.20	.1987	.2027	4.933	.9801	.70	.6442	.8423	1.187	.7648
.21	.2085	.2131	4.692	.9780	.71	.6518	.8595	1.163	.7584
.22	.2182	.2236	4.472	.9759	.72	.6594	.8771	1.140	.7518
.23	.2280	.2341	4.271	.9737	.73	.6669	.8949	1.117	.7452
.24	.2377	.2447	4.086	.9713	.74	.6743	.9131	1.095	.7385
.25	.2474	.2553	3.916	.9689	.75	.6816	.9316	1.073	.7317
.26	.2571	.2660	3.759	.9664	.76	.6889	.9505	1.052	.7248
.27	.2667	.2768	3.613	.9638	.77	.6961	.9697	1.031	.7197
.28	.2764	.2876	3.478	.9611	.78	.7033	.9893	1.011	.7109
.29	.2860	.2984	3.351	.9582	.79	.7104	1.009	.9908	.7038
.30	.2955	.3093	3.233	.9553	.80	.7174	1.030	.9712	.6967
.31	.3051	.3203	3.122	.9523	.81	.7243	1.050	.9520	.6895
.32	.3146	.3314	3.018	.9492	.82	.7311	1.072	.9331	.6822
.33	.3240	.3425	2.920	.9460	.83	.7379	1.093	.9146	.6749
.34	.3335	.3537	2.827	.9428	.84	.7446	1.116	.8964	.6675
.35	.3429	.3650	2.740	.9394	.85	.7513	1.138	.8785	.6600
.36	.3523	.3764	2.657	.9359	.86	.7578	1.162	.8609	.6524
.37	.3616	.3879	2.578	.9323	.87	.7643	1.185	.8437	.5448
.38	.3709	.3994	2.504	.9287	.88	.7707	1.210	.8267	.6372
.39	.3802	.4111	2.433	.9249	.89	.7771	1.235	.8100	.6294
.40	.3894	.4228	2.365	.9211	.90	.7833	1.260	.7936	.6216
.41	.3986	.4346	2.301	.9171	.91	.7895	1.286	.7774	.6137
.42	.4078	.4466	2.239	.9131	.92	.7956	1.313	.7615	.6058
.43	.4169	.4586	2.180	.9090	.93	.8016	1.341	.7458	.5978
.44	.4259	.4708	2.124	.9048	.94	.8076	1.369	.7303	.5898
.45	.4350	.4831	2.070	.9004	.95	.8134	1.398	.7151	.5817
.46	.4439	.4954	2.018	.8961	.96	.8192	1.428	.7001	.5735
.47	.4529	.5080	1.969	.8916	.97	.8249	1.459	.6853	.5653
.48	.4618	.5206	1.921	.8870	.98	.8305	1.491	.6707	.5570
.49	.4706	.5334	1.875	.8823	.99	.8360	1.524	.6563	.5487
.50	.4794	.5463	1.830	.8776	1.00	.8415	1.557	.6421	.5403

Rad	Sin	Tan	Cot	Cos	Rad	Sin	Tan	Cot	Cos
1.00	.8415	1.557	.6421	.5403	1.30	.9636	3.602	.2776	.2675
1.01	.8468	1.592	.6281	.5319	1.31	.9662	3.747	.2669	.2579
1.02	.8521	1.628	.6142	.5234	1.32	.9687	3.903	.2562	.2482
1.03	.8573	1.665	.6005	.5148	1.33	.9711	4.072	.2456	.2385
1.04	.8624	1.704	.5870	.5062	1.34	.9735	4.256	.2350	.2288
1.05	.8674	1.743	.5736	.4976	1.35	.9757	4.455	.2245	.2190
1.06	.8724	1.784	.5604	.4889	1.36	.9779	4.673	.2140	.2092
1.07	.8772	1.827	.5473	.4801	1.37	.9799	4.913	.2035	.1994
1.08	.8820	1.871	.5344	.4713	1.38	.9819	5.177	.1931	.1896
1.09	.8866	1.917	.5216	.4625	1.39	.9837	5.471	.1828	.1798
1.10	.8912	1.965	.5090	.4536	1.40	.9854	5.798	.1725	.1700
1.11	.8957	2.014	.4964	.4447	1.41	.9871	6.165	.1622	.1601
1.12	.9001	2.066	.4840	.4357	1.42	.9887	6.581	.1519	.1502
1.13	.9044	2.120	.4718	.4267	1.43	.9901	7.055	.1417	.1403
1.14	.9086	2.176	.4596	.4176	1.44	.9915	7.602	.1315	.1304
1.15	.9128	2.234	.4475	.4085	1.45	.9927	8.238	.1214	.1205
1.16	.9168	2.296	.4356	.3993	1.46	.9939	8.989	.1113	.1106
1.17	.9208	2.360	.4237	.3902	1.47	.9949	9.887	.1011	.1006
1.18	.9246	2.427	.4120	.3809	1.48	.9959	10.983	.0910	.0907
1.19	.9284	2.498	.4003	.3717	1.49	.9967	12.350	.0810	.0807
1.20	.9320	2.572	.3888	.3624	1.50	.9975	14.101	.0709	.0707
1.21	.9356	2.650	.3773	.3530	1.51	.9982	16.428	.0609	.0608
1.22	.9391	2.733	.3659	.3436	1.52	.9987	19.670	.0508	.0508
1.23	.9425	2.820	.3546	.3342	1.53	.9992	24.498	.0408	.0408
1.24	.9458	2.912	.3434	.3248	1.54	.9995	32.461	.0308	.0308
1.25	.9490	3.010	.3323	.3153	1.55	.9998	48.078	.0208	.0208
1.26	.9521	3.113	.3212	.3058	1.56	.9999	92.620	.0108	.0108
1.27	.9551	3.224	.3102	.2963	1.57	1.0000	1255.8	.0008	.0008
1.28	.9580	3.341	.2993	.2867	1.58	1.0000	−108.65	−.0092	−.0092
1.29	.9608	3.467	.2884	.2771	1.59	.9998	−52.067	−.0192	−.0192
1.30	.9636	3.602	.2776	.2675	1.60	.9996	−34.233	−.0292	−.0292

E.8

Antiderivatives

1. $\int x^n \, dx = \dfrac{x^{n+1}}{n+1} \quad n \ne -1$

$\int \dfrac{dx}{x} = \ln x, x > 0 \quad$ or $\quad \ln |x|, x \ne 0$

2. $\int e^x \, dx = e^x$

3. $\int \sin x \, dx = -\cos x$

4. $\int \cos x \, dx = \sin x$

5. $\int \tan x \, dx = \ln |\sec x| = -\ln |\cos x|$

6. $\int \cot x \, dx = \ln |\sin x| = -\ln |\csc x|$

7. $\int \sec x \, dx = \ln |\sec x + \tan x| = \ln \tan \left(\dfrac{x}{2} + \dfrac{\pi}{4} \right)$

8. $\int \csc x \, dx = \ln |\csc x - \cot x| = \ln \tan \dfrac{x}{2}$

9. $\int \dfrac{dx}{x^2 + a^2} = \dfrac{1}{a} \tan^{-1} \dfrac{x}{a}$

10. $\int \dfrac{dx}{\sqrt{a^2 - x^2}} = \sin^{-1}\dfrac{x}{a}$ $\qquad a > 0$

Expressions Containing ax + b

11. $\int (ax + b)^n\, dx = \dfrac{1}{a(n + 1)}(ax + b)^{n+1}$ $\qquad n \neq -1$

12. $\int \dfrac{dx}{ax + b} = \dfrac{1}{a}\ln |ax + b|$

13. $\int \dfrac{dx}{(ax + b)^2} = -\dfrac{1}{a(ax + b)}$

14. $\int \dfrac{x\, dx}{(ax + b)^2} = \dfrac{b}{a^2(ax + b)} + \dfrac{1}{a^2}\ln |ax + b|$

15. $\int \dfrac{dx}{x(ax + b)} = \dfrac{1}{b}\ln \left|\dfrac{x}{ax + b}\right|$

16. $\int \dfrac{dx}{x^2(ax + b)} = -\dfrac{1}{bx} + \dfrac{a}{b^2}\ln \left|\dfrac{ax + b}{x}\right|$

17. $\int \sqrt{ax + b}\, dx = \dfrac{2}{3a}\sqrt{(ax + b)^3}$

18. $\int x\sqrt{ax + b}\, dx = \dfrac{2(3ax - 2b)}{15a^2}\sqrt{(ax + b)^3}$

19. $\int \dfrac{\sqrt{ax + b}}{x}\, dx = 2\sqrt{ax + b} + b\int \dfrac{dx}{x\sqrt{ax + b}}$

20. $\int \dfrac{dx}{\sqrt{ax + b}} = \dfrac{2\sqrt{ax + b}}{a}$.

21. $\int \dfrac{dx}{x\sqrt{ax + b}} = \dfrac{1}{\sqrt{b}}\ln \left|\dfrac{\sqrt{ax + b} - \sqrt{b}}{\sqrt{ax + b} + \sqrt{b}}\right|$ $\qquad b > 0$

22. $\int \dfrac{dx}{x\sqrt{ax + b}} = \dfrac{2}{\sqrt{-b}}\tan^{-1}\sqrt{\dfrac{ax + b}{-b}}$ $\qquad b < 0$

23. $\int \dfrac{dx}{x^2\sqrt{ax + b}} = -\dfrac{\sqrt{ax + b}}{bx} - \dfrac{a}{2b}\int \dfrac{dx}{x\sqrt{ax + b}}$

24. $\int \sqrt{\dfrac{cx + d}{ax + b}}\, dx = \dfrac{\sqrt{ax + b}\sqrt{cx + d}}{a} + \dfrac{ad - bc}{2a}\int \dfrac{dx}{\sqrt{ax + b}\sqrt{cx + d}}$

Expressions Containing $ax^2 + c$, $ax^n + c$, $x^3 \pm p^2$, and $p^2 - x^2$

25. $\int \dfrac{dx}{p^2 - x^2} = \dfrac{1}{2p}\ln \left|\dfrac{p + x}{p - x}\right|$

26. $\int \dfrac{dx}{ax^2 + c} = \dfrac{1}{\sqrt{ac}}\tan^{-1}\left(x\sqrt{\dfrac{a}{c}}\right)$ $\qquad a$ and $c > 0$

27. $\int \dfrac{dx}{ax^2 + c} = \begin{cases} \dfrac{1}{2\sqrt{-ac}}\ln \left|\dfrac{x\sqrt{a} - \sqrt{-c}}{x\sqrt{a} + \sqrt{-c}}\right| & a > 0,\ c < 0 \\[3mm] \dfrac{1}{2\sqrt{-ac}}\ln \left|\dfrac{\sqrt{c} + x\sqrt{-a}}{\sqrt{c} - x\sqrt{-a}}\right| & a < 0,\ c > 0 \end{cases}$

28. $\int \dfrac{dx}{(ax^2 + c)^n} = \dfrac{1}{2(n - 1)c}\dfrac{x}{(ax^2 + c)^{n-1}} + \dfrac{2n - 3}{2(n - 1)c}\int \dfrac{dx}{(ax^2 + c)^{n-1}}$ $\qquad n > 1$

29. $\int x(ax^2 + c)^n\, dx = \dfrac{1}{2a}\dfrac{(ax^2 + c)^{n+1}}{n + 1}$ $\qquad n \neq -1$

30. $\int \dfrac{x}{ax^2 + c}\, dx = \dfrac{1}{2a}\ln |ax^2 + c|$

31. $\int \sqrt{x^2 \pm p^2} \, dx = \frac{1}{2}[x\sqrt{x^2 \pm p^2} \pm p^2 \ln |x + \sqrt{x^2 \pm p^2}|]$

32. $\int \sqrt{p^2 - x^2} \, dx = \frac{1}{2}\left(x\sqrt{p^2 - x^2} + p^2 \sin^{-1} \frac{x}{p}\right)$

33. $\int \frac{dx}{\sqrt{x^2 \pm p^2}} = \ln |x + \sqrt{x^2 \pm p^2}|$

34. $\int \frac{dx}{\sqrt{p^2 - x^2}} = \sin^{-1} \frac{x}{p}$

Expressions Containing $ax^2 + bx + c$

35. $\int \frac{dx}{ax^2 + bx + c} = \frac{1}{\sqrt{b^2 - 4ac}} \ln \left| \frac{2ax + b - \sqrt{b^2 - 4ac}}{2ax + b + \sqrt{b^2 - 4ac}} \right| \quad b^2 > 4ac$

36. $\int \frac{dx}{ax^2 + bx + c} = \frac{2}{\sqrt{4ac - b^2}} \tan^{-1} \frac{2ax + b}{\sqrt{4ac - b^2}} \quad b^2 < 4ac$

37. $\int \frac{dx}{ax^2 + bx + c} = -\frac{2}{2ax + b} \quad b^2 = 4ac$

38. $\int \frac{dx}{(ax^2 + bx + c)^{n+1}} = \frac{2ax + b}{n(4ac - b^2)(ax^2 + bx + c)^n} + \frac{2(2n - 1)a}{n(4ac - b^2)} \int \frac{dx}{(ax^2 + bx + c)^n}$

39. $\int \frac{x \, dx}{ax^2 + bx + c} = \frac{1}{2a} \ln |ax^2 + bx + c| - \frac{b}{2a} \int \frac{dx}{ax^2 + bx + c}$

40. $\int \frac{dx}{\sqrt{ax^2 + bx + c}} = \frac{1}{\sqrt{a}} \ln |2ax + b + 2\sqrt{a} \sqrt{ax^2 + bx + c}| \quad a > 0$

41. $\int \frac{dx}{\sqrt{ax^2 + bx + c}} = \frac{1}{\sqrt{-a}} \sin^{-1} \frac{-2ax - b}{\sqrt{b^2 - 4ac}} \quad a < 0$

42. $\int \frac{x \, dx}{\sqrt{ax^2 + bx + c}} = \frac{\sqrt{ax^2 + bx + c}}{a} - \frac{b}{2a} \int \frac{dx}{\sqrt{ax^2 + bx + c}}$

43. $\int \sqrt{ax^2 + bx + c} \, dx = \frac{2ax + b}{4a} \sqrt{ax^2 + bx + c} + \frac{4ac - b^2}{8a} \int \frac{dx}{\sqrt{ax^2 + bx + c}}$

Expressions Containing $\sin ax$

44. $\int \sin^2 ax \, dx = \frac{x}{2} - \frac{\sin 2ax}{4a}$

45. $\int \sin^3 ax \, dx = -\frac{1}{a} \cos ax + \frac{1}{3a} \cos^3 ax$

46. $\int \sin^n ax \, dx = -\frac{\sin^{n-1} ax \cos ax}{na} + \frac{n - 1}{n} \int \sin^{n-2} ax \, dx \quad n \text{ positive integer}$

47. $\int \frac{dx}{1 \pm \sin ax} = \mp \frac{1}{a} \tan \left(\frac{\pi}{4} \mp \frac{ax}{2}\right)$

Expressions Containing **cos** *ax*

48. $\displaystyle\int \cos^2 ax\, dx = \frac{x}{2} + \frac{\sin 2ax}{4a}$

49. $\displaystyle\int \cos^3 ax\, dx = \frac{1}{a}\sin ax - \frac{1}{3a}\sin^3 ax$

50. $\displaystyle\int \cos^n ax\, dx = \frac{\cos^{n-1} ax \sin ax}{na} + \frac{n-1}{n}\int \cos^{n-2} ax\, dx$

Expressions Containing Algebraic and Trigonometric Functions

51. $\displaystyle\int x \sin ax\, dx = \frac{1}{a^2}\sin ax - \frac{1}{a}x \cos ax$

52. $\displaystyle\int x \cos ax\, dx = \frac{1}{a^2}\cos ax + \frac{1}{a}x \sin ax$

53. $\displaystyle\int x^n \sin ax\, dx = -\frac{1}{a}x^n \cos ax + \frac{n}{a}\int x^{n-1} \cos ax\, dx$

54. $\displaystyle\int x^n \cos ax\, dx = \frac{1}{a}x^n \sin ax - \frac{n}{a}\int x^{n-1} \sin ax\, dx$ n positive

55. $\displaystyle\int \sin ax \cos bx\, dx = -\frac{\cos(a-b)x}{2(a-b)} - \frac{\cos(a+b)x}{2(a+b)}$

Expressions Containing Exponential and Logarithmic Functions

56. $\displaystyle\int xe^{ax}\, dx = \frac{e^{ax}}{a^2}(ax-1),\ \int xb^{ax}\, dx = \frac{xb^{ax}}{a\ln b} - \frac{b^{ax}}{a^2(\ln b)^2}$ $b > 0$

57. $\displaystyle\int x^n e^{ax}\, dx = \frac{1}{a}x^n e^{ax} - \frac{n}{a}\int x^{n-1}e^{ax}\, dx$ n positive

58. $\displaystyle\int e^{ax} \cos bx\, dx = \frac{e^{ax}}{a^2+b^2}(a \cos bx + b \sin bx)$

59. $\displaystyle\int x^n \ln ax\, dx = x^{n+1}\left[\frac{\ln ax}{n+1} - \frac{1}{(n+1)^2}\right]$ $n \neq -1$

Expressions Containing Trigonometric Inverse Functions

60. $\displaystyle\int \sin^{-1} ax\, dx = x \sin^{-1} ax + \frac{1}{a}\sqrt{1-a^2x^2}$

61. $\displaystyle\int \cos^{-1} ax\, dx = x \cos^{-1} ax - \frac{1}{a}\sqrt{1-a^2x^2}$

62. $\displaystyle\int \csc^{-1} ax\, dx = x \csc^{-1} ax + \frac{1}{a}\ln\left|ax + \sqrt{a^2x^2-1}\right|$

63. $\displaystyle\int \sec^{-1} ax\, dx = x \sec^{-1} ax - \frac{1}{a}\ln\left|ax + \sqrt{a^2x^2-1}\right|$

64. $\displaystyle\int \tan^{-1} ax\, dx = x \tan^{-1} ax - \frac{1}{2a}\ln(1+a^2x^2)$

65. $\displaystyle\int \cot^{-1} ax\, dx = x \cot^{-1} ax + \frac{1}{2a}\ln(1+a^2x^2)$

ANSWERS TO SELECTED ODD-NUMBERED PROBLEMS AND TO GUIDE QUIZZES

CHAPTER 1. THE TWO MAIN CONCEPTS OF CALCULUS

Sec. 1.1. How to Find Speed

1. (*a*) 15 feet per second   (*b*) 10 feet per second
3. (*a*) $16(t_1)^2 - 16(1)^2$ feet   (*b*) $t_1 - 1$ seconds
    (*c*) $16(t_1 + 1)$ feet per second   (*d*) 32 feet per second
5. $2t$ feet per second
9. (*a*) 0.75 feet per second   (*b*) 3 feet per second
    (*c*) 12 feet per second
13. (*a*) 0.2485 feet per second   (*b*) 0.25 feet per second
15. (*a*) 4.1 feet per second per second
    (*b*) $t_1 + 2$ feet per second per second
    (*c*) 4 feet per second per second

Sec. 1.2. How to Find Distance

1. $40 \le d \le 50$
3. (*a*) 5 feet and 1 foot   (*b*) $\frac{15}{4}$ feet and $\frac{7}{4}$ feet
    (*c*) $\frac{91}{27}$ feet and $\frac{55}{27}$ feet
5. $\frac{3}{8}, \frac{5}{8}$
7. (*a*) 4.146 feet, 5.146 feet   (*b*) 4.411 feet, 4.911 feet

Sec. 1.3. Summary: Guide Quiz on Chap. I

1. (*a*) 0.728 feet   (*b*) 3.64 feet per second
3. 29 feet per second
4. (*a*) 2.1875 feet, 5.6875 feet   (*b*) $\frac{8}{3}$ feet, 5 feet

Algebra Review Quiz

1. (*c*)   **2.** (*d*)   **3.** (*c*)
4. (*b*)   **5.** (*c*)   **6.** (*c*)
7. (*c*)   **8.** (*d*)   **9.** (*d*)
10. (*b*)   **11.** (*d*)   **12.** (*b*)
13. (*c*)   **14.** (*b*)   **15.** (*d*)

CHAPTER 2. FUNCTIONS, GRAPHS, AND THE SLOPE OF A LINE

Sec. 2.1. Functions

1. (*a*) 16   (*b*) 16   (*c*) 2   (*d*) 0
3. (*a*) All x (*b*) All x (*c*) $x \ge 0$ (*d*) $x \ne 0$
    (*e*) All x (*f*) $x \ne 0, x \ne -1$
5. (*a*) 5   (*b*) 6.05   (*c*) 1.05   (*d*) 10.5

7. (*a*) 0.2361   (*b*) 0.2485   (*c*) 0.2516
9. (*a*) 0.2   (*b*) -0.5 (*c*) 10   (*d*) 0.4
    (*e*) 0.5   (*f*) -0.2
11. (*a*) 7   (*b*) $(x - 3)(x + 3)$ (*c*) 0.61
    (*d*) 12
13. (*a*) 0.25   (*b*) 1   (*c*) 0.2
    (*d*) -0.05
15. (*a*) 0.24548   (*b*) 0.24695   (*c*) 0.24846
17. (*a*)

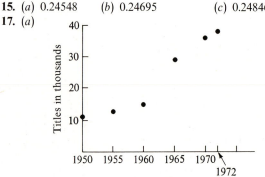

19. $t_1 + t$
21. (*a*) $(1)^2 = (-1)^2$ (*b*) 81   (*c*) $(x^2)^2 = x^4$
23. (*a*) $x \ne 1$ (*b*) $-1, \frac{1}{2}, 2$
25. (*a*), (*b*), (*d*), and (*f*)

Sec. 2.2. Table and Graph

1. (*a*)

x	3	2	1	0	-1	-2	-3
$f(x)$	36	16	4	0	4	16	36

(*b*)

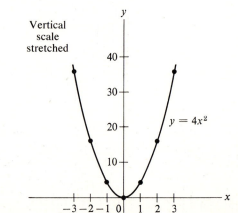

3. (a) $0, \frac{1}{8}, 1, 8, -\frac{1}{8}, -1, -8$
(b)

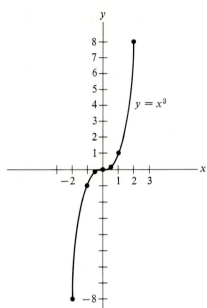

5. The straight line through the points $(0, 0)$ and $(1, -1)$.
7. (a) 0, 3, 6, 9   (b) 0, 3, 6, 9
(c) See diagram in Example 1.
9. The straight line through $(0, 1)$ and $(1, 5)$.
11. The graph in Example 4 ($y = 1/x$) moved to the left 1 unit.
13. (a) $x = 0, x = 2$
(b)

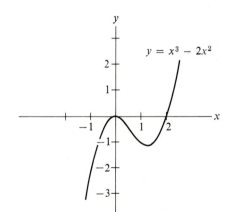

15. 1, 1.21, 0.81. The graph will be like that for Example 5, only in the vicinity of the point $(1, 1)$ instead.
17. The graph in Example 1 ($y = x^2$) moved to the right 3 units and up 4 units.

21.

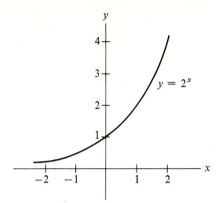

Sec. 2.3. Slope of Line

1. (a) 2   (b) -2 (c) 0   (d) $-\frac{3}{4}$
3. (a) 3, 2   (b) 3, -2 (c) 4, 0   (d) 0, 5
(e) $-4, 0$ (f) 1, 2
5.

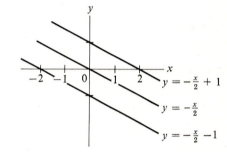

7. (a) $y = 5x - 1$ (b) $y = x + 2$ (c) $y = 4$
9. $y = -x + 6$
11. No.   The line through $(1, -2)$ and $(4, 5)$ has slope $\frac{7}{3}$, but the line through $(4, 5)$ and $(8, 13)$ has slope 2.
13. (a) If $a_1 = a_2$, solve as in Exercise 12 for the intersection.

Sec. 2.4. Summary: Guide Quiz on Chap. 2

1. (a) $x \geq -1$
(b)

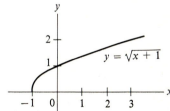

2. $\dfrac{y_2 - y_1}{x_2 - x_1} = \dfrac{ax_2 + b - ax_1 - b}{x_2 - x_1} = a$

3. (a) $-\frac{1}{3}$ (b) -1 (c) 0

4. (*a*)

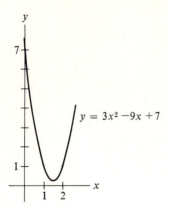

(*b*) No
5. (*a*) 9   (*b*) 5.31
6. (*a*) $2 + h$ (*b*) 8   (*c*) $-5/(1 + h)$
   (*d*) $5 + 3h + h^2$
7. (*a*) The straight line through $(0, -\frac{1}{4})$ and $(2, \frac{3}{4})$.
   (*b*)

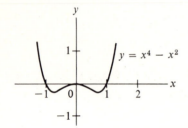

(*c*)

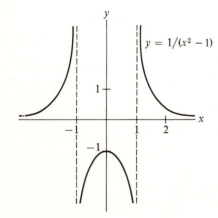

8. (*a*) Yes   (*b*) No   (*c*) Yes

Review Exercises for Chap. 2
1. (*a*) $f(x) = 1/(x + 2)^2$

(*b*)

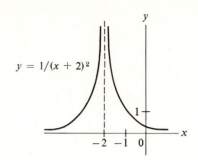

(*c*) $x \neq -2$
3. (*a*)

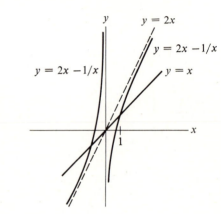

(*b*) $(1, 1)$ and $(-1, -1)$
5. $(a + b)^2 = a^2 + b^2 + 2ab$
7. (*a*) $y = -x/2 + \frac{3}{2}$ (*b*) $y = 7x/2 - 2$
9. (*a*) $|a|$ is large   (*b*) $|a|$ is small   (*c*) $a < 0$
   (*d*) $a > 0$
11. (*a*) 2   (*b*) 2   (*c*) 512   (*d*) 10
13. (*a*) 2   (*b*) 4   (*c*) 5

CHAPTER 3. THE DERIVATIVE

Sec. 3.1. Four Problems

1. (*d*) 8.01   (*e*) 7.99
5. $2x$
7. (*a*) 6.1   (*b*) 6.1   (*c*) 6.01 grams per centimeter
9. (*a*) $2t + 3$ feet per second   (*b*) 5 feet per second
11. (*a*) $6x - 1$ (*b*) $(\frac{1}{6}, \frac{23}{12})$
13. (*b*) Estimate should be near 0   (*c*) 0
15. $\frac{3}{4}$
17. (*a*) 6.1 million dollars per year
   (*b*) 6.01 million dollars per year
   (*c*) 6 million dollars per year
19. (*a*) $(4 - \sqrt{7}, 23 - 8\sqrt{7})$ (*b*) $(-1, 1)$
21. (*a*) Note that the graph passes through $(1, 0)$, $(0, 0)$, and
   $(-1, 0)$.

(b) $3x^2 - 1$ (c) $(\sqrt{3}/3, -2\sqrt{3}/9)$ and $(-\sqrt{3}/3, 2\sqrt{3}/9)$
(d) $(\sqrt{\frac{2}{3}}, -\frac{1}{3}\sqrt{\frac{2}{3}})$ and $(-\sqrt{\frac{2}{3}}, \frac{1}{3}\sqrt{\frac{2}{3}})$

Sec. 3.2. The Derivative of Polynomials

1. (a) 6   (b) -4 (c) 80   (d) 0   (e) 12
   (f) -224
3. (a) 15   (b) 1   (c) 8
5. -22
7. 76 feet per second
9. 811 grams per centimeter
11. (a) $32t + 20$ feet per second   (b) 20 feet per second
13. (a) Use also the information in (b) and (c).
    (b) $x = 0$, $x = \pm\sqrt[4]{5}$ (c) $(1, -4)$ and $(-1, 4)$
15. $y = \frac{1}{2}x - \frac{3}{16}$
17. (a) $8x + 4$ (b) $12x + 1$
19. (a) 4   (b) 3

Sec. 3.3. Derivative of a Function

1. $3/(2\sqrt{x})$
3. $12x + 3$
5. $-2/x^3$
7. (a) 6.1   (b) 6.01   (c) 5.9
9. $2 + 3/(2\sqrt{x})$
11. (a) 1 gram per centimeter   (b) $\frac{1}{2}$ gram per centimeter
13. (a) See Example 4 in Sec. 2.2.   (c) -1 (d) $135°$
15. (a) There are infinitely many such lines.
    (b) The tangent line at $(1, 1)$ meets the graph twice.
    (c) It is the line through that point with slope equal to
        the derivative of the function at the x coordinate of
        the point.
17. (a) Yes   (b) No   (c) No
19. (a) The rate at which water is seeping at time t, in feet
        per hour.   (b) It is slow.

Sec. 3.4. Summary: Guide Quiz on Chap. 3

2. $\frac{13}{4}$
3. (a) 13   (b) -1 (c) 1
4. (a) Distance rocket travels, time interval for that travel,
        average velocity for that time interval
   (b) Increase in population, time interval for that increase,
        average rate of increase
   (c) Mass of segment, length of segment, average density of
        segment
   (d) Length of image of segment, length of segment,
        average magnification of segment
5. $4x^3$

Review Exercises for Chap. 3

1. (a) $5x^4 - 12x^3 + 2$ (b) $-1/x^2$ (c) $1/(2\sqrt{x})$
   (d) $-2/x^3$
3. $4x^3 + 4x$
7. (a) 16.4
   (b) Average rate of growth of profit over first tenth of
        second year
   (c) Slope of line segment joining $(2, 16)$ and $(2.1, 17.64)$
   (d) Average velocity during time interval $(2, 2.1)$
9. (a) 12.61   (b) 11.41   (c) 12
11. (a) $5(x_1^2 + x_1 + 1)$ (b) $-3(x_1 + 1)/x_1^2$ (c) 6
    (d) $4x_1 + 3$
13. (a) The rate at which the value of a car changes
    (b) Depreciation, appreciation, usually former except
        when car is an antique
15. (a) 0, 9, 24   (b) 5, 15   (c) $9 \le t \le 24$
    (d) $5 \le t \le 15$ (e) $0 \le t \le 5, 15 \le t \le 24$ (f) 5, 15
17. (a) 5   (b) 5
19. $y = -x$
21. 32
23. (a) $3x^2 - 3$ (b) $|x| > 1, |x| < 1, |x| = 1$
25. Goes to 0
27. Goes to 0.5
29. Near $\frac{1}{12}$

CHAPTER 4. LIMITS AND CONTINUOUS FUNCTIONS

Sec. 4.1. Exponentiation

1. (a) 8   (b) 4   (c) $\frac{1}{4}$ (d) 16
   (e) $\frac{1}{32}$
3. (a) 0.1   (b) 0.053   (c) 600.06
5. (a) 1   (b) 2   (c) 4   (d) 8
   (e) $\frac{1}{2}$
7. (a) 2^2 (b) 2^{-3} (c) $2^{1/2}$ (d) 2^{-1}
   (e) 2^0 (f) $2^{1/3}$ (g) 2^{-2} (h) $2^{3/2}$
9. (a) $b^{2/3}$ (b) b^{-2} (c) $b^{-1/2}$ (d) $b^{-1/3}$
   (e) b^{-5} (f) $b^{1/6}$
11. (a) 25   (b) 4   (c) 9   (d) 4
21. (a) $0 < x < 1$ (b) $x > 1$
25. (a) 2.25   (b) 2.236
29. (a) 0.9931   (b) 0.1995, 1, 0

Sec. 4.2. The Number e

1. (a) 2.49   (b) 3.05
3. (a) 0.25   (b) 0.30   (c) 0.32
5. (a) 1   (b) $\sqrt{e}$
9. e

Sec. 4.3. Limits

1. $\frac{6}{5}$

3. 0

5. $-\frac{1}{6}$

7. 7

9. 0

11. -1

13. 12

15. $\frac{1}{4}$

17. $\lim_{x\to a} f(x)$ exists for all a.

19. $\lim_{x\to a} f(x)$ exists for all a.

21. $\lim_{x\to a} f(x)$ exists for all a except $a = 1$.
$\lim_{x\to a^+} f(x)$ exists for all a
$\lim_{x\to a^-} f(x)$ exists for all a except $a = 1$.

23.

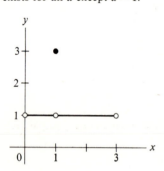

25.

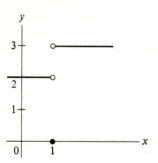

27. (b) Yes   (c) Yes   (d) All a

29. (a) Draw a lot of dots on the lines $y = x$ and $y = -x$.
(b) No   (c) No   (d) Yes   (e) $a = 0$

31. 1

Sec. 4.4. More on Limits and e

1. $\frac{5}{7}$

3. $-\infty$

5. -1

7. $-\infty$

9. 0

11. ∞

13. 0

15. 0

17. ∞

19. ∞

21. 1

23. 0

25. $\sqrt{\frac{2}{3}}$

27. 2.37, 2.44, 2.49, 3.37, 3.16, 3.05

29. 1

31.

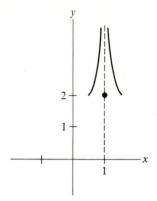

33.

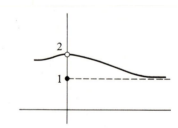

37. 1.271, 1.164, 1.121, yes

39. (a) 30.375, 97.656, 3.052, 0.0226   (b) Goes to 0

41. Goes to 4

Sec. 4.5. Trigonometry

1. (a) $\pi/2$ (b) $\pi/6$ (c) $2\pi/3$ (d) $3\pi/2$ (e) 2π

3. (a) $\frac{5}{3}$ radians   (b) $300°/\pi \doteq 95.5°$

5. (a) $540°/\pi \doteq 171.9°$ (b) $\pi/180$ radians $\doteq 0.0175$ radians

7. (a) Measure an angle of $114.59°$
(b) Lay a diameter along a circumference.

9. (a) $0, \frac{1}{2}, \sqrt{2}/2, \sqrt{3}/2, 1, 0, -1, 0$
(b) See text, where graph is sketched.

13. (a) $\pi/4$ radians   (b) 0.46 radians ($\doteq 26.5°$)
(c) $3\pi/4$ radians   (d) 1.107 radians ($\doteq 63.5°$)
(e) $\pi/3$ radians

15. $\frac{1}{2}, \sqrt{3}/2, \sqrt{3}$

17. (a) $\frac{3}{2}$ (b) $5/\sqrt{2}$ (c) $4\sqrt{3}$

19. (a) 5   (b) $\sqrt{x^2 + y^2}$ (c) $\sqrt{2 - 2\cos\theta}$

27. (a) 2, $\sqrt{2}, 2/\sqrt{3}$, 1

(b)

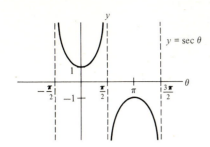

$y = \sec \theta$

31. (a) 0.0166, 0.0171, 0.01736, 0.01743, 0.01745
(b) Goes to $\pi/180 \doteq 0.0175$

Sec. 4.6. Limit of $\sin \theta/\theta$

1. (a) $9\pi/4$ (b) $\theta/2$ (c) 2θ
3. 0.64, 0.90, 0.95
5. 0
7. $\frac{2}{3}$
9. ∞
11. ∞
13. (a) 0, 1, 0, -1, 0, 0.01 (approx.)
(b)

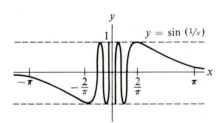

$y = \sin (1/x)$

(d) 0
17. (a) $x \neq 0$ (b) 0.95, 0.64, 0, -0.21, 0
(e)

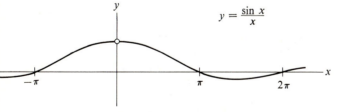

$y = \dfrac{\sin x}{x}$

19. (a) $\lim\limits_{h \to 0} \dfrac{f(x + h) - f(x)}{h}$

(b) Use $\sin (x + h) = \sin x \cos h + \sin h \cos x$.
21. Near $\frac{1}{6}$

Sec. 4.7. Continuous Functions

1. Continuous for all x in $[0, 1]$.

3. $\frac{1}{2}$
5. 1
7. Yes, define $f(0) = 1$.
9. (a) Yes, let $f(0) = e$.   (b) No
11. $\frac{5}{3}$
13. π, 2π, 3π, 4π, 5π
15. $\pi/3$, $5\pi/3$, $7\pi/3$, $11\pi/3$, $13\pi/3$
17. -1, 0, 1
19. Evaluate $x + \sin x$ at 0 and $\pi/2$ and use intermediate-value theorem.
21. Consider the interval $[0, 1]$.
23. Yes
25. For $m = 1$, $X = -1$ or 1, for $1 < m \leq 4$, $X = \sqrt{m}$
27. Only at $x = 0$
29. (a) $f(n) = f(1 + \cdots + 1) = f(1) + \cdots + f(1) = nf(1) = cn$ (n summands)
(b) (n summands) $f(1/n) + \cdots + f(1/n) = nf(1/n) = f(1)$, whence $f(1/n) = c/n$. Show $f(x) = cx$ when x is rational.   (c) Use continuity.

4.8. Summary: Guide Quiz on Exponents

1. (a) 0.125   (b) 2.828   (c) 4   (d) 1
   (e) 0.354   (f) 0.25
2. (a) $3^{1/2}$ (b) 3^{-1} (c) $3^{2/3}$ (d) 3^5
   (e) $3^{3/2}$ (f) 3^{35} (g) $3^{-1/2}$ (h) 3^{10}
3. 1.9
4. (a) b^2 (b) $b^{3/2}$ (c) $b^{5/6}$ (d) $b^{-3/2}$
   (e) $b^{5/6}$ (f) $b^{15/2}$
5. (a) 625, 125   (b) $\sqrt[3]{5}$

Guide Quiz on Trigonometry

1. (a) 114.6°   (b) -0.42, 0.91
2. (a) 5.2 inches   (b) 10.4 square inches
3. (a) See graph in Sec. 4.5
(b) It is the same except shifted to the left $\pi/2$ units.

4.

θ (radians)	$\pi/2$	$2\pi/3$	$3\pi/4$	$-\pi/4$	$-\pi/6$	$9\pi/4$
θ (degrees)	90°	120°	135°	$-45°$	$-30°$	405°
$\cos \theta$	0	$-\frac{1}{2}$	$-\sqrt{2}/2$	$\sqrt{2}/2$	$\sqrt{3}/2$	$\sqrt{2}/2$
$\sin \theta$	1	$\sqrt{3}/2$	$\sqrt{2}/2$	$-\sqrt{2}/2$	$-\frac{1}{2}$	$\sqrt{2}/2$

5. (a) ± 0.8 (b) 53.2°, $-53.2°$
6. (a) $(\sqrt{6} + \sqrt{2})/4$ (b) $(\sqrt{6} - \sqrt{2})/4$
(c) $\sqrt{2} + \sqrt{2}/2$
7. See graph in Sec. 4.5.

Guide Quiz on Chap. 4

1. (a) $\lim_{n \to \infty} (1 + 1/n)^n$ (b) 2.718   (c) e

2. $c^5 - d^5 = (c - d)(c^4 + c^3d + c^2d^2 + cd^3 + d^4) <$
$(c - d)(c^4 + c^4 + c^4 + c^4 + c^4) = 5c^4(c - d)$

3. (a) 2   (b) None   (c) None   (d) 2
(e) $-1, 0, 2, 4, 5$

4. (a) $\frac{1}{2}$ (b) $\frac{3}{2}$ (c) $3\cos(1) - 3 \doteq -1.379$ (d) $\frac{1}{4}$
(e) ∞ (f) e^3

5. (d) $f(x) = \sqrt{x}$ and $x = 4$

6. Approaches 1, approaches 0

7. (a) $0, \sqrt{2}/2$
(b) Intermediate-value theorem, $0 < \frac{1}{2} < \sqrt{2}/2$

8. (a) The graph of $f(x) = |x|$ suffices.
(b) The graph of $f(x)$, where $f(0) = 0, f(x) = 1$ if $x \neq 0$.
(c) The graph of $f(x)$, where $f(x) = 0$ if $x < 0$ and
$f(x) = 1$ if $x \geq 0$.

Review Exercises for Chap. 4

3.

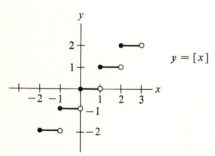

5. (a) False   (b) True   (c) True

7. (a) Multiply top and bottom by $1 + \cos h$.
(d) Within 0.0002, 0.00006, 0.0011

9. Continuous at $-1, 0, 1$; differentiable at 0

11. (a) Goes to ∞ (b) Goes to $-\infty$
(c) Use intermediate-value theorem.

13. (a)

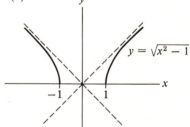

(b) f is not continuous at any integer

15. (a) $0, 1, \frac{1}{2}$

(b) All x

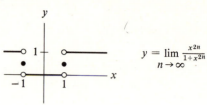

(c) At $x \neq 1, -1$

Review Questions for Chaps. 1–4

1. (a) $\displaystyle\lim_{x_1 \to x} \frac{f(x_1) - f(x)}{x_1 - x}$

(b) Velocity, density, magnification, slope, rate of growth, etc.

3. (a) 7.625   (b) 2.791

5. (a) $\frac{2}{3}$ (b) 6

7. See answer to Exercise 17(e) in Sec. 4.6.

9. (a) $6x^2 + 6x - 12$
(b) Zero for $x = -2, 1$, positive for $x < -2, x > 1$,
negative for $-2 < x < 1$.

11. (a) The slope at $x = 1$ is 3.
(b) Its speed is 3 feet per second at $t = 1$; about 0.03 feet.
(c) The magnification at $x = 1$ is 3; 0.03
(d) The density at $x = 1$ is 3 grams per centimeter; 0.03 grams.

13. (a) $18x^2 - 5$ (b) $12x + 36x^3$ (c) $-5/x^2$

15. (a) $|x| \geq 1$ (b)

(c) Yes

17. (a) $x \neq (2k + 1)\pi/2$, k any integer

19. (a) $\frac{1}{2}$ (b) 4   (c) $-\frac{17}{9}$

21. 0

23. (a) 1   (b) $\frac{1}{3}$

25. (a) $x^3 + 1$ (b) $x^3/3$ (c) $1/x - 18$ (d) $-1/x$
(e) $9x^4/4 + 2x^3 - 3x^2 + 2x - 4$ (f) $10\sqrt{x}$

27. $1/(3x^{2/3})$

29. Suppose that for all $x, f(x) > x$. Then
$x = f(f(x)) > f(x) > x$. Hence, for some $a, f(a) - a \leq 0$.
Similarly, there is b such that $f(b) - b \geq 0$. Apply the
intermediate-value theorem to $g(x) = f(x) - x$.

33. $\frac{3}{2}$

35. No

37. -13 (neither citizen is right).

CHAPTER 5. THE COMPUTATION OF DERIVATIVES

Sec. 5.1. Notations

1. 21

3. $\frac{1}{6}$

5. $24u^3$

7. $15x^4 - x$

9. $-1/u^2$

11. 44

13. $t^3 + t^2 - t$

15. $3u^2$

17. $4s^3 + 6s^2 + 2s$

19. (a) $-32t + 80$ (b) $t < \frac{5}{2}, t > \frac{5}{2}, t = \frac{5}{2}$
   (c) 80 meters per second

21. (a) $u^4/4 + 3$ (b) $2\sqrt{t}$ (c) $3x^2/2 + 5x + 1$
   (d) $5t$

Sec. 5.2. Derivatives of c, sin x, cos x

1. (b) $0 \le x < \pi/2$ and $3\pi/2 < x \le 2\pi$, $\pi/2 < x < 3\pi/2$,
     $x = \pi/2$ and $x = 3\pi/2$ (c) 0 and 2π

3. (a) $\frac{1}{2}$ (b) $-\sqrt{2}/2$ (c) 3   (d) 0

5. (a) 0, 1, 0, -1 centimeters
   (b) 1, 0, 0, -1 centimeters per second
   (c) 1, 0, 0, 1 centimeters per second

11. Use $(\cos [4(x + h)] - \cos 4x)/h$

13. (a) $2 \cos t$ feet per hour   (b) At mean sea level

15. First expand $\sin [(x + h)^2] = \sin (x^2 + 2xh + h^2)$, before looking at usual quotient.

Sec. 5.3 Logarithms

1. (a) $\log_2 32 = 5$ (b) $\log_3 81 = 4$
   (c) $\log_{10} 0.001 = -3$ (d) $\log_5 1 = 0$
   (e) $\log_{1,000} 10 = \frac{1}{3}$ (f) $\log_{49} 7 = \frac{1}{2}$

3. (a) 2, $\frac{1}{2}$ (b) 3, $\frac{1}{3}$ (c) 2, $\frac{1}{2}$

5. (a) $2^x = 7$ (b) $5^s = 2$ (c) $3^{-1} = \frac{1}{3}$ (d) $7^2 = 49$

7. (a) -1 (b) $\frac{3}{2}$ (c) 0

9. (a) 7/6   (b) $-\frac{1}{2}$ (c) $-\frac{1}{6}$

11. 1.1404

13. (c) 0.6309

15. (a)

x	0.01	0.1	1	2	3	4	10	100
$\log_{10} x$	-2	-1	0	0.3010	0.4771	0.6021	1	2
$\log_{10} x/x$	-200	-10	0	0.1505	0.1590	0.1505	0.1	0.02

(b)

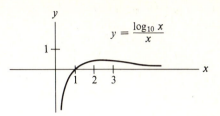

$$y = \frac{\log_{10} x}{x}$$

(c) e (by methods in Chap. 6)

23. (b) One is a scaled version of the other.

25. (b) k

27. Let $\log_2 3 = a/b$, and get $2^a = 3^b$, which is impossible in nonzero integers a and b. (Why?)

29. 1.465

Sec. 5.4. Derivative of Logarithm

1. (a) $(\log_{10} e)/x$ (b) $(\log_2 e)/x$ (c) $(\log_5 e)/x$
   (d) $1/x$ (e) $1/x$

3. (a) $\frac{1}{2}$ (b) $\frac{1}{3}$ (c) 2

5. (a) Similar to graph of $y = \log_{10} x$ in Sec. 5.3, except magnified vertically by a factor $\log_e 10 \doteq 2.3$.
   (b) $45°$

7. (a) $\frac{1}{2}$ (b) 0.04

9. (b) 0.69, 1.10, 1.94

11. (a) $\frac{1}{3}$ (b) 0.02

13. Let $f(x) = 1/x - \cos x$. $f(\pi) = 1/\pi + 1 > 0$,
   $f(2\pi) = 1/2\pi - 1 < 0$.   Use intermediate-value theorem.

Sec. 5.5. Derivative of $f + g$, $f - g$, fg

1. $-\sin x + \cos x$

3. $4/\sqrt{x}$

5. $3x^2 + 5 \cos x$

7. $\cos x$

9. $1 - 1/x^2$

11. $x \cos x$

13. $12x^2 \sin x + 4x^3 \cos x$

15. $3 \cos^2 x - 3 \sin^2 x = 6 \cos^2 x - 3$

17. $5/x - \cos x + x \sin x$

19. $(2 \ln x)/x$

21. $10/x$

23. $x \ln x$

25. $(\ln x)^2$

27. $3(\ln x)^2/x$

31. (a) $x^2/10$ (b) $\ln x^5$ (c) $x^2/2 - 3 \ln x$
   (d) $x^2/2 + 2x - \ln x$ (e) $2 \sin x - 3 \cos x$

Sec. 5.6. Derivative of f/g

1. $1/(3x + 2)^2$

3. $(1 - 2 \ln x)/x^3$

5. $(24x^9 + 8x^7 + 3x^6 - 20x^5 + 5x^4 + 32x^3 - 2x^2 - 24x + 2)/$
$(x^2 + 1)^2$

7. $(\tan x)/x + \ln x \sec^2 x$

9. $(\cos x + 1 - \tan^2 x)/(1 + \sec x)^2$

11. $\ln x \cos 2x + (\sin x \cos x)/x$

13. $(1 - \ln x)/x^2$

15. $(-2x - 15x^4)/(x^2 + 3x^5)^2$

17. $-64/x^9$

19. $(5x \sec^2 x - 15 \tan x)/x^4$

21. $-4 \csc x \cot x - 3 \csc^2 x$

25. (a) $1, 2/\sqrt{3}, \sqrt{2}, 2, -1, -\sqrt{2}, \sqrt{2}$
(b) $x \neq (2k + 1)\pi/2$, k an integer

27. (a) $\sqrt{3}, \sqrt{3}/3, 1, 0, -1$ (b) $\cot x \to +\infty$
(c) $\cot x \to -\infty$
(d)

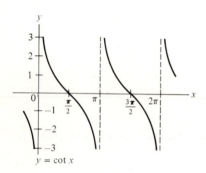

$y = \cot x$

7. $3(1 + \ln 2x)^2/x$

9. $[3(x^2 + 1) \cos 3x - 10x \sin 3x]/(x^2 + 1)^6$

11. $4 \cos 3x \cos 4x - 3 \sin 3x \sin 4x$

13. $(\sec^2 \sqrt{x})/(2\sqrt{x})$

15. $[-5(1 + x^2) \csc^2 5x - 2x \cot 5x]/(1 + x^2)^2$

17. $(3 \cot 3x)/\ln 10$

19. $-6x \csc 3x^2 \cot 3x^2$

21. $x/\sqrt{x^2 - 1}$

23. $x(1 - x^2)^{-3/2}$

25. $x/(3x + 5)$

27. $1/[(x - 3)\sqrt{2x + 3}]$

29. $\sin^3 3x$

31. $\sec 3x$

33. (a) Graph $y = \ln x$ and reflect it across the y axis.

35. (b) 35

Sec. 5.9. Inverse Functions

1.

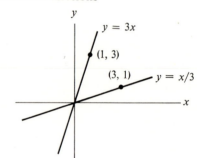

3.

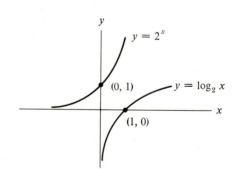

5.

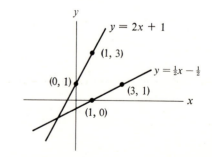

Sec. 5.7. Composite Functions

1. $y = \cos u, u = x^2$

3. $y = u^3, u = \cos x$

5. $y = 3 + u^2, u = \ln x$

7. $y = \sec u, u = 5x$

9. $y = \sin u, u = 1 + \ln x$

11. $y = u^3, u = \sin v, v = 5x$

13. (a) $f \circ g(x) = \sin |x|, g \circ f(x) = |\sin x|$
(b) $f \circ g(x) = (\ln x)^2, g \circ f(x) = \ln (x^2)$
(c) $f \circ g(x) = (x^3)^2 = x^6, g \circ f(x) = (x^2)^3 = x^6$

15. (a) $g'(3) \doteq 2, f'(4) \doteq 3. h'(3) \doteq 6$
(b) $h'(3) = g'(3) \cdot f'(4)$

17. (a) Add 1 to the number, square the result.
(b) Square the number, add 1 to the result.
(c) $(x + 1)^2$ (d) $x^2 + 1$

19. $f(x) = \log x, g(x) = \sin x$

21. (a) x, x (b) x, for $x > 0, x$

Sec. 5.8. Derivative of Composite Function

1. $8(2x^3 - 2x + 5)^3(3x^2 - 1)$

3. $15 \sin^4 3x \cos 3x$

5. $10(1 + 2x)^4 \cos 3x - 3(1 + 2x)^5 \sin 3x$

7. (b) $y = \sqrt[4]{x}$

9. (b) and (c)

11. (a) $\pi/2$ (b) $\pi/6$ (c) 0 (d) $-\pi/6$ (e) $-\pi/3$
(f) $-\pi/2$

15. $-4/(x^2\sqrt{4 - 9x^2})$

17. $10x(1 + x^2)^4 \sin 3x + 3(1 + x^2)^5 \cos 3x$

19. $-3 \sin [\log_{10} (3x + 1)]/[\ln 10(3x + 1)]$

21. $15/(5x + 1) + 12/(6x + 1) - 8/(2x + 1)$

23. The points $(1, -2)$, $(3, 4)$, $(5, 8)$ are on the graph of f.
The points $(-2, 1)$, $(4, 3)$, $(8, 5)$ are on the graph of f^{-1}.

25. $0, \frac{1}{2}, -\pi/2, 2\pi$

29. No, because some fathers have more than one child.

31. $|x|$

Sec. 5.10. Derivative of b^x and x^a

1. $15e^{3x}$

3. $(3 \ln 2)(2^{3x})$

5. $-e^{-x}$

7. $x^2 e^x$

9. $[2x + (2x - 3x^2)e^{3x}]/(1 + e^{3x})^2$

11. $\sqrt{2}x^{\sqrt{2}-1}$

13. 0

15. xe^{3x}

17. $3x \cos 3x$

19. $(5 + 3e^{2x})^{-1}$

21. $(\ln 5x)^2$

23. $(Ae^{kx})' = k \cdot Ae^{kx}$

25. (a) $0, e^{-\pi/2}, 0, -e^{-3\pi/2}, 0, e^{-5\pi/2}, 0, -e^{-7\pi/2}, 0$

(b)

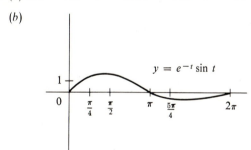

$y = e^{-t} \sin t$

(c) $\pi/4, 5\pi/4$

27. (a) $10x/(x^2 + 3)$ (b) $3e^{3x}/(1 + e^{3x}) + 6/x$
(c) $1/(1 + 2x) + x^2/(1 + x^3)$
(d) $10x/(x^2 - 2) + 5/x - x/(x^2 + 5)$

Sec. 5.11. Derivative of Inverse Trigonometric Functions

1. (a) 0.98 (b) 0.61 (c) -0.88 (d) 0.41
(e) -0.52 (f) 0.93

3. $\sqrt{2}/2$

5. $\sqrt{3}$

7. -1

9. $5/\sqrt{1 - 25x^2}$

11. $3/(1 + 9x^2)$

13. $1/(|x|\sqrt{9x^2 - 1})$

15. $\sqrt{2 - x^2}$

17. $1/(x\sqrt{3x^5 - 1})$

19. $\sqrt{(2 - x)/(1 + x)}$

21. (a) $1/\sqrt{x^2 - 9}$ (b) $1/\sqrt{9 - x^2}$

23. (a) $1/(x\sqrt{2x^2 + 1})$ (b) $1/(|x|\sqrt{2x^2 - 1})$

25. $(\sin^{-1} 2x)^2$

27. (a)

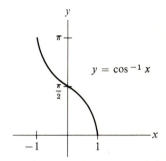

$y = \cos^{-1} x$

Sec. 5.12. Summary: Guide Quiz on Chap. 5 (Computations)

1. $-1/(2x\sqrt{x})$

2. (a) $6(x^3 + x)^5(3x^2 + 1)$
(b) $[6x(\sin 2x)^2 \cos 2x - 4(\sin 2x)^3]/x^5$
(c) $2x^2 + 6x^2 \ln x$ (d) $15 \sec^5 3x \tan 3x$

3. (a) $3x^2/(2\sqrt{1 + x^3})$ (b) $-3 \sin 6x$ (c) $-3e^{-3x}$
(d) $3/\sqrt{1 - 9x^2}$ (e) $2x \cdot 5^{x^2} \ln 5$ (f) $\frac{3}{2}\sqrt{x}$

4. (a) $e^x \sin 2x$ (b) $\sin^{-1} x$ (c) $\tan^{-1} x$
(d) $3 \sec^{-1} 3x \sec 3x \tan 3x + \sec 3x/(|x|\sqrt{9x^2 - 1})$

5. (a) $-\pi/4$ (b) $\pi/3$ (c) $\pi/6$ (d) $2\pi/3$ (e) π
(f) $-\pi/2$

Guide Quiz on Chap. 5 (Concepts)

2. (b) $(\log_{10} e)/x = 1/(x \ln 10)$

3. (a) e^2 (b) $\ln 10$

6. (a) To simplify the derivative of the trig functions
(b) To simplify the derivative of the log functions
(c) To simplify the derivative of the exponential functions

7. (a) The maternal grandmother of x
(b) The paternal grandmother of x
(c) The maternal grandfather of x
(d) Siblings (e) First cousins

8. (a) e^x (b) $\ln x$ (c) $\sqrt[3]{x}$ (d) $x/3$ (e) x^3
(f) $\sin x$

10. (a) 12 (b) $-\frac{1}{2}$ (c) 45 (d) 12 (e) 90

Review Exercises for Chap. 5

1. $12x + 3$
3. $\sin x + \sec x \tan x$
5. $-e^{-3x}(3 \cos 2x + 2 \sin 2x)$
7. (a) $3/x - 10/(1 - 2x)$ (b) $-(\tan x)/3$
9. $1/(9 + x^2)$
11. $(\cos x)/(2\sqrt{\sin x})$
13. $1/\sqrt{x^2 + 1}$
15. $[6(x^2 + x) \sin^2 2x \cos 2x - (2x + 1) \sin^3 2x]/(x^2 + x)^2$
17. $2xe^{-x}/(x^4 + 1) - e^{-x} \tan^{-1} x^2$
19. $6xe^{3x^2}$
21. $2/\sqrt{1 - 4x^2}$
23. $2x \log_{10} e/(x^2 + 1)$
53. $[\ln (1 + 2x)]'$ at $x = 3$, which equals $2/7$.
55. $(\sin \sqrt{x})'$ at $x = 3$, which equals $(\cos \sqrt{3})/(2\sqrt{3})$
57. (a) $-1/(2x^2)$ (b) $1 - 1/x$ (c) $\ln 5x$
59. (a) $3y^2 \, dy/dx$ (b) $-\sin y \, dy/dx$ (c) $e^y \, dy/dx$
    (d) $(-1/y^2) \, dy/dx$
61. $\frac{1}{60}$
63. (a) 2   (b) 3   (c) $\frac{1}{2}$ (d) 1   (e) 3
65. (b) $2x \sin (1/x) - \cos (1/x)$ (c) Look at $\lim_{x \to 0} f'(x)$
67. Deduce that $7^a = 3^b$, which violates unique factorization into primes.

CHAPTER 6. APPLICATIONS OF THE DERIVATIVE

Sec. 6.1. Rolle's Theorem

1. (a) For some X in $(0, 2\pi)$, $-\sin X = 0$.
   (b) π
3. $-1, 0, 1$
5. (a) $\sqrt{2}/2$ (b) $\pi/4, 7\pi/4$ (c) No   (d) No
7. (a) $f(0) = f(2) = 40$
   (b) For some time t in $(0, 2)$, the velocity of the ball must be 0.   (c) $t = 1$
9. No, $f'(x)$ does not exist at $x = 0$.
11. (a) Nothing, f is undefined at $x = -\frac{1}{2}$.   (b) No values
13. Let $g(x) = -f(x)$ and use Theorem 1.
15. (b) No   (c) No, f' does not exist for the integers.
17. The derivative of a sixth-degree polynomial is a fifth-degree polynomial.

Sec. 6.2. Law of the Mean

1. $(1 - \sqrt{7})/3, (1 + \sqrt{7})/3$
3. All x in $(1, 3)$
5. $\pi, 2\pi, 3\pi$
7. $e^{1/(e-1)}$
9. If an interval is magnified by an amount M, then there is some point in the interval where the magnification equals M.
11. $f'(x) = 2$ for some X in $(3, 8)$
13. (a) $1/x$, $1/x$ (b) $\ln \frac{3}{2}$
21. $x - x^2/2 + x^3/3 - \cdots + (-1)^{n+1}x^n/n < \ln (1 + x) < x - x^2/2 + x^3/3 - \cdots (-1)^n x^{n+1}/(n + 1)$, if n is even.

Sec. 6.3. Relative Sizes of e^x, x^a, $\ln x$

1. 0
3. $-\infty$
5. ∞
7. 0
9. 0
11. $(\ln 3)/\ln 2$
13. 0
15. (a) 0.02, 0.003, 0.0004   (b) 0
17. (a) $-1/(3x^3)$ (b) $3x^{4/3}/4$ (c) $\ln x^3$
    (d) $x^4/4 - 2x^3/3$ (e) $5 \tan^{-1} x$ (f) $\frac{3}{2} \ln (1 + x^2)$
    (g) $-1/(3 + 9x)$ (h) $\ln \sqrt[3]{1 + 3x}$ (i) $\frac{1}{2} \sin^{-1} 2x$
23. (b) Only if $x_2 < 0 < x_1$
29. $e^x < 1 + x + x^2/2! + \cdots + x^{n-1}/(n - 1)! + ex^n/n!$
31. 47.647
33. (a) 1.649   (b) 2.718

Sec. 6.4. Growth and Decay

1. (a) 10 grams   (b) $\ln 3 = 1.0986$ (c) 200 percent
3. 70 years
5. 1.435
7. 2064
9. (a) 25 percent   (b) 80 percent
11. (a) $f'(t) = 10^t \ln 10 = f(t) \ln 10$
    (b) No, $10^t = e^{t \ln 10}$
13. (a) $\frac{11}{12}$ (b) -0.087
15. $f(t) = Ae^t$
25. 8.3 percent
27. $y = \frac{1}{2}kt^2 + c$

Sec. 6.5. Graphing

1.

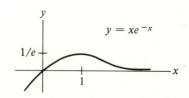

3. A critical point at $(0, 0)$ but no local maxima nor minima.

5.

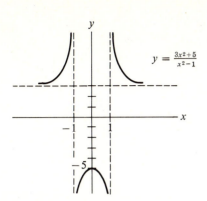

$$y = \frac{3x^2+5}{x^2-1}$$

7.

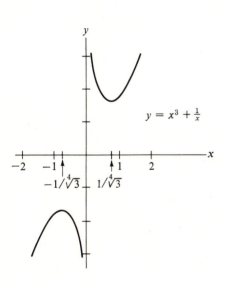

$$y = x^3 + \frac{1}{x}$$

9.

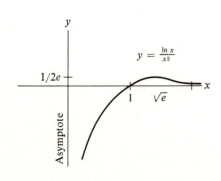

$$y = \frac{\ln x}{x^2}$$

11.

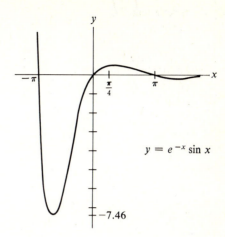

$$y = e^{-x}\sin x$$

13.

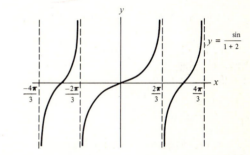

$$y = \frac{\sin}{1+2}$$

15. Shift the graph of $y = x^3$ to the right one unit and up one unit. The point $(1, 1)$ is a critical point and inflection point.

17. The point $((\ln 2)/2, 2\sqrt{2})$ is a global minimum.

19.

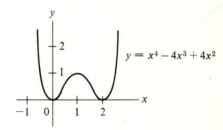

$$y = x^4 - 4x^3 + 4x^2$$

21. Intercepts at $x = 0$ and -2; global minimum at $x = -1/2$;

23.

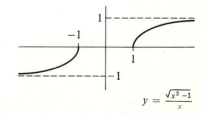

$$y = \frac{\sqrt{x^2-1}}{x}$$

25. Maximum: $1/e$, minimum: 0

27. 3, 0

29. 24, -8

31. $e^{-\pi/4}/\sqrt{2}$, $-e^{-5\pi/4}/\sqrt{2}$

33. (a) $\frac{2}{5}x^{-3/5}$ (b) $3x^2 \sin^{-1} 3x + 3x^3/\sqrt{1 - 9x^2}$
   (c) $e^{-2x}(-2 \cos 5x - 5 \sin 5x)$
   (d) $1/[2(2x + 1)^{1/2}(3x + 2)^{3/2}]$ (e) $-15 \csc^3 5x \cot 5x$
   (f) $x^{\tan^{-1} 2x}[2 \ln x/(1 + 4x^2) + (\tan^{-1} 2x)/x]$

Sec. 6.6. Second Derivative and Motion

1. (a) $f^{(1)}(x) = 3x^2 - 4x + 5$, $f^{(2)}(x) = 6x - 4$, $f^{(3)}(x) = 6$,
   $f^{(4)}(x) = 0$ (b) $f^{(n)}(x) = 0$, $n > 4$

3. $y' = 2/(1 + 4x^2)$, $y'' = -16x/(1 + 4x^2)^2$,
   $y''' = (192x^2 - 16)/(1 + 4x^2)^3$

5. $f^{(1)}(x) = (1 + 2x)^{-1/2}$, $f^{(2)}(x) = -(1 + 2x)^{-3/2}$
   $f^{(3)}(x) = 3(1 + 2x)^{-5/2}$

7. $-2 \csc 2x \cot 2x$, $4 \csc^3 2x + 4 \csc 2x \cot^2 2x$,
   $-40 \csc^3 2x \cot 2x - 8 \csc 2x \cot^3 2x$

9. $e^{-t}(\pi^3 \sin \pi t + 3\pi^2 \cos \pi t - 3\pi \sin \pi t - \cos \pi t)$

15. $y = -16t^2 + 64t$, $v = -32t + 64$

17. $y = -16t^2 + 96$, $\sqrt{6} \doteq 2.45$ seconds

19. The time at which the ball would have been thrown from
   the ground to reach the top of the cliff at $t = 0$ with
   velocity = 64 feet per second.

21. (a) 294.8 feet (b) 86.9 feet (c) 44.5 feet

23. Slows down, approaching 1 mile per second

25. 6.87 miles per second

27. (a) 1.082 miles per second (b) 1.101 miles per second

29. (b) 6 inches (c) $d^2y/dt^2 = -y$ (d) At $y = 0$
   (e) At $y = 6$ or $y = -6$

31. (b) Yes, if the second derivative is always 0, then the first
   derivative must be a constant a; hence the function
   must be $ax + b$.

33. (a) $ake^{-kt} - b$ (b) v approaches $-b$ (c) $-ak^2e^{-kt}$
   (d) Approaches 0

Sec. 6.7. Geometry and sign of f''

1. (a) $3 - 2\sqrt{6}$, 0, $3 + 2\sqrt{6}$ (b) $-1, 5$ (c) 2

3.

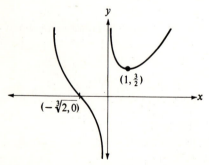

$(1, \frac{3}{2})$

$(-\sqrt[3]{2}, 0)$

5. y' changes sign at $x = 0$; y'' changes sign at $x = \pm 1/\sqrt{3}$

7. Inflection points when $x = \pm\sqrt{2}/2$

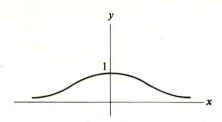

9. y' changes sign when $x = 1$; y'' changes sign when $x = 2$

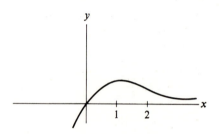

11. Inflection point when $x = 1$.

13. (a) $1\frac{1}{3}, 2\frac{2}{3}, 3\frac{1}{2}, 4\frac{1}{2}$ (b) $x \le 1\frac{1}{3}, 2\frac{2}{3} \le x \le 3\frac{1}{2}, x \ge 4\frac{1}{2}$
   (c) 0, 2, 3, 4 (d) $x \le 0, 2 \le x \le 3, x \ge 4$
   (e) $-1, 1\frac{1}{2}, 2\frac{1}{2}, 3\frac{1}{2}, 4\frac{1}{3}$
   (f) $x \le -1, 1\frac{1}{2} \le x \le 2\frac{1}{2}, 3\frac{1}{2} \le x \le 4\frac{1}{3}$

15. Maxima at $x = \pi/4 + 2\pi k$, minima at $x = 5\pi/4 + 2\pi k$;
   inflection points at $x = 3\pi/4 + \pi k$, where k is any integer.

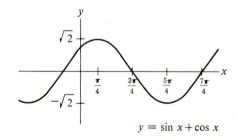

$y = \sin x + \cos x$

17. y' is never 0; y'' is 0 at $x = 1/\sqrt[4]{12}$

19. Relative minimum at $(0, 0)$, relative maximum at $(2, 4e^{-2})$

21. Relative maximum at $(5, 5^5 e^{-5})$

23. Relative minimum at $(-\frac{3}{2}, -\frac{27}{16})$

25. Relative minimum at $(1, 2)$, relative maximum at $(-1, -2)$

27. Relative minimum at $x = (-1 + \sqrt{5})/2$, relative maximum
   at $x = (-1 - \sqrt{5})/2$

29. (a) $x < 1, x > 2$ (b) $1 < x < 2$

31. (a)

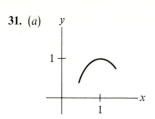

(b)

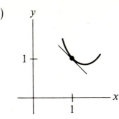

33. Show that $d^2y/dt^2 = 0$ when $y = M/2$.
35. Let $g(t) = f(tx_1 + (1 - t)x_2) - tf(x_1) - (1 - t)f(x_2)$.
Show $g(t) \le 0$ for all t in $[0, 1]$.

Sec. 6.8. Applied Max-Min Problems

1. (c) A tray cannot be constructed if any other x is used.
(d) $3 - \sqrt{3} \doteq 1.268$
3. (a) 1 square inch   (b) 4 inches
5. $\frac{2}{5}$ is squared, while $\frac{3}{5}$ is cubed.
7. (a) $36x - 3x^2$ grams per centimeter
(b) at $x = 6$ centimeters
9. (a) $t = \frac{5}{4}$ seconds   (b) 25 feet
11. (a) $6\sqrt{3}$ thousand dollars
(b) $(\sqrt{11} - 1)/2$ and $(\sqrt{11} + 1)/2$ ($1.16 and $2.16)
13. Width $= r$, height $= \sqrt{3}r$
19. For $s \le \frac{1}{4}$, run the cable in a straight line between the
station and the factory.   For $s > \frac{1}{4}$, run the cable $s - \frac{1}{4}$
miles along the river bank, then across the river to the
factory (or reverse order).
21. $32\pi a^3/81$
23. $r = 100/(3\pi)$, $h = 100/3$
25. $h = \sqrt{8}r$
27. The height should be $\frac{3}{2}$ times the length of the sides.
29. Height = twice diameter, or $r = \sqrt[3]{25/\pi}$, $h = \sqrt[3]{1,600/\pi}$
33. (a) $42D - D^2/3$ or SD (b) 63
35. $13\sqrt{13} \doteq 46.872$ feet

Sec. 6.9. The Differential

1. $dy = 0.6$, $\Delta y = 0.69$
3. $dy = -0.3$, $\Delta y = -0.271$
5. $dy = -0.2$, $\Delta y = -0.223$
7. (a) -0.02
9. (a) $h/2$ (c) 1.105, 1.1, 0.005
11.

dy	Δy	$\Delta y/dy$
27	37	1.370
-13.5	-11.375	0.843
0.3	0.331	1.103
-1.2	-1.141	0.951

13. $dx/(1 + x)$
15. $dx/(1 + x^2)$
17. $3 dx/\sqrt{1 - 9x^2}$
19. $2x2^{x^2} \ln 2\, dx$
21. 10 percent
23. (a) $df = 2x\,\Delta x$, $\Delta f = 2x\,\Delta x + (\Delta x)^2$
(b)

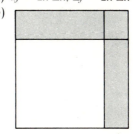

(c)

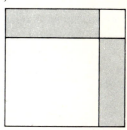

43. (b) $\frac{3}{4}$

Sec. 6.10. L'Hôpital's Rule

1. 1	**3.** 3	**5.** 1
7. 1	**9.** $\frac{1}{2}$	**11.** 1
13. $\frac{2}{3}$	**15.** $\frac{1}{2}$	**17.** $-\infty$
19. 0	**21.** 17	**23.** $\ln \frac{5}{3}$
25. -1	**27.** e^6	**29.** 0
31. Does not exist	**33.** 1	**35.** 1
39. 1		

Sec. 6.11. Summary: Guide Quiz on Chap. 6

3. (a) The graph crosses x axis at $(a, 0)$, decreasing.
(b) There is a maximum at a.
(c) The concavity changes from upward to downward at a.
6. (a) $-\frac{1}{3}\cos 3x + c$
7. $\log_{10} x$, x^{100}, $(1.001)^x$, 2^x
8. (a) $f'(x) = (7 \ln 5 + 8 \ln 6)f(x)$
(b) no, $f(x) = 6^3 e^{(7 \ln 5 + 8 \ln 6)x}$
9. (a) $dA = 2\pi r\, dr$ (b) $D(\pi r^2) = 2\pi r$
10. (a) $2, 5\frac{1}{2}$ (b) $-1, 4, 6$ (c) $-1, 1, 4, 5\frac{1}{2}, 6$
(d) None   (e) -1
13. Global maximum when $x = \sqrt{2}/2$, global minimum when
$x = -\sqrt{2}/2$, inflection points when $x = 0$, $\pm\sqrt{3}/2$.
14. (a) $(-\ln 2)/1,600$ (b) 3,200 years   (c) 3,200 years
(d) 5,313 years
15. 283 grams
16. $(4 - \sqrt{7})/3$ is squared, $(\sqrt{7} - 1)/3$ is cubed.
17. Maximum when $x = 0$, minimum when $r = 0$
18. (a) 8.1
19. (a) 2   (b) 2   (c) 1
20. (a) $-125 \csc 5x \cot^3 5x - 625 \csc^3 5x \cot 5x$
(b) $(4x^7 + 20x^3)/(1 - x^4)^{5/2}$

(c) $(2x^3 - 6x)/(1 + x^2)^3$

(d) $125 \sec 5x \tan^3 5x + 625 \sec^3 5x \tan 5x$

(e) $e^{-\cos 2x}(8 \sin^3 2x + 24 \sin 2x \cos 2x - 8 \sin 2x)$

(f) 30

Guide Quiz on Chaps. 1–6 (Computations)

1. (a) $2, 8, \pi/3$ (b) $-1, 1/\sqrt{3}, \frac{1}{2}$ (c) $\pi/6, -\pi/4, 2\pi/3$
2. (a) $180/\pi \doteq 57.296°$ (b) $\theta r^2/2$
   (c) $\lim_{n \to \infty} (1 + 1/n)^n$ (d) 2.718
3. 40
4. Global maximum when $x = 0$; inflection points when
   $x = \pm\sqrt{2}/2$; x axis is asymptote.
5. (a) $3e^{3x}$ (b) $-5 \ln 10 \cdot 10^{-5x}$
   (c) $3x^2 \cos 3x + 2x \sin 3x$
   (d) $(-5x \sin 5x - 2 \cos 5x)/x^3$
   (e) $\frac{5}{6}x^{-1/6} \sin^{-1} x + x^{5/6}/\sqrt{1 - x^2}$ (f) $-2x \csc^2 x^2$
6. (a) $5x^4 - 2 + 2/(2x + 3)$ (b) $1/(|x|\sqrt{9x^2 - 1})$
   (c) $10(x^7 + x^2)^9(7x^6 + 2x)$ (d) $(2x + 3x^2)/(1 + 3x)^2$
7. (a) $1/\sqrt{x^2 + 25}$ (b) $2\sqrt{6x - x^2}$
8. (a) $3 \tan^{-1} x$ (b) $5 \sin^{-1} x$ (c) $(1/2) \sin^{-1} 2x$
   (d) $x^5/5$ (e) $\sin^{-1} x$ (f) $-1/x$ (g) $\ln x$
   (h) $-\sqrt{1 - x^2}$
9. (a) $x^3 + x^2$ (b) $\sin x$ (c) $\frac{1}{2} \sin 2x$ (d) $2\sqrt{x}$
   (e) $\ln \sqrt{2x + 1}$
10. 4 percent
11. (a), (b), (d), (f)
12. (a), (b)
14. (a) $f(x) = -e^{-x}$ for instance
    (b) If $f''(x) < 0$, then $f(x) < 0$ for some x. *Hint*: Draw a
    tangent line.
    (c) $f(x) = e^x$ for instance

Guide Quiz on Chaps. 1–6 (Concepts)

1. (b) Write $x^4 = e^{4 \ln x}$, etc.
2. (a) To simplify the derivatives of trigonometric functions.
   (b) To simplify the derivatives of exponential and
   logarithmic functions.
3. (a) $2\sqrt{2}/\pi$ (b) $\sqrt{2}/2$ (c) 0 (d) e^3 (e) e^3
   (f) $-\pi$ (g) 2 (h) $-\frac{1}{2}$
4. (a) 1. There can be infinitely many such lines.
      2. Consider $y = x^3$ at $x = 0$; the tangent does cross the
      curve.
5. No, yes, apply the law of the mean to $h(x) = f(x) - g(x)$.
6. $f(3) \geq 3$
7. (b) 0.16 percent, 0.2 percent
8. $a = -1$
9. $y = x^3/6 + ax + b$, a, b constants

Review Exercises for Chaps. 1–6

1. (a) $1/(2\sqrt{x})$ (b) $-1/(2x^{3/2})$
3. $x/(3x + 4)$
5. (a) $e^{\sqrt{x}}/(2\sqrt{x})$ (b) $e^{\sqrt{x}}$
7. (a) $2 \sin 4x$ (b) $6 \sin^2 2x \cos 2x$
9. (a) $-2x/(x^2 + 1)^2$ (b) $(1 - x^2)/(x^2 + 1)^2$
11. (a) $10^x \ln 10$ (b) $(\log_{10} e)/x$
13. (a) $2x \ln x + x$ (b) $(1 - 2 \ln x)/x^3$
15. $x^2/(x + 1)^2$
17. $\sqrt{\frac{15}{2}(5x + 7)}$
19. $1/(3x)$
21. $\ln 10 - \cot x$
23. $\frac{2}{3}$
25. 3
27. $5x^4 \, dx$
29. $5 \sin 5x \, dx$
31. (b) 0.25 (c) 0.236
35. (b) 0.69 (c) 1.6 percent
39. No, yes, (consider the factorization of $f''(x)$ into first- and
    second-degree irreducible polynomials); no.
41. (a) $4\sqrt{3}l/(9 + 4\sqrt{3})$ for triangle and $9l/(9 + 4\sqrt{3})$
    (b) Use all the wire for the square
45. (a) 1.002 billion (b) 0.0225 (c) 2.28 percent
49. $x = e^{-t} + at + b$, a, b constants
51. $d^2y/dx^2 = 6y \, dy/dt = (6y)(3y^2) = 18y^3$
55. Note that the lines $x = 0$, $x = 1$, and $y = 0$ are asymptotes
57. (a) $(-x, y)$ (b) $(-x, -y)$
    (c) Graph in the first quadrant and then reflect across the
    y axis if even, or rotate through $180°$ about the origin
    if odd.
59. (a) 3 (b) Approximately $e^3 - 1 \doteq 19.09$
69. Maximum speed is sustained for 6 blocks and 24 seconds.
73. b
77. (b) implies (a) implies (c).
79. First graph $y = x$ and $y = \sin 2x$; then add the two results
83. (a) $y = \frac{3}{2}x + \frac{19}{4}$
    (b) Note that the line $x = \frac{5}{2}$ is also an asymptote.
85. (a) $y = \pm(b/a)x$
    (b) Top half of hyperbola $x^2/a^2 - y^2/b^2 = 1$.
91. The two X's may be different.
95. It is 1.

CHAPTER 7. THE DEFINITE INTEGRAL

Sec. 7.1. Four Estimates

1. (a) 8.75 (b) 14 (c) more (d) 5, less
3. (a) 8.91 (b) 11.88 (c) 6.48
5. (a) 44.55 grams (b) 59.4 grams (c) 32.4 grams
7. (a) $x^3 + x^2 - x + 5 \ln x - 3/x$ (b) $\frac{1}{3} \ln (x^3 + 1)$

(c) $x^2/2 - 1/x$ (d) $\frac{1}{3}e^{3x}$ (e) $-\frac{1}{2}\cos 2x$
(f) $xe^x - e^x$

9. 8.04, 40.22 grams, 64.35 miles, 8.04 cubic feet
11. (b) 8.9375π cubic feet
13. (a) 306   (b) 319.5   (c) 225   (d) 441
15. 1881.805 ergs
17. (a) 0.719; rectangles are above curve.
    (b) 0.669; rectangles are below curve.

Sec. 7.2. Precise Answers to the Four Problems

1. $\frac{8}{3}$
3. (a) $a^3/3$ (b) $b^3/3$ (c) $(b^3 - a^3)/3$
5. $\frac{7}{3}$ grams
7. less than 0.09
9. (c) $X_i^3(x_i - x_{i-1})$
   (d) $X_1^3(x_1 - x_0) + X_2^3(x_2 - x_1) + \cdots + X_n^3(x_n - x_{n-1})$
15. $b^3/3$ cubic feet
17. Volume of tent $= \frac{1}{3}$ volume of cube $= \frac{1}{3}b^3$
19. (c) Logarithm functions

Sec. 7.3. Summation Notation

1. (a) 6   (b) 20   (c) 14
3. (a) 4   (b) 1   (c) 450
5. (a) $\sum_{i=0}^{100} 2^i$ (b) $\sum_{i=3}^{7} x^i$ (c) $\sum_{i=3}^{102} 1/i$ (d) $\sum_{i=2}^{100} 1/i$
   (e) $\sum_{i=1}^{5} 1/(2i+1)$
   (f) $\sum_{i=1}^{51} 1/(2i-1)^2$ or $\sum_{i=0}^{50} 1/(2i+1)^2$
9. (a) 64   (b) 16   (c) 20
11. (d) 5,050
13. (b) n^2 (c) $2i - 1$ (d) $\sum_{i=1}^{n} (2i-1)$
17. (a) 1, 1.375, 1.625   (b) 1.988, $S_{20} \doteq 1.999$
    (c) It approaches 2.
19. (a) 1.5, 1.083, 0.95, 0.885
    (b) It is not hard to show that S_n decreases as n increases.
        It can be shown that $S_n \to \ln 2$.
    (c) $1/n + 1/(n+1) + \cdots + 1/2n > 1/2n + 1/2n + \cdots + 1/2n = n/2n = \frac{1}{2}$.

Sec. 7.4. The Definite Integral

1. (a) 1   (b) 2   (c) 3
3. (a) Duration of ith time interval
   (b) Speed at sample instant in ith interval
   (c) Approximate distance traveled during ith interval
   (d) Approximate distance traveled from time a to time b
   (e) Distance traveled between times a and b
5. (a) 6   (b) 30   (c) $\frac{152}{3}$
7. (a) $\frac{25}{12}$ (b) $\frac{77}{60}$
11. (a) $\int_0^2 x^3 \, dx$ (b) 4 (million dollars)   (c) $\int_2^3 x^3 \, dx$
    (d) $\frac{65}{4}$ (million dollars)
13. 72

17. (c) 0.835   (d) 1.149
19. (d) 0.8337 and 0.7337

Sec. 7.5. Summary: Guide Quiz on Chap. 7

1. (a) 28   (b) 18,750,000   (c) 90
2. (a) 39   (b) 10   (c) 72   (d) $\frac{31}{5}$
3. (a) $x_0^3(x_1 - x_0) + x_1^3(x_2 - x_1) + x_2^3(x_3 - x_2)$
   (b) $\int_2^4 x^3 \, dx$ (c) Smaller
4. (b) 0.7456   (c) Overestimates   (d) 0.6456
   (e) Underestimates
5. (a) $\int_1^2 1/t \, dt$ (b) $0.645 < \int_1^2 1/t \, dt < 0.746$
7. (a) 0.7456   (b) 0.6456   (c) 0.6919   (d) $\int_1^2 1/x \, dx$

Review Exercises for Chap. 7

1. (a) $\frac{124}{3}$ (b) 12   (c) $\frac{99}{2}$ (d) 4
3. (a) 0.02   (b) $x_0 = 0$, $x_1 = 0.02$, $x_i = 0.02i$
   (c) $X_1 = 0.01$, $X_2 = 0.03$, $X_i = 0.01(2i - 1)$
5. (a) $(2^1 - 2^0) + (2^2 - 2^1) + \cdots + (2^{100} - 2^{99}) = 2^{100} - 1$
   (b) $(2^1 - 2^0) + (2^2 - 2^1) + \cdots + (2^{101} - 2^{100}) = 2^{101} - 1$
   (c) $(\frac{1}{1} - \frac{1}{2}) + (\frac{1}{2} - \frac{1}{3}) + \cdots + (\frac{1}{100} - \frac{1}{101}) = 1 - \frac{1}{101} = \frac{100}{101}$
7. If an object moves c times as fast as another, it will
   travel c times as far during equal amounts of time.
9. (a) $-x + x^2 - x^3 + x^4$ (b) $x + x^2/2 + x^3/3 + x^4/4$
   (c) $-x + x^2/4 - x^3/9 + x^4/16$
11. Let $n = 100$ in Exercise 10.
13. (b) $\frac{15}{32}$ (c) $\frac{1}{3}$ (d) $\frac{1}{3}$
15. (a) 20, 12   (b) 20, 12
17. Distance traveled by the satellite
21. (a) Yes   (b) No   (c) Yes   (d) No   (e) Yes
    (g) Yes.   *Hint*: examine second derivative of $\ln (f + g)$;
        the inequality $(g' - f')^2 \geq 0$ might be of use.

CHAPTER 8. THE FUNDAMENTAL THEOREMS OF CALCULUS

Sec. 8.1. The First Fundamental Theorem

1. (a) $2xe^{x^2}$ (b) x^2 (c) $1/\sqrt{1 - 4x^2}$ (d) x^{-3}
   (e) $1/(x^2 - 1)$ (f) $1/(1 + 4x^2)$ (g) $x \cos x$
   (h) $\sin x$ (i) e^{-x} (j) $\sec x$ (k) $1/\sqrt{1 + 2x}$
   (l) $1/x$
3. $\pi/4$
5. $e^4 - 1$
7. $\frac{5}{72}$
9. $\frac{1}{2}\ln 3$
11. 2
13. 1
15. (a) 1   (c) $(e - 1)/(3e)$
17. (b) $\frac{3}{4}(2^{4/3} - 1)$ (c) $\frac{3}{2}(2^{2/3} - 1)$
19. (c) $4\pi r^3/3$

Sec. 8.2. The Second Fundamental Theorem

1. $\int_0^x \sqrt{1 + t^3}\, dt$
3. $\int_0^x \tan^{-1} t^2\, dt$
5. x^5
7. $\ln x$
9. e^{x^3}
11. $\frac{1}{2}\sqrt{(1 - x)/x}$
13. $-3x^2 + 4x$
15. $f(x^2)2x$
17. $e^{x_1{}^2}$
19. (a) $1/x$

Sec. 8.3. Proof of the Two Fundamental Theorems

1. 0
3. -1
5. (a) $12 \le \int_2^6 f(x)\, dx \le 20$
   (b) The mass is between 12 and 20 grams
7. $1/\sqrt{3}$
9. $2/\ln 2$
11. $x \ln x - x$
15. $2x$
17. (a) Distance traveled during time interval $[t_1, t_2]$.
   (b) Acceleration at time t.

Sec. 8.4. Antiderivatives

1. $\ln x$, $1 + \ln x$, for instance
3. $\sin^{-1} x$
5. $2x^3/3$
7. $2x^{1/2}$
9. $5 \sec^{-1} x$
11. $\frac{1}{2} \sec 2x$
13. $x^5/5$
15. $-\frac{1}{2} \cos 2x$
17. The fundamental theorem of calculus
19. (a) $\sqrt{2}/2$ (b) $\frac{1}{2} \ln 2$ (c) $\frac{3}{2}(2^{2/3} - 1)$
21. $x + C$ for any constant C
23. (a) Function   (b) Number   (c) Number
25. Yes
29. $2k\pi \pm \pi/3$, k an integer
31. (a) Hint: Differentiate $[\int_0^t f(x)\, dx]^2$ and $\int_0^t [f(x)]^3\, dx$.
   (b) $f(x) = x$

Sec. 8.5. Summary: Guide Quiz on Chap. 8

1. (a) True   (b) True   (c) True
2. (a) 9, 45, 72   (b) $\int_0^1 e^{-x^2}\, dx$
3. Both describe the distance traveled.
4. (a) $x^2/2 + 6\sqrt{x}$ (c) $-e^{-3x}/3$
6. (a) $\frac{1}{2} \ln (2x + 3)$ (b) $-\frac{1}{2}(2x + 3)^{-1}$ (c) $\sqrt{2x + 3}$
7. $(4\sqrt{2} - 2 - \ln 8)/3$

8. (a) $3x^2 \sin 2x^3$ (b) $3x^2 \sin 2x^3$
9. (a) 2 (b) 0 (c) 0 (d) $(1 - e)/e$

Guide Quiz on Chaps. 1-8

1. (a) $-1/x^2$ (b) $-1/x$ (c) $\frac{3}{4}$ (d) $3\pi/4$
2. (a) $x/\sqrt{1 + x^2}$ (b) $\sqrt{1 + x^2}$ (c) 1 (d) 1
   (e) 1 foot
3. (a) $5x^4$ (b) $\frac{1}{2}/\sqrt{x}$ (c) $2/(1 - x^2)$ (d) $1/(1 + x^2)$
4. (a) $\frac{1}{2} \ln [(1 + x)/(1 - x)]$ (plus any constant)
   (b) $\tan^{-1} x$ (c) $x^5/5$ (d) $2\sqrt{x}$
5. (a) $\sqrt{9 - x^2}$ (b) $x^2/(2x + 3)^2$ (c) $3 \sec^2 3x - 3$
   (d) $1/(3 + 2x^2)$ (e) $\sin^3 3x$ (f) $-x^2 e^{-2x}$
6. $1 + x/3$
7. $-\frac{1}{4} \sin 2x + c_1 x + c_2$
8. (a) $-3\sqrt{2}/2$ (b) $\frac{65}{4}$
10. Intercepts at $x = 0$; critical numbers $(-1 \pm \sqrt{5})/2$; asymptotes $y = 0$ and $x = -1$.

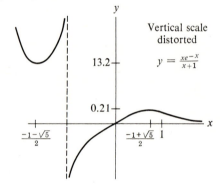

13. (a) $\frac{1}{2}$ (b) $\frac{1}{2}$
14. $200\pi/(\pi + 2)^2$
15. (a) e^{x^2} does not have an elementary antiderivative.
   (b) 1.4537
16. Distance, mass, area of region, volume of solid
17. Velocity, density, equation of curve, magnification, slope, rate of water flow
18. $-\frac{6}{11}(1 - x)^{11/6}$
19. (a) $4x^3/(x^4 + 1)$, $2x/(x^4 + 1)$ (b) $\frac{1}{4} \ln 2$, $\pi/8$

Review Exercises for Chaps. 1-8

1. (a) $e^{-x}(-2x^2 - 4x - 1)/(1 + 2x^2)^2$
   (b) $-7/(9e)$, switching left and right   (c) $-7/(9e)$
3. (a) $2x/(1 + x^2)$ (b) $\frac{1}{2} \ln \frac{5}{2}$
5. (a) $2x^{1/2}$, $-\frac{1}{2}x^{-3/2}$ (b) $-e^{-3x}/3$, $-3e^{-3x}$
   (c) $5 \ln x$, $-5x^{-2}$
7. (a) $-3e^{-3x}/(1 + e^{-3x})$ (b) $5 \cos x$ (c) $5 \cos 5x$
9. (a) $x^{1/3}$ (b) 10^x (c) x^2, $x \ge 0$
11. (a) $\sec x = 1/\cos x$, $\tan x = (\sin x)/\cos x$

25. (a) $\lim_{n \to \infty} (1 + 1/n)^n$ (b) 2.718
29. (a) $20x^3$ (b) $-2x^{-2}$ (c) $6(1 + x)^{-4}$
31. 1.7
33. $-4\sqrt{2}$
35. (a) x^3 (b) $1/(1 + \sqrt{x})$
37. (a) $g(T)$ is the cost of running the machine per unit time
    up to T.   (b) $[f(T) - g(T)]/T$
39. (a) $1 - e^{-b}$ (b) 1
41. (a) Graph oscillates between curves $y = e^{-x}$ and
    $y = -e^{-x}$, touching the top curve when $x = \pi/2 + 2n\pi$
    and the bottom curve when $x = 3\pi/2 + 2n\pi$.   Local
    maxima occur when $x = \pi/4 + 2n\pi$ and local minima
    when $x = 3\pi/4 + 2n\pi$.
    (b) Intercept $x = -1$.   Asymptotes $x = 0$, $x = 3$, and
    $y = 0$.   Critical numbers 1 and -3
49. (a) $a_1 = 0.250$, $a_2 = 0.347$, $a_3 = 0.391$
    (b) $\sum\limits_{k=1}^{n} \dfrac{1}{(1 + k/n)^2} \dfrac{1}{n}$
    (c) $\frac{1}{2}$
51. (b) ce^{kt}
53. $\sqrt{\ln t}$
55. (a) $e^{x^{3/2}} 3x^2$ (b) $-3x^2 \sin x^{3/2}$
57. $\int_0^x e^{-t^2} dt$
59. The same both ways (consider areas of appropriate
    parallelograms)

CHAPTER 9. COMPUTING ANTIDERIVATIVES

Sec. 9.1. Basic Facts

1. $x^4/4$
3. $\frac{3}{4}x^{4/3}$
5. $6\sqrt{x}$
7. $-\frac{5}{2}e^{-2x}$
9. $6 \sec^{-1} x$
11. $\frac{1}{4} \ln (1 + x^4)$
13. $-\ln |1 + \cos x|$
15. $\ln |x + x^2|$
17. $3 \sin x - \sin^{-1} x$
19. $\frac{1}{3}x^3 + \frac{3}{4}x^4$
21. $\frac{2}{7}x^{7/2}$
23. $\ln |x| + 2\sqrt{x}$
25. $\frac{1}{5}x^5 + 2x^3 + 9x$
27. x
29. $y = \frac{1}{168}(2x - 1)^7 + cx + d$
31. 1
33. $\frac{1}{2} \tan^{-1} x^2$
35. both

Sec. 9.2. Substitution

1. $\frac{1}{6}(1 + 3x)^6$
3. $e^{\sin \theta}$
5. $\frac{1}{3}(1 + x^2)^{3/2}$
7. $-\frac{1}{2}(1 + e^{2x})^{-1}$
9. $-e^{1/x}$
11. $-\frac{1}{5}\sqrt{2 - 5t^2}$
13. $-2 \cos \sqrt{x}$
15. $-\cos (\ln x)$
17. $-\frac{1}{4}(x^2 + 1)^{-2}$
19. $\sec \theta$
21. $-\frac{1}{3} \cos 3\theta$
23. $\sqrt{2x + 5}$
25. $-\frac{1}{8}(4x + 3)^{-2}$
27. $-\frac{1}{3}(x^2 + 3x + 5)^{-3}$

29. $-(1 + \sqrt{x})^{-2}$
31. $-e^{-x}$
33. $2e^{2x} + 16x - 8e^{-2x}$
35. $\frac{1}{2} \tan^{-1} x^2$
37. $\frac{1}{2} \sin^{-1} x^2$
39. $\frac{1}{2}x - \frac{1}{12} \sin 6x$
41. $\frac{1}{5}(x^2 - 2)(3 + x^2)^{3/2}$
43. $e^x - e^{-x}$
45. $\frac{1}{2} \tan^{-1} 2x$
47. Not elemenatry
49. $(\ln x)^2$
51. $\frac{1}{3}(x^2 - 1)^{3/2}$

Sec. 9.3. Table of Integrals

1. $\frac{1}{18}(3x + 5)^6$
3. $\frac{1}{4}[\ln |2x - 5| - 5/(2x - 5)]$
5. $\frac{2}{15}\sqrt{(5x - 7)^3}$
7. $\frac{2}{5}\sqrt{5x + 4}$
9. $\sqrt{2} \tan^{-1} \sqrt{(3x - 2)/2}$
11. $\frac{1}{3} \tan^{-1} (x/3)$
13. $\frac{1}{4} \ln |(2 + x)/(2 - x)|$
15. $(1/\sqrt{15}) \tan^{-1} (x\sqrt{5/3})$
17. $\frac{1}{6} \ln (3x^2 + 1)$
19. $\frac{1}{2}[x\sqrt{x^2 - 1} - \ln (x + \sqrt{x^2 - 1})]$
21. $\frac{1}{2}[x\sqrt{x^2 - 3} - 3 \ln (x + \sqrt{x^2 - 3})]$
23. $\frac{1}{4} \ln |2x^2 + x + 3| - (\sqrt{23}/46) \tan^{-1} [(4x + 1)/\sqrt{23}]$
25. $\sin^{-1} [(2x - 3)/\sqrt{29}]$
27. $\sqrt{x^2 + x - 6} + \frac{5}{2} \ln (2x + 1 + 2\sqrt{x^2 + x - 6})$
29. $-\dfrac{\sin^5 2x \cos 2x}{12} - \dfrac{5 \sin^3 2x \cos 2x}{48} - \dfrac{5 \sin 4x}{64} + \dfrac{5x}{16}$
31. $\frac{1}{5} \sin 5x - \frac{1}{15} \sin^3 5x$
33. $\frac{1}{5} \ln |\sec 5x + \tan 5x|$ or $\frac{1}{5} \ln \tan (5x/2 + \pi/4)$
35. $\frac{1}{27}(9x^2 \sin 3x + 6x \cos 3x - 2 \sin 3x)$
37. $-e^{-x}(x^2 + 2x + 2)$
39. $e^{-3x}(-3 \sin 5x - 5 \cos 5x)/34$
41. $x \sin^{-1} 3x + \frac{1}{3}\sqrt{1 - 9x^2}$
43. $x \sec^{-1} 4x - \frac{1}{4} \ln (4x + \sqrt{16x^2 - 1})$
45. $-\frac{1}{3} \tan (\pi/4 - 3x/2)$
47. $\frac{2}{45}(5x^3 + 2)^{3/2}$
49. $\frac{1}{2} \tan^{-1} [(\sin x)/2]$
51. $[(1 + 3x)/6]\sqrt{(1 + 3x)^2 - 5} -$
    $\frac{5}{6} \ln (1 + 3x + \sqrt{(1 + 3x)^2 - 5})$
53. $\frac{1}{4}x^2\sqrt{x^4 + 9} + \frac{9}{4} \ln (x^2 + \sqrt{x^4 + 9})$

Sec. 9.4. Substitution in Definite Integral

1. $[(e + 1)^4 - 16]/4$
3. $\frac{1}{2}$
5. $\frac{1}{6} \ln \frac{10}{3}$
7. $\sqrt{125/3}$
9. $\frac{1}{8}$

11. $\pi/12$

13. $\frac{1}{4}$

15. $6 - 3\sqrt[3]{4}/2$

17. 0

21. When $x = 1$, $y = -7$

23. Jill

Sec. 9.5. Integration by Parts

1. $e^{2x}(2x - 1)/4$

3. $e^{2x}(4x^3 - 6x^2 + 6x - 3)/8$

5. $\sin x - x \cos x$

7. $x \ln (4 + x^2) - 2x + 4 \tan^{-1} (x/2)$

9. $(x \ln x)^2/2 - (x^2/2) \ln x + x^2/4$

11. $e^x(\sin x - \cos x)/2$

13. $(x/3) \tan 3x + \frac{1}{9} \ln |\cos 3x|$

15. $e^{ax}(a \sin bx - b \cos bx)/(a^2 + b^2)$

17. (a) $-\frac{1}{2} \sin x \cos x + x/2$

(b) $-\frac{1}{4} \sin^3 x \cos x - \frac{3}{8} \sin x \cos x + 3x/8$

(c) $-\frac{1}{6} \sin^5 x \cos x - \frac{5}{24} \sin^3 x \cos x - \frac{5}{16} \sin x \cos x +$
$5x/16$

Sec. 9.6. Computing Integrals of $(ax + b)^{-n}, (ax^2 + bx + c)^{-}$, and $x(ax^2 + bx + c)^{-}$

1. $\frac{1}{3} \ln |3x - 4|$

3. $-5/[2(2x + 7)]$

5. $\frac{1}{3} \tan^{-1} (x/3)$

7. $\ln (x^2 + 9) + \tan^{-1} (x/3)$

9. $(1/\sqrt{3}) \tan^{-1} (x/\sqrt{3})$

11. $\frac{1}{4} \ln (2x^2 + 3)$

13. $\frac{1}{4} \ln |(2 - x)/(2 + x)|$

15. $\sqrt{\frac{1}{24}} \ln |(\sqrt{3} - \sqrt{2x})/(\sqrt{3} + \sqrt{2x})|$

17. (a) $2(x + 2)^2 - 8$ (b) $5(x + 1)^2 - 5$
(c) $3(x - 7/6)^2 - \frac{49}{12}$

19. $1/(4\sqrt{2}) \ln |(2x + 1 - \sqrt{2})/(2x + 1) + \sqrt{2})|$

21. $(1/\sqrt{8}) \tan^{-1} [(2x + 1)/\sqrt{2}]$

23. $\frac{1}{4} \ln (2x^2 + 5) - (3/\sqrt{10}) \tan^{-1} (\sqrt{2}x/\sqrt{5})$

25. $(1/\sqrt{10}) \tan^{-1} [(2x + 2)/\sqrt{10}]$

33. $\dfrac{-(13 + 4x)}{332(14x^2 + 8x + 13)} - \dfrac{1}{83\sqrt{166}} \tan^{-1} \dfrac{14x + 4}{\sqrt{166}}$

Sec. 9.7. Partial Fractions

1. $k_1/(x + 1) + k_2/(x + 1)^2 + k_3/(x + 1)^3$

3. $x - 1 + k_1/(x - 1) + k_2/(x + 2)$

5. $(c_1 x + d_1)/(x^2 + x + 1) + (c_2 x + d_2)/(x^2 + x + 1)^2 +$
$(c_3 x + d_3)/(x^2 + x + 1)^3$

7. $x + 3/(x + 1)$

9. $x + \frac{1}{2} + \frac{5}{2}/(2x + 1)$

11. $1/(x + 1) + 3/(x + 1)^2$

13. $(x - 3) + (4x + 15)/(x^2 + 3x + 5)$

15. $x + 2/(x + 2) + 2/(x - 2)$

17. $x + 3 \ln |x - 3|$

19. $\ln |x - 3| + 2 \ln |x + 3|$

21. $2 \ln |x + 2| + 3/(x + 3)$

23. $x^2/2 + x + \frac{27}{5} \ln |x - 3| + \frac{8}{5}(x + 2)$

25. $\frac{1}{2} \ln |(x^2 - 1)/x^2|$

27. $-4 \ln |x - 1| - 2/(x - 1) + 2 \ln (x^2 + 1) + 2 \tan^{-1} x$

29. $x - 4/(x - 2) - 4 \ln |x + 2|$

Sec. 9.8. Rational Functions of $\sin \theta$ and $\cos \theta$

1. $-\frac{1}{5} \sin^4 \theta \cos \theta - \frac{4}{5} \cos \theta - \frac{4}{15} \cos^3 \theta$

3. $\frac{1}{5} \tan^5 \theta$

5. $\frac{1}{7} \cos^7 \theta - (\frac{1}{5}) \cos^5 \theta$

7. $\ln |\sin \theta|$

9. $\frac{1}{3} \tan^3 \theta + \theta - \tan \theta$

11. $-\frac{1}{7} \csc^7 \theta + \frac{1}{5} \csc^5 \theta$

13. $5\theta/2 - \cos 2\theta + \frac{3}{4} \sin 2\theta$

15. $\sin \theta - \frac{1}{3} \sin^3 \theta$

17. $-\frac{1}{6} \cot^6 \theta - \frac{1}{4} \cot^4 \theta$

19. $\tan (\theta/2)$

23. $\frac{1}{2} \sec \theta \tan \theta + \frac{1}{2} \ln |\sec \theta + \tan \theta|$

25. $\sec^2 \theta + 2 \sec \theta \tan \theta + \tan^2 \theta$

27. $\ln |\tan (\theta/2)|$

29. $\displaystyle\int \frac{2(1 - u^2)^3 - 8u(1 + u^2)^2}{(1 + u^2)^3(1 + 2u - u^2)} \, du$

31. $(10/\sqrt{3}) \tan^{-1} [\sqrt{3} \tan (\theta/2)] - \theta$

Sec. 9.9. Trigonometric and Algebraic Substitution

1. $\dfrac{1}{2} \displaystyle\int \dfrac{\cos^2 \theta \, d\theta}{\sin^3 \theta}$

3. $\displaystyle\int \dfrac{\sec^3 \theta \, d\theta}{2\sqrt{2} \tan \theta + 5}$

5. $\displaystyle\int \dfrac{\sec \theta \tan^2 \theta \, d\theta}{}$

7. $\frac{1}{9} \int (3 \cos^3 \theta + \sin \theta) \, d\theta$

9. $\displaystyle\int \dfrac{\sec^3 \theta \, d\theta}{\sqrt{2}(1 + \sec \theta)}$

11. $\sqrt{25 - 4x^2} + 5 \ln \left| \dfrac{\sqrt{25 - 4x^2} - 5}{x} \right|$

13. $\dfrac{x^2 \sqrt{4 + 9x^2}}{27} - \sec \theta + 2 \tan \theta$

15. $(64x^2 - 9)\sqrt{9 - x^2}/(3x^3) + 64 \sin^{-1} (4x/3)$

17. $\frac{1}{4}\ln(4x + \sqrt{16x^2 - 9})$

19. $-(1 - 4x^2)^{5/2}(x^4/36 + x^2/252 + \frac{1}{2,520})$

21. $\frac{1}{3}(9x^2 - 4)^{3/2} - 4\sqrt{9x^2 - 4} + 8\tan^{-1}(\frac{1}{2}\sqrt{9x^2 - 4})$

23. $\frac{1}{16}\int \sin^3\theta\, d\theta/\cos^5\theta$

25. $(2^{11}/3)\int \sec^{12}\theta\, d\theta$

27. $\frac{1}{4}(2\sin\theta + \sin^3\theta)\, d\theta/\cos^9\theta$

31. let $u = \sqrt[n]{(ax + b)/(cx + d)}$

27. $\int (8\cos^3\theta + 1)/(64\cos^6\theta + 5)\, d\theta$

29. $\int 5^7\sqrt{5}\tan^{15}\theta \sec\theta/(5\sec^2\theta + 3 + \sqrt{5}\tan\theta)\, d\theta$

31. $\int \dfrac{\left[\dfrac{12u^2}{(1 - u^2)^2} + \dfrac{1 + u^2}{1 - u^2} + 1\right]\dfrac{2}{1 + u^2}}{2 + \dfrac{2u}{1 - u^2} + \dfrac{1 - u^2}{1 + u^2}}\, du$

33. $(u = \sin\theta)$: $\frac{1}{12}\ln|u - 2| - \frac{1}{24}\ln(u^2 + 2u + 4) + [1/(4\sqrt{3})]\tan^{-1}[(u + 1)/\sqrt{3}]$

35. $-(x^2 + 1)^{3/2}/(3x^3)$

37. $-\ln|3 + \cos x|$

39. $\frac{2}{9}(x^3 - 1)^{3/2}$

41. $\frac{1}{8}x/(4 + x^2) + \frac{1}{16}\tan^{-1}(x/2)$

43. $x/8 - \frac{1}{96}\sin 12x$

45. $\frac{1}{9}\tan^3 3\theta - \frac{1}{3}\tan 3\theta + \theta$

47. $x^2/2 + \ln|x| - \frac{1}{2}x^{-2}$

49. $10^x/\ln 10$

51. $\frac{1}{4}x^2/(x^4 + 1) + \frac{1}{4}\tan^{-1}x^2$

53. $x/2 + \frac{1}{4}\sin 2x$

55. $\frac{1}{3}(x^2 + 4)^{3/2}$

57. $\frac{1}{3}\tan^{-1}x^3$

59. $-\frac{1}{3}\cos x^3$

61. $\frac{1}{5}x^5(\ln x - \frac{1}{5})$

63. $2e^{\sqrt{x}}$

65. $(x - 1)\ln|x^3 - 1| - 3x + \frac{3}{2}\ln(x^2 + x + 1) + \sqrt{3}\tan^{-1}[(2x + 1)/\sqrt{3}]$

67. $-1/(x^2 + 1)^{1/2}$

69. $-2/(x + 1)^{1/2}$

71. $3\tan^{-1}(x + 2)$

73. $\frac{3}{5}x^{5/3} - \frac{3}{4}x^{4/3} + x - \frac{3}{2}x^{2/3} + 3x^{1/3} - 3\ln|1 + x^{1/3}|$

75. $(5x^6 - 6x^4 + 8x^2 - 16)\sqrt{x^2 + 1}/35$

77. $\ln x - \frac{1}{2}\ln(1 + x^2) - (\tan^{-1}x)/x$

79. $e^x(\sin 3x - 3\cos 3x)/10$

81. $\frac{1}{4}x/\sqrt{4 - x^2}$

83. $\frac{1}{8}\ln|(x^2 - 3)/(x^2 + 1)|$

85. $(x - 1)^{2/3}[\frac{3}{8}(x - 1)^2 + \frac{6}{5}(x - 1) + \frac{3}{2}]$

87. $\sqrt{x^2 + 4} - 2\ln|(\sqrt{x^2 + 4} + 2)/x|$

89. $\frac{1}{5}\sec^5\theta$

91. $(-\frac{1}{3})\ln|(3 + \sqrt{x^2 + 9})/x|$

93. $\frac{3}{8}x^{8/3} - \frac{6}{5}x^{5/3} + \frac{3}{2}x^{2/3}$

95. $(-\frac{4}{5})\sin^5 x + \frac{4}{3}\sin^3 x$

97. $e^{2x}/2 - 2x - e^{-2x}/2$

99. $\frac{1}{2}x^2\sin^{-1}x^2 + \frac{1}{2}\sqrt{1 - x^4}$

101. $(-\frac{1}{5})e^{-x} + \frac{1}{25}\ln(5 + e^x) - x/25$

103. $(36x + 14)(3x + 2)^{3/2}/135$

105. $\ln|x - 1| - 2/(x - 1) - \frac{1}{2}/(x - 1)^2$

107. $2\ln|e^x - 1| - x$

109. $\frac{9}{5}x^5 + 2x^3 + x$

111. $x - \frac{1}{3}\ln|x + 1| + \frac{1}{6}\ln(x^2 - x + 1) - (1/\sqrt{3})\tan^{-1}[(2x - 1)/\sqrt{3}]$

Sec. 9.10. Summary: Guide Quiz on Chap. 9

1. $\frac{1}{4}\ln(1 + x^4)$

2. $(x/2)\sqrt{4 - 9x^2} + \frac{2}{3}\sin^{-1}(3x/2)$

3. $\frac{1}{4}\ln|(x - 1)/(x + 1)| - \frac{1}{2}\tan^{-1}x$

4. $\frac{1}{12}\tan^6 2x$

5. $x + \frac{1}{4}\ln|(x - 1)/(x + 1)| - \frac{1}{2}\tan^{-1}x$

6. $\frac{1}{3}\ln(3x + \sqrt{9x^2 + 16})$

7. $\sqrt{x} + \sqrt[4]{x} + \frac{1}{2}\ln|2\sqrt[4]{x} - 1|$

8. $(1/\sqrt{6})\ln|(\sqrt{x + 3} - \sqrt{6})/(\sqrt{x + 3} + \sqrt{6})|$

9. $(\sqrt{3}/6)\ln|(\sqrt{3} + x)/(\sqrt{3} - x)|$

10. $-\frac{1}{2}\cos 2x + \frac{1}{3}\cos^3 2x - \frac{1}{10}\cos^5 2x$

11. $\frac{1}{5}e^x(\cos 2x + 2\sin 2x)$

12. $-\frac{1}{12}\cos 3x/\sin^4 3x - \frac{1}{8}\cos 3x/\sin 3x + \frac{1}{8}\ln|\csc 3x - \cot 3x|$

13. $\frac{1}{2}\tan^{-1}x^2$

14. $\frac{1}{2}\ln|(\sqrt{4 + x^2} - 2)/x|$

15. $\frac{1}{2}x\sqrt{x^2 - 9} + \frac{9}{2}\ln|x + \sqrt{x^2 - 9}|$

16. $(x/8)(-2x^2 + 45)\sqrt{9 - x^2} + (\frac{243}{8})\sin^{-1}(x/3)$

17. $(1/\sqrt{8})\tan^{-1}[(\sqrt{8}\sin x)/(1 + 3\cos x)]$

18. $-\sqrt{x^2 + 25}/(25x)$

Review Exercises for Chap. 9

1. (a) $\int_0^1 x^{3/2}\, dx$ (b) $\frac{2}{5}(4\sqrt{2} - 1)$

3. (a) $\frac{373}{14}$ (b) $\frac{721}{9}$

5. (a) $-1/(2x^2)$ (b) $2\sqrt{x + 1}$ (c) $\frac{1}{5}\ln(1 + 5e^x)$

7. $(-\frac{1}{4})(1 + x^2)^{-2} + \frac{1}{6}(1 + x^2)^{-3}$

9. $\frac{1}{3}(x^3 + 1)\ln(1 + x) + \frac{1}{2}(x + 1)^2 - \frac{1}{9}(x + 1)^3 - (x + 1)$

11. $(1/\sqrt{8})\ln|(\cos\theta - \sqrt{2})/(\cos\theta + \sqrt{2})|$

13. (b) $2\sin\theta$ (c) $\sqrt{8}\sin(\theta/2)$

15. (a) $\frac{6}{11}$

17. $2/(x - 1) + 1/(x^2 + x + 1)$

19. $-1/(x + 1) + x/(x^2 - x + 1)$

21. $x/(x^2 + \sqrt{2}x + 1) - 1/(x^2 - \sqrt{2}x + 1)$

23. $5x + 6 + 1/x - 1/(x + 1)$

25. (a) Odd m (b) Even n (c) m or n odd
  (d) m even or n odd   (e) n even or m odd

113. $\sqrt{2x + 1}$

115. $(\sqrt{17}/34) \ln |(2x^2 - 3 - \sqrt{17})/(2x^2 - 3 + \sqrt{17})|$

117. $\tan^{-1} e^x$

119. $\ln |(x + 2)/(x + 3)|$

121. $2 \ln |x^2 + 5x + 6|$

123. $(2/\sqrt{23}) \tan^{-1} [(4x + 5)/\sqrt{23}]$

125. $(1/\sqrt{73}) \ln |(4x + 5 - \sqrt{73})/(4x + 5 + \sqrt{73})|$

127. $-\cot x$

129. $-(\cot^3 x)/3 - \cot x$

131. $\sin^{-1} (2x - 5)$

133. $x\sqrt{\frac{1}{4} - x^2} + \frac{1}{4} \sin^{-1} 2x$

135. $2\sqrt{x^2 + 1}$

137. $x + \ln |x + 1| + 3 \ln |x - 1| + 4 \tan^{-1} x$

139. $-1/(x + 2) - 3 \ln |x + 1|$

141. $3 \ln |x| + \tan^{-1} 2x$

143. $x^2 + \frac{5}{2} \ln |1 - 3x^2| - \frac{3}{2} \ln |(1 + \sqrt{3}x)/(1 - \sqrt{3}x)|$

145. $\frac{1}{3} \sin^{-1} 3x$

147. (a) $\frac{1}{2} \ln |(x + 1)/(x + 3)|$ (b) $-1/(x + 2)$
  (c) $\tan^{-1} (x + 2)$
  (d) $(1/\sqrt{24}) \ln |(x + 2 - \sqrt{6})/(x + 2 + \sqrt{6})|$

149. (a) $\frac{1}{4} \sin x/\cos^4 x + \frac{3}{8}[\tan x \sec x + \ln |\tan (x/2 + \pi/4)|]$
  (b) $\frac{1}{5} \sec^5 x$ (c) $\frac{1}{2} \sec^2 x$

151. $(x^2 - 2)\sqrt{x^2 + 1}/3$

153. (a) $\int 2(u^2 - 1)^2 \, du$ (b) $\int [(u - 1)^2/\sqrt{u}] \, du$
  (c) $\int 2 \tan^5 \theta \sec \theta \, d\theta$
  (d) $\frac{2}{5}(x + 1)^{5/2} - \frac{4}{3}(x + 1)^{3/2} + 2(x + 1)^{1/2}$

155. (a) $x^2\sqrt{1 + x^2} - \int 2x\sqrt{1 + x^2} \, dx$
  (b) $\int \tan^3 \theta \sec \theta \, d\theta$ (c) $\int (u^2 - 1) \, du$

157. $\frac{2}{5}(1 + x)^{5/2} - \frac{2}{3}(1 + x)^{3/2}$

161. (b) $\frac{1}{2} \sin x - \frac{1}{10} \sin 5x$

CHAPTER 10. COMPUTING AND APPLYING DEFINITE INTEGRALS OVER INTERVALS

Sec. 10.1. Cross-sectional Length

1. (b) $x^2 - x^3$ (c) $y^{1/3} - y^{1/2}$

3. $\sqrt{9 - x^2} + x, -\frac{3}{2}\sqrt{2} \le x \le \frac{3}{2}\sqrt{2}$
  $2\sqrt{9 - x^2}, \frac{3}{2}\sqrt{2} \le x \le 3$

5. $c(y) = 3 - 2y, 0 \le y \le 1.$

7. $e^x - x - 1, 0 \le x \le 2$

9. $\frac{2}{3}x - \frac{2}{3}, 1 \le x \le 3, \frac{16}{3} - \frac{4}{3}x, 3 \le x \le 4$

11. (b) $x + 1 - x^2, \frac{1}{2} - \sqrt{5}/2 \le x \le \frac{1}{2} + \sqrt{5}/2$
  (c) $2\sqrt{y}, 0 \le y \le (3 - \sqrt{5})/2, \sqrt{y} - y + 1,$
  $(3 - \sqrt{5})/2 \le y \le (3 + \sqrt{5})/2$

13. (b) $x/2, 0 \le x \le 1, 1 - x/2, 1 \le x \le 2$
  (c) $1 - y, 0 \le y \le 1$

15. (a) $3y, 0 \le y \le 1,$ (b) $3 - 3y, 0 \le y \le 1,$
  (c) $3y + 3, -1 \le y \le 0$

17. (b) $3\sqrt{4 - x^2}, -2 \le x \le 2$
  (c) $\frac{4}{3}\sqrt{9 - y^2}, -3 \le y \le 3$

Sec. 10.2. Cross-sectional Area

1. (a) s^2, k^2s^2 (b) k^2
  (c) Area is proportional to square of linear dimension.

3. $\int_0^2 \frac{3}{2}x^2 \, dx$

5. (b) $\pi a^2 x^2/h^2$ (c) $\int_0^h (\pi a^2 x^2/h^2) \, dx$

9. (a) $2\sqrt{a^2 - x^2}$ (b) hx/a (c) $2hx\sqrt{a^2 - x^2}/a$
  (d) $\int_0^a (2hx\sqrt{a^2 - x^2}/a) \, dx$ (e) $2ha^2/3$

11. $h\sqrt{a^2 - x^2}$

13. $\pi a^2 h/2$

15. (b) $\frac{1}{2}(a^2 - x^2) \tan \theta$

17. $16\pi\sqrt{9 - y^2}$

Sec. 10.3. Computing Areas and Volumes

1. 25π

3. $\frac{17}{3}$

5. $e^2 - 5$

7. (a) $5\sqrt{5}/6$

9. 6π

11. $\frac{2}{3}$

13. $\frac{3}{2} - 1/\ln 2$

15. $\pi/2 - \pi^2/8$

17. (b) $\frac{32}{3}$

19. $\pi/2$

21. $\pi/4 + \frac{1}{2}$

23. $2\pi/3 - \sqrt{3}/2$

25. $\pi/2$

27. $\pi a^2 h/3$

29. $2a^2 h/3$

31. 144

33. π

35. $422 \pi/5$

37. $\pi/120$

39. $\pi(e - 2)$

41. $\pi/10$

43. $\frac{2}{3}$

Sec. 10.4. Volume by Shell Technique

1. $\int_0^1 2\pi x^3 \, dx$

3. $\int_0^1 2\pi y^3 \, dy$

5. $\int_0^1 2\pi(2 - x)x^2 \, dx$

7. (a) $\pi/2$ (b) $\pi/5$

9. (a) 2π (b) $\pi^2/4$

11. $\pi(e - 1)$

13. (a) $\pi(e^2 + 1)/2$ (b) $\pi(4 - e)$ (c) πe

15. $4\pi a^3/3$

17. $\pi/10$

19. (*a*) 2π (*b*) $\pi^2/4$

Sec. 10.5. Average of a Function

1. $\frac{7}{3}$
3. $1/(e - 1)$
5. $\ln \frac{3}{2}$
7. (*a*) 30 miles per hour   (*b*) $31\frac{1}{9}$ miles per hour
9. (*a*) $\pi r/2$ (*b*) $4r/\pi$ (*c*) r
11. $\frac{45}{8}$
15. $f(a)$

Sec. 10.6. Improper Integrals

1. $\frac{1}{2}$ **11.** π
3. 1   **13.** 0
5. 0   **15.** 0
7. Divergent   **17.** -1
9. Divergent   **19.** Divergent
21. $(\sin 1 + 2 \cos 1)/5$
23. Convergent (see remark preceding exercise)
33. 2,000 mile-pounds

Sec. 10.7. Polar Coordinates

1.

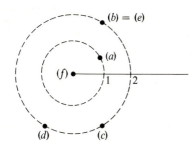

3. (*a*) $(3, 2k\pi + \pi/4)$, k any integer
   (*b*) $(-3, 2k\pi + 5\pi/4)$, k any integer
5. $x^2 + y^2 = y$ **11.** $r = 3/(\cos \theta + 2 \sin \theta)$
7. $4x + 5y = 3$ **13.** $r^2 = 2/\sin 2\theta$
9. $(x^2 + y^2)^{3/2} = 2xy$ **15.** $r = -2 \sec \theta$
17.

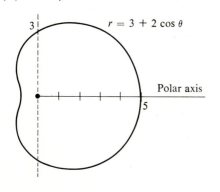

19. An expanding spiral which meets the polar axis at the points $(r, \theta) = (e^{2n\pi}, 0)$, $n = 0, 1, 2, \ldots$
21. See answer to Exercise 23(*a*).
23. (*a*)

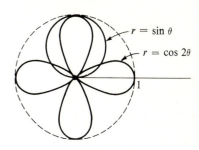

   (*b*) The pole and $(-1, 3\pi/2)$, $(\frac{1}{2}, \pi/6)$, $(\frac{1}{2}, 5\pi/6)$
25.

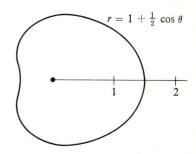

27. The curve has two loops contained between the lines $\theta = \pi/4$ and $\theta = -\pi/4$.
29. (*a*) An ellipse through the points whose rectangular coordinates are $(2, 0)$, $(-\frac{2}{3}, 0)$, $(0, 1)$, and $(0, -1)$.
   (*b*) $3x^2 + 4y^2 - 4x = 4$
31. (*b*) Maximum occurs when $\cos \theta = (-3 + \sqrt{17})/4$

Sec. 10.8. Parametric Equations

1. (*a*)

x	-3	-1	1	3	5
y	-3	-2	-1	0	1

   (*b*) $x - 2y = 3$

3. (*a*)

x	9	4	1	0	1	4	9
y	6	2	0	0	2	6	12

   (*d*) $x^2 + y^2 - 2xy - x = 0$

5. (*a*)

x	2	1.41	0	-1.41	-2	-1.41	0	1.41	2
y	0	2.12	3	2.12	0	-2.12	-3	-2.12	0

   (*c*) An ellipse   (*d*) $x^2/4 + y^2/9 = 1$ (*e*) No
7. (*a*) π (*b*) 2π (*c*) 3π (*d*) 4π

9. $y = x^2$

11. (a) $(1 + e)/(5 + 2\pi)$

13. (a) $x = (1 + \cos\theta)(\cos\theta)$, $y = (1 + \cos\theta)(\sin\theta)$
    (b) 1   (c) 0

15. (b) $k\pi/2$, k an integer   (c) πab

Sec. 10.9. Arc Length and Speed

1. $\frac{59}{24}$

3. 32 feet per second, $32\sqrt{2}$ feet per second, $32\sqrt{5}$ feet per second

5. (a) $(40^{3/2} - 13^{3/2})/27$ (b) $t\sqrt{4 + 9t^2}$
    (c) $(1, 1)$, $(4, 8)$ (d) $y = x^{3/2}$

7. (a) This curve, called a catenary, has the shape of a hanging chain, whose lowest point is $(0, 1)$.
    (b) $(e^b - e^{-b})/2$

9. $\sqrt{5}$

11. (b) $\sqrt{2}\,(e^{2\pi} - 1)$ (c) $\sqrt{2}\,(e^{4\pi} - e^{2\pi})$

13. 3π

15. (b) 10.004 (approx.)   (c) $\int_0^3 \sqrt{1 + x^4}\,dx$
    (d) 14,593 (approx.)   (e) 6.537 (approx.)

17. (a) $\frac{1}{3}\int_1^8 \sqrt{9 + 4x^{-2/3}}\,dx$ (b) $\frac{1}{2}\int_1^4 \sqrt{4 + 9y}\,dy$
    (c) $(40^{3/2} - 13^{3/2})/27$

19. (c) $\int_0^{2\pi} \sqrt{1 + 3\sin^2\theta}\,d\theta$

Sec. 10.10. Area in Polar Coordinates

1. $\pi^3/12$

3. $\pi/[(4 + \pi)(2 + \pi)]$

5. (v) $\frac{1}{2} - \pi/8$

7. (b) $(e^{4\pi} - 1)/4$

9. $18 + 9\pi/4$

11. $\pi/8$

13. $\pi/16$

Sec. 10.11. Area of Surface of Revolution

1. $\int_1^2 2\pi x^3 \sqrt{1 + 9x^4}\,dx$

3. $\int_1^8 2\pi y^{1/3} \sqrt{1 + \frac{1}{9}y^{-4/3}}\,dy$

5. $\int_0^{\pi/2} 2\pi \sin 2\theta \sin\theta \sqrt{\sin^2 2\theta + 4\cos^2 2\theta}\,d\theta$

7. $\pi[e\sqrt{e^2 + 1} - \sqrt{2} + \ln(e + \sqrt{e^2 + 1}) - \ln(1 + \sqrt{2})]$

9. $32\pi/5$

11. $2\pi \int_1^2 \sqrt{4x + 1}\,x^{-3}\,dx$

13. $2\pi \int_1^8 x\sqrt{1 + 16x^{2/3}/9}\,dx$

15. $2\pi \int_1^3 x(x^2/2 + x^{-2}/2)\,dx$

17. $\pi \int_2^6 (1 + \sqrt{x})\sqrt{(4x + 1)/x}\,dx$

19. $2\sqrt{2}\,\pi(2e^\pi + 1)/5$

23. Equal

25. (b) No

27. A section between two concentric circles and two rays from their center

Sec. 10.12. Estimates of Definite Integrals

1. Left-point, 3; right-point, 5; midpoint, $\frac{1}{2}$; trapezoid, 1; Simpson, 0

3. $\frac{5}{6} - \ln 2$; $\ln 2 - \frac{7}{12}$; $\ln 2 - \frac{24}{35}$; $\frac{17}{24} - \ln 2$; $\frac{25}{36} - \ln 2$

5. (a) 0.75   (b) 0.7   (c) 0.6949

7. 0.0008 versus 0.0012; 0.0018 versus 0.0023; 0.00002 versus 0.00003

9. (a) 0.2008   (b) 0.2078   (c) 0.2030

11. (a) 37.326   (b) 40.240   (c) 38.389

13. (a) 0.8834   (b) 0.8655

15. (a) 0.3981   (b) 0.3084

17. (a) $\sqrt{2/e}$ (b) $\sqrt{2/e}/10$

Sec. 10.13. Summary: Guide Quiz on Chap. 10

1. (a) $\int_0^2 (2x - x^2)\,dx$ (b) $\int_0^4 (\sqrt{y} - y/2)\,dy$

2. $\int_{-3}^3 \frac{5}{6}(9 - x^2)\,dx$ or $\int_0^3 \frac{10}{3}x\sqrt{9 - x^2}\,dx$

3. $\int_1^2 (1/x)\sqrt{x^2 + 1}\,dx$

4. $\int_0^{\pi/2} \sin^2(\theta/3)\,d\theta$

5. $\int_0^{\pi/2} \sqrt{\sin^2 2\theta + 4\cos^2 2\theta}\,d\theta$

6. $\int_0^{\pi/2} 6\pi \sin^4 t \cos t\,dt$

7. $\int_0^\pi (2\pi \sin\theta)(2\cos\theta - 2\cos 2\theta) \times$
    $\sqrt{8 + 12\sin^2 2\theta - 8\cos\theta\cos 2\theta + 16\sin\theta\sin 2\theta}\,d\theta$

8. $\int_0^1 \pi y^2\,dy + \int_1^2 \pi y(2 - y)\,dy$

9. $\int_0^{\pi/2} 2\cos^4 t \sin t\,dt$

10. $-3\sqrt{3}/4$

11. (a) $a < -1$ (b) Compare to $\int_0^\infty [1/(2x^4)]\,dx$

12. $\frac{1}{4}[\sin 1 + 2\sin\frac{9}{4} + 2\sin 4 + 2\sin\frac{25}{4} + 2\sin 9 + 2\sin\frac{49}{4} + \sin 16]$;
    $\frac{1}{6}[\sin 1 + 4\sin\frac{9}{4} + 2\sin 4 + 4\sin\frac{25}{4} + 2\sin 9 + 4\sin\frac{49}{4} + \sin 16]$

Review Exercises for Chap. 10

7. 24

9. (a) Halved   (b) Quartered   (c) Quartered   (d) $\frac{1}{16}$

11. 8.883

13. Divergent

17. (b) Infinite

23. $4\pi/3$

25. (b) $\frac{3}{4}$

33. Infinite

35. $\sqrt{2}$

37. (b) $\dfrac{h(t) - h(a)}{g(t) - g(a)}$; $\dfrac{h'(t)}{g'(t)}$ (c) See l'Hôpital's rule

41. $-\frac{1}{25}$

43. (a) $\int_0^1 \sqrt{1 + e^{2x}}\, dx$, $\int_0^1 2\pi e^x \sqrt{1 + e^{2x}}\, dx$, $\int_0^1 2\pi x \sqrt{1 + e^{2x}}\, dx$
(b) The first two are elementary (let $u = e^x$).

45. $0, -1$

47. $-1, 0, 1$

49. (a) $0.15e^3 \doteq 3$ (b) $0.00125e^3 \doteq 0.025$
(c) $0.0025e^3 \doteq 0.050$ (d) $0.00000167e^3 \doteq 0.00003$

51. Finite

53. $x + 2y = 3$

55. (a) $\pi r^2 \sqrt{h}(2 - \sqrt{2})$ (b) $2\pi r^2 \sqrt{h}$

59. *Hint*: Consider the equation $\ln x/x = \ln y/y$, and graph the function $\ln x/x$.

61. $y = \tan (\ln |1 + x| + C)$

CHAPTER 11. ADDITIONAL APPLICATIONS OF THE DERIVATIVE

Sec. 11.1. Implicit Differentiation

1. -4

3. $-\frac{7}{15}$

5. $-\pi/2$

7. $-\frac{3}{5}$

9. $(x^3 - 1)^{3/2} x^5 [9x^2/(2x^3 - 2) + 5/x]$

11. $(\sqrt{\cos x} \sqrt[3]{\ln x})^5 [-\frac{5}{2}\tan x + \frac{5}{3}/(x \ln x)]$

13. $(1 + 3x)^x [3x/(1 + 3x) + \ln (1 + 3x)]$

15. -1

17. (a) 1 (b) 4

19. $(-\frac{21}{10}, \frac{7}{10})$

21. $h/r = 2\sqrt{2}$

23. $h = 2r$

25. Height equals side

27. $\frac{5}{8}$

Sec. 11.2. Related Rates

1. (a) $\frac{3}{4}$ foot per second   (b) $\frac{4}{3}$ feet per second
(c) $9/\sqrt{19}$ feet per second

3. $3 \tan \theta = x$

5. (a) $5/(2\pi)$ yards per hour   (b) $1/(10\pi)$ yards per hour

7. $27\sqrt{2}/4$ square feet per second, increasing

9. $-1, 5$

11. $(\cos t + 2t)/(e^t + 1)$, $[(e^t + 1)(2 - \sin t) - e^t(\cos t + 2t)]/(e^t + 1)^3$

13. (a) -0.0014 feet per second (rounded off)
(b) -7.556 feet per second per second (rounded off)

15. (a) $y = x \tan \theta$;
(b) $\dot{y} = \dot{x} \tan \theta + x \theta \sec^2 \theta$
(c) $\ddot{y} = \ddot{x} \tan \theta + \sec^2 \theta \, [2\dot{x} \dot{\theta} + 2\dot{x} \dot{\theta}^2 \tan \theta + x\ddot{\theta}]$

17. $dy/dx = \frac{1}{2}$, $d^2y/dx^2 = \frac{3}{2}\ddot{x}/(\dot{x})^2$ (inadequate information)

Sec. 11.3. Curvature

1. (a) $5\sqrt{5}/2$ (b) $17\sqrt{17}/2$

3. $e(1 + e^{-2})^{3/2}$

5. 2

7. (a) $4/(e^x + e^{-x})^2$, $(e^x + e^{-x})^2/4$

9. $|\sec x|$

11. Radius of curvature is $3|\cos \theta \sin \theta|$.

13. $-\frac{1}{2}\ln 2$

17. $(a^2 \sin^2 \theta + b^2 \cos^2 \theta)^{3/2}/ab$

23. $2/a$

Sec. 11.4. Newton's Method

3. $4.375, 4.359$

5. (a) $3, 2.259$ (b) $3, 2.259$ (c) $1.917, 1.913$

7. (b) $\frac{6}{7}$

9. (c) 1.146

13. (d) $x_2 \doteq 2.207$ (e) $x_3 \doteq 2.036$

15. (b) $x_2 \doteq 0.3516$ and $x_3 \doteq 0.3589$

Sec. 11.5. Angle Between a Line and Tangent Line

1. $\pi/2$

3. Tangent undefined; angle $\pi/2$

5. $-2\sqrt{2}/3$

7. Undefined (the curves cross at a right angle).

9. $\gamma \doteq 1.964$ radians, $\phi \doteq 2.749$ radians

15. $f(\theta) = a \sin \theta$, where a is a nonzero constant.

19. (b) a

Sec. 11.6. The Hyperbolic Functions

11. (d) 1

17. (a) All x except 0

23. (a) $\cosh 1 \doteq 1.54306$, $\cosh 2 \doteq 3.756$
(b) Correct values: $\cosh 1 \doteq 1.54308$, $\cosh 2 \doteq 3.762$

Sec. 11.7. Summary: Guide Quiz on Chap. 11

1. $-\frac{9}{4}, \frac{113}{32}$

2. $\frac{1}{4}[3 \tan x - \frac{2}{5}/(1 + 2x)]$

3. $h = r$

4. (a) Decreasing, $-5.2\pi/3$ cubic feet per second   (b) No

5. (a) 3 (c) $5\sqrt{10}/3$

7. (a) $x_{i+1} = (3x_i^4 + 7)/(4x_i^3)$ (b) 1.644

9. (a) $\pi - \tan^{-1}(-\frac{1}{2}) \doteq 2.678$ radians   (b) 3.463 radians

10. (b) 1

Review Exercises for Chap. 11

2. (a) $d\phi/ds$

5. $y' = (1 - 2\pi)/(1 + \pi^2)$, $y'' = [(y')^2 - 4\pi y' - 2]/(\pi^2 + 1)$, and plug in the value of y'

7. $(-3 \cos 3t)/(2 \sin 2t)$, $-\frac{3}{4}\cos t/\sin^3 2t$

9. (a) $340/\sqrt{13}$ miles per hour   (b) increasing

11. $-\frac{1}{8}$

13. $(1 + x^2)^3 e^{x^2}(\sin^5 3x)[6x/(1 + x^2) + 2x + 15 \cot 3x]$
15. (a) $\frac{15}{7}$ feet per second   (b) $\frac{15}{7}$ feet per second
17. (a) $25/\sqrt{29}$ feet per second   (b) Decreasing
19. $5\pi/2$ feet per second, $5\sqrt{3}\,\pi/2$ feet per second
21. (a) f' is 0 for at least three values of x in $[1, 4]$.
    (b) f'' is 0 for at least two values of x in $[1, 4]$.
25. $(\dot{x}\ddot{y} - \dot{y}\ddot{x})/(\dot{x})^3$
27. (a) $2^x \ln 2 \sec^{-1} 3x + |x|\, 2^x/\sqrt{9x^2 - 1}$
    (b) $\cot^{-1} 4x/(2\sqrt{x}) - 4\sqrt{x}/(1 + 16x^2)$
    (c) $-e^{-x}(\cos 5x + 5 \sin 5x)$
    (d) $\frac{6}{5}/(1 + 2x) - \frac{9}{4}/(1 + 3x)$
    (e) $[(1 + 2x)^{3/5}/\sin^5 x][\frac{6}{5}/(1 + 2x) - 5 \cot x]$
    (f) $2x^{2x} + 2x^{2x} \ln x$
33. $44°$ and $7°$ (approx.)

CHAPTER 12. PARTIAL DERIVATIVES

Sec. 12.1. Rectangular Coordinates

1.

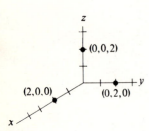

3. (b) $(1, 6, 4)$, $(1, 4, 7)$, $(1, 6, 7)$, $(5, 4, 4)$, $(5, 6, 4)$, $(5, 4, 7)$
    (c) 24   (d) 52   (e) 36
5. (a) 7   (b) 11   (c) $\sqrt{3}$
7. $3\sqrt{37}$
9. (a) $(\pm 7, 0, 0)$ (b) $(0, \pm 7, 0)$ (c) $(0, 0, \pm 7)$
11. Inside

Sec. 12.2. Graphs of Equations

1. The plane parallel to the xz plane that passes through $(0, 1, 0)$
3.

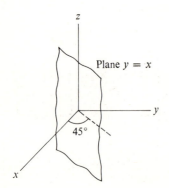

5. The plane parallel to the one in Exercise 3, passing through $(0, 3, 0)$
7.

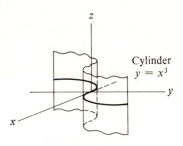

9.

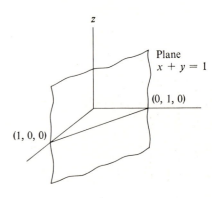

11.

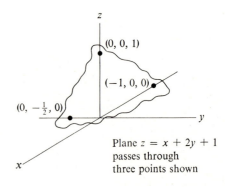

Plane $z = x + 2y + 1$
passes through
three points shown

19. (a)

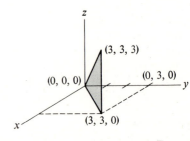

(b) $(0, 0, 0)$, $(3, 3, 3)$, $(3, 3, 0)$ (c) $9\sqrt{2}/2$

Sec. 12.3. Functions and Their Graphs

1.

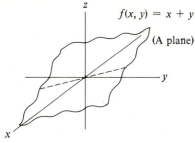

$f(x, y) = x + y$

(A plane)

3. Horizontal plane 3 units above the xy plane.

5.

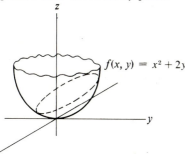

$f(x, y) = x^2 + 2y^2$

7. This is the saddle of Example 3 rotated 45° around the z axis.   (*a*) The x axis and y axis   (*b*) The x axis   (*c*) The y axis   (*d*) A hyperbola   (*e*) A line   (*f*)

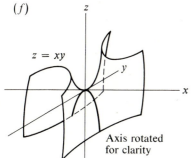

$z = xy$

Axis rotated for clarity

11. (*b*) Top half of sphere of radius 1 and center $(0, 0, 0)$
    (*c*) All (x, y) such that $x^2 + y^2 \le 1$
13. (*a*) A line   (*b*) A line   (*c*) A parabola
15. (*a*) Show that if (x_1, y_1) and (x_2, y_2) are equally distant from $(0, 0)$, then $f(x_1, y_1) = f(x_2, y_2)$.

Sec. 12.4. Partial Derivatives

1. $3x^2y^4$, $4x^3y^3$
3. $1/y$, $-x/y^2$
5. $y^3 \cos (xy^3)$, $3xy^2 \cos (xy^3)$
7. $1 + 1/(2\sqrt{x}) + y/(1 + x^2y^2)$, $3 + x/(1 + x^2y^2)$

9. $e^{x/y}/y$, $-xe^{x/y}/y^2$
11. (*a*) $2xy^4$ (*b*) $4x^2y^3$
13. (*a*) $(y^2 - x^2)/(x^2 + y^2)^2$ (*b*) $-2xy/(x^2 + y^2)^2$
15. 10, 12, -3
17. $12x^2y^7$, $42x^4y^5$, $28x^3y^6$
19. $(2x^2 - y^2)/(x^2 + y^2)^{5/2}$, $(2y^2 - x^2)/(x^2 + y^2)^{5/2}$, $3xy/(x^2 + y^2)^{5/2}$
23. (*a*) $-e^{x^2}$ (*b*) e^{y^2}
25. 1
27. 2
29. 0
31. (*a*) yx^{y-1} (*b*) $x^y \ln x$
33. $\cos (x + 2y + 1) - \cos (x + 2y)$, $2 \cos (x + 2y + 1) - 2 \cos (x + 2y)$
35. $-g(x)$, $g(y)$

Sec. 12.5. The Change, Δf, and df

1. 0.72, 0.7
3. 0.625, 0.5
5. 0.036
7. 17 percent
9. 17 percent
11. -8.1
13. $\frac{13}{30}$
19. (*a*) $u = u(x + \Delta x, y + \Delta y, z + \Delta z) - u(x, y, z)$
$= [u(x + \Delta x, y + \Delta y, z + \Delta z) - u(x + \Delta x, y + \Delta y, z)]$
$+ [u(x + \Delta x, y + \Delta y, z) - u(x + \Delta x, y, z)]$
$+ [u(x + \Delta x, y, z) - u(x, y, z)]$

Sec. 12.6. The Chain Rules

5. 31, 11
7. (*a*) $3r^2 \cos^3 \theta + 2r \sin^2 \theta$ (*b*) θ
9. $3x^2 - y \sin xy + ut$, $-x \sin xy$, tx, ux
13. $bf_1 + df_2$
25. 52 cubic feet per second
27. (*a*) No   (*b*) $x^2y + y^3$

Sec. 12.7. Critical Points

1. (*b*) -18
3. $(1, \frac{1}{3})$
5. $\frac{1}{4}$
7. $\frac{13}{4}$
15. (*b*) $\sqrt{2/3}$

Sec. 12.8. Maximum and Second-Order Partials

1. None
3. None
5. Local minimum at $(-4, -2)$
7. Local minimum at $(0, 2)$

9. Local maximum at $(0, 0)$

11. Local minimum at $(1, -1)$

13. (c)

15. (a)

17. (d)

21. $\sqrt[3]{2} \times \sqrt[3]{2} \times \sqrt[3]{1/4}$

23. 15 of x, 20 of y, 2 of z

25. (a) 12, 15, 613, 1730, 2343 (b) 24/7, 6, 1033.06, 2229.22, 3262.29

Sec. 12.9. Summary: Guide Quiz on Chap. 12

1. (a) Rotate the parabola $z = 2x^2$ in the xz plane around the z axis.

(b)

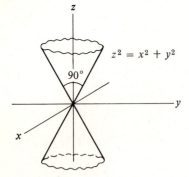

(c) The sphere of radius $\sqrt{6}$ and center $(0, 0, 0)$

(d)

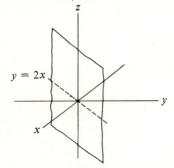

2. (a) $x/\sqrt{x^2 + y^2}$ (b) $2/y - 3/(3y + x)$ (c) $3x^2 y^2 e^{x^3 y}$

(d) $-6 \cos (2x + 3y)$ (e) $6/xy^3$

(f) $6y^2 \sin xy \cos^2 xy - 3y^2 \sin^3 xy$

3. (a) $(ye^y - e^y + 1)/y^2$ (b) $(ye^y - e^y + 1)/y^2$

6. $10\sqrt[3]{25} \times 10\sqrt[3]{25} \times 4\sqrt[3]{25}$

7. (a) 0.2206 (b) 0.2

8. 7 percent

10. $-6 < k < 6$

11. 5.6

Review Exercises for Chap. 12

1. (b) The graph is a saddle, described in answer to Exercise 7 of Sec. 12.3.

3. (a) On the line $x = y$ (b) 0 at $(1, 1)$

5. (a) Neither (b) Relative minimum (c) Neither

(d) Not enough information

7. (a) $(x + y)^3(x - y)^5$, $3(x + y)^2(x - y)^5 + 5(x + y)^3(x - y)^4$, $3(x + y)^2(x - y)^5 - 5(x + y)^3(x - y)^4$

(b) $3u^2 v^5 + 5u^3 v^4$, $3u^2 v^5 - 5u^3 v^4$

9. (a) Perpendicular to xy plane and passing through $(1, 0, 0)$ and $(0, 1, 0)$

(b) Sketch the points $(1, 0, 0)$, $(0, 1, 0)$, $(0, 0, 1)$ first; they determine the plane.

(c) Sketch the points $(2, 0, 0)$, $(0, 3, 0)$, $(0, 0, 4)$ first; they determine the plane.

11. (a)

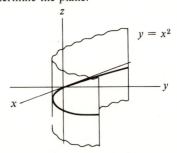

(b)

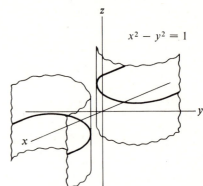

(c)

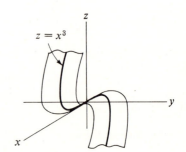

13. (c) $1 - \sqrt{2} < m < 1 + \sqrt{2}$

15. 0 at $(0, 0)$

19. (a) $2xy$ (b) $(3x^2 + 2x)ue^{3x}$

21. A change in t also induces a change in y, which, as well as the induced change in x, influences the change in z.

35. Show that f considered on the line $x + y = k$ is constant.

CHAPTER 13. DEFINITE INTEGRALS OVER PLANE REGIONS

Sec. 13.1. Integral over Plane Region

1. (a) 20 cubic inches   (b) 100 cubic inches
(c) The volume is between 20 and 100 cubic inches

3. (a) It always equals $5A$ (b) $5A$

5. (a) $\frac{25}{4}$

7. Circle

9. An isosceles triangle

13. (a) 0.758   (b) 1.184

15. (b) $\frac{2}{3}$

Sec. 13.2. Describing Plane Region

1. (b) $0 \le x \le 2$, $x/2 \le y \le 1$

3. (b) $0 \le x \le 1$, $x^2 \le y \le x$

5. $-a \le x \le a$, $-\sqrt{a^2 - x^2} \le y \le \sqrt{a^2 - x^2}$

7. A triangle with vertices $(0, 0)$, $(2, 4)$, $(2, 6)$

9. The region bounded by the curves $y = e^x$, $y = 1$, and $x = 1$

11. The region is bounded by parts of two concentric circles and two rays

13. $0 \le y \le 4$, $y/3 \le x \le y/2$ and $4 \le y \le 6$, $y/3 \le x \le 2$

15. $0 \le x \le 1$, $1 \le y \le e^x$

17. (b) $1 \le y \le 2$, $y \le x \le 3y$

Sec. 13.3. Computing $\int_R f(P)\, dA$ by Rectangular Coordinates

1. 16

3. $\frac{3}{2}$

5. $(16\sqrt{2} - 2)/21$

7. $(\pi^2 - 4)/(2\pi^3)$

9. $\frac{16}{3}$

11. 34

13. $2a^3/3$

15. $(\pi^2 + 4)/32$

17. $\frac{1}{2} \ln 2$

19. $\frac{4}{3}$

21. $\ln 2$

23. $1 - \cos 3$

Sec. 13.4. Center of Gravity

1. $(\frac{2}{3}, \frac{1}{3})$

3. $(\frac{4}{3}, 2)$

5. $(0, \frac{12}{5})$

7. $(\frac{3}{4}, \frac{7}{8})$

9. $(3\pi/(36\pi + 16), (48 + 3\pi)/(36\pi + 16))$

11. $(\frac{57}{35}, \frac{75}{14})$

13. No.   Consider the figure in Exercise 1.

15. (a) $\bar{x} = \dfrac{M_1 \bar{x}_1 + M_2 \bar{x}_2}{M_1 + M_2}$, $\bar{y} = \dfrac{M_1 \bar{y}_1 + M_2 \bar{y}_2}{M_1 + M_2}$

Sec. 13.5. Computing $\int_R f(P)\, dA$ by Polar Coordinates

1. $\frac{1}{3}$

3. $\frac{8}{5}$

5. $\int_0^{\pi/2} \left(\int_1^{2 + \cos\theta} r^2\, dr \right) d\theta$

7. $\int_0^{\pi/2} \left(\int_0^{1 + \sin\theta} r^2 \sin\theta\, dr \right) d\theta$

9. $\frac{2}{3}$

11. $3\pi/16$

13. $5\pi/32$

15. (b) $5\pi/16$

17. $32/9\pi$

19. $35\pi/1,024$

21. $3\pi/64$

23. $5\pi a^4/4$

25. $\frac{1}{3}$

27. $\pi/64$

29. $\frac{1}{80} + 3\pi/256$

31. $\pi a^4/8$

33. (a) $f(P) = r \cos\theta$ (b) $f(P) = r$

35. $\frac{1}{4}$

37. $\frac{1}{6}$

Sec. 13.6. Summary: Guide Quiz on Chap. 13

1. (a) $-a \le x \le a$, $0 \le y \le \sqrt{a^2 - x^2}$
(b) $\pi/6 \le \theta \le \pi/4$, $1/\cos\theta \le r \le 2/\cos\theta$

2. (a) $\frac{4}{3}$ (b) $\frac{4}{3}$

3. (a) $2a/3$

4. $\pi a^5/10$

5. (a) $3\pi/64$ (b) $3\pi/64$ grams   (c) $3\pi/64$
(d) $\frac{3}{8}$ degree

6. (a) $2\pi(1 - 101/e^{100})$ cubic feet
(b) $2\pi(1 - 51/e^{50})$ cubic feet

Review Exercises for Chap. 13

1. (a) $-2 \le x \le 2$, $x^2 \le y \le 4$
(b) $0 \le y \le 4$, $-\sqrt{y} \le x \le \sqrt{y}$

3. (a) $x^4/2 + x^3 - x - \frac{1}{2}$ (b) $y^4/2 + y^3 - 3y^2/2$

5. (a) $0 \le x \le a$, $0 \le y \le \sqrt{2a^2 - x^2}$
(b) $0 \le x \le a$, $0 \le y \le a$ and $a \le y \le a\sqrt{2}$
$0 \le x \le \sqrt{2a^2 - y^2}$
(c) $0 \le \theta \le \pi/4$, $0 \le r \le a \sec\theta$, and $0 \le r \le a\sqrt{2}$,
$\pi/4 \le \theta \le \pi/2$

7. $\pi/12$

9. (b) $\int_0^{a/\sqrt{2}} \left(\int_x^{\sqrt{a^2 - x^2}} y\, dy \right) dx$ (c) $a^3\sqrt{2}/6$

11. $\int_0^{a/\sqrt{2}} \left[\int_y^{\sqrt{a^2 - y^2}} (x^2 + y^2)\, dx \right] dy = a^4\pi/16$

13. (a) $\pi/4$ (b) $\frac{2}{3}$ (c) 0   (d) $\frac{4}{3}$

15. (a) $a^4/3$ (b) $a^4/12$

17. (a) $\frac{3}{5}(2^9)$ (b) $\frac{32}{3}$ (c) $2^5/3 + \frac{3}{5}(2^9)$

19. $(\pi^2 + 4)/32$

21. (a) $\pi/4$ (b) $4/9$ (c) $(0, \pi/8)$

23. $2\pi(1 - 2^{-a} a \ln 2 - 2^{-a})/(\ln 2)^2$
25. (a) $Mb^2/6$ (b) $Ma^2/6$ (c) $Ma^2b^2/[6(a^2 + b^2)]$
27. $f(P)$ is the square of the distance from P to the y axis
29. $f(P)$ is the square of the distance from P to the y axis plus the distance to the x axis
31. $f(P)$ is the reciprocal of the distance from P to the pole.
35. (a) By symmetry   (b) $x^2 + y^2 = r^2$ (c) $\pi a^4/2$
    (d) $\pi a^4/4$
37. $3\pi/8 + (\ln 2 - 3)/2$
47. (b) $4\pi/3$; $2\pi - 32/9$ (c) At the edge of town
49. $\pi(a^4 + 2a^2b^2)/2$
51. (a) 2.92   (b) 6.75
53. It is not true that $V = \int_\alpha^\beta A(\theta)\, d\theta$
55. (a) The integrand is nonnegative
57. If f is never 0, it must be always positive or always negative since f is continuous; hence $\int_0^\pi f(\theta)\, d\theta \neq 0$. Assume that there is only one root, a. Then the sign of $f(x) \sin(x - a)$ does not change on $(0, \pi)$. But $\sin(x - a) = \sin x \cos a - \cos x \sin a$, so the $\int_0^\pi f(x) \sin(x - a)\, dx = 0$, which is impossible if the integrand is of constant sign
59. Assume, say, $f(0) > 0$. Since $\int_0^{2\pi} f(\theta)\, d\theta = 0$, there is b in $(0, 2\pi)$ such that $f(b) < 0$. Therefore there are numbers a_1 and a_2 such that $0 < a_1 < b < a_2 < 2\pi$ and $f(a_1) = 0 = f(a_2)$
61. $\int_0^{2\pi} f(\theta)\, d\theta = [g^3(2\pi) - g^3(0)]/3 = 0$. Also, Exercise 56 and integration by parts yields $\int_0^{2\pi} f(\theta) \sin \theta\, d\theta = 0 = \int_0^{2\pi} f(\theta) \cos \theta\, d\theta$. Hence, by Exercise 60, $g^2(\theta)g'(\theta)$ vanishes for at least four distinct numbers; so $g'(\theta)$ must vanish at four points $[g(\theta) \neq 0]$. [The radius is perpendicular to the curve when $g'(\theta) = 0$]

CHAPTER 14. SERIES

Sec. 14.1. Sequences

1. Converges to 0
3. Converges to 0
5. Converges to 0
7. Diverges
9. Converges to 0
11. Converges to 0
13. Diverges
15. Converges to 1
17. 200
19. $\int_1^2 (1/x)\, dx = \ln 2$
21. (a) $\frac{1}{2}, \frac{2}{3}, \frac{3}{4}, \frac{4}{5}$ (b) $n/(n + 1)$
23. (a) $\frac{3}{4} = 0.75$, $\frac{2}{3} \doteq 0.667$, $\frac{5}{8} = 0.625$, $\frac{3}{5} = 0.600$ (What is a simple formula for a_n?)   (b) $\frac{1}{2}$
25. (a) $a_7 \doteq 1.9297$ (b) It approaches 2
27. (a) $a_7 = \frac{85}{128}$ (b) $\frac{2}{3}$
29. Greater than or equal to $1146 \doteq -5/\log 0.99$

Sec. 14.2. Series

1. 2
3. $\frac{1}{3}$
5. Diverges
7. Converges
9. Diverges
11. 10

15. $\sqrt{6}/4 + (\sqrt{6}/2)\sqrt{0.9}/(1 - \sqrt{0.9})$
21. (a) 1, 0.2929, 0.8702, 0.3702, 0.8175, 0.4092

Sec. 14.3. Alternating Series

1. Diverges (nth term does not go to 0).
3. Converges
5. Diverges (since the even partial sums diverge).
11. Diverges (nth term does not go to 0).
13. Converges (geometric series with ratio less than 1).
15. Converges (geometric series)
17. (a) $a_1 = x$, $a_2 = -x^3/6$, $a_3 = x^5/120$
19. $(-1)^{n+1} x^{2n-2}/(2n - 2)!$
21. $a_n = 1/n$, for instance

Sec. 14.4. Integral Test

1. Converges
3. Diverges
5. Diverges
7. Diverges
9. Diverges
11. See Exercise 10
17. (a) $(-\frac{1}{2}) \cos x^2$
    (c) $-x^2 \cos x + 2x \sin x + 2 \cos x$ (e) $(\ln x)^2/2$
    (f) $-(1 + \ln x)/x$

Sec. 14.5. Comparison and Ratio Tests

1. Converges (compare to $\sum_{n=1}^\infty 1/n^2$)
3. Converges by ratio test.
5. Converges by ratio test.
7. Converges by ratio test (remember "e").
9. Converges by limit comparison test (with $\sum_{n=1}^\infty 1/n^2$).
11. Converges by ratio or comparison test.
13. Converges by ratio test.
15. Converges by comparison to $\sum_{n=1}^\infty 2/n^2$.
17. Converges for $0 < x < 1$; diverges for $x \geq 1$.
19. (b) Let $c_n = 1/\sqrt{n}$ and $p_n = 1/n$; let $c_n = 1/n$ and $p_n = 1/n^2$.
23. (a) Use Exercise 19(a).   (b) Let $a_n = (-1)^{n+1}/\sqrt{n}$, for instance.
25. Use comparison test.

Sec. 14.6. Absolute Convergence

1. Converges conditionally
3. Converges absolutely
5. Converges absolutely
7. Converges absolutely for all x, $R = \infty$.
9. Absolutely converges on $(-2, 2)$, conditionally converges for $x = -2$, diverges elsewhere, $R = 2$.
11. Absolutely converges on $(-8, 2)$, diverges elsewhere, $R = 5$.

13. Absolutely converges on $(-1, 1)$, conditionally converges for $x = -1$, diverges elsewhere, $R = 1$.

15. Absolutely converges on $(-2, 0)$, diverges elsewhere, $R = 1$.

17. (a) Nothing   (b) Converges absolutely
(c) Converges absolutely

19. $-1 < x < 7$

Sec. 14.7. Manipulating Power Series

1. $x - x^3/3! + x^5/5!$

3. $1 + x/2 - x^2/8$

5. $x - x^2/2 + x^3/3$

11. $x + x^2 + x^3/3$

13. $1 - 3x/2 + 25x^2/24$

15. ∞

17. $\frac{1}{2} + \sqrt{3}(x - \pi/6)/2 - (x - \pi/6)^2/4 - \sqrt{3}(x - \pi/6)^3/12 + (x - \pi/6)^4/48$

19. $e + e(x - 1) + e(x - 1)^2/2 + e(x - 1)^3/3! + e(x - 1)^4/4!$

21. (a) 2.70833

29. (a) 0.58779

Sec. 14.8. Summary: Guide Quiz on Chap. 14

1. $\sum_{n=1}^{\infty} 1/n$

2. The sum gets arbitrarily large

6. $\sum_{n=1}^{\infty} (-1)^n/n$

8. Diverges (note the relation between S_{2n} and the sum of the harmonic series).

10. (a) Diverges (nth-term test)
(b) Converges (ratio test)
(c) Converges (alternating-series test)
(d) Diverges (integral test)
(e) Diverges [compare to (d)]
(f) Diverges (nth-term test)

11. (a) 0.60417

12. 0.3421 (using first three nonzero terms of series for $\sin \sqrt{x}$)

13. 0.1827

14. (a) -1 (b) $1/e$ (c) $\frac{1}{4}$ (d) $\ln \frac{4}{3}$ (e) $\ln \frac{3}{2}$

16. $x + x^3/3$

17. $\frac{1}{3}$

Review Exercises for Chap. 14

1. Converges to $1/(e - 1)$

3. Diverges by integral test

5. Converges to $-\frac{3}{7}$

7. Converges by alternating-series test

9. Diverges by ratio or nth-term test

11. Diverges by nth-term test

13. Diverges by limit-comparison test with harmonic series

15. Diverges by nth-term test

17. Converges by ratio test

19. Converges by comparison test

21. Converges to $\sin (\pi/2) = 1$

23. Converges only for $x = 0$

25. Absolutely converges for x in $(-1, 1)$, conditionally converges for $x = -1$, divergence elsewhere, $R = 1$.

27. Absolutely converges for x in $(-\frac{2}{3}, 2)$ to $(3x - 2)/(6 - 3x)$, divergence elsewhere, $R = \frac{4}{3}$.

29. Absolutely converges for x in $(-2, 0)$ divergence elsewhere, $R = 1$.

31. $2x - 8x^3/3! + 32x^5/5!$

33. (a) $n \geq 5$ (b) $n \geq 6$ (c) $n \geq 9$ (d) $n \geq 10$

35. -24

37. 0.7475

39. 1.6054

43. 0.1048

65. (a) $a_1 = 1$, $a_2 = 0.625$, $a_3 \doteq 0.518$, $a_4 \doteq 0.469$ (b) $\frac{1}{3}$

CHAPTER 15. TAYLOR'S SERIES AND THE GROWTH OF A FUNCTION

Sec. 15.1. Higher Derivatives and Growth of f

1. $0 \leq f(2) \leq \frac{10}{3}$

9. (b) $\dfrac{x^3}{3(1 + X)^3} \leq \dfrac{(\frac{1}{2})^3}{3 \cdot 1^3} = \dfrac{1}{24}$

11. $\frac{13}{2} \leq g(3) \leq 11$

23. $1 + x + \dfrac{x^2}{2} + \dfrac{x^3}{3} \leq f(x) \leq 1 + x + \dfrac{x^2}{2} + \dfrac{2x^3}{3}$

Sec. 15.2. Taylor's Series

1. $1 - x + x^2/2! - x^3/3! + x^4/4!$

3. $1 + 4(x - 1) + 6(x - 1)^2 + 4(x - 1)^3 + (x - 1)^4$

5. $x - x^3/3! + x^5/5!$

7. $(x - 1) - (x - 1)^2/2 + (x - 1)^3/3$

9. $1 - x + x^2 - x^3 + x^4$

11. $1 - (x - \pi/2)^2/2! + (x - \pi/2)^4/4!$

17. (a) $x - x^3/3$

19. Using $P_4(\frac{2}{3}; 0)$, about 1.9465

21. Using $P_5(2; 0)$, $\frac{1}{15}$

25. (b) $1 + 10x + 45x^2$ (c) $45x^8 + 10x^9 + x^{10}$

27. (c) About $x = 1.37$

29. (a) $1 - 2^2x^2/2! + 2^4x^4/4! - \cdots + (-1)^n 2^{2n} x^{2n}/(2n)! + \cdots$
(b) $1 - x/2! + x^2/4! - x^3/6! + \cdots + (-1)^n x^n/(2n)! + \cdots$

31. (d) The exponential e^{-1/x^2} approaches 0 faster than any power of x (as $x \to 0$).
(e) 0   (f) Only for $x = 0$.

Sec. 15.3. Harmonic Motion

7. (a) c_1 (b) kc_2
9. Consider the point (c_1, c_2) in the xy plane.
11. $A, -A$
13. (a) The curve $y = \sin t$ stretched three-fold vertically.
   (b) The curve $y = \sin 3t$ shrunk by the factor 2π horizontally.   (c) 1

Sec. 15.4. Error in Estimating Integral

7. $6e$
9. (a) $0.005e$ (b) $\frac{19}{45}(0.0001e)$

Sec. 15.5. General Binomial Theorem

5. $1 + \dfrac{x}{2} - \dfrac{x^2}{8} + \dfrac{x^3}{16} - \dfrac{5x^4}{128}$

7. (a) $1 - 2x + 3x^2 - 4x^3 + 5x^4$ (b) $(-1)^n(n+1)$
15. (a) $1 - x^3/2 - x^6/8 - x^9/16$ (b) 0.492
17. (a) Replace x by $\frac{1}{2}$.

   (b) $1 + \dfrac{(r+1)}{2} + \dfrac{(r+2)(r+1)}{2^2 \cdot 2!} +$

   $\dfrac{(r+3)(r+2)(r+1)}{2^3 \cdot 3!} + \dfrac{(r+4)(r+3)(r+2)(r+1)}{2^4 \cdot 4!} +$

Sec. 15.6. Taylor's Series for $f(x, y)$

1. $f_{xxy} = f_{xyx} = f_{yxx} = 140x^4y^6$
   $f_{xyy} = f_{yxy} = f_{yyx} = 210x^4y^5$
   $f_{xxx} = 60x^2y^7, f_{yyy} = 210x^5y^4$
3. $f_{xyy} = f_{yxy} = f_{yyx} = 2/y^3, f_{yyy} = -6x/y^4$, others 0
7. $1 + x + x^2/2 + y^2$
11. (a) $1 + x/3 + y/3$ (b) $\frac{13}{9}$

Sec. 15.7. Summary: Guide Quiz for Chap. 15

2. (a) $1 + x + x^2/2$
3. Approximately 0.2389

4. $\dfrac{\sqrt{3}}{2} + \dfrac{1}{2}\left(x - \dfrac{\pi}{3}\right) - \dfrac{\sqrt{3}/2}{2!}\left(x - \dfrac{\pi}{3}\right)^2 -$

   $\dfrac{\frac{1}{2}}{3!}\left(x - \dfrac{\pi}{3}\right)^3 + \dfrac{\sqrt{3}/2}{4!}\left(x - \dfrac{\pi}{3}\right)^4 + \cdots$

7. $4 + 2x - y + 9x^2/2 - xy + 2y^2 + 5x^3/6 + 7x^2y/2$
   $- xy^2/3 + y^3/2$
9. (a) $(-1)^n\dbinom{n+2}{n}x^n$ (b) $|x| < 1$

Review Exercises for Chap. 15

1. (a) 1.1167   (b) 1.1
3. (a) 1.25   (b) Error less than $1/(7!)$
5. $(x-5)^2 + 11(x-5) + 32$
7. Between 0.144 and 0.288
11. (a) 0 and 48   (b) 24 and 24   (c) 0 and 0
13. (a) 1, 7, 21, 35, 35, 21, 7, 1
   (b) $1 + 7x + 21x^2 + 35x^3 + 35x^4 + 21x^5 + 7x^6 + x^7$
   (c) $a^7 + 7a^6b + 21a^5b^2 + 35a^4b^3 + 35a^3b^4 + 21a^2b^5 + 7ab^6 + b^7$
15. $1 + x/2 - x^2/8 + x^3/16$
19. 1.3495
21. 0.30899
23. 0.005
25. 0.0004
27. 0.042
29. $-(x-\pi) + (x-\pi)^3/6 - (x-\pi)^5/120$
33. (a) $48 + 67(x-2) + 3(x-2)^2 + 4(x-2)^3$
   (b) Using two terms: 54.7, 41.3
35. 0.15595
37. Use Rolle's theorem several times.
39. $f(3) \geq 9/2$
41. (b) About 0.79
   (c) Between $\frac{1}{13}$ and $\frac{1}{26}$ (estimate is too small)
43. $\frac{29}{36}$
45. About 0.364
47. (a) $x^3 - x^9/6 + x^{15}/120 - \cdots$ (b) $\frac{1}{64}$
   (c) Less than $1/(60 \cdot 2^{10})$
49. (a) $20 \cdot 19 \cdot 18 \cdot 17 \cdot 16x^{15}$ (b) $50 \cdot 49 \cdot 48(x-1)^{47}$
   (c) 83!
51. $2 + 3(x-1) + \frac{1}{2}(x-1)^2 - \frac{1}{6}(x-1)^3$
53. $1 + (x+1) + (x+1)^2/2 + (x+1)^3/6 + (x+1)^4/24$
55. (a) $1 + 3x + 3x^2 + x^3$ (b) $1 + 4x + 6x^2 + 4x^3 + x^4$
   (c) $1 + 5x + 10x^2 + 10x^3 + 5x^4 + x^5$
57. (a) 210   (b) 210
61. 3, 2, 5, 12, -1, 10
63. $\frac{1}{3}$
65. (a) $3\cos 100\pi t + 3\sin 100\pi t = 3\sqrt{2}\sin(100\pi t + \pi/4)$
   (b) $3\sqrt{2}$ (c) 50
69. About 1.26
71. $x + x^3/6 + 9x^5/120$.
73. $D^{20}[(\cos 3x + 3\cos x)/4]$ is easy to compute.
79. $3 + 2x + 5x^2/2 + x^3/12$
89. Between 17.5 and 22.5 miles away.
93. Use power series for $\tan^{-1} x$.   (a) 0   (b) 100!
95. Consider its Taylor's series around a number in (a, b).
103. π^2 is irrational.
105. Note the result is true for $k = 1$ and 2. If true for $k = 2n$, then true for $k = 2n + 1$ [by Rolle's theorem and the fact that $(F_{2n+1})' = F_{2n}$]. Also, if true for $k = 2n - 1$,

then true for $k = 2n$.   [Note: $F_{2n} = F_{2n-1} + x^{2n}/(2n)!$.
The graph of $y = F_{2n}(x)$ is concave upward.   If $F_{2n}(x)$
has exactly one root, α, show that $F_{2n-1}(\alpha)$ is equal to 0.
If $F_{2n}(x)$ has two roots, then $F_{2n-1}(\alpha)$ is negative at both,
though 0 between, yet is increasing]

111. $b^2/2$ is correct.   In the second estimate use $e^{-b} \doteq 1 - b + b^2/2$, the third summand being significant in this
problem.

CHAPTER 16. THE MOMENT OF A FUNCTION

Sec. 16.1. Work

1. 280(62:4) foot-pounds
3. 180(62.4) foot-pounds
5. 62.4(18,000) foot-pounds
7. $62.4(\frac{880}{3})$ foot-pounds
9. $(1 + \sqrt{2})(\frac{4}{15})$
11. π
13. 62.4(81π/4) foot-pounds

Sec. 16.2. Force against Dam

1. 62.4(32) pounds
3. 62.4(64π) pounds
5. $62.4(\frac{58}{3})$ pounds
7. $\frac{1}{3} - \pi/4$
9. The woman has about 5 times as much pressure.

Sec. 16.3. Moment of f

1. ln 2
3. 1
5. (ln 2)/2
17. 14π
19. $36\pi(\pi - 1)$
21. (a) $a/3$ (b) $b/3$
23. $\bar{x} = \frac{3}{2}$
25. $\bar{x} = 2/(e^2 - 1)$
35. πla, where a = radius and l = slant height

Sec. 16.4. Summary: Guide Quiz on Chap. 16

1. $2a/(\pi + 2)$
2. 62.4(10 + 15π) foot-pounds
3. 62.4(6 + π) foot-pounds
5. (a) (sin 1)/2   (b) $\pi/2 - 1$

Review Exercises for Chap. 16

1. $\frac{2}{3}, \frac{2}{5}, \frac{2}{7}$
5. The inequality $M_1{}^2 \le M_0 M_2$ reads $[\int_0^a xh(x)\,dx]^2 \le \int_0^a h(x)\,dx \int_0^a x^2h(x)\,dx$, which is the Schwarz inequality in the case $f(x) = \sqrt{h(x)}$, $g(x) = \sqrt{x^2h(x)}$

7. (a) $3W^2M/4$ (b) $W^2M/3$
(c) Mass tends to be farther from x axis than from y axis.
11. (a) If f has no roots in (a, b), then the sign of $f(x)$
remains fixed and $\int_a^b f(x)\,dx$ is either positive or
negative.
(b) If c_1 is the only root, consider $\int_a^b (x - c_1)f(x)\,dx$.
Since the integrand is of fixed sign, the integral is not
0, but equals $\int_a^b xf(x)\,dx - c_1 \int_a^b f(x)\,dx$.
(c) If c_1 and c_2 are the only roots, consider
$\int_a^b (x - c_1)(x - c_2)f(x)\,dx$ and reason as in (b).

CHAPTER 17. MATHEMATICAL MODELS

Sec. 17.1. Probability

1. $\frac{1}{6}$
3. $\frac{2}{9}$
5. $\frac{1}{2}$
7. (a) $\frac{1}{2}$ (b) $\frac{1}{4}$ (c) $\frac{1}{8}$ (d) $1/2^n$
9. $\sum_{k=1}^{\infty} k/2^k$
11. (c) $\sum_{k=1}^{\infty} k2^{k-1}/3^k$
13. $\frac{9}{2}$
15. $\frac{3}{8}$

Sec. 17.2. Probability Distributions

5. 5
7. (b) $\frac{1}{3}$
11. (a) $F(0) = 0$, $F(20) = 0.1$, $F(80) = 0.5$
13. $\int_0^\infty tF'(t)\,dt = \lim_{b\to\infty} \{-t[1 - F(t)]|_0^b + \int_0^b [1 - F(t)]\,dt\}$
$= \int_0^\infty [1 - F(t)]\,dt$
15. (a) The fraction of the original population that reaches
age a is $1 - F(a)$; of this fraction, approximately
$F'(t)\,\Delta t$ have between $t - a$ and $t + \Delta t - a$ years
remaining
(b) Integrate by parts with $u = t - a$, $v = F(t) - 1$

Sec. 17.3. Poisson Model

1. 0.002
3. (a) 0.14   (b) 0.28   (c) 0.28   (d) 0.31
5. (a) Presumably the factory operates continuously at a
constant rate with only a small accident rate
(b) Conditions 1 and 2 are sufficient for the Poisson
distribution to hold
(d) $1/e^2$, $2/e$, $2/e^2$, $4/(3e^2)$, $2/(3e^2)$
7. (a) Assumptions 1 and 2 hold and $k = 1$
(b) $300P_0(1) \doteq 40$ pages
9. (a) 0.223   (b) 0.050   (c) 0.00012
11. $1 - 5e^{-2} \doteq 0.32$
13. $x^2e^{kx}e^{-kx}$
19. Set $k = 1$; then $P_n(x) = x^ne^{-x}/n!$.   Use the fact that
$e^x = e^x[P_0(x) + P_1(x) + \cdots]$

21. $1/k$

Sec. 17.4. Summary: Guide Quiz on Chap. 17

1. (a) $\frac{1}{36}$ (b) 36 times
2. (a) $\frac{1}{9}$ (b) $\frac{64}{729}$
3. $\frac{1}{5}$
4. (a) $e^{-2} \doteq 0.135$ (b) $1 - 5e^{-2} \doteq 0.325$
5. (a) $e^{-3} \doteq 0.050$ (b) $3e^{-3} \doteq 0.15$
   (c) $9e^{-3}/2 \doteq 0.22$ (d) $9e^{-3}/2 \doteq 0.22$ (e) 0.36
6. $\frac{1}{4}$
7. (a) $\frac{1}{6}$ (b) $\frac{5}{6}$ (c) 6

Review Exercises for Chap. 17

1. (a) A measure of change in economic activity.
   (b) Rate of change of the change in economic activity.
   (c) The constant A represents a steady-state equilibrium rate of economic activity. The equation describes how changes from equilibrium influence this activity.
3. (a) 0   (b) 1   (c) Yes; by jumping in time (avoidance).
   (d) At the second trial at least $(0.92)(0.08) + 0.08 = 0.1536$
   (e) Let p_n be the probability of avoidance at the nth stage. Then $\{p_n\}$ is an increasing bounded sequence, hence has a limit $L \le 1$. Since $p_{n+1} \ge 0.92p_n + 0.08$, it follows that $L \ge 0.92L + 0.08$, hence $L \ge 1$. Thus $L = 1$.

CHAPTER 18. DEFINITE INTEGRALS OVER SOLID REGIONS

Sec. 18.1. Integral over Solid Region

1. The mass is between 384 and 1,920 ounces.
3. $20\pi r^3/3$
5. 960 ounces
7. $2\sqrt{6}$ inches
11. Since it lies in a cube of side 10, volume is at most 10^3. (This result can be strengthened)

Sec. 18.2. Description in Rectangular Coordinates

1. $0 \le z \le 1, 0 \le y \le 1 - z, 0 \le x \le 1 - z - y$
3. $-1 \le x \le 1, -\sqrt{1 - x^2} \le y \le \sqrt{1 - x^2}$,
   $0 \le z \le x + y + 2$
5. (a) $0 \le x \le 2, 0 \le y \le 1 - x/2, 0 \le z \le 3x + 4y$
7. $-a \le x \le a, -\sqrt{a^2 - x^2} \le y \le \sqrt{a^2 - x^2}$,
   $-\sqrt{a^2 - x^2 - y^2} \le z \le \sqrt{a^2 - x^2 - y^2}$
9. $-3 \le x \le 5$,
   $2 - \sqrt{16 - (x - 1)^2} \le y \le 2 + \sqrt{16 - (x - 1)^2}$,
   $3 - \sqrt{16 - (x - 1)^2 - (y - 2)^2} \le$
   $z \le 3 + \sqrt{16 - (x - 1)^2 - (y - 2)^2}$

11.

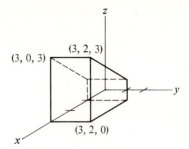

13.

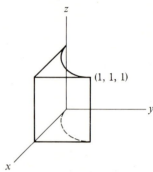

15.

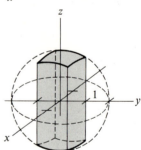

17. The cross section is a square.

Sec. 18.3. Description in Cylindrical or Spherical Coordinates

1. Surface of a right-circular cylinder of radius 1
3. Plane parallel to xy plane
5. Sphere of radius 3
9. $0 \le \theta \le 2\pi, 0 \le r \le a, -\sqrt{a^2 - r^2} \le z \le \sqrt{a^2 - r^2}$
11. (a) $(\sqrt{x^2 + y^2}, \tan^{-1}(y/x), z)$ (There are other descriptions of θ)   (b) $(r \cos \theta, r \sin \theta, z)$
13. (a) $z = 0$ (b) $\theta = \pi/4$ (c) $r = 0$
15. A solid cone.
17. A quadrant of a cylinder, but with tilted top (in the plane $z = x$).
19. The sphere of radius 2, center at the origin.
21. A half plane with the z axis as an edge.
23. The top half of the z axis.

25. $(\rho \sin \phi, \theta, \rho \cos \phi)$

27. $0 \le \theta \le 2\pi, 0 \le \rho \le a, 0 \le \phi \le \pi$ (in any order)

29. $0 \le \theta \le 2\pi, 0 \le \phi \le \pi/6, 0 \le \rho \le a$ (in any order)

31. $0 \le \theta \le 2\pi, 0 \le \phi \le \pi/6, 0 \le \rho \le (3a \sec \phi)/5$

37. $\rho = \sqrt{r^2 + z^2}, \theta, \phi = \tan^{-1}(r/z)$

39. $\rho = \sqrt{x^2 + y^2 + z^2}, \phi = \tan^{-1}\left(\sqrt{x^2 + y^2}/z\right),$
$\theta = \tan^{-1}(y/x)$

Sec. 18.4. Computing $\int_R f(P)\,dV$ with Rectangular Coordinates

1. (a) $\frac{1}{24}$ (b) $\frac{1}{4}$

3. $M/5$

5. $(b^2 + c^2)M/3$

7. 0

9. $\frac{1}{120}$

Sec. 18.5. Computing $\int_R f(P)\,dV$ with Cylindrical or Spherical Coordinates

1. $4\pi a^3/3$

3. $2Ma^2/5$

7. $3Ma^2/2$

9. $\pi a^4/16$

11. $3a/4$

13. $2Ma^2/5$

19. Yes (law of mean is of use)

Sec. 18.6. Summary: Guide Quiz on Chap. 18

2. $3(b^4 - a^4)/4(b^3 - a^3)$

4. (a) $(5, \tan^{-1}\frac{4}{3}, \pi/2)$ (b) $(0, 3 \sin 1, 3 \cos 1)$
(c) $(\sqrt{8}, \pi/4, \pi/4)$

6. (a) $r, \rho^2 \sin \phi$, respectively

8. $a^2/2 + h^2/12$

9. $(e - 1)/4$

10. (c) $M(3a^2/20 + 3h^2/5)$

11. (a) $\rho \cos \phi = 3$ (b) $x^2 + y^2 = 4$ (c) $\rho \sin \phi = 2$
(d) $r^2 + z^2 = 9$

12. a^2

Review Exercises for Chap. 18

15. Use the pythagorean theorem in the form $x^2 + y^2 + z^2 = \rho^2$.

CHAPTER 19. ALGEBRAIC OPERATIONS ON VECTORS

Sec. 19.1. Algebra of Vectors

1. (a)

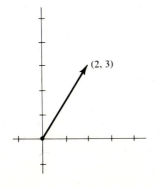

(2, 3)

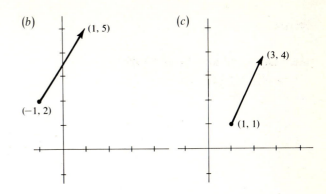

(b)   (1, 5)   (−1, 2)

(c)   (3, 4)   (1, 1)

3. (a) $\overrightarrow{(3, 5)}$ (b) $\overrightarrow{(3, 1)}$

5. (a) $\overrightarrow{(2, 3)}$ (b) $\overrightarrow{(-2, 3)}$ (c) $\overrightarrow{(3, -3)}$

7. (a) 2 and 3   (b) 0 and −3   (c) 0 and 3
(d) −6 and −9

9.

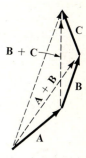

B + C

C

B

$A \times B$

A

11. (a) $\overrightarrow{(1, -2, 4)}$ (b) $\overrightarrow{(3, 4, 2)}$ (c) $\overrightarrow{(-2, -1, 3)}$

13. (a) $\sqrt{59}$ (b) $\sqrt{11}$

15. (a) About 104.4 miles per hour
(b) About 28.3° south of east

17. (a) Tail P, head R (b) Tail Q, head P
(c) Tail R, head Q

Sec. 19.2. Product of Scalar and Vector

3. 3

5. (a) $-4\mathbf{i}$ (b) $5\mathbf{j}$ (c) $\mathbf{0}$

7. (a) $2\mathbf{i}$ (b) $2\mathbf{i}$ (c) $2\mathbf{i}$ (d) $2\mathbf{i} + 2\mathbf{j}$

9. (a) 4   (b) $\mathbf{u}$

11. (a) $\frac{1}{2}\mathbf{i} + (\sqrt{3}/2)\mathbf{j}$

13. $1/\sqrt{10}$ and $3/\sqrt{10}$

15. (b) $\mathbf{i} = \frac{1}{2}\mathbf{u}_1 - (\sqrt{3}/2)\mathbf{u}_2$

17. If $\mathbf{A}$ is $\mathbf{0}$, then $1\mathbf{A} + 0\mathbf{B} + 0\mathbf{C} = \mathbf{0}$.   Similarly if $\mathbf{B} = \mathbf{0}$.   If $\mathbf{B} = x\mathbf{A}$, then $x\mathbf{A} - 1\mathbf{B} + 0\mathbf{C} = \mathbf{0}$. If $\mathbf{A}$ and $\mathbf{B}$ are not parallel, draw a parallelogram with diagonal $\mathbf{C}$ and sides parallel to $\mathbf{A}$ and $\mathbf{B}$. Then the sides are $x\mathbf{A}$ and $y\mathbf{B}$, and $x\mathbf{A} + y\mathbf{B} - 1\mathbf{C} = \mathbf{0}$.

21. (*g*) A straight line.

Sec. 19.3. Dot Product

1. $6\sqrt{2}$
3. 0
5. 0
7. (*c*) Their dot product is 3, which is not 0.
9. $\cos^{-1}\frac{63}{65} = 14.25°$
15. $-\frac{1}{3}$
17. $\sqrt{15}/5$
19. $\frac{1}{3}$
21. $2\sqrt{5}/5$
23. -4
27. $\sqrt{3}$
29. (*d*) There is a profit
31. $(-\sqrt{2}/2)\mathbf{u}_1 + 2\sqrt{2}\,\mathbf{u}_2 - 4\mathbf{u}_3$

Sec. 19.4. Lines and Planes

1. $4x + 5y = 23$
3. $2x + 3y = 23$
5. $2\mathbf{i} - 3\mathbf{j}$
7. $3\mathbf{i} - \mathbf{j}$
9. (*a*) 2   (*b*) $\frac{2}{13}$
11. $\sqrt{14}/14$
17. No
19. $2/\sqrt{29}, -3/\sqrt{29}, 4/\sqrt{29}$, respectively
21. $0, 0, -1$ respectively
23. (*a*) $\frac{2}{7}, -\frac{3}{7}, \frac{6}{7}$, respectively
   (*b*) 73.4°, 115.4°, 31.0°, respectively
25. (*a*) $12\sqrt{61}/61$ (*b*) $5\sqrt{26}/26$ (*c*) 0
27. Let $\mathbf{P} = \overrightarrow{P_0 P_1}$. Then distance is

$$\sqrt{|\mathbf{P}|^2 - (\mathbf{P} \cdot \mathbf{A}/|\mathbf{A}|)^2}$$

31. No
33. Let $\mathbf{A} = \overrightarrow{P_1 P_2}$, $\mathbf{B} = \overrightarrow{P_1 P_3}$, and $\mathbf{C}$ = vector determined as in Exercise 32. Then distance is

$$\frac{|\overrightarrow{P_1 P_0} \cdot \mathbf{C}|}{|\mathbf{C}|}$$

Sec. 19.5. Directional Derivative and Gradient

1. 5
3. $\sqrt{2}/2$
5. -5
7. $(-\frac{3}{125})\mathbf{i} - (\frac{4}{125})\mathbf{j}$
9. $-e\mathbf{i}$
11. $2\mathbf{i} + 3\mathbf{j} + \mathbf{k}$

13. (*a*) 3   (*b*) 2   (*c*) 1   (*d*) -3 (*e*) $6/\sqrt{13}$
19. -3
23. $-\alpha \cos \alpha x \cos \beta y \mathbf{i} + \beta \sin \alpha x \sin \beta y \mathbf{j}$

Sec. 19.6. Determinants

1. -22
3. 22
5. 0
13. (*a*) $(\frac{28}{29}, -\frac{17}{29}, \frac{13}{29})$
17. (*b*) 0
19. (*a*) Area $OCEG = a_1 b_2$, area $OCD = \frac{1}{2} a_1 a_2$,
      area $DEF = \frac{1}{2}(b_2 - a_2)(a_1 - b_1)$, area $OFG = \frac{1}{2} b_1 b_2$

Sec. 19.7. Cross Product

1. $-\mathbf{i}$
3. $-\mathbf{i} + \mathbf{j}$
7. $\sqrt{507}$
9. $-5x + 2y + 9z = 0$
11. The distance is 0

Sec. 19.8. Summary: Guide Quiz on Chap. 19

1. (*a*) -3 (*b*) $\sqrt{6}$ (*c*) $\mathbf{i}/\sqrt{6} + 2\mathbf{j}/\sqrt{6} - \mathbf{k}/\sqrt{6}$
   (*d*) $-\sqrt{6}/2$ (*e*) $-\mathbf{i}/2 - \mathbf{j} + \mathbf{k}/2$ (*f*) $-3/\sqrt{14}$
   (*g*) $-3\sqrt{21}/42$ (*h*) About 109.1°   (*i*) $5\mathbf{i} - 5\mathbf{j} - 5\mathbf{k}$
   (*j*) $-5\mathbf{i} + 5\mathbf{j} + 5\mathbf{k}$ (*k*) $-\mathbf{i}/\sqrt{3} + \mathbf{j}/\sqrt{3} + \mathbf{k}/\sqrt{3}$
   (or its negative)   (*l*) $5\sqrt{3}$
3. $\frac{1}{3}, -\frac{2}{3}, \frac{2}{3}$, respectively
5. (*a*) 8
8. $(\frac{11}{3}, \frac{2}{3}, 1)$
9. (*a*) $\sqrt{14}$ (*b*) $2/\sqrt{14}, 1/\sqrt{14}, 3/\sqrt{14}$, respectively
   (*c*) One is $\mathbf{i}/\sqrt{5} - 2\mathbf{j}/\sqrt{5}$

Review Exercises for Chap. 19

1. (*b*) $a_1 b_1 + a_2 b_2 + a_3 b_3$
3. (*a*) $-2\mathbf{i}$ (*b*) $3\mathbf{j}$ (*c*) $0.72\mathbf{i} + 0.96\mathbf{j}$
   (*d*) $-92\mathbf{i}/41 + 115\mathbf{j}/41$
9. $(-\frac{12}{7}, \frac{6}{7}, -\frac{18}{7})$
15. $8/\sqrt{21}$
17. (*a*) Any numbers proportional to $-21, 5, 17$
   (*b*) $-21/\sqrt{755}, 5/\sqrt{755}, 17/\sqrt{755}$, respectively
   (*c*) 139.8°, 79.5°, 51.8°, respectively
   (*d*) $(\frac{8}{5}, 1, \frac{9}{5})$ is one
19. $x = 1 + 9t, y = 1 + 3t, z = 2 - 5t$
21. Yes
23. (*a*) $(\frac{72}{61}, \frac{48}{61}, \frac{36}{61})$ (*b*) $(-\frac{5}{61}, \frac{78}{61}, \frac{150}{61})$
25. Determine whether or not $Ax_1 + By_1 + C$ and $Ax_2 + By_2 + C$ have the same sign.

27. $\dfrac{|A(x_1 - x_2) + B(y_1 - y_2) + C(z_1 - z_2)|}{\sqrt{A^2 + B^2 + C^2}}$

29. $90° - \cos^{-1} \frac{11}{15} = 47.2°$

31. (b) $(\overrightarrow{p_1, p_2, p_3, p_4, p_5})$ and $(\overrightarrow{x_1, x_2, x_3, x_4, x_5})$

35. One is $4\mathbf{i} + \mathbf{j} - 5\mathbf{k}$

37. 3

53. (a) $\sqrt{x_1{}^2 + x_2{}^2 + \cdots}$ is finite   (b) $\sqrt{3}/3$

55. Their profit (or loss) is 0.

CHAPTER 20. THE DERIVATIVE OF A VECTOR FUNCTION

Sec. 20.1. Derivative of a Vector Function

1. (c) $\mathbf{V}(0) = 32\mathbf{i}$
$\mathbf{V}(1) = 32\mathbf{i} - 32\mathbf{j}$
$\mathbf{V}(2) = 32\mathbf{i} - 64\mathbf{j}$

3. (d) $\sqrt{0.971}$

9. (c) Both are constant.

11. (b) $\mathbf{R}(t) = \cos t^2\mathbf{i} + \sin t^2\mathbf{j}$
$\mathbf{V}(t) = -2t \sin t^2\mathbf{i} + 2t \cos t^2\mathbf{j}$
(c) $|\mathbf{V}(t)| = 2t$.   Speed of particle increases arbitrarily.

13. (b) $\mathbf{R}(25\sqrt{3}/8) = (625\sqrt{3}/2)\mathbf{i} + \frac{1875}{4}\mathbf{j}$
$\mathbf{V}(25\sqrt{3}/8) = 100\mathbf{i}$
(c) $\mathbf{R}(25\sqrt{3}/4) = 625\sqrt{3}\,\mathbf{i}$

17. (b) $\pi/4$

19. (b) $\cos^{-1}(10\sqrt{406}/203) \doteq 6.98°$

Sec. 20.2. Properties of Derivative of Vector Function

1. (b) No

7. *Hint*: Work with each of the components.

9. (a) Yes   (b) No

Sec. 20.3. Acceleration Vector

1. $\mathbf{R}(1) = \mathbf{i} + \mathbf{j}$
$\mathbf{V}(1) = 2\mathbf{i} + 3\mathbf{j}$
$\mathbf{A}(1) = 2\mathbf{i} + 6\mathbf{j}$

3. $\mathbf{R}(0) = \mathbf{0}$, $\mathbf{V}(0) = 4\mathbf{i}$, $\mathbf{A}(0) = -32\mathbf{j}$
$\mathbf{R}(1) = 4\mathbf{i} - 16\mathbf{j}$, $\mathbf{V}(1) = 4\mathbf{i} - 32\mathbf{j}$, $\mathbf{A}(1) = -32\mathbf{j}$
$\mathbf{R}(2) = 8\mathbf{i} - 64\mathbf{j}$, $\mathbf{V}(2) = 4\mathbf{i} - 64\mathbf{j}$, $\mathbf{A}(2) = -32\mathbf{j}$

5. (a) Part of hyperbola $xy = 1$, $x > 0$
(b) $\mathbf{R}(1) = \mathbf{i} + \mathbf{j}$, $\mathbf{V}(1) = \mathbf{i} - \mathbf{j}$, $\mathbf{A}(1) = 2\mathbf{j}$
(c) $\mathbf{V}(t) \to \mathbf{i}$
$\mathbf{A}(t) \to \mathbf{0}$
(d) $|\mathbf{V}(t)| \to 1$ as $t \to \infty$.

7. No

9. (a) Yes
(b) No

11. *Hint*: Differentiate $\mathbf{V} \cdot \mathbf{V} = |\mathbf{V}|^2$

13. Reduces to fact that $ds/ds = 1$, or more specifically,

$$\left(\frac{ds}{ds}\right)^2 = \left(\frac{dx}{ds}\right)^2 + \left(\frac{dy}{ds}\right)^2$$

15. (a) $\mathbf{R} = \cos t^2\mathbf{i} + \sin t^2\mathbf{j}$
$\mathbf{V} = -2t \sin t^2\mathbf{i} + 2t \cos t^2\mathbf{j}$
$\mathbf{A} = (-2 \sin t^2 - 4t^2 \cos t^2)\mathbf{i} + (2 \cos t^2 - 4t^2 \sin t^2)\mathbf{j}$
(b) $\mathbf{R} \cdot \mathbf{A} = -4t^2$

17. 85.5 minutes, 88.7 minutes, 119.5 minutes

19. About 200 miles

21. $\mathbf{R}(t) = \cos t\mathbf{i} + \sin t\mathbf{j}$

23. *Hint*: Compare the computations in this section with those in Sec. 10.6.

Sec. 20.4. The Vectors T and N

1. True

3. (a) Parabola $y = x^2$
(b) $\mathbf{R}(1) = \mathbf{i} + \mathbf{j}$
$\mathbf{V}(1) = \mathbf{i} + 2\mathbf{j}$
$\mathbf{T}(1) = (\mathbf{i} + 2\mathbf{j})/\sqrt{5}$
$\mathbf{N}(1) = (-2\mathbf{i} + \mathbf{j})/\sqrt{5}$

5. (b) $\mathbf{R}(1) = \mathbf{i} + \mathbf{j}$
$\mathbf{V}(1) = 2\mathbf{i} + 3\mathbf{j}$
$\mathbf{T}(1) = (2/\sqrt{13})\mathbf{i} + (3/\sqrt{13})\mathbf{j}$
$\mathbf{N}(1) = (-3/\sqrt{13})\mathbf{i} + (2/\sqrt{13})\mathbf{j}$
(d) $\sqrt{13}$ (e) $22/\sqrt{13}$ (f) $6/\sqrt{13}$

7. $|\mathbf{V} \cdot \mathbf{T}| = \text{speed}$, $\mathbf{V} \cdot \mathbf{N} = 0$.

9. 3

11. (a) $\mathbf{V}(1) = 32\mathbf{i} - 32\mathbf{j}$
(b) $\mathbf{V}(1) = 32\sqrt{2}\mathbf{T}(1) + 0\mathbf{N}(1)$
(c) $\mathbf{A}(1) = (32/\sqrt{2})\mathbf{T}(1) + (32/\sqrt{2})\mathbf{N}(1)$

Sec. 20.5. Components of Acceleration Vector

1. $A_T = 3$, $A_N = \frac{16}{5}$

3. $A_T = 0$, $A_N = \frac{1}{3}$

5. $A_T = 10\pi$, $A_N = 20\pi^2$

9. (b) Slowing down, $A_T < 0$
(d) $A_T = -\frac{3}{13}$, $A_N = \frac{41}{13}$

11. $r = 3$

19. (b) $v = |\mathbf{V}| = \sqrt{13}$, $d^2s/dt^2 = \mathbf{A} \cdot \mathbf{T} = -6\sqrt{13}$
(c) $13\sqrt{13}/17$

Sec. 20.6. Level Curves and Surfaces

1. Circle of radius 13, center $(0, 0)$

3. Hyperbola $xy = -1$

5. $10\mathbf{i} + 24\mathbf{j}$

7. $-\mathbf{i} + \mathbf{j}$

13. $4\mathbf{i} + 6\mathbf{j} + 8\mathbf{k}$

15. $18\mathbf{i} + 3\mathbf{j} + 2\mathbf{k}$

17. (b) (The parabola $z = x^2$, rotated around z axis)

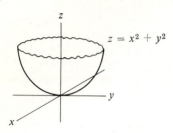

$z = x^2 + y^2$

(c) $2\mathbf{i} + 2\mathbf{j} - \mathbf{k}$

19. (d) Stream lines of air flow.

21. (b) $\cos\theta = 3\sqrt{11}$.

23. $\cos^{-1}(57/\sqrt{161 \cdot 21}) \doteq 11.4°$

Sec. 20.7. Surface Integrals

1. $\frac{1}{3}(5^{3/2} - 1)$

5. $\frac{91}{6}$

7. $162\pi/\sqrt{2}$

9. (a) $4\pi/3$ (b) $4\pi/3$

13. (a) Symmetry of sphere in x, y, and z.   (b) $4\pi a^4$
    (c) $4\pi a^4/3$ (d) $20\pi a^4/3$

15. $8\pi a^4 M/3$, if mass of surface is M

Sec. 20.8. Lagrange Multipliers

1. 2

3. (a) 1 square inch   (b) 4 inches

5. $h = \sqrt{8}r$

7. $(\frac{3}{7}, \frac{6}{7}, \frac{9}{7})$

9. $\frac{1}{27}$

11. $(0, \frac{1}{2}, 1)$

13. (a) $1/n^n$ (b) Let $x_i = \sqrt{a_i}/\sqrt{\sum_{i=1}^{n} a_i}$

15. $\sqrt{2}$

Sec. 20.9. Summary: Guide Quiz on Chap. 20

1. (a) First sketch the parabola $y = x^2$ in the xy plane of an xyz coordinate system. Then raise it 1 unit, to be in the plane $z = 1$.
    (b) $\Delta\mathbf{R} = 0.1\mathbf{i} + 0.21\mathbf{j} + 0\mathbf{k}$ (c) $\mathbf{i} + 2.1\mathbf{j} + 0\mathbf{k}$

3. (a) $\frac{2}{3}\mathbf{i} + \frac{2}{3}\mathbf{j} - \frac{1}{3}\mathbf{k}$ (b) $(1/\sqrt{14})(\mathbf{i} + 2\mathbf{j} + 3\mathbf{k})$
    (c) $\frac{1}{3}(x\mathbf{i} + y\mathbf{j} + z\mathbf{k})$

4. (a) 12   (b) $12/\sqrt{14}$

5. $\cos^{-1}(6/\sqrt{42})$

6. (a) $8x + 3y = 11$ (b) $2x + y + 2z = 5$

7. (a) $\mathbf{G}'(t)$ is parallel. Let $\mathbf{T} = \mathbf{G}'/|\mathbf{G}'|$. Then $d\mathbf{T}/dt$ is perpendicular.

(b) ∇f.

(c) If surfaces are $f(x, y, z) = 0$ and $g(x, y, z) = 0$, both ∇f and ∇g are perpendicular to the curve, while $\nabla f \times \nabla g$ is parallel to it.

(d) $\mathbf{k} \times \mathbf{G}'(t)$ for instance.

8. (a) Slowing down   (c) $A_T = -\sqrt{13}/13$, $A_N = 5\sqrt{13}/13$
    (d) $r = 13\sqrt{13}/5$ (e) $\frac{3}{2}$ (f) $-\frac{5}{2}$

9. (a)

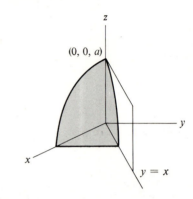

$(0, 0, a)$

$y = x$

(b) a^2

10. See section on Lagrange multipliers.

11. (c) $(\frac{1}{2}, \frac{1}{4}, \frac{3}{8})$

Review Exercises for Chap. 20

1. (a) $2\mathbf{i} + 3\mathbf{j} - \mathbf{k}$
    (b) Any vector perpendicular to the one in (a), for example, $3\mathbf{i} - 2\mathbf{j}$.

3. (b) $\sqrt{3}$

5. (b) $A_T = \dfrac{\sqrt{2}(e^2 - e^{-2})}{2\sqrt{e^2 + e^{-2}}}$, $A_N = \sqrt{\dfrac{2}{e^2 + e^{-2}}}$

7. (a) $\mathbf{V}(1) = 3\mathbf{i} + e\mathbf{j}$
    $\mathbf{A}(1) = 6\mathbf{i} + e\mathbf{j}$
    (b) $v = \sqrt{9 + e^2}$, $A_T = (18 + e^2)/\sqrt{9 + e^2}$,
    $A_N = 3e/\sqrt{9 + e^2}$, $r = (9 + e^2)^{3/2}/(3e)$

9. (b) $\cos^{-1} 0.6 \doteq 53°$

15. For instance, $6\mathbf{i} + 3\mathbf{j} - 2\mathbf{k}$

17. (b) $\cos^{-1}[31/(\sqrt{26}\sqrt{37})] \doteq 0°$

19. 0

23. (c) $\mathbf{i} + 3\mathbf{j} + 4\mathbf{k}$ (d) $\sqrt{26}$

25. $(2 + \frac{4}{3}t)\mathbf{i} + (5 + \frac{8}{3}t)\mathbf{j} + (7 + \frac{8}{3}t)\mathbf{k}$

31. (a), (b) $\sqrt{6}$

CHAPTER 21. INTEGRALS OF SCALAR AND VECTOR FIELDS

Sec. 21.1. Vector and Scalar Fields

1. 9

3. $y \cos xy - 2xy \sin xy^2$

5. Vector field

7. **0**

9. $-\mathbf{i} + y^2\mathbf{j} + (1 - 2yz)\mathbf{k}$

13. (b) $\mathbf{F} = 2x\mathbf{i} - y\mathbf{j} - z\mathbf{k}$

Sec. 21.2. Line Integrals

1. 2

3. $\frac{19}{28}$

5. $\frac{3}{2}$

7. -10

9. (a) $\pi/4$ (b) $9\pi/4$

11. $\frac{4}{3}$

13. 1

15. (a) $\frac{2}{3}$ (b) $\frac{2}{3}$ (c) $\frac{3}{4}$ (d) $\frac{1}{2}$

Sec. 21.3. Integral of Normal Component

1. $0, 18\pi$

3. $0, 7\pi a^2$

5. (b) Leaving   (c) 3π

7. It tends to measure the flow of fluid along the curve C—which might be thought of as "circulation."

19. $2ab$

21. ab

Sec. 21.4. Conservative Fields

1. (a) 8   (b) 0   (c) $e^8 - 1$

15. x^2yz

17. (a) x^2y^2

(b) For instance, the triangle whose vertices are $(0, 0)$, $(1, 0)$, $(1, 1)$.

Sec. 21.5. Summary: Guide Quiz on Chap. 21

1. (a) $\frac{22}{3}$ (b) 2

2. $\dfrac{b \cos t\mathbf{i} + a \sin t\mathbf{j}}{\sqrt{b^2 \cos^2 t + a^2 \sin^2 t}}$

5. (a) $\nabla f = 3x^2y^2z^4\mathbf{i} + 2x^3yz^4\mathbf{j} + 4x^3y^2z^3\mathbf{k}$

(b) $6xy^2z^4 + 2x^3z^4 + 12x^3y^2z^2$ (c) 0

8. $\frac{5}{6}$

Review Exercises for Chap. 21

1. $-\frac{7}{2}$

3. (a) Yes   (b) No

5. No

7. (a) Conservative   (b) Not conservative
   (c) Conservative

11. (a) $-\frac{3}{2}$ (b) No

13. 2π

15. (a) They meet at $(1, 3, 0)$ (b) $\cos^{-1}(9/\sqrt{798}) \doteq 71.4°$

19. (c) Yes   (d) 2π (e) By part (d),
    (f) z axis not in domain.

21. The integrand must be 0, hence $f_x = 0$, etc.   Thus f is constant.

23. Let Z = speed of plane relative to ground, and θ = angle of tangent line.   Then $V^2 = W^2 + Z^2 - 2WZ \cos \theta$, hence $Z = W \cos \theta \pm \sqrt{V^2 - W^2 \sin^2 \theta}$.   Choose "+" for "worst" case. Then

$$\int_C dS/Z = \int_C \frac{(W \cos \theta + \sqrt{V^2 - W^2 \sin^2 \theta})}{W^2 - V^2} \, ds$$

$$= \frac{1}{W^2 - V^2}\int_C \sqrt{V^2 - W^2 \sin^2 \theta} \, ds$$

$$> \int_C \frac{ds}{\sqrt{V^2 - W^2}} > \int_C \frac{ds}{V}.$$

CHAPTER 22. GREEN'S THEOREM AND ITS GENERALIZATIONS

Sec. 22.1. Green's Theorem

1. 5π

3. $\frac{5}{28}$

Sec. 22.2. Proof of Green's Theorem

5. (c) $\frac{9}{40}$

9. (a), (b) $3\pi/2$

15. (a) Check if formula yields a positive or negative number.

Sec. 22.3. Functions from Plane to Plane

1. (a) $(0, 0)$, $(4, 0)$, $(2, 3)$

5. (a) $x = 0$ (b) parabola $x = y - y^2$

7. (a) $(0, 0)$, $(1, 1)$, $(\sqrt{2}, 2)$, $(\sqrt{3}, 3)$, $(2, 4)$
   (c) $\sqrt{y} = x$, a parabola

9. (a) $(2, 1)$, $(3, 1)$, $(-2, -1)$, $(-3, -1)$ (b) Clockwise

11. (a) $(-1, -1)$, $(1, -2)$, $(3, -3)$, $(5, -4)$

15. (b) *Hint:* A conic section in the uv plane has the equation $Au^2 + Buv + Cv^2 + Du + Ev + F = 0$.

Sec. 22.4. Magnification and Jacobian

1. (a) -6 (b) -6

3. (a) 6   (b) 9

9. (b) Counterclockwise   (c) 3, yes

11. (b) $\frac{4}{3}$

13. (d) $\frac{1}{3}$

19. $\frac{1}{60}$

Sec. 22.5. Divergence Theorem

3. Use divergence theorem

5. By divergence theorem, its absolute value is not larger than 5(surface area).

7. (*a*) By divergence theorem
(*b*) Continuity of integrand is assumed.

9. Use divergence theorem.

11. If $g(P_0) = a > 0$, then there would be a region R throughout which $g(P) \geq a/2$, say. Then $\int_R g(P)\, dV \neq 0$, indeed is at least $(a/2)(\text{area of } R)$.

13. (*a*) Symmetry   (*b*) $4\pi a^5/5$ (*c*) $4\pi a^6/5$

17. Use the divergence theorem.

19. Use a "fluid-loss" argument.

Sec. 22.6. Stokes' Theorem

5. -4π

7. 1. The divergence theorem helps.

9. No

13. 0

15. (*b*) Parallel to z axis   (*c*) **curl F** = **k**

Sec. 22.7. Summary: Guide Quiz on Chap. 22

6. 0

7. (*a*) $(2, 3), (1, -2), (-2, -3), (-1, 2)$
(*b*) An ellipse   (*c*) 7π (*d*) $35\pi/4$.

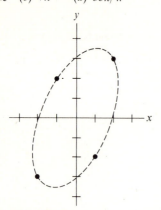

Review Exercises for Chap. 22

7. 2

9. 2

21. The center of gravity is at the origin.

29. (*d*) Volume $= \pm \begin{vmatrix} a_1 & a_2 & a_3 \\ b_1 & b_2 & b_3 \\ c_1 & c_2 & c_3 \end{vmatrix}$

35. (*e*) A rectangle

CHAPTER 23. THE INTERCHANGE OF LIMITS

Sec. 23.2. The Derivative $d(\int_a^b f(x, y)\, dx)/dy$

3. $2/\pi$

5. $f(u, u) + \int_0^u f_u(u, x)\, dx$

Sec. 23.3. Power Series

3. This follows directly from Exercise 2. It may also be proved in the fashion of Theorem 1.

Sec. 23.4. Interchange of Limits

17. Use the equation

$$\frac{1}{1^2} + \frac{1}{2^2} + \frac{1}{3^2} + \cdots =$$

$$\frac{1}{1^2} + \frac{1}{3^2} + \frac{1}{5^2} + \cdots + \frac{1}{4}\left(\frac{1}{1^2} \quad \frac{1}{3^2} \quad \frac{1}{5^2} + \cdots\right)$$

APPENDIX A

The Real Numbers

1. (*a*) Equality

3. Rational: (*a*), (*d*), (*e*), (*f*). The text proves that $\sqrt{2}$ and π are irrational.

5. (*a*) $62{,}395/9{,}990$ (*b*) $20{,}162/990$

Completeness of the Real Numbers

1. Yes, the set of negative irrational numbers, for instance

3. (*a*) 10   (*b*) 10

APPENDIX B

Sec. B.1

1. (*c*) 13

3. (*a*) $(x - 1)^2 + (y + 1)^2 = 25$ (*c*) No

5. (*a*) $(x - 3)^2 + y^2 = 9$

7. (*c*) $\sqrt{A^2/4 + B^2/4 - C}$

9. No

Sec. B.2

1. (*a*) $-2, 3$ (*b*) $1, 1$ (*c*) $-1, 5$ (*d*) $6, 0$

3. (*a*) $(4, 0)$ (*b*) $(0, 3)$

5. (*a*) $y = -x/2 + 3$ (*b*) $(0, 3)$

7. (*a*) $y = -3x/2 + 8$ (*c*) Yes

Sec. B.3

1. Foci at $(\pm 2\sqrt{6}, 0)$

3. Place two stakes 3 feet from the center at points on the line through the center parallel to the long side of the garden

5. Foci at $(0, \pm\sqrt{2})$

7. Focus at $(-\frac{3}{2}, 0)$ directrix $x = \frac{3}{2}$

Sec. B.4

3. (a) $\frac{4}{3}$ (b) $\frac{3}{4}$ (c) 1   (d) $\frac{4}{3}$

5. (b) $r = 2/(1 + \cos \theta)$ (c) $r = -4 \cos \theta/\sin^2 \theta$

7. (a) Multiply by r and write in terms of x and y

APPENDIX C

Sec. C.I

9. (a) The sequence $a_n = (-1)^n$ would have a limit, say 0, for $\varepsilon = 2$ would suffice for all N

(b) The same sequence as (a) approaches, say, 8, with $N = 1$ and $\varepsilon = 10$

APPENDIX D

Sec. D.I

5. (a) $(x - 2)(x^2 + 2x + 4)$

(b) $(x - \sqrt[3]{4})(x^2 + \sqrt[3]{4}x + \sqrt[3]{16})$

7. $a = -1, b = 5$

9. (a) $2/(x - 2) + 8/(x - 2)^2 + 9/(x - 2)^3$

(b) $1/(x + 1) - 1/(x + 1)^2$ (c) $1/(x - 3)^2 + x/(x^2 + 1)$

11. 20 in either case

Sec. D.2

1. (a) $(2x - 1)(x + 2)^2$ (b) Prime

(c) $2[x + (1 - \sqrt{57})/4][x + (1 + \sqrt{57})/4]$

3. (b) $\dfrac{x/2 - \frac{5}{2}}{x^2 + x + 1} + \dfrac{-x/2 + \frac{7}{2}}{x^2 - x + 1}$

7. $\dfrac{\frac{1}{23}}{x - 2} - \dfrac{4x/23 + \frac{10}{23}}{4x^2 + 2x + 3}$

9. (a) $x^2 + \dfrac{1 - x}{1 - x + x^2}$ (b) $1 - \dfrac{2}{x^2 + 1} + \dfrac{1}{(x^2 + 1)^2}$

INDEX